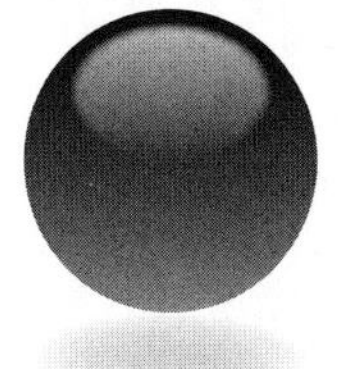

Student Solutions Manual to Accompany Atkins' Physical Chemistry

TENTH EDITION

Charles Trapp

Marshall Cady

Carmen Giunta

OXFORD UNIVERSITY PRESS

OXFORD
UNIVERSITY PRESS

Great Clarendon Street, Oxford, OX2 6DP,
United Kingdom

Oxford University Press is a department of the University of Oxford.
It furthers the University's objective of excellence in research, scholarship,
and education by publishing worldwide. Oxford is a registered trade mark of
Oxford University Press in the UK and in certain other countries

Impression: 1

British Library Cataloguing in Publication Data
Data available

ISBN 978-0-19-870800-1

Printed in Great Britain by
Ashford Colour Press Ltd, Gosport, Hampshire

Published in the United States and Canada by
W.H. Freeman and Company
41 Madison Avenue
New York, NY 10010
www.whfreeman.com

ISBN-13: 978-1-4641-2449-5

ISBN-10: 1-4641-2449-3

Table of contents

Preface

This manual provides detailed solutions to all the end-of-chapter (a) Exercises, and to the odd-numbered Discussion questions and Problems. Solutions to Exercises and Problems carried over from previous editions have been reworked, modified, or corrected when needed.

The solutions to some of the Exercises and many of the Problems in this edition relied more heavily on the mathematical, graphical, and molecular modelling software that is now generally accessible to physical chemistry students. This is particularly true for some of the new Problems which specifically request the use of such software for their solutions. For many of the solutions in this manual we have used Excel™ for spreadsheet calculations and graphing, and Mathcad™ for mathematical calculations and the plotting of the results. When a quantum chemical calculation or molecular modelling process has been called for, we have often provided the solution with PC Spartan Pro because of its common availability. However, the majority of the Exercises and many of the Problems can still be solved with a modern hand-held scientific calculator.

In general, we have adhered rigorously to the rules for significant figures in displaying the final answers. However, when intermediate answers are shown, they are often given with one more figure than would be justified by the data. These excess digits are usually indicated with an overline.

We have carefully checked the solutions in this manual for errors and expect that most have been eliminated, but would be grateful to any readers who bring any remaining ones to our attention.

We warmly thank our publishers, especially Jonathan Crowe and Alice Mumford, as well as Sian Jenkins, for their patience in guiding this complex, detailed project to completion. We also thank Peter Atkins and Julio de Paula for the opportunity to participate in the development of their outstanding *Physical Chemistry* text.

C. T.
M. C.
C. G.

Preface

Foundations

A Matter

Answers to discussion questions

DA.1 The **nuclear atomic model** consists of atomic number Z protons concentrated along with all atomic neutrons within the nucleus, an extremely small central region of the atom. Z electrons occupy **atomic orbitals**, which are voluminous regions of the atom that describe where electrons are likely to be found with no more than two electrons in any orbital. The electrostatic attraction binds the negatively charged electrons to the positively charged nucleus and the so-called strong interaction binds the protons and neutrons within the nucleus.

The atomic orbitals are arranged in shells around the nucleus, each shell being characterized by a **principal quantum number**, $n = 1, 2, 3, 4 \ldots$. A shell consists of n^2 individual orbitals, which are grouped together into n subshells. The **subshells**, and the orbitals they contain, are denoted s, p, d, and f. For all neutral atoms other than hydrogen, the subshells of a given shell have slightly different energies.

The **atomic number, Z**, is the number of protons in an atom. These protons are located within the nucleus. The **nucleon number, A**, which is also commonly called the **mass number**, is the total number of protons and neutrons in an atom. Theses nucleons are located within the nucleus.

DA.3 A single bond is a shared pair of electrons between adjacent atoms within a molecule while a multiple bond involves the sharing of either two pairs of electrons (a double bond) or three pairs of electrons (a triple bond).

DA.5 The solid phase of matter has a shape that is independent of the container it occupies. It has a density compatible with the close proximity of either its constituent elemental atoms or its constituent molecules and, consequently, it has low compressibility. Constituent atoms, or molecules, are held firmly at specific lattice sites by relatively strong, net forces of attraction between neighbouring constituents. Solids may be characterized by terms such as brittle, ductile, tensile strength, toughness, and hardness.

A liquid adopts the shape of the part of the container that it occupies; it can flow under the influence of gravitational attraction to occupy any shape. Like a solid, it has a density caused by the close proximity of either its constituent elemental atoms or its constituent molecules and it has low compressibility. Liquids can flow because the constituent atoms or molecules have enough average kinetic energy to overcome the attractive forces between neighbouring constituents, thereby making it possible for them to slip past immediate neighbours. This causes constituents to be placed randomly in contrast to the orderly array in crystals. Liquids are characterized by terms such as surface tension, viscosity, and capillary action. Liquids within a vertical, narrow tube exhibit a meniscus that is either concave-up or concave-down depending upon the nature of the attractive or repulsive forces between the liquid and the material of the tube.

Gases have no fixed shape or volume. They expand to fill the container volume. The constituent molecules move freely and randomly. A perfect gas has a total molecular volume that is negligibly small compared to the container volume and, because of the relatively large average distance between molecules, intermolecular attractive forces are negligibly small. Gases are compressible.

Solutions to exercises

EA.1(a)

	Example	Element	Ground-state electronic configuration
(a)	Group 2	Ca, calcium	$[Ar]4s^2$
(b)	Group 7	Mn, manganese	$[Ar]3d^54s^2$
(c)	Group 15	As, arsenic	$[Ar]3d^{10}4s^2 4p^3$

EA.2(a) (a) chemical formula and name: $MgCl_2$, magnesium chloride; ions: Mg^{2+} and Cl^-; oxidation numbers of the elements: magnesium, + 2; chlorine, –1

(b) chemical formula and name: FeO, iron(II) oxide; ions: Fe^{2}+ and O^{2-}; oxidation numbers of the elements: iron, + 2; oxygen, –2

(c) chemical formula and name: Hg_2Cl_2, mercury(I) chloride; ions: Cl^- and Hg_2^{2+} (a polyatomic ion); oxidation numbers of the elements: mercury, + 1; chlorine, –1

EA.3(a) (a) Methylamine, a gas $H_3C–NH_2$

(b) Acetoxime, a solid $(H_3C)_2C=NOH$

(c) Acetonitrile, a liquid $H_3CC\equiv N$

EA.4(a) (a) Sulfite anion, SO_3^{2-}

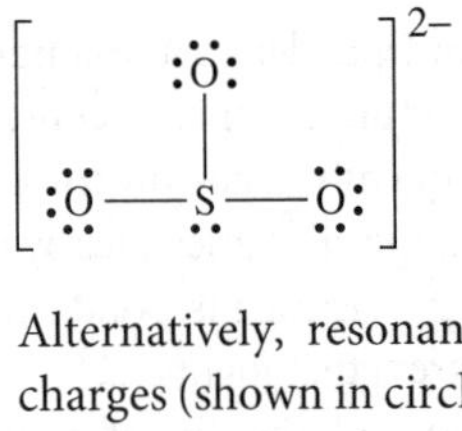

Alternatively, resonance structures may be drawn and, if desired, formal charges (shown in circles below) may be indicated.

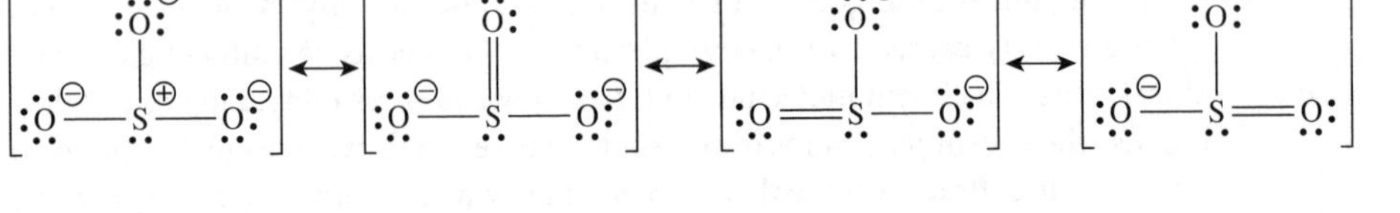

(b) Xenon tetrafluoride, XeF_4

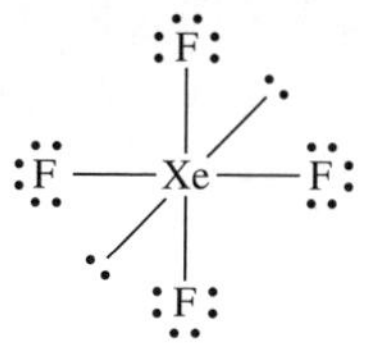

:F: — Xe — :F: (four F atoms bonded to Xe, two lone pairs on Xe)

(c) White phosphorus, P_4

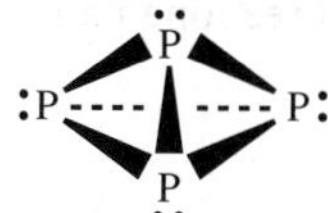

EA.5(a) Beryllium dichloride :Cl — Be — Cl:

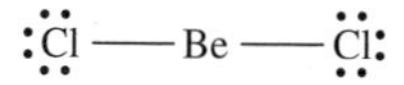

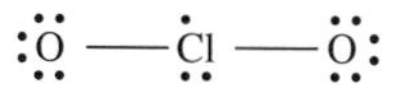

Chlorine dioxide :O — Cl — O:

Nitrogen monoxide N═O:

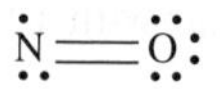

EA.6(a) Molecular and polyatomic ion shape are predicted by drawing the Lewis structure and applying the concepts of VSEPR theory.

(a) PCl_3 Lewis structure:

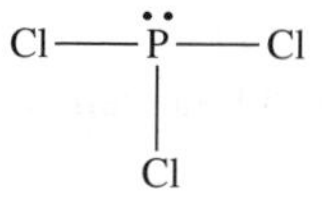

Cl — P — Cl, with P also bonded to Cl (lone pair on P)

Orientations caused by repulsions between one lone pair and three bonding pairs:

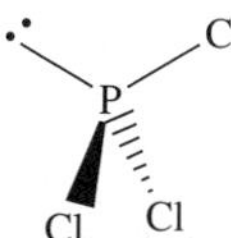

Molecular shape: trigonal pyramidal and bond angles somewhat smaller than 109.5°

(b) PCl_5 Lewis structure:

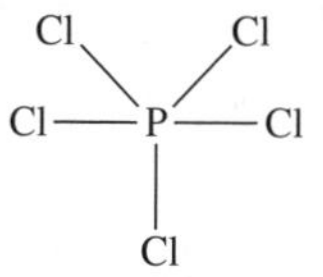

Cl — P — Cl, with P also bonded to three further Cl atoms

Orientations caused by repulsions between five bonding pairs (no lone pair):

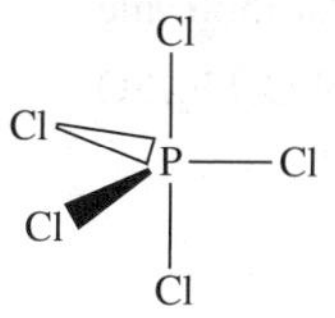

Cl (axial), Cl, Cl, Cl (equatorial) — P — Cl (axial)

Molecular shape: trigonal bipyramidal with equatorial bond angles of 120° and axial bond angles of 90°

(c) XeF_2 Lewis structure:

Orientations caused by repulsions between three lone pair and two bonding pairs:

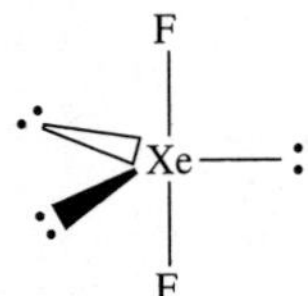

Molecular shape: linear with a 180° bond angle.

(d) XeF_4 Lewis structure:

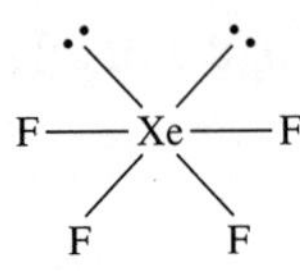

Orientations caused by repulsions between two lone pairs and four bonding pairs:

Molecular shape: square planar with a 90° bond angles.

EA.7(a) (a) $\overset{\delta^+}{C}-\overset{\delta^-}{Cl}$

(b) P—H

Nonpolar or weakly polar.

(c) $\overset{\delta^+}{N}-\overset{\delta^-}{O}$

EA.8(a) (a) CO_2 is a linear, nonpolar molecule.

(b) SO_2 is a bent, polar molecule.

(c) N_2O is linear, polar molecule.

(d) SF_4 has a see-saw configuration and it is a polar molecule.

EA.9(a) In the order of increasing dipole moment: CO_2, N_2O, SF_4, SO_2

EA.10(a) (a) Mass is an extensive property.

(b) Mass density is an intensive property.

(c) Temperature is an intensive property.

(d) Number density is an intensive property.

EA.11(a) (a) $n = \frac{m}{M} = 25.0\ \text{g}\left(\frac{1\ \text{mol}}{46.069\ \text{g}}\right) = \boxed{0.543\ \text{mol}}$ [A.3]

(b) $N = nN_A = 0.543\ \text{mol}\left(\frac{6.0221\times10^{23}\ \text{molecules}}{\text{mol}}\right) = \boxed{3.27\times10^{23}\ \text{molecules}}$

EA.12(a) (a) $m = nM = 10.0\ \text{mol}\left(\frac{18.015\ \text{g}}{\text{mol}}\right) = \boxed{180.\ \text{g}}$ [A.3]

(b) $\text{weight} = F_{\text{gravity on Earth}} = mg_{\text{Earth}}$

$$= (180.\ \text{g})\times(9.81\ \text{m}\,\text{s}^{-2})\times\left(\frac{1\ \text{kg}}{1000\ \text{g}}\right) = 1.77\ \text{kg}\,\text{m}\,\text{s}^{-2} = \boxed{1.77\ \text{N}}$$

EA.13(a) $p = \frac{F}{A} = \frac{mg}{A}$

$$= \frac{(65\ \text{kg})\times(9.81\ \text{m}\,\text{s}^{-2})}{150\ \text{cm}^2}\left(\frac{1\ \text{cm}^2}{10^{-4}\ \text{m}^2}\right) = 4.3\times10^4\ \text{N}\,\text{m}^{-2} = 4.3\times10^4\ \text{Pa}\left(\frac{1\ \text{bar}}{10^5\ \text{Pa}}\right)$$

$$= \boxed{0.43\ \text{bar}}$$

EA.14(a) $0.43\ \text{bar}\left(\frac{1\ \text{atm}}{1.01325\ \text{bar}}\right) = \boxed{0.42\ \text{atm}}$

EA.15(a) (a) $1.45\ \text{atm}\left(\frac{1.01325\times10^5\ \text{Pa}}{1\ \text{atm}}\right) = \boxed{1.47\times10^5\ \text{Pa}}$

(b) $1.45\ \text{atm}\left(\frac{1.01325\ \text{bar}}{1\ \text{atm}}\right) = \boxed{1.47\ \text{bar}}$

EA.16(a) $T/\text{K} = \theta/{}^\circ\text{C} + 273.15 = 37.0 + 273.15 = 310.2$ [A.4]

$$\boxed{T = 310.2\ \text{K}}$$

EA.17(a) To devise an equation relating the Fahrenheit and Celsius scales requires that consideration be given to both the degree size and a common reference point in each scale. Between the normal freezing point of water and its normal boiling point there is a degree scaling of 100 °C and 180 °F. Thus, the scaling ratio is 5 °C per 9 °F. A convenient reference point is provided by the normal freezing point of water, which is 0 °C and 32 °F. So to calculate the Celsius temperature from a given Fahrenheit temperature, 32 must be subtracted from the Fahrenheit temperature (θ_F) followed by scaling to the Celsius temperature (θ) with the ratio 5 °C/9 °F.

$$\boxed{\theta/°\mathrm{C} = \tfrac{5}{9}\times(\theta_F/°\mathrm{F} - 32) \quad \text{or} \quad \theta_F/°\mathrm{F} = \tfrac{9}{5}\times\theta/°\mathrm{C} + 32}$$

$$\theta_F/°\mathrm{F} = \tfrac{9}{5}\times\theta/°\mathrm{C} + 32 = \tfrac{9}{5}\times 78.5 + 32 = 173$$

$$\boxed{\theta_F = 173\ °\mathrm{F}}$$

EA.18(a) $110\ \mathrm{kPa}\times\left(\dfrac{(7.0+273.15)\ \mathrm{K}}{(20.0+273.15)\ \mathrm{K}}\right) = \boxed{105\ \mathrm{kPa}}$

EA.19(a) $pV = nRT\ [\mathrm{A.5}] = \dfrac{mRT}{M}$

$$M = \frac{mRT}{pV} = \frac{\rho RT}{p} \quad \text{where } \rho \text{ is the mass density } [\mathrm{A.2}]$$

$$= \frac{(3.710\ \mathrm{kg\,m^{-3}})(8.314\ \mathrm{J\,K^{-1}\,mol^{-1}})(773.15\ \mathrm{K})}{93.2\times10^3\ \mathrm{Pa}} = 0.256\ \mathrm{kg\,mol^{-1}} = 256\ \mathrm{g\,mol^{-1}}$$

The molecular mass is eight times as large as the atomic mass of sulfur ($32.06\ \mathrm{g\ mol^{-1}}$) so the molecular formula is $\boxed{S_8}$.

EA.20(a) $n = 22\ \mathrm{g}\times\left(\dfrac{1\ \mathrm{mol}}{30.07\ \mathrm{g}}\right) = 0.73\ \mathrm{mol}\ [\mathrm{A.3}]$

$$p = \frac{nRT}{V}\ [\mathrm{A.5}] = \frac{(0.73\ \mathrm{mol})(8.314\ \mathrm{J\,K^{-1}\,mol^{-1}})(298.15\ \mathrm{K})}{1000\ \mathrm{cm^3}}\left(\frac{\mathrm{cm^3}}{10^{-6}\ \mathrm{m^3}}\right)$$

$$= 1.8\times10^6\ \mathrm{Pa} = \boxed{1.8\ \mathrm{MPa}}$$

EA.21(a) $n_{N_2} = 1.0$ mole and $n_{H_2} = 2.0$ mole

$$p_{N_2} = \frac{n_{N_2}RT}{V}\ [\mathrm{A.5}] = \frac{(1.0\ \mathrm{mol})(8.314\ \mathrm{J\,K^{-1}\,mol^{-1}})(278.15\ \mathrm{K})}{10.0\ \mathrm{dm^3}}\left(\frac{\mathrm{dm^3}}{10^{-3}\ \mathrm{m^3}}\right) = \boxed{2.3\times10^5\ \mathrm{Pa}}$$

Since there are twice as many moles of hydrogen as nitrogen, the hydrogen partial pressure must be twice as large.

$$p_{H_2} = \boxed{4.6\times10^5\ \mathrm{Pa}} \qquad p = p_{N_2} + p_{H_2}\ [1.6] = \boxed{6.9\times10^5\ \mathrm{Pa}}$$

B Energy

Answers to discussion questions

DB.1 In the classical physics of Newton the force $\boldsymbol{F}$ acting on a body so as to cause acceleration $\boldsymbol{a}$ equals the mass m of the object multiplied by its acceleration: $\boldsymbol{F}=m\boldsymbol{a}$ where the bold faced symbols represent vector quantities. Coulomb's law describes the particularly important electrostatic force between two point charges Q_1 and Q_2 separated by the distance r:

$$F=\frac{Q_1Q_2}{4\pi\varepsilon_0 r^2} \text{ in a vacuum } (\varepsilon_0 \text{ is the vacuum permittivity})$$

and

$$F=\frac{Q_1Q_2}{4\pi\varepsilon_r\varepsilon_0 r^2} \text{ in a medium that has the relative permittivity } \varepsilon_r \text{ (formerly, dielectric constant)}$$

Convention assigns a negative value to the Coulomb force when it is attractive and a positive value when it is repulsive. The SI unit of force is the newton (N) and 1 N = 1 kg m s^{-2}.

An infinitesimal amount of **work**, dw, done on a body when it experiences an infinitesimal displacement, d$\boldsymbol{s}$, is defined by the scalar product (sometimes called the dot-product): $\mathrm{d}w=-\boldsymbol{F}\cdot\mathrm{d}\boldsymbol{s}$ where $\boldsymbol{F}$ is now the force that opposes the displacement. When the opposing force and displacement lie along the same direction, the z direction for example, the infinitesimal work done on the body simplifies to: $\mathrm{d}w=-F\mathrm{d}z$ where F is the magnitude of the opposing force. The SI unit of work is the joule (J) and 1 J = 1 N m = 1 Pa m^3 = 1 kg m^2 s^{-2}.

Continuing the example of displacement along the z direction alone, integration of the infinitesimal work between the initial position 'i' and the final position 'f' gives the total work, w, done on the object: $w=-\int_i^f F\mathrm{d}z$. Provided that the opposing force is constant over the displacement, the integral simplifies to give: $w=-F\times(z_f-z_i)=-F\Delta z$ where $\Delta z=z_f-z_i$.

Energy is the capacity to do work. It is a property and the SI unit of energy, like the unit of work, is the joule. The **law of conservation of energy** states that the total energy E of an isolated system is conserved; that is, the total energy of an isolated system is a constant. Even in an isolated system, however, energy can be transferred from one location to another and transformed from one form to another. The transfers and transformations involve heat, work, gravitational potential energy, Coulomb potential energy, and electromagnetic radiation.

DB.3 The Second Law of thermodynamics states that any spontaneous (that is, natural) change in an isolated system is accompanied by an increase in the entropy of the system. This tendency is commonly expressed by saying that the natural direction of change is accompanied by dispersal of energy from a localized region to a less organized form.

Yes, the entropy of a non-isolated system can decrease during a spontaneous process but for this to happen the total entropy ($S_{\text{system}} + S_{\text{environment}}$) must increase.

DB.5 Kinetic molecular theory, a model for a perfect gas, assumes that the molecules, imagined as particles of negligible size, are in ceaseless, random motion and do not interact except during their brief collisions.

Solutions to exercises

EB.1(a) $a = \mathrm{d}\upsilon/\mathrm{d}t = g$ so $\mathrm{d}\upsilon = g\,\mathrm{d}t$. The acceleration of free fall is constant near the surface of the Earth.

$$\int_{\upsilon=0}^{\upsilon(t)} \mathrm{d}\upsilon = \int_{t=0}^{t=t} g\,\mathrm{d}t$$

$$\upsilon(t) = gt$$

(a) $$\upsilon(1.0\ \mathrm{s}) = (9.81\ \mathrm{m\,s^{-2}})\times(1.0\ \mathrm{s}) = \boxed{9.81\ \mathrm{m\,s^{-1}}}$$

$$E_\mathrm{k} = \tfrac{1}{2}m\upsilon^2 = \tfrac{1}{2}(0.0010\ \mathrm{kg})\times(9.81\ \mathrm{m\,s^{-1}})^2 = \boxed{48\ \mathrm{mJ}}$$

(b) $$\upsilon(3.0\ \mathrm{s}) = (9.81\ \mathrm{m\,s^{-2}})\times(3.0\ \mathrm{s}) = \boxed{29.4\ \mathrm{m\,s^{-1}}}$$

$$E_\mathrm{k} = \tfrac{1}{2}m\upsilon^2 = \tfrac{1}{2}(0.0010\ \mathrm{kg})\times(29.4\ \mathrm{m\,s^{-1}})^2 = \boxed{0.43\ \mathrm{J}}$$

EB.2(a) The terminal velocity occurs when there is a balance between the force exerted by the electric field and the force of frictional drag. It will be in the direction of the field and have the magnitude s_{terminal}.

$$ze\mathcal{E} = 6\pi\eta R s_{\text{terminal}}$$

$$\boxed{s_{\text{terminal}} = \frac{ze\mathcal{E}}{6\pi\eta R}}$$

EB.3(a) $x(t) = A\sin(\omega t) + B\cos(\omega t)$

$$\begin{aligned}\frac{\mathrm{d}^2x}{\mathrm{d}t^2} &= A\omega\frac{\mathrm{d}\cos(\omega t)}{\mathrm{d}t} - B\omega\frac{\mathrm{d}\sin(\omega t)}{\mathrm{d}t}\\ &= A\omega\times(-\omega\sin(\omega t)) - B\omega\times(\omega\cos(\omega t))\\ &= -\omega^2(A\sin(\omega t) + B\cos(\omega t)) = -\omega^2 x \quad \text{where } \omega = (k/m)^{1/2}\end{aligned}$$

$$m\frac{\mathrm{d}^2x}{\mathrm{d}t^2} = -m\omega^2 x = -m\times\left(\frac{k}{m}\right)x = -kx$$

This confirms that $x(t)$ satisfies the harmonic oscillator equation of motion (see *Brief illustration* B.2).

EB.4(a) $w = \frac{1}{2}kx^2$ where $x = R - R_e$ is the displacement from equilibrium [*Brief illustation* B.3]

(a) $w = \frac{1}{2}(450\ \text{N m}^{-1}) \times (10 \times 10^{-12}\ \text{m})^2 = 2.25 \times 10^{-20}\ \text{N m} = \boxed{2.25 \times 10^{-20}\ \text{J}}$

(b) $w = \frac{1}{2}(450\ \text{N m}^{-1}) \times (20 \times 10^{-12}\ \text{m})^2 = 9.00 \times 10^{-20}\ \text{N m} = \boxed{9.00 \times 10^{-20}\ \text{J}}$

EB.5(a) $E_k = e\Delta\phi$

$$\tfrac{1}{2}mv^2 = e\Delta\phi \quad \text{or} \quad v = \left(\frac{2e\Delta\phi}{m}\right)^{1/2}$$

$$v = \left(\frac{2(1.6022 \times 10^{-19}\ \text{C}) \times (100 \times 10^3\ \text{V})}{9.10938 \times 10^{-31}\ \text{kg}}\right)^{1/2} = 1.88 \times 10^8 \left(\frac{\text{C V}}{\text{kg}}\right)^{1/2} = 1.88 \times 10^8 \left(\frac{\text{J}}{\text{kg}}\right)^{1/2}$$

$$= 1.88 \times 10^8 \left(\frac{\text{kg m}^2\ \text{s}^{-2}}{\text{kg}}\right)^{1/2} = \boxed{1.88 \times 10^8\ \text{m s}^{-1}}$$

$$E = E_k = e\Delta\phi = e \times (100\ \text{kV}) = \boxed{100\ \text{keV}}$$

EB.6(a) The work needed to separate two ions to infinity is identical to the Coulomb potential drop that occurs when the two ions are brought from an infinite separation, where the interaction potential equals zero, to a separation of r.

In a vacuum:

$$w = -V = -\left(\frac{Q_1Q_2}{4\pi\varepsilon_0 r}\right)[\text{B.14}] = -\left(\frac{(e)\times(-e)}{4\pi\varepsilon_0 r}\right) = \frac{e^2}{4\pi\varepsilon_0 r}$$

$$= \frac{(1.6022 \times 10^{-19}\ \text{C})^2}{4\pi(8.85419 \times 10^{-12}\ \text{J}^{-1}\ \text{C}^2\ \text{m}^{-1}) \times (200 \times 10^{-12}\ \text{m})} = \boxed{1.15 \times 10^{-18}\ \text{J}}$$

In water:

$$w = -V = -\left(\frac{Q_1Q_2}{4\pi\varepsilon r}\right) = -\left(\frac{(e)\times(-e)}{4\pi\varepsilon r}\right) = \frac{e^2}{4\pi\varepsilon r} = \frac{e^2}{4\pi\varepsilon_r\varepsilon_0 r}\ [\text{B.15}] \text{ where } \varepsilon_r = 78 \text{ for water at 25 °C}$$

$$= \frac{(1.6022 \times 10^{-19}\ \text{C})^2}{4\pi(78) \times (8.85419 \times 10^{-12}\ \text{J}^{-1}\ \text{C}^2\ \text{m}^{-1}) \times (200 \times 10^{-12}\ \text{m})} = \boxed{1.48 \times 10^{-20}\ \text{J}}$$

EB.7(a) We will model a solution by assuming that the LiH pair consists of the two point charge ions Li^+ and H^-. The electric potential will be calculated along the line of the ions.

$$\phi = \phi_{\text{Li}^+} + \phi_{\text{H}^-}\ [\text{B.17}] = \frac{e}{4\pi\varepsilon_0 r_{\text{Li}^+}} + \frac{(-e)}{4\pi\varepsilon_0 r_{\text{H}^-}}\ [2.16] = \frac{e}{4\pi\varepsilon_0}\left(\frac{1}{r_{\text{Li}^+}} - \frac{1}{r_{\text{H}^-}}\right)$$

$$\phi = \frac{1.6022 \times 10^{-19}\ \text{C}}{4\pi(8.85419 \times 10^{-12}\ \text{J}^{-1}\ \text{C}^2\ \text{m}^{-1})}\left(\frac{1}{200 \times 10^{-12}\ \text{m}} - \frac{1}{150 \times 10^{-12}\ \text{m}}\right)$$

$$= -2.40\ \text{J C}^{-1} = -2.40\ \text{C V C}^{-1} = \boxed{-2.40\ \text{V}}$$

EB.8(a) We will assume that the electric circuit has a negligibly small heat capacity so that all of the electrically generated heat energy is received by the water.

$$\Delta U_{H_2O} = \text{energy dissipated by the electric circuit}$$
$$= I\Delta\phi\Delta t \text{ [B.20]}$$
$$= (2.23\text{ A})\times(15.0\text{ V})\times(720\text{ s}) = 24.1\times10^3\text{ C s}^{-1}\text{ V s} = \boxed{24.1\text{ kJ}}$$

$$\Delta U_{H_2O} = (nC\Delta T)_{H_2O} \text{ [B.21]}$$

$$\Delta T = \frac{\Delta U_{H_2O}}{(nC)_{H_2O}} = \frac{\Delta U_{H_2O}}{(mC/M)_{H_2O}} = \frac{24.1\times10^3\text{ J}}{(200\text{ g})\times(75.3\text{ J K}^{-1}\text{ mol}^{-1})/(18.02\text{ g mol}^{-1})}$$
$$= 28.8\text{ K} = \boxed{28.8\text{ °C}}$$

EB.9(a) $\Delta T = \frac{\Delta U}{C}\text{ [B.21]} = \frac{100.\text{ J}}{3.67\text{ J K}^{-1}} = \boxed{27.2\text{ K or }27.2\text{ °C}}$

EB.10(a) $n = 100.\text{ g}\times\left(\frac{1\text{ mol}}{207.2\text{ g}}\right) = 0.483\text{ mol}$

$$\Delta U = C\Delta T\text{ [B.21]} = nC_m\Delta T$$
$$= (0.483\text{ mol})(26.44\text{ J K}^{-1}\text{ mol}^{-1})(10.0\text{ K}) = \boxed{128\text{ J}}$$

EB.11(a) $C_s = C_m/M = (111.46\text{ J K}^{-1}\text{ mol}^{-1})\times\left(\frac{1\text{ mol}}{46.069\text{ g}}\right) = \boxed{2.4194\text{ J K}^{-1}\text{ g}^{-1}}$

EB.12(a) $C_m = C_s M = (4.18\text{ J K}^{-1}\text{ g}^{-1})\times\left(\frac{18.02\text{ g}}{\text{mol}}\right) = \boxed{75.3\text{ J K}^{-1}\text{ mol}^{-1}}$

EB.13(a) Dividing eqn B.23 by the number of moles n and using the molar properties (intensive) H_m, U_m, and V_m yields the equation:

$$H_m - U_m = pV_m$$

Substitution of the perfect gas equation of state $pV_m = RT$ yields:

$$H_m - U_m = RT = (8.3145\text{ J mol}^{-1}\text{ K}^{-1})\times(1\,000\text{ K}) = \boxed{8.3145\text{ kJ mol}^{-1}}$$

EB.14(a) $S_{H_2O(g)} > S_{H_2O(l)}$

EB.15(a) $S_{Fe(3\,000\text{ K})} > S_{Fe(300\text{ K})}$

EB.16(a) In a state of dynamic equilibrium, which is the character of all chemical equilibria, the forward and reverse reactions are occurring at the same rate and there is no net tendency to change in either direction. Examples:

$2\,SO_2(g) + O_2(g) \rightleftharpoons 2\,SO_3(g)$ Addition of oxygen shifts the equilibrium to the right. An increase in pressure also shifts it to the right so as to reduce the number of moles of gas.

$CaCO_3(s) \rightleftharpoons CaO(s) + CO_2(g)$ Addition of carbon dioxide shifts the equilibrium to the left. An increase in pressure also shifts it of the left so as to reduce the number of moles of gas.

$CO(g) + H_2O(g) \rightleftharpoons CO_2(g) + H_2(g)$ Addition of carbon monoxide shifts the equilibrium to the right. An increase in pressure has no effect on the equilibrium because there are equal numbers of moles of gas on left and right.

EB.17(a) $\dfrac{N_i}{N_j} = \mathrm{e}^{-(\varepsilon_i-\varepsilon_j)/kT} = \mathrm{e}^{-\Delta\varepsilon_{ij}/kT}$ [B.25a]

(a) $$\frac{N_2}{N_1} = \mathrm{e}^{-(1.0\ \mathrm{eV})\times(1.602\times10^{-19}\ \mathrm{JeV^{-1}})/\{(1.381\times10^{-23}\ \mathrm{JK^{-1}})\times(300\ \mathrm{K})\}} = 1.6\times10^{-17}$$

(b) $$\frac{N_2}{N_1} = \mathrm{e}^{-(1.0\ \mathrm{eV})\times(1.602\times10^{-19}\ \mathrm{JeV^{-1}})/\{(1.381\times10^{-23}\ \mathrm{JK^{-1}})\times(3000\ \mathrm{K})\}} = 0.021$$

EB.18(a) $$\lim_{T\to0}\left(\frac{N_{\mathrm{upper}}}{N_{\mathrm{lower}}}\right) = \lim_{T\to0}\left(\mathrm{e}^{-\Delta\varepsilon/kT}\right)\ [2.25\mathrm{a}] = \mathrm{e}^{-\infty} = 0$$

In the limit of the absolute zero of temperature all particles occupy the lower state. The upper state is empty.

EB.19(a)

$$\Delta\varepsilon = \varepsilon_{\mathrm{upper}} - \varepsilon_{\mathrm{lower}} = \tilde{\nu}hc = (2\,500\ \mathrm{cm^{-1}})(6.626\times10^{-34}\ \mathrm{Js})(3.000\times10^{10}\ \mathrm{cms^{-1}}) = 4.970\times10^{-20}\ \mathrm{J}$$

$$\frac{N_{\mathrm{upper}}}{N_{\mathrm{lower}}} = \mathrm{e}^{-\Delta\varepsilon/kIT}\ [\mathrm{B.25a}] = \mathrm{e}^{-(4.970\times10^{-20}\ \mathrm{J})/\{(1.381\times10^{-23}\ \mathrm{JK^{-1}})\times(293\ \mathrm{K})\}} = \boxed{4.631\times10^{-6}}$$

The ratio $N_{\mathrm{upper}}/N_{\mathrm{lower}}$ is so small that the population of the upper level is approximately zero.

EB.20(a) Molecules can survive for long periods without undergoing chemical reaction at low temperatures when few molecules have the requisite speed and corresponding kinetic energy to promote excitation and bond breakage during collisions.

EB.21(a) $\upsilon_{\mathrm{mean}} \propto (T/M)^{1/2}$ [B.26]

$$\frac{\upsilon_{\mathrm{mean}}(T_2)}{\upsilon_{\mathrm{mean}}(T_1)} = \frac{(T_2/M)^{1/2}}{(T_1/M)^{1/2}} = \left(\frac{T_2}{T_1}\right)^{1/2}$$

$$\frac{\upsilon_{\mathrm{mean}}(313\ \mathrm{K})}{\upsilon_{\mathrm{mean}}(273\ \mathrm{K})} = \left(\frac{313\ \mathrm{K}}{273\ \mathrm{K}}\right)^{1/2} = \boxed{1.07}$$

EB.22(a) $\upsilon_{\mathrm{mean}} \propto (T/M)^{1/2}$ [B.26]

$$\frac{\upsilon_{\mathrm{mean}}(M_2)}{\upsilon_{\mathrm{mean}}(M_1)} = \frac{(T/M_2)^{1/2}}{(T/M_1)^{1/2}} = \left(\frac{M_1}{M_2}\right)^{1/2}$$

$$\frac{\upsilon_{\mathrm{mean}}(\mathrm{N_2})}{\upsilon_{\mathrm{mean}}(\mathrm{CO_2})} = \left(\frac{44.0\ \mathrm{g\,mol^{-1}}}{28.0\ \mathrm{g\,mol^{-1}}}\right)^{1/2} = \boxed{1.25}$$

EB.23(a) A gaseous argon atom has three translational degrees of freedom (the components of motion in the x, y, and z directions). Consequently, the equipartition theorem assigns a mean energy of $\frac{3}{2}kT$ to each atom. The molar internal energy is

$$U_\mathrm{m} = \tfrac{3}{2}N_\mathrm{A}kT = \tfrac{3}{2}RT = \tfrac{3}{2}(8.3145\ \mathrm{J\,mol^{-1}\,K^{-1}})(298\ \mathrm{K}) = 3.72\ \mathrm{kJ\,mol^{-1}}$$

$$U = nU_\mathrm{m} = mM^{-1}U_\mathrm{m} = (5.0\ \mathrm{g})\left(\frac{1\ \mathrm{mol}}{39.95\ \mathrm{g}}\right)\left(\frac{3.72\ \mathrm{kJ}}{\mathrm{mol}}\right) = \boxed{0.47\ \mathrm{kJ}}$$

EB.24(a) (a) A gaseous, linear, carbon dioxide molecule has three quadratic translational degrees of freedom (the components of motion in the x, y, and z directions) but it has only two rotational quadratic degrees of freedom because there is no rotation along the internuclear line. There is a total of five quadratic degrees of freedom for the molecule. Consequently, the equipartition theorem assigns a mean energy of $\frac{5}{2}kT$ to each molecule. The molar internal energy is

$$U_\mathrm{m} = \tfrac{5}{2}N_\mathrm{A}kT = \tfrac{5}{2}RT = \tfrac{5}{2}(8.3145\ \mathrm{J\,mol^{-1}\,K^{-1}})(293\ \mathrm{K}) = 6.09\ \mathrm{kJ\,mol^{-1}}$$

$$U = nU_\mathrm{m} = mM^{-1}U_\mathrm{m} = (10.0\ \mathrm{g})\left(\frac{1\ \mathrm{mol}}{44.0\ \mathrm{g}}\right)\left(\frac{6.09\ \mathrm{kJ}}{\mathrm{mol}}\right) = \boxed{1.38\ \mathrm{kJ}}$$

(b) A gaseous, nonlinear, methane molecule has three quadratic translational degrees of freedom (the components of motion in the x, y, and z directions) and three quadratic rotational degrees of freedom. Consequently, the equipartition theorem assigns a mean energy of $\frac{6}{2}kT$ to each molecule. The molar internal energy is

$$U_\mathrm{m} = \tfrac{6}{2}N_\mathrm{A}kT = 3RT = 3(8.3145\ \mathrm{J\,mol^{-1}\,K^{-1}})(293\ \mathrm{K}) = 7.31\ \mathrm{kJ\,mol^{-1}}$$

$$U = nU_\mathrm{m} = mM^{-1}U_\mathrm{m} = (10.0\ \mathrm{g})\left(\frac{1\ \mathrm{mol}}{16.04\ \mathrm{g}}\right)\left(\frac{7.31\ \mathrm{kJ}}{\mathrm{mol}}\right) = \boxed{4.56\ \mathrm{kJ}}$$

EB.25(a) See Exercise B.23(a) for the description of the molar internal energy of argon.

$$C_{V,\mathrm{m}} = \frac{\partial U_\mathrm{m}}{\partial T} = \frac{\partial(\frac{3}{2}RT)}{\partial T} = \tfrac{3}{2}R = \tfrac{3}{2}(8.3145\ \mathrm{J\,mol^{-1}\,K^{-1}}) = \boxed{12.47\ \mathrm{J\,mol^{-1}\,K^{-1}}}$$

EB.26(a) See Exercise B.24(a) for the description of the molar internal energies of carbon dioxide and methane.

(a) $U_\mathrm{m} = \frac{5}{2}RT$ for carbon dioxide

$$C_{V,\mathrm{m}} = \frac{\partial U_\mathrm{m}}{\partial T} = \frac{\partial(\frac{5}{2}RT)}{\partial T} = \tfrac{5}{2}R = \tfrac{5}{2}(8.3145\ \mathrm{J\,mol^{-1}\,K^{-1}}) = \boxed{20.79\ \mathrm{J\,mol^{-1}\,K^{-1}}}$$

(b) $U_{\mathrm{m}} = 3RT$ for methane

$$C_{V,\mathrm{m}} = \frac{\partial U_{\mathrm{m}}}{\partial T} = \frac{\partial(3RT)}{\partial T} = 3R = 3\left(8.3145\ \mathrm{J\,mol^{-1}\,K^{-1}}\right) = \boxed{24.94\ \mathrm{J\,mol^{-1}\,K^{-1}}}$$

C Waves

Answer to discussion question

DC.1 The study of wave motion includes the examination of water waves, sound waves, electromagnetic waves (microwave, infrared, visible, ultraviolet, etc.), radio waves, seismic waves, and de Broglie waves. Mechanical waves, generated by a source that causes particulate displacements within an elastic medium, are **transverse** when the particles experience displacements perpendicular to the direction of propagation while they are **longitudinal** when the particles are displaced parallel to the direction of wave propagation. Water molecules are displaced in ocean waves. Gas molecules are displaced in sound waves. Differential equations are used to describe the wave motion in space and time.

Solutions to exercises

EC.1(a) $$c_{\mathrm{H_2O}} = \frac{c}{n_{\mathrm{r}}}\ [\mathrm{C.4}] = \frac{3.00\times10^{8}\ \mathrm{m\,s^{-1}}}{1.33} = \boxed{2.26\times10^{8}\ \mathrm{m\,s^{-1}}}$$

EC.2(a) $$\lambda = \frac{1}{\tilde{\nu}}\ [\mathrm{C.5}] = \frac{1}{2\,500\ \mathrm{cm^{-1}}}\left(\frac{10^{6}\ \mu\mathrm{m}}{10^{2}\ \mathrm{cm}}\right) = \boxed{4.00\ \mu\mathrm{m}}$$

$$\nu = \frac{c}{\lambda}\ [\mathrm{C.1}] = \frac{3.00\times10^{8}\ \mathrm{m\,s^{-1}}}{4.00\times10^{-6}\ \mathrm{m}} = 7.50\times10^{13}\ \mathrm{s^{-1}} = \boxed{7.50\times10^{13}\ \mathrm{Hz}}$$

Integrated activities

IF.1 (a) Plots of the Maxwell–Boltzmann distribution at T = 200 K, 500 K, 1 000 K, and 2 000 K with M = 100 g mol^{-1} are calculated in the following Mathcad Prime 2 worksheet. The worksheet looks much the same in all versions of Mathcad. How would you change the worksheet so as to plot the distribution at the fix temperature T = 300 K for a range of molar masses such as 20 g mol^{-1}, 50 g mol^{-1}, and 100 g mol^{-1}?

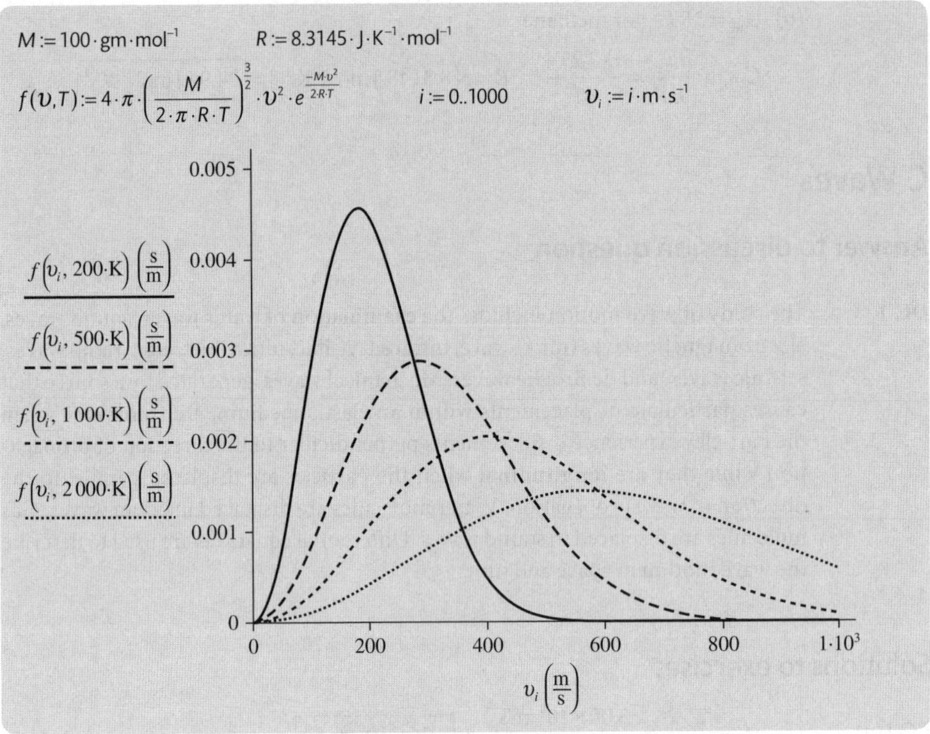

(b) The fraction of molecules with speeds in the range 100 m s^{-1} to 200 m s^{-1} at 300 K and 1000 K is calculated with a numerical integration of the Maxwell–Boltzmann distribution over the desired range of speeds at fixed temperature. The following integration calculations are a continuation of the above Mathcad worksheet. Note that the integration over all possible speeds (0–∞ m s^{-1}) equals 1 because the sum of all fractions equals 1.

$$fraction(T) := \int_{100 \cdot m \cdot s^{-1}}^{200 \cdot m \cdot s^{-1}} f(\upsilon, T)\,dv$$

$fraction(300 \cdot K) = 0.281$ $\qquad$ $fraction(1000 \cdot K) = 0.066$

$$\int_{0 \cdot m \cdot s^{-1}}^{2000 \cdot m \cdot s^{-1}} f(\upsilon, 300 \cdot K)\,d\upsilon = 1$$

1 The properties of gases

1A The perfect gas

Answer to discussion question

D1A.1 An equation of state is an equation that relates the variables that define the state of a system to each other. Boyle, Charles, and Avogadro established these relations for gases at low pressures (perfect gases) by appropriate experiments. Boyle determined how volume varies with pressure ($V \propto 1/p$), Charles how volume varies with temperature ($V \propto T$), and Avogadro how volume varies with amount of gas ($V \propto n$). Combining all of these proportionalities into one we find

$$V \propto \frac{nT}{p}$$

Inserting the constant of proportionality, R, yields the perfect gas equation

$$V = \frac{RnT}{p} \quad \text{or} \quad pV = nRT$$

Solutions to exercises

E1A.1(a) The perfect gas law [1A.5] is $pV = nRT$

Solving for the pressure gives $p = \frac{nRT}{V}$

The amount of xenon is $n = \frac{131\,\text{g}}{131\,\text{g}\,\text{mol}^{-1}} = 1.00\,\text{mol}$

$$p = \frac{(1.00\ \text{mol}) \times (0.0821\ \text{dm}^3\ \text{atm}\ \text{K}^{-1}\ \text{mol}^{-1}) \times (298.15\ \text{K})}{1.0\ \text{dm}^3} = \boxed{24\ \text{atm}}$$

So $\boxed{\text{no}}$, the sample would not exert a pressure of 20 atm, but 24 atm if it were a perfect gas.

E1A.2(a) Boyle's law [1A.4a] applies.

$$pV = \text{constant} \quad \text{so} \quad p_f V_f = p_i V_i$$

Solve for the initial pressure:

$$p_i = \frac{V_f}{V_i} \times p_f$$

$V_f = 4.65\ \text{dm}^3$, $V_i = (4.65+2.20)\ \text{dm}^3 = 6.85\ \text{dm}^3$, $p_f = 5.04$ bar

Therefore,

(a) $p_i = \left(\dfrac{4.65\ \text{dm}^3}{6.85\ \text{dm}^3}\right) \times (5.04\ \text{bar}) = \boxed{3.42\ \text{bar}}$

(b) Since 1 atm = 1.013 bar, $p_i = 3.42\ \text{bar} \times \dfrac{1\ \text{atm}}{1.013\ \text{bar}} = \boxed{3.38\ \text{atm}}$

E1A.3(a) The perfect gas law, $pV = nRT$ [1A.5], can be rearranged to $\dfrac{p}{T} = \dfrac{nR}{V} = \text{constant}$, if n and V are constant. Hence, $\dfrac{p_f}{T_f} = \dfrac{p_i}{T_i}$ or, solving for p_f, $p_f = \dfrac{T_f}{T_i} \times p_i$

Note that the internal pressure is the pump pressure **plus** atmospheric pressure.

$$p_i = 24\ \text{lb in}^{-2} + 14.7\ \text{lb in}^{-2} = 38.\overline{7}\ \text{lb in}^{-2}, \quad T_i = 268\ \text{K}(-5\ °\text{C}), \quad T_f = 308\ \text{K}(35\ °\text{C})$$

$$\text{so} \quad p_f = \frac{308\ \text{K}}{268\ \text{K}} \times 38.\overline{7}\ \text{lb in}^{-2} = 44.\overline{5}\ \text{lb in}^{-2}$$

Therefore, $p(\text{pump}) = 44.\overline{5}\ \text{lb in}^{-2} - 14.7\ \text{lb in}^{-2} = \boxed{30\ \text{lb in}^{-2}}$

Complications are those factors that destroy the constancy of V or n, such as the change in volume of the tyre, the change in rigidity of the material from which it is made, and loss of pressure by leaks and diffusion.

E1A.4(a) The perfect gas law in the form $p = \dfrac{nRT}{V}$ [1A.5] is appropriate. T and V are given; n must be calculated.

$$n = \frac{0.255\ \text{g}}{20.18\ \text{g mol}^{-1}} = 1.26 \times 10^{-2}\ \text{mol}, \quad T = 122\ \text{K}, \quad V = 3.00\ \text{dm}^3$$

Therefore, upon substitution,

$$p = \frac{(1.26 \times 10^{-2}\ \text{mol}) \times (0.08206\ \text{dm}^3\ \text{atm K}^{-1}\ \text{mol}^{-1}) \times (122\ \text{K})}{3.00\ \text{dm}^3} = \boxed{4.20 \times 10^{-2}\ \text{atm}}$$

E1A.5(a) Boyle's law [1A.4a] in the form $p_fV_f=p_iV_i$ is solved for V_f:

$$V_f = \frac{p_i}{p_f} \times V_i$$

The initial pressure is 1.0 atm. The final pressure is the pressure of the air plus the hydrostatic pressure of the water

$$p_f = p_{ex} + \rho gh \ [1A.1] = p_i + \rho gh = 1.0\,\text{atm} + \rho gh$$

where $\rho gh = 1.025\,\text{g cm}^{-3} \times \frac{1\,\text{kg}}{10^3\,\text{g}} \times \left(\frac{1\,\text{cm}}{10^{-2}\,\text{m}}\right)^3 \times 9.81\,\text{m s}^{-2} \times 50\,\text{m} = 5.0\bar{3} \times 10^5\,\text{Pa}$

Hence, $p_f = (1.0\bar{1} \times 10^5\,\text{Pa}) + (5.0\bar{3} \times 10^5\,\text{Pa}) = 6.0\bar{4} \times 10^5\,\text{Pa}$

$$V_f = \frac{1.0\bar{1} \times 10^5\,\text{Pa}}{6.0\bar{4} \times 10^5\,\text{Pa}} \times 3.0\,\text{m}^3 = \boxed{0.50\,\text{m}^3}$$

E1A.6(a) The pressure in the apparatus is given by

$$p = p_{ex} + \rho g \Delta h \ [1A.1]$$

where, using the fact that 760 Torr = 1 atm = 1.013×10^5 Pa,

$$p_{ex} = 770\,\text{Torr} \times \left(\frac{1.013 \times 10^5\,\text{Pa}}{760\,\text{Torr}}\right) = 1.026 \times 10^5\,\text{Pa}$$

and $\rho g \Delta h = 0.99707\,\text{g cm}^{-3} \times \left(\frac{1\,\text{kg}}{10^3\,\text{g}}\right) \times \left(\frac{1\,\text{cm}}{10^{-2}\,\text{m}}\right)^3 \times 9.806\,\text{m s}^{-2} \times (-0.100\,\text{m})$

$$= -978\,\text{Pa}$$

so $p = 1.026 \times 10^5\,\text{Pa} - 978\,\text{Pa} = 1.017 \times 10^5\,\text{Pa} = \boxed{102\,\text{kPa}}$

E1A.7(a) Rearrange the perfect gas equation [1A.5] to give $R = \frac{pV}{nT}$

Everything on the right hand side is related to measured quantities.

$$V = (20.000\,\text{dm}^3) \times \left(\frac{10^{-1}\,\text{m}}{1\,\text{dm}}\right)^3 = 2.0000 \times 10^{-2}\,\text{m}^3$$

$$n = \frac{m}{M} = \frac{0.25132\,\text{g}}{4.00260\,\text{g mol}^{-1}} = 0.062789\,\text{mol}$$

The gas pressure is calculated as the force per unit area that a column of water of height 206.402 cm exerts on the gas due to its weight. (See Example 1A.1 and eqn 1A.1.)

$$p=\rho gh=(0.99707\text{ g cm}^{-3})\times(206.402\text{ cm})\times(9.8067\text{ m s}^{-2})\times\left(\frac{1\text{ kg}}{10^3\text{ g}}\right)\times\left(\frac{1\text{ cm}}{10^{-2}\text{ m}}\right)^2$$

$$=2.0182\times10^4\text{ Pa}$$

So $$R=\frac{(2.0182\times10^4\text{ Pa})\times(2.0000\times10^{-2}\text{ m}^3)}{(0.062789\text{ mol})\times(773.15\text{ K})}=\boxed{8.3147\text{ J K}^{-1}\text{ mol}^{-1}}$$

The accepted value is $R=8.3145\text{ J K}^{-1}\text{mol}^{-1}$. Although gas volume data should be extrapolated to zero pressure for the best value of R, helium is close to being a perfect gas under the conditions here, and thus a value of R close to the accepted value is obtained.

E1A.8(a) Since $p<1\text{atm}$, the approximation that the vapour is a perfect gas is adequate. We introduce the molar mass into the perfect gas law [1A.5] by

$$pV=nRT=\frac{m}{M}RT$$

Upon rearrangement

$$M=\frac{m}{V}\left(\frac{RT}{p}\right)=\rho\left(\frac{RT}{p}\right)=(3.710\text{ kg m}^{-3})\times\frac{(8.3145\text{ Pa m}^3\text{ K}^{-1}\text{ mol}^{-1})\times(773\text{ K})}{9.32\times10^4\text{ Pa}}$$

$$=0.256\text{ kg mol}^{-1}=256\text{ g mol}^{-1}$$

This molar mass must be an integral multiple of the molar mass of atomic sulfur; hence

$$\text{number of S atoms}=\frac{256\text{g mol}^{-1}}{32.1\text{g mol}^{-1}}=8$$

The formula of the vapour is then $\boxed{S_8}$

E1A.9(a) The partial pressure of the water vapour in the room is 60% of the equilibrium vapour pressure, which must be looked up in a handbook like the *CRC* or other resource such as the NIST Chemistry WebBook.

$$p=(0.60)\times(0.0356\text{bar})=0.021\bar{4}\text{ bar}=2.1\bar{4}\times10^3\text{ Pa}$$

Assuming that the perfect gas equation [1A.5] applies, with $n=\frac{m}{M}$, $pV=\frac{m}{M}RT$ or

$$m=\frac{pVM}{RT}=\frac{(2.14\times10^3\text{ Pa})\times(400\text{m}^3)\times(18.02\text{g mol}^{-1})}{(8.3145\text{J K}^{-1}\text{ mol}^{-1})\times(300\text{K})}$$

$$=6.2\times10^3\text{ g}=\boxed{6.2\text{kg}}$$

E1A.10(a) (a) For simplicity assume a container of volume 1 m^3. Then the total mass is

$$m_T = n_{N_2}M_{N_2} + n_{O_2}M_{O_2} = 1146\text{ g} \qquad (1)$$

Assuming that air is a perfect gas, $p_T V = n_T RT$, where n_T is the total amount of gas

$$n_T = \frac{p_T V}{RT} = \frac{(0.987\text{ bar})\times(10^5\text{ Pa bar}^{-1})\times(1\text{ m}^3)}{(8.3145\text{ Pa m}^3\text{ K}^{-1}\text{ mol}^{-1})\times(300\text{ K})} = 39.6\text{ mol} \qquad (2)$$

$$n_T = n_{N_2} + n_{O_2} = 39.6\text{ mol}$$

Equations (1) and (2) are simultaneous equations for the amounts of gas and may be solved for them. Inserting n_{O_2} from (2) into (1) we get

$$n_{N_2}\times(28.0136\text{ g mol}^{-1}) + (39.6\text{ mol} - n_{N_2})\times(31.9988\text{ g mol}^{-1}) = 1146\text{ g}$$

$$(126\bar{6} - 1146)\text{g} = (3.9852\text{ g mol}^{-1})\times n_{N_2}$$

$$n_{N_2} = 30.1\bar{6}\text{ mol}$$

$$n_{O_2} = n_T - n_{N_2} = (39.6 - 30.1\bar{6})\text{ mol} = 9.4\bar{1}\text{ mol}$$

The mole fractions are $x_{N_2} = \dfrac{30.1\bar{6}\text{ mol}}{39.6\text{ mol}} = \boxed{0.762}$ $x_{O_2} = \dfrac{9.4\bar{1}\text{ mol}}{39.6\text{ mol}} = \boxed{0.238}$

The partial pressures are

$$p_{N_2} = (0.762)\times(0.987\text{ bar}) = \boxed{0.752\text{ bar}}$$

and $p_{O_2} = (0.238)\times(0.987\text{ bar}) = \boxed{0.235\text{ bar}}$

The sum checks, (0.752+0.235) bar=0.987 bar.

(b) The simplest way to solve this part is to realize that n_T, p_T, and m_T remain the same as in part (a) as these are experimentally determined quantities. However, the simultaneous equations that need to be solved are modified as follows:

$$m_T = n_{N_2}M_{N_2} + n_{O_2}M_{O_2} + n_{Ar}M_{Ar} = 1146\text{ g} \qquad (1')$$

$$n_T = n_{N_2} + n_{O_2} + n_{Ar} = 39.6\text{ mol} \qquad (2')$$

Since x_{Ar}=0.010, n_{Ar}=0.40 mol, that means

$$n_{O_2} = 39.6\text{ mol} - n_{N_2} - n_{Ar} = 39.2\text{ mol} - n_{N_2}$$

and $$(n_{N_2})\times(28.0136\text{ g mol}^{-1}) + (39.2\text{ mol} - n_{N_2})\times(31.9988\text{ g mol}^{-1})$$
$$= 1146\text{ g} - (0.396\text{ mol})\times(39.948\text{ g mol}^{-1})$$

Solving the equations yields

$$n_{N_2} = 30.9\bar{4}\ \text{mol and } x_{N_2} = \boxed{0.782}$$

and $n_{O_2} = 8.23\ \text{mol and } x_{O_2} = \boxed{0.208}$

The partial pressures are:

$$p_{Ar} = (0.010)\times(0.987\ \text{bar}) = \boxed{0.0099\ \text{bar}}$$

$$p_{N_2} = (0.782)\times(0.987\ \text{bar}) = \boxed{0.772\ \text{bar}}$$

and $p_{O_2} = (0.208)\times(0.987\ \text{bar}) = \boxed{0.205\ \text{bar}}$

E1A.11(a) This exercise uses the formula, $M = \frac{\rho RT}{p}$, which was developed and used in Exercise 1A.8(a). Substituting the data,

$$M = \frac{(1.23\times10^3\ \text{g m}^{-3})\times(8.3145\ \text{Pa m}^3\ \text{K}^{-1}\ \text{mol}^{-1})\times(330\ \text{K})}{20\times10^3\ \text{Pa}} = \boxed{169\ \text{g mol}^{-1}}$$

E1A.12(a) The easiest way to solve this exercise is to assume a convenient sample mass, say 1.000 g, then calculate the volume at each temperature, plot the volume against the Celsius temperature, and extrapolate to $V=0$. Draw up the following table

θ/°C	ρ/(g dm^{-3})	V/dm^3
-85	1.877	0.5328
0	1.294	0.7728
100	0.946	1.057

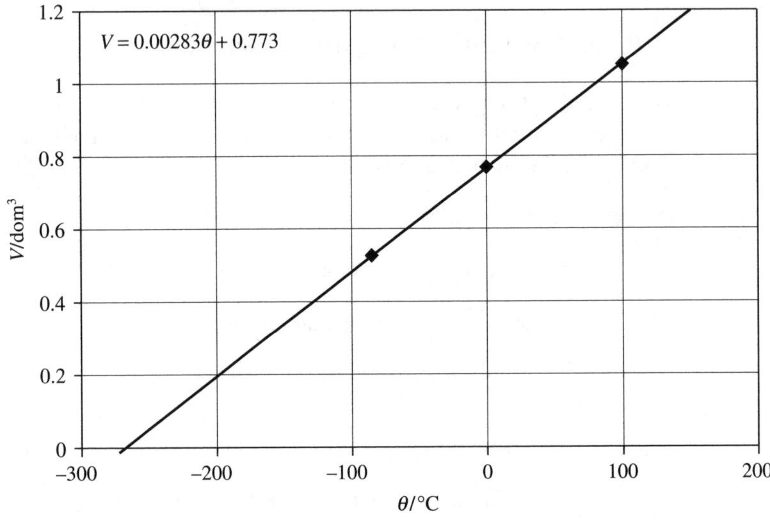

Figure 1A.1

V is plotted against θ in Fig. 1A.1. The extrapolation gives a value for absolute zero close to −273 °C. Alternatively, one could use an equation for V as a linear function of θ (the equation of the best-fit straight line through the data, also shown in Fig. 1A.1)

$$V = V_0 + \alpha\theta.$$

Charles's law says that V is directly proportional to absolute temperature [1A.4b]

$$V \propto T$$

so the volume of a perfect gas would vanish at zero absolute temperature. Solve for θ at absolute zero:

$$V = 0 = V_0 + \alpha\theta \quad \text{so} \quad \theta = -\frac{V_0}{\alpha} = -\frac{0.773\ \text{dm}^3}{0.00283\ \text{dm}^3\ {}^\circ\text{C}^{-1}} = \boxed{-273\ {}^\circ\text{C}}$$

E1A.13(a) $n = n(\text{H}_2) + n(\text{N}_2) = 2.0\ \text{mol} + 1.0\ \text{mol} = 3.0\ \text{mol} \quad x_\text{J} = \dfrac{n_\text{J}}{n}$ [1A.9]

(a) $x(\text{H}_2) = \dfrac{2.0\ \text{mol}}{3.0\ \text{mol}} = \boxed{0.67} \qquad x(\text{N}_2) = \dfrac{1.0\ \text{mol}}{3.0\ \text{mol}} = \boxed{0.33}$

(b) The perfect gas law is assumed to hold for each component individually as well as for the mixture as a whole. Hence, $p_\text{J} = n_\text{J}\dfrac{RT}{V}$

$$\frac{RT}{V} = \frac{(0.08206\ \text{dm}^3\ \text{atm mol}^{-1}\ \text{K}^{-1}) \times (273.15\ \text{K})}{22.4\ \text{dm}^3} = 1.00\ \text{atm mol}^{-1}$$

$$p(\text{H}_2) = (2.0\ \text{mol}) \times (1.00\ \text{atm mol}^{-1}) = \boxed{2.0\ \text{atm}}$$

$$p(\text{N}_2) = (1.0\ \text{mol}) \times (1.00\ \text{atm mol}^{-1}) = \boxed{1.0\ \text{atm}}$$

(c) $p = p(\text{H}_2) + p(\text{N}_2)\,[1\text{A}.10] = (2.0 + 1.0)\ \text{atm} = \boxed{3.0\ \text{atm}}$

Solutions to problems

P1A.1 Since the Neptunians know about perfect gas behaviour, we may assume that they will write $pV = nRT$ at both temperatures. We may also assume that they will establish the relationship between their temperature scales to be

$$\theta / {}^\circ\text{N} = T/(\text{absolute units}) + \text{absolute zero}/{}^\circ\text{N}$$

analogous to

$$\theta / {}^\circ\text{C} = T/K + \text{absolute zero}/{}^\circ\text{C}$$

Thus $pV(T_1) = 28.0\,\text{dm}^3\,\text{atm} = nRT_1 = nR\times(T_1 + 0\,°\text{N})$

and $pV(T_2) = 40.0\,\text{dm}^3\,\text{atm} = nRT_2 = nR\times(T_1 + 100\,°\text{N})$.

So $T_1 = \dfrac{28.0\,\text{dm}^3\,\text{atm}}{nR}$ and $T_1 + 100\,°\text{N} = \dfrac{40.0\,\text{dm}^3\,\text{atm}}{nR}$.

Dividing yields

$$\frac{T_1 + 100\,°\text{N}}{T_1} = \frac{40.0\,\text{dm}^3\,\text{atm}}{28.0\,\text{dm}^3\,\text{atm}} = 1.43$$

Thus $T_1 + 100\,°\text{N} = 1.43T_1$ so $T_1 = 233$ absolute units.

So absolute zero $= \theta - T = (0 - 233)\,°\text{N} = \boxed{-233\,°\text{N}}$.

Comment. To facilitate communication with Earth students we have converted the Neptunians' units of the pV product to units familiar to humans, namely dm^3 atm. However, we see from the solution that only the ratio of pV products is required, and that will be the same in any civilization.

Question. If the Neptunians' unit of volume is the venus (V), their unit of pressure is the poseidon (P), their unit of amount is the nereid (n), and their unit of absolute temperature is the titan (T), what is the value of the Neptunians' gas constant (R) in units of V, P, n, and T?

P1A.3 The value of absolute zero can be expressed in terms of α by using the requirement that the volume of a perfect gas becomes zero at the absolute zero of temperature. Hence

$$0 = V_0[1 + \alpha\theta(\text{abs. zero})]$$

Then $\theta(\text{abs. zero}) = -\dfrac{1}{\alpha}$

All gases become perfect in the limit of zero pressure, so the best value of α and, hence, of θ(abs. zero) is obtained by extrapolating α to zero pressure. This is done in Fig. 1A.2. Using the extrapolated value,

$$\alpha = 3.6637\times10^{-3}\,°\text{C}^{-1},$$

or $\theta(\text{abs. zero}) = -\dfrac{1}{3.6637\times10^{-3}\,°\text{C}^{-1}} = \boxed{-272.95\,°\text{C}}$

which is close to the accepted value of −273.15 °C.

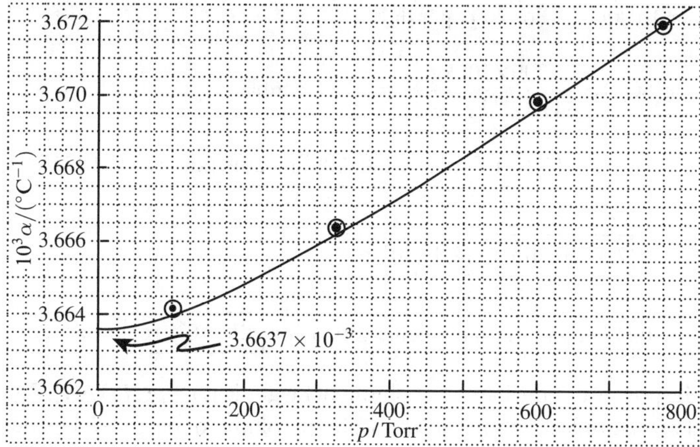

Figure 1A.2

P1A.5 At constant n and V, $\dfrac{p}{T} = \dfrac{nR}{V} = \text{constant}$. Hence, $\dfrac{p}{T} = \dfrac{p_3}{T_3}$, where p is the measured pressure at temperature T, and p_3 and T_3 are the triple point pressure and temperature, respectively. Rearranging, $p = \left(\dfrac{p_3}{T_3}\right)T$.

(a) If we think of p and T as representing changes from the triple point, we can write

$$p_3 + \Delta p = \left(\frac{p_3}{T_3}\right)(T + \Delta T) \quad \text{so} \quad \Delta p = \left(\frac{p_3}{T_3}\right)T + \left(\frac{p_3}{T_3}\right)\Delta T - p_3 = \left(\frac{p_3}{T_3}\right)\Delta T$$

Thus the change in pressure, Δp, is proportional to the change in temperature, ΔT:

$$\Delta p = \frac{6.69\ \text{kPa}}{273.16\ \text{K}} \times (\Delta T) = 0.0245\ \text{kPa K}^{-1} \times (\Delta T)$$

and a temperature change of 1 K corresponds to a pressure change of $\boxed{0.0245\,\text{kPa}}$.

(b) $p = \left(\dfrac{T}{T_3}\right)p_3 = \left(\dfrac{373.16\text{K}}{373.16\text{K}}\right) \times (6.69\,\text{kPa}) = \boxed{9.14\,\text{kPa}}$

(c) Since $\dfrac{p}{T}$ is a constant at constant n and V, it always has the value 0.0245 kPa K^{-1}.

Hence Δp near 373 K is the same as Δp near 273 K for a given temperature change, namely $\boxed{0.0245\,\text{kPa}}$ for a temperature change of 1.00 K.

P1A.7 $1\,\text{t} = 10^3\,\text{kg}$ For 200 t per day.

$$n(\text{SO}_2) = \frac{200\times10^3\,\text{kg}}{64\times10^{-3}\,\text{kg mol}^{-1}} = 3.1\times10^6\,\text{mol}$$

$$V = \frac{nRT}{p} = \frac{(3.1\times10^6\,\text{mol})\times(0.08206\,\text{dm}^3\,\text{atm mol}^{-1}\,\text{K}^{-1})\times1073\,\text{K}}{1.0\,\text{atm}}$$

$$= \boxed{2.8\times10^8\,\text{dm}^3} = \boxed{2.8\times10^5\,\text{m}^3}$$

For 300 t per day, similar calculation yields $\boxed{4.1\times10^8\,\text{dm}^3} = \boxed{4.1\times10^5\,\text{m}^3}$.

P1A.9 The pressure at the base of a column of height H is $p = \rho g H$ [1A.1]. But the pressure at any altitude h within the atmospheric column of height H depends only on the air above it; therefore

$$p = \rho g(H-h) \quad \text{and} \quad \text{d}p = -\rho g\text{d}h$$

Since $\rho = \dfrac{pM}{RT}$ [Exercise 1A.8(a)], $\text{d}p = \dfrac{pMg\text{d}h}{RT}$, implying that $\dfrac{\text{d}p}{p} = -\dfrac{Mg\text{d}h}{RT}$

This relation integrates to $p = p_0\text{e}^{-Mgh/RT}$

For air, $M = 29\,\text{g mol}^{-1}$, and at 298 K

$$\frac{Mg}{RT} \approx \frac{(29\times10^{-3}\,\text{kg mol}^{-1})\times(9.81\,\text{m s}^{-2})}{(8.3145\,\text{J K}^{-1}\,\text{mol}^{-1})\times(298\,\text{K})} = 1.1\bar{5}\times10^{-4}\,\text{m}^{-1}\,[1\,\text{J} = 1\,\text{kg m}^2\,\text{s}^{-2}]$$

(a) $h = 15$ cm

$$p = p_0\times\text{e}^{(-0.15\,\text{m})\times(1.1\bar{5}\times10^{-4}\,\text{m}^{-1})} = 0.9999\overline{8}p_0$$

That is, the pressure at the top is indistinguishable from the pressure at the bottom. To find the small pressure difference, expand the exponential function:

$$\text{e}^{-x} = 1 - x + \dots \quad \text{so} \quad \frac{p}{p_0} = \text{e}^{-Mgh/RT} \approx 1 - \frac{Mgh}{RT}$$

Thus the relative pressure drop is

$$\frac{p_0 - p}{p_0} \approx \frac{Mgh}{RT} = 1.1\bar{5}\times10^{-4}\,\text{m}^{-1}\times0.15\,\text{m} = \boxed{1.7\times10^{-5}}$$

(b) $h = 11\,\text{km} = 1.1\times10^4\,\text{m}$

$$p = p_0\times\text{e}^{(-1.1\times10^4)\times(1.1\bar{5}\times10^{-4}\,\text{m}^{-1})} = 0.28p_0 \text{ so } \frac{p_0 - p}{p_0} = \boxed{0.72}$$

P1A.11 Refer to Fig. 1A.3.

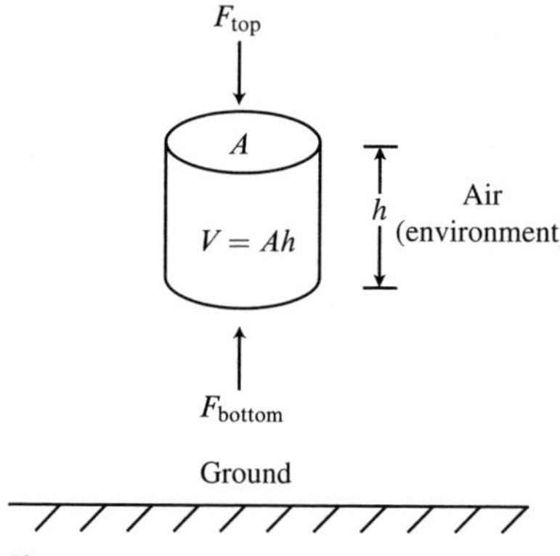

Figure 1A.3

The buoyant force on the cylinder is

$$F_{buoy} = F_{bottom} - F_{top} = A(p_{bottom} - p_{top})$$

According to the barometric formula (See Problem 1A.9),

$$p_{top} = p_{bottom}e^{-Mgh/RT}$$

where M is the molar mass of the environment (air). Since h is small, the exponential can be expanded in a Taylor series around $h=0$:

$$e^{-x} = 1 - x + \ldots$$

Keeping the first-order term only yields

$$p_{top} = p_{bottom}\left(1 - \frac{Mgh}{RT}\right)$$

The buoyant force becomes

$$F_{buoy} = Ap_{bottom}\left(1 - 1 + \frac{Mgh}{RT}\right) = Ah\left(\frac{p_{bottom}M}{RT}\right)g$$
$$= \left(\frac{p_{bottom}VM}{RT}\right)g = nMg\left[n = \frac{p_{bottom}V}{RT}\right]$$

n is the number of moles of the environment (air) displaced by the balloon, and $nM = m$, the mass of the displaced environment. Thus $F_{buoy} = mg$. The net force is the difference between the buoyant force and the weight of the balloon. Thus

$$F_{net} = mg - m_{balloon}g = (m - m_{balloon})g$$

This is Archimedes' principle.

P1A.13 Use the barometric formula [*Impact* 1.1] to relate the partial pressures of nitrogen and oxygen:

$$p = p_0 e^{-hMg/RT}.$$

The pressures of N_2 and O_2 would be, respectively,

$$p_N = p_{0N} e^{-hM_N g/RT} \quad \text{and} \quad p_O = p_{0O} e^{-hM_O g/RT}.$$

Divide the two expressions and solve for h:

$$\frac{p_N}{p_O} = \frac{p_{0N} e^{-hM_N g/RT}}{p_{0O} e^{-hM_O g/RT}} = \frac{p_{0N}}{p_{0O}} e^{h(M_O - M_N)g/RT}$$

$$h = \frac{RT}{g(M_O - M_N)} \ln\left(\frac{p_N}{p_O} \times \frac{p_{0O}}{p_{0N}}\right)$$

$$= \frac{(8.3145\,\text{J}\,\text{K}^{-1}\,\text{mol}^{-1}) \times (298\,\text{K})}{(9.807\,\text{m}\,\text{s}^{-2}) \times (32.00 - 28.02) \times 10^{-3}\,\text{kg}\,\text{mol}^{-1}} \ln\left(\frac{90}{10} \times \frac{20}{80}\right)$$

$$= 5.1 \times 10^4\,\text{m} = \boxed{51\,\text{km}}$$

At this altitude, the partial pressure of N_2 would be

$$p_N = (0.80\,\text{atm}) \times \exp\left(\frac{-(5.1 \times 10^4\,\text{m}) \times (28.02 \times 10^{-3}\,\text{kg}\,\text{mol}^{-1}) \times (9.807\,\text{m}\,\text{s}^{-2})}{(8.3145\,\text{J}\,\text{K}^{-1}) \times (298\,\text{K})}\right)$$

$$= 0.0027\,\text{atm}$$

By hypothesis, this accounts for 90% of the atmosphere, so the total pressure would be

$$p_{\text{total}} = \frac{p_N}{0.90} = \boxed{0.0029\,\text{atm}}$$

Comment. This treatment takes no account of the effect of entropy **against** the tendency of gases to fractionate under the influence of gravity. See Section 5A.2 for a treatment of the thermodynamics of mixing. The effect of gravity alone, as suggested in this problem, is to encourage heavier gases to be more concentrated at lower altitudes and lighter ones higher; however, the effect of entropy is to encourage the mixing of gases.

1B The kinetic model

Answer to discussion question

D1B.1 The three assumptions on which the kinetic model is based are given in Section 1B.1

1. The gas consists of molecules in random motion obeying the laws of classical mechanics.
2. Molecules are small in the sense that their diameters are much smaller than the average distance travelled between collisions.
3. The molecules interact only through brief elastic collisions.

None of these assumptions is strictly true; however, many of them are good approximations under a wide range of conditions including conditions of ambient temperature and pressure. In particular,

1. Molecules are subject to laws of quantum mechanics (See Chapter 7); however, for all but the lightest gases at low temperatures, non-classical effects are not important.
2. Obviously with increasing pressure, the average distance between molecules will decrease, eventually becoming comparable to the dimensions of the molecules themselves—if the gas remains a gas at all.
3. We learned about intermolecular interactions in general chemistry—hydrogen bonding, interactions of dipole moments, and the like. Those interactions operate when molecules are separated by small distances (i.e. not just when they are in contact, but not when they are far apart). So when assumption 2 breaks down, so does assumption 3, because the molecules are often close enough together to interact even when not colliding.

Solutions to exercises

E1B.1(a) The mean speed is [1B.8]

$$v_{\text{mean}} = \left(\frac{8RT}{\pi M}\right)^{1/2}$$

The mean translational kinetic energy is

$$\langle E_{\text{k}}\rangle = \langle \tfrac{1}{2}mv^2\rangle = \tfrac{1}{2}m\langle v^2\rangle = \tfrac{1}{2}mv_{\text{rms}}^2 = \frac{m}{2}\left(\frac{3RT}{M}\right)[1\text{B}.3] = \frac{3kT}{2}$$

The ratios of species 1 to species 2 at the same temperature are

$$\frac{v_{\text{mean},1}}{v_{\text{mean},2}} = \left(\frac{M_2}{M_1}\right)^{1/2} \quad \text{and} \quad \frac{\langle E_{\text{k}}\rangle_1}{\langle E_{\text{k}}\rangle_2} = 1$$

(a) $$\frac{v_{\text{mean},\text{H}_2}}{v_{\text{mean},\text{Hg}}} = \left(\frac{200.6}{2\times 1.008}\right)^{1/2} = \boxed{9.975}$$

(b) The mean translation kinetic energy is independent of molecular mass and depends upon temperature alone! Consequently, because the mean translational kinetic energy for a gas is proportional to T, the ratio of mean translational kinetic energies for gases at the same temperature always equals 1.

E1B.2(a) The root mean square speed [1B.3] is

$$v_{\mathrm{rms}}=\left(\frac{3RT}{M}\right)^{1/2}$$

For H_2

$$v_{\mathrm{rms}}=\left(\frac{3(8.3145\,\mathrm{J\,K^{-1}\,mol^{-1}})(20+273)\,\mathrm{K}}{2\times1.008\times10^{-3}\,\mathrm{kg\,mol^{-1}}}\right)^{1/2}=\boxed{1.90\times10^{3}\,\mathrm{m\,s^{-1}}}=\boxed{1.90\,\mathrm{km\,s^{-1}}}$$

For O_2

$$v_{\mathrm{rms}}=\left(\frac{3(8.3145\,\mathrm{J\,K^{-1}\,mol^{-1}})(20+273)\,\mathrm{K}}{2\times15.9994\times10^{-3}\,\mathrm{kg\,mol^{-1}}}\right)^{1/2}=\boxed{478\,\mathrm{m\,s^{-1}}}$$

Comment. Note that M must be given in units compatible with those of R, or must be converted. In practice, expressing both R and M in MKS units ($\mathrm{J\,mol^{-1}\,K^{-1}}$ and $\mathrm{kg\,mol^{-1}}$, respectively) leads to a result also in MKS units ($\mathrm{m\,s^{-1}}$).

E1B.3(a) The Maxwell–Boltzmann distribution of speeds [1B.4] is

$$f(v)=4\pi\left(\frac{M}{2\pi RT}\right)^{3/2}v^{2}\mathrm{e}^{-Mv^{2}/2RT}$$

and the fraction of molecules that have a speed between v and $v+\mathrm{d}v$ is $f(v)\mathrm{d}v$. The fraction of molecules to have a speed in the range between v_1 and v_2 is, therefore, $\int_{v_1}^{v_2}f(v)\mathrm{d}v$. If the range is relatively small, however, such that $f(v)$ is nearly constant over that range, the integral may be approximated by $f(v)\Delta v$, where $f(v)$ is evaluated anywhere within the range and $\Delta v=v_2-v_1$. Thus, we have

$$\int_{v_1}^{v_2}f(v)\mathrm{d}v\approx f(v)\Delta v=4\pi\left(\frac{2\times14.007\times10^{-3}\,\mathrm{kg\,mol^{-1}}}{2\pi(8.3145\,\mathrm{J\,K^{-1}\,mol^{-1}})(400\,\mathrm{K})}\right)^{3/2}(205\,\mathrm{m\,s^{-1}})^{2}$$
$$\times\exp\left(-\frac{(2\times14.007\times10^{-3}\,\mathrm{kg\,mol^{-1}})(205\,\mathrm{m\,s^{-1}})^{2}}{2(8.3145\,\mathrm{J\,K^{-1}\,mol^{-1}})(400\,\mathrm{K})}\right)\times(210-200)\,\mathrm{m\,s^{-1}}$$
$$=\boxed{0.00687},\text{nearly }0.7\%$$

E1B.4(a) The most probable, mean, and mean relative speeds are, respectively

$$v_{\mathrm{mp}}=\left(\frac{2RT}{M}\right)^{1/2}\ [1\mathrm{B}.9]\quad v_{\mathrm{mean}}=\left(\frac{8RT}{\pi M}\right)^{1/2}\ [1\mathrm{B}.8]\quad v_{\mathrm{rel}}=\left(\frac{8RT}{\pi\mu}\right)^{1/2}\ [1\mathrm{B}.10\mathrm{b}]$$

The temperature is $T=(20+273)\ \mathrm{K}=293\ \mathrm{K}$. For CO_2 the molar mass is

$$M=(12.011+2\times15.9994)\times10^{-3}\ \mathrm{kg\,mol^{-1}}=44.010\times10^{-3}\ \mathrm{kg\,mol^{-1}}$$

so $$v_{\mathrm{mp}}=\left(\frac{2(8.3145\ \mathrm{J\,K^{-1}\,mol^{-1}})(293\ \mathrm{K})}{44.010\times10^{-3}\ \mathrm{kg\,mol^{-1}}}\right)^{1/2}=\boxed{333\ \mathrm{m\,s^{-1}}}$$

and $$v_{\mathrm{mean}}=\left(\frac{8(8.3145\ \mathrm{J\,K^{-1}\,mol^{-1}})(293\ \mathrm{K})}{\pi(44.010\times10^{-3}\ \mathrm{kg\,mol^{-1}})}\right)^{1/2}=\boxed{375\ \mathrm{m\,s^{-1}}}$$

For many purposes, air can be considered as a gas with an average molar mass of 29.0 g mol^{-1}. In that case, the reduced molar mass [1B.10b] is

$$\mu=\frac{M_{\mathrm{A}}M_{\mathrm{B}}}{M_{\mathrm{A}}+M_{\mathrm{B}}}=\frac{(29.0\ \mathrm{g\,mol^{-1}})(44.0\ \mathrm{g\,mol^{-1}})}{(29.0+44.0)\ \mathrm{g\,mol^{-1}}}=17.5\ \mathrm{g\,mol^{-1}}=17.5\times10^{-3}\ \mathrm{kg\,mol^{-1}}$$

and $$v_{\mathrm{rel}}=\left(\frac{8(8.3145\ \mathrm{J\,K^{-1}\,mol^{-1}})(293\ \mathrm{K})}{\pi(17.5\times10^{-3}\ \mathrm{kg\,mol^{-1}})}\right)^{1/2}=\boxed{596\ \mathrm{m\,s^{-1}}}$$

Comment. One computes the average molar mass of air just as one computes the average molar mass of an isotopically mixed element. For elements, this amounts to taking an average of the various isotopes weighted by the abundance of those isotopes. For air, the species that have different masses are not just different isotopes but different chemical species (N_2, O_2, Ar, etc.), but taking a weighted average is the same procedure. By examining this approximation for a relative speed of a molecule in air, we realize that in fact that CO_2 would have slightly lower mean speeds relative to N_2 and relative to O_2. But then again, we make a similar approximation already when we use a mean molar mass to represent CO_2 rather than isotopically specific molar masses for, say, $^{12}C^{16}O_2$.

E1B.5(a) (a) $$v_{\mathrm{mean}}=\left(\frac{8RT}{\pi M}\right)^{1/2}[1\mathrm{B}.8]=\left(\frac{8(8.3145\ \mathrm{J\,K^{-1}\,mol^{-1}})(298\ \mathrm{K})}{\pi(2\times14.007\times10^{-3}\ \mathrm{kg\,mol^{-1}})}\right)^{1/2}=\boxed{475\ \mathrm{m\,s^{-1}}}$$

(b) The mean free path [1B.13] is

$$\lambda=\frac{kT}{\sigma p}=\frac{kT}{\pi d^2 p}=\frac{(1.381\times10^{-23}\ \mathrm{J\,K^{-1}})(298\ \mathrm{K})}{\pi(395\times10^{-12}\ \mathrm{m})^2(1.0\ \mathrm{atm})}\times\frac{1\ \mathrm{atm}}{1.013\times10^5\ \mathrm{Pa}}=\boxed{8.3\times10^{-8}\ \mathrm{m}}$$

At atmospheric pressure, the mean free path is roughly 100 nm, large compared to the collision diameter, but small compared to the laboratory scale. Note that we converted from atm to Pa because 1 J = 1 Pa m^3.

(c) The collision frequency is related to the mean free path and relative mean speed by [1B.12]

$$\lambda=\frac{v_{\mathrm{rel}}}{z}\quad\text{so}\quad z=\frac{v_{\mathrm{rel}}}{\lambda}=\frac{2^{1/2}v_{\mathrm{mean}}}{\lambda}[1\mathrm{B}.10\mathrm{a}]$$

$$z=\frac{2^{1/2}(475\ \mathrm{m\,s^{-1}})}{8.3\times10^{-8}\ \mathrm{m}}=\boxed{8.1\times10^{9}\ \mathrm{s^{-1}}}$$

E1B.6(a) The volume and radius of a spherical container are related by

$$V = 4\pi r^3/3 \quad \text{so} \quad r = (3V/4\pi)^{1/3}.$$

Thus $d=2r=2(3\times 100\ \text{cm}^3/4\pi)^{1/3}=5.76\ \text{cm}=5.76\times 10^{-2}\ \text{m}$

The mean free path [1B.13] is

$$\lambda = \frac{kT}{\sigma p}$$

Solve this expression for the pressure and set λ equal to the container diameter *d*:

$$p = \frac{kT}{\sigma\lambda} = \frac{(1.381\times 10^{-23}\ \text{J K}^{-1})(293\ \text{K})}{0.36\times(10^{-9}\ \text{m})^2(5.76\times 10^{-2}\ \text{m})} = 0.19\bar{5}\ \text{J m}^{-3} = \boxed{0.19\bar{5}\ \text{Pa}}$$

E1B.7(a) The mean free path [1B.13] is

$$\lambda = \frac{kT}{\sigma p} = \frac{(1.381\times 10^{-23}\ \text{J K}^{-1})(217\ \text{K})}{0.43\times(10^{-9}\ \text{m})^2(0.050\ \text{atm})(1.013\times 10^5\ \text{Pa atm}^{-1})} = \boxed{1.4\times 10^{-6}\ \text{m}}$$

Solutions to problems

P1B.1 The time in seconds for a disk to rotate 360° is the inverse of the frequency. The time for it to advance 2° is $\left(\frac{2°}{360°}\right)\times\frac{1}{\nu}$. This is the time required for slots in neighbouring disks to coincide along the atomic beam. For an atom to pass through all neighbouring slots it must have the speed

$$v_x = \frac{\text{disk spacing}}{\text{alignment time}} = \frac{1.0\ \text{cm}}{\left(\frac{2°}{360°}\right)\times\frac{1}{\nu}} = 180\nu\ \text{cm} = 180\times(\nu/\text{Hz})\ \text{cm s}^{-1}$$

Hence, the distributions of the *x*-component of velocity are

ν/Hz	20	40	80	100	120
v_x/cm s^{-1}	3 600	7 200	14 400	18 000	21 600
I_{exp} (40 K)	0.846	0.513	0.069	0.015	0.002
I_{exp} (100 K)	0.592	0.485	0.217	0.119	0.057

Theoretically, the velocity distribution in the *x*-direction [1B.5] is

$$f(v_x) = \left(\frac{m}{2\pi kT}\right)^{1/2} e^{-mv_x^2/2kT} = \left(\frac{M}{2\pi RT}\right)^{1/2} e^{-Mv_x^2/2RT}$$

Therefore, as $I \propto f, I \propto T^{-1/2}e^{-mv_x^2/2kT}$.

Since $\dfrac{mv_x^2}{2kT}=\dfrac{(83.80\times10^{-3}\,\text{kg mol}^{-1})\{1.80\times(\nu/\text{Hz})\text{m s}^{-1}\}^2}{2(8.3145\,\text{J K}^{-1}\text{mol}^{-1})T}=\dfrac{1.63\times10^{-2}(\nu/\text{Hz})^2}{T/\text{K}}$

we can write $I\propto(T/\text{K})^{-1/2}\,\text{e}^{-1.63\times10^{-2}\times(\nu/\text{Hz})^2/(T/\text{K})}$ and draw up the following table, obtaining the constant of proportionality by fitting I to the value at T=40 K, ν=80 Hz.

ν/Hz	20	40	80	100	120
I_{calc} (40 K)	0.80	0.49	(0.069)	0.016	0.003
I_{calc} (100 K)	0.56	0.46	0.209	0.116	0.057

The calculated values are in fair agreement with the experimental data.

P1B.3 The most probable speed of a gas molecule corresponds to the condition that the Maxwell distribution (eqn 1B.4) be a maximum (it has no minimum); hence we find it by setting the first derivative of the function to zero and solve for the value of v at which this condition holds.

$$f(v)=4\pi\left(\frac{M}{2\pi RT}\right)^{3/2}v^2\text{e}^{-Mv^2/2RT}=\text{const}\times v^2\text{e}^{-Mv^2/2RT}$$

$$\frac{\text{d}f(v)}{\text{d}v}=0\quad\text{when}\quad\left(2-\frac{Mv^2}{RT}\right)=0$$

So, $$v_{\text{mp}}=\boxed{\left(\frac{2RT}{M}\right)^{1/2}}$$ which is eqn 1B.9.

The average translational kinetic energy is

$$\langle E_k\rangle=\langle\tfrac{1}{2}mv^2\rangle=\tfrac{1}{2}m\langle v^2\rangle$$

The average is obtained by substituting the distribution (eqn 1B.4) into eqn 1B.7:

$$\langle v^2\rangle=\int_0^\infty v^2f(v)\text{d}v=4\pi\left(\frac{M}{2\pi RT}\right)^{3/2}\int_0^\infty v^4\text{e}^{-Mv^2/2RT}\text{d}v$$

The integral evaluates to $\dfrac{3\pi^{1/2}}{8}\left(\dfrac{M}{2RT}\right)^{-5/2}$ [Integral G.6]

Then $\langle v^2\rangle=4\pi\left(\dfrac{M}{2\pi RT}\right)^{3/2}\times\dfrac{3\pi^{1/2}}{8}\left(\dfrac{M}{2RT}\right)^{-5/2}=\dfrac{3RT}{M}$ as implied in eqn 1B.3.

Thus $\langle\varepsilon\rangle=\dfrac{m}{2}\langle v^2\rangle=\left(\dfrac{m}{2}\right)\left(\dfrac{3RT}{M}\right)=\tfrac{3}{2}kT$

where we have used $M=mN_A$ and $R=kN_A$.

P1B.5 Call the initial mean velocity a; then in the emerging beam

$$\langle v_x \rangle = K\int_0^a v_x f(v_x)\mathrm{d}v_x$$

where $f(v_x)$ is eqn 1B.5 and K is a constant which ensures that the distribution in the emergent beam is also normalized. That is,

$$1 = K\int_0^a f(v_x)\mathrm{d}v_x = K\left(\frac{m}{2\pi kT}\right)^{1/2}\int_0^a \mathrm{e}^{-mv_x^2/2kT}\,\mathrm{d}v_x$$

This integral cannot be evaluated analytically but it can be related to the error function by defining

$$\chi^2 = \frac{mv_x^2}{2kT} \quad \text{so that} \quad \mathrm{d}v_x = \left(\frac{2kT}{m}\right)^{1/2}\mathrm{d}\chi$$

Then $$1 = K\left(\frac{m}{2\pi kT}\right)^{1/2}\left(\frac{2kT}{m}\right)^{1/2}\int_0^b \mathrm{e}^{-\chi^2}\,\mathrm{d}\chi \quad \left[b = (m/2kT)^{1/2}\times a\right]$$

$$= \frac{K}{\pi^{1/2}}\int_0^b \mathrm{e}^{-\chi^2}\,\mathrm{d}\chi = \tfrac{1}{2}K\mathrm{erf}(b)$$

where erf(z) is the error function [Integral G.6]: $\mathrm{erf}(z) = \left(2/\pi^{1/2}\right)\int_0^z \mathrm{e}^{-\chi^2}\mathrm{d}\chi$.

Therefore, $K = \dfrac{2}{\mathrm{erf}(b)}$.

The mean velocity of the emerging beam is

$$\langle v_x \rangle = K\left(\frac{m}{2\pi kT}\right)^{1/2}\int_0^a v_x \mathrm{e}^{-mv_x^2/2kT}\mathrm{d}v_x = K\left(\frac{m}{2\pi kT}\right)^{1/2}\left(\frac{-kT}{m}\right)\int_0^a \frac{\mathrm{d}}{\mathrm{d}v_x}\left(\mathrm{e}^{-mv_x^2/2kT}\mathrm{d}v_x\right)$$

$$= -K\left(\frac{kT}{2m\pi}\right)^{1/2}\left(\mathrm{e}^{-ma^2/2kT} - 1\right)$$

Now use the expression for $a = \langle v_x \rangle_{\text{initial}}$, namely, as will be shown below, $a = (2kT/m\pi)^{1/2}$. This expression for the average magnitude of the one-dimensional velocity in the x direction may be obtained from

$$\langle v_x \rangle_{\text{initial}} = 2\int_0^\infty v_x f(v_x)\mathrm{d}v_x = 2\int_0^\infty v_x\left(\frac{m}{2\pi kT}\right)^{1/2}\mathrm{e}^{-mv_x^2/2kT}\,\mathrm{d}v_x$$

$$= \left(\frac{m}{2\pi kT}\right)^{1/2}\left(\frac{2kT}{m}\right) = \left(\frac{2kT}{m\pi}\right)^{1/2}$$

It may also be obtained very quickly by setting $a = \infty$ in the expression for $\langle v_x \rangle$ in the emergent beam with $\mathrm{erf}(b) = \mathrm{erf}(\infty) = 1$. Substituting $a = (2kT/m\pi)^{1/2}$ into the expression for the emergent beam yields $\mathrm{e}^{-ma^2/2kT} = \mathrm{e}^{-1/\pi}$ and $\mathrm{erf}(b) = \mathrm{erf}(1/\pi^{1/2})$. Therefore

$$\langle v_x \rangle = \left(\frac{2kT}{m\pi}\right)^{1/2} \times \frac{1 - \mathrm{e}^{-1/\pi}}{\mathrm{erf}\left(1/\pi^{1/2}\right)}.$$

From tables of the error function, or from readily available software,

$$\mathrm{erf}\left(1/\pi^{1/2}\right)=\mathrm{erf}(0.56)=0.57 \qquad \text{and} \qquad e^{-1/\pi}=0.73$$

Therefore, $\langle v_x\rangle = \boxed{0.47\langle v_x\rangle_{\text{initial}}}$

P1B.7 The root mean square speed, v_{rms}, of a perfect gas is given by eqn 1B.3

$$v_{\text{rms}}=\left(\frac{3kT}{m}\right)^{1/2}$$

Consider a range of speeds Δv around v_{rms} and nv_{rms}:

$$\frac{f(nv_{\text{rms}})}{f(v_{\text{rms}})}=\frac{(nv_{\text{rms}})^2 e^{-Mn^2v_{\text{rms}}^2/2RT}}{v_{\text{rms}}^2 e^{-mv_{\text{rms}}^2/2RT}}=n^2e^{-(n^2-1)mv_{\text{rms}}^2/2RT}=\boxed{n^2e^{3(1-n^2)/2}}$$

Therefore,

$$\frac{f(3v_{\text{rms}})}{f(v_{\text{rms}})}=9\times e^{-3\times 8/2}=\boxed{5.53\times10^{-5}} \quad \text{and} \quad \frac{f(4v_{\text{rms}})}{f(v_{\text{rms}})}=16\times e^{-3\times15/2}=\boxed{2.71\times10^{-9}}$$

P1B.9 Newton's gravitational force law: $F=\dfrac{Gmm'}{r^2}$ where G is the gravitational constant and r is the centre-to-centre distance between masses m and m'. In this problem $r\geq R$ where R is a planet radius. The minimum work w required to move an object of mass m from a position near the planet's surface to infinity is

$$w=\int_R^\infty F\,\mathrm{d}r=Gmm'\int_R^\infty\frac{1}{r^2}\mathrm{d}r=-Gmm'\times\left[\frac{1}{r}\right]_{r=R}^{r=\infty}=\left(\frac{Gm'}{R^2}\right)m=mg_{\text{planet}}R$$

where $g_{\text{planet}}=\dfrac{Gm'}{R^2}$ is the gravitational acceleration of a planet and

$$\frac{g_{\text{planet}}}{g_{\text{Earth}}}=\left(\frac{m}{R^2}\right)_{\text{planet}}\Big/\left(\frac{m}{R^2}\right)_{\text{Earth}}.$$

Using data provided in the problem:

$$g_{\text{Mars}}=g_{\text{Earth}}\times\left(\frac{m}{R^2}\right)_{\text{Mars}}\Big/\left(\frac{m}{R^2}\right)_{\text{Earth}}=g_{\text{Earth}}\times\left(\frac{m_{\text{Mars}}}{m_{\text{Earth}}}\right)\times\left(\frac{R_{\text{Earth}}}{R_{\text{Mars}}}\right)^2$$

$$=\left(9.81\,\mathrm{m\,s^{-2}}\right)\times(0.108)\times\left(\frac{6.37}{3.38}\right)^2=3.76\,\mathrm{m\,s^{-2}}$$

The escape speed v_{esc} is determined by the minimum kinetic energy that provides the energy w.

$$\tfrac{1}{2}mv_{\text{esc}}^2=(gR)_{\text{planet}}m \qquad \text{so that} \qquad v_{\text{esc}}=(2gR)_{\text{planet}}^{2}$$

(a) The escape speed for Earth:

$$v_{\text{esc}}=\left\{2\left(9.81\,\mathrm{m\,s^{-2}}\right)\left(6.37\times10^6\,\mathrm{m}\right)\right\}^{1/2}=\boxed{11.2\,\mathrm{km\,s^{-1}}}$$

(b) The escape speed for Mars:

$$v_{esc} = \{2(3.76\,\text{m s}^{-2})(3.38\times10^6\,\text{m})\}^{1/2} = \boxed{5.04\,\text{km s}^{-1}}$$

Since $v_{mean} = (8RT/\pi M)^{1/2}$ [1B.8], the temperature at which the mean gas speed corresponds to the escape speed is given by $T = \pi M v_{esc}^2/8R$ and computed temperature values for hydrogen, helium, and oxygen are summarized in the following table.

$10^{-3}\,T$/K	H_2	He	O_2	
Earth	11.9	23.7	190	$v_{mean} = 12.2\,\text{km s}^{-1}$
Mars	2.4	4.8	38	$v_{mean} = 5.0\,\text{km s}^{-1}$

In order to calculate the proportion *P* of molecules that have speeds exceeding the escape velocity, we must integrate the Maxwell distribution [1B.4] from v_{esc} to infinity. *P* is a function of *M*, *T*, and v_{esc}.

$$P(M,T,v_{esc}) = \int_{v_{esc}}^{\infty} f(v)\text{d}v = 4\pi\left(\frac{m}{2\pi RT}\right)^{3/2}\int_{v_{esc}}^{\infty} v^2 e^{-Mv^2/2RT}\,\text{d}v$$

The integral of this expression has no analytical solution but it is easily evaluated numerically on a scientific calculator or with computer software. Avoid unit errors by using SI units throughout. Here is a Mathcad setup for the computations along with the desired calculations:

$$R := 8.3145 \quad P(M,T,v_{esc}) := 4\cdot\pi\cdot\left(\frac{M}{2\cdot\pi\cdot R\cdot T}\right)^{1/5}\cdot\int_{v_{esc}}^{\infty} v^2\cdot e^{\frac{-M\cdot v^2}{2\cdot R\cdot T}}\,dv$$

For gases of the Earth's atmosphere, $v_{esc} = 11.2\times10^3\,\text{m s}^{-1}$:

Hydrogen at 240 K: $P(0.002016, 240, 11.2\times10^3) = 2.498\times10^{-27}$

Hydrogen at 1500 K: $P(0.002016, 1500, 11.2\times10^3) = 1.487\times10^{-4}$

Helium at 240 K: $P(0.004, 240, 11.2\times10^3) = 3.210\times10^{-54}$

Helium at 1500 K: $P(0.004, 1500, 11.2\times10^3) = 9.516\times10^{-9}$

Oxygen at 240 K: $P(0.032, 240, 11.2\times10^3) = 0.000\times10^{0}$

Oxygen at 1500 K: $P(0.032, 1500, 11.2\times10^3) = 1.888\times10^{-69}$

For gases of the Mars atmosphere, $v_{esc} = 5.04\times10^{3}\ \mathrm{m\,s^{-1}}$:

Hydrogen at 240 K: $P\left(0.002016,\ 240,\ 5.04\times10^{3}\right) = 1.122\times10^{-5}$

Hydrogen at 1500 K: $P\left(0.002016,\ 1500,\ 5.04\times10^{3}\right) = 2.502\times10^{-1}$

Helium at 240 K: $P\left(0.004,\ 240,\ 5.04\times10^{3}\right) = 5.093\times10^{-11}$

Helium at 1500 K: $P\left(0.004,\ 1500,\ 5.04\times10^{3}\right) = 4.307\times10^{-2}$

Oxygen at 240 K: $P\left(0.032,\ 240,\ 5.04\times10^{3}\right) = 5.659\times10^{-88}$

Oxygen at 1500 K: $P\left(0.032,\ 1500,\ 5.04\times10^{3}\right) = 4.246\times10^{-14}$

Based on these numbers alone, it would appear that H_2 and He would be depleted from the atmosphere of both Earth and Mars only after many (millions?) years; that the rate on Mars, though still slow, would be many orders of magnitude larger than on Earth; that O_2 would be retained on Earth indefinitely; and that the rate of O_2 depletion on Mars would be very slow (billions of years?), though not totally negligible. The temperatures of both planets may have been higher in past times than they are now.

P1B.11 On a spreadsheet or other mathematical software, make a column of velocity values and then several columns for $f(v)$ [1B.4] at different temperatures. Figure 1B.1 shows $f(v)$ plotted against v for temperatures of 200, 300, 400, 600, 800, 1000, 1500, and 2000 K. Each curve is labelled with the numerical value of T/K.

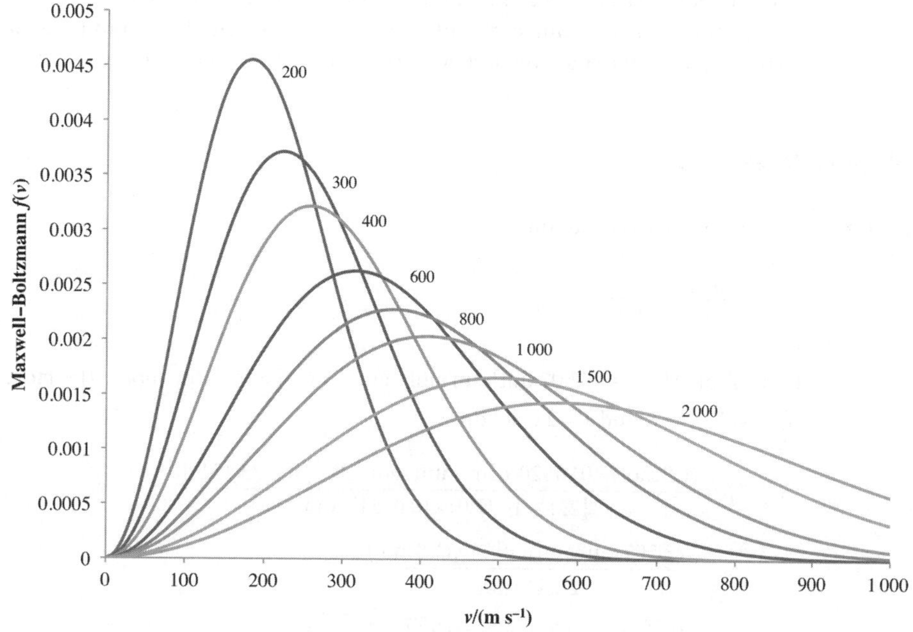

Figure 1B.1

1C Real gases

Answers to discussion questions

D1C.1 Consider three temperature regions:

(1) $T < T_B$. At very low pressures, all gases show a compression factor, $Z \approx 1$. At high pressures, all gases have $Z > 1$, signifying that they have a molar volume greater than a perfect gas, which implies that repulsive forces are dominant. At intermediate pressures, most gases show $Z < 1$, indicating that attractive forces reducing the molar volume below the perfect value are dominant.

(2) $T \approx T_B$. $Z \approx 1$ at low pressures, slightly greater than 1 at intermediate pressures, and significantly greater than 1 only at high pressures. There is a balance between the attractive and repulsive forces at low to intermediate pressures, but the repulsive forces predominate at high pressures where the molecules are very close to each other.

(3) $T > T_B$. $Z > 1$ at all pressures because the frequency of collisions between molecules increases with temperature.

D1C.3 The van der Waals equation 'corrects' the perfect gas equation for both attractive and repulsive interactions between the molecules in a real gas. See *Justification* 1.1 for a fuller explanation.

The Berthelot equation accounts for the volume of the molecules in a manner similar to the van der Waals equation but the term representing molecular attractions is modified to account for the effect of temperature. Experimentally one finds that the van der Waals a decreases with increasing temperature. Theory (see Chapter 16) also suggests that intermolecular attractions can decrease with temperature. This variation of the attractive interaction with temperature can be accounted for in the equation of state by replacing the van der Waals a with a/T.

Solutions to exercises

E1C.1(a) The van der Waals equation [1C.5a] is

$$p = \frac{nRT}{V - nb} - \frac{an^2}{V^2}$$

From Table 1C.3, $a = 5.507\ \mathrm{dm^6\,atm\,mol^{-2}}$ and $b = 6.51 \times 10^{-2}\ \mathrm{dm^3\,mol^{-1}}$. Therefore,

(i) $T = 273.15$ K and $V = 22.414\ \mathrm{dm^3}$

$$p = \frac{(1.0\,\mathrm{mol}) \times (0.08206\ \mathrm{dm^3\ atm\ mol^{-1}\ K^{-1}}) \times (273.15\mathrm{K})}{[22.414 - (1.0) \times (0.0651)]\ \mathrm{dm^3}} - \frac{(5.507\,\mathrm{dm^6\,atm\,mol^{-2}}) \times (1.0\,\mathrm{mol})^2}{(22.414\,\mathrm{dm^3})^2}$$

$$= 1.0\overline{03}\ \mathrm{atm} - 0.011\ \mathrm{atm} = 0.9\overline{92} = \boxed{1.0\ \mathrm{atm}}$$

(ii) $T = 1000$ K and $V = 100\ \text{cm}^3$

$$p = \frac{(1.0\,\text{mol})\times(0.08206\ \text{dm}^3\ \text{atm mol}^{-1}\ \text{K}^{-1})\times(1000\,\text{K})}{(0.100-0.0651)\text{dm}^3} - \frac{(5.507\,\text{dm}^6\,\text{atm mol}^{-2})\times(1.0\,\text{mol})^2}{(0.100\,\text{dm}^3)^2}$$

$$= 2.3\bar{5}\times10^3\ \text{atm} - 5.5\bar{1}\times10^2\ \text{atm} = \boxed{1.8\times10^3\ \text{atm}}$$

E1C.2(a) The conversions needed are as follows:

$1\ \text{atm} = 1.013\times10^5\,\text{Pa}, 1\,\text{Pa} = 1\,\text{kg}\,\text{m}^{-1}\,\text{s}^{-2}, 1\,\text{dm}^6 = (10^{-1}\,\text{m})^6 = 10^{-6}\,\text{m}^6, 1\,\text{dm}^3 = 10^{-3}\,\text{m}^3.$

Therefore,

$$a = 0.751\ \text{atm dm}^6\ \text{mol}^{-2}\times\frac{1.013\times10^5\ \text{kg m}^{-1}\ \text{s}^{-2}}{1\ \text{atm}}\times\frac{10^{-6}\ \text{m}^6}{\text{dm}^6}$$

$$= \boxed{7.61\times10^{-2}\ \text{kg m}^5\ \text{s}^{-2}\ \text{mol}^{-2}}$$

and $b = 0.0226\,\text{dm}^3\,\text{mol}^{-1}\times\dfrac{10^{-3}\,\text{m}^3}{\text{dm}^3} = \boxed{2.26\times10^{-5}\,\text{m}^3\,\text{mol}^{-1}}$

E1C.3(a) The compression factor Z is [1C.1]

$$Z = \frac{V_\text{m}}{V_\text{m}^\circ} = \frac{pV_\text{m}}{RT}$$

V_m is the actual molar volume, V_m° is the perfect gas molar volume. $V_\text{m}^\circ = \dfrac{RT}{p}$.

Since V_m is 12 per cent smaller than that of a perfect gas, $V_\text{m} = 0.88V_\text{m}^\circ$, and

(a) $Z = \dfrac{0.88V_\text{m}^\circ}{V_\text{m}^\circ} = \boxed{0.88}$

(b) $V_\text{m} = \dfrac{ZRT}{p} = \dfrac{(0.88)\times(0.08206\ \text{dm}^3\ \text{atm mol}^{-1}\ \text{K}^{-1})\times(250\ \text{K})}{15\ \text{atm}} = \boxed{1.2\,\text{dm}^3\ \text{mol}^{-1}}$

Since $V_\text{m} < V_\text{m}^\circ$ $\boxed{\text{attractive}}$ forces dominate.

E1C.4(a) The amount of gas is first determined from its mass; then the van der Waals equation [1C.5a] is used to determine its pressure at the working temperature. The initial conditions of 300 K and 100 atm are in a sense superfluous information.

$$n = \frac{92.4\times10^3\ \text{g}}{28.02\ \text{g mol}^{-1}} = 3.30\times10^3\ \text{mol}$$

$$V = 1.000\ \text{m}^3 = 1.000\times10^3\,\text{dm}^3$$

$$p=\frac{nRT}{V-nb}-\frac{an^2}{V^2}=\frac{(3.30\times10^3\,\text{mol})\times(0.08206\,\text{dm}^3\ \text{atm}\,\text{mol}^{-1}\,\text{K}^{-1})\times(500\,\text{K})}{(1.000\times10^3\,\text{dm}^3)-(3.30\times10^3\,\text{mol})\times(0.037\,\text{dm}^3\,\text{mol}^{-1})}$$

$$-\frac{(1.352\,\text{dm}^6\,\text{atm}\,\text{mol}^{-2})\times(3.30\times10^3\,\text{mol})^2}{(1.000\times10^3\,\text{dm}^3)^2}$$

$$=(155-14.7)\,\text{atm}=\boxed{140\,\text{atm}}$$

E1C.5(a) (a) According to the perfect gas law

$$p=\frac{nRT}{V}=\frac{(10.0\,\text{mol})\times(0.08206\,\text{dm}^3\ \text{atm}\,\text{mol}^{-1}\,\text{K}^{-1})\times(300\,\text{K})}{4.860\,\text{dm}^3}=\boxed{50.7\,\text{atm}}$$

(b) According to the van der Waals equation [1C.5a]

$$p=\frac{nRT}{V-nb}-a\left(\frac{n}{V}\right)^2=\frac{(10.0\,\text{mol})\times(0.08206\,\text{dm}^3\ \text{atm}\,\text{mol}^{-1}\,\text{K}^{-1})\times(300\,\text{K})}{(4.860\,\text{dm}^3)-(10.0\,\text{mol})\times(0.06380\,\text{dm}^3\,\text{mol}^{-1})}$$

$$-(5.489\,\text{dm}^6\,\text{atm}\,\text{mol}^{-2})\times\left(\frac{10.0\,\text{mol}}{4.860\,\text{dm}^3}\right)^2$$

$$=58.3-23.2=\boxed{35.1\,\text{atm}}$$

The compression factor is calculated from eqn 1C.2. That equation requires a value for the pressure, p, in particular the actual pressure as determined experimentally. This pressure is neither the perfect gas pressure nor the van der Waals pressure. However, on the assumption that the van der Waals equation provides a value for the pressure close to the experimental value, we can estimate the compression factor as follows

$$Z=\frac{pV}{nRT}=\frac{(35.1\,\text{atm})\times(4.860\,\text{dm}^3)}{(10.0\,\text{mol})\times(0.08206\,\text{dm}^3\ \text{atm}\,\text{mol}^{-1}\,\text{K}^{-1})\times(300\,\text{K})}=\boxed{0.692}$$

E1C.6(a) Equations 1C.6 are solved for b and a, respectively, and yield

$$b=V_c/3 \quad \text{and} \quad a=27b^2p_c=3V_c^2p_c.$$

Substituting the critical constants

$$b=\frac{98.7\,\text{cm}^3\,\text{mol}^{-1}}{3}=32.9\,\text{cm}^3\,\text{mol}^{-1}=0.0329\,\text{dm}^3\,\text{mol}^{-1}$$

and $a=3\times(98.7\times10^{-3}\,\text{dm}^3\,\text{mol}^{-1})^2\times(45.6\,\text{atm})=1.33\,\text{dm}^6\,\text{atm}\,\text{mol}^{-2}$

But this problem is overdetermined. We have another piece of information

$$T_c=\frac{8a}{27Rb}$$

If we use T_c along with V_c as above, we would arrive at the same value of b along with

$$a=\frac{27RbT_c}{8}=\frac{9RV_cT_c}{8}$$

$$=\frac{9(0.08206\ \text{dm}^3\ \text{atm mol}^{-1}\ \text{K}^{-1})(98.7\times10^{-3}\ \text{dm}^3\ \text{mol}^{-1})(190.6\ \text{K})}{8}$$

$$=1.74\ \text{dm}^6\ \text{atm mol}^{-2}$$

Or we could use T_c along with p_c. In that case, we can solve the pair of equations for a and b by first setting the two expressions for a equal to each other:

$$a=27b^2p_c=\frac{27RbT_c}{8}$$

Solving the resulting equation for b yields

$$b=\frac{RT_c}{8p_c}=\frac{(0.08206\ \text{dm}^3\ \text{atm mol}^{-1}\ \text{K}^{-1})(190.6\ \text{K})}{8(45.6\ \text{atm})}=0.0429\ \text{dm}^3\ \text{mol}^{-1}$$

and then

$$a=27(0.0429\ \text{dm}^3\ \text{mol}^{-1})^2(45.6\ \text{atm})=2.26\ \text{dm}^6\ \text{atm mol}^{-2}$$

These results are summarized in the following table

Using	$a/\text{dm}^6\ \text{atm mol}^{-2}$	$b/\text{dm}^3\ \text{mol}^{-1}$
V_c & p_c	1.33	0.0329
V_c & T_c	1.74	0.0329
p_c & T_c	2.26	0.0429

One way of selecting best values for these parameters would be to take the mean of the three determinations, namely $a=\boxed{1.78\ \text{dm}^6\ \text{atm mol}^{-2}}$ and $b=\boxed{0.0362\ \text{dm}^3\ \text{mol}^{-1}}$. By interpreting b as the excluded volume of a mole of spherical molecules, we can obtain an estimate of molecular size. The centres of spherical particles are excluded from a sphere whose radius is the diameter of those spherical particles (i.e. twice their radius); that volume times the Avogadro constant is the molar excluded volume b

$$b=N_A\left(\frac{4\pi(2r)^3}{3}\right)\quad\text{so}\quad r=\frac{1}{2}\left(\frac{3b}{4\pi N_A}\right)^{1/3}$$

$$r=\frac{1}{2}\left(\frac{3(0.0362\ \text{dm}^3\ \text{mol}^{-1})}{4\pi(6.022\times10^{23}\ \text{mol}^{-1})}\right)^{1/3}=1.22\times10^{-9}\ \text{dm}=\boxed{0.122\ \text{nm}}$$

E1C.7(a) The Boyle temperature, T_B, is the temperature at which the virial coefficient $B=0$. In order to express T_B in terms of a and b, the van der Waals equation [1C.5b] must be recast into the form of the virial equation.

$$p=\frac{RT}{V_m-b}-\frac{a}{V_m^2}$$

Factoring out $\frac{RT}{V_m}$ yields $p=\frac{RT}{V_m}\left\{\frac{1}{1-b/V_m}-\frac{a}{RTV_m}\right\}$

So long as $b/V_m<1$, the first term inside the brackets can be expanded using

$$(1-x)^{-1}=1+x+x^2+\ldots,$$

which gives

$$p=\frac{RT}{V_m}\left\{1+\left(b-\frac{a}{RT}\right)\times\left(\frac{1}{V_m}\right)+\cdots\right\}$$

We can now identify the second virial coefficient as $B=b-\frac{a}{RT}$

At the Boyle temperature

$$B=0=b-\frac{a}{RT_B} \quad \text{so} \quad T_B=\frac{a}{bR}=\frac{27T_c}{8}$$

(a) From Table 1C.3, $a=6.260\ \text{dm}^6\ \text{atm mol}^{-2}$ and $b=0.0542\ \text{dm}^3\ \text{mol}^{-1}$. Therefore,

$$T_B=\frac{6.260\ \text{dm}^6\ \text{atm mol}^{-2}}{(5.42\times10^{-2}\ \text{dm}^3\ \text{mol}^{-1})\times(0.08206\ \text{dm}^3\ \text{atm mol}^{-1}\ \text{K}^{-1})}=\boxed{1.41\times10^3\ \text{K}}$$

(b) As in Exercise 1C.6(a),

$$b=N_A\left(\frac{4\pi(2r)^3}{3}\right) \quad \text{so} \quad r=\frac{1}{2}\left(\frac{3b}{4\pi N_A}\right)^{1/3}$$

$$r=\frac{1}{2}\left(\frac{3(0.0542\ \text{dm}^3\ \text{mol}^{-1})}{4\pi(6.022\times10^{23}\ \text{mol}^{-1})}\right)^{1/3}=1.39\times10^{-9}\ \text{dm}=\boxed{0.139\ \text{nm}}$$

E1C.8(a) The reduced temperature and pressure of hydrogen are calculated from the relations

$$T_r=\frac{T}{T_c} \quad \text{and} \quad p_r=\frac{p}{p_c} \quad [1C.8]$$

$$T_r=\frac{298\ \text{K}}{33.23\ \text{K}}=8.97\ [T_c=33.23\ \text{K, Table 1C.2}]$$

$$p_r=\frac{1.0\ \text{atm}}{12.8\ \text{atm}}=0.078\ [p_c=12.8\ \text{atm, Table 1C.2}]$$

Hence, the gases named will be in corresponding states at $T=8.97T_c$ and at $p=0.078p_c$.

(a) For ammonia, $T_c=405.5$ K and $p_c=111.3$ atm, so

$$T=8.97(405.5\,\text{K})=\boxed{3.64\times10^3\,\text{K}}$$

$$p=0.078(111.3\,\text{atm})=\boxed{8.7\,\text{atm}}$$

(b) For xenon, $T_c=289.75$ K and $p_c=58.0$ atm, so

$$T=8.97(289.75\,\text{K})=\boxed{2.62\times10^3\,\text{K}}$$

$$p=0.078(58.0\,\text{atm})=\boxed{4.5\,\text{atm}}$$

(c) For helium, $T_c=5.21$ K and $p_c=2.26$ atm, so

$$T=8.97(5.2\,\text{K})=\boxed{47\,\text{K}}$$

$$p=0.078(2.26\,\text{atm})=\boxed{0.18\,\text{atm}}$$

E1C.9(a) The van der Waals equation [1C.5b] is solved for b, which yields

$$b=V_\text{m}-\frac{RT}{\left(p+\dfrac{a}{V_\text{m}^2}\right)}$$

Substituting the data

$$b=5.00\times10^{-4}\,\text{m}^3\,\text{mol}^{-1}-\frac{(8.3145\,\text{J K}^{-1}\,\text{mol}^{-1})\times(273\text{K})}{\left\{(3.0\times10^6\,\text{Pa})+\left(\dfrac{0.50\,\text{m}^6\,\text{Pa mol}^{-2}}{(5.00\times10^{-4}\,\text{m}^3\,\text{mol}^{-1})^2}\right)\right\}}$$

$$=\boxed{0.46\times10^{-4}\,\text{m}^3\,\text{mol}^{-1}}$$

$$Z=\frac{pV_\text{m}}{RT}[1\text{C}.2]=\frac{(3.0\times10^6\,\text{Pa})\times(5.00\times10^{-4}\,\text{m}^3\,\text{mol}^{-1})}{(8.3145\,\text{J K}^{-1}\,\text{mol}^{-1})\times(273\text{K})}=\boxed{0.66}$$

Comment. The definition of Z involves the actual pressure, volume, and temperature and does not depend upon the equation of state used to relate these variables.

Solutions to problems

P1C.1 (a) According to the perfect gas law

$$V_\text{m}=\frac{RT}{p}=\frac{(0.08206\,\text{dm}^3\,\text{atm mol}^{-1}\,\text{K}^{-1})\times(350\,\text{K})}{2.30\,\text{atm}}=\boxed{12.5\,\text{dm}^3\,\text{mol}^{-1}}$$

(b) From the van der Waals equation [1C.5b] $p=\dfrac{RT}{V_m-b}-\dfrac{a}{V_m^2}$,

we obtain $V_m=\dfrac{RT}{p+\dfrac{a}{V_m^2}}+b$.

Then, with a and b from Table 1C.3

$$V_m\approx\frac{(0.08206\,\text{dm}^3\ \text{atm mol}^{-1}\ \text{K}^{-1})\times(350\,\text{K})}{(2.30\,\text{atm})+\left(\dfrac{6.260\,\text{dm}^3\ \text{atm mol}^{-2}}{(12.5\,\text{dm}^3\ \text{mol}^{-1})^2}\right)}+5.42\times10^{-2}\ \text{dm}^3\ \text{mol}^{-1}$$

$$\approx\frac{28.7\,\text{dm}^3\ \text{mol}^{-1}}{2.34}+5.42\times10^{-2}\ \text{dm}^3\ \text{mol}^{-1}\approx\boxed{12.3\,\text{dm}^3\ \text{mol}^{-1}}$$

Substitution of 12.3 $\text{dm}^3\,\text{mol}^{-1}$ into the denominator of the first expression again results in $V_m=12.3\ \text{dm}^3\,\text{mol}^{-1}$, so the cycle of approximation may be terminated.

P1C.3 As indicated by the parts of eqn 1C.3, the compression factor of a gas may be expressed as either a virial expansion in p or in $\left(\dfrac{1}{V_m}\right)$. The virial form of the van der Waals equation is derived in Exercise 1C.7(a) and is $p=\frac{RT}{V_m}\left\{1+\left(b-\frac{a}{RT}\right)\times\left(\frac{1}{V_m}\right)+\cdots\right\}$

Rearranging, $Z=\dfrac{pV_m}{RT}=1+\left(b-\frac{a}{RT}\right)\times\left(\dfrac{1}{V_m}\right)+\cdots$

On the assumption that the perfect gas expression for V_m is adequate for the second term in this expansion, we can readily obtain Z as a function of p.

$$Z=1+\left(\frac{1}{RT}\right)\times\left(b-\frac{a}{RT}\right)p+\cdots$$

(a) $T=T_c=126.3$ K

$$V_m=\left(\frac{RT}{p}\right)\times Z=\frac{RT}{p}+\left(b-\frac{a}{RT}\right)+\cdots$$

$$=\frac{(0.08206\,\text{dm}^3\ \text{atm mol}^{-1}\ \text{K}^{-1})\times(126.3\text{K})}{10.0\,\text{atm}}$$

$$+\left\{(0.0391\,\text{dm}^3\text{mol}^{-1})-\left(\frac{1.390\,\text{dm}^6\text{atm mol}^{-2}}{(0.08206\,\text{dm}^3\ \text{atm mol}^{-1}\ \text{K}^{-1})\times(126.3\,\text{K})}\right)\right\}$$

$$=(1.036-0.095)\,\text{dm}^3\,\text{mol}^{-1}=\boxed{0.941\,\text{dm}^3\,\text{mol}^{-1}}$$

$$Z=\left(\frac{p}{RT}\right)\times(V_m)=\frac{(10.0\,\text{atm})\times(0.941\text{dm}^3\,\text{mol}^{-1})}{(0.08206\,\text{dm}^3\ \text{atm mol}^{-1}\ \text{K}^{-1})\times(126.3\text{K})}=0.908$$

(b) The Boyle temperature is the temperature at which the second virial coefficient is zero, hence $Z=1$ (correct to the first power in p), and the gas is close to perfect. However, if we assume that N_2 is a van der Waals gas, when the second virial coefficient is zero

$$\left(b-\frac{a}{RT_B}\right)=0, \quad \text{or} \quad T_B=\frac{a}{bR}$$

$$T_B=\frac{1.390\,\text{dm}^6\,\text{atm mol}^{-2}}{(0.0391\,\text{dm}^3\,\text{mol}^{-1})\times(0.08206\,\text{dm}^3\,\text{atm mol}^{-1}\,\text{K}^{-1})}=433\,\text{K}$$

The experimental value (Table 1C.2) is 327.2 K. Two possible resolutions of the discrepancy spring to mind.

1. Terms beyond the first power in p should not be dropped in the expansion for Z.
2. Nitrogen is only approximately a van der Waals gas.

When $Z=1$, $V_m=\dfrac{RT}{p}$, and using $T_B=327.2\,\text{K}$

$$V_m=\frac{(0.08206\,\text{dm}^3\,\text{atm mol}^{-1}\,\text{K}^{-1})\times 327.2\,\text{K}}{10.0\,\text{atm}}=\boxed{2.69\,\text{dm}^3\,\text{mol}^{-1}}$$

This is the ideal value of V_m. Using the experimental value of T_B and inserting this value into the expansion for V_m above, we have

$$V_m=\frac{0.08206\,\text{dm}^3\,\text{atm mol}^{-1}\,\text{K}^{-1}\times 327.2\,\text{K}}{10.0\,\text{atm}}$$

$$+\left\{0.0391\,\text{dm}^3\,\text{mol}^{-1}-\left(\frac{1.390\,\text{dm}^6\,\text{atm mol}^{-2}}{0.08206\,\text{dm}^3\,\text{atm mol}^{-1}\,\text{K}^{-1}\times 327.2\,\text{K}}\right)\right\}$$

$$=(2.68\bar{5}-0.013)\,\text{dm}^3\,\text{mol}^{-1}=\boxed{2.67\,\text{dm}^3\,\text{mol}^{-1}}$$

and $$Z=\frac{V_m}{V_m^{\circ}}=\frac{2.67\,\text{dm}^3\,\text{mol}^{-1}}{2.69\,\text{dm}^3\,\text{mol}^{-1}}=0.995\approx 1$$

(c) See Section 2D.2(a) and Table 2D.2 for inversion temperature: $T=T_I=621$ K

$$V_m=\frac{0.08206\,\text{dm}^3\,\text{atm mol}^{-1}\,\text{K}^{-1}\times 621\,\text{K}}{10.0\,\text{atm}}$$

$$+\left\{0.0391\,\text{dm}^3\,\text{mol}^{-1}-\left(\frac{1.390\,\text{dm}^6\,\text{atm mol}^{-2}}{0.08206\,\text{dm}^3\,\text{atm K}^{-1}\,\text{mol}^{-1}\times 621\,\text{K}}\right)\right\}$$

$$=(5.09\bar{6}+0.012)\,\text{dm}^3\,\text{mol}^{-1}=\boxed{5.11\,\text{dm}^3\,\text{mol}^{-1}}$$

and $$Z=\frac{5.11\,\text{dm}^3\,\text{mol}^{-1}}{5.10\,\text{dm}^3\,\text{mol}^{-1}}=1.002\approx 1$$

Based on the values of T_B and T_I given in Tables 1C.2 and 2D.2 and assuming that N_2 is a van der Waals gas, the calculated value of Z is closest to 1 at the

$\boxed{\text{inversion temperature}}$, but the difference from the value at T_B is less than the accuracy of the method.

P1C.5 (a) $V_m = \dfrac{V}{n} = V \times \dfrac{M}{m} = \dfrac{M}{\rho} = \dfrac{18.02\,\text{g mol}^{-1}}{133.2\,\text{g dm}^{-3}} = \boxed{0.1353\,\text{dm}^3\,\text{mol}^{-1}}$

(b) $Z = \dfrac{pV_m}{RT}\,[1\text{C}.2] = \dfrac{(327.6\,\text{atm})\times(0.1353\,\text{dm}^3\,\text{mol}^{-1})}{(0.08206\,\text{dm}^3\,\text{atm}\,\text{K}^{-1}\,\text{mol}^{-1})\times(776.4\,\text{K})} = \boxed{0.6957}$

(c) Two expansions for Z based on the van der Waals equation are given in Problem 1C.3. They are

$$Z = 1 + \left(b - \frac{a}{RT}\right)\times\left(\frac{1}{V_m}\right) + \cdots$$

$$= 1 + \left\{(0.03049\,\text{dm}^3\,\text{mol}^{-1}) - \left(\frac{5.464\,\text{dm}^6\,\text{atm}\,\text{mol}^{-2}}{(0.08206\,\text{dm}^3\,\text{atm}\,\text{K}^{-1}\,\text{mol}^{-1})\times(776.4\,\text{K})}\right)\right\}$$

$$\times \frac{1}{0.1353\,\text{dm}^3\,\text{mol}^{-1}} = 1 - 0.4085 = 0.5915$$

and

$$Z = 1 + \left(\frac{1}{RT}\right)\times\left(b - \frac{a}{RT}\right)\times(p) + \cdots$$

$$= 1 + \frac{1}{(0.08206\,\text{dm}^3\,\text{atm}\,\text{K}^{-1}\,\text{mol}^{-1})\times(776.4\,\text{K})}$$

$$\times\left\{(0.03049\,\text{dm}^3\,\text{mol}^{-1}) - \left(\frac{5.464\,\text{dm}^6\,\text{atm}\,\text{mol}^{-2}}{(0.08206\,\text{dm}^3\,\text{atm}\,\text{K}^{-1}\,\text{mol}^{-1})\times(776.4\,\text{K})}\right)\right\}\times 327.6\,\text{atm}$$

$$= 1 - 0.2842 \approx \boxed{0.7158}$$

In this case the expansion in p gives a value close to the experimental value; the expansion in $\dfrac{1}{V_m}$ is not as good. However, when terms beyond the second are included the results from the two expansions for Z converge.

P1C.7 $V_c = 2b, \quad T_c = \dfrac{a}{4bR}$ [Table 1C.4]

Hence, with V_c and T_c from Table 1C.2,

$$b = \tfrac{1}{2}V_c = \tfrac{1}{2}\times(118.8\,\text{cm}^3\,\text{mol}^{-1}) = \boxed{59.4\,\text{cm}^3\,\text{mol}^{-1}}$$

$$a = 4bRT_c = 2RT_cV$$

$$= (2)\times(0.08206\,\text{dm}^3\,\text{atm}\,\text{mol}^{-1}\,\text{K}^{-1})\times(289.75\,\text{K})\times(118.8\times10^{-3}\,\text{dm}^3\,\text{mol}^{-1})$$

$$= \boxed{5.649\,\text{dm}^6\,\text{atm}\,\text{mol}^{-2}}$$

Hence

$$p=\frac{RT}{V_m-b}e^{-a/RTV_m}$$
$$=\frac{(8.206\times10^{-2}\ \mathrm{dm^3\ atm\ K^{-1}\ mol^{-1}})\times(298\ \mathrm{K})}{(1.0\ \mathrm{dm^3\ mol^{-1}})-(59.4\times10^{-3}\ \mathrm{dm^3\ mol^{-1}})}$$
$$\times\exp\left(\frac{-(5.649\ \mathrm{dm^6\ atm\ mol^{-2}})}{(0.08206\ \mathrm{dm^3\ atm\ mol^{-1}\ K^{-1}})\times(298\ \mathrm{K})\times(1.0\ \mathrm{dm^3\ mol^{-1}})}\right)$$
$$=26.\bar{0}\ \mathrm{atm}\times e^{-0.23\bar{1}}=\boxed{21\ \mathrm{atm}}$$

P1C.9 This expansion has already been given in the solution to Exercise 1C.7(a); the result is

$$p=\frac{RT}{V_m}\left(1+\left[b-\frac{a}{RT}\right]\frac{1}{V_m}+\frac{b^2}{V_m^2}+\cdots\right)$$

Compare this expansion with the virial expansion [1C.3b]

$$p=\frac{RT}{V_m}\left(1+\frac{B}{V_m}+\frac{C}{V_m^2}+\cdots\right)$$

and hence find $\boxed{B=b-\frac{a}{RT}}$ and $\boxed{C=b^2}$

Since $C=1200\ \mathrm{cm^6\ mol^{-2}}$, $b=C^{1/2}=\boxed{34.6\ \mathrm{cm^3\ mol^{-1}}}$

$$a=RT(b-B)=(8.206\times10^{-2})\times(273\ \mathrm{dm^3\ atm\ mol^{-1}})\times(34.6+21.7)\ \mathrm{cm^3\ mol^{-1}}$$
$$=(22.4\bar{0}\ \mathrm{dm^3\ atm\ mol^{-1}})\times(56.3\times10^{-3}\ \mathrm{dm^3 mol^{-1}})=\boxed{1.26\ \mathrm{dm^6\ atm\ mol^{-2}}}$$

P1C.11 The critical point corresponds to a point of zero slope which is simultaneously a point of inflection in a plot of pressure versus molar volume. A critical point exists if there are values of p, V, and T that result in a point which satisfies these conditions.

$$p=\frac{RT}{V_m}-\frac{B}{V_m^2}+\frac{C}{V_m^3}$$

$$\left.\begin{aligned}\left(\frac{\partial p}{\partial V_m}\right)_T&=-\frac{RT}{V_m^2}+\frac{2B}{V_m^3}-\frac{3C}{V_m^4}=0\\ \left(\frac{\partial^2 p}{\partial V_m^2}\right)_T&=\frac{2RT}{V_m^3}-\frac{6B}{V_m^4}+\frac{12C}{V_m^5}=0\end{aligned}\right\}\text{at the critical point}$$

That is, at the critical point

$$\left.\begin{aligned}-RT_cV_c^2+2BV_c-3C=0\\ RT_cV_c^2-3BV_c+6C=0\end{aligned}\right\}$$

which solve to $V_c = \boxed{\dfrac{3C}{B}}$, $T_c = \boxed{\dfrac{B^2}{3RC}}$

Now use the equation of state to find p_c

$$p_c = \frac{RT_c}{V_c} - \frac{B}{V_c^2} + \frac{C}{V_c^3} = \left(\frac{RB^2}{3RC}\right)\times\left(\frac{B}{3C}\right) - B\left(\frac{B}{3C}\right)^2 + C\left(\frac{B}{3C}\right)^3 = \boxed{\frac{B^3}{27C^2}}$$

It follows that $Z_c = \dfrac{p_c V_c}{RT_c} = \left(\dfrac{B^3}{27C^2}\right)\times\left(\dfrac{3C}{B}\right)\times\left(\dfrac{1}{R}\right)\times\left(\dfrac{3RC}{B^2}\right) = \boxed{\dfrac{1}{3}}$

P1C.13 For a real gas we may use the virial expansion in terms of p [1C.3a]

$$p = \frac{nRT}{V}(1+B'p+\text{L}) = \rho\frac{RT}{M}(1+B'p+\cdots) \quad \left[\frac{n}{V} = \frac{\rho}{M}\right]$$

which rearranges to $\dfrac{p}{\rho} = \dfrac{RT}{M} + \dfrac{RTB'}{M}p + \cdots$

Therefore, the limiting slope of a plot of $\dfrac{p}{\rho}$ against p is $\dfrac{B'RT}{M}$.

Draw up the following table. Bear in mind that 1 kPa $= 10^3$ kg m^{-1} s^{-2}.

p/kPa	12.223	25.20	36.97	60.37	85.23	101.3
ρ/(kg m^{-3})	0.225	0.456	0.664	1.062	1.468	1.734
$\dfrac{p/\rho}{10^3\ \text{m}^2\ \text{s}^{-2}}$	54.3	55.3	55.7	56.8	58.1	58.4

$\dfrac{p}{\rho}$ is plotted in Fig. 1C.1.

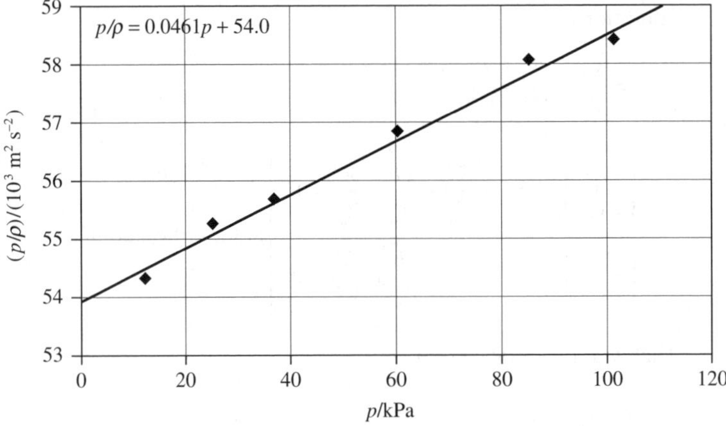

Figure 1C.1

A straight line fits the data rather well. From Fig. 1C.1, the slope is

$$\frac{B'RT}{M} = 0.0461\times10^3\ \text{m}^2\ \text{s}^{-2}\ \text{kPa}^{-1} = 0.0461\ \text{m}^3\ \text{kg}^{-1}$$

and the y-intercept is

$$\frac{RT}{M} = 5.40\times10^4\ \text{m}^2\ \text{s}^{-2}$$

hence $\quad B' = \dfrac{4.61\times10^{-2}\ \text{kg}^{-1}\ \text{m}^3}{5.40\times10^4\ \text{m}^2\ \text{s}^{-2}} = 0.854\times10^{-6}\ \text{Pa}^{-1}$

or $\quad B' = (0.854\times10^{-6}\ \text{Pa}^{-1})\times(1.0133\times10^5\ \text{Pa atm}^{-1}) = \boxed{0.0866\ \text{atm}^{-1}}$.

Comparing equations 1C.3a and 1C.3b, we see

$$B = RTB' = (0.08206\ \text{dm}^3\ \text{atm mol}^{-1}\ \text{K}^{-1})\times(298\text{K})\times(0.0866\ \text{atm}^{-1})$$
$$= \boxed{2.12\ \text{dm}^3\ \text{mol}^{-1}}$$

Comment. The relation between B' and B is part of Problem 1C.12. However, since these terms involve only the first power in p and $1/V_m$, respectively, the relationship between them can be obtained simply by substituting the ideal-gas relationship between p and $1/V_m$, namely $p=RT/V_m$. Deviations from this relationship would find their way into higher powers of p and $1/V_m$.

P1C.15 The critical temperature is that temperature above which the gas cannot be liquefied by the application of pressure alone. Below the critical temperature, two phases, liquid and gas, may coexist at equilibrium; and in the two-phase region there is more than one molar volume corresponding to the same conditions of temperature and pressure. Therefore, any equation of state that can even approximately describe this situation must allow for more than one real root for the molar volume at some values of T and p, but as the temperature is increased above T_c, allows only one real root. Thus, appropriate equations of state must be equations of odd degree in V_m.

The equation of state for gas A may be rewritten $V_m^2 - \dfrac{RT}{p}V_m - \dfrac{RTb}{p} = 0$, which is a quadratic and never has just one real root. Thus, this equation can never model critical behaviour. It could possibly model in a very crude manner a two-phase situation, since there are some conditions under which a quadratic has two real positive roots, but not the process of liquefaction.

The equation of state of gas B is a first-degree equation in V_m; because such an equation has only one root, it can never model critical behaviour, the process of liquefaction, or the existence of a two-phase region.

A cubic equation is the equation of lowest degree which can show a cross-over from more than one real root to just one real root as the temperature increases. The van der Waals equation is a cubic equation in V_m.

P1C.17 The two masses represent the same volume of gas under identical conditions, and therefore, the same number of molecules (Avogadro's principle) and moles, n. Thus, the masses can be expressed as

$$nM_{\mathrm{N}} = 2.2990\,\mathrm{g}$$

for 'chemical nitrogen' and

$$n_{\mathrm{Ar}}M_{\mathrm{Ar}} + n_{\mathrm{N}}M_{\mathrm{N}} = n\{x_{\mathrm{Ar}}M_{\mathrm{Ar}} + (1-x_{\mathrm{Ar}})M_{\mathrm{N}}\} = 2.3102\,\mathrm{g}$$

for 'atmospheric nitrogen'. Dividing the latter expression by the former yields

$$\frac{x_{\mathrm{Ar}}M_{\mathrm{Ar}}}{M_{\mathrm{N}}} + (1-x_{\mathrm{Ar}}) = \frac{2.3102}{2.2990} \quad \text{so} \quad x_{\mathrm{Ar}}\left(\frac{M_{\mathrm{Ar}}}{M_{\mathrm{N}}} - 1\right) = \frac{2.3102}{2.2990} - 1$$

$$\text{and } x_{\mathrm{Ar}} = \frac{\dfrac{2.3102}{2.2990} - 1}{\dfrac{M_{\mathrm{Ar}}}{M_{\mathrm{N}}} - 1} = \frac{\dfrac{2.3102}{2.2990} - 1}{\dfrac{39.95\,\mathrm{g\,mol^{-1}}}{28.013\,\mathrm{g\,mol^{-1}}} - 1} = \boxed{0.011}$$

Comment. This value for the mole fraction of argon in air is close to the modern value of 0.9. (See S. Y. Park *et al.*, *Metrologia* **41**, 387 (2004).)

P1C.19 (a) and (b) Plots of pressure vs. volume are shown at the two temperatures in Fig. 1C.2(a).

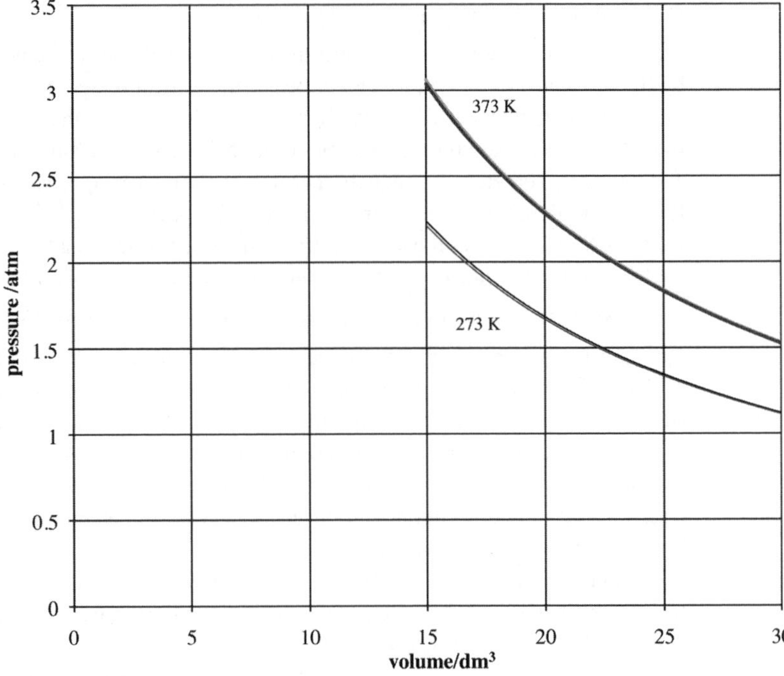

Figure 1C.2(a)

In fact, at each temperature, two nearly superimposed curves are plotted, one using the perfect gas law [1A.5] and one the van der Waals equation [1C.5a] and parameters for CO_2 from Table 1C.3. At these volumes, the perfect gas pressure is very slightly greater, by about 1% or less.

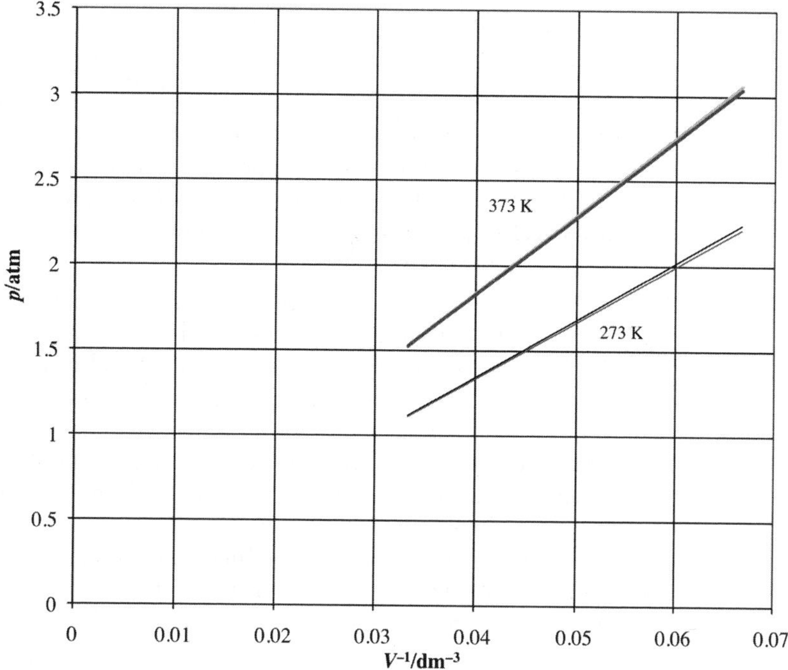

Figure 1C.2(b)

The same pressures are plotted against $1/V$ in Fig. 1C.2(b). According to the perfect gas law, the plot should be linear. In fact, the plots of van der Waals pressure vs. $1/V$ are also very nearly linear under these conditions.

Question. Try the same exercise with volumes smaller by a factor of 10. Are the van der Waals and perfect gas curves significantly different from each other?

P1C.21 The compression factor [1C.2] is

$$Z = \frac{pV_{\mathrm{m}}}{RT}$$

Substituting the van der Waals equation [1C.5b] for p yields

$$Z = \frac{V_{\mathrm{m}}}{RT}\left(\frac{RT}{V_{\mathrm{m}}-b} - \frac{a}{V_m^2}\right) = \frac{V_{\mathrm{m}}}{V_{\mathrm{m}}-b} - \frac{a}{RTV_{\mathrm{m}}}$$

To find a minimum, find the value of V_{m} that makes the derivative of Z vanish:

$$\frac{\mathrm{d}Z}{\mathrm{d}V_{\mathrm{m}}} = 0 = \frac{1}{V_{\mathrm{m}}-b} - \frac{V_{\mathrm{m}}}{(V_{\mathrm{m}}-b)^2} + \frac{a}{RTV_{\mathrm{m}}^2} = \frac{-b}{(V_{\mathrm{m}}-b)^2} + \frac{a}{RTV_{\mathrm{m}}^2}$$

This is a quadratic equation in V_m, which can be rearranged into the following standard form

$$(a-bRT)V_m^2-2abV_m+ab^2=0$$

This can be solved using the quadratic formula:

$$V_m=\frac{2ab\pm\{4a^2b^2-4(a-bRT)ab^2\}^{1/2}}{2(a-bRT)}=\boxed{\frac{b\{1\pm(bRT/a)^{1/2}\}}{1-bRT/a}}$$

Note that this expression for the position of the minimum depends on the van der Waals parameters *a* and *b* as well as the temperature. For most of the gases listed in Table 1C.3, $bRT/a<1$ at 298 K. (Exceptions include H_2, He, and Ne.) If $bRT/a<1$, then both roots of V_m are positive. In that case, however, one would correspond to a maximum and one to a minimum.

Whether or not there is a physically reasonable minimum is quite another matter from whether or not there is a mathematical minimum in *Z*. See, for example, Fig. 1C.3, which shows plots of *Z* vs. V_m at $T=298$ K, as calculated from the van der Waals equation for Ne, Ar, and NH_3 (with parameters taken from Table 1C.3). As noted above, $bRT/a>1$ for Ne, so V_m has only one positive root. That root, however, occurs at $V_m<b$, thus leading to a negative value of the pressure—not physically reasonable. NH_3 has two positive roots, one of which occurs at $V_m<b$; however, even the root with $V_m>b$ yields a value of $Z<0$—not physically reasonable. That is, *Z* for NH_3 has a minimum at a physically reasonable value of V_m, but the compression factor there is not reasonable. Finally, Ar has a physically reasonable minimum in *Z* at a physically reasonable value of V_m.

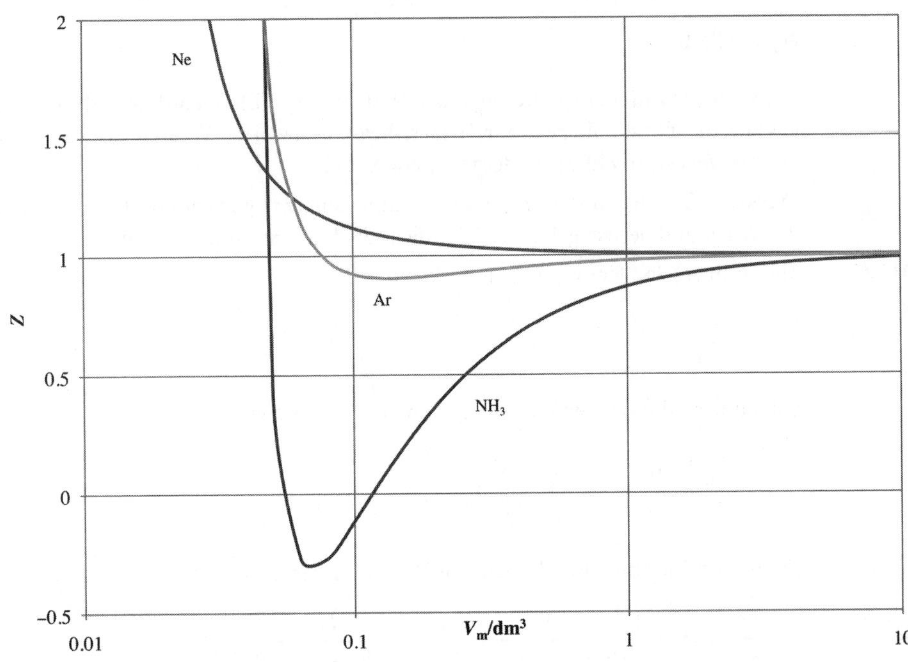

Figure 1C.3

2 The First Law

Assume all gases are perfect unless stated otherwise. Unless otherwise stated, thermochemical data are for 298.15 K.

2A Internal energy

Answers to discussion questions

D2A.1 In physical chemistry, the universe is considered to be divided into two parts: the system and its surroundings. In thermodynamics, the system is the object of interest which is separated from its surroundings, the rest of the universe, by a boundary. The characteristics of the boundary determine whether the system is open, closed, or isolated. An open system has a boundary that permits the passage of both matter and energy. A closed system has a boundary that allows the passage of energy but not of matter. Closed systems can be either adiabatic or diathermic. The former do not allow the transfer of energy as a result of a temperature difference, but the latter do. An isolated system is one with a boundary that allows neither the transfer of matter nor energy between the system and the surroundings.

In thermodynamics, the state of a system is characterized by state functions which are thermodynamic properties that are independent of the previous history of the system. In quantum mechanics, the state of a system is characterized by a time-independent wave function.

D2A.3 See Table 2A.1, which lists four varieties of work: expansion, surface expansion, extension, and electrical. There is also work associated with processes in magnetic and gravitational fields which we will not describe in detail.

Solutions to exercises

E2A.1(a) See *Foundations B* for a brief description of the equipartition theorem. For a more complete explanation of the basis of the theorem and when it can be used see Chapter 15 where the following equation is developed and its application explained.

$$C_{V,\mathrm{m}} = \tfrac{1}{2}(3 + \nu_{\mathrm{R}}^{*} + 2\nu_{\mathrm{V}}^{*})R$$

with a mode active if $T > \theta_M$

(i) $\nu_R^* = 2,\ \nu_V^* \approx 1$; hence $C_{V,m} = \frac{1}{2}(3+2+2\times1)R = \boxed{\frac{7}{2}R}$ [experimental = 3.4R]

Hence the energy is

$$\frac{7}{2}RT = \frac{7}{2}\times 8.314\ \text{J K}^{-1}\ \text{mol}^{-1}\times 298.15\ \text{K} = 8\,671\ \text{J mol}^{-1} = \boxed{8.671\ \text{kJ mol}^{-1}}$$

Note that I_2 has quite a low vibrational wavenumber, so

$$\theta_V = \frac{hc\tilde{\nu}}{k} = \frac{(6.626\times10^{-34}\ \text{J s})(2.998\times10^{10}\ \text{cm s}^{-1})(214\ \text{cm}^{-1})}{1.381\times10^{-23}\ \text{J K}^{-1}} = 308\ \text{K}$$

The temperature specified for this exercise is less than this, but only slightly. When the temperature is close to the vibrational temperature, the mode is closer to active than inactive. Hence, applying the equipartition theorem may not be adequate in this case.

(ii) $\nu_R^* = 3,\ \nu_V^* \approx 0$; hence $C_{V,m} = \frac{1}{2}(3+3+0)R = \boxed{3R}$ [experimental = 3.2R]

Hence the energy is

$$E = 3RT = 3\times 8.314\ \text{J K}^{-1}\ \text{mol}^{-1}\times 298.15\ \text{K} = 7\,436\ \text{J mol}^{-1} = \boxed{7.436\ \text{kJ mol}^{-1}}$$

(iii) $\nu_R^* = 3,\ \nu_V^* \approx 4$; hence $C_{V,m} = \frac{1}{2}(3+3+2\times4)R = \boxed{7R}$ [experimental = 8.8R]

Hence the energy is

$$E = 7RT = 8\times 8.314\ \text{J K}^{-1}\ \text{mol}^{-1}\times 298.15\ \text{K} = \boxed{17.35\ \text{kJ mol}^{-1}}$$

Note that data from the book by G. H. Herzberg (*Molecular Spectra and Molecular Structure II. Infrared and Raman Spectra of Polyatomic Molecules,* D. van Nostrand Company, Inc., Princeton, NJ, 1945, pp 364–365) give low vibrational wavenumbers for four modes of benzene, and we have included these above. There are 26 more vibrational modes we have neglected, taking them to be inactive. Slight activity from these modes accounts for the difference of about 1.8R in the heat capacity between the experimental value and our estimate.

E2A.2(a) (i) pressure, (ii) temperature, and (iv) enthalpy are state functions.

E2A.3(a) This is an expansion against a constant external pressure; hence $w = -p_{ex}\Delta V$ [2A.6]

$$p_{ex} = (1.0\ \text{atm})\times(1.013\times10^5\ \text{Pa atm}^{-1}) = 1.0\bar{1}\times10^5\ \text{Pa}$$

The change in volume is the cross-sectional area times the linear displacement:

$$\Delta V = (50\ \text{cm}^2)\times(15\ \text{cm})\times\left(\frac{1\ \text{m}}{100\ \text{cm}}\right)^3 = 7.5\times10^{-4}\ \text{m}^3$$

so $\quad w = -(1.01\times10^5\ \text{Pa})\times(7.5\times10^{-4}\ \text{m}^3) = \boxed{-75\ \text{J}}$ as $1\ \text{Pa m}^3 = 1\ \text{J}$

E2A.4(a) For all cases $\Delta U = 0$, since the internal energy of a perfect gas depends only on temperature. From the definition of enthalpy, $H = U + pV$, so $\Delta H = \Delta U + \Delta(pV) = \Delta U + \Delta(nRT)$ (perfect gas). Hence, $\Delta H = 0$ as well, at constant temperature for all processes in a perfect gas.

(i) $\boxed{\Delta U = \Delta H = 0}$

$$w = -nRT\ln\left(\frac{V_f}{V_i}\right) \text{ [2A.9]}$$

$$= -(1.00\text{ mol})\times(8.314\text{ J K}^{-1}\text{ mol}^{-1})\times(293\text{ K})\times\ln\left(\frac{30.0\text{ dm}^3}{10.0\text{ dm}^3}\right)$$

$$= -2.68\times10^3\text{ J} = \boxed{-2.68\text{ kJ}}$$

$$q = \Delta U - w \text{ [First Law]} = 0 + 2.68\text{ kJ} = \boxed{+2.68\text{ kJ}}$$

(ii) $\boxed{\Delta U = \Delta H = 0}$

$$w = -p_{ex}\Delta V \text{ [2A.6]} \quad \Delta V = (30.0 - 10.0)\text{ dm}^3 = 20.0\text{ dm}^3$$

p_{ex} can be computed from the perfect gas law $pV = nRT$.

$$\text{so } p_{ex} = p_f = \frac{nRT}{V_f} = \frac{(1.00\text{ mol})\times(0.08206\text{ dm}^3\text{ atm K}^{-1}\text{ mol}^{-1})\times(293\text{ K})}{30.0\text{ dm}^3} = 0.801\text{ atm}$$

$$w = -(0.801\text{ atm})\times\left(\frac{1.013\times10^5\text{ Pa}}{1\text{ atm}}\right)\times(20.0\text{ dm}^3)\times\left(\frac{1\text{ m}^3}{10^3\text{ dm}^3}\right)$$

$$= -1.62\times10^3\text{ Pa m}^3 = -1.62\times10^3\text{ J} = \boxed{-1.62\text{ kJ}}$$

$$q = \Delta U - w = 0 + 1.62\text{ kJ} = \boxed{+1.62\text{ kJ}}$$

(iii) $\boxed{\Delta U = \Delta H = 0}$

Free expansion is expansion against no force, so $\boxed{w = 0}$ and $q = \Delta U - w = 0 - 0 = \boxed{0}$

Comment. An isothermal free expansion of a perfect gas is also adiabatic.

E2A.5(a) For a perfect gas at constant volume

$$\frac{p}{T} = \frac{nR}{V} = \text{constant,} \quad \text{hence,} \quad \frac{p_1}{T_1} = \frac{p_2}{T_2}$$

$$p_2 = \left(\frac{T_2}{T_1}\right)\times p_1 = \left(\frac{400\text{ K}}{300\text{ K}}\right)\times(1.00\text{ atm}) = \boxed{1.33\text{ atm}}$$

$$\Delta U = nC_{V,\text{m}}\Delta T \;[2.16\text{b}] = (n)\times\left(\tfrac{3}{2}R\right)\times(400\,\text{K}-300\,\text{K})$$
$$= (1.00\,\text{mol})\times\left(\tfrac{3}{2}\right)\times(8.314\,\text{J K}^{-1}\,\text{mol}^{-1})\times(100\,\text{K})$$
$$= 1.25\times10^{3}\,\text{J} = \boxed{+1.25\,\text{kJ}}$$

$\boxed{w=0}$ [constant volume] $\quad q = \Delta U - w$ [First Law] $= 1.25\,\text{kJ} - 0 = \boxed{+1.25\,\text{kJ}}$

E2A.6(a) (i) $w = -p_{\text{ex}}\Delta V \qquad [2\text{A}.6]$

$$p_{\text{ex}} = (200\,\text{Torr})\times(133.3\,\text{Pa Torr}^{-1}) = 2.66\bar{6}\times10^{4}\,\text{Pa}$$

$$\Delta V = 3.3\,\text{dm}^3 = 3.3\times10^{-3}\,\text{m}^3$$

Therefore, $w = (-2.66\bar{6}\times10^{4}\,\text{Pa})\times(3.3\times10^{-3}\,\text{m}^3) = \boxed{-88\,\text{J}}$

(ii) $w = -nRT\ln\left(\dfrac{V_\text{f}}{V_\text{i}}\right)$ [2A.9]

$$n = \frac{4.50\,\text{g}}{16.04\,\text{g mol}^{-1}} = 0.280\bar{5}\,\text{mol},\quad RT = 2.577\,\text{kJ mol}^{-1},$$

$$V_\text{i} = 12.7\,\text{dm}^3,\; V_\text{f} = 16.0\,\text{dm}^3$$

$$w = -(0.280\bar{5}\,\text{mol})\times(2.577\,\text{kJ mol}^{-1})\times\ln\left(\frac{16.0\,\text{dm}^3}{12.7\,\text{dm}^3}\right) = \boxed{-167\,\text{J}}$$

Solutions to problems

P2A.1 $w = -\int_{V_1}^{V_2} p\,\text{d}V \quad$ with $\quad p = \dfrac{nRT}{V-nb} - \dfrac{n^2a}{V^2} \quad$ [Topic 1C]

Therefore,

$$w = -nRT\int_{V_1}^{V_2}\frac{\text{d}V}{V-nb} + n^2a\int_{V_1}^{V_2}\frac{\text{d}V}{V^2} = \boxed{-nRT\,\ln\left(\frac{V_2-nb}{V_1-nb}\right) - n^2a\left(\frac{1}{V_2}-\frac{1}{V_1}\right)}$$

This expression can be interpreted more readily if we assume $V >> nb$, which is certainly valid at all but the highest pressures. Then using the first term of the Taylor series expansion,

$$\ln(1-x) = -x - \frac{x^2}{2}\ldots \quad \text{for } |x| << 1$$

$$\ln(V-nb) = \ln V + \ln\left(1-\frac{nb}{V}\right) \approx \ln V - \frac{nb}{V}$$

and, after substitution

$$w \approx -nRT\ln\left(\frac{V_2}{V_1}\right)+n^2bRT\left(\frac{1}{V_2}-\frac{1}{V_1}\right)-n^2a\left(\frac{1}{V_2}-\frac{1}{V_1}\right)$$

$$\approx -nRT\ln\left(\frac{V_2}{V_1}\right)-n^2(a-bRT)\left(\frac{1}{V_2}-\frac{1}{V_1}\right)$$

$$\approx +w_0-n^2(a-bRT)\left(\frac{1}{V_2}-\frac{1}{V_1}\right)=\text{perfect gas value + van der Waals correction.}$$

w_0, the perfect gas value, is negative in expansion and positive in compression. Considering the correction term, in expansion $V_2>V_1$, so $\left(\frac{1}{V_2}-\frac{1}{V_1}\right)<0$. If attractive forces predominate, $a>bRT$ and the work done *by* the van der Waals gas is less in magnitude (less negative) than the perfect gas—the gas cannot easily expand. If repulsive forces predominate, $bRT>a$ and the work done *by* the van der Waals gas is greater in magnitude than the perfect gas—the gas easily expands. In the numerical calculations, consider a doubling of the initial volume.

(a) $w_0=-nRT\ln\left(\frac{V_f}{V_i}\right)=(-1.0\,\text{mol}^{-1})\times(8.314\,\text{J K}^{-1}\,\text{mol}^{-1})\times(298\,\text{K})\times\ln\left(\frac{2.0\,\text{dm}^3}{1.0\,\text{dm}^3}\right)$

$$w_0=-1.7\bar{2}\times10^3\,\text{J}=\boxed{-1.7\,\text{kJ}}$$

(b) $w=w_0-(1.0\,\text{mol})^2\times[0-(5.11\times10^{-2}\,\text{dm}^3\,\text{mol}^{-1})\times(8.314\,\text{J K}^{-1}\text{mol}^{-1})\times(298\,\text{K})]$

$$\times\left(\frac{1}{2.0\,\text{dm}^3}-\frac{1}{1.0\,\text{dm}^3}\right)=(-1.7\bar{2}\times10^3\,\text{J})-(63\,\text{J})=-1.7\bar{8}\times10^3\,\text{J}=\boxed{-1.8\,\text{kJ}}$$

(c) $w=w_0-(1.0\,\text{mol})^2\times(4.2\,\text{dm}^6\,\text{atm mol}^{-2})\times\left(\frac{1}{2.0\,\text{dm}^3}-\frac{1}{1.0\,\text{dm}^3}\right)$

$$w=w_0+2.1\,\text{dm}^3\,\text{atm}$$

$$=(-1.7\bar{2}\times10^3\,\text{J})+(2.1\,\text{dm}^3\,\text{atm})\times\left(\frac{1\,\text{m}}{10\,\text{dm}}\right)^3\times\left(\frac{1.01\times10^5\,\text{Pa}}{1\,\text{atm}}\right)$$

$$=(-1.7\bar{2}\times10^3\,\text{J})+(0.21\times10^3\,\text{J})=\boxed{-1.5\,\text{kJ}}$$

Schematically, the indicator diagrams for the cases (a), (b), and (c) would appear as in Fig. 2A.6 of the text. For case (b) the pressure is always greater than the perfect gas pressure and for case (c) always less. Therefore,

$$\int_{V_1}^{V_2} p\,\mathrm{d}V(\text{c})<\int_{V_1}^{V_2} p\,\mathrm{d}V(\text{a})<\int_{V_1}^{V_2} p\,\mathrm{d}V(\text{b})$$

and we see that $w(\text{b})>w(\text{a})>w(\text{c})$. See Fig. 2A.1.

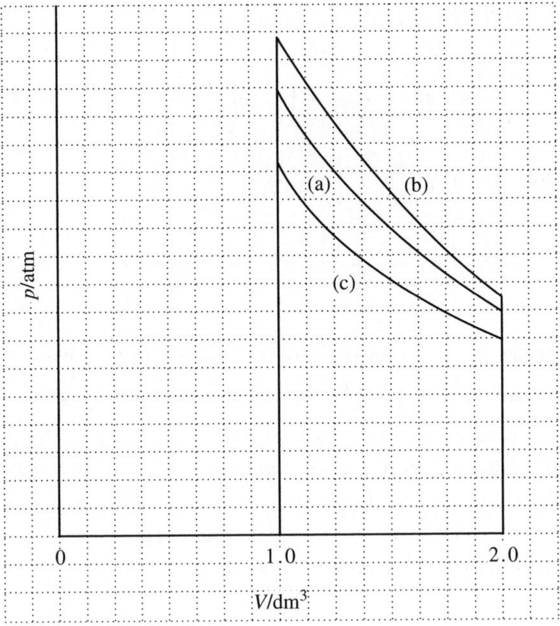

Figure 2A.1

P2A.3 $w = -\int_{V_1}^{V_2} p\,\mathrm{d}V$

Inserting $\dfrac{V}{n} = V_{\mathrm{m}}$ into the virial equation for p we obtain

$$p = nRT\left(\frac{1}{V} + \frac{nB}{V^2} + \frac{n^2C}{V^3} + \ldots\right)$$

Therefore, $$w = -nRT\int_{V_1}^{V_2}\left(\frac{1}{V} + \frac{nB}{V^2} + \frac{n^2C}{V^3} + \ldots\right)\mathrm{d}V$$

$$w = -nRT\ln\frac{V_2}{V_1} + n^2RTB\left(\frac{1}{V_2} - \frac{1}{V_1}\right) + \frac{1}{2}n^3RTC\left(\frac{1}{V_2^2} - \frac{1}{V_2^2}\right) + \ldots$$

For $n = 1$ mol: $nRT = (1.0\,\mathrm{mol}) \times (8.314\,\mathrm{J\,K^{-1}\,mol^{-1}}) \times (273\,\mathrm{K}) = 2.2\bar{7}\,\mathrm{kJ}$

From Table 1.C1, $B = -21.7\ \mathrm{cm^3\,mol^{-1}}$ and $C = 1200\ \mathrm{cm^6\,mol^{-2}}$, so

$$n^2BRT = (1.0\ \mathrm{mol}) \times (-21.7\,\mathrm{cm^3\,mol^{-1}}) \times (2.2\bar{7}\,\mathrm{kJ}) = -49.\bar{3}\,\mathrm{kJ\,cm^3}$$
$$\tfrac{1}{2}n^3CRT = \tfrac{1}{2}(1.0\ \mathrm{mol})^2 \times (1200\ \mathrm{cm^6\,mol^{-2}}) \times (2.2\bar{7}\,\mathrm{kJ}) = +1362\,\mathrm{kJ\,cm^6}$$

Therefore,

(a) $$w = -2.2\bar{7}\,\mathrm{kJ}\ln 2 - (49.\bar{3}\,\mathrm{kJ}) \times \left(\frac{1}{1\,000} - \frac{1}{500}\right) + (1\,3\overline{62}\,\mathrm{kJ}) \times \left(\frac{1}{1\,000^2} - \frac{1}{500^2}\right)$$
$$= (-1.5\bar{7}) + (0.049) - (4.1\times10^{-3})\,\mathrm{kJ} = -1.5\bar{2}\,\mathrm{kJ} = \boxed{-1.5\,\mathrm{kJ}}$$

(b) A perfect gas corresponds to the first term of the expansion of p, so

$$w = -1.5\bar{7}\,\text{kJ} = \boxed{-1.6\,\text{kJ}}.$$

P2A.5 One major limitation of Hooke's law is that it applies to displacements from a single equilibrium value of the end-to-end distance. In fact, if a DNA molecule or any other macromolecular chain that is susceptible to strong nonbonding intramolecular interactions is disturbed sufficiently from one equilibrium configuration, it is likely to settle into a different equilibrium configuration, a so-called 'local minimum' in potential energy. Hooke's law is a good approximation for systems that have a single equilibrium configuration corresponding to a single minimum in potential energy. Another limitation is the assumption that it is just as easy (or as difficult) to move the ends away from each other in any direction. In fact, the intramolecular interactions would be quite different depending on whether one were displacing an end along the chain or outward from the chain. (See Fig. 2A.2)

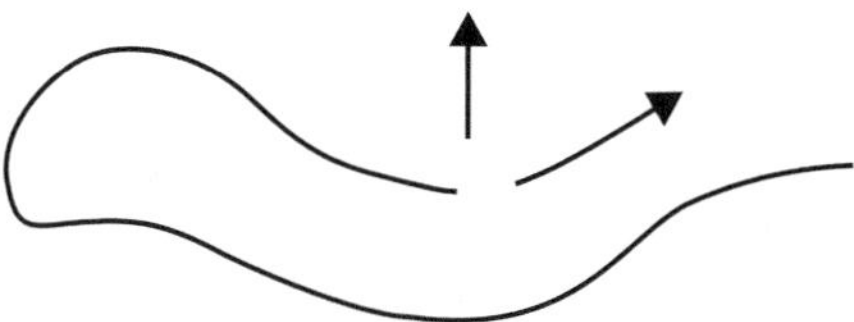

Figure 2A.2

Work is $dw = -Fdx = +k_F\, x\, dx$. This integrates to

$$w = \int_0^{x_f} k_F x\,dx = \tfrac{1}{2}k_F x^2\Big|_0^{x_f} = \boxed{\tfrac{1}{2}k_F x_f^2} \text{ (See Fig. 2A.3.)}$$

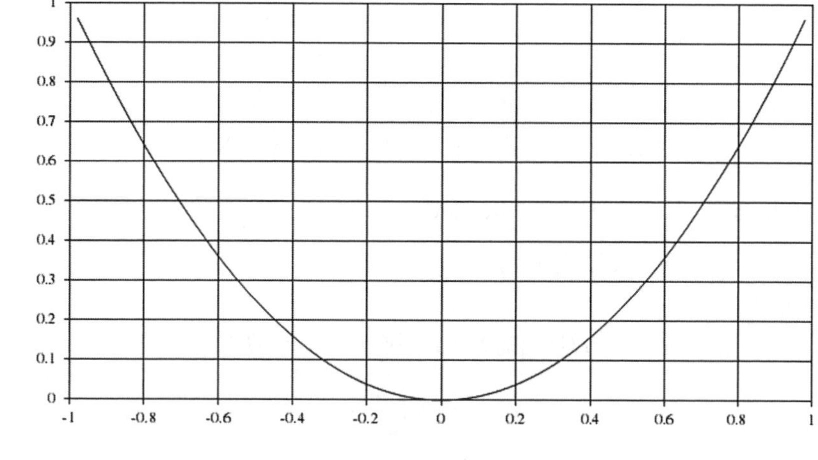

Figure 2A.3

P2A.7 (a) For $\nu << 1$, the natural log can be expanded: $\ln(1+\nu) \approx \nu$ and $\ln(1-\nu) \approx -\nu$. Therefore

$$|F| = \frac{kT}{2l}\ln\left(\frac{1+\nu}{1-\nu}\right) = \frac{kT}{2l}[\ln(1+\nu) - \ln(1-\nu)]$$
$$\approx \frac{kT}{2l}[\nu - (-\nu)] = \frac{\nu kT}{l} = \frac{nkT}{Nl} = \frac{xkT}{Nl^2}$$

(b) Figure 2A.3 in Problem 2A.6 already suggested what the derivation in part (a) confirms: that the one-dimensional chain model and Hooke's law have the same behaviour for small displacements.

Part (a) allows us to identify $\frac{kT}{Nl^2}$ as the Hooke's law force constant.

2B Enthalpy

Answer to discussion question

D2B.1 The difference results from the definition $H = U + pV$; hence $\Delta H = \Delta U + \Delta(pV)$. As $\Delta(PV)$ is not usually zero, except for isothermal processes in a perfect gas, the difference between ΔH and ΔU is a nonzero quantity. As shown in Topic 2B of the text, ΔH can be interpreted as the heat associated with a process at constant pressure and ΔU as the heat at constant volume.

Solutions to exercises

E2B1(a) $C_p = \frac{q_p}{\Delta T}[2\text{B}.7] = \frac{229\,\text{J}}{2.55\,\text{K}} = 89.8\,\text{J}\,\text{K}^{-1}$

so $C_{p,\text{m}} = (89.8\,\text{J}\,\text{K}^{-1})/(3.0\,\text{mol}) = \boxed{30\,\text{J}\,\text{K}^{-1}\,\text{mol}^{-1}}$.

For a perfect gas $C_{p,\text{m}} - C_{V,\text{m}} = R$ [2B.9]

$$C_{V,\text{m}} = C_{p,\text{m}} - R = (30 - 8.3)\,\text{J}\,\text{K}^{-1}\,\text{mol}^{-1} = \boxed{22\,\text{J}\,\text{K}^{-1}\,\text{mol}^{-1}}$$

E2B.2(a) (i) $q = \Delta H$, since pressure is constant

$$\Delta H = \int_{T_\text{i}}^{T_\text{f}} \text{d}H, \quad \text{d}H = nC_{p,\text{m}}\text{d}T$$

$$\text{d}(H/\text{J}) = \{20.17 + 0.3665(T/\text{K})\}\text{d}(T/\text{K})$$

$$\Delta(H/\text{J}) = \int_{T_\text{i}}^{T_\text{f}}(H/\text{J}) = \int_{298}^{373}\{20.17 + 0.3665(T/\text{K})\}\text{d}(T/\text{K})$$
$$= (20.17)\times(373-298) + \left(\frac{0.3665}{2}\right)\times\left(\frac{T}{\text{K}}\right)^2\Bigg|_{298}^{373}$$
$$= (1.51\bar{3}\times10^3) + (9.22\bar{2}\times10^3) = 1.07\times10^4$$

$$q = \Delta H = \boxed{1.07\times10^4\ \text{J}} = \boxed{+10.7\,\text{kJ}} \text{ for 1.00 mol}$$

$$w = -p_{\text{ex}}\Delta V \quad [2\text{A}.6] \quad \text{where} \quad p_{\text{ex}} = p$$

$$w = -p\Delta V = -\Delta(pV)\,[\text{constant pressure}] = -\Delta(nRT)[\text{perfect gas}] = -nR\Delta T$$
$$= (-1.00\,\text{mol})\times(8.314\,\text{J}\,\text{K}^{-1}\,\text{mol}^{-1})\times(373\,\text{K} - 298\,\text{K})$$
$$= \boxed{-0.624\times10^3\ \text{J}} = \boxed{-0.624\,\text{kJ}}$$

$$\Delta U = q + w = (10.7\,\text{kJ}) - (0.624\,\text{kJ}) = \boxed{+10.1\,\text{kJ}}$$

(ii) The energy and enthalpy of a perfect gas depend on temperature alone; hence it does not matter whether the temperature change is brought about at constant volume or constant pressure; ΔH and ΔU are the same.

$$\Delta H = \boxed{+10.7\ \text{kJ}}, \quad \Delta U = \boxed{+10.1\ \text{kJ}}$$

Under constant volume, $w = \boxed{0}$

$$q = \Delta U - w = \boxed{+10.1\ \text{kJ}}$$

E2B.3(a) $q_p = C_p\Delta T\,[2\text{B}.7] = nC_{p,\text{m}}\Delta T = (3.0\,\text{mol})\times(29.4\,\text{J}\,\text{K}^{-1}\,\text{mol}^{-1})\times(25\,\text{K}) = \boxed{+2.2\ \text{kJ}}$

$\Delta H = q_p\ [2\text{B}.2] = \boxed{+2.2\ \text{kJ}}$

$$\Delta U = \Delta H - \Delta(pV)[\text{from } H \equiv U + pV] = \Delta H - \Delta(nRT)[\text{perfect gas}] = \Delta H - nR\Delta T$$
$$= (2.2\,\text{kJ}) - (3.0\,\text{mol})\times(8.314\,\text{J}\,\text{K}^{-1}\,\text{mol}^{-1})\times(25\,\text{K}) = (2.2\,\text{kJ}) - (0.62\,\text{kJ}) = \boxed{+1.6\,\text{kJ}}$$

Solutions to problems

P2B.1 The change in the molar enthalpy of formation of SO_2 as the temperature increases from 300 K to 1500 K is calculated as follows:

$$\Delta(\Delta_{\text{f}}H^{\ominus}) = \int_{300}^{1500} C_{p,\text{m}}^{\ominus}(T)\,\text{d}T$$

$C_{p,\text{m}}^{\ominus}(T)$ can be obtained by fitting the heat capacity data to an equation of the form of eqn 2B.8. When this is done using mathematical software we find

$$C_{p,\text{m}}^{\ominus}(T)/(\text{J}\,\text{K}^{-1}\,\text{mol}^{-1}) = 48.01 + 6.535\times10^{-3}\,T - 9.294\times10^{5}\,T^{-2}$$

Inserting this expression and performing the integration we obtain

$$\Delta(\Delta_{\text{f}}H^{\ominus}) = \boxed{62.2\,\text{kJ}\,\text{mol}^{-1}}$$

P2B.3 Since the volume is fixed, $\boxed{w=0}$

Since $\Delta U = q$ at constant volume, $\boxed{\Delta U = +2.35\ \text{kJ}}$

$$\Delta H = \Delta U + \Delta(pV) = \Delta U + V\Delta p \quad [\Delta V = 0]$$

From the van der Waals equation [1C.5a]

$$p = \frac{RT}{V_\text{m} - b} - \frac{a}{V_\text{m}^2} \quad \text{so} \quad \Delta p = \frac{R\Delta T}{V_\text{m} - b} \quad [\Delta V_m = 0 \text{ at constant volume}]$$

Therefore, $\Delta H = \Delta U + \dfrac{RV\Delta T}{V_\text{m} - b}$

From the data,

$$V_m = \frac{15.0\,\text{dm}^3}{2.0\ \text{mol}} = 7.5\,\text{dm}^3\,\text{mol}^{-1},\ \Delta T = (341 - 300)\,\text{K} = 41\,\text{K}$$

$$V_\text{m} - b = (7.5 - 4.3\times 10^{-2})\ \text{dm}^3\ \text{mol}^{-1} = 7.4\bar{6}\,\text{dm}^3\ \text{mol}^{-1}$$

$$\frac{RV\Delta T}{V_\text{m} - b} = \frac{(8.314\,\text{J}\,\text{K}^{-1}\ \text{mol}^{-1})\times(15.0\,\text{dm}^3)\times(41\,\text{K})}{7.4\bar{6}\ \text{dm}^3\,\text{mol}^{-1}} = 0.68\ \text{kJ}$$

Therefore, $\Delta H = (2.35\ \text{kJ}) + (0.68\ \text{kJ}) = \boxed{+3.03\ \text{kJ}}$

2C Thermochemistry

Answer to discussion question

D2C.1 When a system is subjected to constant pressure conditions, and only expansion work can occur, the energy supplied as heat is the change in enthalpy of the system. Thus enthalpy changes in the system can be determined by measuring the amount of heat supplied. The calorimeters used for measuring these heats and hence the enthalpy changes of the systems being studied are constant pressure calorimeters. A simple example is a thermally insulated vessel (a coffee cup calorimeter) open to the atmosphere: the heat released in the reaction is determined by measuring the change in temperature of the contents. For a combustion reaction a constant-pressure flame calorimeter may be used where a certain amount of substance burns in a supply of oxygen and the rise in temperature is monitored. Isothermal titration calorimetry is also used, as explained in Section 2C.4(b). The most sophisticated way of measuring enthalpy changes is to use a differential scanning calorimeter, which is described in detail in Section 2C.4(a).

Solutions to exercises

E2C.1(a) At constant pressure

$$q = \Delta H = n\Delta_{\text{vap}}H^{\ominus} = (0.75\text{ mol})\times(30.0\text{ kJ mol}^{-1}) = \boxed{22.5\text{ kJ}}$$

and $w = -p\Delta V \approx -pV_{\text{vapour}} = -nRT = -(0.75\text{ mol})\times(8.3145\text{ J K}^{-1}\text{ mol}^{-1})\times(250\text{ K})$

$$w = -1.6\times10^{3}\text{ J} = \boxed{-1.6\text{ kJ}}$$

$$\Delta U = w + q = -1.6 + 22.5\text{ kJ} = \boxed{20.9\text{ kJ}}$$

Comment. Because the vapour is treated as a perfect gas, the specific value of the external pressure provided in the statement of the exercise does not affect the numerical value of the answer.

E2C.2(a) The reaction is

$$C_6H_5C_2H_5(l) + \tfrac{21}{2}\,O_2(g) \rightarrow 8\,CO_2(g) + 5\,H_2O(l)$$

$$\begin{aligned}\Delta_c H^{\ominus} &= 8\Delta_f H^{\ominus}(CO_2,g) + 5\Delta_f H^{\ominus}(H_2O,l) - \Delta_f H^{\ominus}(C_6H_5C_2H_5,\,l)\\ &= [(8)\times(-393.51)+(5)\times(-285.83)-(-12.5)]\text{ kJ mol}^{-1}\\ &= \boxed{-4564.7\text{ kJ mol}^{-1}}\end{aligned}$$

E2C.3(a) First $\Delta_f H[(CH_2)_3,g]$ is calculated, and then that result is used to calculate $\Delta_r H$ for the isomerization

$$(CH_2)_3(g) + \tfrac{9}{2}\,O_2(g) \rightarrow 3\,CO_2(g) + 3\,H_2O(l) \quad \Delta_c H = -2\,091\text{ kJ mol}^{-1}$$

$$\begin{aligned}\Delta_f H[(CH_2)_3,g] &= -\Delta_c H + 3\Delta_f H(CO_2,g) + 3\Delta_f H(H_2O,g)\\ &= [+2\,091 + (3)\times(-393.51) + (3)\times(-285.83)]\text{ kJ mol}^{-1}\\ &= \boxed{+53\text{ kJ mol}^{-1}}\end{aligned}$$

$$(CH_2)_3(g) \rightarrow C_3H_6(g) \quad \Delta_r H = ?$$

$$\begin{aligned}\Delta_r H &= \Delta_f H(C_3H_6,g) - \Delta_f H[(CH_2)_3,g]\\ &= (20.42 - 53)\text{ kJ mol}^{-1} = \boxed{-33\text{ kJ mol}^{-1}}\end{aligned}$$

E2C.4(a) Because $\Delta_f H^{\ominus}(H^{+},aq) = 0$ the whole of $\Delta_f H^{\ominus}(HCl,\,aq)$ is ascribed to $\Delta_f H^{\ominus}(Cl^{-},aq)$. Therefore, $\Delta_f H^{\ominus}(Cl^{-},aq) = \boxed{-167\text{ kJ mol}^{-1}}$

E2C.5(a) For naphthalene the reaction is $C_{10}H_8(s) + 12\,O_2(g) \rightarrow 10\,CO_2(g) + 4\,H_2O(l)$
A bomb calorimeter gives $q_V = n\Delta_c U^{\ominus}$ rather than $q_p = n\Delta_c H^{\ominus}$; thus we need

$$\Delta_c U^{\ominus} = \Delta_c H^{\ominus} - \Delta n_g RT\,[2B.4], \quad \Delta n_g = -2\text{ mol}$$

$$\Delta_c H^\ominus = -5\,157\,\text{kJ mol}^{-1} \text{ [Table 2C.1]}$$

$$\Delta_c U^\ominus = (-5\,157\,\text{kJ mol}^{-1}) - (-2)\times(8.3\times10^{-3}\,\text{kJ K}^{-1}\,\text{mol}^{-1})\times(298\,\text{K})$$
$$= \boxed{-5\,152\,\text{kJ mol}^{-1}}$$

$$|q| = |q_V| = |n\Delta_c U^\ominus| = \left(\frac{120\times10^{-3}\,\text{g}}{128.18\,\text{g mol}^{-1}}\right)\times(5\,152\,\text{kJ mol}^{-1}) = 4.82\bar{3}\,\text{kJ}$$

$$C = \frac{|q|}{\Delta T} = \frac{4.82\bar{3}\,\text{kJ}}{3.05\,\text{K}} = \boxed{1.58\,\text{kJ K}^{-1}}$$

When phenol is used the reaction is

$$C_6H_5OH(s) + \frac{15}{2}\,O_2(g) \rightarrow 6\,CO_2(g) + 3\,H_2O(l)$$

$$\Delta_c H^\ominus = -3\,054\,\text{kJ mol}^{-1} \text{ [Table 2C.1]}$$

$$\Delta_c U^\ominus = \Delta_c H^\ominus - \Delta n_g RT,\ \Delta n_g = -\frac{3}{2}$$
$$= (-3\,054\,\text{kJ mol}^{-1}) + (\tfrac{3}{2})\times(8.314\times10^{-3}\,\text{kJ K}^{-1}\,\text{mol}^{-1})\times(298\,\text{K})$$
$$= -3\,050\,\text{kJ mol}^{-1}$$

$$|q| = \frac{150\times10^{-3}\,\text{g}}{94.12\,\text{g mol}^{-1}}\times(3\,050\,\text{kJ mol}^{-1}) = 4.86\,\text{kJ}$$

$$\Delta T = \frac{|q|}{C} = \frac{4.86\,\text{kJ}}{1.58\,\text{kJ K}^{-1}} = \boxed{+3.08\,\text{K}}$$

Comment. In this case, $\Delta_c U^\ominus$ and $\Delta_c H^\ominus$ differed by about 0.1 per cent. Thus, to within 3 significant figures, it would not have mattered if we had used $\Delta_c H^\ominus$ instead of $\Delta_c U^\ominus$, but for very precise work it would.

E2C.6(a) (i) reaction(3) = (−2) × reaction(1) + reaction(2) and $\Delta n_g = -1$

The enthalpies of reactions are combined in the same manner as the equations (Hess's law).

$$\Delta_r H^\ominus(3) = (-2)\times\Delta_r H^\ominus(1) + \Delta_r H^\ominus(2)$$
$$= [(-2)\times(-184.62) + (-483.64)]\,\text{kJ mol}^{-1}$$
$$= \boxed{-114.40\,\text{kJ mol}^{-1}}$$

$$\Delta_r U = \Delta_r H - \Delta n_g RT\,[2B.4] = (-114.40\,\text{kJ mol}^{-1}) - (-1)\times(2.48\,\text{kJ mol}^{-1})$$
$$= \boxed{-111.92\,\text{kJ mol}^{-1}}$$

(ii) $\Delta_f H^\ominus$ refers to the formation of one mole of the compound, hence

$$\Delta_f H(\mathrm{J}) = \frac{\Delta_r H(\mathrm{J})}{\nu_\mathrm{J}}$$

$$\Delta_f H^\ominus(\mathrm{HCl,g}) = \frac{-184.62}{2}\,\mathrm{kJ\,mol^{-1}} = \boxed{-92.31\,\mathrm{kJ\,mol^{-1}}}$$

$$\Delta_f H^\ominus(\mathrm{H_2O,g}) = \frac{-483.64}{2}\,\mathrm{kJ\,mol^{-1}} = \boxed{-241.82\,\mathrm{kJ\,mol^{-1}}}$$

E2C.7(a) $\Delta_r H^\ominus = \Delta_r U^\ominus + \Delta n_g RT$ [2B.4]; $\Delta n_g = +2$

$$= (-1\,373\,\mathrm{kJ\,mol^{-1}}) + 2\times(2.48\,\mathrm{kJ\,mol^{-1}}) = \boxed{-1\,368\,\mathrm{kJ\,mol^{-1}}}$$

Comment. As a number of these exercises have shown, the use of $\Delta_r H$ as an approximation for $\Delta_r U$ is often valid.

E2C.8(a) (a) $\Delta_r H^\ominus = \sum_{\text{Products}} \nu\Delta_f H^\ominus - \sum_{\text{Reactants}} \nu\Delta_f H^\ominus$ [2C.5]

$$\Delta_r H^\ominus(298\,\mathrm{K}) = [(-110.53) - (-241.82)]\,\mathrm{kJ\,mol^{-1}} = \boxed{+131.29\,\mathrm{kJ\,mol^{-1}}}$$

$$\Delta_r U^\ominus(298\,\mathrm{K}) = \Delta_r H^\ominus(298\,\mathrm{K}) - \Delta n_g RT \text{ [2B.4]}$$

$$= (131.29\,\mathrm{kJ\,mol^{-1}}) - (1)\times(2.48\,\mathrm{kJ\,mol^{-1}}) = \boxed{+128.81\,\mathrm{kJ\,mol^{-1}}}$$

(b) $\Delta_r H^\ominus(478\,K) = \Delta_r H^\ominus(298\,K) + (T_2 - T_1)\Delta_r C_p^\ominus$ [Example 2C.2]

$$\Delta_r C_p^\ominus = C_{p,\mathrm{m}}^\ominus(\mathrm{CO,g}) + C_{p,\mathrm{m}}^\ominus(\mathrm{H_2,g}) - C_{p,\mathrm{m}}^\ominus(\mathrm{C,gr}) - C_{p,\mathrm{m}}^\ominus(\mathrm{H_2O,g})$$
$$= (29.14 + 28.82 - 8.53 - 33.58)\times 10^{-3}\,\mathrm{kJ\,K^{-1}\,mol^{-1}}$$
$$= 15.85\times 10^{-3}\,\mathrm{kJ\,K^{-1}\,mol^{-1}}$$

$$\Delta_r H^\ominus(478\,\mathrm{K}) = (131.29\,\mathrm{kJ\,mol^{-1}}) + (15.85\times 10^{-3}\,\mathrm{kJ\,K^{-1}\,mol^{-1}})\times(180\,\mathrm{K})$$
$$= (131.29 + 2.85)\,\mathrm{kJ\,mol^{-1}} = \boxed{+134.14\,\mathrm{kJ\,mol^{-1}}}$$

$$\Delta_r U^\ominus(478\,\mathrm{K}) = \Delta_r H^\ominus(478\,\mathrm{K}) - (1)\times(8.31\times 10^{-3}\,\mathrm{kJ\,K^{-1}\,mol^{-1}})\times(478\,\mathrm{K})$$
$$= (134.14 - 3.97)\,\mathrm{kJ\,mol^{-1}} = \boxed{+130.17\,\mathrm{kJ\,mol^{-1}}}$$

Comment. The differences in both $\Delta_r H^\ominus$ and $\Delta_r U^\ominus$ between the two temperatures are small and justify the use of the approximation that $\Delta_r C_p^\ominus$ is a constant.

E2C.9(a) For the reaction $\mathrm{CH_4(g) + 2\,O_2(g) \rightarrow CO_2(g) + 2\,H_2O(g)}$

$$\Delta_r H^\ominus = \Delta_f H^\ominus(\mathrm{CO_2,g}) + 2\times\Delta_f H^\ominus(\mathrm{H_2O,g}) - \Delta_f H^\ominus(\mathrm{CH_4,g})$$

In order to calculate the enthalpy of reaction at 500 K we first calculate its value at 298 K using data in Tables 2C.1 and 2C.2.

$$\Delta_r H^\ominus(298\,\text{K}) = -393.51\,\text{kJ mol}^{-1} + 2\times(-241.82\,\text{kJ mol}^{-1}) - (-74.81\,\text{kJ mol}^{-1})$$
$$= -802.34\,\text{kJ mol}^{-1}$$

Then using data on the heat capacities of all the reacting substances we can calculate the change in enthalpy, ΔH, of each substance as the temperature increases from 298 K to 500 K. The enthalpy of reaction at 500 K could be obtained by adding all these enthalpy changes to the enthalpy of reaction at 298 K. This process is shown below:

$$\Delta_r H^\ominus(500\,\text{K}) = \Delta_r H^\ominus(298\,\text{K}) + \Delta H(\text{CO}_2,\text{g}) + 2\times\Delta H(\text{H}_2\text{O},\text{g}) - \Delta H(\text{CH}_4,\text{g})$$
$$-2\times\Delta H(\text{O}_2,\text{g})$$

However, we need not calculate each individual ΔH values. It is more efficient to proceed as follows using eqn 2C.7a:

$$\Delta_r H^\ominus(500\,\text{K}) = \Delta_r H^\ominus(298\,\text{K}) + \int_{298\,\text{K}}^{500\,\text{K}} \Delta_r C_p^\ominus \,\text{d}T \quad [2\text{C}.7\text{a}]$$

We will express the temperature dependence of the heat capacities in the form of the equation given in Problem 2C.7 as data for the heat capacities of the substances involved in this reaction are available only in that form. They are not available for all the substances in the form of the equation of Table 2B.1.

If $C_{p,\text{m}}^\ominus = \alpha + \beta T + \gamma T^2$ then

$$\Delta_r C_p^\ominus = \Delta\alpha + \Delta\beta T + \Delta\gamma T^2 \text{ with } \Delta\alpha = \sum_J \nu_J \alpha_J \text{ etc.}$$

Using the data given in Problem 2C.7, we calculate

$$\Delta\alpha = (26.86 + 2\times30.36 - 14.16 - 2\times25.72)\,\text{J K}^{-1}\,\text{mol}^{-1} = 21.98\,\text{J K}^{-1}\,\text{mol}^{-1}$$

$$\Delta\beta = (6.97 + 2\times9.61 - 75.50 - 2\times12.98)\,\text{mJ K}^{-2}\,\text{mol}^{-1} = -75.27\,\text{mJ K}^{-2}\,\text{mol}^{-1}$$

$$\Delta\gamma = [-0.820 + 2\times1.184 - (-17.99) - 2\times(-3.862)]\,\mu\text{J K}^{-3}\,\text{mol}^{-1}$$
$$= 27.262\,\mu\text{J K}^{-3}\,\text{mol}^{-1}$$

Integrating eqn. 2C.7a between T_1(298 K) and T_2(500 K) we obtain

$$\Delta_r H^\ominus(500\,\text{K}) = \Delta_r H^\ominus(298\,\text{K}) + \Delta\alpha(500\,\text{K} - 298\,\text{K}) + \tfrac{1}{2}\Delta\beta[(500\,\text{K})^2 - (298\,\text{K})^2]$$
$$+\tfrac{1}{3}\Delta\gamma[(500\,\text{K})^3 - (298\,\text{K})^3] = \boxed{-803.07\,\text{kJ mol}^{-1}}$$

E2C.10 (a) Since enthalpy is a state function, $\Delta_r H$ for the process (see Fig. 2C.1)

$$\text{Mg}^{2+}(\text{g}) + 2\,\text{Cl}(\text{g}) + 2\,\text{e}^- \rightarrow \text{MgCl}_2(\text{aq})$$

is independent of path; therefore the change in enthalpy for the path on the left is equal to the change in enthalpy for the path on the right. All numerical values are in kJ mol^{-1}.

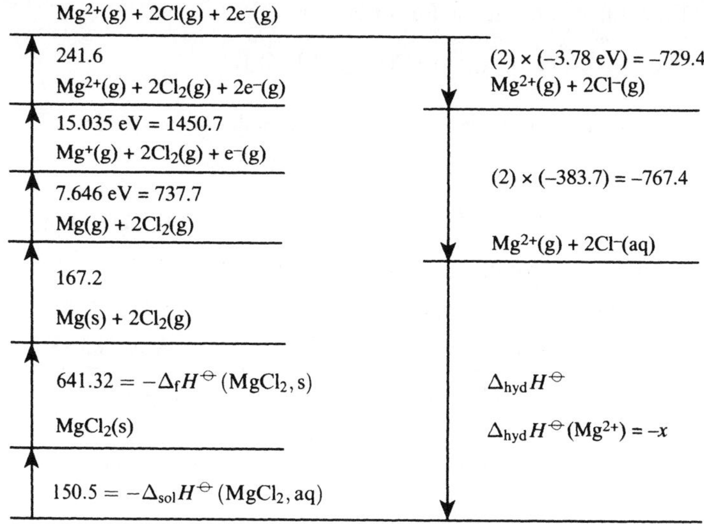

Figure 2C.1

The cycle is the distance traversed upward along the left plus the distance traversed downward on the right. The sum of these distances is zero. Note that $E_{ea} = -\Delta_{eg}H$. Therefore, following the cycle up the left and down the right and using kJ units,

$$-(-150.5)-(-641.32)+(167.2)+(241.6)+(737.7+1450.7)$$
$$+2\times(-364.\bar{7})+2\times(-383.7)+\Delta_{hyd}H(\mathrm{Mg}^{2+})=0$$

which yields $\Delta_{hyd}H(\mathrm{Mg}^{2+}) = \boxed{-1\,892\,\mathrm{kJ\,mol^{-1}}}$

Solutions to problems

P2C.1 The calorimeter is a constant-volume instrument, as described in the text (Section 2A.4); therefore $\Delta U = q_V$.

The calorimeter constant is determined from the data for the combustion of benzoic acid

$$\Delta U = \left(\frac{0.825\ \mathrm{g}}{122.12\ \mathrm{g\,mol^{-1}}}\right)\times(-3\,251\ \mathrm{kJ\,mol^{-1}}) = -21.9\bar{6}\,\mathrm{kJ}$$

Since $\Delta T = 1.940\ \mathrm{K}$, $C = \dfrac{|q|}{\Delta T} = \dfrac{21.9\bar{6}\ \mathrm{kJ}}{1.940\ \mathrm{K}} = 11.3\bar{2}\ \mathrm{kJ\,K^{-1}}$

For D-ribose, $\Delta U = -C\Delta T = -(11.3\bar{2}\ \mathrm{kJ\,K^{-1}})\times(0.910\ \mathrm{K})$

$$\text{Therefore, } \Delta_r U = \frac{\Delta U}{n} = -(11.3\bar{2}\ \mathrm{kJ\,K^{-1}})\times(0.910\ \mathrm{K})\times\left(\frac{150.13\ \mathrm{g\,mol^{-1}}}{0.727\ \mathrm{g}}\right)$$
$$= -2\,12\bar{7}\ \mathrm{kJ\,mol^{-1}}$$

The combustion reaction for D-ribose is

$$C_5H_{10}O_5(s)+5\,O_2(g)\rightarrow 5\,CO_2(g)+5\,H_2O(l)$$

Since there is no change in the number of moles of gas, $\Delta_r H=\Delta_r U$ [2B.4]
The enthalpy of formation is obtained from the sum

	$\Delta H/(\text{kJ mol}^{-1})$
$5\,CO_2(g)+5\,H_2O(l)\rightarrow C_5H_{10}O_5(s)+5\,O_2(g)$	2127
$5\,C(s)+5\,O_2(g)\rightarrow 5\,CO_2(g)$	$5\times(-393.51)$
$5\,H_2(g)+\frac{5}{2}\,O_2(g)\rightarrow 5\,H_2O(l)$	$5\times(-285.83)$
$5\,C(s)+5\,H_2(g)+\frac{5}{2}\,O_2(g)\rightarrow C_5H_{10}O_5(s)$	$-1\,270$

$$\Delta_f H=\boxed{-1\,270\ \text{kJ mol}^{-1}}$$

P2C.3 Data: methane–octane normal alkane combustion enthalpies

species	CH_4	C_2H_6	C_3H_8	C_4H_{10}	C_5H_{12}	C_6H_{14}	C_8H_{18}
$\Delta_c H/(\text{kJ mol}^{-1})$	−890	−1 560	−2 220	−2 878	−3 537	−4 163	−5 471
$M/(\text{g mol}^{-1})$	16.04	30.07	44.10	58.13	72.15	86.18	114.23

Suppose that $\Delta_c H=kM^n$. There are two methods by which a regression analysis can be used to determine the values of k and n. If you have a software package that can perform a 'power fit' of the type $Y=aX^b$, the analysis is direct using $Y=\Delta_c H$ and $X=M$. Then, $k=a$ and $n=b$. Alternatively, taking the logarithm yields another equation—one of linear form

$$\ln|\Delta_c H|=\ln|k|+n\ln M \qquad \text{where } k<0$$

This equation suggests a linear regression fit of $\ln(\Delta_c H)$ against $\ln M$ (Fig. 2C.2). The intercept is $\ln k$ and the slope is n. Linear regression fit

$$\ln|k|=4.2112,\quad \text{standard deviation}=0.0480;\quad k=-e^{4.2112}=\boxed{-67.44}$$

$$\boxed{n=0.9253},\quad \text{standard deviation}=0.0121$$

$$R=1.000$$

This is a good regression fit; essentially all of the variation is explained by the regression.

For decane the experimental value of $\Delta_c H$ equals −6 772.5 kJ mol^{-1} *(CRC Handbook of Chemistry and Physics)*. The predicted value is

$$\Delta_c H=kM^n=-67.44(142.28)^{(0.9253)}\ \text{kJ mol}^{-1}=\boxed{-6\,625.5\ \text{kJ mol}^{-1}}$$

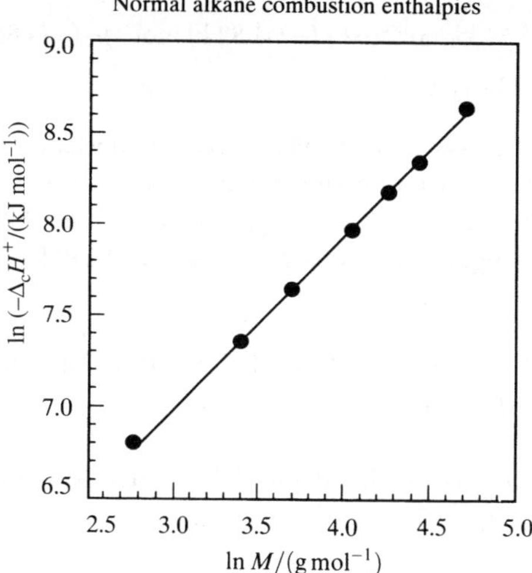

Figure 2C.2

$$\text{Per cent error of prediction} = \left|\frac{-6\,772.5-(-6\,625.5)}{-6\,625.5}\right| \times 100$$

$$\text{Per cent error of prediction} = \boxed{2.17 \text{ per cent}}$$

P2C.5 We must relate the formation of $DyCl_3$

$$\mathrm{Dy(s)+1.5\,Cl_2(g)\rightarrow DyCl_3(s)}$$

to the three reactions for which we have information. This reaction can be seen as a sequence of reaction (2), three times reaction (3), and the reverse of reaction (1), so

$$\Delta_f H^\ominus(\mathrm{DyCl_3,s}) = \Delta_r H^\ominus(2) + 3\Delta_r H^\ominus(3) - \Delta_r H^\ominus(1),$$

$$\Delta_f H^\ominus(\mathrm{DyCl_3,s}) = [-699.43 + 3(-158.31) - (-180.06)]\ \mathrm{kJ\ mol^{-1}}$$

$$= \boxed{-994.30\ \mathrm{kJ\ mol^{-1}}}$$

P2C.7 This problem is essentially the same as Exercise 2C.9(a). Here however we are asked to estimate the enthalpy of combustion at 350 K rather than 500 K.

For the reaction $\mathrm{CH_4(g) + 2\,O_2(g) \rightarrow CO_2(g) + 2\,H_2O(g)}$

$$\Delta_r H^\ominus = \Delta_f H^\ominus(\mathrm{CO_2,g}) + 2\times\Delta_f H^\ominus(\mathrm{H_2O,g}) - \Delta_f H^\ominus(\mathrm{CH_4,g})$$

In order to calculate the enthalpy of reaction at 500 K, we first calculate its value at 298 K using data in Tables 2C.2 and 2C.3.

$$\Delta_r H^\ominus(298\ \text{K}) = -393.51\ \text{kJ mol}^{-1} + 2\times(-241.82\ \text{kJ mol}^{-1}) - (-74.81\ \text{kJ mol}^{-1})$$
$$= -802.34\ \text{kJ mol}^{-1}$$

Then using data on the heat capacities of all the reacting substances we can calculate the change in enthalpy, ΔH, of each substance as the temperature increases from 298 K to 350 K. The enthalpy of reaction at 350 K could be obtained by adding all these enthalpy changes to the enthalpy of reaction at 298 K. This process is shown below:

$$\Delta_r H^\ominus(350\ \text{K}) = \Delta_r H^\ominus(298\ \text{K}) + \Delta H(\text{CO}_2,\text{g}) + 2\times\Delta H(\text{H}_2\text{O},\text{g}) - \Delta H(\text{CH}_4,\text{g})$$
$$-2\times\Delta H(\text{O}_2,\text{g})$$

However, we need not calculate each individual ΔH values. It is more efficient to proceed as follows using eqn 2C.7a:

$$\Delta_r H^\ominus(350\ \text{K}) = \Delta_r H^\ominus(298\ \text{K}) + \int_{298\text{K}}^{500\text{K}} \Delta_r C_p^\ominus \text{d}T \quad [2\text{C}.7\text{a}]$$

If $C_{p,\text{m}}^\ominus = \alpha + \beta T + \gamma T^2$ then

$\Delta_r C_p^\ominus = \Delta\alpha + \Delta\beta T + \Delta\gamma T^2$ with $\Delta\alpha = \sum_J \nu_J \alpha_J$ etc.

Using the data given in the problem, we calculate

$$\Delta\alpha = (26.86 + 2\times30.36 - 14.16 - 2\times25.72)\ \text{J K}^{-1}\ \text{mol}^{-1} = 21.98\ \text{J K}^{-1}\ \text{mol}^{-1}$$

$$\Delta\beta = (6.97 + 2\times9.61 - 75.50 - 2\times12.98)\ \text{mJ K}^{-2}\ \text{mol}^{-1} = -75.27\ \text{mJ K}^{-2}\ \text{mol}^{-1}$$

$$\Delta\gamma = [-0.820 + 2\times1.184 - (-17.99) - 2\times(-3.862)]\ \mu\text{J K}^{-3}\ \text{mol}^{-1}$$
$$= 27.262\ \mu\text{J K}^{-3}\ \text{mol}^{-1}$$

Integrating eqn 2C.7a between T_1(298 K) and T_2(350 K) we obtain

$$\Delta_r H^\ominus(350\ \text{K}) = \Delta_r H^\ominus(298\ \text{K}) + \Delta\alpha(350\ \text{K} - 298\ \text{K}) + \tfrac{1}{2}\Delta\beta[(350\ \text{K})^2 - (298\ \text{K})^2]$$
$$+\frac{1}{3}\Delta\gamma[(350\ \text{K})^3 - (298\ \text{K})^3] = \boxed{-802.31\ \text{kJ mol}^{-1}}$$

We note that in this example there was very little change in the enthalpy of combustion as the temperature increased from 298 K to 350 K. This is due to the coincidental cancellation of the enthalpy changes from the terms involving $\Delta\alpha$ and $\Delta\beta$.

P2C.9 The needed data are the enthalpy of vaporization and heat capacity of water, available in the *Resource section.*

$$C_{p,\text{m}}(\text{H}_2\text{O,l}) = 75.3\ \text{J K}^{-1}\ \text{mol}^{-1} \quad \Delta_{\text{vap}}H^\ominus(\text{H}_2\text{O}) = 44.0\ \text{kJ mol}^{-1}$$

$$n(\text{H}_2\text{O}) = \frac{65\ \text{kg}}{0.018\ \text{kg mol}^{-1}} = 3.6\times10^3\ \text{mol}$$

From $\Delta H = nC_{p,\mathrm{m}}\Delta T$ we obtain

$$\Delta T = \frac{\Delta H}{nC_{p,\mathrm{m}}} = \frac{1.0\times10^4\ \mathrm{kJ}}{(3.6\times10^3\ \mathrm{mol})\times(0.0753\ \mathrm{kJ\ K^{-1}\ mol^{-1}})} = \boxed{+37\ \mathrm{K}}$$

From $\Delta H = n\Delta_{\mathrm{vap}}H^{\ominus} = \dfrac{m}{M}\Delta_{\mathrm{vap}}H^{\ominus}$

$$m = \frac{M\times\Delta H}{\Delta_{\mathrm{vap}}H^{\ominus}} = \frac{(0.018\ \mathrm{kg\ mol^{-1}})\times(1.0\times10^4\ \mathrm{kJ})}{44.0\ \mathrm{kJ\ mol^{-1}}} = \boxed{4.09\ \mathrm{kg}}$$

Comment. This estimate would correspond to about 30 glasses of water per day, which is much higher than the average consumption. The discrepancy may be a result of our assumption that evaporation of water is the main mechanism of heat loss.

2D State functions and exact differentials

Answer to discussion question

D2D.1 One can use the general expression for π_T given in *Justification* 2D.1 (and proved in Topic 3D) to derive its specific form for a van der Waals gas, as given in Exercise 2D.2(a), that is, $\pi_T = a/V_{\mathrm{m}}^2$. For an isothermal expansion in a van der Waals gas $\mathrm{d}U_{\mathrm{m}} = (a/V_{\mathrm{m}})^2$. Hence, $\Delta U_{\mathrm{m}} = -a(1/V_{\mathrm{m},2} - 1/V_{\mathrm{m},1})$. See this derivation in the solution to Exercise 2D.2(a). This formula corresponds to what one would expect for a real gas. As the molecules get closer and closer the molar volume gets smaller and smaller and the energy of attraction gets larger and larger.

Solutions to exercises

E2D.1(a) See Exercise 2D.2(a) and Problem 2D.9. The internal pressure of a van der Waals gas is $\pi_T = a/V_{\mathrm{m}}^2$.

The molar volume can be estimated from the perfect gas equation.

$$V_{\mathrm{m}} = \frac{RT}{p} = \frac{0.08206\ \mathrm{dm^3\ atm\ K^{-1}\ mol^{-1}}\times400\ \mathrm{K}}{1.00\ \mathrm{bar}\times\left(\dfrac{1.000\ \mathrm{atm}}{1.013\ \mathrm{bar}}\right)} = 33.26\ \mathrm{dm^3\ mol^{-1}}$$

$$\pi_T = \frac{a}{V_{\mathrm{m}}^2} = \frac{5.464\ \mathrm{atm\ dm^6\ mol^{-2}}}{(33.26\ \mathrm{dm^3\ mol^{-1}})^2} = 4.96\times10^{-3}\ \mathrm{atm} = \boxed{5.03\ \mathrm{mbar}}$$

E2D.2(a) The internal energy is a function of temperature and volume, $U_{\mathrm{m}} = U_{\mathrm{m}}(T,V_{\mathrm{m}})$, so

$$\mathrm{d}U_{\mathrm{m}} = \left(\frac{\partial U_{\mathrm{m}}}{\partial T}\right)_{V_{\mathrm{m}}}\mathrm{d}T + \left(\frac{\partial U_{\mathrm{m}}}{\partial V_{\mathrm{m}}}\right)_T\mathrm{d}V_{\mathrm{m}} \qquad [\pi_T = (\partial U_{\mathrm{m}}/\partial V_{\mathrm{m}})_T]$$

For an isothermal expansion $\mathrm{d}T = 0$; hence

$$\mathrm{d}U_\mathrm{m} = \left(\frac{\partial U_\mathrm{m}}{\partial V_\mathrm{m}}\right)_T \mathrm{d}V = \pi_T\,\mathrm{d}V_\mathrm{m} = \frac{a}{V_\mathrm{m}^2}\mathrm{d}V_\mathrm{m}$$

$$\Delta U_\mathrm{m} = \int_{V_{\mathrm{m},1}}^{V_{\mathrm{m},2}} \mathrm{d}U_\mathrm{m} = \int_{V_{\mathrm{m},2}}^{V_{\mathrm{m},1}} \frac{a}{V_\mathrm{m}^2}\mathrm{d}V_\mathrm{m} = a\int_{1.00\ \mathrm{dm^3\ mol^{-1}}}^{20.0\ \mathrm{dm^3\ mol^{-1}}} \frac{\mathrm{d}V_\mathrm{m}}{V_\mathrm{m}^2} = -\frac{a}{V_\mathrm{m}}\Bigg|_{1.00\ \mathrm{dm^3\ mol^{-1}}}^{20.0\ \mathrm{dm^3\ mol^{-1}}}$$

$$= -\frac{a}{20.0\ \mathrm{dm^3\ mol^{-1}}} + \frac{a}{1.00\ \mathrm{dm^3\ mol^{-1}}} = \frac{19.0a}{20.0\ \mathrm{dm^3\ mol^{-1}}}$$

$$= 0.950\ a\ \mathrm{mol\ dm^{-3}}$$

From Table 1C.3, $a = 1.352\ \mathrm{dm^6\ atm\ mol^{-1}}$

$$\Delta U_\mathrm{m} = (0.950\ \mathrm{mol\ dm^{-3}}) \times (1.352\ \mathrm{dm^6\ atm\ mol^{-2}})$$

$$= (1.28\bar{4}\ \mathrm{dm^3\ atm\ mol^{-1}}) \times \left(\frac{1\ \mathrm{m}}{10\ \mathrm{dm}}\right)^3 \times \left(\frac{1.013\times10^5\ \mathrm{Pa}}{\mathrm{atm}}\right) = \boxed{+130.\bar{1}\ \mathrm{J\ mol^{-1}}}$$

$w = -\int p\,\mathrm{d}V_\mathrm{m}$ where $p = \dfrac{RT}{V_\mathrm{m} - b} - \dfrac{a}{V_\mathrm{m}^2}$ for a van der Waals gas. Hence,

$$w = -\int\left(\frac{RT}{V_\mathrm{m} - b}\right)\mathrm{d}V_\mathrm{m} + \int\frac{a}{V_\mathrm{m}^2}\mathrm{d}V_\mathrm{m} = -q + \Delta U_\mathrm{m}$$

Therefore,

$$q = \int_{1.00\mathrm{dm^3\ mol^{-1}}}^{20.0\mathrm{dm^3\ mol^{-1}}}\left(\frac{RT}{V_\mathrm{m} - b}\right)\mathrm{d}V_\mathrm{m} = RT\ln(V_\mathrm{m} - b)\Big|_{1.00\mathrm{dm^3\ mol^{-1}}}^{20.0\mathrm{dm^3\ mol^{-1}}}$$

$$= (8.314\mathrm{J K^{-1} mol^{-1}}) \times (298\mathrm{K}) \times \ln\left(\frac{20.0 - 3.9\times10^{-2}}{1.00 - 3.9\times10^{-2}}\right) = \boxed{+7.52\times10^3\ \mathrm{J mol^{-1}}}$$

and $w = -q + \Delta U_\mathrm{m} = -(7.52\times10^3\ \mathrm{J mol^{-1}}) + (130.\bar{1}\ \mathrm{J\ mol^{-1}}) = \boxed{-7.39\times10^3\mathrm{J mol^{-1}}}$

E2D.3(a) $\alpha = \left(\frac{1}{V}\right)\left(\frac{\partial V}{\partial T}\right)_p$ [2D.6]; $\quad \alpha_{320} = \left(\frac{1}{V_{320}}\right)\left(\frac{\partial V}{\partial T}\right)_{p,320}$

$$\left(\frac{\partial V}{\partial T}\right)_p = V_{300}(3.9\times10^{-4}/\mathrm{K} + 2.96\times10^{-6}T/\mathrm{K}^2)$$

$$\left(\frac{\partial V}{\partial T}\right)_{p,320} = V_{300}(3.9\times10^{-4}/\mathrm{K} + 2.96\times10^{-6}\times320/\mathrm{K}) = 1.34\times10^{-3}\mathrm{K}^{-1}V_{300}$$

$$V_{320} = V_{300}\{(0.75) + (3.9\times10^{-4})\times(320) + (1.48\times10^{-6})\times(320)^2\} = (V_{300})\times(1.02\bar{6})$$

$$\text{so}\quad \alpha_{320} = \left(\frac{1}{V_{320}}\right)\left(\frac{\partial V}{\partial T}\right)_{p,320} = \left(\frac{1}{1.02\bar{6}V_{300}}\right)\times(1.34\times10^{-3}\,\mathrm{K}^{-1}\,V_{300})$$

$$\alpha_{320} = \frac{1.34\times10^{-3}\,\mathrm{K}^{-1}}{1.02\bar{6}} = \boxed{1.31\times10^{-3}\,\mathrm{K}^{-1}}$$

Comment. Knowledge of the density at 300 K is not required to solve this exercise, but it would be required to obtain numerical values of the volumes at the two temperatures.

E2D.4(a) The isothermal compressibility is

$$\kappa_T = -\left(\frac{1}{V}\right)\left(\frac{\partial V}{\partial p}\right)_T \text{ [2D.7]} \quad \text{so} \quad \left(\frac{\partial V}{\partial p}\right)_T = -\kappa_T V$$

At constant temperature

$$\mathrm{d}V = \left(\frac{\partial V}{\partial p}\right)_T \mathrm{d}p \quad \text{so} \quad \mathrm{d}V = -\kappa_T V\,\mathrm{d}p \quad \text{or} \quad \frac{\mathrm{d}V}{V} = -\kappa_T\,\mathrm{d}p$$

Substituting $V = \dfrac{m}{\rho}$ yields $\mathrm{d}V = -\dfrac{m}{\rho^2}\mathrm{d}\rho\,;\ \dfrac{\mathrm{d}V}{V} = -\dfrac{\mathrm{d}\rho}{\rho} = -\kappa_T \mathrm{d}p$

Therefore, $\dfrac{\delta\rho}{\rho} \approx \kappa_T \delta p$

For $\dfrac{\delta\rho}{\rho} = 0.10\times10^{-2} = 1.0\times10^{-3},\ \delta p \approx \dfrac{1.0\times10^{-3}}{\kappa_T} = \dfrac{1.0\times10^{-3}}{4.96\times10^{-5}\,\mathrm{atm}^{-1}} = \boxed{2.0\times10^{3}\ \mathrm{atm}}$

E2D.5(a) The isothermal Joule–Thomson coefficient is

$$\mu_T = \left(\frac{\partial H_\mathrm{m}}{\partial p}\right)_T = -\mu C_{p,\mathrm{m}} \text{ [2D.14]} = (-0.25\,\mathrm{K\,atm}^{-1})\times(29\,\mathrm{J\,K}^{-1}\,\mathrm{mol}^{-1})$$

$$= \boxed{-7.2\,\mathrm{J\,atm}^{-1}\,\mathrm{mol}^{-1}}$$

$$\mathrm{d}H = n\left(\frac{\partial H_\mathrm{m}}{\partial p}\right)_T \mathrm{d}p = -n\mu C_{p,\mathrm{m}}\,\mathrm{d}p$$

$$\Delta H = \int_{p_1}^{p_2}(-n\mu C_{p,\mathrm{m}})\mathrm{d}p = -n\mu C_{p,\mathrm{m}}(p_2 - p_1) \quad [\mu \text{ and } C_p \text{ are constant}]$$

$$\Delta H = -(10.0\,\mathrm{mol})\times(+7.2\,\mathrm{J\,atm}^{-1}\,\mathrm{mol}^{-1})\times(-85\,\mathrm{atm}) = +6.1\,\mathrm{kJ}$$

$$\text{so} \quad q(\text{supplied}) = +\Delta H = \boxed{+6.1\,\mathrm{kJ}}$$

Solutions to problems

P2D.1 The coefficient of thermal expansion is

$$\alpha=\frac{1}{V}\left(\frac{\partial V}{\partial T}\right)_p\approx\frac{\Delta V}{V\Delta T}\quad\text{so}\quad\Delta V\approx\alpha V\Delta T$$

This change in volume is equal to the change in height (sea level rise, Δh) times the area of the ocean (assuming that area remains constant). We will use α of pure water, although the oceans are complex solutions. For a 2 °C rise in temperature

$$\Delta V=(2.1\times10^{-4}\,\text{K}^{-1})\times(1.37\times10^{9}\,\text{km}^3)\times(2.0\text{K})=5.8\times10^{5}\,\text{km}^3$$

so $$\Delta h=\frac{\Delta V}{A}=1.6\times10^{-3}\,\text{km}=\boxed{1.6\text{m}}$$

Since the rise in sea level is directly proportional to the rise in temperature, $\Delta T=1\,°\text{C}$ would lead to $\Delta h=\boxed{0.80\text{m}}$ and $\Delta T=3.5\,°\text{C}$ would lead to $\Delta h=\boxed{2.8\text{m}}$

Comment. More detailed models of climate change predict somewhat smaller rises, but the same order of magnitude.

P2D.3 Using the Euler's chain relation and the reciprocal identity [*Mathematical background* 2]

$$\left(\frac{\partial p}{\partial T}\right)_V=-\left(\frac{\partial p}{\partial V}\right)_T\left(\frac{\partial V}{\partial T}\right)_p$$

Substituting into the given expression for

$$C_p-C_V=-T\left(\frac{\partial p}{\partial V}\right)_T\left(\frac{\partial V}{\partial T}\right)_p^2$$

Using the reciprocal identity again

$$C_p-C_V=-\frac{T\left(\frac{\partial V}{\partial T}\right)_p^2}{\left(\frac{\partial V}{\partial p}\right)_T}$$

For a perfect gas, $pV=nRT$, so

$$\left(\frac{\partial V}{\partial T}\right)_p^2=\left(\frac{nR}{p}\right)^2\quad\text{and}\quad\left(\frac{\partial V}{\partial p}\right)_T=-\frac{nRT}{p^2}$$

so $$C_p-C_V=\frac{-T\left(\frac{nR}{p}\right)^2}{-\frac{nRT}{p^2}}=\boxed{nR}$$

P2D.5 $p = \dfrac{nRT}{V-nb} - \dfrac{n^2a}{V^2}$ [van der Waals eqn]

Hence $\boxed{T = \left(\dfrac{p}{nR}\right)\times(V-nb) + \left(\dfrac{na}{RV^2}\right)\times(V-nb)}$

$$\boxed{\left(\frac{\partial T}{\partial p}\right)_V = \frac{V-nb}{nR}} = \frac{V_m - b}{R} = \frac{1}{\left(\frac{\partial p}{\partial T}\right)_V}$$

For Euler's chain relation, we need to show that $\left(\dfrac{\partial T}{\partial p}\right)_V\left(\dfrac{\partial p}{\partial V}\right)_T\left(\dfrac{\partial V}{\partial T}\right)_p = -1$

Hence, in addition to $\left(\dfrac{\partial T}{\partial p}\right)_V$ we need $\left(\dfrac{\partial p}{\partial V}\right)_T$ and $\left(\dfrac{\partial V}{\partial T}\right)_p = \dfrac{1}{\left(\frac{\partial T}{\partial V}\right)_p}$

$$\left(\frac{\partial p}{\partial V}\right)_T = \frac{-nRT}{(V-nb)^2} + \frac{2n^2a}{V^3}$$

which can be found from

$$\left(\frac{\partial T}{\partial V}\right)_p = \left(\frac{p}{nR}\right) + \left(\frac{na}{RV^2}\right) - \left(\frac{2na}{RV^3}\right)\times(V-nb)$$

$$\left(\frac{\partial T}{\partial V}\right)_p = \left(\frac{T}{V-nb}\right) - \left(\frac{2na}{RV^3}\right)\times(V-nb)$$

Therefore,

$$\left(\frac{\partial T}{\partial p}\right)_V\left(\frac{\partial p}{\partial V}\right)_T\left(\frac{\partial V}{\partial T}\right)_p = \frac{\left(\frac{\partial T}{\partial p}\right)_V\left(\frac{\partial p}{\partial V}\right)_T}{\left(\frac{\partial T}{\partial V}\right)_p}$$

$$= \frac{\left(\frac{V-nb}{nR}\right)\times\left(\frac{-nRT}{(V-nb)^2} + \frac{2n^2a}{V^3}\right)}{\left(\frac{T}{V-nb}\right) - \left(\frac{2na}{RV^3}\right)\times(V-nb)} = \frac{\left(\frac{-T}{V-nb}\right) + \left(\frac{2na}{RV^3}\right)\times(V-nb)}{\left(\frac{T}{V-nb}\right) - \left(\frac{2na}{RV^3}\right)\times(V-nb)}$$

$$= -1$$

P2D.7 $c_s=\left(\frac{RT\gamma}{M}\right)^{1/2}$, $p=\rho\frac{RT}{M}$, so $\frac{RT}{M}=\frac{p}{\rho}$; hence $\boxed{c_s=\left(\frac{\gamma p}{\rho}\right)^{1/2}}$

For argon, $\gamma=\frac{5}{3}$, so $c=\left(\frac{(8.314\,\mathrm{J\,K^{-1}\,mol^{-1}})\times(298\,\mathrm{K})\times\frac{5}{3}}{39.95\times10^{-3}\,\mathrm{kg\,mol^{-1}}}\right)^{1/2}=\boxed{322\ \mathrm{m\,s^{-1}}}$

P2D.9 $\mu\equiv\left(\frac{\partial T}{\partial p}\right)_H$ [2D.13]

Use of Euler's chain relation [*MB* 2] yields

$$\mu=-\frac{\left(\frac{\partial H_\mathrm{m}}{\partial p}\right)_T}{C_{p,\mathrm{m}}}\ \text{[\textit{Justification} 2D.2]}$$

$$\left(\frac{\partial H_\mathrm{m}}{\partial p}\right)_T=\left(\frac{\partial U_\mathrm{m}}{\partial p}\right)_T+\left[\frac{\partial(pV_\mathrm{m})}{\partial p}\right]_T=\left(\frac{\partial U_\mathrm{m}}{\partial V_\mathrm{m}}\right)_T\left(\frac{\partial V_\mathrm{m}}{\partial p}\right)_T+\left[\frac{\partial(pV_\mathrm{m})}{\partial p}\right]_T$$

Use the virial expansion of the van der Waals equation in terms of p. (See the solution to Problem 1.9.) Now let us evaluate some of these derivatives.

$$\left(\frac{\partial U_\mathrm{m}}{\partial V_\mathrm{m}}\right)_T=\left(\frac{\partial U}{\partial V}\right)_T=\pi_T=\frac{a}{V_\mathrm{m}^2}\ \text{[Exercise 2D.1(a)]}$$

$$pV_\mathrm{m}=RT\left[1+\frac{1}{RT}\left(b-\frac{a}{RT}\right)p+\dots\right]$$

$$\left[\frac{\partial(pV_\mathrm{m})}{\partial p}\right]_T\approx b-\frac{a}{RT},\quad\left(\frac{\partial V_\mathrm{m}}{\partial p}\right)_T\approx-\frac{RT}{p^2}$$

Substituting $\left(\frac{\partial H_\mathrm{m}}{\partial p}\right)_T\approx\left(\frac{a}{V_m^2}\right)\times\left(-\frac{RT}{p^2}\right)+\left(b-\frac{a}{RT}\right)\approx\frac{-aRT}{(pV_\mathrm{m})^2}+\left(b-\frac{a}{RT}\right)$

Since $\left(\frac{\partial H_\mathrm{m}}{\partial p}\right)_T$ is in a sense a correction term, that is, it approaches zero for a perfect gas, little error will be introduced by the approximation, $(pV_\mathrm{m})^2=(RT)^2$.

Thus $\left(\frac{\partial H_\mathrm{m}}{\partial p}\right)_T\approx\left(-\frac{a}{RT}\right)+\left(b-\frac{a}{RT}\right)=\left(b-\frac{2a}{RT}\right)$ and $\mu=\frac{\left(\frac{2a}{RT}-b\right)}{C_{p,\mathrm{m}}}$

P2D.11 We compute μ from

$$\mu=-\frac{1}{C_p}\left(\frac{\partial H}{\partial p}\right)_T$$

and we estimate $\left(\frac{\partial H}{\partial p}\right)_T$ from the enthalpy and pressure data. We are given both enthalpy and heat capacity data on a mass basis rather than a molar basis; however, the masses will cancel, so we need not convert to a molar basis.

(a) At 300 K

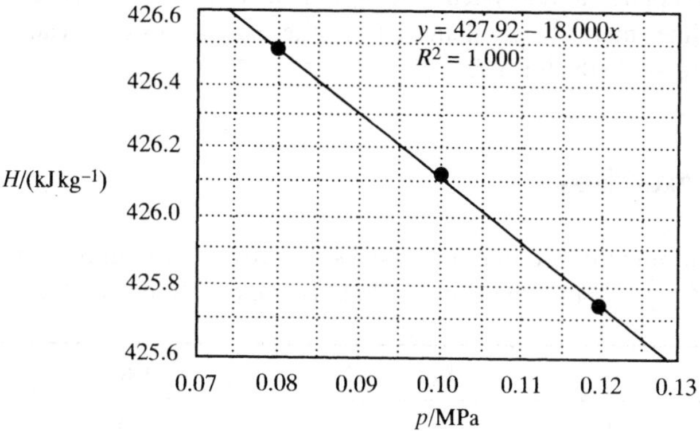

Figure 2D.1(a).

The regression analysis gives the slope as $-18.0\,\mathrm{J\,g^{-1}\,MPa^{-1}} \approx \left(\frac{\partial H}{\partial p}\right)_T$, see Fig. 2D.1(a),

so $$\mu = -\frac{-18.0\,\mathrm{kJ\,kg^{-1}\,MPa^{-1}}}{0.7649\,\mathrm{kJ\,kg^{-1}\,K^{-1}}} = \boxed{23.5\,\mathrm{K\,MPa^{-1}}}$$

(b) At 350 K

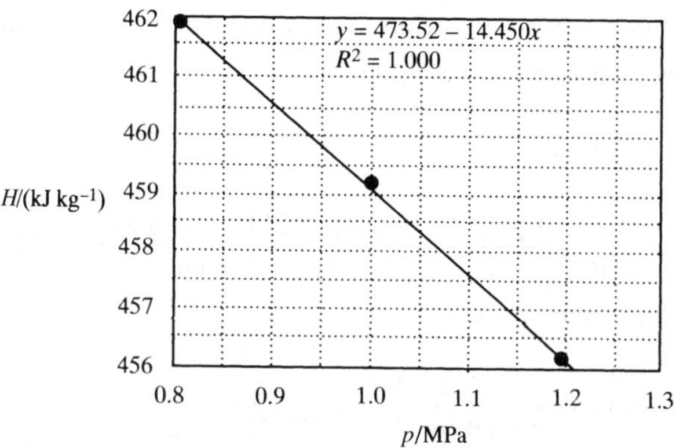

Figure 2D.1(b).

The regression analysis gives the slope as $-14.5\,\mathrm{J\,g^{-1}\,MPa^{-1}} \approx \left(\frac{\partial H}{\partial p}\right)_T$, see Fig. 2D.1(b),

so $$\mu = -\frac{-14.5\,\mathrm{kJ\,kg^{-1}\,MPa^{-1}}}{1.0392\,\mathrm{kJ\,kg^{-1}\,K^{-1}}} = \boxed{14.0\,\mathrm{K\,MPa^{-1}}}$$

2E Adiabatic changes

Answer to discussion question

D2E.1 See Fig. 2E.2 of the text and the Interactivity associated with that figure. Heat is not supplied to the system during an adiabatic expansion; therefore, the pressure declines more steeply than in an isothermal expansion because the temperature decreases in the former.

Solutions to exercises

E2E.1(a) Assuming that all rotational modes are active we can draw up the following table for $C_{V,\mathrm{m}}$, $C_{p,\mathrm{m}}$ and γ with and without active vibrational modes.

	$C_{V,\mathrm{m}}$	$C_{p,\mathrm{m}}$	γ	Exptl	
$NH_3(v_V^*=0)$	3R	4R	1.33	1.31	closer
$NH_3(v_V^*=6)$	9R	10R	1.11		
$CH_4(v_V^*=0)$	3R	4R	1.33	1.31	closer
$CH_4(v_V^*=9)$	12R	13R	1.08		

The experimental values are obtained from Tables 2C.2 and 2C.4 assuming $C_{p,\mathrm{m}} = C_{V,\mathrm{m}} + R$. It is clear from the comparison in the above table that the vibrational modes are not active. This is confirmed by the experimental vibrational wavenumbers (see Herzberg, *Molecular Spectra and Molecular Structure II. Infrared and Raman Spectra of Polyatomic Molecules*, D. Van Nostrand Company, Inc., Princeton, NJ, 1945) all of which are much greater than kT at 298 K.

E2E.2(a) For reversible adiabatic expansion

$$T_\mathrm{f} = T_\mathrm{i}\left(\frac{V_\mathrm{i}}{V_\mathrm{f}}\right)^{1/c} \text{ [2E.2a]}$$

where $c = \dfrac{C_{V,\mathrm{m}}}{R} = \dfrac{C_{p,\mathrm{m}} - R}{R} = \dfrac{(20.786 - 8.3145)\,\mathrm{J\,K^{-1}\,mol^{-1}}}{8.3145\,\mathrm{J\,K^{-1}\,mol^{-1}}} = 1.500$,

so the final temperature is

$$T_\mathrm{f} = (273.15\,\mathrm{K}) \times \left(\frac{1.0\,\mathrm{dm^3}}{3.0\,\mathrm{dm^3}}\right)^{1/1.500} = \boxed{13\bar{1}\,\mathrm{K}}$$

E2E.3(a) In an adiabatic process, the initial and final pressures are related by (eqn. 2E.3)

$$p_\mathrm{f} V_\mathrm{f}^\gamma = p_\mathrm{i} V_\mathrm{i}^\gamma$$

where $\gamma = \dfrac{C_p}{C_V} = \dfrac{C_V + nR}{C_V} = \dfrac{20.8\,\mathrm{J\,K^{-1}} + (1.0\,\mathrm{mol})(8.31\,\mathrm{J\,K^{-1}\,mol^{-1}})}{20.8\,\mathrm{J\,K^{-1}}} = 1.40$

Find V_i from the perfect gas law:

$$V_i = \frac{nRT_i}{p_i} = \frac{(1.0\text{ mol})\times(8.31\text{ J K}^{-1}\text{ mol}^{-1})\times(300\text{ K})}{4.25\text{ atm}}\times\frac{1\text{ atm}}{1.013\times10^5\text{ Pa}}$$

$$V_i = 5.7\bar{9}\times10^{-3}\text{ m}^3$$

so $$V_f = V_i\left(\frac{p_i}{p_f}\right)^{1/\gamma} = (5.7\bar{9}\times10^{-3}\text{ m}^3)\times\left(\frac{4.25\text{ atm}}{2.50\text{ atm}}\right)^{1/1.40} = \boxed{0.0084\bar{6}\text{ m}^3}$$

Find the final temperature from the perfect gas law:

$$T_f = \frac{p_f V_f}{nR} = \frac{(2.50\text{ atm})\times(0.0084\bar{6}\text{ m}^3)}{(1.0\text{ mol})\times(8.31\text{ J K}^{-1}\text{ mol}^{-1})}\times\frac{1.013\times10^5\text{ Pa}}{1\text{ atm}}$$

$$T_f = \boxed{25\bar{7}\text{ K}}$$

Adiabatic work is (eqn 2E.1)

$$w_{ad} = C_V\Delta T = (20.8\text{ J K}^{-1})\times(25\bar{7}-300)\text{ K} = \boxed{-0.89\times10^3\text{ J}}$$

E2E.4(a) Reversible adiabatic work is

$$w = C_V\Delta T\ [2E.1] = n(C_{p,m}-R)\times(T_f - T_i)$$

where the temperatures are related by

$$T_f = T_i\left(\frac{V_i}{V_f}\right)^{1/c}\ [2E.2a] \qquad \text{where} \quad c = \frac{C_{V,m}}{R} = \frac{C_{p,m}-R}{R} = 3.463$$

So $$T_f = [(27.0+273.15)\text{K}]\times\left(\frac{500\times10^{-3}\text{ dm}^3}{3.00\text{ dm}^3}\right)^{1/3.463} = 179\text{K}$$

and $$w = \left(\frac{2.45\text{ g}}{44.0\text{ g mol}^{-1}}\right)\times[(37.11-8.3145)\text{J K}^{-1}\text{mol}^{-1}]\times(179-300)\text{K} = \boxed{-194\text{J}}$$

E2E.5(a) For reversible adiabatic expansion

$$p_f V_f^\gamma = p_i V_i^\gamma\ [2E.3] \quad \text{so } p_f = p_i\left(\frac{V_i}{V_f}\right)^\gamma = (67.4\text{kPa})\times\left(\frac{0.50\text{ dm}^3}{2.0\text{ dm}^3}\right)^{1.4} = \boxed{9.7\text{kPa}}$$

Solution to problem

P2E.1 The temperatures and volumes in reversible adiabatic expansion are related by eqn 2E.2a:

$$T_f = T_i\left(\frac{V_i}{V_f}\right)^{1/c} \qquad \text{where} \qquad c = \frac{C_{V,m}}{R}$$

From eqn 2E.3, we can relate the pressures and volumes:

$$p_f = p_i\left(\frac{V_i}{V_f}\right)^{\gamma} \qquad \text{where} \qquad \gamma = \frac{C_{p,m}}{C_{V,m}}$$

We are looking for $C_{p,m}$, which can be related to c and γ:

$$c\gamma = \left(\frac{C_{V,m}}{R}\right)\times\left(\frac{C_{p,m}}{C_{V,m}}\right) = \frac{C_{p,m}}{R}$$

Solving both relationships for the ratio of volumes, we have:

$$\left(\frac{p_f}{p_i}\right)^{1/\gamma} = \frac{V_i}{V_f} = \left(\frac{T_f}{T_i}\right)^{c} \qquad \text{so} \qquad \frac{p_f}{p_i} = \left(\frac{T_f}{T_i}\right)^{c\gamma}$$

Therefore

$$C_{p,m} = R\frac{\ln\left(\frac{p_f}{p_i}\right)}{\ln\left(\frac{T_f}{T_i}\right)} = (8.314\,\mathrm{J\,K^{-1}\,mol^{-1}})\times\left(\frac{\ln\left(\frac{81.840\ \mathrm{kPa}}{202.94\ \mathrm{kPa}}\right)}{\ln\left(\frac{248.44\ \mathrm{K}}{298.15\ \mathrm{K}}\right)}\right) = \boxed{41.40\,\mathrm{J\,K^{-1}\,mol^{-1}}}$$

Integrated activities

I2.1 A state function is a thermodynamic property, the value of which is independent of the history of the system. Examples of state functions are the properties of pressure, temperature, internal energy, enthalpy as well as the properties of entropy, Gibbs energy, and Helmholtz energy to be discussed fully in Chapter 3. The differentials of state functions are exact differentials. Hence, we can use the mathematical properties of exact differentials to draw far-reaching conclusions about the relations between physical properties and establish connections that were unexpected but turn out to be very significant. One practical importance of these results is that we can obtain the value of a property we require from the combination of measurements of other properties without actually having to measure the required property itself, the measurement of which might be very difficult.

I2.3 (a) See the solution to Problem 2A.1. There we developed the following equations:

$$w=-\int_{V_1}^{V_2} p\,\mathrm{d}V \quad \text{with} \quad p=\frac{nRT}{V-nb}-\frac{n^2a}{V^2} \quad \text{[van der Waals equation]}$$

Therefore, $w=-nRT\int_{V_1}^{V_2}\frac{\mathrm{d}V}{V-nb}+n^2a\int_{V_1}^{V_2}\frac{\mathrm{d}V}{V^2}=\boxed{-nRT\ln\left(\frac{V_2-nb}{V_1-nb}\right)-n^2a\left(\frac{1}{V_2}-\frac{1}{V_1}\right)}$

This expression can be interpreted more readily if we assume $V>>nb$, which is certainly valid at all but the highest pressures. Then using the first term of the Taylor series expansion,

$$\ln(1-x)=-x-\frac{x^2}{2}\ldots \quad \text{for } |x|<<1$$

$$\ln(V-nb)=\ln V+\ln\left(1-\frac{nb}{V}\right)\approx \ln V-\frac{nb}{V}$$

and, after substitution

$$w\approx -nRT\ln\left(\frac{V_2}{V_1}\right)+n^2bRT\left(\frac{1}{V_2}-\frac{1}{V_1}\right)-n^2a\left(\frac{1}{V_2}-\frac{1}{V_1}\right)$$

$$\approx -nRT\ln\left(\frac{V_2}{V_1}\right)-n^2(a-bRT)\left(\frac{1}{V_2}-\frac{1}{V_1}\right)$$

$$\approx +w_0-n^2(a-bRT)\left(\frac{1}{V_2}-\frac{1}{V_1}\right)=\text{perfect gas value + van der Waals correction.}$$

w_0 is the perfect gas value

$$w_0=-nRT\ln\left(\frac{V_\mathrm{f}}{V_\mathrm{i}}\right)=(-1.0\,\mathrm{mol}^{-1})\times(8.314\,\mathrm{J\,K^{-1}\,mol^{-1}})\times(298\,\mathrm{K})\times\ln\left(\frac{3.0\,\mathrm{dm}^3}{1.0\,\mathrm{dm}^3}\right)$$

$$w_0=-2.7\bar{2}\times10^3\,\mathrm{J}=\boxed{-2.7\,\mathrm{kJ}}$$

The van der Waals correction term is

$$-n^2(a-bRT)\left(\frac{1}{V_2}-\frac{1}{V_1}\right)$$

For CO_2(g) $a=3.610$ atm dm^6 mol^{-2} and $b=0.0429$ dm^3 mol^{-1}
Converting to SI units, these values become

$$a=0.3657\ \mathrm{J\,m^3mol^{-2}} \text{ and } b=4.29\times10^{-5}\ \mathrm{m^3mol^{-1}}$$

Inserting these SI values into the van der Waals correction term above, we obtain, after carrying out the calculation for 1.0 mol of gas, 173 J = 0.173 kJ. So the contribution to the total work due to van der Waals interactions is small compared to the perfect gas contribution. If the pressure of the gas were higher, then the van der Waals correction term would be greater.

The total work then of the isothermal reversible expansion of the van der Waals gas under the stated conditions is

$$w = 2.7\overline{2}\ \text{kJ} + 0.17\overline{3}\ \text{kJ} = 2.8\overline{9}\ \text{kJ} = \boxed{2.9\ \text{kJ}}\ (2\ \text{sig figs})$$

(b) See Fig. 2E.2 and eqn 2E.3 of the text and the applet on the W H Freeman website (whfreeman.com) entitled *Pressure changes during the adiabatic expansion of gases.*

$$p_\text{f} = \left(\frac{V_\text{i}}{V_\text{f}}\right)^{\gamma} p_\text{i}\ [2\text{E}.3].$$

Larger γ results in larger p_f when the relative volume $\left(\frac{V_\text{i}}{V_\text{f}}\right)$ is greater than unity. The opposite is true when the relative volume is less than unity.

Use the graph in the applet on the website to visually demonstrate this dependence.

3 The Second and Third Laws

Assume that all gases are perfect and that data refer to 298.15 K unless otherwise stated.

3A Entropy

Answers to discussion questions

D3A.1 We must remember that the Second Law of thermodynamics states only that the total entropy of both the system (here, the molecules organizing themselves into cells) and the surroundings (here, the medium) must increase in a naturally occurring process. It does not state that entropy must increase in a portion of the universe that interacts with its surroundings. In this case, the cells grow by using chemical energy from their surroundings (the medium) and in the process the increase in the entropy of the medium outweighs the decrease in entropy of the system. Hence, the Second Law is not violated.

D3A.3 For a thorough discussion of the relationship between the various formulations of the Second Law, see Sections 3A.2 and 3A.3. We will not repeat all of that discussion here and will merely summarize the main points.

There are two equivalent statements of the Second Law that are based on directly observable processes:

1. The **Kelvin statement:** No process is possible in which the sole result is the absorption of heat from a reservoir and its complete conversion into work.
2. The **Clausius statement:** No process is possible in which the sole result is the transfer of energy from a cooler to a hotter body.

It can be shown that these statements are equivalent and that they both lead to the existence of a state function of the system, S, called the entropy, defined through the relation $dS = \frac{dq_{rev}}{T}$. dS can be shown to be an exact differential, that is, $\oint dS = 0$. Hence, S is a property of all systems. In Section 3A.3(c), it is shown that the differential of the entropy as defined above leads to the Clausius inequality, that is $dS \geq \frac{dq}{T}$, where q represents the actual heat associated with a real, necessarily irreversible, process. We now suppose that the system is isolated from its surroundings, so that $dq = 0$. The Clausius inequality then implies that $dS \geq 0$, and we conclude that in an isolated system the entropy cannot decrease when a spontaneous change occurs. Since the universe as a whole can be considered an isolated system, this implies that $\Delta S_{\text{universe}} = \Delta S_{\text{sys}} + \Delta S_{\text{surr}} \geq 0$, which is another version of the Second Law.

Solutions to exercises

E3A.1(a) All spontaneous processes are irreversible processes, which implies through eqn 3A.12, the Clausius inequality, that $\Delta S_{\text{tot}} = \Delta S_{\text{sys}} + \Delta S_{\text{surr}} > 0$, for all spontaneous processes. In this case, $\Delta S_{\text{tot}} = 0$. Therefore, the process is $\boxed{\text{not spontaneous}}$.

E3A.2(a) Efficiency, η, is $\dfrac{\text{work performed}}{\text{heat absorbed}} = \dfrac{|w|}{q_{\text{h}}} = \dfrac{3.00\text{ kJ}}{10.00\text{ kJ}} = 0.300$. For an ideal heat engine we have $\eta_{\text{rev}} = 1 - \dfrac{T_{\text{c}}}{T_{\text{h}}}$ [3A.10] $= 0.300 = 1 - \dfrac{T_{\text{c}}}{273.16\text{ K}}$. Solving for T_{c}, we obtain $\boxed{T_{\text{c}} = 191.2\text{ K}}$ as the temperature of the organic liquid.

E3A.3(a) Assume that the block is so large that its temperature does not change significantly as a result of the heat transfer. Then

$$\Delta S = \int_{\text{i}}^{\text{f}} \frac{\text{d}q_{\text{rev}}}{T}\ [3\text{A}.2] = \frac{1}{T}\int_{\text{i}}^{\text{f}} \text{d}q_{\text{rev}}\ [\text{constant } T] = \frac{q_{\text{rev}}}{T}$$

(a) $\Delta S = \dfrac{100\times10^{3}\text{ J}}{273.15\text{K}} = \boxed{366\text{ J K}^{-1}}$

(b) $\Delta S = \dfrac{100\times10^{3}\text{ J}}{323.15\text{ K}} = \boxed{309\text{J K}^{-1}}$

E3A.4(a) $\boxed{I_2(\text{g})}$ will have the higher standard molar entropy at 298 K, primarily because ΔS_{fus} and ΔS_{vap} are greater for I_2. At 298 K, I_2 is a solid in its standard state.

E3A.5(a)

$$\Delta S = nR\ln\left(\frac{V_{\text{f}}}{V_{\text{i}}}\right)\ [3\text{A}.14]$$

$$= \left(\frac{15\text{ g}}{44\text{ g mol}^{-1}}\right) \times 8.314\text{ J K}^{-1}\text{ mol}^{-1} \times \ln\left(\frac{3.0}{1.0}\right) = \boxed{3.1\text{ J K}^{-1}}$$

E3A.6(a) Trouton's rule in the form $\Delta_{\text{vap}}H^{\ominus} = T_{\text{b}} \times 85\text{ J K}^{-1}\text{ mol}^{-1}$ can be used to obtain approximate enthalpies of vaporization. For benzene

$$\Delta_{\text{vap}}H^{\ominus} = (273.2 + 80.1)\text{ K} \times 85\text{ J K}^{-1}\text{ mol}^{-1} = \boxed{30.0\text{ kJ mol}^{-1}}$$

E3A.7(a) $S_{\text{m}}(T_{\text{f}}) = S_{\text{m}}(T_{\text{i}}) + \displaystyle\int_{T_{\text{i}}}^{T_{\text{f}}} \frac{C_{V,\text{m}}}{T}\text{d}T$ [3A.20]

If we assume that neon is a perfect gas then $C_{V,m}$ may be taken to be constant and given by

$$C_{V,m} = C_{p,m} - R; \qquad C_{p,m} = 20.786\ \text{J K}^{-1}\ \text{mol}^{-1}\ [\text{Table 2C.2}]$$
$$= (20.786 - 8.314)\ \text{J K}^{-1}\ \text{mol}^{-1}$$
$$= 12.472\ \text{J K}^{-1}\ \text{mol}^{-1}$$

Integrating, we obtain

$$S_m(500\ \text{K}) = S_m(298\text{K}) + C_{V,m} \ln \frac{T_f}{T_i}$$
$$= (146.22\ \text{J K}^{-1}\ \text{mol}^{-1}) + (12.472\ \text{J K}^{-1}\ \text{mol}^{-1}) \ln\left(\frac{500\ \text{K}}{298\ \text{K}}\right)$$
$$= (146.22 + 6.45)\ \text{J K}^{-1}\ \text{mol}^{-1} = \boxed{152.67\ \text{J K}^{-1}\ \text{mol}^{-1}}$$

E3A.8(a) Since entropy is a state function, ΔS may be calculated from the most convenient path, which in this case corresponds to constant-pressure heating followed by constant-temperature compression.

$$\Delta S = nC_{p,m} \ln\left(\frac{T_f}{T_i}\right)\ [\text{3A.20, at } p_i] + nR \ln\left(\frac{V_f}{V_i}\right)\ [\text{3A.14, at } T_f].$$

Since pressure and volume are inversely related (Boyle's law), $\frac{V_f}{V_i} = \frac{p_i}{p_f}$. Hence,

$$\Delta S = nC_{p,m} \ln\left(\frac{T_f}{T_i}\right) - nR \ln\left(\frac{p_f}{p_i}\right) = (3.00\ \text{mol}) \times \frac{5}{2} \times (8.314\ \text{J K}^{-1}\ \text{mol}^{-1}) \times \ln\left(\frac{398\ \text{K}}{298\ \text{K}}\right)$$
$$-(3.00\ \text{mol}) \times (8.314\ \text{J K}^{-1}\ \text{mol}^{-1}) \times \ln\left(\frac{5.00\ \text{atm}}{1.00\ \text{atm}}\right)$$
$$= (18.0\bar{4} - 40.1\bar{4})\ \text{J K}^{-1} = -22.1\ \text{J K}^{-1}$$

Though ΔS (system) is negative, the process can still occur spontaneously if ΔS (total) is positive.

E3A.9(a) Since the container is isolated, the heat flow is zero and therefore $\boxed{\Delta H = 0}$; since the masses of the blocks are equal, the final temperature must be their mean temperature, 25 °C. Specific heat capacities are heat capacities per gram and are related to the molar heat capacities by

$$C_s = \frac{C_m}{M}\ [C_{p,m} \approx C_{V,m} = C_m]$$

So $\quad nC_m = mC_s\,[nM = m]$

$$\Delta S = mC_s \ln\left(\frac{T_f}{T_i}\right)\ [\text{3A.20}]$$

$$\Delta S_1 = (1.00\times10^3\,\text{g})\times(0.385\,\text{J K}^{-1}\,\text{g}^{-1})\times\ln\left(\frac{298\,\text{K}}{323\,\text{K}}\right) = -31.0\,\text{J K}^{-1}$$

$$\Delta S_2 = (1.00\times10^3\,\text{g})\times(0.385\,\text{J K}^{-1}\,\text{g}^{-1})\times\ln\left(\frac{298\,\text{K}}{273\,\text{K}}\right) = 33.7\,\text{J K}^{-1}$$

$$\Delta S_{\text{tot}} = \Delta S_1 + \Delta S_2 = \boxed{+2.7\,\text{J K}^{-1}}.$$

Comment. The positive value of ΔS_{tot} corresponds to a spontaneous process. The heat capacity of each block is $C=mC_s$ where C_s is the specific heat capacity.

So,

$$\Delta H(\text{individual}) = mC_s\Delta T = 10.0\times10^3\,\text{g}\times0.449\,\text{J K}^{-1}\,\text{g}^{-1}\times(\pm37.\bar{5}\,\text{K}) = \pm168\,\text{kJ}$$

These two enthalpy changes add up to zero: $\boxed{\Delta H_{\text{tot}} = 0}$

$$\Delta S = mC_s\ln\left(\frac{T_f}{T_i}\right);\quad 100\,^\circ\text{C} = 373.2\,\text{K};\ 25\,^\circ\text{C} = 298.2\,\text{K};\ 62.\bar{5}\,^\circ\text{C} = 335.\bar{7}\,\text{K}$$

E3A.10(a) $\Delta S(\text{gas}) = nR\ln\dfrac{V_f}{V_i}\,[3\text{A}.14] = \left(\dfrac{14\,\text{g}}{28.02\,\text{g mol}^{-1}}\right)\times(8.314\,\text{J K}^{-1}\,\text{mol}^{-1})\times(\ln 2)$

$= \boxed{+2.9\,\text{J K}^{-1}}$

(a) $\Delta S(\text{surroundings}) = \boxed{-2.9\,\text{J K}^{-1}}$ [overall zero entropy production]

$\Delta S(\text{total}) = \boxed{0}$ [reversible process]

(b) $\Delta S(\text{gas}) = \boxed{+2.9\,\text{J K}^{-1}}$ [S a state function]

$\Delta S(\text{surroundings}) = \boxed{0}$ [surroundings do not change]

$\Delta S(\text{total}) = \boxed{+2.9\,\text{J K}^{-1}}$

(c) $\Delta S(\text{gas}) = \boxed{0}$ [$q_{\text{rev}} = 0$]

$\Delta S(\text{surroundings}) = \boxed{0}$ [no heat transferred to surroundings]

$\Delta S(\text{total}) = \boxed{0}$

E3A.11(a) (a) $\Delta_{\text{vap}}S = \dfrac{\Delta_{\text{vap}}H}{T_b} = \dfrac{29.4\times10^3\,\text{J mol}^{-1}}{334.88\,\text{K}} = \boxed{+87.8\,\text{J K}^{-1}\,\text{mol}^{-1}}$

(b) If the vaporization occurs reversibly, $\Delta S_{\text{tot}} = 0$, so $\Delta S_{\text{surr}} = \boxed{-87.8\,\text{J K}^{-1}\,\text{mol}^{-1}}$

E3A.12(a) $\Delta S = nC_p(\text{H}_2\text{O, s})\ln\dfrac{T_f}{T_i} + n\dfrac{\Delta_{\text{fus}}H}{T_{\text{fus}}} + nC_p(\text{H}_2\text{O, l})\ln\dfrac{T_f}{T_i} + n\dfrac{\Delta_{\text{vap}}H}{T_{\text{vap}}} + nC_p(\text{H}_2\text{O, g})\ln\dfrac{T_f}{T_i}$

$$n = \frac{10.0\,\text{g}}{18.02\,\text{g mol}^{-1}} = 0.555\,\text{mol}$$

$$\Delta S = 0.555\text{ mol}\times 38.02\text{ J K}^{-1}\text{mol}^{-1}\times\ln\frac{273.15}{263.15} + 0.555\text{ mol}\times\frac{6.008\text{ kJ mol}^{-1}}{273.15\text{ K}}$$
$$+0.555\text{ mol}\times 75.291\text{ J K}^{-1}\text{ mol}^{-1}\times\ln\frac{373.15}{273.15}$$
$$+0.555\text{ mol}\times\frac{44.016\text{ kJ mol}^{-1}}{373.15\text{ K}} + 0.555\text{ mol}\times 33.58\text{ J K}^{-1}\text{ mol}^{-1}\times\ln\frac{388.15}{373.15}$$

$$\boxed{\Delta S = 92.2\text{ J K}^{-1}}$$

Comment. This calculation has been based on the assumption that the heat capacities remain constant over the range of temperatures involved and that the enthalpy of vaporization at 298.15 K given in Table 3A.2 can be applied to the vaporization at 373.15 K. Neither one of these assumptions is strictly valid. Therefore, the calculated value is only approximate.

Solutions to problems

P3A.1 The isotherms correspond to T=constant, and the reversibly traversed adiabats correspond to S=constant. Thus we can represent the cycle as in Fig. 3A.1.

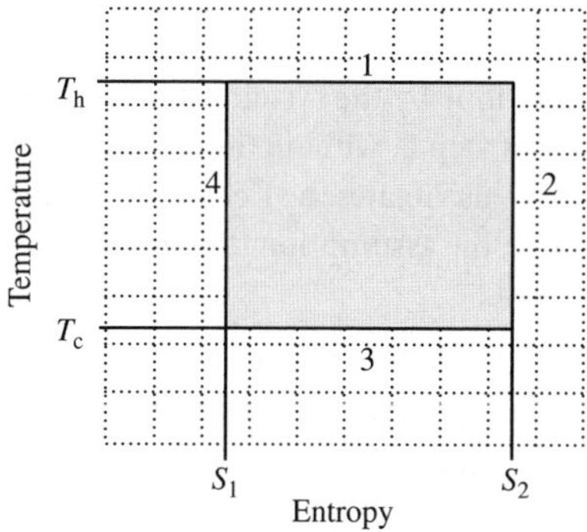

Figure 3A.1

In this figure, paths 1, 2, 3, and 4 correspond to the four stages of the Carnot cycle listed in the text following Fig. 3A.7. The area within the rectangle is

$$\text{Area} = \int_{\text{cyclic}} T\,\mathrm{d}S = (T_h - T_c)\times(S_2 - S_1) = (T_h - T_c)\Delta S = (T_h - T_c)nR\ln\frac{V_B}{V_A}$$

(isothermal expansion from V_A to V_B, stage 1). But,

$$w(\text{cycle}) = \eta q_h = \left(\frac{T_h - T_c}{T_h}\right)nRT_h\ln\frac{V_B}{V_A}\ [\text{text Fig. 3A.7}] = nR(T_h - T_c)\ln\frac{V_B}{V_A}.$$

Therefore, the area is equal to the net work done in the cycle.

P3A.3 Paths A and B in Fig. 3A.2 are the reversible adiabatic paths which are assumed to cross at state 1. Path C (dashed) is an isothermal path which connects the adiabatic paths at states 2 and 3. Now go round the cycle ($1 \rightarrow 2$, step 1; $2 \rightarrow 3$, step 2; $3 \rightarrow 1$, step 3).

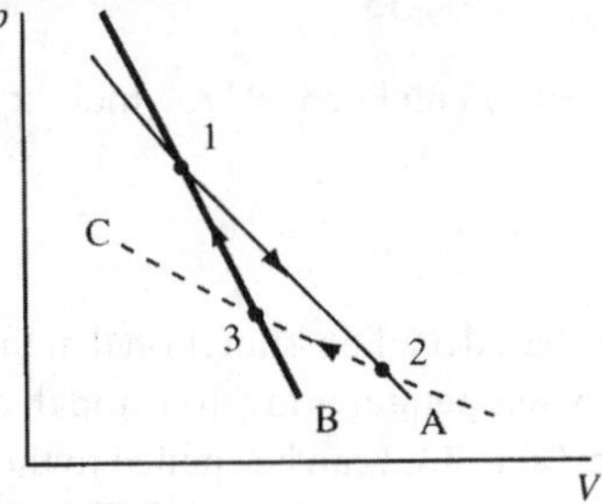

Figure 3A.2

Step 1 $\Delta U_1 = q_1 + w_1 = w_1$ [q_1=0, adiabatic]

Step 2 $\Delta U_2 = q_2 + w_2 = 0$ [isothermal step, energy depends on temperature only]

Step 3 $\Delta U_3 = q_3 + w_3 = w_3$ [q_3=0, adiabatic]

For the cycle $\Delta U = 0 = w_1 + q_2 + w_2 + w_3$ or $w(\text{net}) = w_1 + w_2 + w_3 = -q_2$

But, $\Delta U_1 = -\Delta U_3$ [$\Delta T_1 = -\Delta T_2$]; hence $w_1 = -w_3$, and $w(\text{net}) = w_2 = -q_2$, or $-w(\text{net}) = q_2$

A net amount of work has been done by the system from heat obtained from a heat reservoir at the temperature of step 2, without at the same time transferring heat from a hot to a cold reservoir. This violates the Kelvin statement of the Second Law of thermodynamics. Therefore, the assumption that the two adiabatic reversible paths may intersect is disproved.

Question. May any adiabatic paths intersect, reversible or not?

Alternative solution not requiring the system to be a perfect gas

Note that step 2 above effectively requires the system to be a perfect gas. The following solution is more general:

Suppose that two adiabats cross at point 1. Consider the isotherm at T crossing both adiabats at points 2 and 3. We now can define a quasi-steady closed cycle along the two adiabats and the isotherm.

By definition of state variables:

$$\Delta U = \oint dU = 0$$

$$\Delta S = \oint dS = 0$$

Using the First Law:

$$\Delta U = \oint dq + \oint dw = q + w$$

$$\Rightarrow q = -w$$

and the Second Law for reversible processes we have:

$$\Delta S = \oint \frac{dq}{T} = \frac{q}{T}$$

$$\Rightarrow q = 0$$

As T is finite we find:

$$w = -q = 0$$

As the work corresponds to the surface area of our closed cycle we conclude that the two adiabats coincide (are a single curve).

P3A.5 In case (a), the electric heater converts 1.00 kJ of electrical energy into heat, providing $\boxed{1.00\text{ kJ}}$ of energy as heat to the room. (The Second Law places no restriction on the complete conversion of work to heat—only on the reverse process.) In case (b), we want to find the heat deposited in the room $|q_h|$:

$$|q_h| = |q_c| + |w| \quad \text{where} \quad \frac{|q_c|}{|w|} = c = \frac{T_c}{T_h - T_c}$$

$$\text{so} \quad |q_c| = \frac{|w|T_c}{T_h - T_c} = \frac{1.00\text{ kJ} \times 260\text{ K}}{(295-260)\text{ K}} = 7.4\text{ kJ}$$

The heat transferred to the room is $|q_h| = |q_c| + |w| = 7.4\text{ kJ} + 1.00\text{ kJ} = \boxed{8.4\text{ kJ}}$. Most of the thermal energy the heat pump deposits into the room comes from outdoors. Difficult as it is to believe on a cold winter day, the intensity of thermal energy (that is, the absolute temperature) outdoors is a substantial fraction of that indoors. The work put into the heat pump is not simply converted to heat, but is 'leveraged' to transfer additional heat from outdoors.

P3A.7

$$\Delta S_m = \int_{T_1}^{T_2} \frac{C_{p,m}\text{d}T}{T}\ [3\text{A}.19] = \int_{T_1}^{T_2}\left(\frac{a+bT}{T}\right)\text{d}T = a\ln\left(\frac{T_2}{T_1}\right) + b(T_2 - T_1)$$

$$a = 91.47\text{ J K}^{-1}\text{ mol}^{-1}, \qquad b = 7.5\times10^{-2}\text{ J K}^{-2}\text{ mol}^{-1}$$

$$\Delta S_m = (91.47\text{ J K}^{-1}\text{ mol}^{-1}) \times \ln\left(\frac{300\text{ K}}{273\text{ K}}\right) + (0.075\text{ J K}^{-2}\text{ mol}^{-1})\times(27\text{ K})$$

$$= \boxed{10.\bar{7}\text{ J K}^{-1}\text{ mol}^{-1}}$$

P3A.9

	q	w	$\Delta U = \Delta H$	ΔS	ΔS_{sur}	ΔS_{tot}
Path (a)	2.74 kJ	−2.74 kJ	0	9.13 J K^{-1}	−9.13 J K^{-1}	0
Path (b)	1.66 kJ	−1.66 kJ	0	9.13 J K^{-1}	−5.53 J K^{-1}	3.60 J K^{-1}

Path (a)

$$w = -nRT\ln\left(\frac{V_f}{V_i}\right)\ [2\text{A}.9] = -nRT\ln\left(\frac{p_i}{p_f}\right)\ [\text{Boyle's law}]$$

$$= -(1.00\text{ mol})\times(8.314\text{ J K}^{-1}\text{ mol}^{-1})\times(300\text{ K})\times\ln\left(\frac{3.00\text{ atm}}{1.00\text{ atm}}\right) = -2.74\times10^{3}\text{ J}$$

$$= \boxed{-2.74\text{ kJ}}$$

$$\Delta H = \Delta U = \boxed{0} \text{ [isothermal process in perfect gas]}$$

$$q = \Delta U - w = 0 - (-2.74\ \text{kJ}) = \boxed{+2.74\ \text{kJ}}$$

$$\Delta S = \frac{q_{\text{rev}}}{T} \text{ [isothermal]} = \frac{2.74\times10^{3}\ \text{J}}{300\ \text{K}} = \boxed{+9.13\ \text{J K}^{-1}}$$

$$\Delta S_{\text{tot}} = \boxed{0} \text{ [reversible process]}$$

$$\Delta S_{\text{tot}} = \Delta S_{\text{sur}} = \Delta S_{\text{tot}} - \Delta S = 0 - 9.13\ \text{J K}^{-1} = \boxed{-9.13\ \text{J K}^{-1}}$$

Path (b)

$$w = -p_{\text{ex}}(V_{\text{f}} - V_{\text{i}}) = -p_{\text{ex}}\left(\frac{nRT}{p_{\text{f}}} - \frac{nRT}{p_{\text{i}}}\right) = -nRT\left(\frac{p_{\text{ex}}}{p_{\text{f}}} - \frac{p_{\text{ex}}}{p_{\text{i}}}\right)$$

$$= -(1.00\ \text{mol})\times(8.314\ \text{J K}^{-1})\times(300\ \text{K})\times\left(\frac{1.00\ \text{atm}}{1.00\ \text{atm}} - \frac{1.00\ \text{atm}}{3.00\ \text{atm}}\right)$$

$$= -1.66\times10^{3}\ \text{J} = \boxed{-1.66\ \text{kJ}}$$

$$\Delta H = \Delta U = \boxed{0} \text{ [isothermal process in perfect gas]}$$

$$q = \Delta U - w = 0 - (-1.66\ \text{kJ}) = \boxed{+1.66\ \text{kJ}}$$

$$\Delta S = \frac{q_{\text{rev}}}{T} \text{ [isothermal]} = \frac{2.74\times10^{3}\ \text{J}}{300\text{K}} = \boxed{+9.13\ \text{J K}^{-1}}$$

(Note: One can arrive at this by using q from *Path (a)* as the reversible path, or one can simply use ΔS from *Path (a)*, realizing that entropy is a state function.)

$$\Delta S_{\text{sur}} = \frac{q_{\text{sur}}}{T_{\text{sur}}} = \frac{-q}{T_{\text{sur}}} = \frac{-1.66\times10^{3}\ \text{J}}{300\ \text{K}} = \boxed{-5.53\ \text{J K}^{-1}}$$

$$\Delta S_{\text{tot}} = \Delta S + \Delta S_{\text{sur}} = (9.13 - 5.53)\ \text{J K}^{-1} = \boxed{+3.60\ \text{J K}^{-1}}$$

P3A.11 $\Delta S = \boxed{nC_{p,\text{m}} \ln\frac{T_{\text{f}}}{T_{\text{h}}} + nC_{p,\text{m}} \ln\frac{T_{\text{f}}}{T_{\text{c}}}}$ [3A.20] [T_{f} is the final temperature, $T_{\text{f}} = \frac{1}{2}(T_{\text{h}} + T_{\text{c}})$]

In the present case, $T_{\text{f}} = \frac{1}{2}(500\ \text{K} + 250\ \text{K}) = 375\ \text{K}$

$$\Delta S = nC_{p,\text{m}} \ln\frac{T_{\text{f}}^{2}}{T_{\text{h}}T_{\text{c}}} = nC_{p,\text{m}} \ln\frac{(T_{\text{h}} + T_{\text{c}})^{2}}{4T_{\text{h}}T_{\text{c}}} = \left(\frac{500\text{g}}{63.54\ \text{g cm}^{-3}}\right)\times(24.4\ \text{J K}^{-1}\ \text{mol}^{-1})$$

$$\times \ln\left(\frac{375^{2}}{500\times250}\right) = \boxed{+22.6\ \text{J K}^{-1}}$$

P3A.13 Taking the hint, we have

$$\Delta_{trs}S^{\ominus}(25°C) = \Delta S_i + \Delta S_{ii} + \Delta S_{iii}$$

We are not given the heat capacity of either the folded or unfolded protein, but if we let $C_{p,m}$ be the heat capacity of the folded protein, the heat capacity of the unfolded protein is $C_{p,m} + 6.28\ \text{kJ K}^{-1}\,\text{mol}^{-1}$. So for the heating and cooling steps, we have:

$$\Delta S_i = C_p \ln\left(\frac{T_f}{T_i}\right) = C_{p,m} \ln\left(\frac{348.7\ \text{K}}{298.2\ \text{K}}\right)\ [3A.20]$$

and $\Delta S_{iii} = (C_{p,m} + 6.28\ \text{kJ K}^{-1}\,\text{mol}^{-1})\ln\left(\frac{298.2\ \text{K}}{348.7\ \text{K}}\right)$, so

$$\Delta S_i + \Delta S_{iii} = C_{p,m} \ln\left(\frac{348.7\ \text{K}}{298.2\ \text{K}}\right) + \left(C_{p,m} + 6.28\ \text{kJ K}^{-1}\,\text{mol}^{-1}\right)\ln\left(\frac{298.2\ \text{K}}{348.7\ \text{K}}\right)$$

$$= \left(6.28\ \text{kJ K}^{-1}\,\text{mol}^{-1}\right)\ln\left(\frac{298.2\ \text{K}}{348.7\ \text{K}}\right) = -0.983\ \text{kJ K}^{-1}\,\text{mol}^{-1}$$

For the transition itself, use eqn 3A.17:

$$\Delta S_{ii} = \frac{\Delta_{trs}H^{\ominus}}{T_{trs}} = \frac{509\ \text{kJ mol}^{-1}}{348.2\ \text{K}} = 1.46\bar{0}\ \text{kJ K}^{-1}\,\text{mol}^{-1}$$

Hence, $\Delta_{trs}S^{\ominus} = (1.46\bar{0} - 0.983)\ \text{kJ K}^{-1}\,\text{mol}^{-1} = 0.47\bar{7}\ \text{kJ K}^{-1}\,\text{mol}^{-1} = \boxed{47\bar{7}\ \text{J K}^{-1}\,\text{mol}^{-1}}$

3B The measurement of entropy

Answer to discussion question

D3B.1 Because solutions of cations cannot be prepared in the absence of anions and *vice versa*, in order to assign numerical values to the entropies of ions in solution, we arbitrarily assign the value of zero to the standard entropy of H^+ ions in water at all temperatures, i.e. $S^{\ominus}(H^+, aq) = 0$. With this choice, the entropies of ions in water are values relative to the hydrogen ion in water; hence they may be either positive or negative. Ion entropies vary as expected on the basis that they are related to the degree to which the ions order the water molecules around them in solution. Small, highly charged ions induce local structure in the surrounding water, and the disorder of the solution is decreased more than for the case of large, singly charged ions.

Solutions to exercises

E3B.1(a) In each case, $S_m = R \ln s$, where s is the number of orientations of about equal energy that the molecule can adopt. Therefore,

(a) $S_m = R\ln 3 = 8.3145\,\text{J K}^{-1}\,\text{mol}^{-1} \times \ln 3 = \boxed{9.13\,\text{J K}^{-1}\,\text{mol}^{-1}}$

(b) $S_m = R\ln 5 = 8.3145\ \text{J K}^{-1}\ \text{mol}^{-1} \times \ln 5 = \boxed{13.4\ \text{J K}^{-1}\ \text{mol}^{-1}}$

(c) $S_m = R\ln 6 = 8.3145\ \text{J K}^{-1}\ \text{mol}^{-1} \times \ln 6 = \boxed{14.9\ \text{J K}^{-1}\ \text{mol}^{-1}}$

E3B.2(a) In each case, $\Delta_r S^\ominus = \sum_{\text{Products}} \nu S_m^\ominus - \sum_{\text{Reactants}} \nu S_m^\ominus$ [3B.2]

with S_m values obtained from Tables 2C.1 and 2C.2.

(a) $\Delta_r S = 2S_m^\ominus(\text{CH}_3\text{COOH,l}) - 2S_m^\ominus(\text{CH}_3\text{CHO,g}) - S_m^\ominus(\text{O}_2\text{,g})$

$= [(2\times 159.8) - (2\times 250.3) - 205.14]\ \text{J K}^{-1}\ \text{mol}^{-1} = \boxed{-386.1\ \text{J K}^{-1}\ \text{mol}^{-1}}$

(b) $\Delta_r S^\ominus = 2S_m^\ominus(\text{AgBr,s}) + S_m^\ominus(\text{Cl}_2\text{,g}) - 2S_m^\ominus(\text{AgCl,s}) - S_m^\ominus(\text{Br}_2\text{,l})$

$= [(2\times 107.1) + (223.07) - (2\times 96.2) - (152.23)]\ \text{J K}^{-1}\ \text{mol}^{-1}$

$= \boxed{+92.6\ \text{J K}^{-1}\ \text{mol}^{-1}}$

(c) $\Delta_r S^\ominus = S_m^\ominus(\text{HgCl}_2\text{,s}) - S_m^\ominus(\text{Hg,l}) - S_m^\ominus(\text{Cl}_2\text{,g})$

$= [146.0 - 76.02 - 223.07]\ \text{J K}^{-1}\ \text{mol}^{-1} = \boxed{-153.1\ \text{J K}^{-1}\ \text{mol}^{-1}}$

Solutions to problems

P3B.1 $S_m^\ominus(T) = S_m^\ominus(298\ \text{K}) + \Delta S$

$$\Delta S = \int_{T_1}^{T_2} C_{p,m}\frac{dT}{T} = \int_{T_1}^{T_2}\left(\frac{a}{T} + b + \frac{c}{T^3}\right) dT = a\ln\frac{T_2}{T_1} + b(T_2 - T_1) - \frac{1}{2}c\left(\frac{1}{T_2^2} - \frac{1}{T_1^2}\right)$$

(a) $S_m^\ominus(373\ \text{K}) = (192.45\ \text{J K}^{-1}\ \text{mol}^{-1}) + (29.75\ \text{J K}^{-1}\ \text{mol}^{-1}) \times \ln\left(\frac{373}{298}\right)$

$+ (25.10\times 10^{-3}\ \text{J K}^{-2}\ \text{mol}^{-1}) \times (75.0\ \text{K})$

$+ (\tfrac{1}{2}) \times (1.55\times 10^5\ \text{J K}^{-1}\ \text{mol}^{-1}) \times \left(\frac{1}{(373.15)^2} - \frac{1}{(298.15)^2}\right)$

$= \boxed{200.7\ \text{J K}^{-1}\ \text{mol}^{-1}}$

(b) $S_m^\ominus(773\ \text{K}) = (192.45\ \text{J K}^{-1}\ \text{mol}^{-1}) + (29.75\ \text{J K}^{-1}\ \text{mol}^{-1}) \times \ln\left(\frac{773}{298}\right)$

$+ (25.10\times 10^{-3}\ \text{J K}^{-2}\ \text{mol}^{-1}) \times (475\ \text{K})$

$+ (\tfrac{1}{2}) \times (1.55\times 10^5\ \text{J K}^{-1}\ \text{mol}^{-1}) \times \left(\frac{1}{773^2} - \frac{1}{298^2}\right)$

$= \boxed{232.0\ \text{J K}^{-1}\ \text{mol}^{-1}}$

P3B.3

$$\Delta_r H^\ominus = \sum_{\text{products}} \nu_J \Delta_f H^\ominus(J) - \sum_{\text{reactants}} \nu_J \Delta_f H^\ominus(J) \; [2C.5]$$

$$\Delta_r H^\ominus(298\text{ K}) = 1\times\Delta_f H^\ominus(CO,g) + 1\times\Delta_f H^\ominus(H_2O,g) - 1\times\Delta_f H^\ominus(CO_2,g)$$

$$= \{-110.53 - 241.82 - (-393.51)\}\text{ kJ mol}^{-1} = \boxed{+41.16\text{ kJ mol}^{-1}}$$

$$\Delta_r S^\ominus = \sum_{\text{products}} \nu_J S_m^\ominus(J) - \sum_{\text{reactants}} \nu_J S_m^\ominus(J) \; [3B.2]$$

$$\Delta_r S^\ominus(298\text{ K}) = 1\times S_m^\ominus(CO,g) + 1\times S_m^\ominus(H_2O,g) - 1\times S_m^\ominus(CO_2,g) - 1\times S_m^\ominus(H_2,g)$$

$$= (197.67 + 188.83 - 213.74 - 130.684)\text{ kJ mol}^{-1} = \boxed{+42.08\text{ J K}^{-1}\text{ mol}^{-1}}$$

$$\Delta_r H^\ominus(398\text{ K}) = \Delta_r H^\ominus(298\text{ K}) + \int_{298\text{ K}}^{398\text{ K}} \Delta_r C_p \mathrm{d}T \; [2C.7a]$$

$$= \Delta_r H^\ominus(298\text{ K}) + \Delta_r C_p \Delta T \; [\text{heat capacities constant}]$$

$$\Delta_r C_p = 1\times C_{p,m}(CO,g) + 1\times C_{p,m}(H_2O,g) - 1\times C_{p,m}(CO_2,g) - 1\times C_{p,m}(H_2,g)$$

$$= (29.14 + 33.58 - 37.11 - 28.824)\text{ J K}^{-1}\text{ mol}^{-1} = -3.21\text{ J K}^{-1}\text{ mol}^{-1}$$

$$\Delta_r H^\ominus(398\text{ K}) = (41.16\text{ kJ mol}^{-1}) + (-3.21\text{ JK}^{-1}\text{ mol}^{-1})\times(100\text{ K}) = \boxed{+40.84\text{ kJ mol}^{-1}}$$

For each substance in the reaction

$$\Delta S_m = C_{p,m}\ln\left(\frac{T_f}{T_i}\right) = C_{p,m}\ln\left(\frac{398\text{ K}}{298\text{ K}}\right) \quad [3A.20]$$

Thus

$$\Delta_r S^\ominus(398\text{ K}) = \Delta_r S^\ominus(298\text{ K}) + \sum_{\text{products}} \nu_J C_{p,m}(J)\ln\left(\frac{T_f}{T_i}\right) - \sum_{\text{reactants}} \nu_J C_{p,m}(J)\ln\left(\frac{T_f}{T_i}\right)$$

$$= \Delta_r S^\ominus(298\text{K}) + \Delta_r C_p \ln\left(\frac{398\text{K}}{298\text{K}}\right)$$

$$= (42.01\text{JK}^{-1}\text{ mol}^{-1}) + (-3.21\text{J K}^{-1}\text{ mol}^{-1})\ln\left(\frac{398\text{K}}{298\text{K}}\right)$$

$$= (42.01 - 0.93)\text{J K}^{-1}\text{ mol}^{-1} = \boxed{+41.08\text{ J K}^{-1}\text{ mol}^{-1}}$$

Comment. Both $\Delta_r H^\ominus$ and $\Delta_r S^\ominus$ changed little over 100 K for this reaction. This is not an uncommon result.

P3B.5 Draw up the following table and proceed as in Problems 3B.2 and 3B.4:

T/K	14.14	16.33	20.03	31.15	44.08	64.81
$(C_{p,m}/T)$ solidus $(\text{J K}^{-2}\text{ mol}^{-1})$	0.671	0.778	0.908	1.045	1.063	1.024

T/K	100.90	140.86	183.59	225.10	262.99	298.06
$(C_{p,m}/T)$ solidus $(\text{J K}^{-2}\text{ mol}^{-1})$	0.942	0.861	0.787	0.727	0.685	0.659

Plot $C_{p,m}$ against T (Fig. 3B.1(a)) and $C_{p,m}/T$ against T (Fig. 3B.1(b)), extrapolating to $T=0$ with $C_{p,m}=aT^3$ fitted at $T=14.14\,K$, which gives $a=3.36\,mJ\,K^{-1}\,mol^{-1}$.

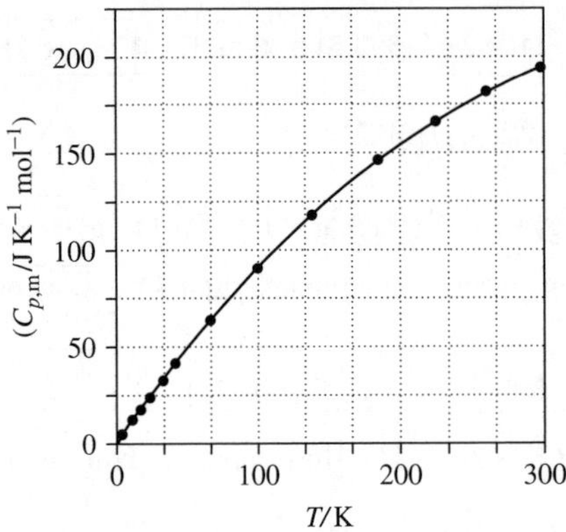

Figure 3B.1(a)

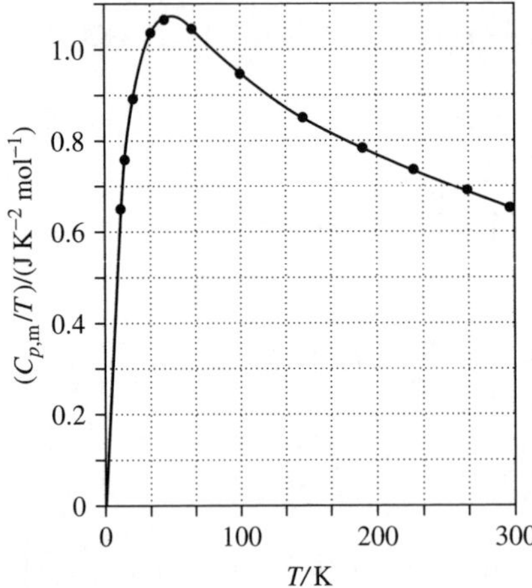

Figure 3B.1(b)

Integration by determining the area under the curve then gives

$$H_m^{\ominus}(298\,K)-H_m^{\ominus}(0)=\int_0^{298\,K} C_{p,m}\,dT=\boxed{34.4\ kJ\,mol^{-1}}$$

$$S_m(298\ K)=S_m(0)+\int_0^{298\ K}\frac{C_{p,m}}{T}dT=S_m(0)+\boxed{243\ J\,K^{-1}\,mol^{-1}}$$

P3B.7 Einstein solid: $C_{V,m}(T)=3Rf_E(T)\quad f_E(T)=\left(\frac{\theta_E}{T}\right)^2\left(\frac{e^{\theta/2T}}{e^{\theta/T}-1}\right)^2$

$$S_m=\int_0^T (C_{V,m}/T)dT=3R\int_0^T [f_E(T)/T]dT$$

$$=3R\int_0^T\left(\frac{\theta_E}{T}\right)^2\left(\frac{1}{T}\right)\left\{\frac{e^{\theta/T}}{(1-e^{\theta/T})^2}\right\}dT$$

Set $x=\theta_E/T$, then $dT=-T^2dx/\theta_E$; hence

$\frac{S_m}{3R}=\int_{\theta_E/T}^{\infty}\left[\frac{xe^x}{(1-e^x)^2}\right]dx$. This integral can be evaluated analytically. The result is

$$\frac{S_m}{3R}=\left[\frac{x}{e^x-1}-\ln(1-e^{-x})\right] \text{ with } x\equiv\frac{\theta_E}{T}$$

A more instructive alternative derivation starting from the partition function follows, but you will need to read ahead to Chapter 15 where partition functions are introduced. The partition function of one dimension of vibration of an harmonic oscillator is

$$q^V=\frac{1}{1-e^{-\beta hc\tilde{\nu}}}\ [\text{Chapter 15}]=\frac{1}{1-e^{-\theta_E/T}}$$

The energy is calculated from

$$U_m-U_m(0)=kT^2\left[\frac{\partial \ln Q}{\partial T}\right]_V \quad \text{with} \quad Q^V=(q^V)^N$$

Performing the differentiation we obtain

$$U_m-U_m(0)=R\theta_E\left(\frac{1}{e^{\theta_E/T}-1}\right).$$ We have used $N=N_A$ and $N_Ak=R$

The molar entropy is calculated from

$$S_m=\frac{U_m-U_m(0)}{T}+k\ln Q\ [\text{Chapter 15}]=\frac{U_m-U_m(0)}{T}+R\ln q^V$$

Substituting for $U_m-U_m(0)$ and q^V and multiplying by 3 for the 3 dimensions of vibration in a crystal we obtain $\frac{S_m}{3R}=\left[\frac{x}{e^x-1}-\ln(1-e^{-x})\right]$ with $x\equiv\frac{\theta_E}{T}$. Note that this is the same expression obtained above. We now plot this result against $y\equiv\frac{1}{x}=\frac{T}{\Theta_E}$, Fig. 3B.2(a).

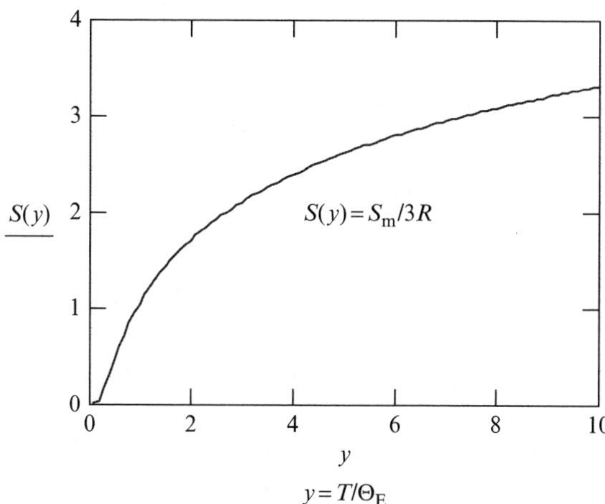

Figure 3B.2(a)

Debye solid: $C_{V,\mathrm{m}}(T)=3Rf_{\mathrm{D}}(T)\quad f_{\mathrm{D}}(T)=3\left(\dfrac{T}{\theta_{\mathrm{D}}}\right)^3\displaystyle\int_0^{\theta_{\mathrm{D}}/T}\frac{x^4\mathrm{e}^x}{(\mathrm{e}^x-1)^2}\mathrm{d}x$

$$S_{\mathrm{m}}=\int_0^T(C_{V,\mathrm{m}}/T)\mathrm{d}T=3R\int_0^T[f_{\mathrm{D}}(T)/T]\mathrm{d}T$$

The function $f_{\mathrm{D}}(T)$ cannot be integrated analytically, only numerically, the integration yielding a table of numbers, one for each value of $\frac{\theta_{\mathrm{D}}}{T}$. The entropy requires a second integration which would be the area under a plot of $C_{V,\mathrm{m}}$ against T. This double integration can certainly be accomplished rather easily with available software such as Mathcad and we do this below. However, we could again proceed as we did in the alternative derivation of the entropy equation for the Einstein solid, starting from the partition function for the Debye solid. This derivation is complex, but in essence is conceptually the same as what we did above for the case of the Einstein solid. The derivation finally yields an equation for the entropy that involves only a single numerical integration.

$$\frac{S}{3R}=\frac{3}{u^3}\int_0^u\left[\frac{x}{\mathrm{e}^x-1}-\ln(1-\mathrm{e}^{-x})\right]x^2\mathrm{d}x \text{ with } u\equiv\frac{\theta_{\mathrm{D}}}{T}$$

In the graph below, we plot this function against $y=\dfrac{1}{u}=\dfrac{T}{\Theta_{\mathrm{D}}}$

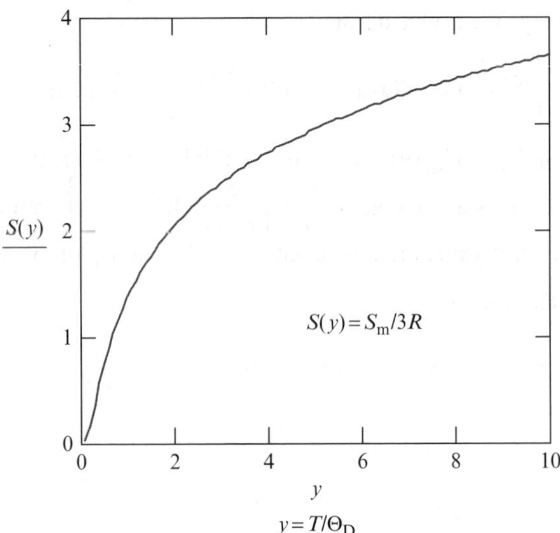

Figure 3B.2(b)

The entropy can be calculated by double integration of the equation.

$\dfrac{S_{\mathrm{m}}}{3R}=\displaystyle\int_0^y 3y^2\left[\int_0^{\frac{1}{y}}\frac{x^4\mathrm{e}^x}{(\mathrm{e}^x-1)^2}\mathrm{d}x\right]\mathrm{d}y$. The plot obtained in Fig. 3B.2(c) is identical to the one above, Fig. 3B.2(b). Also note the Debye plots are not much different from the Einstein plot. It is a little steeper at low values of y.

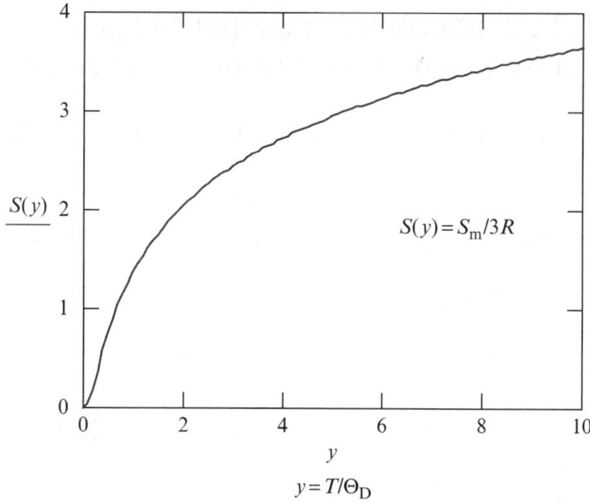

Figure 3B.2(c)

3C Concentrating on the system

Answer to discussion question

D3C.1 These criteria of spontaneity are obtained from a combination of the First Law of thermodynamics with the Second Law in the form of the Clausius inequality, $TdS \geq dq$. In its most general form this fundamental relation may be written as

$$-dU - p_{ex}dV + dw_{add} + TdS \geq 0$$

where we divided the work into pressure–volume work and additional work. Specific conditions may now be applied to this relation.

Under conditions of constant temperature and volume, with no additional work, the relation reduces to $dA \leq 0$, where A is defined as $U - TS$.

Under conditions of constant temperature and pressure, with no additional work, the relation reduces to $dG \leq 0$, where G is defined as $U + pV - TS = H - TS$.

In these relations, choosing the inequality provides the criteria for *spontaneous change*. Choosing the equal sign gives us the criteria for *equilibrium* under the conditions specified.

Solutions to exercises

E3C.1(a) In each case we use

$$\Delta_r G^{\ominus} = \Delta_r H^{\ominus} - T\Delta_r S^{\ominus} \text{ [3C.17]}$$

along with

$$\Delta_r H^{\ominus} = \sum_{\text{Products}} \nu\Delta_f H^{\ominus} - \sum_{\text{Reactants}} \nu\Delta_f H^{\ominus} \text{ [2C.5]}$$

(a) $\Delta_r H^\ominus = 2\Delta_f H^\ominus(CH_3COOH, l) - 2\Delta_f H^\ominus(CH_3CHO, g)$
$= [2\times(-484.5) - 2\times(-166.19)]\ \text{kJ mol}^{-1} = -636.6\bar{2}\ \text{kJ mol}^{-1}$

$\Delta_r G^\ominus = -636.6\bar{2}\ \text{kJ mol}^{-1} - (298.15\ \text{K})\times(-386.1\ \text{J K}^{-1}\ \text{mol}^{-1}) = \boxed{-521.5\ \text{kJ mol}^{-1}}$

(b) $\Delta_r H^\ominus = 2\Delta_f H^\ominus(AgBr, s) - 2\Delta_f H^\ominus(AgCl, s)$
$= [2\times(-100.37) - 2\times(-127.07)]\ \text{kJ mol}^{-1} = +53.40\ \text{kJ mol}^{-1}$

(c) $\Delta_r H^\ominus = \Delta_f H^\ominus(HgCl_2, s) = -224.3\ \text{kJ mol}^{-1}$

$\Delta_r G^\ominus = -224.3\ \text{kJ mol}^{-1} - (298.15\ \text{K})\times(-153.1\ \text{J K}^{-1}\ \text{mol}^{-1}) = \boxed{-178.7\ \text{kJ mol}^{-1}}$

E3C.2(a) $\Delta_r G^\ominus = \Delta_r H^\ominus - T\Delta_r S^\ominus$ [3C.17] $\Delta_r H^\ominus = \sum_{\text{Products}} \nu\Delta_f H^\ominus - \sum_{\text{Reactants}} \nu\Delta_f H^\ominus$ [2C.5]

$\Delta_r S^\ominus = \sum_{\text{Products}} \nu S_m^\ominus - \sum_{\text{Reactants}} \nu S_m^\ominus$ [3B.2]

$\Delta_r H^\ominus = 2\Delta_f H^\ominus(H_2O,l) - 4\Delta_f H^\ominus(HI,g) = \{2\times(-285.83) - 4\times(+26.48)\}\ \text{kJ mol}^{-1}$
$= -677.58\ \text{kJ mol}^{-1}$

$\Delta_r S = 2S_m^\ominus(I_2, s) + 2S_m^\ominus(H_2O, l) - 4S_m^\ominus(HI, g) - S_m^\ominus(O_2, g)$
$= [(2\times116.135) + (2\times69.91) - (4\times206.59) - (205.14)]\ \text{J K}^{-1}\ \text{mol}^{-1}$
$= -659.41\ \text{J K}^{-1}\ \text{mol}^{-1} = -0.65941\ \text{kJ K}^{-1}\ \text{mol}^{-1}$

$\Delta_r G^\ominus = -677.58\ \text{kJ mol}^{-1} - (298.15\ \text{K})\times(-0.65941\ \text{k J K}^{-1}\ \text{mol}^{-1}) = \boxed{-480.98\ \text{kJ mol}^{-1}}$

Question. Repeat the calculation based on $\Delta_f G$ data of Table 2C.2. What difference, if any, is there from the value above?

E3C.3(a) $CH_4(g) + 2\,O_2(g) \rightarrow CO_2(g) + 2\,H_2O(l)$

$\Delta_r G^\ominus = \sum_{\text{Products}} \nu\Delta_f G^\ominus - \sum_{\text{Rectants}} \nu\Delta_f G^\ominus$ [3C.18]

$\Delta_r G^\ominus = \Delta_f G^\ominus(CO_2,g) + 2\Delta_f G^\ominus(H_2O,l) - \Delta_f G^\ominus(CH_4,g)$
$= \{-394.36 + (2\times-237.13) - (-50.72)\}\ \text{kJ mol}^{-1} = -817.90\ \text{kJ mol}^{-1}$

Therefore, the maximum non-expansion work is $\boxed{817.90\ \text{kJ mol}^{-1}}$ since $|w_{\text{add}}| = |\Delta G|$

E3C.4(a) In each case $\Delta_r G^\ominus = \sum_{\text{Products}} \nu\Delta_f G^\ominus - \sum_{\text{Reactants}} \nu\Delta_f G^\ominus$ [3C.18]

with $\Delta_f G^\ominus(J)$ values from Tables 2C.1 and 2C.2.

(a) $\Delta_r G^\ominus = 2\Delta_f G^\ominus(CH_3COOH,l) - 2\Delta_f G^\ominus(CH_3CHO,g)$
$= [2\times(-389.9) - 2\times(-128.86)]\ \text{kJ mol}^{-1} = \boxed{-522.1\ \text{kJ mol}^{-1}}$

(b) $\Delta_r G^\ominus = 2\Delta_f G^\ominus(\text{AgBr,s}) - 2\Delta_f G^\ominus(\text{AgCl,s}) = [2\times(-96.90) - 2\times(-109.79)]\text{ kJ mol}^{-1}$

$= \boxed{+25.78\text{ kJ mol}^{-1}}$

(c) $\Delta_r G^\ominus = \Delta_f G^\ominus(\text{HgCl}_2\text{,s}) = \boxed{-178.6\text{ kJ mol}^{-1}}$

Comment. In each case these values of $\Delta_r G^\ominus$ agree closely with the calculated values in Exercise 3C.1(a).

E3C.5(a) The formation reaction for ethyl acetate is

$$4\,\text{C(s)} + 4\,\text{H}_2\text{(g)} + \text{O}_2\text{(g)} \rightarrow \text{CH}_3\text{COOC}_2\text{H}_5\text{(l)}$$

$$\Delta_f G^\ominus = \Delta_f H^\ominus - T\Delta_f S^\ominus \text{ [3C.17]}$$

$\Delta_f H^\ominus$ is to be obtained from $\Delta_c H^\ominus$ for ethyl acetate and data from Tables 3C.1 and 3C.2.

Thus

$$\text{CH}_3\text{COOC}_2\text{H}_5\text{(l)} + 5\,\text{O}_2\text{(g)} \rightarrow 4\,\text{CO}_2\text{(g)} + 4\,\text{H}_2\text{O(l)}$$

$$\Delta_c H^\ominus = 4\Delta_f H^\ominus(\text{CO}_2\text{,g}) + 4\Delta_f H^\ominus(\text{H}_2\text{O,l}) - \Delta_f H^\ominus(\text{CH}_3\text{COOC}_2\text{H}_5\text{(l)})$$

$$\Delta_f H^\ominus(\text{CH}_3\text{COOC}_2\text{H}_5\text{(l)}) = 4\Delta_f H^\ominus(\text{CO}_2\text{,g}) + 4\Delta_f H^\ominus(\text{H}_2\text{O,l}) - \Delta_c H^\ominus$$
$$= [4\times(-393.51) + 4\times(-285.83) - (-2231]\text{ kJ mol}^{-1}$$
$$= -486.\bar{4}\text{ kJ mol}^{-1}$$

$$\Delta_r S^\ominus = \sum_{\text{Products}} \nu S_m^\ominus - \sum_{\text{Reactants}} \nu S_m^\ominus \text{ [3B.2]}$$

$$\Delta_f S = S_m^\ominus(\text{CH}_3\text{COOC}_2\text{H}_5\text{(l)}) - 4S_m^\ominus(\text{C,s}) - 4S_m^\ominus(\text{H}_2\text{,g}) - S_m^\ominus(\text{O}_2\text{,g})$$
$$= [259.4 - (4\times 5.740) - (4\times 130.68) - (205.14)]\text{ J K}^{-1}\text{ mol}^{-1}$$
$$= -491.4\text{ J K}^{-1}\text{ mol}^{-1}$$

Hence $\Delta_r G^\ominus = -486.\bar{4}\text{ kJ mol}^{-1} - (298.15\text{ K})\times(-491.\bar{4}\text{ J K}^{-1}\text{ mol}^{-1}) = \boxed{-340\text{ kJ mol}^{-1}}$

Solutions to problems

P3C.1 First, determine the final state in each section. In section B, the volume was halved at constant temperature, so the pressure was doubled: $p_{\text{B,f}} = 2p_{\text{B,i}}$. The piston ensures that the pressures are equal in both chambers, so $p_{\text{A,f}} = 2p_{\text{B,i}} = 2p_{\text{A,i}}$. From the perfect gas law

$$\frac{T_{\text{A,f}}}{T_{\text{A,i}}} = \frac{p_{\text{A,f}}V_{\text{A,f}}}{p_{\text{A,i}}V_{\text{A,i}}} = \frac{(2p_{\text{A,i}})\times(3.00\text{ dm}^3)}{(p_{\text{A,i}})\times(2.00\text{ dm}^3)} = 3.00 \quad \text{so} \quad T_{\text{A,f}} = 900\text{ K}.$$

(a) $\Delta S_A = nC_{V,m}\ln\left(\frac{T_{\text{A,f}}}{T_{\text{A,i}}}\right)\text{[3A.20]} + nR\ln\left(\frac{V_{\text{A,f}}}{V_{\text{A,i}}}\right)$ [3A.14]

$$\Delta S_A = (2.0\,\text{mol})\times(20\,\text{J K}^{-1}\,\text{mol}^{-1})\times\ln 3.00 + (2.00\,\text{mol})\times(8.314\,\text{J K}^{-1}\,\text{mol}^{-1})\times\ln\left(\frac{3.00\,\text{dm}^3}{2.00\,\text{dm}^3}\right) = \boxed{50.7\,\text{J K}^{-1}}$$

$$\Delta S_B = nR\ln\left(\frac{V_{B,f}}{V_{B,i}}\right) = (2.00\,\text{mol})\times(8.314\,\text{J K}^{-1}\,\text{mol}^{-1})\times\ln\left(\frac{1.00\,\text{dm}^3}{2.00\,\text{dm}^3}\right) = \boxed{-11.5\,\text{J K}^{-1}}$$

(b) The Helmholtz free energy is defined as $A = U - TS$. Because section B is isothermal, $\Delta U = 0$ and $\Delta(TS) = T\Delta S$, so

$$\Delta A_B = -T_B\Delta S_B = -(300\text{ K})\left(-11.5\text{ J K}^{-1}\right) = 3.46\times10^3\text{ J} = \boxed{+3.46\text{ kJ}}$$

In section A, we cannot compute $\Delta(TS)$, so we cannot compute ΔU. ΔA $\boxed{\text{is indeterminate}}$ in both magnitude and sign. We know that in a perfect gas, U depends only on temperature; moreover, $U(T)$ is an increasing function of T, for $\dfrac{\partial U}{\partial T} = C$ (heat capacity), which is positive; since $\Delta T > 0, \Delta U > 0$ as well. But $\Delta(TS) > 0$ too, since both the temperature and the entropy increase.

(c) Likewise, under constant-temperature conditions

$$\Delta G = \Delta H - T\Delta S$$

In section B, $\Delta H_B = 0$ (constant temperature, perfect gas), so

$$\Delta G_B = -T_B\Delta S_B = -(300\text{K})\times(-11.5\,\text{J K}^{-1}) = \boxed{3.46\times10^3\text{ J}}$$

ΔG_A is $\boxed{\text{indeterminate}}$ in both magnitude and sign.

(d) $\Delta S(\text{total system}) = \Delta S_A + \Delta S_B = (50.7 - 11.5)\,\text{J K}^{-1} = \boxed{+39.2\,\text{J K}^{-1}}$

If the process has been carried out reversibly as assumed in the statement of the problem we can say

$$\Delta S(\text{system}) + \Delta S(\text{surroundings}) = 0$$

Hence, $\Delta S(\text{surroundings}) = \boxed{-39.2\,\text{J K}^{-1}}$

Question. Can you design this process such that heat is added to section A reversibly?

P3C.3 (a) At constant temperature,

$$\Delta_r G = \Delta_r H - T\Delta_r S \quad \text{so} \quad \Delta_r S = \frac{\Delta_r H - \Delta_r G}{T}$$

and $\Delta_r S = \dfrac{[-20-(-31)]\ \text{kJ mol}^{-1}}{310\ \text{K}} = +0.035\ \text{kJ K}^{-1}\ \text{mol}^{-1} = \boxed{+35\ \text{J K}^{-1}\ \text{mol}^{-1}}$

(b) The power density P is

$$P = \frac{|\Delta_r G|n}{V}$$

where n is the number of moles of ATP hydrolyzed per second

$$n = \frac{N}{N_A} = \frac{10^6\ \text{s}^{-1}}{6.02\times10^{23}\ \text{mol}^{-1}} = 1.6\bar{6}\times10^{-18}\ \text{mol s}^{-1}$$

and V is the volume of the cell

$$V = \frac{4}{3}\pi r^3 = \frac{4}{3}\pi(10\times10^{-6}\,\text{m})^3 = 4.1\bar{9}\times10^{-15}\ \text{m}^3$$

Thus $P = \dfrac{|\Delta_r G|n}{V} = \dfrac{(31\times10^3\ \text{J mol}^{-1})\times(1.6\bar{6}\times10^{-18}\ \text{mol s}^{-1})}{4.1\bar{9}\times10^{-15}\ \text{m}^3} = \boxed{12\ \text{W m}^{-3}}$

This is orders of magnitude less than the power density of a computer battery, which is about

$$P_{\text{battery}} = \frac{15\ \text{W}}{100\ \text{cm}^3}\times\left(\frac{100\ \text{cm}}{1\ \text{m}}\right)^3 = \boxed{1.5\times10^4\ \text{W m}^{-3}}$$

(c) Simply make a ratio of the magnitudes of the free energies

$$\frac{14.2\ \text{kJ (mol glutamine)}^{-1}}{31\ \text{kJ (mol ATP)}^{-1}} = \boxed{0.46\frac{\text{mol ATP}}{\text{mol glutamine}}}$$

3D Combining the First and Second Laws

Answer to discussion question

D3D.1 The relation $(\partial G/\partial T)_p = -S$ [3D.8] shows that the Gibbs function of a system decreases with T at constant p in proportion to the magnitude of its entropy. This makes good sense when one considers the definition of G, which is $G = U + pV - TS$. Hence, G is expected to decrease with T in proportion to S when p is constant. Furthermore, an increase in temperature causes entropy to increase according to

$$\Delta S = \int_i^f \mathrm{d}q_{\text{rev}}/T$$

The corresponding increase in molecular disorder causes a decline in the Gibbs energy. (Entropy is always positive.)

Solutions to exercises

E3D.1(a) $\Delta G = nRT\ln\left(\frac{p_f}{p_i}\right)$ [3D.14] $= nRT\ln\left(\frac{V_i}{V_f}\right)$ [Boyle's law]

$$\Delta G = (2.5\times10^{-3}\ \text{mol})\times(8.314\ \text{J K}^{-1}\ \text{mol}^{-1})\times(300\ \text{K})\times\ln\left(\frac{42}{600}\right) = \boxed{-17\ \text{J}}$$

E3D.2(a) $\left(\frac{\partial G}{\partial T}\right)_p = -S$ [3D.8]; hence $\left(\frac{\partial G_f}{\partial T}\right)_p = -S_f$, and $\left(\frac{\partial G_i}{\partial T}\right)_p = -S_i$

$$\Delta S = S_f - S_i = -\left(\frac{\partial G_f}{\partial T}\right)_p + \left(\frac{\partial G_i}{\partial T}\right)_p = -\left(\frac{\partial (G_f - G_i)}{\partial T}\right)_p$$

$$= -\left(\frac{\partial \Delta G}{\partial T}\right)_p = -\frac{\partial}{\partial T}\left(-85.40\ \text{J} + 36.5\ \text{J}\times\frac{T}{\text{K}}\right) = \boxed{-36.5\ \text{J K}^{-1}}$$

E3D.3(a) We will assume that the volume and molar volume of octane change little over the range of pressures given and that, therefore, eqn 3D.13, which applies to incompressible substances, can be used to solve this exercise. The change in Gibbs energy for this sample is then given by

$$\Delta G = nV_m\Delta p\ [3D.13] = V\Delta p$$

$$\Delta G = (1.0\ \text{dm}^3)\times\left(\frac{1\text{m}^3}{10^3\ \text{dm}^3}\right)\times(99\ \text{atm})\times(1.013\times10^5\ \text{Pa atm}^{-1})$$

$$= 10\ \text{kPa m}^3 = \boxed{+10\ \text{kJ}}$$

In order to calculate the change in Gibbs energy per mole, we calculate the molar volume

$$V_m = \frac{M}{\rho(\text{density})} = \frac{114.23\ \text{g mol}^{-1}}{0.703\ \text{g cm}^{-3}}\times\frac{10^{-6}\ \text{m}^3}{\text{cm}^3} = 1.625\times10^{-4}\ \text{m}^3$$

$$\Delta G_m = V_m\Delta p\ [3D.13] = 1.625\times10^{-4}\ \text{m}^3\ \text{mol}^{-1}\times(99\ \text{atm})\times(1.013\times10^5\ \text{Pa atm}^{-1})$$

$$= \boxed{1.6\ \text{kJ mol}^{-1}}$$

E3D.4(a) $\Delta G_m = RT\ln\frac{p_f}{p_i}$ [3D.15] $= (8.314\ \text{J K}^{-1}\ \text{mol}^{-1})\times(298\text{K})\times\ln\left(\frac{100.0}{1.0}\right)$

$$= \boxed{+11\ \text{kJ mol}^{-1}}$$

Solutions to problems

P3D.1 The Gibbs–Helmholtz equation [3D.9] may be recast into an analogous equation involving ΔG and ΔH, since

$$\left(\frac{\partial \Delta G}{\partial T}\right)_p = \left(\frac{\partial G_f}{\partial T}\right)_p - \left(\frac{\partial G_i}{\partial T}\right)_p$$

and $\Delta H = H_\mathrm{f} - H_\mathrm{i}$

Thus, $\left(\frac{\partial}{\partial T}\frac{\Delta_\mathrm{r}G^\ominus}{T}\right)_p = -\frac{\Delta_\mathrm{r}H^\ominus}{T^2}$

$$\mathrm{d}\left(\frac{\Delta_\mathrm{r}G^\ominus}{T}\right) = \left(\frac{\partial}{\partial T}\frac{\Delta_\mathrm{r}G^\ominus}{T}\right)_p \mathrm{d}T[\text{constant pressure}] = -\frac{\Delta_\mathrm{r}H^\ominus}{T^2}\mathrm{d}T$$

$$\Delta\left(\frac{\Delta_\mathrm{r}G^\ominus}{T}\right) = -\int_{T_\mathrm{c}}^{T}\frac{\Delta_\mathrm{r}H^\ominus\,\mathrm{d}T}{T^2}$$

$$\approx -\Delta_\mathrm{r}H^\ominus\int_{T_\mathrm{c}}^{T}\frac{\mathrm{d}T}{T^2} = \Delta_\mathrm{r}H^\ominus\left(\frac{1}{T}-\frac{1}{T_\mathrm{c}}\right) \quad [\Delta_\mathrm{r}H^\ominus \text{ assumed constant}]$$

Therefore, $\frac{\Delta_\mathrm{r}G^\ominus(T)}{T} - \frac{\Delta_\mathrm{r}G^\ominus(T_\mathrm{c})}{T_\mathrm{c}} \approx \Delta_\mathrm{r}H^\ominus\left(\frac{1}{T}-\frac{1}{T_\mathrm{c}}\right)$

and so $\Delta_\mathrm{r}G^\ominus(T) = \frac{T}{T_\mathrm{c}}\Delta_\mathrm{r}G^\ominus(T_\mathrm{c}) + \left(1-\frac{T}{T_\mathrm{c}}\right)\Delta_\mathrm{r}H^\ominus(T_\mathrm{c})$

$$= \tau\Delta_\mathrm{r}G^\ominus(T_\mathrm{c}) + (1-\tau)\Delta_\mathrm{r}H^\ominus(T_\mathrm{c}) \quad \text{where} \quad \tau = \frac{T}{T_\mathrm{c}}$$

For the reaction

$$2\,\mathrm{CO(g)} + \mathrm{O_2(g)} \rightarrow 2\,\mathrm{CO_2(g)}$$

$$\Delta_\mathrm{r}G^\ominus(T_\mathrm{c}) = 2\Delta_\mathrm{f}G^\ominus(\mathrm{CO_2,g}) - 2\Delta_\mathrm{f}G^\ominus(\mathrm{CO,g})$$
$$= [2\times(-394.36) - 2\times(-137.17)]\ \mathrm{kJ\,mol^{-1}} = -514.38\ \mathrm{kJ\,mol^{-1}}$$

$$\Delta_\mathrm{r}H^\ominus(T_\mathrm{c}) = 2\Delta_\mathrm{f}H^\ominus(\mathrm{CO_2,g}) - 2\Delta_\mathrm{f}H^\ominus(\mathrm{CO,g})$$
$$= [2\times(-393.51) - 2\times(-110.53)]\ \mathrm{kJ\,mol^{-1}} = -565.96\ \mathrm{kJ\,mol^{-1}}$$

Therefore, since $\tau = \frac{375}{298.15} = 1.25\bar{8}$

$$\Delta_\mathrm{r}G^\ominus(375\,\mathrm{K}) = \{(1.25\bar{8})\times(-514.38) + (1-1.25\bar{8})\times(-565.96)\}\ \mathrm{kJ\,mol^{-1}}$$
$$= \boxed{-501\ \mathrm{kJ\,mol^{-1}}}$$

P3D.3 $w_\mathrm{add,max} = \Delta_\mathrm{r}G$ [3C.16]

$$\Delta_\mathrm{r}G^\ominus(37\,^\circ\mathrm{C}) = \tau\,\Delta_\mathrm{r}G^\ominus(T_\mathrm{c}) + (1-\tau)\Delta_\mathrm{r}H^\ominus(T_\mathrm{c}) \quad \left[\text{Problem 3D.2, } \tau = \frac{T}{T_\mathrm{c}}\right]$$

$$= \left(\frac{310\,\mathrm{K}}{298.15\,\mathrm{K}}\right)\times(-6333\ \mathrm{kJ\,mol^{-1}}) + \left(1-\frac{310\,\mathrm{K}}{298.15\,\mathrm{K}}\right)\times(-5797\ \mathrm{kJ\,mol^{-1}})$$
$$= -6354\ \mathrm{kJ\,mol^{-1}}$$

The difference is

$$\Delta_r G^{\ominus}(37°\text{C}) - \Delta_r G^{\ominus}(T_c) = \{-6354-(-6333)\}\ \text{kJ mol}^{-1} = \boxed{-21\ \text{kJ mol}^{-1}}$$

Therefore an additional $21\ \text{kJ mol}^{-1}$ of non-expansion work may be done at the higher temperature.

Comment. As shown by Problem 3D.1, increasing the temperature does not necessarily increase the maximum non-expansion work. The relative magnitude of $\Delta_r G^{\ominus}$ and $\Delta_r H^{\ominus}$ is the determining factor.

P3D.5 $H \equiv U + pV$

$$dH = dU + p\,dV + V\,dp = T\,dS - p\,dV\ [3.43] + p\,dV + V\,dp = T\,dS + V\,dp$$

Since H is a state function, dH is exact, and it follows that

$$\left(\frac{\partial H}{\partial S}\right)_p = T \quad \text{and} \quad \boxed{\left(\frac{\partial V}{\partial S}\right)_p = \left(\frac{\partial T}{\partial p}\right)_S}$$

Similarly, $A \equiv U - TS$

$$dA = dU - T\,dS - S\,dT = T\,dS - p\,dV\ [3.43] - T\,dS - S\,dT = -p\,dV - S\,dT$$

Since dA is exact,

$$\boxed{\left(\frac{\partial S}{\partial V}\right)_T = \left(\frac{\partial p}{\partial T}\right)_V}$$

P3D.7 If $S = S(T,p)$

then $\quad dS = \left(\frac{\partial S}{\partial T}\right)_p dT + \left(\frac{\partial S}{\partial p}\right)_T dp$

$$T\,dS = T\left(\frac{\partial S}{\partial T}\right)_p dT + T\left(\frac{\partial S}{\partial p}\right)_T dp$$

Use $\quad \left(\frac{\partial S}{\partial T}\right)_p = \left(\frac{\partial S}{\partial H}\right)_p \left(\frac{\partial H}{\partial T}\right)_p = \frac{1}{T} \times C_p \quad \left[\left(\frac{\partial H}{\partial S}\right)_p = T, \text{Problem 3D.5}\right]$

$$\left(\frac{\partial S}{\partial p}\right)_T = -\left(\frac{\partial V}{\partial T}\right)_p \quad \text{[Maxwell relation]}$$

Hence, $\quad T\,dS = C_p\,dT - T\left(\frac{\partial V}{\partial T}\right)_p dp = \boxed{C_p\,dT - \alpha TV\,dp}$

For reversible, isothermal compression, $T\,dS = dq_{rev}$ and $dT = 0$; hence

$$dq_{rev} = -\alpha TV\,dp$$

$$q_{rev} = \int_{p_i}^{p_f} -\alpha TV\,dp = \boxed{-\alpha TV \Delta p}\ [\alpha \text{ and } V \text{ assumed constant}]$$

For mercury

$$q_{\text{rev}} = (-1.82\times10^{-4}\ \text{K}^{-1})\times(273\ \text{K})\times(1.00\times10^{-4}\ \text{m}^{-3})\times(1.0\times10^{8}\ \text{Pa}) = \boxed{-0.50\ \text{kJ}}$$

P3D.9 It is most efficient to first explore how the molar Gibbs energy depends on pressure using the full van der Waals expression, and then in turn to set a to zero and finally both a and b to zero in the expression and repeat the calculation. We use Mathcad[*] to perform the calculation and plot the graphs. We also use data for CO_2 as our sample gas.

The general solution of this problem requires that we develop a function that can calculate the molar volume for given values of pressure, temperature, and van der Waals constants.

Misc. definitions:

$R := 8.31447\cdot \text{J}\cdot\text{K}^{-1}\cdot\text{mol}^{-1}$ $\text{dm} := 10^{-1}\cdot\text{m}$ $\text{kJ} := 10^{3}\cdot\text{J}$ $\text{TOL} := 10^{-3}$

Function for calculation of the molar volume:

V_m estimate for Given/Find solve block: $V := 0.1\cdot\text{dm}^3\cdot\text{mol}^{-1}$

Given $p = \dfrac{R\cdot T}{V-b} - \dfrac{a}{V^2}$ $V_m(p,T,a,b) := \text{Find}(V)$

$a := 3.610\,\text{atm}\,\text{dm}^6\,\text{mol}^{-2}$ $b := 4.29\cdot10^{-2}\cdot\text{dm}^3\cdot\text{mol}^{-1}$

To see the effect of first reducing a to zero and then the effect of reducing b to zero we define a_1 and b_1 as follows.

$a_1 := 0\ \text{atm}\ \text{dm}^6\ \text{mol}^{-2}$

$b_1 := 0\,\text{dm}^3\,\text{mol}^{-1}$

$V_m(250\text{bar}, 325\text{K}, a, b) = 7.004\times10^{-2}\ \text{dm}^3\,\text{mol}^{-}$

$V_m(250\text{bar}, 325\text{K}, a_1, b) = 1.510\times10^{-1}\,\text{dm}^3\,\text{mol}^{-}$

$V_m(250\text{bar}, 325\text{K}, a_1, b_1) = 1.081\times10^{-1}\,\text{dm}^3\,\text{mol}^{-}$

We note that when $b=0$, for this choice of pressure and temperature, there is no real solution for the molar volume.

We can now use eqn 3D.12b to calculate the molar Gibbs energy change when pressure changes at constant temperature. Let $\Delta G_m(p,T) = G_m(p,T) - G_m(1\,\text{bar},T)$, then we calculate the following integrals. The first integral has both a and b nonzero, the second has $a=0$, the third $b=0$.

$$\Delta G_{ma}(p,T) := \int_{1\cdot\text{bar}}^{p} V_m(p,T,a_1,b)\,dp$$

$$\Delta G_{mb}(p,T) := \int_{1\cdot\text{bar}}^{p} V_m(p,T,a,b_1)\,dp$$

$$\Delta G_m(p,T) := \int_{1\cdot\text{bar}}^{p} V_m(p,T,a,b)\,dp$$

This last integral cannot be evaluated over the complete range of pressures from 1 bar to 300 bar because the molar volume is not real over all of this range. Let us then calculate the following integral which has both *a* and *b* equal zero, which is the case of the perfect gas.

$$\Delta G_{id}(p,T) := \int_{1\cdot bar}^{p} V_m(p,T,a_1,b_1)\,dp$$

For our carbon dioxide example, the Gibbs energy change that occurs upon increasing the pressure from 1 bar to 300 bar at 325 K is:

$$\Delta G_m(300bar, 325K) = 1.283 \times 10^1 kJ \cdot mol^{-1}$$
$$\Delta G_{ma}(300bar, 325K) = 1.670 \times 10^1 kJ \cdot mol^{-1}$$
$$\Delta G_{id}(300bar, 325K) = 1.541 \times 10^1 kJ \cdot mol^{-1}$$

We plot the change of the Gibbs energy over a chosen pressure range at the particular temperature of 325 K. See Fig. 3D.1(a).

$p := 1\cdot bar, 2\cdot bar \ldots 300\cdot bar$

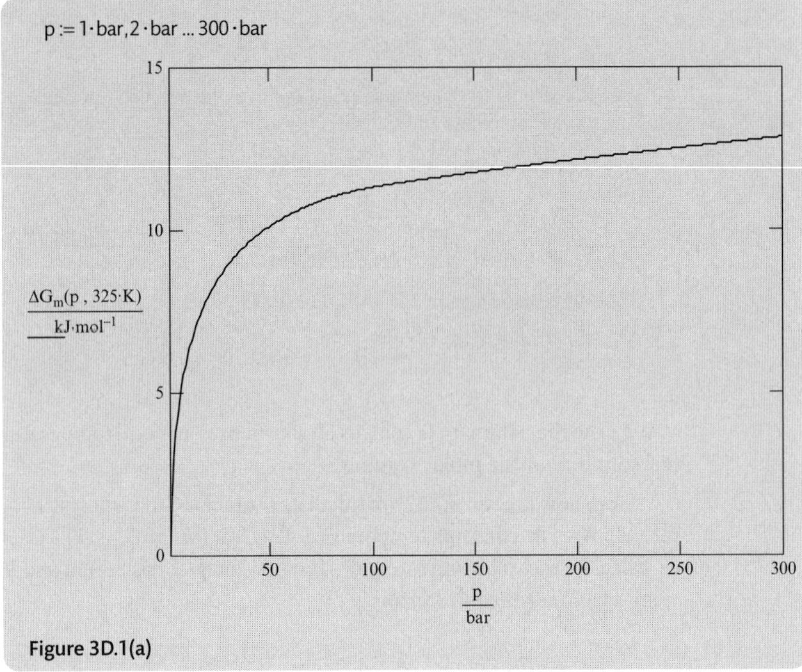

Figure 3D.1(a)

To see the effect of reducing *a* to zero, and of both *a* and *b* to zero (the perfect gas), we plot on the same graph below, Fig. 3D.1(b), ΔG_m, ΔG_{ma}(with $a=0$), and ΔG_{id}(both *a* and *b* equal zero).

Examination of Fig. 3D.1(b) shows that when intermolecular attractions are absent, but repulsions present (the top curve), the Gibbs energy increases most rapidly with pressure. When both attractions and repulsions are present (the bottom curve, same as Fig. 3D.1(a)), the Gibbs energy increases least rapidly with pressure, at least for the case of carbon dioxide, but since the rate of increase is dependent upon the magnitude of a and b, so it would not necessarily be the same for a different gas.

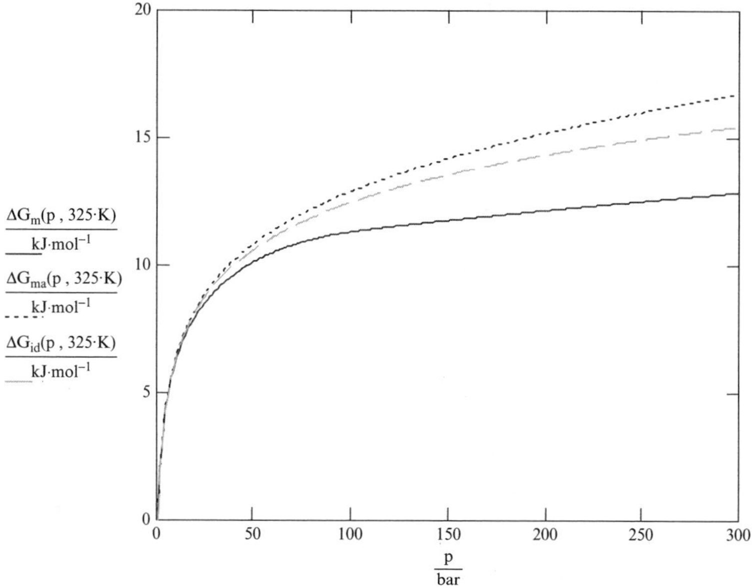

Figure 3D.1(b)

Integrated activities

I3.1 Consider the temperature as a function of pressure and enthalpy: $T=T(p,H)$

so $$dT=\left(\frac{\partial T}{\partial p}\right)_H dp+\left(\frac{\partial T}{\partial H}\right)_p dH$$

The Joule–Thomson expansion is a constant-enthalpy process (Section 2D.3). Hence,

$$dT=\left(\frac{\partial T}{\partial p}\right)_H dp=\mu dp$$

$$\Delta T=\int_{p_i}^{p_f}\mu dp=\mu\Delta p \quad [\mu \text{ is constant}]$$

$$=(0.21\text{K atm}^{-1})\times(1.00\text{ atm}-100\text{ atm})=\boxed{-21\text{ K}}$$

$$T_f=T_i+\Delta T=(373-21)\text{ K}=352\text{ K}\,[\text{Mean } T=363\text{ K}]$$

Consider the entropy as a function of temperature and pressure: $S=S(T,p)$

Therefore, $$dS=\left(\frac{\partial S}{\partial T}\right)_p dT+\left(\frac{\partial S}{\partial p}\right)_T dp$$

$$\left(\frac{\partial S}{\partial T}\right)_p = \frac{C_p}{T} \quad \left(\frac{\partial S}{\partial p}\right)_T = -\left(\frac{\partial V}{\partial T}\right)_p \quad \text{[Maxwell relation]}$$

For $V_\mathrm{m} = \frac{RT}{p}(1+Bp)$

$$\left(\frac{\partial V_\mathrm{m}}{\partial T}\right)_p = \frac{R}{p}(1+Bp)$$

Then $\mathrm{d}S_\mathrm{m} = \frac{C_{p,\mathrm{m}}}{T}\mathrm{d}T - \frac{R}{p}(1+Bp)\mathrm{d}p$

or $\mathrm{d}S_\mathrm{m} = \frac{C_{p,\mathrm{m}}}{T}\mathrm{d}T - \frac{R}{p}\mathrm{d}p - RB\mathrm{d}p$

Upon integration

$$\Delta S_\mathrm{m} = \int_1^2 \mathrm{d}S_\mathrm{m} = C_{p,\mathrm{m}} \ln\left(\frac{T_2}{T_1}\right) - R\ln\left(\frac{p_2}{p_1}\right) - RB(p_2 - p_1)$$

$$= \tfrac{5}{2}R\ln\left(\frac{352}{373}\right) - R\ln\left(\frac{1}{100}\right) - R\left(-\frac{0.525\ \text{atm}^{-1}}{363}\right) \times (-99\ \text{atm})$$

$$= \boxed{+35.9\ \mathrm{J\,K^{-1}\,mol^{-1}}}$$

I3.3 (a) See the solution to Problem 3D.4 where it is established that for a van der Waals gas

$$\left(\frac{\partial S}{\partial V}\right)_T = \frac{R}{V_\mathrm{m} - b}. \text{ Then}$$

$$\Delta S_\mathrm{m} = \int_{V_{\mathrm{m}1}}^{V_{\mathrm{m}2}} \left(\frac{R}{V_\mathrm{m} - b}\right)\mathrm{d}V_\mathrm{m} = R\ln\left(\frac{V_{m2} - b}{V_{\mathrm{m}1} - b}\right) = R\ln\left(\frac{10 - 0.0429}{1 - 0.0429}\right)$$

$$\Delta S_\mathrm{m} = 2.3421\,R = 2.3421 \times 8.3145\ \mathrm{J\,K^{-1}\,mol^{-1}} = \boxed{19.473\ \mathrm{J\,K^{-1}\,mol^{-1}}}$$

Note that this change in entropy treating CO_2 as a van der Waals gas is only slightly larger than treating it as a perfect gas for the same change of volume at this temperature.

(b) Use eqn 3A.20, substitute the given equation for C and integrate. The final equation for ΔS_m becomes

$$\Delta S_\mathrm{m} = a\ln\frac{T_2}{T_1} + b(T_2 - T_1) - \frac{1}{2}c\left(\frac{1}{T_2^2} - \frac{1}{T_1^2}\right)$$

Then using various values of T_1, T_2, a, b, and c, we can construct the table below. In all cases we will assume $S_m(0)=0$. We have also assumed that there have been no phase changes, only temperature changes within the phase which is unspecified (solid, liquid, or gas).

Calc	T_1/K	T_2/K	a/J K^{-1} mol^{-1}	b/mJ K^{-2} mol^{-1}	c/kJ K mol^{-1}	ΔS_m/J K^{-1} mol^{-1}
1	273.15	373.15	44.22	8.79	−862	11.993
2		373.15	34.22	8.79	−862	8.873
3		373.15	54.22	8.79	−862	15.112
4		473.15	44.22	8.79	−862	22.201
5		373.15	44.22	−8.79	−862	10.235
6		473.15	44.22	−8.79	−862	22.536
7		373.15	44.22	3.79	−862	11.493
8		373.15	44.22	−3.79	−862	10.375
9		373.15	44.22	8.79	0	14.674
10		473.15	44.22	8.79	0	26.052
11		373.15	44.22	8.79	+862	17.355
12		473.15	44.22	8.79	+862	29.904

Comparing calculations 1 to 4, 5 to 6, 9 to 10, and 11 to 12 shows that no matter what the values of the coefficients a, b, and c, increasing the temperature increases the entropy. That is to be expected. Comparing calculation 1 to calculations 2 and 3 shows how the parameter a affects the entropy; greater a results in greater entropy, as expected from the form of the entropy equation above. Comparison of calculation 1 to calculations 5, 7, and 8 shows how the parameter b affects the entropy. The entropy is not highly dependent on the parameter b; even changing the sign of b does not greatly alter the entropy. However, changing the sign of parameter c results in substantial change in the entropy as can be seen by comparing calculation 1 to calculations 9 and 11, or by comparing calculation 4 to calculations 10 and 12.

(c) See Section 3D.2(c), and eqn 3D.7, which leads to eqn 3D.8.

$$\left(\frac{\partial G}{\partial p}\right)_T = V \,[\text{3D.8}]$$

Thus the physical significance of this partial derivative is simply the volume of the system. The volume of the system is determined by p and T, but at constant T only by p. The manner in which V depends on p is determined by the nature of the system and the equation of state that describes that system. For a perfect gas

$$V = \frac{nRT}{p},$$

which at constant T is simply Boyle's law. We can plot this V against p as shown in Fig. I3.1 below.

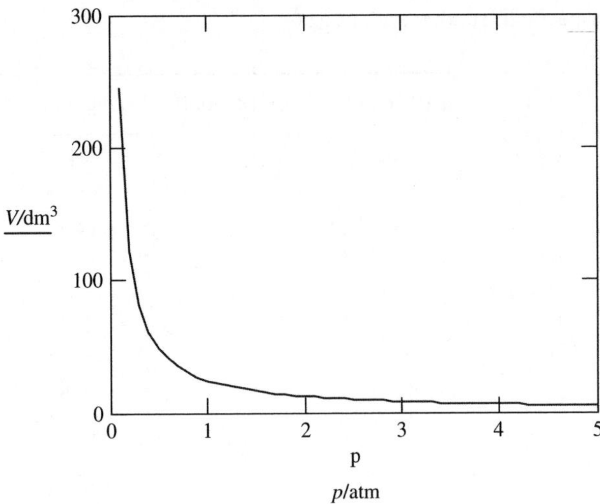

Figure I3.1 (*V* against *p* for a perfect gas)

(d) The starting point for the calculation is eqn 3D.18. To evaluate the integral, we need an analytical expression for Z, which can be obtained from the equation of state.

The integral in eqn 3D.18 that we require is

$$\ln\phi = \int_0^p \left(\frac{Z-1}{p}\right) dp$$

Now go to the whfreeman.com web site and access the applet entitled *Fugacity of Gases*. In this applet, the fugacity coefficient for a van der Waals gas is calculated as a function of the reduced volume, V_r, for various values of the reduced temperature, T_r.

The fugacity coefficient is obtained from the integration

$$\ln(\phi) = \int_0^P \frac{Z-1}{P} dp$$

To carry out this integration, we note that

$$Z(V_r, T_r) = \frac{3 \cdot V_r}{3 \cdot V_r - 1} - \frac{9}{8 \cdot T_r V_r^2}$$

$$p_r(V_r, T_r) = \frac{8T_r}{3 \cdot V_r - 1} - \frac{3}{V_r^2}$$

To change the variable of integration from **p** to $\mathbf{V_r}$ we need

$$dp_r / dV_r = \frac{d}{dV_r}\left(\frac{8\cdot T_r}{3\cdot V_r - 1} - \frac{3}{V_r^2}\right) = -6\frac{(4\cdot T_r V_r^3 - 9\cdot V_r^2 + 6\cdot V_r - 1)}{\left[(3\cdot V_r - 1)^2\cdot V_r^3\right]}$$

The expression **{(Z − 1)/p} dp** may be written

$$\frac{\frac{3\cdot V_r}{3\cdot V_r - 1} - \frac{9}{8\cdot T_r\cdot V_r^2} - 1}{\frac{8\cdot T_r}{3\cdot V_r - 1} - \frac{3}{V_r^2}} \cdot \left[-6\cdot\frac{(4\cdot T_r\cdot V_r^3 - 9\cdot V_r^2 + 6\cdot V_r - 1)}{\left[(3\cdot V_r - 1)^2\cdot V_r^3\right]}\right] dV_r$$

The factor multiplying $\mathbf{dV_r}$ simplifies to

$$\frac{-3}{4}\cdot\frac{(-27\cdot V_r + 9 + 8\cdot T_r\cdot V_r^2)}{\left[(3\cdot V_r - 1)^2\cdot T_r\cdot V_r^3\cdot(8\cdot T_r\cdot V_r^2 - 9\cdot V_r + 3)\right]}\cdot(4\cdot T_r\cdot V_r^3 - 9\cdot V_r^2 + 6\cdot V_r - 1)$$

Therefore, the integral we require is

$$I(V_r,T_r) = \int_{V_r}^{\infty}\frac{3}{4}\cdot\frac{(-27\cdot V_r + 9 + 8\cdot T_r\cdot V_r^2)}{\left[(3\cdot V_r - 1)^2\cdot T_r\cdot V_r^3\cdot(8\cdot T_r\cdot V_r^2 - 9\cdot V_r + 3)\right]}\cdot(4\cdot T_r\cdot V_r^3 - 9\cdot V_r^2 + 6\cdot V_r - 1)\,dV_r$$

The fugacity coefficient is

$$\phi(V_r, T_r) = \exp(-I(V_r, T_r))$$

where **I** is the integral evaluated above.

To graph this equation set the reduced temperature ($\mathbf{T_r}$) and hit **New Plot**. Up to five plots can be displayed at one time. The **Clear** button will remove all plots. To see the parameters for each plot hit the **Legend on/off** button. The **Redraw** button will refresh the graph. This is useful when the function domain has been changed. To see the value of each plot at a given point, move your cursor to the desired location then click and hold.

Examining the plots we see that the fugacity coefficient is always less than one for all values of V_r greater than 0.8 and that the fugacity coefficient decreases as T_r increases above 1.0 for all values of V_r.

4 Physical transformations of pure substances

4A Phase diagrams of pure substances

Answers to discussion questions

D4A.1 Chemical potential is the single function that governs phase stability. The phase whose chemical potential is least under a set of given conditions is the most stable. Conditions under which two or more phases have equal chemical potentials are conditions under which those phases are in equilibrium. Understanding how chemical potential varies with physical conditions such as temperature, pressure, and (in later chapters) composition allows one to compute chemical potentials for various phases and to map out the conditions for stability of those phases and for equilibrium between them.

D4A.3 For two phases to be in equilibrium, the chemical potentials of each component must be equal in the two phases. In a one-component system, this means that the chemical potential of that one component must be the same in all phases that are in equilibrium. The chemical potential is a function of two variables, say p and T (and not of composition in a one-component system). Thus, if we had four phases α, β, γ, and δ in equilibrium, we would have

$$\mu(p,T;\alpha)=\mu(p,T;\beta)=\mu(p,T;\gamma)=\mu(p,T;\delta)$$

This is a set of three independent equations in only two variables (p and T), which are not compatible. See *Justification* 4A.1.

Solutions to exercises

E4A.1(a) The phase rule (eqn 4A.1) relates the number of phases (P), components (C), and degrees of freedom (F) of a thermodynamic system:

$$F=C-P+2.$$

Restricting to pure substances ($C=1$) and rearranging for phases gives

$$P=3-F.$$

Areas in the phase diagram have two degrees of freedom; one can vary pressure and temperature independently (within limits) and stay within the area. Thus, $F = 2$ and $P = 1$ in areas. Lines have one degree of freedom; one can vary pressure or temperature, but to stay on the line the value of the other is determined by the line. Thus, $F = 1$ and $P = 2$ on lines. Points on the phase diagram have zero degrees of freedom; one can vary neither pressure nor temperature and remain on a given point. Thus, $F = 0$ and $P = 3$ on points.

(a) is in an area, so there is a $\boxed{\text{single phase}}$. (b) and (d) are on lines, so there are $\boxed{\text{two phases}}$ present. (c) is a point, so there are $\boxed{\text{three phases}}$ present.

E4A.2(a) For pure substances (one-component systems), the chemical potential is the molar Gibbs energy:

$$\mathrm{d}G = (\mu_2 - \mu_1)\mathrm{d}n$$

so $$\Delta G = (\mu_2 - \mu_1)n = (+7.1\ \mathrm{kJ\,mol^{-1}})(0.10\times10^{-3}\ \mathrm{mol}) = +0.71\times10^{-3}\ \mathrm{kJ} = \boxed{0.71\ \mathrm{J}}.$$

E4A.3(a) Use the phase rule (eqn 4A.1)

$$F = C - P + 2$$

to solve for the number of phases:

$$P = C - F + 2 = 2 - F + 2 = 4 - F \le \boxed{4}$$

The maximum number of phases in equilibrium occurs when the number of degrees of freedom is at a minimum, namely zero; that number is four.

4B Thermodynamic aspects of phase transitions

Answers to discussion questions

D4B.1 See Section 4B.1(a). The mathematical reason can be seen in eqn 4B.1,

$$\left(\frac{\partial \mu}{\partial T}\right)_p = -S_\mathrm{m}$$

Because $S_\mathrm{m} > 0$ for all pure substances, the slope of the change in chemical potential with respect to change in temperature is negative: chemical potential decreases with increasing temperature. The physical reason is also entropic. Recall that the Gibbs function weighs two competing criteria for stability, namely lower energy and greater entropy (disorder). The higher the temperature, the greater the weight on the entropic term—or rather the lower the relative weight of the energetic term as higher-energy states become more populated at higher temperature.

D4B.3 A differential scanning calorimeter can be used to identify phase transitions, whether first-order, second-order, or lambda transitions. (See Section 4B.3 for classification of phase transitions.) Differential scanning calorimetry (DSC) measures constant-pressure heat capacity (C_p) as a function of temperature by electrically heating a sample and comparing it to a reference sample. (See Section 2C.4(a) for more information on DSC.) Phase transitions are characterized by discontinuities in C_p as a function of temperature. Familiar first-order phase transitions such as fusion and evaporation exhibit latent enthalpies—non zero changes in enthalpy at constant temperature. Thus, these phase changes have infinite spikes in C_p. Second-order transitions have discontinuities in C_p albeit not infinite ones. Either way, such changes in C_p will stand out against an otherwise constant or slowly changing plot of C_p vs. temperature.

Solutions to exercises

E4B.1(a) The difference between the definition of normal and standard transition temperatures is the pressure at which the transition takes place: normal refers to exactly 1 atm (101 325 Pa), while standard refers to exactly 1 bar (exactly 10^5 Pa). At the standard melting temperature and pressure, the liquid and solid phases are in equilibrium, so their chemical potentials are equal:

$$\mu_{\text{liquid}}(T_{\text{std}}, p_{\text{std}}) = \mu_{\text{solid}}(T_{\text{std}}, p_{\text{std}})$$

The same can be said at the normal melting temperature and pressure:

$$\mu_{\text{liquid}}(T_{\text{norm}}, p_{\text{norm}}) = \mu_{\text{solid}}(T_{\text{norm}}, p_{\text{norm}})$$

Equations 4B.1 and 4B.2 show how the chemical potential changes with temperature and pressure, so for small changes we can write

$$\mathrm{d}\mu = \left(\frac{\partial \mu}{\partial T}\right)_p \mathrm{d}T + \left(\frac{\partial \mu}{\partial p}\right)_T \mathrm{d}p = -S_{\mathrm{m}}\mathrm{d}T + V_{\mathrm{m}}\mathrm{d}p$$

Assuming that the differences between standard and normal melting point are small enough, we can equate the differences in the chemical potentials of the two phases:

$$\Delta\mu_{\text{solid}} = -S_{\text{m,solid}}\Delta T + V_{\text{m,solid}}\Delta p = -S_{\text{m,liquid}}\Delta T + V_{\text{m,liquid}}\Delta p = \Delta\mu_{\text{liquid}},$$

where Δp is defined as $p_{\text{norm}} - p_{\text{std}}$. Rearrange to isolate ΔT:

$$(S_{\text{m,liquid}} - S_{\text{m,solid}})\Delta T = (V_{\text{m,liquid}} - V_{\text{m,solid}})\Delta p,$$

$$(\Delta_{\text{fus}}S)\Delta T = (V_{\text{m,liquid}} - V_{\text{m,solid}})\Delta p \quad \text{so} \quad \Delta T = \frac{(V_{\text{m,liquid}} - V_{\text{m,solid}})\Delta p}{\Delta_{\text{fus}}S}.$$

The molar volume of each phase is the molar mass divided by the density. We need to look up the density of ice and of liquid water near the melting temperature. We also need to find $\Delta_{\text{fus}}S$ or to use eqn 3A.17.

$$\Delta T=\left(\frac{1}{\rho_{\text{liquid}}}-\frac{1}{\rho_{\text{solid}}}\right)\frac{M\Delta p}{\Delta_{\text{fus}}S}=\left(\frac{1}{\rho_{\text{liquid}}}-\frac{1}{\rho_{\text{solid}}}\right)\frac{MT_{\text{f}}\Delta p}{\Delta_{\text{fus}}H}$$

$$=\left(\frac{1\,\text{cm}^3}{1.000\,\text{g}}-\frac{1\,\text{cm}^3}{0.915\,\text{g}}\right)\frac{(18.02\,\text{g}\,\text{mol}^{-1})(273.15\,\text{K})(1\,325\,\text{Pa})}{6\,008\,\text{J}\,\text{mol}^{-1}}\times\frac{1\,\text{m}^3}{10^6\,\text{cm}^3}$$

$$=\boxed{-1.0\times10^{-4}\,\text{K}}$$

That is, the normal melting temperature is 0.1 mK lower than the standard melting temperature.

E4B.2(a) Equation 4B.1 shows how the chemical potential changes with temperature

$$\text{d}\mu=\left(\frac{\partial\mu}{\partial T}\right)_p\text{d}T=-S_{\text{m}}\text{d}T$$

so $$\Delta\mu=-\int S_{\text{m}}\,\text{d}T=-S_{\text{m}}\Delta T=-69.91\,\text{J}\,\text{K}^{-1}\,\text{mol}^{-1}\times(100-25)\,\text{K}$$

$$=\boxed{5.2\times10^3\,\text{J}\,\text{mol}^{-1}}=\boxed{5.2\,\text{kJ}\,\text{mol}^{-1}}$$

Note: we assumed that the entropy is constant over the temperature range.

E4B.3(a) Equation 4B.2 shows how the chemical potential changes with pressure

$$\text{d}\mu=\left(\frac{\partial\mu}{\partial p}\right)_T\text{d}p=V_{\text{m}}\text{d}p=\frac{M}{\rho}\text{d}p$$

so $$\Delta\mu=\int\frac{M}{\rho}\text{d}p=\frac{M}{\rho}\Delta p=\frac{63.55\,\text{g}\,\text{mol}^{-1}}{8.960\,\text{g}\,\text{cm}^{-3}}\times(10\times10^6-100\times10^3)\,\text{Pa}\times\frac{1\,\text{m}^3}{10^6\,\text{cm}^3}$$

$$=\boxed{70\,\text{J}\,\text{mol}^{-1}}$$

Note: we assumed that the sample is incompressible.

E4B.4(a) The effect on vapour pressure of a change in applied external pressure on a liquid is given by eqn 4B.3:

$$p=p^*\text{e}^{V_{\text{m}}(\text{l})\Delta P/RT}.$$

For water, the molar volume at 20 °C (based on density from the *CRC Handbook*) is

$$V_{\text{m}}=\frac{M}{\rho}=\frac{18.02\,\text{g}\,\text{mol}^{-1}}{0.9982\,\text{g}\,\text{cm}^{-3}}=18.05\,\text{cm}^3\,\text{mol}^{-1}$$

so $$\frac{V_{\text{m}}(\text{l})\Delta P}{RT}=\frac{18.05\,\text{cm}^3\,\text{mol}^{-1}\times(20\times10^6-1.0\times10^5)\,\text{Pa}}{8.3145\,\text{J}\,\text{K}^{-1}\,\text{mol}^{-1}\times293\,\text{K}}\times\frac{1\,\text{m}^3}{10^6\,\text{cm}^3}=0.147$$

and $$p=p^*\text{e}^{V_{\text{m}}(\text{l})\Delta P/RT}=(2.34\,\text{kPa})\text{e}^{0.147}=\boxed{2.71\,\text{kPa}}.$$

E4B.5(a) Use the Clapeyron equation (eqn 4B.5a)

$$\frac{dp}{dT}=\frac{\Delta_{trs}S}{\Delta_{trs}V}$$

Assume that $\Delta_{fus}S$ and $\Delta_{fus}T$ are independent of temperature:

$$\Delta_{fus}S=\Delta_{fus}V\times\left(\frac{dp}{dT}\right)\approx\Delta_{fus}V\times\frac{\Delta p}{\Delta T}$$

$$\Delta_{fus}S=\left[(163.3-161.0)\times10^{-6}\ \text{m}^3\ \text{mol}^{-1}\right]\times\left(\frac{(100-1)\times(1.013\times10^5\ \text{Pa})}{(351.26-350.75)\ \text{K}}\right)$$

$$=\boxed{+45.2\bar{3}\ \text{J K}^{-1}\ \text{mol}^{-1}}$$

At the melting temperature

$$\Delta_{fus}H=T_f\Delta_{fus}S=(350.75\ \text{K})\times(45.23\ \text{J K}^{-1}\ \text{mol}^{-1})=\boxed{+16\ \text{kJ mol}^{-1}}$$

E4B.6(a) On the assumption that the vapour is a perfect gas and that $\Delta_{vap}H$ is independent of temperature, we may write [4B.11]

$$p=p^*e^{-\chi},\quad \chi=\left(\frac{\Delta_{vap}H}{R}\right)\times\left(\frac{1}{T}-\frac{1}{T^*}\right),\quad \ln\frac{p^*}{p}=\chi$$

$$\frac{1}{T}=\frac{1}{T^*}+\frac{R}{\Delta_{vap}H}\ln\frac{p^*}{p}$$

$$=\frac{1}{297.25\text{K}}+\left(\frac{8.3145\text{J K}^{-1}\ \text{mol}^{-1}}{28.7\times10^3\ \text{J mol}^{-1}}\right)\ln\frac{53.3\ \text{kPa}}{70.0\ \text{kPa}}=3.28\bar{5}\times10^{-3}\ \text{K}^{-1}$$

Hence $T=\boxed{304\ \text{K}}=\boxed{31°\text{C}}$.

E4B.7(a) Integrating the Clausius-Clapeyron equation [4B.10] yields an expression for $\ln p$:

$$\int d\ln p=\int\frac{\Delta_{vap}H}{RT^2}dT$$

so $$\ln p=\text{constant}-\frac{\Delta_{vap}H}{RT}$$

Therefore, $\Delta_{vap}H=(2501.8\ \text{K})\times R=(2501.8\ \text{K})\times(8.3145\ \text{J K}^{-1}\ \text{mol}^{-1})$

$$=\boxed{+20.80\ \text{kJ mol}^{-1}}$$

E4B.8(a) (a) The indefinitely integrated form of eqn 4B.10 is used as in Exercise 4B.7(a).

$$\ln p=\text{constant}-\frac{\Delta_{vap}H}{RT},\quad\text{or}\quad \log p=\text{constant}-\frac{\Delta_{vap}H}{2.303\ RT}$$

Therefore,

$$\Delta_{\text{vap}}H = (2.303)\times(1\,780\,\text{K})\times R = (2.303)\times(1\,780\,\text{K})\times(8.3145\,\text{J K}^{-1}\,\text{mol}^{-1})$$

$$= \boxed{+34.08\,\text{kJ mol}^{-1}}$$

(b) The normal boiling point corresponds to $p = 1.000$ atm $= 760$ Torr,

so $\quad \log 760 = 7.960 - \dfrac{1\,780\,\text{K}}{T_\text{b}}$

and $\quad T_\text{b} = \boxed{350.5\,\text{K}}$.

E4B.9(a) $\quad \Delta T \approx \dfrac{\Delta_{\text{fus}}V}{\Delta_{\text{fus}}S}\times\Delta p\ [\text{4B.5a and Exercise 4B.5(a)}]$

$$\approx \frac{T_\text{f}\Delta_{\text{fus}}V}{\Delta_{\text{fus}}H}\Delta p = \frac{T_\text{f}M\Delta p}{\Delta_{\text{fus}}H}\times\Delta\left(\frac{1}{\rho}\right)\quad [V_\text{m} = M/\rho]$$

$$\approx \frac{278.6\,\text{K}\times 999\,\text{atm}\times 78.11\,\text{g mol}^{-1}}{10.59\times10^{3}\,\text{J mol}^{-1}}\times\left(\frac{1}{0.879\,\text{g cm}^{-3}} - \frac{1}{0.891\,\text{g cm}^{-3}}\right)$$

$$\times\left(\frac{1.013\times10^{5}\,\text{Pa}}{1\,\text{atm}}\right)\times\left(\frac{1\,\text{m}^3}{10^6\,\text{cm}^3}\right)$$

$$\approx 3.2\,\text{K}$$

Therefore, at 1 000 atm, $T_\text{f} = (278.6 + 3.2)\,\text{K} = \boxed{281.8\,\text{K}}$ or $\boxed{8.7\,°\text{C}}$.

E4B.10(a) The rate of loss of mass of water may be expressed as

$$\frac{\text{d}m}{\text{d}t} = \frac{\text{d}}{\text{d}t}(nM) \quad \text{where} \quad n = \frac{q}{\Delta_{\text{vap}}H}$$

Thus $\quad \dfrac{\text{d}n}{\text{d}t} = \dfrac{\text{d}q/\text{d}t}{\Delta_{\text{vap}}H} = \dfrac{(1.2\times10^{3}\,\text{W m}^{-2})\times(50\,\text{m}^2)}{44.0\times10^{3}\,\text{J mol}^{-1}} = 1.4\,\text{mol s}^{-1}$

and $\quad \dfrac{\text{d}m}{\text{d}t} = (1.4\,\text{mol s}^{-1})\times(18.02\,\text{g mol}^{-1}) = \boxed{25\,\text{g s}^{-1}}$

E4B.11(a) Assume perfect gas behaviour. The volume is 75 m^3.

$$n = \frac{pV}{RT} = \frac{m}{M} \quad \text{so} \quad m = \frac{pVM}{RT}$$

(a) $\quad m = \dfrac{(3.2\times10^{3}\,\text{Pa})\times(75\,\text{m}^3)\times(18.02\,\text{g mol}^{-1})}{(8.3145\,\text{J K}^{-1}\,\text{mol}^{-1})\times(298\,\text{K})} = \boxed{1.7\times10^{3}\,\text{g}}$ water

(b) $\quad m = \dfrac{(13.1\times10^{3}\,\text{Pa})\times(75\,\text{m}^3)\times(78.11\,\text{g mol}^{-1})}{(8.3145\,\text{J K}^{-1}\,\text{mol}^{-1})\times(298\,\text{K})} = \boxed{31\times10^{3}\,\text{g}}$ benzene

(c) $m=\dfrac{(0.23\ \text{Pa})\times(75\ \text{m}^3)\times(200.59\ \text{g mol}^{-1})}{(8.3145\ \text{J K}^{-1}\ \text{mol}^{-1})\times(298\ \text{K})}=\boxed{1.4\ \text{g}}$ mercury

Comment. Note that an ordinary reagent bottle of benzene would evaporate completely before saturating the air of the laboratory with benzene vapour.

Question. Assuming all the mercury vapour breathed remains in the body, how long would it take to accumulate 1.4 g? Make reasonable assumptions about the volume and frequency of a breath.

E4B.12(a) The Clausius–Clapeyron equation [4B.10] integrates to the form [4B.11] which may be rewritten as

$$\ln\left(\frac{p_2}{p_1}\right)=\frac{\Delta_{\text{vap}}H}{R}\times\left(\frac{1}{T_1}-\frac{1}{T_2}\right)$$

(a) $\Delta_{\text{vap}}H=R\ln\left(\dfrac{p_2}{p_1}\right)\times\left(\dfrac{1}{T_1}-\dfrac{1}{T_2}\right)^{-1}$

$$=(8.3145\ \text{J K}^{-1}\ \text{mol}^{-1})\times\ln\left(\frac{5.3\ \text{kPa}}{1.3\ \text{kPa}}\right)\times\left(\frac{1}{359.0\ \text{K}}-\frac{1}{392.5\ \text{K}}\right)^{-1}$$

$$=\boxed{+4.9\times10^4\ \text{J mol}^{-1}}=\boxed{+49\ \text{kJ mol}^{-1}}$$

(b) Rearrange the equation above to isolate a temperature

$$\frac{R}{\Delta_{\text{vap}}H}\ln\left(\frac{p_2}{p_1}\right)+\frac{1}{T_2}=\frac{1}{T_1}$$

The normal boiling point (T_1 in this equation) corresponds to a vapour pressure of 101.3 kPa. Using the data at 119.3 °C (T_2), we have

$$\left(\frac{8.3145\ \text{J K}^{-1}\ \text{mol}^{-1}}{+4.9\times10^4\ \text{J mol}^{-1}}\right)\ln\left(\frac{5.3\ \text{kPa}}{101.3\ \text{kPa}}\right)+\frac{1}{392.5\ \text{K}}=2.0\bar{5}\times10^{-3}\ \text{K}^{-1}=\frac{1}{48\bar{8}\ \text{K}}$$

Thus, the estimated boiling point is $48\bar{8}\ \text{K}=\boxed{21\bar{5}\ °\text{C}}$

The accepted value is 218 °C.

(c) At the boiling point

$$\Delta_{\text{vap}}S=\frac{\Delta_{\text{vap}}H}{T_b}\approx\frac{+49\times10^3\ \text{J mol}^{-1}}{48\bar{8}\ \text{K}}=\boxed{+101\ \text{J K}^{-1}\ \text{mol}^{-1}}$$

E4B.13(a)

$$\Delta T\approx\frac{T_f M\Delta p}{\Delta_{\text{fus}}H}\times\Delta\left(\frac{1}{\rho}\right)\ \text{[Exercise 4B.9(a)]}$$

$$\Delta T\approx\left(\frac{(273.15\ \text{K})\times(49\times10^5\ \text{Pa})\times(18.0\ \text{g mol}^{-1})}{6.01\times10^3\ \text{J mol}^{-1}}\right)$$

$$\times\left(\frac{1}{1.00\ \text{g cm}^{-3}}-\frac{1}{0.92\ \text{g cm}^{-3}}\right)\times\frac{1\ \text{m}^3}{10^6\ \text{cm}^3}=-0.35\ \text{K}$$

$T_f(50\ \text{bar})=(273.15-0.35)\ \text{K}=\boxed{272.80\ \text{K}}$

E4B.14(a) $\Delta_{vap}H = \Delta_{vap}U + \Delta_{vap}(pV) = 40.656\ \text{kJ mol}^{-1}$

$$\Delta_{vap}(pV) = p\Delta_{vap}V = p(V_{gas} - V_{liq}) \approx pV_{gas} = RT\ [\text{perfect gas}]$$

$$\Delta_{vap}(pV) \approx (8.3145\ \text{J K}^{-1}\ \text{mol}^{-1}) \times (373.2\ \text{K}) = 3102\ \text{J mol}^{-1}$$

$$\text{Fraction} = \frac{\Delta_{vap}(pV)}{\Delta_{vap}H} = \frac{3.102\ \text{kJ mol}^{-1}}{40.656\ \text{kJ mol}^{-1}} = \boxed{0.0763} = 7.63\ \text{per cent}$$

Solutions to problems

P4B.1 At the triple point, T_3, the vapour pressures of liquid and solid are equal, hence

$$10.5916 - \frac{1871.2\text{K}}{T_3} = 8.3186 - \frac{1425.7\text{K}}{T_3};\quad T_3 = \boxed{196.0\text{K}}$$

$$\log(p_3/\text{Torr}) = \frac{-1871.2\text{K}}{196.0\text{K}} + 10.5916 = 1.044\bar{7};\quad p_3 = \boxed{11.1\text{Torr}}$$

P4B.3 (a) $$\frac{dp}{dT} = \frac{\Delta_{vap}S}{\Delta_{vap}V}[4\text{B}.5\text{a}] = \frac{\Delta_{vap}H}{T_b\Delta_{vap}V}\ [4\text{B}.9]$$

$$= \frac{14.4\times10^3\ \text{J mol}^{-1}}{(180\ \text{K})\times(14.5\times10^{-3} - 115\times10^{-6})\text{m}^3\ \text{mol}^{-1}} = \boxed{+5.56\times10^3\ \text{Pa K}^{-1}}$$

(b) $$\frac{dp}{dT} = \frac{\Delta_{vap}H}{RT^2}\times p\left[4.11, \text{with}\ \text{d}\ln p = \frac{dp}{p}\right]$$

$$= \frac{(14.4\times10^3\ \text{J mol}^{-1})\times(1.013\times10^5\ \text{Pa})}{(8.3145\ \text{J K}^{-1}\ \text{mol}^{-1})\times(180\text{K})^2} = +5.41\times10^3\ \text{Pa K}^{-1}$$

The percentage error is $\boxed{2.6\ \text{per cent}}$.

P4B.5 (a) $$\left(\frac{\partial\mu(\text{l})}{\partial p}\right)_T - \left(\frac{\partial\mu(\text{s})}{\partial p}\right)_T = V_m(\text{l}) - V_m(\text{s})\ [4\text{B}.12] = M\Delta\left(\frac{1}{\rho}\right)$$

$$= (18.02\ \text{g mol}^{-1})\times\left(\frac{1}{1.000\ \text{g cm}^{-3}} - \frac{1}{0.917\ \text{g cm}^{-3}}\right)$$

$$= \boxed{-1.63\ \text{cm}^3\ \text{mol}^{-1}}$$

(b) $$\left(\frac{\partial \mu(\mathrm{g})}{\partial p}\right)_T - \left(\frac{\partial \mu(\mathrm{l})}{\partial p}\right)_T = V_{\mathrm{m}}(\mathrm{g}) - V_{\mathrm{m}}(\mathrm{l})$$

$$= (18.02\ \mathrm{g\,mol^{-1}}) \times \left(\frac{1}{0.598\ \mathrm{g\,dm^{-3}}} - \frac{1}{0.958\times10^{3}\ \mathrm{g\,dm^{-3}}}\right)$$

$$= \boxed{+30.1\ \mathrm{dm^3\ mol^{-1}}}$$

(c) $\mu(\mathrm{g},1.2\ \mathrm{atm}) - \mu(\mathrm{l},1.2\ \mathrm{atm}) = \mu(\mathrm{g},1.2\ \mathrm{atm}) - \mu(\mathrm{g},1\ \mathrm{atm}) - \{\mu(\mathrm{l},1.2\ \mathrm{atm}) - \mu(\mathrm{l},1\ \mathrm{atm})\}$

because $\mu(\mathrm{g},1\ \mathrm{atm}) = \mu(\mathrm{l},1\ \mathrm{atm})$

Thus $\mu(\mathrm{g},1.2\ \mathrm{atm}) - \mu(\mathrm{l},1.2\ \mathrm{atm}) = \Delta\mu(\mathrm{g}) - \Delta\mu(\mathrm{l})$

where $\Delta\mu$ is the difference in chemical potential of a given phase at 1.2 atm compared to that at 1 atm.

$$\Delta\mu \approx \left(\frac{\partial \mu}{\partial p}\right)_T \Delta p = V_{\mathrm{m}}\Delta p \quad [4\mathrm{B}.2]$$

so $\mu(\mathrm{g},1.2\ \mathrm{atm}) - \mu(\mathrm{l},1.2\ \mathrm{atm}) = \{V_{\mathrm{m}}(\mathrm{g}) - V_{\mathrm{m}}(\mathrm{l})\}\Delta p \approx V_{\mathrm{m}}(\mathrm{g})\Delta p$

$$\Delta\mu(\mathrm{g},1.2\ \mathrm{atm}) - \Delta\mu(\mathrm{g},1.2\ \mathrm{atm}) \approx (30.1\times10^{-3}\ \mathrm{m^3\ mol^{-1}}) \times (0.2\ \mathrm{atm}) \times \frac{1.013\times10^{5}\ \mathrm{Pa}}{1\ \mathrm{atm}}$$

$$\approx \boxed{+6\times10^{2}\ \mathrm{J\,mol^{-1}}}$$

Since $\mu(\mathrm{g},1.2\ \mathrm{atm}) > \mu(\mathrm{l},1.2\ \mathrm{atm})$, the gas tends to condense into a liquid.

P4B.7 The amount (moles) of water evaporated is $n_{\mathrm{g}} = \dfrac{p_{\mathrm{H_2O}}V}{RT}$

The heat leaving the water is $q = n_{\mathrm{g}}\Delta_{\mathrm{vap}}H$

The temperature change of the water is $\Delta T = \dfrac{-q}{n_{\mathrm{liq}}C_{p,\mathrm{m}}} = \dfrac{-p_{\mathrm{H_2O}}V\Delta_{\mathrm{vap}}H}{RTn_{\mathrm{liq}}C_{p,\mathrm{m}}}$.

Therefore

$$\Delta T = \frac{-(3.17\ \mathrm{kPa}) \times (50.0\,\mathrm{dm^3}) \times (44.0\times10^{3}\ \mathrm{J\,mol^{-1}})}{(8.3145\ \mathrm{kPa\,dm^3\ K^{-1}\ mol^{-1}}) \times (298\ \mathrm{K}) \times (75.5\,\mathrm{J\,K^{-1}\ mol^{-1}}) \times \left(\dfrac{250\ \mathrm{g}}{18.02\,\mathrm{g\,mol^{-1}}}\right)}$$

$$= -2.7\ \mathrm{K}$$

The final temperature will be about $\boxed{22\ °\mathrm{C}}$.

P4B.9 (a) T_{b} is the temperature at which the vapour pressure is 1 atm (760 Torr);

$$T_{\mathrm{b}} = \boxed{227.5\ °\mathrm{C}}.$$

(b) Integrating the Clausius–Clapeyron equation [4B.10] yields an expression for $\ln p$:

$$\ln p = \mathrm{constant} - \frac{\Delta_{\mathrm{vap}}H}{RT}$$

Therefore, plot $\ln p$ against $1/T$ and identify $-\Delta_{vap}H/R$ as the slope of the plot. Construct the following table

θ/°C	57.4	100.4	133.0	157.3	203.5	227.5
T/K	330.6	373.6	406.2	430.5	476.7	500.7
1 000 K/T	3.02	2.68	2.46	2.32	2.10	2.00
ln (p/Torr)	0.00	2.30	3.69	4.61	5.99	6.63

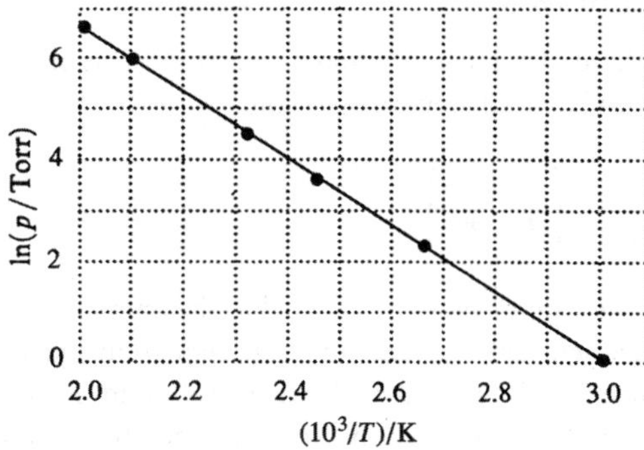

Figure 4B.1

The points are plotted in Fig. 4B.1. The slope is -6.4×10^3 K, so $\Delta_{vap}H =$ $\boxed{+53\ \mathrm{kJ\,mol^{-1}}}$.

P4B.11 $\left(\frac{\partial \Delta G}{\partial p}\right)_T = \left(\frac{\partial G_\beta}{\partial p}\right)_T - \left(\frac{\partial G_\alpha}{\partial p}\right)_T = V_\beta - V_\alpha$ [analogous to 4B.2]

Therefore, if $V_\alpha = V_\beta$, ΔG is independent of pressure. In general, $V_\alpha \neq V_\beta$, so that ΔG is nonzero, though small, since $V_\beta - V_\alpha$ is small.

P4B.13 The amount (moles) of gas bubbled through liquid is $\frac{PV}{RT}$

(P is the initial pressure of the gas and the pressure of the emerging gaseous mixture).

The amount (moles) of vapour carried away is $\frac{m}{M}$

Mole fraction of vapour in the gaseous mixture is

$$x = \frac{\frac{m}{M}}{\frac{m}{M}+\frac{PV}{RT}} = \frac{MRT}{mRT+PVM}\times\frac{m}{M} = \frac{mRT}{mRT+PVM}$$

The partial pressure of the vapour (i.e. the vapour pressure) is

$$p = xP = \frac{mRT}{mRT+PVM}\times P = \frac{P\left(\frac{mRT}{PVM}\right)}{\left(\frac{mRT}{PVM}\right)+1} = \frac{mPA}{mA+1},\quad \text{where } A = \frac{RT}{PVM}$$

Inserting the data yields

For geraniol, $M = 154.2\ \mathrm{g\,mol^{-1}}$, $T = 383$ K, $V = 5.00\ \mathrm{dm^3}$, $P = 1.00$ atm, and $m = 0.32$ g

$M = 154.2\ \text{g mol}^{-1}$, $T = 383$ K, $V = 500\ \text{dm}^3$, $p = 1.00$ atm, and $m = 0.32$ g, so

$$A = \frac{(0.08206\ \text{dm}^3\ \text{atm}\,\text{mol}^{-1}\,\text{K}^{-1}) \times (383\ \text{K})}{(1.00\ \text{atm}) \times (5.00\ \text{dm}^3) \times (154.2\ \text{g}\,\text{mol}^{-1})} = 0.0408\ \text{g}^{-1}$$

Therefore

$$p = \frac{(0.32\ \text{g}) \times (760\ \text{Torr}) \times (0.0408\ \text{g}^{-1})}{(0.32\ \text{g}) \times (0.0408\ \text{g}^{-1}) + 1} = \boxed{9.8\ \text{Torr}}$$

P4B.15 Equation 4B.11 for the dependence of vapour pressure on temperature comes from integrating the Clausius–Clapeyron equation.

$$p = p^* \text{e}^{-\chi} \quad \chi = \frac{\Delta_{\text{vap}} H}{R} \times \left(\frac{1}{T} - \frac{1}{T^*} \right)$$

Boiling occurs when the vapour pressure is equal to the ambient pressure at altitude h and ambient temperature T, so we can set this equation equal to $p = p_0 \text{e}^{-a/H}$. Let us choose the normal boiling point as the reference conditions in eqn 4B.11; that means $T^* = T_\text{b}$ and p^* is exactly 1 atm. Note that the other temperature T in eqn 4B.11 is the boiling temperature at altitude a (call it T_a), not the ambient temperature. Let us choose exactly 1 atm as the reference pressure p_0 in the barometric formula. Thus

$$\exp\left(-\frac{a}{H} \right) = \exp\left\{ -\frac{\Delta_{\text{vap}} H}{R} \times \left(\frac{1}{T_a} - \frac{1}{T_\text{b}} \right) \right\}$$

It follows that $\dfrac{1}{T_a} = \dfrac{1}{T_\text{b}} + \dfrac{aR}{H \Delta_{\text{vap}} H}$

For water at 3 000 m,

$$\frac{1}{T_a} = \frac{1}{373\ \text{K}} + \frac{(3\,000\ \text{m})(8.3145\ \text{J}\,\text{K}^{-1}\ \text{mol}^{-1})}{(8\,000\ \text{m})(40.7 \times 10^3\ \text{J}\,\text{mol}^{-1})}$$

$$= \frac{1}{373\ \text{K}} + \frac{7.66 \times 10^{-5}}{\text{K}} = \frac{2.76 \times 10^{-3}}{\text{K}}$$

Hence, $T_a = \boxed{363\ \text{K}}$ (90 °C).

P4B.17 The Clapeyron equation is inadequate because both V and S are continuous through a second-order transition. Thus, ΔV and ΔS vanish, resulting in an indeterminate form 0/0 . Just as with l'Hospital's rule in calculus, examining derivatives helps resolve the indeterminate form.

(1) $V = V(T,p)$

so $$\text{d}V = \left(\frac{\partial V}{\partial T} \right)_p \text{d}T + \left(\frac{\partial V}{\partial p} \right)_T \text{d}p$$

$$\left(\frac{\partial V}{\partial T} \right)_p = \alpha V\ [\text{2D.6}], \quad \left(\frac{\partial V}{\partial p} \right)_T = -\kappa_T V\ [\text{2D.7}]$$

hence, $\text{d}V = \alpha V\,\text{d}T - \kappa_T V\,\text{d}p$.

This equation applies to both phases 1 and 2, and since V is continuous through a second-order transition

$$\alpha_1\,\mathrm{d}T-\kappa_{T,1}\,\mathrm{d}p=\alpha_2\,\mathrm{d}T-\kappa_{T,2}\,\mathrm{d}p$$

Solving for $\dfrac{\mathrm{d}p}{\mathrm{d}T}$ yields $\boxed{\dfrac{\mathrm{d}p}{\mathrm{d}T}=\dfrac{\alpha_2-\alpha_1}{\kappa_{T,2}-\kappa_{T,1}}}$

(2) $S_\mathrm{m}=S_\mathrm{m}(T,p)$

so $\mathrm{d}S_\mathrm{m}=\left(\dfrac{\partial S_\mathrm{m}}{\partial T}\right)_p\mathrm{d}T+\left(\dfrac{\partial S_\mathrm{m}}{\partial p}\right)_T\mathrm{d}p.$

To evaluate the derivative $\partial S_\mathrm{m}/\partial T$, consider $\mathrm{d}S$ at constant p:

$$\mathrm{d}S=\frac{\mathrm{d}q_\mathrm{rev}}{T}=\frac{\mathrm{d}H}{T}=\frac{C_p\mathrm{d}T}{T}\quad\text{so}\quad\left(\frac{\partial S_\mathrm{m}}{\partial T}\right)_p=\frac{C_{p,\mathrm{m}}}{T}$$

Also $\left(\dfrac{\partial S_\mathrm{m}}{\partial p}\right)_T=-\left(\dfrac{\partial V_\mathrm{m}}{\partial T}\right)_p$ [Maxwell relation; Table 3D.1] $=\alpha V_\mathrm{m}$

Thus $\mathrm{d}S_\mathrm{m}=\dfrac{C_{p,\mathrm{m}}}{T}\mathrm{d}T-\alpha V_\mathrm{m}\,\mathrm{d}p$

This relation applies to both phases. For second-order transitions both S_m and V_m are continuous through the transition, so that

$$\frac{C_{p,\mathrm{m}1}}{T}\mathrm{d}T-\alpha_1V_\mathrm{m}\,\mathrm{d}p=\frac{C_{p,\mathrm{m}2}}{T}\mathrm{d}T-\alpha_2V_\mathrm{m}\,\mathrm{d}p$$

Solving for $\dfrac{\mathrm{d}p}{\mathrm{d}T}$ yields $\boxed{\dfrac{\mathrm{d}p}{\mathrm{d}T}=\dfrac{C_{p,\mathrm{m}2}-C_{p,\mathrm{m}1}}{TV_\mathrm{m}(\alpha_2-\alpha_1)}}$

Integrated activities

I4.1 The equations describing the coexistence curves for the three states are

(a) Solid–liquid boundary

$$p=p^*+\frac{\Delta_\mathrm{fus}H}{\Delta_\mathrm{fus}V}\ln\frac{T}{T^*}\quad[4B.7]$$

(b) Liquid–vapour boundary

$$p=p^*\mathrm{e}^{-\chi},\quad \chi=\frac{\Delta_\mathrm{vap}H}{R}\times\left(\frac{1}{T}-\frac{1}{T^*}\right)\quad[4B.11]$$

(c) Solid–vapour boundary

$$p=p^*\mathrm{e}^{-\chi},\quad \chi=\frac{\Delta_\mathrm{sub}H}{R}\times\left(\frac{1}{T}-\frac{1}{T^*}\right)\quad[\text{analogous to 4B.11}]$$

We need $\Delta_{sub}H = \Delta_{fus}H + \Delta_{vap}H = 41.4\ \text{kJ mol}^{-1}$

$$\Delta_{fus}V = M \times \left(\frac{1}{\rho(\text{l})} - \frac{1}{\rho(\text{s})}\right) = \left(\frac{78.11\ \text{g mol}^{-1}}{\text{g cm}^{-3}}\right) \times \left(\frac{1}{0.879} - \frac{1}{0.891}\right)$$

$$= +1.20\ \text{cm}^3\ \text{mol}^{-1}$$

After insertion of these numerical values into the above equations, we obtain

(a) $$p = p^* + \left(\frac{10.6 \times 10^3\ \text{J mol}^{-1}}{1.20 \times 10^{-6}\ \text{m}^3\ \text{mol}^{-1}}\right) \ln \frac{T}{T^*}$$

$$= p^* + 8.86 \times 10^9\ \text{Pa} \times \ln \frac{T}{T^*} = p^* + (6.64 \times 10^7\ \text{Torr}) \ln \frac{T}{T^*}\quad [1\ \text{Torr} = 133.3\ \text{Pa}]$$

This line is plotted as a in Fig. I4.1, starting at the triple point (p^*, T^*) = (36 Torr, 5.50 °C (278.65 K)).

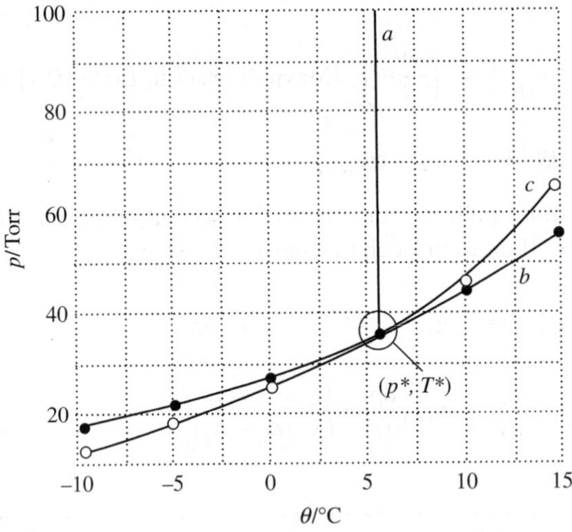

Figure I4.1

(b) $$\chi = \left(\frac{30.8 \times 10^3\ \text{J mol}^{-1}}{8.3145\ \text{J K}^{-1}\ \text{mol}^{-1}}\right) \times \left(\frac{1}{T} - \frac{1}{T^*}\right) = (370\bar{4}\ \text{K}) \times \left(\frac{1}{T} - \frac{1}{T^*}\right)$$

$$p = p^* \text{e}^{-370\bar{4}\ \text{K} \times (1/T - 1/T^*)}$$

This equation is plotted as curve b in Fig. I4.1, starting from the triple point (p^*, T^*).

(c) $$\chi = \left(\frac{41.4 \times 10^3\ \text{J mol}^{-1}}{8.3145\ \text{J K}^{-1}\ \text{mol}^{-1}}\right) \times \left(\frac{1}{T} - \frac{1}{T^*}\right) = (497\bar{9}\ \text{K}) \times \left(\frac{1}{T} - \frac{1}{T^*}\right)$$

$$p = p^* \text{e}^{-497\bar{9}\ \text{K} \times (1/T - 1/T^*)}$$

These points are plotted as curve c in Fig. I4.1, starting from the triple point (p^*, T^*). The lighter lines in Fig. I4.1 represent extensions of lines b and c into regions where the liquid and solid states respectively are not stable.

I4.3 (a) $\Delta G_m = (n-4)\Delta_{hb}H_m - (n-2)T_m\Delta_{hb}S_m$ (1)

Enthalpy and entropy terms give a Gibbs energy change of $\Delta G = \Delta H - T\Delta S$ for a constant temperature process. The enthalpy term is justified by $n-4$ independent hydrogen bonds for which each requires $\Delta_{hb}H_m$ of heat to break during melting dissociation. The entropy term is justified by n–2 highly ordered, but independent, structures for which each experiences an entropy increase of $\Delta_{hb}H_m$ during the melting process.

(b) $\Delta_{trs}S = \dfrac{\Delta_{trs}H}{T_{trs}}$ [3A.17] yields $T_{trs} = \dfrac{\Delta_{trs}H}{\Delta_{trs}S}$ which here becomes

$$T_m = \frac{(n-4)\Delta_{hb}H_m}{(n-2)\Delta_{hb}S_m}$$

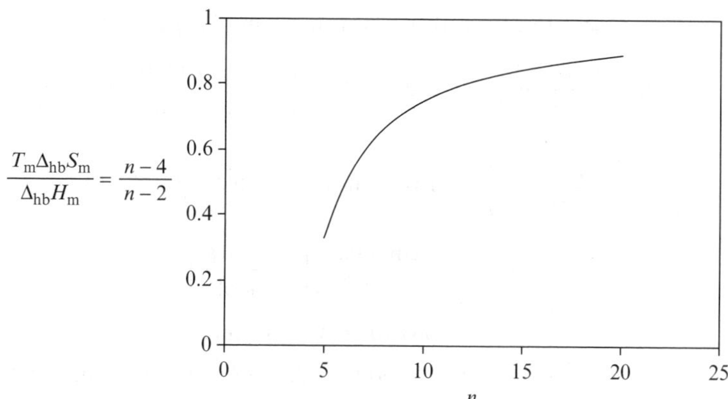

Figure I4.2

(c) See Fig. I4.2.

Consider $$\frac{1}{T_m}\frac{dT_m}{dn} = \frac{\Delta_{hb}H_m}{T_m\Delta_{hb}S_m}\frac{d(T_m\Delta_{hb}S_m/\Delta_{hb}H_m)}{dn} = \left(\frac{n-2}{n-4}\right)\frac{d}{dn}\left(\frac{n-4}{n-2}\right)$$

$$= \left(\frac{n-2}{n-4}\right)\left(\frac{2}{(n-2)^2}\right) = \frac{2}{(n-4)(n-2)}$$

This expression will be less than 1% when $\frac{2}{(n-4)(n-2)} < 0.01$ or when n equals, or is larger than the value given by $n^2 - 6n + 8 = 200$. The positive root of this quadratic is $n = \boxed{17}$. T_m changes by about 1% or less upon addition of another amino acid residue when the polypeptide consists of 17 or more residues.

I4.5 $\text{C(graphite)} \rightleftharpoons \text{C(diamond)} \quad \Delta_r G = 2.8678\ \text{kJ mol}^{-1}$ at $T = 25\ °\text{C}$

We want the pressure at which $\Delta_r G = 0$; above that pressure the reaction will be spontaneous. Equation 3D.8 determines the rate of change of $\Delta_r G$ with p at constant T.

$$(1) \quad \left(\frac{\partial \Delta_r G}{\partial p}\right)_T = \Delta_r V = (V_D - V_G)M$$

where M is the molar mass of carbon; V_D and V_G are the specific volumes of diamond and graphite, respectively. $\Delta_r G(p)$ may be expanded in a Taylor series around the pressure $p_0 = 100$ kPa.

$$(2)\quad \Delta_r G(p) = \Delta_r G(p_0) + \left(\frac{\partial \Delta_r G(p_0)}{\partial p}\right)_T (p-p_0) + \frac{1}{2}\left(\frac{\partial^2 \Delta_r G(p_0)}{\partial p^2}\right)_T (p-p_0)^2 + \dots$$

We will neglect the third and higher-order terms in $(p - p_0)$; the derivative of the first-order term can be calculated from eqn 1. An expression for the derivative of the second-order term can be derived from eqn 1.

$$(3)\quad \left(\frac{\partial^2 \Delta_r G}{\partial p^2}\right)_T = \left\{\left(\frac{\partial V_D}{\partial p}\right)_T - \left(\frac{\partial V_G}{\partial p}\right)_T\right\} M = (V_G \kappa_{T,G} - V_D \kappa_{T,D})M \; [2D.7]$$

Calculating the derivatives of eqns 1 and 2 at p_0,

$$(4)\quad \left(\frac{\partial \Delta_r G(p_0)}{\partial p}\right)_T = (0.284 - 0.444) \times \left(\frac{\text{cm}^3}{\text{g}}\right) \times \left(\frac{12.01\,\text{g}}{\text{mol}}\right) = -1.92\ \text{cm}^3\ \text{mol}^{-1}$$

$$(5)\quad \left(\frac{\partial^2 \Delta_r G(p_0)}{\partial p^2}\right)_T = \left\{0.444\left(3.04\times10^{-8}\right) - 0.284\left(0.187\times10^{-8}\right)\right\}$$

$$\times\left(\frac{\text{cm}^3\ \text{kPa}^{-1}}{\text{g}}\right)\times\left(\frac{12.01\,\text{g}}{\text{mol}}\right)$$

$$= 1.56\times10^{-7}\ \text{cm}^3\ \text{kPa}^{-1}\ \text{mol}^{-1}$$

It is convenient to convert the value of $\Delta_r G(p_0)$ to the units $\text{cm}^3\ \text{kPa mol}^{-1}$

$$(6)\quad \Delta_r G(p_0) = 2.8678\times10^3\ \text{J mol}^{-1} \times \left(\frac{10^6\ \text{cm}^3}{\text{m}^3}\right) \times \left(\frac{1\ \text{kPa}}{10^3\ \text{Pa}}\right)$$

$$= 2.8678\times10^6\ \text{cm}^3\ \text{kPa mol}^{-1}$$

Setting $\chi = p - p_0$, and dividing through by $\text{cm}^3\ \text{mol}^{-1}$, eqns 2 and 3–6 give

$$2.8678\times10^6\ \text{kPa} - 1.92\chi + (7.80\times10^{-8}\ \text{kPa}^{-1})\chi^2 = 0$$

when $\Delta_r G(p) = 0$. One real root of this equation is

$$\chi = 1.60\times10^6\ \text{kPa} = p - p_0$$

or $\quad p = 1.60\times10^6\ \text{kPa} - 100\ \text{kPa} = 1.60\times10^6\ \text{kPa} = \boxed{1.60\times10^4\ \text{bar}}$.

Above this pressure the reaction is spontaneous. The other real root is much higher, 2.3×10^7 kPa.

Question. What interpretation might you give to the other real root?

5 Simple mixtures

5A The thermodynamic description of mixtures

Answers to discussion questions

D5A.1 A partial molar quantity X_J is defined analogous to eqn 5A.1 as the partial derivative of a state function with respect to the amount of a component J, holding all other variables constant:

$$X_J = \left(\frac{\partial X}{\partial n_J}\right)_{p,T,n'}$$

A physical answer would note that how a state function varies with the amount of one component in a mixture depends on what else is in the mixture because it depends on the chemical environment, defined by the properties and proportions of all of the components in a mixture. How the volume of water depends on the amount of water is surely different for water mixed with ethanol compared to water mixed with sodium chloride! A mathematical answer would add that what is held constant in the derivative that defines partial molar quantities is the *amount* of every other component, not the *proportion* of other components: as the *amount* of component A increases while keeping the *amount* of B constant, the *proportion* of B (mole fraction, for example) decreases.

D5A.3 *Perfect* gases spontaneously mix in all proportions. There are, however, conceivable circumstances under which two *real* gases might not mix spontaneously. Consider allowing two gases initially at the same pressure p to mix (so that mixing them would not change the pressure). Consider carrying out this process at constant temperature as well. Then mixing would be spontaneous if $\Delta_{mix}G < 0$. Consider entropic and enthalpic contributions:

$$\Delta_{mix}G = \Delta_{mix}H - T\Delta_{mix}S$$

Now $\Delta_{mix}S > 0$ would be positive, so mixing would always be favoured entropically. The only circumstances under which mixing might not be spontaneous would be if $\Delta_{mix}H > T\Delta_{mix}S$, that is if interactions between the species were so energetically unfavourable as to outweigh the entropic term. We might consider such unfavourable interactions unlikely in gases; after all, $\Delta_{mix}H = 0$ for perfect gases. At the same time, we know that there are liquids for which such unfavourable interactions prevent mixing at least in some proportions and at some temperatures (the proverbial oil and water). Consider two such species at a temperature above the critical temperature of both and at a pressure high enough to make their densities more typical of liquids than gases. Under such circumstances, the species are technically gases, although the term supercritical fluid is more informative [Section 1C.1(c)]. One might find examples of immiscibility among such supercritical fluids.

D5A.5 Raoult's law (eqn 5A.21) defines the behaviour of ideal solutions. Like ideal gases, what makes the behaviour ideal can be expressed in terms of intermolecular interactions. Unlike ideal gases, however, the interactions in an ideal solution cannot be neglected. Instead, ideal behaviour amounts to having the same interactions among molecules of the mixture's different components as molecules of each component have with other molecules of that same component. In shorthand, ideal behaviour consists of A–B interactions being the same as A–A and B–B interactions. If that is the case, then the cohesive forces that would keep a molecule in the liquid phase would be the same in the solution as in a pure liquid, and the vapour pressure of a component will differ from that of a pure liquid only in proportion to its abundance (mole fraction). Thus, we expect Raoult's law to be valid for mixtures of components that have very similar chemical structures. Similar structures imply both similar intermolecular interactions (governed largely by polarity) and similar sizes (implying that the mole fraction is a good approximation to the relative proportion of the surface area occupied by each component—a factor that is relevant to rates of evaporation and condensation).

In an ideal dilute solution, on the other hand, Raoult's law holds for the solvent in the limit as x_A approaches 1, not because A–B interactions are like A–A interactions, but because there are so many more A–A interactions than A–B interactions that A–A interactions dominate the behaviour of the solvent. For the solute, on the other hand, there are many more A–B interactions than B–B interactions in the limit as x_B approaches zero. Thus, only one kind of interaction (A–B) is important in determining the affinity of the solute for the solution.

Solutions to exercises

E5A.1(a) The partial molar volume is

$$V_J = \left(\frac{\partial V}{\partial n_J}\right)_{p,T,n'} [5A.1] = \left(\frac{dv}{dx}\right)\left(\frac{dV}{dv}\right)\left(\frac{dx}{dn_J}\right)$$

Right away we see that $V_A = \boxed{0}$ because V is independent of n_A $[dx/dn_A = 0]$

$$V_B = (35.6774 - 2\times 0.45923x + 3\times 0.017325x^2)\left(\frac{\text{cm}^3}{\text{mol}}\right)$$

$$= \boxed{(35.6774 - 0.91846x + 0.051975x^2)\ \text{cm}^3\ \text{mol}^{-1}}$$

E5A.2(a) Let A stand for water and B for NaCl(aq)

$$V_J = \left(\frac{\partial V}{\partial n_J}\right)_{p,T,n'} [5A.1] = \left(\frac{dv}{dx}\right)\left(\frac{dV}{dv}\right)\left(\frac{\partial x}{\partial n_J}\right)_{n'}$$

Now $x = \frac{b}{b^{\ominus}} = \frac{n_B}{n_A M_A b^{\ominus}}$ so $\left(\frac{\partial x}{\partial n_B}\right)_{n_A} = \frac{1}{n_A M_A b^{\ominus}}$

and $V_B = \left(16.62 + \frac{3}{2}\times 1.77 x^{1/2} + 2\times 0.12x\right)\frac{\text{cm}^3}{n_A M_A b^{\ominus}}$

Evaluate this expression for $b = 0.100$ mol kg^{-1} ($x = 0.100$), recalling that the original expression for v applies for 1.000 kg of water (i.e. for $n_A M_A = 1.000$ kg). The result is $V_B = \boxed{17.5\ \text{cm}^3\ \text{mol}^{-1}}$.

For a solution consisting of 0.100 mol NaCl and 1.000 kg water, corresponding to 55.49 mol H_2O, the total volume is given both by

$$V = \{1003 + 16.62\times 0.100 + 1.77\times(0.100)^{3/2} + 0.12\times(0.100)^2\}\ \text{cm}^3 = 1004.\bar{7}\ \text{cm}^3$$

and by

$$V = V_A n_A + V_B n_B\ [5A.3].$$

Therefore, $V_A = \frac{V - V_B n_B}{n_A} = \frac{1004.\bar{7}\ \text{cm}^3 - (17.5\ \text{cm}^3)\times(0.100\ \text{mol})}{55.49\ \text{mol}} = \boxed{18.\overline{07}\ \text{cm}^3\ \text{mol}^{-1}}$

Comment. V_A is essentially the same as the molar volume of pure water at 25 °C. How does the partial molar volume of NaCl(aq) in this solution compare to molar volume of pure solid NaCl?

E5A.3(a) Use the Gibbs–Duhem equation [5A.13], replacing infinitesimal changes in chemical potential ($d\mu_J$) with small finite changes ($\delta\mu_J$)

$$\delta\mu_B \approx -\frac{n_A}{n_B}\delta\mu_A = -\frac{0.10 n_B}{n_B}\times(+12\ \text{J mol}^{-1}) = \boxed{-1.2\ \text{J mol}^{-1}}$$

E5A.4(a) The Gibbs energy of mixing perfect gases is

$$\Delta_{\text{mix}}G = nRT(x_A \ln x_A + x_B \ln x_B)[5A.16] = pV(x_A \ln x_A + x_B \ln x_B)[\text{perfect gas law}]$$

Because the compartments are of equal size, each contains half of the gas; therefore,

$$\begin{aligned}\Delta_{\text{mix}}G &= (pV)\times\left(\tfrac{1}{2}\ln\tfrac{1}{2} + \tfrac{1}{2}\ln\tfrac{1}{2}\right) = -pV\ln 2 \\ &= (-1.0)\times(1.013\times10^5\ \text{Pa})\times(5.0\times10^{-3}\ \text{m}^3)\times(\ln 2) \\ &= -3.5\times10^2\ \text{J} = \boxed{-0.35\ \text{kJ}}\end{aligned}$$

The entropy of mixing perfect gases is

$$\Delta_{\text{mix}}S = -nR(x_A \ln x_A + x_B \ln x_B)[5A.17] = \frac{-\Delta_{\text{mix}}G}{T} = \frac{+350\ \text{kJ}}{298\ \text{K}} = \boxed{+1.2\ \text{J K}^{-1}}$$

E5A.5(a) $\Delta_{mix}S = -nR\sum_J x_J \ln x_J$ [5A.17]

We need mole fractions:

$$x_J = \frac{n_J}{\sum_J n_J}$$

Since we have mass percentages, 100.0 g is a convenient sample size. The amounts of each component are

$$n_{N_2} = 75.5\text{ g} \times \frac{1\text{ mol}}{2\times 14.007\text{ g}} = 2.69\bar{5}\text{ mol}$$

$$n_{O_2} = 23.2\text{ g} \times \frac{1\text{ mol}}{2\times 16.00\text{ g}} = 0.725\text{ mol}$$

$$n_{Ar} = 1.3\text{ g} \times \frac{1\text{ mol}}{39.95\text{ g}} = 0.033\text{ mol}$$

The mole fractions are

$$x_{N_2} = \frac{n_{N_2}}{n_{N_2} + n_{O_2} + n_{Ar}} = \frac{2.69\bar{5}\text{ mol}}{(2.69\bar{5}+0.725+0.033)\text{ mol}} = 0.781$$

Similarly, $x_{O_2} = 0.210$ and $x_{Ar} = 0.0094$

Once we have mole fractions, the convenient sample size is for a total of one mole of gas:

$$\begin{aligned}\Delta_{mix}S &= -R\sum_J x_J \ln x_J \\ &= -R\{(0.781\ln 0.781)+(0.210\ln 0.210)+(0.0094\ln 0.0094)\} \\ &= 0.565R = \boxed{+4.70\text{ J K}^{-1}\text{ mol}^{-1}}\end{aligned}$$

E5A.6(a) Let B refer to benzene and M to methylbenzene. Because the two components are structurally similar, we assume Raoult's law [5A.21] applies.

$$p_{total} = p_B + p_M = x_B p_B{}^* + x_M p_M{}^*$$

We need the mole fractions that correspond to equal masses.

$$x_B = \frac{n_B}{n_B + n_M} = \frac{m_B / M_B}{m_B / M_B + m_M / M_M}$$

Because $m_B = m_M$, the sample masses cancel. (Alternatively, one can assume a convenient sample size, for example exactly one gram of each component.)

$$x_B = \frac{1/M_B}{1/M_B + 1/M_M} = \frac{(78.11\text{ g mol}^{-1})^{-1}}{(78.11\text{ g mol}^{-1})^{-1} + (93.13\text{ g mol}^{-1})^{-1}} = 0.544$$

and $x_M = 0.466$

Thus $p_{total} = (0.544)(10\,kPa)+(0.466)(2.8\,kPa) = \boxed{6.7\,kPa}$

E5A.7(a) Let A denote acetone and C chloroform. The total volume of the solution is

$$V = n_A V_A + n_C V_C$$

V_A and V_C are given; hence we need to determine n_A and n_C in 1.000 kg of the solution with the stated mole fraction. The total mass of the sample is

(a) $m = n_A M_A + n_C M_C$

We also know that

$$x_A = \frac{n_A}{n_A + n_C}$$

implying that $(x_A - 1)n_A + x_A n_C = 0$,

and hence that

(b) $-x_C n_A + x_A n_C = 0$.

Thus, we have two equations in two unknowns. On solving (a) and (b), we find

$$n_A = \left(\frac{x_A}{x_C}\right) \times n_C \quad \text{and} \quad n_C = \frac{m x_C}{x_A M_A + x_C M_C}$$

Since $x_C = 0.4693$, $x_A = 1 - x_C = 0.5307$,

$$n_C = \frac{(0.4693)\times(1000\,g)}{[(0.5307)\times(58.08)+(0.4693)\times(119.37)]\,g\,mol^{-1}} = 5.404\ mol,$$

and $n_A = \left(\frac{0.5307}{0.4693}\right)\times(5.404)\,mol = 6.111\,mol$

The total volume, $V = n_A V_A + n_C V_C$, is therefore

$$V = (6.111\,mol)\times(74.166\,cm^3\,mol^{-1}) + (5.404\,mol)\times(80.235\,cm^3\,mol^{-1})$$

$$= \boxed{886.8\,cm^3}$$

E5A.8(a) Let W denote water and E ethanol. The total volume of the solution is

$$V = n_W V_W + n_E V_E$$

We are given V_W, we need to determine n_W and n_E in order to solve for V_E, for

$$V_E = \frac{V - n_W V_W}{n_E}$$

Take 100 cm^3 of solution as a convenient sample. The mass of this sample is

$$m = \rho V = (0.914\,g\,cm^{-3})\times(100\,cm^3) = 91.4\,g.$$

Half of this mass is water and half ethanol, so the moles of each component are

$$n_W = \frac{(0.50)\times(91.4\,g)}{18.02\,g\,mol^{-1}} = 2.5\,mol \quad \text{and} \quad n_E = \frac{(0.50)\times(91.4\,g)}{46.07\,g\,mol^{-1}} = 0.99\,mol.$$

$$V_E = \frac{V - n_W V_W}{n_E} = \frac{100\,cm^3 - (2.5\,mol)\times(17.4\,cm^3\,mol^{-1})}{0.99\,mol} = \boxed{56\,cm^3\,mol^{-1}}$$

E5A.9(a) Henry's law is [5A.23] $p_B = x_B K_B$, so check whether p_B/x_B is equal to a constant (K_B)

x	0.005	0.012	0.019
p/kPa	32.0	76.9	121.8
(p/kPa)/x	6.4×10^3	6.4×10^3	6.4×10^3

Hence, $K_B = \boxed{6.4\times10^3\,kPa}$

E5A.10(a) Refer to *Brief illustration* 5A.4 and use the Henry's law constant from Table 5A.1. Henry's law in terms of molal concentration is $p_B = b_B K_B$. We can find the molality of nitrogen in benzene if we know the partial pressure of nitrogen in air. From Example 1A.3, we see that the mole fraction of nitrogen in air is 0.780, so its partial pressure (assuming the standard pressure of 1 bar = 100 kPa) is 78.0 kPa. So the molal solubility of nitrogen in benzene saturated with air at 25 °C and 1 bar is

$$b_{N_2} = \frac{p_{N_2}}{K_{N_2}} = \frac{78.0\,kPa}{1.87\times10^4\,kPa\,kg\,mol^{-1}} = 4.17\times10^{-3}\,mol\,kg^{-1}$$

To find the molar solubility, we assume that the density of the solution is the same as that of pure benzene, given at a nearby temperature (20 °C) in Table 0.1:

$$[N_2] = b_{N_2}\rho_{benzene} = 4.17\times10^{-3}\,mol\,kg^{-1}\times0.879\,kg\,dm^{-3} = \boxed{3.67\times10^{-3}\,mol\,dm^{-3}}$$

E5A.11(a) With concentrations expressed in molalities, Henry's law [5A.23] becomes $p_B = b_B K_B$. Solving for b_B, the solubility, we have $b_B = \dfrac{p_B}{K}$.

(a) $p_B = 0.10\,atm = 10.1\,kPa$

$$b = \frac{10.1\,kPa}{3.01\times10^3\,kPa\,kg\,mol^{-1}} = \boxed{3.4\times10^{-3}\,mol\,kg^{-1}}$$

(b) $p_B = 1.00\,atm = 101.3\,kPa$

$$b = \frac{101.3\,kPa}{3.01\times10^3\,kPa\,kg\,mol^{-1}} = \boxed{3.37\times10^{-2}\,mol\,kg^{-1}}$$

E5A.12(a) As in Exercise 5A.11(a), we have $b_B = \frac{p_B}{K} = \frac{5.0\times101.3\text{ kPa}}{3.01\times10^3\text{ kPa kg mol}^{-1}} = 0.17\text{ mol kg}^{-1}$

Hence, the molality of the solution is about 0.17 mol kg^{-1}. Since molalities and molar concentrations (molarities) for dilute aqueous solutions are numerically approximately equal, the molar concentration is about $\boxed{0.17\text{ mol dm}^{-3}}$.

Solutions to problems

P5A.1 Refer to Example 5A.2. Let A refer to water, the solvent, and B to the solute, the salt. The Gibbs–Duhem equation (analogous to 5A.12a) is

$$n_A dV_A + n_B dV_B = 0 = n_A dv_A + n_B dv_B \text{ [divide by cm}^3\text{ mol}^{-1}]$$

Therefore

$$dv_A = -\left(\frac{n_B}{n_A}\right) dv_B$$

and, integrating

$$v_A = v_A^* - \int_0^{v_B}\left(\frac{n_B}{n_A}\right) dv_B = v_A^* - \frac{19.121}{2}\int_0^{b/b^\ominus}\left(\frac{n_B}{n_A}\right) x^{-1/2} dx$$

But $\frac{n_B}{n_A} = xb^\ominus M_A$ [Example 5A.2]

$$\text{so } v_A = v_A^* - \frac{19.121}{2} b^\ominus M_A \int_0^{b/b^\ominus} x^{+1/2}\, dx = v_A^* - \frac{19.121}{2}\times\frac{2}{3} b^\ominus M_A x^{3/2}$$

$$= \boxed{18.079 - 0.11482x^{3/2}}$$

In the last step, we need M_A and $b^\ominus$ to be in compatible units: $M_A = 1.8015\times10^{-2}$ kg mol^{-1}.

P5A.3 $p_A = y_A p$ and $p_B = y_B p$ [1A.8]. Hence, draw up the following table

p_A/kPa	0	1.399	3.566	5.044	6.996	7.940	9.211	10.105	11.287	12.295
x_A	0	0.0898	0.2476	0.3577	0.5194	0.6036	0.7188	0.8019	0.9105	1
y_A	0	0.0410	0.1154	0.1762	0.2772	0.3393	0.4450	0.5435	0.7284	1

p_B/kPa	0	4.209	8.487	11.487	15.462	18.243	23.582	27.334	32.722	36.066
x_B	0	0.0895	0.1981	0.2812	0.3964	0.4806	0.6423	0.7524	0.9102	1
y_B	0	0.2716	0.4565	0.5550	0.6607	0.7228	0.8238	0.8846	0.9590	1

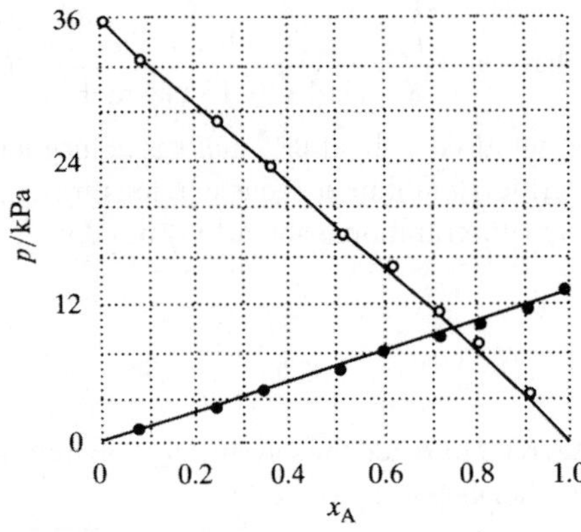

Figure 5A.1

The data are plotted in Fig. 5A.1. We can assume, at the lowest concentrations of both A and B, that Henry's law (eqn 5A.23) will hold. The Henry's law constants are then given by

$$K_A = \frac{p_A}{x_A} = \boxed{15.58 \text{ kPa}} \text{ from the point at } x_A = 0.0898$$

$$K_B = \frac{p_B}{x_B} = \boxed{47.03 \text{ kPa}} \text{ from the point at } x_B = 0.0895$$

P5A.5 The 97% saturated haemoglobin in the lungs releases oxygen in the capillary until the haemoglobin is 75% saturated. 100 cm^3 of blood in the lung containing 15.0 g of Hb at 97% saturated with O_2 binds

$$1.34 \text{ cm}^3 \text{ g}^{-1} \times 15.0 \text{g} = 20.1 \text{ cm}^3 \text{ O}_2.$$

The same 100 cm^3 of blood in capillaries would contain

$$20.1 \text{ cm}^3 \text{ O}_2 \times \frac{75\%}{97\%} = 15.5 \text{ cm}^3$$

Therefore, about (20.1 – 15.5) cm^3 or $\boxed{4.6 \text{ cm}^3}$ of O_2 is given up in the capillaries to body tissue.

5B The properties of solutions

Answers to discussion questions

D5B.1 A regular solution has excess entropy of zero, but an excess enthalpy that is non-zero and dependent on composition, perhaps in the manner of eqn 5B.6. We can think of a regular solution as one in which the different molecules of the solution are distributed randomly, as in an ideal solution, but have different energies of interaction with each other. Real solutions that are not regular may also have a non-random distribution of molecules. This is particularly likely in the case of ionic solutions, in which ions of one charge are more likely to be surrounded by ions of the opposite charge than of the same charge. (See Topic 5F for treatment of ionic solutions.)

D5B.3 Colligative properties depend on the identity of the solvent and on the concentration—but not the identity—of the solute. Thus osmometry (and other colligative properties, for that matter) can be used to determine the concentration of a solute of unknown molar mass. This information can be combined with information about composition by mass (e.g., of the masses of solute and solvent used to prepare the solution) to determine the molar mass. In a way, the fact that colligative properties are independent of the identity of the solute that allows them to provide an independent measure of the solute concentration—a measure that does not depend on the properties of the solute.

Solutions to exercises

E5B.1(a) In Exercise 5A.10(a), the Henry's law constant was determined for concentrations expressed in mole fractions; $K_B = 6.4 \times 10^3$ kPa. Thus the concentration must be converted from molality to mole fraction.

$$m(\text{GeCl}_4) = 1\,000\text{ g, corresponding to } n(\text{GeCl}_4) = \frac{1\,000\text{ g}}{214.39\text{ g mol}^{-1}} = 4.664\text{ mol}$$

$$\text{Therefore } x_{\text{HCl}} = \frac{0.10\text{ mol}}{(0.10\text{ mol}) + (4.664\text{ mol})} = 0.021$$

The pressure is

$$p_{\text{HCl}} = K_B x_{\text{HCl}}\,[5\text{A}.23] = (0.021) \times (6.4 \times 10^3\text{ kPa}) = \boxed{1.3 \times 10^2\text{ kPa}}.$$

E5B.2(a) We assume that the solvent, benzene, is ideal and obeys Raoult's law [5A.21]. As usual, let A denote the solvent (benzene) but let's avoid using B and call the solute O.

$$p_A = x_A p_A{}^* \quad \text{and} \quad x_A = \frac{n_A}{n_A + n_O}$$

$$\text{Hence } p_A = \frac{n_A p_A^*}{n_A + n_O}\text{; which solves to}$$

$$n_O = \frac{n_A (p_A^* - p_A)}{p_A}$$

Then, since $n_O = \frac{m_O}{M_O}$, where m_O is the mass of O present,

$$M_O = \frac{m_O p_A}{n_A (p_A^* - p_A)} = \frac{m_O M_A p_A}{m_A (p_A^* - p_A)}$$

From the data

$$M_O = \frac{(19.0\text{ g}) \times (78.11\text{ g mol}^{-1}) \times (51.5\text{ kPa})}{(500\text{ g}) \times (53.3 - 51.5)\text{ kPa}} = \boxed{85\text{ g mol}^{-1}}$$

E5B.3(a) Let B denote the compound and A the solvent, CCl_4. $K_f = 30\ \mathrm{K\,kg\,mol^{-1}}$ [Table 5B.1].

$$M_B = \frac{m_B}{n_B}$$

$$n_B = m_A b_B \quad \text{where} \quad b_B = \frac{\Delta T}{K_f} \ [5B.13]$$

$$\text{Thus } M_B = \frac{m_B K_f}{m_A \Delta T} = \frac{(100\ \mathrm{g}) \times (30\ \mathrm{K\,kg\,mol^{-1}})}{(0.750\ \mathrm{kg}) \times (10.5\ \mathrm{K})} = \boxed{3.8 \times 10^2\ \mathrm{g\,mol^{-1}}}$$

E5B.4(a) From the osmotic pressure, compute the concentration, and from the concentration the freezing point. According to the van 't Hoff equation [5B.16], the osmotic pressure is

$$\Pi = [B]RT \quad \text{so} \quad [B] = \frac{\Pi}{RT} = \frac{n_B}{V_{soln}}$$

The expression for freezing point depression [5B.13] includes the molality b rather than the molarity [B]. In dilute solutions, the two concentration measures are readily related:

$$b = \frac{n_B}{m_A} \approx \frac{n_B}{V_{soln}\rho_{soln}} = \frac{[B]}{\rho_{soln}} = \frac{\Pi}{RT\rho_{soln}}$$

The freezing point depression is

$$\Delta T = K_f b \approx \frac{K_f \Pi}{RT\rho_{soln}} \quad \text{where } K_f = 1.86\ \mathrm{K\,mol^{-1}\,kg} \text{ [Table 5B.1]}$$

The density of a dilute aqueous solution is approximately that of water:

$$\rho \approx 1.0\ \mathrm{g\,cm^{-3}} = 1.0 \times 10^3\ \mathrm{kg\,m^{-3}}$$

$$\text{So } \Delta T \approx \frac{(1.86\ \mathrm{K\,kg\,mol^{-1}}) \times (120 \times 10^3\ \mathrm{Pa})}{(8.3145\ \mathrm{J\,K^{-1}\,mol^{-1}}) \times (300\ \mathrm{K}) \times (10^3\ \mathrm{kg\,m^{-3}})} = 0.089\ \mathrm{K}$$

Therefore, the solution will freeze at about $\boxed{-0.09\ ^\circ\mathrm{C}}$.

Comment. Osmotic pressures are typically large: even dilute solutions with small freezing point depressions have large osmotic pressures.

E5B.5(a) $\Delta_{mix} G = nRT \sum_J x_J \ln x_J$ [5A.16] and

$$\Delta_{mix} S = -nR \sum_J x_J \ln x_J \,[5A.17] = \frac{-\Delta_{mix} G}{T}$$

$n = 2.50\ \mathrm{mol}$ and $x_{hexane}\ 0.50/2.50 = 0.20$; $x_{heptane} = 2.00/2.50 = 0.800$

Therefore,

$$\Delta_{mix} G = (2.50\ \mathrm{mol})(8.3145\ \mathrm{J\,K^{-1}\,mol^{-1}})(298\ \mathrm{K})\{(0.20 \ln 0.20) + (0.800 \ln 0.800)\}$$

$$= -3.10 \times 10^3\ \mathrm{J} = \boxed{-3.10\ \mathrm{kJ}}$$

and $\Delta_{\text{mix}}S = \dfrac{-\Delta_{\text{mix}}G}{T} = \dfrac{+3.10\times10^3}{298\text{ K}} = \boxed{+10.4\text{ J K}^{-1}}$

For an ideal solution, $\Delta_{\text{mix}}H = \boxed{0}$, just as it is for a mixture of perfect gases. It can be demonstrated from

$$\Delta_{\text{mix}}H = \Delta_{\text{mix}}G + T\Delta_{\text{mix}}S = \Delta_{\text{mix}}G + T\left(\frac{-\Delta_{\text{mix}}G}{T}\right) = 0$$

E5B.6(a) (a) Hexane and heptane form nearly ideal solutions; therefore.

$$\Delta_{\text{mix}}S = -nRT(x_{\text{A}}\ln x_{\text{A}} + x_{\text{B}}\ln x_{\text{B}})[5\text{A}.17]$$

We need to differentiate eqn 5A.17 with respect to x_{A} and look for the value of x_{A} at which the derivative is zero. Since $x_{\text{B}} = 1 - x_{\text{A}}$, we need to differentiate

$$\Delta_{\text{mix}}S = -nRT\{x_{\text{A}}\ln x_{\text{A}} + (1-x_{\text{A}})\ln(1-x_{\text{A}})\}$$

This gives $\left(\text{using}\,\dfrac{\mathrm{d}\ln x}{\mathrm{d}x} = \dfrac{1}{x}\right)$

$$\frac{\mathrm{d}\Delta_{\text{mix}}S}{\mathrm{d}x_{\text{A}}} = -nR\{\ln x_{\text{A}} + 1 - \ln(1-x_{\text{A}}) - 1\} = -nR\ln\frac{x_{\text{A}}}{1-x_{\text{A}}}$$

which is zero when $x_{\text{A}} = \boxed{\tfrac{1}{2}}$. Hence, the maximum entropy of mixing occurs for the preparation of a mixture that contains equal mole fractions of the two components.

(b) Because entropy of mixing is maximized when $n_{\text{hexane}} = n_{\text{heptane,}}$

$$\frac{m_{\text{hexane}}}{M_{\text{hexane}}} = \frac{m_{\text{heptane}}}{M_{\text{heptane}}}$$

This makes the mass ratio

$$\frac{m_{\text{hexane}}}{m_{\text{heptane}}} = \frac{M_{\text{hexane}}}{M_{\text{heptane}}} = \frac{86.17\,\text{g mol}^{-1}}{100.20\,\text{g mol}^{-1}} = \boxed{0.8600}$$

E5B.7(a) The ideal solubility in terms of mole fraction is given by eqn 5B.15, where B stands for the solute (anthracene):

$$\ln x_{\text{B}} = \frac{\Delta_{\text{fus}}H}{R}\times\left(\frac{1}{T_{\text{f}}} - \frac{1}{T}\right)$$

$$= \left(\frac{28.8\times10^3\text{ J mol}^{-1}}{8.3145\text{ J K}^{-1}\text{ mol}^{-1}}\right)\times\left(\frac{1}{490\text{K}} - \frac{1}{298\text{K}}\right) = -4.55$$

Therefore, $x_{\text{B}} = \text{e}^{-4.35} = 0.0105$.

$$x_B = \frac{n_B}{n_A + n_B} \approx \frac{n_B}{n_A} \left[i.e., \frac{n(\text{anthracene})}{n(\text{benzene})} \right]$$

(The approximation is justified because x_B is small.) Therefore, in 1 kg of benzene,

$$n_{\text{anth}} = x_{\text{anth}} \times n_{\text{benz}} \approx x_{\text{anth}} \times \frac{m_{\text{benz}}}{M_{\text{benz}}} = (0.0105) \times \left(\frac{1\,000\ \text{g}}{78.11\ \text{g mol}^{-1}} \right) = 0.135\ \text{mol}$$

The molality of the solution is therefore $\boxed{0.135\ \text{mol kg}^{-1}}$. Since $M = 178\ \text{g mol}^{-1}$, 0.135 mol corresponds to $\boxed{24.0\ \text{g anthracene}}$ per kg of benzene.

E5B.8(a) The best value of the molar mass is obtained from values of the data extrapolated to zero concentration, since it is under this condition that the van 't Hoff equation [5B.16] applies.

$\Pi V = n_B RT$ so $\Pi = \frac{mRT}{MV} = \frac{cRT}{M}$ where $c = m/V$.

But the osmotic pressure is also equal to the hydrostatic pressure

$\Pi = \rho g h$ [1A.1] so $h = \left(\frac{RT}{\rho g M} \right) c$

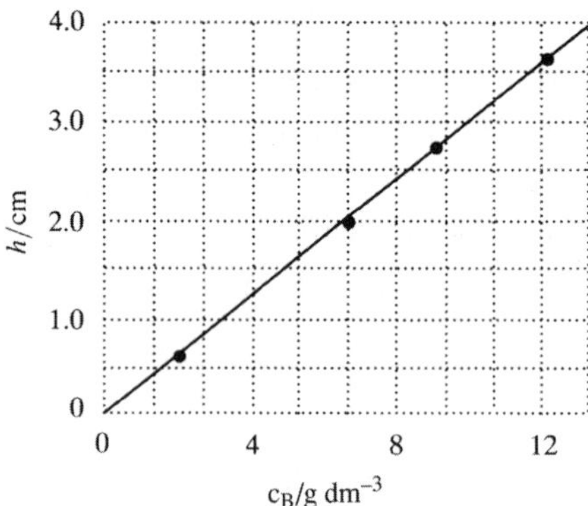

Figure 5B.1

Hence, plot h against c and identify the slope as $\frac{RT}{\rho g M}$. Figure 5B.1 shows the plot of the data. The slope of the line is 0.29 cm/(g dm^{-3}), so

$$\frac{RT}{\rho g M} = \frac{0.29\ \text{cm}}{\text{g dm}^{-3}} = 0.29\ \text{cm dm}^3\ \text{g}^{-1} = 2.9 \times 10^{-3}\ \text{m}^4\ \text{kg}^{-1}$$

Therefore,

$$M = \frac{RT}{(\rho g) \times (2.9 \times 10^{-3}\ \text{m}^4\ \text{kg}^{-1})}$$

$$= \frac{(8.3145\ \text{J K}^{-1}\ \text{mol}^{-1}) \times (298\ \text{K})}{(1.004 \times 10^3\ \text{kg m}^{-3}) \times (9.81\ \text{m s}^{-2}) \times (2.9 \times 10^{-3}\ \text{m}^4\ \text{kg}^{-1})} = \boxed{87\ \text{kg mol}^{-1}}$$

E5B.9(a) In an ideal dilute solution the solvent (CCl_4, A) obeys Raoult's law [5A.21] and the solute (Br_2, B) obeys Henry's law [5A.23]; hence

$$CCl_4\text{: } p_A = x_A p^* = (0.950)\times(33.85\,\text{Torr}) = \boxed{32.2\,\text{Torr}}.$$

$$Br_2\text{: } p_B = x_B K_B = (0.050)\times(122.36\,\text{Torr}) = \boxed{6.1\,\text{Torr}}.$$

$$p_{\text{total}} = (32.2+6.1)\,\text{Torr} = \boxed{38.3\,\text{Torr}}.$$

The composition of the vapour in equilibrium with the liquid is

$$CCl_4\text{: } y_A = \frac{p_A}{p_{\text{total}}} = \frac{32.2\,\text{Torr}}{38.3\,\text{Torr}} = \boxed{0.840}$$

$$Br_2\text{: } y_B = \frac{p_B}{p_{\text{total}}} = \frac{6.1\,\text{Torr}}{38.3\,\text{Torr}} = \boxed{0.160}$$

E5B.10(a) Let subscript 1 denote the methylbenzene and 2 the dimethylbenzene. Assume that the structurally similar liquids obey Raoult's law [5A.21]. The partial pressures of the two liquids sum to 0.50 atm = 0.50 atm × 101.3 kPa atm^{-1} = 50.7 kPa.

$$p_1 + p_2 = p = x_1 p_1^* + x_2 p_1^* = x_1 p_1^* + (1-x_1)p_2^*$$

Solve for x_1:

$$x_1 = \frac{p - p_2^*}{p_1^* - p_2^*} = \frac{(50.7-20.0)\,\text{kPa}}{(53.3-20.0)\,\text{kPa}} = \boxed{0.92}$$

and $x_2 = 1 - 0.92 = \boxed{0.08}$.

The vapour phase mole fractions are given by eqn 1A.8:

$$y_1 = \frac{p_1}{p} = \frac{x_1 p_1^*}{p} = \frac{(0.92)\times 53.3\,\text{kPa}}{50.7\,\text{kPa}} = \boxed{0.97}$$

and $y_2 = \dfrac{x_2 p_2^*}{p} = \dfrac{(0.08)\times 20.0\,\text{kPa}}{50.7\,\text{kPa}} = \boxed{0.03}$

E5B.11(a) The partial vapour pressures are given by Raoult's law [5A.21]:

$$p_A = x_A p_A^* \quad \text{and} \quad p_B = x_B p_B^* = (1-x_B)p_B^*.$$

Equation 1A.8 relates these vapour pressures to the vapour-phase mole fractions:

$$y_A = \frac{p_A}{p_{\text{total}}} = \frac{x_A p_A^*}{x_A p_A^* + (1-x_A)p_B^*}$$

Solve for x_A:

$$x_A p_A^* + (1-x_A)p_B^* = \frac{x_A p_A^*}{y_A}$$

$$x_A\left(p_A^* - p_B^* - \frac{p_A^*}{y_A}\right) = -p_B^*$$

$$x_A = \frac{p_B^*}{p_B^* + \frac{p_A^*}{y_A} - p_A^*} = \frac{52.0\text{ kPa}}{\left(52.0 + \frac{76.7}{0.350} - 76.7\right)\text{kPa}} = \boxed{0.267}$$

and $x_B = 1 - x_A = 1 - 0.267 = \boxed{0.733}$.

The total vapour pressure is

$$p_{\text{total}} = x_A p_A^* + x_B p_B^* = 0.267 \times 76.7\text{ kPa} + 0.733 \times 52.0\text{ kPa} = \boxed{58.6\text{ kPa}}$$

E5B.12(a) (a) If the solution is ideal, then the partial vapour pressures are given by Raoult's law [5A.21]:

$$p_A{}^\circ = x_A p_A^* = 0.6589 \times 127.6\text{ kPa} = 84.1\text{ kPa}$$

and $p_B{}^\circ = x_B p_B^* = (1-x_B)p_B^* = (1-0.6589) \times 50.60\text{ kPa} = 17.26\text{ kPa}$.

(Note the use of the symbol ° to emphasize that these are idealized quantities; we do not yet know if they are the actual partial vapour pressures.) At the normal boiling temperature, the partial vapour pressures must add up to 1 atm (101.3 kPa). These ideal partial vapour pressures **do** add up to 101.3 kPa, so the $\boxed{\text{solution is ideal}}$.

(b) Because the solution is ideal, the ideal partial vapour pressures are the actual partial vapour pressures. We can use them to find the vapour phase composition:

$$y_A = \frac{p_A}{p_{\text{total}}} = \frac{84.1\text{ kPa}}{101.3\text{ kPa}} = \boxed{0.830} \quad \text{and} \quad y_B = \frac{p_B}{p_{\text{total}}} = \frac{17.26\text{ kPa}}{101.3\text{ kPa}} = \boxed{0.1703}$$

E5B.13(a) (a) If the solution is ideal, then the partial vapour pressures are given by Raoult's law [5A.21]:

$p_{DE} = x_{DE}p_{DE}{}^* = 0.60 \times 22.9\text{ kPa} = 13.7\text{ kPa}$

and $p_{DP} = x_{DP}p_{DP}{}^* = 0.40 \times 17.1\text{ kPa} = 6.8\text{ kPa}$

The total pressure is

$$p_{\text{total}} = p_{DE} + p_{DP} = (13.7 + 6.8)\text{kPa} = \boxed{20.6\text{ kPa}}.$$

(b) The composition of the vapour is given by

$$y_{DE} = \frac{p_{DE}}{p_{\text{total}}} = \frac{13.7\text{ kPa}}{20.6\text{ kPa}} = \boxed{0.668} \quad \text{and} \quad y_{DP} = \frac{p_{DP}}{p_{\text{total}}} = \frac{6.8\text{ kPa}}{20.6\text{ kPa}} = \boxed{0.332}$$

Solutions to problems

P5B.1 $\Delta T = \dfrac{RT_f^{*2} x_B}{\Delta_{fus}H}$ [5B.12] and $x_B \approx \dfrac{n_B}{n_A} = \dfrac{n_B M_A}{1000\,\text{g}}$

Hence $\Delta T = \dfrac{n_B M_A RT_f^{*2}}{\Delta_{fus}H \times 1000\,\text{g}} = \dfrac{bM_A RT_f^{*2}}{\Delta_{fus}H}$ [b is molality of solution]

$$= b \times \left(\frac{(0.06005\,\text{kg mol}^{-1}) \times (8.3145\,\text{J K}^{-1}\,\text{mol}^{-1}) \times (290\,\text{K})^2}{11.4 \times 10^3\,\text{J mol}^{-1}} \right)$$

$$= 3.68\,\text{K} \times b/(\text{mol kg}^{-1})$$

Thus the apparent molality (based on measured freezing point depression) is

$$b_{app} = \frac{\Delta T}{3.68\,\text{K}}\,\text{mol kg}^{-1} = \nu b,$$

where b is the actual molality and ν may be interpreted as the number of ions in solution per one formula unit of KF. The apparent molar mass of KF can be determined from the apparent molality by the relation

$$M_{B,app} = \frac{b}{b_{app}} \times M_B = \frac{1}{\nu} \times M_B = \frac{1}{\nu} \times (58.1\,\text{g mol}^{-1})$$

where M_B is the actual molar mass of KF. We can draw up the following table from the data.

$b/(\text{mol kg}^{-1})$	0.015	0.037	0.077	0.295	0.602
$\Delta T/\text{K}$	0.115	0.295	0.470	1.381	2.67
$b_{app}/(\text{mol kg}^{-1})$	0.0312	0.0801	0.128	0.375	0.725
$\nu = b_{app}/b$	2.08	2.16	1.66	1.27	1.20
$M_{B,app}/(\text{g mol}^{-1})$	27.9	26.8	35.1	45.7	48.3

A possible explanation is that the dissociation of KF into ions is complete at the lower concentrations but incomplete at the higher concentrations. Values of ν greater than 2 are hard to explain, but they could be a result of the approximations involved in obtaining eqn 5B.12. See the original reference for further information about the interpretation of the data.

P5B.3 Let subscript c stand for cyclohexane and p for pentadecane. The partial molar volume of cyclohexane is

$$V_c = \left(\frac{\partial V}{\partial n_c} \right)_{n_p} \quad [5A.1]$$

A similar expression holds for V_p. V_c can be evaluated graphically by plotting V against n_c (holding n_p constant) and finding the slope at the desired point. In a similar manner, V_p can be evaluated by plotting V against n_p. To find V_c, V is needed at a variety of n_c while holding n_p constant, say at exactly 1 mol; conversely, to find V_p, V is needed at a variety of n_p while holding n_c constant. The mole fraction in this system is

$$x_c = \frac{n_c}{n_c + n_p} \quad \text{so} \quad n_c = \frac{x_c n_p}{1 - x_c} \quad \text{and} \quad n_p = \frac{(1 - x_c) n_c}{x_c}.$$

From n_c and n_p, the mass of the sample can be calculated, and the volume can be calculated from

$$V = \frac{m}{\rho} = \frac{n_c M_c + n_p M_p}{\rho}.$$

Draw up the following table, using $M_c = 84.16\ \mathrm{g\,mol^{-1}}$ and $M_p = 212.41\ \mathrm{g\,mol^{-1}}$. Note that the central columns are the given data. The columns on either side hold one component constant at 1 mol and compute the amount of the other component and the volume.

n_c/mol (n_p=1)	V/cm^3	x_c	ρ/g cm^{-3}	n_p/mol (n_c=1)	V/cm^3
2.295	529.4	0.6965	0.7661	0.4358	230.7
3.970	712.2	0.7988	0.7674	0.2519	179.4
9.040	1264	0.9004	0.7697	0.1106	139.9

These values are plotted in Fig. 5.2(a) and (b).

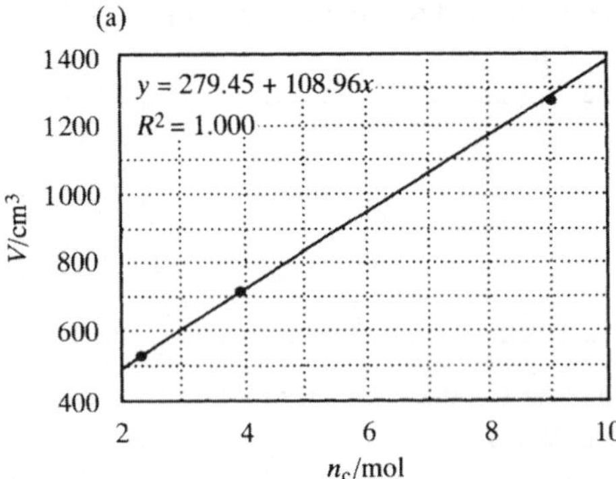

Figure 5B.2(a)

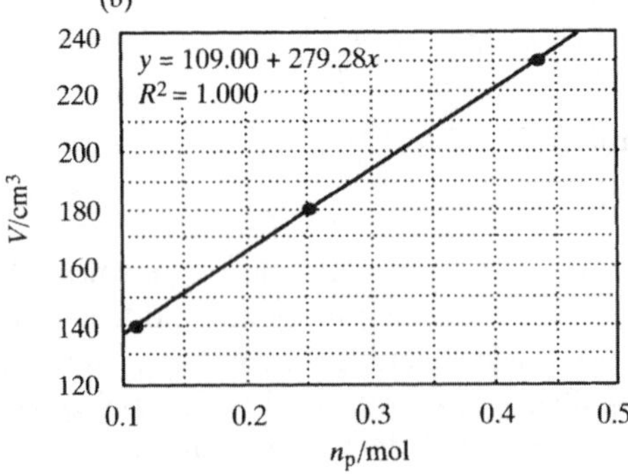

Figure 5B.2(b)

These plots show no curvature, so in this case, perhaps due to the limited number of data points, the partial molar volumes are constant over the range of concentrations.

$$V_c = \boxed{109.0\,\text{cm}^3\,\text{mol}^{-1}} \quad \text{and} \quad V_p = \boxed{279.3\,\text{cm}^3\,\text{mol}^{-1}}$$

P5B.5 $S = S_0 e^{\tau/T}$ may be written in the form $\ln S = \ln S_0 + \frac{\tau}{T}$, which indicates that a plot of $\ln S$ against $1/T$ should be linear with slope τ and intercept $\ln S_0$. Linear regression analysis gives $\tau = \boxed{165\,\text{K}}$ (standard deviation = 2 K)

$\ln\{S_0/\text{g solute }(100\text{ g solvent})^{-1}\} = 2.990$, standard deviation 0.007; $R = \boxed{0.99978}$ so the linear regression explains 99.98 per cent of the variation.

$$S_0 = e^{2.990}\,\text{mol dm}^{-3} = \boxed{19.89\,\text{g solute}(100\,\text{g solvent})^{-1}},$$

or, converting to molality

$$S_0 = \frac{19.89\text{ g solute}}{0.1000\text{ kg solvent}} \times \frac{1\text{ mol solute}}{158.17\text{ g solute}} = \boxed{1.257\text{ mol kg}^{-1}}$$

Equation 5B.15 gives the theoretical solubility in mole fraction units:

$$x_B = \exp\left\{\frac{\Delta_{\text{fus}}H}{R}\left(\frac{1}{T_f} - \frac{1}{T}\right)\right\} = \exp\left(\frac{\Delta_{\text{fus}}H}{RT_f}\right)\exp\left(-\frac{\Delta_{\text{fus}}H}{RT}\right)$$

Comparing the theoretical expression to the empirical $S = S_0 e^{\tau/T}$ is not straightforward because of the different units. For dilute solutions, mole fraction and molality are directly proportional:

$$x_B = \frac{n_B}{n_A + n_B} \approx \frac{n_B}{n_A} \quad \text{and} \quad b = \frac{n_B}{m_A} = \frac{n_B}{n_A M_A} \approx \frac{x_B}{M_A}.$$

We identify S_0 with the temperature-independent part of eqn 5B.15 and constants:

$$S_0 = \boxed{\frac{1}{M_A}\exp\left(\frac{\Delta_{\text{fus}}H}{RT^*}\right)}.$$

Comparing the temperature-dependent portion, we would identify $\tau = \boxed{-\Delta_{\text{fus}}H/R}$. This identification is problematic. The empirical τ is positive, but so is $\Delta_{\text{fus}}H$. The empirical solubility decreases with increasing temperature, but the theoretical solubility increases. That the ideal expression fails to capture the solubility behaviour of a reasonably concentrated ionic solution should not be **too** surprising in light of Topic 5F.

P5B.7 (a) The $\Delta_{\text{mix}}G(x_{\text{Pb}})$ curves show that at 1 500 K lead and copper are totally miscible. They mix to form a homogeneous solution no matter what the relative amounts may be. However, the curve at 1 100 K displays two distinct minima, and we expect two partially miscible phases.

(i) $F = C - P + 2 = 2 - 2 + 2 = \boxed{2}$ at 1 500 K

(ii) $F = C - P + 2 = 2 - 1 + 2 = \boxed{3}$ at 1 100 K

(b)

(i) When a homogeneous, equilibrium mixture with $x_{\text{Pb}} = 0.1$ is cooled from 1 500 K to 1 100 K, no phase separation occurs. The solution composition does not change.

(ii) When a homogeneous, equilibrium mixture with $x_{\text{Pb}} = 0.7$ is cooled slowly, two partially miscible phases appear somewhere between 1 300 and 1 100 K. The separation occurs because the composition lies between two minima on the $\Delta_{\text{mix}}G$ curve at 1 100 K and phase separation lowers the total Gibbs energy. The composition of the two phases is determined by the equilibrium criterion $\mu_i(\alpha) = \mu_i(\beta)$ between the α and β phases. Since the chemical potential is the tangent of the $\Delta_{\text{mix}}G$ curve, we conclude that the straight line that is tangent to $\Delta_{\text{mix}}G(x)$ at two values of x (a double tangent) determines the composition of the two partially miscible phases. The two minima in the 1 100 K data appear to have the same value of $\Delta_{\text{mix}}G$, namely $-1\ \text{kJ mol}^{-1}$, so a horizontal line is tangent to both at the compositions $x_{\text{Pb}}(\alpha) = \boxed{0.19}$ and $x_{\text{Pb}}(\beta) = \boxed{0.82}$. (Note that the tangent points and the minima coincide in this case, but they need not. If the minima fall at different values of $\Delta_{\text{mix}}G$, then the line tangent to both of them is not horizontal and the points of tangency need not be the minima.) The relative amounts of the two phases is determined by the lever rule [5C.6].

$$\frac{n_\alpha}{n_\beta} = \frac{l_\beta}{l_\alpha} = \frac{0.82 - 0.70}{0.70 - 0.19} = \boxed{0.24}$$

(c) Solubility at 1 100 K is determined by the positions of the two minima in the $\Delta_{\text{mix}}G$ curve. The maximum amount of lead that can be dissolved in copper yields a mixture that has $x_{\text{Pb}} = 0.19$; any more lead produces a second phase. So the solubility of lead in copper at this temperature is $\boxed{x_{\text{Pb}} = 0.19}$. The other minimum in the curve occurs at $x_{\text{Pb}} = 0.82$; any ***less*** lead produces a second phase. This lead-rich composition corresponds to the maximum solubility of copper in lead, namely $x_{\text{Cu}} = \boxed{0.18}$.

Question. Express these solubilities in g solute per 100 g solvent.

P5B.9 Start with eqn 5B.11

$$\mu_A^\star(\text{s}) = \mu_A^\star(\text{l}) + RT \ln x_A.$$

Rearrange, analogous to *Justification* 5B.1, to

$$\ln x_A = \frac{\mu_A^\star(\text{s}) - \mu_A^\star(\text{l})}{RT} = \frac{-\Delta_{\text{fus}}G}{RT}$$

So $\dfrac{\mathrm{d}\ln x_A}{\mathrm{d}T} = -\dfrac{1}{R} \times \dfrac{\mathrm{d}}{\mathrm{d}T}\left(\dfrac{\Delta_{\text{fus}}G}{T}\right) = \boxed{\dfrac{\Delta_{\text{fus}}H}{RT^2}}$ [Gibbs–Helmholtz equation, 3D.10].

Integration yields

$$\int_1^{x_A} \mathrm{d}\ln x_A = \int_{T^*}^{T} \frac{\Delta_{\text{fus}}H\,\mathrm{d}T}{RT^2} \approx \frac{\Delta_{\text{fus}}H}{R}\int_{T^*}^{T}\frac{\mathrm{d}T}{T^2}$$

or $\boxed{\ln x_A = \dfrac{-\Delta_{\text{fus}}H}{R} \times \left(\dfrac{1}{T} - \dfrac{1}{T^*}\right)}$

The approximations $\ln x_A = \ln(1-x_B) \approx -x_B$ and $T \approx T^*$ then lead to eqns 5B.12 here, just as (in *Justification* 5B.1) they lead to eqn 5B.9.

$$\ln\left(\frac{a_B}{r}\right) = \varphi - \varphi(0) + \int_0^r \left(\frac{\varphi - 1}{r}\right) dr$$

P5B.11 We use eqn 5B.18 in the form given in Example 5B.2 with $\Pi = \rho g h$, then

$$\frac{\Pi}{c} = \frac{RT}{M}\left(1 + \frac{B}{M}c\right) = \frac{RT}{M} + \frac{RTB}{M^2}c$$

where c is the mass concentration of the polymer. Therefore plot Π/c against c. The intercept gives RT/M and the slope gives RT/M^2.

The transformed data to plot are given in the table

$c/(\text{mg cm}^{-3})$	1.33	2.10	4.52	7.18	9.87
$(\Pi/c)/(\text{N m}^{-2}\,\text{mg}^{-1}\,\text{cm}^3)$	22.6	24.3	29.2	34.3	39.5

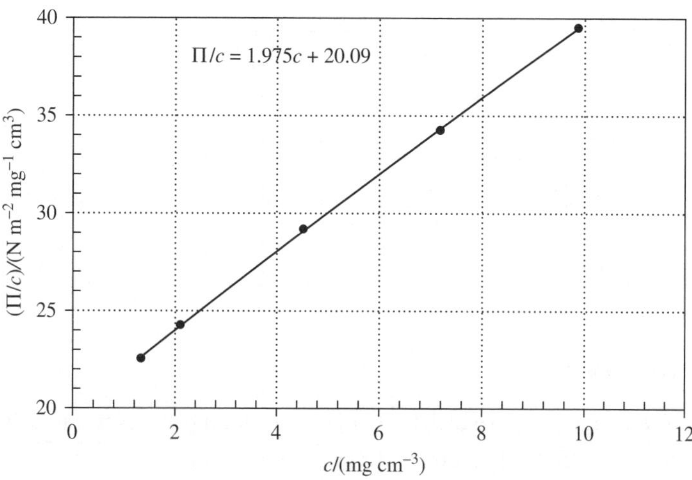

Figure 5B.3

The plot is shown in Fig. 5B.3. The intercept is 29.1 $\text{N m}^{-2}/(\text{mg cm}^{-3})$. The slope is 1.974 $\text{N m}^{-2}/(\text{mg cm}^{-3})^2$. Therefore

$$M = \frac{RT}{29.1\,\mathrm{N\,m^{-2}/(mg\,cm^{-3})}}$$

$$= \frac{8.3145\,\mathrm{J\,K^{-1}\,mol^{-1}} \times 303.15\,\mathrm{K}}{20.09\,\mathrm{N\,m^{-2}/(mg\,cm^{-3})}} \times \left(\frac{1\,\mathrm{g}}{10^3\,\mathrm{mg}}\right) \times \left(\frac{10^6\,\mathrm{cm^3}}{1\,\mathrm{m^3}}\right)$$

$$= 1.26 \times 10^5\ \mathrm{g\,mol^{-1}} = \boxed{1.26 \times 10^5\,\mathrm{g\,mol^{-1}}}$$

$$B = \frac{M}{RT} \times 1.974\,\mathrm{N\,m^{-2}/(mg\,cm^{-3})^2}$$

$$= \frac{M}{\left(\frac{RT}{M}\right)} \times 1.974\,\mathrm{N\,m^{-2}/(mg\,cm^{-3})^2}$$

$$= \frac{1.25\bar{5} \times 10^5\ \mathrm{g\,mol^{-1}} \times 1.974\,\mathrm{N\,m^{-2}/(mg\,cm^{-3})^2}}{20.09\,\mathrm{N\,m^{-2}/(mg\,cm^{-3})}}$$

$$= 1.23 \times 10^4\,\mathrm{g\,mol^{-1}/(mg\,cm^{-3})}$$

$$= 1.23 \times 10^7\ \mathrm{g\,mol^{-1}/(g\,cm^{-3})} = \boxed{1.23 \times 10^4\ \mathrm{dm^3\,mol^{-1}}}$$

P5B.13 The model for excess enthalpy is [5B.6]

$$H^{\mathrm{E}} = n\xi RTx_{\mathrm{A}}x_{\mathrm{B}}$$

Molar excess enthalpy is plotted against composition for $\xi = 1$ at several temperatures in Fig. 5B.4 (curves labelled by T/K).

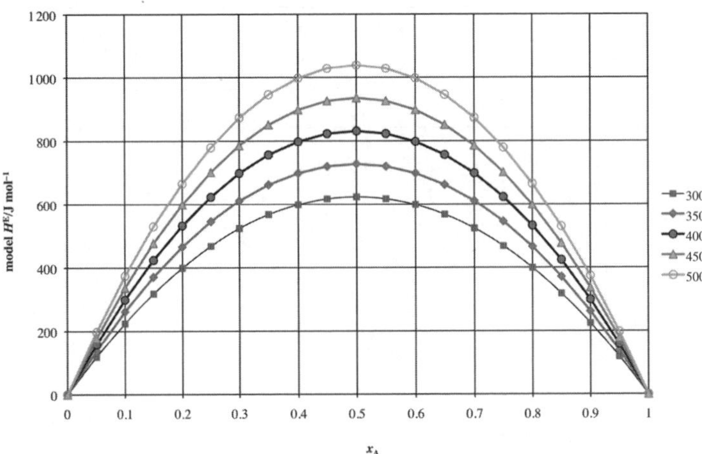

Figure 5B.4

The composition at which the temperature dependence is strongest is the composition at which the function is at its maximum, namely $\boxed{x_{\mathrm{A}} = x_{\mathrm{B}} = 0.5}$.

P5B.15 In a plot of h/c vs. c, the slope is $\dfrac{RTB}{\rho g M^2}$, so

$$B = \text{slope} \times \frac{\rho g M^2}{RT} = \frac{\text{slope}}{y\ \text{intercept}} \times M$$

The data to be plotted are in Example 5B.2, and the plot is in Fig. 5B.15 of the main text. If we use the raw h and c data from the example and run the linear regression on h/c without rounding, the slope is 0.0758 cm and the intercept is $0.202\ \text{cm g}^{-1}\,\text{dm}^3$. Therefore

$$M = \frac{(8.3145\ \text{J K}^{-1}\ \text{mol}^{-1})(298\ \text{K})}{(980\ \text{g dm}^{-3})(9.81\ \text{m s}^{-2})} \times \frac{100\ \text{cm m}^{-1}}{0.202\ \text{cm g}^{-1}\ \text{dm}^3} = 127\ \text{kg mol}^{-1} = 1.27\times10^5\ \text{g mol}^{-1}$$

[Note that the non-MKS units of g, dm^3, and cm cancel, leaving the molar mass in MKS units of kg mol^{-1}.] The virial coefficient, then, is

$$B = \frac{0.202\ \text{cm g}^{-1}\ \text{dm}^3}{0.0758\ \text{cm}} \times (1.27\times10^5\ \text{g mol}^{-1}) = \boxed{4.78\times10^4\ \text{dm}^3\ \text{mol}^{-1}}$$

5C Phase diagrams of binary systems

Answers to discussion questions

D5C.1 (a) See Fig. 5C.1(a). Congruent melting is evident from the presence of liquid AB wherever liquids are present (along with whichever of A or B happens to be in excess).

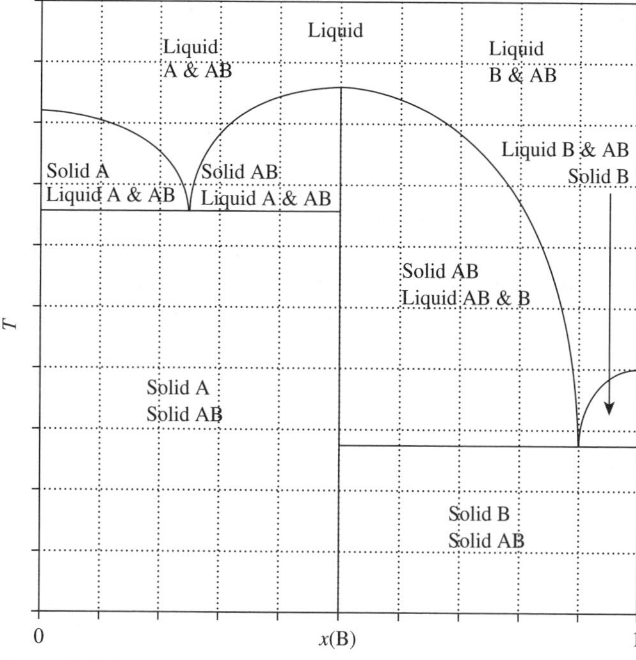

Figure 5C.1(a)

(b) See Fig. 5C.1(b). Incongruent melting is reflected in the absence of liquid AB_2. When solid AB_2 melts, it decomposes.

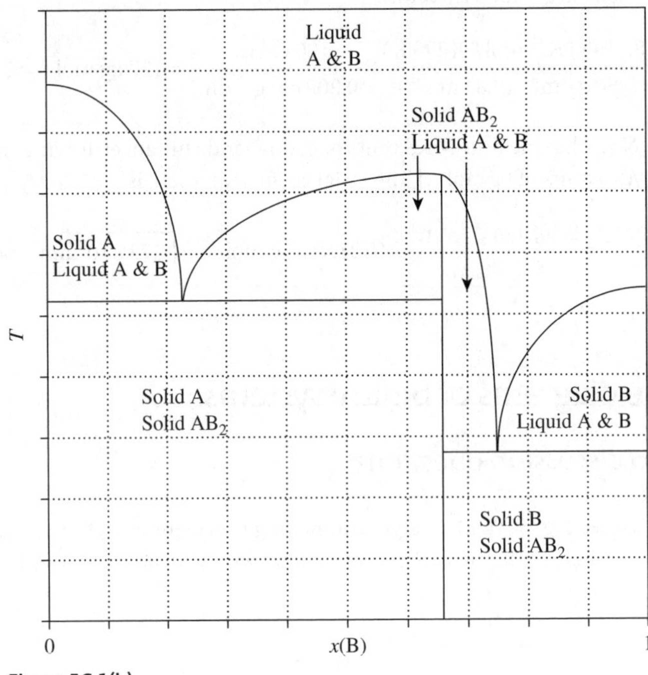

Figure 5C.1(b)

(c) See Fig. 5C.1(c). Vapours are always miscible and these liquids are miscible, so both the vapour region at high temperature and the liquid region at low temperature are two-component single-phase regions. In between, the liquid–vapour coexistence regions are two-phase regions, each of which has both components. The curve that bounds the coexistence regions at high temperature is the vapour composition curve, while the curve that bounds the region at low temperatures is the liquid boiling temperature. Figure 5C.1(c) depicts a high-boiling azeotrope; however, a low-boiling azeotrope could illustrate what the question describes.

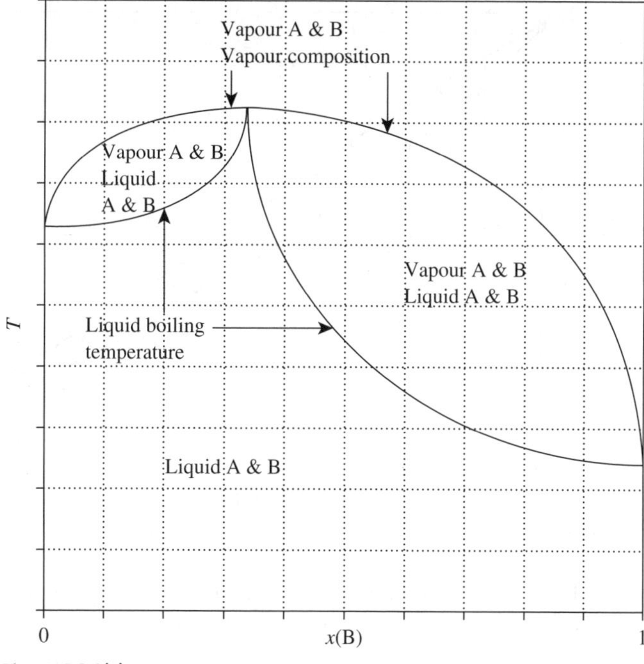

Figure 5C.1(c)

D5C.3 The principal factor is the shape of the two-phase liquid–vapour region in the phase diagram (usually a temperature–composition diagram). The closer the liquid and vapour lines are to each other, the more steps of the sort described in Section 5C.2(a) are needed to move from a given mixed composition to an acceptable enrichment in one of the components. (See Fig. 5C.10 of the main text.) But the presence of an azeotrope could prevent the desired degree of separation from being achieved. Incomplete miscibility of the components at specific concentrations could also affect the number of plates required.

Solutions to exercises

E5C.1(a) Add the boiling point of methylbenzene to the table at $x_M = y_M = 1$ and the boiling point of octane at $x_M = y_M = 0$. Plot the boiling temperatures against liquid mole fractions and the same boiling temperatures against vapour mole fractions on the same plot.

The phase diagram is shown in Fig. 5C.2. The phase boundary curves are polynomial fits to the data points.

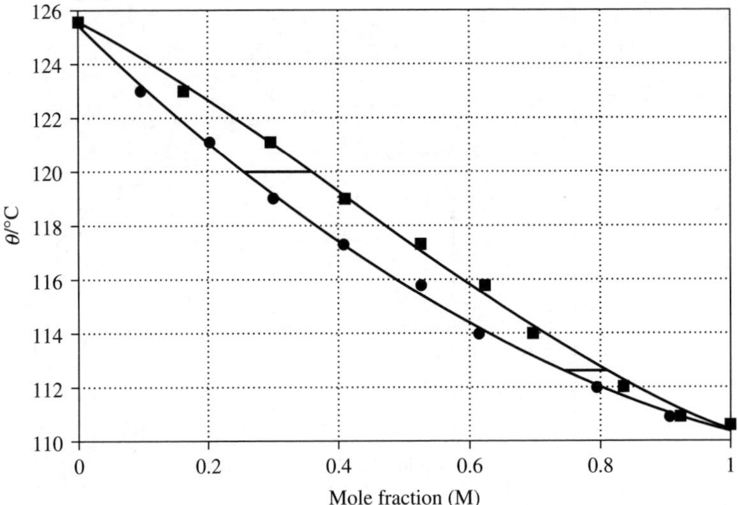

Figure 5C.2

(a) Find $x_M = 0.25$ on the lower curve and draw a horizontal tie line to the upper curve. The mole fraction at that point is $\boxed{y_M = 0.36}$.

(b) Find $x_M = 0.75$ (i.e. $x_O = 0.25$) on the lower curve and draw a horizontal tie line to the upper curve. The mole fraction at that point is $\boxed{y_M = 0.80 \text{ (i.e. } y_O = 0.20)}$.

E5C.2(a) The phase diagram is shown in Fig. 5C.3.

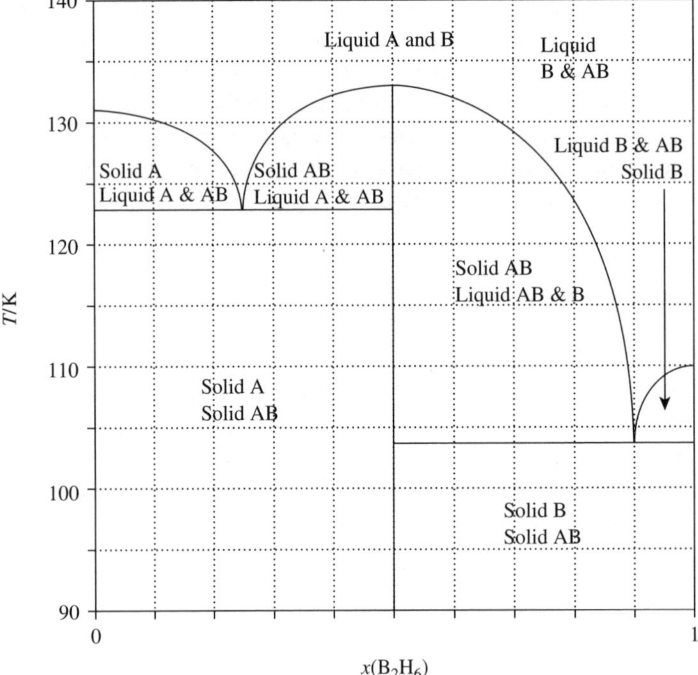

Figure 5C.3

E5C.3(a) Refer to the figure given with the exercise. At the lowest temperature shown on the phase diagram, there are two liquid phases, a water-rich phase (x_B=0.07) and a methylpropanol-rich phase (x_B=0.88); the latter phase is about 10 times as abundant as the former (lever rule). On heating, the compositions of the two phases change, the water-rich phase increasing significantly in methylpropanol and the methylpropanol-rich phase more gradually increasing in water. (Note how the composition of the left side of the diagram changes more with temperature than the right.) The relative proportions of the phases continue to be given by the lever rule. Just before the isopleth intersects the phase boundary, the methylpropanol-rich phase (x_B=0.8) is in equilibrium with a vanishingly small water-rich phase (x_B=0.38). Then the phases merge, and the single-phase region is encountered with x_B=0.8.

E5C.4(a) The feature that indicates incongruent melting (Section 5C.4(c)) is circled in Fig. 5C.4. The incongruent melting point is marked as $T_1 \approx 350\,°C$. The composition of the eutectic is $x_B \approx \boxed{0.25}$ and its melting point is labelled $T_2 = \boxed{193\,°C}$.

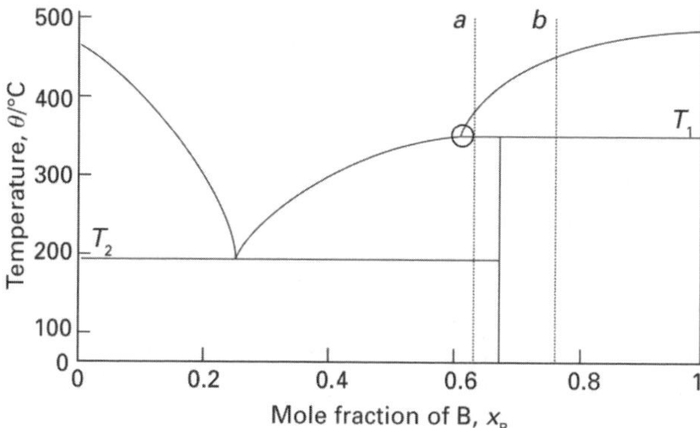

Figure 5C.4

E5C.5(a) The cooling curves are shown in Fig. 5C.5(b). Note the breaks (abrupt change in slope) at temperatures corresponding to points a_1, a_2, b_1, and b_2 of the phase diagram (Fig. 5C.5(a)). Also note the eutectic halt at b_3.

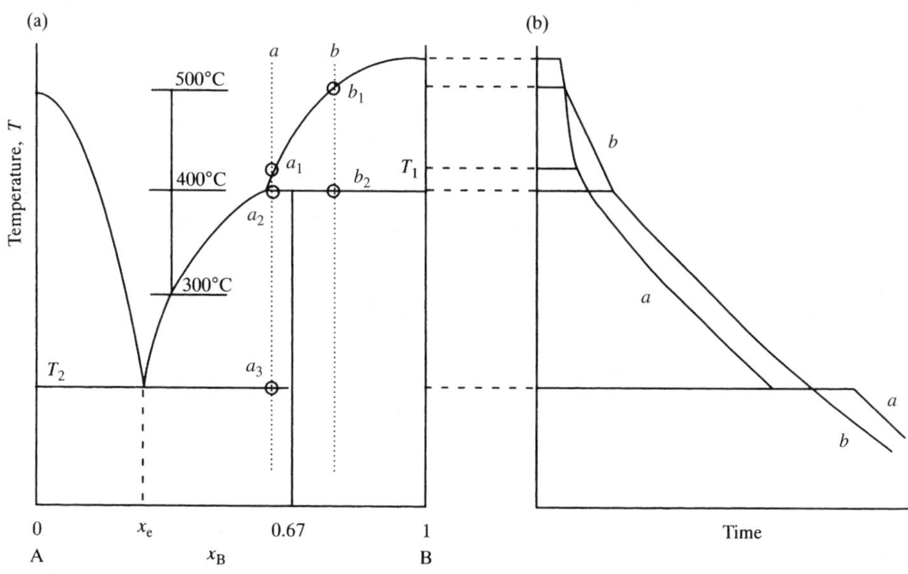

Figure 5C.5

E5C.6(a) Refer to Fig. 5C.6. Dotted horizontal lines have been drawn at the relevant temperatures.

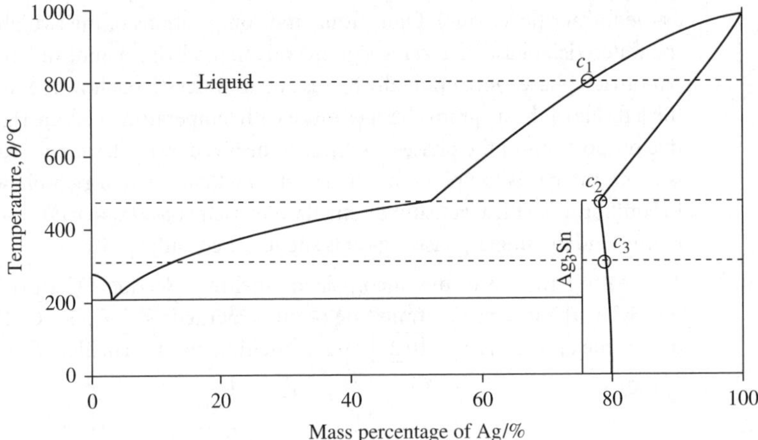

Figure 5C.6

(a) The solubility of silver in tin at 800 °C is determined by the point c_1. (At higher proportions of silver, the system separates into two phases, a liquid and a solid phase rich in silver.) The point c_1 corresponds to 76 per cent silver by mass.

(b) The compound Ag_3Sn decomposes at this temperature. Three phases are in equilibrium here: a liquid containing atomic Ag and Sn about 52 per cent Ag by mass; a solid solution of Ag_3Sn in Ag; and solid Ag_3Sn. See point c_2.

(c) At point c_3, two phases coexist: solid Ag_3Sn and a solid solution of the compound and metallic silver. Because this point is close to the Ag_3Sn composition, the solid solution is mainly Ag_3Sn—at least when measured in mass terms. Let us express the composition of the solid solution as a ratio of moles of compound (n_c) to moles of atomic silver (n_a). We can relate these quantities to the silver mass fraction c_{Ag} by employing the definition of mass fraction, namely the mass of silver (from the compound and from atomic silver) over the total sample mass:

$$c_{Ag} = \frac{m_{Ag}}{m_{Ag} + m_{Sn}} = \frac{(3n_c + n_a)M_{Ag}}{(3n_c + n_a)M_{Ag} + n_c M_{Sn}}.$$

Rearrange this relationship, collecting terms in n_c on one side and n_a on the other:

$$n_c\{3M_{Ag}(c_{Ag} - 1) + M_{Sn}c_{Ag}\} = n_a M_{Ag}(1 - c_{Ag}).$$

The mole ratio of compound to atomic silver is given by

$$\frac{n_c}{n_a} = \frac{M_{Ag}(1 - c_{Ag})}{3M_{Ag}(c_{Ag} - 1) + M_{Sn}c_{Ag}}.$$

At 300 °C, $c_{\text{Ag}}=0.78$ (point c_3 on the coexistence curve), so

$$\frac{n_\text{c}}{n_\text{a}}=\frac{(107.9\text{ g mol}^{-1})\times(1-0.78)}{3\times(107.9\text{ g mol}^{-1})\times(0.78-1)+(118.7\text{ g mol}^{-1})\times0.78}=\boxed{1.11}$$

At 460 °C, $c_{\text{Ag}}=0.77$ (point c_2 on the coexistence curve), so

$$\frac{n_\text{c}}{n_\text{a}}=\frac{(107.9\text{ g mol}^{-1})\times(1-0.77)}{3\times(107.9\text{ g mol}^{-1})\times(0.77-1)+(118.7\text{ g mol}^{-1})\times0.77}=\boxed{1.46}$$

E5C.7(a) (a) See Fig. 5C.7(a) and (b).

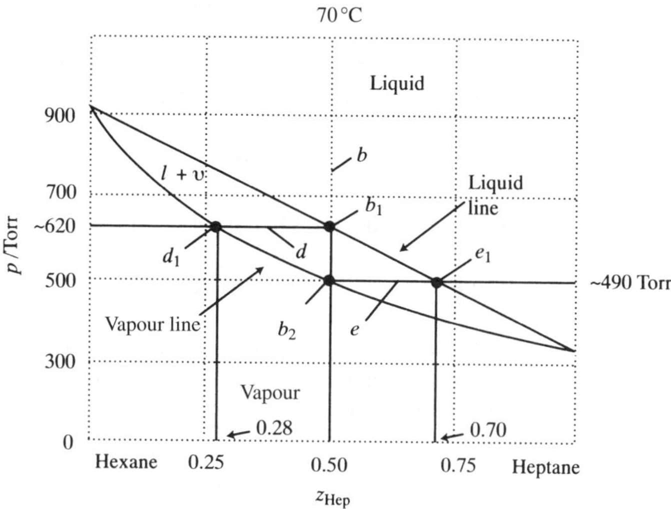

Figure 5C.7(a)

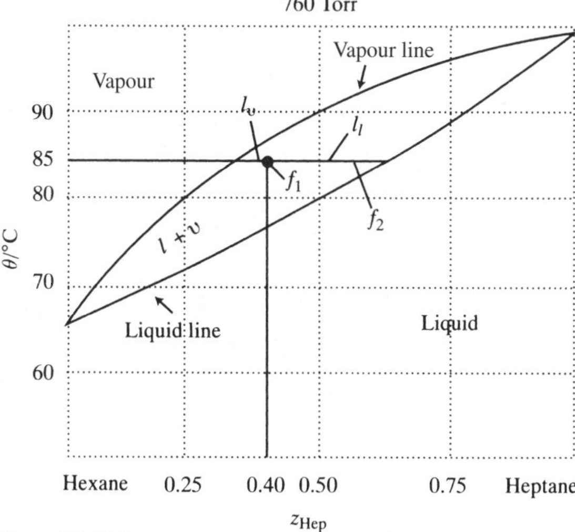

Figure 5C.7(b)

(b) Follow line b in Fig. 5C.7(a) down to the liquid line which intersects at point b_1. The vapour pressure at b_1 is $\boxed{620\ \text{Torr}}$.

(c) Follow line b in Fig. 5C.7(a) down to the vapour line which intersects at point b_2. The vapour pressure at b_2 is $\boxed{490\ \text{Torr}}$. From points b_1 to b_2, the system changes from essentially all liquid to essentially all vapour.

(d) Consider tie line d; point b_1 gives the mole fractions of the liquid, which are

$$x_{\text{Hep}} = 0.50 = 1 - x_{\text{Hex}}, \quad \text{so} \quad x_{\text{Hex}} = \boxed{0.50}.$$

Point d_1 gives the mole fractions in the vapour which are

$$y_{\text{Hep}} = 0.28 = 1 - y_{\text{Hex}}, \quad \text{so} \quad y_{\text{Hex}} = \boxed{0.72}.$$

The initial vapour is richer in the more volatile component, hexane.

(e) Consider tie line e; point b_2 gives the mole fractions in the vapour, which are

$$y_{\text{Hep}} = 0.50 = 1 - y_{\text{Hex}}, \quad \text{so} \quad y_{\text{Hex}} = \boxed{0.50}.$$

Point e_1 gives the mole fractions in the liquid, which are

$$x_{\text{Hep}} = 0.70 = 1 - x_{\text{Hex}}, \quad \text{so} \quad x_{\text{Hex}} = \boxed{0.30}$$

(f) Consider tie line in Fig. 5C.7(b). The section, l_l, from point f_1 to the liquid line gives the relative amount of vapour; the section, l_v, from point f_1 to the vapour line gives the relative amount of liquid. That is

$$n_v l_v = n_l l_l \,[5\text{C}.6] \quad \text{or} \quad \frac{n_v}{n_l} = \frac{l_l}{l_v} \approx \frac{6}{1}$$

Since the total amount is 2 mol, $n_v = 1.6$ mol and $n_l = 0.4$ mol.

E5C.8(a) The phase diagram is drawn in Fig. 5C.8.

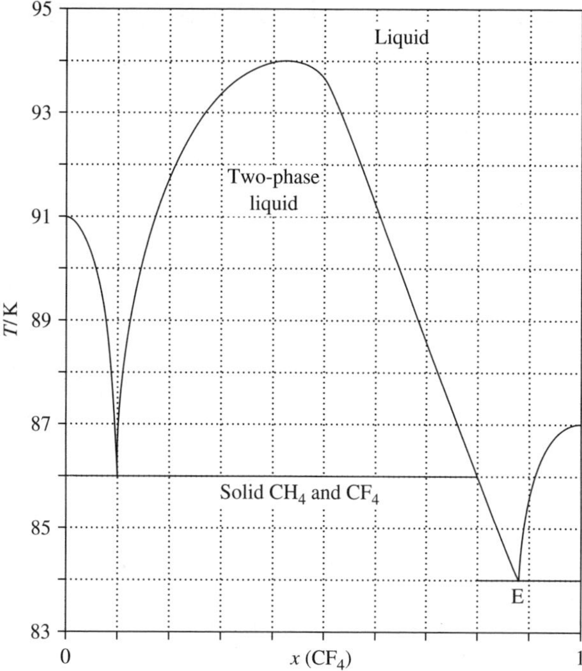

Figure 5C.8

E5C.9(a) The cooling curves are sketched in Fig. 5C.9. Note the breaks and halts. The breaks correspond to changes in the rate of cooling due to the freezing out of a solid which releases its heat of fusion and thus slows down the cooling process. The halts correspond to the existence of three phases and hence no variance until one of the phases disappears.

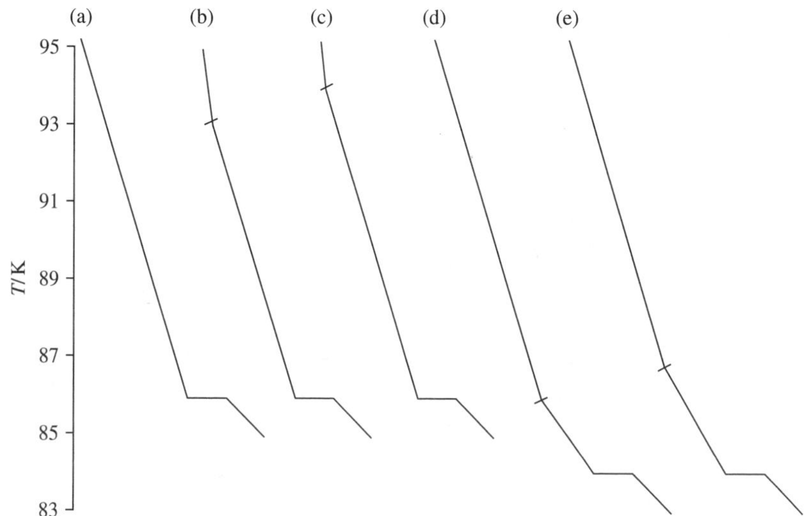

Figure 5C.9

E5C.10(a) The phase diagram is sketched in Fig. 5C.10.

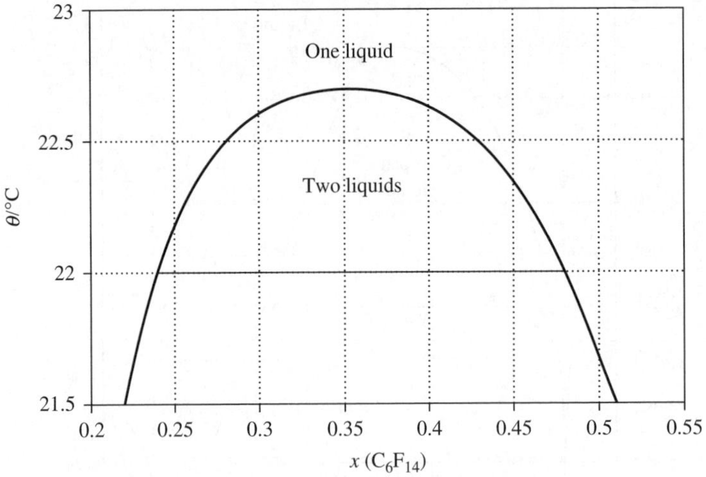

Figure 5C.10

(a) The mixture has a single liquid phase at all compositions.

(b) Adding C_6F_{14} amounts to moving from left to right on the phase diagram on the 22.0 °C line. When the overall composition reaches $x(C_6F_{14})=0.24$ the mixture separates into two liquid phases of compositions, the more abundant phase having $x=0.24$ and an initially minuscule phase of $x=0.48$. The compositions of the two phases remain the same but their relative amounts change as more C_6F_{14} is added until the overall composition reaches $x=0.48$. At that point, the phase having $x=0.24$ vanishes, and from there on the mixture forms a single liquid phase.

Solutions to problems

P5C.1 (a) The data, including that for pure chlorobenzene, are plotted in Fig. 5C.11.

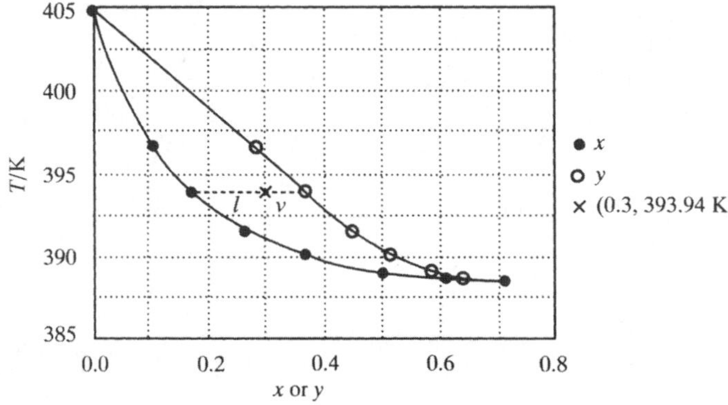

Figure 5C.11

(b) The smooth curve through the (x, T) data crosses $x=0.300$ at $\boxed{391.0\,\text{K}}$, the boiling point of the mixture.

(c) We need not interpolate data, for 393.94 K is a temperature for which we have experimental data. The mole fraction of 1-butanol in the liquid phase is 0.1700 and in the vapour phase 0.3691. According to the lever rule, the proportions of the two phases are in an inverse ratio of the distances their mole fractions are from the composition point in question. That is

$$\frac{n_{\text{liq}}}{n_{\text{vap}}}=\frac{v}{l}=\frac{0.3691-0.300}{0.300-0.1700}=\boxed{0.532}$$

P5C.3 The phase diagram is shown in Fig. 5C.12(a). The values of x_S corresponding to the three compounds are: (1) P_4S_3, 0.43; (2) P_4S_7, 0.64; (3) P_4S_{10}, 0.71.

The diagram has four eutectics labelled e_1, e_2, e_3, and e_4; eight two-phase liquid–solid regions, t_1 through t_8; and four two-phase solid regions, S_1, S_2, S_3, and S_4. The composition and physical state of the regions are as follows:

l: liquid S and P

S_1: solid P and solid P_4S_3

S_2: solid P_4S_3 and solid P_4S_7

S_3: solid P_4S_7 and solid P_4S_{10}

S_4: solid P_4S_{10} and solid S

t_1: liquid P and S and solid P

t_2: liquid P and S and solid P_4S_3

t_3: liquid P and S and solid P_4S_3

t_4: liquid P and S and solid P_4S_7

t_5: liquid P and S and solid P_4S_7

t_6: liquid P and S and solid P_4S_{10}

t_7: liquid P and S and solid P_4S_{10}

t_8: liquid P and S and solid S

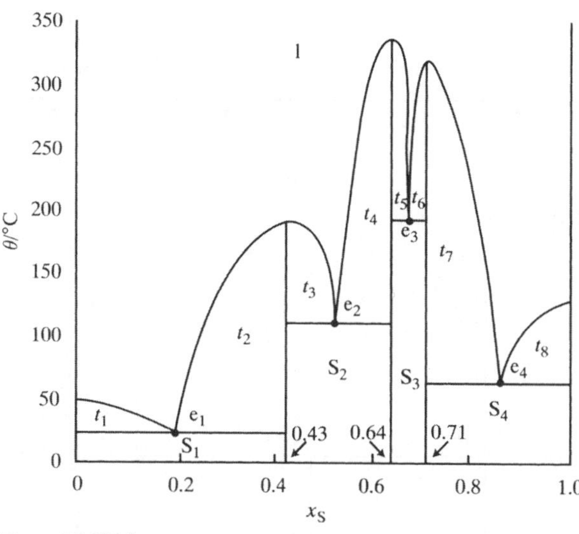

Figure 5C.12(a)

A break in the cooling curve (Fig. 5.17(b)) occurs at point $b_1 \approx 125\,°\text{C}$ as a result of solid P_4S_3 forming; a eutectic halt occurs at point $e_1 \approx 20\,°\text{C}$.

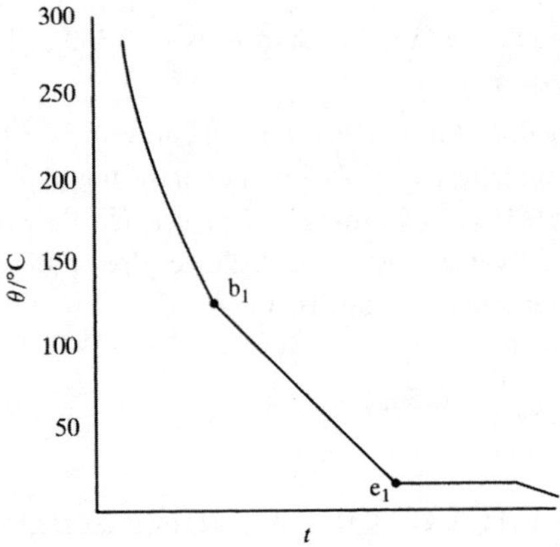

Figure 5C.12(b)

P5C.5 See Fig. 5C.13(a). The number of distinct chemical species (as opposed to components) and phases present at the indicated points are, respectively

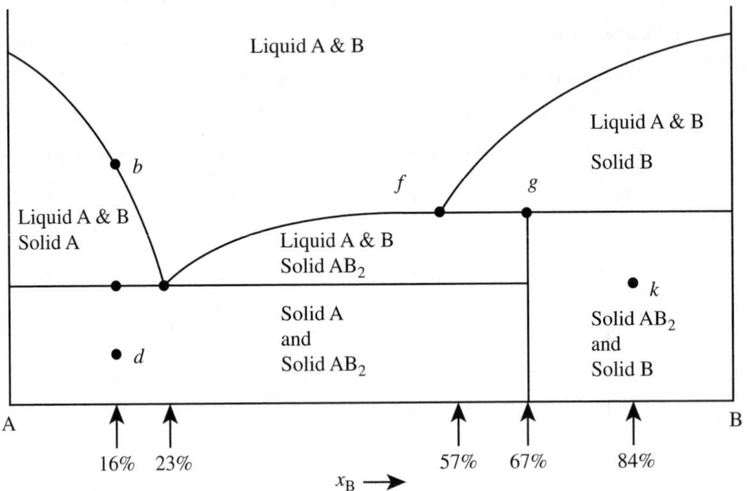

Figure 5C.13(a)

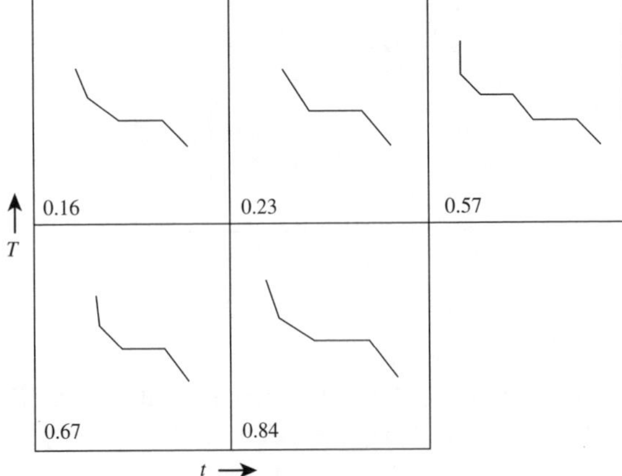

Figure 5C.13(b)

b(3,2), *d*(2,2), *e*(4,3), *f*(4,3), *g*(4,3), *k*(2,2)

[Liquid A and solid A are here considered distinct species.]

The cooling curves are shown in Fig. 5C.13(b).

P5C.7 (a) **eutectics:** $x_{Si}=0.402$ at 1 268 °C $x_{Si}=0.694$ at 1 030 °C

congruent melting compounds: Ca_2Si $T_f=1\,314$ °C

CaSi $T_f=1\,324$ °C

incongruent melting compound: $CaSi_2$ $T_f=1040$ °C (melts into CaSi(s) and Si-rich liquid with x_{Si} around 0.69)

(b) At 1 000 °C the phases in equilibrium are $\boxed{Ca_2Si \text{ and a Ca-rich liquid } (x_{Si}=0.13)}$. The lever rule (eqn 5C.6) gives the relative amounts:

$$\frac{n_{Ca_2Si}}{n_{liq}}=\frac{l_{liq}}{l_{Ca_2Si}}=\frac{0.2-0.13}{0.333-0.2}=\boxed{0.5}$$

(c) Si(s) begins to appear at about 1 300 °C. Further cooling causes more Si(s) to freeze out of the melt so that the melt becomes more concentrated in Ca. There is a eutectic at $x_{Si}=0.694$ and 1 030 °C. Just before the eutectic is reached (i.e. at a temperature just higher than 1 030 °C), the lever rule says that the relative amounts of the Si(s) and liquid ($x_{Si}=0.694$) phases are:

$$\frac{n_{Si}}{n_{liq}}=\frac{l_{liq}}{l_{Si}}=\frac{0.80-0.694}{1.0-0.80}=\boxed{0.53}$$

At the eutectic temperature a third phase appears, $CaSi_2$(s). As the melt cools at this temperature, both Si(s) and $CaSi_2$(s) freeze out of the melt while the composition of the melt remains constant. At a temperature slightly below 1 030 °C, all the melt will have frozen to Si(s) and $CaSi_2$(s) with the relative amounts:

$$\frac{n_{Si}}{n_{CaSi_2}}=\frac{l_{CaSi_2}}{l_{Si}}=\frac{0.80-0.667}{1.0-0.80}=\boxed{0.67}$$

Plotting the mole fraction of Si(s) and of $CaSi_2$(s) vs. that of the eutectic liquid is a convenient way to show relative amounts of the three phases as the eutectic liquid freezes. See Fig. 5C.14. Equations for the graph are derived from the law of conservation of atoms. For silicon atoms,

$$nz_{Si}=n_{liq}w_{Si}+n_{Si}x_{Si}+n_{CaSi_2}y_{Si}$$

where n = total number of moles of all phases

w_{Si} = Si fraction in eutectic liquid = 0.694

x_{Si} = Si fraction in Si(s) = 1.000

y_{Si} = Si fraction in $CaSi_2$(s) = 0.667

z_{Si} = Si fraction in melt = 0.800

This equation may be rewritten in mole fractions of each phase by dividing by n:

$$z_{Si}=(\text{mol fraction liq})\,w_{Si}+(\text{mol fraction Si})\,x_{Si}+(\text{mol fraction CaSi}_2)\,y_{Si}$$

Since (mol fraction liq) + (mol fraction Si) + (mol fraction $CaSi_2$) = 1 or (mol fraction $CaSi_2$) = 1 – (mol fraction liq) – (mol fraction Si) we may write:

$$z_{Si} = (\text{mol fraction liq})w_{Si} + (\text{mol fraction Si})x_{Si}$$
$$+[1-(\text{mol fraction liq} + \text{mol fraction Si})]y_{Si}$$

Solving for mol fraction Si:

$$\text{mol fraction Si} = \frac{(z_{Si} - y_{Si}) - (w_{Si} - y_{Si})(\text{mol fraction liq})}{x_{Si} - y_{Si}}$$

and mol fraction $CaSi_2 = 1-(\text{mol fraction liq} + \text{mol fraction Si})$

These two equations are used to prepare plots of the mol fraction of Si and mol fraction of $CaSi_2$ against the mol fraction of the melt in the range 0–0.65.

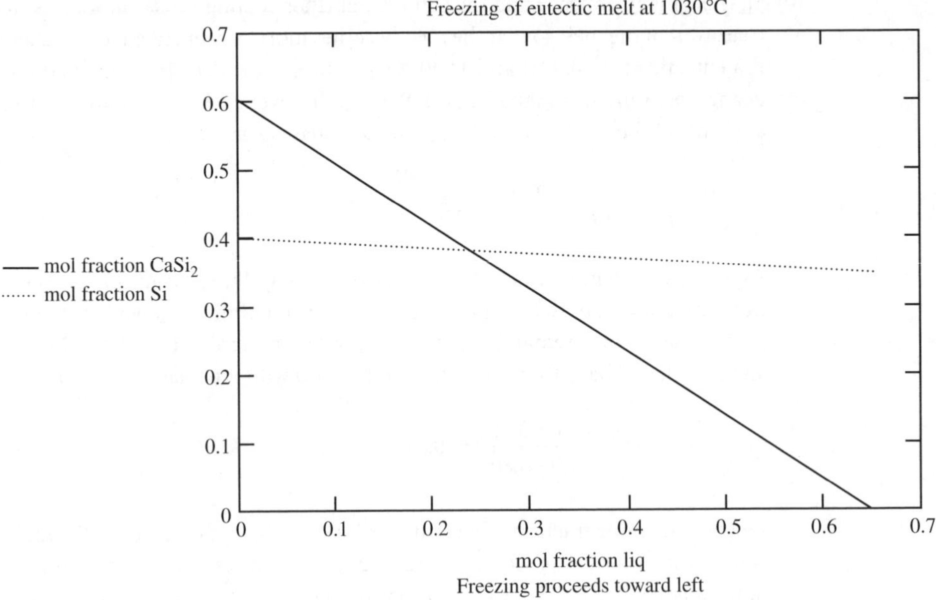

Figure 5C.14

P5C.9 Equation 5C.4 is

$$y_A = \frac{x_A p_A^*}{p_B^* + (p_A^* - p_B^*)x_A}$$

To see if this can be expressed as a function of the ratio p_A^* / p_B rather than of each pressure separately, divide both numerator and denominator by p_B^*:

$$\boxed{y_A = \frac{x_A p_A^* / p_B^*}{1 + (p_A^* / p_B^* - 1)x_A}}$$

The plot of y_A vs. x_A at several values of the vapour pressure ratio is shown in Fig. 5C.2 of the main text.

P5C.11 Follow the derivation in Section 5C.3(b) to arrive at an expression that gives the compositions at which $\Delta_{mix}G$ is minimized. That expression is [5C.7]

$$\ln\frac{x_A}{1-x_A} = -\xi(1-2x_A) \quad \text{where} \quad H^E = \xi nRTx_Ax_B \text{ [5B.6]}$$

While ξ is a convenient dimensionless parameter for modelling how the excess enthalpy varies with composition, it masks the temperature dependence. Enthalpy changes often vary slowly with temperature, and we can explore the consequences of treating H^E as roughly temperature independent (a reasonable assumption if the intermolecular forces involved are roughly temperature independent). So let

$$H^E = 4nH^{max}x_Ax_B$$

where H^{max} is the maximum molar excess enthalpy of mixing (which occurs when $x_A=x_B=0.5$). In terms of this new parameter, the old parameter ξ is equal to $4H^{max}/RT$. Substituting for ξ into eqn 5C.3(b) yields

$$\ln\frac{x_A}{1-x_A} = -\frac{4H^{max}(1-2x_A)}{RT}$$

Therefore, minima in the Gibbs function of mixing occur at temperatures

$$T_{min} = \frac{4H^{max}(2x_A-1)}{R\ln\dfrac{x_A}{1-x_A}}$$

The dimensionless temperature RT_{min}/H^{max} is plotted against composition in Fig. 5C.15.

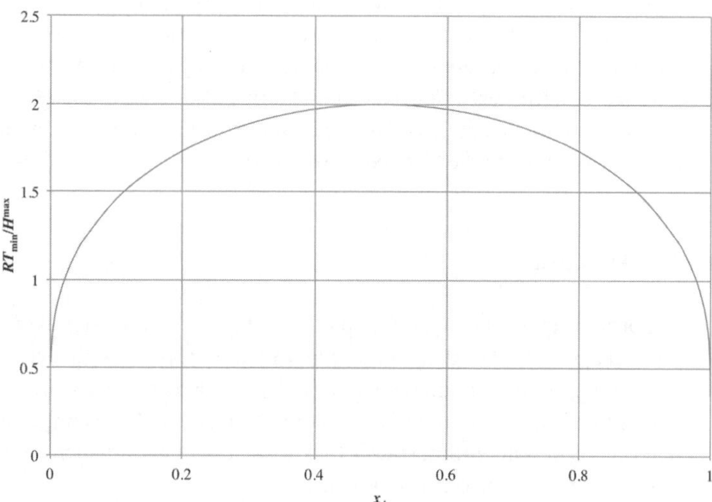

Figure 5C.15

For temperatures less than $2H^{\max}/R$ there are minima at two compositions, corresponding to phase separation. At temperatures greater than $2H^{\max}/R$, there is no phase separation that would lower the Gibbs energy compared to a single mixed phase. Thus $T=2H^{\max}/R$ corresponds to an upper critical temperature for this model of mixing.

5D Phase diagrams of ternary systems

Answers to discussion questions

D5D.1 The phase rule [5A.1] for three components ($C=3$) implies that the degrees of freedom $F=5-P$. If we use two of those degrees of freedom to fix temperature and pressure, then the remaining degrees of freedom equal $3-P$. The maximum number of phases in equilibrium at any given temperature and pressure, therefore, is $\boxed{3}$, for then there would be no remaining degrees of freedom (fixed T, p, and three components' proportions).

D5D.3 The phase rule [5A.1] for four components ($C=4$) implies that the degrees of freedom $F=6-P$. If we use two of those degrees of freedom to fix temperature and pressure, then the remaining degrees of freedom equal $4-P$. Those four degrees of freedom would be the proportions of the four components. A regular tetrahedron would seem to be the object that could depict four mole fractions constrained to sum to one. After all, it has four vertices, each of which could represent one of the four pure components (analogous to the three vertices of an equilateral triangle). And it has four faces, each of which *is* an equilateral triangle. Points on any face would represent a three-component system (the component represented by the opposite vertex being missing); the faces, then, are exactly the ternary phase diagrams discussed in Topic 5D. What is not immediately obvious, however, is that any point in the interior of the tetrahedron can represent compositions of four components constrained to sum to a constant (that is, four mole fractions constrained to sum to 1). It is not obvious, but it is true that for any interior point of a regular tetrahedron, the sum of its distances from the four faces is a constant: $d_1+d_2+d_3+d_4=\text{constant}$, where the four d_i represent the distances of the same point from the four sides of the tetrahedron. This fact is an extension of Viviani's theorem, the result that $d_1+d_2+d_3=\text{constant}$ for an equilateral triangle, where the three d_i represent the distances of the same point from the three edges of the triangle.

Solutions to exercises

E5D.1(a) The ordered triples (x_A, x_B, x_C) are plotted in Fig. 5D.1. The vertices of the triangular phase diagram are labelled for the component that is pure at that vertex. For example, the top of the diagram is pure A: (1, 0, 0). As a reminder, at the edge opposite a labelled vertex, that component's mole fraction is zero. For example, the base of the diagram represents compositions $(0, x_B, x_C)$. The point at which all mole fractions are the same is in the centre of the diagram, $\left(\frac{1}{3},\frac{1}{3},\frac{1}{3}\right)$.

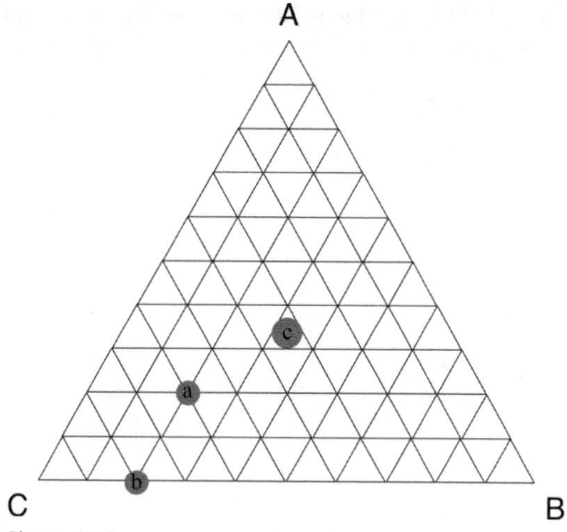

Figure 5D.1

E5D.2(a) Note that the compositions are given in mass percentages, not mole percentages, so we simply convert to moles before plotting. Assume a convenient sample size, such as 100 g, making the numerical values of the mass percentages equal to masses in grams:

(a) NaCl: $25\ \text{g} \times \dfrac{1\ \text{mol}}{58.44\ \text{g}} = 0.42\overline{8}\ \text{mol}$

$Na_2SO_4 \cdot 10H_2O$: $25\ \text{g} \times \dfrac{1\ \text{mol}}{322.21\ \text{g}} = 0.077\overline{6}\ \text{mol}$

H_2O: $50\ \text{g} \times \dfrac{1\ \text{mol}}{18.016\ \text{g}} = 2.7\overline{8}\ \text{mol}$

To get mole fractions, divide these amounts by the total amount of 3.28 mol:

$$x(\text{NaCl}) = \frac{0.42\overline{8}\ \text{mol}}{3.28\ \text{mol}} = 0.13$$

$$x(\text{Na}_2\text{SO}_4 \cdot 10\,\text{H}_2\text{O}) = 0.024 \quad x(\text{H}_2\text{O}) = 0.85$$

This composition is point a in Fig. 5D.2.

(b) We want to plot a line representing equal masses of salt with varying amounts of water. One point on that line has no water. Compute the mole fractions that correspond to the amounts of salt computed in part (a) with no water. In that case, the total amount is:

$$n_{\text{total}} = (0.42\overline{8} + 0.077\overline{6})\ \text{mol} = 0.50\overline{5}\ \text{mol}$$

$$\text{so} \quad x(\text{NaCl}) = \frac{0.42\overline{8}\ \text{mol}}{0.50\overline{5}\ \text{mol}} = 0.85 \quad \text{and} \quad x(\text{Na}_2\text{SO}_4 \cdot 10\,\text{H}_2\text{O}) = 0.15$$

Plot this point on the edge opposite the vertex labelled H_2O, close to the vertex labelled NaCl. The other extreme has the salts in the same proportions, but in amounts negligible compared to that of water, so the other end of this line lies at the vertex labelled H_2O. The line is labelled b on Fig. 5D.2, and note that it goes through point (a) as it must.

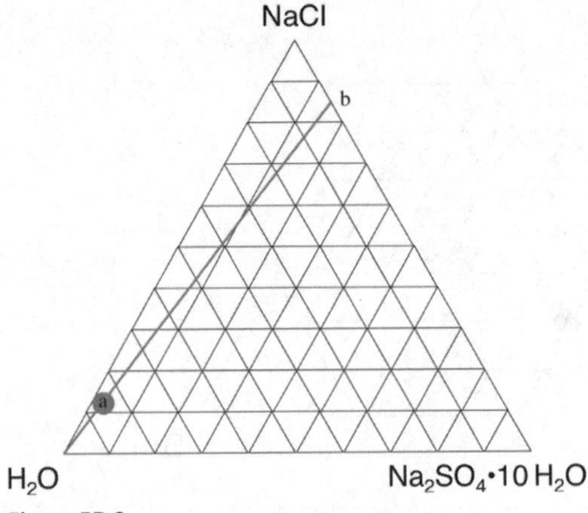

Figure 5D.2

E5D.3(a) First convert to moles and find composition by mole fraction.

$$H_2O\text{ (W): } 2.3\text{ g}\times\frac{1\text{ mol}}{18.016\text{ g}}=0.12\bar{8}\text{ mol}$$

$$CHCl_3\text{ (C): } 9.2\text{ g}\times\frac{1\text{ mol}}{119.4\text{ g}}=0.077\text{ mol}$$

$$CH_3COOH\text{ (A): } 3.1\text{ g}\times\frac{1\text{ mol}}{60.05\text{ g}}=0.052\text{ mol}$$

To get mole fractions, divide these amounts by the total amount of 0.256 mol:

$$x_W=\frac{0.12\bar{8}\text{ mol}}{0.256\text{ mol}}=0.50 \quad x_C=0.30 \quad x_A=0.20$$

This point is plotted in Fig. 5D.3; it is very close to the label a_2 in the original Fig. 5D.4 of the main text. $\boxed{\text{Two phases}}$ are present, since our point lies beneath the phase-boundary arc. Our point is sufficiently close to the tie line labelled $a_2'a_2''$ that we use that tie line to answer questions about the compositions and abundances of the phases. One phase, richest in water, is represented by point a_2', where $(x_W, x_C, x_A)=\boxed{(0.58,0.20,0.22)}$; the other, richest in chloroform, is represented by point a_2'', $\boxed{(0.07,0.82,0.11)}$. Our test point lies about five times further from a_2'' than from a_2', so the $\boxed{\text{water-rich phase is about five times as abundant}}$ as the chloroform-rich phase there.

(a) If water is added to our mixture, the composition changes from our point along the line connecting it to the vertex labelled W. That system would have two phases until the overall composition was about (0.80, 0.12, 0.08); after that, the phases coalesce into a single phase that would increasingly resemble pure water.

(b) If acetic acid is added to our mixture, the composition changes from our point along the line connecting it to the vertex labelled A. That system would have two phases until the overall composition was about (0.43, 0.26, 0.31); after that, the phases coalesce into a single phase that would increasingly resemble pure acetic acid. Note, however, that the single phase region is reached much more quickly by adding acetic acid than by adding water. As noted in Section 5D.2(a) of the text, acetic acid helps both other components dissolve.

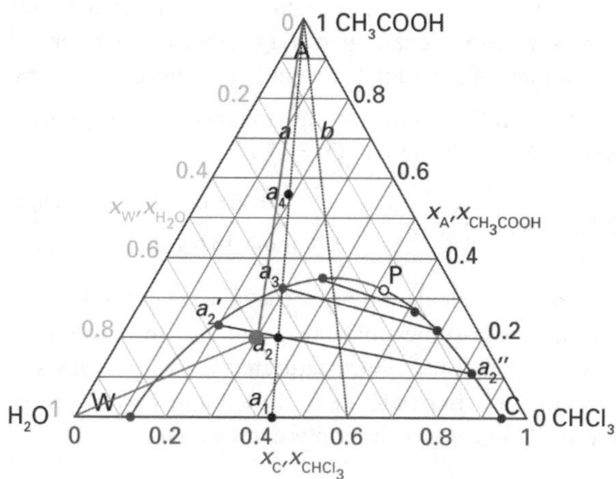

Figure 5D.3

E5D.4(a) The phase diagram showing the three given compositions is shown in Fig. 5D.4.

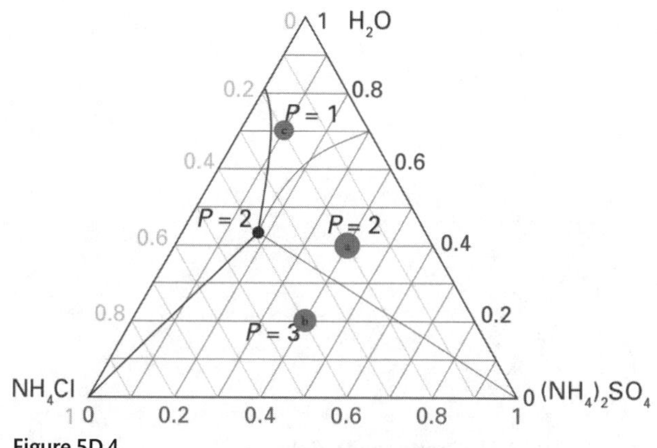

Figure 5D.4

Point (a) is in a two-phase region, point (b) in a three-phase region, and point (c) in a single-phase region. Note that the single-phase region has an abundance of water, which is what allows the other two components to mix (i.e. by dissolving in water).

E5D.5(a) (a) The solubility of NH_4Cl in water would be determined by the point along the water–NH_4Cl edge (i.e. no $(NH_4)_2SO_4$) that intersects a boundary of the water-rich, single-phase region. That point has $x(NH_4Cl)=0.19$ and $x(H_2O)=0.81$. We must convert these mole fractions into molarity:

$$[NH_4Cl]=\frac{n_{NH_4Cl}}{V_{solution}}\approx\frac{0.19\,\text{mol}}{0.81\,\text{mol}\times 18.0\,\text{g mol}^{-1}}\times\frac{1000\,\text{g}}{1\,\text{dm}^3}=\boxed{13\,\text{mol dm}^{-3}}$$

Comment. The calculation bases the volume of the solution on the amount of solvent and assumes that the volume of the saturated solution has the same density as water. Since saturated NH_4Cl has a density more like 1 075 g dm^{-3}, a better estimate of the solubility would be 14 mol dm^{-3}.

(b) The corresponding point on the water–$(NH_4)_2SO_4$ edge (no NH_4Cl) has $x((NH_4)_2SO_4)=0.30$ and $x(H_2O)=0.70$.

$$[(NH_4)_2SO_4]=\frac{n_{(NH_4)_2SO_4}}{V_{solution}}\approx\frac{0.30\,\text{mol}}{0.70\,\text{mol}\times 18.0\,\text{g mol}^{-1}}\times\frac{1\,000\,\text{g}}{1\,\text{dm}^3}=\boxed{24\,\text{mol dm}^{-3}}$$

Comment. The calculation bases the volume of the solution on the amount of solvent and assumes that the volume of the saturated solution has the same density as water. Since saturated $(NH_4)_2SO_4$ has a density more like 1 250 g dm^{-3}, a better estimate of the solubility would be 29 mol dm^{-3}; however, even 24 mol dm^{-3} is a serious overestimate.

Solutions to problems

P5D.1 The phase diagram is shown in Fig. 5D.5.

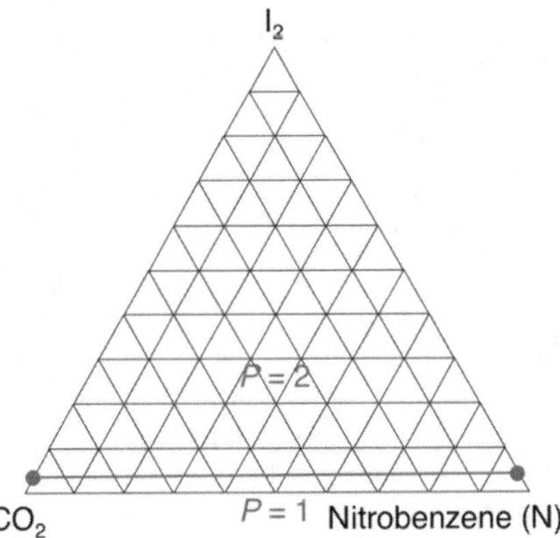

Figure 5D.5

A straight line joins the points along the I_2–CO_2 edge and I_2–nitrobenzene edge that represent saturated binary solutions of I_2 in each solvent. Below that nearly horizontal line, there is a single phase. Above it are two phases, solid I_2 in equilibrium with a liquid saturated with I_2.

P5D.3 (a) Refer to Fig. 5D.6.

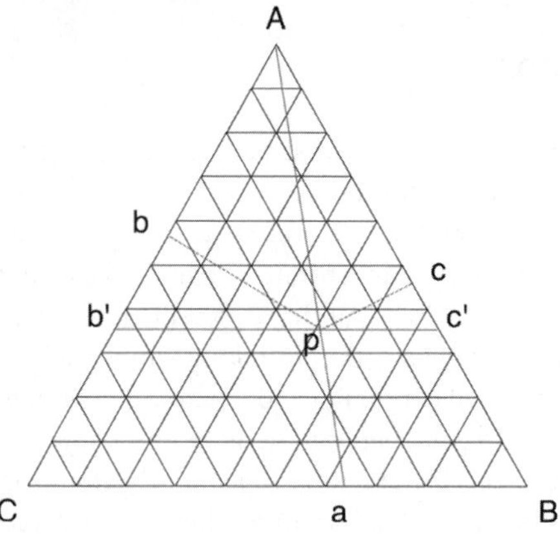

Figure 5D.6

We are asked to show that at any point p on a line joining apex A to a point a on the opposite edge, the proportions of components B and C are constant and equal to the initial ratio x_B/x_C when no A is present. At any point p on line Aa, x_B is equal to the length of pb, where pb is the line segment from p perpendicular to the baseline opposite apex B, and x_C is equal to the length of pc, where pc is the line segment from p perpendicular to the baseline opposite apex C. In other words, we are to prove that

$$\frac{\text{pb}}{\text{pc}} = \frac{\text{aB}}{\text{aC}}$$

where each pair of letters refers to the length of the corresponding line segment. We can do so by drawing the line segment b′c′ through p parallel to the baseline CB. Because b′c′ is parallel to CB, triangle AaC is similar to triangle Apb′ and triangle AaB is similar to Apc′. (The triangles are similar because the corresponding angles in them are identical). Because the triangles are similar, the proportions of corresponding lengths are equal. In particular

$$\frac{\text{pb}'}{\text{pc}'} = \frac{\text{aB}}{\text{aC}}$$

But we need pb/pc, not pb′/pc′. But if we can prove that triangles pb′b and pc′c are similar, then we would have

$$\frac{\text{pb}}{\text{pc}} = \frac{\text{pb}'}{\text{pc}'} = \frac{\text{aB}}{\text{aC}}$$

In fact, pb′b and pc′c **are** similar because angle pbb′ = angle pcc′ (both being perpendicular by construction) and angle pb′b = angle pc′c (both being 60°, since b′c′ is paralell to the baseline of the original equilateral triangle). Two triangles are similar if they have two pairs of identical angles (because if two are identical the third must also be identical, since the angles of a triangle add up to 180°). Therefore, the perpendicular distances pb and pc are in the same proportion as the distances from p along the parallel b′c′, which in turn are in the same proportion as the distances from a along the original baseline.

5E Activities

Answers to discussion questions

D5E.1 Activities are 'effective concentrations' in the sense that we say that real (non-ideal) mixtures behave 'as if' their concentration had a different value, and we call that different value the activity. In practice, we are really saying that real (non-ideal) mixtures that actually have a certain concentration behave like ***ideal*** mixtures that have a different concentration, and we call that different concentration the activity. Thus the activity replaces concentration in relationships such as Raoult's law [5A.21], Henry's law [5A.23], and expressions for chemical potential [5E.1, 5E.9].

What accounts for the difference between activity and concentration is non-ideal interactions between species. The nature of ideal behaviour for solvents and solutes is discussed in Discussion question 5A.5. Non-ideal interactions are not mysterious. Intermolecular interactions between, say, polar and nonpolar molecules are well-understood to be different from interactions between different species of similar structure. And the interactions among ions are so strong that their deviations from ideal behaviour warrant a separate topic (5F).

D5E.3 The main way to measure activities given in this topic is by measuring partial vapour pressures, as implied in the definitions of solvent activity [5E.2] and solute activity [5E.10]. In practice, measurements of properties that depend on the chemical potential also offer ways of determining activity through chemical potential relationships [5E.1, 5E.9].

Solutions to exercises

E5E.1(a) For A (Raoult's law basis; concentration in mole fraction):

$$a_A = \frac{p_A}{p_A^*}\,[5E.2] = \frac{250\ \text{Torr}}{300\ \text{Torr}} = \boxed{0.833}; \qquad \gamma_A = \frac{a_A}{x_A}\,[5E.4] = \frac{0.833}{0.90} = \boxed{0.93}$$

For B (Henry's law basis; concentration in mole fraction):

$$a_B = \frac{p_B}{K_B}\,[5E.10] = \frac{25\ \text{Torr}}{200\ \text{Torr}} = \boxed{0.125}; \qquad \gamma_B = \frac{a_B}{x_B}\,[5E.11] = \frac{0.125}{0.10} = \boxed{1.2\overline{5}}$$

For B (Henry's law basis; concentration in molality), the activity coefficient remains the same, but the activity is

$$a_B = \gamma_B \times \frac{b_B}{b^{\ominus}}\,[5E.14] = 0.12\overline{5} \times 2.22 = \boxed{2.8}$$

Comment. The two methods for the 'solute' B give different values for the activities. This is reasonable since the reference states are different. However, the activity coefficient is the same: it is the ratio between activity and a concentration unit, and it incorporates all deviations from an ideality defined by Henry's law.

Question. What are the activity and activity coefficient of B in the Raoult's law basis?

E5E.2(a) From eqn 1A.8 (partial pressures) and y_A we can compute the partial pressures:

$$y_A = \frac{p_A}{p_A + p_M} = \frac{p_A}{101.3\,\text{kPa}} = 0.516$$

So $p_A = 101.3\text{ kPa} \times 0.516 = 52.3\text{ kPa}$

and $p_M = 101.3\text{ kPa} - 52.3\text{ kPa} = 49.0\text{ kPa}$

$$a_A = \frac{p_A}{p_A^*}\,[5E.2] = \frac{52.3\,\text{kPa}}{105\,\text{kPa}} = \boxed{0.498} \quad \text{and} \quad a_M = \frac{p_M}{p_M^*} = \frac{49.0\,\text{kPa}}{73.5\,\text{kPa}} = \boxed{0.667}$$

$$\gamma_A = \frac{a_A}{x_A}\,[5E.4] = \frac{0.498}{0.400} = \boxed{1.24} \quad \text{and} \quad \gamma_M = \frac{a_M}{x_M} = \frac{0.668}{0.600} = \boxed{1.11}$$

E5E.3(a) The biological standard state is defined as pH 7, which implies $a_{H^+} = 10^{-7}$. All other activities in the biological standard state are unity, just as in the chemical standard state; in the chemical standard state, $a_{H^+} = 1 = 10^0$ as well (which implies pH 0). As a result, the biological standard molar Gibbs function for H^+ is lower than that of the chemical standard by 7 RT ln 10 [5E.16], which is equal to 39.96 kJ mol^{-1} at 25 °C [*Brief illustration* 5E.3]. For the given reaction, the standard Gibbs energy is

$$\Delta G^{\ominus} = 2\Delta_f G^{\ominus}(B) + 2\Delta_f G^{\ominus}(H^+) - \Delta_f G^{\ominus}(A)$$

The biological standard is

$$\Delta G^{\oplus} = 2\Delta_f G^{\oplus}(B) + 2\Delta_f G^{\oplus}(H^+) - \Delta_f G^{\oplus}(A)$$
$$= 2\Delta_f G^{\ominus}(B) + 2\{\Delta_f G^{\ominus}(H^+) - 39.96\text{ kJ mol}^{-1}\} - \Delta_f G^{\ominus}(A)$$

Comparing the two, we have

$$\Delta G^{\oplus} - \Delta G^{\ominus} = -2 \times 39.96\text{ kJ mol}^{-1} = \boxed{-79.92\text{ kJ mol}^{-1}}$$

E5E.4(a) The partial pressures of both components are given by eqn 5E.19

$$p_J = p_J^* x_J e^{\xi(1-x_J)^2}$$

The total pressure is the sum of the two partial pressures. The vapour-pressure diagram is plotted in Fig. 5E.1.

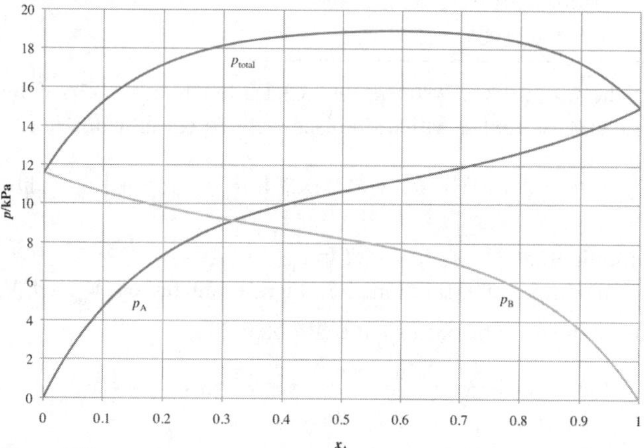

Figure 5E.1

Comment. The figure shows that upon adding the other component to either pure component, the vapour pressure rises (as is evident from the fact that the total pressure increases as one moves from either edge of the graph). This is consistent with the physical interpretation given in Section 5E.3: positive ξ corresponds to endothermic mixing, reflecting unfavourable interactions between the components.

Solutions to problems

P5E.1 The activity of a solvent is

$$a_A = \frac{p_A}{p_A^*} \text{ [5E.2]} = x_A \gamma_A \text{ [5E.4]}$$

so the activity coefficient is

$$\gamma_A = \frac{p_A}{x_A p_A^*} = \frac{y_A p}{x_A p_A^*}$$

where the last equality applies the eqn 1A.8 (partial pressures) to the vapour phase. Since the data are given in terms of trichloromethane mole fractions, the last equation applies directly, changing subscript A to T. For the other component:

$$\gamma_E = \frac{y_E p}{x_E p_E^*} = \frac{(1-y_T)p}{(1-x_T)p_E^*}$$

Substituting the data yields the following table of results.

p/kPa	x_T	y_T	γ_T	γ_E
23.40	0.000	0.000	–	1.000
21.75	0.129	0.065	0.418	0.998
20.25	0.228	0.145	0.490	1.031
18.75	0.353	0.285	0.576	1.023
18.15	0.511	0.535	0.723	0.920
20.25	0.700	0.805	0.885	0.725
22.50	0.810	0.915	0.966	0.497
26.30	1.000	1.000	1.00	–

P5E.3 Retrace the argument in *Justification* 5B.3, leading to eqn 5B.16 of the text. Exactly the same process applies with a_A in place of x_A. At equilibrium

$$\mu_A^*(p) = \mu_A(a_A, p+\Pi) = \mu_A^*(p+\Pi) + RT\ln a_A = \mu_A^*(p) + \int_p^{p+\Pi} V_m \mathrm{d}p + RT\ln a_A$$

which implies that $\int_p^{p+\Pi} V_m \mathrm{d}p = -RT\ln a_A$.

For an incompressible solution, the integral evaluates to ΠV_m, so $\Pi V_m = -RT\ln a_A$.

In terms of the osmotic coefficient ϕ (Problem 5E.2)

$$\Pi V_m = r\phi RT \quad \text{where} \quad r = \frac{x_B}{x_A} = \frac{n_B}{n_A} \quad \phi = -\frac{x_A}{x_B}\ln a_A = -\frac{1}{r}\ln a_A$$

For a dilute solution, $V=nV_m \approx n_A V_m$

Hence, $\Pi V = n_B \phi RT$

and therefore, with $[B]=\frac{n_B}{V}$, $\boxed{\Pi = \phi[B]RT}$.

5F The activities of ions

Answers to discussion questions

D5F.1 The coulombic (electrostatic) interactions of the ions in solution with each other and with the solvent are responsible for the deviation of their activity coefficients from the ideal value of 1. The electrostatic ion–ion interaction is the stronger of the two and is fundamentally responsible for the deviation. Because of this interaction there is a build up of charge of opposite sign around any given ion in the overall electrically neutral solution. In addition, polar solvent molecules can orient their dipole moments to interact favourably with ions. The energy, and hence, the chemical potential of any given ion is lowered as a result of the existence of this ionic atmosphere. (This is true in sufficiently dilute solutions, at least, where the ionic atmospheres are not forced too close together.) The lowering of the chemical potential below its ideal value is identified with a negative value of $RT\ln\gamma_\pm$ and a value of $\gamma_\pm$ less than unity.

D5F.3 The term B in the Truesdell–Jones equation [5F.11a] and the Davies equation [5F.11b] can be interpreted as an indicator of the distance of closest approach of the ions. (Note that B does not have units of distance, so it cannot literally be the distance of closest approach.) It and the parameter C in the latter equation, however, are best thought of as empirical parameters to be fit to data.

Solutions to exercises

E5F.1(a) The definition of ionic strength is

$$I=\frac{1}{2}\sum_i\left(\frac{b_i}{b^\ominus}\right)z_i^2 \quad [5F.9]$$

and if b is the molal concentration of an M_pX_q salt, the molal concentrations of the ions are

$$b_M = p\times b \quad \text{and} \quad b_X = q\times b.$$

Hence $I=\frac{1}{2}(pz_+^2+qz_-^2)\left(\frac{b}{b^\ominus}\right)$

For KCl (or any other compound of monovalent ions)

$$I=\frac{1}{2}(1\times1+1\times1)\left(\frac{b}{b^\ominus}\right)=\left(\frac{b}{b^\ominus}\right).$$

For $CuSO_4$ (or any other compound of divalent ions)

$$I=\frac{1}{2}(1\times2^2+1\times2^2)\left(\frac{b}{b^{\ominus}}\right)=4\left(\frac{b}{b^{\ominus}}\right).$$

Thus, for this mixture

$$I=I(\mathrm{KCl})+I(\mathrm{CuSO_4})=\left(\frac{b(\mathrm{KCl})}{b^{\ominus}}\right)+4\left(\frac{b(\mathrm{CuSO_4})}{b^{\ominus}}\right)$$

$$=0.10+4\times0.20=\boxed{0.90}$$

Comment. Note that the strength of a solution of more than one electrolyte may be calculated by summing the ionic strengths of each electrolyte considered as a separate solution, as in the solution to this exercise, by summing the product $\frac{1}{2}\left(\frac{b_i}{b^{\ominus}}\right)z_i^2$ for each individual ion as in the definition of I [5F.9].

E5F.2(a) The original KNO_3 solution has an ionic strength of 0.150. (For compounds of monovalent ions, the ionic strength is numerically equal to the molal concentration, as shown in Exercise 5F.1(a).) Therefore, the ionic strengths of the added salts must be 0.100.

(a) For $Ca(NO_3)_2$ $I=\frac{1}{2}(1\times2^2+2\times1^2)\left(\frac{b}{b^{\ominus}}\right)$ [5F.9] $=3\left(\frac{b}{b^{\ominus}}\right)$

Therefore, the solution should be made $0.100\ \mathrm{mol\,kg^{-1}}/3=0.0333\ \mathrm{mol\,kg^{-1}}$ in $Ca(NO_3)_2$. The mass that should be added to 500 g of the solution is therefore

$$(0.500\ \mathrm{kg})\times(0.0333\ \mathrm{mol\,kg^{-1}})\times(164\ \mathrm{g\,mol^{-1}})=\boxed{2.73\ \mathrm{g}}.$$

(b) For NaCl, the ionic strength is numerically equal to the molality; therefore, the solution should be made $0.100\ \mathrm{mol\,kg^{-1}}$ in NaCl. The mass that should be added to 500 g of the solution is therefore

$$(0.500\ \mathrm{kg})\times(0.100\ \mathrm{mol\,kg^{-1}})\times(58.44\ \mathrm{g\,mol^{-1}})=\boxed{2.92\ \mathrm{g}}.$$

E5F.3(a) The solution is dilute, so use the Debye–Hückel limiting law.

$$\log\gamma_{\pm}=-|z_+z_-|AI^{1/2}\ [5\mathrm{F}.8]$$

$$I=\frac{1}{2}\sum_i\left(\frac{b_i}{b^{\ominus}}\right)z_i^2\ [5\mathrm{F}.9]$$

$$=\frac{1}{2}\{(0.010\times2^2)+(2\times0.010\times1^2)+(0.030\times1^2)+(0.030\times1^2)\}=\boxed{0.060}$$

$$\mathrm{CaCl_2}:\log\gamma_{\pm}=-2\times1\times0.509\times(0.060)^{1/2}=-0.25\quad\text{so}\quad\gamma_{\pm}=\boxed{0.56}.$$

The activities of the ions are

$$a(\mathrm{Ca^{2+}})=\gamma_+b_+/b^{\ominus}\approx\gamma_{\pm}b_+/b^{\ominus}=0.56\times0.010=\boxed{0.0056}$$

and $a(\text{Cl}^-)=\gamma_- b_-/b^{\ominus} \approx \gamma_\pm b_-/b^{\ominus} = 0.56\times(2\times 0.010) = \boxed{0.011}$

(The ≈ sign is a reminder that the activities of the separate ions are not available, so we use the same mean ionic activity coefficient for both ions.)

Question. What are the activity coefficients and activities of NaF in the same solution?

E5F.4(a) The extended Debye–Hückel law [5F.11a] is

$$\log\gamma_\pm = -\frac{A|z_+z_-|I^{1/2}}{1+BI^{1/2}}$$

Solve for B.

$$B=-\left(\frac{1}{I^{1/2}}+\frac{A|z_+z_-|}{\log\gamma_\pm}\right)=-\left(\frac{1}{(b/b^{\ominus})^{1/2}}+\frac{0.509}{\log\gamma_\pm}\right)$$

Draw up the following table

$b/(\text{mol kg}^{-1})$	5.0×10^{-3}	10.0×10^{-3}	20.0×10^{-3}
$\gamma_\pm$	0.930	0.907	0.879
B	2.01	2.01	2.02

The values of B are constant, illustrating that the extended law fits these activity coefficients with $B=\boxed{2.01}$.

Solution to problem

P5F.1 According to the Debye–Hückel limiting law

$$\log\gamma_\pm = -0.509|z_+z_-|I^{1/2}\,[5\text{F}.8] = -0.509\left(\frac{b}{b^{\ominus}}\right)^{1/2}\,[5\text{F}.9]$$

We draw up the following table for the given data and have extended it to the limit of zero concentration, where the solution must become ideal.

$b/(\text{mmol kg}^{-1})$	0.0	1.0	2.0	5.0	10.0	20.0
$I^{1/2}$	0.0	0.032	0.045	0.071	0.100	0.141
$\log\gamma_\pm$(D-H)	0	−0.0161	−0.0228	−0.0360	−0.0509	−0.0720
$\log\gamma_\pm$(exp)	0	−0.0155	−0.0214	−0.0327	−0.0446	−0.0599
$\gamma_\pm$(D-H)	1	0.964	0.949	0.920	0.889	0.847
$\gamma_\pm$(exp)	1	0.9649	0.9519	0.9275	0.9024	0.8712

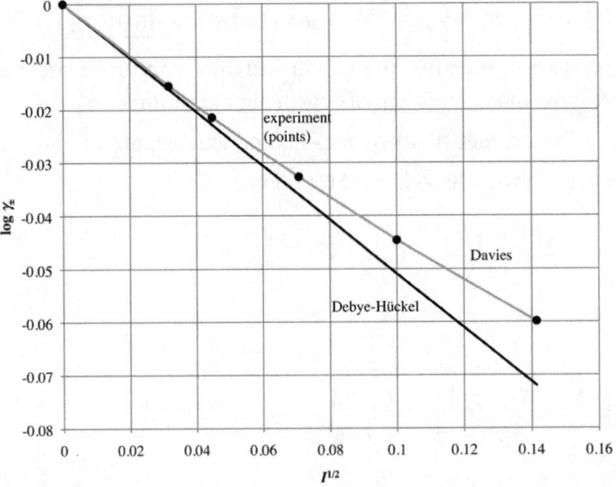

Figure 5F.1

The experimental points and the line representing the Debye–Hückel limiting law are plotted against $I^{1/2}$ in Fig. 5F.1. Note that the limiting law is not a good representation of the experimental points in that its line lies below all of the given experimental points. The data support the limiting law in the sense that they coincide in the limit of very low concentration (about 1 mmol kg^{-1} or less). A sufficiently good value of B in the extended Debye–Hückel law may be obtained by assuming that the constant A in the extended law is the same as A in the limiting law and taking $C=0$.

$$\log \gamma_{\pm} = -\frac{A|z_+z_-|I^{1/2}}{1+BI^{1/2}} \; [\text{5F.11b with } C=0] = -\frac{AI^{1/2}}{1+BI^{1/2}}$$

Rearrange to solve for B:

$$B = -\frac{A}{\log \gamma_{\pm}} - \frac{1}{I^{1/2}}$$

Use the experimental data and a spreadsheet to determine B from every data point. We see that B is relatively constant except for the 1.0 mmol kg^{-1} data point, so we take $B=1.42$ (average excluding that data point) to compute $\log \gamma_{\pm}$ from the Davies equation:

Thus $\log \gamma_{\pm} = -\dfrac{0.509I^{1/2}}{1+1.42I^{1/2}}$

We extend the above table:

b/(mmol kg^{-1})	0.0	1.0	2.0	5.0	10.0	20.0
B		1.18	1.41	1.43	1.41	1.43
$\log \gamma_{\pm}$(Davies)	0	−0.0154	−0.0214	−0.0327	−0.0446	−0.0599
$\gamma_{\pm}$(Davies)	1	0.965	0.952	0.927	0.902	0.871

The curve representing the Davies equation is also plotted in Fig. 5F.1, and that curve goes through all of the data points, clearly representing an improvement in fit over the limiting law.

Comment. Saying that the limiting law is 'not a good representation' of the experimental data is, perhaps, a bit harsh when one considers that the activity coefficients agree to within 2-3% over the range of the data.

Comment. Sufficiently general curve-fitting software could be used to find optimal values for both B and C in the Davies equation. The approach used here demonstrates that the Davies equation improves the fit by finding an optimal value for B having selected an arbitrary value for C.

Integrated activities

I5.1 (a) On a Raoult's law basis, $a=\dfrac{p}{p^*}$ [5E.2] and $a=\gamma x$ [5E.4], so $\gamma=\dfrac{p}{xp^*}$. On a Henry's law basis, $a=\dfrac{p}{K}$ [5E.10], so $\gamma=\dfrac{p}{xK}$. The vapour pressures of the pure components are given in the table of data and are $p_{\mathrm{I}}^*=47.12$ kPa and $p_{\mathrm{A}}^*=37.38$ kPa.

(b) The Henry's law constant for iodoethane is determined by plotting the data and extrapolating the low concentration data to $x_{\mathrm{I}}=1$. (The Henry's law constant for ethyl acetate can also be determined by extrapolating the low-x_{A} data to $x_{\mathrm{A}}=1$, i.e. $x_{\mathrm{I}}=0$.) The data are plotted in Fig. I5.1, and both Henry's law extrapolations are shown as dotted lines. The values obtained are $K_{\mathrm{I}}=64.4$ kPa and $K_{\mathrm{A}}=56.1$ kPa. (Note that the figure displays dashed lines, corresponding to Raoult's law in addition to the dotted Henry's law lines and the solid experimental curves.)

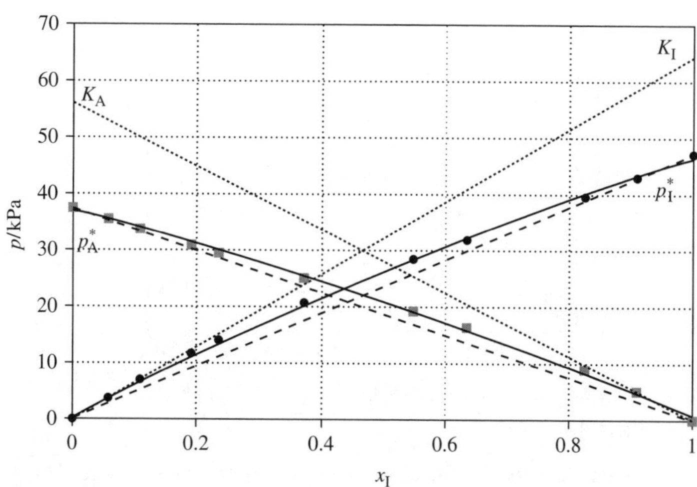

Figure I5.1

Then draw up the following table based on the partial pressures given in the data.

x_I	0	0.0579	0.1095	0.1918	0.2353	0.3718
p_I/kPa	0	3.73	7.03	11.7	14.05	20.72
p_A/kPa	37.38[†]	35.48	33.64	30.85	29.44	25.05
γ_I(R)	–	1.367	1.362	1.295	1.267	1.183 [$p_I/x_Ip_I^*$]
γ_A(R)	1.000	1.008	1.011	1.021	1.030	1.067 [$p_A/x_Ap_A^*$]
γ_I(H)	1.000	1.000	0.997	0.947	0.927	0.865 [p_I/x_IK_I]

x_I	0.5478	0.6349	0.8253	0.9093	1
p_I/kPa	28.44	31.88	39.58	43.00	47.12[‡]
p_A/kPa	19.23	16.39	8.88	5.09	0
γ_I(R)	1.102	1.066	1.018	1.004	1
γ_A(R)	1.138	1.201	1.360	1.501	–
γ_I(H)	0.806	0.779	0.744	0.734	0.731

[†]The value of p_A^*; [‡] the value of p_I^*.

Question. In this problem both I and A were treated as solvents, but only I as a solute. Extend the table by including a row for γ_A(H).

I5.3 The Henry's law constant of the solute (here CO_2) is the slope of a plot of p_B versus x_B in the limit of zero x_B. The partial pressures of CO_2 are almost but not quite equal to the total pressures reported:

$$p_{CO_2} = py_{CO_2} = p(1-y_{cyc})$$

CO_2 partial pressures are shown in the table below and plotted in Fig. I5.2.

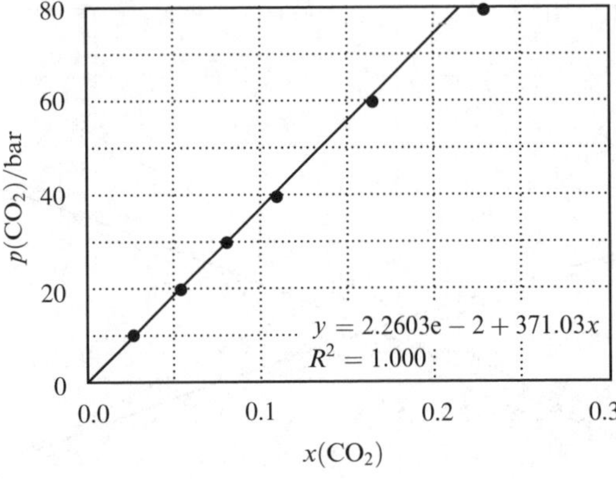

Figure I5.2

Linear regression of the low-pressure points gives $K_B = \boxed{371\,\text{bar}}$.

The activity of a solute is

$$a_B = \frac{p_B}{K_B}\,[5E.10] = x_B\gamma_B\,[5E.11]$$

so the activity coefficient is

$$\gamma_B = \frac{p_B}{x_B K_B} = \frac{y_B p}{x_B K_B} = \frac{(1-y_A)p}{(1-x_A)K_B}$$

where the last equality applies eqn 1A.8. A spreadsheet applied this equation to the above data to yield

p/bar	p_{CO_2} / bar	y_{cyc}	x_{cyc}	γ_{CO_2}
10.0	9.7	0.0267	0.9741	1.01
20.0	19.7	0.0149	0.9464	0.99
30.0	29.7	0.0112	0.9204	1.00
40.0	39.6	0.009 47	0.892	0.99
60.0	59.5	0.008 35	0.836	0.98
80.0	79.3	0.009 21	0.773	0.94

I5.5 Begin with the Gibbs–Duhem equation [5A.12a] and divide through by total moles:

$$x_A d\mu_A + x_B d\mu_B = 0.$$

Dividing through by dx_A:

$$x_A \frac{d\mu_A}{dx_A} + x_B \frac{d\mu_B}{dx_A} = 0$$

The mole fractions add up to a constant (namely 1), so $dx_B = -dx_A$, and we can write

$$x_A \frac{d\mu_A}{dx_A} - x_B \frac{d\mu_B}{dx_B} = 0$$

$$(*) \quad \frac{d\mu_A}{d\ln x_A} = \frac{d\mu_B}{d\ln x_B} \quad \left[d\ \ln x = \frac{dx}{x}\right].$$

Substitute the chemical potential of a real gas, namely

$$\mu = \mu^{\ominus} + RT\ln\frac{f}{p^{\ominus}} \quad \text{[5A.14a with fugacity in place of pressure]},$$

into the starred equation, bearing in mind that the derivatives in it are really partial derivatives at constant p and T:

$$\left(\frac{\partial \ln f_A}{\partial \ln x_A}\right)_{p,T} = \left(\frac{\partial \ln f_B}{\partial \ln x_B}\right)_{p,T}.$$

Now, go back to the starred equation and substitute into it the expression for chemical potential of a solvent in a solution, namely

$$\mu_A = \mu_A^* + RT\ln a_A \text{[5E.1]} = \mu_A^* + RT\ln\frac{p_A}{p_A^*} \text{[5E.2]},$$

yielding $\left(\dfrac{\partial \ln p_{\mathrm{A}}}{\partial \ln x_{\mathrm{A}}}\right)_{p,T} = \left(\dfrac{\partial \ln p_{\mathrm{B}}}{\partial \ln x_{\mathrm{B}}}\right)_{p,T}$.

If A satisfies Raoult's law, we can write $p_{\mathrm{A}}=x_{\mathrm{A}}p_{\mathrm{A}}^{*}$, which implies that

$$\left(\frac{\partial \ln p_{\mathrm{A}}}{\partial \ln x_{\mathrm{A}}}\right)_{p,T} = \frac{\partial \ln x_{\mathrm{A}}}{\partial \ln x_{\mathrm{A}}} + \frac{\partial \ln p_{\mathrm{A}}^{*}}{\partial \ln x_{\mathrm{A}}} = 1+0=1$$

which requires $\left(\dfrac{\partial \ln p_{\mathrm{B}}}{\partial \ln x_{\mathrm{B}}}\right)_{p,T} = 1$ as well. Both Raoult's law ($p_{\mathrm{B}}=x_{\mathrm{B}}p_{\mathrm{B}}^{*}$) and Henry's law ($p_{\mathrm{B}}=x_{\mathrm{B}}K_{\mathrm{B}}$) satisfy this requirement, as can be determined by differentiation. Indeed, Henry's law can be considered the more general of the two; Raoult's law can be considered a special case of Henry's law in which $K_{\mathrm{B}}=p_{\mathrm{B}}^{*}$.

I5.7 $\mathrm{A(s)} \rightleftharpoons \mathrm{A(l)}$

$\mu_{\mathrm{A}}^{*}(\mathrm{s}) = \mu_{\mathrm{A}}^{*}(\mathrm{l}) + RT\ln a_{\mathrm{A}}$

so $\Delta_{\mathrm{fus}}G=\mu_{\mathrm{A}}^{*}(\mathrm{l})-\mu_{\mathrm{A}}^{*}(\mathrm{s})=-RT\ln a_{\mathrm{A}}$.

Hence, $\ln a_{\mathrm{A}} = \dfrac{-\Delta_{\mathrm{fus}}G}{RT}$

Differentiate with respect to T and use the Gibbs–Helmholtz equation [3D.11]:

$$\frac{\mathrm{d}\ln a_{\mathrm{A}}}{\mathrm{d}T} = -\frac{1}{R}\frac{\mathrm{d}}{\mathrm{d}T}\left(\frac{\Delta_{\mathrm{fus}}G}{T}\right) = \frac{\Delta_{\mathrm{fus}}H}{RT^{2}}$$

The freezing point depression is $\Delta T=T_{\mathrm{f}}-T$, so $\mathrm{d}\Delta T=-\mathrm{d}T$.

and $\dfrac{\mathrm{d}\ln a_{\mathrm{A}}}{\mathrm{d}\Delta T} = \dfrac{-\Delta_{\mathrm{fus}}H}{RT^{2}} \approx \dfrac{-\Delta_{\mathrm{fus}}H}{RT_{\mathrm{f}}^{2}}$

But $K_{\mathrm{f}} = \dfrac{RT_{\mathrm{f}}^{2}M_{\mathrm{A}}}{\Delta_{\mathrm{fus}}H}$ [from 5B.12 and 5B.13]

Therefore,

$$\frac{\mathrm{d}\ln a_{\mathrm{A}}}{\mathrm{d}\Delta T} = \frac{-M_{\mathrm{A}}}{K_{\mathrm{f}}} \quad \text{and} \quad \mathrm{d}\ln a_{\mathrm{A}} = \frac{-M_{\mathrm{A}}\mathrm{d}\Delta T}{K_{\mathrm{f}}}.$$

According to the Gibbs–Duhem equation [5A.12a]

$$n_{\mathrm{A}}\,\mathrm{d}\mu_{\mathrm{A}} + n_{\mathrm{B}}\mathrm{d}\mu_{\mathrm{B}} = 0.$$

which implies that

$$n_{\mathrm{A}}\mathrm{d}\ln a_{\mathrm{A}} + n_{\mathrm{B}}\,\mathrm{d}\ln a_{\mathrm{B}} = 0 \; [\mu = \mu^{*} + RT\ln a],$$

and hence that $\mathrm{d}\ln a_{\mathrm{A}} = -\dfrac{n_{\mathrm{B}}}{n_{\mathrm{A}}}\mathrm{d}\ln a_{\mathrm{B}}$.

Hence, $\dfrac{\mathrm{d}\ln a_{\mathrm{B}}}{\mathrm{d}\Delta T}=\dfrac{n_{\mathrm{A}}M_{\mathrm{A}}}{n_{\mathrm{B}}K_{\mathrm{f}}}=\dfrac{1}{b_{\mathrm{B}}K_{\mathrm{f}}}$ $\quad$ [for $n_{\mathrm{A}}M_{\mathrm{A}}=1\,\mathrm{kg}$]

We know from the Gibbs–Duhem equation that

$$x_{\mathrm{A}}\mathrm{d}\mu_{\mathrm{A}}+x_{\mathrm{B}}\mathrm{d}\mu_{\mathrm{B}}=0.$$

and hence that $\int \mathrm{d}\ln a_{\mathrm{A}}=-\int \dfrac{x_{\mathrm{B}}}{x_{\mathrm{A}}}\mathrm{d}\ln a_{\mathrm{B}}=\ln a_{\mathrm{A}}$.

The osmotic coefficient was defined in Problem 5E.2 as

$$\phi=-\frac{1}{r}\ln a_{\mathrm{A}}=-\frac{x_{\mathrm{A}}}{x_{\mathrm{B}}}\ln a_{\mathrm{A}}$$

Therefore,

$$\phi=\frac{x_{\mathrm{A}}}{x_{\mathrm{B}}}\int\frac{x_{\mathrm{B}}}{x_{\mathrm{A}}}\mathrm{d}\ln a_{\mathrm{B}}=\frac{1}{b}\int_0^b b\mathrm{d}\ln a_{\mathrm{B}}=\frac{1}{b}\int_0^b b\mathrm{d}\ln\gamma b=\frac{1}{b}\int_0^b b\mathrm{d}\ln b+\frac{1}{b}\int_0^b b\mathrm{d}\ln\gamma$$
$$=1+\frac{1}{b}\int_0^b b\mathrm{d}\ln\gamma$$

From the Debye–Hückel limiting law [5F.8]
$\ln\gamma=-A'b^{1/2}$ [$A'=A\ln 10=2.303A$]

Hence, $\mathrm{d}\ln\gamma=-\dfrac{1}{2}A'b^{-1/2}\,\mathrm{d}b$

$$\phi=1+\frac{1}{b}\left(-\frac{1}{2}A'\right)\int_0^b b^{1/2}\,\mathrm{d}b=1-\frac{1}{2}\left(\frac{A'}{b}\right)\times\frac{2}{3}b^{3/2}=1-\frac{1}{3}A'b^{1/2}$$

Comment. For the depression of the freezing point in the solution of a 1,1-electrolyte

$$\ln a_{\mathrm{A}}=\frac{-\Delta_{\mathrm{fus}}G}{RT}+\frac{\Delta_{\mathrm{fus}}G}{RT^{*}}$$

and hence $-r\phi=\dfrac{-\Delta_{\mathrm{fus}}H}{R}\left(\dfrac{1}{T}-\dfrac{1}{T^{*}}\right)$

Therefore, $\phi=\dfrac{\Delta_{\mathrm{fus}}Hx_{\mathrm{A}}}{Rx_{\mathrm{B}}}\left(\dfrac{1}{T}-\dfrac{1}{T^{*}}\right)=\dfrac{\Delta_{\mathrm{fus}}Hx_{\mathrm{A}}}{Rx_{\mathrm{B}}}\left(\dfrac{T^{*}-T}{TT^{*}}\right)\approx\dfrac{\Delta_{\mathrm{fus}}Hx_{\mathrm{A}}\Delta T}{Rx_{\mathrm{B}}T^{*2}}$

$$\approx\frac{\Delta_{\mathrm{fus}}H\Delta T}{\nu Rb_{\mathrm{B}}T^{*2}M_{\mathrm{A}}}$$

where ν is the number of ions per formula unit (here 2). Therefore, since $K_{\mathrm{f}}=\dfrac{MRT^{*2}}{\Delta_{\mathrm{fus}}H}$

$$\phi=\frac{\Delta T}{2b_{\mathrm{B}}K_{\mathrm{f}}}$$

I5.9 (a) Implicit in the model we are exploring is that at most one small molecule A can bind to each macromolecule, so the species are limited to free A, free M, and MA (no MA_2, etc.). In that case, the average number bound is also equal to the fraction of macromolecules that are bound to small molecules. Therefore, $[MA]=\nu[M]$, where [M] is the total concentration of M, bound and unbound. Similarly, $[M]_{free}=(1-\nu)[M]$. Therefore

$$K=\frac{[MA]}{[M]_{free}[A]_{free}}=\frac{\nu[M]}{(1-\nu)[M][A]_{free}}=\frac{\nu}{(1-\nu)[A]_{out}}$$

where we use $[A]_{free}=[A]_{out}$ in the last step.

(b) At the risk of introducing more notation, let us rewrite the binding equilibrium in terms of individual binding sites, S:

$$S+A\rightleftharpoons SA,$$

for which the equilibrium constant is

$$K=\frac{[SA]}{[S]_{free}[A]_{free}}$$

The concentration of bound sites is the average number per site times the number of sites per macromolecule times the total concentration of macromolecules:

$$[SA]=(\nu/N)N[M]=\nu[M].$$

Similarly, the concentration of free sites is

$$[S]_{free}=(1-\nu/N)N[M]=(N-\nu)[M].$$

Thus $K=\dfrac{\nu[M]}{(N-\nu)[M][A]_{out}}$,

which rearranges to the Scatchard equation

$$KN-\nu K=\frac{\nu}{[A]_{out}}.$$

(c) The requisite plot is of $\nu/[EB]_{out}$ vs. ν. If the Scatchard equation applies, the points should fall on a straight line with slope $-K$ and y-intercept KN. Use

$$\nu=\frac{[EB]_{bound}}{[M]}\quad\text{and}\quad[EB]_{bound}=[EB]_{in}-[EB]_{out}$$

to draw up the following table:

$[EB]_{out}/(\mu mol\,dm^{-3})$	0.042	0.092	0.204	0.526	1.150
$[EB]_{bound}/(\mu mol\,dm^{-3})$	0.250	0.498	1.000	2.005	3.000
ν	0.250	0.498	1.000	2.005	3.000
$(\nu/[EB]_{out})/(dm^3\,\mu mol^{-1})$	5.95	5.41	4.90	3.81	2.61

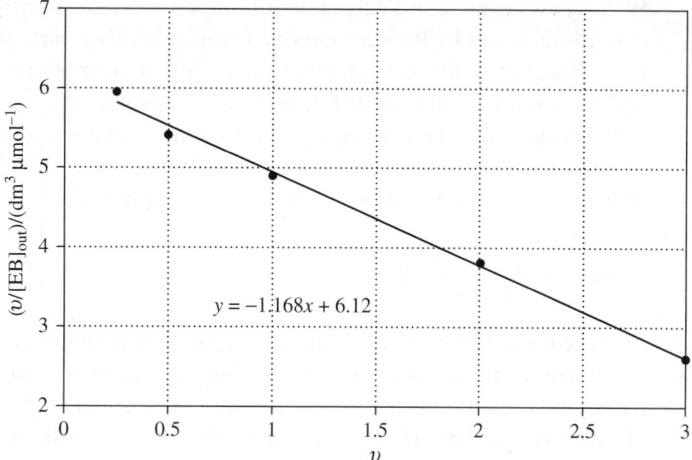

Figure I5.3

A plot of $v/[\mathrm{EB}]_{\mathrm{out}}$ against v is shown in Fig. I5.3.

The slope is $-1.168\ \mathrm{dm^3\ \mu mol^{-1}} = -1.168\times10^6\ \mathrm{dm^3\ mol^{-1}}$, hence $K = \boxed{1.168\times10^6}$. (Note that K is conventionally defined as a dimensionless quantity when concentrations are given in $\mathrm{mol\ dm^{-3}}$; however, one can specify units and say $K = 1.167\ \mathrm{dm^3\ \mu mol^{-1}}$. The intercept at $v = 0$ is

$$KN = 6.12\ \mathrm{dm^3 \mu mol^{-1}} \quad \text{so} \quad N = \boxed{5.23},$$

and this is the average number of binding sites per oligonucleotide. The close fit of the data to a straight line indicates that the identical and independent sites model $\boxed{\text{is applicable}}$.

I5.11 $\mathrm{PX}_v(\mathrm{s}) \rightleftharpoons \mathrm{P}^{v+}(\mathrm{aq}) + v\mathrm{X}^-(\mathrm{aq})$

This process is a solubility equilibrium described by a solubility constant K_s.

$$K_\mathrm{s} = a_\mathrm{P} \times a_\mathrm{X}^v$$

Introducing activity coefficients and concentrations, b, we obtain

$$K_\mathrm{s} = b_\mathrm{P} \times b_\mathrm{X}^v \times \gamma_\pm^{v+1}$$

At low to moderate ionic strengths we can use the Debye–Hückel limiting law [5F.8] as a good approximation for $\gamma_\pm$:

$$\log \gamma_\pm = -|z_- z_+| A I^{1/2}.$$

Addition of a salt, such as $(\mathrm{NH_4})_2\mathrm{SO_4}$, causes I to increase, $\log \gamma_\pm$ to become more negative, and $\gamma_\pm$ to decrease. However, K_s is a true equilibrium constant and remains unchanged. Therefore, the concentration of P^{v+} increases and the protein solubility increases proportionately.

We may also explain this effect in terms of Le Chatelier's principle. As the ionic strength increases by the addition of an inert electrolyte such as $(NH_4)_2SO_4$, the ions of the protein that are in solution attract one another less strongly, so that the equilibrium is shifted in the direction of increased solubility.

The explanation of the salting out effect is somewhat more complicated and can be related to the failure the Debye–Hückel limiting law at higher ionic strengths. At high ionic strengths we may write the Davies equation [5F.11b]

$$\log \gamma_{\pm} = \frac{-|z_+z_-|AI^{1/2}}{1+BI^{1/2}} + CI.$$

At low concentrations of inert salt, $I^{1/2} > I$, the first term dominates, $\gamma_{\pm}$ decreases with increasing I, and salting in occurs; however, at high concentrations, $I > I^{1/2}$, the second term dominates, $\gamma_{\pm}$ increases with increasing I, and salting out occurs. The Le Chatelier's principle explanation is that the water molecules are tied up by ion–dipole interactions and become unavailable for solvating the protein, thereby leading to decreased solubility.

6 Chemical equilibrium

6A The equilibrium constant

Answer to discussion question

D6A.1 The position of equilibrium is always determined by the condition that the **reaction quotient**, defined by the expression

$$Q = \frac{\text{activities of products}}{\text{activities of reactants}} = \prod_J a_J^{\nu_J} \text{ [6A.12] with the } \pm \text{ convention}$$

for stoichiometric numbers,

must equal the **thermodynamic equilibrium constant, *K***, a ratio that depends upon temperature but is independent of both composition and pressure. Mixing in of an additional amount of reactant or product to an equilibrium system destroys that equality and the reaction system spontaneously responds by shifting in such a way as to restore the equality. That implies that some of the added reactant or product must be removed by the reacting system and the amounts of other components will also be affected. These adjustments restore the concentrations to new equilibrium values but the value of the equilibrium constant is unchanged.

Chemical additions to an equilibrium system increases Gibbs energy, a condition that provides the impetus to spontaneously return to the equilibrium thermodynamic state, the state of minimum Gibbs energy, for which

$\Delta_r G = 0$ [6A.3, the **condition for equilibrium**] and,

$$\text{therefore, } \Delta_r G^\ominus = -RT\ln Q_{eq} = -RT\ln K \text{ [6A.14].}$$

If the mixing of reactants and products gives $\Delta_r G < 0$, reactant activities will spontaneously diminish to increase product activities until $\Delta_r G = 0$. If the mixing gives $\Delta_r G > 0$, product activities will spontaneously diminish to increase reactant activities until $\Delta_r G = 0$. For the general reaction

$$a\text{A} + b\text{B} \rightleftharpoons c\text{C} + d\text{D} \quad Q = a_C^c a_D^d / a_A^a a_B^b \text{ [6A.12]}$$

we say that, when the mixing gives $\Delta_r G < 0$, the reaction proceeds spontaneously to the right (*forward* reaction) until equilibrium is achieved but, when $\Delta_r G > 0$, the reaction proceeds spontaneously to the left (*reverse* direction). If reactant or product is added to an equilibrium mixture, the reaction spontaneously shifts in the direction that lowers the Gibbs energy of the reaction mixture (see text Fig. 6A.1, remember that $\Delta_r G$ is the *slope* of G plotted against the extent of reaction [6A.1]). The reaction spontaneously shifts to the right upon addition of reactant to an equilibrium mixture; left upon addition of product. We must also remember that thermodynamics says nothing about the rate at which the reaction occurs or shifts. A spontaneous reaction may occur very rapidly, infinitely slowly, or at any intermediate speed.

Solutions to exercises

E6A.1(a) $\mathrm{A} \rightarrow 2\,\mathrm{B}$

$n_J = n_J(0) + \nu_J \Delta\xi$ where ξ is the extent of reaction; ν_J is negative for reactants and positive for products.

$$n_A = 1.5\ \text{mol} - 1\times(0.60\ \text{mol}) = \boxed{0.9\ \text{mol}}$$

$$n_B = 0 + 2\times(0.60\ \text{mol}) = \boxed{1.2\ \text{mol}}$$

E6A.2(a) $\mathrm{A} \rightarrow 2\,\mathrm{B} \quad \Delta_r G = -6.4\ \text{kJ mol}^{-1}$

$$\Delta_r G = \left(\frac{\partial G}{\partial \xi}\right)_{p,T} \quad [6A.1]$$

With the approximation that $\left(\frac{\partial G}{\partial \xi}\right)_{p,T} \simeq \frac{\Delta G}{\Delta \xi}$, which is valid when $\Delta\xi$ is very small, we find that

$$\Delta G \simeq \Delta_r G \times \Delta\xi$$
$$\simeq (-6.4\ \text{kJ mol}^{-1})\times(+0.10\ \text{mol}) = \boxed{-0.64\ \text{kJ}}$$

E6A.3(a) $N_2(g) + 3\,H_2(g) \rightarrow 2\,NH_3(g) \quad \Delta_r G^{\ominus} = -32.9\ \text{kJ mol}^{-1}$

$$\Delta_r G = \Delta_r G^{\ominus} + RT\ln Q \quad [6A.10]$$
$$= (-32.9\ \text{kJ mol}^{-1}) + (2.4790\ \text{kJ mol}^{-1})\times \ln Q \quad \text{at } 298.15\ \text{K}$$

The above equation is used to calculate $\Delta_r G$ values at the given Q values in the following table.

Part	Q	ln Q	$\Delta_r G$ / kJ mol^{-1}
(i)	0.010	−4.605	−44.3
(ii)	1.00	0	−32.9
(iii)	10.0	2.303	−27.2
(iv)	100 000	11.513	−4.36
(v)	1 000 000	13.816	+ 1.35

The above equation also indicates that a plot of ln Q against $\Delta_r G$ should be linear so points iv and v, which straddle $\Delta_r G = 0$, can be used to perform a linear interpolation to find K from our equilibrium knowledge that $\ln Q = \ln Q_{\text{equilibrium}} = \ln K$ when $\Delta_r G = 0$. Performing the linear interpolation:

$$\begin{aligned}\ln Q &= \ln Q_{\text{iv}} + \left(\frac{\ln Q_{\text{v}} - \ln Q_{\text{iv}}}{\Delta_r G_{\text{v}} - \Delta_r G_{\text{iv}}}\right) \times (\Delta_r G - \Delta_r G_{\text{iv}}) \\ &= 11.513 + \left(\frac{13.816 - 11.513}{1.35 - (-4.36)}\right) \times \left(\frac{\Delta_r G}{\text{kJ mol}^{-1}} - (-4.36)\right) \\ &= 11.513 + 0.4033 \times \left(\frac{\Delta_r G}{\text{kJ mol}^{-1}} + 4.36\right)\end{aligned}$$

Thus,

$$\begin{aligned}\ln K &= 11.513 + 0.4033 \times (0 + 4.36) \\ &= 13.27\end{aligned}$$

$$K = e^{13.27} = \boxed{5.80 \times 10^5} \text{ from a two-point interpolation}$$

The two-point interpolation is in agreement with the result given by eqn 6A.8:

$$\begin{aligned}K &= e^{-\Delta_r G^\ominus / RT} \quad [6A.8] \\ &= e^{-(-32.9 \times 10^3\ \text{J mol}^{-1}) / \{(8.3145\ \text{J mol}^{-1}\ \text{K}^{-1}) \times (298.15\ \text{K})\}} \\ &= 5.81 \times 10^5\end{aligned}$$

E6A.4(a) $2\,H_2O(g) \rightleftharpoons 2\,H_2(g) + O_2(g)$ $T = 2\,257$ K, $p = 1\text{ bar} = p^\ominus$, $\alpha = 0.0177$ at equilibrium.

We draw up the following equilibrium table (Example 6A.2).

	$H_2O(g)$	$H_2(g)$	$O_2(g)$
Amount at equilibrium	$(1-\alpha)n$	αn	$\frac{1}{2}\alpha n$
Mole fraction	$\frac{1-\alpha}{1+\frac{1}{2}\alpha}$	$\frac{\alpha}{1+\frac{1}{2}\alpha}$	$\frac{\frac{1}{2}\alpha}{1+\frac{1}{2}\alpha}$
Partial pressure	$\frac{(1-\alpha)p}{1+\frac{1}{2}\alpha}$	$\frac{\alpha p}{1+\frac{1}{2}\alpha}$	$\frac{\frac{1}{2}\alpha p}{1+\frac{1}{2}\alpha}$

$$K = \left(\prod_J a_J^{\nu_J}\right)_{\text{equilibrium}} [6A.13] = \left(\prod_J (p_J / p^\ominus)^{\nu_J}\right)_{\text{equilibrium}} \quad \text{(perfect gas assumption)}$$

$$\begin{aligned}&= \frac{(p_{H_2}/p^\ominus)^2 (p_{O_2}/p^\ominus)}{(p_{H_2O}/p^\ominus)^2} = \frac{p_{H_2}^2 p_{O_2}}{p_{H_2O}^2 p^\ominus} = \frac{\left(\frac{\alpha p}{1+\frac{1}{2}\alpha}\right)^2 \left(\frac{\frac{1}{2}\alpha p}{1+\frac{1}{2}\alpha}\right)}{\left(\frac{(1-\alpha)p}{1+\frac{1}{2}\alpha}\right)^2 p^\ominus} \\ &= \frac{p}{2p^\ominus}\left\{\frac{\alpha^3}{(1-\alpha)^2 \times (1+\frac{1}{2}\alpha)}\right\} = \frac{1}{2}\left\{\frac{(0.0177)^3}{(1-0.0177)^2 \times (1+\frac{1}{2} \times 0.0177)}\right\} \\ &= \boxed{2.85 \times 10^{-6}}\end{aligned}$$

E6A.5(a) (i) $N_2O_4(g) \rightleftharpoons 2\,NO_2(g)$ $T = 298.15\,\text{K}, p = 1\,\text{bar} = p^{\ominus}, \alpha = 0.1846$ at equilibrium. We draw up the following equilibrium table (Example 6A.2).

	$N_2O_4(g)$	$NO_2(g)$
Amount at equilibrium	$(1-\alpha)n$	$2\alpha n$
Mole fraction	$\dfrac{1-\alpha}{1+\alpha}$	$\dfrac{2\alpha}{1+\alpha}$
Partial pressure	$\dfrac{(1-\alpha)p}{1+\alpha}$	$\dfrac{2\alpha p}{1+\alpha}$

$$K = \left(\prod_J a_J^{\nu_J}\right)_{\text{equilibrium}} [6A.13] = \left(\prod_J (p_J/p^{\ominus})^{\nu_J}\right)_{\text{equilibrium}} \quad \text{(perfect gas assumption)}$$

$$= \frac{(p_{NO_2}/p^{\ominus})^2}{p_{N_2O_4}/p^{\ominus}} = \frac{p_{NO_2}^{\,2}}{p_{N_2O_4}\,p^{\ominus}} = \frac{\left(\dfrac{2\alpha p}{1+\alpha}\right)^2}{\left(\dfrac{(1-\alpha)p}{1+\alpha}\right)p^{\ominus}}$$

$$= \frac{4p}{p^{\ominus}}\left\{\frac{\alpha^2}{(1-\alpha)\times(1+\alpha)}\right\} = 4\left\{\frac{(0.1846)^2}{(1-0.1846)\times(1+0.1846)}\right\}$$

$$= \boxed{0.141} \text{ at } 25\,°\text{C}$$

(ii) $$\ln K_2 = \ln K_1 - \frac{\Delta_r H^{\ominus}}{R}\left(\frac{1}{T_2} - \frac{1}{T_1}\right) \quad [6B.22]$$

$$\ln K_{100\,°\text{C}} = \ln K_{25\,°\text{C}} - \frac{\Delta_r H^{\ominus}}{R}\left(\frac{1}{373.15\,\text{K}} - \frac{1}{298.15\,\text{K}}\right)$$

$$= \ln(0.141) - \left(\frac{+56.2\times10^3\,\text{J mol}^{-1}}{8.3145\,\text{J K}^{-1}\,\text{mol}^{-1}}\right)\times\left(\frac{1}{373.15\,\text{K}} - \frac{1}{298.15\,\text{K}}\right) = 2.60$$

$$K_{100\,°\text{C}} = e^{2.60} = \boxed{13.5}$$

As expected, the temperature increase causes a shift to the right when the reaction is endothermic.

E6A.6(a) $PbO(s) + CO(g) \rightleftharpoons Pb(s) + CO_2(g)$

(i) Using data tables of the text *Resource section* at 25 °C, we find

$$\Delta_r G^{\ominus} = \Delta_f G^{\ominus}(\text{Pb, s}) + \Delta_f G^{\ominus}(\text{CO}_2\text{, g}) - \Delta_f G^{\ominus}(\text{PbO, s, red}) - \Delta_f G^{\ominus}(\text{CO, g})$$

$$= (-394.36\,\text{kJ mol}^{-1}) - (-188.93\,\text{kJ mol}^{-1}) - (-137.17\,\text{kJ mol}^{-1})$$

$$= \boxed{-68.26\,\text{kJ mol}^{-1}}$$

$$\begin{aligned}\Delta_r H^\ominus &= \Delta_f H^\ominus(\text{Pb, s}) + \Delta_f H^\ominus(\text{CO}_2\text{, g}) - \Delta_f H^\ominus(\text{PbO, s, red}) - \Delta_f H^\ominus(\text{CO, g})\\ &= (-393.51\,\text{kJ mol}^{-1}) - (-218.99\,\text{kJ mol}^{-1}) - (-110.53\,\text{kJ mol}^{-1})\\ &= -63.99\,\text{kJ mol}^{-1} \quad \text{[Used in part (ii)]}\end{aligned}$$

$$\ln K = \frac{-\Delta_r G^\ominus}{RT}\,[6\text{A}.14] = \frac{+68.26\times10^3\,\text{J mol}^{-1}}{(8.3145\,\text{J K}^{-1}\,\text{mol}^{-1})\times(298.15\,\text{K})} = 27.54$$

$$K = \text{e}^{27.54} = \boxed{9.13\times10^{11}}$$

$$\text{(ii)}\quad \ln K_2 = \ln K_1 - \frac{\Delta_r H^\ominus}{R}\left(\frac{1}{T_2} - \frac{1}{T_1}\right) \quad [6\text{A}.22]$$

$$\begin{aligned}\ln K_{400\,\text{K}} &= \ln K_{298.15\,\text{K}} - \frac{\Delta_r H^\ominus}{R}\left(\frac{1}{400\,\text{K}} - \frac{1}{298.15\,\text{K}}\right)\\ &= \ln\left(9.13\times10^{11}\right) - \left(\frac{-63.99\times10^3\ \text{J mol}^{-1}}{8.3145\ \text{J K}^{-1}\ \text{mol}^{-1}}\right)\times\left(\frac{1}{400\ \text{K}} - \frac{1}{298.15\ \text{K}}\right) = 21.0\end{aligned}$$

$$K_{400\,\text{K}} = \text{e}^{21.0} = \boxed{1.32\times10^9}$$

As expected, the temperature increase causes a shift to the left when the reaction is exothermic.

$$\Delta_r G^\ominus = -RT\ln K \quad [6\text{A}.14]$$

$$\Delta_r G^\ominus_{400\,\text{K}} = -\left(8.3145\ \text{J K}^{-1}\ \text{mol}^{-1}\right)\times(400\ \text{K})\ln\left(1.32\times10^9\right) = \boxed{-69.8\ \text{kJ mol}^{-1}}$$

E6A.7(a) $H_2CO(g) \rightarrow CO(g) + H_2(g)$ For this gas phase reaction $\Delta\nu = \sum_J \nu_J = 1+1-1 = 1$

$$K = \left(\prod_J a_J^{\nu_J}\right)_{\text{equilibrium}} [6\text{A}.13] = \left(\prod_J (\gamma_J p_J/p^\ominus)^{\nu_J}\right)_{\text{equilibrium}} = \left(\prod_J \gamma_J^{\nu_J}\right)_{\text{equilibrium}} \left(\prod_J (p_J/p^\ominus)^{\nu_J}\right)_{\text{equilibrium}}$$

$$= K_\gamma K_p \quad \text{where} \quad K_\gamma = \left(\prod_J \gamma_J^{\nu_J}\right)_{\text{equilibrium}} \quad \text{and } K_p = \left(\prod_J (p_J/p^\ominus)^{\nu_J}\right)_{\text{equilibrium}}$$

Let us assume that the gases are perfect gases. Then, the activity coefficients equal 1 and $K_\gamma = 1$. Additionally, $p_J = n_J RT/V = [J]RT$.. Substitution gives

$$K = K_p = \left(\prod_J ([J]RT/p^\ominus)^{\nu_J}\right)_{\text{equilibrium}} = \left(\prod_J ([J]/c^\ominus)^{\nu_J}\right)_{\text{equilibrium}} \left(\prod_J (c^\ominus RT/p^\ominus)^{\nu_J}\right)_{\text{equilibrium}}$$

$$= K_c\left(c^\ominus RT/p^\ominus\right)^{\Delta\nu} \quad \text{where} \quad K_c = \left(\prod_J ([J]/c^\ominus)^{\nu_J}\right)_{\text{equilibrium}} \quad [6.17(\text{a})]$$

$$\boxed{K = K_c \times (c^{\ominus} RT / p^{\ominus})} \quad \text{because } \Delta\nu = 1 \text{ for this reaction}$$

Since $c^{\ominus} R / p^{\ominus} = 0.0831451\ \text{K}^{-1}$, this expression may be written in the form

$$K = (0.0831451\ \text{K}^{-1}) \times K_c T$$

E6A.8(a) Draw up the following equilibrium table for the reaction equation: $2\,\text{A} + \text{B} \rightleftharpoons 3\,\text{C} + 2\,\text{D}$.

	A	B	C	D	Total
Initial amounts/mol	1.00	2.00	0	1.00	4.00
Stated change/mol			+0.90		
Implied change/mol	−0.60	−0.30	+0.90	+0.60	
Equilibrium amounts/mol	0.40	1.70	0.90	1.60	4.60
Mole fractions	0.087	0.370	0.196	0.348	1.001

(i) The mole fractions are given in the table.

(ii) $K_x = \prod_{\text{J}} x_{\text{J}}^{\nu_{\text{J}}}$

$$K_x = \frac{(0.196)^3 \times (0.348)^2}{(0.087)^2 \times (0.370)} = 0.32\bar{6} = \boxed{0.33}$$

(iii) $p_{\text{J}} = x_{\text{J}} p, \quad p = 1\ \text{bar}, \quad p^{\ominus} = 1\ \text{bar}$

Assuming that the gases are perfect, $a_{\text{J}} = p_{\text{J}} / p^{\ominus}$, hence

$$K = \frac{(p_{\text{C}} / p^{\ominus})^3 \times (p_{\text{D}} / p^{\ominus})^2}{(p_{\text{A}} / p^{\ominus})^2 \times (p_{\text{B}} / p^{\ominus})}$$

$$= \frac{x_{\text{C}}^3 x_{\text{D}}^2}{x_{\text{A}}^2 x_{\text{B}}} \times \left(\frac{p}{p^{\ominus}}\right)^2 = K_x \quad \text{when} \quad p = 1.00\ \text{bar} = \boxed{0.33}$$

(iv) $\Delta_r G^{\ominus} = -RT \ln K = -(8.3145\ \text{J K}^{-1}\ \text{mol}^{-1}) \times (298\ \text{K}) \times (\ln 0.32\bar{6}) = \boxed{+2.8\ \text{kJ mol}^{-1}}$

E6A.9(a) borneol $\rightleftharpoons$ isoborneol

$$\Delta_r G = \Delta G^{\ominus} + RT \ln Q \ [\text{6A.10}] \quad \text{where} \quad Q = \frac{p_{\text{isoborneol}}}{p_{\text{borneol}}} \ [\text{perfect gas assumption.}]$$

$$p_{\text{borneol}} = x_{\text{borneol}} p = \frac{0.15\ \text{mol}}{0.15\ \text{mol} + 0.30\ \text{mol}} \times 600\ \text{Torr} = 200\ \text{Torr}$$

$$p_{\text{isoborneol}} = p - p_{\text{borneol}} = 600\ \text{Torr} - 200\ \text{Torr} = 400\ \text{Torr}$$

$$Q = \frac{400\ \text{Torr}}{200\ \text{Torr}} = 2.00$$

$$\Delta_r G = (+9.4 \text{ kJ mol}^{-1}) + (8.314 \text{ J K}^{-1} \text{ mol}^{-1}) \times (503 \text{ K}) \times (\ln 2.00) = \boxed{+12.3 \text{ kJ mol}^{-1}}$$

This mixture reacts spontaneously to the left under these conditions.

E6A.10(a) $\tfrac{1}{2} N_2(g) + \tfrac{3}{2} H_2(g) \rightarrow NH_3(g) \quad \Delta_f G^\ominus = -16.5 \text{ kJ mol}^{-1}$

$$\Delta_f G = \Delta_f G^\ominus + RT \ln Q \text{ [6A.10]} \quad \text{where} \quad Q = \prod_J a_J^{\nu_J} \text{ [6A.12b]}$$

$$= \Delta_f G^\ominus + RT \ln \frac{p_{NH_3}/p^\ominus}{(p_{N_2}/p^\ominus)^{1/2}(p_{H_2}/p^\ominus)^{3/2}} \text{ [perfect gas assumption]}$$

$$= (-16.5 \text{ kJ mol}^{-1}) + (8.3145 \times 10^{-3} \text{ kJ K}^{-1} \text{ mol}^{-1}) \times (298.15 \text{ K}) \times \ln\left(\frac{4}{3^{1/2} \times 1^{3/2}}\right)$$

$$= \boxed{-14.4 \text{ kJ mol}^{-1}}$$

Since $\Delta_f G < 0$, the spontaneous direction of reaction is $\boxed{\text{toward the ammonia product}}$.

E6A.11(a) $CaF_2(s) \rightleftharpoons Ca^{2+}(aq) + 2F^-(aq) \quad K_s = 3.9 \times 10^{-11}$

$$\Delta_r G^\ominus = -RT \ln K_s$$

$$= -(8.3145 \text{ J K}^{-1} \text{ mol}^{-1}) \times (298.15 \text{ K}) \times \ln(3.9 \times 10^{-11}) = +59.4 \text{ kJ mol}^{-1}$$

$$= \Delta_f G^\ominus(CaF_2, aq) - \Delta_f G^\ominus(CaF_2, s)$$

$$\Delta_f G^\ominus(CaF_2, aq) = \Delta_r G^\ominus + \Delta_f G^\ominus(CaF_2, s)$$

$$= (59.4 - 1167) \text{ kJ mol}^{-1} = \boxed{-1108 \text{ kJ mol}^{-1}}$$

Solutions to problems

P6A.1 $I_2(s) + Br_2(g) \rightarrow 2\,IBr(g) \quad K = 0.164 \text{ at } 25°C$

(a) $\Delta_r G^\ominus = -RT \ln K$ [6A.14]

$$= -(8.3145 \text{ J K}^{-1} \text{ mol}^{-1}) \times (298 \text{ K}) \times (\ln 0.164) = 4.48 \times 10^3 \text{ J mol}^{-1}$$

$$= \boxed{+4.48 \text{ kJ mol}^{-1}}$$

(b) Draw up the following equilibrium table

	I_2	Br_2	IBr
Amounts	–	$(1-\alpha)n$	$2\alpha n$
Mole fractions	–	$\dfrac{(1-\alpha)}{(1+\alpha)}$	$\dfrac{2\alpha}{(1+\alpha)}$
Partial pressure $p_J = x_J p$	–	$\dfrac{(1-\alpha)p}{(1+\alpha)}$	$\dfrac{2\alpha p}{(1+\alpha)}$

$$K=\prod_J a_J^{\nu_J}\ [6A.13]=\frac{(p_{IBr}/p^{\ominus})^2}{p_{Br_2}/p^{\ominus}}\ \text{[perfect gases]}$$

$$=\frac{\{(2\alpha)^2 p/p^{\ominus}\}}{(1-\alpha)\times(1+\alpha)}=\frac{(4\alpha^2 p/p^{\ominus})}{1-\alpha^2}=0.164$$

With $p=0.164$ atm, $4\alpha^2=1-\alpha^2 \quad \alpha^2=1/5 \quad \alpha=0.447$

$$p_{IBr}=\frac{2\alpha}{1+\alpha}\times p=\frac{(2)\times(0.447)}{1+0.447}\times(0.164\text{ atm})=\boxed{0.101\text{ atm}}$$

(c) The equilibrium table needs to be modified as follows

$$p=p_{I_2}+p_{Br_2}+p_{IBr},\quad p_{Br_2}=x_{Br_2}p,\quad p_{IBr}=x_{IBr}p,\quad p_{I_2}=x_{I_2}p$$

with $x_{Br_2}=\dfrac{(1-\alpha)n}{(1+\alpha)n+n_{I_2}}$ [$n=$ amount of Br_2 introduced into the container]

and $x_{IBr}=\dfrac{2\alpha n}{(1+\alpha)n+n_{I_2}}$.

K is constructed as above [6A.13], but with these modified partial pressures. In order to complete the calculation additional data are required, namely, the amount of Br_2 introduced, n, and the equilibrium vapour pressure of $I_2(s)$. n_{I_2} can be calculated from a knowledge of the volume of the container at equilibrium which is most easily determined by successive approximations since p_{I_2} is small.

What is the partial pressure of IBr(g) if 0.0100 mol of Br_2(g) is introduced into the container? The partial pressure of $I_2(s)$ at 25°C is 0.305 Torr.

P6A.3 Draw up the following table for the reaction: $H_2(g)+I_2(g)\rightleftharpoons 2\,HI(g) \quad K=870.$

	H_2	I_2	HI	Total
Initial amounts/mol	$n_{H_2,0}=0.300$	$n_{I_2,0}=0.400$	$n_{HI,0}=0.200$	$n_0=0.900$
Change/mol	$-z$	$-z$	$+2z$	
Equilibrium amounts/mol	$n_{H_2,0}-z$	$n_{I_2,0}-z$	$n_{HI,0}+2z$	$n_0=0.900$
Mole fraction	$(n_{H_2,0}-z)/n$	$(n_{I_2,0}-z)/n$	$(n_{HI_2,0}+2z)/n$	1

$$K=\frac{(p_{HI}/p^{\ominus})^2}{(p_{H_2}/p^{\ominus})(p_{I_2}/p^{\ominus})}\ [\text{perfect gases}, p_J=x_Jp]$$

$$=\frac{(x_{HI})^2}{(x_{H_2})(x_{I_2})}=\frac{(n_{HI,0}+2z)^2}{(n_{H_2,0}-z)(n_{I_2,0}-z)}$$

$$(K-4)z^2-\{K(n_{H_2,0}+n_{I_2})+4n_{HI,0}\}z+Kn_{H_2,0}n_{I_2,0}-n_{HI,0}^2=0$$

$$866z^2-609.80z+104.36=0$$

$$z=\frac{609.80\pm\sqrt{(-609.80)^2-4\times(866)\times(104.36)}}{2\times(866)}=0.411 \text{ or } 0.293$$

$$=0.293 \text{ because } z \text{ cannot exceed } n_{H_2,0} \text{ so we reject the 0.411 value.}$$

The final composition is therefore $\boxed{0.007 \text{ mol } H_2}$, $\boxed{0.107 \text{ mol } I_2}$, and $\boxed{0.786 \text{ mol HI}}$.

P6A.5 We draw up the following table using the stoichiometry $A+3B\rightarrow 2C$ and $\Delta n_J = \nu_J\xi$

	A	B	C	Total
Initial amounts/mol	1	3	0	4
Change, Δn_J / mol	$-\xi$	-3ξ	$+2\xi$	
Equilibrium amounts/mol	$1-\xi$	$3(1-\xi)$	2ξ	$2(2-\xi)$
More fraction	$\frac{1-\xi}{2(2-\xi)}$	$\frac{3(1-\xi)}{2(2-\xi)}$	$\frac{\xi}{2-\xi}$	1

$$K=\frac{(p_C/p^\ominus)^2}{(p_A/p^\ominus)(p_B/p^\ominus)^3}=\frac{x_C^2}{x_A x_B^3}\times\left(\frac{p^\ominus}{p}\right)^2=\frac{\xi^2}{(2-\xi)^2}\times\frac{2(2-\xi)}{1-\xi}\times\frac{2^3(2-\xi)^3}{3^3(1-\xi)^3}\times\left(\frac{p^\ominus}{p}\right)^2$$

$$=\frac{16(2-\xi)^2\xi^2}{27(1-\xi)^4}\times\left(\frac{p^\ominus}{p}\right)^2$$

Since K is independent of the pressure

$$\frac{(2-\xi)^2\xi^2}{(1-\xi)^4}=a^2\left(\frac{p}{p^\ominus}\right)^2 \quad \text{where} \quad a^2=\frac{27}{16}K, \text{ a constant}$$

Therefore, $(2-\xi)\xi=a\left(\frac{p}{p^\ominus}\right)\times(1-\xi)^2$

$$\left(1+\frac{ap}{p^\ominus}\right)\xi^2-2\left(1+\frac{ap}{p^\ominus}\right)\xi+\frac{ap}{p^\ominus}=0$$

which solves to $\boxed{\xi=1-\left(\frac{1}{1+ap/p^\ominus}\right)^{1/2}}$.

We choose the root with the negative sign because ξ lies between 0 and 1. The variation of ξ with p is shown in Fig. 6A.1.

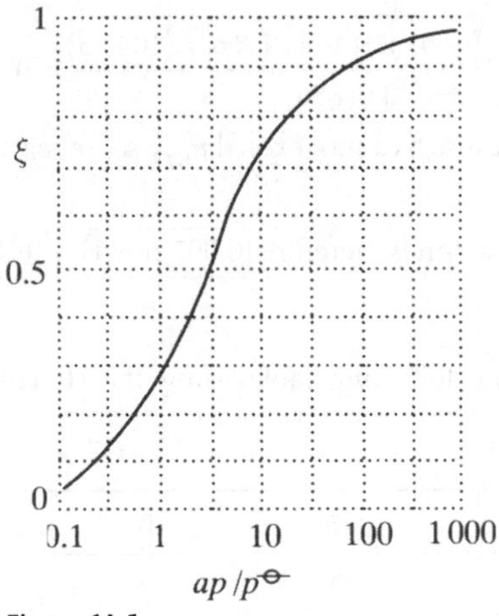

Figure 6A.1

6B The response of equilibria to the conditions

Answers to discussion questions

D6B.1 The definition of the thermodynamic equilibrium constant involves activities rather than pressures:

$$K=\left(\prod_{\mathrm{J}} a_{\mathrm{J}}^{\nu_{\mathrm{J}}}\right)_{\text{equilibrium}} \quad [6\text{A}.13] \text{ with the } \pm \text{ convention for stoichiometric numbers}$$

At low pressures, the activities of gases may be replaced with their partial pressures with little error, but at high pressures that is not a good approximation. The difference between the equilibrium constant expressed in activities and the constant expressed in pressures is dependent upon two factors: the stoichiometry of the reaction and the magnitude of the partial pressures. Thus there is no one answer to this question. For the example of the ammonia synthesis reaction, in a range of pressures where the activity coefficients are greater than one, an increase in pressure results in a greater shift to the product side than would be predicted by the constant expressed in partial pressures.

The activity coefficients of real gases give an indication of dominant intermolecular forces. Coefficients greater than 1 are observed when repulsions dominate; coefficients less than one dominate when attractions dominate. In the limit of zero pressure all gases behave as perfect gases with activity coefficients equal to 1 and, therefore, $a_{\mathrm{J(gas)}}=\gamma_{\mathrm{J}}p_{\mathrm{J}}/p^{\ominus}=p_{\mathrm{J}}/p^{\ominus}$ in the perfect gas case only. Thus, we conclude that intermolecular forces cause the thermodynamic equilibrium constant to respond differently to changes in pressure from the equilibrium constant expressed in terms of partial pressures.

For an exothermic reaction, such as the ammonia synthesis, an increase in temperature will shift the reaction to the reactant side, but the relative shift is independent of the activity coefficients. The ratio $\ln(K_2/K_1)$ depends on $\Delta_r H^\ominus$. See eqn 6B.4.

D6B.3 The physical dependence of the equilibrium constant on temperature, as predicted by the van 't Hoff =equation, $\frac{\mathrm{d}\ln K}{\mathrm{d}T}=\frac{\Delta_r H^\ominus}{RT^2}$ [6B.2(a)], can be seen when the expression $\Delta_r G^\ominus = \Delta_r H^\ominus - T\Delta_r S^\ominus$ is written in the form $R\ln K = -\Delta_r H^\ominus/T + \Delta_r S^\ominus$. When the reaction is exothermic and the temperature is raised, and hence K decreases, since T occurs in the denominator, the reaction shifts to favour the reactants. When the reaction is endothermic, increasing T makes ln K less negative, or K more positive and products are favoured. Another factor of importance when the reaction is endothermic is the increasing entropy of their reacting system resulting in a more positive ln K, favouring products.

The molecular basis of the van 't Hoff equation stems from the Boltzmann distribution of molecules over the available energy levels. When the temperature of an equilibrium system is changed, the Boltzmann distribution adjusts. For an endothermic reaction a temperature increase causes an increased population of the higher energy states of products at the expense of the population of the lower energy states of reactants and the reaction shifts to the right. The converse applies to a temperature increase for an exothermic reaction and the reaction shifts to the left.

Solutions to exercises

E6B.1(a) At 1 280 K, $\Delta_r G^\ominus = +33\times10^3\,\mathrm{J\,mol^{-1}}$; thus

$$\ln K_1(1\,280\,\mathrm{K}) = -\frac{\Delta_r G^\ominus}{RT}\,[6\mathrm{A}.14] = -\frac{33\times10^3\,\mathrm{J\,mol^{-1}}}{(8.3145\,\mathrm{J\,K^{-1}\,mol^{-1}})\times(1\,280\,\mathrm{K})} = -3.1\bar{0}$$

$$K_1 = \boxed{0.045}$$

$$\ln K_2 = \ln K_1 - \frac{\Delta_r H^\ominus}{R}\left(\frac{1}{T_2}-\frac{1}{T_1}\right)\,[6\mathrm{B}.4]$$

We look for the temperature T_2 that corresponds to $\ln K_2 = \ln(1) = 0$. This is the crossover temperature. Solving for T_2 from eqn 6B.4 with ln K_2=0, we obtain

$$\frac{1}{T_2} = \frac{R\ln K_1}{\Delta_r H^\ominus} + \frac{1}{T_1} = \left(\frac{(8.3145\,\mathrm{J\,K^{-1}\,mol^{-1}})\times(-3.1\bar{0})}{224\times10^3\,\mathrm{J\,mol^{-1}}}\right) + \left(\frac{1}{1\,280\,\mathrm{K}}\right) = 6.6\bar{6}\times10^{-4}\,\mathrm{K^{-1}}$$

$$T_2 = \boxed{15\bar{0}\bar{0}\,\mathrm{K}}$$

E6B.2(a) $2\,C_3H_6(g) \rightarrow C_2H_4(g) + C_4H_8(g) \quad T = 400\,\mathrm{K}$

$$\ln K = A + \frac{B}{T} + \frac{C}{T^2} \quad \text{where} \quad A = -1.04,\ B = -1\,088\,\mathrm{K},\ C = 1.51\times10^5\,\mathrm{K^2}$$

$$\Delta_r G^\ominus = -RT\ln K\ [6A.14] = RT\times\left(A+\frac{B}{T}+\frac{C}{T^2}\right)$$

$$= -(8.3145\ \mathrm{J\,K^{-1}\,mol^{-1}})\times(400\ \mathrm{K})\times\left(-1.04-\frac{1088\ \mathrm{K}}{400\ \mathrm{K}}+\frac{1.51\times10^5\ \mathrm{K}^2}{(400\ \mathrm{K})^2}\right)$$

$$= +9.37\ \mathrm{kJ\,mol^{-1}}$$

$$\Delta_r H^\ominus = -R\frac{\mathrm{d}\ln K}{\mathrm{d}(1/T)}\ [6B.2(b)]$$

$$= -R\frac{\mathrm{d}}{\mathrm{d}(1/T)}\left(A+\frac{B}{T}+\frac{C}{T^2}\right) = -R\times\left(B+\frac{2C}{T}\right)$$

$$= -(8.3145\ \mathrm{J\,K^{-1}\,mol^{-1}})\times\left\{(-1088\ \mathrm{K})+2\times\frac{1.51\times10^5\ \mathrm{K}^2}{400\ \mathrm{K}}\right\} = \boxed{+2.77\ \mathrm{kJ\,mol^{-1}}}$$

$$\Delta_r G^\ominus = \Delta_r H^\ominus - T\Delta_r S^\ominus$$

$$\Delta_r S^\ominus = \frac{\Delta_r H^\ominus - \Delta_r G^\ominus}{T} = \frac{2.77\ \mathrm{kJ\,mol^{-1}} - 9.37\ \mathrm{kJ\,mol^{-1}}}{400\ \mathrm{K}} = \boxed{-16.5\ \mathrm{J\,K^{-1}\,mol^{-1}}}$$

E6B.3(a) $H_2CO(g) \rightarrow CO(g) + H_2(g)$ For this gas phase reaction $\Delta\nu = \sum_J \nu_J = 1$

$$K = \left(\prod_J a_J^{\nu_J}\right)_{\text{equilibrium}}\ [6A.13] = \left(\prod_J (\gamma_J p_J/p^\ominus)^{\nu_J}\right)_{\text{equilibrium}} = \left(\prod_J \gamma_J^{\nu_J}\right)_{\text{equilibrium}}\left(\prod_J (p_J/p^\ominus)^{\nu_J}\right)_{\text{equilibrium}}$$

$$= K_\gamma K_p \quad \text{where} \quad K_\gamma = \left(\prod_J \gamma_J^{\nu_J}\right)_{\text{equilibrium}} \quad \text{and } K_p = \left(\prod_J (p_J/p^\ominus)^{\nu_J}\right)_{\text{equilibrium}}$$

Let us assume that the gases are perfect gases. Then, the activity coefficients equal 1 and $K_\gamma = 1$. Additionally, $p_J = x_J p$. Substitution gives

$$K = K_p = \left(\prod_J (x_J p/p^\ominus)^{\nu_J}\right)_{\text{equilibrium}} = \left(\prod_J x_J^{\nu_J}\right)_{\text{equilibrium}}\left(\prod_J (p/p^\ominus)^{\nu_J}\right)_{\text{equilibrium}}$$

$$= K_x(p/p^\ominus)^{\Delta\nu} \quad \text{where} \quad K_x = \left(\prod_J x_J^{\nu_J}\right)_{\text{equilibrium}}$$

For this reaction:

$$K = K_x\times(p/p^\ominus) \quad \text{or} \quad K_x = K\times(p/p^\ominus)^{-1} \qquad \text{because } \Delta\nu = 1$$

Percentage change:

$$K_x = \frac{K_x(1\ \mathrm{bar}) - K_x(2\ \mathrm{bar})}{K_x(1\ \mathrm{bar})}\times100\%$$

$$= \frac{K\times(1\ \mathrm{bar}/p^\ominus)^{-1} - K\times(2\ \mathrm{bar}/p^\ominus)^{-1}}{K\times(1\ \mathrm{bar}/p^\ominus)^{-1}}\times100\% \quad \text{(Remember that } K \text{ is independent of pressure.)}$$

$$= \frac{1-\tfrac{1}{2}}{1}\times100\% = \boxed{50\%}$$

E6B.4(a) borneol(g) $\rightleftharpoons$ isoborneol(g) $K = 0.106$ at 503 K; $M = 154.24\ \text{g mol}^{-1}$

Let z be the number of moles of isoborneol that convert to borneol during the establishment of equilibrium.

	borneol	isoborneol	Total
Initial amounts/mol	$n_{\text{borneol}}=7.50/154.24=0.0486$	$n_{\text{iso}}=14.0/154.24=0.0908$	$n=0.1394$
Equilibrium amounts/mol	$n_{\text{borneol}}+z$	$n_{\text{iso}}-z$	$n=0.1394$
Equilibrium mole fractions	$(n_{\text{borneol}}+z)/n$	$(n_{\text{iso}}-z)/n$	

$$K=\left(\frac{p_{\text{isoborneol}}}{p_{\text{borneol}}}\right)_{\text{equilibrium}} \quad \text{(perfect gas assumption)}$$

$$=\left(\frac{x_{\text{isoborneol}}}{x_{\text{borneol}}}\right)_{\text{equilibrium}} \quad (\text{because } p_{\text{A}}=x_{\text{A}}p)$$

$$=\frac{n_{\text{iso}}-z}{n_{\text{borneol}}+z}$$

$$z=\frac{n_{\text{iso}}-n_{\text{borneol}}K}{1+K}=\frac{0.0908-0.0486\times 0.106}{1+0.106}=0.0774$$

Thus, the equilibrium mole fractions are

$$x_{\text{borneol}}=\frac{n_{\text{borneol}}+z}{n}=\frac{0.0486+0.0774}{0.1394}=\boxed{0.9039}$$

$$x_{\text{iso}}=\frac{n_{\text{borneol}}-z}{n}=\frac{0.0908-0.0774}{0.1394}=\boxed{0.0961}$$

E6B.5(a) $\ln\frac{K_2}{K_1}=-\frac{\Delta_r H^\ominus}{R}\left(\frac{1}{T_2}-\frac{1}{T_1}\right)$ [6B.4]

$$\Delta_r H^\ominus = R\times\left(\frac{1}{T_1}-\frac{1}{T_2}\right)^{-1}\times\ln\left(\frac{K_2}{K_1}\right)$$

$$=\left(8.3145\ \text{J K}^{-1}\ \text{mol}^{-1}\right)\times\left(\frac{1}{298\ \text{K}}-\frac{1}{308\ \text{K}}\right)^{-1}\times\ln\left(\frac{K_2}{K_1}\right)$$

$$=\left(76.31\ \text{kJ mol}^{-1}\right)\times\ln\left(\frac{K_2}{K_1}\right)$$

(a) $K_2/K_1=2.00$

$$\Delta_r H^\ominus=\left(76.31\ \text{kJ mol}^{-1}\right)\times\ln(2.00)=\boxed{52.89\ \text{kJ mol}^{-1}}$$

(b) $K_2/K_1=0.500$

$$\Delta_r H^\ominus=\left(76.31\ \text{kJ mol}^{-1}\right)\times\ln(0.500)=\boxed{-52.89\ \text{kJ mol}^{-1}}$$

E6B.6(a) The decomposition reaction is $CaCO_3(s) \rightleftharpoons CaO(s) + CO_2(g)$.

For the purposes of this exercise we may assume that the required temperature is that temperature at which $K=1$ at a pressure of 1 bar. For $K=1$, $\ln K = 0$ and $\Delta_r G^\ominus = 0$. Then,

$$\Delta_r G^\ominus = \Delta_r H^\ominus - T\Delta_r S^\ominus = 0$$

$$T = \frac{\Delta_r H^\ominus}{\Delta_r S^\ominus}$$

We now estimate that the values of both $\Delta_r H^\ominus$ and $\Delta_r S^\ominus$ are not too different then the values at 25 °C and calculate each with standard values found in the text *Resource section.*

$$\Delta_r H^\ominus = \sum_J \nu_J \Delta_{f,J} H^\ominus = \{(-635.09) - (393.51) - (-1\,206.9)\}\ \text{kJ mol}^{-1} = +178.3\ \text{kJ mol}^{-1}$$

$$\Delta_r S^\ominus = \sum_J \nu_J S_J^\ominus = \{(39.75) + (213.74) - (92.9)\}\ \text{J K}^{-1}\ \text{mol}^{-1} = +160.6\ \text{J K}^{-1}\ \text{mol}^{-1}$$

$$T : \frac{178.3\times 10^3\ \text{J mol}^{-1}}{160.6\ \text{J K}^{-1}\ \text{mol}^{-1}} = \boxed{1110\ \text{K}}\ (837\ °\text{C})$$

E6B.7(a) $A_2B(s) \rightleftharpoons A_2(g) + B(g)$ Vapour pressures: $p_{367°C} = 208$ kPa and $p_{477°C} = 547$ kPa

The gases originate from the dissociation of the solid alone so $p_{A_2} = p_B = \tfrac{1}{2}p$. The equilibrium constants and standard Gibbs energy of reaction at these two temperatures (367 °C and 477 °C) are:

$$\begin{aligned} K &= (p_{A_2}/p^\ominus)\times(p_B/p^\ominus)\ \text{[perfect gas assumption]} \\ &= (\tfrac{1}{2}p/p^\ominus)\times(\tfrac{1}{2}p/p^\ominus) \\ &= \tfrac{1}{4}(p/p^\ominus)^2 \end{aligned}$$

$$K_{367°C} = \tfrac{1}{4}(208\ \text{kPa}/100\ \text{kPa})^2 = 1.08$$

$$K_{477°C} = \tfrac{1}{4}(547\ \text{kPa}/100\ \text{kPa})^2 = 7.48$$

$$\Delta_r G^\ominus = -RT\ln K\ \text{[6A.14]}$$

$$\Delta_r G^\ominus_{367°C} = -(8.3145\ \text{J K}^{-1}\text{mol}^{-1})\times(640.15\ \text{K})\times\ln(1.08) = -0.410\ \text{kJ mol}^{-1}$$

$$\Delta_r G^\ominus_{477°C} = -(8.3145\ \text{J K}^{-1}\text{mol}^{-1})\times(750.15\ \text{K})\times\ln(7.48) = -12.6\ \text{kJ mol}^{-1}$$

Assuming that the enthalpy of reaction is independent of temperature:

$$\ln\frac{K_2}{K_1} = -\frac{\Delta_r H^\ominus}{R}\left(\frac{1}{T_2} - \frac{1}{T_1}\right)\ \text{[6B.4]}$$

$$\Delta_r H^\ominus = R\times\left(\frac{1}{T_1}-\frac{1}{T_2}\right)^{-1}\times\ln\left(\frac{K_2}{K_1}\right)$$

$$=(8.3145\ \mathrm{J\ K^{-1}\ mol^{-1}})\times\left(\frac{1}{640.15\ \mathrm{K}}-\frac{1}{750.15\ \mathrm{K}}\right)^{-1}\times\ln\left(\frac{7.48}{1.08}\right)$$

$$=\boxed{70.2\ \mathrm{kJ\ mol^{-1}}}$$

The standard reaction entropy at these two temperatures (367 °C and 477 °C) are:

$$\Delta_r G^\ominus = \Delta_r H^\ominus - T\Delta_r S^\ominus$$

$$\Delta_r S^\ominus = \frac{\Delta_r H^\ominus - \Delta_r G^\ominus}{T}$$

$$\Delta_r S^\ominus_{367\,°\mathrm{C}} = \frac{(70.2-(-0.410))\ \mathrm{kJ\ mol^{-1}}}{640.15\ \mathrm{K}}$$

$$=110\ \mathrm{J\ mol^{-1}\ K^{-1}}$$

$$\Delta_r S^\ominus_{477\,°\mathrm{C}} = \frac{(70.2-(-12.6))\ \mathrm{kJ\ mol^{-1}}}{750.15\ \mathrm{K}}$$

$$=110\ \mathrm{J\ mol^{-1}\ K^{-1}}$$

Thus, we see that $\Delta_r S^\ominus = \boxed{110\ \mathrm{kJ\ mol^{-1}\ K^{-1}}}$ is independent of temperature in this temperature range. In the following we calculate the reaction properties at 422 °C (695.15 K).

$$\Delta_r G^\ominus_{422\,°\mathrm{C}} = \Delta_r H^\ominus - T_{422\,°\mathrm{C}}\Delta_r S^\ominus = (70.2-695.15\times 0.110)\ \mathrm{kJ\ mol^{-1}} = \boxed{-6.3\ \mathrm{kJ\ mol^{-1}}}$$

$$K_{422\,°\mathrm{C}} = \mathrm{e}^{-\Delta_r G^\ominus_{422\,°\mathrm{C}}/RT} = \mathrm{e}^{-(-6.3\times10^3\ \mathrm{J\ mol^{-1}})/\{(8.3145\times 695.15)\ \mathrm{J\ mol^{-1}}\}} = \boxed{3.0}$$

Solutions to problems

P6B.1 $CH_4(g) \rightleftharpoons C(s) + 2\,H_2(g)$

This reaction is the reverse of the formation reaction. Consequently,

$$\Delta_r G^\ominus = -\Delta_f G^\ominus = -(\Delta_f H^\ominus - T\Delta_f S^\ominus)$$

$$=-\{-74.85\ \mathrm{kJ\ mol^{-1}} - (298\ \mathrm{K})\times(-80.67\ \mathrm{J\ K^{-1}\ mol^{-1}})\}$$

$$=+50.81\ \mathrm{kJ\ mol^{-1}}$$

(a) $K = \mathrm{e}^{-\Delta_r G^\ominus/RT}$ [6A.14]

$$=\mathrm{e}^{-(5.081\times10^4\ \mathrm{J\ mol^{-1}})/(8.3145\ \mathrm{J\ K^{-1}\ mol^{-1}}\times 298\ \mathrm{K})}$$

$$=\boxed{1.24\times10^{-9}}$$

(b) $\Delta_r H^\ominus = -\Delta_f H^\ominus = 74.85\ \text{kJ mol}^{-1}$

$$\ln K(50\,°\text{C}) = \ln K(298\ \text{K}) - \frac{\Delta_r H^\ominus}{R}\left(\frac{1}{323\ \text{K}} - \frac{1}{298\ \text{K}}\right)\ [6\text{B}.4]$$

$$= -20.508 - \left(\frac{7.4850\times10^4\ \text{J mol}^{-1}}{8.3145\ \text{J K}^{-1}\ \text{mol}^{-1}}\right)\times\left(-2.59\bar{7}\times10^{-4}\right) = -18.17\bar{0}$$

$$K(50\,°\text{C}) = \boxed{1.29\times10^{-8}}$$

(c) Draw up the equilibrium table

	$CH_4(g)$	$H_2(g)$
Amounts	$(1-\alpha)n$	$2\alpha n$
Gas mole fractions	$\frac{(1-\alpha)}{(1+\alpha)}$	$\frac{2\alpha}{(1+\alpha)}$
Partial pressures	$\left(\frac{1-\alpha}{1+\alpha}\right)p$	$\left(\frac{2\alpha}{1+\alpha}\right)p$

$$K = \prod_j a_j^{\nu_j}\ [6\text{A}.13] = \frac{(p_{\text{H}_2}/p^\ominus)^2}{p_{\text{CH}_4}/p^\ominus} = \frac{\left(\left(\frac{2\alpha}{1+\alpha}\right)\frac{p}{p^\ominus}\right)^2}{\left(\frac{1-\alpha}{1+\alpha}\right)\frac{p}{p^\ominus}}$$

$$= \frac{4\alpha^2 p/p^\ominus}{1-\alpha^2} \simeq 4\alpha^2 p/p^\ominus \quad (\text{Since } K \ll 1,\ \text{we expect } \alpha \ll 1.)$$

$$\alpha = \left(\frac{K}{4p/p^\ominus}\right)^{1/2} = \left(\frac{1.24\times10^{-9}}{4\times(0.010)}\right)^{1/2} = \boxed{1.8\times10^{-4}}$$

(d) Le Chatelier's principle provides the answers. As pressure increases, α decreases, since the more compact state (less moles of gas) is favoured at high pressures. As temperature increases the side of the reaction which can absorb heat is favoured. Since $\Delta_r H^\ominus$ is positive, that is the right-hand side, hence α increases. This can also be seen from the results of parts (a) and (b), K increased from 25°C to 50°C, implying that α increased.

P6B.3 $CO_2(g) \rightleftharpoons CO(g) + \frac{1}{2}\,O_2(g)$

Draw up the following equilibrium table and recognize that, since $\alpha \ll 1$, α may be neglected when compared to 1 within mole fraction factors.

	CO_2	CO	O_2
Amounts	$(1-\alpha)n$	αn	$\frac{1}{2}\alpha n$
Mole fractions	$\frac{1-\alpha}{1+\frac{1}{2}\alpha}$	$\frac{\alpha}{1+\frac{1}{2}\alpha}$	$\frac{\frac{1}{2}\alpha}{1+\frac{1}{2}\alpha}$
Approximate mole fractions	1	α	$\frac{1}{2}\alpha$

$$K=\prod_J a_J^{\nu_J}\ [6A.13]=\frac{(p_{CO}/p^{\ominus})(p_{O_2}/p^{\ominus})^{1/2}}{p_{CO_2}/p^{\ominus}}\ [\text{perfect gases}]$$

$$=\frac{x_{CO}x_{O_2}^{1/2}}{x_{CO_2}}(p/p^{\ominus})^{1/2}\ [p_J=x_J p]$$

$$=\frac{x_{CO}x_{O_2}^{1/2}}{x_{CO_2}}\ [p=p^{\ominus}]$$

$$\simeq\alpha(\tfrac{1}{2}\alpha)^{1/2}$$

$$\simeq(\tfrac{1}{2}\alpha^3)^{1/2} \qquad \text{(i)}$$

$$\Delta_r G^{\ominus}=-RT\ln K\ [6A.14] \qquad \text{(ii)}$$

The calculated values of K and $\Delta_r G^{\ominus}$ are given in the table below. From any two pairs of K and T, may be calculated.

$$\ln K_2=\ln K_1-\frac{\Delta_r H^{\ominus}}{R}\left(\frac{1}{T_2}-\frac{1}{T_1}\right)\ [6B.4]$$

Solving for $\Delta_r H^{\ominus}$.

$$\Delta_r H^{\ominus}=\frac{R\ln\left(\frac{K_2}{K_1}\right)}{\left(\frac{1}{T_1}-\frac{1}{T_2}\right)}=\frac{(8.3145\ \text{J K}^{-1}\ \text{mol}^{-1})\times\ln\left(\frac{7.23\times10^{-6}}{1.22\times10^{-6}}\right)}{\left(\frac{1}{1\,395\text{K}}-\frac{1}{1\,498\text{K}}\right)}=\boxed{300.\ \text{kJ mol}^{-1}}$$

$$\Delta_r S^{\ominus}=\frac{\Delta_r H^{\ominus}-\Delta_r G^{\ominus}}{T} \qquad \text{(iii)}$$

K, $\Delta_r G^{\ominus}$, and $\Delta_r S^{\ominus}$ are calculated with equations (i), (ii), and (iii).

T/K	1 395	1 443	1 498
$\alpha/10^{-4}$	1.44	2.50	4.71
$K/10^{-6}$	1.22	2.80	7.23
$\Delta_r G^{\ominus}$/kJ mol^{-1}	158	153	147
$\Delta_r S^{\ominus}$/J K^{-1} mol^{-1}	102	102	102

Comment. $\Delta_r S^{\ominus}$ is essentially constant over this temperature range but it is much different from its value at 25 °C. $\Delta_r H^{\ominus}$, however, is only slightly different.

Question. What are the values of $\Delta_r H^{\ominus}$ and $\Delta_r S^{\ominus}$ at 25 °C for this reaction?

P6B.5 The equilibrium we need to consider is $A_2(g) \rightleftharpoons 2\,A(g)$ where A is acetic acid. It is convenient to express the equilibrium constant in terms of α, the degree of dissociation of the dimer, which is the predominant species at low temperatures. Consequently, we let n represent the amount of dimer present should all the contained mass be dimerized. Thus, $n = m_{\text{total}} / M_{A_2}$ with $M_{A_2} = 120.1\ \text{g mol}^{-1}$. We take the experimental masses to be: $m_{\text{total}} = 0.0463$ g at 437 K and $m_{\text{total}} = 0.0379$ g at 471 K.

	A	A_2	Total
At equilibrium	$2\alpha n$	$(1-\alpha)n$	$(1+\alpha)n$
Mole fraction	$\dfrac{2\alpha}{1+\alpha}$	$\dfrac{1-\alpha}{1+\alpha}$	1
Partial pressure	$\dfrac{2\alpha p}{1+\alpha}$	$\left(\dfrac{1-\alpha}{1+\alpha}\right)p$	p

The equilibrium constant for the dissociation is

$$K = \frac{(p_A / p^{\ominus})^2}{p_{A_2} / p^{\ominus}} = \frac{p_A^2}{p_{A_2} p^{\ominus}} = \frac{4\alpha^2 (p/p^{\ominus})}{1-\alpha^2}$$

We also know that

$$pV = n_{\text{total}} RT = (1+\alpha)nRT, \text{ implying that } \alpha = \frac{pV}{nRT} - 1 = \frac{pVM_{A_2}}{m_{\text{total}} RT} - 1$$

In the T=437 K experiment,

$$\alpha_{437\,\text{K}} = \frac{pVM_{A_2}}{m_{\text{total}} RT} - 1 = \frac{(101.9\ \text{kPa}) \times (21.45 \times 10^{-3}\ \text{dm}^3) \times (120.1\ \text{g mol}^{-1})}{(0.0519\ \text{g}) \times (8.3145\ \text{kPa dm}^3\ \text{K}^{-1}\ \text{mol}^{-1}) \times (437\ \text{K})} - 1 = 0.392$$

$$K_{437\,\text{K}} = \frac{4 \times (0.392)^2 \times (101.9\ \text{kPa}/100.0\ \text{kPa})}{1-(0.392)^2} = \boxed{0.740} \quad [p^{\ominus} = 1\ \text{bar} = 100.0\ \text{kPa}]$$

In the T=471 K experiment,

$$\alpha_{471\,\text{K}} = \frac{pVM_{A_2}}{m_{\text{total}} RT} - 1 = \frac{(101.9\ \text{kPa}) \times (21.45 \times 10^{-3}\ \text{dm}^3) \times (120.1\ \text{g mol}^{-1})}{(0.0380\ \text{g}) \times (8.3145\ \text{kPa dm}^3\ \text{K}^{-1}\ \text{mol}^{-1}) \times (471\ \text{K})} - 1 = 0.764$$

$$K_{471\,\text{K}} = \frac{4 \times (0.764)^2 \times (101.9\ \text{kPa}/100.0\ \text{kPa})}{1-(0.764)^2} = \boxed{5.71} \quad [p^{\ominus} = 1\ \text{bar} = 100.0\ \text{kPa}]$$

The enthalpy of dissociation is

$$\Delta_r H^{\ominus} = \frac{R \ln(K_2/K_1)}{\left(\dfrac{1}{T_1} - \dfrac{1}{T_2}\right)} \text{[6B.4]} = \frac{(8.3145\ \text{J mol}^{-1}\ \text{K}^{-1}) \ln(5.71/0.740)}{\left(\dfrac{1}{437\ \text{K}} - \dfrac{1}{471\ \text{K}}\right)} = +103\ \text{kJ mol}^{-1}$$

The enthalpy of dimerization is the negative of this value, or $\boxed{-103\ \text{kJ mol}^{-1}}$ (i.e. per mole of dimer) and the equilibrium constant for the dimerization is the inverse of the dissociation constants calculated above.

P6B.7 The reaction is $\text{Si(s)} + \text{H}_2\text{(g)} \rightleftharpoons \text{SiH}_2\text{(g)}$.

The equilibrium constant is

$$K = \exp\left(-\Delta_f G^\ominus / RT\right)\ [6A.14] = \exp\left(-\Delta_f H^\ominus / RT\right)\exp\left(\Delta_f S^\ominus / RT\right)$$

Let h be the uncertainty in $\Delta_f H^\ominus$ so that the high value is the low value plus h. The K based on the low value is

$$K_{\text{low }H} = \exp\left(\frac{-\Delta_f H^\ominus_{\text{low}}}{RT}\right)\exp\left(\frac{\Delta_f S^\ominus}{R}\right) = \exp\left(\frac{-\Delta_f H^\ominus_{\text{high}}}{RT}\right)\exp\left(\frac{h}{RT}\right)\exp\left(\frac{\Delta_f S^\ominus}{R}\right)$$

$$= \exp\left(\frac{h}{RT}\right)K_{\text{high }H}.$$

So, $$\frac{K_{\text{low }H}}{K_{\text{high }H}} = \exp\left(\frac{h}{RT}\right)$$

(a) At 298 K, $$\frac{K_{\text{low }H}}{K_{\text{high }H}} = \exp\left(\frac{(289-243)\ \text{kJ mol}^{-1}}{(8.3145\times10^{-3}\ \text{kJ K}^{-1}\ \text{mol}^{-1})\times(298\ \text{K})}\right) = \boxed{1.2\times10^{8}}.$$

(b) At 700 K, $$\frac{K_{\text{low }H}}{K_{\text{high }H}} = \exp\left(\frac{(289-243)\ \text{kJ mol}^{-1}}{(8.3145\times10^{-3}\ \text{kJ K}^{-1}\ \text{mol}^{-1})\times(700\ \text{K})}\right) = \boxed{2.7\times10^{3}}.$$

P6B.9 $a\text{A(g)} + b\text{B(g)} \rightleftharpoons c\text{C(g)} + d\text{D(g)}$ For this gas phase reaction $\Delta\nu = \sum_{\text{J}} \nu_{\text{J}} = c + d - a - b$

$$K = \left(\prod_{\text{J}} a_{\text{J}}^{\nu_{\text{J}}}\right)_{\text{equilibrium}} [6A.13] = \left(\prod_{\text{J}} (\gamma_{\text{J}} p_{\text{J}} / p^\ominus)^{\nu_{\text{J}}}\right)_{\text{equilibrium}} = \left(\prod_{\text{J}} \gamma_{\text{J}}^{\nu_{\text{J}}}\right)_{\text{equilibrium}} \left(\prod_{\text{J}} (p_{\text{J}} / p^\ominus)^{\nu_{\text{J}}}\right)_{\text{equilibrium}}$$

$$= K_\gamma K_p \quad \text{where} \quad K_\gamma = \left(\prod_{\text{J}} \gamma_{\text{J}}^{\nu_{\text{J}}}\right)_{\text{equilibrium}} \quad \text{and } K_p = \left(\prod_{\text{J}} (p_{\text{J}} / p^\ominus)^{\nu_{\text{J}}}\right)_{\text{equilibrium}}$$

Let us assume that the gases are perfect gases. Then, the activity coefficients equal 1 and $K_\gamma = 1$. Additionally, $p_{\text{J}} = n_{\text{J}} RT / V = [\text{J}]RT$. Substitution gives

$$K = K_p = \left(\prod_{\text{J}} ([\text{J}]RT / p^\ominus)^{\nu_{\text{J}}}\right)_{\text{equilibrium}} = \left(\prod_{\text{J}} ([\text{J}]/c^\ominus)^{\nu_{\text{J}}}\right)_{\text{equilibrium}} \left(\prod_{\text{J}} (c^\ominus RT / p^\ominus)^{\nu_{\text{J}}}\right)_{\text{equilibrium}}$$

$$= K_c \left(c^\ominus RT / p^\ominus\right)^{\Delta\nu} \quad \text{where} \quad K_c = \left(\prod_{\text{J}} ([\text{J}]/c^\ominus)^{\nu_{\text{J}}}\right)_{\text{equilibrium}}$$

Taking the natural logarithm, differentiating with respect to T, and application of the van 't Hoff equation yield a workable equation for finding the temperature dependence of K_c.

$$\ln(K) = \ln(K_c) + \Delta\nu \ln(c^{\ominus}RT/p^{\ominus})$$

$$\frac{\mathrm{d}\ln(K_c)}{\mathrm{d}T} = \frac{\mathrm{d}\ln(K)}{\mathrm{d}T} - \frac{\Delta\nu}{T}$$

$$= \frac{\Delta_r H^{\ominus}}{RT^2} - \frac{\Delta\nu}{T} \quad \text{[6B.2(a), the van 't Hoff equation]}$$

We now assume that $\Delta_r H^{\ominus}$ is constant in the temperature range of interest and integrate between a reference temperature T_{ref} and the temperature T of interest to find that

$$\ln\left(\frac{K_c(T)}{K_c(T_{ref})}\right) = \frac{\Delta_r H^{\ominus}}{R} \times \left(\frac{1}{T_{ref}} - \frac{1}{T}\right) - \Delta\nu \times \ln\left(\frac{T}{T_{ref}}\right)$$

Exponentiation yields the desired relation.

$$\boxed{K_c(T) = K_c(T_{ref}) \times \left(\frac{T_{ref}}{T}\right)^{\Delta\nu} \mathrm{e}^{\frac{\Delta_r H^{\ominus}}{R} \times \left(\frac{1}{T_{ref}} - \frac{1}{T}\right)}} \quad \text{where } K_c(T_{ref}) = K(T_{ref}) \times \left(\frac{p^{\ominus}}{c^{\ominus}RT_{ref}}\right)^{\Delta\nu}$$

6C Electrochemical cells

Answers to discussion questions

D6C.1 On a very basic level we observe that a concentration gradient establishes a chemical potential gradient and it is the potential gradient that can generate an electric current. The extreme example is provided by the **electrolyte concentration cell**, which is by definition a galvanic cell consisting of two electrodes of the same metal in different concentration of the same salt of that metal. The concentration cell having two M^+ (aq)/M electrodes is $M(s)|M^+(aq,L)||M^+(aq,R)|M(s)$ and the net reaction is $M^+(aq,R) \rightarrow M^+(aq,L)$. For such a cell $E^{\ominus}_{cell} = 0$ and the Nernst equation gives the cell potential:

$$E_{cell} = E^{\ominus}_{cell} - \frac{RT}{\nu F}\ln Q \quad [6C.4]$$

$$= -\frac{RT}{F}\ln\frac{a_{M^+(L)}}{a_{M^+(R)}}$$

With $[M^+(R)] = 10 \times [M^+(L)]$ and an assumption that $\gamma_{M^+(L)} \simeq \gamma_{M^+(R)}$:

$$E_{cell} = -(25.693\ \text{mV}) \times \left(\ln\frac{1}{10}\right) = 59.16\ \text{mV}$$

The cell potential is half as large for the $M(s)|M^{2+}(aq,L)||M^{2+}(aq,R)|M(s)$ cell.

D6C.3 In electrochemistry, the non-expansion work is identified with maximum electrical work, the Gibbs energy of the cell reaction $\Delta_r G$, and the definition of the cell potential E_{cell} through the relations $w_{e,max} = \Delta_r G$ [3C16b] $= -\nu FE_{cell}$ [6C.2] at constant temperature and pressure. Maximum work is produced when a change occurs reversibly. It follows that, to draw thermodynamic conclusions from measurements of the work that a cell can do, we must ensure that the cell is operating reversibly. Furthermore, the relations require that the cell is operating at a specific, constant composition. The conditions of reversibility and constant composition are achieved by measuring the cell potential when it is balanced by an exactly opposing external potential so that there is no current flow, the zero-current condition. In effect, the cell reaction is poised for change, but not actually changing.

Solutions to exercises

E6C.1(a) The cell notation specifies the right and left electrodes. Note that for proper cancellation we must equalize the number of electrons in half-reactions being combined. To calculate the standard cell potential we have used $E^{\ominus}_{cell} = E^{\ominus}_R - E^{\ominus}_L$, with standard electrode potentials from the data table.

		$E^{\ominus}$
(i)	R: $2\,Ag^+(aq) + 2\,e^- \rightarrow 2\,Ag(s)$	+0.80 V
	L: $Zn^{2+}(aq) + 2\,e^- \rightarrow Zn(s)$	−0.76 V
	Overall (R − L): $2\,Ag^+(aq) + Zn(s) \rightarrow 2\,Ag(s) + Zn^{2+}(aq)$	+1.56 V
(ii)	R: $2\,H^+(aq) + 2\,e^- \rightarrow H_2(g)$	0
	L: $Cd^{2+}(aq) + 2\,e^- \rightarrow Cd(s)$	−0.40 V
	Overall (R − L): $Cd(s) + 2\,H^+(aq) \rightarrow Cd^{2+}(aq) + H_2(g)$	+0.40 V
(iii)	R: $Cr^{3+}(aq) + 3\,e^- \rightarrow Cr(s)$	−0.74 V
	L: $3\,[Fe(CN)_6]^{3-}(aq) + 3\,e^- \rightarrow 3\,[Fe(CN)_6]^{4-}(aq)$	+0.36 V
	Overall (R − L): $Cr^{3+}(aq) + 3\,[Fe(CN)_6]^{4-}(aq) \rightarrow Cr(s) + 3\,[Fe(CN)_6]^{3-}(aq)$	−1.10 V

Comment. Those cells for which $E^{\ominus}_{cell} > 0$ may operate as spontaneous galvanic cells under standard conditions. Those for which $E^{\ominus}_{cell} < 0$ may operate as non-spontaneous electrolytic cells. Recall that $E^{\ominus}_{cell}$ informs us of the spontaneity of a cell under standard conditions only. For other conditions we require E_{cell}.

E6C.2(a) The conditions (concentrations, etc.) under which these reactions occur are not given. For the purposes of this exercise we assume standard conditions. The specification of the right and left electrodes is determined by the direction of the reaction as written. As always, in combining half-reactions to form an overall cell reaction we must write the half-reactions with equal number of electrons to ensure proper cancellation. We first identify the half-reactions, and then set up the corresponding cell.

		$E^\ominus$
(i)	R: $2\,Cu^{2+}(aq) + 2\,e^- \rightarrow Cu(s)$	+0.34 V
	L: $Zn^{2+}(aq) + 2\,e^- \rightarrow Zn(s)$	−0.76 V
	Hence the cell is	
	$Zn(s)\,|\,ZnSo_4(aq)\,\|\,CuSO_4(aq)\,|\,Cu(s)$	$\boxed{+1.10\text{ V}}$
(ii)	R: $AgCl(s) + e^- \rightarrow Ag(s) + Cl^-(aq)$	+0.22 V
	L: $H^+(aq) + e^- \rightarrow \tfrac{1}{2}H_2(g)$	0
	and the cell is	
	$Pt|H_2(g)|H^+(aq)|AgCl(s)|Ag(s)$	
	or $Pt|H_2(g)|HCl(aq)|AgCl(s)|Ag(s)$	$\boxed{+0.22\text{ V}}$
(iii)	R: $O_2(g) + 4\,H^+(aq) + 4\,e^- \rightarrow 2\,H_2O(l)$	+1.23 V
	L: $4\,H^+(aq) + 4\,e^- \rightarrow 2\,H_2(g)$	0
	and the cell is	
	$Pt|H_2(g)|H^+(aq)\,|\,O_2(g)\,|\,Pt$	$\boxed{+1.23\text{ V}}$

Comment. All of these cells have $E^\ominus_{\text{cell}} > 0$, corresponding to a spontaneous cell reaction under standard conditions. If $E^\ominus_{\text{cell}}$ had turned out to be negative, the spontaneous reaction would have been the reverse of the one given, with the right and left electrodes of the cell also reversed.

E6C.3(a) $Ag|AgBr(s)|KBr(aq,0.050\text{ mol kg}^{-1})||Cd(NO_3)_2(aq,0.010\text{ mol kg}^{-1})|Cd(s)$

(i) R: $Cd^{2+}(aq) + 2\,e^- \rightarrow Cd(s)$ $\quad E^\ominus = -0.40\text{ V}$

L: $2\,AgBr(s) + 2\,e^- \rightarrow 2Ag(s) + 2\,Br^-(aq)$ $\quad E^\ominus = +0.0713\text{ V}$

R − L: $\boxed{Cd^{2+}(aq) + 2Br^-(aq) + 2Ag(s) \rightarrow Cd(s) + 2AgBr(s)}$ $\quad E^\ominus_{\text{cell}} = -0.47\text{ V}$

The cell reaction is not spontaneous toward the right under standard conditions because $E^\ominus_{\text{cell}} < 0$.

(ii) The Nernst equation for the above cell reaction is:

$$E_{\text{cell}} = E^\ominus_{\text{cell}} - \frac{RT}{\nu F}\ln Q \text{ [6C.4]} \quad \text{where} \quad \nu = 2$$

$$Q = \frac{1}{a_{Cd^{2+}}a^2_{Br^-}} = \frac{1}{\gamma_{Cd^{2+}}\gamma^2_{Br^-}} \times \frac{(b^\ominus)^3}{b_{Cd^{2+}}b^2_{Br^-}} = \frac{1}{\gamma_{\pm,R}\gamma^2_{\pm,L}} \times \frac{(b^\ominus)^3}{b_{Cd^{2+}}b^2_{Br^-}}$$

$b_{Cd^{2+}} = 0.010\text{ mol kg}^{-1}$ for the right-hand electrode and $b_{Br^-} = 0.050\text{ mol kg}^{-1}$ for the left-hand electrode.

(iii) The ionic strength and mean activity coefficient at the right-hand electrode are:

$$I_R = \tfrac{1}{2}\sum_i z_i^2 (b_i / b^{\ominus}) \text{ [5F.9]} = \tfrac{1}{2}\{4(0.010)+1(.020)\} = 0.030$$

$$\log\gamma_{\pm,R} = -|z_+z_-|AI^{1/2} \text{ [5F.8]} = -2\times(0.509)\times(0.030)^{1/2} = -0.176$$

$$\gamma_{\pm,R} = 0.667$$

The ionic strength and mean acitivity coefficient at the left-hand electrode are:

$$I_L = \tfrac{1}{2}\sum_i z_i^2 (b_i / b^{\ominus}) \text{ [5F.9]} = \tfrac{1}{2}\{1(0.050)+1(0.050)\} = 0.050$$

$$\log\gamma_{\pm,L} = -|z_+z_-|AI^{1/2} \text{ [5F.8]} = -1\times(0.509)\times(0.050)^{1/2} = -0.114$$

$$\gamma_{\pm,L} = 0.769$$

Therefore,

$$Q = \left(\frac{1}{(0.667)\times(0.769)^2}\right)\times\left(\frac{1}{(0.010)\times(0.050)^2}\right) = 1.01\times10^5$$

and

$$E_{\text{cell}} = -0.47\text{ V} - \left(\frac{25.693\times10^{-3}\text{ V}}{2}\right)\ln(1.01\times10^5)$$

$$= \boxed{-0.62\text{ V}}$$

Solutions to problems

P6C.1 (a) The cell reaction is

$$H_2(g) + \tfrac{1}{2}O_2(g) \rightarrow H_2O(l)$$

$$\Delta_r G^{\ominus} = \Delta_f G^{\ominus}(H_2O, l) = -237.13\text{ kJ mol}^{-1} \text{ [data tables]}$$

$$E^{\ominus}_{\text{cell}} = -\frac{\Delta_r G^{\ominus}}{\nu F}\text{[6C.2]} = \frac{+237.13\text{ kJ mol}^{-1}}{(2)\times(96.485\text{ kC mol}^{-1})} = \boxed{+1.23\text{ V}}$$

(b) $C_4H_{10}(g) + \tfrac{13}{2}O_2(g) \rightarrow 4\,CO_2(g) + 5\,H_2O(l)$

$$\Delta_f G^{\ominus} = 4\Delta_f G^{\ominus}(CO_2, g) + 5\Delta_f G^{\ominus}(H_2O, l) - \Delta_f G^{\ominus}(C_4H_{10}, g)$$

$$= [(4)\times(-394.36)+(5)\times(-237.13)-(-17.03)]\text{ kJ mol}^{-1} \text{ [data tables]}$$

$$= -2746.06\text{ kJ mol}^{-1}$$

In this reaction, the number of electrons transferred, ν, is not immediately apparent as in part (a). To find ν we break the cell reaction down into half-reactions as follows

$$\text{R: } \tfrac{13}{2}\,O_2(g)+26\,e^-+26\,H^+(aq)\rightarrow 13\,H_2O(l)$$
$$\text{L: } 4\,CO_2(g)+26\,e^-+26\,H^+(aq)\rightarrow C_4H_{10}(g)+8\,H_2O(1)$$
$$\text{R}-\text{L: } C_4H_{10}(g)+\tfrac{13}{2}\,O_2(g)\rightarrow 4\,CO_2(g)+5\,H_2O(l)$$

Hence, $v=26$.

Therefore, $E=\dfrac{-\Delta G^{\ominus}}{vF}=\dfrac{+2\,746.06\text{ kJ mol}^{-1}}{(26)\times(96.485\text{ kC mol}^{-1})}=\boxed{+1.09\text{ V}}$

P6C.3 $\log(s_A/s^{\ominus})=-\left(a+\dfrac{b}{T/\text{K}}\right)$

where $a=5.39$ and $b=768$ for H_2 while $a=5.98$ and $b=980$ for CO

Let 'A' represent either H_2 or CO. The equilibrium to be considered is

$$\text{A(g,1bar)}\rightleftharpoons\text{A(soln)}\quad K=\frac{(c/c^{\ominus})}{(p/p^{\ominus})}=\frac{s}{s^{\ominus}}=10^{-(a+b\times\text{K}/T)}\quad\text{where } s^{\ominus}=1\text{ mol cm}^{-3}\text{ bar}^{-1}$$

$$\begin{aligned}\Delta_r H^{\ominus}&=-R\times\frac{\mathrm{d}\ln K}{\mathrm{d}(1/T)}\ [6B.2(b)]=-R\ln(10)\times\frac{\mathrm{d}\log(K)}{\mathrm{d}(1/T)}\\&=-R\ln(10)\times\frac{\mathrm{d}\log(s/s^{\ominus})}{\mathrm{d}(1/T)}\\&=-R\ln(10)\times\frac{\mathrm{d}(-(a+b\times\text{K}/T))}{\mathrm{d}(1/T)}\\&=R\ln(10)\times b\times\text{K}\end{aligned}$$

$$\begin{aligned}\Delta_r H^{\ominus}(H_2)&=(8.3145\times10^{-3}\text{ kJ mol}^{-1}\text{ K}^{-1})\times\ln(10)\times(768\text{ K})\\&=\boxed{+14.7\text{ kJ mol}^{-1}}\end{aligned}$$

$$\begin{aligned}\Delta_r H^{\ominus}(\text{CO})&=(8.3145\times10^{-3}\text{ kJ mol}^{-1}\text{ K}^{-1})\times\ln(10)\times(980\text{ K})\\&=\boxed{+18.8\text{ kJ mol}^{-1}}\end{aligned}$$

6D Electrode potentials

Answer to discussion question

D6D.1 Construct a cell using a standard hydrogen electrode and an electrode designed around the redox couple of interest. The cell potential E is measured with a high impedance voltmeter under zero current conditions. When using SHE as a reference electrode, E_{cell} is the desired half-reaction potential [6D.1 and 6D.3]. Should the redox couple have one or more electroactive species (i) that are solvated with concentration b_i, E_{cell} must be measured over a range of b_i values.

The Nernst equation [6C.4], with Q being the cell reaction quotient, is the starting point for analysis of the $E_{cell}(b_i)$ data.

$$E_{cell} = E^{\ominus}_{cell} - \frac{RT}{\nu F}\ln Q \quad [6C.4]$$

It would seem that substitution of E_{cell} and Q values would allow the computation of the standard redox potential $E^{\ominus}_{cell}$ for the couple. However, a problem arises because the calculation of Q requires not only knowledge of the concentrations of the species involved in the cell reaction but also of their activity coefficients. These coefficients are not usually available, so the calculation cannot be directly completed. However, at very low concentrations, the Debye–Hückel limiting law for the coefficients holds. The procedure then is to substitute the Debye–Hückel law for the activity coefficients into the specific form of the Nernst equation for the cell under investigation and carefully examine the equation to determine what kind of plot to make of the $E_{cell}(b_i)$ data so that extrapolation of the plot to zero concentration, where the Debye–Hückel law is valid, gives a plot intercept that equals $E^{\ominus}_{cell}$. See Section 6D.1(a) for the details of this procedure and an example for which the relevant graph involves a plot of $E_{cell} + (2RT/F)\ln b$ against $b^{1/2}$ (see eqn 6D.4).

Solutions to exercises

E6D.1(a) In each case the equilibrium constant is calculated with the expression $\ln K = \frac{\nu F E^{\ominus}_{cell}}{RT}$ [6C.5].

(i) $Sn(s) + Sn^{4+}(aq) \rightleftharpoons 2\ Sn^{2+}(aq)$

R: $Sn^{4+} + 2\ e^- \rightarrow Sn^{2+}(aq)$ $+0.15$ V

L: $Sn^{2+}(aq) + 2\ e^- \rightarrow Sn(s)$ -0.14 V

$E^{\ominus}_{cell} = +0.29$ V

$$\ln K = \frac{\nu F E^{\ominus}_{cell}}{RT}\ [6C.5] = \frac{(2)\times(0.29\ \text{V})}{25.693\ \text{mV}} = 22.\bar{6}$$

$$K = e^{22.\bar{6}} = \boxed{6.5\times10^{9}}$$

(ii) $Sn(s) + 2\ AgCl(s) \rightleftharpoons SnCl_2(aq) + 2\ Ag(s)$

R: $2\ AgCl(s) + 2\ e^- \rightarrow 2\ Ag(s) + 2Cl^-(aq)$ $+0.22$ V

L: $Sn^{2+}(aq) + 2\ e^- \rightarrow Sn(s)$ -0.14 V

$E^{\ominus}_{cell} = +0.36$ V

$$\ln K = \frac{\nu F E^{\ominus}_{cell}}{RT}\ [6C.5] = \frac{(2)\times(0.36\ \text{V})}{25.693\ \text{mV}} = +2\bar{8}.0$$

$$K = e^{2\bar{8}.0} = \boxed{1.4\times10^{12}}$$

E6D.2(a) Ag|AgI(s)|AgI(aq)|Ag $E^{\ominus}$

R: $Ag^+(aq) + e^- \rightarrow Ag(s)$ +0.7996 V

L: $AgI(s) + e^- \rightarrow Ag(s) + I^-(aq)$ −0.1513 V

Overall: $Ag^+(aq) + I^-(aq) + AgI(s)$ +0.9509 V $\nu = 1$

(i) $\ln K = \dfrac{\nu F E^{\ominus}_{\text{cell}}}{RT}$ [6C.5] $= \dfrac{1\times(+0.9509\ \text{V})}{25.693\ \text{mV}} = 37.01$

$K = e^{37.01} = 1.18\times10^{16}$

The solubility equilibrium is written as the reverse of the cell reaction. Therefore, the solubility product of AgI(s) is $K_{\text{sp}} = K^{-1} = 1/1.18\times10^{16} = \boxed{8.47\times10^{-17}}$.

(ii) The solubility product of AgI(s) is very small. Consequently, the molar solubility, *s*, of silver iodide must also be very low and we can reasonably take the activity coefficients of the aqueous ions to equal 1.

$$K_{\text{sp}} = [Ag^+][I^-]/(c^{\ominus})^2 = s^2/(c^{\ominus})^2$$

$$s = K_{\text{sp}}^{1/2}c^{\ominus} = (8.47\times10^{-17})^{1/2}\ \text{mol dm}^{-3} = \boxed{9.20\times10^{-9}\ \text{mol dm}^{-3}\ \text{or}\ 2.16\ \mu\text{g dm}^{-3}}$$

Solution to problem

P6D.1 $Pt|H_2(g,p^{\ominus})|HCl(aq,b)|Hg_2Cl_2(s)|Hg(l)$

$\tfrac{1}{2}Hg_2Cl_2(s) + \tfrac{1}{2}H_2(g) \rightarrow Hg(l) + HCl(aq)$ and $\nu = 1$

$$E_{\text{cell}} = E^{\ominus}_{\text{cell}} - \frac{RT}{F}\ln a_{H^+}a_{Cl^-}\ \ [\text{6C.4},\ p_{H_2(g)} = p^{\ominus},\ \text{activities of pure liquids and solids equal 1}]$$

$$a_{H^+} = \gamma_+ b_+ = \gamma_+ b;\quad a_{Cl^-} = \gamma_- b_- = \gamma_- b\quad [b = b/b^{\ominus}\ \text{here and below}]$$

$$a(H^+)a(Cl^-) = \gamma_+\gamma_- b^2 = \gamma_\pm^2 b^2\ \ [\text{5F.3}]$$

$$E_{\text{cell}} = E^{\ominus}_{\text{cell}} - \frac{2RT}{F}\ln b - \frac{2RT}{F}\ln\gamma_\pm \qquad \text{(a)}$$

Converting from natural logarithms to common logarithms (base 10) in order to introduce the Debye–Hückel expression, we obtain

$$\begin{aligned} E_{\text{cell}} &= E^{\ominus}_{\text{cell}} - \frac{(2.303)\times2RT}{F}\log b - \frac{(2.303)\times2RT}{F}\log\gamma_\pm \\ &= E^{\ominus}_{\text{cell}} - (0.1183\ \text{V})\log b - (0.1183\ \text{V})\log\gamma_\pm \\ &= E^{\ominus}_{\text{cell}} - (0.1183\ \text{V})\log b - (0.1183\ \text{V})\times(-|z_+z_-|AI^{1/2})\ \ [\text{5F.8}] \\ &= E^{\ominus}_{\text{cell}} - (0.1183\ \text{V})\log b + (0.1183\ \text{V})\times A\times b^{1/2}\ \ [\text{5F.9},\ I = b] \end{aligned}$$

Rearranging,

$$E_{\text{cell}} + (0.1183\ \text{V})\log b = E^{\ominus}_{\text{cell}} + \text{constant} \times b^{1/2}$$

Therefore, a plot of E_{cell} + (0.1183 V) log b against $b^{1/2}$ should be linear and the intercept at $b=0$ is $E^{\ominus}_{\text{cell}}$. Draw up the following table.

b/(mmol kg^{-1})	1.6077	3.0769	5.0403	7.6938	10.9474
E_{cell}/V	0.60080	0.56825	0.54366	0.52267	0.50532
$(b/b^{\ominus})^{1/2}$	0.04010	0.05547	0.07100	0.08771	0.1046
$E_{\text{cell}}/\text{V} + 0.1183 \times \log b$	0.27029	0.27109	0.27186	0.27260	0.27337

The linear least-squares fit of the plot gives a slope of 0.047533 V mol$^{-1/2}$ kg$^{1/2}$ and an intercept of 0.26843 V with $R^2=0.998856$, which indicates a satisfactory linear relation. Thus, $E^{\ominus}_{\text{cell}} = \boxed{+0.26843\ \text{V}}$.

For the activity coefficients we obtain from equation (a) above:

$$\ln\gamma_{\pm} = \frac{E^{\ominus}_{\text{cell}} - E_{\text{cell}}}{2RT/F} - \ln\frac{b}{b^{\ominus}} = \frac{0.26843 - E_{\text{cell}}/V}{0.05139} - \ln\frac{b}{b^{\ominus}}$$

and we draw up the following table, which summarizes calculated values of mean activity coefficients.

b/(mmol kg^{-1})	1.6077	3.0769	5.0403	7.6938	10.9474
E_{cell}/V	0.60080	0.56825	0.54366	0.52267	0.50532
$\ln\gamma_{\pm}$	−0.03465	−0.05038	−0.06542	−0.07993	−0.09500
$\gamma_{\pm}$	0.9659	0.9509	0.9367	0.9232	0.9094

Integrated activities

I6.1 If we knew $\Delta_r H^{\ominus}$ for the reaction $Cl_2O(g) + H_2O(g) \rightarrow 2\,HOCl(g)$, we could calculate $\Delta_f H^{\ominus}$(HOCl) from

$$\Delta_r H^{\ominus} = 2\,\Delta_f H^{\ominus}(\text{HOCl,g}) - \Delta_f H^{\ominus}(\text{Cl}_2\text{O,g}) - \Delta_f H^{\ominus}(\text{H}_2\text{O,g})$$

We can find $\Delta_r H^{\ominus}$ if we know $\Delta_r G^{\ominus}$ and $\Delta_r S^{\ominus}$, since

$$\Delta_r G^{\ominus} = \Delta_r H - T\Delta_r S$$

And we can find $\Delta_r G^{\ominus}$ from the equilibrium constant.

$$\begin{aligned}\Delta_r G^{\ominus} &= -RT\ln K \\ &= -(8.3145\times10^{-3}\ \text{kJ K}^{-1}\ \text{mol}^{-1})\times(298\ \text{K})\ln\left(8.2\times10^{-2}\right) \\ &= 6.2\ \text{kJ mol}^{-1}\end{aligned}$$

$$\begin{aligned}\Delta_r H^\ominus &= \Delta_r G^\ominus + T\Delta_r S^\ominus \\ &= 6.2\ \text{kJ mol}^{-1} + (298\ \text{K}) \times (16.38 \times 10^{-3}\ \text{kJ K}^{-1}\ \text{mol}^{-1}) \\ &= 11.1\ \text{kJ mol}^{-1}\end{aligned}$$

Finally,

$$\begin{aligned}\Delta_f H^\ominus(\text{HOCl,g}) &= \tfrac{1}{2}\{\Delta_r H^\ominus + \Delta_f H^\ominus(\text{Cl}_2\text{O,g}) + \Delta_f H^\ominus(\text{H}_2\text{O,g})\} \\ &= \tfrac{1}{2}\{11.1 + 77.2 + (-241.82)\}\ \text{kJ mol}^{-1} \\ &= \boxed{-76.8\ \text{kJ mol}^{-1}}\end{aligned}$$

I6.3 $Hg_2Cl_2(s) + Zn(s) \rightarrow 2\,Hg(l) + ZnCl_2(aq)$ and $\nu = 2$

(a)
$$\begin{aligned}Q &= a_{Zn^{2+}} a_{Cl^-}^2 \\ &= \{\gamma_+ b_{Zn^{2+}}/b^\ominus\} \times \gamma_-^2 \{b_{Cl^-}/b^\ominus\}^2 \quad \text{where } b_{Zn^{2+}} = b,\ b_{Cl^-} = 2b, \text{ and } \gamma_+\gamma_-^2 = \gamma_\pm^3 \\ &= 4\gamma_\pm^3 b^3 \quad [b \equiv b/b^\ominus \text{ here and below}]\end{aligned}$$

$$\begin{aligned}E_{\text{cell}} &= E_{\text{cell}}^\ominus - \frac{RT}{\nu F}\ln Q \quad [6\text{C}.4] = E_{\text{cell}}^\ominus - \frac{RT}{\nu F}\ln(4\gamma_\pm^3 b^3) \\ &= \boxed{E_{\text{cell}}^\ominus - \frac{RT}{\nu F}\{\ln(4b^3) + \ln(\gamma_\pm^3)\}}\end{aligned}$$

(b) $E_{\text{cell}}^\ominus = E_{Hg_2^{2+}/Hg}^\ominus - E_{Zn^{2+}/Zn}^\ominus = +0.2676\ \text{V} - (-0.7628\ \text{V}) = \boxed{1.0304\ \text{V}}$

(c) $\Delta_r G = -\nu F E_{\text{cell}} = -(2) \times (9.6485 \times 10^4\ \text{C mol}^{-1}) \times (1.2272\ \text{V}) = \boxed{-236.81\ \text{kJ mol}^{-1}}$

$$\Delta_r G^\ominus = -\nu F E_{\text{cell}}^\ominus = -(2) \times (9.6485 \times 10^4\ \text{C mol}^{-1}) \times (1.0304\ \text{V}) = \boxed{-198.84\ \text{kJ mol}^{-1}}$$

$$\ln K = -\frac{\Delta_r G^\ominus}{RT} = \frac{1.9884 \times 10^5\ \text{J mol}^{-1}}{(8.3145\ \text{J K}^{-1}\ \text{mol}^{-1}) \times (298.15\ \text{K})} = 80.211 \quad \text{so} \quad K = \boxed{6.84 \times 10^{34}}$$

(d) From part (a) at 25 °C:

$$\begin{aligned}\ln(\gamma_\pm^3) &= \frac{\nu F}{RT}(E_{\text{cell}}^\ominus - E_{\text{cell}}) - \ln(4b^3) \\ &= \left(\frac{2}{25.693 \times 10^{-3}\ \text{V}}\right) \times (1.0304 - 1.2272)\text{V} - \ln(4 \times (0.0050)^3) \\ &= -0.8107\end{aligned}$$

$$\gamma_\pm = (e^{-0.8107})^{1/3} = \boxed{0.7632}$$

(e)

$$\begin{aligned}I &= \tfrac{1}{2}\sum_i z_i^2 (b_i/b^\ominus) \quad [5\text{F}.9] \quad \text{where} \quad b_{Zn^{2+}} = b = 0.0050\ \text{mol kg}^{-1},\ b_{Cl^-} = 2b = 0.0100\ \text{mol kg}^{-1} \\ &= \tfrac{1}{2} \times (4 \times 0.0050 + 0.0100) = 0.0150\end{aligned}$$

$$\log \gamma_{\pm} = -|z_- z_+| A I^{1/2} \text{ [5F.8]} = -(2)\times(0.509)\times(0.015)^{1/2} = -0.12\bar{5}$$

$$\gamma_{\pm} = 10^{-0.12\bar{5}} = \boxed{0.75}$$

This compares remarkably well to the value obtained from experimental data in part (d).

(f) $$\Delta_r S = -\left(\frac{\partial \Delta_r G}{\partial T}\right)_p = \nu F\left(\frac{\partial E}{\partial T}\right)_p \text{ [6C.6]}$$

$$= (2)\times(9.6485\times10^4 \text{ C mol}^{-1})\times(-4.52\times10^{-4} \text{ V K}^{-1})$$

$$= \boxed{-87.2 \text{ J K}^{-1} \text{ mol}^{-1}}$$

$$\Delta_r H = \Delta_r G + T\Delta_r S = (-236.81 \text{ kJ mol}^{-1}) + (298.15 \text{ K})\times(-87.2 \text{ J K}^{-1} \text{ mol}^{-1})$$

$$= \boxed{-262.4 \text{ kJ mol}^{-1}}$$

I6.5 $\text{Ag(s)}|\text{AgX(s)}|\text{MX}(b_1)|\text{M}_x\text{Hg(s)}|\text{MX}(b_2)|\text{AgX(s)}|\text{Ag(s)}$ with b_2=0.09141 mol kg^{-1} for all (b_1,E_{cell}) data pairs. We are to use the data pair (b_1,E_{cell})=(1.350 mol kg^{-1}, 0.1336 V) with a given functional form for the mean activity coefficient as a reference for the calculation of the mean activity coefficients for the remainder of the data pairs. The parameters provided for the Davies equation [5F.11(b)] for the mean activity coefficient are to be used for the reference only. The 1:1 salt of the cell solutions means that $|z_+ z_-| = 1$ and $I = b/b^{\ominus}$.

$$\log(\gamma_{\pm,\text{ref}}) = -\frac{A|z_+ z_-| I_{\text{ref}}^{1/2}}{1+BI_{\text{ref}}^{1/2}} + CI_{\text{ref}}$$

$$= -\frac{1.461\times(1.350)^{1/2}}{1+1.70\times(1.350)^{1/2}} + 0.20\times(1.350) = -0.3006$$

$$\gamma_{\pm,\text{ref}} = 10^{-0.3006} = \boxed{0.5005} \quad \text{and} \quad a_{\text{ref}} = \gamma_{\pm,\text{ref}} b_{1,\text{ref}}/b^{\ominus} = 0.5005\times(1.35) = 0.6757$$

This concentration cell consists of two back-to-back cells each of the type $\text{M}_x\text{Hg(s)}|\text{MX}(b)|\text{AgX(s)}|\text{Ag(s)}$. The right-hand cell is described by the following R and L halves.

$$\text{R:}\quad \text{AgX(s)} + \text{e}^- \rightarrow \text{Ag(s)} + \text{X}^-(b_2)$$

$$\text{L:}\quad \text{M}^+(b_2) + \text{e}^- \xrightarrow{\text{Hg}} \text{M}_x\text{Hg(s)} \quad (\text{reduction of M}^+ \text{ and formation of amalgam})$$

$$\text{R}-\text{L:}\quad \text{M}_x\text{Hg(s)} + \text{AgX(s)} \xrightarrow{\text{Hg}} \text{Ag(s)} + \text{M}^+(b_2) + \text{X}^-(b_2) \quad \text{and} \quad \nu = 1$$

Similarly, the left-hand cell is described by R and L halves that give

$$\text{R}-\text{L:}\quad \text{Ag(s)} + \text{M}^+(b_1) + \text{X}^-(b_1) \xrightarrow{\text{Hg}} \text{M}_x\text{Hg(s)} + \text{AgX(s)}$$

Adding the R – L halves gives the net cell reaction:
$M^+(b_1) + X^-(b_1) \rightarrow M^+(b_2) + X^-(b_2)$

The data pair $(b_1, E_{cell}) = (0.09141\ \text{mol kg}^{-1}, 0.0000\ \text{V})$ has $b_1 = b_2$. It is the cell at equilibrium with $Q_{eq} = 1$, because left and right half-cell activities are equal, and $E^{\ominus}_{cell} = 0$, which simplifies the Nernst equation [6C.4].

$$E_{cell} = -\frac{RT}{F}\ln Q = -\frac{RT}{F}\ln\left(\frac{a(M^+(b_2))\times a(X^-(b_2))}{a(M^+(b_1))\times a(X^-(b_1))}\right)$$
$$= -\frac{RT}{F}\ln\left(\frac{\gamma_\pm^2(b_2)\times b_2^2}{\gamma_\pm^2(b_1)\times b_1^2}\right)$$
$$= \frac{2RT\ \ln 10}{F}\log\left(\frac{\gamma_\pm(b_1)\times b_1}{\gamma_\pm(b_2)\times b_2}\right)$$

Evaluating this with the reference data pair gives $\gamma_\pm(b_2)$.

$$E_{cell,ref} = \frac{2RT\ \ln 10}{F}\log\left(\frac{\gamma_\pm(b_{1,ref})\times b_{1,ref}}{\gamma_\pm(b_2)\times b_2}\right)$$

$$\log(\gamma_\pm(b_2)) = \log\left(\frac{\gamma_\pm(b_{1,ref})\times b_{1,ref}}{b_2}\right) - \frac{F}{2RT\ \ln 10}E_{cell,ref}$$
$$= \log\left(\frac{0.6757}{0.09141}\right) - \frac{0.1336\ \text{V}}{2\times(0.059160\ \text{V})} = -0.2604$$

$$\gamma_\pm(b_2) = 10^{-0.2604} = 0.5490 \quad \text{and} \quad a(b_2) = \gamma_\pm(b_2)\times b_2/b^{\ominus}$$
$$= 0.5490\times 0.09141 = 0.05018$$

Have the value of $a(b_2)$, we can return to the working equation for E_{cell} and use it to calculate $\gamma_\pm(b_1)$ at each (b_1, E_{cell}) data pair. First, solve for $\log(\gamma_\pm(b_1))$.

$$\log(\gamma_\pm(b_1)) = \frac{F}{2RT\ \ln 10}E_{cell} - \log\left(\frac{b_1}{\gamma_\pm(b_2)\times b_2}\right)$$
$$= \frac{E_{cell}}{2\times(0.059160\ \text{V})} - \log\left(\frac{b_1/b^{\ominus}}{0.05018}\right)$$

Then draw up the following table.

b_1 / (mol kg^{-1})	0.0555	0.09141	0.1652	0.2171	1.040	1.350
E / V	−0.0220	0.0000	0.0263	0.0379	0.01156	0.1336
$\log(\gamma_\pm)$	−0.2296	−0.2605	−0.2952	−0.3158	−0.3395	−0.3007
$\gamma_\pm$	0.5893	0.5490	0.5068	0.4833	0.4576	0.5005

I6.7 (a) ATP hydrolysis at physiological pH, $ATP(aq)+H_2O(l) \rightarrow ADP(aq)+P_i^-(aq) + H_3O^+(aq)$, converts two reactant moles into three product moles. The increased number of chemical species present in solution increases the disorder of the system by increasing the number of molecular rotational, vibrational, and translational degrees of freedom. This is an effective increase in the number of available molecular states and an increase in entropy.

(b) At physiological pH, the oxygen atoms of ATP are deprotonated, negatively charged, and the molecule is best represented as ATP^{4-}. The electrostatic repulsions between the highly charged oxygen atoms of ATP^{4-} is expected to give it an exergonic hydrolysis free energy by making the hydrolysis enthalpy negative. Also, the deprotonated phosphate species, $P_i(aq)$, produced in the hydrolysis ATP has more resonance structures than ATP^{4-}. Resonance lowers the energy of the dissociated phosphate making the hydrolysis enthalpy more negative and contributing to the exergonicity of the hydrolysis.

The electrostatic repulsion between the highly charged oxygen atoms of ATP^{4-} is a hypothesis that is consistent with the observation that protonated ATP, H_4ATP, has an exergonic hydrolysis free energy of smaller magnitude because the negative repulsions of oxygen atoms are not present. Likewise for $MgATP^{2-}$ because the Mg^{2+} ion lies between negatively charged oxygen atoms, thereby, reducing repulsions and stabilizing the ATP molecule.

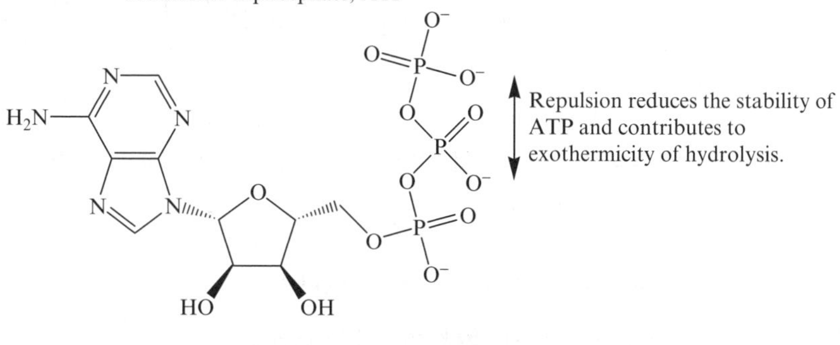

Adenosine triphosphate, ATP^{4-}

Repulsion reduces the stability of ATP and contributes to exothermicity of hydrolysis.

I6.9

$$C_6H_{12}O_6(aq) + 6\,O_2(g) \rightarrow 6\,CO_2(g) + 6\,H_2O(l)$$

$$\Delta_r G^{\oplus} = \Delta_r G^{\ominus} = -2\,880 \text{ kJ mol}^{-1} \text{ (i.e. pH independent)}$$

$$ATP(aq) + H_2O(l) \rightarrow ADP(aq) + P_i^-(aq) + H_3O^+(aq) \quad \Delta_r G^{\oplus} = -31\,\text{kJ mol}^{-1}$$

Refer to *Impact on Biochemistry* I6.1 for information necessary to the solution of this problem.

(a) If we assume that each mole of ATP formed during the aerobic breakdown of glucose produces about −31 kJ mol^{-1} and 38 moles of ATP are produced per mole of glucose consumed, then

$$\text{percentage efficiency} = \frac{38 \times (-31\ \text{kJ mol}^{-1})}{-2\,880\ \text{kJ mol}^{-1}} \times 100\% = \boxed{41\%}.$$

(b) For the oxidation of glucose under the biological conditions of $T = 310$ K, $p_{CO_2} = 5.3 \times 10^{-2}$ atm, $p_{O_2} = 0.132$ atm, and $[\text{glucose}] = 5.6 \times 10^{-12}$ mol dm^{-3} we have

$$\Delta_r G = \Delta_r G^{\ominus} + RT \ln Q = \Delta_r G^{\ominus} + RT \ln\left(\frac{(p_{CO_2}/p^{\ominus})^6 \times 1\ \text{M}}{[\text{glucose}] \times (p_{O_2}/p^{\ominus})^6}\right)$$

$$= -2\,880\ \text{kJ mol}^{-1} + (8.3145\ \text{J K}^{-1}\ \text{mol}^{-1}) \times (310\ \text{K}) \ln\left(\frac{0.053^6}{5.6 \times 10^{-12} \times 0.132^6}\right)$$

$$= -2\,827\ \text{kJ mol}^{-1}$$

This is 2% less exergonic than the standard value of −2 880 kJ mol^{-1}.

For the ATP hydrolysis under the given conditions of $T = 310$ K, $p_{CO_2} = 5.3 \times 10^{-2}$ atm, $p_{O_2} = 0.132$ atm, $[\text{glucose}] = 5.6 \times 10^{-12}$ mol dm^{-3}, $[\text{ATP}] = [\text{ADP}] = [\text{P}_i] = 1.0 \times 10^{-4}$ mol dm^{-3}, and pH = 7.4 we have

$$\Delta_r G = \Delta_r G^{\oplus} + RT \ln Q^{\oplus} = \Delta_r G^{\oplus} + RT \ln\left(\frac{[\text{ADP}] \times [\text{P}_i^-]}{[\text{ATP}] \times (1\ \text{M})} \times \frac{[\text{H}^+]}{10^{-7}\ \text{M}}\right)$$

$$\left[\text{i.e. } [\text{H}^+]^{\oplus} = 10^{-7}\ \text{M}, [\text{J}]^{\oplus} = 1\ \text{M}\right]$$

$$= -31\ \text{kJ mol}^{-1} + (8.3145\ \text{J K}^{-1}\ \text{mol}^{-1}) \times (310\ \text{K}) \ln(1.0 \times 10^{-4} \times 10^{-7.4} / 10^{-7})$$

$$= -57\ \text{kJ mol}^{-1}$$

With this value for $\Delta_r G$ the efficiency becomes

$$\text{efficiency} = \frac{38 \times (-57\ \text{kJ mol}^{-1})}{-2\,827\ \text{kJ mol}^{-1}} = \boxed{77\%}$$

(c) The theoretical limit of the diesel engine is

$$\varepsilon = 1 - \frac{T_c}{T_h} = 1 - \frac{873\ \text{K}}{1923\ \text{K}} = \boxed{55\%}$$

With an efficiency of 75% of the theoretical limit the energy conversion is reduced to 41% in a heat engine. We see that the biological efficiency (part (b), 77%) under the conditions given is greater than that of the diesel engine. What limits the efficiency of the diesel engine, or any heat engine, is that heat engines must convert heat ($q \approx \Delta_c H$) into useful work ($w_{\text{add,max}} = \Delta_r G$). Because of the Second Law, a substantial fraction of that heat is wasted. The biological process involves $\Delta_r G$ directly and does not go through a heat step.

I6.11 The half-reactions involved are:

$$\text{R: } \text{cyt}_{\text{ox}} + \text{e}^- \rightarrow \text{cyt}_{\text{red}} \qquad E^\ominus_{\text{cyt}}$$

$$\text{L: } \text{D}_{\text{ox}} + \text{e}^- \rightarrow \text{D}_{\text{red}} \qquad E^\ominus_{\text{D}}$$

The overall cell reaction is:

$$\text{R}-\text{L: } \text{cyt}_{\text{ox}} + \text{D}_{\text{red}} \rightleftharpoons \text{cyt}_{\text{red}} + \text{D}_{\text{ox}} \qquad E^\ominus_{\text{cell}} = E^\ominus_{\text{cyt}} - E^\ominus_{\text{D}}$$

(a) The Nernst equation for the cell reaction is

$$E_{\text{cell}} = E^\ominus_{\text{cell}} - \frac{RT}{F}\ln\frac{[\text{cyt}_{\text{red}}][\text{D}_{\text{ox}}]}{[\text{cyt}_{\text{ox}}][\text{D}_{\text{red}}]}$$

At equilibrium, $E_{\text{cell}} = 0$; therefore

$$\ln\left(\frac{[\text{cyt}_{\text{red}}]_{\text{eq}}[\text{D}_{\text{ox}}]_{\text{eq}}}{[\text{cyt}_{\text{ox}}]_{\text{eq}}[\text{D}_{\text{red}}]_{\text{eq}}}\right) = \frac{F}{RT}\left(E^\ominus_{\text{cyt}} - E^\ominus_{\text{D}}\right)$$

$$\ln\left(\frac{[\text{D}_{\text{ox}}]_{\text{eq}}}{[\text{D}_{\text{red}}]_{\text{eq}}}\right) = \ln\left(\frac{[\text{cyt}_{\text{ox}}]_{\text{eq}}}{[\text{cyt}_{\text{red}}]_{\text{eq}}}\right) + \frac{F}{RT}\left(E^\ominus_{\text{cyt}} - E^\ominus_{\text{D}}\right)$$

Therefore a plot of $\ln\left([\text{D}_{\text{ox}}]_{\text{eq}}/[\text{D}_{\text{red}}]_{\text{eq}}\right)$ against $\ln\left([\text{cyt}_{\text{ox}}]_{\text{eq}}/[\text{cyt}_{\text{red}}]_{\text{eq}}\right)$ is linear with a slope of one and an intercept of $F\left(E^\ominus_{\text{cyt}} - E^\ominus_{\text{D}}\right)/RT$.

(b) A plot of $\ln\left([\text{D}_{\text{ox}}]_{\text{eq}}/[\text{D}_{\text{red}}]_{\text{eq}}\right)$ against $\ln\left([\text{cyt}_{\text{ox}}]_{\text{eq}}/[\text{cyt}_{\text{red}}]_{\text{eq}}\right)$ does appear to be linear but a linear regression fit has a slope slightly different than 1. It is 1.0116, which makes it difficult to assess the uncertainty of the final $E^\ominus_{\text{cyt}}$ computation, so another and simpler analysis is desired. Solving the above working equation for $E^\ominus_{\text{cyt}}$ gives

$$E^\ominus_{\text{cyt}} = E^\ominus_{\text{D}} + \frac{RT}{F}\ln\left(\frac{[\text{D}_{\text{ox}}]_{\text{eq}}}{[\text{D}_{\text{red}}]_{\text{eq}}}\bigg/\frac{[\text{cyt}_{\text{ox}}]_{\text{eq}}}{[\text{cyt}_{\text{red}}]_{\text{eq}}}\right)$$

$$= 0.237\ \text{V} + \left(25.693\times10^{-3}\ \text{V}\right)\ln\left(\frac{[\text{D}_{\text{ox}}]_{\text{eq}}}{[\text{D}_{\text{red}}]_{\text{eq}}}\bigg/\frac{[\text{cyt}_{\text{ox}}]_{\text{eq}}}{[\text{cyt}_{\text{red}}]_{\text{eq}}}\right)$$

Thus, $E^\ominus_{\text{cyt}}$ can be calculated for each experiment and the mean determined for the set. Uncertainties are then due to experimentation alone without introducing a computational uncertainty, and the standard deviation of the mean represents that uncertainty. Draw up the following table.

$[D_{ox}]_{eq}/[D_{red}]_{eq}$	0.00279	0.00843	0.0257	0.0497	0.0748	0.238	0.534
$[cyt_{ox}]_{eq}/[cyt_{red}]_{eq}$	0.0106	0.0230	0.0894	0.197	0.335	0.809	1.39
$\ln\left(\frac{[D_{ox}]_{eq}}{[D_{red}]_{eq}}\Big/\frac{[cyt_{ox}]_{eq}}{[cyt_{red}]_{eq}}\right)$	−1.335	−1.004	−1.247	−1.377	−1.499	−1.224	−0.957
$E^{\ominus}_{cyt}/V$	0.2023	0.2109	0.2046	0.2012	0.1981	0.2052	0.2122

Taking the mean gives $E^{\ominus}_{cyt} = \boxed{+0.205\ V}$ with a standard deviation of 0.005 V.

I6.13 $\frac{1}{2}N_2(g)+\frac{3}{2}H_2(g)\rightarrow NH_3(g) \qquad \Delta\nu=\sum_J \nu_J=-1$

This is the ammonia formation reaction for which we find the standard reaction thermodynamic functions in the text data tables.

$$\Delta_r H^{\ominus}(298\ K)=-46.11\ kJ \quad \text{and} \quad \Delta_r S^{\ominus}(298\ K)=-99.38\ J\ K^{-1}$$

Use text data tables to establish functions for the constant pressure heat capacity of reactants and products. Define a function $\Delta_r C_p^{\ominus}(T)$ that makes it possible to calculate $\Delta_r C_p$ at 1 bar and any temperature (eqns 2C.7b and 2B.8). Define functions that make it possible to calculate the reaction enthalpy and entropy at 1 bar and any temperature (eqns 2C.7a and 3A.19 applied to chemical reaction equations).

$$\Delta_r H^{\ominus}(T)=\Delta_r H^{\ominus}(298\ K)+\int_{298.15\,K}^{T}\Delta_r C_p^{\ominus}(T)\mathrm{d}T$$

$$\Delta_r S^{\ominus}(T)=\Delta_r S^{\ominus}(298\ K)+\int_{298.15\,K}^{T}\frac{\Delta_r C_p^{\ominus}(T)}{T}\mathrm{d}T$$

(i) For a perfect gas reaction mixture $\Delta_r H$ is independent of pressure at constant temperature. Consequently, $\Delta_r H(T,p)=\Delta_r H^{\ominus}(T)$. The pressure dependence of the reaction entropy may be evaluated with the expression:

$$\begin{aligned}\Delta_r S(T,p)&=\Delta_r S^{\ominus}(T)+\sum_{\text{Products–Reactants},i}\nu_i\int_{1\,bar}^{p}\left(\frac{\partial S_{m,i}}{\partial p}\right)_T \mathrm{d}p\\
&=\Delta_r S^{\ominus}(T)-\sum_{\text{Products–Reactants},i}\nu_i\int_{1\,bar}^{p}\left(\frac{\partial V_{m,i}}{\partial T}\right)_p \mathrm{d}p \quad \text{(Table 3.5)}\\
&=\Delta_r S^{\ominus}(T)-\sum_{\text{Products–Reactants},i}\nu_i\int_{1\,bar}^{p}\frac{R}{p}\mathrm{d}p\\
&=\Delta_r S^{\ominus}(T)-\sum_{\text{Products–Reactants},i}\nu_i\int_{1\,bar}^{p}\frac{R}{p}\mathrm{d}p\\
&=\Delta_r S^{\ominus}(T)-\left[\sum_{\text{Products–Reactants},i}\nu_i\right]R\ln\left(\frac{p}{1\ bar}\right)=\Delta_r S^{\ominus}(T)-(-1)\,R\ln\left(\frac{p}{1\ bar}\right)\\
&=\Delta_r S^{\ominus}(T)+R\ln\left(\frac{p}{1\ bar}\right)\end{aligned}$$

The above two equations make it possible to calculate $\Delta_r G(T,p)$.

$$\Delta_r G(T,p) = \Delta_r H(T,p) - T\Delta_r S(T,p)$$

Once the above functions have been defined on a scientific calculator or with mathematical software on a computer, the root function may be used to evaluate pressure where $\Delta_r G(T,p) = -500$ J at a given temperature.

(a) and (b) perfect gas mixture:

For $T = (450+273.15)\,\text{K} = 723.15$ K, $\text{root}(\Delta_r G(723.15\text{ K}, p) + 500\text{ J}) = \boxed{156.5\text{ bar}}$

For $T = (400+273.15)\,\text{K} = 673.15$ K, $\text{root}(\Delta_r G(673.15\text{ K}, p) + 500\text{ J}) = \boxed{81.8\text{ bar}}$

(ii) For a van der Waals gas mixture $\Delta_r H$ does depend upon pressure. The calculation equation is:

$$\Delta_r H(T,P) = \Delta_r H^{\ominus}(T) + \sum_{\text{Products–Reactants,i}} \nu_i \int_{1\text{bar}}^{p} \left(\frac{\partial H_{\text{m,i}}}{\partial p}\right)_T \text{d}p$$

[The sum involves the i = 1, 2, 3 gases (NH_3, N_2, or H_2)]

$$= \Delta_r H^{\ominus}(T) + \sum_{\text{Products–Reactants,i}} \nu_i \int_{1\text{ bar}}^{p} \left[V_{\text{m,i}} - T\left(\frac{\partial V_{\text{m,i}}}{\partial T}\right)_p\right] \text{d}p$$

[The equation of this substitution is proven below*.]

where $(\partial V_{\text{m,i}}/\partial T)_p = R(V_{\text{m,i}} - b_i)^{-1}(RT(V_{\text{m,i}} - b_i)^{-2} - 2a_i V_{\text{m,i}}^{-3})^{-1}$ for each gas i (NH_3, N_2, or H_2) and $V_{\text{m,i}}(T,p) = \text{root}\left(p - \frac{RT}{V_{\text{m,i}} - b_i} + \frac{a_i}{V_{\text{m,i}}^2}\right)$

The functional equation for $\Delta_r S$ calculations is:

$$\Delta_r S(T,p) = \Delta_r S^{\ominus}(T) - \sum_{\text{Products–Reactants,i}} \nu_i \int_{1\text{bar}}^{p} \left(\frac{\partial V_{\text{m,i}}}{\partial T}\right)_p \text{d}p$$

where $(\partial V_{\text{m,i}}/\partial T)_p$ and $V_{\text{m,i}}(T,p)$ are calculated as described above. As usual,

$$\Delta_r G(T,p) = \Delta_r H(T,p) - T\Delta_r S(T,p)$$

(a) and (b) van der Waals gas mixture:

For $T = 723.15$ K, $\text{root}(\Delta_r G(723.15\text{ K}, p) + 500\text{ J}) = \boxed{132.5\text{ bar}}$

For $T = 673.15$ K, $\text{root}(\Delta_r G(673.15\text{ K}, p) + 500\text{ J}) = \boxed{73.7\text{ bar}}$

(c) $\Delta_r G(T,p)$ isotherms (see Fig. I6.1) $\boxed{\text{confirm}}$ Le Chatelier's principle. Along an isotherm, $\Delta_r G$ decreases as pressure increases. This corresponds to a shift to the right in the reaction equation and reduces the stress by shifting to the side that has fewer total moles of gas. Additionally the reaction is exothermic, so Chatelier's principle predicts a shift to the left with an increase in temperature. The isotherms confirm this as an increase in $\Delta_r G$ as temperature is increased at constant pressure.

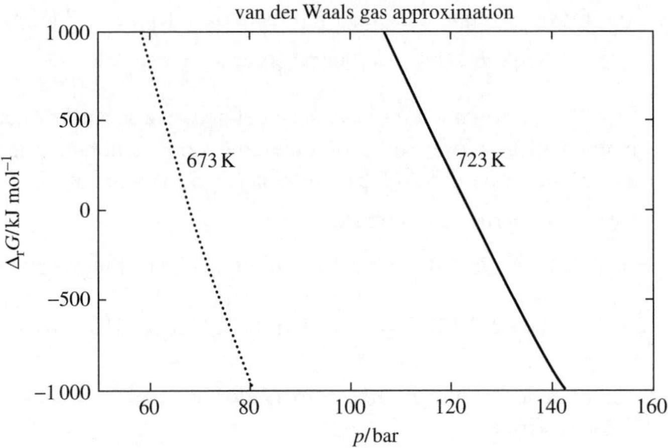

Figure I6.1

* There are many thermodynamic equations that are very useful when deriving desired computation equations. One of them (used above) is: $\left(\frac{\partial H}{\partial p}\right)_T = -T\left(\frac{\partial V}{\partial T}\right)_p + V$. To prove this relationship, first use an identity of partial derivatives that involves a change of variable

$$\left(\frac{\partial H}{\partial p}\right)_T = \left(\frac{\partial H}{\partial S}\right)_p \left(\frac{\partial S}{\partial p}\right)_T + \left(\frac{\partial H}{\partial p}\right)_S$$

We will be able to identify some of these terms if we examine an expression for dH analogous to the fundamental equation of thermodynamics [3D.1]. From the definition of enthalpy [2B.1], we have:

$$\mathrm{d}H = \mathrm{d}U + p\mathrm{d}V + V\mathrm{d}p = T\mathrm{d}S - p\mathrm{d}V \text{ [3D.1]} + p\mathrm{d}V + V\mathrm{d}p = T\mathrm{d}S + V\mathrm{d}p$$

Compare this expression to the exact differential of H considered as a function of S and p:

$$\mathrm{d}H = \left(\frac{\partial H}{\partial S}\right)_p \mathrm{d}S + \left(\frac{\partial H}{\partial p}\right)_S \mathrm{d}p$$

Thus, $\left(\frac{\partial H}{\partial S}\right)_p = T$, $\left(\frac{\partial H}{\partial p}\right)_S = V$ [dH exact]

Substitution yields $\left(\frac{\partial H}{\partial p}\right)_T = T\left(\frac{\partial S}{\partial p}\right)_T + V = \boxed{-T\left(\frac{\partial V}{\partial T}\right)_p + V}$ [Maxwell relation]

7 Introduction to quantum theory

7A The origins of quantum mechanics

Answers to discussion questions

D7A.1 At the end of the nineteenth century and the beginning of the twentieth, there were many experimental results on the properties of matter and radiation that could not be explained on the basis of established physical principles and theories. Here we list only some of the most significant.

(1) The photoelectric effect revealed that electromagnetic radiation, classically considered to be a wave, also exhibits the particle-like behaviour of photons. Each photon is a discrete unit, or quantum, of energy that is absorbed during collisions with electrons. Photons are never partially absorbed. They either completely give up their energy or they are not absorbed. The energy of a photon can be calculated if either the radiation frequency or wavelength is known: $E_{\text{photon}} = h\nu = hc/\lambda$.

(2) Absorption and emission spectra indicated that atoms and molecules can only absorb or emit discrete packets of energy (i.e. photons). This means that an atom or molecule has specific, allowed energy levels and we say that their energies are quantized. During a spectroscopic transition the atom or molecule gains or loses the energy ΔE by either absorption of a photon or emission of a photon, respectively. Thus, spectral lines must satisfy the **Bohr frequency condition:** $\Delta E = h\nu$.

(3) Neutron and electron diffraction studies indicated that these particles also possess wave-like properties of constructive and destructive interference. The joint particle and wave character of matter and radiation is called **wave-particle duality**. The **de Broglie relation,** $\lambda_{\text{de Broglie}} = h/p$, connects the wave character of a particle ($\lambda_{\text{de Broglie}}$) with its particulate momentum (p).

(4) The concept of classical physics that all energies are permitted incorrectly predicts that the ideal emitter of electromagnetic radiation, an object called a **black body**, which emits and absorbs all wavelengths of radiation uniformly, should emit an infinite amount of high-frequency radiation at all temperatures. This **ultraviolet catastrophe** sharply contrasts with the experimental evidence that at a given temperature the emissions from a pinhole in an empty container have a maximum intensity at intermediate frequencies and little-to-no intensity at both low and high frequencies. Max Planck (in 1900) correctly accounted for the experimental observations by postulating that the permitted

energies of an electromagnetic oscillator of frequency ν are **quantized** as integer multiples of $h\nu$.

(5) Classical physics also fails to explain why the heat capacities of monatomic solids such as copper metal vary from zero in the limit of the absolute zero of temperature to a value of about $3R$ at high temperature. In order to explain these observations Einstein (in 1905) found it necessary to postulate that each atom of the solid oscillates about its equilibrium position with a frequency ν and that the oscillation energy is restricted to quantized values that must equal an integer multiple of $h\nu$.

D7A.3 The heat capacities of monatomic solids are primarily a result of the energy acquired by vibrations of the atoms about their equilibrium positions. If this energy can be acquired continuously, we expect that the **equipartition of energy principle** should apply. This principle states that for each direction of motion and for each kind of energy (potential and kinetic) the associated energy should be ½kT. Hence, for three directions and both kinds of motion, a total of $3kT$, which gives a heat capacity of $3k$ per atom, or $3R$ per mole, independent of temperature. But the experiments show a temperature dependence. The heat capacity falls steeply below $3R$ at low temperatures. Einstein showed that by allowing the energy of the atomic oscillators to be quantized according to Planck's formula $E = nh\nu = nhc/\lambda$ where the **quantum number *n*** can equal zero or any positive integer, rather than the continuous variability of energy prescribed by classical physics, this temperature dependence could be explained. The physical reason is that at low temperatures only a few atomic oscillators have enough energy to populate the higher quantized levels; at higher temperatures more of them can acquire the energy to become active.

Solutions to exercises

E7A.1(a) $\Delta E = h\nu = h/T \quad [T = \text{period} = 1/\nu] \quad \text{and} \quad \Delta E_\text{m} = N_\text{A}\Delta E$

(a) For $T = 1.0$ fs

$$\Delta E = (6.626\times10^{-34}\ \text{J s})/(1.0\times10^{-15}\ \text{s}) = \boxed{6.6\times10^{-19}\ \text{J}}$$

$$\Delta E_\text{m} = (6.022\times10^{23}\ \text{mol}^{-1})\times(6.6\times10^{-19}\ \text{J}) = \boxed{4.0\times10^{2}\ \text{kJ mol}^{-1}}$$

(b) For $T = 10$ fs

$$\Delta E = (6.626\times10^{-34}\ \text{J s})/(10\times10^{-15}\ \text{s}) = \boxed{6.6\times10^{-20}\ \text{J}}$$

$$\Delta E_\text{m} = (6.022\times10^{23}\ \text{mol}^{-1})\times(6.6\times10^{-20}\ \text{J}) = \boxed{40\ \text{kJ mol}^{-1}}$$

(c) For $T = 1.0$ s

$$\Delta E = (6.626\times10^{-34}\ \text{J s})/(1.0\ \text{s}) = \boxed{6.6\times10^{-34}\ \text{J}}$$

$$\Delta E_\text{m} = (6.022\times10^{23}\ \text{mol}^{-1})\times(6.6\times10^{-34}\ \text{J}) = \boxed{4.0\times10^{-13}\ \text{kJ mol}^{-1}}$$

E7A.2(a) $E = h\nu = \frac{hc}{\lambda}$ [Example 7A.2] $= \frac{(6.626\times10^{-34}\ \mathrm{J\,s})\times(2.998\times10^{8}\ \mathrm{m\,s^{-1}})}{\lambda}$

$$= \frac{1.986\times10^{-25}\ \mathrm{J\,m}}{\lambda}$$

$$E_\mathrm{m} = N_\mathrm{A}E = \frac{N_\mathrm{A}hc}{\lambda} = \frac{0.1196\ \mathrm{J\,m}}{\lambda}$$

We can therefore draw up the following table.

λ/nm	E/aJ	E_m/(kJ mol^{-1})
(a) 600	331	199
(b) 550	361	218
(c) 400	497	299

E7A.3(a) Upon absorption of a photon by a free hydrogen atom, the law of conservation of energy requires that the acquired kinetic energy, E_k, of the atom equal the energy of the absorbed photon: $E_\mathrm{k} = E_\mathrm{photon} = \tfrac{1}{2}m_\mathrm{H}\upsilon^2$. The values of E_photon are calculated in Exercise 7A.2(a) so the atom is accelerated to the speed

$$\upsilon = \sqrt{\frac{2E_\mathrm{photon}}{m_\mathrm{H}}} = \sqrt{\frac{2N_\mathrm{A}E_\mathrm{photon}}{M_\mathrm{H}}}$$

$$= \sqrt{\frac{2\times(6.022\times10^{23}\ \mathrm{mol^{-1}})\times E_\mathrm{photon}}{1.0079\times10^{-3}\ \mathrm{kg\,mol^{-1}}}} = (3.457\times10^{13}\ \mathrm{m\,s^{-1}})\times\sqrt{E_\mathrm{photon}/\mathrm{J}}$$

We can therefore draw up the following table.

λ/nm	E_photon/aJ	υ/(km s^{-1})
(a) 600	331	19.9
(b) 550	361	20.8
(c) 400	497	24.4

E7A.4(a) The total energy emitted in time Δt is $P\Delta t$ where P is the power of the glow-worm emission. The energy of each emitted photon $E_\mathrm{photon} = h\nu = hc/\lambda$ [C.3]. The total number of photons emitted in an interval Δt is then the total energy divided by the energy per photon.

$$N = \frac{P\Delta t}{E_\mathrm{photon}} = \frac{P\Delta t\lambda}{hc}$$

Assuming that de Broglie's relation applies to each photon, and recognizing that the law of conservation of linear momentum specifies that the loss of a photon imparts an equivalent momentum (in the opposite direction) to the glow-worm, the total momentum imparted to the glow-worm in time Δt is

$$p = Np_{\text{photon}} = \frac{Nh}{\lambda} \text{ [7A.14]} = \frac{P\Delta t\lambda}{hc} \times \frac{h}{\lambda} = \frac{P\Delta t}{c}$$

Since $p = (mv)_{\text{glow-worm}}$, the final speed of the glow-worm is

$$\begin{aligned} v &= \frac{P\Delta t}{cm_{\text{glow-worm}}} \\ &= \frac{(0.10\text{ W})\times(10\text{ y})\times(3.1557\times10^{7}\text{ s y}^{-1})}{(2.9979\times10^{8}\text{ m s}^{-1})\times(0.0050\text{ kg})} = \boxed{21\text{ m s}^{-1}} \end{aligned}$$

Comment. Note that the answer is independent of the wavelength of the radiation emitted: the greater the wavelength the smaller the photon momentum, but the greater the number of photons emitted.

Question. If this glow-worm eventually turns into a firefly which glows for 1 s intervals while flying with a speed of 0.1 m s^{-1} what additional speed does the 1 s glowing impart to the firefly? Ignore any frictional effects of air.

E7A.5(a) The total energy emitted in time Δt is $P\Delta t$ where P is the power of the emission. The energy of each emitted photon is given by $E_{\text{photon}} = h\nu = hc/\lambda$ [C.3]. The total number of photons emitted in an interval Δt is then the total energy divided by the energy per photon.

$$\begin{aligned} N &= \frac{P\Delta t}{E_{\text{photon}}} = \frac{P\Delta t\lambda}{hc} \\ &= \frac{(1.00\text{ s})\times(550\times10^{-9}\text{ m})}{(6.626\times10^{-34}\text{ J s})\times(2.998\times10^{8}\text{ m s}^{-1})}P = (2.77\times10^{18})\times P/\text{W} \end{aligned}$$

(a) When $P = 1.0$ W, $N = \boxed{2.77\times10^{18}}$.

(b) When $P = 100$ W, $N = \boxed{2.77\times10^{20}}$.

E7A.6(a) $E_{\text{k}} = h\nu - \Phi = \dfrac{hc}{\lambda} - \Phi$ [7A.13] and, since $E_{\text{k}} = \tfrac{1}{2}m_{\text{e}}v^2$, $v = \sqrt{2E_{\text{k}}/m_{\text{e}}}$

$\Phi = 2.14\text{ eV} = (2.14)\times(1.602\times10^{-19}\text{ J}) = 3.43\times10^{-19}\text{ J}$

(a) For $\lambda = 700$ nm

$$\begin{aligned} E_{\text{k}} &= \frac{(6.626\times10^{-34}\text{ J s})\times(2.998\times10^{8}\text{ m s}^{-1})}{700\times10^{-9}\text{ m}} - 3.43\times10^{-19}\text{ J} \\ &= 2.84\times10^{-19}\text{ J} - 3.43\times10^{-19}\text{ J} \end{aligned}$$

Inspection of the above equation reveals that the photon energy of the radiation is less than the work function and we conclude that $\boxed{\text{no electron ejection}}$ occurs.

(b) For $\lambda = 300$ nm

$$E_k = \frac{(6.626\times10^{-34}\ \text{J s})\times(2.998\times10^{8}\ \text{m s}^{-1})}{300\times10^{-9}\ \text{m}} - 3.43\times10^{-19}\ \text{J}$$
$$= 6.62\times10^{-19}\ \text{J} - 3.43\times10^{-19}\ \text{J}$$
$$= \boxed{3.19\times10^{-19}\ \text{J}}$$
$$\upsilon = \sqrt{2\times(3.19\times10^{-19}\ \text{J})/(9.109\times10^{-31}\ \text{kg})} = \boxed{837\ \text{km s}^{-1}}$$

E7A.7(a) $E_{\text{binding}} = E_{\text{photon}} - E_k = h\nu - \tfrac{1}{2}m_e\upsilon^2 = \dfrac{hc}{\lambda} - \tfrac{1}{2}m_e\upsilon^2$

$$E_{\text{binding}} = \frac{hc}{\lambda} - \tfrac{1}{2}m\upsilon^2$$
$$= \frac{(6.626\times10^{-34}\ \text{J s})\times(2.998\times10^{8}\ \text{m s}^{-1})}{150\times10^{-12}\ \text{m}} - \tfrac{1}{2}(9.13\times10^{-31}\ \text{kg})\times(2.14\times10^{7}\ \text{m s}^{-1})^2$$
$$= (1.12\times10^{-15}\ \text{J})\times\left(\frac{1\ \text{eV}}{1.602\times10^{-19}\ \text{J}}\right) = \boxed{6.96\ \text{keV}}\ \text{with the relativistic mass correction.}$$

Note: The photoelectron is moving at 7.1% of the speed of light. So, in order to calculate a more accurate value of the binding energy, it would be necessary to use the relativistic mass in place of the rest mass.

$$m = \frac{m_e}{\left(1-(\upsilon/c)^2\right)^{1/2}} = \frac{9.109\times10^{-31}\ \text{kg}}{\left(1-(2.14\times10^{7}\ \text{m s}^{-1}/2.998\times10^{8}\ \text{m s}^{-1})^2\right)^{1/2}} = 9.13\times10^{-31}\ \text{kg}$$

$$E_{\text{binding}} = \frac{hc}{\lambda} - \tfrac{1}{2}m\upsilon^2$$
$$= \frac{(6.626\times10^{-34}\ \text{J s})\times(2.998\times10^{8}\ \text{m s}^{-1})}{150\times10^{-12}\ \text{m}} - \tfrac{1}{2}(9.13\times10^{-31}\ \text{kg})\times(2.14\times10^{7}\ \text{m s}^{-1})^2$$
$$= (1.12\times10^{-15}\ \text{J})\times\left(\frac{1\ \text{eV}}{1.602\times10^{-19}\ \text{J}}\right) = \boxed{6.96\ \text{keV}}\ \text{with the relativistic mass correction.}$$

The relativistic mass correction did not make a difference in this exercise.

E7A.8(a) The de Broglie relation is $\lambda = \dfrac{h}{p}$ [7A.14] $= \dfrac{h}{m\upsilon}$.

Hence, $$\upsilon = \frac{h}{m_e\lambda} = \frac{6.626\times10^{-34}\ \text{J s}}{(9.109\times10^{-31}\ \text{kg})\times(100\times10^{-12}\ \text{m})} = \boxed{7.27\times10^{6}\ \text{m s}^{-1}}$$

The kinetic energy acquired by an electron upon acceleration through a voltage of $\mathcal{E}$ equals $e\mathcal{E}$.

$$E_k = \frac{m_e v^2}{2} = e\mathcal{E}$$

Solving for the potential difference gives

$$\mathcal{E} = \frac{m_e v^2}{2e} = \frac{(9.109\times10^{-31}\ \text{kg})\times(7.27\times10^{6}\ \text{m s}^{-1})^2}{2\times(1.602\times10^{-19}\ \text{C})} = \boxed{150\ \text{V}}$$

E7A.9(a) The de Broglie relation is $\lambda = \frac{h}{p}$ [7A.14] $= \frac{h}{mv}$.

Hence, $v = \frac{h}{m_e\lambda} = \frac{6.626\times10^{-34}\ \text{J s}}{(9.109\times10^{-31}\ \text{kg})\times(0.030\ \text{m})} = \boxed{0.024\ \text{m s}^{-1}}$,

which is very slow!

E7A.10(a) $\lambda = \frac{h}{p}[7A.14] = \frac{h}{mv} = \frac{h}{m_e\alpha c} = \frac{\alpha^{-1}h}{m_e c} = \frac{137\times(6.626\times10^{-34}\ \text{J s})}{(9.109\times10^{-31}\ \text{kg})\times(2.998\times10^{8}\ \text{m s}^{-1})}$

$= \boxed{332\ \text{pm}}$

Comment. One wavelength of the matter wave of an electron with this velocity just fits in the first Bohr orbit. The velocity of the electron in the first Bohr classical orbit is thus $\frac{1}{137}c$.

Question. What is the wavelength of an electron with velocity approaching the speed of light? Such velocities can be achieved with particle accelerators.

E7A.11(a) The de Broglie wavelength is $\lambda = \frac{h}{p} = \frac{h}{mv}$ [7A.14]

(a) $\lambda = \frac{6.626\times10^{-34}\ \text{J s}}{(1.0\times10^{-3}\ \text{kg})\times(1.0\times10^{-2}\ \text{m s}^{-1})} = \boxed{6.6\times10^{-29}\ \text{m}}$

(b) $\lambda = \frac{6.626\times10^{-34}\ \text{J s}}{(1.0\times10^{-3}\ \text{kg})\times(1.00\times10^{5}\ \text{m s}^{-1})} = \boxed{6.6\times10^{-36}\ \text{m}}$

(c) $\lambda = \frac{6.626\times10^{-34}\ \text{J s}}{(4.003)\times(1.6605\times10^{-27}\ \text{kg})\times(1\,000\ \text{m s}^{-1})} = \boxed{99.7\ \text{pm}}$

Comment. The wavelengths in (a) and (b) are smaller than the dimensions of any known particle, whereas that in (c) is comparable to atomic dimensions.

Question. For stationary particles, $v = 0$, corresponding to an infinite de Broglie wavelength. What meaning can be ascribed to this result?

Solutions to problems

P7A.1 A cavity approximates an ideal black body; hence the Planck distribution applies

$$\rho=\frac{8\pi hc}{\lambda^5}\left(\frac{1}{e^{hc/\lambda kT}-1}\right) \text{[7A.6]}$$

Since the wavelength range is small (5 nm) we may write as a good approximation

$$\Delta E=\rho\Delta\lambda, \quad \lambda\approx 652.5\,\text{nm}$$

$$\frac{hc}{\lambda k}=\frac{(6.626\times10^{-34}\text{ J s})\times(2.998\times10^{8}\text{ m s}^{-1})}{(652.5\times10^{-9}\text{ m})\times(1.381\times10^{-23}\text{ J K}^{-1})}=2.205\times10^{4}\text{ K}$$

$$\frac{8\pi hc}{\lambda^5}=\frac{8\pi\times(6.626\times10^{-34}\text{ J s})\times(2.998\times10^{8}\text{ m s}^{-1})}{(652.5\times10^{-9}\text{ m})^5}=4.221\times10^{7}\text{ J m}^{-4}$$

$$\Delta E=(4.221\times10^{7}\text{ J m}^{-4})\times\left(\frac{1}{e^{(2.205\times10^4\text{ K})/T}-1}\right)\times(5\times10^{-9}\text{ m})=\frac{0.21\bar{1}\text{ J m}^{-3}}{e^{(2.205\times10^4\text{ K})/T}-1}$$

(a) $T=298\text{ K},\ \Delta E=\dfrac{0.21\bar{1}\text{ J m}^{-3}}{e^{(2.205\times10^4\text{ K})/(298\text{ K})}-1}=\boxed{1.6\times10^{-33}\text{ J m}^{-3}}$

(b) $T=3273\text{ K},\ \Delta E=\dfrac{0.21\bar{1}\text{ J m}^{-3}}{e^{(2.205\times10^4\text{ K})/(3273\text{ K})}-1}=\boxed{2.5\times10^{-4}\text{ J m}^{-3}}$

Comment. The energy density in the cavity does not depend on the volume of the cavity, but the total energy in any given wavelength range does, as well as the total energy over all wavelength ranges.

Question. What is the total energy in this cavity within the range 650–655 nm at the stated temperatures?

P7A.3 The Planck distribution is $\rho(\lambda,T)=\dfrac{8\pi hc}{\lambda^5\left(e^{hc/\lambda kT}-1\right)}$ [7A.6].

We look for the value $\lambda=\lambda_{max}$ at which ρ is a maximum. At this point $d\rho/d\lambda=0$ so we find the derivative and evaluate it at $\lambda=\lambda_{max}$.

$$\begin{aligned}\frac{d\rho}{d\lambda}&=\frac{d}{d\lambda}\left\{8\pi hc\lambda^{-5}\left(e^{hc/\lambda kT}-1\right)^{-1}\right\}\\&=8\pi hc\left\{\left(e^{hc/\lambda kT}-1\right)^{-1}\frac{d\lambda^{-5}}{d\lambda}+\lambda^{-5}\frac{d}{d\lambda}\left(e^{hc/\lambda kT}-1\right)^{-1}\right\}\\&=8\pi hc\left\{-5\lambda^{-6}\left(e^{hc/\lambda kT}-1\right)^{-1}+\lambda^{-5}\left(e^{hc/\lambda kT}-1\right)^{-2}\left(\frac{hc}{\lambda^2kT}\right)e^{\frac{hc}{\lambda kT}}\right\}\\&=8\pi hc\lambda^{-7}\left(e^{hc/\lambda kT}-1\right)^{-1}\left\{-5\lambda+\left(\frac{hc}{kT}\right)e^{hc/\lambda kT}\left(e^{hc/\lambda kT}-1\right)^{-1}\right\}\end{aligned}$$

Thus, at $\lambda=\lambda_{max}$

$$-5\lambda_{max}+\left(\frac{hc}{kT}\right)e^{hc/\lambda_{max}kT}\left(e^{hc/\lambda_{max}kT}-1\right)^{-1}=0 \text{ or } \lambda_{max}=\left(\frac{hc}{5kT}\right)e^{hc/\lambda_{max}kT}\left(e^{hc/\lambda_{max}kT}-1\right)^{-1}$$

This is a transcendental equation which can be written in the form

$$x_{\max}e^{x_{\max}}\left(e^{x_{\max}}-1\right)^{-1}=5 \quad \text{where} \quad x_{\max}=\frac{hc}{\lambda_{\max}kT}$$

Using the root function or the numeric solver of a scientific calculator to solve for $x_{\max}$, we find $x_{\max}=4.965$. Thus, we find the approximation $\boxed{\lambda_{\max}T\cong \tfrac{1}{5}hc/k}$, which is Wien's law that $\lambda_{\max}T$ is a constant.

P7A.5 We begin by reviewing the Stefan–Boltzmann law, derived in Problem 7A.6, that relates the total energy density $\mathcal{E}$ of black-body radiation to the temperature T of the body:

$$\mathcal{E}=\int_0^\infty \rho(\lambda)\,d\lambda=8\pi hc\int_0^\infty\frac{d\lambda}{\lambda^5\left(e^{hc/\lambda kT}-1\right)}\ [7A.6]=\left(\frac{4}{c}\right)\sigma T^4$$

where $\sigma=\dfrac{2\pi^5k^4}{15h^3c^2}=5.671\times10^{-8}\ \text{W m}^{-2}\ \text{K}^{-4}$ is the Stefan–Boltzmann constant. The power emitted by a region of surface divided by the area of the surface is called the excitance M and it can be shown (see note below for derivation) that according to the Stefan–Boltzmann law: $M=\tfrac{c}{4}\mathcal{E}=\sigma T^4$.

Let I_{solar} ($=343\ \text{W m}^{-2}$) be the solar energy flux at the top of the Earth's atmosphere and let a ($=0.30$) be the albedo, the fraction of the solar flux that is reflected back into space by the Earth or the atmosphere. The solar energy absorbed by the Earth is $(1-a)I_{\text{solar}}$. The radiation energy emitted by the Earth is the Earth's excitance. Since the average temperature of the Earth changes but little over tens of thousands of years or more, we surmise that there is a steady-state balance between the energy absorbed by the Earth and the radiation energy emitted (and lost into space) by the Earth. That is

$$(1-a)I_{\text{solar}}=\sigma T^4$$

$$T=\left(\frac{(1-a)I_{\text{solar}}}{\sigma}\right)^{1/4}=\left(\frac{(1-0.30)(343\ \text{W m}^{-2})}{5.671\times10^{-8}\ \text{W m}^{-2}\ \text{K}^{-4}}\right)^{1/4}$$

$$=\boxed{25\bar{5}\ \text{K or}\ 1\bar{8}\ {}^\circ\text{C}}$$

Wien's law, which is derived in Problem 7A.3, relates the temperature to the wavelength of the most intense radiation.

$$\lambda_{\max}=\frac{hc}{5kT}$$

$$=\frac{(6.626\times10^{-34}\ \text{J s})\times(2.998\times10^{8}\ \text{m s}^{-1})}{5\times(1.381\times10^{-23}\ \text{J K}^{-1})\times(25\bar{5}\ \text{K})}$$

$$=\boxed{11\ \mu\text{m}}$$

Note: The derivation of the Stefan–Boltzmann law, $M = \frac{c}{4}\mathcal{E} = \sigma T^4$, is aided by examination of Fig. 7A.1. We consider the hemisphere to be a non-reflecting black-body chamber of radius $r_{max} = c\,\Delta t$ where Δt is the period over which emitted energy flux is observed normal to a window at the origin. The chamber interior has an isotropic total energy density for which $\mathcal{E}$ is identical at all points. Furthermore, r_{max} has been chosen with the recognition that, because of the unique value of the speed of light, only infinitesimal volume elements for which $r \le r_{max}$ will contribute to the observed energy flux in the period Δt; r being the distance from an infinitesimal volume element in the interior of the hemisphere to the observation point (coordinate origin). The total energy emission from an infinitesimal element is $\mathcal{E}\,dxdydz$, total energy flux at the distance r is $\mathcal{E}\,dxdydz/(4\pi r^2\Delta t)$, and the projection of the flux upon the observation direction is $\mathcal{E}\cos(\theta)\,dxdydz/(4\pi r^2\Delta t)$. We need to add up all the contributions from volume elements close enough to contribute to the observation. The elements must be such that $0 \le r \le r_{max}$ while $0 \le \phi \le 2\pi$. The range of θ is found by recognition that $0 \le z \le r_{max}$ and, since $z = r\cos(\theta)$, $\theta = \cos^{-1}(z/r_{max})$. This implies that $0 \le \theta \le \pi/2$, which is most easily seen by examination of Fig. 7A.1. We now perform the continuous sum (integration) over all the volume elements that contribute to the flux observation M.

$$M = \iiint \frac{\mathcal{E}\cos\theta}{4\pi r^2 \Delta t}\mathrm{d}x\mathrm{d}y\mathrm{d}z = \frac{\mathcal{E}}{4\pi\Delta t}\iiint \frac{\mathcal{E}\cos(\theta)}{4\pi r^2}\mathrm{d}x\mathrm{d}y\mathrm{d}z$$

$$= \frac{\mathcal{E}}{4\pi\Delta t}\int_0^{c\Delta t}\int_0^{\pi/2}\int_0^{2\pi}\frac{\cos(\theta)}{r^2}r^2\sin(\theta)\mathrm{d}r\mathrm{d}\theta\mathrm{d}\phi$$

$$= \frac{\mathcal{E}}{4\pi\Delta t}\int_0^{c\Delta t}\int_0^{\pi/2}\int_0^{2\pi}\cos(\theta)\sin(\theta)\mathrm{d}r\mathrm{d}\theta\mathrm{d}\phi$$

$$= -\frac{\mathcal{E}}{4\pi\Delta t}\times\int_0^{c\Delta t}\mathrm{d}r\times\int_0^{\pi/2}\cos(\theta)\mathrm{d}(\cos(\theta))\times\int_0^{2\pi}\mathrm{d}\phi = -\frac{\mathcal{E}}{4\pi\Delta t}\times(c\Delta t)\times\left(-\frac{1}{2}\right)\times(2\pi)$$

$$= \boxed{\frac{c\mathcal{E}}{4}}$$

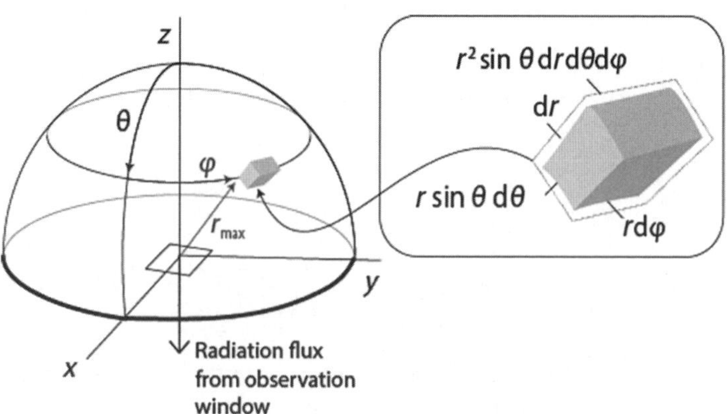

Figure 7A.1.

P7A.7 (a) With a little manipulation, a small-wavelength approximation of the Planck distribution can be derived that has the same form as Wien's formula. First examine the Planck distribution,

$$\rho_{\text{Planck}} = \frac{8\pi hc}{\lambda^5(e^{hc/\lambda kT}-1)},$$

for small-wavelength behaviour. The factor λ^{-5} gets large as λ gets small, but the factor $1/\left(e^{hc/\lambda kT}-1\right)$ gets small even faster. Focus on the latter factor, and try to express it in terms of a single decaying exponential (as in Wien's formula), at least in the small-λ limit. Multiplying it by one in the form of $e^{-hc/\lambda kT}/e^{-hc/\lambda kT}$, yields $e^{-hc/\lambda kT}/\left(1-e^{-hc/\lambda kT}\right)$, where $e^{-hc/\lambda kT}$ is small, so let us call it ε. The factor, then, becomes $\varepsilon/(1-\varepsilon)$, which can be expressed as a power series in ε as $\varepsilon(1+\varepsilon\cdots)$. For sufficiently small wavelengths, then, the Planck distribution may be approximated as:

$$\rho_{\text{Planck}} \approx \frac{8\pi hc\varepsilon}{\lambda^5} = \frac{8\pi hc e^{-hc/\lambda kT}}{\lambda^5}$$

This has the same form as Wien's formula:

$$\rho_{\text{Wien}} = \frac{a}{\lambda^5}e^{-b/\lambda kT}$$

Comparing the two formulas gives the values of the Wien constants: $a = \boxed{8\pi hc}$ and $b = \boxed{hc}$.

(b) The wavelength at which the Planck distribution, $\rho(\lambda,T) = \dfrac{8\pi hc}{\lambda^5\left(e^{hc/\lambda kT}-1\right)}$ [7A.6], is a maximum is found by setting the derivative of the distribution function with respect to wavelength equal to zero and solving the resulting transcendental equation with the numeric solver of a scientific calculator (see Problem 7A.3). This gives

$$\lambda_{\max}T = \frac{hc}{4.965\,k} = 2.898\times10^{-3}\ \text{m K} \qquad \text{[prediction of the Planck distribution]}$$

Following the same procedure with the Wien distribution gives a very similar result.

$$\frac{d\rho_{\text{Wein}}}{d\lambda} = \frac{d}{d\lambda}\left(\frac{8\pi hc}{\lambda^5}e^{-hc/\lambda kT}\right) = 8\pi hc\left\{-\frac{5}{\lambda^6}e^{-hc/\lambda kT} + \left(\frac{1}{\lambda^5}\right)\left(\frac{hc}{\lambda^2 kT}\right)e^{-hc/\lambda kT}\right\}$$
$$= 8\pi hc\lambda^{-7}e^{-hc/\lambda kT}\left\{-5\lambda + \left(\frac{hc}{kT}\right)\right\}$$

This derivative equals zero when the distribution is a maximum at $\lambda = \lambda_{\max}$ and inspection of the factors reveals that this occurs when

$$-5\lambda_{\max}+\left(\frac{hc}{kT}\right)=0 \quad \text{or} \quad \lambda_{\max}T=\frac{hc}{5k} \qquad \text{[Prediction of the Wien distribution]}$$

Thus, the maximum of the Wien distribution agrees well with that of the Planck distribution. The only difference being a factor of 1/4.965 in the expression for $\lambda_{\max} = T$ in the Planck distribution and a factor of 1/5 for the Wien distribution.

The Stefan–Boltzmann law gives the energy density $\mathcal{E}$ as a function of temperature. According to the Planck distribution (see Problem 7A.6), the energy density of black-body radiation is

$$\mathcal{E}_{\text{Planck}}=\int_0^\infty \rho_{\text{Planck}}(\lambda)\text{d}\lambda=\left(\frac{4}{c}\right)\sigma T^4$$

where $\sigma=\dfrac{2\pi^5k^4}{15h^3c^2}=5.671\times10^{-8}\ \text{W m}^{-2}\,\text{K}^{-4}$ is the Stefan–Boltzmann constant.

The energy density of the Wien distribution is

$$\mathcal{E}_{\text{Wien}}=\int_0^\infty \rho_{\text{Wien}}(\lambda)\,\text{d}\lambda=8\pi hc\int_0^\infty \frac{\text{e}^{-hc/\lambda kT}\,\text{d}\lambda}{\lambda^5}$$

Let $x=\dfrac{hc}{\lambda kT}$. Then, $\text{d}x=-\dfrac{hc}{\lambda^2kT}\text{d}\lambda$ or $\text{d}\lambda=-\dfrac{\lambda^2kT}{hc}\text{d}x$.

$$\mathcal{E}_{\text{Wien}}=8\pi kT\int_0^\infty\frac{\text{e}^{-x}\lambda^2\text{d}x}{\lambda^5}=8\pi kT\int_0^\infty\frac{\text{e}^{-x}\text{d}x}{\lambda^3}=8\pi kT\left(\frac{kT}{hc}\right)^3\int_0^\infty x^3\text{e}^{-x}\,\text{d}x$$

$$=8\pi kT\left(\frac{kT}{hc}\right)^3\times(6)\quad\text{[standard integral]}$$

$$=\boxed{\left(\frac{4}{c}\right)\sigma_{\text{Wien}}T^4}\quad \text{where } \sigma_{\text{Wien}}=\frac{12\pi k^4}{h^3c^2}=5.239\times10^{-8}\ \text{kg s}^{-3}\,\text{K}^{-4}$$

Comparing the energy densities of the Planck and Wien distribution, we see that both predict that the energy density is proportional to T^4. However, the Wien distribution predicts a constant of proportionality that is low by about 8%.

P7A.9 Confirmation that θ_E has the dimensions of temperature:

$$\theta_\text{E}=\frac{h\nu}{k}\quad\text{so}\quad[\theta_\text{E}]_{\text{SI unit}}=\frac{(\text{J s})\times(\text{s}^{-1})}{\text{J K}^{-1}}=\text{K}$$

In terms of θ_E the Einstein equation [7A.11] for the heat capacity of solids is

$$C_{V,\text{m}}=3R\left(\frac{\theta_\text{E}}{T}\right)^2\times\left(\frac{\text{e}^{\theta_\text{E}/2T}}{\text{e}^{\theta_\text{E}/T}-1}\right)^2\quad\left[\text{7A.9, classical value}=\lim_{T\to\infty}C_{V,\text{m}}=3R\right]$$

The Einstein expression for $C_{V,\mathrm{m}}$ approaches the classical value when $T \gg \theta_\mathrm{E}$ or when $h\nu/kT \ll 1$ as demonstrated in the text (Section 7A.1(b)). The criterion for classical behaviours is therefore that $\boxed{T \gg \theta_\mathrm{E}}$.

$$\theta_\mathrm{E} = \frac{h\nu}{k} = \frac{(6.626\times10^{-34}\ \mathrm{J\,Hz^{-1}})\times\nu}{1.381\times10^{-23}\ \mathrm{J\,K^{-1}}} = (4.798\times10^{-11}\ \mathrm{K})\times(\nu/\mathrm{Hz})$$

(a) For diamond, $\nu = 46.5\times10^{12}$ Hz,

$$\theta_\mathrm{E} = (4.798\times10^{-11}\ \mathrm{K})\times(46.5\times10^{12}) = \boxed{22\overline{3}1\ \mathrm{K}}$$

$$\text{At 298 K,}\quad \frac{C_{V,\mathrm{m}}}{3R} = \left(\frac{22\overline{3}1\ \mathrm{K}}{298\ \mathrm{K}}\right)^2 \times \left(\frac{\mathrm{e}^{22\overline{3}1/(2\times298)}}{\mathrm{e}^{22\overline{3}1/298}-1}\right)^2 = \boxed{0.0315}$$

(b) For copper, $\nu = 7.15\times10^{12}$ Hz, $\theta_\mathrm{E} = (4.798\times10^{-11}\ \mathrm{K})\times(7.15\times10^{12}) = \boxed{343\ \mathrm{K}}$

$$\text{At 298 K,}\quad \frac{C_{V,\mathrm{m}}}{3R} = \left(\frac{343\ \mathrm{K}}{298\ \mathrm{K}}\right)^2 \times \left(\frac{\mathrm{e}^{343/(2\times298)}}{\mathrm{e}^{343/298}-1}\right)^2 = \boxed{0.897}$$

Comment. For many metals the classical value is approached at room temperature; consequently, the failure of classical theory became apparent only after methods for achieving temperatures well below 25 °C were developed in the latter part of the nineteenth century.

7B Dynamics of microscopic systems

Answers to discussion questions

D7B.1 In quantum mechanics, all dynamical properties of a physical system have associated with them a corresponding operator. The system itself is described by a wavefunction. The observable properties of the system can be obtained in one of two ways from the wavefunction depending upon whether or not the wavefunction is an eigenfunction of the operator for the observable.

When the function representing the state of the system is an eigenfunction of the operator $\hat{\Omega}$, we solve the eigenvalue equation (eqn 7C.2b) $\hat{\Omega}\psi = \omega\psi$ in order to obtain the observable values, ω, of the dynamical property. The time-independent Schrödinger equation, $\hat{H}\psi = E\psi$ [7B.1 and 7C.1a], is an especially important example; the wavefunction ψ is an eigenfunction of the system hamiltonian operator while the system energy E is an eigenvalue. When ψ is not an eigenfunction of $\hat{\Omega}$, we can only find the average or expectation value of the dynamical property by performing the integration shown in eqn 7C.11: $\langle\Omega\rangle = \int\psi^*\hat{\Omega}\psi\,\mathrm{d}\tau$.

D7B.3 The detailed discussion of the constraints that the Born probability interpretation places upon acceptable wavefunctions, found in Section 7B.2(b), is a must read. In summary form the wavefunction must not be infinite over a non-infinitesimal region, must be single-valued, must be continuous, and must have a continuous slope.

Solutions to exercises

E7B.1(a) The time-independent, **stationary state** wavefunction in three-dimensional space is a function of x, y, and z so we write $\psi(x,y,z)$ or $\psi(\boldsymbol{r})$. The infinitesimal space element is $d\tau = dxdydz$ with each variable ranging from $-\infty$ to $+\infty$. For systems with spherical symmetry it is best to work in the spherical polar coordinates r, θ, and ϕ where the variables have the ranges $0 \le r \le \infty$, $0 \le \theta \le \pi$, and $0 \le \phi \le 2\pi$; the infinitesimal space element is $d\tau = r^2 \sin\theta\, dr\, d\theta\, d\phi$.

Remarkably, when the potential energy term of the hamiltonian is either zero or a constant value throughout space, the time-independent wavefunction does not depend upon the particle mass! Mass does appear in both the kinetic energy operator and eigenvalues of operators that contain the kinetic energy operator. Similarly, electrical charge does not appear in the time-independent wavefunction in this particular case.

It is reasonable to expect that in some special cases the probability densities in each of the three independent directions should be mutually independent. This implies that the probability density for the time-independent wavefunction $\psi(\boldsymbol{r})$ should be the product of three probability densities, one for each coordinate: $|\psi(\boldsymbol{r})|^2 \propto |X(x)|^2 \times |Y(y)|^2 \times |Z(z)|^2$. Subsequently, we see that the wavefunction is the product of three independent wavefunctions in such a special case and we write $\psi(\boldsymbol{r}) \propto X(x) \times Y(y) \times Z(z)$. Such a wavefunction is said to exhibit the **separation of variables.**

E7B.2(a) An isolated, freely moving hydrogen atom is expected to have a translational, time-independent wavefunction that is a function of the centre-of-mass coordinates x_{cm}, y_{cm}, and z_{cm} so we write $\psi_{cm}(x_{cm},y_{cm},z_{cm})$ or $\psi_{cm}(\boldsymbol{r}_{cm})$ with each variable ranging from $-\infty$ to $+\infty$. The infinitesimal space element for the centre of mass variables is $d\tau_{cm} = dx_{cm}dy_{cm}dz_{cm}$. We expect that there are special cases for which the translational wavefunction exhibits the separation of variables: $\psi_{cm}(\boldsymbol{r}_{cm}) \propto X_{cm}(x_{cm}) \times Y(y_{cm}) \times Z_{cm}(z_{cm})$.

In general we expect that the total wavefunction of the isolated hydrogen atom is the product of the centre-of-mass wavefunction and an electronic wavefunction, $\psi_{el}(\boldsymbol{r})$ or $\psi_{el}(x,y,z)$, that originates from the interaction between the electron and proton:

$$\psi_{total} = \psi_{cm}(\boldsymbol{r}_{cm}) \times \psi_{el}(\boldsymbol{r}) = \psi_{cm}(\boldsymbol{r}_{cm}) \times \psi_{el}(x,y,z)$$

with the origin of x, y, and z at the centre-of-mass, variables ranging from $-\infty$ to $+\infty$, and the infinitesimal space element being $d\tau = dxdydz$. The electronic wavefunction does not exhibit the separation of Cartesian variables because the electrostatic potential between the electron and nucleus is proportional to $1/r$, which cannot be written as a sum of separate terms in the variables x, y, and z. But the hydrogen atom, as viewed from the centre-of-mass, has spherical symmetry which suggests that the electronic wavefunction variables may be separated when using the spherical polar coordinates r, θ, ϕ where $0 \le r \le \infty$, $0 \le \theta \le \pi$, $0 \le \phi \le 2\pi$, and

$d\tau = r^2 \sin\theta\, dr\, d\theta\, d\phi$. The separated functions are symbolized as $R(r)$, $\Theta(\theta)$, and $\Phi(\phi)$ and the electronic wavefunction becomes: $\psi_{el} = R(r)\times\Theta(\theta)\times\Phi(\phi)$.

E7B.3(a) The normalized wavefunction has the form $\psi(\phi) = Ne^{i\phi}$ where N is the normalization constant.

$$\int_0^{2\pi} \psi^*\psi \; d\varphi = 1 \; [7B.4c]$$

$$N^2\int_0^{2\pi} e^{-i\phi}e^{i\phi}\; d\phi = N^2\int_0^{2\pi} d\phi = 2\pi N^2 = 1$$

$$\boxed{N = \left(\frac{1}{2\pi}\right)^{1/2}}$$

E7B.4(a) $\psi(\phi) = (1/2\pi)^{1/2}\, e^{i\phi}$ so $|\psi(\phi)|^2 = (1/2\pi)e^{i\phi}e^{-i\phi} = 1/2\pi$.

Thus, the probability of finding the atom in an infinitesimal volume element at any angle is $\boxed{(1/2\pi)d\phi}$.

E7B.5(a) The normalized wavefunction is $\psi = \left(\frac{1}{2\pi}\right)^{½} e^{i\phi}$.

$$\text{Probability that } \pi/2 \le \phi \le 3\pi/2 = \int_{\pi/2}^{3\pi/2} \psi^*\psi\; d\phi$$

$$= \left(\frac{1}{2\pi}\right)\int_{\pi/2}^{3\pi/2} e^{-i\phi}e^{i\phi}\; d\phi = \left(\frac{1}{2\pi}\right)\int_{\pi/2}^{3\pi/2} d\phi$$

$$= \left(\frac{1}{2\pi}\right)\phi\Big|_{\phi=\pi/2}^{\phi=3\pi/2} = \left(\frac{1}{2\pi}\right)\times\left(\frac{3\pi}{2} - \frac{\pi}{2}\right)$$

$$= \boxed{\frac{1}{2}}$$

Solutions to problems

P7B.1 We require $\int \psi^*\psi\, d\tau = 1$, and so write $\psi = Nf$ and find normalization constant N for the given function f.

(a) $$\int \psi^*\psi\, d\tau = N^2\int_0^L \sin^2\frac{n\pi x}{L}dx$$

$$= \frac{1}{2}N^2\int_0^L \left(1 - \cos\frac{2n\pi x}{L}\right)dx \text{ [trigonometric identity]}$$

$$= \frac{1}{2}N^2\left(x - \frac{L}{2n\pi}\sin\frac{2n\pi x}{L}\right)\Bigg|_0^L$$

$$= \frac{L}{2}N^2 = 1 \quad \text{if } \boxed{N = \left(\frac{2}{L}\right)^{1/2}}$$

(b) $\int \psi^*\psi \,d\tau = N^2\int_{-L}^{L} c^2\,dx = 2N^2c^2L = 1$ if $\boxed{N = \dfrac{1}{c(2L)^{1/2}}}$

(c) $\int \psi^*\psi \,d\tau = \int_0^{\infty}\int_0^{\pi}\int_0^{2\pi} \psi^*\psi r^2 \sin\theta\,dr\,d\theta\,d\phi = N^2\int_0^{\infty} e^{-2r/a}r^2\,dr\int_0^{\pi}\sin\theta\,d\theta\int_0^{2\pi}d\phi$

$$= N^2\left(\frac{a^3}{4}\right)\times(2)\times(2\pi) = 1 \quad \text{if} \quad \boxed{N = \frac{1}{(\pi a^3)^{1/2}}}$$

(d) $\int \psi^*\psi \,d\tau = \int_0^{\infty}\int_0^{\pi}\int_0^{2\pi} \psi^*\psi r^2 \sin\theta\,dr\,d\theta\,d\phi$

$$= N^2\int_0^{\infty} r^2\times r^2 e^{-r/a}\,dr\int_0^{\pi}\sin^3\theta\,d\theta\int_0^{2\pi}\cos^2\phi\,d\phi \qquad [x = r\cos\phi\sin\theta]$$

$$= N^2 4!a^5\times\frac{4}{3}\times\pi = 32\pi a^5 N^2 = 1 \quad \text{if} \quad \boxed{N = \frac{1}{(32\pi a^5)^{1/2}}}$$

We have used $\int \sin^3\theta\,d\theta = -\tfrac{1}{3}(\cos\theta)(\sin^2\theta+2)$ and $\int_0^{2\pi}\cos^2\phi\,d\phi = \int_0^{2\pi}\sin^2\phi\,d\phi$, as found in tables of integrals. Also, $\int_0^{2\pi}(\cos^2\phi+\sin^2\phi)\,d\phi = \int_0^{2\pi}d\phi = 2\pi$.

P7B.3 The normalization constant for this wavefunction is

$$N = \left(\frac{1}{\int_0^{\infty} e^{-ax}\times e^{-ax}\,dx}\right)^{1/2} [7\text{B}.2] = \left(\frac{1}{\int_0^{\infty} e^{-2ax}\,dx}\right)^{1/2} = \left(\frac{-2a}{e^{-2ax}\big|_{x=0}^{\infty}}\right)^{1/2} = (2a)^{1/2}$$

so the normalized wavefunction is $\psi = (2a)^{1/2}e^{-ax}$.

$$P = \int_b^{\infty}\psi^2\,dx = (2a)\int_b^{\infty}e^{-2ax}\,dx = e^{-2ab} \quad \text{where} \quad ab = (2\text{ m}^{-1})(1\text{ m}) = 2$$
$$= e^{-4} = \boxed{0.0183}$$

P7B.5 The normalized wavefunction is

$$\psi(r) = \left(\frac{1}{\pi a_0^3}\right)^{1/2} e^{-r/a_0} \quad \text{and} \quad \psi(r)^2 = \frac{1}{\pi a_0^3}e^{-2r/a_0} \quad \text{where } a_0 = 53\text{ pm}$$

The probability P that the particle will be found in the region is the integral summation of all the infinitesimal probabilities of finding the particle within the region: $P = \int_{\text{region}}\psi(r)^2\,d\tau$. In this problem we are to integrate over the region of a sphere for which the radius (b = 1.0 pm = a_0/53) is a very small fraction of the characteristic wavefunction distance a_0. Consequently, we may assume that ψ does not vary much within the sphere. Then, the probability is given by

$$P=\int_{\text{region}}\psi(r)^2\,\mathrm{d}\tau \sim \psi(a)^2\int_{\text{region}}\mathrm{d}\tau$$

where a is the position of the centre of the sphere

$$\sim\left(\frac{4\pi b^3}{3}\right)\left(\frac{1}{\pi a_0^3}\right)\mathrm{e}^{-2a/a_0}$$

$$\sim\left(\frac{4}{3}\right)\left(\frac{b}{a_0}\right)^3\mathrm{e}^{-2a/a_0}$$

(a) $b = 1.0\ \text{pm} = a_0/53$ is the radius of a sphere that is centred on the origin ($a = 0$).

$$P(0,b)=\frac{4}{3\times(53)^3}\mathrm{e}^{-0}=\boxed{9.0\times10^{-6}}$$

(b) $b = 1.0\ \text{pm} = a_0/53$ is the radius of a sphere that is centred at $a = a_0$.

$$P=\left(\frac{4}{3\times(53)^3}\right)\mathrm{e}^{-2}=\boxed{1.2\times10^{-6}}$$

P7B.7 The most probable location occurs when the probability density, $|\psi|^2$, is a maximum. Thus, we wish to find the value $x = x_{\text{max}}$ such that $\mathrm{d}|\psi|^2/\mathrm{d}x = 0$.

$$\psi(x)=Nx\mathrm{e}^{-x^2/2a^2}$$

$$|\psi^2|=N^2x^2\mathrm{e}^{-x^2/a^2}$$

$$\mathrm{d}|\psi^2|/\mathrm{d}x=N^2\left\{2x\mathrm{e}^{-x^2/a^2}-\left(2x^3/a^2\right)\mathrm{e}^{-x^2/a^2}\right\}=2N^2x\left\{1-x^2/a^2\right\}\mathrm{e}^{-x^2/a^2}$$

The above derivative equals zero when the factor $1-x^2/a^2$ equals zero so we conclude that $\boxed{x_{\text{max}}=a}$.

7C The principles of quantum theory

Answers to discussion questions

D7C.1 According to eqns 7C.5, 7C.11, and 7C.12, the expectation value for the kinetic energy of a particle equals the expectation value of the 'curvature' operator, $\mathrm{d}^2/\mathrm{d}x^2$:

$$\langle E_{\mathrm{k}}\rangle=-\frac{\hbar^2}{2m}\left\langle\frac{\mathrm{d}^2}{\mathrm{d}x^2}\right\rangle$$

This illustrates the important concept that a wavefunction that is sharply curved will have a higher expectation value for kinetic energy than a wavefunction of low curvature. Thus, we expect that, since total energy is conserved and particulate energy is the sum of kinetic and potential terms, the shape of a wavefunction is an indicator of the interplay between kinetic and potential energy. To clarify this association, carefully examine text Figs. 7C.1–7C.3 along with the associated text discussion.

D7C.3 If the wavefunction describing the linear momentum of a particle is precisely known, the particle has a definite state of linear momentum; but then according to the uncertainty principle (eqn 7C.13a), the position of the particle is completely unknown. Conversely, if the position of a particle is precisely known, its linear momentum cannot be described by a single wavefunction. Rather, the wavefunction is a superposition of many wavefunctions, each corresponding to a different value for the linear momentum. All knowledge of the linear momentum of the particle is lost when its position is specified exactly. In the limit of an infinite number of superposed wavefunctions, the wavepacket illustrated in text Fig. 7C.6 turns into the sharply spiked packet shown in text Fig. 7C.5. But the requirement of the superposition of an infinite number of momentum wavefunctions in order to locate the particle means a complete lack of knowledge of the momentum.

Solutions to exercises

E7C.1(a) The classical expression for the potential energy of a harmonic oscillator is given by $E_p = {}^1/_2 k_f x^2$ where k_f is the force constant. To construct the potential energy operator, $\hat{V}$, we replace the position x in the classical expression with the position operator identity $\hat{x} = x\times$. This rather simple procedure yields: $\boxed{\hat{V} = \tfrac{1}{2}k_f x^2}$.

E7C.2(a) Let f and g be functions of x and examine the integral $\int_{-\infty}^{\infty} f^* \left(-\frac{\hbar^2}{2m}\frac{d^2}{dx^2}\right) g\, dx$.

Integrate successively by parts and use the fact that these functions must be well behaved at the boundaries (i.e. they equal zero at infinity in either direction).

$$\int_{-\infty}^{\infty} f^* \left(-\frac{\hbar^2}{2m}\frac{d^2}{dx^2}\right) g\, dx = \left(-\frac{\hbar^2}{2m}\right)\int_{-\infty}^{\infty} f^* \left(\frac{d^2}{dx^2}\right) g\, dx = \left(-\frac{\hbar^2}{2m}\right)\int_{-\infty}^{\infty} f^* \left(\frac{d}{dx}\right)\frac{dg}{dx}\, dx$$

$$= \left(-\frac{\hbar^2}{2m}\right)\times\left\{\left[f^*\frac{dg}{dx}\right]_{-\infty}^{\infty} - \int_{-\infty}^{\infty}\frac{df^*}{dx}\times\frac{dg}{dx}\, dx\right\}$$

$$= -\left(-\frac{\hbar^2}{2m}\right)\times\left\{\int_{-\infty}^{\infty}\frac{df^*}{dx}\times\frac{dg}{dx}\, dx\right\}$$

$$= -\left(-\frac{\hbar^2}{2m}\right)\times\left\{\left[g\frac{df^*}{dx}\right]_{-\infty}^{\infty} - \int_{-\infty}^{\infty} g\times\frac{d^2 f^*}{dx^2}\, dx\right\}$$

$$= \left(-\frac{\hbar^2}{2m}\right)\times\left\{\int_{-\infty}^{\infty} g\times\frac{d^2 f^*}{dx^2}\, dx\right\}$$

$$= \int_{-\infty}^{\infty} g\left(-\frac{\hbar^2}{2m}\frac{d^2}{dx^2}\right) f^*\, dx = \left\{\int_{-\infty}^{\infty} g^*\left(-\frac{\hbar^2}{2m}\frac{d^2}{dx^2}\right) f\, dx\right\}^*$$

Thus, $\int_{-\infty}^{\infty} f^* \left(-\frac{\hbar^2}{2m}\frac{d^2}{dx^2}\right) g\, dx = \left\{\int_{-\infty}^{\infty} g^*\left(-\frac{\hbar^2}{2m}\frac{d^2}{dx^2}\right) f\, dx\right\}^*$

This is exactly the criteria that a hermitian operator must satisfy [7C.7] so we conclude that the kinetic energy operator is hermitian.

E7C.3(a) ψ_i and ψ_j are orthogonal if $\int \psi_i^* \psi_j \, d\tau = 0$ [7C.8]. Where $n \neq m$ and both n and m are integers, explicit integration gives

$$\begin{aligned}
&\int_0^L \sin(n\pi x/L) \times \sin(m\pi x/L)\,dx \\
&= \left[\frac{\sin(\pi(n-m)x/L)}{2\pi(n-m)/L} - \frac{\sin(\pi(n+m)x/L)}{2\pi(n+m)/L}\right]_{x=0}^{x=L} \\
&= \frac{\sin(\pi(n-m))}{2\pi(n-m)/L} - \frac{\sin(\pi(n+m))}{2\pi(n+m)/L} - \left\{\frac{\sin(0)}{2\pi(n-m)/L} - \frac{\sin(0)}{2\pi(n+m)/L}\right\} \\
&= 0 \quad \text{because the sine of an integer multiple of } \pi \text{ equals zero.}
\end{aligned}$$

Thus, the functions $\sin(n\pi x/L)$ and $\sin(m\pi x/L)$ are orthogonal in the region $0 \leq x \leq L$.

Alternatively, successively integrate by parts.

$$\begin{aligned}
&\int_0^L \sin(n\pi x/L) \times \sin(m\pi x/L)\,dx \\
&= \left[\sin(n\pi x/L) \times \left(\frac{-\cos(m\pi x/L)}{m\pi/L}\right)\right]_{x=0}^{x=L} \\
&\quad - \int_0^L (n\pi/L)(\cos(n\pi x/L)) \times \left(\frac{-\cos(m\pi x/L)}{m\pi/L}\right) dx \quad \text{[integration by parts]} \\
&= \left(\frac{n}{m}\right) \int_0^L \cos(n\pi x/L) \times \cos(m\pi x/L)\,dx \quad [\text{use } \sin(n\pi) = 0 \text{ for multiples of } \pi] \\
&= \left(\frac{n}{m}\right) \times \left\{\left[\cos(n\pi x/L) \times \left(\frac{\sin(m\pi x/L)}{m\pi/L}\right)\right]_{x=0}^{x=L} - \int_0^L (n\pi/L)(-\sin(n\pi x/L)) \times \left(\frac{\sin(m\pi x/L)}{m\pi/L}\right) dx\right\} \\
&\quad \text{[integration by parts]} \\
&= \left(\frac{n}{m}\right)^2 \times \left\{\int_0^L \sin(n\pi x/L) \times \sin(m\pi x/L)\,dx\right\} \quad [\text{use } \sin(m\pi) = 0 \text{ for multiples of } \pi]
\end{aligned}$$

Thus, $\left(1 - \left(\frac{n}{m}\right)^2\right) \times \left\{\int_0^L \sin(n\pi x/L) \times \sin(m\pi x/L)\,dx\right\} = 0$

and we conclude that the integral necessarily equals zero when $n \neq m$.

E7C.4(a) ψ_i and ψ_j are orthogonal if $\int \psi_i^* \psi_j \, d\tau = 0$ [7C.8].

$$\begin{aligned}
\int_0^{2\pi} \left(e^{i\phi}\right)^* \times e^{2i\phi}\,d\varphi &= \int_0^{2\pi} e^{-i\phi} \times e^{2i\phi}\,d\varphi = \int_0^{2\pi} e^{i\phi}\,d\phi = \frac{1}{i} e^{i\phi} \Big|_{\phi=0}^{\phi=2\pi} \\
&= \frac{1}{i}\left(e^{2\pi i} - e^0\right) = \frac{1}{i}\left(e^{2\pi i} - 1\right) = \frac{1}{i}\left(\cos(2\pi) - i\sin(2\pi) - 1\right) = \frac{1}{i}(1 - 0 - 1) = 0
\end{aligned}$$

(The Euler identity $e^{ai} = \cos(a) - i\sin(a)$ has been used in the math manipulations.)

Thus, the functions $e^{i\phi}$ and $e^{2i\phi}$ are orthogonal in the region $0 \le \phi \le 2\pi$

E7C.5(a) The normalized form of this wavefunction is:

$$\psi(x) = \left(\frac{2}{L}\right)^{1/2} \sin(2\pi x / L)$$

The expectation value of the electron position is:

$$\langle x \rangle = \int_0^L \psi^* x \psi \,\mathrm{d}x \ [7.2] = \left(\frac{2}{L}\right)\int_0^L x \sin^2\left(\frac{2\pi x}{L}\right)\mathrm{d}x$$

$$= \left(\frac{2}{L}\right) \times \left[\frac{x^2}{4} - \frac{x\sin\left(\frac{4\pi x}{L}\right)}{8\pi / L} - \frac{\cos\left(\frac{4\pi x}{L}\right)}{8(2\pi / L)^2}\right]_{x=0}^{x=L}$$

$$= \left(\frac{2}{L}\right) \times \left[\frac{L^2}{4} - \frac{L\sin(4\pi)}{8\pi / L} - \frac{\cos(4\pi)}{8(2\pi / L)^2} - \left\{\frac{0^2}{4} - \frac{0 \times \sin(0)}{8\pi / L} - \frac{\cos(0)}{8(2\pi / L)^2}\right\}\right]$$

$$= \left(\frac{2}{L}\right) \times \left(\frac{L^2}{4}\right)$$

$$= \boxed{\frac{L}{2}}$$

E7C.6(a) The normalized form of this wavefunction is:

$$\psi(x) = \left(\frac{2}{L}\right)^{1/2} \sin(\pi x / L)$$

$$\frac{\mathrm{d}\psi}{\mathrm{d}x} = \left(\frac{2}{L}\right)^{1/2} \left(\frac{\pi}{L}\right)\cos(\pi x / L)$$

The expectation value of the electron momentum is:

$$\langle p_x \rangle = \int_0^L \psi^* \hat{p}_x \psi \,\mathrm{d}x \ [7C.11] = \int_0^L \psi^* \left(\frac{\hbar}{\mathrm{i}}\frac{\mathrm{d}}{\mathrm{d}x}\right)\psi \,\mathrm{d}x = \left(\frac{\hbar}{\mathrm{i}}\right)\int_0^L \psi^* \left(\frac{\mathrm{d}\psi}{\mathrm{d}x}\right)\mathrm{d}x$$

$$= \left(\frac{\hbar}{\mathrm{i}}\right)\left(\frac{2}{L}\right)^{1/2}\left(\frac{\pi}{L}\right)\int_0^L \psi^* \cos\left(\frac{\pi x}{L}\right)\mathrm{d}x = \left(\frac{\hbar}{\mathrm{i}}\right)\left(\frac{2}{L}\right)\left(\frac{\pi}{L}\right)\int_0^L \sin\left(\frac{\pi x}{L}\right)\cos\left(\frac{\pi x}{L}\right)\mathrm{d}x$$

$$= \left(\frac{h}{\mathrm{i}L^2}\right)\int_0^L \sin\left(\frac{\pi x}{L}\right)\cos\left(\frac{\pi x}{L}\right)\mathrm{d}x$$

$$= \left(\frac{h}{\mathrm{i}L^2}\right) \times \left[\frac{\sin^2\left(\frac{\pi x}{L}\right)}{2\pi / L}\right]_{x=0}^{x=L} = \left(\frac{h}{\mathrm{i}L^2}\right) \times \left(\frac{\sin^2(\pi)}{2\pi / L} - \frac{\sin^2(0)}{2\pi / L}\right) = \left(\frac{h}{\mathrm{i}L^2}\right) \times (0+0) = \boxed{0}$$

E7C.7(a) The minimum uncertainty in position and momentum is given by the uncertainty principle in the form $\Delta p\Delta q \geq \tfrac{1}{2}\hbar$ [7C.13a] with the choice of the equality. The uncertainty in momentum is $\Delta p = m\Delta\upsilon$ so:

$$\Delta\upsilon_{\min} = \frac{\hbar}{2m\Delta q} = \frac{1.055\times10^{-34}\,\mathrm{J\,s}}{(2)\times(0.500\,\mathrm{kg})\times(1.0\times10^{-6}\,\mathrm{m})} = \boxed{1.1\times10^{-28}\,\mathrm{m\,s^{-1}}}$$

The minimum uncertainty in position of a 5.0 g mass is

$$\Delta q_{\min} = \frac{\hbar}{2m\Delta\upsilon} = \frac{1.055\times10^{-34}\,\mathrm{J\,s}}{(2)\times(0.0050\,\mathrm{kg})\times(1.0\times10^{-5}\,\mathrm{m\,s^{-1}})} = \boxed{1.1\times10^{-27}\,\mathrm{m}}$$

Comment. These uncertainties are extremely small; thus, the ball and bullet are effectively classical particles.

Question. If the ball were stationary (no uncertainty in position) the uncertainty in speed would be infinite. Thus, the ball could have a very high speed, contradicting the fact that it is stationary. What is the resolution of this apparent paradox?

E7C.8(a) The desired uncertainty in the proton momentum is

$$\begin{aligned}\Delta p &= 1.00\times10^{-4}\,p = 1.00\times10^{-4}\,m_{\mathrm{p}}\upsilon \\ &= \left(1.00\times10^{-4}\right)\times\left(1.673\times10^{-27}\,\mathrm{kg}\right)\times\left(0.45\times10^{6}\,\mathrm{m\,s^{-1}}\right) = 7.5\overline{3}\times10^{-26}\,\mathrm{kg\,m\,s^{-1}}\end{aligned}$$

Thus, the minimum uncertainty in position must be

$$\Delta x = \frac{\hbar}{2\Delta p}\,[7\mathrm{C}.13\mathrm{a}] = \frac{1.055\times10^{-34}\,\mathrm{J\,s}}{2\times\left(7.5\overline{3}\times10^{-26}\,\mathrm{kg\,m\,s^{-1}}\right)} = \boxed{700\,\mathrm{pm}}$$

E7C.9(a) The quantity $\left[\hat{\Omega}_1,\hat{\Omega}_2\right] = \hat{\Omega}_1\hat{\Omega}_2 - \hat{\Omega}_2\hat{\Omega}_1$ [7C.15] is referred to as the commutator of the operators $\hat{\Omega}_1$ and $\hat{\Omega}_2$. In obtaining the commutator it is necessary to realize that the operators operate on functions; thus, we find expressions for $\left[\hat{\Omega}_1,\hat{\Omega}_2\right]\psi(x) = \hat{\Omega}_1\hat{\Omega}_2\psi(x) - \hat{\Omega}_2\hat{\Omega}_1\psi(x)$.

(a) $$\left[\frac{\mathrm{d}}{\mathrm{d}x},\frac{1}{x}\right]\psi = \frac{\mathrm{d}}{\mathrm{d}x}\times\left(\frac{1}{x}\psi\right) - \frac{1}{x}\frac{\mathrm{d}}{\mathrm{d}x}\psi = \left(-\frac{1}{x^2}\right)\psi + \frac{1}{x}\frac{\mathrm{d}}{\mathrm{d}x}\psi - \frac{1}{x}\frac{\mathrm{d}}{\mathrm{d}x}\psi = \left(-\frac{1}{x^2}\right)\psi$$

Thus, $$\left[\frac{\mathrm{d}}{\mathrm{d}x},\frac{1}{x}\right] = \boxed{-\frac{1}{x^2}}$$

(b) $$\left[\frac{\mathrm{d}}{\mathrm{d}x},x^2\right]\psi = \frac{\mathrm{d}}{\mathrm{d}x}\times\left(x^2\psi\right) - x^2\frac{\mathrm{d}}{\mathrm{d}x}\psi = (2x)\psi + x^2\frac{\mathrm{d}}{\mathrm{d}x}\psi - x^2\frac{\mathrm{d}}{\mathrm{d}x}\psi = (2x)\psi$$

Thus, $$\left[\frac{\mathrm{d}}{\mathrm{d}x},x^2\right] = \boxed{2x}$$

Solutions to problems

P7C.1 Time-independent Schrödinger equation of a single particle in one dimension:

$$\left(-\frac{\hbar^2}{2m}\frac{d^2}{dx^2}+\hat{V}\right)\psi = E\psi \text{ [7B.1]}$$

(a) $\left(-\frac{\hbar^2}{2m_e}\frac{d^2}{dx^2}-\frac{e^2}{4\pi\varepsilon_0 x}\right)\psi = E\psi$

(b) $\left(-\frac{\hbar^2}{2m}\frac{d^2}{dx^2}\right)\psi = E\psi$

(c) $F = c$ (a constant) implies that $V = -cx$ because $F = -dV/dx$.

$$\left(-\frac{\hbar^2}{2m}\frac{d^2}{dx^2}-cx\right)\psi = E\psi$$

P7C.3 Form $\hat{\Omega}f$ in each case. If the result is ωf where ω is a constant, then f is an eigenfunction of the operator $\hat{\Omega}$ and ω is the eigenvalue [7C.2a, b, and c]. We check whether f is an eigenfunction of the $\hat{\Omega} = d/dx$ operator.

(a) $\frac{d}{dx}f = \frac{d}{dx}e^{ikx} = ike^{ikx} = ikf$

$\boxed{\text{Yes}}$, the function is an eigenfunction with the eigenvalue ik.

(b) $\frac{d}{dx}f = \frac{d}{dx}k = 0 = 0\times f$

$\boxed{\text{Yes}}$, the function is an eigenfunction with the eigenvalue 0.

(c) $\frac{d}{dx}f = \frac{d}{dx}kx = k$

No, the function is not an eigenfunction because $k \neq \text{constant}\times f$.

(e) $\frac{d}{dx}f = \frac{d}{dx}e^{-ax^2} = -2axe^{-ax^2} = -2axf$

No, the function is not an eigenfunction because $-2ax \neq$ constant.

P7C.5 Form $\hat{\Omega}f$ in each case of Problem 7C.3 (and of Problem 7C.4). If the result is ωf where ω is a constant, then f is an eigenfunction of the operator $\hat{\Omega}$ and ω is the eigenvalue [7C.2a, b, and c]. We first check whether f is an eigenfunction of the $\hat{\Omega} = d/dx$ operator after which we check whether f is an eigenfunction of the $\hat{\Omega} = d^2/dx^2$ operator.

(a) $f = e^{ikx}$

$$\frac{d}{dx}f = \frac{d}{dx}e^{ikx} = ike^{ikx} = \text{constant}\times f$$

$$\frac{d^2}{dx^2}f = \frac{d}{dx}\left(ike^{ikx}\right) = (ik)^2 e^{ikx} = \text{constant}\times f$$

$\boxed{\text{Yes}}$, f is an eigenfunction of d^2/dx^2 with eigenvalue is $\boxed{-k^2}$.

(b) $f = k$

$$\frac{d}{dx}f = \frac{d}{dx}k = 0 = \text{constant} \times f$$

$$\frac{d^2}{dx^2}f = \frac{d}{dx}(0) = 0 = \text{constant} \times f \quad \boxed{\text{Yes}}, f \text{ is a trivial eigenfunction of } d^2/dx^2.$$

(c) $f = kx$

$$\frac{d}{dx}f = \frac{d}{dx}(kx) = k \neq \text{constant} \times f$$

$$\frac{d^2}{dx^2}f = \frac{d}{dx}(k) = 0 = \text{constant} \times f \quad \boxed{\text{Yes}}, f \text{ is a trivial eigenfunction of } d^2/dx^2.$$

(d) $f = e^{-ax^2}$

$$\frac{d}{dx}f = \frac{d}{dx}e^{-ax^2} = -2axe^{-ax^2} \neq \text{constant} \times f$$

$$\frac{d^2}{dx^2}f = \frac{d}{dx}\left(-2axe^{-ax^2}\right) = -2a\left(1 - 2ax^2\right)e^{-ax^2} \neq \text{constant} \times f \quad \boxed{\text{No}},$$

f is not an eigenfunction of d^2/dx^2.

(i) In addition to being an eigenfunction of d/dx the functions of $\boxed{\text{(a) and (b)}}$ are also eigenfunctions of d/dx^2.

(ii) The function of $\boxed{\text{(c)}}$, not being eigenfunction of d/dx, is a (trivial) eigenfunction of d/dx^2.

The problem may be repeated with the functions of Problem 7C.4 to get the following results.

(a) $f = x^3 - kx$

$$\frac{d}{dx}f = \frac{d}{dx}\left(x^3 - kx\right)$$

$$= 3x^2 - k \neq \text{constant} \times f \quad \text{No}, f \text{ is not an eigenfunction of } d/dx.$$

$$\frac{d^2}{dx^2}f = \frac{d}{dx}\left(3x^2 - k\right)$$

$$= 6x \neq \text{constant} \times f \quad \text{No}, f \text{ is not an eigenfunction of } d^2/dx^2.$$

(b) $f = \cos kx$

$$\frac{d}{dx}f = \frac{d}{dx}(\cos kx)$$

$$= -k\sin kx \neq \text{constant} \times f \quad \text{No}, f \text{ is not an eigenfunction of } d/dx.$$

$$\frac{d^2}{dx^2}f = \frac{d}{dx}(-k\sin kx) = -k^2\cos kx = \text{constant} \times f$$

Yes, f is an eigenfunction of d^2/dx^2. Its eigenvalue is $-k^2$.

(c) $f = x^2 + 3x - 1$

$$\frac{d}{dx}f = \frac{d}{dx}\left(x^2 + 3x - 1\right)$$

$$= 2x + 3 \neq \text{constant} \times f \quad \text{No, } f \text{ is not an eigenfunction of d/d}x.$$

$$\frac{d^2}{dx^2}f = \frac{d}{dx}(2x+3) = 2 \neq \text{constant} \times f \quad \text{No, } f \text{ is not an eigenfunction of d}^2/\text{d}x^2.$$

Of these Problem 7C.4 functions, only $f = \cos kx$ is an eigenfunction of d^2/dx^2. It is not, however, an eigenfunction of d/dx.

P7C.7 (a) The function e^{+ikx} is an eigenfunction of the linear momentum operator, $\hat{p}_x = \frac{\hbar}{i}\frac{d}{dx}$ [7C.3]. It has the eigenvalues $+k\hbar$:

$$\hat{p}e^{+ikx} = \frac{\hbar}{i}\frac{d}{dx}e^{+ikx} = \left(\frac{\hbar}{i}\right) \times (ik)e^{+ikx} = +k\hbar e^{+ikx}$$

Consequently, the particle has the linear momentum $+k\hbar$.

(b) The wavefunction $\psi = N\cos kx$ is not an eigenfunction of the linear momentum operator so we find the expectation value for linear momentum with eqn 7C.11.

$$\langle p_x \rangle = \int_{-\infty}^{\infty} \psi^* \hat{p}_x \psi \, dx \ [7C.11] = N^2 \int_{-\infty}^{\infty} \cos kx \left(\frac{\hbar}{i}\frac{d}{dx}\right) \cos kx dx$$

$$= N^2\left(\frac{\hbar}{i}\right)\int_{-\infty}^{\infty} \cos kx \left(\frac{d}{dx}\right)\cos kx dx = -kN^2\left(\frac{\hbar}{i}\right)\int_{-\infty}^{\infty} \cos kx \times \sin kx dx$$

$$= -kN^2\left(\frac{\hbar}{i}\right)\lim_{\chi\to\infty}\left[\frac{\sin^2 kx}{2k}\right]_{x=-\chi}^{x=\chi} \quad \text{[standard integral]}$$

$$= -kN^2\left(\frac{\hbar}{i}\right)\lim_{\chi\to\infty}\left[\frac{\sin^2(k\chi)}{2k} - \frac{\sin^2(-k\chi)}{2k}\right]$$

$$= -kN^2\left(\frac{\hbar}{i}\right)\lim_{\chi\to\infty}\left[\frac{\sin^2(k\chi)}{2k} - \frac{\sin^2(k\chi)}{2k}\right] = \boxed{0}$$

(c) $\psi = Ne^{-ax^2}$

$$\frac{d}{dx}\psi = N\frac{d}{dx}e^{-ax^2} = -2aNxe^{-ax^2} = -2ax\psi$$

The wavefunction is not an eigenfunction of the linear momentum operator so we find the expectation value for linear momentum with eqn 7C.11.

$$\langle p_x \rangle = \int_{-\infty}^{\infty} \psi^* \hat{p}_x \psi \, dx \ [7C.11] = N^2 \int_{-\infty}^{\infty} e^{-ax^2}\left(\frac{\hbar}{i}\frac{d}{dx}\right)e^{-ax^2} dx$$

$$= -2aN^2\left(\frac{\hbar}{i}\right)\int_{-\infty}^{\infty} xe^{-2ax^2} dx$$

The integrand of the above integral is an odd function so, when it is integrated around its centre of symmetry at $x=0$, the integral equals zero. Thus, $\langle p_x\rangle = \boxed{0}$.

P7C.9 $\psi = a^{1/2}\mathrm{e}^{-ax/2}$

$$\langle x\rangle = \int_0^\infty \psi^* x\psi\,\mathrm{d}x = a\int_0^\infty x\mathrm{e}^{-ax}\,\mathrm{d}x \quad \text{[Integrate by parts]}$$

$$= a\left\{\left[x\times\left(-\frac{1}{a}\mathrm{e}^{-ax}\right)\right]_0^\infty - \int_0^\infty\left(-\frac{1}{a}\mathrm{e}^{-ax}\right)\mathrm{d}x\right\} = \int_0^\infty \mathrm{e}^{-ax}\,\mathrm{d}x$$

$$= \left[-\frac{1}{a}\mathrm{e}^{-ax}\right]_0^\infty = -\frac{1}{a}\left[\mathrm{e}^{-\infty}-\mathrm{e}^{-0}\right] = -\frac{1}{a}[0-1] = \boxed{\frac{1}{a}}$$

P7C.11 In each case the normalization constant $N^{-2} = \int|\psi|^2\,\mathrm{d}\tau$ [7B.2] must be evaluated by analytically determining the integral over the whole space of the wavefunction. The normalization integrals are best evaluated using the spherical coordinates (r,θ,ϕ) for which $0\le r\le\infty$, $0\le\theta\le\pi$, and $0\le\phi\le 2\pi$. It is helpful to recognize that, when a wavefunction has the separation of variables form $\psi(r,\theta,\phi) = R(r)\times\Theta(\theta)\times\Phi(\phi)$, the integral over the space of all variables is

$$N^{-2} = \int|\psi|^2\,\mathrm{d}\tau = \int_{r=0}^\infty\int_{\theta=0}^\pi\int_{\phi=0}^{2\pi}|\psi|^2 r^2\sin\theta\,\mathrm{d}r\,\mathrm{d}\theta\,\mathrm{d}\phi \quad \left[\text{Note that } \mathrm{d}\tau = r^2\sin\theta\,\mathrm{d}r\,\mathrm{d}\theta\,\mathrm{d}\phi.\right]$$

$$= \int_{r=0}^\infty\int_{\theta=0}^\pi\int_{\phi=0}^{2\pi} r^2\times\left(R(r)\times\Theta(\theta)\times\Phi(\phi)\right)^2\sin\theta\,\mathrm{d}r\,\mathrm{d}\theta\,\mathrm{d}\phi$$

$$= \int_{r=0}^\infty r^2\times R(r)^2\,\mathrm{d}r\times\int_{\theta=0}^\pi \sin(\theta)\times\Theta(\theta)^2\,\mathrm{d}\theta\times\int_{\phi=0}^{2\pi}\Phi(\phi)^2\,\mathrm{d}\phi$$

In the special case (i) for which $\Theta(\theta)=1$ and $\Phi(\phi)=1$:

$$\int_{\theta=0}^\pi \sin(\theta)\times\Theta(\theta)^2\,\mathrm{d}\theta\times\int_{\phi=0}^{2\pi}\Phi(\phi)^2\,\mathrm{d}\phi = \int_{\theta=0}^\pi\sin(\theta)\mathrm{d}\theta\times\int_{\phi=0}^{2\pi}\mathrm{d}\phi = [-\cos\theta]_{\theta=0}^\pi\times[\phi]_{\phi=0}^{2\pi} = 4\pi$$

and the normalization integral is

$$N^{-2} = 4\pi\int_{r=0}^\infty r^2\times R(r)^2\,\mathrm{d}r$$

In the special case (ii) for which $\Theta(\theta)=\sin\theta$ and $\Phi(\phi)=\cos\phi$:

$$\int_{\theta=0}^\pi \sin(\theta)\times\Theta(\theta)^2\,\mathrm{d}\theta\times\int_{\phi=0}^{2\pi}\Phi(\phi)^2\,\mathrm{d}\phi = \int_{\theta=0}^\pi\sin^3(\theta)\mathrm{d}\theta\times\int_{\phi=0}^{2\pi}\cos^2(\phi)\mathrm{d}\phi$$

$$= \left[-\cos\theta + \tfrac{1}{3}\cos^3\theta\right]_{\theta=0}^\pi\times\left[\tfrac{\phi}{2}+\tfrac{1}{4}\sin2\phi\right]_{\phi=0}^{2\pi}$$

$$= \tfrac{4}{3}\pi$$

and the normalization integral is

$$N^{-2} = \tfrac{4}{3}\pi\int_{r=0}^\infty r^2\times R(r)^2\,\mathrm{d}r$$

(a) (i) The unnormalized wavefunction is $\psi=\left(2-\dfrac{r}{a_0}\right)e^{-r/a_0}$ and

$$\psi^2=\left(2-\frac{r}{a_0}\right)^2 e^{-2r/a_0}$$

$$\begin{aligned}N^{-2}&=4\pi\int_{r=0}^{\infty}r^2\times R(r)^2\,dr\\&=4\pi\int_{r=0}^{\infty}r^2\left(2-\frac{r}{a_0}\right)^2 e^{-2r/a_0}\,dr\\&=4\pi a_0^3\int_{\chi=0}^{\infty}\chi^2(2-\chi)^2e^{-2\chi}\,d\chi\quad\text{where}\quad\chi=r/a_0\\&=4\pi a_0^3\int_{\chi=0}^{\infty}\{4\chi^2-4\chi^3+\chi^4\}e^{-2\chi}\,d\chi\end{aligned}$$

[Use the standard integral $\int_0^{\infty}\chi^n e^{-a\chi}\,d\chi=n!/a^{n+1}$.]

$$\begin{aligned}&=4\pi a_0^3\{4\times2!/2^{2+1}-4\times3!/2^{3+1}+4!/2^{4+1}\}\\&=\pi a_0^3\end{aligned}$$

Hence, $\boxed{N=(\pi a_0^3)^{-1/2}}$ and the normalized wavefunction is

$$\psi=\left(\frac{1}{\pi a_0^3}\right)^{1/2}\left(2-\frac{r}{a_0}\right)e^{-r/a_0}.$$

(i) The unnormalized wavefunction is $\psi=r\sin\theta\cos\varphi e^{-r/2a_0}$ and

$$\psi^2=r^2\sin^2\theta\cos^2\varphi e^{-r/a_0}.$$

$$\begin{aligned}N^{-2}&=\tfrac{4}{3}\pi\int_{r=0}^{\infty}r^2\times R(r)^2\,dr\\&=\tfrac{4}{3}\pi\int_{r=0}^{\infty}r^2\times\{re^{-r/2a_0}\}^2\,dr\\&=\tfrac{4}{3}\pi a_0^5\int_{\chi=0}^{\infty}\chi^4e^{-\chi}\,d\chi\quad\text{where}\quad\chi=r/a_0\\&=(\tfrac{4}{3}\pi a_0^5)\times(4!)\quad[\text{Use the standard integral }\int_0^{\infty}\chi^ne^{-a\chi}\,d\chi=n!/a^{n+1}.]\\&=32\pi a_0^5\end{aligned}$$

Hence, $\boxed{N=(32\pi a_0^5)^{-1/2}}$ and the normalized wavefunction is $\psi=\left(\dfrac{1}{32\pi a_0^5}\right)^{1/2}$ $r\sin\theta\cos\varphi e^{-r/2a_0}$.

(b) Since normalization constants do not affect orthogonality, we use the unnormalized wavefunctions to examine the integral $\int \psi_1\psi_2 \mathrm{d}\tau$. The wavefunctions are orthogonal if the integral proves to equal zero.

$$\int \psi_1\psi_2 \mathrm{d}\tau = \int \left\{\left(2-\frac{r}{a_o}\right)\mathrm{e}^{-\frac{r}{a_o}}\right\}\left\{r\sin\theta\cos\phi\,\mathrm{e}^{-\frac{r}{2a_o}}\right\}\mathrm{d}\tau$$

$$= \int_{r=0}^{\infty}\int_{\theta=0}^{\pi}\int_{\phi=0}^{2\pi}\left\{\left(2-\frac{r}{a_o}\right)\mathrm{e}^{-\frac{r}{a_o}}\right\}\left\{r\sin\theta\cos\phi\,\mathrm{e}^{-\frac{r}{2a_o}}\right\}r^2\sin\theta\,\mathrm{d}r\,\mathrm{d}\theta\,\mathrm{d}\phi$$

$$= \int_{r=0}^{\infty}\left\{\left(2r^3-\frac{r^4}{a_o}\right)\mathrm{e}^{-\frac{3r}{2a_o}}\right\}\mathrm{d}r \times \int_{\theta=0}^{\pi}\sin^2\theta\,\mathrm{d}\theta \times \int_{\phi=0}^{2\pi}\cos\phi\;\mathrm{d}\phi$$

The integral on the far right equals zero:

$\int_{\phi=0}^{2\pi}\cos\phi\,\mathrm{d}\phi = \sin\phi\,|_0^{2\pi} = \sin(2\pi)-\sin(0) = 0-0 = 0$

Hence, $\int \psi_1\psi_2 \mathrm{d}\tau = 0$ and we conclude that the wavefunctions are orthogonal.

(c) The normalized excited state wavefunctions for the hydrogen atom are

(i) $\psi = \left(\frac{1}{\pi a_0^3}\right)^{½}\left(2-\frac{r}{a_0}\right)\mathrm{e}^{-r/a_0}$ and (ii) $\psi = \left(\frac{1}{32\pi a_0^5}\right)^{½} r\sin\theta\cos\phi\,\mathrm{e}^{-r/2a_0}$

$$\langle r\rangle = 4\pi\int_{r=0}^{\infty} r^3 \times R(r)^2\,\mathrm{d}r$$

$$= 4\pi\int_{r=0}^{\infty} r^3 \times \left\{\left(\frac{1}{\pi a_0^3}\right)^{½}\left(2-\frac{r}{a_0}\right)\mathrm{e}^{-r/a_0}\right\}^2 \mathrm{d}r$$

$$= 4a_0\int_{\chi=0}^{\infty}\chi^3 \times \left\{(2-\chi)\mathrm{e}^{-\chi}\right\}^2\mathrm{d}\chi \quad \text{where} \quad \chi = r/a_0$$

$$= 4a_0\int_{\chi=0}^{\infty}\left\{4\chi^3-4\chi^4+\chi^5\right\}\mathrm{e}^{-2\chi}\,\mathrm{d}\chi$$

[Use the standard integral $\int_0^{\infty}\chi^n\mathrm{e}^{-a\chi}\,\mathrm{d}\chi = n!/a^{n+1}$.]

$$= 4a_0\left\{4\times 3!/2^{3+1} - 4\times 4!/2^{4+1} + 5!/2^{5+1}\right\}$$

$$= \boxed{1.5a_0}$$

$$\langle r^2\rangle = 4\pi\int_{r=0}^{\infty} r^4 \times R(r)^2\,\mathrm{d}r$$

$$= 4\pi\int_{r=0}^{\infty} r^4 \times \left\{\left(\frac{1}{\pi a_0^3}\right)^{½}\left(2-\frac{r}{a_0}\right)\mathrm{e}^{-r/a_0}\right\}^2 \mathrm{d}r$$

$$= 4a_0^2\int_{\chi=0}^{\infty}\chi^4 \times \left\{(2-\chi)\mathrm{e}^{-\chi}\right\}^2\mathrm{d}\chi \quad \text{where} \quad \chi = r/a_0$$

$$= 4a_0^2\int_{\chi=0}^{\infty}\left\{4\chi^4-4\chi^5+\chi^6\right\}\mathrm{e}^{-2\chi}\,\mathrm{d}\chi$$

[Use the standard integral $\int_0^{\infty}\chi^n\mathrm{e}^{-a\chi}\,\mathrm{d}\chi = n!/a^{n+1}$.]

$$= 4a_0^2\left\{4\times 4!/2^{4+1} - 4\times 5!/2^{5+1} + 6!/2^{6+1}\right\}$$

$$= \boxed{4.5a_0^2}$$

(ii) $\langle r\rangle = \frac{4}{3}\pi\int_{r=0}^{\infty} r^3\times R(r)^2\,\mathrm{d}r$

$$= \tfrac{4}{3}\pi\int_{r=0}^{\infty} r^3\times\left[\left\{\left(\frac{1}{32\pi a_0^5}\right)^{1/2} r\mathrm{e}^{-r/2a_0}\right\}\right]^2 \mathrm{d}r$$

$$= \tfrac{a_0}{24}\int_{\chi=0}^{\infty}\chi^5\mathrm{e}^{-\chi}\,\mathrm{d}\chi \quad \text{where} \quad \chi = r/a_0$$

$$= \left(\tfrac{a_0}{24}\right)\times(5!) \quad [\text{Use the standard integral } \int_0^{\infty}\chi^n\mathrm{e}^{-a\chi}\,\mathrm{d}\chi = n!/a^{n+1}.]$$

$$= \boxed{5a_0}$$

$$\langle r^2\rangle = \tfrac{4}{3}\pi\int_{r=0}^{\infty} r^4\times R(r)^2\,\mathrm{d}r$$

$$= \tfrac{4}{3}\pi\int_{r=0}^{\infty} r^4\times\left[\left\{\left(\frac{1}{32\pi a_0^5}\right)^{1/2} r\mathrm{e}^{-r/2a_0}\right\}\right]^2 \mathrm{d}r$$

$$= \tfrac{a_0^2}{24}\int_{\chi=0}^{\infty}\chi^6\mathrm{e}^{-\chi}\,\mathrm{d}\chi \quad \text{where} \quad \chi = r/a_0$$

$$= \left(\tfrac{a_0^2}{24}\right)\times(6!) \quad [\text{Use the standard integral } \int_0^{\infty}\chi^n\mathrm{e}^{-a\chi}\,\mathrm{d}\chi = n!/a^{n+1}.]$$

$$= \boxed{30a_0^2}$$

P7C.13 For the hermitian operator $\hat{\Omega}$:

$$\langle\Omega^2\rangle = \int\psi^*\hat{\Omega}^2\psi d\tau = \int\psi^*\hat{\Omega}\left(\hat{\Omega}\psi\right)d\tau = \left\{\int\left(\hat{\Omega}\psi\right)^*\hat{\Omega}\psi d\tau\right\}^* \text{ [7C.7]}$$

The integrand on the far right is a function times its complex conjugate, which must always be a real, positive number. When this type of integrand is integrated over real space, the result is always a real, positive number. Thus, the expectation value of the square of a hermitian operator is always positive.

P7C.15 $\hat{x} = x\times$ and $\hat{p}_x = \dfrac{\hbar}{\mathrm{i}}\dfrac{\mathrm{d}}{\mathrm{d}x}$ [7C.3]

$$[\hat{x},\hat{p}_x]\psi = (\hat{x}\hat{p}_x - \hat{p}_x\hat{x})\psi = \hat{x}\hat{p}_x\psi - \hat{p}_x\hat{x}\psi$$

$$= \hat{x}\hat{p}_x\psi - (\hat{p}_x\hat{x})\psi - \hat{x}(\hat{p}_x\psi) \quad \text{[rule for differentiation of a product]}$$

$$= -(\hat{p}_x\hat{x})\psi = -\left(\frac{\hbar}{\mathrm{i}}\frac{\mathrm{d}}{\mathrm{d}x}x\right)\psi = -\frac{\hbar}{\mathrm{i}}\psi$$

$$= \mathrm{i}\hbar\psi$$

Thus, $\boxed{[\hat{x},\hat{p}_x] = \mathrm{i}\hbar}$ for all wavefunctions because the derivation does not depend upon a specific wavefunction. Since the commutator of position and momentum operators does not equal zero, it is not possible to simultaneously measure both x and p_x with arbitrary precision. They are complimentary observables for which for which the most general form of the Heisenberg uncertainty principle is $\Delta x\Delta p_x \geq \tfrac{1}{2}\left|\langle[\hat{x},\hat{p}_x]\rangle\right|$ [7C.17].

P7C.17 (a) In the momentum representation $\hat{p}_x = p_x \times$, consequently

$$[\hat{x},\hat{p}_x]\phi = [\hat{x},p_x\times]\phi = \hat{x}p_x\times\phi - p_x\times\hat{x}\phi = i\hbar\phi \text{ [7C.16]}$$

Suppose that the position operator has the form $\hat{x} = a\dfrac{d}{dp_x}$ where a is a complex number. Then,

$$a\frac{d}{dp_x}(p_x\times\phi) - p_x\times\left(a\frac{d}{dp_x}\phi\right) = i\hbar\phi$$

$$\frac{d}{dp_x}(p_x\times\phi) - p_x\times\left(\frac{d}{dp_x}\phi\right) = \frac{i\hbar}{a}\phi$$

$$\frac{dp_x}{dp_x}\phi + p_x\frac{d\phi}{dp_x} - p_x\frac{d\phi}{dp_x} = \frac{i\hbar}{a}\phi \quad \text{[rule for differentiation of a product]}$$

$$\phi = \frac{i\hbar}{a}\phi$$

This is true when $a = i\hbar$. We conclude that $\hat{x} = i\hbar\dfrac{d}{dp_x}$ in the momentum representation.

(b) The fact that integration is the inverse of differentiation suggests the guess that in the momentum representation

$$\hat{x}^{-1}\phi = \left(i\hbar\frac{d}{dp_x}\right)^{-1}\phi = \left(\frac{1}{i\hbar}\int_{-\infty}^{p_x} dp_x\right)\phi = \frac{1}{i\hbar}\int_{-\infty}^{p_x}\phi\, dp_x$$

where the symbol $\int_{-\infty}^{p_x} dp_x$ is understood to be an integration operator which uses any function on its right side as an integrand. To validate the guess that $\hat{x}^{-1} = \dfrac{1}{i\hbar}\int_{-\infty}^{p_x} dp_x$ we need to confirm the operator relation $\hat{x}^{-1}\hat{x} = \hat{x}\hat{x}^{-1} = \hat{1}$. Using the Leibnitz's rule for differentiation of integrals:

$$\hat{x}\hat{x}^{-1}\phi = \left(i\hbar\frac{d}{dp_x}\right)\left(\frac{1}{i\hbar}\int_{-\infty}^{p_x} dp_x\right)\phi = \left(\frac{d}{dp_x}\right)\left(\int_{-\infty}^{p_x}\phi\, dp_x\right)$$

$$= \left(\int_{-\infty}^{p_x}\frac{d\phi}{dp_x}\, dp_x\right) + \phi(p_x)\lim_{c\to-\infty}\frac{dc}{dp_x} - \phi(-\infty)\frac{dp_x}{dp_x} = \left(\int_{-\infty}^{p_x}\frac{d\phi}{dp_x}\, dp_x\right) - \phi(-\infty)$$

Since $\phi(-\infty)$ must equal zero, we find that

$$\hat{x}\hat{x}^{-1}\phi = \int_{-\infty}^{p_x}\frac{d\phi}{dp_x}\, dp_x = \int_{-\infty}^{p_x} d\phi = \phi(p_x) - \phi(-\infty) = \phi(p_x) = \phi$$

from which we conclude that $\hat{x}\hat{x}^{-1} = \hat{1}$. Likewise,

$$\hat{x}^{-1}\hat{x}\phi = \left(\frac{1}{i\hbar}\int_{-\infty}^{p_x} dp_x\right)\left(i\hbar\frac{d}{dp_x}\right)\phi = \int_{-\infty}^{p_x} d\phi = \phi(p_x) - \phi(-\infty) = \phi(p_x) = \phi$$

Integrated activities

I7.1 (a) $CH_4(g) \rightarrow C(\text{graphite}) + 2\,H_2(g)$

$$\Delta_r G^{\ominus} = -\Delta_f G^{\ominus}(CH_4) = -(-50.72\ \text{kJ mol}^{-1}) = 50.72\ \text{kJ mol}^{-1} \quad \text{at } T_{\text{ref}} = 298.15\ \text{K}$$

$$\Delta_r H^{\ominus} = -\Delta_f H^{\ominus}(CH_4) = -(-74.81\ \text{kJ mol}^{-1}) = 74.81\ \text{kJ mol}^{-1} \quad \text{at } T_{\text{ref}} = 298.15\ \text{K}$$

We want to find the temperature at which $\Delta_r G^{\ominus}(T) = 0$. Below this temperature methane is stable with respect to decomposition into the elements. Above this temperature it is unstable. Assuming that the heat capacities are basically independent of temperature.

$$\Delta_r C_p^{\ominus}(T) \simeq \Delta_r C_p^{\ominus}(T_{\text{ref}}) = \{8.527 + 2(28.824) - 35.31\}\ \text{J K}^{-1}\ \text{mol}^{-1} = 30.865\ \text{J K}^{-1}\ \text{mol}^{-1}$$

$$\Delta_r H^{\ominus}(T) = \Delta_r H^{\ominus}(T_{\text{ref}}) + \int_{T_{\text{ref}}}^{T} \Delta_r C_p^{\ominus}(T)\text{d}T\ [2\text{C.7a}] = \Delta_r H^{\ominus}(T_{\text{ref}}) + \Delta_r C_p^{\ominus} \times (T - T_{\text{ref}})$$

A modification of the Gibbs–Helmholtz equation [3D.10] for chemical reactions can be integrated at constant pressure $(p^{\ominus})$ with use of the properties listed above. The result is a computational equation for $\Delta_r G^{\ominus}(T)$.

$$\left(\frac{\partial}{\partial T}\left(\frac{\Delta_r G^{\ominus}}{T}\right)\right)_p = -\frac{\Delta_r H^{\ominus}}{T^2} \quad [3\text{D.10}]$$

$$\int_{T_{\text{ref}}}^{T} \text{d}\left(\frac{\Delta_r G^{\ominus}}{T}\right) = -\int_{T_{\text{ref}}}^{T} \frac{\Delta_r H^{\ominus}}{T^2}\text{d}T$$

$$\frac{\Delta_r G^{\ominus}(T)}{T} = \frac{\Delta_r G^{\ominus}(T_{\text{ref}})}{T_{\text{ref}}} - \int_{T_{\text{ref}}}^{T} \left\{\frac{\Delta_r H^{\ominus}(T_{\text{ref}}) + \Delta_r C_p^{\ominus} \times (T - T_{\text{ref}})}{T^2}\right\}\text{d}T$$

$$= \frac{\Delta_r G^{\ominus}(T_{\text{ref}})}{T_{\text{ref}}} - \{\Delta_r H^{\ominus}(T_{\text{ref}}) - \Delta_r C_p^{\ominus} \times T_{\text{ref}}\}\int_{T_{\text{ref}}}^{T} \frac{1}{T^2}\text{d}T - \Delta_r C_p^{\ominus}\int_{T_{\text{ref}}}^{T} \frac{1}{T}\text{d}T$$

$$= \frac{\Delta_r G^{\ominus}(T_{\text{ref}})}{T_{\text{ref}}} + \{\Delta_r H^{\ominus}(T_{\text{ref}}) - \Delta_r C_p^{\ominus} \times T_{\text{ref}}\} \times \left\{\frac{1}{T} - \frac{1}{T_{\text{ref}}}\right\} - \Delta_r C_p^{\ominus} \ln\left(\frac{T}{T_{\text{ref}}}\right)$$

$$= \{170.1\ \text{J K}^{-1}\ \text{mol}^{-1}\} + \{6.561 \times 10^4\ \text{JK}^{-1}\text{mol}^{-1}\} \times \left\{\frac{1}{T} - \frac{1}{298.15\ \text{K}}\right\}$$

$$- \{30.865\ \text{JK}^{-1}\text{mol}^{-1}\}\ln\left(\frac{T}{298.15\ \text{K}}\right)$$

The temperature at which $\Delta_r G^{\ominus}(T) = 0$ can be found by finding the root of the right hand side of the above equation on a scientific calculator (the numeric solver can also be used). We find the temperature to be 811 K and conclude that above $\boxed{811\ \text{K}}$ methane is unstable. Alternatively, the above relation can be used to prepare a plot of $\Delta_r G^{\ominus}(T)/T$ against T. Inspection of the plot for the point at which $\Delta_r G^{\ominus}(T)/T = 0$ yields the temperature at which methane becomes unstable.

(b) Wien's law, which is derived in Problem 7A.3, relates the temperature to the wavelength of the most intense radiation.

$$\lambda_{\max} = \frac{hc}{5kT}$$
$$= \frac{(6.626\times10^{-34}\ \mathrm{J\,s})\times(2.998\times10^{8}\ \mathrm{m\,s^{-1}})}{5\times(1.381\times10^{-23}\ \mathrm{J\,K^{-1}})\times(1\,000\ \mathrm{K})}$$
$$= \boxed{2.88\ \mu\mathrm{m}}$$

(c) We begin by reviewing the Stefan–Boltzmann law, derived in Problem 7A.6, that relates the total energy density $\mathcal{E}$ of black-body radiation to the temperature T of the body:

$$\mathcal{E} = \int_0^\infty \rho(\lambda)\,\mathrm{d}\lambda = 8\pi hc\int_0^\infty \frac{\mathrm{d}\lambda}{\lambda^5\left(\mathrm{e}^{hc/\lambda kT}-1\right)}\ [7.8] = \left(\frac{4}{c}\right)\sigma T^4$$

where $\sigma = \dfrac{2\pi^5 k^4}{15h^3c^2} = 5.671\times10^{-8}\ \mathrm{W\,m^{-2}\,K^{-4}}$ is the Stefan–Boltzmann constant.

Consequently, the ratio of the energy density of a star at a 1000 K surface to the energy density of the Sun's 6 000 K surface is

$$\left(\frac{T_{\text{star}}}{T_{\text{Sun}}}\right)^4 = \left(\frac{1\,000\ \mathrm{K}}{6\,000\ \mathrm{K}}\right)^4 = \boxed{7.72\times10^{-4}}$$

(d) We find the fraction of the total energy density by integration of the Planck distribution over the visible range of radiation (between about 700 nm (red) and 420 nm (violet)) followed by division by the total energy density.

$$\mathcal{E}_{\text{visible}} = \int_{420\ \mathrm{nm}}^{700\ \mathrm{nm}} \rho(\lambda)\,\mathrm{d}\lambda = 8\pi hc\int_{420\ \mathrm{nm}}^{700\ \mathrm{nm}} \frac{\mathrm{d}\lambda}{\lambda^5\left(\mathrm{e}^{hc/\lambda kT}-1\right)}\quad [7\mathrm{A}.7]$$

At a high temperature the integration range is small compared to the total range of emitted wavelengths so we estimate the above integrand to be a constant equal to the value of the Planck distribution at the mid-point of the visible range (560 nm).

$$\text{Integrand} = \frac{1}{(560\times10^{-9}\ \mathrm{m})^5\times\left(\mathrm{e}^{(6.626\times10^{-34}\ \mathrm{J\,s})\times(2.998\times10^{8}\ \mathrm{m\,s^{-1}})/\{(560\times10^{-9}\ \mathrm{m})\times\times(1.381\times10^{-23}\ \mathrm{J\,K^{-1}})\times(1000\ \mathrm{K})\}}-1\right)}$$
$$= 1.270\times10^{20}\ \mathrm{m^{-5}}$$

$$\mathcal{E}_{\text{visible}} = 8\pi hc\times(\text{integrand})\times\Delta\lambda$$
$$= 8\pi\times(6.626\times10^{-34}\ \mathrm{J\,s})\times(2.998\times10^{8}\ \mathrm{m\,s^{-1}})\times(1.270\times10^{20}\ \mathrm{m^{-5}})$$
$$\times(700-420)\times10^{-9}\ \mathrm{m}$$
$$= 1.78\times10^{-10}\ \mathrm{J\,m^{-3}}$$

$$\begin{aligned}\text{Fraction of total energy density} &= \frac{\mathcal{E}_{\text{visible}}}{\mathcal{E}_{\text{total}}} \\ &= \frac{\mathcal{E}_{\text{visible}}}{\frac{4}{c}\sigma T^4} \\ &= \frac{1.78\times10^{-10}\ \text{J m}^{-3}}{\left(\frac{4}{2.998\times10^{8}\ \text{m s}^{-1}}\right)\times\left(5.671\times10^{-8}\ \text{W m}^{-2}\,\text{K}^{-4}\right)\times\left(1\,000\ \text{K}\right)^4} \\ &= \boxed{2.35\times10^{-7}}\end{aligned}$$

Very little of the brown dwarf's radiation is in the visible. It doesn't shine brightly.

8 The quantum theory of motion

8A Translation

Answers to discussion questions

D8A.1 In quantum mechanics, particles are said to have wave characteristics. The fact of the existence of the particle then requires that the wavelengths of the waves representing it be such that the wave does not experience destructive interference upon reflection by a barrier or in its motion around a closed loop. This requirement restricts the wavelength to values $\lambda = 2/n \times L$, where L is the length of the path and n is a positive integer. Then using the relations $\lambda = h/p$ and $E = p^2/2m$, the energy is quantized at $E = n^2h^2/8mL^2$. This derivation applies specifically to the particle in a box, the derivation is similar for the particle on a ring; the same principles apply. (See Section 8C.1).

D8A.3 The physical origin of tunnelling is related to the probability density of the particle, which according to the Born interpretation is the square of the wavefunction that represents the particle. This interpretation requires that the wavefunction of the system be everywhere continuous, even at barriers. Therefore, if the wavefunction is non zero on one side of a barrier it must be non zero on the other side of the barrier and this implies that the particle has tunnelled into the barrier. The transmission probability depends upon the mass of the particle (specifically $m^{1/2}$, through eqns 8A.21): the greater the mass the smaller the probability of tunnelling. Electrons and protons have small masses, molecular groups large masses; therefore, tunnelling effects are more observable in process involving electrons and protons.

Solutions to exercises

E8A.1(a) If the wavefunction is an eigenfunction of an operator, the corresponding eigenvalue is the value of corresponding observable [Section 7C.1(b)]. Applying the linear momentum operator $\hat{p} = \frac{\hbar}{\mathrm{i}}\frac{\mathrm{d}}{\mathrm{d}x}$ [7C.3] to the wavefunction yields

$$\hat{p}\psi = \frac{\hbar}{\mathrm{i}}\frac{\mathrm{d}}{\mathrm{d}x}\psi = \frac{\hbar}{\mathrm{i}}\frac{\mathrm{d}}{\mathrm{d}x}\mathrm{e}^{\mathrm{i}kx} = \hbar k\mathrm{e}^{\mathrm{i}kx}$$

so the wavefunction is an eigenfunction of the linear momentum; thus, the value of the linear momentum is the eigenvalue

$$\hbar k = 1.0546\times10^{-34}\,\mathrm{J\,s}\times3\times(10^{-9}\,\mathrm{m})^{-1} = \boxed{3\times10^{-25}\,\mathrm{kg\,m\,s^{-1}}}$$

Similarly, applying the kinetic energy operator $\hat{E}_\mathrm{k} = -\dfrac{\hbar^2}{2m}\dfrac{\mathrm{d}^2}{\mathrm{d}x^2}$ [7C.5] to the wavefunction yields

$$\hat{E}_\mathrm{k}\psi = -\frac{\hbar^2}{2m}\frac{\mathrm{d}^2}{\mathrm{d}x^2}\psi = -\frac{\hbar^2}{2m}\frac{\mathrm{d}^2}{\mathrm{d}x^2}\mathrm{e}^{ikx} = \frac{\hbar^2k^2}{2m}\mathrm{e}^{ikx}$$

so the wavefunction is an eigenfunction of this operator as well; thus, its value is the eigenvalue

$$\frac{\hbar^2k^2}{2m} = \frac{\{1.0546\times10^{-34}\,\mathrm{J\,s}\times3\times(10^{-9}\,\mathrm{m})^{-1}\}^2}{2\times9.11\times10^{-31}\,\mathrm{kg}} = \boxed{5\times10^{-20}\,\mathrm{J}}$$

E8A.2(a) The wavefunction for the particle is [8A.2 with $B=0$ because the particle is moving toward positive x]

$$\psi_k = \boxed{A\mathrm{e}^{ikx}}$$

The index k is given by the relationship

$$E_k = \frac{\hbar^2k^2}{2m} = 20\,\mathrm{J}$$

$$\text{so } k = \frac{\sqrt{2mE_k}}{\hbar} = \frac{\left(2(2.0\times10^{-3}\,\mathrm{kg})(20\,\mathrm{J})\right)^{1/2}}{1.0546\times10^{-34}\,\mathrm{J\,s}} = \boxed{2.7\times10^{33}\,\mathrm{m^{-1}}}$$

E8A.3(a) $E = \dfrac{n^2h^2}{8m_\mathrm{e}L^2}$ [8A.6b]

$$\frac{h^2}{8m_\mathrm{e}L^2} = \frac{(6.626\times10^{-34}\,\mathrm{J\,s})^2}{(8)\times(9.11\times10^{-31}\,\mathrm{kg})\times(1.0\times10^{-9}\,\mathrm{m})^2} = 6.02\times10^{-20}\,\mathrm{J}$$

The conversion factors required are

$$1\,\mathrm{eV} = 1.602\times10^{-19}\,\mathrm{J};\quad 1\,\mathrm{cm^{-1}} = 1.986\times10^{-23}\,\mathrm{J};\quad 1\,\mathrm{eV} = 96.485\,\mathrm{kJ\,mol^{-1}}$$

(i) $E_2 - E_1 = (4-1)\dfrac{h^2}{8m_\mathrm{e}L^2} = \dfrac{3h^2}{8m_\mathrm{e}L^2} = (3)\times(6.02\times10^{-20}\,\mathrm{J})$

$$= \boxed{1.81\times10^{-19}\,\mathrm{J}},\ \boxed{1.13\,\mathrm{eV}},\ \boxed{9100\,\mathrm{cm^{-1}}},\ \boxed{109\,\mathrm{kJ\,mol^{-1}}}$$

(ii) $E_6 - E_5 = (36-25)\dfrac{h^2}{8m_\mathrm{e}L^2} = \dfrac{11h^2}{8m_\mathrm{e}L^2} = (11)\times(6.02\times10^{-20}\,\mathrm{J})$

$$= \boxed{6.6\times10^{-19}\,\mathrm{J}},\ \boxed{4.1\,\mathrm{eV}},\ \boxed{33\,000\,\mathrm{cm^{-1}}},\ \boxed{400\,\mathrm{kJ\,mol^{-1}}}$$

Comment. The energy level separations increase as n increases.

Question. For what value of n is $E_{n+1} - E_n$ for the system of this exercise equal to the ionization energy of the hydrogen atom, which is 13.6 eV?

E8A.4(a) The wavefunctions are [8A.6a]

$$\psi_n = \left(\frac{2}{L}\right)^{1/2} \sin\left(\frac{n\pi x}{L}\right)$$

The required probability is

$$P = \int \psi^* \psi \, dx = \frac{2}{L}\int \sin^2\left(\frac{n\pi x}{L}\right) dx \approx \frac{2\Delta x}{L}\sin^2\left(\frac{n\pi x}{L}\right)$$

where $\Delta x = 0.02L$ and the function is evaluated at $x = 0.50\,L$.

(i) For $n=1$ $\quad P = \left(\frac{2}{L}\right) \times 0.02L \times \sin^2\left(\frac{\pi}{2}\right) = \boxed{0.04}$

(ii) For $n=2$ $\quad P = \left(\frac{2}{L}\right) \times 0.02L \times \sin^2 \pi = 0$ so $P \approx \boxed{0}$

E8A.5(a) The wavefunction for a particle in a square-well potential is [8A.6a]

$$\psi_1 = \left(\frac{2}{L}\right)^{1/2} \sin\left(\frac{n\pi x}{L}\right)$$

and the momentum operator is [7C.3]

$$\hat{p} = \frac{\hbar}{\mathrm{i}}\frac{\mathrm{d}}{\mathrm{d}x}$$

so $$\langle p \rangle = \int_0^L \psi_1^* \hat{p} \psi_1 \, \mathrm{d}x = \frac{2\hbar}{\mathrm{i}L}\int_0^L \sin\left(\frac{n\pi x}{L}\right)\frac{\mathrm{d}}{\mathrm{d}x}\sin\left(\frac{n\pi x}{L}\right)\mathrm{d}x$$

$$= \frac{2\pi\hbar}{\mathrm{i}L^2}\int_0^L \sin\left(\frac{n\pi x}{L}\right)\cos\left(\frac{n\pi x}{L}\right)\mathrm{d}x = \boxed{0} \text{ for all } n \text{ including } n=1$$

$$\hat{p}^2 = -\hbar^2 \frac{\mathrm{d}^2}{\mathrm{d}x^2}$$

$$\langle p^2 \rangle = -\frac{2\hbar^2}{L}\int_0^L \sin\left(\frac{n\pi x}{L}\right)\frac{\mathrm{d}^2}{\mathrm{d}x^2}\sin\left(\frac{n\pi x}{L}\right)\mathrm{d}x = \left(\frac{2\hbar^2}{L}\right)\times\left(\frac{n\pi}{L}\right)^2\int_0^L \sin^2 ax\,\mathrm{d}x \left[a = \frac{n\pi}{L}\right]$$

$$= \left(\frac{2\hbar^2}{L}\right)\times\left(\frac{n\pi}{L}\right)^2\left(\frac{1}{2}x - \frac{1}{4a}\sin 2ax\right)\Bigg|_0^L = \left(\frac{2\hbar^2}{L}\right)\times\left(\frac{n\pi}{L}\right)^2\times\left(\frac{L}{2}\right) = \frac{n^2h^2}{4L^2}$$

So for $n=1$: $\langle p^2 \rangle = \boxed{\dfrac{h^2}{4L^2}}$

Comment. The expectation value of $\hat{p}$ is zero because on average the particle moves to the left as often as the right.

E8A.6(a) The wavefunction is

$$\psi_n = \left(\frac{2}{L}\right)^{1/2} \sin\left(\frac{n\pi x}{L}\right) \quad [8A.6a]$$

Hence $\langle x\rangle = \int \psi^* x\psi\,dx = \frac{2}{L}\int_0^L x\sin^2\left(\frac{n\pi x}{L}\right)dx$

Use Integral T.11 from the *Resource section*

$$\int x\sin^2 ax\,dx = \frac{x^2}{4} - \frac{x\sin 2ax}{4a} - \frac{\cos 2ax}{8a^2}$$

$$\text{so } \langle x\rangle = \frac{2}{L}\left[\frac{x^2}{4} - \frac{Lx}{4n\pi}\sin\left(\frac{2n\pi x}{L}\right) - \frac{L^2}{8(n\pi)^2}\cos\left(\frac{2n\pi x}{L}\right)\right]_0^L = \boxed{\frac{L}{2}} \text{ for all } n.$$

$$\langle x^2\rangle = \frac{2}{L}\int_0^L x^2\sin^2\left(\frac{n\pi x}{L}\right)dx$$

Use Integral T.12 from the *Resource section*

$$\int x^2\sin^2 ax\,dx = \frac{x^3}{6} - \left(\frac{x^2}{4a} - \frac{1}{8a^3}\right)\sin 2ax - \frac{x\cos 2ax}{4a^2}$$

$$\text{so } \langle x^2\rangle = \frac{2}{L}\left[\frac{x^3}{6} - \left(\frac{Lx^2}{4n\pi} - \frac{L^3}{(2n\pi)^3}\right)\sin\left(\frac{2n\pi x}{L}\right) - \frac{L^2 x}{(2n\pi)^2}\cos\left(\frac{2n\pi x}{L}\right)\right]_0^L$$

$$= \frac{2}{L}\left(\frac{L^3}{6} - \frac{L^3}{(2n\pi)^2}\right) = L^2\left(\frac{1}{3} - \frac{1}{2n^2\pi^2}\right)$$

For $n=1$, $\langle x^2\rangle = \boxed{L^2\left(\frac{1}{3} - \frac{1}{2\pi^2}\right)}$

E8A.7(a) The zero-point energy is the ground-state energy, that is, with $n=1$:

$$E = \frac{n^2h^2}{8m_eL^2}[8A.6b] = \frac{h^2}{8m_eL^2}$$

Set this equal to the rest energy m_ec^2 and solve for L:

$$m_ec^2 = \frac{h^2}{8m_eL^2} \quad \text{so} \quad L = \boxed{\frac{h}{8^{1/2}m_ec} = \frac{\lambda_C}{8^{1/2}}}$$

In absolute units, the length is

$$L = \frac{6.63\times10^{-34}\,\text{J s}}{8^{1/2}\times(9.11\times10^{-31}\,\text{kg})\times(3.00\times10^8\,\text{m s}^{-1})} = 8.58\times10^{-13}\,\text{m} = 0.858\,\text{pm}$$

In terms of the Compton wavelength of an electron, $\lambda_C = \frac{h}{m_ec}, L = \frac{\lambda_C}{2\sqrt{2}}$.

E8A.8(a) $\psi_3 = \left(\dfrac{2}{L}\right)^{1/2} \sin\left(\dfrac{3\pi x}{L}\right)$ [8A.6a]

$$P(x) \propto \psi_3^2 \propto \sin^2\left(\frac{3\pi x}{L}\right)$$

The maxima and minima in $P(x)$ correspond to $\dfrac{\mathrm{d}P(x)}{\mathrm{d}x} = 0$.

$$\frac{\mathrm{d}P(x)}{\mathrm{d}x} \propto \sin\left(\frac{3\pi x}{L}\right)\cos\left(\frac{3\pi x}{L}\right) \propto \sin\left(\frac{6\pi x}{L}\right) \quad [2\sin\alpha\cos\alpha = \sin 2\alpha]$$

$\sin\theta = 0$ when $\theta = \left(\dfrac{6\pi x}{L}\right) = n'\pi$, $n' = 0, 1, 2, \ldots$, which corresponds to $x = \dfrac{n'L}{6}, n' \leq 6$.

$n' = 0$, 2, 4, and 6 correspond to minima in P, leaving $n' = 1$, 3, and 5 for the maxima, that is

$$x = \boxed{\frac{L}{6}, \frac{L}{2}, \text{and } \frac{5L}{6}}$$

Comment. Maxima in ψ^2 correspond to maxima *and* minima in ψ itself, so one can also solve this exercise by finding all points where $\dfrac{\mathrm{d}\psi}{\mathrm{d}x} = 0$..

E8A.9(a) In the original box [8A.6b]

$$E_1 = \frac{n^2h^2}{8mL^2}$$

In the longer box

$$E_2 = \frac{n^2h^2}{8m(1.1L)^2}$$

So $\quad \Delta E = \dfrac{n^2h^2}{8mL^2}\left(\dfrac{1}{1.1^2} - 1\right)$

and the relative change is

$$\frac{\Delta E}{E_1} = \frac{1}{1.1^2} - 1 = -0.174 = \boxed{-17.4\%}$$

E8A.10(a) The energy is [8A.6b]

$$E_n = \frac{n^2h^2}{8mL^2}$$

so the difference between neighbouring levels is

$$\Delta E_n = E_{n+1} - E_1 = \frac{\{(n+1)^2 - n^2\}h^2}{8mL^2} = \frac{(2n+1)h^2}{8mL^2}.$$

Set this difference equal to the thermal energy $kT/2$ and solve for n:

$$\frac{(2n+1)h^2}{8mL^2}=\frac{kT}{2},$$

$$\text{so } n=\boxed{\frac{2kTmL^2}{h^2}-\frac{1}{2}}$$

E8A.11(a) The energy levels are given by [8A.15b]

$$E_{n_1,n_2}=\left(\frac{n_1^2}{L_1^2}+\frac{n_2^2}{L_2^2}\right)\frac{h^2}{8m}=\left(\frac{n_1^2}{1}+\frac{n_2^2}{4}\right)\frac{h^2}{8mL^2}$$

$$E_{2,2}=\left(\frac{2^2}{1}+\frac{2^2}{4}\right)\frac{h^2}{8mL^2}=\frac{5h^2}{8mL^2}$$

We are looking for another state that has the same energy. By inspection we note that the first term in parentheses in $E_{2,2}$ works out to be 4 and the second 1; we can arrange for those values to be reversed:

$$E_{1,4}=\left(\frac{1^2}{1}+\frac{4^2}{4}\right)\frac{h^2}{8mL^2}=\frac{5h^2}{8mL^2}$$

So in this box, the state $\boxed{n_1=1,\ n_2=4}$ is degenerate to the state $n_1=2$, $n_2=2$. The question notes that degeneracy frequently accompanies symmetry, and suggests that one might be surprised to find degeneracy in a box with unequal lengths. Symmetry is a matter of degree. This box is less symmetric than a square box, but it is more symmetric than boxes whose sides have a non-integer or irrational ratio. Every state of a square box except those with $n_1=n_2$ is degenerate (with the state that has n_1 and n_2 reversed). Only a few states in this box are degenerate. In this system, a state (n_1, n_2) is degenerate with a state $(n_2/2, 2n_1)$ as long as the latter state (a) exists (*i.e.*, $n_2/2$ must be an integer) and (b) is distinct from (n_1, n_2). A box with incommensurable sides, say, L and $2^{1/2}L$, would have no degenerate levels.

E8A.12(a) $E_{n_1,n_2,n_3}=\dfrac{(n_1^2+n_2^2+n_3^2)h^2}{8m_eL^2}$ [8A.16b]

$$E_{1,1,1}=\frac{3h^2}{8mL^2},\qquad 3E_{1,1,1}=\frac{9h^2}{8mL^2}$$

Hence, we require the values of n_1, n_2, and n_3 that make

$$n_1^2+n_2^2+n_3^2=9$$

Therefore, $(n_1, n_2, n_3)=(1, 2, 2)$, $(2, 1, 2)$, and $(2, 2, 1)$ and the degeneracy is $\boxed{3}$.

E8A.13(a) The transmission probability [8A.23a] depends on the energy of the tunneling particle relative to the barrier height ($\varepsilon=E/V=1.5\,\text{eV}/(2.0\,\text{eV})=0.75$), on the width of the barrier ($L=100$ pm), and on the decay parameter of the wavefunction inside the barrier (κ), where [8A.20]

$$\kappa=\frac{\{2m(V-E)\}^{1/2}}{\hbar}=\frac{\{2\times 9.11\times 10^{-31}\,\text{kg}\times(2.0-1.5)\,\text{eV}\times 1.602\times 10^{-19}\,\text{J eV}^{-1}\}^{1/2}}{1.0546\times 10^{-34}\,\text{J s}}$$

$$=3.\overline{6}\times 10^{9}\,\text{m}^{-1}$$

We note that $\kappa L = 3.\bar{6}\times10^{9}\,\mathrm{m^{-1}}\times100\times10^{-12}\,\mathrm{m} = 0.3\bar{6}$ is not large compared to 1, so we must use eqn 8A.23a for the transmission probability.

$$T=\left\{1+\frac{(e^{\kappa L}-e^{-\kappa L})^2}{16\varepsilon(1-\varepsilon)}\right\}^{-1}=\left\{1+\frac{(e^{0.3\bar{6}}-e^{-0.3\bar{6}})^2}{16\times0.75\times(1-0.75)}\right\}^{-1}=\boxed{0.8}$$

Solutions to problems

P8A.1 $E=\dfrac{n^2h^2}{8mL^2}$ [8A.6b], $E_2-E_1=\dfrac{3h^2}{8mL^2}$

For O_2, $m=32.00\times(1.6605\times10^{-27}\,\mathrm{kg})=5.314\times10^{-26}\,\mathrm{kg}$,

so $$E_2-E_1=\frac{(3)\times(6.626\times10^{-34}\,\mathrm{J\,s})^2}{(8)\times(5.314\times10^{-26}\,\mathrm{kg})\times(5.0\times10^{-2}\,\mathrm{m})^2}=\boxed{1.24\times10^{-39}\,\mathrm{J}}$$

We set $E=\dfrac{n^2h^2}{8mL^2}=\dfrac{kT}{2}$ and solve for n:

$$n=\frac{2L(kTm)^{1/2}}{h}$$

$$=\frac{2\times(5.0\times10^{-2}\,\mathrm{m})\times\{(1.381\times10^{-23}\,\mathrm{J\,K^{-1}})\times(300\,\mathrm{K})\times5.314\times10^{-26}\,\mathrm{kg})\}^{1/2}}{6.626\times10^{-34}\,\mathrm{J\,s}}$$

$$=\boxed{2.2\times10^{9}}$$

At this level,

$$E_n-E_{n-1}=\{n^2-(n-1)^2\}\times\frac{h^2}{8mL^2}=(2n-1)\times\frac{h^2}{8mL^2}\approx(2n)\times\frac{h^2}{8mL^2}$$

From above $\dfrac{h^2}{8mL^2}=\dfrac{E_2-E_1}{3}=4.13\times10^{-40}\,\mathrm{J}$,

so $$E_n-E_{n-1}=2\times(2.2\times10^{9})\times(4.13\times10^{-40}\,\mathrm{J})\approx\boxed{1.8\times10^{-30}\,\mathrm{J}}\quad[\text{or }1.1\,\mu\mathrm{J\,mol^{-1}}]$$

P8A.3 (a) Suppose that a particle moves classically at the constant speed v. It starts at $x=0$ at $t=0$ and at $t=\tau$ is at position $x=L$. $v=\dfrac{L}{\tau}$ and $x=vt$.

$$\langle x\rangle=\frac{1}{\tau}\int_{t=0}^{\tau}x\,\mathrm{d}t=\frac{1}{\tau}\int_{t=0}^{\tau}vt\,\mathrm{d}t=\frac{v}{\tau}\int_{t=0}^{\tau}t\,\mathrm{d}t=\frac{v}{2\tau}t^2\Big|_{t=0}^{\tau}$$

$$=\frac{v\tau^2}{2\tau}=\frac{v\tau}{2}=\boxed{\frac{L}{2}}$$

$$\langle x^2\rangle=\frac{1}{\tau}\int_{t=0}^{\tau}x^2\,\mathrm{d}t=\frac{v^2}{\tau}\int_{t=0}^{\tau}t^2\,\mathrm{d}t=\frac{v^2}{3\tau}t^3\Big|_{t=0}^{\tau}=\frac{(v\tau)^2}{3}=\frac{L^2}{3}$$

so $\langle x^2\rangle^{1/2}=\boxed{\dfrac{L}{3^{1/2}}}$

(b) $\psi_n = \left(\frac{2}{L}\right)^{1/2} \sin\left(\frac{n\pi x}{L}\right)$ for $0 \le x \le L$ [8A.7a]

$$\langle x \rangle_n = \int_{x=0}^{L} \psi_n^* x \psi_n \, dx = \frac{2}{L}\int_0^L x \sin^2\left(\frac{n\pi x}{L}\right) dx$$

$$= \frac{2}{L}\left\{\frac{x^2}{4} - \frac{x\sin\left(\frac{2n\pi x}{L}\right)}{4(n\pi/L)} - \frac{\cos\left(\frac{2n\pi x}{L}\right)}{8(n\pi/L)^2}\right\}\Bigg|_{x=0}^{x=L} = \frac{2}{L}\left(\frac{L^2}{4}\right) = \boxed{\frac{L}{2}}$$

This agrees with the classical result for all values of n.

$$\langle x^2 \rangle_n = \int_{x=0}^{L} \psi_n^* x^2 \psi_n \, dx = \frac{2}{L}\int_{x=0}^{L} x^2 \sin^2\left(\frac{n\pi x}{L}\right) dx$$

$$= \frac{2}{L}\left\{\frac{x^3}{6} - \left(\frac{x^2}{4(n\pi/L)} - \frac{1}{8(n\pi/L)^3}\right)\sin\left(\frac{2n\pi x}{L}\right) - \frac{x\cos\left(\frac{2n\pi x}{L}\right)}{8(n\pi/L)^2}\right\}\Bigg|_{x=0}^{x=L}$$

$$= \frac{2}{L}\left(\frac{L^3}{6} - \frac{L}{8(n\pi/L)^2}\right) = \frac{L^2}{3} - \frac{1}{4(n\pi/L)^2}$$

so $$\langle x^2 \rangle_n^{1/2} = \boxed{\left(\frac{L^2}{3} - \frac{1}{4(n\pi/L)^2}\right)^{1/2}}$$

This agrees with the classical result in the limit of large quantum numbers:

$$\lim_{n\to\infty} \langle x^2 \rangle_n^{1/2} = \frac{L}{3^{1/2}}$$

P8A.5 The rate of tunnelling is proportional to the transmission probability, so a ratio of tunnelling rates is equal to the corresponding ratio of transmission probabilities (given in eqn 8A.23a). The desired factor is T_1/T_2, where the subscripts denote the tunnelling distances in nanometres:

$$\frac{T_1}{T_2} = \frac{1 + \frac{(e^{\kappa L_2} - e^{-\kappa L_2})^2}{16\varepsilon(1-\varepsilon)}}{1 + \frac{(e^{\kappa L_1} - e^{-\kappa L_1})^2}{16\varepsilon(1-\varepsilon)}}.$$

If $$\frac{(e^{\kappa L_2} - e^{-\kappa L_2})^2}{16\varepsilon(1-\varepsilon)} >> 1,$$

then $$\frac{T_1}{T_2} \approx \frac{(e^{\kappa L_2} - e^{-\kappa L_2})^2}{(e^{\kappa L_1} - e^{-\kappa L_1})^2} \approx e^{2\kappa(L_2 - L_1)} = e^{2(7/\text{nm})(2.0-1.0)\text{nm}} = \boxed{1.2\times 10^6}.$$

That is, the tunnelling rate increases about a million-fold. Note: if the first approximation does not hold, we need more information, namely $\varepsilon = E/V$. If the first approximation is valid, then the second is also likely to be valid, namely that the negative exponential is negligible compared to the positive one.

P8A.7 (a) The wavefunctions in each region (see Fig. 8A.1(a)) are (eqns 8A.18, 8A.20, and 8A.21):

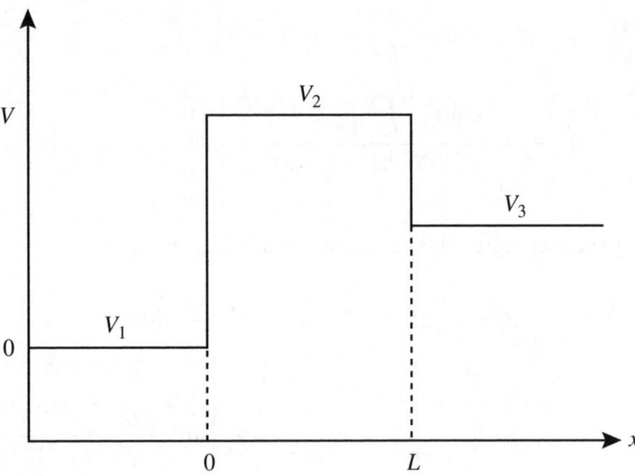

Figure 8A.1(a)

$$\psi_1(x)=e^{ik_1x}+B_1e^{-ik_1x}$$
$$\psi_2(x)=A_2e^{k_2x}+B_2e^{-k_2x}$$
$$\psi_3(x)=A_3e^{ik_3x}$$

With the above choice of $A_1=1$ the transmission probability is simply $T=|A_3|^2$. The wavefunction coefficients are determined by the criteria that both the wavefunctions and their first derivatives with respect to x be continuous at potential boundaries

$$\psi_1(0)=\psi_2(0) \quad \text{and} \quad \psi_2(L)=\psi_3(L)$$

$$\frac{d\psi_1(0)}{dx}=\frac{d\psi_2(0)}{dx} \quad \text{and} \quad \frac{d\psi_2(L)}{dx}=\frac{d\psi_3(L)}{dx}$$

These criteria establish the algebraic relationships:

$$1+B_1-A_2-B_2=0$$
$$(-ik_1-k_2)A_2+(-ik_1+k_2)B_2+2ik_1=0$$
$$A_2e^{k_2L}+B_2e^{-k_2L}-A_3e^{ik_3L}=0$$
$$A_2k_2e^{k_2L}-B_2k_2e^{-k_2L}-iA_3k_3e^{ik_3L}=0$$

Solving the simultaneous equations for A_3 gives

$$A_3=\frac{4k_1k_2\,e^{ik_3L}}{(ia+b)e^{k_2L}-(ia-b)e^{-k_2L}}$$

where $a=k_2^{\,2}-k_1k_3$ and $b=k_1k_2+k_2k_3$

Since $\sinh(z) = (e^{z} - e^{-z})/2$ or $e^{z} = 2\sinh(z) + e^{-z}$, substitute $e^{k_2 L} = 2\sinh(k_2 L) + e^{-k_2 L}$ giving

$$A_3 = \frac{2k_1 k_2 e^{ik_3 L}}{(ia+b)\sinh(k_2 L) + b\,e^{-k_2 L}}$$

the transmission coefficient is

$$\boxed{\begin{array}{l} T = |A_3|^2 = A_3 \times A_3^{\star} = \dfrac{4k_1^2 k_2^2}{(a^2+b^2)\sinh^2(k_2 L) + b^2} \\ \text{where } a^2+b^2 = (k_1^2+k_2^2)(k_2^2+k_3^2) \text{ and } b^2 = k_2^2(k_1+k_3)^2 \end{array}}$$

(b) In the special case for which $V_1 = V_3 = 0$, eqns 8A.18 and 8A.21 require that $k_1 = k_3$. Additionally,

$$\left(\frac{k_1}{k_2}\right)^2 = \frac{E}{V_2 - E} = \frac{\varepsilon}{1-\varepsilon} \text{ where } \varepsilon = E/V_2.$$

$$a^2 + b^2 = (k_1^2 + k_2^2)^2 = k_2^4\left\{1 + \left(\frac{k_1}{k_2}\right)^2\right\}^2$$

$$b^2 = 4k_1^2 k_2^2$$

$$\frac{a^2+b^2}{b^2} = \frac{k_2^2\left\{1+\left(\frac{k_1}{k_2}\right)^2\right\}^2}{4k_1^2}$$

$$= \frac{1}{4\varepsilon(1-\varepsilon)}$$

$$T = \frac{b^2}{b^2 + (a^2+b^2)\sinh^2(k_2 L)} = \frac{1}{1 + \left(\frac{a^2+b^2}{b^2}\right)\sinh^2(k_2 L)}$$

$$= \left\{1 + \frac{\sinh^2(k_2 L)}{4\varepsilon(1-\varepsilon)}\right\}^{-1}$$

$$= \left\{1 + \frac{(e^{k_2 L} - e^{-k_2 L})^2}{16\varepsilon(1-\varepsilon)}\right\}^{-1}$$

This proves eqn 8A.23a where $V_1 = V_3 = 0$

In the limit of a high and wide barrier, $k_2 L >> 1$. This implies both that $e^{-k_2 L}$ is negligibly small compared to $e^{k_2 L}$ and that 1 is negligibly small compared to $e^{2k_2 L}/\{16\varepsilon(1-\varepsilon)\}$. The previous equation simplifies to [8A.23b]

$$T = 16\varepsilon(1-\varepsilon)e^{-2k_2 L}$$

(c) The specified graph is shown in Fig. 8A.1(b).

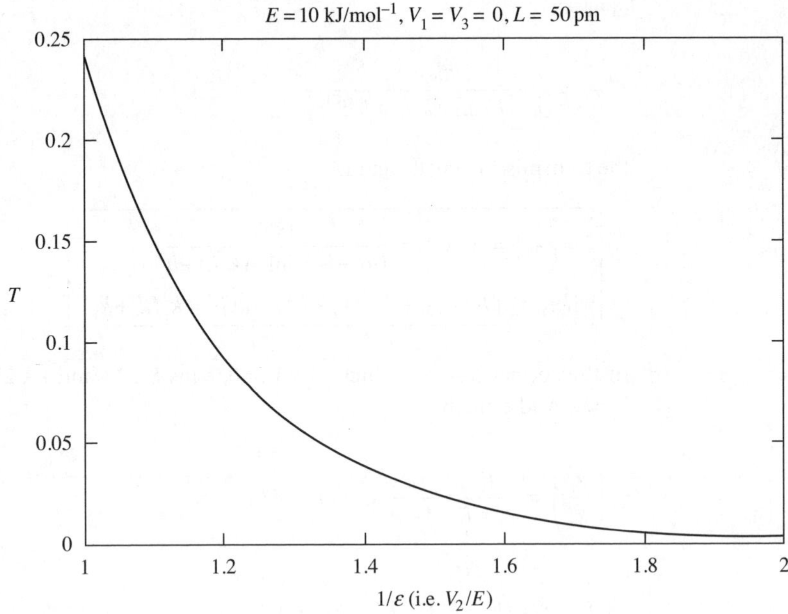

Figure 8A.1(b)

8B Vibrational motion

Answers to discussion questions

D8B.1 The separation of vibrational energy levels of the harmonic oscillator is uniform, $\hbar\omega$ (eqn. 8B.5), where $\omega=(k_f/m)^{1/2}$ (eqn. 8B.4), regardless of the level. Thus the energy separation varies in direct proportion to the square root of the force constant and in inverse proportion to the square root of the mass.

D8B.3 The physical reason is the uncertainty principle (Section 7C.3). If a particle is at least partly localized, its position is not completely uncertain, and therefore its momentum, and hence its kinetic energy, cannot be exactly zero. The potential energy of the harmonic oscillator at least partially localizes the oscillator in the neighbourhood of the equilibrium position, so zero momentum (and therefore zero kinetic energy) is precluded. Furthermore, the confinement is not perfect, which makes the particle's potential energy also differ from zero.

Solutions to exercises

E8B.1(a) $E=\left(v+\frac{1}{2}\right)\hbar\omega,\quad \omega=\left(\frac{k_f}{m}\right)^{1/2}$ [8B.4]

The zero-point energy corresponds to $v=0$; hence

$$E_0 = \frac{1}{2}\hbar\omega = \frac{1}{2}\hbar\left(\frac{k_f}{m}\right)^{1/2} = \left(\frac{1}{2}\right)\times(1.055\times10^{-34}\,\text{J s})\times\left(\frac{155\,\text{N m}^{-1}}{2.33\times10^{-26}\,\text{kg}}\right)^{1/2}$$

$$= \boxed{4.30\times10^{-21}\,\text{J}}$$

E8B.2(a) The difference in adjacent energy levels is

$$\Delta E = E_{v+1} - E_v = \hbar\omega\ [8\text{B}.5] = \hbar\left(\frac{k_f}{m}\right)^{1/2}\ [8\text{B}.4]$$

Hence

$$k_f = m\left(\frac{\Delta E}{\hbar}\right)^2 = (1.33\times10^{-25}\,\text{kg})\times\left(\frac{4.82\times10^{-21}\,\text{J}}{1.055\times10^{-34}\,\text{J s}}\right)^2 = 278\,\text{kg s}^{-2} = \boxed{278\,\text{N m}^{-1}}$$

E8B.3(a) The requirement for a transition to occur is that $\Delta E(\text{system}) = E(\text{photon})$,

so $\Delta E(\text{system}) = \hbar\omega\ [8\text{B}.5] = E(\text{photon}) = h\nu = \dfrac{hc}{\lambda}$

Therefore, $\dfrac{hc}{\lambda} = \dfrac{h\omega}{2\pi} = \left(\dfrac{h}{2\pi}\right)\times\left(\dfrac{k_f}{m}\right)^{1/2}$ [8B.4]

$$\lambda = 2\pi c\left(\frac{m}{k_f}\right)^{1/2} = (2\pi)\times(2.998\times10^8\,\text{m s}^{-1})\times\left(\frac{1.0078\times1.6605\times10^{-27}\,\text{kg}}{855\,\text{N m}^{-1}}\right)^{1/2}$$

$$= 2.64\times10^{-6}\,\text{m} = \boxed{2.64\,\mu\text{m}}$$

E8B.4(a) The frequency of a harmonic oscillator is [8B.4]

$$\omega = \left(\frac{k_f}{m}\right)^{1/2}.$$

^{1}H (H) and ^{2}H (D) are isotopes, so we expect that the force constant is the same in H_2 and D_2. They differ in mass. So the frequencies are inversely proportional to the square root of the mass:

$$\omega_D = \omega_H\left(\frac{m_H}{m_D}\right)^{1/2}$$

But the appropriate mass is not the mass of the molecule but its "effective mass" [8B.7]

$$\mu = \frac{m_1 m_2}{m_1 + m_2} = \frac{m}{2}\ [m_1 = m_2 = m]$$

For H_2: $\mu = \dfrac{m}{2} = \dfrac{1.0078\times(1.6605\times10^{-27}\,\text{kg})}{2} = \boxed{8.3673\times10^{-28}\,\text{kg}}$

For D_2: $\mu = \dfrac{m}{2} = \dfrac{2.0141\times(1.6605\times10^{-27}\,\text{kg})}{2} = \boxed{1.6722\times10^{-27}\,\text{kg}}$

$$\omega_{\mathrm{D}} = 131.9\,\mathrm{THz} \times \left(\frac{8.3673\times10^{-28}\,\mathrm{kg}}{1.6722\times10^{-27}\,\mathrm{kg}}\right)^{1/2} = \boxed{93.3\,\mathrm{THz}}$$

E8B.5(a) (a) From introductory physics, we have

$$\omega = \left(\frac{g}{l}\right)^{1/2}$$

$$\Delta E = \hbar\omega\ [8B.5] = (1.055\times10^{-34}\,\mathrm{J\,s})\times\left(\frac{9.81\,\mathrm{m\,s^{-2}}}{1.0\,\mathrm{m}}\right)^{1/2}$$

$$= \boxed{3.3\times10^{-34}\,\mathrm{J}}$$

(b) $\Delta E = h\nu = (6.626\times10^{-34}\,\mathrm{J\,Hz^{-1}})\times(5\,\mathrm{Hz}) = \boxed{3.3\times10^{-33}\,\mathrm{J}}$

E8B.6(a) The zero-point energy is

$$E_0 = \frac{1}{2}\hbar\omega[8B.6] = \frac{\hbar}{2}\left(\frac{k}{\mu}\right)^{1/2}\ [8B.4]$$

$$\mu = 34.9688\times(1.6605\times10^{-27}\,\mathrm{kg})/2 = 2.9033\times10^{-26}\,\mathrm{kg}$$

where we have used eqn 8B.7 for two equal masses, as in Exercise 8B.4(a).

$$\text{so } E_0 = \left(\frac{1.0546\times10^{-34}\,\mathrm{J\,s}}{2}\right)\times\left(\frac{329\,\mathrm{N\,m^{-1}}}{2.9033\times10^{-26}\,\mathrm{kg}}\right)^{1/2} = \boxed{5.61\times10^{-21}\,\mathrm{J}}$$

E8B.7(a) The harmonic oscillator wavefunctions have the form [8B.8]

$$\psi_v(x) = N_v H_v(y)\exp\left(-\tfrac{1}{2}y^2\right) \quad \text{with} \quad y = \frac{x}{\alpha} \quad \text{and} \quad \alpha = \left(\frac{\hbar^2}{mk_{\mathrm{f}}}\right)^{1/4}$$

The exponential function approaches zero only as x approaches $\pm\infty$, so the nodes of the wavefunction are the nodes of the Hermite polynomials.

$$H_4(y) = 16y^4 - 48y^2 + 12 = 0\ \ [\text{Table 8B.1}]$$

Dividing through by 4 and letting $z = y^2$, we have a quadratic equation

$$4z^2 - 12z + 3 = 0$$

$$\text{so } z = \frac{-b\pm\sqrt{b^2-4ac}}{2a} = \frac{12\pm\sqrt{12^2-4\times4\times3}}{2\times4} = \frac{3\pm\sqrt{6}}{2}$$

Evaluating the result numerically yields $z = 0.275$ or 2.72, so $y = \pm0.525$ or ±1.65. Therefore $x = \boxed{\pm0.525\alpha \text{ or } \pm1.65\alpha}$.

Comment. Numerical values could also be obtained graphically by plotting $H_4(y)$.

E8B.8(a) The most probable displacements are the values of x that maximize ψ^2. As noted in Exercise 8A.8(a), maxima in ψ^2 correspond to maxima *and* minima in ψ itself, so one can solve this exercise by finding all points where $\frac{d\psi}{dx}=0$. The wavefunction is [8B.10]

$$\psi_1 = 2N_1 y\exp\left(-\tfrac{1}{2}y^2\right) \quad \text{with} \quad y=\frac{x}{\alpha} \quad \text{and} \quad \alpha=\left(\frac{\hbar^2}{mk_f}\right)^{1/4}$$

$$\frac{d\psi_1}{dx}=\frac{dy}{dx}\frac{d\psi_1}{dy}=\frac{2N_1}{\alpha}\left\{\exp\left(-\tfrac{1}{2}y^2\right)-y^2\exp\left(-\tfrac{1}{2}y^2\right)\right\}=0$$

Dividing through by constants and the exponential functions yields

$$1-y^2=0 \quad \text{so} \quad y=\pm 1 \quad \text{and} \quad x=\boxed{\pm\alpha}.$$

E8B.9(a) Example 8B.4 analyses the classical turning points of the harmonic oscillator. In terms of the dimensionless variable y, the turning points are $y_{tp}=\pm(2v+1)^{1/2}$. The probability of extension beyond the classical turning point is

$$P=\int_{x_{tp}}^{\infty}\psi_v^2\,dx=\alpha N_v^2\int_{y_{tp}}^{\infty}\{H_v(y)\}^2 e^{-y^2}\,dy$$

For $v=1$, $H_1(y)=2y$ and $N_1=\left(\dfrac{1}{2\alpha\pi^{1/2}}\right)^{1/2}$

$$P=4\alpha N_1^2\int_{3^{1/2}}^{\infty}y^2e^{-y^2}\,dy$$

Use integration by parts:

$$\int u\,dv=uv-\int v\,du$$

where $u=y,\quad dv=ye^{-y^2}dy$

so $du=dy,\quad v=-\tfrac{1}{2}e^{-y^2}$

and $$P=-2\alpha N_1^2\left(ye^{-y^2}\Big|_{3^{1/2}}^{\infty}-\int_{3^{1/2}}^{\infty}e^{-y^2}\,dy\right)$$

$$=\pi^{-1/2}\left(3^{1/2}e^{-3}+\int_{3^{1/2}}^{\infty}e^{-y^2}\,dy\right)$$

The remaining integral can be expressed in terms of the error function.

$$\operatorname{erf} z=1-\frac{2}{\pi^{1/2}}\int_z^{\infty}e^{-y^2}\,dy$$

so $$\int_{3^{1/2}}^{\infty}e^{-y^2}\,dy=\frac{\pi^{1/2}(1-\operatorname{erf}3^{1/2})}{2}$$

Finally, using $\operatorname{erf}3^{1/2}=0.986$,

$$P=\pi^{-1/2}\left(3^{1/2}e^{-3}+\frac{\pi^{1/2}(1-\operatorname{erf}3^{1/2})}{2}\right)=\boxed{0.056}$$

Comment. This is the probability of an extension greater than the positive classical turning point. There is an equal probability of a compression smaller than the negative classical turning point, so the total probability of finding the oscillator in a classically forbidden region is $\boxed{0.112}$.

Comment. Note that molecular parameters such as m and k do not enter into the calculation.

Solutions to problems

P8B.1 $\omega=\left(\dfrac{k_\mathrm{f}}{\mu}\right)^{1/2}$ [8B.4]

Also, $\omega=2\pi\nu=\dfrac{2\pi c}{\lambda}=2\pi c\tilde{\nu}$

Therefore $k_\mathrm{f}=\omega^2\mu=4\pi^2c^2\tilde{\nu}^2\mu=\dfrac{4\pi^2c^2\tilde{\nu}^2m_1m_2}{m_1+m_2}$.

We draw up the following table using isotope masses from the *Resource section.*

	$^1H^{35}Cl$	$^1H^{81}Br$	$^1H^{127}I$	$^{12}C^{16}O$	$^{14}N^{16}O$
$\tilde{\nu}/m^{-1}$	299 000	265 000	231 000	217 000	190 400
$10^{27}\,m_1/kg$	1.6735	1.6735	1.6735	19.926	23.253
$10^{27}\,m_2/kg$	58.066	134.36	210.72	26.560	26.560
$k/(N\,m^{-1})$	516	412	314	1902	1595

Therefore, the order of stiffness, is $\boxed{\mathrm{HI}<\mathrm{HBr}<\mathrm{HCl}<\mathrm{NO}<\mathrm{CO}}$.

P8B.3 Assuming that one can identify the CO peak in the infrared spectrum of the CO–myoglobin complex, taking infrared spectra of each of the isotopic variants of CO–myoglobin complexes can show which atom binds to the haem group and determine the C≡O force constant. Compare isotopic variants to $^{12}C^{16}O$ as the standard; when an isotope changes but the vibrational frequency does not, then the atom whose isotope was varied is the atom that binds to the haem. See table below, which includes predictions of the wavenumber of all isotopic variants compared to that of $\tilde{\nu}$ ($^{12}C^{16}O$). (As usual, the better the experimental results agree with the whole set of predictions, the more confidence one would have with the conclusion.)

Wavenumber for isotopic variant	If O binds	If C binds
$\tilde{\nu}(^{12}C^{18}O)=$	$\tilde{\nu}(^{12}C^{16}O)^{\dagger}$	$(16/18)^{1/2}\tilde{\nu}(^{12}C^{16}O)$
$\tilde{\nu}(^{13}C^{16}O)=$	$(12/13)^{1/2}\tilde{\nu}(^{12}C^{16}O)$	$\tilde{\nu}(^{12}C^{16}O)^{\dagger}$
$\tilde{\nu}(^{13}C^{18}O)=$	$(12/13)^{1/2}\tilde{\nu}(^{12}C^{16}O)$	$(16/18)^{1/2}\tilde{\nu}(^{12}C^{16}O)$

† That is, no change compared to the standard.

The wavenumber is related to the force constant as follows [8B.4]:

$$\omega = 2\pi c\tilde{\nu} = \left(\frac{k_\mathrm{f}}{m}\right)^{1/2} \quad \text{so} \quad k_\mathrm{f} = m(2\pi c\tilde{\nu})^2.$$

Hence $k_\mathrm{f} = (m/m_\mathrm{u})(1.66\times10^{-27}\,\mathrm{kg})[(2\pi)(2.998\times10^{10}\,\mathrm{cm\,s^{-1}})\tilde{\nu}(^{12}\mathrm{C}^{16}\mathrm{O})]^2$,

and $k_\mathrm{f}/(\mathrm{kg\,s^{-1}}) = (5.89\times10^{-5})(m/m_\mathrm{u})[\tilde{\nu}(^{12}\mathrm{C}^{16}\mathrm{O})/\mathrm{cm^{-1}}]^2$

Here m is the mass of the atom that is not bound in atomic mass units, i.e. $12m_\mathrm{u}$ if O is bound and $16m_\mathrm{u}$ if C is bound. (Of course, one can compute k_f from any of the isotopic variants, and take k_f to be a mean derived from all the relevant data.)

P8B.5 $\langle E_\mathrm{k}\rangle = \int_{-\infty}^{+\infty} \psi^* \hat{E}_\mathrm{k}\psi\,\mathrm{d}x$

$$\hat{E}_\mathrm{k} = -\frac{\hbar^2}{2m}\frac{\mathrm{d}^2}{\mathrm{d}x^2}[7\mathrm{C}.5] = -\frac{\hbar^2}{2m\alpha^2}\frac{\mathrm{d}^2}{\mathrm{d}y^2} = -\frac{\hbar\omega}{2}\frac{\mathrm{d}^2}{\mathrm{d}y^2}, \quad \left[x=\alpha y, \alpha^2 = \frac{\hbar}{m\omega}\right]$$

which implies that

$$\hat{E}_\mathrm{k}\psi = -\frac{\hbar\omega}{2}\left(\frac{\mathrm{d}^2\psi}{\mathrm{d}y^2}\right)$$

We then use $\psi = NH\mathrm{e}^{-y^2/2}$, and obtain

$$\frac{\mathrm{d}^2\psi}{\mathrm{d}y^2} = N\frac{\mathrm{d}^2}{\mathrm{d}y^2}(H\mathrm{e}^{-y^2/2}) = N\{H'' - 2yH' - H + y^2H\}\mathrm{e}^{-y^2/2}$$

From Table 8B.1

$$H_v'' - 2yH_v' = -2vH_v$$

$$y^2H_v = y\left(\tfrac{1}{2}H_{v+1} + vH_{v-1}\right) = \tfrac{1}{2}\left(\tfrac{1}{2}H_{v+2} + (v+1)H_v\right) + v\left(\tfrac{1}{2}H_v + (v-1)H_{v-2}\right)$$
$$= \tfrac{1}{4}H_{v+2} + v(v-1)H_{v-2} + \left(v+\tfrac{1}{2}\right)H_v$$

Hence, $\dfrac{\mathrm{d}^2\psi}{\mathrm{d}y^2} = N\left[\tfrac{1}{4}H_{v+2} + v(v-1)H_{v-2} - \left(v+\dfrac{1}{2}\right)H_v\right]\mathrm{e}^{-y^2/2}$

Therefore,

$$\langle E_\mathrm{k}\rangle = N_v^2\left(-\frac{\hbar\omega}{2}\right)\int_{-\infty}^{+\infty} H_v\left\{\frac{1}{4}H_{v+2} + v(v-1)H_{v-2} - \left(v+\frac{1}{2}\right)H_v\right\}\mathrm{e}^{-y^2}\,\mathrm{d}x[\mathrm{d}x = \alpha\mathrm{d}y]$$

$$= \alpha N_v^2\left(-\tfrac{1}{2}\hbar\omega\right)\left\{0 + 0 - \left(v+\tfrac{1}{2}\right)\pi^{1/2}2^v v!\right\} \text{ [Hermite polynomials orthogonal]}$$

$$= \boxed{\frac{1}{2}\left(v+\frac{1}{2}\right)\hbar\omega} \quad \left[N_v^2 = \frac{1}{\alpha\pi^{1/2}2^v v!}, \text{Example 8B.2}\right]$$

P8B.7 The turning points in terms of the displacement x are (using Example 8B.4 and eqns 8B.4 and 8B.8)

$$x_\mathrm{tp} = \pm\left(\frac{2E}{k}\right)^{1/2} = \pm\left(\frac{2(v+\tfrac{1}{2})\hbar\omega}{k}\right)^{1/2} = \pm\left(\frac{(2v+1)\hbar}{(km)^{1/2}}\right)^{1/2} = \pm(2v+1)^{1/2}\alpha$$

In terms of the dimensionless variable $y=x/\alpha$, $y_{tp}=\pm(2v+1)^{1/2}$

The probability of extension beyond one classical turning point for a harmonic oscillator with quantum number v is

$$P_v=\int_{|x_{tp}|}^{\infty}\psi_v^2\,dx=\alpha N_v^2\int_{(2v+1)^{1/2}}^{\infty}\{H_v(y)\}^2e^{-y^2}\,dy=\frac{1}{\pi^{1/2}2^v v!}\int_{(2v+1)^{1/2}}^{\infty}\{H_v(y)\}^2e^{-y^2}\,dy$$

This expression can be evaluated for an arbitrary integer v. Mathematical software packages such as Maple, Mathcad, and Mathematica have the Hermite polynomials built in, so the above expression can be entered in a relatively straightforward manner. Here is what it looks like in Mathcad:

```
v := 0..10

N2(v) := 1 / (π^(1/2) 2^v v!)

P(v) := N2(v) ∫_{√(2v+1)}^{∞} (Her(v,y))^2 exp(−y^2) dy
```

The table of results is

v	0	1	2	3	4	5	6	7	8	9	10
$P(v)$ %	7.90	5.61	4.79	4.31	3.98	3.73	3.54	3.38	3.25	3.14	3.04

where $P(v)$ is expressed as a percentage (i.e. as $P(v)/100$).

The points are plotted (as probabilities, not percentages) in Fig. 8B.1.

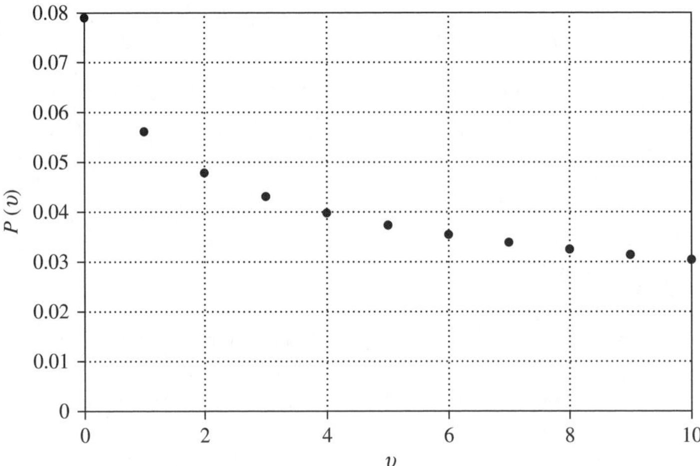

Figure 8B.1

Comment. $P(v)$ decreases with increasing v, albeit very slowly after the first few states. According to the correspondence principle, the quantum result reduces to the classical result in the limit of very high quantum numbers. So the probability of finding a displacement beyond the classical turning point must vanish in the limit of very high v.

P8B.9 Mathematical software can animate the real part or the imaginary part of $\Psi(x,t)$, or you may wish to have it display $|\Psi(x,t)|^2$. Try a "pure" state, that is, let $c=1$ for one value of m_l and 0 for all others. This "packet" does not spread; in fact, $|\Psi(x,t)|^2$ does not change, which is one reason why pure states are sometimes called stationary states. Also try making all the coefficients in the sum equal (all 1, for example). Whatever your choice of coefficients, the pattern will repeat with a period T that makes all the time-dependent factors equal to the exponential of ($2\pi i\times$an integer). Because the energy is

$$E_v = \hbar\omega(v+\tfrac{1}{2}),$$

then the exponent is equal to $2\pi i\times$an integer when

$$\frac{iE_v t}{\hbar} = \frac{i\hbar\omega(v+\frac{1}{2})t}{\hbar} = 2\pi i\times\text{integer} \quad \text{so} \quad T = \frac{2\pi}{\omega(v+\frac{1}{2})} = \frac{4\pi}{\omega(2v+1)}.$$

Thus a component of the packet returns to its initial value when $t=T$ and at intervals of T thereafter. For a harmonic oscillator, $T=\frac{4\pi}{\omega}$ is the period for the ground state and it is **a** period for any other state (because $\frac{4\pi}{\omega}$ is an integral multiple of $\frac{4\pi}{\omega(2v+1)}$ for all positive integers v.

Figure 8B.2 shows the evolution of a wavepacket composed of the first eleven harmonic oscillator wavefunctions equally weighted. That is, we have set all coefficients equal to 1:

$$\Psi(x,t) = \sum_{v=0}^{10} \psi_v(x)e^{-iE_v t/\hbar}.$$

(Note: the resulting time-dependent wavefunction is not normalized; however, $|\Psi(x,t)|^2$ is proportional to the time-dependent probability of finding the oscillator 'at' (x,t) .)

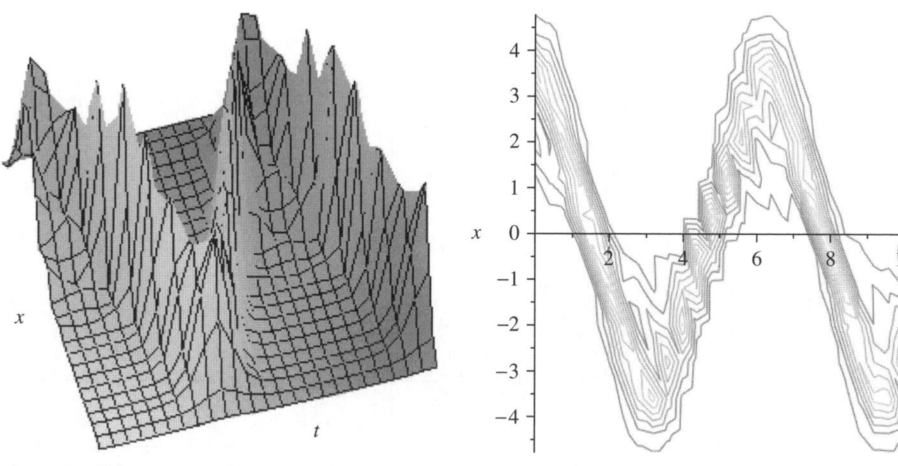

Figure 8B.2(a)

Figure 8B.2(b)

Figure 8B.2(a) shows a three-dimensional plot of the evolution of $|\Psi(x,t)|^2$ as a function of x and t, and Fig. 8B.2(b) shows a contour plot of the same function. In both plots, x is graphed in units of α (i.e. x is really x/α) and t in units of $1/\omega$ (i.e. t is really ωt). Notice that this wavepacket is fairly localized. At time 0 it is most likely found between $x=3$ and 4. As the wavepacket moves, it hangs together more or less. That is, its height and breadth do not change by much. Notice that the track repeats itself after a period of t (or rather ωt) $=2\pi$.

Figure 8B.3 shows four snapshots of the wavepacket's spatial distribution, at $t=0, 1, 4$, and 6, during which time we can see that the peak moves across the graph and almost all the way back again.

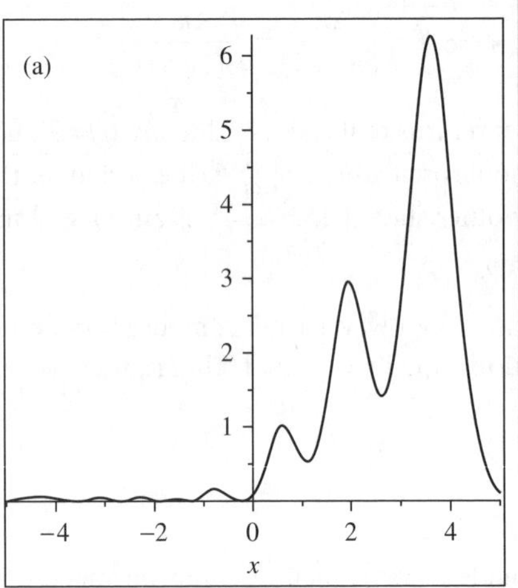

Figure 8B.3(a)

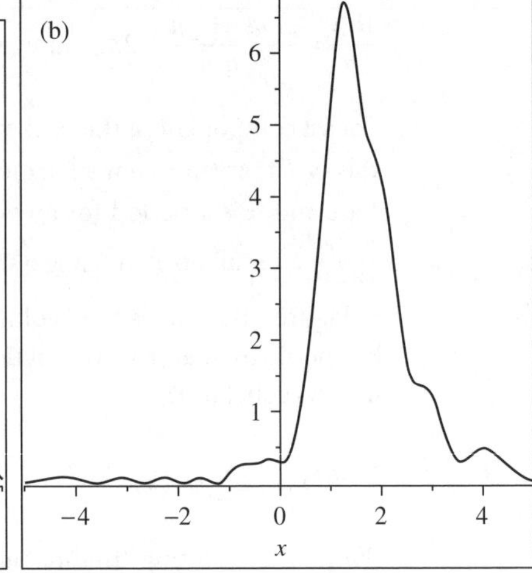

Figure 8B.3(b)

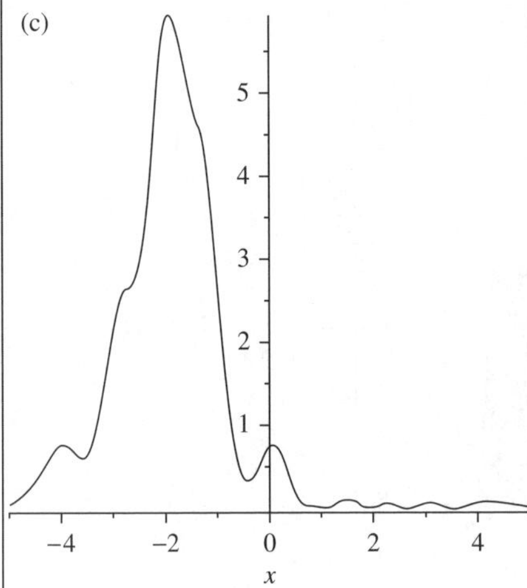

Figure 8B.3(c)

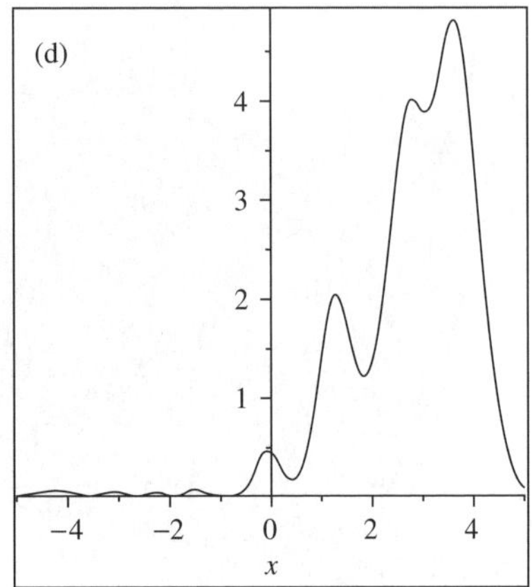

Figure 8B.3(d)

P8B.11 As expressed in the problem, the potential energy function assumes that ϕ is defined as we would expect; that is, $\phi=0$ corresponds to an eclipsed conformation. Thus, $\phi=0$ is **not** a stable equilibrium point, and small displacements from this point are **not** harmonic; in fact, $\phi=0$ is a position of unstable equilibrium, and small displacements from it would grow larger. We must express the potential energy in terms of displacements from a stable equilibrium position. One such equilibrium position is the staggered conformation directly opposite $\phi=0$, namely $\phi=\pi$. So let the displacement $x=\phi-\pi$. So, in terms of x, the potential energy function is $V=-V_0\cos 3x$. Conventionally, the potential energy in harmonic motion is measured with respect to that stable equilibrium position. Note that the potential energy at the stable equilibrium position is $V=-V_0$. We can redefine the potential energy function to measure energy relative to the stable equilibrium by letting

$$V'=V_0+V=V_0-V_0\cos 3x=V_0(1-\cos 3x).$$

Use the first two terms of the Taylor series expansion of cosine:

$$V'=V_0(1-\cos 3x)\approx V_0\left(1-1+\frac{(3x)^2}{2}\right)=\frac{9V_0}{2}x^2$$

The Schrödinger equation becomes

$$-\frac{\hbar^2}{2I}\frac{\partial^2\psi}{\partial x^2}+\frac{9V_0}{2}x^2\psi=E\psi \quad \text{[8C.9b with a non zero potential]}$$

This has the form of the Schrödinger equation for the harmonic oscillator wavefunction (eqn 8B.3). The difference in adjacent energy levels is:

$$E_1-E_0=\hbar\omega \text{ [8B.5]} \qquad \text{where} \quad \omega=\left(\frac{9V_0}{I}\right)^{1/2} \text{ [adapting 8B.4]}$$

If the displacements are sufficiently large, the potential energy does not rise as rapidly with the angle as would a harmonic potential (i.e. the cosine potential energy is not well approximated by the first few terms of its expansion). Each successive energy level would become lower than that of a harmonic oscillator, so the energy levels would become progressively closer together.

Question. The next term in the Taylor series for the potential energy is $-\frac{27V_0}{8}x^4$. Treat this as a perturbation to the harmonic oscillator wavefunction and compute the first-order correction to the energy.

8C Rotational motion

Answers to discussion questions

D8C.1 In quantum mechanics, particles are said to have wave characteristics. The fact of the existence of the particle then requires that the wavelengths of the waves representing it be such that the wave does not experience destructive interference in its motion around a closed loop. This means that a whole number of wavelengths must fit on the circumference of the ring [8C.3]:

$$n\lambda = 2\pi r$$

making the wavelength quantized. This in turn requires the angular momentum and energy to be quantized (eqns 8C.4 and 8C.5 respectively). Mathematically, this comes out in the cyclic boundary condition discussed in Section 8C.1(b), which restricts the constant m_l in the wavefunction $\psi(\phi)=e^{im_l\phi}$ to be an integer. That way, the wavefunction has the same value at ϕ and at $\phi+2\pi$—as it must since these two 'different' angles represent the same angular position. The constant m_l must be an integer because that ensures $e^{2\pi i m_l}=1$.

D8C.3 A vector can be used to represent angular momentum. The vector's length represents the magnitude of the angular momentum, and its direction is along the axis of rotation. In classical mechanics, both the length and direction are well defined and continuously variable. In quantum mechanics, however, the length is quantized. The orientation of the vector is both restricted and indefinite in that one component of the vector (call it the z component) is quantized, while the other two components are indefinite. The result is that the classical mechanical vector is replaced by a set of cones that represent vectors allowed by quantum constraints. (See Section 8C.2(e).) Each cone is the set of vectors of a given (quantized) length and definite z component; the cone is generated by taking one such vector and 'sweeping' it through all possible directions in the xy plane. (The 'sweeping' does not correspond to any sort of motion. At this stage, the model describes time-independent rotational states.)

Solutions to exercises

E8C.1(a) The magnitude of angular momentum is [8C.21a]

$$\langle l^2\rangle^{1/2}=\{l(l+1)\}^{1/2}\hbar=(1\times 2)^{1/2}\hbar=\boxed{2^{1/2}\hbar}$$

Possible projections on to an arbitrary axis are [8C.21b]

$$\langle l_z\rangle = m_l\hbar$$

where $m_l=0$ or ± 1. So possible projections include $\boxed{0, \pm\hbar}$

E8C.2(a) Normalization requires [7B.4a]

$$\int \psi^*\psi\, d\tau = 1$$

That is, using the unnormalized wavefunction [8C.11]

$$\int_0^{2\pi} N^2 e^{-im_l\phi}e^{im_l\phi}\,d\phi = N^2\int_0^{2\pi} d\phi = 2\pi N^2 = 1$$

Thus $N^2=\dfrac{1}{2\pi}\qquad N=\boxed{\left(\dfrac{1}{2\pi}\right)^{1/2}}$

E8C.3(a) The energy levels of a particle on a ring are given by [8C.6b]

$$E=\frac{m_l^2\hbar^2}{2I}=\frac{m_l^2\hbar^2}{2mr^2},\quad m_l=0,\pm1,\pm2,...$$

The minimum excitation energy is the energy difference between the ground state (m_l=0) and the first excited level (m_l=±1). So the minimum excitation energy is

$$\Delta E=\frac{(1^2-0^2)\hbar^2}{2mr^2}=\frac{(1.0546\times10^{-34}\,\text{J s})^2}{2\times(1.6726\times10^{-27}\,\text{kg})\times(100\times10^{-12}\,\text{m})^2}=\boxed{3.32\times10^{-22}\,\text{J}}.$$

E8C.4(a) The energy levels are [8C.20]

$$E=\frac{l(l+1)\hbar^2}{2I},\quad l=0,1,2,\ldots$$

The minimum energy to start it rotating is the minimum excitation energy, the energy to take it from the motionless l=0 to the rotating l=1 state:

$$\Delta E=E_1=\frac{1\times2\times(1.0546\times10^{-34}\,\text{J s})^2}{2\times(5.27\times10^{-47}\,\text{kg m}^2)}=\boxed{2.11\times10^{-22}\,\text{J}}$$

E8C.5(a) The energy levels are [8C.20]

$$E=\frac{l(l+1)\hbar^2}{2I},\quad l=0,1,2,\ldots$$

So the excitation energy is

$$\Delta E=E_2-E_1=\frac{(2\times3-1\times2)\times(1.0546\times10^{-34}\,\text{J s})^2}{2\times(5.27\times10^{-47}\,\text{kg m}^2)}=\boxed{4.22\times10^{-22}\,\text{J}}$$

E8C.6(a) The energy levels are [8C.20]

$$E=\frac{l(l+1)\hbar^2}{2I}\quad l=0,1,2,\ldots$$

So the minimum energy ***allowed*** for this system is zero—but that corresponds to rest, not rotation. So the minimum energy of rotation occurs for the state that has l=1. The angular momentum in that state is [8C.21a]

$$\{l(l+1)\}^{1/2}\hbar=\boxed{2^{1/2}\hbar}=2^{1/2}\times(1.0546\times10^{-34}\,\text{J s})=\boxed{1.49\times10^{-34}\,\text{J s}}.$$

Comment. Note that the moment of inertia does not enter into the result. Thus the minimum angular momentum is the same for a molecule of CH_4 as for a molecule of C_{60} as for a football.

E8C.7(a) The diagrams are drawn by forming a vector of length $\{l(l+1)\}^{1/2}$ and with a projection m_j on the z-axis. (See Fig. 8C.1). Each vector represents the edge of a cone around the z-axis (that for m_j=0 represents the side view of a disc perpendicular to z).

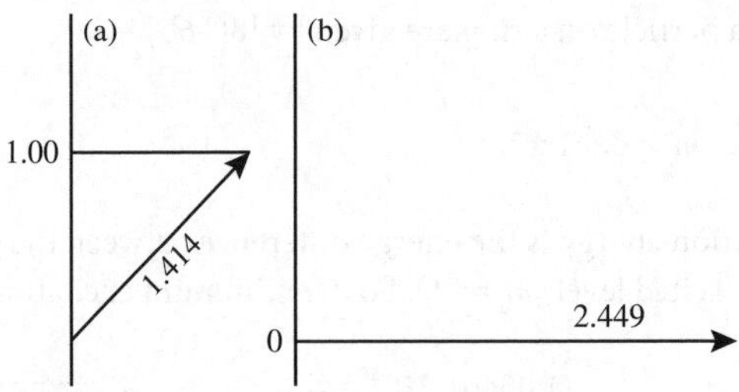

Figure 8C.1

E8C.8(a) The rotational energy depends only on the quantum number l [8C.20], but there are distinct states for every allowed value of m_l, which can range from $-l$ to l in integer steps. For $l=3$, possible values of $m_l=-3, -2, -1, 0, 1, 2, 3$. There are 7 such values, so the degeneracy is $\boxed{7}$.

Solutions to problems

P8C.1 The angular momentum states are defined by the quantum number $m_l=0, \pm1, \pm2$, etc. The energy of state m_l is [8C.6b]

$$E_{m_l}=\frac{m_l^2\hbar^2}{2I}$$

and the angular momentum is [8C.4]

$$J_z=m_l\hbar$$

(a) If there are 22 electrons, two in each of the lowest 11 states, then the highest occupied states are $m_l=\pm5$, so,

$$J_z=\pm5\hbar=\pm5\times(1.0546\times10^{-34}\,\mathrm{J\,s})=\pm5.275\times10^{-34}\,\mathrm{J\,s}.$$

and $E_{\pm5}=\dfrac{25\hbar^2}{2I}$.

The moment of inertia of an electron on a ring of radius 440 pm is

$$I=mr^2=(9.11\times10^{-31}\,\mathrm{kg})\times(440\times10^{-12}\,\mathrm{m})^2=1.76\times10^{-49}\,\mathrm{kg\,m^2}.$$

Hence $E_{\pm5}=\dfrac{25\times(1.055\times10^{-34}\,\mathrm{J\,s})^2}{2\times(1.76\times10^{-49}\,\mathrm{kg\,m^2})}=\boxed{7.89\times10^{-19}\,\mathrm{J}}$.

(b) The lowest unoccupied energy level is $m_l=\pm6$, which has energy

$$E_{\pm6}=\frac{36\times(1.055\times10^{-34}\,\mathrm{J\,s})^2}{2\times(1.76\times10^{-49}\,\mathrm{kg\,m^2})}=1.14\times10^{-18}\,\mathrm{J}.$$

Radiation that would induce a transition between these levels must have a frequency such that

$$h\nu=\Delta E \quad \text{so} \quad \nu=\frac{\Delta E}{h}=\frac{(11.4-7.89)\times10^{-19}\,\mathrm{J}}{6.626\times10^{-34}\,\mathrm{J\,s}}=\boxed{5.2\times10^{14}\,\mathrm{Hz}}.$$

This corresponds to a wavelength of about 570 nm, a wave of visible light.

P8C.3 In each case, if the function is an eigenfunction of the operator, the eigenvalue is also the expectation value; if it is not an eigenfunction we form [7C.11]

$$\langle\Omega\rangle = \int \psi^* \hat{\Omega}\psi \,\mathrm{d}\tau$$

(a) $\hat{l}_z \mathrm{e}^{\mathrm{i}\phi} = \frac{\hbar}{\mathrm{i}}\frac{\mathrm{d}}{\mathrm{d}\phi}\mathrm{e}^{\mathrm{i}\phi}$ [8C.13b] $= \hbar\mathrm{e}^{\mathrm{i}\phi}$; hence $l_z = \boxed{+\hbar}$

(b) $\hat{l}_z \mathrm{e}^{-2\mathrm{i}\phi} = \frac{\hbar}{\mathrm{i}}\frac{\mathrm{d}}{\mathrm{d}\phi}\mathrm{e}^{-2\mathrm{i}\phi} = -2\hbar\mathrm{e}^{-2\mathrm{i}\phi}$; hence $l_z = \boxed{-2\hbar}$

(c) $\langle l_z\rangle \propto \int_0^{2\pi}\cos\phi\left(\frac{\hbar}{\mathrm{i}}\frac{\mathrm{d}}{\mathrm{d}\phi}\cos\phi\right)\mathrm{d}\phi \propto -\frac{\hbar}{\mathrm{i}}\int_0^{2\pi}\cos\phi\sin\phi\,\mathrm{d}\phi = \boxed{0}$

(d) $\langle l_z\rangle = N^2\int_0^{2\pi}(\cos\chi\mathrm{e}^{\mathrm{i}\phi}+\sin\chi\mathrm{e}^{-\mathrm{i}\phi})^*\left(\frac{\hbar}{\mathrm{i}}\frac{\mathrm{d}}{\mathrm{d}\phi}\right)\times(\cos\chi\mathrm{e}^{\mathrm{i}\phi}+\sin\chi\mathrm{e}^{-\mathrm{i}\phi})\mathrm{d}\phi$

$$= \frac{\hbar}{\mathrm{i}}N^2\int_0^{2\pi}(\cos\chi\mathrm{e}^{-\mathrm{i}\phi}+\sin\chi\mathrm{e}^{\mathrm{i}\phi})\times(\mathrm{i}\cos\chi\mathrm{e}^{\mathrm{i}\phi}-\mathrm{i}\sin\chi\mathrm{e}^{-\mathrm{i}\phi})\mathrm{d}\phi$$
$$= \hbar N^2\int_0^{2\pi}(\cos^2\chi-\sin^2\chi+\cos\chi\sin\chi[\mathrm{e}^{2\mathrm{i}\phi}-\mathrm{e}^{-2\mathrm{i}\phi}])\mathrm{d}\phi$$
$$= \hbar N^2(\cos^2\chi-\sin^2\chi)\times(2\pi) = 2\pi\hbar N^2\cos 2\chi$$

We must evaluate the normalization constant:

$$N^2\int_0^{2\pi}(\cos\chi\mathrm{e}^{\mathrm{i}\phi}+\sin\chi\mathrm{e}^{-\mathrm{i}\phi})^*(\cos\chi\mathrm{e}^{\mathrm{i}\phi}+\sin\chi\mathrm{e}^{-\mathrm{i}\phi})\mathrm{d}\phi = 1$$
$$1 = N^2\int_0^{2\pi}(\cos^2\chi+\sin^2\chi+\cos\chi\sin\chi[\mathrm{e}^{2\mathrm{i}\phi}+\mathrm{e}^{-2\mathrm{i}\phi}])\mathrm{d}\phi$$
$$= 2\pi N^2(\cos^2\chi+\sin^2\chi) = 2\pi N^2 \quad \text{so } N^2 = \frac{1}{2\pi}$$

Therefore

$$\langle l_z\rangle = \boxed{\hbar\cos 2\chi}\,[\chi \text{ is a parameter}]$$

For the kinetic energy we use $\hat{E}_\mathrm{k} = \frac{\hat{J}_z^2}{2I}$[8C.1] $= -\frac{\hbar^2}{2I}\frac{\mathrm{d}^2}{\mathrm{d}\phi^2}$[8C.1b]

(a) $\hat{E}_\mathrm{k}\mathrm{e}^{\mathrm{i}\phi} = -\frac{\hbar^2}{2I}(\mathrm{i}^2\mathrm{e}^{\mathrm{i}\phi}) = \frac{\hbar^2}{2I}\mathrm{e}^{\mathrm{i}\phi}$; hence $\langle E_\mathrm{k}\rangle = \boxed{\frac{\hbar^2}{2I}}$

(b) $\hat{E}_\mathrm{k}\mathrm{e}^{-2\mathrm{i}\phi} = -\frac{\hbar^2}{2I}(2\mathrm{i})^2\mathrm{e}^{-2\mathrm{i}\phi} = \frac{4\hbar^2}{2I}\mathrm{e}^{-2\mathrm{i}\phi}$; hence $\langle E_\mathrm{k}\rangle = \boxed{\frac{2\hbar^2}{I}}$

(c) $\hat{E}_\mathrm{k}\cos\phi = -\frac{\hbar^2}{2I}(-\cos\phi) = \frac{\hbar^2}{2I}\cos\phi$; hence $\langle E_\mathrm{k}\rangle = \boxed{\frac{\hbar^2}{2I}}$

(d) $\hat{E}_\mathrm{k}(\cos\chi\mathrm{e}^{\mathrm{i}\phi}+\sin\chi\mathrm{e}^{-\mathrm{i}\phi}) = -\frac{\hbar^2}{2I}(-\cos\chi\mathrm{e}^{\mathrm{i}\phi}-\sin\chi\mathrm{e}^{-\mathrm{i}\phi}) = \frac{\hbar^2}{2I}(\cos\chi\mathrm{e}^{\mathrm{i}\phi}+\sin\chi\mathrm{e}^{-\mathrm{i}\phi})$

and hence $\langle E_\mathrm{k}\rangle = \boxed{\frac{\hbar^2}{2I}}$

Comment. All of these functions are eigenfunctions of the kinetic energy operator, which is also the total energy or hamiltonian operator, since the potential energy is zero for this system.

P8C.5

$$E=\frac{l(l+1)\hbar^2}{2I}[8C.20]=\frac{l(l+1)\hbar^2}{2\mu R^2} \quad [I=\mu R^2]$$

$$=\left(\frac{l(l+1)\times(1.055\times10^{-34}\,\mathrm{J\,s})^2}{(2)\times(1.6605\times10^{-27}\,\mathrm{kg})\times(160\times10^{-12}\,\mathrm{m})^2}\right)\times\left(\frac{1}{1.008}+\frac{1}{126.90}\right)$$

Therefore,

$$E=l(l+1)\times(1.31\times10^{-22}\,\mathrm{J})$$

The energies may be expressed in terms of equivalent frequencies with

$$\nu=\frac{E}{h}=(1.509\times10^{33}\,\mathrm{J^{-1}\,s^{-1}})E.$$

Hence, the energies and equivalent frequencies are

l	0	1	2	3
$10^{22}\,E/\mathrm{J}$	$\boxed{0}$	$\boxed{2.62}$	$\boxed{7.86}$	$\boxed{15.72}$
ν/GHz	0	396	1 188	2 376

P8C.7 Call the integral I:

$$I=\int_0^{\pi}\int_0^{2\pi}Y_{3,3}^{*}Y_{3,3}\sin\theta\,\mathrm{d}\theta\,\mathrm{d}\phi=\left(\frac{1}{64}\right)\times\left(\frac{35}{\pi}\right)\int_0^{\pi}\sin^6\theta\sin\theta\,\mathrm{d}\theta\int_0^{2\pi}\mathrm{d}\phi \quad [\text{Table 8C.1}]$$

Integration over $\mathrm{d}\phi$ yields a factor of 2π. Noting that $\sin\theta\,\mathrm{d}\theta=\mathrm{d}\cos\theta$, and that $\sin^2\theta=1-\cos^2\theta$, the integral becomes

$$I=\left(\frac{1}{64}\right)\times\left(\frac{35}{\pi}\right)\times(2\pi)\int_{-1}^{1}(1-\cos^2\theta)^3\,\mathrm{d}\cos\theta$$

Letting $x=\cos\theta$ and expanding the integrand, we have

$$I=\frac{35}{32}\int_{-1}^{1}(1-3x^2+3x^4-x^6)\mathrm{d}x=\frac{35}{32}\left(x-x^3+\tfrac{3}{5}x^5-\tfrac{1}{7}x^7\right)\Big|_{-1}^{1}=\frac{35}{32}\times\frac{32}{35}=\boxed{1}$$

P8C.9

$$\hat{\mathbf{l}}=\hat{\mathbf{r}}\times\hat{\mathbf{p}}=\begin{vmatrix}\mathbf{i} & \mathbf{j} & \mathbf{k}\\ \hat{x} & \hat{y} & \hat{z}\\ \hat{p}_x & \hat{p}_y & \hat{p}_z\end{vmatrix} \quad [\text{see any book treating the vector product of vectors}]$$

$$=\mathbf{i}(\hat{y}\hat{p}_z-\hat{z}\hat{p}_y)+\mathbf{j}(\hat{z}\hat{p}_x-\hat{x}\hat{p}_z)+\mathbf{k}(\hat{x}\hat{p}_y-\hat{y}\hat{p}_x)$$

Therefore,

$$\hat{l}_x = (\hat{y}\hat{p}_z - \hat{z}\hat{p}_y) = \boxed{\frac{\hbar}{\mathrm{i}}\left(y\frac{\partial}{\partial z} - z\frac{\partial}{\partial y}\right)}$$

$$\hat{l}_y = (\hat{z}\hat{p}_x - \hat{x}\hat{p}_z) = \boxed{\frac{\hbar}{\mathrm{i}}\left(z\frac{\partial}{\partial x} - x\frac{\partial}{\partial z}\right)}$$

$$\hat{l}_z = (\hat{x}\hat{p}_y - \hat{y}\hat{p}_x) = \boxed{\frac{\hbar}{\mathrm{i}}\left(x\frac{\partial}{\partial y} - y\frac{\partial}{\partial x}\right)}$$

We have used $\hat{p}_x = \frac{\hbar}{\mathrm{i}}\frac{\partial}{\partial x}$, etc. The commutator of $\hat{l}_x$ and $\hat{l}_y$ is $(\hat{l}_x\hat{l}_y - \hat{l}_y\hat{l}_x)$. We note that the operations always imply operation on a function. We form

$$\hat{l}_x\hat{l}_y f = -\hbar^2\left(y\frac{\partial}{\partial z} - z\frac{\partial}{\partial y}\right)\left(z\frac{\partial}{\partial x} - x\frac{\partial}{\partial z}\right)f$$

$$= -\hbar^2\left(yz\frac{\partial^2 f}{\partial z\partial x} + y\frac{\partial f}{\partial x} - yx\frac{\partial^2 f}{\partial z^2} - z^2\frac{\partial^2 f}{\partial y\partial x} + zx\frac{\partial^2 f}{\partial z\partial y}\right)$$

and $\hat{l}_y\hat{l}_x f = -\hbar^2\left(z\frac{\partial}{\partial x} - x\frac{\partial}{\partial z}\right)\left(y\frac{\partial}{\partial z} - z\frac{\partial}{\partial y}\right)f$

$$= -\hbar^2\left(zy\frac{\partial^2 f}{\partial x\partial z} - z^2\frac{\partial^2 f}{\partial x\partial y} - xy\frac{\partial^2 f}{\partial z^2} + xz\frac{\partial^2 f}{\partial z\partial y} + x\frac{\partial f}{\partial y}\right)$$

Since multiplication and differentiation are each commutative, the results of the operation $\hat{l}_x\hat{l}_y$ and $\hat{l}_y\hat{l}_x$ differ only in one term. For $\hat{l}_y\hat{l}_x f$, $x\frac{\partial f}{\partial y}$ replaces $y\frac{\partial f}{\partial x}$. Hence, the commutator of the operations, $(\hat{l}_x\hat{l}_y - \hat{l}_y\hat{l}_x)$ is $-\hbar^2\left(y\frac{\partial}{\partial x} - x\frac{\partial}{\partial y}\right)$ or $\boxed{-\frac{\hbar}{\mathrm{i}}\hat{l}_z.}$

Similarly $(\hat{l}_y\hat{l}_z - \hat{l}_z\hat{l}_y) = -\frac{\hbar}{\mathrm{i}}\hat{l}_x$ and $(\hat{l}_z\hat{l}_x - \hat{l}_x\hat{l}_z) = -\frac{\hbar}{\mathrm{i}}\hat{l}_y$.

P8C.11 We are to show that $[\widehat{l^2},\hat{l}_z] = [\hat{l}_x^2 + \hat{l}_y^2 + \hat{l}_z^2,\hat{l}_z] = [\hat{l}_x^2,\hat{l}_z] + [\hat{l}_y^2,\hat{l}_z] + [\hat{l}_z^2,\hat{l}_z] = 0$

The three commutators are:

$$[\hat{l}_z^2,\hat{l}_z] = \hat{l}_z^2\hat{l}_z - \hat{l}_z\hat{l}_z^2 = \hat{l}_z^3 - \hat{l}_z^3 = 0$$

$$[\hat{l}_x^2,\hat{l}_z] = = \hat{l}_x^2\hat{l}_z - \hat{l}_z\hat{l}_x^2 = \hat{l}_x^2\hat{l}_z - \hat{l}_x\hat{l}_z\hat{l}_x + \hat{l}_x\hat{l}_z\hat{l}_x - \hat{l}_z\hat{l}_x^2$$

$$= \hat{l}_x(\hat{l}_x\hat{l}_z - \hat{l}_z\hat{l}_x) + (\hat{l}_x\hat{l}_z - \hat{l}_z\hat{l}_x)\hat{l}_x = \hat{l}_x[\hat{l}_x,\hat{l}_z] + [\hat{l}_x,\hat{l}_z]\hat{l}_x$$

$$= \hat{l}_x(-\mathrm{i}\hbar\hat{l}_y) + (-\mathrm{i}\hbar\hat{l}_y)\hat{l}_x = -\mathrm{i}\hbar(\hat{l}_x\hat{l}_y + \hat{l}_y\hat{l}_x) \quad [8C.27]$$

$$[\hat{l}_y^2,\hat{l}_z] = \hat{l}_y^2\hat{l}_z - \hat{l}_z\hat{l}_y^2 = \hat{l}_y^2\hat{l}_z - \hat{l}_y\hat{l}_z\hat{l}_y + \hat{l}_y\hat{l}_z\hat{l}_y - \hat{l}_z\hat{l}_y^2$$

$$= \hat{l}_y(\hat{l}_y\hat{l}_z - \hat{l}_z\hat{l}_y) + (\hat{l}_y\hat{l}_z - \hat{l}_z\hat{l}_y)\hat{l}_y = \hat{l}_y[\hat{l}_y,\hat{l}_z] + [\hat{l}_y,\hat{l}_z]\hat{l}_y$$

$$= \hat{l}_y(\mathrm{i}\hbar\hat{l}_x) + (\mathrm{i}\hbar\hat{l}_x)\hat{l}_y = \mathrm{i}\hbar(\hat{l}_y\hat{l}_x + \hat{l}_x\hat{l}_y) \quad [8C.27]$$

Therefore, $[\widehat{l^2},\hat{l}_z] = -\mathrm{i}\hbar(\hat{l}_x\hat{l}_y + \hat{l}_y\hat{l}_x) + \mathrm{i}\hbar(\hat{l}_x\hat{l}_y + \hat{l}_y\hat{l}_x) + 0 = 0$

We may also conclude that $[\widehat{l^2},\hat{l}_x] = 0$ and $[\widehat{l^2},\hat{l}_y] = 0$ because $\hat{l}_x$, $\hat{l}_y$, and $\hat{l}_z$ occur symmetrically in $\widehat{l^2}$.

Integrated activities

I8.1 Macroscopic synthesis and material development always contains elements of molecular randomness. Crystal structures are never perfect. A product of organic synthesis is never absolutely free of impurities, although impurities may be at a level that is lower than measurement techniques make possible. Alloys are grainy and slightly non-homogeneous within any particular grain. Furthermore, the random distribution of atomic/molecular positions and orientations within, and between, macroscopic objects causes the conversion of energy to non-useful heat during manufacturing processes. Production efficiencies are difficult to improve. Nanometre technology on the 1 nm to 100 nm scale may resolve many of these problems. Self-organization and production processes by nanoparticles and nanomachines may be able to exclude impurities and greatly improve homogeneity by effective examination and selection of each atom/molecule during nanosynthesis and nanoproduction processes. Higher efficiencies of energy usage may be achievable as nanomachines produce idealized materials at the smaller sizes and pass their products to larger nanomachines for production of larger scale materials.

The directed, non-random, use of atoms and molecules by nanotechniques holds the promise for the production of smaller transistors and wires for the electronics and computer industries. Unusual material strengths, optical properties, magnetic properties, and catalytic properties may be achievable. Higher efficiencies of photo-electronic conversion would be a boon to mankind. There is hope that science will devise nanoparticles that destroy pathogens and repair tissues. See *Impact* I8B.1 for discussion of nano-quantum dots that have unusual optical and magnetic properties.

I8.3 The energy of any given molecule is

$$E=\frac{n^2h^2}{8mL^2}\quad [\text{8A.16b with } n^2=n_1{}^2+n_2{}^2+n_3{}^2 \text{ and equal lengths}]$$

So the internal energy of a sample of N molecules is

$$U=NE=\frac{Nn^2h^2}{8mL^2}=\frac{Nn^2h^2}{8mV^{2/3}}$$

(a) In the last step we used $V=L^3$, because we are interested in how the energy changes with volume. Consider an adiabatic change of volume, that is, a change in which no heat enters or leaves the sample. In that case, the change in energy is entirely work (First Law with $q=0$). Differentiate the expression for U:

$$\mathrm{d}w=\left(\frac{\partial U}{\partial V}\right)_{\text{adiabatic}}\mathrm{d}V=\boxed{-\frac{Nn^2h^2}{12mV^{5/3}}\mathrm{d}V}\quad(1)$$

(b) Note that we did not have to specify the energy level to obtain the above result. The lowest energy level is $n_1=n_2=n_3=1$, so $n^2=3$; however, the derivation applies what follows applies to any allowed energy level. So $\boxed{\text{no, the work would not be different}}$.

(c) Expansion work has the form $dw = -p_{ex}dV$ [2A.5a]. Can these expressions be reconciled, and if so, under what conditions? First, note that the expression that multiplies dV in equation (1) refers to the sample, so if it is some sort of pressure, it must be the sample pressure, and not an arbitrary external pressure. Hence, if the expressions can be reconciled, it must be for *reversible* adiabatic expansion or compression (in which the internal and external pressures are equal). The expression that multiplies dV can be expressed as

$$\frac{Nn^2h^2}{12mV^{5/3}} = \frac{2}{3}\frac{N}{V}E.$$

In fact, the kinetic model of gases (Topic 1B) says that the pressure of a gas is equal to $\frac{2}{3}\frac{N}{V}E$ where E is the average kinetic energy of the gas molecules—completely consistent with interpreting it as the average particle-in-a-box energy. To summarize, reversible adiabatic work for a gas of particle-in-a-box molecules is $dw = -pdV$, where the pressure is

$$p = \frac{Nn^2h^2}{12mL^5} = \frac{2}{3}\frac{N}{V}E$$

In expansion, the volume increases, meaning that the box gets bigger. Equation 8A.16b tells us that the kinetic energy decreases, even as the quantum numbers remain constant. This is also consistent with what we know of adiabatic expansion and the kinetic model of gases: the temperature of the sample drops on expansion, and temperature is related to the kinetic energy ($T \propto v^2 \propto E$).

(d) In isothermal expansion, energy must enter the system as heat to maintain the temperature. We can interpret this influx of heat as an increase in quantum numbers (an excitation of the molecules) that offsets the falling energy levels.

I8.5 (a) Expectation values for $\langle x \rangle$, $\langle x^2 \rangle$, $\langle p \rangle$, and $\langle p^2 \rangle$ were evaluated in Exercises 8A.5 and 8A.6.

$$\langle x \rangle = \frac{L}{2} \text{ for all } n$$

$$\langle x^2 \rangle = L^2\left(\frac{1}{3} - \frac{1}{2n^2\pi^2}\right)$$

$$\Delta x = \left[L^2\left(\frac{1}{3} - \frac{1}{2n^2\pi^2}\right) - \frac{L^2}{4}\right]^{1/2} = \boxed{L\left(\frac{1}{12} - \frac{1}{2\pi^2 n^2}\right)^{1/2}}$$

$$\langle p \rangle = 0 \text{ for all } n$$

$$\langle p^2 \rangle = \frac{n^2h^2}{4L^2}$$

$$\Delta p = \left(\frac{n^2h^2}{4L^2}\right)^{1/2} = \boxed{\frac{nh}{2L}}$$

$$\Delta p \Delta x = \frac{nh}{2L} \times L\left(\frac{1}{12} - \frac{1}{2\pi^2 n^2}\right)^{1/2} = \frac{nh}{2\sqrt{3}}\left(1 - \frac{1}{24\pi^2 n^2}\right)^{1/2} > \frac{\hbar}{2}$$

(b) $\langle x\rangle = 0$ for all v [8B.12a, or by symmetry]

and $\langle x^2\rangle = \left(v+\frac{1}{2}\right)\times\left(\frac{\hbar^2}{mk_f}\right)^{1/2}$ [8B.12] $=\left(v+\frac{1}{2}\right)\times\frac{\hbar}{\omega m}$ [8B.4]

so $\Delta x = \boxed{\left\{\left(v+\frac{1}{2}\right)\frac{\hbar}{\omega m}\right\}^{1/2}}$

$\langle p\rangle = 0$ [by symmetry, or by noting that the integrand is an odd function of x]

and $\langle p^2\rangle = 2m\langle E_k\rangle = 2m\int_{-\infty}^{+\infty}\psi^*\hat{E}_k\psi\,dx$

$$\hat{E}_k = -\frac{\hbar^2}{2m}\frac{d^2}{dx^2}\ [7C.5] = -\frac{\hbar^2}{2m\alpha^2}\frac{d^2}{dy^2} = -\frac{\hbar\omega}{2}\frac{d^2}{dy^2},\quad \left[x=\alpha y,\ \alpha^2=\frac{\hbar}{m\omega}\right]$$

which implies that

$$\hat{E}_k\psi = -\frac{\hbar\omega}{2}\left(\frac{d^2\psi}{dy^2}\right)$$

We then use $\psi = NHe^{-y^2/2}$, and obtain

$$\frac{d^2\psi}{dy^2} = N\frac{d^2}{dy^2}(He^{-y^2/2}) = N\{H'' - 2yH' - H + y^2H\}e^{-y^2/2}$$

From Table 8B.1

$$H_v'' - 2yH_v' = -2vH_v$$

$$y^2H_v = y\left(\tfrac{1}{2}H_{v+1} + vH_{v-1}\right) = \tfrac{1}{2}\left(\tfrac{1}{2}H_{v+2} + (v+1)H_v\right) + v\left(\tfrac{1}{2}H_v + (v-1)H_{v-2}\right)$$
$$= \tfrac{1}{4}H_{v+2} + v(v-1)H_{v-2} + \left(v+\tfrac{1}{2}\right)H_v$$

Hence, $\frac{d^2\psi}{dy^2} = N\left[\tfrac{1}{4}H_{v+2} + v(v-1)H_{v-2} - \left(v+\frac{1}{2}\right)H_v\right]e^{-y^2/2}$

Therefore,

$$\langle E_k\rangle = N_v^2\left(-\frac{\hbar\omega}{2}\right)\int_{-\infty}^{+\infty}H_v\left\{\frac{1}{4}H_{v+2} + v(v-1)H_{v-2} - \left(v+\frac{1}{2}\right)H_v\right\}e^{-y^2}dx\,[dx=\alpha dy]$$

$$= \alpha N_v^2\left(-\frac{1}{2}\hbar\omega\right)\left\{0+0-\left(v+\frac{1}{2}\right)\pi^{1/2}2^v v!\right\}\ \text{[Hermite polynomials orthogonal]}$$

$$= \frac{1}{2}\left(v+\frac{1}{2}\right)\hbar\omega\quad\left[N_v^2 = \frac{1}{\alpha\pi^{1/2}2^v v!},\ \text{Example 8B.2}\right]$$

and $\langle p^2\rangle = \left(v+\frac{1}{2}\right)m\hbar\omega$

$$\Delta p = \boxed{\left\{\left(v+\frac{1}{2}\right)\hbar\omega m\right\}^{1/2}}$$

$$\Delta p\Delta x = \left(v+\frac{1}{2}\right)\hbar \geq \frac{\hbar}{2}$$

Comment. Both results show a consistency with the uncertainty principle in the form $\Delta p\Delta q \geq \frac{\hbar}{2}$ as given in eqn 7C.13.

9 Atomic structure and spectra

9A Hydrogenic atoms

Answers to discussion questions

D9A.1 The Schrödinger equation for the hydrogen atom is a six-dimensional partial differential equation, three dimensions for each particle in the atom. One cannot directly solve a multidimensional differential equation; it must be broken down into one-dimensional equations. This is the separation of variables procedure. The choice of coordinates is critical in this process. The separation of the Schrödinger equation can be accomplished in a set of coordinates that are natural to the system, but not in others. These natural coordinates are those directly related to the description of the motion of the atom. The atom as a whole (centre of mass) can move from point to point in three-dimensional space. The natural coordinates for this kind of motion are the Cartesian coordinates of a point in space. The internal motion of the electron with respect to the proton is most naturally described with spherical polar coordinates. So the six-dimensional Schrödinger equation is first separated into two three-dimensional equations, one for the motion of the centre of mass, the other for the internal motion. The separation of the centre of mass equation and its solution is fully discussed in Chapter 8. The equation for the internal motion is separable into three one-dimensional equations, one in the angle ϕ, another in the angle θ, and a third in the distance r. The solutions of these three one-dimensional equations can be obtained by standard techniques and were already well known long before the advent of quantum mechanics. Another choice of coordinates would not have resulted in the separation of the Schrödinger equation just described. For the details of the separation procedure, see Chapter 8 and in particular Section 9A.1(a).

D9A.3 (a) A boundary surface for a hydrogenic orbital is drawn so as to contain most (say 90%) of the probability density of an electron in that orbital. Its shape varies from orbital to orbital because the electron density distribution is different for different orbitals.

(b) The radial distribution function gives the probability that the electron will be found anywhere within a shell of radius r around the nucleus. It gives a better picture of where the electron is likely to be found with respect to the nucleus than the probability density which is the square of the wavefunction.

Solutions to exercises

E9A.1(a) The energies are $E=-\dfrac{hc\tilde{R}_H}{n^2}$ [9A.9 with 9A.14], and the orbital degeneracy g of an energy level of principal quantum number n is

$$g=\sum_{l=0}^{n-1}(2l+1)=1+3+5+\cdots+2n-1=\frac{(1+2n-1)n}{2}=n^2$$

(i) $E=-hc\tilde{R}_H$ implies that $n=1$, so $\boxed{g=1}$ [the 1s orbital].

(ii) $E=-\dfrac{hc\tilde{R}_H}{9}$ implies that $n=3$, so $\boxed{g=9}$ (3s orbital, the three 3p orbitals, and the five 3d orbitals).

(iii) $E=-\dfrac{hc\tilde{R}_H}{25}$ implies that $n=5$, so $\boxed{g=25}$ (the 5s orbital, the three 5p orbitals, the five 5d orbitals, the seven 5f orbitals, the nine 5g orbitals).

E9A.2(a) $R_{1,0}=Ne^{-r/a_0}$

$$\int_0^\infty R^2r^2\mathrm{d}r=1=\int_0^\infty N^2r^2e^{-2r/a_0}\mathrm{d}r=N^2\times\frac{2!}{\left(\frac{2}{a_0}\right)^3}=1\quad\left[\int_0^\infty x^n\mathrm{e}^{-ax}\,\mathrm{d}x=\frac{n!}{a^{n+1}}\right]$$

$$N^2=\frac{4}{a_0^3},\quad \boxed{N=\frac{2}{a_0^{3/2}}}$$

Thus,

$$R_{1,0}=2\left(\frac{1}{a_0}\right)^{3/2}\mathrm{e}^{-r/a_0},$$

which agrees with Table 9A.1.

E9A.3(a) $R_{2,0}\propto\left(2-\dfrac{\rho}{2}\right)\mathrm{e}^{-\rho/4}$ with $\rho\equiv\dfrac{2r}{a_0}$ [Table 9A.1]

Note: for convenience ρ defined here is twice ρ defined in Table 9A.1.

$$\frac{\mathrm{d}R}{\mathrm{d}r}=\frac{2}{a_0}\frac{\mathrm{d}R}{\mathrm{d}\rho}=\frac{2}{a_0}\left(-\frac{1}{2}-\frac{1}{2}+\frac{1}{8}\rho\right)\mathrm{e}^{-\rho/4}=0\quad\text{when }\rho=8$$

Hence, the wavefunction has an extremum at $r=\boxed{4a_0}$. Since $2-\dfrac{\rho}{2}<0,\psi<0$ and the extremum is a minimum (more formally: $\dfrac{\mathrm{d}^2\psi}{\mathrm{d}r^2}>0$ at $\rho=8$). The second extremum is at $\boxed{r=0.}$ It is not a minimum and in fact is a physical maximum, though not one that can be obtained by differentiation. To see that it is a maximum substitute $\rho=0$ into $R_{2,0}$.

E9A.4(a) Here we use the probability density function ψ^2, rather than the radial distribution function P, since we are seeking the probability at a point, namely $\psi^2\,\mathrm{d}\tau$.

The probability density varies as

$$\psi^2 = \frac{1}{\pi a_0^3}\mathrm{e}^{-2r/a_0} \text{ [From 9A.17, the ground state is assumed.]}$$

Therefore, the maximum value is at $r = 0$ and ψ^2 is 50 per cent of the maximum when

$$\mathrm{e}^{-2r/a_0} = 0.50$$

implying that $r = -\frac{1}{2}a_0 \ln 0.50$, which is at $\boxed{r = 0.35a_0}$ [18 pm]

E9A.5(a) The radial nodes correspond to $R_{3,0} = 0$. $R_{3,0} \propto 6 - 2\rho + \frac{1}{9}\rho^2$ (Table 9A.1); the radial nodes occur at

$$6 - 6\rho + \rho^2 = 0, \quad \text{or} \quad \rho = 1.27 \text{ and } 4.73.$$

Since $r = \frac{n\rho a_0}{2}$, the radial nodes occur at $\boxed{101\,\text{pm and } 376\,\text{pm}}$

E9A.6(a) This exercise has already been solved in Problem 8B.12 by use of the virial theorem. Here we will solve it by straightforward integration.

$$\psi_{1,0,0} = R_{1,0}Y_{0,0} = \left(\frac{1}{\pi a_0^3}\right)^{1/2}\mathrm{e}^{-r/a_0} \text{ [Tables 8C.1 and 9A.1]}$$

The potential energy operator is

$$V = -\frac{Ze^2}{4\pi\varepsilon_0}\times\left(\frac{1}{r}\right) = -k\left(\frac{1}{r}\right)$$

$$\langle V\rangle = -k\left\langle\frac{1}{r}\right\rangle\left[k = \frac{e^2}{4\pi\varepsilon_0}\right] = -k\int_0^\infty\int_0^\pi\int_0^{2\pi}\left(\frac{1}{\pi a_0^3}\right)\mathrm{e}^{-r/a_0}\left(\frac{1}{r}\right)\mathrm{e}^{-r/a_0}r^2\mathrm{d}r\sin\theta\,\mathrm{d}\theta\,\mathrm{d}\phi$$

$$= -k\times(4\pi)\times\left(\frac{1}{\pi a_0^3}\right)\int_0^\infty r\mathrm{e}^{-2r/a_0}\,\mathrm{d}r = -k\times\left(\frac{4}{a_0^3}\right)\times\left(\frac{a_0^2}{4}\right) = -k\left(\frac{1}{a_0}\right)$$

$$\left[\text{We have used} \int_0^\pi \sin\theta\,\mathrm{d}\theta = 2, \int_0^{2\pi}\mathrm{d}\theta = 2\pi, \text{and} \int_0^\infty x^n\mathrm{e}^{-ax}\,\mathrm{d}x = \frac{n!}{a^{n+1}}\right]$$

Hence,

$$\langle V\rangle = -\frac{e^2}{4\pi\varepsilon_0 a_0} = \boxed{2E_{1s}}$$

The kinetic energy operator is $-\dfrac{\hbar^2}{2\mu}\nabla^2$ [9A.5], hence

$$\langle E_K\rangle \equiv \langle T\rangle = \int \psi_{1s}^*\left(-\frac{\hbar^2}{2\mu}\right)\nabla^2\psi_{1s}\,d\tau$$

$$\nabla^2\psi_{1s} = \frac{1}{r}\frac{\partial^2(r\psi_{1s})}{\partial r^2} + \frac{1}{r^2}\Lambda^2\psi_{1s} \text{ [Problem 8C.6]}$$

$$= \left(\frac{1}{\pi a_0^3}\right)^{1/2} \times \left(\frac{1}{r}\right) \times \left(\frac{d^2}{dr^2}\right) re^{-r/a_0}$$

$[\Lambda^2\psi_{1s} = 0,\ \psi_{1s}$ contains no angular variables$]$

$$= \left(\frac{1}{\pi a_0^3}\right)^{1/2}\left[-\left(\frac{2}{a_0 r}\right)+\left(\frac{1}{a_0^2}\right)\right]e^{-r/a_0}$$

$$\langle T\rangle = -\left(\frac{\hbar^2}{2\mu}\right)\times\left(\frac{1}{\pi a_0^3}\right)\int_0^\infty\left[-\left(\frac{2}{a_0 r}\right)+\left(\frac{1}{a_0^2}\right)\right]e^{-2r/a_0}r^2\,dr \times \int_0^\pi \sin\theta\,d\theta\int_0^{2\pi}d\phi$$

$$= -\left(\frac{2\hbar^2}{\mu a_0^3}\right)\int_0^\infty\left[-\left(\frac{2r}{a_0}\right)+\left(\frac{r^2}{a_0^2}\right)\right]e^{-2r/a_0}\,dr = -\left(\frac{2\hbar^2}{\mu a_0^3}\right)\times\left(-\frac{a_0}{4}\right) = \frac{\hbar^2}{2\mu a_0^2}$$

$$= \boxed{-E_{1s}}$$

Hence, $\langle T\rangle + \langle V\rangle = 2E_{1s} - E_{1s} = E_{1s}$

Comment. E_{1s} may also be written as

$$E_{1s} = -\frac{\mu e^4}{32\pi^2\varepsilon_0^2\hbar^2}$$

Question. Are the three different expressions for E_{1s} given in this exercise all equivalent?

E9A.7(a) $P_{2s} = 4\pi r^2\psi_{2s}^2$

$$\psi_{2s} = \left(\frac{1}{4\pi}\right)\frac{1}{2\sqrt{2}}\left(\frac{Z}{a_0}\right)^{3/2}\times\left(2-\frac{\rho}{2}\right)e^{-\rho/4} \quad \left[\rho = \frac{2Zr}{a_0}\right]$$

[Note: ρ defined here is $n \times \rho$, as defined in Table 9A.1.]

$$P_{2s} = \left(\frac{1}{4\pi}\right)\left(\frac{a_0\rho}{2Z}\right)^2\times\left(\frac{1}{8}\right)\times\left(\frac{Z}{a_0}\right)^3\left(2-\frac{\rho}{2}\right)^2 e^{-\rho/2}$$

$$P_{2s} = k\rho^2\left(2-\frac{\rho}{2}\right)^2 e^{-\rho/2} \quad \left[k = \frac{Z}{128\pi a_0} = \text{constant}\right]$$

The most probable value of r, or equivalently, ρ is where

$$\frac{\mathrm{d}}{\mathrm{d}\rho}\left\{\rho^2\left(2-\frac{\rho}{2}\right)^2 \mathrm{e}^{-\rho/2}\right\}=0$$

$$\propto\left\{2\rho\left(2-\frac{\rho}{2}\right)^2-2\rho^2\left(2-\frac{\rho}{2}\right)-\rho^2\left(2-\frac{\rho}{2}\right)^2\right\}\mathrm{e}^{-\rho/2}=0$$

$$\propto \rho(\rho-4)(\rho^2-12\rho+16)=0 \qquad [\mathrm{e}^{-\rho/2} \text{ is never zero, except as } \rho\to\infty]$$

Thus, $\rho^*=0, \rho^*=4, \rho^*=6\pm2\sqrt{5}$

The principal (outermost) maximum is at $\rho^*=6\pm2\sqrt{5}$

Hence, $r^*=(6+2\sqrt{5})\dfrac{a_0}{2Z}=\boxed{5.24\dfrac{a_0}{Z}}$

E9A.8(a) The most probable radius occurs when the radial distribution function is a maximum. At this point the derivative of the function w/r/t either r or ρ equals zero.

$$\left(\frac{\mathrm{d}R_{21}}{\mathrm{d}\rho}\right)_{\max}=0=\left(\frac{\mathrm{d}\left(\rho\mathrm{e}^{-\rho/2}\right)}{\mathrm{d}\rho}\right)_{\max} \text{[Table 9A.1]}=\left(1-\frac{\rho}{2}\right)\mathrm{e}^{-\rho/2}$$

The function is a maximum when $\rho=2$. Since $\rho=(2Z/na_0)r$ and $n=2$, this corresponds to $\boxed{r=2a_0/Z}$.

E.9A.9(a) Identify l and use angular momentum $=\{l(l+1)\}^{1/2}\hbar$.

(i) $l=0$, so angular momentum $=0$

(ii) $l=0$, so angular momentum $=0$

(iii) $l=2$, so angular momentum $=\sqrt{6}\hbar$

The total number of nodes is equal to $n-1$ and the number of angular nodes is equal to l; hence the number of radial nodes is equal to $n-l-1$. We can draw up the following table

	1s	3s	3d
n, l	1,0	3,0	3,2
Angular nodes	0	0	2
Radial nodes	0	2	0

E9A.10(a) See Fig. 9A.5 of the text to see the radial node of the 3p orbital. Its position can be determined by setting $(4-\rho)=0$ [see Table 9A.1]. Hence, $\rho=4$ and $r=\boxed{6a_0/Z}$.

For the angular nodes, see Fig. 9A.15 of the text. The number of angular nodes is the value of the quantum number l which for p orbitals is 1. Hence, each of the three p orbitals has one angular node. To locate the angular nodes look for the value of θ that makes the wavefunction zero.

p_0 orbital: see eqn 9A.20a and Fig. 9A.15. The nodal plane is the $\boxed{xy\,\text{plane}}$ and $\boxed{\theta=\pi/2}$ is an angular node.

p_x orbital: see eqns 9A.21 and 9A.22 and Fig. 9A.15. The nodal plane is the $\boxed{yz}$ plane and $\boxed{\theta=0}$ is an angular node.

p_y orbital: see eqns 9A.21 and 9A.22 and Fig. 9A.15. The nodal plane is the $\boxed{\text{xz}}$ plane and $\boxed{\theta=0}$ is an angular node.

Solutions to problems

P9A.1 Consider $\psi_{2p_z}=\psi_{2,1,0}$ which extends along the z-axis. The most probable point along the z-axis is where the radial function has its maximum value (for ψ^2 is also a maximum at that point). From Table 9A.1 we know that

$$R_{21}\propto \rho e^{-\rho/2}$$

and so $\dfrac{dR}{d\rho}=\left(1-\dfrac{1}{2}\rho\right)e^{-\rho/4}=0$ when $\rho=2$.

Therefore, $r^*=\dfrac{2a_0}{Z}$, and the point of maximum probability lies at

$$z=\pm\frac{2a_0}{Z}=\boxed{\pm106\,\text{pm}}$$

Comment. Since the radial portion of a 2p function is the same, the same result would have been obtained for all of them. The direction of the most probable point would, however, be different.

P9A.3 (a) We must show that $\int|\psi_{3p_x}|^2 d\tau=1$. The integrations are most easily performed in spherical coordinates.

$$\int|\psi_{3p_x}|^2 d\tau=\int_0^{2\pi}\int_0^{\pi}\int_0^{\infty}\left|\psi_{3p_x}\right|^2 r^2\sin(\theta)dr\,d\theta\,d\phi$$

$$=\int_0^{2\pi}\int_0^{\pi}\int_0^{\infty}\left|R_{31}(\rho)\left\{\frac{Y_{1-1}-Y_{11}}{\sqrt{2}}\right\}\right|^2 r^2\sin(\theta)dr\,d\theta\,d\phi$$

[Table 9A.1, eqns 9A.20, 9A.21, & 9A.22]

where $\rho=2r/a_0, r=\rho a_0/2, dr=(a_0/2)d\rho$. [Note that for convenience ρ defined here is $n\times\rho$ as defined in Table 9A.1, with $n=3$ here.]

$$=\frac{1}{2}\int_0^{2\pi}\int_0^{\pi}\int_0^{\infty}\left(\frac{a_0}{2}\right)^3\left|\left[\left(\frac{1}{27(6)^{1/2}}\right)\left(\frac{1}{a_0}\right)^{3/2}\left(4-\frac{1}{3}\rho\right)\rho e^{-\rho/6}\right]\right.$$

$$\left.\times\left[\left(\frac{3}{8\pi}\right)^{1/2}2\sin(\theta)\cos(\phi)\right]\right|^2\rho^2\sin(\theta)d\rho\,d\theta\,d\phi$$

$$=\frac{1}{46\,656\pi}\int_0^{2\pi}\int_0^{\pi}\int_0^{\infty}\left|\left(4-\frac{1}{3}\rho\right)\rho e^{-\rho/6}\sin(\theta)\cos(\phi)\right|^2\rho^2\sin(\theta)d\rho\,d\theta\,d\phi$$

$$=\frac{1}{46\,656\pi}\underbrace{\int_0^{2\pi}\cos^2(\phi)d\phi}_{\pi}\underbrace{\int_0^{\pi}\sin^3(\theta)d\theta}_{4/3}\underbrace{\int_0^{\infty}\left(4-\frac{1}{3}\rho\right)^2\rho^4e^{-\rho/3}d\rho}_{34992}$$

$=1$ Thus, ψ_{3p_x} is normalized to 1.

We must also show that $\int \psi_{3\mathrm{p}_x}\psi_{3\mathrm{d}_{xy}}\,\mathrm{d}\tau = 0$

Using Tables 8C.1 and 9A.1, we find that

$$\psi_{3\mathrm{p}_x} = \frac{1}{54(2\pi)^{1/2}}\left(\frac{1}{a_0}\right)^{3/2}\left(4-\frac{1}{3}\rho\right)\rho\mathrm{e}^{-\rho/6}\sin(\theta)\cos(\phi)$$

$$\psi_{3\mathrm{d}_{xy}} = R_{32}\left\{\frac{Y_{22}-Y_{2-2}}{\sqrt{2}\mathrm{i}}\right\}$$

$$= \frac{1}{32(2\pi)^{1/2}}\left(\frac{1}{a_0}\right)^{3/2}\rho^2\mathrm{e}^{-\rho/6}\sin^2(\theta)\sin(2\phi)$$

where $\rho = 2r/a_0, r = \rho a_0/2, \mathrm{d}r = (a_0/2)\mathrm{d}\rho$.

$$\int \psi_{3\mathrm{p}_x}\psi_{3\mathrm{d}_{xy}}\,\mathrm{d}\tau = \text{constant}\times\int_0^\infty \rho^5\mathrm{e}^{-\rho/3}\,\mathrm{d}\rho\underbrace{\int_0^{2\pi}\cos(\phi)\sin(2\phi)\mathrm{d}\phi}_{0}\int_0^{\pi}\sin^3(\theta)\mathrm{d}\theta$$

Since the integral equals zero, $\psi_{3\mathrm{p}_x}$ and $\psi_{3d_{xy}}$ are orthogonal.

(b) Radial nodes are determined by finding the ρ values, $(\rho = 2r/a_0)$, [ρ here is defined as in Table 9A.1], for which the radial wavefunction equals zero. These values are the roots of the polynomial portion of the wavefunction. For the 3s orbital, $6-6\rho+\rho^2=0$ [Table 9A.1], when

$$\boxed{\rho_{\text{node}} = 3+\sqrt{3} \quad \text{and} \quad \rho_{\text{node}} = 3-\sqrt{3}}$$

The 3s orbital has these two spherically symmetrical nodes. There is no node at $\rho = 0$ so we conclude that there is a finite probability of finding a 3*s* electron at the nucleus.

For the 3_{p_x} orbital, $(4-\rho)(\rho)=0$, when $\boxed{\rho_{\text{node}}=0 \quad \text{and} \quad \rho_{\text{node}}=4}$. There is a zero probability of finding a $3\mathrm{p}_x$ electron at the nucleus.

For the $3\mathrm{d}_{xy}$ orbital $\boxed{\rho_{\text{node}}=0}$ is the only radial node.

(c) $\langle r\rangle_{3s} = \int |R_{10}Y_{00}|^2 r\,\mathrm{d}\tau = \int |R_{10}Y_{00}|^2 r^3\sin(\theta)\mathrm{d}r\,\mathrm{d}\theta\,\mathrm{d}\phi$

$$= \int_0^\infty R_{10}^2 r^3\,\mathrm{d}r\underbrace{\int_0^{2\pi}\int_0^{\pi}|Y_{00}|^2\sin(\theta)\mathrm{d}\theta\,\mathrm{d}\phi}_{1}$$

$$= \frac{a_0}{3\,888}\underbrace{\int_0^\infty (6-2\rho+\rho^2/9)^2\rho^3\mathrm{e}^{-\rho/3}\,\mathrm{d}\rho}_{52\,488}$$

[Note that for convenience ρ defined here is again $n \times \rho$ as defined in Table 9A.1, with $n = 3$ here.]

$$\boxed{\langle r \rangle_{3s} = \frac{27a_0}{2}}$$

(d) The plot in Fig. 9A.1(a) shows that the 3s orbital has larger values of the radial distribution function for $r < a_0$. This penetration of inner core electrons of multi-electron atoms means that a 3s electron experiences a larger effective nuclear charge and, consequently, has a lower energy than either a 3p or $3d_{xy}$ electron. This reasoning also leads us to conclude that a $3p_x$ electron has less energy than a $3d_{xy}$ electron.

$E_{3s} < E_{3p_x} < E_{3d_{xy}}$.

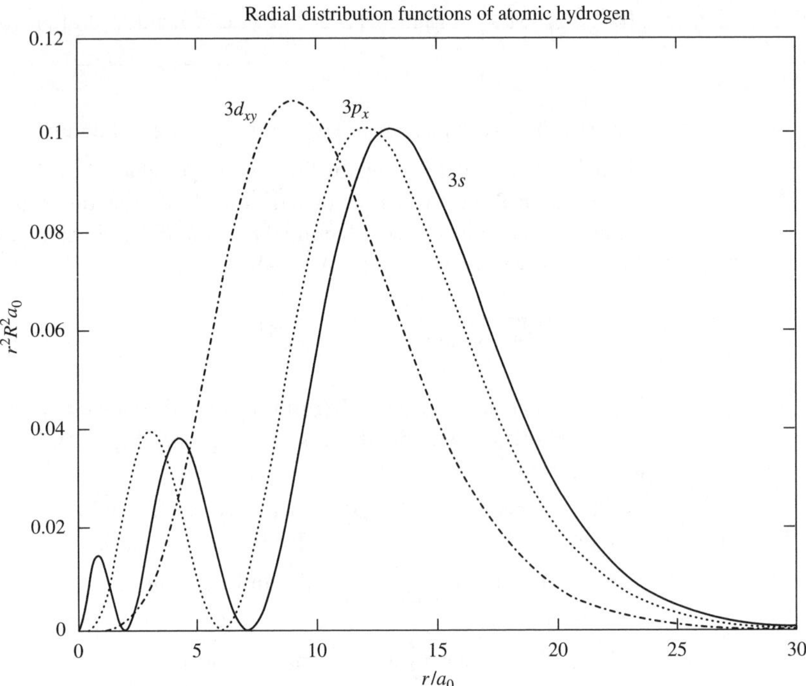

Figure 9A.1(a)

(e) Polar plots with $\theta = 90°$, as shown in Fig. 9A.1(b).

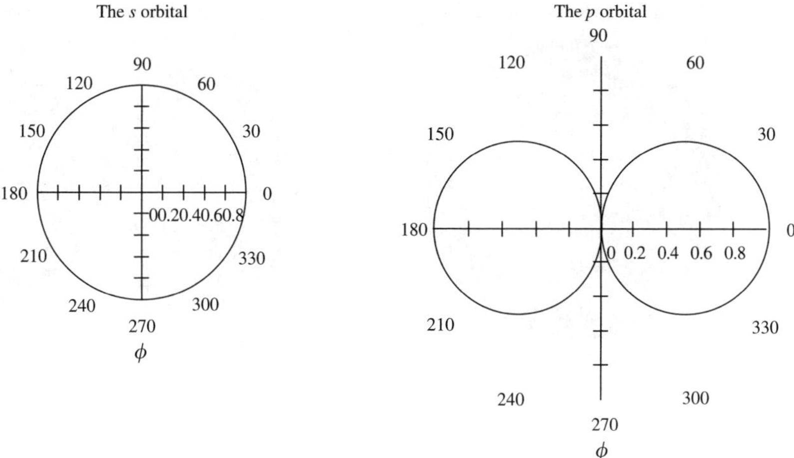

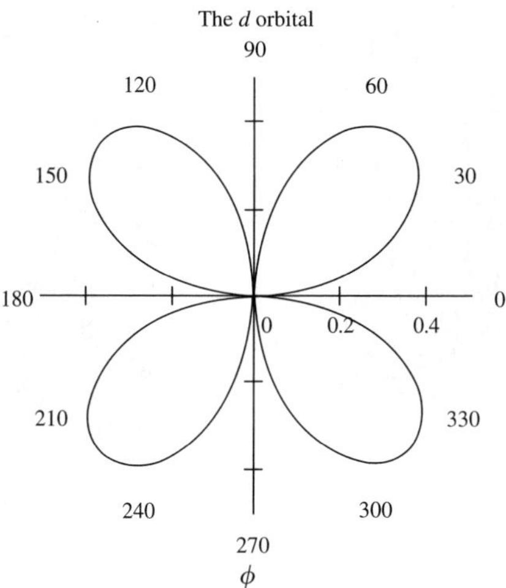

Figure 9A.1(b)

Boundary surface are shown overleaf in Fig. 9A.1(c).

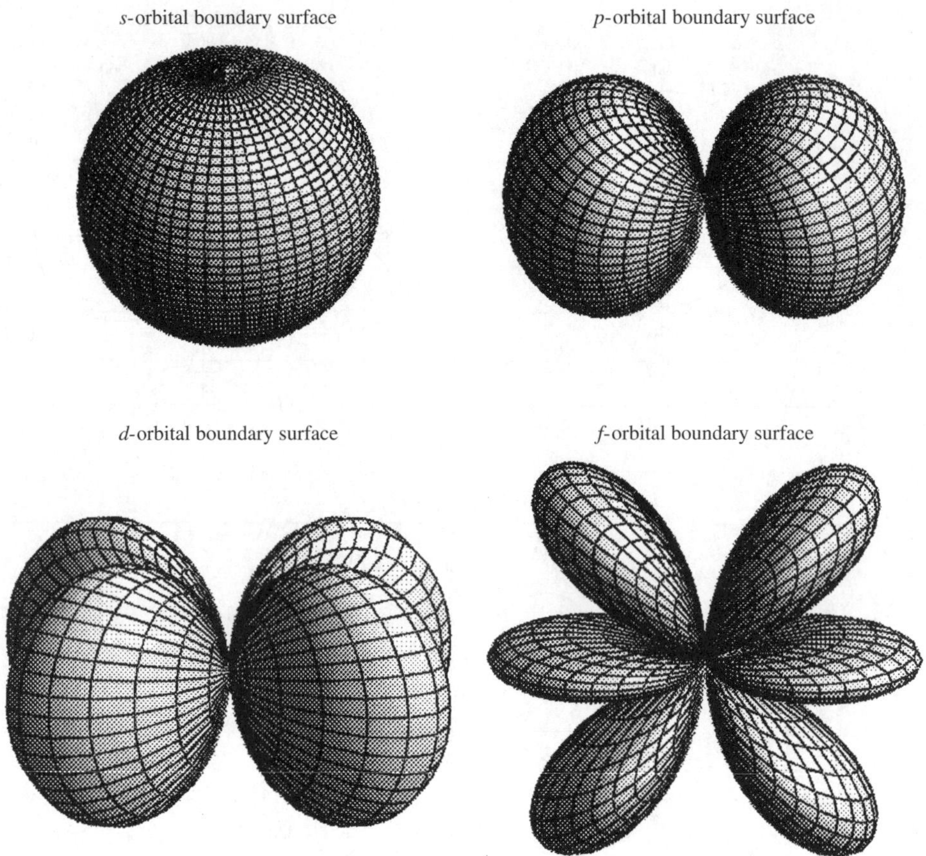

Figure 9A.1(c)

P9A.5 The general rule to use in deciding commutation properties is that operators having no variable in common will commute with each other. We first consider the commutation of $\hat{l}_z$ with the Hamiltonian. This is most easily solved in spherical polar coordinates.

$$\hat{l}_z = \frac{\hbar}{\mathrm{i}}\frac{\partial}{\partial\phi}$$

$$H = -\frac{\hbar^2}{2\mu}\nabla^2 + V \quad V = -\frac{Ze^2}{4\pi\varepsilon_0 r}$$

Since V has no variable in common with $\hat{l}_z$, this part of the Hamiltonian and $\hat{l}_z$ commute.

$$\nabla^2 = \text{terms in } r \text{ only} + \text{terms in } \theta \text{ only} + \frac{1}{r^2\sin^2\theta}\frac{\partial^2}{\partial\phi^2}$$

The terms in r only and θ only necessarily commute with $\hat{l}_z(\phi \text{ only})$. The final term in ∇^2 contains $\dfrac{\partial^2}{\partial\phi^2}$ which commutes with $\dfrac{\partial}{\partial\phi}$, since an operator necessarily commutes with itself. By symmetry we can deduce that if H commutes with $\hat{l}_z$ it must also commute with $\hat{l}_x$ and $\hat{l}_y$ since they are related to each other by a simple transformation of coordinates. This proves useful in establishing the commutation of l^2 and H. We form

$$\hat{l}^2=\hat{l}\cdot\hat{l}=(i\hat{l}_x+j\hat{l}_y+k\hat{l}_Z)\cdot(i\hat{l}_x+j\hat{l}_y+k\hat{l}_z)=\hat{l}_x^2+\hat{l}_y^2+\hat{l}_z^2$$

If H commutes with each of $\hat{l}_x,\hat{l}_y$, and $\hat{l}_z$ it must commute with $\hat{l}_x^2,\hat{l}_y^2$, and $\hat{l}_z^2$. Therefore it also commutes with $\hat{l}^2$. Thus H commutes with both $\hat{l}^2$ and $\hat{l}_z$.

Comment. As described at the end of Topic 7C, the physical properties associated with non-commuting operators cannot be simultaneously known with precision. However, since $H,\hat{l}^2$, and $\hat{l}_z$ commute we may simultaneously have exact knowledge of the energy, the total orbital angular momentum, and the projection of the orbital angular momentum along an arbitrary axis.

P9A.7

$$\langle r^m\rangle_{nl}=\int r^m|\psi_{nl}|^2\,\mathrm{d}\tau=\int_0^\infty\int_0^{2\pi}\int_0^\pi r^{m+2}|R_{nl}Y_{l0}|^2\sin(\theta)\mathrm{d}\theta\mathrm{d}\phi\mathrm{d}r$$

$$=\int_0^\infty r^{m+2}|R_{nl}|^2\,\mathrm{d}r\int_0^{2\pi}\int_0^\pi|Y_{l0}|^2\sin(\theta)\mathrm{d}\theta\mathrm{d}\phi=\int_0^\infty r^{m+2}|R_{nl}|^2\,\mathrm{d}r$$

With $r=(na_0/2Z)\rho$ and $m=-1$ the expectation value is

$$\langle r^{-1}\rangle_{nl}=\left(\frac{na_0}{2Z}\right)^2\int_0^\infty\rho|R_{nl}|^2\,\mathrm{d}\rho$$

(a) $$\langle r^{-1}\rangle_{1s}=\left(\frac{a_0}{2Z}\right)^2\left\{2\left(\frac{Z}{a_0}\right)^{3/2}\right\}^2\int_0^\infty\rho\,\mathrm{e}^{-\rho}\mathrm{d}\rho \qquad \text{[Table 9A.1]}$$

$$=\boxed{\frac{Z}{a_0}} \quad \text{because} \quad \int_0^\infty\rho\,\mathrm{e}^{-\rho}\mathrm{d}\rho=1$$

(b) $$\langle r^{-1}\rangle_{2s}=\left(\frac{a_0}{Z}\right)^2\left\{\frac{1}{8^{1/2}}\left(\frac{Z}{a_0}\right)^{3/2}\right\}^2\int_0^\infty\rho(2-\rho)^2\,\mathrm{e}^{-\rho}\mathrm{d}\rho \qquad \text{[Table 9A.1]}$$

$$=\frac{Z}{8a_0}(2) \quad \text{because} \quad \int_0^\infty\rho(2-\rho)^2\,\mathrm{e}^{-\rho}\mathrm{d}\rho=2$$

$$\langle r^{-1}\rangle_{2s}=\boxed{\frac{Z}{4a_0}}$$

(c) $$\langle r^{-1}\rangle_{2p} = \left(\frac{a_0}{Z}\right)^2 \left\{\frac{1}{24^{1/2}}\left(\frac{Z}{a_0}\right)^{3/2}\right\}^2 \int_0^\infty \rho^3\, e^{-\rho}\, d\rho \qquad \text{[Table 9A.1]}$$

$$= \frac{Z}{24a_0}(6) \quad \text{because} \quad \int_0^\infty \rho^3\, e^{-\rho}\, d\rho = 6$$

$$\langle r^{-1}\rangle_{2p} = \boxed{\frac{Z}{4a_0}}$$

The general formula for a hydrogenic orbital is $\langle r^{-1}\rangle_{nl} = \dfrac{Z}{n^2 a_0}$.

(d) No. Consider, for example, the case of the 1s orbital. The average value r is $\langle r\rangle_{1s} = \dfrac{3a_0}{2Z}$, hence $\dfrac{1}{\langle r\rangle_{1s}} = \dfrac{2Z}{3a_0}$. We saw above in part (a) that $\langle r^{-1}\rangle_{1s} = \boxed{\dfrac{Z}{a_0}}$. Similarly, $\langle 1/r\rangle \neq 1/\langle r\rangle$ for the other orbitals as well.

P9A.9 (a) The trajectory is defined, which is not allowed according to quantum mechanics.

(b) The angular momentum of a three-dimensional system is given by $\{l(l+1)\}^{1/2}\hbar$, not by $n\hbar$. In the Bohr model, the ground state possesses orbital angular momentum ($n\hbar$, with $n = 1$), but the actual ground state has no angular momentum ($l = 0$). Moreover, the distribution of the electron is quite different in the two cases. The two models can be distinguished experimentally by (a) showing that there is zero orbital angular momentum in the ground state (by examining its magnetic properties) and (b) examining the electron distribution (such as by showing that the electron and the nucleus do come into contact, Chapter 14).

P9A.11 The wavenumber of a spectroscopic transition is related to the difference in the relevant energy levels. For a one-electron atom or ion, the relationship is

$$hc\tilde{\nu} = \Delta E = \frac{Z^2\mu_{He}e^4}{32\pi^2\varepsilon_0^2\hbar^2 n_1^2} - \frac{Z^2\mu_{He}e^4}{32\pi^2\varepsilon_0^2\hbar^2 n_2^2} = \frac{Z^2\mu_{He}e^4}{32\pi^2\varepsilon_0^2\hbar^2}\left(\frac{1}{n_2^2} - \frac{1}{n_1^2}\right)$$

Solving for $\tilde{\nu}$, using the definition $\hbar = h/2\pi$ and the fact that $Z = 2$ for He, yields

$$\tilde{\nu} = \frac{\mu_{He}e^4}{2\varepsilon_0^2 h^3 c}\left(\frac{1}{n_2^2} - \frac{1}{n_1^2}\right)$$

Note that the wavenumbers are proportional to the reduced mass, which is very close to the mass of the electron for both isotopes. In order to distinguish between them, we need to carry lots of significant figures in the calculation.

$$\tilde{\nu}=\frac{\mu_{\text{He}}(1.60218\times10^{-19}\,\text{C})^4}{2(8.85419\times10^{-12}\,\text{J}^{-1}\text{C}^2\text{m}^{-1})^2\times(6.62607\times10^{-34}\,\text{Js})^3\times(2.99792\times10^{10}\,\text{cm s}^{-1})}\times\left(\frac{1}{n_2^2}-\frac{1}{n^2}\right)$$

$$\tilde{\nu}/\text{cm}^{-1}=4.81870\times10^{35}(\mu_{\text{He}}/\text{kg})\left(\frac{1}{n_2^2}-\frac{1}{n_1^2}\right)$$

The reduced masses for the ^{4}He and ^{3}He nuclei are

$$\mu=\frac{m_{\text{e}}m_{\text{nuc}}}{m_{\text{e}}+m_{\text{nuc}}}$$

where $m_{\text{nuc}}=4.00260m_{\text{u}}$ for ^{4}He and 3.01603 m_{u} for ^{3}He, or, in kg

$$^4\text{He}\ m_{\text{nuc}}=(4.00260m_{\text{u}})\times(1.66054\times10^{-27}\,\text{kg}\,m_{\text{u}}^{-1})=6.64648\times10^{-27}\,\text{kg}$$

$$^3\text{He}\ m_{\text{nuc}}=(3.01603m_{\text{u}})\times(1.66054\times10^{-27}\,\text{kg}\,m_{\text{u}}^{-1})=5.00824\times10^{-27}\,\text{kg}$$

The reduced masses are

$$^4\text{He}\ \mu=\frac{(9.10939\times10^{-31}\,\text{kg})\times(6.64648\times10^{-27}\,\text{kg})}{(9.10939\times10^{-31}+6.64648\times10^{-27})\text{kg}}=9.10814\times10^{-31}\,\text{kg}$$

$$^3\text{He}\ \mu=\frac{(9.10939\times10^{-31}\,\text{kg})\times(5.00824\times10^{-27}\,\text{kg})}{(9.10939\times10^{-31}+5.00824\times10^{-27})\text{kg}}=9.10773\times10^{-31}\,\text{kg}$$

Finally, the wavenumbers for $n=3\rightarrow n=2$ are:

$$^4\text{He}\ \tilde{\nu}=(4.81870\times10^{35})\times(9.10814\times10^{-31})\times(1/4-1/9)\text{cm}^{-1}=\boxed{60\,957.4\,\text{cm}^{-1}}$$

$$^3\text{He}\ \tilde{\nu}=(4.81870\times10^{35})\times(9.10773\times10^{-31})\times(1/4-1/9)\text{cm}^{-1}=\boxed{60\,954.7\,\text{cm}^{-1}}$$

The wavenumbers for $n=2\rightarrow n=1$ are

$$^4\text{He}\ \tilde{\nu}=(4.81870\times10^{35})\times(9.10814\times10^{-31})\times(1/1-1/4)\text{cm}^{-1}=\boxed{329\,170\,\text{cm}^{-1}}$$

$$^3\text{He}\ \tilde{\nu}=(4.81870\times10^{35})\times(9.10773\times10^{-31})\times(1/1-1/4)\text{cm}^{-1}=\boxed{329\,155\,\text{cm}^{-1}}$$

9B Many-electron atoms

Answers to discussion questions

D9B.1 See Section 9B.2 of the text and any general chemistry book, for example, Sections 1.10–1.13 of P. Atkins and L. Jones, *Chemical Principles*, 2nd ed. W. H. Freeman, and Co., New York (2002).

D9B.3 In the crudest form of the orbital approximation, the many-electron wavefunctions for atoms are represented as a simple product of one-electron wavefunctions. At a somewhat more sophisticated level, the many electron wavefunctions are written as linear combinations of such simple product functions that explicitly satisfy the Pauli exclusion principle. Relatively good one-electron functions are generated by the Hartree–Fock self-consistent field method described in Section 9B.3. If we place no restrictions on the form of the one-electron functions, we reach the Hartree–Fock limit which gives us the best value of the calculated energy within the orbital approximation. The orbital approximation is based on the disregard of significant portions of the electron–electron interaction terms in the many-electron Hamiltonian, so we cannot expect that it will be quantitatively accurate. By abandoning the orbital approximation, we could in principle obtain essentially exact energies; however, there are significant conceptual advantages to retaining the orbital approximation. Increased accuracy can be obtained by reintroducing the neglected electron–electron interaction terms and including their effects on the energies of the atom.

Solutions to exercises

E9B.1(a) Sc: $[Ar]4s^2 3d^1$

Ti: $[Ar]4s^2 3d^2$

V: $[Ar]4s^2 3d^3$

Cr: $[Ar]4s^2 3d^4$ or $[Ar]4s^1 3d^5$ (most probable)

Mn: $[Ar]4s^2 3d^5$

Fe: $[Ar]4s^2 3d^6$

Co: $[Ar]4s^2 3d^7$

Ni: $[Ar]4s^2 3d^8$

Cu: $[Ar]4s^2 3d^9$ or $[Ar]4s^1 3d^{10}$ (most probable)

Zn: $[Ar]4s^2 3d^{10}$

E9B.2(a) (i) $1s^2 2s^2 2p^6 3s^2 3p^6 3d^8 = \boxed{[Ar]3d^8}$

(ii) All subshells except 3*d* are filled and hence have no net spin. Applying Hund's rule to $3d^8$ shows that there are two unpaired spins. The paired spins do not contribute to the net spin, hence we consider only $s_1 = \frac{1}{2}$ and $s_2 = \frac{1}{2}$. The Clebsch–Gordan series produces

$$S = s_1 + s_2, \ldots, |s_1 - s_2|, \quad \text{hence } \boxed{S = 1, 0}$$

$$M_S = -S, -S+1, \ldots, S$$

$$\text{For } S=1, \quad \boxed{M_S = -1, 0, +1}$$

$$S=0, \quad \boxed{M_S = 0}$$

Solutions to problems

P9B.1 If we assume that the innermost electron is a hydrogen-like 1s orbital we may write

$$r^* = \frac{a_0}{Z} = \frac{52.92\,\text{pm}}{126} = \boxed{0.420\,\text{pm}}$$

This result clearly shows that relativistic effects cannot be ignored. If one now calculates the velocity of the electron in this orbit one obtains the value 2.75×10^8 m s^{-1} which is close to the velocity of light.

Question. Calculate this velocity from the Bohr theory of the atom.

P9B.3 Electronic configurations of neutral, 4th period transition atoms in the ground state are summarized in the following table along with observed, positive oxidation states. The most common, positive oxidation states are indicated with bright boxing.

Group	3	4	5	6	7	8	9	10	11	12
Oxidation state	Sc	Ti	V	Cr	Mn	Fe	Co	Ni	Cu	Zn
0	$3d4s^2$	$3d^24s^2$	$3d^34s^2$	$3d^54s$	$3d^54s^2$	$3d^64s^2$	$3d^74s^2$	$3d^84s^2$	$3d^{10}4s$	$3d^{10}4s^2$
+1			☺	☺	☺		☺	☺	☺	
+2		☺	☺	☺	☺	☺	☺	☺	☺	☺
+3	☺	☺	☺	☺	☺	☺	☺	☺		
+4		☺	☺	☺	☺	☺	☺	☺		
+5			☺	☺	☺	☺				
+6				☺	☺	☺				
+7					☺					

Toward the middle of the first transition series (Cr, Mn, and Fe) elements exhibit the widest ranges of oxidation states. This phenomenon is related to the availability of both electrons and orbitals favourable for bonding. Elements to the left (Sc and Ti) of the series have few electrons and relatively low effective nuclear charge leaves d orbitals at high energies that are relatively unsuitable for bonding. To the far right (Cu and Zn) effective nuclear charge may be higher but there are few, if any, orbitals available for bonding. Consequently, it is more difficult to produce a range of compounds that promote a wide range of oxidation states for elements at either end of the series. At the middle and right of the series the +2 oxidation state is very commonly observed because normal reactions can provide the requisite ionization energies for the removal of 4s electrons. The readily available +2 and +3 oxidation states of Mn, Fe, and the +1 and +2 oxidation states of Cu make these cations useful in electron transfer processes occurring in chains of specialized protein within biological cells. The special size and charge of the Zn^{2+} cation makes it useful for the function of some enzymes. The tendency of Fe^{2+} and Cu^{+} to bind oxygen proves very useful in haemoglobin and electron transport (respiratory) chain, respectively.

9C Atomic spectra

Answer to discussion questions

D9C.1 When hydrogen atoms are excited into higher energy levels they emit energy at a series of discrete wavenumbers which show up as a series of lines in the spectrum. The lines in the Lyman series, for which the emission of energy corresponds to the final state $n_1 = 1$, are in the ultraviolet region of the electromagnetic spectrum. The Balmer series, for which $n_1 = 2$, is in the visible region. The Paschen series, $n_1 = 3$, is in the infrared. The Brackett series, $n_1 = 4$, is also in the infrared. The Pfund series, $n_1 = 5$, is in the far infrared. The Humphrey series, $n_1 = 6$, is in the very far infrared.

D9C.3 An electron has a magnetic moment and magnetic field due to its orbital angular momentum. It also has a magnetic moment and magnetic field due to its spin angular momentum. There is an interaction energy between magnetic moments and magnetic fields. That between the spin magnetic moment and the magnetic field generated by the orbital motion is called spin–orbit coupling. The energy of interaction is proportional to the scalar product of the two vectors representing the spin and orbital angular momenta and hence depends upon the orientation of the two vectors. See Fig. 9C.4. The total angular momentum of an electron in an atom is the vector sum of the orbital and spin angular momenta, as illustrated in Fig. 9C.5 and expressed in eqn 9C.5. The spin–orbit coupling results in a splitting of the energy levels associated with atomic terms as shown in Figs 9C.6 and 9C.7. This splitting shows up in atomic spectra as a fine structure as illustrated in Fig. 9C.7

Solutions to exercises

E9C.1(a) Equation 9A.1 implies that the shortest wavelength corresponds to $n_2 = \infty$, and the longest to $n_2 = 2$. Solve eqn 9A.1 for λ.

$$\lambda = \frac{(1/n_1^2 - 1/n_2^2)^{-1}}{\tilde{R}_{\mathrm{H}}}$$

Shortest: $$\lambda = \frac{(1/1^2 - 1/\infty^2)^{-1}}{109\,677\ \mathrm{cm}^{-1}} = \boxed{9.118 \times 10^{-6}\ \mathrm{cm}}$$

Longest: $$\lambda = \frac{(1/1^2 - 1/2^2)^{-1}}{109\,677\ \mathrm{cm}^{-1}} = \boxed{1.216 \times 10^{-5}\ \mathrm{cm}}$$

E9C.2(a) For atoms N, eqn 9A.9 may be rewritten in terms of the Rydberg constant $\tilde{R}_{\rm N}$ as

$$E_n = -\frac{Z^2\mu_{\rm N}hc\tilde{R}_{\rm He^+}}{m_{\rm e}n^2} \approx -\frac{Z^2hc\tilde{R}_\infty}{n^2} \quad [9A.15]$$

where to within 0.01% the ratio $\mu_{\rm N}/m_{\rm e}$ is unity. Equation 9A.1 can then be rewritten as

$$\tilde{\nu} = Z^2\tilde{R}_\infty\left(\frac{1}{n_1^2}-\frac{1}{n_2^2}\right) \qquad \lambda = \frac{1}{\tilde{\nu}} \qquad \nu = \frac{c}{\lambda}$$

$$\tilde{\nu} = 4\times 109\,737\ \text{cm}^{-1}\left(\frac{1}{1^2}-\frac{1}{2^2}\right) = \boxed{3.292\times10^5\ \text{cm}^{-1}} \qquad \lambda = \boxed{3.038\times10^{-6}\,\text{cm}}$$

$$\nu = \frac{2.9978\times10^{10}\ \text{cm s}^{-1}}{3.0376\times10^{-6}\ \text{cm}} = \boxed{9.869\times10^{15}\ \text{s}^{-1}}$$

E9C.3(a) This is essentially the photoelectric effect [Topic 7A] with the ionization energy of the ejected electron being the work function Φ.

$$h\nu = \frac{1}{2}m_{\rm e}v^2 + I$$

$$\begin{aligned} I &= h\nu - \frac{1}{2}m_e v^2 = (6.626\times10^{-34}\ \text{J Hz}^{-1})\times\left(\frac{2.998\times10^8\ \text{m s}^{-1}}{58.4\times10^{-9}\ \text{m}}\right) \\ &\quad -(\tfrac{1}{2})\times(9.109\times10^{-31}\ \text{kg})\times(1.59\times10^6\ \text{m s}^{-1})^2 \\ &= 2.25\times10^{-18}\ \text{J, corresponding to } \boxed{14.0\text{eV}} \end{aligned}$$

E9C.4(a) The selection rules for a many-electron atom are given by the set [9C.8]. For a single-electron transition these amount to $\Delta n =$ any integer; $\Delta l = \pm 1$. Hence

(i) $2\text{s}\rightarrow 1\text{s};\ \Delta l = 0,$ $\boxed{\text{forbidden}}$

(ii) $2\text{p}\rightarrow 1\text{s};\ \Delta l = -1,$ $\boxed{\text{allowed}}$

(iii) $3\text{d}\rightarrow 2\text{p};\ \Delta l = -1,$ $\boxed{\text{allowed}}$

E9C.5(a) We use the Clebsch–Gordan series [9C.5] in the form

$j = l+s, l+s-1, \ldots, |l-s|$ [lower-case for a single electron]

(i) $l = 2,\quad s = \frac{1}{2};\quad \text{so } j = \boxed{\frac{5}{2},\frac{3}{2}}$

(ii) $l = 3,\quad s = \frac{1}{2};\quad \text{so } j = \boxed{\frac{7}{2},\frac{5}{2}}$

E9C.6(a) The Clebsch–Gordan series for $\boxed{l=1}$ and $s = \frac{1}{2}$ leads to $j = \frac{3}{2}$ and $\frac{1}{2}$.

E9C.7(a) The letter D indicates that $L = 2$, the superscript 1 is the value of $2S+1$, so $S = 0$ and the subscript 2 is the value of J. Hence, $\boxed{L=2, S=0, J=2}$

E9C.8(a) Use the Clebsch–Gordan series in the form

$$S' = s_1+s_2, s_1+s_2-1, \ldots, |s_1-s_2|$$

and

$$S = S' + s_1, S' + s_1 - 1, \ldots, |S' - s_1|$$

in succession. The multiplicity is $2S+1$.

(i) $S = \frac{1}{2} + \frac{1}{2}, \frac{1}{2} - \frac{1}{2} = \boxed{1,0}$ with multiplicities $\boxed{3,1}$ respectively

(ii) $S' = 1,0$; then $S = \boxed{\frac{3}{2}, \frac{1}{2}}$ [from 1], $\boxed{\text{and} \frac{1}{2}}$ [from 0], with multiplicities $\boxed{4,2,2}$

E9C.9(a) These electrons are not equivalent (different subshells), hence all the terms that arise from the vector model and the Clebsch–Gordan series are allowed (Example 9C.2).

$$L = l_1 + l_2, \ldots, |l - l_2| \text{ [9C.5]} = 2 \text{ only}$$

$$S = s_1 + s_2, \ldots, |s_1 - s_2| = 1,0$$

The allowed terms are then ^{3}D and ^{1}D. The possible values of J are given by

$$J = L + S, \ldots, |L - S| \text{[9C.7]} = 3, 2, 1 \text{ for } {}^3\text{D and 2 for } {}^1\text{D}$$

The allowed complete term symbols are then

$$\boxed{{}^3\text{D}_3, {}^3\text{D}_2, {}^3\text{D}_1, {}^1\text{D}_2}$$

The $\boxed{{}^3\text{D set of terms are the lower in energy}}$ [Hund's rule]

Comment. Hund's rule in the form given in the text does not allow the energies of the triplet terms to be distinguished. Experimental evidence indicates that $^3\text{D}_1$ is lowest.

E9C.10(a) Use the Clebsch–Gordan series in the form

$$J = L + S, L + S - 1, \ldots, |L - S|$$

The number of states (M_J values) is $2J+1$ in each case.

(i) $L = 0, S = 0$; hence $\boxed{J = 0}$ and there is only $\boxed{1 \text{ state}}$ ($M_J = 0$).

(ii) $L = 1, S = \frac{1}{2}$; hence $\boxed{J = \frac{3}{2}, \frac{1}{2}}$ $({}^2\text{P}_{3/2}, {}^2\text{P}_{1/2})$ with $\boxed{4, 2 \text{ states respectively}}$.

(iii) $L = 1, S = 1$; hence $\boxed{J = 2,1,0}$ $({}^3\text{P}_2, {}^3\text{P}_1, {}^3\text{P}_0)$ with $\boxed{5,3,1 \text{ states respectively}}$.

E9C.11(a) Closed shells and subshells do not contribute to either L or S and thus are ignored in what follows.

(i) Li[He]2s^1: $S = \frac{1}{2}$, $L = 0$; $J = \frac{1}{2}$, so the only term is $\boxed{{}^2\text{S}_{1/2}}$

(ii) Na[Ne]3p^1: $S = \frac{1}{2}$, $L = 1$; $J = \frac{3}{2}, \frac{1}{2}$, so the terms are $\boxed{{}^2\text{P}_{3/2} \text{ and } {}^2\text{P}_{1/2}}$

E9C.12(a) See eqn. 9C.8 for the selection rules. (i) $\boxed{\text{allowed}}$, (ii) $\boxed{\text{forbidden}}$, (iii) $\boxed{\text{allowed}}$.

Solutions to problems

P9C.1 All lines in the hydrogen spectrum fit the Rydberg formula

$$\frac{1}{\lambda}=\tilde{R}_{\mathrm{H}}\left(\frac{1}{n_1^2}-\frac{1}{n_2^2}\right)\quad\left[9\mathrm{A}.1,\text{ with }\tilde{\nu}=\frac{1}{\lambda}\right]\quad \tilde{R}_{\mathrm{H}}=109\,677\,\mathrm{cm}^{-1}$$

Find n_1 from the value of $\lambda_{\max}$, which arises from the transition $n_1+1\rightarrow n_1$

$$\frac{1}{\lambda_{\max}\tilde{R}_{\mathrm{H}}}=\frac{1}{n_1^2}-\frac{1}{(n_1+1)^2}=\frac{2n_1+1}{n_1^2(n_1+1)^2}$$

$$\lambda_{\max}\tilde{R}_{\mathrm{H}}=\frac{n_1^2(n_1+1)^2}{2n_1+1}=(12\,368\times10^{-9}\,\mathrm{m})\times(109\,677\times10^2\,\mathrm{m}^{-1})=135.65$$

Since $n_1=1,2,3$, and 4 have already been accounted for, try $n_1=5,6,\ldots$. With $n_1=6$ we get $\dfrac{n_1^2(n_1+1)^2}{2n_1+1}=136$. Hence, the Humphreys series is $\boxed{n_2\rightarrow 6}$ and the transitions are given by

$$\frac{1}{\lambda}=(109\,677\,\mathrm{cm}^{-1})\times\left(\frac{1}{36}-\frac{1}{n_2^2}\right),\quad n_2=7,8,\cdots$$

and occur at 12 372 nm, 7 503 nm, 5 908 nm, 5 129 nm, . . ., 3 908 nm (at $n_2=15$), converging to 3 282 nm as $n_2\rightarrow\infty$, in agreement with the quoted experimental result.

P9C.3 A Lyman series corresponds to $n_1=1$; hence

$$\tilde{\nu}=\tilde{R}_{\mathrm{Li}^{2+}}\left(1-\frac{1}{n^2}\right),\quad n=2,3,\cdots\quad\left[\tilde{\nu}=\frac{1}{\lambda}\right]$$

Therefore, if the formula is appropriate, we expect to find that $\tilde{\nu}\left(1-\dfrac{1}{n^2}\right)^{-1}$ is a constant ($\tilde{R}_{\mathrm{Li}^{2+}}$).

We therefore draw up the following table.

n	2	3	4
$\tilde{\nu}/\mathrm{cm}^{-1}$	740 747	877 924	925 933
$\tilde{\nu}\left(1-\frac{1}{n^2}\right)^{-1}/\mathrm{cm}^{-1}$	987 663	987 665	987 662

Hence, the formula does describe the transitions, and $\boxed{\tilde{R}_{\mathrm{Li}^{2+}}=987\,663\,\mathrm{cm}^{-1}}$. The Balmer transitions lie at

$$\tilde{\nu}=\tilde{R}_{\mathrm{Li}^{2+}}\left(\frac{1}{4}-\frac{1}{n^2}\right)\quad n=3,4,\ldots$$

$$=(987\,663\,\mathrm{cm}^{-1})\times\left(\frac{1}{4}-\frac{1}{n^2}\right)=\boxed{137\,175\,\mathrm{cm}^{-1}},\boxed{185\,187\,\mathrm{cm}^{-1}},\ldots$$

The ionization energy of the ground-state ion is given by

$$\tilde{\nu}=\tilde{R}_{\mathrm{Li}^{2+}}\left(1-\frac{1}{n^2}\right),\quad n\rightarrow\infty$$

and hence corresponds to

$$\tilde{\nu}=987\,663\ \mathrm{cm}^{-1},\quad \text{or}\quad \boxed{122.5\ \mathrm{eV}}$$

P9C.5 The 7p configuration has just one electron outside a closed subshell. That electron has $l = 1$, $s = 1/2$, and $j = 1/2$ or $3/2$, so the atom has $L = 1$, $S = 1/2$, and $J = 1/2$ or $3/2$. The term symbols are $\boxed{{}^2\mathrm{P}_{1/2} \text{ and } {}^2\mathrm{P}_{3/2}}$, of which the former has the lower energy. The 6d configuration also has just one electron outside a closed subshell; that electron has $l=2$, $s=1/2$, and $j=3/2$ or $5/2$, so the atom has $\mathrm{L}=2$, $S=1/2$, and $J=3/2$ or $5/2$. The term symbols are $\boxed{{}^2\mathrm{D}_{3/2} \text{ and } {}^2\mathrm{D}_{5/2}}$, of which the former has the lower energy. According to the simple treatment of spin–orbit coupling, the energy is given by

$$E_{l,s,j}=\frac{1}{2}hc\tilde{A}[j(j+1)-l(l+1)-s(s+1)]$$

where A is the spin–orbit coupling constant. So

$$E({}^2\mathrm{P}_{1/2})=\frac{1}{2}hc\tilde{A}\left[\frac{1}{2}(1/2+1)-1(1+1)-\frac{1}{2}(1/2+1)\right]=-hc\tilde{A}$$

and $$E({}^2\mathrm{D}_{3/2})=\frac{1}{2}hc\tilde{A}\left[\frac{3}{2}(3/2+1)-2(2+1)-\frac{1}{2}(1/2+1)\right]=-\frac{3}{2}hc\tilde{A}$$

This approach would predict the ground state to be $\boxed{{}^2\mathrm{D}_{3/2}}$

Comment. The computational study cited finds the ${}^2\mathrm{P}_{1/2}$ level to be lowest, but the authors caution that the error of similar calculations on Y and Lu is comparable to the computed difference between levels.

P9C.7 $\tilde{R}_{\mathrm{H}}=k\mu_{\mathrm{H}},\ \tilde{R}_{\mathrm{D}}=k\mu_{\mathrm{D}},\ \tilde{R}_{\infty}=k\mu$ [9A.15]

where $\tilde{R}_{\infty}$ corresponds to an infinitely heavy nucleus, with $\mu=m_{\mathrm{e}}$.

Since $\mu=\dfrac{m_{\mathrm{e}}m_{\mathrm{N}}}{m_{\mathrm{e}}+m_{\mathrm{N}}}$ [N = p or d]

$$\tilde{R}_{\mathrm{H}}=k\mu_{\mathrm{H}}=\frac{km_{\mathrm{e}}}{1+\frac{m_{\mathrm{e}}}{m_{\mathrm{p}}}}=\frac{\tilde{R}_{\infty}}{1+\frac{m_{\mathrm{e}}}{m_{\mathrm{p}}}}$$

Likewise, $\tilde{R}_{\mathrm{D}}=\dfrac{\tilde{R}_{\infty}}{1+\frac{m_{\mathrm{e}}}{m_{\mathrm{d}}}}$ where m_{p} is the mass of the proton and m_{d} the mass of the deuteron. The two lines in question lie at

$$\frac{1}{\lambda_{\mathrm{H}}}=\tilde{R}_{\mathrm{H}}\left(1-\frac{1}{4}\right)=\frac{3}{4}\tilde{R}_{\mathrm{H}}\quad \frac{1}{\lambda_{\mathrm{D}}}=\tilde{R}_{\mathrm{D}}\left(1-\frac{1}{4}\right)=\frac{3}{4}\tilde{R}_{\mathrm{D}}$$

and hence

$$\frac{\tilde{R}_{\mathrm{H}}}{\tilde{R}_{\mathrm{D}}}=\frac{\lambda_{\mathrm{D}}}{\lambda_{\mathrm{H}}}=\frac{\tilde{\nu}_{\mathrm{H}}}{\tilde{\nu}_{\mathrm{D}}}$$

Then, since

$$\frac{\tilde{R}_H}{\tilde{R}_D}=\frac{1+\frac{m_e}{m_d}}{1+\frac{m_e}{m_p}},\quad m_d=\frac{m_e}{\left(1+\frac{m_e}{m_p}\right)\frac{\tilde{R}_H}{\tilde{R}_D}-1}$$

we can calculate m_d from

$$m_d=\frac{m_e}{\left(1+\frac{m_e}{m_p}\right)\frac{\lambda_D}{\lambda_H}-1}=\frac{m_e}{\left(1+\frac{m_e}{m_p}\right)\frac{\tilde{\nu}_H}{\tilde{\nu}_D}-1}$$

$$=\frac{9.10939\times10^{-31}\,\text{kg}}{\left(1+\frac{9.1039\times10^{-31}\,\text{kg}}{1.67262\times10^{-27}\,\text{kg}}\right)\times\left(\frac{82\,259.098\,\text{cm}^{-1}}{82\,281.476\,\text{cm}^{-1}}\right)-1}=\boxed{3.3429\times10^{-27}\,\text{kg}}$$

Since $I=\tilde{R}_\infty hc$,

$$\frac{I_D}{I_H}=\frac{\tilde{R}_D}{\tilde{R}_H}=\frac{\tilde{\nu}_D}{\tilde{\nu}_H}=\frac{82\,281.476\text{ cm}^{-1}}{82\,259.098\text{ cm}^{-1}}=\boxed{1.000272}$$

P9C.9 (a) The splitting of adjacent energy levels is related to the difference in wavenumber of the spectral lines as follows:

$$hc\Delta\tilde{\nu}=\Delta E=\mu_B\mathcal{B}\text{, so }\Delta\tilde{\nu}=\frac{\mu_B\mathcal{B}}{hc}=\frac{(9.274\times10^{-24}\text{ J T}^{-1})(2\text{ T})}{(6.626\times10^{-34}\text{ J s})(2.998\times10^{10}\text{ cm s}^{-1})}$$

$$\Delta\tilde{\nu}=\boxed{0.9\text{ cm}^{-1}}.$$

(b) Transitions induced by absorbing visible light have wavenumbers in the tens of thousands of reciprocal centimetres, so normal Zeeman splitting is $\boxed{\text{small}}$ compared to the difference in energy of the states involved in the transition. Take a wavenumber from the middle of the visible spectrum as typical:

$$\tilde{\nu}=\frac{1}{\lambda}=\frac{1}{600\text{ nm}}\left(\frac{10^9\text{ nm m}^{-1}}{10^2\text{ cm m}^{-1}}\right)=1.7\times10^4\text{ cm}^{-1}$$

Or take the Balmer series as an example, as suggested in the problem; the Balmer wavenumbers are (eqn 9A.1):

$$\tilde{\nu}=\tilde{R}_H\left(\frac{1}{2^2}-\frac{1}{n^3}\right).$$

The smallest Balmer wavenumber is

$$\tilde{\nu}=(109\,677\text{ cm}^{-1})\times(1/4-1/9)=15\,233\text{ cm}^{-1}$$

and the upper limit is

$$\tilde{\nu}=(109\,677\text{ cm}^{-1})\times(1/4-0)=27\,419\text{ cm}^{-1}$$

P9C.11 (a) The speed distribution in the molecular beam is related to the speed distribution within the chamber by a factor of $v\cos\theta$, as shown in Fig. 9C.1. Since an integration over all possible θ must be performed, the $\cos\theta$ factor may be absorbed into the constant of proportionality.

$$f_{\text{beam}}(v) = Cvf_{\text{chamber}}(v) \quad \text{where } C \text{ is to be determined}$$

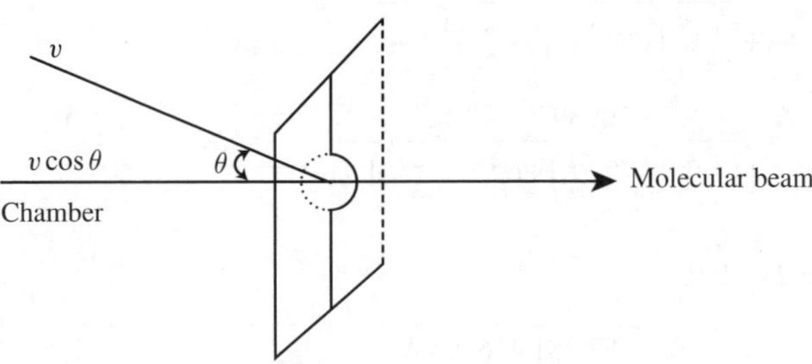

Figure 9C.1

By normalization over the possible beam speeds $(0 < v_{\text{beam}} < \infty)$

$$f_{\text{beam}} = Cv\left(v^2 e^{-(mv^2/2kT)}\right)$$
$$= Cv^3 e^{-(mv^2/2kT)}$$

$$\int_{v=0}^{\infty} f_{\text{beam}}\,dv = 1 = C\int_{v=0}^{\infty} v^3 e^{-(mv^2/2kT)}\,dv = C\left\{\frac{1}{2(m/2kT)^2}\right\}$$

$$C = 2(m/2kT)^2$$

$$\langle v^2\rangle = \int_{v=0}^{\infty} v^2 f_{\text{beam}}(v)\,dv = C\int v^5 e^{-(mv^2/2kT)}\,dv$$
$$= C\left\{\frac{1}{(m/2kT)^3}\right\} = 2\frac{(m/2kT)^2}{(m/2kT)^3}$$
$$= \frac{4kT}{m}$$

$$\langle E_{\text{K}}\rangle = \frac{m}{2}\langle v^2\rangle = \frac{m}{2}\left(\frac{4kT}{m}\right) = \boxed{2kT}$$

(b) $$\Delta x = \left(\frac{2\mu_{\text{B}}L^2}{4E_{\text{K}}}\right)\frac{d\mathcal{B}}{dz}$$

$$\frac{d\mathcal{B}}{dz} = \frac{4E_K\Delta x}{2\mu_B L^2} = \frac{4(2kT)\Delta x}{2\mu_B L^2}$$

or $$= \frac{4kT\,\Delta x}{\mu_B L^2}$$

$$= \frac{4(1.3807\times10^{-23}\ \mathrm{J\,K^{-1}})\times(1\,000\ \mathrm{K})\times(1.00\times10^{-3}\ \mathrm{m})}{(9.27402\times10^{-24}\ \mathrm{J\,T^{-1}})\times(50\times10^{-2}\ \mathrm{m})^2}$$

$$\frac{d\mathcal{B}}{dz} = \boxed{23.8\ \mathrm{T\,m^{-1}}}$$

Integrated activities

I9.1 (a) ${}^2S_{1/2} \rightarrow {}^2P_{3/2}$ (or ${}^2P_{1/2}$)

(b) See the solution to Exercise 9C.2(a).

For atoms N, eqn 9A.9 may be rewritten in terms of the Rydberg constant $\tilde{R}_N$ as

$$E_n = -\frac{Z^2\mu_A hc\tilde{R}_{He^+}}{m_e n^2} \approx -\frac{Z^2hc\tilde{R}_\infty}{n^2}$$

where to within 0.01% the ratio μ_A/m_e is unity. Equation 9A.1 can then be rewritten as

$$\tilde{\nu} = Z^2\tilde{R}_\infty\left(\frac{1}{n_1^2}-\frac{1}{n_2^2}\right) \quad \lambda = \frac{1}{\tilde{\nu}} \quad \nu = \frac{c}{\lambda}$$

$$\tilde{\nu} = 4\times109\,737\ \mathrm{cm^{-1}}\left(\frac{1}{1^2}-\frac{1}{4^2}\right) = \boxed{4.115\times10^5\ \mathrm{cm^{-1}}} \quad \lambda = \boxed{2.430\times10^{-6}\,\mathrm{cm}}$$

$$\nu = \frac{2.9978\times10^{10}\ \mathrm{cm\,s^{-1}}}{2.430\times10^{-6}\ \mathrm{cm}} = \boxed{1.234\times10^{16}\ \mathrm{s^{-1}}}$$

(c) See Example 9A.2. The mean radius is given by $\langle r\rangle = \int_0^\infty r^3R_{n,l}^2\mathrm{d}r$. After substituting eqn 9A.12 for $R_{n,l}$ and integrating, the general expression for the mean radius of a hydrogenic orbital with quantum numbers l and n is obtained (see eqn 10.19 of Atkins and De Paula, *Physical Chemistry*, 8th edition).

$$\langle r_{n,l}\rangle = n^2\left\{1+\frac{1}{2}\left(1-\frac{l(l+1)}{n^2}\right)\right\}\frac{a_0}{Z}$$

$$\langle r_{4,1}\rangle = \boxed{\frac{23}{2}a_0}$$

Hence, since $\langle r_{1,0}\rangle = \boxed{\frac{3}{4}a_0}$, the mean radius has increased by $\boxed{\frac{43}{4}a_0}$

10 Molecular structure

10A Valence-bond theory

Answers to discussion questions

D10A.1 The Born–Oppenheimer approximation treats the nuclei of the multi-particle system of electrons and nuclei as if they were fixed. The dependence of energy on nuclear positions is then obtained by solving the Schrödinger equation at many different (fixed) nuclear geometries. Molecular potential energy curves and surfaces are the plots of molecular energy (computed under the Born–Oppenheimer approximation) vs. nuclear coordinates.

D10A.3 See Section 10A.2(b) for details on carbon-atom hybridization. The carbon atoms in alkanes are sp^3-hybridized. This explains the nearly tetrahedral bond angles about the carbon atoms in alkanes. The double-bonded carbon atoms in alkenes are sp^2-hybridized. This explains the bond angles of approximately 120° about these atoms. The simultaneous overlap of sp^2-hybridized orbitals and unhybridized p orbitals in C=C double bonds explains the resistance of such bonds to torsion and the co-planarity of the atoms attached to those atoms. The triple-bonded carbon atoms in alkynes are sp-hybridized. This explains the 180° bond angles about these atoms. The arrangement of sp-hybridized orbitals and unhybridized p orbitals in C≡C triple bonds explains the resistance of such bonds to torsion as well. The central carbon atom in allene is also sp-hybridized. Each of its C=C double bonds involves one of its sp-hybrids and one unhybridized p orbital. Its two unhybridized p orbitals are oriented along perpendicular axes, which accounts for the two CH_2 groups to which it is attached lying in different planes.

D10A.5 Resonance refers to the superposition of the wavefunctions representing different electron distributions in the same nuclear framework. The wavefunction resulting from the superposition is called a resonance hybrid. Resonance allows us to understand the true electron distribution (or rather a good approximation to it) in terms of contributions of various limiting structures readily pictured and understood in the context of chemical concepts. For example, we can understand polar covalent bonds in terms of greater or lesser amounts of covalent and ionic character in a resonance hybrid wavefunction, or we can understand the properties of a structure that does not correspond to a single Lewis structure as a hybrid of component wavefunctions, each of which *does* correspond to a single Lewis structure.

Solutions to exercises

E10A.1(a) Let s represent the H1s atomic orbital and p_z an F2p orbital centred on the F nucleus. Then the spatial portion of wavefunction (for the bonding electrons only) would be

$$s(1)p_z(2)+s(2)p_z(1)$$

The spatial factor is symmetric with respect to interchange of the electrons' labels, so the spin factor must be antisymmetric:

$$\alpha(1)\beta(2)-\alpha(2)\beta(1).$$

So the total (unnormalized) wavefunction for this bond would be

$$\psi=\boxed{\{s(1)p_z(2)+s(2)p_z(1)\}\times\{\alpha(1)\beta(2)-\alpha(2)\beta(1)\}}$$

Comment. This simple VB function embodies a perfectly covalent H–F bond. A slightly better wavefunction, particularly for the distribution of the nonbonding valence electrons (i.e. lone pairs of fluorine) would have an sp^3 hybrid rather than a p_z orbital.

E10.2A(a) Let s represent the H1s atomic orbital and p_z an F2p orbital centred on the F nucleus. (As in Exercise 10.1(a), an sp^3 hybrid would be even better than the p_z.) Call the wavefunction that describes a perfectly covalent H–F bond ψ_{VB}:

$$\psi_{VB}=\{s(1)p_z(2)+s(2)p_z(1)\}\times\sigma_-$$

where σ_- is the spin wavefunction that corresponds to paired spins

$$\sigma_-=\alpha(1)\beta(2)-\alpha(2)\beta(1)$$

The wavefunction that corresponds to H^+F^- is

$$\psi_{H^+F^-}=p_z(1)p_z(2)\times\sigma_-$$

and the wavefunction that corresponds to H^-F^+ is

$$\psi_{H^-F^+}=s(1)s(2)\times\sigma_-$$

The wavefunction corresponding to the resonance hybrid is

$$\boxed{\psi=a\psi_{VB}+b\psi_{H^-F^+}+c\psi_{H^+F^-}}$$

where a, b, and c are coefficients determined by the variation principle and normalization.

Question. Based on your knowledge of the properties of H and F, which coefficient would you expect to have the larger magnitude, b or c?

E10A.3(a) The valence bond description of P_2 is similar to that of N_2, a triple bond. The three bonds are a σ from the overlap of sp hybrid orbitals and two π bonds from the overlap of unhybridized 3p orbitals.

In the tetrahedral P_4 molecule there are six single P–P bonds of roughly 200 kJ mol^{-1} bond enthalpy each. So the total bonding enthalpy is roughly 1200 kJ mol^{-1}. In the transformation

$$P_4 \rightarrow 2P_2$$

there is a loss of about 800 kJ mol^{-1} in σ-bond enthalpy. This loss is not likely to be made up by the formation of 4 P–P π bonds. Period 3 atoms, such as P, are too large to get close enough to each other to form strong π bonds.

E10A.4(a) All of the carbon atoms are sp^2 hybridized. The C–H σ bonds are formed by the overlap of Csp^2 orbitals with H1s orbitals. The C–C σ bonds are formed by the overlap of Csp^2 orbitals. The C–C π bonds are formed by the overlap of C2p orbitals. This description predicts double bonds between carbon atoms 1 & 2 and 3 & 4, but (unlike a simple molecular-orbital description of the bonding) no double-bond character between carbon atoms 2 & 3.

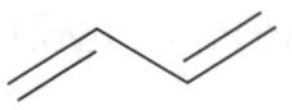

E10A.5(a) $h_1 = s + p_x + p_y + p_z \qquad h_2 = s - p_x - p_y + p_z$

We need to evaluate

$$\int h_1 h_2 \, d\tau = \int (s + p_x + p_y + p_z)(s - p_x - p_y + p_z) \, d\tau$$

We assume that the atomic orbitals are normalized atomic orbitals which are mutually orthogonal. We expand the integrand, noting that all cross terms integrate to zero because of orthogonality. The remaining terms integrate to one, because of normalization, yielding

$$\int h_1 h_2 \, d\tau = \int s^2 \, d\tau - \int p_x^2 \, d\tau - \int p_y^2 \, d\tau + \int p_z^2 \, d\tau = 1 - 1 - 1 + 1 = 0$$

E10A.6(a) Normalization requires

$$\int \psi^* \psi \, d\tau = 1 \qquad \text{where} \qquad \psi = Nh = N(s + 2^{1/2}p)$$

Solve for the normalization constant N:

$$1 = N^2 \int (s + 2^{1/2}p)^*(s + 2^{1/2}p) d\tau = N^2 \int (|s|^2 + 2^{1/2}p^*s + 2^{1/2}s^*p + 2|p|^2) d\tau = 3N^2$$

In the last step, we used the fact that the s and p orbitals are each normalized to see that the squared terms integrate to 1 and the fact that they are orthogonal to each other to see that the cross terms integrate to 0. Thus

$\boxed{N = 3^{-1/2}}$ and $\boxed{\psi = 3^{-1/2}(\mathrm{s} + 2^{1/2}\,\mathrm{p})}$

Solutions to problems

P10A.1 Substitute explicit forms of the orbitals into the expressions, using 2s and 2p orbitals. First use eqns 9A.22 to write the px and py orbitals in terms of the complex hydrogenic orbitals defined in eqn. 9A.21:

$$\psi = \frac{1}{3^{1/2}}\left(\mathrm{s} + \frac{1}{2^{1/2}} \times \frac{\mathrm{p}_{+1} - \mathrm{p}_{-1}}{2^{1/2}} + \frac{3^{1/2}\mathrm{i}}{2^{1/2}} \times \frac{\mathrm{p}_{+1} + \mathrm{p}_{-1}}{2^{1/2}}\right)$$

Then substitute the radial wavefunctions from Table 9A.1 and spherical harmonics from Table 8C.1:

$$\psi = \frac{Z^{3/2}\mathrm{e}^{-\rho/2}}{3^{1/2}a^{3/2}}\left(\frac{1}{8^{1/2}} \times \frac{2-\rho}{2\pi^{1/2}} - \frac{\rho}{2\times 24^{1/2}} \times \frac{3^{1/2}\sin\theta}{(8\pi)^{1/2}} \times \{(\mathrm{e}^{+\mathrm{i}\phi} + \mathrm{e}^{-\mathrm{i}\phi}) + 3^{1/2}\mathrm{i}(\mathrm{e}^{+\mathrm{i}\phi} - \mathrm{e}^{-\mathrm{i}\phi})\}\right)$$

Pull out some common factors and combine the imaginary exponentials into sine and cosine:

$$\psi = \boxed{\frac{Z^{3/2}\mathrm{e}^{-\rho/2}}{(24\pi)^{1/2}a^{3/2}}\left(\frac{2-\rho}{2} + \frac{\rho\sin\theta}{8^{1/2}} \times (-\cos\phi + 3^{1/2}\sin\phi)\right)}$$

ρ is defined in eqn 9A.11a, and it is proportional to r. The direction of maximum amplitude is the direction, defined by the angles θ and ϕ, in which ψ is a maximum. To find maxima, take derivatives with respect to angles and set them equal to zero. Actually, the θ dependence can be seen by inspection or by setting $\partial\psi/\partial\theta = 0$. Either way, we note that the maximum is at
$\theta = \pi/2$, or in the xy plane.

$$\frac{\partial\psi}{\partial\phi} = 0 = \frac{Z^{3/2}\mathrm{e}^{-\rho/2}}{(24\pi)^{1/2}a^{3/2}}\left(\frac{\rho\sin\theta}{8^{1/2}} \times (\sin\phi + 3^{1/2}\cos\phi)\right)$$

which requires

$$\sin\phi = -3^{1/2}\cos\phi \quad \text{or} \quad \phi = \tan^{-1}(-3^{1/2}) = \boxed{120°} \text{ or } -60°$$

P10A.3 One approach is to construct the explicit forms of the hybrid orbitals and find the values of ϕ that maximize the squares of their magnitudes. We need the component unhybridized orbitals, put together from the radial and angular functions listed in Tables 9A.1 and 8C.1. This approach is perhaps best applied to sp^2 hybrids, because the form of these hybrids has different numerical coefficients in front of the unhybridized p orbitals, not just different signs.

$$2\mathrm{s} = R_{20}Y_{00} = \frac{1}{8^{1/2}}\left(\frac{Z}{a}\right)^{3/2} \times (2-\rho)\mathrm{e}^{-\rho/2} \times \left(\frac{1}{4\pi}\right)^{1/2} = \left(\frac{1}{32\pi}\right)^{1/2}\left(\frac{Z}{a}\right)^{3/2}(2-\rho)\mathrm{e}^{-\rho/2}$$

Use eqns 9A.22 to write the p_x and p_y orbitals in terms of the complex hydrogenic orbitals defined in eqn 9A.21:

$$\begin{aligned} 2\mathrm{p}_x &= -\frac{1}{2^{1/2}}R_{21}(Y_{1,+1} - Y_{1,-1}) \\ &= -\frac{1}{2^{1/2}} \times \frac{1}{24^{1/2}}\left(\frac{Z}{a}\right)^{3/2} \rho\mathrm{e}^{-\rho/2} \times \left(\frac{3}{8\pi}\right)^{1/2} \sin\theta(-\mathrm{e}^{+\mathrm{i}\phi} - \mathrm{e}^{-\mathrm{i}\phi}) \\ &= \left(\frac{1}{32\pi}\right)^{1/2}\left(\frac{Z}{a}\right)^{3/2} \rho\mathrm{e}^{-\rho/2}\sin\theta\cos\phi \end{aligned}$$

$$\begin{aligned} 2\mathrm{p}_y &= \frac{\mathrm{i}}{2^{1/2}}R_{21}(Y_{1,+1} + Y_{1,-1}) \\ &= \frac{\mathrm{i}}{2^{1/2}} \times \frac{1}{24^{1/2}}\left(\frac{Z}{a}\right)^{3/2} \rho\mathrm{e}^{-\rho/2} \times \left(\frac{3}{8\pi}\right)^{1/2} \sin\theta(-\mathrm{e}^{+\mathrm{i}\phi} + \mathrm{e}^{-\mathrm{i}\phi}) \\ &= \left(\frac{1}{32\pi}\right)^{1/2}\left(\frac{Z}{a}\right)^{3/2} \rho\mathrm{e}^{-\rho/2}\sin\theta\sin\phi \end{aligned}$$

where $\rho = \dfrac{2Zr}{2a} = \dfrac{Zr}{a}$ [9A.11a]

In forming each hybrid, we neglect the factor $\left(\dfrac{1}{32\pi}\right)^{1/2}\left(\dfrac{Z}{a}\right)^{3/2}\mathrm{e}^{-\rho/2}$ common to each component; an angle-independent multiplicative term cannot influence the angle at which the hybrid is maximal.

Next, form the hybrids, using eqn 10A.7:

$$h_1 = \mathrm{s} + 2^{1/2}\mathrm{p}_y = (2-\rho) + 2^{1/2}(\rho\sin\theta\sin\phi) = 2 + \rho(2^{1/2}\sin\theta\sin\phi - 1)$$

$$\begin{aligned} h_2 &= \mathrm{s} + \left(\frac{3}{2}\right)^{1/2}\mathrm{p}_x - \left(\frac{1}{2}\right)^{1/2}\mathrm{p}_y = (2-\rho) + \left(\frac{3}{2}\right)^{1/2}\rho\sin\theta\cos\phi - \left(\frac{1}{2}\right)^{1/2}\rho\sin\theta\sin\phi \\ &= 2 + \rho\left(\frac{3^{1/2}\cos\phi - \sin\phi}{2^{1/2}}\sin\theta - 1\right) \end{aligned}$$

and
$$\begin{aligned} h_3 &= \mathrm{s} - \left(\frac{3}{2}\right)^{1/2}\mathrm{p}_x - \left(\frac{1}{2}\right)^{1/2}\mathrm{p}_y = (2-\rho) - \left(\frac{3}{2}\right)^{1/2}\rho\sin\theta\cos\phi - \left(\frac{1}{2}\right)^{1/2}\rho\sin\theta\sin\phi \\ &= 2 - \rho\left(\frac{3^{1/2}\cos\phi + \sin\phi}{2^{1/2}}\sin\theta + 1\right) \end{aligned}$$

To find the angle ϕ at which the hybrids have maximum probability, differentiate with respect to ϕ, and set the derivative equal to zero, because positive maxima and negative minima in the hybrid orbitals correspond to maxima in the probability. Differentiation would work even for h_1, but it is unnecessary there. One can see by inspection that the function is maximized when sin ϕ is maximal, namely at $\phi=\pi/2$ (90°). It should come as no surprise that this orbital points along the positive y-axis.

$$\frac{\partial h_2}{\partial\phi}=0=\left(\frac{\rho\sin\theta}{2^{1/2}}\right)(-3^{1/2}\sin\phi-\cos\phi)$$

so $\dfrac{\sin\phi}{\cos\phi}=-\dfrac{1}{3^{1/2}}=\tan\phi$ or $\phi=5\pi/6$ or $11\pi/6$ (150° or 330°).

Of these, $11\pi/6$ (330°) is a maximum and $5\pi/6$ a minimum. The larger amplitude is at the maximum ($11\pi/6$ or 330°). Finally

$$\frac{\partial h_3}{\partial\phi}=0=-\left(\frac{\rho\sin\theta}{2^{1/2}}\right)(-3^{1/2}\sin\phi+\cos\phi)$$

so $\dfrac{\sin\phi}{\cos\phi}=\dfrac{1}{3^{1/2}}=\tan\phi$ or $\phi=\pi/6$ or $7\pi/6$ (30° or 210°).

Of these, $7\pi/6$ (210°) is a maximum and corresponds to the larger amplitude. The angles 90°, 210°, and 330° are 120° apart.

So much for the case $\lambda=2$. The cases $\lambda=1$ and 3 can be treated from the functional forms of the hybrids and the facts that the unhybridized p orbitals are maximally positive along positive Cartesian axes and maximally negative along the negative Cartesian axes. Since s orbitals have constant positive values for all angles, we can figure out the direction of the hybrid orbitals by figuring out the direction in which the component p orbitals reinforce the s contribution.

The sp hybrids are

$$h_1=\mathrm{s}+\mathrm{p}_z \text{ and } h_2=\mathrm{s}-\mathrm{p}_z \quad [10\text{A}.8]$$

Therefore, h_1 points in the same direction as the positive lobe of the p_z orbital (i.e. the positive z axis) and h_2 points in the same direction as the negative lobe of the p_z orbital (i.e. the negative z-axis, where we have positive s and negative p_z). So equivalent sp orbitals point in directions that differ by 180°.

Finally, consider the sp^3 orbitals [10A.5]

$$h_1=\mathrm{s}+\mathrm{p}_x+\mathrm{p}_y+\mathrm{p}_z \quad h_2=\mathrm{s}-\mathrm{p}_x-\mathrm{p}_y+\mathrm{p}_z$$

$$h_3=\mathrm{s}-\mathrm{p}_x+\mathrm{p}_y-\mathrm{p}_z \quad h_4=\mathrm{s}+\mathrm{p}_x-\mathrm{p}_y-\mathrm{p}_z$$

As explained in *Justification* 10A.2, we know that h_1 points to the Cartesian point (+1, +1, +1), because the three p orbitals have equal weight and it is their positive lobes that reinforce the positive s contribution. By similar reasoning, h_2 points to (–1,–1, + 1), h_3 to (–1, + 1,–1), and h_4 to (+ 1,–1,–1). As illustrated in Fig. 10A.1, these points fall on the vertices of a regular tetrahedron.

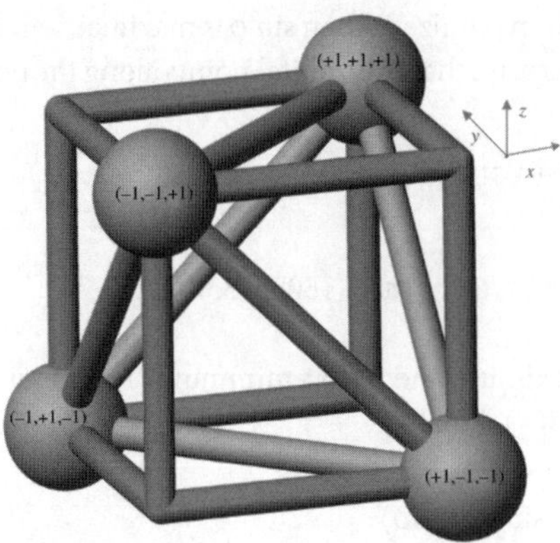

Figure 10A.1

As noted in the text, the angle between the corresponding hybrid orbitals is, therefore, the tetrahedral angle, arcos(−1/3) = 109.47°. Rearranging this equation, we see that

$$-1/3=\cos 109.47° \text{ or } -1/(\cos 109.47°)=3$$

Similarly

$$-1/\cos 120°=2 \text{ and } -1/\cos 180°=1$$

so for these sp$^\lambda$ hybrids, equivalent orbitals are separated by an angle θ such that

$$\lambda=-1/\cos\theta$$

A plot of λ vs. θ is shown in Fig. 10A.2A. The angle between two orbitals can only range from 0 to 180°; however, λ is negative for $0 < \theta < 180°$, and λ approaches infinity as θ approaches 90° from the right.

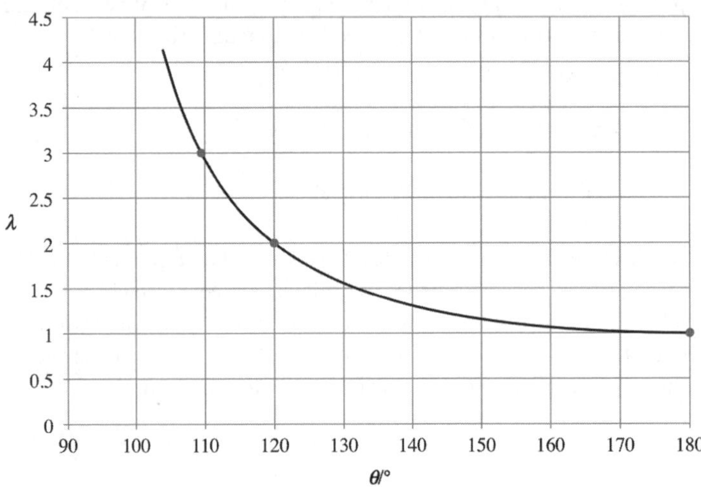

Figure 10A.2

10B Principles of molecular orbital theory

Answer to discussion question

D10B.1 As stated in Section 10B.1(b), the cause of reduced energy of bonding molecular orbitals compared to atomic orbitals is not entirely clear, although what is clear is that bonding character correlates strongly with molecular orbitals that have an accumulation of electron density between nuclei due to overlap and constructive interference of their component atomic orbitals. A simple and plausible explanation of this correlation is that enhanced electron probability between nuclei lowers potential energy by putting electrons in a position where they can be attracted to two nuclei at the same time; however, the source of the reduced energy may be more complicated.

Solutions to exercises

E10B.1(a)
$$\int\psi^2\,\mathrm{d}\tau = N^2\int(\psi_A+\lambda\psi_B)^2\,\mathrm{d}\tau = N^2\int(\psi_A^2+\lambda^2\psi_B^2+2\lambda\psi_A\psi_B)\mathrm{d}\tau = 1$$
$$= N^2(1+\lambda^2+2\lambda S)\quad\left[\int\psi_A\psi_B\,\mathrm{d}\tau = S\right]$$

Hence $\boxed{N=\left(\dfrac{1}{1+2\lambda S+\lambda^2}\right)^{1/2}}$

E10B.2(a) Let $\psi_1 = N(0.145A + 0.844B)$ and $\psi_2 = aA + bB$

First, let us normalize ψ_1:

$$\int\psi_1{}^*\psi_1\,\mathrm{d}\tau = 1 = N^2\int(0.145A+0.844B)^*(0.145A+0.844B)\mathrm{d}\tau,$$

$$0.0210\int|A|^2\,\mathrm{d}\tau+0.712\int|B|^2\,\mathrm{d}\tau+0.122\int B^*A\mathrm{d}\tau+0.122\int A^*B\mathrm{d}\tau=\frac{1}{N^2}$$

The first two integrals are 1 due to normalization and the latter two are the overlap integral, $S = 0.250$. So

$$0.210+0.712+2\times0.122\times0.250=\frac{1}{N^2}=0.795,$$

so $\boxed{N=1.12}$ which makes $\boxed{\psi_1=0.163A+0.947B}$

Orthogonality of the two molecular orbitals requires

$$\int\psi_1{}^*\psi_2\,\mathrm{d}\tau=0=N\int(0.145A+0.844B)^*(aA+bB)\mathrm{d}\tau$$

Dividing by N yields

$$0.145a\int|A|^2\,d\tau+0.844b\int|B|^2\,d\tau+0.844a\int B^*A\,d\tau+0.145b\int A^*B\,d\tau=0,$$

$$0.145a+0.844b+0.844a\times0.250+0.145b\times0.250=0=0.356a+0.880b$$

Normalization of ψ_2 requires

$$\int\psi_2{}^*\psi_2\,d\tau=1=\int(aA+bB)^*(aA+bB)d\tau,$$

$$a^2\int|A|^2\,d\tau+b^2\int|B|^2\,d\tau+ab\int B^*A\,d\tau+ab\int A^*B\,d\tau=1$$

So $a^2+b^2+2ab\times0.250=1=a^2+b^2+0.500ab$

We have two equations in the two unknown coefficients a and b. Solve the first equation for a in terms of b:

$$a=-0.880b/0.356=-2.47b$$

Substitute this result into the second (quadratic) equation:

$$1=(-2.47b)^2+b^2+0.500(-2.47b)b=5.88b^2$$

So $\boxed{b=0.412}$, $\boxed{a=-1.02}$, and $\boxed{\psi_2=-1.02A+0.412B}$

E10B.3(a) $E_H=E_1=-hc\tilde{R}_H$ [9A.14]

Draw up the following table using the data in the question and using

$$\frac{e^2}{4\pi\varepsilon_0R}=\frac{e^2}{4\pi\varepsilon_0a_0}\times\frac{a_0}{R}=\frac{e^2}{4\pi\varepsilon_0\times(4\pi\varepsilon_0\hbar^2/m_ee^2)}\times\frac{a_0}{R}$$

$$=\frac{m_ee^4}{16\pi^2\varepsilon_0^2\hbar^2}\times\frac{a_0}{R}=E_h\times\frac{a_0}{R}\quad\left[E_h\equiv\frac{m_ee^4}{16\pi^2\varepsilon_0^2\hbar^2}=2hc\tilde{R}_H\right]$$

so that $\dfrac{e^2}{4\pi\varepsilon_0R}\times\dfrac{1}{E_h}=\dfrac{a_0}{R}$

R/a_0	0	1	2	3	4	∞
$\frac{e^2}{4\pi\varepsilon_0R}\times\frac{1}{E_h}=\frac{a_0}{R}$	∞	1	0.500	0.333	0.250	0
$(V_1+V_2)/E_h$	2.000	1.465	0.843	0.529	0.342	0
$(E-E_H)/E_h$	∞	0.212	−0.031	−0.059	−0.038	0

The points are plotted in Fig. 10B.1.

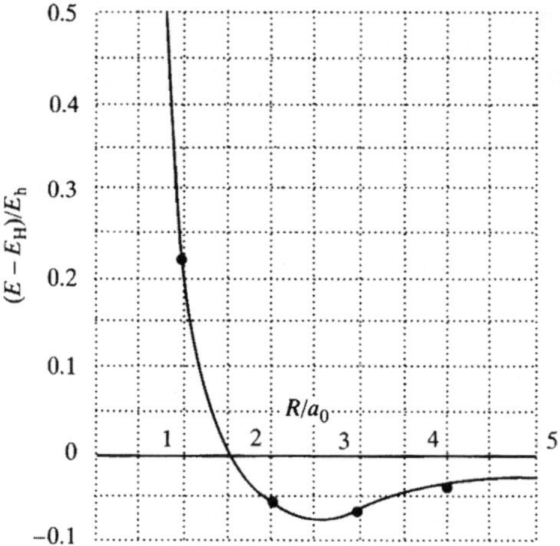

Figure 10B.1

The minimum occurs at $R = 2.5a_0$, so $R = 130$ pm. At that bond length

$$E - E_H = -0.07E_h = -1.91\,\text{eV}$$

Hence, the dissociation energy is predicted to be about $\boxed{1.9\,\text{eV}}$ and the equilibrium bond length about $\boxed{130\,\text{pm}}$.

E10B.4(a) Figure 10.B2 shows sketches of the orbitals in question. By inspection, one can see that the bonding orbital is odd, $\boxed{\text{u}}$, with respect to inversion, and the antibonding orbital is even, $\boxed{\text{g}}$.

bonding

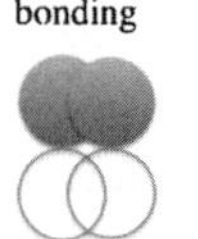

antibonding

Figure 10B.2

Solutions to problems

P10B.1 The energy of repulsion is given by Coulomb's law:

$$E = \frac{e^2}{4\pi\varepsilon_0 R} = \frac{(1.602\times10^{-19}\,\text{C})^2}{4\pi\times8.854\times10^{-12}\,\text{J}^{-1}\,\text{C}^2\,\text{m}^{-1}\times74.1\times10^{-12}\,\text{m}} = 3.11\times10^{-18}\,\text{J}$$

The repulsion energy of a mole of H_2 is

$$E_{\text{molar}} = N_A E = 6.022\times10^{23}\,\text{mol}^{-1}\times3.11\times10^{-18}\,\text{J} = \boxed{1.87\times10^6\,\text{J}\,\text{mol}^{-1} = 1.87\,\text{MJ}\,\text{mol}^{-1}}$$

The gravitational energy is

$$E_{\text{grav}} = \frac{-Gm_1m_2}{R} = \frac{-6.673\times10^{-11}\ \text{m}^3\ \text{kg}^{-1}\ \text{s}^{-2}\times(1.672\times10^{-27}\ \text{kg})^{-2}}{74.1\times10^{-12}\ \text{m}} = -2.52\times10^{-54}\ \text{J}$$

The gravitational energy is many, many orders of magnitude smaller than the energy of electrostatic repulsion. This is an illustration of the fact that gravitational effects tend to be negligible on the atomic scale.

P10B.3 The hamiltonian is

$$\hat{H} = -\frac{\hbar^2}{2m}\nabla^2 - \frac{e^2}{4\pi\varepsilon_0}\left(\frac{1}{r_\text{A}}+\frac{1}{r_\text{B}}-\frac{1}{R}\right) = -\frac{\hbar^2}{2m}\nabla^2 - j_0\left(\frac{1}{r_\text{A}}+\frac{1}{r_\text{B}}-\frac{1}{R}\right) \quad [10\text{B}.1]$$

where we have introduced the abbreviation j_0 for the collection of constants $e^2/4\pi\varepsilon_0$. The expectation value of this operator is

$$\langle H\rangle = \int\psi^*\hat{H}\psi\,\text{d}\tau,$$

where $\psi_\pm = N_\pm(A\pm B)$ [10B.2] and $N_\pm = \dfrac{1}{\{2(1+S)\}^{1/2}}$ [Example 10B.1]

Thus, $\langle H\rangle = \int\psi_\pm{}^*\left\{-\dfrac{\hbar^2}{2m}\nabla^2 - j_0\left(\dfrac{1}{r_\text{A}}+\dfrac{1}{r_\text{B}}-\dfrac{1}{R}\right)\right\}\psi_\pm\,\text{d}\tau$

Pull out the nuclear repulsion term

$$\begin{aligned}\langle H\rangle &= \int\psi_\pm{}^*\left(-\frac{\hbar^2}{2m}\nabla^2 - \frac{j_0}{r_\text{A}} - \frac{j_0}{r_\text{B}}\right)\psi_\pm\,\text{d}\tau + \frac{j_0}{R}\int\psi_\pm{}^*\psi_\pm\,\text{d}\tau \\ &= N_\pm^2\int(A\pm B)\left(-\frac{\hbar^2}{2m}\nabla^2 - \frac{j_0}{r_\text{A}} - \frac{j_0}{r_\text{B}}\right)(A\pm B)\,\text{d}\tau + \frac{j_0}{R}\end{aligned}$$

Expand the remaining integral into four integrals, letting $\hat{H}_\text{H}$ represent the hamiltonian operator of a hydrogen atom. Note that

$$\hat{H}_\text{H}A = -\frac{\hbar^2}{2m}\nabla^2 A - \frac{j_0}{r_\text{A}}A = E_{\text{H1s}}A \quad\text{and}\quad \hat{H}_\text{H}B = -\frac{\hbar^2}{2m}\nabla^2 B - \frac{j_0}{r_\text{B}}B = E_{\text{H1s}}B$$

So $$\begin{aligned}\langle H\rangle &= N_\pm^2\int A\left(\hat{H}_\text{H} - \frac{j_0}{r_\text{B}}\right)A\,\text{d}\tau \pm N_\pm^2\int A\left(\hat{H}_\text{H} - \frac{j_0}{r_\text{A}}\right)B\,\text{d}\tau \\ &\quad \pm N_\pm^2\int B\left(\hat{H}_\text{H} - \frac{j_0}{r_\text{B}}\right)A\,\text{d}\tau + N_\pm^2\int B\left(\hat{H}_\text{H} - \frac{j_0}{r_\text{A}}\right)B\,\text{d}\tau + \frac{j_0}{R}\end{aligned}$$

and $$\begin{aligned}\langle H\rangle &= N_\pm^2\int A\left(E_{\text{H1s}} - \frac{j_0}{r_\text{B}}\right)A\,\text{d}\tau \pm N_\pm^2\int A\left(E_{\text{H1s}} - \frac{j_0}{r_\text{A}}\right)B\,\text{d}\tau \\ &\quad \pm N_\pm^2\int B\left(E_{\text{H1s}} - \frac{j_0}{r_\text{B}}\right)A\,\text{d}\tau + N_\pm^2\int B\left(E_{\text{H1s}} - \frac{j_0}{r_\text{A}}\right)B\,\text{d}\tau + \frac{j_0}{R}\end{aligned}$$

Pull out the terms involving E_{His}:

$$\langle H\rangle = E_{\text{H1s}}N_{\pm}^2\int(A^2 \pm 2AB + B^2)\text{d}\tau$$
$$- j_0N_{\pm}^2\int\frac{A^2}{r_{\text{B}}}\text{d}\tau \mp j_0N_{\pm}^2\int\frac{AB}{r_{\text{A}}}\text{d}\tau \mp j_0N_{\pm}^2\int\frac{AB}{r_{\text{B}}}\text{d}\tau - j_0N_{\pm}^2\int\frac{B^2}{r_{\text{A}}}\text{d}\tau + \frac{j_0}{R}$$

What multiplies E_{His} is the normalization integral, which equals 1. Now examine the remaining four integrals. The first and fourth of these are equal to each other; likewise for the second and third. (This can be demonstrated by interchanging the labels A and B, which are arbitrary.) Thus

$$\langle H\rangle = E_{\text{H1s}} - 2N_{\pm}^2\left(j_0\int\frac{A^2}{r_{\text{B}}}\text{d}\tau \pm j_0\int\frac{AB}{r_{\text{A}}}\text{d}\tau\right) + \frac{j_0}{R}$$
$$= E_{\text{H1s}} - 2N_{\pm}^2(j \pm k) + \frac{j_0}{R} = E_{\text{H1s}} - \frac{j \pm k}{1 \pm S} + \frac{j_0}{R}$$

where we have used the definitions of j and k in eqn 10B.5. The expectation values work out to eqns 10B.4 and 10B.7:

$$\langle H\rangle_+ = \boxed{E_{\text{H1s}} - \frac{j+k}{1+S} + \frac{j_0}{R}} = E_{1\sigma} \quad \text{and} \quad \langle H\rangle_- = \boxed{E_{\text{H1s}} - \frac{j-k}{1-S} + \frac{j_0}{R}} = E_{2\sigma}$$

P10B.5 (a) Start from eqn 10B.4 (whose derivation is covered in Problem 10B.3).

$$E_{1\sigma} = E_{\text{H1s}} - \frac{j+k}{1+S} + \frac{j_0}{R}$$

where we define the Coulomb integral j and the resonance integral k as in eqns 10B.5 and use the relationship of the normalization constant N to the overlap integral S as defined in Example 10B.1. We must write the integrals S, j, and k explicitly to evaluate them. Use

$$A = \frac{\text{e}^{-r_{\text{A}}/a_0}}{(\pi a_0^3)^{1/2}} \quad \text{and} \quad B = \frac{\text{e}^{-r_{\text{B}}/a_0}}{(\pi a_0^3)^{1/2}} \quad [\textit{Brief illustration } 10\text{B}.1]$$

Define a coordinate system centred on nucleus A, so that r_{A} is a coordinate and

$$r_{\text{B}} = \{x^2 + y^2 + (z-R)^2\}^{1/2}[\textit{Brief illustration}10\text{B}.1] = (x^2 + y^2 + z^2 + R^2 - 2zR)^2$$
$$r_{\text{B}} = (r_{\text{A}}^2 + R^2 - 2zR)^{1/2} = (r_{\text{A}}^2 + R^2 - 2r_{\text{A}}R\cos\theta)^{1/2}$$

The volume element is

$$\text{d}\tau = r_{\text{A}}^2 \sin\theta \,\text{d}r_{\text{A}}\,\text{d}\theta\,\text{d}\phi$$

Thus, the overlap integral is

$$S=\int AB\mathrm{d}\tau=\int_0^{2\pi}\int_0^{\pi}\int_0^{\infty}\frac{\mathrm{e}^{-r_\mathrm{A}/a_0}}{(\pi a_0^3)^{1/2}}\frac{\mathrm{e}^{-r_\mathrm{B}/a_0}}{(\pi a_0^3)^{1/2}}r_\mathrm{A}^2\sin\theta\,\mathrm{d}r_\mathrm{A}\,\mathrm{d}\theta\,\mathrm{d}\phi$$

Integration over $\mathrm{d}\phi$ yields a factor of 2π. Numerical integration over $\mathrm{d}\theta$ and $\mathrm{d}r_\mathrm{A}$ can be done by approximating the differentials by small but finite $\delta\theta$ and δr_A and summing over a series of θ_j and $r_{\mathrm{A},i}$:

$$S=\frac{2}{a_0^3}\int_0^{\pi}\int_0^{\infty}\mathrm{e}^{-r_\mathrm{A}/a_0}\mathrm{e}^{-r_\mathrm{B}/a_0}r_\mathrm{A}^2\sin\theta\,\mathrm{d}r_\mathrm{A}\,\mathrm{d}\theta\approx 2\delta\theta\frac{\delta r_\mathrm{A}}{a_0}\sum_j\sum_i\mathrm{e}^{-r_{\mathrm{A},i}/a_0}\mathrm{e}^{-r_{\mathrm{B},ij}/a_0}\left(\frac{r_{\mathrm{A},i}}{a_0}\right)^2\sin\theta_j$$

Similar treatment of the other integrals yields

$$j\approx 2\delta\theta\frac{\delta r_\mathrm{A}}{a_0}\times\frac{j_0}{a_0}\sum_j\sum_i\frac{\mathrm{e}^{-2r_{\mathrm{A},i}/a_0}}{\left(r_{\mathrm{B},ij}/a_0\right)}\left(\frac{r_{\mathrm{A},i}}{a_0}\right)^2\sin\theta_j$$

and
$$k\approx 2\delta\theta\frac{\delta r_\mathrm{A}}{a_0}\times\frac{j_0}{a_0}\sum_j\sum_i\frac{\mathrm{e}^{-r_{\mathrm{A},i}/a_0}\mathrm{e}^{-r_{\mathrm{B},ij}/a_0}}{\left(r_{\mathrm{B},ij}/a_0\right)}\left(\frac{r_{\mathrm{A},i}}{a_0}\right)^2\sin\theta_j$$

To evaluate the integrals, express the distances in units of a_0. S is dimensionless. j and k have dimensions of energy; they have units of j_0/a_0. In the following table and plots, the numerical values of the integrals were computed using θ_j from 0 to π in increments that correspond to 10° and $r_{\mathrm{A},i}$ from 0 to $5a_0$ in increments of $0.25a_0$.

R/a_0	S_num	$\frac{j}{j_0/a_0}$	$\frac{k}{j_0/a_0}$	S_anal	$\frac{j_\mathrm{anal}}{j_0/a_0}$	$\frac{k_\mathrm{anal}}{j_0/a_0}$
1.00	0.8524	0.7227	0.7220	0.8584	0.7293	0.7358
1.25	0.7875	0.6460	0.6287	0.7939	0.6522	0.6446
1.50	0.7183	0.5781	0.5408	0.7252	0.5837	0.5578
1.75	0.6480	0.5192	0.4606	0.6553	0.5240	0.4779
2.00	0.5789	0.4684	0.3890	0.5865	0.4725	0.4060
2.25	0.5125	0.4249	0.3262	0.5204	0.4284	0.3425
2.50	0.4501	0.3876	0.2718	0.4583	0.3906	0.2873
2.75	0.3924	0.3555	0.2253	0.4009	0.3581	0.2397
3.00	0.3397	0.3279	0.1858	0.3485	0.3300	0.1991
3.25	0.2922	0.3038	0.1525	0.3013	0.3057	0.1648
3.50	0.2498	0.2829	0.1245	0.2592	0.2845	0.1359
3.75	0.2122	0.2644	0.1012	0.2219	0.2660	0.1117
4.00	0.1791	0.2482	0.0817	0.1893	0.2496	0.0916

A plot of S vs. R is shown in Fig. 10B.3(a) and a plot of j and k vs. R is shown in Fig. 10B.3(b). Note that the numerical results at this level of approximation agree well but not identically with the analytical ones. The disagreement becomes more noticeable at larger R, which is not surprising given that all sums were truncated at $r_A = 5a_0$. The disagreement is more pronounced in k than in S or j. The numerical results can be made arbitrarily close to the analytical by using smaller increments and extending the range of r_A.

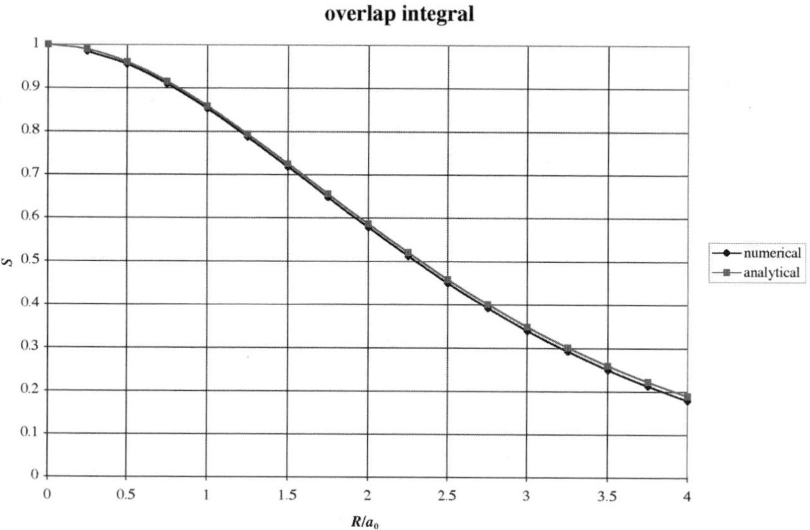

Figure 10B.3(a)

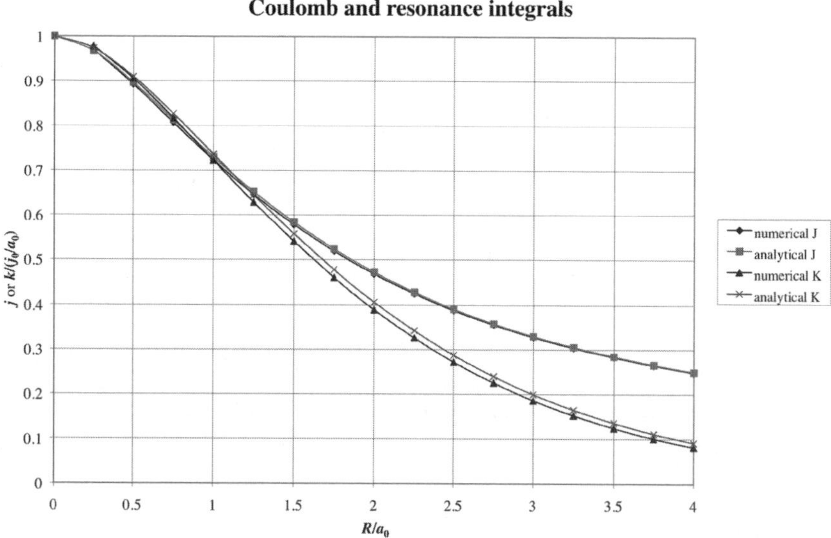

Figure 10B.3(b)

(b) Use eqn 10B.4 to compute the total energy, whether from numerical or analytical values of the integrals. Results are shown in the table below and in Fig. 10B.3(c).

R/a_0	$\dfrac{E_{\text{num}}}{j_0/a_0}$	$\dfrac{E_{\text{anal}}}{j_0/a_0}$
1.00	−0.280	−0.288
1.25	−0.413	−0.423
1.50	−0.485	−0.495
1.75	−0.523	−0.534
2.00	−0.543	−0.554
2.25	−0.552	−0.563
2.50	−0.555	−0.565
2.75	−0.554	−0.563
3.00	−0.550	−0.559
3.25	−0.545	−0.554
3.50	−0.540	−0.548
3.75	−0.535	−0.542
4.00	−0.530	−0.537

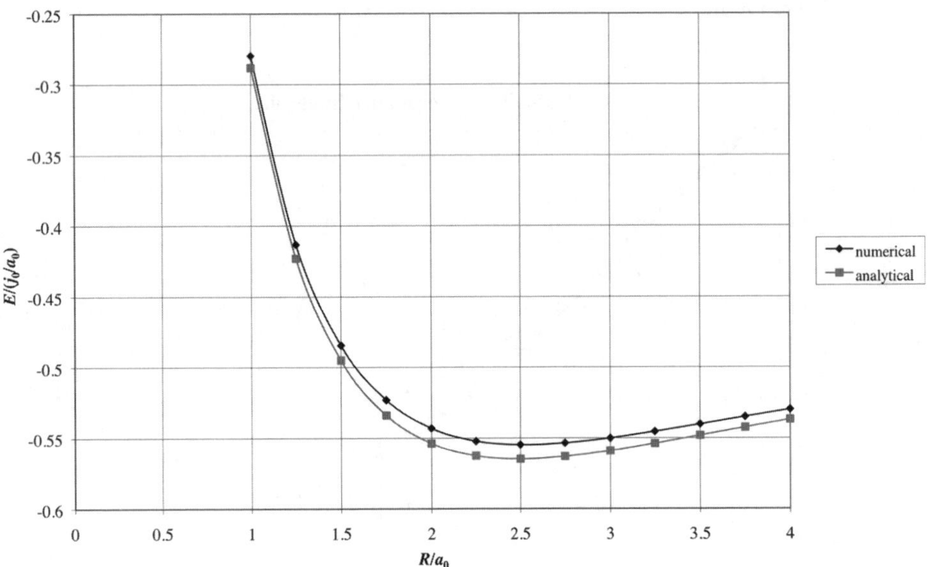

Figure 10B.3(c)

The minimum in both total energy curves lies at $R = \boxed{2.5a_0 = 1.3\times10^{-10}\,\text{m}}$.

The minimum total energy is $E_{\text{min}} = \boxed{-0.555j_0/a_0 = -15.1\,\text{eV}}$ (numerical) and $\boxed{0.565j_0/a_0 = -15.4\,\text{eV}}$ (analytical). The bond dissociation energy is

$$D_e = E_H - E_{\text{min}} = (-0.500+0.555)j_0/a_0 = \boxed{0.055j_0/a_0 = 1.5\,\text{eV}}\text{(numerical)}$$

$$D_e = (-0.500+0.565)j_0/a_0 = \boxed{0.065j_0/a_0 = 1.8\,\text{eV}}\text{(analytical)}$$

Comment. Dissociation energies are typically (relatively) small differences between larger quantities (electron binding energies). Even small errors in the latter lead to relatively large errors in the former. In this case, the numerical binding energy differed by about 2% from the analytical value; the difference in D_e was closer to 20%.

10C Homonuclear diatomic molecules

Answers to discussion questions

D10C.1 σ bonding with a $d_{x^2-y^2}$ orbital is shown in Fig. 10C.1(a).

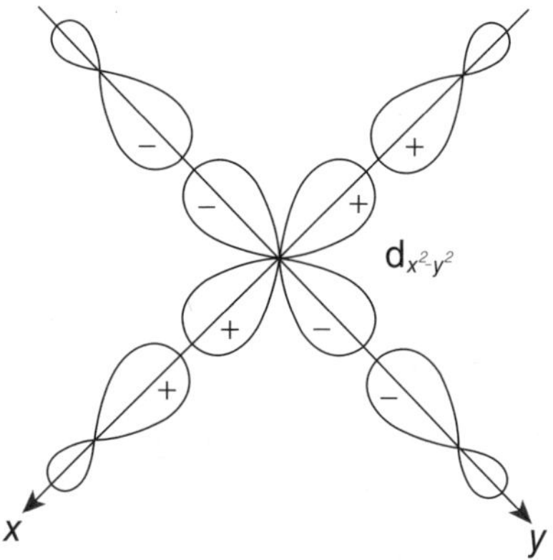

Figure 10C.1(a)

The σ antibonding orbital looks the same but with the p orbital lobes pointed in the opposite direction.

σ bonding with the d_{z^2} orbital is shown in Fig. 10C.1(b).

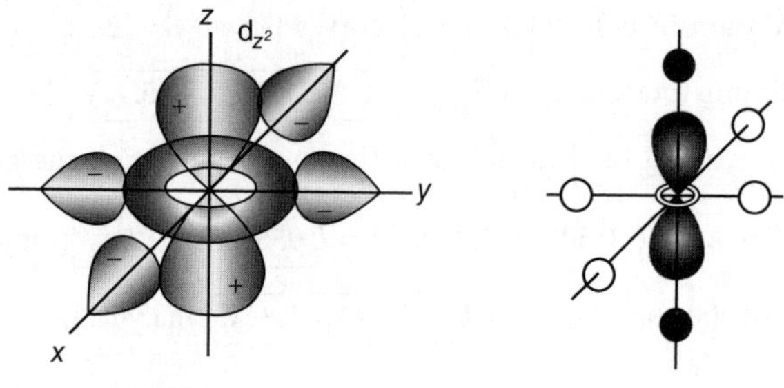

Figure 10C.1(b)

In this figure, only one of the p-orbital lobes is shown in each p orbital. p orbitals with positive lobes may also approach this orbital along the + and − z-direction as indicated in the smaller diagram on the right. The antibonding diagrams are similar, but with the signs of the p orbital lobes reversed.

For π bonding (see Fig. 10C.1(c)), only the d_{xy}, d_{yz}, and d_{xz} orbitals undergo π bonding with p orbitals on neighbouring atoms. The bonding arrangement is pictured here with the d_{xy} orbital. The diagrams for the d_{yz} and d_{xz} orbitals are similar. The antibonding diagrams have the signs of the p orbital lobes reversed.

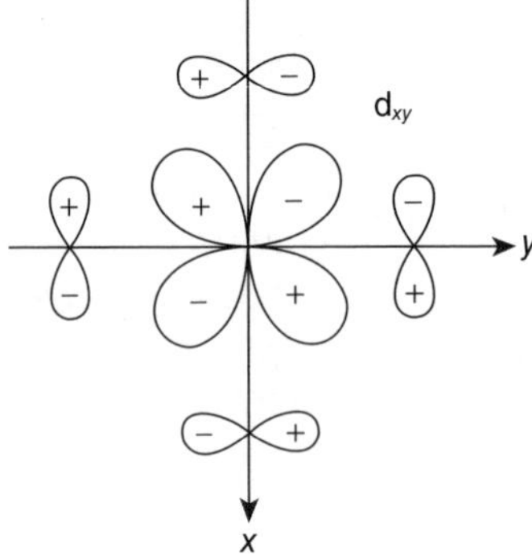

Figure 10C.1(c)

D10C.3 The Born–Oppenheimer approximation treats the nuclei of the multi-particle system of electrons and nuclei as if they were fixed. The dependence of energy on nuclear positions is then obtained by solving the Schrödinger equation at many different (fixed) nuclear geometries. In molecular orbital theory, specifically, one forms molecular orbitals from linear combinations of atomic orbitals within a fixed framework of nuclei, and then at a slightly different fixed framework, and so on until one finds the framework of nuclei that yields the lowest-energy wavefunction.

D10C.5 All other things being equal, bond strength and bond length are inversely related: the shorter the bond, the stronger the bond. (For example, think of carbon–carbon single, double, and triple bonds and their bond dissociation enthalpies. Or consider the lengths and strengths of the single bond in Cl_2, Br_2, and I_2.) Atomic orbital overlap is similarly related to bond length (or internuclear distance): the smaller the internuclear distance, the greater the overlap and therefore the stronger the bond.

Solutions to exercises

E10C.1(a) Refer to Fig. 10C.11 of the text. Place two of the valence electrons in each orbital starting with the lowest-energy orbital, until all valence electrons are used up. Apply Hund's rule to the filling of degenerate orbitals.

(a) Li_2 (2 electrons) $\boxed{1\sigma_g^2, b=1}$

(b) Be_2 (4 electrons) $\boxed{1\sigma_g^2 1\sigma_u^2, b=0}$

(c) C_2 (8 electrons) $\boxed{1\sigma_g^2 1\sigma_u^2 1\pi_u^4, b=2}$

E10C.2(a) Refer to Fig. 10C.11 of the text.

$$\begin{aligned} &B_2\text{(6 electrons)}: \quad 1\sigma_g^2 1\sigma_u^2 1\pi_u^2 \quad b=1 \\ &C_2\text{(8 electrons)}: \quad 1\sigma_g^2 1\sigma_u^2 1\pi_u^4 \quad b=2 \end{aligned}$$

The bond orders of B_2 and C_2 are respectively 1 and 2; so $\boxed{C_2}$ should have the greater bond dissociation enthalpy. The experimental values are approximately 4 eV and 6 eV, respectively.

E10C.3(a) Refer to Figure 10C.11 of the text.

$$\begin{aligned} &F_2^+\text{ (13 electrons)}: 1\sigma_g^2 1\sigma_u^2 2\sigma_g^2 1\pi_u^4 1\pi_g^3, b=1.5 \\ &F_2^+\text{ (14 electrons)}: 1\sigma_g^2 1\sigma_u^2 2\sigma_g^2 1\pi_u^4 1\pi_g^4, b=1 \end{aligned}$$

The bond orders of $F_2{}^+$ and F_2 are respectively 1.5 and 1; so $\boxed{F_2^+}$ should have the greater bond dissociation energy.

E10C.4(a) Refer to Fig. 10C.11 of the text.

$$\begin{aligned} &Li_2\text{(2 electrons)}: && 1\sigma_g^2 && \boxed{b=1} \\ &Be_2\text{(4 electrons)}: && 1\sigma_g^2 1\sigma_u^2 && \boxed{b=0} \\ &B_2\text{(6 electrons)}: && 1\sigma_g^2 1\sigma_u^2 1\pi_u^2 && \boxed{b=1} \\ &C_2\text{(8 lectrons)}: && 1\sigma_g^2 1\sigma_u^2 1\pi_u^4 && \boxed{b=2} \\ &N_2\text{(10 electrons)}: && 1\sigma_g^2 1\sigma_u^2 1\pi_u^4 2\sigma_g^2 && \boxed{b=3} \\ &O_2\text{(12 electrons)}: && 1\sigma_g^2 1\sigma_u^2 2\sigma_g^2 1\pi_u^4 1\pi_g^2 && \boxed{b=2} \\ &F_2\text{(14 electrons)}: && 1\sigma_g^2 1\sigma_u^2 2\sigma_g^2 1\pi_u^4 1\pi_g^4 && \boxed{b=1} \end{aligned}$$

E10C.5(a) Refer to Fig. 10C.11 of the text. The configurations are shown below. The HOMO is the rightmost orbital in the configuration and it has a box around it.

$$\mathrm{Li_2^+}\text{(1 electron)}: \quad \boxed{1\sigma_g}^1$$

$$\mathrm{Be_2^+}\text{(3 electrons)}: \quad 1\sigma_g^2\boxed{1\sigma_u}^1$$

$$\mathrm{B_2^+}\text{(5 electrons)}: \quad 1\sigma_g^2 1\sigma_u^2\boxed{1\pi_u}^1$$

$$\mathrm{C_2^+}\text{(7 electrons)}: \quad 1\sigma_g^2 1\sigma_u^2\boxed{1\pi_u}^3$$

$$\mathrm{N_2^+}\text{(9 electrons)}: \quad 1\sigma_g^2 1\sigma_u^2 1\pi_u^4\boxed{2\sigma_g}^1$$

$$\mathrm{O_2^+}\text{(11 electrons)}: \quad 1\sigma_g^2 1\sigma_u^2 2\sigma_g^2 1\pi_u^4\boxed{1\pi_g}^1$$

$$\mathrm{F_2^+}\text{(13 electrons)}: \quad 1\sigma_g^2 1\sigma_u^2 2\sigma_g^2 1\pi_u^4\boxed{1\pi_g}^3$$

$$\mathrm{Li_2^-}\text{(3 electrons)}: \quad 1\sigma_g^2\boxed{1\sigma_u}^1$$

$$\mathrm{Be_2^-}\text{(5 electrons)}: \quad 1\sigma_g^2 1\sigma_u^2\boxed{1\pi_u}^1$$

$$\mathrm{B_2^-}\text{(7 electrons)}: \quad 1\sigma_g^2 1\sigma_u^2\boxed{1\pi_u}^3$$

$$\mathrm{C_2^-}\text{(9 electrons)}: \quad 1\sigma_g^2 1\sigma_u^2 1\pi_u^4\boxed{2\sigma_g}^1$$

$$\mathrm{N_2^-}\text{(11 electrons)}: \quad 1\sigma_g^2 1\sigma_u^2 2\sigma_g^2 1\pi_u^4\boxed{1\pi_g}^1$$

$$\mathrm{O_2^-}\text{(13 electrons)}: \quad 1\sigma_g^2 1\sigma_u^2 2\sigma_g^2 1\pi_u^4\boxed{1\pi_g}^3$$

$$\mathrm{F_2^-}\text{(15 electrons)}: \quad 1\sigma_g^2 1\sigma_u^2 2\sigma_g^2 1\pi_u^4 1\pi_g^4\boxed{2\sigma_u}^1$$

Comment. Note that each diatomic cation is isoelectronic with the diatomic anion of the next element, and has the same configuration and LUMO.

E10C.6(a) Energy is conserved, so when the photon is absorbed, its energy is transferred to the electron. Part of that energy overcomes the binding energy (ionization energy) and the remainder is manifest as the now freed electron's kinetic energy.

$$E_{\text{photon}} = I + E_{\text{kinetic}}$$

$$\text{so } E_{\text{kinetic}} = E_{\text{photon}} - I = \frac{hc}{\lambda} - I = \frac{(6.626\times10^{-34}\ \text{J s})\times(2.998\times10^{8}\ \text{m s}^{-1})}{(100\times10^{-9}\ \text{m})\times(1.602\times10^{-19}\ \text{J eV}^{-1})} - 12.0\ \text{eV}$$
$$= 0.4\overline{0}\ \text{eV} = 6.\overline{4}\times10^{-20}\ \text{J}$$

The speed is obtained from the kinetic energy:

$$E_{\text{kinetic}} = \frac{mv^2}{2} \quad \text{so} \quad v = \sqrt{\frac{2E_{\text{kinetic}}}{m}} = \sqrt{\frac{2\times6.\overline{4}\times10^{-20}\ \text{J}}{9.11\times10^{-31}\ \text{kg}}} = \boxed{4\times10^{5}\ \text{m s}^{-1}}$$

E10C.7(a) The expression in eqn 10C.4 cannot be solved analytically for R, but solving it numerically or graphically is fairly easy. In a spreadsheet, plot S as a function of ZR/a_0, and look for the value of ZR/a_0 where $S = 0.20$. One can change the scale of the plot and the spacing of grid points to find the graphical solution to arbitrary precision.

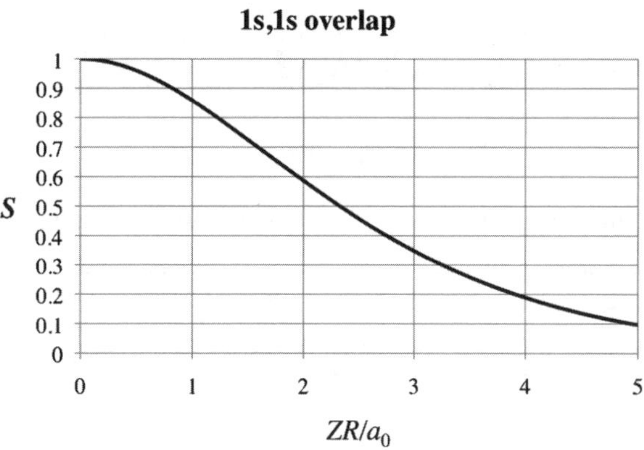

Figure 10C.2

By inspection of the plot (Fig. 10.C2), $S = 0.20$ at approximately $ZR/a_0 = 3.9$. Or one could inspect the spreadsheet in the neighbourhood of those points:

ZR/a_0	3.80	3.85	3.90	3.95	4.00
S	0.215	0.208	0.202	0.195	0.189

We can see that the value of S crosses $S = 0.20$ between the values of $ZR/a_0 = 3.90$ and 3.95. So to two significant figures, $ZR/a_0 = 3.9$, agreeing with the graphical estimate.

(a) For H_2, $Z = 1$, so $R = 3.9a_0/Z = 3.9\times5.29\times10^{-11}\ \text{m}/1 = \boxed{2.1\times10^{-10}\ \text{m} = 0.21\ \text{nm}}$

(b) For He_2, $Z = 2$, so $R = 3.9a_0/Z = 3.9\times5.29\times10^{-11}\ \text{m}/2 = \boxed{1.0\times10^{-10}\ \text{m} = 0.10\ \text{nm}}$

Solutions to problems

P10C.1 Think of the p orbital starting far from the s orbital and approaching it. At long distances, there is no overlap, and as the distance decreases, the positive overlap increases—but only up to a point. Eventually, some of the 'other side' of the p orbital (i.e. the region of negative amplitude) begins to overlap the s orbital. When the centres of the orbitals coincide, the region of positive overlap cancels the negative region. Thus, the overlap goes to zero at very large **and** very small separations, as shown in Fig. 10C.3(a).

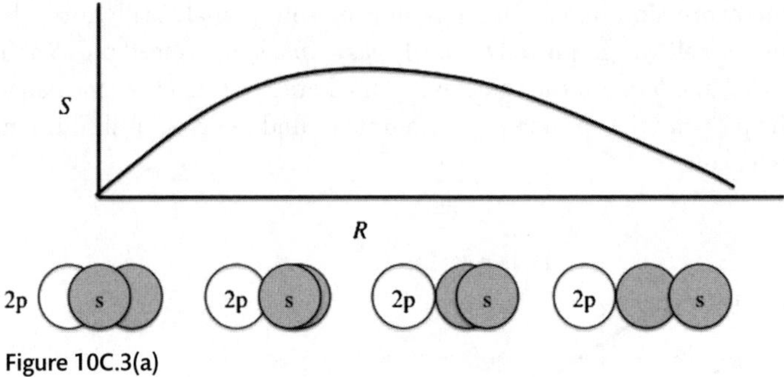

Figure 10C.3(a)

Now use the equation given in the problem to draw up the following table:

R/a_0	0	1	2	3	4	5	6	7	8	9	10
S	0	0.429	0.588	0.523	0.379	0.241	0.141	0.078	0.041	0.021	0.01

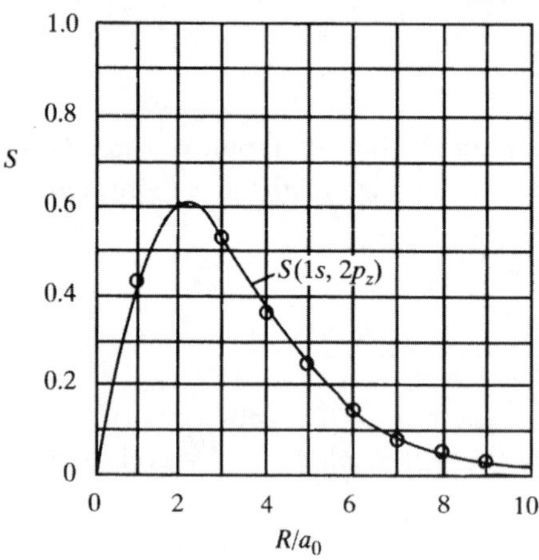

Figure 10C.3(b)

The points are plotted in Fig. 10C.3(b). One can find the maximum overlap graphically from the plot, or numerically from a spreadsheet. Or one can approach the matter analytically by setting $\mathrm{d}S/\mathrm{d}R$ equal to zero; however, the resulting expression is a cubic equation, which is itself most readily solved numerically. The maximum overlap occurs at $R = \boxed{2.1a_0}$.

P10C.3 (a) & (b) Let $\psi = c_A\psi_A + c_B\psi_B$

Normalization of ψ requires

$$\int|\psi|^2\mathrm{d}\tau = 1 = \int|c_A\psi_A + c_B\psi_B|^2\mathrm{d}\tau = \int(c_A\psi_A + c_B\psi_B)^*(c_A\psi_A + c_B\psi_B)\mathrm{d}\tau$$

$$1 = |c_A|^2\int|\psi_A|^2\mathrm{d}\tau + |c_B|^2\int|\psi_B|^2\mathrm{d}\tau + c_A{}^*c_B\int\psi_A{}^*\psi_B\mathrm{d}\tau c + c_B{}^*c_A\int\psi_B{}^*\psi_A\mathrm{d}\tau$$

In this expression, the first two integrals are 1 because the **atomic** orbitals are normalized. The last two integrals are zero because the atomic orbitals are orthogonal. (Remember, overlap is neglected by hypothesis.) Thus, normalization of ψ requires

$$1=|c_A|^2+|c_B|^2$$

If we choose the two coefficients to be cosine and sine of the same parameter θ (note: not the spherical polar coordinate θ), then the squares of the coefficients sum to one as required. Strictly speaking, the parameter θ can range from 0 to π if the wavefunction is defined as above (with a plus sign), because we must allow for the possibility that the coefficients c_A and c_B have different signs; alternatively, if we allow the wavefunction to be

$$\psi=c_A\psi_A \pm c_B\psi_B$$

then we can restrict θ to 0 to $\pi/2$. (Note: There is no loss of generality in forcing one of the coefficients to be positive, as long as we allow the coefficients to have different signs.)

(c) In homonuclear diatomics, the squares of the coefficients are equal by symmetry, so $\cos^2\theta=\sin^2\theta \quad \cos\theta=\pm\sin\theta \quad \tan\theta=\pm 1 \quad \theta=\tan^{-1}(\pm 1)$

or $\theta=\boxed{\pi/4 \text{ or } 3\pi/4}$

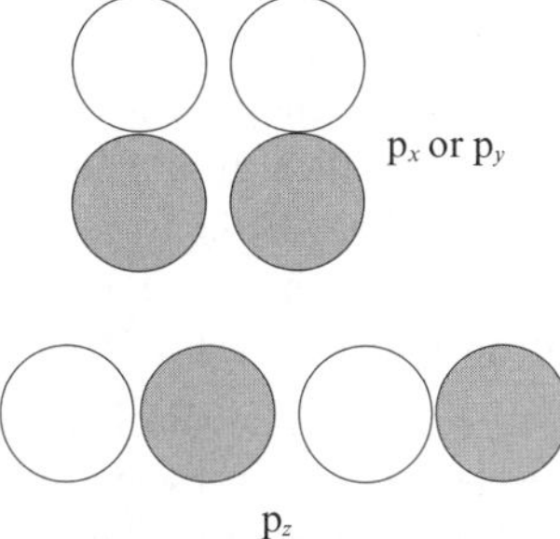

Figure 10C.4

Which combination is bonding and which is antibonding depends on what the atomic orbitals are, as illustrated in Fig. 10C.4. Constructive interference between the atomic orbitals is characteristic of bonding MOs, and that is sometimes achieved by adding the atomic orbital functions (as in s orbitals or p_x or p_y orbitals) and sometimes by subtracting one atomic orbital function from the other (as in p_z orbitals). Adding the atomic orbitals corresponds to equal coefficients: $\theta=\pi/4$; subtraction corresponds to equal and opposite coefficients: $\theta=3\pi/4$.

10D Heteronuclear diatomic molecules

Answers to discussion questions

D10D.1 Both the Pauling and Mulliken methods for measuring the attracting power of atoms for electrons seem to make good chemical sense. If we look at eqn. 10D.2 (the Pauling scale), we see that if $D(\text{A–B})$ were equal to $\frac{1}{2}[D(\text{A–A}) + D(\text{B–B})]$ the calculated electronegativity difference would be zero, as expected for completely non-polar bonds. Hence, any increased strength of the A–B bond over the average of the A–A and B–B bonds can reasonably be thought of as being due to the polarity of the A–B bond, which in turn is due to the difference in electronegativity of the atoms involved. Therefore, this difference in bond strengths can be used as a measure of electronegativity difference. To obtain numerical values for individual atoms, a reference state (atom) for electronegativity must be established. The value for fluorine is arbitrarily set at 4.0.

The Mulliken scale [10D.3] may be more intuitive than the Pauling scale because we are used to thinking of ionization energies and electron affinities as measures of the electron attracting powers of atoms. The choice of factor $\frac{1}{2}$, however, is arbitrary, though reasonable, and no more arbitrary than the specific form of eqn 10D.2 that defines the Pauling scale.

D10D.3 It can be proven that if an arbitrary wavefunction is used to calculate the energy of a system, the value calculated is never less than the true energy. This is the variation principle. This principle allows an enormous amount of latitude in constructing wavefunctions. We can continue modifying the wavefunctions in nearly any arbitrary manner (consistent with the minimal constraints of 'well-behaved' functions mentioned in Section 7B.2(b)) until we find a set that we feel provides an energy close to the true minimum in energy. Thus we can construct wavefunctions containing many parameters and then minimize the energy with respect to those parameters. These parameters may be interpretable in terms of chemically or physically significant quantities—but they need not be. Examples of the mathematical steps involved are illustrated in Section 10D.2.

D10D.5 Carbon is an essential building block of complex biological structures. It can form covalent bonds with many other elements, such as hydrogen, nitrogen, oxygen, sulfur, and, more importantly, other carbon atoms. As a consequence, such networks as long carbon–carbon chains (as in lipids) and chains of peptide links can form readily. Furthermore, carbon atoms can form chains and rings containing single, double, or triple C–C bonds. Such a variety of bonding options leads to the intricate molecular architectures of proteins, nucleic acids, and cell membranes. But the balance of bond strengths is critical to biology: bonds need to be sufficiently strong to maintain the structure of the cell and yet need to be susceptible to dissociation and rearrangement during biochemical reactions.

Intermediate values of ionization energy and electron affinity make carbon's bonding primarily covalent rather than ionic. A low ionization energy would favour cation formation (particularly in interactions with elements whose electron affinity is high) and a high electron affinity would favour anion formation (particularly in interactions with elements whose ionization energy is low). Intermediate value of these quantities also means that the energy of the valence atomic orbitals in carbon is not terribly different from the valence orbital energies of other non-metals; this in turn implies that the molecular orbitals carbon can form with them are significantly different in energy from the component atomic orbitals. That is, the atomic orbitals of carbon can interact with those of both moderately electronegative and electropositive elements to produce molecular orbitals with considerable bonding character—leading to strong covalent bonds. The fact that carbon has four valence electrons is intimately connected to its ability to form four bonds—a connection that has been rationalized by pre-quantum notions such as the octet rule as well as by quantum chemical theories as valence-bond and molecular orbital theory. The ability of carbon to make four bonds is in turn directly related to the incredible variety and complexity of organic chemistry and biochemistry. That carbon can make four bonds allows it to form long chains and rings (requiring two bonds for all but the terminal carbons for the C–C σ skeleton) and still have valences left over for a variety of functional groups. Finally, the strength of C–C bonds relative to C–N bonds and C–O bonds permits large organic molecules to persist in an atmosphere of oxygen and nitrogen.

Solutions to exercises

E10D.1(a) Refer to Fig. 10C.11 of the main text. Note that CO and CN^- are isoelectronic with N_2; note, however, that the σ and π orbitals no longer have u or g symmetry, so they are simply labelled consecutively.

(a) CO (10 electrons) $\boxed{1\sigma^2 2\sigma^2 1\pi^4 3\sigma^2}$

(b) NO (11 electrons) $\boxed{1\sigma^2 2\sigma^2 3\sigma^2 1\pi^4 2\pi^1}$

(c) CN^- (10 electrons) $\boxed{1\sigma^2 2\sigma^2 1\pi^4 3\sigma^2}$

E10D.2(a) We can use a version of Fig. 10C.12 of the text, but with the energy levels of F lower than those of Xe as in Fig. 10D.1.

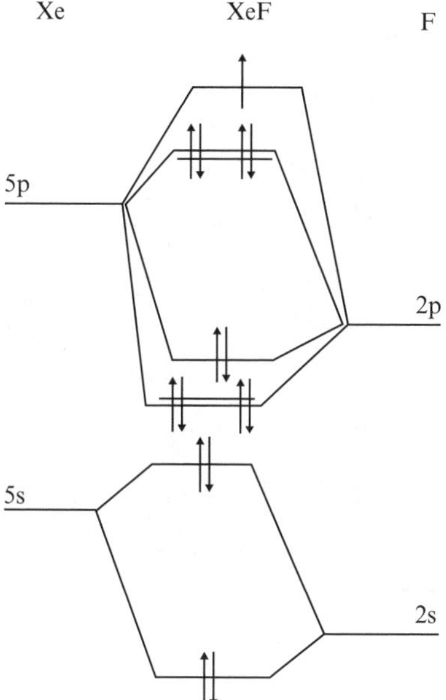

Figure 10D.1

For XeF we insert 15 valence electrons. The bond order is increased when XeF^+ is formed from XeF, because an electron is removed from an antibonding orbital. Therefore, we predict that XeF^+ has a shorter bond length than XeF.

E10D.3(a) Form the electron configurations and find the bond order. Note that NO^+ is isoelectronic with N_2, so for it we expect 1π to lie below 3σ, and vice versa for NO^-, which is isoelectronic with O_2. See Fig. 10C.11 of the main text.

$$NO^+ (10\text{ electrons}) \quad 1\sigma^2 2\sigma^2 1\pi^4 3\sigma^2 \qquad b=3$$

$$NO^- (12\text{ electrons}) \quad 1\sigma^2 2\sigma^2 3\sigma^2 1\pi^4 2\pi^2 \quad b=2$$

Based on the electron configurations, we would expect $\boxed{NO^+}$ to have the stronger and therefore the shorter bond.

E10D.4(a) Draw up the following table using data from Table 10D.1:

element	Li	Be	B	C	N	O	F	Ne
χ_M	1.28	1.99	1.83	2.67	3.08	3.22	4.43	4.60
$(\chi_M)^{1/2}$	1.13	1.41	1.35	1.63	1.75	1.79	2.10	2.14
χ_P (from table)	0.98	1.57	2.04	2.55	3.04	3.44	3.98	
χ_P (from formula)	0.16	0.53	0.46	0.84	1.00	1.05	1.47	1.53

A plot (Fig. 10D.2(a)) of the Pauling electronegativities (actual and from the formula) vs. the square root of the Mulliken electronegativities shows that the formula does a poor job. The formula consistently underestimates the Pauling electronegativity, and it underestimates the rise in electronegativity across the period.

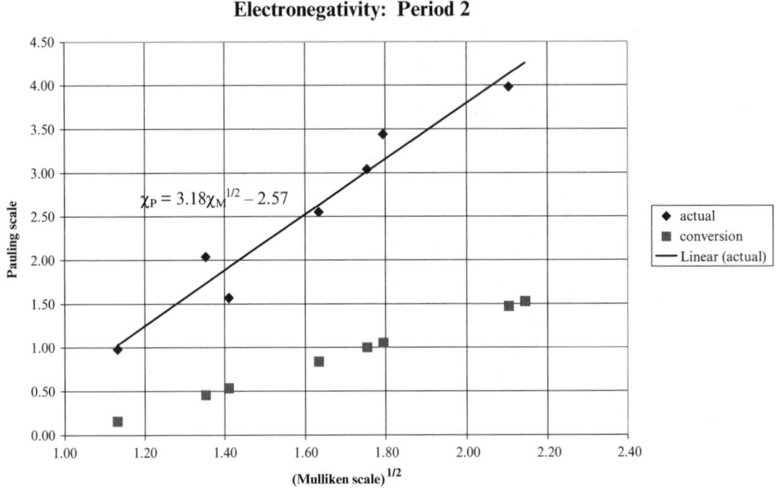

Figure 10D.2(a)

The problem is that the Mulliken electronegativities given in the table have already been scaled to the range of Pauling electronegativities ***using this very conversion***. That is, the Mulliken electronegativities given in Table 10D.1 are not defined by eqn 10D.3, but rather by eqn 10D.3 and then eqn 10D.4. Using Mulliken electronegativities based on eqn 10D.3 alone (the 'intrinsic Mulliken electronegativities' also known as 'Mulliken *a* parameters' from Steven G. Bratsch, 'Revised Mulliken electronegativities: I. Calculation and conversion to Pauling units,' *J. Chem. Educ.* **1988**, *65*, 34–41) leads to the following table.

element	Li	Be	B	C	N	O	F	Ne
χ_M (Bratsch *a*)	3.01	4.65	6.37	8.15	10.00	12.55	15.30	15.71
$(\chi_M)^{1/2}$	1.73	2.16	2.52	2.85	3.16	3.54	3.91	3.96
χ_P (from table)	0.98	1.57	2.04	2.55	3.04	3.44	3.98	
χ_P (from formula)	0.97	1.54	2.04	2.48	2.90	3.41	3.91	

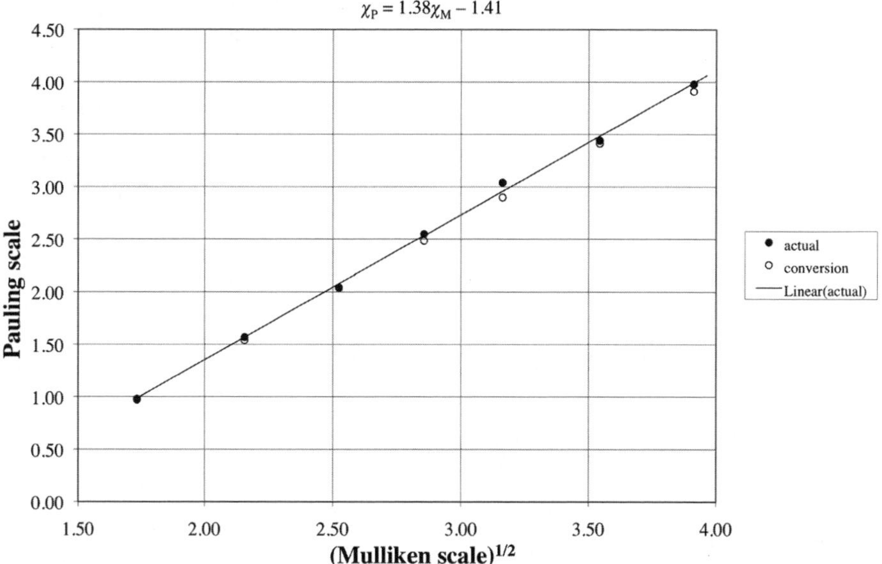

Figure 10D.2(b)

A plot (Fig. 10D.2(b)) of actual Pauling electronegativities (from Table 10D.1; filled circles) vs. the square root of the intrinsic Mulliken electronegativities (from Table 5 of Bratsch's article) yields an excellent regression line with an equation very close to eqn 10D.4. The open circles on the same plot are the result of applying eqn 10D.4 to the intrinsic Mulliken electronegativities from Bratsch; these are essentially the 'scaled' Mulliken electronegativities. As you can see, they fall very close to the best-fit line of the actual Pauling electronegativity values.

E10D.5(a) Following *Brief Illustration* 10D.1, we draw up a table of ionization energies and electron affinities from Tables 9B.2 and 9B.3. The mean of those two energies is the estimated orbital energy for each atom.

	I/eV	E_{ea}/eV	$\frac{1}{2}(I+E_{ea})$/eV
H	13.6	0.75	$\boxed{7.2}$
Cl	13.0	3.6	$\boxed{8.3}$

E10D.6(a) In the zero overlap approximation, the molecular orbital energies are given by eqn 10D.8c:

$$E_{\pm}=\tfrac{1}{2}(\alpha_A+\alpha_B)\pm\tfrac{1}{2}(\alpha_A-\alpha_B)\left\{1+\left(\frac{2\beta}{\alpha_A-\alpha_B}\right)^2\right\}^{1/2}$$

Taking $\beta=-1.0$ eV (a typical value), we have

$$E_{\pm}/\text{eV}=\frac{1}{2}(-7.2-8.3)\pm\frac{1}{2}(-7.2+8.3)\left\{1+\left(\frac{2(-1)}{-7.2+8.3}\right)^2\right\}^{1/2}=\boxed{-6.6\text{ or }-8.9}$$

E10D.7(a) If overlap cannot be neglected, then the molecular orbital energies are given by eqn 10D.8a:

$$E_{\pm}=\frac{\alpha_A+\alpha_B-2\beta S\pm\{(\alpha_A+\alpha_B-2\beta S)^2-4(1-S^2)(\alpha_A\alpha_B-\beta^2)\}^{1/2}}{2(1-S^2)}$$

Taking $\beta=-1.0$ eV (a typical value), we have

$$\frac{E_{\pm}}{\text{eV}}=\frac{-7.2-8.3+2(1.0)(0.20)\pm\{(-7.2-8.3+2(1.0)(0.20))^2-4(1-0.20^2)(7.2\times8.3-1.0^2)\}^{1/2}}{2(1-0.20^2)}$$

so $E_{\pm}=\boxed{-5.0\text{ or }-10.7\text{ eV}}$

Solutions to problems

P10D.1 If overlap is not neglected, the determinant is

$$\begin{vmatrix}\alpha_A-E & \beta-SE\\ \beta-SE & \alpha_B-E\end{vmatrix}$$

and expanding it yields eqn 10D.7:

$$0=(1-S^2)E^2+(2\beta S-\alpha_A-\alpha_B)E+\alpha_A\alpha_B-\beta^2$$

This is a quadratic equation in E where $a=1-S^2$, $b=2\beta S-\alpha_A-\alpha_B$, and $c=\alpha_A\alpha_B-\beta^2$. The solution is

$$E_{\pm}=\frac{-b\pm\sqrt{b^2-4ac}}{2a}$$

Let us evaluate b^2-4ac:

$$\begin{aligned}b^2-4ac&=4\beta^2S^2-4\alpha_A\beta S-4\alpha_B\beta S+\alpha_A^2+2\alpha_A\alpha_B+\alpha_B^2\\&\quad-4\alpha_A\alpha_B+4\beta^2+4\alpha_A\alpha_B S^2-4\beta^2S^2\\&=-4\alpha_A\beta S-4\alpha_B\beta S+\alpha_A^2+\alpha_B^2-2\alpha_A\alpha_B+4\beta^2+4\alpha_A\alpha_B S^2.\end{aligned}$$

The terms can be gathered into the following factors:

$$b^2-4ac=(\alpha_A-\alpha_B)^2+4(\beta+\alpha_A S)(\beta+\alpha_B S)$$

$$\text{So }E_{\pm}=\frac{\alpha_A+\alpha_B-2\beta S}{2(1-S^2)}\pm\frac{\{(\alpha_A-\alpha_B)^2+4(\beta+\alpha_A S)(\beta+\alpha_B S)\}^{1/2}}{2(1-S^2)}$$

$$=\boxed{\frac{\alpha_A+\alpha_B-2\beta S}{2(1-S^2)}\pm\frac{\alpha_A-\alpha_B}{2(1-S^2)}\left(1+\frac{4(\beta+\alpha_A S)(\beta+\alpha_B S)}{(\alpha_A-\alpha_B)^2}\right)^{1/2}}$$

Comment. The effect of overlap is to mix the Coulomb and resonance contributions, producing terms such as $\beta+S\alpha$ and $\alpha-S\beta$ in place of 'pure' α or β terms. Even if overlap is not negligible, one can consider the situation if the atomic orbitals are energetically far apart. In this case, the condition is that

$$(\alpha_A-\alpha_B)^2\gg4(\beta+\alpha_A S)(\beta+\alpha_B S)$$

If so, the square root term can be expanded

$$E_{\pm}=\frac{\alpha_A+\alpha_B-2\beta S}{2(1-S^2)}\pm\frac{\alpha_A-\alpha_B}{2(1-S^2)}\left(1+\frac{4(\beta+\alpha_A S)(\beta+\alpha_B S)}{2(\alpha_A-\alpha_B)^2}-\dots\right)$$

$$\text{so }E_{+}\approx\boxed{\frac{\alpha_A-\beta S}{1-S^2}+\frac{(\beta+\alpha_A S)(\beta+\alpha_B S)}{(\alpha_A-\alpha_B)(1-S^2)}}$$

$$\text{and }E_{-}\approx\boxed{\frac{\alpha_B-\beta S}{1-S^2}-\frac{(\beta+\alpha_A S)(\beta+\alpha_B S)}{(\alpha_A-\alpha_B)(1-S^2)}}$$

P10D.3 At first blush, we simply have three terms rather than two. But the fact that two of the atomic orbitals belong to the same nucleus alters matters. Overlap, for instance, is zero for atomic orbitals on the same nucleus, because the orbitals are orthogonal. Also the resonance integral for different atomic orbitals on the same nucleus (which we would denote by β_{BC}) vanishes, as it must for orbitals that have zero overlap.

Thus $(\alpha_A - E)c_A + (\beta_{AB} - ES_{AB})c_B + (\beta_{AC} - ES_{AC})c_C = 0$

$(\beta_{AB} - ES_{AB})c_A + (\alpha_B - E)c_B = 0$

and $(\beta_{AC} - ES_{AC})c_A + (\alpha_C - E)c_C = 0$

(i) First consider the case $S = 0$ (without, however, making the resonance integrals also vanish). The secular determinant—call it $f(E)$—is

$$f(E) = \begin{vmatrix} \alpha_A - E & \beta_{AB} & \beta_{AC} \\ \beta_{AB} & \alpha_B - E & 0 \\ \beta_{AC} & 0 & \alpha_C - E \end{vmatrix}$$

Solve for the energies by expanding the secular determinant and setting it equal to zero. Let us expand using the second row:

$$f(E) = \beta_{AB}\{\beta_{AB}(\alpha_C - E)\} - (\alpha_B - E)\{(\alpha_A - E)(\alpha_C - E) - \beta_{AC}^2\} = 0$$

$$0 = E^3 - (\alpha_A + \alpha_B + \alpha_C)E^2 + (\alpha_A\alpha_B + \alpha_A\alpha_C + \alpha_B\alpha_C - \beta_{AB}^2 - \beta_{AC})^2 E + \alpha_B\beta_{AC}^2 + \alpha_C\beta_{AB}^2 - \alpha_A\alpha_B\alpha_C$$

Substituting in the parameters, we have

$$0 = E^3 + (26.0\text{ eV})E^2 + (221\text{ eV}^2)E - 614\text{ eV}^3$$

This cubic equation can be solved numerically and/or graphically. Fig. 10D.3(a) displays a plot of $f(E)$ vs. E/eV. The roots are $\boxed{E/\text{eV} = -10.7, -8.7, \text{and } -6.6}$.

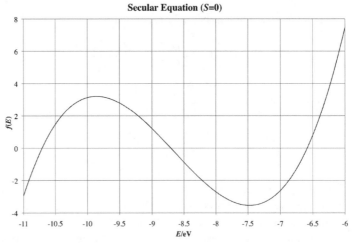

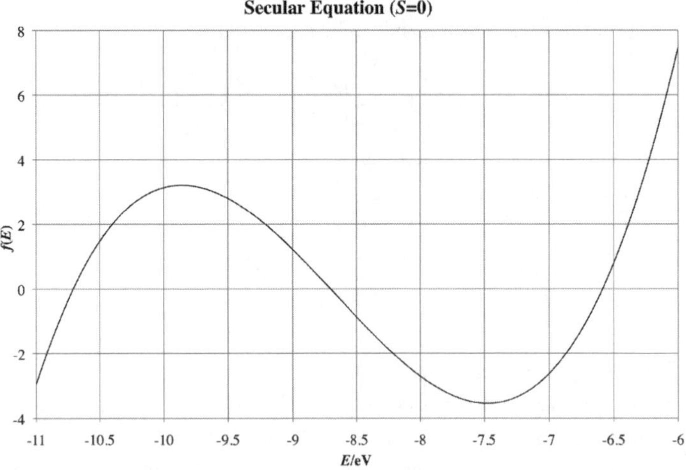

Figure 10D.3(a)

(ii) Now consider $S = 0.2$. The secular determinant—call it $g(E)$—is

$$g(E)=\begin{vmatrix} \alpha_A - E & \beta_{AB} - ES & \beta_{AC} - ES \\ \beta_{AB} - ES & \alpha_B - E & 0 \\ \beta_{AC} - ES & 0 & \alpha_C - E \end{vmatrix}$$

Again, we expand using the second row:

$$g(E)=(\beta_{AB}-ES)\{(\beta_{AB}-ES)(\alpha_C-E)\}-(\alpha_B-E)\{(\alpha_A-E)(\alpha_C-E)-(\beta_{AC}-ES)^2\}=0$$

$$\begin{aligned} 0 &= E^3(1-2S^2) \\ &+\{S^2(\alpha_B+\alpha_C)+2S(\beta_{AB}+\beta_{AC})-\alpha_A-\alpha_B-\alpha_C\}E^2 \\ &+\{\alpha_A\alpha_B+\alpha_A\alpha_C+\alpha_B\alpha_C-\beta_{AB}^2-\beta_{AC}^2-2S(\alpha_B\beta_{AC}+\alpha_C\beta_{AB})\}E \\ &+\alpha_B\beta_{AC}^2+\alpha_C\beta_{AB}^2-\alpha_A\alpha_B\alpha_C \end{aligned}$$

Substituting in the parameters, we have

$$0=0.92E^3+(24.5\,\text{eV})E^2+(214\,\text{eV}^2)E-614\,\text{eV}^3$$

This cubic equation can also be solved numerically and/or graphically. Before we do so, however, we note that the coefficients are not markedly different from those considered in part (i) above; therefore, we do not anticipate very different roots of this secular equation. Figure 10D.3(b) displays a plot of $g(E)$ vs. E/eV. The roots are $\boxed{E/\text{eV}=-10.8,-8.9,\text{and}-6.9}$.

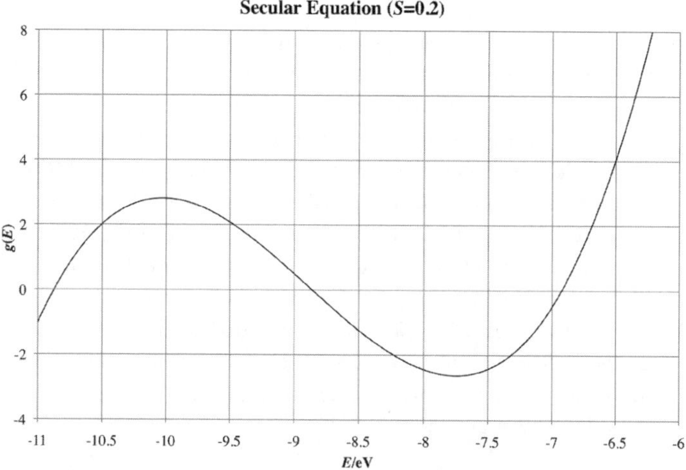

Figure 10D.3(b)

10E Polyatomic molecules

Answers to discussion questions

D10E.1 The Hückel method parametrizes, rather than calculates, the energy integrals that arise in molecular orbital theory. In the simplest version of the method, the overlap integral is also parametrized. The energy integrals, α and β, are always considered to be adjustable parameters; their numerical values emerge only at the end of the calculation by comparison to experimental energies. The simple form of the method has three other rather drastic approximations, listed in Section 10E.1 of the text, which eliminate many terms from the secular determinant and make it easier to solve. Ease of solution was important in the early days of quantum chemistry before the advent of computers. Without the use of these approximations, calculations on polyatomic molecules would have been difficult to accomplish.

The simple Hückel method is usually applied only to the calculation of π-electron energies in conjugated organic systems. It is based on the assumption of the separability of the σ- and π-electron systems in the molecule. This is a very crude approximation and works best when the energy level pattern is determined largely by the symmetry of the molecule. (See Chapter 11.)

D10E.3 See Section 9B.3, but modified for molecules (multiple nuclei). A Hartree–Fock (HF) wavefunction for a closed-shell species that has N_e electrons is a Slater determinant [9B.5] of $N_e/2$ distinct one-electron molecular orbitals (MOs), ψ_m, each doubly occupied. The HF equations [9B.11] for the MOs are a set of one-electron Schrödinger equations. Each HF hamiltonian (analogous to eqn 9B.10) contains a term for the kinetic energy of the electron, terms for the attraction of the electron to each of the species' nuclei, and terms for the repulsion of the electron from each of the other electrons (both direct Coulombic repulsion and electron-exchange terms that represent spin-correlation effects). The Coulomb and exchange portions of each HF hamiltonian needs the MOs occupied by the *other* electrons of the species: in other words, to solve for any MO, we need to know the other MOs. The HF method circumvents this difficulty by using initial guesses for the MOs as part of the hamiltonians, using the hamiltonians to find improved estimates of the MOs, and putting those MOs back into the hamiltonians for another round of solutions; this iterative procedure is followed until the MOs converge. Because iteration continues until self-consistency is achieved, the method is said to provide a **self-consistent field** (SCF). Finally, the form of the MOs is typically a linear combination of atomic orbitals (LCAO-MO, eqn 10E.1). Solving for the MOs, then, amounts to finding the optimal coefficients of the LCAO-MOs by means of the variation principle. The atomic orbitals in an LCAO-MO are called the basis set. Selection of the basis set (i.e. how many functions and of what form) can influence the amount of computational time needed to obtain an HF wavefunction as well as the quality of the result.

D10E.5 In *ab initio* methods an attempt is made to evaluate all integrals that appear in the secular determinant. Approximations are still employed, but these are mainly associated with the construction of the wavefunctions involved in the integrals. In semi-empirical methods, many of the integrals are expressed in terms of spectroscopic data or physical properties. Semi-empirical methods exist at several levels. At some levels, in order to simplify the calculations, many of the integrals are set equal to zero. Density functional theory (DFT) is different from the Hartree-Fock self-consistent field (HF-SCF) methods, the *ab initio* methods, in that DFT focuses on the electron density while HF-SCF methods focus on the wavefunction. However, they both attempt to evaluate integrals from first principles, so DFT methods are in that sense *ab initio* methods: both are iterative self-consistent methods in that the calculations are repeated until the energy and wavefunctions (HF-SCF) or energy and electron density (DFT) are unchanged to within some acceptable tolerance.

Solutions to exercises

E10E.1(a) In setting up the secular determinant we use the Hückel approximations outlined in Section 10E.1:

$$\text{(a)}\quad \begin{vmatrix} \alpha-E & \beta & 0 \\ \beta & \alpha-E & \beta \\ 0 & \beta & \alpha-E \end{vmatrix}$$

$$\text{(b)}\quad \begin{vmatrix} \alpha-E & \beta & \beta \\ \beta & \alpha-E & \beta \\ \beta & \beta & \alpha-E \end{vmatrix}$$

The atomic orbital basis is $1s_A$, $1s_B$, $1s_C$ in each case; in linear H_3 we ignore A, C overlap because A and C are not neighbouring atoms; in cyclic H_3 we include it because they are.

E10E.2(a) We use the molecular orbital energy level diagram in Fig. 10E.4 of the main text. As usual, we fill the orbitals starting with the lowest-energy orbital, obeying the Pauli principle and Hund's rule. We then write

(i) $C_6H_6^-$ (7 electrons): $\boxed{a_{2u}^2 e_{1g}^4 e_{2u}^1}$

$$E_\pi = 2(\alpha+2\beta)+4(\alpha+\beta)+(\alpha-\beta) = \boxed{7\alpha+7\beta}$$

(ii) $C_6H_6^+$ (5 electrons): $\boxed{a_{2u}^2 e_{1g}^3}$

$$E_\pi = 2(\alpha+2\beta)+3(\alpha+\beta) = \boxed{5\alpha+7\beta}$$

E10E.3(a) The π-bond formation energy is the difference between the π-electron binding energy and the Coulomb energies α [10E.12]:

$$E_{bf} = E_\pi - N_C\alpha$$

The delocalization energy is the difference between the energy of the delocalized π system and the energy of isolated π bonds:

$$E_{\text{delocal}} = E_{\pi} - N_{\text{C}}(\alpha+\beta)$$

(i) $E_{\pi} = 7\alpha + 7\beta$ [Exercise 10E.2(a)]

so $E_{\text{bf}} = 7\alpha + 7\beta - 7\alpha = \boxed{7\beta}$

and $E_{\text{delocal}} = 7\alpha + 7\beta - 7(\alpha+\beta) = \boxed{0}$

Comment. With an odd number of π electrons, we do not have a whole number of π bonds in the formula for delocalization energy. In effect, we compare the π-electron binding energy to the energy of 3.5 isolated π bonds—whatever that means. The result is that the benzene anion has none of the 'extra' stabilization we associate with aromaticity.

(ii) $E_{\pi} = 5\alpha + 7\beta$ [Exercise 10E.2(a)]

so $E_{\text{bf}} = 5\alpha + 7\beta - 5\alpha = \boxed{7\beta}$

and $E_{\text{delocal}} = 5\alpha + 7\beta - 5(\alpha+\beta) = \boxed{2\beta}$

E10E.4(a) The structures are numbered to match the row and column numbers shown in the determinants:

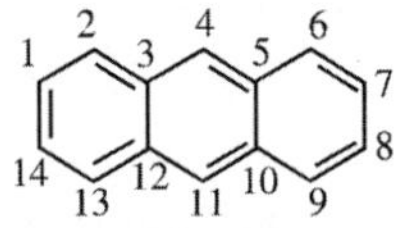

1 2 3 4 5 6 7 8 9 10 11 12 13 14
anthracene

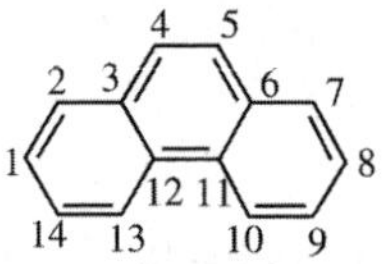

1 2 3 4 5 6 7 8 9 10 11 12 13 14
phenanthrene

(i) The secular determinant of anthracene in the Hückel approximation is:

	1	2	3	4	5	6	7	8	9	10	11	12	13	14
1	$\alpha-E$	β	0	0	0	0	0	0	0	0	0	0	0	β
2	β	$\alpha-E$	β	0	0	0	0	0	0	0	0	0	0	0
3	0	β	$\alpha-E$	β	0	0	0	0	0	0	0	β	0	0
4	0	0	β	$\alpha-E$	β	0	0	0	0	0	0	0	0	0
5	0	0	0	β	$\alpha-E$	β	0	0	0	β	0	0	0	0
6	0	0	0	0	β	$\alpha-E$	β	0	0	0	0	0	0	0
7	0	0	0	0	0	β	$\alpha-E$	β	0	0	0	0	0	0
8	0	0	0	0	0	0	β	$\alpha-E$	β	0	0	0	0	0
9	0	0	0	0	0	0	0	β	$\alpha-E$	β	0	0	0	0
10	0	0	0	0	β	0	0	0	β	$\alpha-E$	β	0	0	0
11	0	0	0	0	0	0	0	0	0	β	$\alpha-E$	β	0	0
12	0	0	β	0	0	0	0	0	0	0	β	$\alpha-E$	β	0
13	0	0	0	0	0	0	0	0	0	0	0	β	$\alpha-E$	β
14	β	0	0	0	0	0	0	0	0	0	0	0	β	$\alpha-E$

(ii) The secular determinant of phenanthrene in the Hückel approximation is:

	1	2	3	4	5	6	7	8	9	10	11	12	13	14
1	α–E	β	0	0	0	0	0	0	0	0	0	0	0	β
2	β	α–E	β	0	0	0	0	0	0	0	0	0	0	0
3	0	β	α–E	β	0	0	0	0	0	0	0	β	0	0
4	0	0	β	α–E	β	0	0	0	0	0	0	0	0	0
5	0	0	0	β	α–E	β	0	0	0	0	0	0	0	0
6	0	0	0	0	β	α–E	β	0	0	0	β	0	0	0
7	0	0	0	0	0	β	α–E	β	0	0	0	0	0	0
8	0	0	0	0	0	0	β	α–E	β	0	0	0	0	0
9	0	0	0	0	0	0	0	β	α–E	β	0	0	0	0
10	0	0	0	0	0	0	0	0	β	α–E	β	0	0	0
11	0	0	0	0	0	β	0	0	0	β	α–E	β	0	0
12	0	0	β	0	0	0	0	0	0	0	β	α–E	β	0
13	0	0	0	0	0	0	0	0	0	0	0	β	α–E	β
14	β	0	0	0	0	0	0	0	0	0	0	0	β	α–E

E10E.5(a) The secular determinants from Exercise 10E.4(a) can be diagonalized with the assistance of general-purpose mathematical software. Alternatively, programs specifically designed for Hückel calculations (such as the Simple Huckel Molecular Orbital Theory Calculator at the University of Calgary, **http://www.chem.ucalgary.ca/SHMO/** or Hückel software in *Explorations in Physical Chemistry*, 2nd Ed. by Julio de Paula, Valerie Walters, and Peter Atkins, **http://ebooks.bfwpub.com/explorations.php**) can be used. In both molecules, 14 π electrons fill seven orbitals.

(i) In anthracene, the energies of the filled orbitals are $\alpha + 2.414\beta$, $\alpha + 2\beta$, $\alpha + 1.414\beta$ (doubly degenerate), $\alpha+\beta$ (doubly degenerate), and $\alpha+0.414\beta$, so the total π-electron binding energy is $\boxed{14\alpha+19.314\beta}$.

(ii) For phenanthrene, the energies of the filled orbitals are $\alpha+2.435\beta$, $\alpha+1.951\beta$, $\alpha+1.516\beta$, $\alpha+1.3060\beta$, $\alpha+1.142\beta$, $\alpha+0.769\beta$, and $\alpha+0.605\beta$, so the total π-electron binding energy is $\boxed{14\alpha+19.448\beta}$.

E10E.6(a) HeH^+ is a two-electron molecule. The hamiltonian for the electrons in HeH^+ is

$$\hat{H}=-\frac{\hbar^2}{2m}\nabla_1^2-\frac{\hbar^2}{2m}\nabla_2^2-\frac{e^2}{4\pi\varepsilon_0 r_{1\mathrm{H}}}-\frac{e^2}{4\pi\varepsilon_0 r_{2\mathrm{H}}}-\frac{2e^2}{4\pi\varepsilon_0 r_{1\mathrm{He}}}-\frac{2e^2}{4\pi\varepsilon_0 r_{2\mathrm{He}}}+\frac{e^2}{4\pi\varepsilon_0 r_{12}}$$

The first two terms represent kinetic energy of the electrons, the next two attraction of the electrons to the H nucleus, the next two attraction of the electrons to the He nucleus, and the last electron–electron repulsion. There is an additional term that does not enter into the Schrödinger equation of the electrons but into the total energy as a function of internuclear separation, namely a nuclear–nuclear repulsion term $\dfrac{2e^2}{4\pi\varepsilon_0 R}$.

Solutions to problems

P10E.1 In the simple Hückel approximation

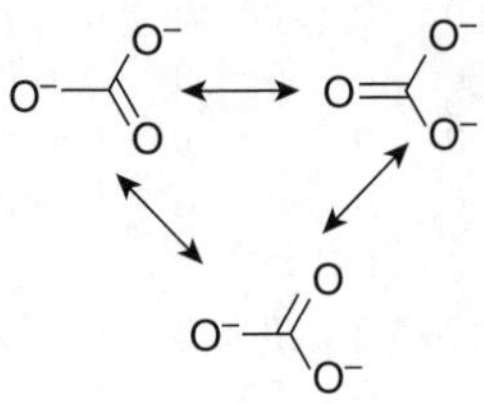

$$\begin{vmatrix} \alpha_O - E & 0 & 0 & \beta \\ 0 & \alpha_O - E & 0 & \beta \\ 0 & 0 & \alpha_O - E & \beta \\ \beta & \beta & \beta & \alpha_C - E \end{vmatrix} = 0$$

$$(E-\alpha_O)^2 \times \{(E-\alpha_O)\times(E-\alpha_C)-3\beta^2\}=0$$

Therefore, the factors yield

$$E-\alpha_O=0(\text{twice}) \text{ so } \boxed{E=\alpha_O},$$

and $(E-\alpha_O)\times(E-\alpha_C)-3\beta^2=0=E^2-E(\alpha_O+\alpha_C)+\alpha_O\alpha_C-3\beta^2$

$$\text{so } E_\pm = \frac{\alpha_O+\alpha_C \pm \sqrt{(\alpha_O+\alpha_C)^2-4(\alpha_O\alpha_C-3\beta^2)}}{2} = \frac{\alpha_O+\alpha_C \pm \sqrt{(\alpha_O-\alpha_C)^2+12\beta^2}}{2}$$

$$= \boxed{\frac{1}{2}\left(\alpha_O+\alpha_C \pm (\alpha_O-\alpha_C)\sqrt{1+\frac{12\beta^2}{(\alpha_O-\alpha_C)^2}}\right)}$$

Because Coulomb integrals approximate ionization energies in magnitude, α_O is more negative than α_C. Therefore, E_+ corresponds to the lowest-energy orbital and E_- to the highest-energy orbital; the degenerate orbitals at $E=\alpha_O$ fall in between.

The π energies in the absence of resonance are derived for just one of the three structures, i.e. for a structure containing a single localized π bond.

$$\begin{vmatrix} \alpha_O - E & \beta \\ \beta & \alpha_C - E \end{vmatrix} = 0$$

Expanding the determinant and solving for E gives

$$(E-\alpha_O)\times(E-\alpha_C)-\beta^2=0=E^2-E(\alpha_O+\alpha_C)+\alpha_O\alpha_C-\beta^2,$$

so $$E_{\text{local}\pm} = \frac{1}{2}\left(\alpha_\text{O} + \alpha_\text{C} \pm (\alpha_\text{O} - \alpha_\text{C})\sqrt{1 + \frac{4\beta^2}{(\alpha_\text{O} - \alpha_\text{C})^2}}\right)$$

There are two π electrons in the system, so the delocalization energy is

$$2E_+ - 2E_{\text{local}+} = \boxed{(\alpha_\text{O} - \alpha_\text{C})\left(\sqrt{1 + \frac{12\beta^2}{(\alpha_\text{O} - \alpha_\text{C})^2}} - \sqrt{1 + \frac{4\beta^2}{(\alpha_\text{O} - \alpha_\text{C})^2}}\right)}$$

If $12\beta^2 << (\alpha_\text{O} - \alpha_\text{C})^2$, we can use $(1 + x)^{1/2} \approx 1 + x/2$, so the delocalization energy is

$$\approx (\alpha_\text{O} - \alpha_\text{C})\left(1 + \frac{12\beta^2}{2(\alpha_\text{O} - \alpha_\text{C})^2} - 1 - \frac{4\beta^2}{2(\alpha_\text{O} - \alpha_\text{C})^2}\right) = \boxed{\frac{4\beta^2}{\alpha_\text{O} - \alpha_\text{C}}}$$

P10E.3 We use the Hückel approximation, neglecting overlap integrals.

The secular determinant of ethene is $\begin{vmatrix} \alpha - E & \beta \\ \beta & \alpha - E \end{vmatrix}$

Mathematical software (such as the Simple Huckel Molecular Orbital Theory Calculator at Canada's University of Calgary, http://www.chem.ucalgary.ca/SHMO/) diagonalizes the hamiltonian matrix to

$$\boldsymbol{E} = \begin{pmatrix} \alpha + \beta & 0 \\ 0 & \alpha - \beta \end{pmatrix}$$

and the matrix that achieves the diagonalization is

$$\boldsymbol{c} = \begin{pmatrix} 0.707 & 0.707 \\ 0.707 & -0.707 \end{pmatrix}$$

The secular determinant of butadiene is $\begin{vmatrix} \alpha - E & \beta & 0 & 0 \\ \beta & \alpha - E & \beta & 0 \\ 0 & \beta & \alpha - E & \beta \\ 0 & 0 & \beta & \alpha - E \end{vmatrix}$

The hamiltonian matrix is diagonalized to

$$\boldsymbol{E} = \begin{pmatrix} \alpha + 1.618\beta & 0 & 0 & 0 \\ 0 & \alpha + 0.618\beta & 0 & 0 \\ 0 & 0 & \alpha - 0.618\beta & 0 \\ 0 & 0 & 0 & \alpha - 1.618\beta \end{pmatrix}$$

and the matrix that achieves the diagonalization is

$$\boldsymbol{c}=\begin{pmatrix} 0.372 & 0.602 & 0.602 & 0.372 \\ 0.602 & 0.372 & -0.372 & -0.602 \\ 0.602 & -0.372 & -0.372 & 0.602 \\ 0.372 & -0.602 & 0.602 & -0.372 \end{pmatrix}$$

The secular determinant of hexatriene is $\begin{vmatrix} \alpha-E & \beta & 0 & 0 & 0 & 0 \\ \beta & \alpha-E & \beta & 0 & 0 & 0 \\ 0 & \beta & \alpha-E & \beta & 0 & 0 \\ 0 & 0 & \beta & \alpha-E & \beta & 0 \\ 0 & 0 & 0 & \beta & \alpha-E & \beta \\ 0 & 0 & 0 & 0 & \beta & \alpha-E \end{vmatrix}$

The hamiltonian matrix is diagonalized to

$$\boldsymbol{E}=\begin{pmatrix} \alpha+1.802\beta & 0 & 0 & 0 & 0 & 0 \\ 0 & \alpha+1.247\beta & 0 & 0 & 0 & 0 \\ 0 & 0 & \alpha+0.445\beta & 0 & 0 & 0 \\ 0 & 0 & 0 & \alpha-0.445\beta & 0 & 0 \\ 0 & 0 & 0 & 0 & \alpha-1.247\beta & 0 \\ 0 & 0 & 0 & 0 & 0 & \alpha-1.802\beta \end{pmatrix}$$

and the matrix that achieves the diagonalization is

$$\boldsymbol{c}=\begin{pmatrix} 0.232 & 0.418 & 0.521 & 0.521 & 0.418 & 0.232 \\ 0.418 & 0.521 & 0.232 & -0.232 & -0.521 & -0.418 \\ 0.521 & 0.232 & -0.418 & -0.418 & 0.232 & 0.521 \\ 0.521 & -0.232 & -0.418 & 0.418 & 0.232 & -0.521 \\ 0.418 & -0.521 & 0.232 & 0.232 & -0.521 & 0.418 \\ 0.232 & -0.418 & 0.521 & -0.521 & 0.418 & -0.232 \end{pmatrix}$$

The secular determinant of octatetraene is

$$\begin{vmatrix} \alpha-E & \beta & 0 & 0 & 0 & 0 & 0 & 0 \\ \beta & \alpha-E & \beta & 0 & 0 & 0 & 0 & 0 \\ 0 & \beta & \alpha-E & \beta & 0 & 0 & 0 & 0 \\ 0 & 0 & \beta & \alpha-E & \beta & 0 & 0 & 0 \\ 0 & 0 & 0 & \beta & \alpha-E & \beta & 0 & 0 \\ 0 & 0 & 0 & 0 & \beta & \alpha-E & \beta & 0 \\ 0 & 0 & 0 & 0 & 0 & \beta & \alpha-E & \beta \\ 0 & 0 & 0 & 0 & 0 & 0 & \beta & \alpha-E \end{vmatrix}$$

The hamiltonian matrix is diagonalized to

$$\boldsymbol{E}=\begin{pmatrix} \alpha+1.879\beta & 0 & 0 & 0 & 0 & 0 & 0 & 0 \\ 0 & \alpha+1.532\beta & 0 & 0 & 0 & 0 & 0 & 0 \\ 0 & 0 & \alpha+\beta & 0 & 0 & 0 & 0 & 0 \\ 0 & 0 & 0 & \alpha+0.347\beta & 0 & 0 & 0 & 0 \\ 0 & 0 & 0 & 0 & \alpha-0.347\beta & 0 & 0 & 0 \\ 0 & 0 & 0 & 0 & 0 & \alpha-\beta & 0 & 0 \\ 0 & 0 & 0 & 0 & 0 & 0 & \alpha-1.532\beta & 0 \\ 0 & 0 & 0 & 0 & 0 & 0 & 0 & \alpha-1.879\beta \end{pmatrix}$$

and the matrix that achieves the diagonalization is

$$\boldsymbol{c}=\begin{pmatrix} 0.161 & 0.303 & 0.408 & 0.464 & 0.464 & 0.408 & 0.303 & 0.161 \\ 0.303 & 0.464 & 0.408 & 0.161 & -0.161 & -0.408 & -0.464 & -0.303 \\ 0.408 & 0.408 & 0 & -0.408 & -0.408 & 0 & 0.408 & 0.408 \\ 0.464 & 0.161 & -0.408 & -0.303 & 0.303 & 0.408 & -0.161 & -0.464 \\ 0.464 & -0.161 & -0.408 & 0.303 & 0.303 & -0.408 & -0.161 & 0.464 \\ 0.408 & -0.408 & 0 & 0.408 & -0.408 & 0 & 0.408 & -0.408 \\ 0.303 & -0.464 & 0.408 & -0.161 & -0.161 & 0.408 & -0.464 & 0.303 \\ 0.161 & -0.303 & 0.408 & -0.464 & 0.464 & -0.408 & 0.303 & -0.161 \end{pmatrix}$$

The columns of the $\boldsymbol{c}$ matrices are coefficients of the atomic orbitals in each molecular orbital. The first column represents the lowest-energy molecular orbital. In each molecule, all of the coefficients in the first column are positive, meaning that the lowest-energy MO has no nodes. Note also that the coefficients rise from one end of the molecule to the middle and then decrease to the other end; thus the lowest-energy MO extends over all of the carbon atoms of the chain. Note that the number of sign changes within a given column increases as one moves from the lowest-energy MO (leftmost column) to the highest-energy (rightmost column). Each time the sign of a coefficient changes, there is a node in the wavefunction. So, for example, the second column in octatetraene has four positive and four negative coefficients, but only one sign change and one node. The eighth column also has four positive and four negative coefficients, but the sign changes between every atom; there are seven nodes. The number of nodes increases with the energy of the MO.

P10E.5 In butadiene, the HOMO is bonding with respect to the two bonds on the ends (the ones conventionally depicted as double bonds) and antibonding with respect to the central bond (the one conventionally depicted as a single bond). The situation is reversed in the LUMO. Thus, a HOMO to LUMO transition weakens the terminal C=C bonds and strengthens the central C–C bond. (See Fig. 10E.1(a) based on Fig. 10E.2 of the main text.)

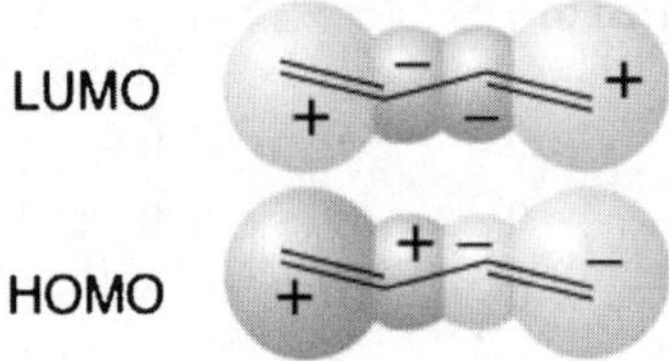

Figure 10E.1(a)

In benzene, the situation is more complicated because there are two degenerate HOMOs and two degenerate LUMOs. (See Fig. 10E.1(b) based on Fig. 10E.4 of the main text.) We will analyse these orbitals in terms of the top and bottom halves as shown in the figure. The HOMO on the left (of Fig. 10E.1(b)) is bonding ***between*** the two halves, but antibonding ***within*** each half; the HOMO on the right is completely bonding ***within*** each half but antibonding ***between*** the halves. The LUMO on the left is also bonding ***between*** the halves, but completely antibonding ***within*** each half (more antibonding than the HOMO on the left); the LUMO on the right is antibonding ***between*** the halves, and antibonding ***within*** each half as well. A transition from the right-hand HOMO to the right-hand LUMO or left-hand HOMO to left-hand LUMO would not affect the bonding ***between*** the halves, and it would destabilize the bonding ***within*** each half. A transition from the left-hand HOMO to the right-hand LUMO would weaken the bonding ***between*** the halves, leaving the bonding ***within*** each half unchanged. A transition from the right-hand HOMO to the left-hand LUMO would strengthen the bonding ***between*** the halves and destabilize the bonding between each atom ***within*** each half.

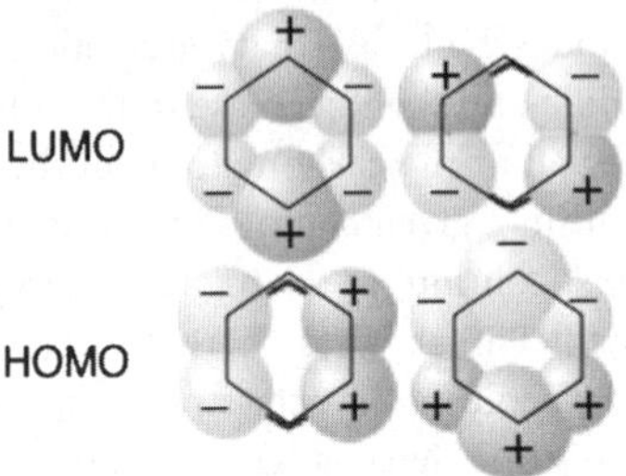

Figure 10E.1(b)

P10E.7 The standard reduction potential is related to the tendency of an atom, ion, or molecule to accept an electron: the more positive the reduction potential, the greater that tendency. For most chemical species, the most energetically favourable orbital to accept an additional electron is the LUMO. (We say most because odd electron species or species with degenerate HOMOs can accept an electron at an even lower energy level than the completely unoccupied LUMO. Still, the principle applies if we extend the idea to the lowest energy level that is less than completely filled.) The lower the energy level of the accepted electron in its new species, the lower the barrier to the transfer, the greater the tendency for the transfer, and the greater the reduction potential. Thus the **standard potential increases as the LUMO decreases**.

P10E.9 The secular determinant for a cyclic species HN^m has the form

$$\begin{array}{c} \begin{array}{cccccccc} 1 & 2 & 3 & \dots & \dots & \dots & N-1 & N \end{array} \\ \begin{vmatrix} x & 1 & 0 & \dots & \dots & \dots & 0 & 1 \\ 1 & x & 1 & \dots & \dots & \dots & 0 & 0 \\ 0 & 1 & x & 1 & \dots & \dots & 0 & 0 \\ 0 & 0 & 1 & x & 1 & \dots & 0 & 0 \\ \vdots & \vdots & \vdots & \vdots & \vdots & \vdots & \vdots & \vdots \\ \vdots & \vdots & \vdots & \vdots & \vdots & \vdots & \vdots & 1 \\ 1 & 0 & 0 & 0 & 0 & \dots & 1 & x \end{vmatrix} \end{array}$$

where $x = \dfrac{\alpha - E}{\beta}$ or $E = \alpha - \beta x$

Expanding the determinant, finding the roots of the polynomial, and solving for the total binding energy yields the following table. Note that $\alpha < 0$ and $\beta < 0$.

Species	Number of e^-	Permitted x (roots)	Total binding energy
H_4	4	$-2, 0, 0, 2$	$4\alpha + 4\beta$
H_5^+	4	$-2, \frac{1}{2}(1-\sqrt{5}), \frac{1}{2}(1-\sqrt{5}), \frac{1}{2}(1+\sqrt{5}), \frac{1}{2}(1+\sqrt{5})$	$4\alpha + (3+\sqrt{5})\beta$
H_5	5	$-2, \frac{1}{2}(1-\sqrt{5}), \frac{1}{2}(1-\sqrt{5}), \frac{1}{2}(1+\sqrt{5}), \frac{1}{2}(1+\sqrt{5})$	$5\alpha + \frac{1}{2}(5+3\sqrt{5})\beta$
H_5^-	6	$-2, \frac{1}{2}(1-\sqrt{5}), \frac{1}{2}(1-\sqrt{5}), \frac{1}{2}(1+\sqrt{5}), \frac{1}{2}(1+\sqrt{5})$	$6\alpha + (2+2\sqrt{5})\beta$
H_6	6	$-2, -1, -1, 1, 1, 2$	$6\alpha + 8\beta$
H_7^+	6	$-2, -1.248, -1.248, 0.445, 0.445, 1.802, 1.802$	$6\alpha + 8.992\beta$

Thus, for example, the energy levels in H_5 are

$$\alpha + 2\beta,\ \alpha + \frac{\beta}{2}\left(\sqrt{5}-1\right) \approx \alpha + 0.618\beta \text{ (twice)}, \alpha - \frac{\beta}{2}\left(1+\sqrt{5}\right) \approx \alpha - 1.618\beta \text{ (twice)}$$

and the total binding energy for H_5 is

$$2(\alpha+2\beta) + 3\left\{\alpha + \frac{\beta}{2}\left(\sqrt{5}-1\right)\right\} = 5\alpha + \frac{\beta}{2}\left(3\sqrt{5}+5\right) \approx 5\alpha + 5.854\beta$$

Now we estimate $\Delta_r U$ based on the difference in binding energy between products and reactants in the following reactions.

$$H_4 \rightarrow 2H_2 \qquad \Delta_r U = 4(\alpha+\beta)-(4\alpha+4\beta)=0$$

$$H_5^+ \rightarrow H_2 + H_3^+ \qquad \Delta_r U = 2(\alpha+\beta)+(2\alpha+2\beta)-(4\alpha+5.236\beta) = 0.764\beta < 0$$

The above $\Delta_r U$ values indicate that H_4 and H_5^+ can fall apart without an energy penalty.

$$H_5^- \rightarrow H_2 + H_3^- \qquad \Delta_r U = 2(\alpha+\beta)+(4\alpha+2\beta)-(6\alpha+6.472\beta) = -2.472\beta > 0$$

$$H_6 \rightarrow 3H_2 \qquad \Delta_r U = 6(\alpha+\beta)-(6\alpha+8\beta) = -2\beta > 0$$

$$H_7^+ \rightarrow 2H_2 + H_3^+ \qquad \Delta_r U = 4(\alpha+\beta)+(2\alpha+4\beta)-(6\alpha+8.992\beta) = -0.992\beta > 0$$

The $\Delta_r U$ values for H_5^-, H_6, and H_7^+ suggest that they are stable.

	Satisfies Hückel 4*n* + 2 low energy rule?	
Species	**4*n* + 2 electrons?**	**Stable?**
H_4, 4e$^-$	No	No
H_5^+, 4e$^-$	No	No
H_5^-, 6e$^-$	Yes	Yes
H_6, 6e$^-$	Yes	Yes
H_7^+, 6e$^-$	Yes	Yes

Hückel's 4n + 2 rule successfully predicts the stability of hydrogen rings.

P10E.11 (a) Ethanol

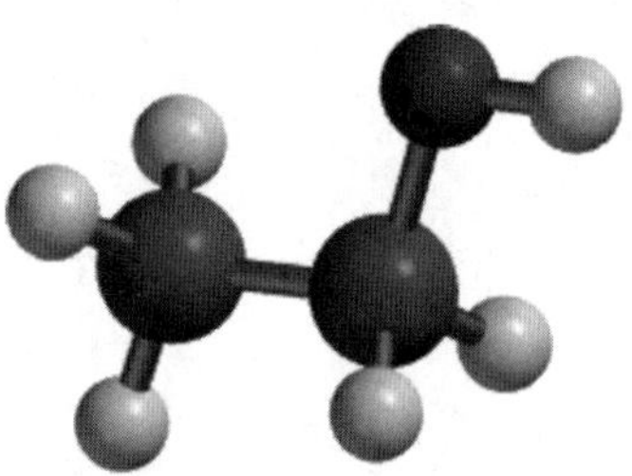

Ethanol	AM1*	PM3*	exp
C_1–C_2/pm	151.2	151.8	153.0
C_1–O/pm	142.0	141.0	142.5
C_1–H/pm	112.4	110.8	110
C_2–H/pm	111.6	109.7	109
O–H/pm	96.4	94.7	97.1
C_2–C_1–O/°	107.34	107.81	107.8
C_1–O–H/°	106.65	106.74	
H–C_1–H/°	108.63	106.98	
H–C_2–H/°	108.24	107.24	
$\Delta_f H^\ominus(g)$ / kJ mol^{-1}	−262.18	−237.87	−235.10
Dipole/D	1.55	1.45	1.69

*Spartan '10™

Figure 10E.2(a)

Both methods give fair agreement with experiment but PM3 gives a better estimate of formation enthalpy. The HOMO and LUMO are shown in Fig. 10E.2(a).

(b 1,4-dichlorobenzene, D_{2h}

1,4-dichlorobenzene	AM1*	PM3*	exp
C_1–C_2/pm	139.9	139.4	138.8
C_2–C_3/pm	139.3	138.9	138.8
C_1–Cl/pm	169.9	168.5	173.9
C_2–H/pm	110.1	109.6	
Cl–C_1–C_2/°	119.71	119.42	
C_1–C_2–H/°	120.40	120.06	
Cl–C_1–C_2–H/°	0.00	0.00	
$\Delta_f H^\ominus(g)$ / kJ mol^{-1}	33.36	42.31	24.6
dipole/D	0.00	0.00	0.0

*Spartan '10™

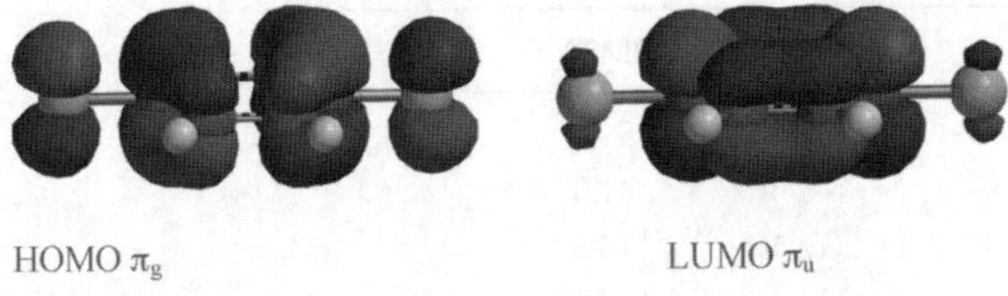

HOMO π_g LUMO π_u

Figure 10E.2(b)

Both methods give poor agreement with the experimental formation enthalpy but both methods are in broad agreement with experimental structural values. The HOMO and LUMO are shown in Fig. 10E.2 (b).

P10E.13 (a) The table displays PM3 calculations of the LUMO, HOMO, and the predicted $\pi^* \leftarrow \pi(\Delta E)$ energies for the all trans alternate double bond polyene series: ethene, butadiene, hexatriene, octatetraene, and decapentaene. Figure 10E.3(a) plots the transition energies against the number of alternate double bonds for HF/6-31G*, PM3, and DF/B3LYP/6-31G* calculations with Spartan '10™. It is seen that the DF/B3LYP/6-31G* calculations are in very good agreement with experimentation while the HF/6-31G* are in very poor agreement.

PM3 calculations* for trans polyene series

Species	E_{LUMO} / eV*	E_{HOMO} / eV*	ΔE / eV	ΔE / cm^{-1}	$\tilde{\nu}_{obs}$ / cm^{-1}
C_2H_4	1.2282	−10.6411	11.8693	95730	61500
C_4H_6	0.2634	−9.4671	9.7305	78480	46080
C_6H_8	−0.2494	−8.8993	8.6499	69770	39750
C_8H_{10}	−0.5568	−8.5767	8.0199	64680	32900
$C_{10}H_{12}$	−0.7556	−8.3755	7.6199	61460	–

*Spartan '10™

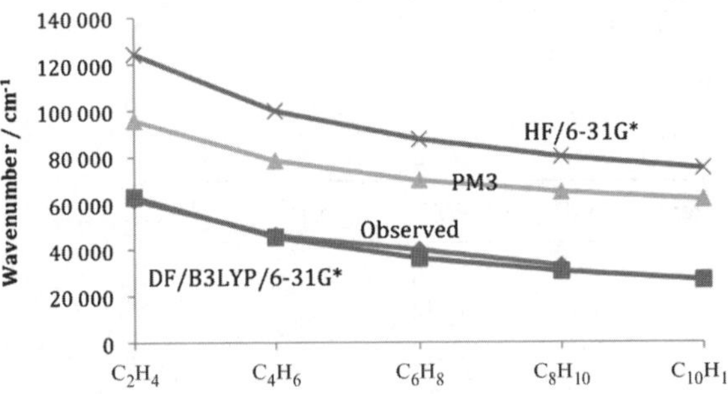

Figure 10E.3(a)

(b) Figure 10E.3(a) shows that the PM3 transition calculations can be 'calibrated' to agree with spectroscopic measurements by the subtraction of about 32 000 cm^{-1}. This however has no practical application; having the experimental absorptions for the first four members of the homologous series, it is far better to extrapolate the experimental data to the fifth member rather than to extrapolate a 'calibrated' set of computations. In fact, extrapolations are extremely risky and only small extrapolations are generally considered. Figure 10E.3(a) is a demonstration that DF/B3LYP/6-31G* calculations are in far better agreement with spectroscopic measurement than are PM3 and HF computations. So we expect that the direct DF calculation of the decapentaene transition to be very reliable. Also a study of transitions for molecules that cannot be easily examined experimentally might better be undertaken with the DF method. A plot of calculated minus observed wavenumbers against observed, shown in Fig. 10E.3(b) leads to identical conclusions. The PM3 calculations are about 32 000 cm^{-1} higher than both the DF calculations and the observed values and the linear regression fits of the Fig. 10E.3(b) plots show low correlation coefficients. Accepting the linearity, we can use the regression fits to extrapolate the decapentaene transition. The PM3 transition calculation for decapentaene is 61 460 cm^{-1} while the DF/B3LYP/6-31G* calculation is $\boxed{26\,780\ \mathrm{cm^{-1}}}$. Using PM3 decapentaene calculation and solving the regression fit for the extrapolated observation gives:

$$\text{predicted transition of PM3 regression fit} = \frac{61\,460 - 27\,046}{1.1123} \times \mathrm{cm^{-1}} = 30\,940\ \mathrm{cm^{-1}}$$

Doing this for the DF calculation gives:

$$\text{predicted transition of DF regression fit} = \frac{26\,780 + 7\,911}{1.1452} \times \mathrm{cm^{-1}} = 30\,290\ \mathrm{cm^{-1}}$$

These are reasonable predictions for the decapentaene transition but the direct DF/B3LYP/6-31G* calculation of 26 780 cm^{-1} is expected to be better.

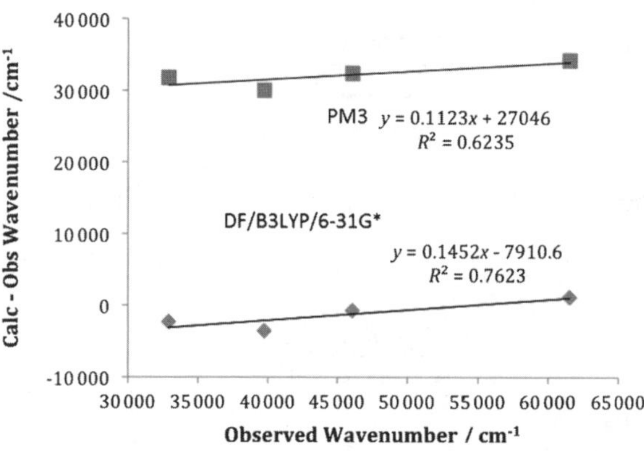

Figure 10E.3(b)

(c) Quantum calculations often involve approximations (AO approximations, Gaussian estimates of the AOs, integral neglect, etc.), which provide information about property trends for a homologous series of substances and reduce both computational time and cost. It is apparent that each computational method used in this problem adequately reflects the transition trends (see

Fig. 10E.3(b)). However, there are instances for which we wish to have some highly accurate knowledge of a property and studies, illustrated by this problem, help us identify the desired computational method. Such studies also encourage us toward the development of yet more reliable quantum calculations.

Integrated activities

I10.1 Ethene: The MO diagram is to show the formation of a σ bond and a π bond between two sp^2 hybridized CH_2 fragments. The fragments are shown, along with their associated p_z orbitals, at the far left and far right of the Fig. I10.1(a) MO diagram with a total of four electrons. Knowing that a σ bond and a π bond form between the carbon atoms, we place the π MO lower in energy than the σ* MO. Would the ethene molecule exist if the order of the energies of the π and σ* orbitals were reversed?

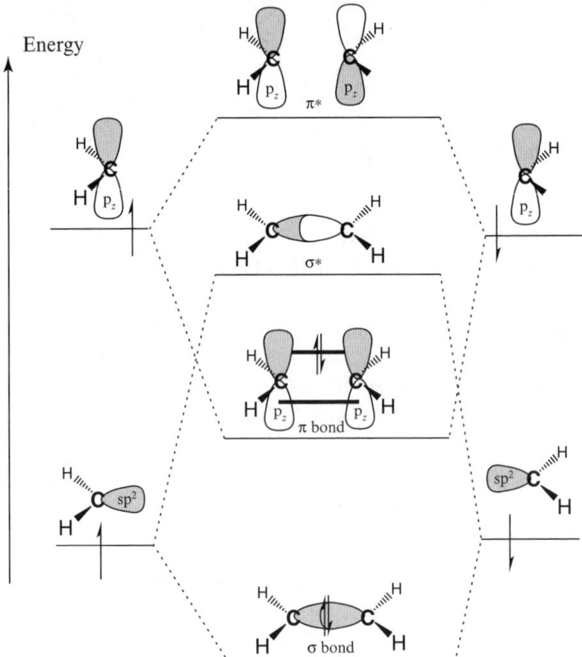

Figure I10.1(a)

Ethyne: The MO diagram is to show the formation of a σ bond and two π bonds between two sp hybridized CH fragments. The fragments are shown, along with their associated p_x and p_y orbitals, at the far left and far right of the Fig. I10.1(b) MO diagram with a total of six electrons. Knowing that a σ bond and two π bonds form between the carbon atoms, we place the π MO lower in energy than the σ*

MO. Would the ethyne molecule exist if the order of the energies of the π and σ* orbitals were reversed?

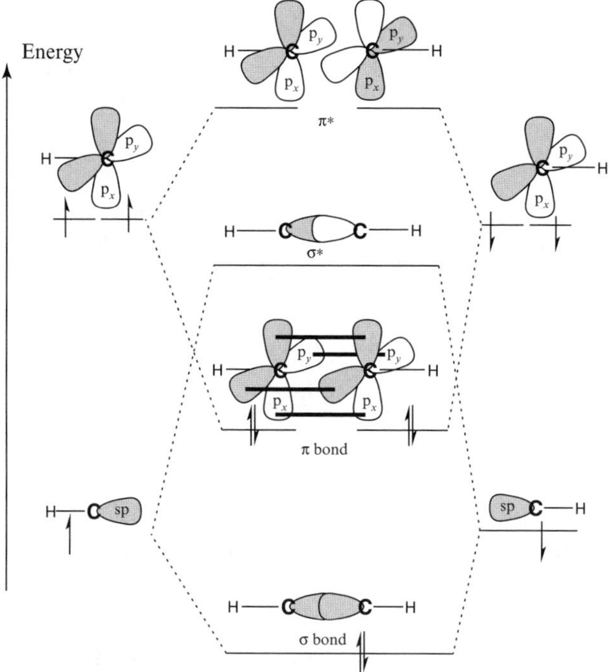

Figure I10.1(b)

I10.3 (a) The table below defines six 1,4-benzoquinone molecules and displays reduction potentials ($E^{\ominus}$) and computed LUMO energies calculated with both the PM3 and DF/B3LYP/6-31G* methods.

Species	R_2	R_3	R_5	R_6	$E^{\ominus}$/V	E_{LUMO}/eV*	E_{LUMO}/eV§
1	H	H	H	H	0.078	−1.7063	−3.5377
2	CH_3	H	H	H	0.023	−1.6512	−3.3897
3	CH_3	H	CH_3	H	−0.067	−1.5826	−3.2417
4	CH_3	CH_3	CH_3	H	−0.165	−1.3581	−3.1114
5	CH_3	CH_3	CH_3	CH_3	−0.260	−1.2149	−3.0063
6	CH_3	CH_3	CH_3O	CH_3O		−1.43414	−2.9922

* Spartan '10™ PM3
§ Spartan '10™ DF/B3LYP/6-31G*

The calculated E_{LUMO} values for species 1-5 are plotted against $E^{\ominus}$ measurements in Fig. I10.2. E_{LUMO} values calculated with both the PM3 and DF/B3LYP/6-31G* methods are seen to be linear in $E^{\ominus}$ and the regression fits are displayed within the figure. Absolute values differ between the two computational methods but it is clear that the two methods produce identical series trends: as $E^{\ominus}$ increases, E_{LUMO} values decrease at the approximate rate of 1.5 eV per V.

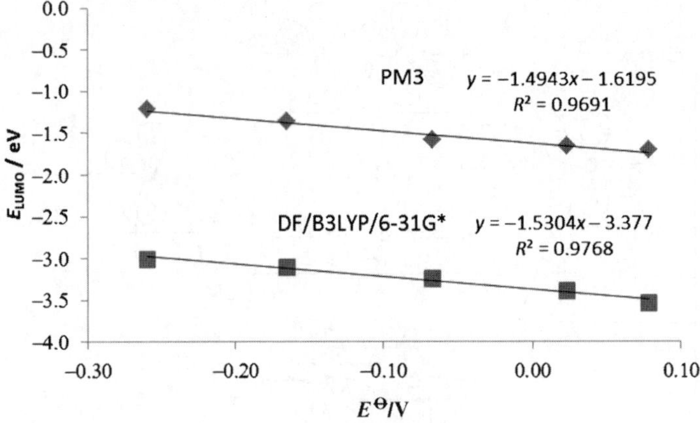

Figure I10.2

(b) We estimate the value of $E^{\ominus}$ for species 6, our model for ubiquinone, using the linear regression fit of the DF/B3LYP/6-31G* computations.

$$E^{\ominus}(\text{species 6}) = -\{(E_{\text{LUMO}}/\text{eV}+3.377)/1.5304\}\text{V}$$
$$= -\{(-2.9922+3.377)/1.5304\}\text{V} = \boxed{-0.251\text{ V}}$$

(c) Species 4 is the working model for plastoquinone. Using the linear regression fit of the DF/B3LYP/6-31G* computations so that we can make a comparison with the calculation of part (b), its one-electron reducing potential is:

$$E^{\ominus}(\text{species 4}) = -\{(E_{\text{LUMO}}/\text{eV}+3.377)/1.5304\}\text{V}$$
$$= -\{(-3.1114+3.377)/1.5304\}\text{V} = \boxed{-0.174\text{ V}}$$

The species 6 analogue of ubiquinone has a more negative reduction potential than the species 4 analogue of plastoquinone so we conclude that ubiquinone is a better oxidizing agent than plastoquinone or, conversely, plastoquinone is a better reducing agent than ubiquinone.

(d) Coenzyme Q (i.e. ubiquinone) acts as an oxidizing agent when oxidizing NADH and $FADH_2$ in respiration, the overall oxidation of glucose by oxygen. Plastoquinone, on the other hand, acts as a reducing agent by reducing oxidized plastocyanin in photosynthesis. Respiration involves the oxidation of glucose by oxygen, while photosynthesis involves a reduction to glucose and oxygen. It stands to reason that the better oxidizing agent, ubiquinone, is employed in oxidizing glucose in respiration, while the better reducing agent is used to meet the reduction demands in photosynthesis. (Note, however, that both species are recycled to their original forms: reduced ubiquinone is oxidized by iron (III) and oxidized plastoquinone is reduced by water.)

I10.5 (a) The transitions occur for photons whose energies are equal to the difference in energy between the highest occupied and lowest unoccupied orbital energies:

$$E_{\text{photon}} = E_{\text{LUMO}} - E_{\text{HOMO}}$$

If N is the number of carbon atoms in these species, then the number of π electrons is also N. These N electrons occupy the first $N/2$ orbitals, so orbital number $N/2$ is the HOMO and orbital number $1^+ N/2$ is the LUMO. Writing the photon energy in terms of the wavenumber, substituting the given energy expressions with this identification of the HOMO and LUMO gives

$$hc\tilde{\nu} = \left(\alpha + 2\beta\cos\frac{(\frac{1}{2}N+1)\pi}{N+1}\right) - \left(\alpha + 2\beta\cos\frac{\frac{1}{2}N\pi}{N+1}\right)$$
$$= 2\beta\left(\cos\frac{(\frac{1}{2}N+1)\pi}{N+1} - \cos\frac{\frac{1}{2}N\pi}{N+1}\right).$$

Solving for β yields

$$\beta = \frac{hc\tilde{\nu}}{2\left(\cos\frac{(\frac{1}{2}N+1)\pi}{N+1} - \cos\frac{\frac{1}{2}N\pi}{N+1}\right)}$$

Draw up the following table

Species	N	$\tilde{\nu}$ / cm^{-1}	estimated β/eV
C_2H_4	2	61 500	-3.813
C_4H_6	4	46 080	-4.623
C_6H_8	6	39 750	-5.538
C_8H_{10}	8	32 900	-5.873

(b) The total energy of the π electron system is the sum of the energies of occupied orbitals weighted by the number of electrons that occupy them. In C_8H_{10}, each of the first four orbitals is doubly occupied, so

$$E_\pi = 2\sum_{k=1}^{4} E_k = 2\sum_{k=1}^{4}\left(\alpha + 2\beta\cos\frac{k\pi}{9}\right) = 8\alpha + 4\beta\sum_{k=1}^{4}\cos\frac{k\pi}{9} = 8\alpha + 9.518\beta$$

The delocalization energy is the difference between this quantity and that of four isolated double bonds:

$$E_{\text{deloc}} = E_\pi - 8(\alpha+\beta) = 8\alpha + 9.518\beta - 8(\alpha+\beta) = \boxed{1.518\beta}$$

Using the estimate of β from part (a) yields $E_{\text{deloc}} = \boxed{8.913\text{ eV}}$

(c) Draw up the following table, in which the orbital energy decreases as we go down. For the purpose of comparison, we express orbital energies as $(E_k - \alpha)/\beta$. Recall that β is negative (as is α for that matter), so the orbital with the greatest value of $(E_k - \alpha)/\beta$ has the lowest energy.

	energy	Coefficients					
orbital	**$(E_k-\alpha)/\beta$**	**1**	**2**	**3**	**4**	**5**	**6**
6	−1.8019	0.2319	−0.4179	0.5211	−0.5211	0.4179	−0.2319
5	−1.2470	0.4179	−0.5211	0.2319	0.2319	−0.5211	0.4179
4	−0.4450	0.5211	−0.2319	−0.4179	0.4179	0.2319	−0.5211
3	0.4450	0.5211	0.2319	−0.4179	−0.4179	0.2319	0.5211
2	1.2470	0.4179	0.5211	0.2319	−0.2319	−0.5211	−0.4179
1	1.8019	0.2319	0.4179	0.5211	0.5211	0.4179	0.2319

The orbitals are shown schematically in Fig. I10.3, with each vertical pair of lobes representing a p orbital on one of the carbons in hexatriene. Shaded lobes represent one sign of the wavefunction (say positive) and unshaded lobes the other sign. Where adjacent atoms have atomic orbitals of the same sign, the resulting molecular orbital is bonding with respect to those atoms; where adjacent atoms have a different sign, there is a node between the atoms and the resulting molecular orbital is antibonding with respect to them. The lowest energy orbital is totally bonding (no nodes between atoms) and the highest energy orbital is totally antibonding (nodes between each adjacent pair). Note that the orbitals have increasing antibonding character as their energy increases. The size of each atomic p orbital is proportional to the magnitude of the coefficient of that orbital in the molecular orbital. So, for example, in orbitals 1 and 6, the largest lobes are in the centre of the molecule, so electrons that occupy those orbitals are more likely to be found near the centre of the molecule than on the ends. In the ground state of the molecule, there are two electrons in each of orbitals 1, 2, and 3, with the result that the probability of finding a π electron in hexatriene is uniform over the entire molecule.

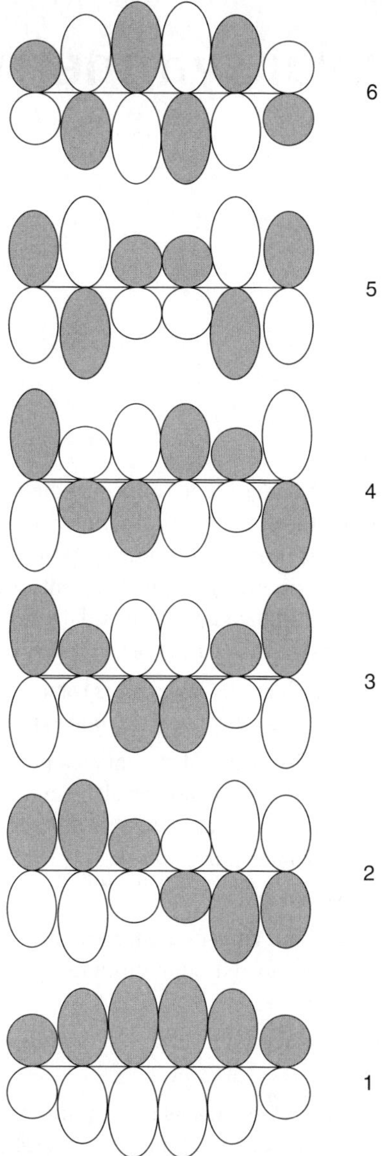

Figure I10.3

11 Molecular symmetry

11A Symmetry elements

Answer to discussion questions

D11A.1 The point group to which a molecule belongs is determined by the **symmetry operations** (actions in space) that move the point atoms of a molecule but result in a visually unchanged picture of the molecule in space. These operations may include rotations reflections and inversions. Specific symmetry elements are associated with each symmetry operations and the student is well advised to practice identifying the symmetry elements possessed by a molecule. The **symmetry elements** are:

(a) The n-fold axis of symmetry C_n in which the molecule is rotated 360°/n about a symmetry axis. The C_n of greatest n is called the **principal axis.**

(b) Reflection in a mirror plane σ. Mirror planes that contain a principal axis are denoted σ_v 'vertical' mirror planes. A vertical mirror plane that bisects two C_2 axes is denoted a σ_d 'dihedral' plane. A mirror plane that is perpendicular to a principal axis is denoted a σ_h 'horizontal' mirror plane.

(c) Inversion i through a centre of symmetry.

(d) The n-fold improper rotation S_n about an n-fold improper rotation axis. This is a rotation through 360°/n followed by reflection through a plane that is perpendicular to the axis of rotation.

(e) The identity E in which there is no action in space. Every molecule possesses the E symmetry element.

To assign a molecule to a point group, first identify all the symmetry elements possessed by the molecule. For example, inspection of the NH_3 molecular shape (ignore the lone pair, consider only the positions of the point atoms) reveals that it possesses the symmetry elements E, C_3, and three vertical mirror planes ($3\sigma_v$). This is the collection of elements for the C_{3v} point group so NH_3 belongs to the group C_{3v}.

So, identification of a molecule's point group requires the examination of a molecular model, which may be a plastic model, a drawing, or a mental picture, for its symmetry elements. The point group assignment corresponds to the identified collection of symmetry elements. The flow diagram in Fig. 11A.7 of the text often simplifies the assignment process.

D11A.3 The permanent dipole moment is a fixed property of a molecule and as a result it must remain unchanged through any symmetry operation of the molecule. Recall that the dipole moment is a vector quantity; therefore both its magnitude and direction must be unaffected by the operation. That can only be the case if the dipole moment is coincident with *all* of the symmetry elements of the molecule. Hence molecules belonging to point groups containing symmetry elements that do not satisfy this criterion can be eliminated. Molecules with a centre of symmetry cannot possess a dipole moment because any vector is changed through inversion. Molecules with more than one C_n axis cannot be polar since a vector cannot be coincident with more than one axis simultaneously. If the molecule has a plane of symmetry, the dipole moment must lie in the plane; if it has more than one plane of symmetry, the dipole moment must lie in the axis of intersection of these planes. A molecule can also be polar if it has one plane of symmetry and no C_n. Examination of the character tables at the end of the data section shows that the only point groups that satisfy these restrictions are C_s, C_n, and C_{nv}.

Solutions to exercises

E11A.1(a) Chloromethane belongs to the point group C_{3v}. The elements, other than the $\boxed{\text{identity } E}$, are a $\boxed{C_3 \text{ axis}}$ and $\boxed{\text{three vertical mirror planes } \sigma_v}$. The symmetry axis passes through the C and Cl nuclei. The mirror planes are defined by the three C–CH planes. The C_3 principal axis and one of the σ_v mirror planes are shown in Fig. 11A.1.

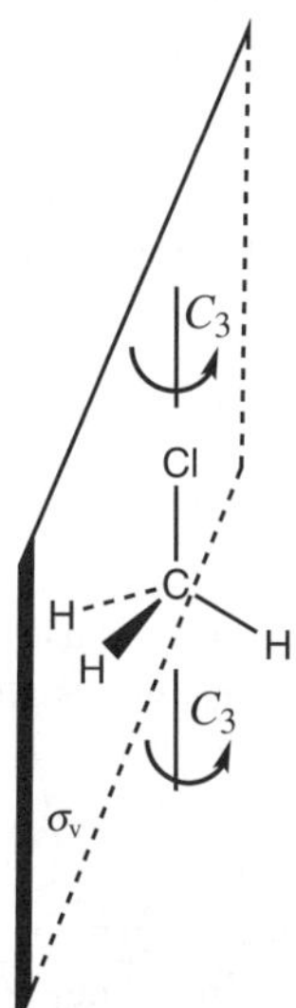

Figure 11A.1

E11A.2(a) Naphthalene belongs to the point group $\boxed{D_{2h}}$ and it has the symmetry elements shown overleaf in Fig. 11A.2. There are $\boxed{3\,C_2 \text{ axes}}$, $\boxed{\text{a centre of inversion}}$, and $\boxed{3\,\sigma_h \text{ mirror planes}}$.

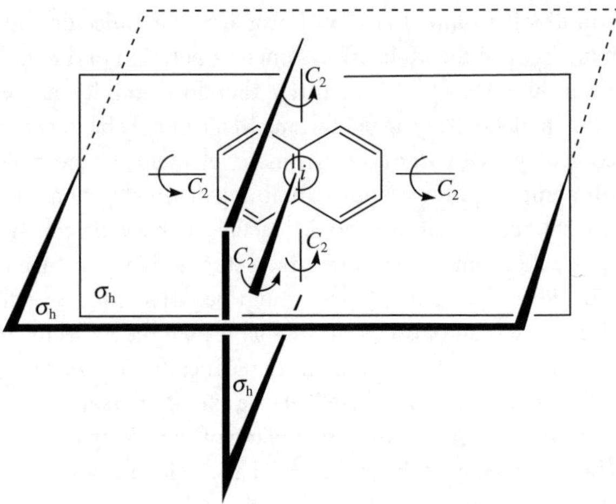

Figure 11A.2

E11A.3(a) Sketch a figure of the object, identify symmetry elements, and use the flow diagram in Fig. 11A.7 of the text when it simplifies the group assignment.

(i) Sphere: an infinite number of symmetry axes; therefore R_3

(ii) Isosceles triangles: E, C_2, σ_v, and σ'_v; therefore C_{2v}

(iii) Equilateral triangle: $\underbrace{\underbrace{E,\ C_3,\ C_2}_{D_3},\ \sigma_h}_{\boxed{D_{3h}}}$

(iv) Cylinder: E, C_∞, C_2, σ_h; therefore $\boxed{D_{\infty h}}$

E11A.4(a) Make a sketch of the molecule, identify symmetry elements, and use the flow diagram in Fig. 11A.7 of the text when it simplifies the point group assignment.

(i) NO_2: E, C_2, σ_v, σ'_v; $\boxed{C_{2v}}$

(ii) N_2O (linear N–N–O molecule): E, C_∞, C_2, σ_v; $\boxed{C_{\infty v}}$

(iii) $CHCl_3$: E, C_3, $3\sigma_v$; $\boxed{C_{3v}}$

(iv) $CH_2=CH_2$: E, C_2, $2C'_2$, i, σ_h; $\boxed{C_{2h}}$

E11A.5(a) Make a sketch of the molecule, identify symmetry elements, and use the flow diagram in Fig. 11A.7 of the text when it simplifies the point group assignment.

(a) *cis*-CHCl=CHCl: E, C_2, σ_v, σ'_v; $\boxed{C_{2v}}$

(b) *trans*-CHCl=CHCl; E, C_2, σ_h, i; $\boxed{C_{2h}}$

E11A.6(a) Only molecules belonging to the groups C_n, C_{nv}, and C_s may be polar [Section 11A.3(a)]; hence of the molecules listed only (i) $\boxed{\text{pyridine}}$ (C_{2v}) and (ii) $\boxed{\text{nitroethane}}$ (C_s) are polar.

E11A.7(a) The parent of the dichloronaphthalene isomers is shown to the right. Care must be taken when determining possible isomers because naphthalene is a flat molecule that belongs to the point group D_{2h}, as discussed in Exercise 11A.2(a). It has an inversion centre, mirror plane, and rotational axes that cause superficially distinct visual images to actually be the same molecule viewed from different angles. For example, Fig. 11A.3 structures are all 1,3-dichloronaphthalene. By drawing figures that avoid the redundancy caused by the symmetry elements you will find a total of ten dichloronaphthalene isomers.

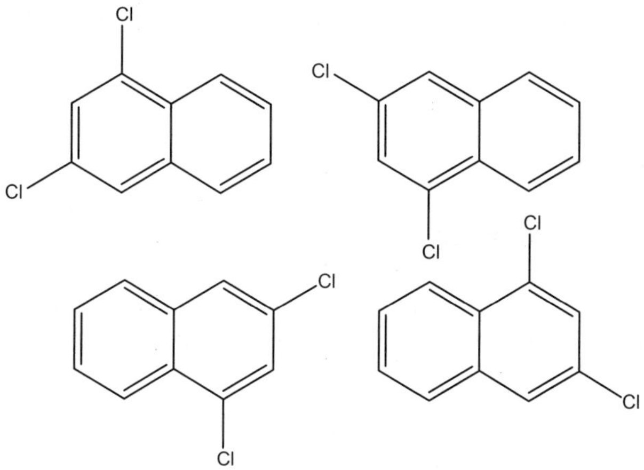

Figure 11A.3

The names and point groups of the ten isomers are summarized in the following table.

Isomers and point groups of *m,n*-dichloronaphthalene

m,n	1,2	1,3	1,4	1,5	1,6	1,7	1,8	2,3	2,6	2,7
Point group	C_s	C_s	C_{2v}	C_{2h}	C_s	C_s	C_{2v}	C_{2v}	C_{2h}	C_{2v}

E11A.8(a) In each molecule we must look for an improper rotation axis, perhaps in the disguised form $S_1 = \sigma$ or $S_2 = i$ (Section 11A.3(b)). If present, the molecule cannot be chiral. D_{2h} contains $\boxed{i}$ and C_{3h} contains $\boxed{\sigma_h}$; therefore, molecules belonging to these point groups cannot be chiral and cannot be optically active.

Solutions to problems

P11A.1 (a) Staggered CH_3CH_3: $E,\ C_3,\ C_2,\ i,\ 2S_6,\ 3\sigma_d$; $\boxed{D_{3d}}$

(b) Chair C_6H_{12}: $E,\ C_3,\ C_2,\ i,\ 2S_6,\ 3\sigma_d$; $\boxed{D_{3d}}$

Boat C_6H_{12}: $E,\ C_2,\ \sigma_v,\ \sigma_v'$; $\boxed{C_{2v}}$

(c) B_2H_6: $E,\ C_2,\ 2C_2',\ i,\ \sigma_h,\ 2\sigma_v'$; $\boxed{D_{2h}}$

(d) $[Co(en)_3]^{3+}$: $E,\ 2C_3,\ 3C_2$; $\boxed{D_3}$

(e) Crown S_8: $E,\ C_4,\ C_2,\ 4C_2',4\sigma_d,\ 2S_8$; $\boxed{D_{4d}}$

Only boat C_6H_{12} may be polar, since all the others are *D* point groups. Only $[Co(en)_3]^{3+}$ belongs to a group without an improper rotation axis ($S_1 = \sigma$), and hence is chiral.

P11A.3 The most distinctive symmetry operation is the $\boxed{S_4}$ axis through the central atom and aromatic nitrogens on both ligands. That axis is also a $\boxed{C_2}$ axis. The group is $\boxed{S_4}$.

11B Group theory

Answers to discussion questions

D11B.1 Within the context of quantum theory and molecular symmetry a **group** is a collection of transformations (*R*, *S*, *T*, etc.) that satisfy these criteria:

1. One of the transformations is the identity (*E*).
2. For every transformation *R*, the inverse transformation R^{-1} is included in the collection so that the combination RR^{-1} is equivalent to the identity: $RR^{-1} = E$.
3. The product *RS* is equivalent to a single member of the collection of transformations.
4. Multiple transformations obey the associative rule: $R(ST) = (RS)T$.

D11B.3 Figure 11B.1 uses the C_{3v} point group to illustrate structural features of a character table. The central columns are labelled with the symmetry operations, or **symmetry classes**, of the group. Each operation is prefixed with the number of operations in the class, called the **degeneracy of the class**. For example, the degeneracy of the C_3 class is 2, which corresponds to clockwise and anticlockwise rotations of the principal axis, shown in Fig. 11B.2. There are 3 vertical reflection planes in the C_{3v} group; these are also shown in Fig. 11B.2. The **group order**, *h*, is the sum of class degeneracies. It is the total number of symmetry operations of the group. The first column of Fig. 11B.1 shows that the C_{3v} group contains three **symmetry species**. They are the **Mulliken symbols** used to label the irreducible representations of the group. The symmetry species A and B are one-dimensional representations, E is the symbol for a two-dimensional representation, and T is the symbol of a three-dimensional representation. It is useful to remember that E species levels are doubly degenerate while T species levels are triply degenerate.

Number of symmetry species = number of classes [11B.4]

Also, the sum of the squares of the dimension d_i of all symmetry species *i* equals the group order.

$$\sum_i d_i^{\,2} = h \quad [11B.5]$$

For the C_{3v} group the dimensions are 1, 1, and 2 for the A_1, A_2, and E irreducible representations, respectively. Adding the squares gives $1^2 + 1^2 + 2^2 = 6$ and we see that the sum of the squares equals h.

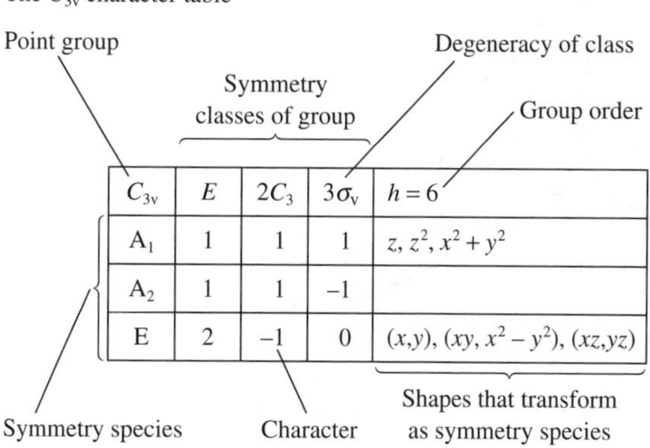

C_{3v}	E	$2C_3$	$3\sigma_v$	$h = 6$
A_1	1	1	1	z, z^2, x^2+y^2
A_2	1	1	−1	
E	2	−1	0	$(x,y), (xy, x^2-y^2), (xz,yz)$

Figure 11B.1

The **characters** below a symmetry class of a point group are the traces (the sum of diagonal elements) of the matrix representations of the class operation(s) for a symmetry species. When the character under the principal axis is +1 for a one-dimensional symmetry species, the Mulliken symbol of the species is A; if the character is −1, the Mulliken symbol of the species is B. All characters of the symmetry species A_1 are +1. Characters of +1 are said to be **symmetric** with respect to the specific operation of the character; characters of −1 are said to be **antisymmetric** with respect to the operation. The subscript 1 or 2 is added to the Mulliken symbol when a species is either symmetric or antisymmetric under a σ_v operation, respectively. The subscripts g and u (gerade and ungerade) indicate that a species is either symmetric or antisymmetric under the i operation. Single primes or double primes are added to the species symbol to indicate symmetry or antisymmetry under the σ_h operation.

The last column of the character table catalogues the symmetry species to which useful functions belong. Any function that is proportional to one of the summarized functions belongs to the same symmetry species. For example, a quick examination of the C_{3v} group reveals that the basic functions z, z^2, and $x^2 + y^2$ belong to the A_1 **irrep** (irreducible representation) so we immediately recognize that the p_z, d_{z^2}, and $d_{x^2+y^2}$ atomic orbitals all belong to the A_1 species. The d_{xy} and $d_{x^2-y^2}$ orbitals **jointly** belong to the E symmetry species of the C_{3v} group; they are said to **span** the representation and to be **basis functions** for the representation.

We take a moment to show that x and y jointly span the E irreducible representation of the C_{3v} group. The operations indicated in Fig. 11B.2 are useful to this effort.

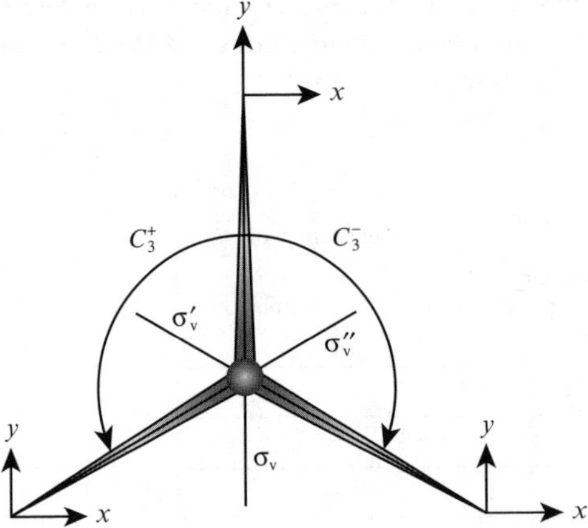

Figure 11B.2

By inspection we deduce that the matrix representation of σ_v is

$$D(\sigma_v)=\begin{pmatrix} -1 & 0 \\ 0 & 1 \end{pmatrix}$$

because this matrix transforms (x,y) to $(-x,y)$ in accord with the requirements of the σ_v reflection shown in Fig. 11B.2. The trace of this matrix is $-1+1=0$, which is the character of the σ_v operation of the E symmetry species. The character of the C_3^+ operation is more difficult to deduce. The general form of the two dimensional rotation of angle ϕ is

$$D(C_\phi)=\begin{pmatrix} \cos\phi & \sin\phi \\ -\sin\phi & \cos\phi \end{pmatrix}$$

so the matrix for a 120° ($2\pi/3$ radian) rotation of the principal axis, which is the central z-axis in Fig. 11B.2, is

$$D(C_3^+)=\begin{pmatrix} \cos(2\pi/3) & \sin(2\pi/3) \\ -\sin(2\pi/3) & \cos(2\pi/3) \end{pmatrix}=\begin{pmatrix} -\tfrac{1}{2} & \sqrt{3}/2 \\ -\sqrt{3}/2 & -\tfrac{1}{2} \end{pmatrix}$$

The trace of this matrix is $-\tfrac{1}{2}-\tfrac{1}{2}=-1$, which is the character of the C_3 operation of the E symmetry species. Thus, x and y jointly span E.

D11B.5 The letters and subscripts of a symmetry species provide information about the symmetry behaviour of the species. An A or a B is used to denote a one-dimensional representation; A is used if the character under the principal rotation is +1 (symmetric behaviour), and B is used if the character is −1 (antisymmetric behaviour). Subscripts are used to distinguish the irreducible representations if there is more than one of the same type: A_1 is reserved for the representation with the character 1 under all operations. E denotes a two-dimensional irreducible representation and T a three-dimensional irreducible representation. These labels are called **Mulliken**

symbols. For groups with an inversion centre, a subscript g (gerade) indicates symmetric behaviour under the inversion operation; a subscript u (ungerade) indicates antisymmetric behaviour. A horizontal mirror is assigned ′ or ″ superscripts if the behaviour is symmetric or antisymmetric, respectively, under the σ_h operation.

Solutions to exercises

E11B.1(a) Since the p_z orbitals are perpendicular to the molecular plane, we recognize that the set of p_z orbitals on each atom of BF_3 experience the σ_h change $(p_B, p_{F1}, p_{F2}, p_{F3})\boldsymbol{D}(\sigma_h) = (-p_B, -p_{F1}, -p_{F2}, -p_{F3})$. Consequently, we find by inspection that

$$\boldsymbol{D}(\sigma_h) = \begin{pmatrix} -1 & 0 & 0 & 0 \\ 0 & -1 & 0 & 0 \\ 0 & 0 & -1 & 0 \\ 0 & 0 & 0 & -1 \end{pmatrix}$$

E11B.2(a) The matrix representations of the operations σ_h and C_3 are deduced in Exercises 11B.1(a) and 11B.1(b). According to the precepts of group theory, the successive application of these operations yields another member of the D_{3h} group to which BF_3 belongs and, in fact, by definition the operation $\sigma_h C_3$ should yield the S_3 symmetry operation. The matrix representation of S_3 can be found by matrix multiplication of the component operations.

$$\boldsymbol{D}(\sigma_h)\boldsymbol{D}(C_3) = \begin{pmatrix} -1 & 0 & 0 & 0 \\ 0 & -1 & 0 & 0 \\ 0 & 0 & -1 & 0 \\ 0 & 0 & 0 & -1 \end{pmatrix} \begin{pmatrix} 1 & 0 & 0 & 0 \\ 0 & 0 & 1 & 0 \\ 0 & 0 & 0 & 1 \\ 0 & 1 & 0 & 0 \end{pmatrix} = \begin{pmatrix} -1 & 0 & 0 & 0 \\ 0 & 0 & -1 & 0 \\ 0 & 0 & 0 & -1 \\ 0 & -1 & 0 & 0 \end{pmatrix} = \boldsymbol{D}(S_3)$$

The result may be checked by matrix operation on the p_z orbital vector where the effort should yield $(p_B, p_{F1}, p_{F2}, p_{F3})\boldsymbol{D}(S_3) = (-p_B, -p_{F3}, -p_{F1}, -p_{F2})$ and, as expected:

$$(p_B, p_{F1}, p_{F2}, p_{F3}) \begin{pmatrix} -1 & 0 & 0 & 0 \\ 0 & 0 & -1 & 0 \\ 0 & 0 & 0 & -1 \\ 0 & -1 & 0 & 0 \end{pmatrix} = (-p_B, -p_{F3}, -p_{F1}, -p_{F2})$$

E11B.3(a) Consider the equilateral triangle $P_1P_2P_3$, which belongs to the D_{3h} point group (text Fig. 11A.8). The three C_2 axes and the three σ_v mirror planes of this triangle are shown overleaf in Fig. 11B.3.

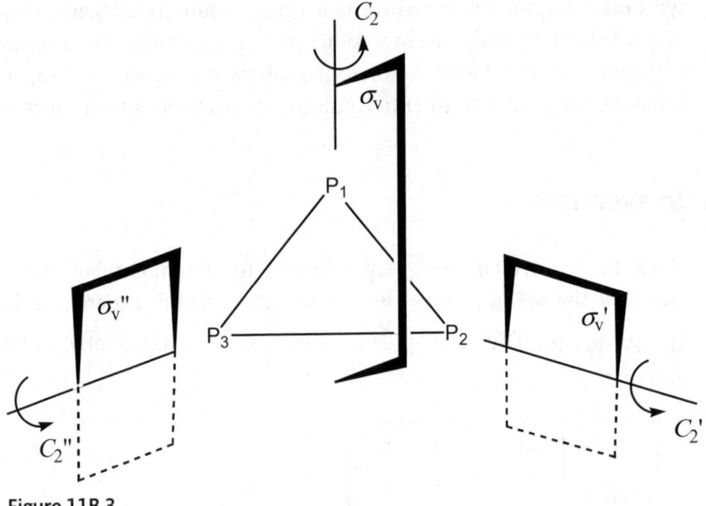

Figure 11B.3

The C_2 and C_2' axes belong to the same class if there is a member S of the group such that $C_2' = S^{-1}C_2S$ [11B.1] where S^{-1} is the inverse of S. We will work with $S = \sigma_v''$, an operator for which $S^{-1} = \sigma_v''$. By comparison of the action of $S^{-1}C_2S$ upon the vector (P_1,P_2,P_3) with the action of C_2' upon the same vector we can determine whether or not the equality of eqn 11B.1 holds. If it does, C_2 and C_2' axes belong to the same class.

$$\begin{aligned} S^{-1}C_2S(P_1,P_2,P_3) &= \sigma_v''C_2\sigma_v''(P_1,P_2,P_3) \\ &= \sigma_v''C_2(P_2,P_1,P_3) \\ &= \sigma_v''(P_2,P_3,P_1) \\ &= (P_3,P_2,P_1) \qquad \text{(i)} \\ C_2'(P_1,P_2,P_3) &= (P_3,P_2,P_1) \qquad \text{(ii)} \end{aligned}$$

Equations (i) and (ii) indicate that $C_2' = S^{-1}C_2S$ where $S = \sigma_v''$ and we conclude that C_2 and C_2' belong to the same class. By either using the same argument or seeing the necessities of symmetry we find that C_2 and C_2'' also belong to the same class. Consequently, C_2, C_2', and C_2'' all belong to the same class.

E11B.4(a) Because the largest character is 3 in the column headed E in the O_h character table, we know that the maximum orbital degeneracy is 3.

E11B.5(a) Benzene belongs to the D_{6h} point group. Because the largest character is 2 in the column headed E in the D_{6h} character table, we know that the maximum orbital degeneracy is 2.

Solutions to problems

P11B.1 Note that $R^2 = E$ for all the operations of the groups, that $ER = RE = R$ always, and that $RR' = R'R$ for this group. Since $C_2\sigma_h = i$, $\sigma_h i = C_2$, and $iC_2 = \sigma_h$ we can draw up the following group multiplication table

$R\downarrow R'\rightarrow$	E	C_2	σ_h	i
E	E	C_2	σ_h	i
C_2	C_2	E	i	σ_h
σ_h	σ_h	i	E	C_2
i	i	σ_h	C_2	E

The $\boxed{trans\text{-CHCl=CHCl}}$ molecule belongs to the group C_{2h}.

P11B.3 Refer to Fig. 11A.3 of the text. Place 1s orbitals h_1 and h_2 on the H atoms and s, p_x, p_y, and p_z on the O atom. The *z*-axis is the C_2 axis; *x* lies perpendicular to σ_v'; *y* lies perpendicular to σ_v. Then draw up the following table of the effect of the operations on the basis

	E	C_2	σ_v	σ_v'
h_1	h_1	h_2	h_2	h_1
h_2	h_2	h_1	h_1	h_2
s	s	s	s	s
p_x	p_x	$-p_x$	p_x	$-p_x$
p_y	p_y	$-p_y$	$-p_y$	p_y
pz	p_z	p_z	p_z	p_z

Each operation column is the changed basis that results from the application of the operation *R* upon the basis (h_1, h_2, s, p_x, p_y, p_z). We want the 6 × 6 matrix representative of each *R* such that the changes are expressed in the form: the columns headed by each operations *R* in the form (changed basis) = (original basis)*D*(*R*). We use the methods set out in Section 11B.2.

(i) Under *E* the change $(h_1, h_2, s, p_x, p_y, p_z) \leftarrow (h_1, h_2, s, p_x, p_y, p_z)$ is reproduced by:

$$\boldsymbol{D}(E)=\begin{bmatrix} 1 & 0 & 0 & 0 & 0 & 0 \\ 0 & 1 & 0 & 0 & 0 & 0 \\ 0 & 0 & 1 & 0 & 0 & 0 \\ 0 & 0 & 0 & 1 & 0 & 0 \\ 0 & 0 & 0 & 0 & 1 & 0 \\ 0 & 0 & 0 & 0 & 0 & 1 \end{bmatrix}$$

(ii) Under C_2 the change $(h_2, h_1, s, -p_x, -p_y, p_z) \leftarrow (h_1, h_2, s, p_x, p_y, p_z)$ is reproduced by

$$D(C_2)=\begin{bmatrix} 0 & 1 & 0 & 0 & 0 & 0 \\ 1 & 0 & 0 & 0 & 0 & 0 \\ 0 & 0 & 1 & 0 & 0 & 0 \\ 0 & 0 & 0 & -1 & 0 & 0 \\ 0 & 0 & 0 & 0 & -1 & 0 \\ 0 & 0 & 0 & 0 & 0 & 1 \end{bmatrix}$$

(iii) Under σ_v the change $(h_2, h_1, s, p_x, -p_y, p_z) \leftarrow (h_1, h_2, s, p_x, p_y, p_z)$ is reproduced by

$$D(\sigma_v)=\begin{bmatrix} 0 & 1 & 0 & 0 & 0 & 0 \\ 1 & 0 & 0 & 0 & 0 & 0 \\ 0 & 0 & 1 & 0 & 0 & 0 \\ 0 & 0 & 0 & 1 & 0 & 0 \\ 0 & 0 & 0 & 0 & -1 & 0 \\ 0 & 0 & 0 & 0 & 0 & 1 \end{bmatrix}$$

(iv) Under σ_v' the change $(h_1, h_2, s, -p_x, p_y, p_z) \leftarrow (h_1, h_2, s, p_x, p_y, p_z)$ is reproduced by

$$D(\sigma_v')=\begin{bmatrix} 1 & 0 & 0 & 0 & 0 & 0 \\ 0 & 1 & 0 & 0 & 0 & 0 \\ 0 & 0 & 1 & 0 & 0 & 0 \\ 0 & 0 & 0 & -1 & 0 & 0 \\ 0 & 0 & 0 & 0 & 1 & 0 \\ 0 & 0 & 0 & 0 & 0 & 1 \end{bmatrix}$$

(a) Confirm that $C_2\sigma_v = \sigma_v'$ by multiplication of their matrix representations.

$$D(C_2)D(\sigma_v)=\begin{bmatrix} 0 & 1 & 0 & 0 & 0 & 0 \\ 1 & 0 & 0 & 0 & 0 & 0 \\ 0 & 0 & 1 & 0 & 0 & 0 \\ 0 & 0 & 0 & -1 & 0 & 0 \\ 0 & 0 & 0 & 0 & -1 & 0 \\ 0 & 0 & 0 & 0 & 0 & 1 \end{bmatrix}\begin{bmatrix} 0 & 1 & 0 & 0 & 0 & 0 \\ 1 & 0 & 0 & 0 & 0 & 0 \\ 0 & 0 & 1 & 0 & 0 & 0 \\ 0 & 0 & 0 & 1 & 0 & 0 \\ 0 & 0 & 0 & 0 & -1 & 0 \\ 0 & 0 & 0 & 0 & 0 & 1 \end{bmatrix}$$

$$=\begin{bmatrix} 1 & 0 & 0 & 0 & 0 & 0 \\ 0 & 1 & 0 & 0 & 0 & 0 \\ 0 & 0 & 1 & 0 & 0 & 0 \\ 0 & 0 & 0 & -1 & 0 & 0 \\ 0 & 0 & 0 & 0 & 1 & 0 \\ 0 & 0 & 0 & 0 & 0 & 1 \end{bmatrix}=D(\sigma_v')$$

(b) Confirm that $\sigma_v\sigma_v' = C_2$ by multiplication of their matrix representations.

$$D(\sigma_v)D(\sigma_v') = \begin{bmatrix} 0 & 1 & 0 & 0 & 0 & 0 \\ 1 & 0 & 0 & 0 & 0 & 0 \\ 0 & 0 & 1 & 0 & 0 & 0 \\ 0 & 0 & 0 & 1 & 0 & 0 \\ 0 & 0 & 0 & 0 & -1 & 0 \\ 0 & 0 & 0 & 0 & 0 & 1 \end{bmatrix}\begin{bmatrix} 1 & 0 & 0 & 0 & 0 & 0 \\ 0 & 1 & 0 & 0 & 0 & 0 \\ 0 & 0 & 1 & 0 & 0 & 0 \\ 0 & 0 & 0 & -1 & 0 & 0 \\ 0 & 0 & 0 & 0 & 1 & 0 \\ 0 & 0 & 0 & 0 & 0 & 1 \end{bmatrix}$$

$$= \begin{bmatrix} 0 & 1 & 0 & 0 & 0 & 0 \\ 1 & 0 & 0 & 0 & 0 & 0 \\ 0 & 0 & 1 & 0 & 0 & 0 \\ 0 & 0 & 0 & -1 & 0 & 0 \\ 0 & 0 & 0 & 0 & -1 & 0 \\ 0 & 0 & 0 & 0 & 0 & 1 \end{bmatrix} = D(C_2)$$

(a) The **character** χ (chi), of an operation in a particular matrix representation is the sum of the diagonal elements of the representative of that operation. For the matrix representations the characters of the representation Γ are:

R	E	C_2	σ_v	σ_v'
Γ	6	0	2	4

(b) Compare the set of characters found in part (a) with the set of characters for each of the four irreps in the C_{2v} character table (shown below).

C_{2v}	E	C_2	σ_v	σ_v'
A_1	1	1	1	1
A_2	1	1	−1	−1
B_1	1	−1	1	−1
B_2	1	−1	−1	1

We quickly see that there is no match with the irreps A_1, A_2, B_1, and B_2. Consequently, the representation of this basis is reducible to a combination of these irreps.

(c) Using the inspection method of trial and error, we find in short order that 3× the characters of A_1 plus the characters of B_1 plus 2× the characters of B_2 yields the character set of part (a). The sums are detailed in the following table. Therefore, the representation of this basis set is reduced to $\boxed{\Gamma = 3A_1 + B_1 + 2B_2}$.

	E	C_2	σ_v	σ_v'
$3A_1$	3	3	3	3
B_1	1	−1	1	−1
$2B_2$	2	−2	−2	2
$3A_1 + B_1 + 2B_2$	6	0	2	4

P11B.5 The basis set consists of four 1s hydrogen orbitals, which is written as $\boldsymbol{f}=(\mathrm{A,B,C,D})$ when positioned as shown in Fig. 11B.4. When reflected in the mirror plane σ_{dAB} of the figure, the basis vector becomes (B,A,C,D). The order of the T_{d} group is 24 so there are 24 matrices to find. In addition to E there are $8C_3$ operations; clockwise and anticlockwise C_3 axes along the C–H$_{\mathrm{A}}$ bond, shown in the figure, are labeled $C_{3\mathrm{A}}^{+}$ and $C_{3\mathrm{A}}^{-}$. Likewise, the clockwise and anticlockwise C_3 axes along the C–H$_{\mathrm{B}}$ bond, shown in the figure, are labelled $C_{3\mathrm{B}}^{+}$ and $C_{3\mathrm{B}}^{-}$. There are $3C_2$ operations; the one that bisects the H$_{\mathrm{A}}$–C–H$_{\mathrm{B}}$ angle is labelled $C_{2\mathrm{AB}}$. There are 6 S_4. The clockwise and anticlockwise S_4 axes that bisect the H$_{\mathrm{A}}$–C–H$_{\mathrm{B}}$ angle are labelled $S_{4\mathrm{AB}}^{+}$ and $S_{4\mathrm{AB}}^{-}$. Finally, there are 6 σ_{d} mirror planes; the one, shown in the figure, that bisects the H$_{\mathrm{A}}$–C–H$_{\mathrm{B}}$ angle is labelled σ_{dAB}.

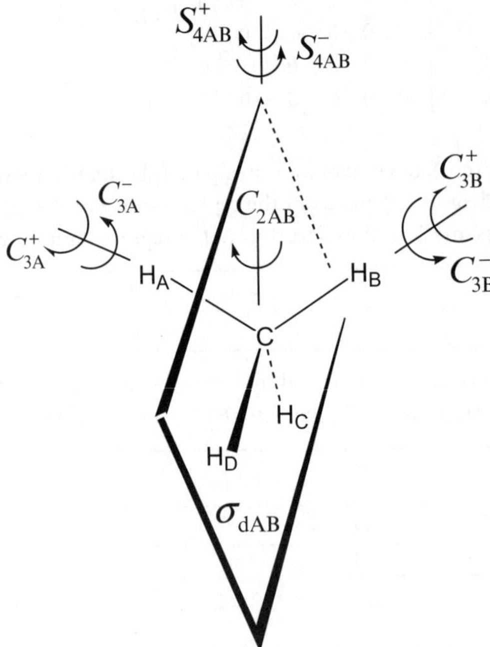

Figure 11B.4

$$E\boldsymbol{f}=(\mathrm{A,B,C,D})=\boldsymbol{f}\begin{pmatrix}1&0&0&0\\0&1&0&0\\0&0&1&0\\0&0&0&1\end{pmatrix}=\boldsymbol{f}\boldsymbol{D}(E);\ \chi=4$$

$$C_{3\mathrm{A}}^{+}\boldsymbol{f}=(\mathrm{A,C,D,B})=\boldsymbol{f}\begin{pmatrix}1&0&0&0\\0&0&0&1\\0&1&0&0\\0&0&1&0\end{pmatrix}=\boldsymbol{f}\boldsymbol{D}(C_{3\mathrm{A}}^{+});\ \chi=1$$

$$C_{3A}^{-}\boldsymbol{f}=(\text{A,D,B,C})=\boldsymbol{f}\begin{pmatrix}1&0&0&0\\0&0&1&0\\0&0&0&1\\0&1&0&0\end{pmatrix}=\boldsymbol{fD}(C_{3A}^{-});\ \chi=1$$

$$C_{3B}^{+}\boldsymbol{f}=(\text{D,B,A,C})=\boldsymbol{f}\begin{pmatrix}0&0&1&0\\0&1&0&0\\0&0&0&1\\1&0&0&0\end{pmatrix}=\boldsymbol{fD}(C_{3B}^{+});\ \chi=1$$

$$C_{3B}^{-}\boldsymbol{f}=(\text{C,B,D,A})=\boldsymbol{f}\begin{pmatrix}0&0&0&1\\0&1&0&0\\1&0&0&0\\0&0&1&0\end{pmatrix}=\boldsymbol{fD}(C_{3B}^{-});\ \chi=1$$

$$C_{3C}^{+}\boldsymbol{f}=(\text{B,D,C,A})=\boldsymbol{f}\begin{pmatrix}0&0&0&1\\1&0&0&0\\0&0&1&0\\0&1&0&0\end{pmatrix}=\boldsymbol{fD}(C_{3C}^{+});\ \chi=1$$

$$C_{3C}^{-}\boldsymbol{f}=(\text{D,A,C,B})=\boldsymbol{f}\begin{pmatrix}0&1&0&0\\0&0&0&1\\0&0&1&0\\1&0&0&0\end{pmatrix}=\boldsymbol{fD}(C_{3C}^{-});\ \chi=1$$

$$C_{3D}^{+}\boldsymbol{f}=(\text{C,A,B,D})=\boldsymbol{f}\begin{pmatrix}0&1&0&0\\0&0&1&0\\1&0&0&0\\0&0&0&1\end{pmatrix}=\boldsymbol{fD}(C_{3D}^{+});\ \chi=1$$

$$C_{3D}^{-}\boldsymbol{f}=(\text{B,C,A,D})=\boldsymbol{f}\begin{pmatrix}0&0&1&0\\1&0&0&0\\0&1&0&0\\0&0&0&1\end{pmatrix}=\boldsymbol{fD}(C_{3D}^{-});\ \chi=1$$

$$C_{2AB}\boldsymbol{f}=(\text{B,A,D,C})=\boldsymbol{f}\begin{pmatrix}0&1&0&0\\1&0&0&0\\0&0&0&1\\0&0&1&0\end{pmatrix}=\boldsymbol{fD}(C_{2AB});\ \chi=0$$

$$C_{2BC}\boldsymbol{f}=(\mathrm{D,C,B,A})=\boldsymbol{f}\begin{pmatrix}0&0&0&1\\0&0&1&0\\0&1&0&0\\1&0&0&0\end{pmatrix}=\boldsymbol{fD}(C_{2BC});\ \chi=0$$

$$C_{2AC}\boldsymbol{f}=(\mathrm{C,D,A,B})=\boldsymbol{f}\begin{pmatrix}0&0&1&0\\0&0&0&1\\1&0&0&0\\0&1&0&0\end{pmatrix}=\boldsymbol{fD}(C_{2AC});\ \chi=0$$

$$S_{4AB}^{+}\boldsymbol{f}=(\mathrm{D,C,A,B})=\boldsymbol{f}\begin{pmatrix}0&0&1&0\\0&0&0&1\\0&1&0&0\\1&0&0&0\end{pmatrix}=\boldsymbol{fD}(S_{4AB}^{+});\ \chi=0$$

$$S_{4AB}^{-}\boldsymbol{f}=(\mathrm{C,D,B,A})=\boldsymbol{f}\begin{pmatrix}0&0&0&1\\0&0&1&0\\1&0&0&0\\0&1&0&0\end{pmatrix}=\boldsymbol{fD}(S_{4AB}^{-});\ \chi=0$$

$$S_{4BC}^{+}\boldsymbol{f}=(\mathrm{B,D,A,C})=\boldsymbol{f}\begin{pmatrix}0&0&1&0\\1&0&0&0\\0&0&0&1\\0&1&0&0\end{pmatrix}=\boldsymbol{fD}(S_{4BC}^{+});\ \chi=0$$

$$S_{4BC}^{-}\boldsymbol{f}=(\mathrm{C,A,D,B})=\boldsymbol{f}\begin{pmatrix}0&1&0&0\\0&0&0&1\\1&0&0&0\\0&0&1&0\end{pmatrix}=\boldsymbol{fD}(S_{4BC}^{-});\ \chi=0$$

$$S_{4AC}^{+}\boldsymbol{f}=(\mathrm{B,C,D,A})=\boldsymbol{f}\begin{pmatrix}0&0&0&1\\1&0&0&0\\0&1&0&0\\0&0&1&0\end{pmatrix}=\boldsymbol{fD}(S_{4AC}^{+});\ \chi=0$$

$$S_{4AC}^{-}\boldsymbol{f}=(\mathrm{D,A,B,C})=\boldsymbol{f}\begin{pmatrix}0&1&0&0\\0&0&1&0\\0&0&0&1\\1&0&0&0\end{pmatrix}=\boldsymbol{fD}(S_{4AC}^{-});\ \chi=0$$

$$\sigma_{\text{dAB}} f = (\text{B,A,C,D}) = f\begin{pmatrix} 0 & 1 & 0 & 0 \\ 1 & 0 & 0 & 0 \\ 0 & 0 & 1 & 0 \\ 0 & 0 & 0 & 1 \end{pmatrix} = f\boldsymbol{D}(\sigma_{\text{dAB}}); \chi = 2$$

$$\sigma_{\text{dAC}} f = (\text{C,B,A,D}) = f\begin{pmatrix} 0 & 0 & 1 & 0 \\ 0 & 1 & 0 & 0 \\ 1 & 0 & 0 & 0 \\ 0 & 0 & 0 & 1 \end{pmatrix} = f\boldsymbol{D}(\sigma_{\text{dAC}}); \chi = 2$$

$$\sigma_{\text{dAD}} f = (\text{D,B,C,A}) = f\begin{pmatrix} 0 & 0 & 0 & 1 \\ 0 & 1 & 0 & 0 \\ 0 & 0 & 1 & 0 \\ 1 & 0 & 0 & 0 \end{pmatrix} = f\boldsymbol{D}(\sigma_{\text{dAD}}); \chi = 2$$

$$\sigma_{\text{dBC}} f = (\text{A,C,B,D}) = f\begin{pmatrix} 1 & 0 & 0 & 0 \\ 0 & 0 & 1 & 0 \\ 0 & 1 & 0 & 0 \\ 0 & 0 & 0 & 1 \end{pmatrix} = f\boldsymbol{D}(\sigma_{\text{dBC}}); \chi = 2$$

$$\sigma_{\text{dBD}} f = (\text{A,D,C,B}) = f\begin{pmatrix} 1 & 0 & 0 & 0 \\ 0 & 0 & 0 & 1 \\ 0 & 0 & 1 & 0 \\ 0 & 1 & 0 & 0 \end{pmatrix} = f\boldsymbol{D}(\sigma_{\text{dBD}}); \chi = 2$$

$$\sigma_{\text{dCD}} f = (\text{A,B,D,C}) = f\begin{pmatrix} 1 & 0 & 0 & 0 \\ 0 & 1 & 0 & 0 \\ 0 & 0 & 0 & 1 \\ 0 & 0 & 1 & 0 \end{pmatrix} = f\boldsymbol{D}(\sigma_{\text{dCD}}); \chi = 2$$

Question. Find the representations for the 1s orbital basis at the corners of a regular trigonal bipyramid.

P11B.7 Representation 1: $\boldsymbol{D}(C_3)\boldsymbol{D}(C_2) = 1\times 1 = 1 = \boldsymbol{D}(C_6)$

and from the C_{6v} character table this is either A_1 or A_2. Hence, either $\boldsymbol{D}(\sigma_v) = \boldsymbol{D}(\sigma_d)$

$= \boxed{+1 \text{ or } -1}$, respectively.

Representation 2: $\boldsymbol{D}(C_3)\boldsymbol{D}(C_2) = 1\times(-1) = -1 = \boldsymbol{D}(C_6)$

and from the C_{6v} character table this is either B_1 or B_2. Hence, either

$\boldsymbol{D}(\sigma_v) = -\boldsymbol{D}(\sigma_d) = \boxed{+1}$ or $\boldsymbol{D}(\sigma_v) = -\boldsymbol{D}(\sigma_d) = \boxed{-1}$, respectively.

P11B.9 f orbitals: $f_{z(5z^2-3r^2)}$, $f_{y(5y^2-3r^2)}$, $f_{x(5x^2-3r^2)}$, $f_{z(x^2-y^2)}$, $f_{y(x^2-z^2)}$, $f_{x(z^2-y^2)}$, f_{xyz}

(a) C_{2v}. The functions x^2, y^2, and z^2 belong to the irreducible species A_1 so they are invariant under all symmetry operations of the group. Thus, $f_{z(5z^2-3r^2)}$ transforms as $z(A_1)$, $f_{y(5y^2-3r^2)}$ as $y(B_2)$, $f_{x(5x^2-3r^2)}$ as $x(B_1)$, and likewise for $f_{z(x^2-y^2)}$, $f_{y(x^2-z^2)}$, and $f_{x(z^2-y^2)}$. The function f*xyz* transforms as $B_1 \times B_2 \times A_1 = A_2$. Therefore, in group C_{2v}, $f \rightarrow \boxed{2A_1 + A_2 + 2B_1 + 2B_2}$.

(b) C_{3v}. In C_{3v}, both z and z^2 transform as A_1; r^2 is also invariant, and hence $f_{z(5z^2-3r^2)}$ transforms as $z^3(A_1)$. ($f_{z(x^2-y^2)}$, f_{xyz}) transform as (x^2-y^2, xy) so this pair is a joint basis for E (see point group character table). The set $\{f_{x(5x^2-3r^2)}, f_{y(5y^2-3r^2)}, f_{x(z^2-y^2)}, f_{y(x^2-z^2)}\}$ remain to be accounted for in the C_{3v} environment. We will now examine each of these under the class operations: E, $2C_3$, and $3\sigma_v$. Only one representation of each multi-member class need be examined (C_3^+ with C_3 along the z axis, and σ_v with σ_v defined by the yz plane). The irreps spanned by these four are found by the decomposition of the characters $\chi(E)$, $2\chi(C_3^+)$, and $3\chi(\sigma_v)$. $\boldsymbol{D}(E)$ is a 4×4 unit matrix so $\chi(E) = 4$ and we turn to the task of finding $\chi(C_3^+)$ and $\chi(\sigma_v)$.

The z-axis is invariant under both C_3^+ and σ_v and it is important to recognize that (see Discussion question 11B.1):

$$C_3^+(x,y)\rightarrow\left(-\frac{1}{2}x+\frac{1}{2}\sqrt{3}\,y,-\frac{1}{2}\sqrt{3}\,x-\frac{1}{2}y\right) \text{ and } \sigma_v(x,y)\rightarrow(-x,y).$$

Thus

$$C_3^+\left(x\left(5x^2-3r^2\right)\right)\rightarrow C_3^+(x)\times\left\{5\left(-\frac{1}{2}x+\frac{1}{2}\sqrt{3}\,y\right)^2-3r^2\right\}$$

$$=C_3^+(x)\times\left\{5\left(\frac{1}{4}x^2-\frac{1}{2}\sqrt{3}\,xy+\frac{3}{4}y^2\right)-3r^2\right\}$$

$$C_3^+\left(x\left(z^2-y^2\right)\right)\rightarrow C_3^+(x)\times\left\{z^2-\left(-\frac{1}{2}\sqrt{3}x-\frac{1}{2}y\right)^2\right\}=-C_3^+(x)\times\left\{\frac{3}{4}x^2+\frac{1}{2}\sqrt{3}\,xy+\frac{1}{4}y^2-z^2\right\}$$

The xy terms within the above expressions are removed by forming the combination orbital

$$f_1 = f_{x(5x^2-3r^2)} - 5f_{x(z^2-y^2)}$$

$$C_3^+(f_1)\rightarrow C_3^+(x)\times\left\{5\left(x^2+y^2\right)-3r^2-5z^2\right\}$$

All terms in the factor $\left\{5\left(x^2+y^2\right)-3r^2-5z^2\right\}$ belong to A_1 so the character depends upon $C_3^+(x)$ alone.

$$C_3^+(f_1)\rightarrow C_3^+(x)=-\frac{1}{2}x+\frac{1}{2}\sqrt{3}\,y$$

In the same way we define the combination orbital

$$f_2 = f_{y(5y^2-3r^2)} + 5f_{y(x^2-z^2)}$$

and find that

$$C_3^+(f_2)\rightarrow C_3^+(y)\times\left\{5\left(x^2+y^2\right)-3r^2-5z^2\right\}\rightarrow C_3^+(y)=-\frac{1}{2}\sqrt{3}x-\frac{1}{2}y.$$

Thus

$$C_3^+(f_1,f_2)\rightarrow\left(-\frac{1}{2}x+\frac{1}{2}\sqrt{3}\,y,-\frac{1}{2}\sqrt{3}x-\frac{1}{2}y\right)$$

which has the matrix form

$$(\mathrm{f}_1,\mathrm{f}_2)\boldsymbol{D}(C_3^+)\rightarrow(xy)\begin{pmatrix}-\tfrac{1}{2} & -\tfrac{1}{2}\sqrt{3}\\ \tfrac{1}{2}\sqrt{3} & -\tfrac{1}{2}\end{pmatrix}$$

and we conclude that $\chi(C_3^+)=-\tfrac{1}{2}-\tfrac{1}{2}=-1$.

Now, consider the effect of σ_v on each of the four orbitals.

$$\sigma_v\left(\mathrm{f}_{x(5x^2-3r^2)}\right)\rightarrow-\mathrm{f}_{x(5x^2-3r^2)}\quad\text{because } \sigma_v(x)\rightarrow-x$$

$$\sigma_v\left(\mathrm{f}_{y(5y^2-3r^2)}\right)\rightarrow\mathrm{f}_{y(5y^2-3r^2)}\quad\text{because } \sigma_v(y)\rightarrow y$$

$$\sigma_v\left(\mathrm{f}_{x(z^2-y^2)}\right)\rightarrow-\mathrm{f}_{x(z^2-y^2)}$$

$$\sigma_v\left(\mathrm{f}_{y(x^2-z^2)}\right)\rightarrow\mathrm{f}_{y(x^2-z^2)}$$

The matrix form of these transformations is

$$\left(\mathrm{f}_{x(5x^2-3r^2)},\mathrm{f}_{y(5y^2-3r^2)},\mathrm{f}_{x(z^2-y^2)},\mathrm{f}_{y(x^2-z^2)}\right)\boldsymbol{D}(\sigma_v)$$

$$=\left(\mathrm{f}_{x(5x^2-3r^2)},\mathrm{f}_{y(5y^2-3r^2)},\mathrm{f}_{x(z^2-y^2)},\mathrm{f}_{y(x^2-z^2)}\right)\begin{pmatrix}-1&0&0&0\\0&1&0&0\\0&0&-1&0\\0&0&0&1\end{pmatrix}$$

and we conclude that $\chi(\sigma_v)=-1+1-1+1=0$. In summary, we have found that these four orbitals have the characters $\chi(E)=4$, $\chi(C_3^+)=-1$, and $\chi(\sigma_v)=0$. The species of the four is indicated by the total membership of each symmetry operation $\{\chi(E),2\chi(C_3^+),3\chi(\sigma_v)\}=\{4,-2,0\}$ for which inspection of the character table immediately yields the decomposition 2E. Finally, we conclude that the set of seven f orbitals in C_{3v} spans $\boxed{\mathrm{A}_1+3\mathrm{E}}$.

(c) T_d. Make the inspired guess that the f orbitals are a basis of dimension 3 + 3 + 1, suggesting the decomposition T + T + A. Is the A representation A_1 or A_2? We see from the character table that the effect of S_4 discriminates between A_1 and A_2. Under S_4, $x\rightarrow y, y\rightarrow -x, z\rightarrow -z$, and so $\mathrm{S}_4\{xyz\}\rightarrow xyz$. The character is $\chi=1$, and so fxyz spans A_1. Likewise, $(x^3, y^3, z^3)\rightarrow(y^3,-x^3,-z^3)$ and $\chi=0+0-1=-1$. Hence, the trio

$$\left(\mathrm{f}_{x(5x^2-3r^2)},\mathrm{f}_{y(5y^2-3r^2)},\mathrm{f}_{z(5z^2-3r^2)}\right)$$

spans T_2. Finally

$$\mathrm{S}_4\{x(z^2-y^2),\ y(x^2-z^2),\ z(x^2-y^2)\}\rightarrow\{y(z^2-x^2),-x(z^2-y^2),-z(y^2-x^2)\}$$

resulting in $\chi = 1$, indicating that $\left(f_{x(z^2-y^2)}, f_{y(x^2-z^2)}, f_{z(x^2-y^2)}\right)$ belongs to T_1. Therefore, in T_d, $f \rightarrow \boxed{A_1 + T_1 + T_2}$.

(d) O_h. Anticipate an A + T + T decomposition as in the other cubic group. Since x, y, and z all have odd parity, all the irreducible representatives will be ungerade (u). Under S_4, $xyz \rightarrow xyz$ (as in (c)), and so the representation is $\chi = 1$, which causes f_{xyz} to belong to A_{2u} (see the character table). Under S_4, as before, and $\chi = -1$, indicating that $\left(f_{x(5x^2-3r^2)}, f_{y(5y^2-3r^2)}, f_{z(5z^2-3r^2)}\right)$ belongs to T_{1u}. In the same way, the remaining three functions, $\left(f_{x(z^2-y^2)}, f_{y(x^2-z^2)}, f_{z(x^2-y^2)}\right)$, span T_{2u}:

$$S_4\{x(z^2-y^2), y(z^2-x^2), z(x^2-y^2)\} \rightarrow \{y(z^2-x^2), -x(z^2-y^2), -z(y^2-x^2)\} \quad \chi = +1$$

$$C_4\{x(z^2-y^2), y(z^2-x^2), z(x^2-y^2)\} \rightarrow \{-y(z^2-x^2), x(z^2-y^2), -z(y^2-x^2)\} \quad \chi = -1$$

Hence, in O_h, $f \rightarrow \boxed{A_{2u} + T_{1u} + T_{2u}}$.

The f orbitals will cluster into sets according to their irreducible representations. Thus, (a) $f \rightarrow A_1 + T_1 + T_2$ in T_d symmetry, and there is one non-degenerate orbital and two sets of triply degenerate orbitals. (b) In O_h symmetry, $f \rightarrow A_{2u} + T_{1u} + T_{2u}$, and the pattern of splitting (but not the order of energies) is the same as that in T_d.

11C Applications of symmetry

Answer to discussion question

D11C.1 Character tables provide a way to: (a) assign symmetry symbols for orbitals; (b) know whether overlap integrals are nonzero; (c) determine what atomic orbitals can contribute to a LCAO-MO; (d) determine the maximum orbital degeneracy of a molecule; and (e) determine whether a transition is allowed.

Solutions to exercises

E11C.1(a) The p_x orbital spans B_1 of the C_{2v} point group while z and p_z span A_1. Following the Section 11C.1(a) procedure for deducing the symmetry species spanned by the product f_1f_2 and hence to see whether it does indeed span A_1, we write a table of the characters of each function and multiply the rows.

C_{2v}	E	C_2	σ_v	σ_v'
p_x	1	−1	1	−1
z	1	1	1	1
p_z	1	1	1	1
p_xzp_z	1	−1	1	−1

The characters of the product p_xzp_z are those of B_1 alone, so the integrand does not span A_1. It follows that the integral must be zero.

E11C.2(a) For a C_{3v} molecule, x and y span E while z spans A_1. Thus, the x and y components of the dipole moment [11C.6] have transition integrands that span $A_2 \times E \times A_1$ for the $A_1 \rightarrow A_2$ transition. By inspection of the C_{3v} character table we find the **decomposition of the direct product** to be: $A_2 \times E \times A_1 = E$. The integrand spans E alone. Since it does not span A_1, the x and y components of the transition integral must be zero. The transition integrand for the z component spans $A_2 \times A_1 \times A_1 = A_2$ for the $A_1 \rightarrow A_2$ transition. Consequently, the z component of the transition integral must also equal zero and we conclude that the transition is forbidden.

Should these considerations prove confusing, write a table with columns headed by the three components of the electric dipole moment operator, μ.

Component of μ:	x	y	z
A_1	1 1 1	1 1 1	1 1 1
$\Gamma(\mu)$	2 -1 0	2 -1 0	1 1 1
A_2	1 1 -1	1 -1 -1	1 1 -1
$A_1\,\Gamma(\mu)\,A_2$	2 -1 0	2 -1 0	1 1 -1
	E	E	A_2

Since A_1 is not present in any product, the transition dipole moment must be zero.

E11C.3(a) We first determine how x and y individually transform under the operations of the C_{4v} group. Using these results we determine how the products xy transforms. The transform of xy is the product of the transforms of x and y. Under each operation the functions transform as follows.

	E	C_2	C_4	σ_v	σ_d
x	x	$-x$	y	x	$-y$
y	y	$-y$	$-x$	$-y$	$-x$
xy	xy	xy	$-xy$	$-xy$	xy
χ	1	1	-1	-1	1

From the C_{4v} character table, we see that this set of characters belongs of B_2.

E11C.4(a) Recall that $p_x \propto x, p_y \propto y, p_z \propto z, d_{xy} \propto xy, d_{xz} \propto xz, d_{yz} \propto yz, d_{z^2} \propto z^2, d_{x^2-y^2} \propto x^2 - y^2$. Additionally, when the functions f_1 and f_2 of an overlap integral are bases for irreducible representations of a group, the integral must vanish if they are different symmetry species; if they are the same symmetry species, then the integral may be nonzero (Section 11C.1). Since the combination $p_x(A) - p_x(B)$ of the two O atoms (with x perpendicular to the plane) spans A_2, the orbital on N must span A_2 for a nonzero overlap.

Now refer to the C_{2v} character table. The s orbital spans A_1 and the p orbitals of the central N atom span $A_1(p_z)$, $B_1(p_x)$, and $B_2(p_y)$. Therefore, no N orbitals span A_2, and hence $p_x(A) - p_x(B)$ is a nonbonding combination. If d orbitals are available, as they are in S of the SO_2 molecule, we could form a molecular orbital with d_{xy} which is a basis for A_2.

E11C.5(a) The electric dipole moment operator transforms as $x(B_1)$, $y(B_2)$, and $z(A_1)$ of the C_{2v} character table. Transitions are allowed if $\int \psi_f^* \mu \psi_i \, d\tau$ is nonzero (Example 11C.5

of text), and hence are forbidden unless $\Gamma_f \times \Gamma(\mu) \times \Gamma_i$ contains A_1. Since $\Gamma_i = A_1$ this requires $\Gamma_f \times \Gamma(\mu) = A_1$. Since $B_1 \times B_1 = A_1$ and $B_2 \times B_2 = A_1$ and $A_1 \times A_1 = A_1$ x-polarized light may cause a transition to a B_1 term, y-polarized light to a B_2 term, and z-polarized light to an A_1 term.

E11C.6(a)

C_{4v}, $h=8$	E	C_2	$2C_4$	$2\sigma_v$	$2\sigma_d$
A_1	1	1	1	1	1
A_2	1	1	1	−1	−1
B_1	1	1	−1	1	−1
B_2	1	1	−1	−1	1
E	2	−2	0	0	0

$$n(\Gamma) = \frac{1}{h}\sum_R \chi^{(\Gamma)}(R)\chi(R) \quad [11C.2] \quad \text{where } \chi(R) = (5,1,1,3,1)$$

$$n(A_1) = \tfrac{1}{8}\{1(1\times5)+1(1\times1)+2(1\times1)+2(1\times3)+2(1\times1)\} = 2$$

$$n(A_2) = \tfrac{1}{8}\{1(1\times5)+1(1\times1)+2(1\times1)+2(-1\times3)+2(-1\times1)\} = 0$$

$$n(B_1) = \tfrac{1}{8}\{1(1\times5)+1(1\times1)+2(-1\times1)+2(1\times3)+2(-1\times1)\} = 1$$

$$n(B_2) = \tfrac{1}{8}\{1(1\times5)+1(1\times1)+2(-1\times1)+2(-1\times3)+2(1\times1)\} = 0$$

$$n(E) = \tfrac{1}{8}\{1(2\times5)+1(-2\times1)+2(0\times1)+2(0\times3)+2(0\times1)\} = 1$$

Thus, this set of basis functions spans $\boxed{2A_1 + B_1 + E}$.

E11C.7(a) (i) The point group of benzene is D_{6h}. In D_{6h} μ spans $E_{1u}(x, y)$ and $A_{2u}(z)$, and the ground term is A_{1g}. Then, using $A_{2u} \times A_{1g} = A_{2u}$, $E_{1u} \times A_{1g} = E_{1u}$, $A_{2u} \times A_{2u} = A_{1g}$, and $E_{1u} \times E_{1u} = A_{1g} + A_{2g} + E_{2g}$, we conclude that the upper term is $\boxed{\text{either } E_{1u} \text{ or } A_{2u}}$.

(ii) Naphthalene belongs to D_{2h}. In D_{2h} itself, the components of μ span $B_{3u}(x)$ $B_{2u}(y)$ and $B_{1u}(z)$ and the ground term is A_g. Hence, since $A_g \times \Gamma = \Gamma$ in this group, the upper terms are $\boxed{B_{3u}\,(x\text{-polarized})}$, $\boxed{B_{2u}\,(y\text{-polarized})}$, and $\boxed{B_{1u}\,(z-\text{polarized})}$.

E11C.8(a) See Fig. 11C.1, which shows that we assume that θ is the angle to the σ_h plane of group C_s and, therefore, it is a dimension that can exhibit a symmetrical integration interval from one side of the plane to the other side in a C_s object like methanol, bromochloromethane, and O=N–Cl. We consider the integral

$$I = \int_{-a}^{a} f_1 f_2 \mathrm{d}\theta = \int_{-a}^{a} \sin\theta\cos\theta\,\mathrm{d}\theta$$

and hence draw up the following table for the effect of operations in the group C_s.

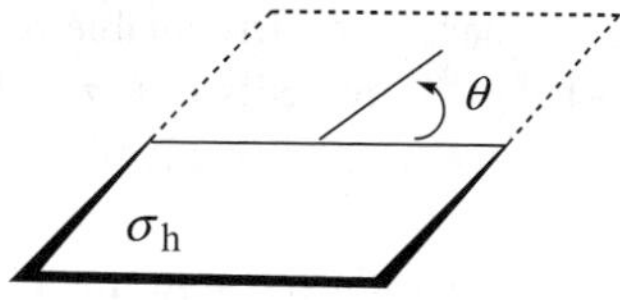

Figure 11C.1

	E	σ_h
$f_1 = \sin\theta$	$\sin\theta$	$-\sin\theta$
$f_2 = \cos\theta$	$\cos\theta$	$\cos\theta$

In terms of characters:

	E	σ_h	Symmetry species
f_1	1	−1	A″
f_2	1	1	A′
f_1f_2	1	−1	A″

Since the product does not span the totally symmetric species A′ the integral is necessarily $\boxed{\text{zero}}$.

Solutions to problems

P11C.1 A quick rule for determining the character without first having to set up the matrix representation is to count 1 each time a basis function is left unchanged by the operation, because only these functions give a nonzero entry on the diagonal of the matrix representative. In some cases there is a sign change, $(\ldots -f\ldots) \leftarrow (\ldots f\ldots)$; then -1 occurs on the diagonal, and so count -1. The character of the identity is always equal to the dimension of the basis since each function contributes 1 to the trace.

E: all four orbitals are left unchanged; hence $\chi = 4$

C_3: One orbital is left unchanged; hence $\chi = 1$

C_2: No orbitals are left unchanged; hence $\chi = 0$

σ_d: Two orbitals are left unchanged; hence $\chi = 2$

S_4: No orbitals are left unchanged; hence $\chi = 0$

The character set 4, 1, 0, 2, 0 spans $\boxed{A_1 + T_2}$. Inspection of the character table of the group T_d shows that an s orbital spans A_1 and that the three p orbitals on the C atom span T_2. Hence, the $\boxed{\text{s and p}}$ orbitals of the C atom may form molecular orbitals with the four H1s orbitals. In T_d, the d orbitals of the central atom span $E + T_2$ (character table, final column), and so only the T_2 set $\boxed{(d_{xy}, d_{yz}, d_{zx})}$ may contribute to molecular orbital formation with the H orbitals.

P11C.3 Consider the integral $\int_{-a}^{a}(3x^2-1)\mathrm{d}x = 3\int_{-a}^{a}x^2\,\mathrm{d}x - \int_{-a}^{a}1\,\mathrm{d}x = 3I_2 - I_1$. Integral I_1 has the unit integrand, which spans the totally symmetric irreducible symmetric species in all point groups and is, consequently, nonzero. The integrand of I_2 is the product $x \times x$, which also spans the totally symmetric irreducible representation in all point groups because

$$n(A_1) = \frac{1}{h}\sum_R \chi^{(A_1)}(R)\chi(R) = \frac{1}{h}\sum_R \chi(R) = 1 \quad [11C.2]$$

whenever $\chi(R)$ are the characters of the square of an irreducible representation (e.g., $\Gamma_i \times \Gamma_i$). Thus, the integral $\int_{-a}^{a}(3x^2-1)\mathrm{d}x$ is not necessarily zero because neither I_1 nor I_2 are necessarily zero. The integral will be accidently zero when the integration interval causes $3I_2$ to equal I_1.

Since the above conclusion is valid for all point groups, in the following parts we simply use eqn 11C.2 to show that I_2 is not necessarily zero in each point group because the integrand spans A_1.

(a) A cube, O_h: x spans T_{1u} and x^2 spans $T_{1u} \times T_{1u}$ so the number of times that the integrand of I_2 spans A_{1g} is:

$$n(A_{1g}) = \frac{1}{48}\{1\times9+8\times0+6\times1+6\times1+3\times1+1\times9+6\times1+8\times0+3\times1+6\times1\} = 1$$

(b) A tetrahedron, T_d: x spans T_2 and x^2 spans $T_2 \times T_2$ so the number of times that the integrand of I_2 spans A_1 is:

$$n(A_1) = \frac{1}{24}\{1\times9+8\times0+3\times1+6\times1+6\times1\} = 1$$

(c) A hexagonal prism, D_{6h}: x spans E_{1u} and x^2 spans $E_{1u} \times E_{1u}$ so the number of times that the integrand of I_2 spans A_{1g} is:

$$n(A_{1g}) = \frac{1}{24}\left\{\begin{matrix}1\times4+2\times1+2\times1+1\times4+3\times0+3\times0\\ +1\times4+2\times1+2\times1+1\times4+3\times0+3\times0\end{matrix}\right\} = 1$$

P11C.5 The p_z orbitals of the fluorine atoms in XeF_4 are shown in Fig. 11C.2. (The orbital has a positive wavefunction sign in shaded lobes and a negative wavefunction sign in unshaded lobes.) These orbitals may form π molecular orbitals with the xenon atom provided that a nonzero overlap integral exists.

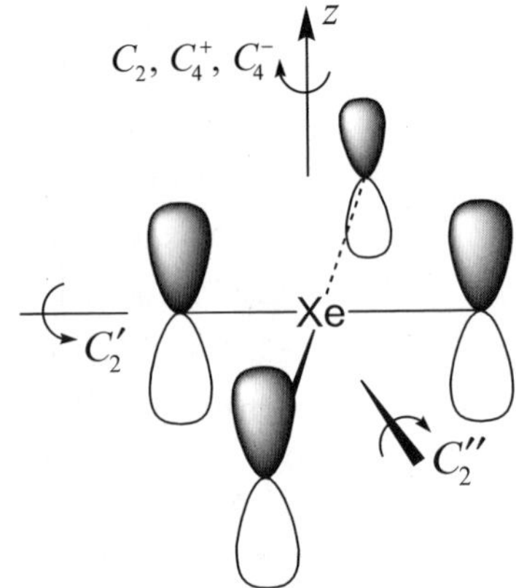

Figure 11C.3

To find the symmetry species spanned by the fluorine p_z orbitals, we use a quick rule for determining the character of the basis set under each symmetry operation of the group: count 1 each time a basis function is left unchanged by the operation, because only these functions give a nonzero entry on the diagonal of the matrix representative. In some cases there is a sign change, $(\ldots -f \ldots) \leftarrow (\ldots f \ldots)$; then 1 occurs on the diagonal, and so count 1. The character of the identity is always equal to the dimension of the basis since each function contributes 1 to the trace. Although XeF_4 belongs to the D_{4h} group, we will use the D_4 subgroup for convenience. Here is a summary of the characters exhibited by the fluorine orbital under each symmetry operation:

D_4	E	C_2	$2C_4$	$2C_2'$	$2C_2''$
fluorine p_z orbitals	4	0	0	−2	0

Inspection of the D_4 character table shows that the fluorine p_z orbitals span $A_2 + B_2 + E$. Further inspection of the D_4 character table reveals that p_z belongs to A_2, d_{xy} belongs to B_2, and both the $\{p_x, p_y\}$ set and the $\{d_{xz}, d_{yz}\}$ set belong to E. Consequently, only these orbitals of the central atom may possibly have nonzero overlap with the fluorine p_z orbitals. We must now use the procedure of Section 11C.2(b) to find the four **symmetry-adapted linear combinations (SALC)** of the fluorine p_z orbitals. Using clockwise labelling of the Fp_z orbitals, we write a table that summarizes the effect of each operation on each orbital. (The symbol A is used to represent the orbital $p_z(A)$ at fluorine atom A, etc.)

D_4	A	B	C	D
E	A	B	C	D
C_2	C	D	A	B
C_4^+	D	A	B	C
C_4^-	B	C	D	A
C_2'(A–C)	–A	–D	–C	–B
C_2'(B–D)	–C	–B	–A	–D
C_2''(A–Xe–B)	–B	–A	–D	–C
C_2''(B–Xe–C)	–D	–C	–B	–A

To generate the B_2 combination, we take the characters for B_2 (1,1,–1,–1,–1,–1,1,1) and multiply column 3 and sum the terms. Then, divide by the order of the group (8):

$$p_1(B_2) = \tfrac{1}{8}\{C+A-B-D+C+A-D-B\} = \tfrac{1}{4}\{p_z(A)-p_z(B)+p_z(C)-p_z(D)\}$$

To generate the A_2 combination, we take the characters for A_2 and multiply column 1:

$$p_2(A_2) = \tfrac{1}{4}\{p_z(A)+p_z(A)+p_z(C)+p_z(D)\}$$

To generate the two E combinations, we take the characters for E and multiply column 1 for one of them and multiply column 2 for the other:

$$p_3(E) = \tfrac{1}{4}\{p_z(A)-p_z(C)\}$$
$$p_4(E) = \tfrac{1}{4}\{p_z(B)-p_z(D)\}$$

Other column multiplications yield these same combinations or zero.

Figure 11C.3(a) overleaf shows that the $p_z(A_2)$ orbital of the central atom does have nonzero overlap with p_2. Figure 11C.3(b) shows that $d_{xz}(E)$ has a nonzero overlap with p_3. However, Fig. 11C.3(c) shows that, because of the balance between constructive and destructive interference, there is zero overlap between $d_{xy}(B_2)$ and p_1. Thus, there is no orbital of the central atom that forms a nonzero overlap with the p_1 combination so p_1 is nonbonding.

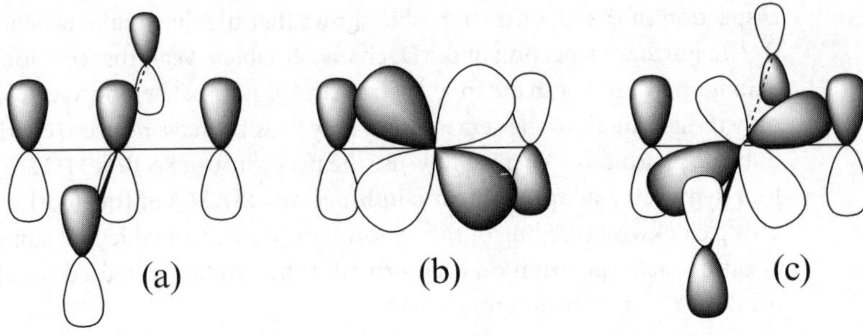

Figure 11C.3

12 Rotational and vibrational spectra

Note: The masses of nuclides are listed in Table 0.2 of the *Resource section*.

12A General features of molecular spectroscopy

Answers to discussion questions

D12A.1 The classical concept of a 'selection rule' is that, for the molecule to be able to interact with the electromagnetic field and absorb or create a photon of frequency ν, it must possess, at least transiently, a dipole oscillating at that frequency. A **gross selection rule** specifies the general features that a molecule must have if it is to have a spectrum of a given kind and **specific selection rules** express the allowed transitions in terms of the changes in quantum numbers. Chapter 12 explores the quantum mechanical foundations of these rules, which we summarize as they apply to spectroscopy.

For both microwave and infrared spectroscopy, the allowed transitions depend on the existence of an oscillating dipole moment which can stir the electromagnetic field into oscillation (and vice versa for absorption). For microwave rotational spectroscopy, this implies that the molecule must have a permanent dipole moment, which is equivalent to an oscillating dipole when the molecule is rotating. See Figs 12C.1 and 12C.2 of the text. In the case of infrared vibrational spectroscopy, the physical basis of the gross selection rule is that the molecule have a structure that allows for the existence of an oscillating dipole moment when the molecule vibrates. That is, the electric dipole of a molecule must change when the atoms are displaced relative to one another. Polar molecules necessarily satisfy this requirement, but nonpolar molecules may also have a fluctuating dipole moment upon vibration. See Fig. 12D.3.

For rotational Raman spectroscopy the gross selection rule is that the molecule must be anisotropically polarizable, which is to say that its polarizability, α, depends upon the direction of the electric field relative to the molecule. Nonspherical rotors satisfy this condition. Therefore, linear and symmetric rotors are rotationally Raman active. See Fig. 12C.6.

For vibrational Raman spectroscopy the gross selection rule is that the polarizability of the molecule must change as the molecule vibrates. All diatomic molecules satisfy this condition as the molecules swell and contract during a vibration, the control of the nuclei over the electrons varies, and the molecular polarizability changes. Hence both homonuclear and heteronuclear diatomics are vibrationally Raman active. In polyatomic molecules it is usually quite difficult to judge by inspection whether or not the molecule is anisotropically polarizable; hence group theoretical methods are relied on for judging the Raman activity of the various normal modes of vibration. See Sections 12D.5 and 12E.4.

D12A.3 Textbook schematics of absorption, emission, and Raman spectrometers are found in Figs 12A.6, 12A.14, and 12A.15, respectively. A fundamental difference in the experimental arrangements is in the relative direction of the radiation exiting from the sample with respect to the incident radiation upon the sample. In absorption spectroscopy the angle between the two is 180°; that is, they are in line with each other and the analysed radiation is said to be **transmitted**. Identical beams incident upon the sample and reference cells are compared to eliminate small absorptions within the sample spectrum from components present in both cells, a solvent for example. In emission spectrometry the angle is 90°, which reduces the problem of distinguishing the relatively low intensity emission from the extremely intense incident radiation. Extremely low intensity Raman scattered radiation is collected with a curved mirror from angles in back-scattered directions and directed to the detector. This arrangement reduces problems caused by the high intensity incident radiation and increases the intensity of interest at the detector.

Solutions to exercises

E12A.1(a) The ratio of Einstein coefficients A/B is

$$\frac{A}{B}=\frac{8\pi h\nu^3}{c^3}[12\text{A}.9]$$

The frequency is

$$\nu=\frac{c}{\lambda} \quad \text{so} \quad \frac{A}{B}=\frac{8\pi h}{\lambda^3}$$

(a) $$\frac{A}{B}=\frac{8\pi(6.626\times10^{-34}\,\text{J s})}{(70.8\times10^{-12}\,\text{m})^3}=\boxed{0.0469\,\text{J m}^{-3}\,\text{s}}$$

(b) $$\frac{A}{B}=\frac{8\pi h}{\lambda^3}=\frac{8\pi(6.626\times10^{-34}\,\text{J s})}{(500\times10^{-9}\,\text{m})^3}=\boxed{1.33\times10^{-13}\,\text{J m}^{-3}\,\text{s}}$$

(c) $$\frac{A}{B}=\frac{8\pi h}{\lambda^3}=8\pi h\tilde{\nu}^3=8\pi\left(6.626\times10^{-34}\,\text{J s}\right)\left(3\,000\,\text{cm}^{-1}\times(10^{-2}\,\text{m}^{-1}/1\,\text{cm}^{-1})\right)^3$$

$$=\boxed{4.50\times10^{-28}\,\text{J m}^{-3}\,\text{s}}$$

E12A.2(a) $\log\frac{I}{I_0}=-\log\frac{I_0}{I}=-\varepsilon[\text{J}]L$ [12A.13 and 12A.14]

$$=(-723\,\text{dm}^3\,\text{mol}^{-1}\,\text{cm}^{-1})\times(4.25\times10^{-3}\,\text{mol dm}^{-3})\times(0.250\,\text{cm})$$
$$=-0.768$$

Hence, $\frac{I}{I_0}=10^{-0.768}=0.171$, and the reduction in intensity is $\boxed{82.9 \text{ per cent}}$.

E12A.3(a) $\log\frac{I}{I_0}=-\log\frac{I_0}{I}=-\varepsilon[\text{J}]L$ [12A.13 and 12A.14]

Hence, $\varepsilon=-\frac{1}{[\text{J}]L}\log\frac{I}{I_0}=-\frac{\log(0.181)}{(1.39\times10^{-4}\,\text{mol dm}^{-3})\times(1.00\,\text{cm})}$

$$=\boxed{5.34\times10^3\,\text{dm}^3\,\text{mol}^{-1}\,\text{cm}^{-1}}$$

E12A.4(a) $\log T=-A=-\varepsilon[\text{J}]L$ [12A.12-14]

$$[\text{J}] = -\frac{1}{\varepsilon L}\log T=\frac{-\log(1-0.385)}{(386\,\text{dm}^3\,\text{mol}^{-1}\,\text{cm}^{-1})\times(0.500\,\text{cm})}$$
$$= \boxed{1.09\,\text{mmol dm}^{-3}}$$

E12A.5(a) $\varepsilon=-\frac{1}{[\text{J}]L}\log\frac{I}{I_0}$ [12A.13 and 12A.14] with $L=0.20\,\text{cm}$

We use this formula to draw up the following table.

$[Br_2]$/mol dm^{-3}	0.0010	0.0050	0.0100	0.0500	
I/I_0	0.814	0.356	0.127	3.0×10^{-5}	
ε/(dm^3 mol^{-1} cm^{-1})	447	449	448	452	mean: $44\bar{9}$

Hence, the molar absorption coefficient is $\varepsilon = \boxed{450\,\text{dm}^3\,\text{mol}^{-1}\,\text{cm}^{-2}}$.

E12A.6(a) $\varepsilon=-\frac{1}{[\text{J}]L}\log\frac{I}{I_0}$ [12A.13, 12A.14] $=\frac{-1}{(0.010\,\text{mol dm}^{-3})\times(0.20\,\text{cm})}\log(0.48)$

$$=\boxed{15\bar{9}\,\text{dm}^3\,\text{mol}^{-1}\,\text{cm}^{-1}}$$

$$T=\frac{I}{I_0}=10^{-[\text{J}]\varepsilon L}\ \text{[12A.12-14]}$$

$$=10^{(-0.010\,\text{mol dm}^{-3})\times(15\bar{9}\,\text{dm}^3\,\text{mol}^{-1}\text{cm}^{-1})\times(0.40\,\text{cm})}=10^{-0.63\bar{6}}=0.23,\ \text{or}\ \boxed{23 \text{ per cent}}$$

E12A.7(a) Our computation uses the less murky molar absorption coefficient of $6.2\times10^{-5}\,\mathrm{dm^3\,mol^{-1}\,cm^{-1}}$. For water,

$$[\mathrm{H_2O}]\approx\frac{1.00\ \mathrm{kg\,dm^{-3}}}{18.02\ \mathrm{g\,mol^{-1}}}=55.5\ \mathrm{mol\,dm^{-3}}$$

$$\log\frac{I}{I_0}=-\varepsilon[\mathrm{J}]L \quad \text{so} \quad L=-\frac{1}{\varepsilon[\mathrm{J}]}\log\frac{I}{I_0} \quad [12\mathrm{A}.13 \text{ and } 12\mathrm{A}.14]$$

(a) $$L=-\frac{1}{(6.2\times10^{-5}\,\mathrm{dm^3\,mol^{-1}\,cm^{-1}})\times(55.5\,\mathrm{mol\,dm^{-3}})}\log\frac{1}{2}$$

$$=87\,\mathrm{cm}=\boxed{0.87\,\mathrm{m}}$$

(b) $$L=-\frac{1}{(6.2\times10^{-5}\,\mathrm{dm^3\,mol^{-1}\,cm^{-1}})\times(55.5\,\mathrm{mol\,dm^{-3}})}\log\frac{1}{10}$$

$$=2.9\times10^{2}\,\mathrm{cm}=\boxed{2.9\,\mathrm{m}}$$

E12A.8(a) $$\mathcal{A}=\int_{\text{band}}\varepsilon(\tilde{\nu})\,\mathrm{d}\tilde{\nu}\ [12\mathrm{A}.15]=\int_{\tilde{\nu}_\mathrm{i}}^{\tilde{\nu}_\mathrm{f}}\varepsilon(\tilde{\nu})\,\mathrm{d}\tilde{\nu}$$

Since $\tilde{\nu}=\lambda^{-1}$ and $\tilde{\nu}/\mathrm{cm^{-1}}=10^7/(\lambda/\mathrm{nm})$, the initial, peak, and final wavenumbers of the absorption band are:

$\tilde{\nu}_\mathrm{i}/\mathrm{cm^{-1}}=10^7/(300)=3.3\times10^4$, $\tilde{\nu}_\mathrm{peak}/\mathrm{cm^{-1}}=10^7/(270)=3.7\times10^4$, and $\tilde{\nu}_\mathrm{f}/\mathrm{cm^{-1}}=10^7/(220)=4.5\times10^4$.

The positions of the wavenumber end points and peak (max) of the band are schematically presented in Fig. 12A.1. It is apparent that, because of the relative position of the peak, the molar absorption coefficient is not symmetrically distributed around the peak wavenumber. The distribution is skewed toward higher wavenumbers. However, a reasonable estimate of the area under the curve may be approximated by adding the areas of triangle 1 and triangle 2 shown as dashed lines in the figure.

$$\mathcal{A}=\int_{\tilde{\nu}_\mathrm{i}}^{\tilde{\nu}_\mathrm{f}}\varepsilon(\tilde{\nu})\,\mathrm{d}\tilde{\nu}=\text{area 1}+\text{area 2}$$

$$=\frac{1}{2}(\tilde{\nu}_\mathrm{peak}-\tilde{\nu}_\mathrm{i})\varepsilon_\mathrm{peak}+\frac{1}{2}(\tilde{\nu}_\mathrm{f}-\tilde{\nu}_\mathrm{peak})\varepsilon_\mathrm{peak}=\frac{1}{2}(\tilde{\nu}_\mathrm{f}-\tilde{\nu}_\mathrm{i})\varepsilon_\mathrm{peak}$$

$$=\frac{1}{2}\times(1.2\times10^4\ \mathrm{cm^{-1}})\times2.21\times10^4\ \mathrm{dm^3\ mol^{-1}\ cm^{-1}}$$

$$=\boxed{1.3\times10^8\ \mathrm{dm^3\ mol^{-1}\ cm^{-2}}}$$

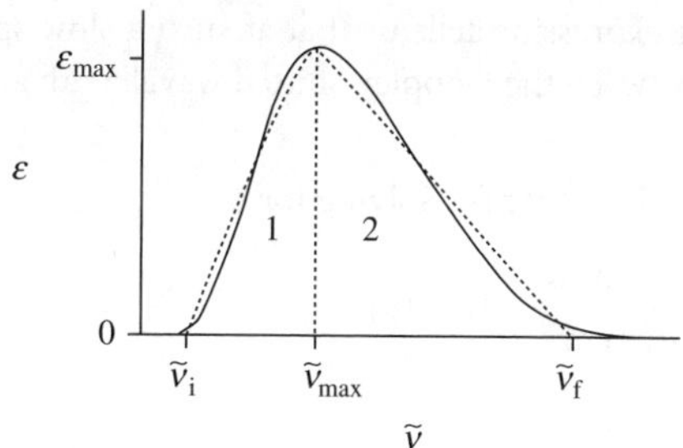

Figure 12A.1

E12A.9(a) We will make the assumption that the absorption curve can be approximated by an isosceles triangle, as shown in Fig. 12A.2. The base of the absorption band triangle is twice the half-width at half-height, $\Delta\tilde{\nu}_{1/2}$, because the small triangle of base $\Delta\tilde{\nu}_{1/2}$ is similar to the absorption band triangle but has half the height. Thus, the area of the absorption band, which equals the integrated absorption coefficient [12A.15], is given by

$$\mathcal{A} = \tfrac{1}{2}\times\varepsilon_{max}\times 2\Delta\tilde{\nu}_{1/2}\ [\text{area} = \tfrac{1}{2}\times\text{height}\times\text{base}] = \varepsilon_{max}\Delta\tilde{\nu}_{1/2} = (5\,000\ \text{cm}^{-1})\times\varepsilon_{max}$$

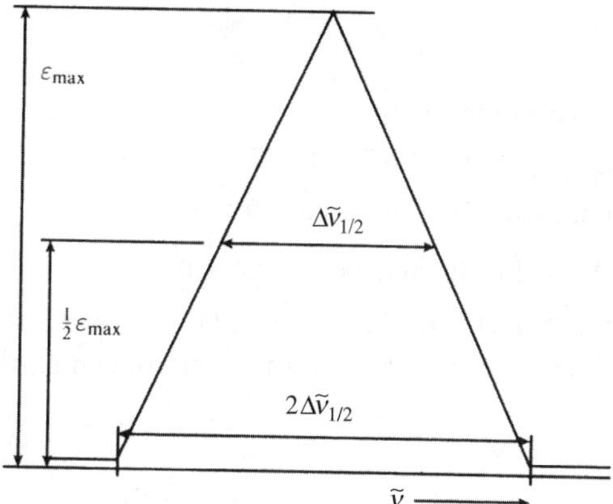

Figure 12A.2

(a) $\mathcal{A} = (5\,000\ \text{cm}^{-1})\times(1\times10^{4}\ \text{dm}^{3}\ \text{mol}^{-1}\ \text{cm}^{-1}) = \boxed{5\times10^{7}\ \text{dm}^{3}\ \text{mol}^{-1}\ \text{cm}^{-2}}$

(b) $\mathcal{A} = (5\,000\ \text{cm}^{-1})\times(5\times10^{2}\ \text{dm}^{3}\ \text{mol}^{-1}\ \text{cm}^{-1}) = \boxed{2.5\times10^{6}\ \text{dm}^{3}\ \text{mol}^{-1}\ \text{cm}^{-2}}$

E12A.10(a) $\nu_{approach} = \left(\dfrac{1+s/c}{1-s/c}\right)^{1/2}\nu$ [12A.16a]

or $\lambda_{approach} = \left(\dfrac{1-s/c}{1+s/c}\right)^{1/2}\lambda$ where $c = 2.9979\times10^{8}\ \text{m s}^{-1} = 1.0793\times10^{9}\ \text{km h}^{-1}$

$$= \left(\frac{1-60/1.0793\times10^{9}}{1+60/1.0793\times10^{9}}\right)^{1/2}\lambda$$

$$= \boxed{0.9999\times\lambda}$$

The coefficient in the above expression tells us that at such a slow speed there is an insignificant difference between the Doppler-shifted wavelength and the wavelength of the traffic light.

E12A.11(a) $\delta E \approx \hbar/\tau$ so, since $E = h\nu$, $\delta\nu = (2\pi\tau)^{-1}$. Solving for τ:

$$\tau = (2\pi\delta\nu)^{-1} = (2\pi c\delta\tilde{\nu})^{-1} = \frac{5.309\,\text{ps}}{\delta\tilde{\nu}/\text{cm}^{-1}} \quad [12\text{A}.19]$$

(a) $\tau = \dfrac{5.309\,\text{ps}}{0.20} = \boxed{27\,\text{ps}}$

(b) $\tau = \dfrac{5.309\,\text{ps}}{2.0} = \boxed{2.7\,\text{ps}}$

E12A.12(a) $\tau = \dfrac{5.309\,\text{ps}}{\delta\tilde{\nu}/\text{cm}^{-1}}$ [12A.19, see Exercise 12A.11(a)]

(a) $\tau \approx 1.0 \times 10^{-13}\,\text{s} = 0.1\,\text{ps}$, implying that $\boxed{\delta\tilde{\nu} = 53\,\text{cm}^{-1}}$.

(b) $\tau \approx 100 \times (1 \times 10^{-13}\,\text{s}) = 10\,\text{ps}$, implying that $\boxed{\delta\tilde{\nu} = 0.53\,\text{cm}^{-1}}$.

Solutions to problems

P12A.1 Fraction transmitted to the retina is

$(1-0.30) \times (1-0.25) \times (1-0.09) \times 0.57 = 0.272$

The number of photons focused on the retina in 0.1 s is

$0.272 \times (40\,\text{mm}^2) \times (0.1\,\text{s}) \times (4 \times 10^3\,\text{mm}^{-2}\,\text{s}^{-1}) = \boxed{4.4 \times 10^3}$.

P12A.3 The derivation is similar to that of the Beer–Lambert law in text *Justification* 12A.1 but let $\mathrm{d}x$ be the infinitesimal thickness of the layer containing the absorbing species J. Then,

$$\mathrm{d}I = -\kappa[\text{J}]I\mathrm{d}x$$

$$\frac{\mathrm{d}I}{I} = -\kappa[\text{J}]\mathrm{d}x$$

$$\int_{I_0}^{I}\frac{\mathrm{d}I}{I} = -\kappa\int_0^L[\text{J}]\mathrm{d}x = -\kappa[\text{J}]_0\int_0^L \mathrm{e}^{-x/\lambda}\,\mathrm{d}x$$

$$\ln(I)\Big|_{I=I_0}^{I=I} = \kappa\lambda[\text{J}]_0\,\mathrm{e}^{-x/\lambda}\Big|_{x=0}^{x=L}$$

$$-\ln\frac{I_0}{I} = -\kappa\lambda[\text{J}]_0\left(1-\mathrm{e}^{-L/\lambda}\right)$$

$$(\ln 10)\log\frac{I_0}{I} = \kappa\lambda[\text{J}]_0\left(1-\mathrm{e}^{-L/\lambda}\right)$$

Let $A = \log\dfrac{I_0}{I}$ [12A.13] and $\varepsilon' = \kappa\lambda/(\ln 10)$. Then,

$$\boxed{A = \varepsilon'[\text{J}]_0\left(1-\mathrm{e}^{-L/\lambda}\right)}$$

In the case for which $L/\lambda >> 1, e^{-L/\lambda} = 0$, $\boxed{A = \varepsilon'[J]_0}$, and the absorbance is independent of the cell length.

In the case for which $L/\lambda << 1, e^{-L/\lambda} \sim 1 + (-L/\lambda) + (-L/\lambda)^2/2$ (Taylor series truncated after 2nd order).

Thus,

$$A = \varepsilon'[J](1 - e^{-L/\lambda})$$
$$A = \varepsilon'[J](1 - 1 + L/\lambda - (L/\lambda)^2/2)$$
$$A = \frac{\varepsilon' L[J]}{\lambda}(1 - L/2\lambda)$$
$$A = \varepsilon L[J] \times (1 - L/2\lambda) \text{ when } L/\lambda << 1$$

P12A.5 Suppose that non-absorbing species B (pyridine in this problem) is progressively added to a solution of light absorbing species A (I_2 in this problem). Furthermore, suppose that A and B form a complex of light absorbing species AB. Suppose that all absorbance measurements are made at the equilibrium $A + B \rightleftharpoons AB$. By the conservation of mass the sum of the equilibrium concentrations of the absorbing species, $[A]_e + [AB]_e$, is a constant for all additions of B. That is, the sum of concentrations of all species that contain A must equal the concentration of A present before any B is added, $[A]_0$. The absorbance at any wavelength is

$$\begin{aligned} A_\lambda &= \varepsilon_{A,\lambda} L[A]_e + \varepsilon_{AB,\lambda} L[AB]_e \\ &= \varepsilon_{A,\lambda} L([A]_0 - [AB]_e) + \varepsilon_{AB,\lambda} L[AB]_e \\ &= \varepsilon_{A,\lambda} L[A]_0 + (\varepsilon_{AB,\lambda} - \varepsilon_{A,\lambda}) L[AB]_e \end{aligned}$$

The first term to the right in the above equation is a constant for all additions of B; the second term is generally not constant and, consequently, A_λ generally varies with the addition of B. However, at any wavelength for which $\varepsilon_{AB,\lambda} = \varepsilon_{A,\lambda}$ the second term vanishes making A_λ a constant at all additions of B. The wavelength at which this happens is called the **isosbestic point**. It is characterized by the equilibrium between absorbing species and $\varepsilon_{AB,\lambda} = \varepsilon_{A,\lambda}$.

P12A.7 Gaussian distribution: $\varepsilon(\tilde{\nu}) = \varepsilon_{max} e^{-\frac{1}{2}\left(\frac{\tilde{\nu}-\mu}{\sigma}\right)^2}$ where μ is the mean of $\tilde{\nu}$ and σ is the standard deviation of the distribution. Dividing by ε_{max}, taking the natural logarithm, and solving for $\tilde{\nu} - \mu$ gives

$$\tilde{\nu} - \mu = \pm\left(2\ln\frac{\varepsilon_{max}}{\varepsilon}\right)^{1/2}\sigma$$

The width of the distribution at half-height, $\Delta\tilde{\nu}_{1/2}$, equals $2|\tilde{\nu} - \mu|$ evaluated at $\varepsilon = \varepsilon_{max}/2$. Thus,

$$\Delta\tilde{\nu}_{1/2} = 2(2\ln 2)^{1/2}\sigma \quad \text{or} \quad \sigma = \frac{\Delta\tilde{\nu}_{1/2}}{2(2\ln 2)^{1/2}}$$

We can now evaluate the integrated absorption coefficient, $\mathcal{A}$, in terms of ε_{max} and $\Delta\tilde{\nu}_{1/2}$.

Let $x = \dfrac{\tilde{\nu}-\mu}{\sigma}$ and $dx = \dfrac{1}{\sigma}d\tilde{\nu}$ and $\varepsilon = \varepsilon_{max}e^{-\frac{1}{2}x^2}$

Then $\mathcal{A} = \int_{-\infty}^{\infty}\varepsilon\, d\tilde{\nu}$ [40.10] $= \varepsilon_{max}\sigma\int_{-\infty}^{\infty}e^{-\frac{1}{2}x^2}\,dx = (2\pi)^{1/2}\varepsilon_{max}\sigma$ [standard integral]

$$= \boxed{\frac{1}{2}\left(\frac{\pi}{\ln 2}\right)^{1/2}\varepsilon_{max}\Delta\tilde{\nu}_{1/2}}$$

$$= 1.064467\,\varepsilon_{max}\Delta\tilde{\nu}_{1/2}$$

The Gaussian distribution is symmetric about the mean value of $\tilde{\nu}$, μ, which is the value of $\tilde{\nu}$ at the peak of the distribution. The absorption band of text Fig. 12.2 does not quite have this symmetry. It appears to be a skewed slightly toward the higher wavenumbers. Never the less, we estimate $\mathcal{A}$ by assuming that it can be approximated as a single Gaussian characterized by ε_{max} and $\Delta\tilde{\nu}_{1/2}$ values that are coarsely read off text Fig. 12.2.

Coarse estimate:

$$\mathcal{A} = 1.064467\,\varepsilon_{max}\Delta\tilde{\nu}_{1/2} = 1.064467\times(10\text{ dm}^3\text{ mol}^{-1}\text{ cm}^{-1})\times(5.4\times10^3\text{ cm}^{-1})$$

$$= \boxed{5.7\times10^4\text{ dm}^3\text{ mol}^{-1}\text{ cm}^{-2}}$$

Let us now suppose that the slightly non-Gaussian shape exhibited by text Fig. 12.2 results from two separate absorption lines each of which has a molar absorption coefficient that is a Gaussian function of wavenumber. Figure 12.2 is then an 'apparent' molar absorption coefficient that is the sum of two independent Gaussians each characterized by an amplitude A, a mean value μ, and a standard deviation σ. That's a total of six parameters to be adjusted to fit the data of the figure. We label the three parameters of the predominate Gaussian, the one with the lower mean wavenumber, with a '1'; the low amplitude, higher mean distribution is labelled with a '2'. A lot of $(\tilde{\nu}, \varepsilon)$, data pairs are needed to determine precise values of the parameters so we expanded Fig. 12.2 and used Photoshop to read a total of 20 data pairs, several of which are displayed in the following Mathcad Prime 2 worksheet. Calling the sum of the two Gaussians G_{sum}, the worksheet uses guess values for the 6 parameters to calculate the difference $\varepsilon_{obs} - G_{sum}$ at each of the 20 $\tilde{\nu}_{obs}$, the difference is squared and summed over all data pairs, which the worksheet calls the 'sum of the squared errors' *SSE*. The idea is to systematically adjust the six parameters so as to minimize *SSE*. Mathcad performs the minimization process with the 'minerr()' function within a solve block. The symbol '*v*' is used to represent wavenumber within the worksheet.

$$\upsilon = \begin{bmatrix} 2.2\cdot 10^4 \\ 2.249\cdot 10^4 \\ 2.293\cdot 10^4 \\ \vdots \end{bmatrix} \text{cm}^{-1} \qquad \varepsilon = \begin{bmatrix} 0.412 \\ 0.7 \\ 1.317 \\ \vdots \end{bmatrix} \text{dm}^3\cdot\text{mol}^{-1}\cdot\text{cm}^{-1}$$

$$G_{sum}(\upsilon,\mu_1,\Delta\upsilon_1,A_1,\mu_2,\Delta\upsilon_2,A_2) := A_1\cdot e^{\frac{-1}{2}\left(\frac{\upsilon-\mu_1}{\Delta\upsilon_1}\right)^2} + A_2\cdot e^{\frac{-1}{2}\left(\frac{\upsilon-\mu_2}{\Delta\upsilon_2}\right)^2}$$

$$SSE(\mu_1,\Delta\upsilon_1,A_1,\mu_2,\Delta\upsilon_2,A_2) := \sum_{i=0}^{19}\left(\varepsilon_i - G_{sum}(\upsilon_i,\mu_1,\Delta\upsilon_1,A_1,\mu_2,\Delta\upsilon_2,A_2)\right)^2$$

Guess Values

$\mu_1 := 26\cdot 10^3\cdot\text{cm}^{-1}$ $\qquad\mu_2 := 29\cdot 10^3\cdot\text{cm}^{-1}$

$\Delta\upsilon_1 := 2\cdot 10^3\cdot\text{cm}^{-1}$ $\qquad\Delta\upsilon_2 := 2\cdot 10^3\cdot\text{cm}^{-1}$

$A_1 := 8\cdot\text{dm}^3\cdot\text{mol}^{-1}\cdot\text{cm}^{-1}$ $\qquad A_2 := 3\cdot\text{dm}^3\cdot\text{mol}^{-1}\cdot\text{cm}^{-1}$

Constraints

$SSE(\mu_1,\Delta\upsilon_1,A_1,\mu_2,\Delta\upsilon_2,A_2) = 0$

Solver

$Parameters := \textbf{minerr}(\mu_1,\Delta\upsilon_1,A_1,\mu_2,\Delta\upsilon_2,A_2)$

$\mu_1 := Parameters_0 \qquad \mu_1 = (2.59\cdot 10^4)\frac{1}{\text{cm}}$

$\Delta\upsilon_1 := Parameters_1 \qquad \Delta\upsilon_1 = (1.597\cdot 10^3)\frac{1}{\text{cm}}$

$A_1 := Parameters_2 \qquad A_1 = 6.931\frac{\text{dm}^3}{\text{mol}\cdot\text{cm}}$

$\mu_2 := Parameters_3 \qquad \mu_2 = (2.856\cdot 10^4)\frac{1}{\text{cm}}$

$\Delta\upsilon_2 := Parameters_4 \qquad \Delta\upsilon_2 = (2.234\cdot 10^3)\frac{1}{\text{cm}}$

$A_2 := Parameters_5 \qquad A_2 = 5.131\frac{\text{dm}^3}{\text{mol}\cdot\text{cm}}$

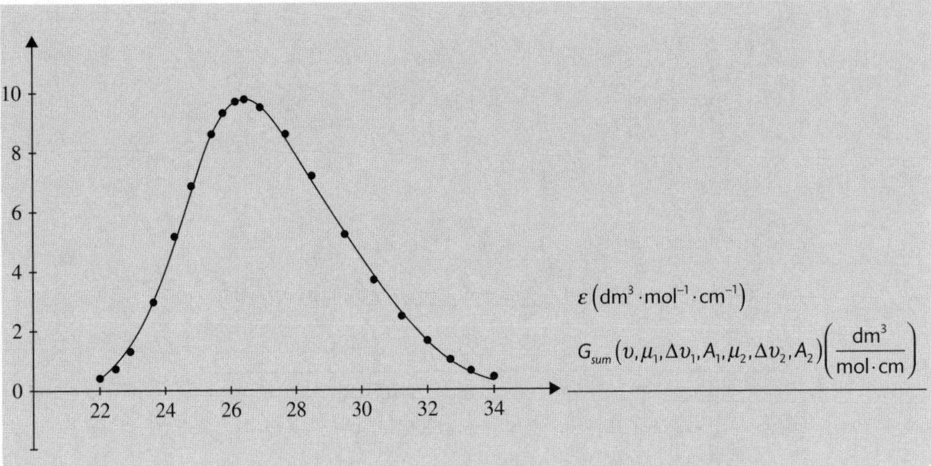

The plot shows that the sum of two Gaussians with adjusted parameters fits the data very nicely. The values of the six parameters are listed just above the plot. The following worksheet section uses the fitted function to calculate the integrated absorption coefficient with eqn 12A.15. The earlier, coarse estimate is seen to be rather close to the more precise calculation.

$$integrated\_absorption\_coeff := \int_{0\cdot cm^{-1}}^{10^6\cdot cm^{-1}} G_{sum}(\upsilon,\mu_1,\Delta\upsilon_1,A_1,\mu_2,\Delta\upsilon_2,A_2)\,d\upsilon$$

$$integrated\_absorption\_coeff = (5.649\cdot 10^4)\,dm^3\cdot mol^{-1}\cdot cm^{-2}$$

P12A.9 (a) Compute the ratios ν_{star}/ν for all three lines. We are given wavelength data, so we can use:

$$\frac{\nu_{star}}{\nu}=\frac{\lambda}{\lambda_{star}}.$$

The ratios are:

$$\frac{438.392\,\text{nm}}{438.882\,\text{nm}}=0.998884,\quad \frac{440.510\,\text{nm}}{441.000\,\text{nm}}=0.998889,\text{ and }\frac{441.510\,\text{nm}}{442.020\,\text{nm}}=0.998846.$$

The frequencies of the stellar lines are all less than those of the stationary lines, so we infer that the star is $\boxed{\text{receding}}$ from Earth. The Doppler effect follows:

$$\nu_{receding}=\nu f\quad\text{where } f=\left(\frac{1-s/c}{1+s/c}\right)^{1/2}\text{, so}$$

$$f^2(1+s/c)=(1-s/c),\quad (f^2+1)s/c=1-f^2,\quad s=\frac{1-f^2}{1+f^2}c$$

Our average value of f is 0.998873. (Note: the uncertainty is actually greater than the significant figures here imply, and a more careful analysis would treat uncertainty explicitly.) So the speed of recession with respect to the Earth is:

$$s=\left(\frac{1-0.997747}{1+0.997747}\right)c=\boxed{1.128\times10^{-3}\ \mathrm{c}}=\boxed{3.381\times10^{5}\ \mathrm{m\ s^{-1}}}$$

(b) One could compute the star's radial velocity with respect to the Sun if one knew the Earth's speed with respect to the Sun along the Sun–star vector at the time of the spectral observation. This could be estimated from quantities available through astronomical observation: the Earth's orbital velocity times the cosine of the angle between that velocity vector and the Earth–star vector at the time of the spectral observation. (The Earth–star direction, which is observable by earth-based astronomers, is practically identical to the Sun–star direction, which is technically the direction needed.) Alternatively, repeat the experiment half a year later. At that time, the Earth's motion with respect to the Sun is approximately equal in magnitude and opposite in direction compared to the original experiment. Averaging f values over the two experiments would yield f values in which the Earth's motion is effectively averaged out.

P12A.11 Transform eqn 12A.18 to the standard Gaussian form with the transformations $\sigma(T)=\left(kT/mc^2\right)^{1/2}$, which is the standard deviation, and $x(\nu_{\text{obs}})=(\nu_{\text{obs}}-\nu)/\nu$. σ^2 is the ratio of thermal energy to total, relativistic energy while x is a linear function of ν_{obs}. The study of dI/dν_{obs} and the effects of varying T are equivalent to study of dI/dx and the effects of varying σ. Here's a Mathcad Prime 2 study.

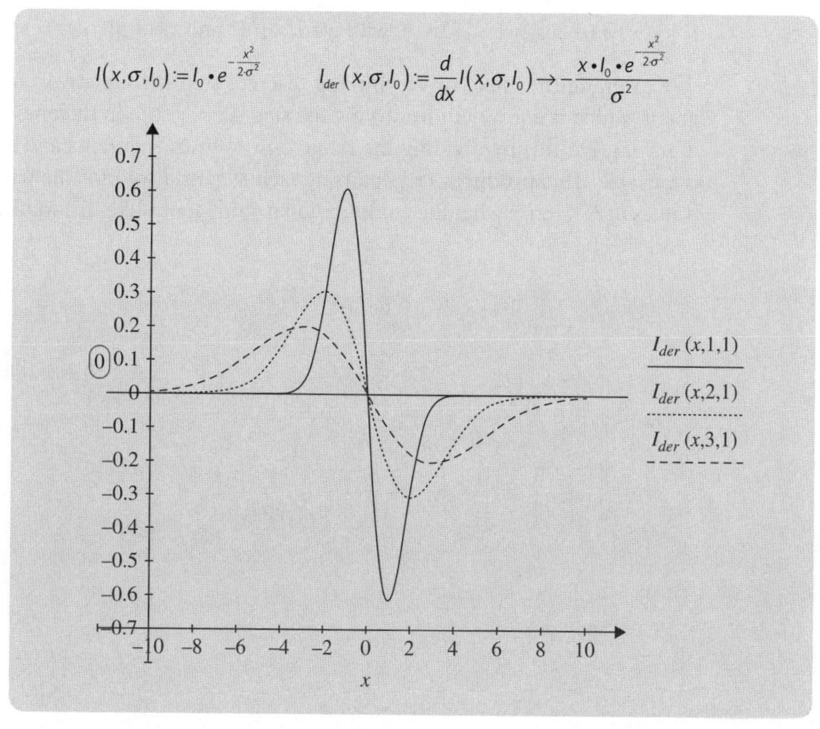

The plot indicates that as the standard deviation increases from 1 to 3, which is equivalent to a systematic increase in T, the extrema separate and the distribution of signal from the phase-sensitive detector broadens. An analytic determination of the dependence of the extrema separation upon σ involves finding dI_{der}/dx, setting it equal to zero, and solving for x.

$$\frac{d}{dx}I_{der}(x,\sigma,I_0)\xrightarrow{\text{solve},x}\begin{bmatrix}\sigma\\-\sigma\end{bmatrix}$$

Thus, the separation is $2\sigma(T)=\boxed{2(kT/mc^2)^{1/2}}$.

To finish the study, we calculate the standard deviation exhibited by an atomic hydrogen gas at the temperature of the surface of the sun (6 000 K).

$$\sigma_{T=6000\text{ K}}=\left(\frac{kT}{mc^2}\right)^{1/2}=\left(\frac{RT}{Mc^2}\right)^{1/2}$$
$$=\left(\frac{(8.314\text{ J K}^{-1}\text{ mol}^{-1})\times(6\,000\text{ K})}{(1.0\times10^{-3}\text{ kg mol}^{-1})\times(3.00\times10^{8}\text{ m s}^{-1})^2}\right)^{1/2}$$
$$=2.4\times10^{-5}$$

P12A.13 For a monochromatic beam of wavenumber $\tilde{\nu}$, the ratio of the output of a Michelson interferometer to the entering intensity I_0 is given by eqn 12A.20 where p is the path length difference imposed by the position of mirror M_1 in text Fig. 12A.9.

$$y(x)=1+\cos 2\pi x\quad [\text{12A.20 with } y\equiv I(x)/I_0 \text{ and } x\equiv\tilde{\nu}p]$$

We examine the effect of varying the discrete step increment Δp of the instrument, which is equivalent to varying the step $\Delta x=\tilde{\nu}\cdot\Delta p$, in the one-period range $0\le x\le 1$. Do this by dividing the range into N equal segments and treating N as a variable. This produces (x,y) pairs at each step and we plot the pairs for select values of N. Here's what the computational study looks like in a Mathcad Prime 2 worksheet:

```
f(N) :=  for i ε 0..N
             x_i ← i/N
             y_i ← 1+cos(2·π·x_i)
         return [x
                 y]

x(N) := f(N)_0,0
y(N) := f(N)_1,0
```

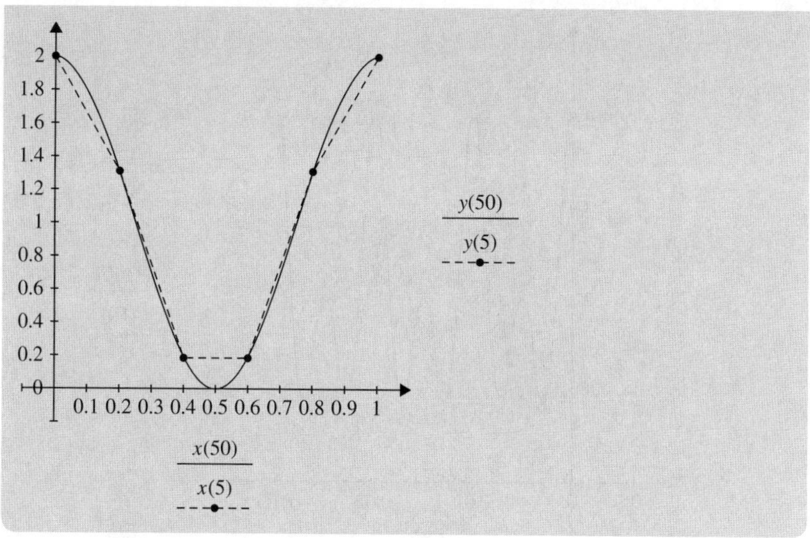

The plot clearly shows that when a large number of data pairs N are used, the Michelson interferometer output replicates the continuous variation of eqn 12A.20 over the whole range. Even the selection of a small number of data pairs like $N=$ 5 provides a good representation of the interferogram but one wonders whether important information is lost in regions between sampled points. To explore this question, we create a Prime 2 worksheet of an interferogram as a function of N when there are two spectrum lines of wavenumbers 100 cm^{-1} and 200 cm^{-1} of equal intensity (e.g. 1; the interferometer parameter p has the unit cm).

```
f(N) := | for i ε 0..N
        |   p_i ← (i/N)·(2·π/200)
        |   y_i ← (1+cos(2·π·100p_i))+(1+cos(2·π·200·p_i))
        | return [ p ]
        |        [ y ]
```

$p(N) := f(N)_{0,0}$

$y(N) := f(N)_{1,0}$

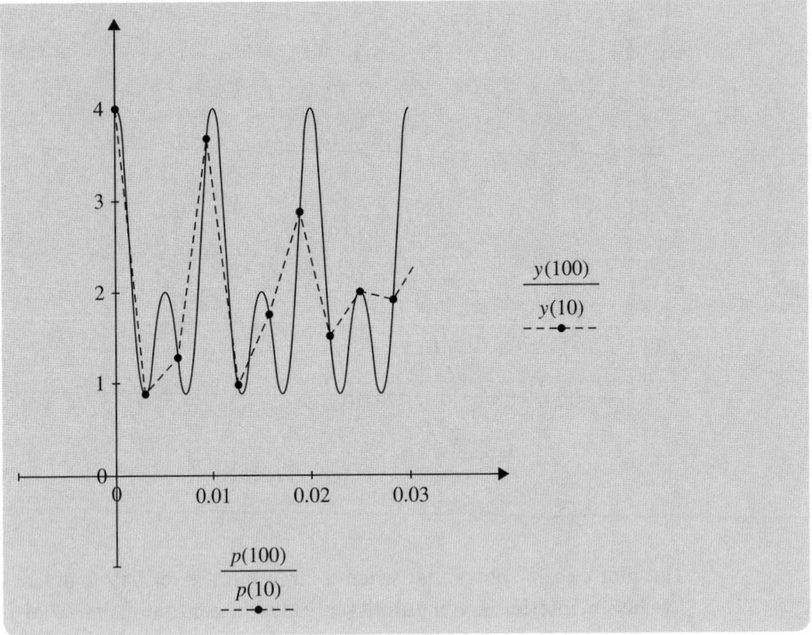

Once again, large values of N reproduce the interferogram of the continuous variation described by eqn 12A.21. However, major features are omitted from the interferogram, or distorted, when either the range of p is smaller than a cycle in p or the discrete step value Δp is too large to represent the interferogram detail within a cycle. The plot of the data set $(p(10),y(10))$ is clearly a poor representation when the range is $0 \le p \le 0.03\,\mathrm{cm}$ when the input consists of $100\,\mathrm{cm}^{-1}$ and $200\,\mathrm{cm}^{-1}$ lines. We explore the effect of the distortions by performing a discrete Fourier transform on the interferogram of the two-line spectrum with the discrete, incremental form of eqn 12A.22 (see text Example 12.2). The worksheet uses the symbol v to represent wavenumber and the plot divisions of 225 and 25 are inserted to normalize peaks to 1.

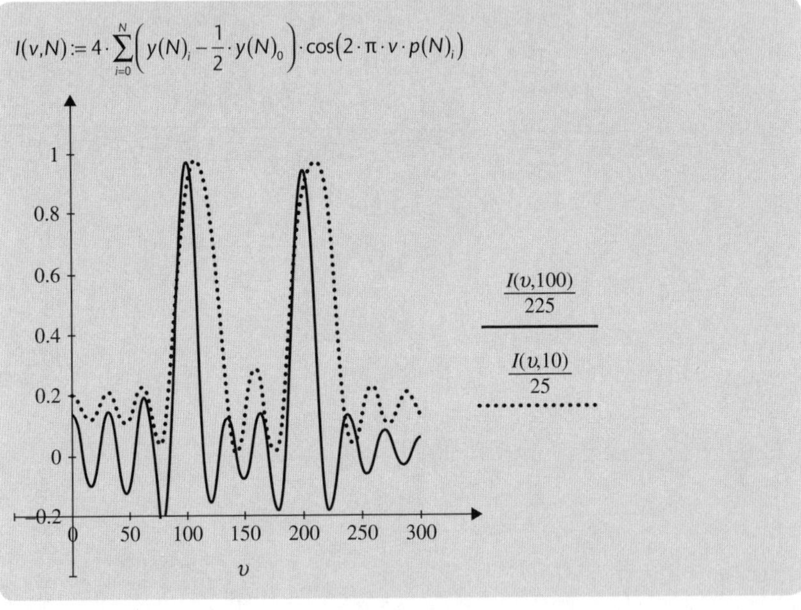

When N is large, the above plot clearly shows the two lines at correct wavenumbers of $100\,cm^{-1}$ and $200\,cm^{-1}$. At the small value of N, the lines have been distorted to incorrectly high wavenumber values. This distortion problem is called **aliasing**, a problem that can also give spurious lines and must be avoided in spectroscopic work.

12B Molecular rotation

Answers to discussion questions

D12B.1 Symmetric rotor: the energy depends on J and K^2 [eqn 12B.13], hence each level except the $K=0$ level is doubly degenerate. In addition, states of a given J have a component of their angular momentum along an external, laboratory-fixed axis, characterized by the quantum number M_J, that can take on $2J+1$ values. The quantum number M_J does not affect the energy, consequently all $2J+1$ orientations of the molecule have the same energy. It follows that a symmetric rotor level is $2(2J+1)$-fold degenerate for $K\neq 0$, and $2J+1$ degenerate for $K=0$.

Linear rotor: A linear rotor has K fixed at 0, but there are still $2J+1$ values of M_J, so the degeneracy is $2J+1$.

Spherical rotor: A spherical rotor can be regarded as a version of a symmetric rotor in which $A=B$, and consequently, the energy is independent of the $2J+1$ values that K can assume. Hence, there is a simultaneous degeneracy of $2J+1$ in both K and M_J, resulting in a total degeneracy of $(2J+1)^2$.

Asymmetric rotor: In an asymmetric rotor there is no longer a preferred direction in the molecule which carries out a simple rotation about J, thus for each value of J, there is only a degeneracy of $2J+1$, as in a linear rotor.

If a decrease in rigidity affects the symmetry of the molecule, the rotational degeneracy could be affected also. For example, if a wobble resulted in a spherical rotor being transformed into an asymmetric rotor the rotational degeneracy of the molecule would change.

Solutions to exercises

E12B.1(a) Ozone is an asymmetric rotor similar to H_2O; hence we follow the method of Example 12B.1 and use $I=\sum_i m_i x_i^2$ [12B.1] to calculate the moment of inertia, I_C about the C_2 axis.

$$I=m_O(x_O^2+0+x_O^2)=2m_O x_O^2=2m_O(R\sin\phi)^2,$$

where the bond angle is denoted 2ϕ and the bond length R. Substitution of the data gives

$$I = 2\times 15.9949 m_u \times 1.66054\times 10^{-27}\ \text{kg}/m_u \times (1.28\times 10^{-10}\ \text{m}\times \sin 58.5°)^2$$

$$= \boxed{6.33\times 10^{-46}\ \text{kg m}^2}$$

The corresponding rotational constant is

$$\tilde{A} = \frac{\hbar}{4\pi c I_c} = \frac{1.05447\times 10^{-34}\ \text{J s}}{4\pi\times 2.998\times 10^{8}\ \text{m s}^{-1}\times 6.33\times 10^{-46}\ \text{kg m}^2} = 44.21\ \text{m}^{-1} = \boxed{0.4421\ \text{cm}^{-1}}$$

E12B.2(a) In order to conform to the symbols used in the first symmetric rotor figure of Table 12B, we will use the molecular formula BA_4. $I_{\parallel}$ is along a bond and $I_{\perp}$ is perpendicular to both $I_{\parallel}$ and a molecular face that does not contain $I_{\parallel}$ (see the spherical rotor of text Fig. 12B.3). For our molecule, $R' = R$, $m_C = m_A$, and the equations of Table 12.1 simplify to

$$I_{\parallel}/m_A R^2 = 2(1-\cos\theta)$$

$$I_{\perp}/m_A R^2 = 1-\cos\theta + (m_A + m_B)\times(1+2\cos\theta)/m$$

$$+\left\{(3m_A + m_B) + 6m_A\left[\tfrac{1}{3}(1+2\cos\theta)\right]^{1/2}\right\}/m \text{ where } m = 4m_A + m_B.$$

The $I_{\parallel}/m_A R^2$ moment of inertia ratio does not depend upon specific atomic masses so the plot of this ratio against θ, shown in Fig. 12B.1 describes the angular dependence of all molecules having the formula BA_4. However, the $I_{\perp}/m_A R^2$ moment of inertia ratio does have a specific atomic mass dependency so we will plot its angular dependence for CH_4, an important fuel and powerful greenhouse gas. The computational equation is

$$I_{\perp}/m_A R^2 = 1-\cos\theta + 13(1+2\cos\theta)/16 + \left\{15 + 6\left[\tfrac{1}{3}(1+2\cos\theta)\right]^{1/2}\right\}/16$$

and its plot is also found in Fig. 12B.1. As θ increases, atoms move away from the axis of $I_{\parallel}$, which causes the moment of inertia around this axis to increase. Atoms move toward the axis of $I_{\perp}$ as θ increases, thereby decreasing the moment of inertia around this axis.

Comment. Figure 12B.1 suggests that $I_{\parallel}(\theta_{\text{tetra}}) = I_{\perp}(\theta_{\text{tetra}})$ for all tetrahedral molecules where θ_{tetra} is the tetrahedral angle (approx. 109.471°). Can you provide an analytic proof that this is true? Hint: The exact value of θ_{tetra} is $\pi/2 + \sin^{-1}(1/3)$ in radians where $\sin^{-1}()$ is the arcsin() function.

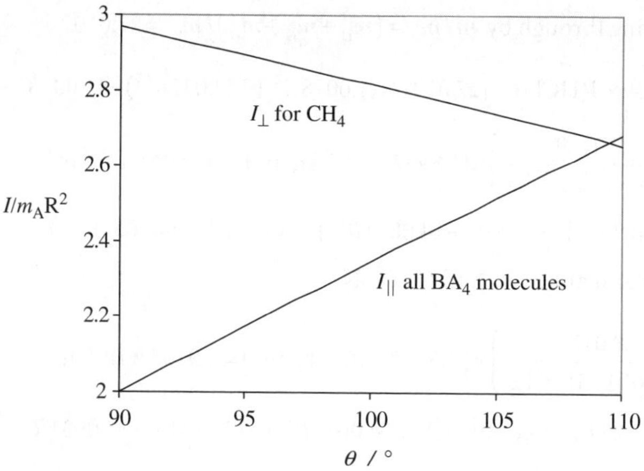

Figure 12B.1

E12B.3(a) (a) asymmetric; (b) oblate symmetric; (c) spherical; (d) prolate symmetric.

E12B.4(a) The determination of two unknowns requires data from two independent experiments and the equation which relates the unknowns to the experimental data. In this exercise two independently determined values of B for two isotopically different HCN molecules are used to obtain the moments of inertia of the molecules and from these, by use of the equation for the moment of inertia of linear triatomic rotors (Table 12.1), the interatomic distances R_{HC} and R_{CN} are calculated.

Rotational constants, which are usually expressed in wavenumbers (cm^{-1}), are sometimes expressed in frequency units (Hz). The conversion between the two is

$$B/\text{Hz} = c \times \tilde{B}/\text{cm}^{-1} \; [c \text{ in cm s}^{-1}]$$

Thus, $B(\text{in Hz}) = \dfrac{\hbar}{4\pi I}$ and $I = \dfrac{\hbar}{4\pi B}$

Let, ${}^1\text{H} = \text{H}$, ${}^2\text{H} = \text{D}$, $R_{HC} = R_{DC} = R$, $R_{CN} = R'$. Then

$$I(\text{HCN}) = \frac{1.05457 \times 10^{-34}\,\text{J s}}{(4\pi) \times (4.4316 \times 10^{10}\,\text{s}^{-1})} = 1.8937 \times 10^{-46}\,\text{kg m}^2$$

$$I(\text{DCN}) = \frac{1.05457 \times 10^{-34}\,\text{J s}}{(4\pi) \times (3.6208 \times 10^{10}\,\text{s}^{-1})} = 2.3178 \times 10^{-46}\,\text{kg m}^2$$

and from Table 12.1 with isotope masses from the *Data section.*

$$I(\text{HCN}) = m_H R^2 + m_N R'^2 - \frac{(m_H R - m_N R')^2}{m_H + m_C + m_N}$$

$$I(\text{HCN}) = \left[(1.0078R^2) + (14.0031R'^2) - \left(\frac{(1.0078R - 14.0031R')^2}{1.0078 + 12.0000 + 14.0031}\right)\right] m_u$$

Multiplying through by $m/m_u = (m_H + m_C + m_N)/m_u = 27.0109$

$$27.0109 \times I(\text{HCN}) = \{27.0109 \times (1.0078R^2 + 14.0031R'^2) - (1.0078R - 14.0031R')^2\}m_u$$

or $$\left(\frac{27.0109}{1.66054 \times 10^{-27}\ \text{kg}}\right) \times (1.8937 \times 10^{-46}\ \text{kg m}^2) = 3.0804 \times 10^{-18}\ \text{m}^2$$

$$= \{27.0109 \times (1.0078R^2 + 14.0031R'^2) - (1.0078R - 14.0031R')^2\} \qquad \text{(a)}$$

In a similar manner we find for DCN

$$\left(\frac{28.0172}{1.66054 \times 10^{-27}\ \text{kg}}\right) \times (2.3178 \times 10^{-46}\ \text{kg m}^2) = 3.9107 \times 10^{-18}\ \text{m}^2$$

$$= \left\{28.0172 \times (2.0141R^2 + 14.0031R'^2) - (2.0141R - 14.0031R')^2\right\} \qquad \text{(b)}$$

Thus there are two simultaneous quadratic equations (a) and (b) to solve for R and R'. These equations are most easily solved by readily available computer programs or by successive approximations. The results are

$$R = 1.065 \times 10^{-10}\ \text{m} = \boxed{106.5\ \text{pm}} \quad \text{and} \quad R' = 1.156 \times 10^{-10}\ \text{m} = \boxed{115.6\ \text{pm}}$$

These values are easily verified by direct substitution into the equations and agree well with the accepted values $R_{HC} = 1.064 \times 10^{-10}$ m and $R_{CN} = 1.156 \times 10^{-10}$ m.

E12B.5(a) The centrifugal distortion constant is given by

$$\tilde{D}_J = \frac{4\tilde{B}^3}{\tilde{\nu}^2} \quad \text{[12B.17, see also Integrated activity I12.2]}$$

$$\tilde{D}_J = \frac{4 \times (6.511\ \text{cm}^{-1})^3}{(2308\ \text{cm}^{-1})^2} = \boxed{2.073 \times 10^{-4}\ \text{cm}^{-1}}$$

$$\tilde{D}_J \propto \frac{1}{m_H^3} \text{ and } \frac{\tilde{D}_J(^2\text{H})}{\tilde{D}_J(^1\text{H})} = \frac{m_{^1\text{H}}^3}{m_{^2\text{H}}^3} = \frac{(1.0078\ m_u)^3}{(2.0140\ m_u)^3} = \boxed{0.1253}$$

$$\text{Therefore, } \tilde{D}_J \propto \frac{1}{m_H^3} \text{ and } \frac{\tilde{D}_J(^2\text{H})}{\tilde{D}_J(^1\text{H})} = \frac{m_{^1\text{H}}^3}{m_{^2\text{H}}^3} = \frac{(1.0078\ m_u)^3}{(2.0140\ m_u)^3} = \boxed{0.1253}$$

We have assumed that the internuclear distance remains constant upon substitution.

Solutions to problems

P12B.1 The centre of mass of a diatomic molecule lies at a distance x from atom A and is such that the masses on either side of it balance

$$m_A x = m_B(R - x)$$

and hence it is at

$$x = \frac{m_B}{m}R \quad m = m_A + m_B$$

The moment of inertia of the molecule is

$$I = m_A x^2 + m_B (R-x)^2 \text{ [12B.1]} = \frac{m_A m_B^2 R^2}{m^2} + \frac{m_B m_A^2 R^2}{m^2} = \frac{m_A m_B}{m} R^2$$

$$= \boxed{m_{eff} R^2} \text{ since } m_{eff} = \frac{m_A m_B}{m_A + m_B}$$

12C Rotational spectroscopy

Answers to discussion questions

D12C.1 The gross selection rules tell us which are the allowed spectroscopic transitions. For both microwave and infrared spectroscopy, the allowed transitions depend on the existence of an oscillating dipole moment which can stir the electromagnetic field into oscillation (and vice versa for absorption). For microwave rotational spectroscopy, this implies that the molecule must have a permanent dipole moment, which is equivalent to an oscillating dipole when the molecule is rotating. See Fig. 12C.1 of the text. In the case of infrared vibrational spectroscopy, the physical basis of the gross selection rule is that the molecules have a structure that allows for the existence of an oscillating dipole moment when the molecule vibrates. Polar molecules necessarily satisfy this requirement, but nonpolar molecules may also have a fluctuating dipole moment upon vibration. See Fig. 12D.3 of the text.

D12C.3 $^1H^{12}C{\equiv}^{12}C^1H$ is a linear molecule. It has hydrogen nuclei with half integral spin and carbon nuclei with zero spin. Rotation by 180° produces an equivalent form. This rotation is equivalent to the interchange of identical hydrogen nuclei. Since hydrogen is a fermion, the overall rotational wave function must be antisymmetric with respect to exchange of identical particles. The overall statistics are Fermi statistics. This leads to the result that symmetric hydrogen nuclear spin functions must be paired with antisymmetrical spatial wave functions for rotation and vice versa. Consequently $^1H^{12}C{\equiv}^{12}C^1H$ will exhibit ortho and para forms just as for H_2. The (odd J)/(even J) statistical weight ratio is 3/1.

$^2H^{12}C{\equiv}^{12}C^2H$ is a linear molecule. It has deuterium nuclei with integral spin and carbon nuclei with zero spin. Rotation by 180° produces an equivalent form. This rotation is equivalent to the interchange of identical deuterium nuclei. Since deuterium is a boson, the overall rotational wave function must be symmetric with respect to exchange of identical particles. The overall statistics are Bose statistics. This leads to the result that symmetric deuterium nuclear spin functions must be paired with symmetrical spatial wave functions for rotation and also leads to the result that antisymmetric deuterium nuclear spin functions must be paired with antisymmetrical spatial wave functions for rotation. Consequently

$^2H^{12}C{\equiv}^{12}C^2H$ will exhibit ortho and para forms. The (odd J)/(even J) statistical weight ratio is 3/6.

From the point of view of the effect of nuclear statistics on the occupation of rotational energy levels the situation in $^1H^{13}C{\equiv}^{13}C^1H$ is similar to the case of $^2H^{12}C{\equiv}^{12}C^2H$. $^1H^{13}C{\equiv}^{13}C^1H$ will exhibit ortho and para forms. The (odd J)/(even J) statistical weight ratio is 6/10.

Solutions to exercises

E12C.1(a) Polar molecules show a pure rotational absorption spectrum. Therefore, select the polar molecules based on their well-known structures. Alternatively, determine the point groups of the molecules and use the rule that only molecules belonging to C_n, C_{nv}, and C_s may be polar, and in the case of Cn and Cnv, that dipole must lie along the rotation axis. Hence the polar molecules are

(ii) HCl **(iv)** CH_3Cl **(v)** CH_2Cl_2

Their point group symmetries are

(ii) $C_{\infty v}$ **(iv)** C_{3v} **(v)** C_{2v}

E12C.2(a) NO is a linear rotor, and we assume there is little centrifugal distortion; hence

$$\tilde{F}(J) = \tilde{B}J(J+1) \text{ [12B.9]}$$

with $\tilde{B}=\dfrac{\hbar}{4\pi cI}$, $I=m_{\text{eff}}R^2$ [12B.15], and

$$m_{\text{eff}} = \frac{m_N m_O}{m_N+m_O} \text{[nuclide masses from the } \textit{Resource section}]$$

$$= \left(\frac{(14.003m_u)\times(15.995m_u)}{(14.003m_u)+(15.995m_u)}\right)\times(1.6605\times10^{-27}\text{ kg}m_u^{-1})=1.240\times10^{-26}\text{ kg}$$

Then, $I=(1.240\times10^{-26}\text{ kg})\times(1.15\times10^{-10}\text{ m})^2 = 1.64\bar{0}\times10^{-46}\text{ kg m}^2$

and $\tilde{B}=\dfrac{1.0546\times10^{-34}\text{ J s}}{(4\pi)\times(2.998\times10^8\text{ m s}^{-1})\times(1.64\bar{0}\times10^{-46}\text{ kg m}^2)}=170.\bar{7}\text{ m}^{-1}=1.70\bar{7}\text{cm}^{-1}$

The wavenumber of the $J=3\leftarrow2$ transition is

$$\tilde{\nu}=2\tilde{B}(J+1)\text{[12C.8a]}=6\tilde{B}[J=2]=(6)\times(1.70\bar{7}\text{ cm}^{-1})=10.2\bar{4}\text{cm}^{-1}$$

The frequency is

$$\nu=\tilde{\nu}c=(10.2\bar{4}\text{ cm}^{-1})\times\left(\frac{10^2\text{ m}^{-1}}{1\text{cm}^{-1}}\right)\times(2.998\times10^8\text{ ms}^{-1})=\boxed{3.07\times10^{11}\text{ Hz}}$$

When centrifugal distortion is taken into account the frequency decreases, as can be seen by considering eqn 12.8b.

Question. What is the percentage change in these calculated values if centrifugal distortion is included?

E12C.3(a) The wavenumber of the transition is related to the rotational constant by

$$hc\tilde{\nu} = \Delta E = hc\tilde{B}[J(J+1)-(J-1)J] = 2hc\tilde{B}J \text{ [12B.10]}$$

where J refers to the upper state (J=3). The rotational constant is related to molecular structure by

$$\tilde{B} = \frac{\hbar}{4\pi cI} \text{ [12B.7]}$$

where I is moment of inertia. Putting these expressions together yields

$$\tilde{\nu} = 2\tilde{B}J = \frac{\hbar J}{2\pi cI} \quad \text{so} \quad I = \frac{\hbar J}{2\pi c\tilde{\nu}} = \frac{(1.0546\times10^{-34}\text{ J s})\times(3)}{2\pi(2.998\times10^{10}\text{ cm s}^{-1})\times(63.56\text{cm}^{-1})}$$
$$= 2.642\times10^{-47}\text{ kg m}^2$$

The moment of inertia is related to the bond length by

$$I = m_{\text{eff}}R^2 \text{ so } R = \sqrt{\frac{I}{m_{\text{eff}}}}$$

$$m_{\text{eff}}^{-1} = m_{\text{H}}^{-1} + m_{\text{Cl}}^{-1} = \frac{(1.0078\ m_{\text{u}})^{-1} + (34.9688\ m_{\text{u}})^{-1}}{1.66054\times10^{-27}\text{ kg } m_{\text{u}}^{-1}} = 6.1477\times10^{26}\text{ kg}^{-1}$$

and $R = \sqrt{(6.1477\times10^{26}\text{ kg}^{-1})\times(2.642\times10^{-47}\text{ kg m}^2)} = 1.274\times10^{-10}\text{ m} = \boxed{127.4\text{pm}}$

E12C.4(a) If the spacing of lines is constant, the effects of centrifugal distortion are negligible. Hence we may use for the wavenumbers of the transitions

$$\tilde{F}(J) - \tilde{F}(J-1) = 2\tilde{B}J \text{ [12B.10]}$$

Since J=1, 2, 3 …, the spacing of the lines is $2\tilde{B}$

$$12.604\text{ cm}^{-1} = 2\tilde{B}$$

$$\tilde{B} = 6.302\text{ cm}^{-1} = 6.302\times10^{2}\text{ m}^{-1}$$

$$I = \frac{\hbar}{4\pi c\tilde{B}} = m_{\text{eff}}R^2$$

$$\frac{\hbar}{4\pi c} = \frac{1.0546\times10^{-34}\text{ J s}}{(4\pi)\times(2.9979\times10^{8}\text{ m s}^{-1})} = 2.7993\times10^{-44}\text{ kg m}$$

$$I = \frac{2.7993\times10^{-44}\text{ kg m}}{6.302\times10^{2}\text{m}^{-1}} = \boxed{4.442\times10^{-47}\text{ kg m}^2}$$

$$m_{\text{eff}} = \frac{m_{\text{Al}} m_{\text{H}}}{m_{\text{Al}} + m_{\text{H}}}$$

$$= \left(\frac{(26.98)\times(1.008)}{(26.98)+(1.008)}\right) m_{\text{u}} \times \left(1.6605\times10^{-27}\,\text{kg}\,m_{\text{u}}^{-1}\right) = 1.613\bar{6}\times10^{-27}\,\text{kg}$$

$$R = \left(\frac{I}{m_{\text{eff}}}\right)^{1/2} = \left(\frac{4.442\times10^{-47}\,\text{kg m}^2}{1.6136\times10^{-27}\,\text{kg}}\right)^{1/2} = 1.659\times10^{-10}\,\text{m} = \boxed{165.9\,\text{pm}}$$

E12C.5(a) See eqn 12C.9 and Problem 12C.9. The most highly populated rotational level is given by

$$J_{\text{max}} \approx \left(\frac{kT}{2hc\tilde{B}}\right)^{1/2} - \frac{1}{2}\ [12\text{C}.9]$$

For Cl_2 after substituting for the constants this expression becomes

$$J_{\text{max}} \approx \left(\frac{T/\text{K}}{0.7021}\right)^{1/2} - \frac{1}{2}$$

(a) At 25°C = 298.15 K, $J_{\text{max}} \approx \left(\frac{298.15/\text{K}}{0.7021}\right)^{1/2} - \frac{1}{2} \approx \boxed{20}$

(b) At 100°C = 373.15 K, $J_{\text{max}} \approx \left(\frac{373.15/\text{K}}{0.7021}\right)^{1/2} - \frac{1}{2} \approx \boxed{23}$

Answers are rounded off to the nearest integer.

E12C.6(a) We select those molecules with an anisotropic polarizability. A practical rule to apply is that spherical rotors do not have anisotropic polarizabilities. Therefore $\boxed{\text{(iii) CH}_4\ \text{is inactive}}$. $\boxed{\text{(i), (ii), and (iv) are active}}$.

E12C.7(a) The Stokes lines appear at

$$\tilde{\nu}(J+2 \leftarrow J) = \tilde{\nu}_i - 2\tilde{B}(2J+3)\ [12\text{C}.15] \quad \text{with } J=0,\ \tilde{\nu} = \tilde{\nu}_i - 6\tilde{B}$$

Since $\tilde{B} = 1.9987\,\text{cm}^{-1}$ (Table 12D.1), the Stokes line appears at

$$\tilde{\nu} = (20487) - (6)\times(1.9987\,\text{cm}^{-1}) = \boxed{20\,475\,\text{cm}^{-1}}$$

E12C.8(a) The separation of lines is $4\tilde{B}$ [eqns 12C.15], so $\tilde{B} = 0.2438\,\text{cm}^{-1}$. Then we use

$$R = \left(\frac{\hbar}{4\pi m_{\text{eff}} c\tilde{B}}\right)^{1/2}$$

with $m_{\text{eff}} = \frac{1}{2}m(^{35}\text{Cl}) = \left(\frac{1}{2}\right)\times(34.9688 m_{\text{u}}) = 17.4844 m_{\text{u}}$

Therefore,

$$R = \left(\frac{1.05457\times10^{-34}\,\text{J s}}{(4\pi)\times(17.4844)\times(1.6605\times10^{-27}\,\text{kg})\times(2.9979\times10^{10}\,\text{cm s}^{-1})\times(0.2438\,\text{cm}^{-1})}\right)^{1/2}$$

$$= 1.989\times10^{-10}\,\text{m} = \boxed{198.9\,\text{pm}}$$

E12C.9(a) For diatomic molecules we can use eqn 12.16 to determine the statistical ratio of weights of populations. For chlorine-35 $I=3/2$, hence

$$\text{Ratio of (odd J/even J) weights of populations is } \frac{I+1}{I}=\boxed{\frac{5}{3}}$$

Solutions to problems

P12C.1 Rotational line separations are $2\tilde{B}$ (in wavenumber units), $2B$ (in frequency units), and $(2\tilde{B})^{-1}$ in wavelength units. Hence the transitions are separated by $\boxed{596\text{ GHz}}$, $\boxed{19.9\text{ cm}^{-1}}$, and $\boxed{0.503\text{ mm}}$.

Ammonia is a symmetric rotor (Section 12B.2(b)) and we know that

$$\tilde{B}=\frac{\hbar}{4\pi cI_{\perp}}\quad[12\text{B}.14]$$

and from Table 12B.1,

$$I_{\perp}=m_{\text{A}}R^2(1-\cos\theta)+\left(\frac{m_{\text{A}}m_{\text{B}}}{m}\right)R^2(1+2\cos\theta)$$

$m_{\text{A}}=1.6735\times10^{-27}\text{ kg}, m_{\text{B}}=2.3252\times10^{-26}\text{ kg}$, and $m=2.8273\times10^{-26}\text{ kg}$ with $R=101.4\text{ pm}$ and $\theta=106°78'$, which gives

$$\begin{aligned}I_{\perp}&=\left(1.6735\times10^{-27}\text{ kg}\right)\times\left(101.4\times10^{-12}\text{ m}\right)^2\times\left(1-\cos 106°78'\right)\\&\quad+\left(\frac{\left(1.6735\times10^{-27}\right)\times\left(2.3252\times10^{-26}\right)\text{kg}^2}{2.8273\times10^{-26}\text{ kg}}\right)\\&\quad\times\left(101.4\times10^{-12}\text{ m}\right)^2\times\left(1+2\cos 106°78'\right)\\&=2.815\bar{8}\times10^{-47}\text{ kg m}^2\end{aligned}$$

Therefore,

$$\tilde{B}=\frac{1.05457\times10^{-34}\text{ J s}}{(4\pi)\times\left(2.9979\times10^{8}\text{ m s}^{-1}\right)\times\left(2.815\bar{8}\times10^{-47}\text{ kg m}^2\right)}=994.1\text{ m}^{-1}=\boxed{9.941\text{ cm}^{-1}}$$

which is in accord with the data.

P12C.3 $\tilde{\nu}=2\tilde{B}(J+1)\,[12\text{C}.8\text{a}]=2\tilde{B}$

Hence, $\tilde{B}(^1\text{HCl})=10.4392\text{ cm}^{-1}$, $\tilde{B}(^2\text{HCl})=5.3920\text{ cm}^{-1}$

$$\tilde{B}=\frac{\hbar}{4\pi cI}\;[12.7]\qquad I=m_{\text{eff}}R^2\;[\text{Table 12B.1}]$$

$$R^2=\frac{\hbar}{4\pi cm_{\text{eff}}\tilde{B}}\qquad\frac{\hbar}{4\pi c}=2.79927\times10^{-44}\text{ kg m}$$

$$\begin{aligned}m_{\text{eff}}(\text{HCl})&=\left(\frac{(1.007825m_{\text{u}})\times(34.96885m_{\text{u}})}{(1.007825m_{\text{u}})+(34.96885m_{\text{u}})}\right)\times(1.66054\times10^{-27}\text{ kg}\,m_{\text{u}}^{-1})\\&=1.62665\times10^{-27}\text{ kg}\end{aligned}$$

$$m_{\text{eff}}(\text{DCl}) = \left(\frac{(2.0140m_\text{u})\times(34.96885m_\text{u})}{(2.0140m_\text{u})+(34.96885m_\text{u})}\right)\times(1.66054\times10^{-27}\ \text{kg}\,m_\text{u}^{-1})$$

$$= 3.1622\times10^{-27}\ \text{kg}$$

$$R^2(\text{HCl}) = \frac{2.79927\times10^{-44}\ \text{kg m}}{(1.62665\times10^{-27}\ \text{kg})\times(1.04392\times10^{3}\ \text{m}^{-1})} = 1.64848\times10^{-20}\ \text{m}^2$$

$$R(\text{HCl}) = 1.28393\times10^{-10}\ \text{m} = \boxed{128.393\,\text{pm}}$$

$$R^2(^2\text{HCl}) = \frac{2.79927\times10^{-44}\ \text{kg m}}{(3.1622\times10^{-27}\ \text{kg})\times(5.3920\times10^{2}\ \text{m}^{-1})} = 1.6417\times10^{-20}\ \text{m}^2$$

$$R(^2\text{HCl}) = 1.2813\times10^{-10}\ \text{m} = \boxed{128.13\,\text{pm}}$$

Thus the bond lengths are $\boxed{\text{slightly different}}$.

Comment. Since the effects of centrifugal distortion have not been taken into account, the number of significant figures in the calculated values of R above should be no greater than 4, despite the fact that the data are precise to 6 figures.

P12C.5 From the equation for a linear rotor in Table 12B.1 it is possible to show that $Im = m_\text{a}m_\text{c}(R+R')^2 + m_\text{a}m_\text{b}R^2 + m_\text{b}m_\text{c}R'^2$.

Thus,

$$I(^{16}\text{O}^{12}\text{C}^{32}\text{S}) = \left(\frac{m(^{16}\text{O})m(^{32}\text{S})}{m(^{16}\text{O}^{12}\text{C}^{32}\text{S})}\right)\times(R+R')^2 + \left(\frac{m(^{12}\text{C})\{m(^{16}\text{O})R^2 + m(^{32}\text{S})R'^2\}}{m(^{16}\text{O}^{12}\text{C}^{32}\text{S})}\right)$$

$$I(^{16}\text{O}^{12}\text{C}^{34}\text{S}) = \left(\frac{m(^{16}\text{O})m(^{34}\text{S})}{m(^{16}\text{O}^{12}\text{C}^{34}\text{S})}\right)\times(R+R')^2 + \left(\frac{m(^{12}\text{C})\{m(^{16}\text{O})R^2 + m(^{34}\text{S})R'^2\}}{m(^{16}\text{O}^{12}\text{C}^{34}\text{S})}\right)$$

$m(^{16}\text{O}) = 15.9949m_\text{u}$, $m(^{12}\text{C}) = 12.0000m_\text{u}$, $m(^{32}\text{S}) = 31.9721m_\text{u}$, and $m(^{34}\text{S}) = 33.9679m_\text{u}$.

Hence,

$$I(^{16}\text{O}^{12}\text{C}^{32}\text{S})/m_\text{u} = (8.5279)\times(R+R')^2 + (0.20011)\times(15.9949R^2 + 31.9721R'^2)$$

$$I(^{16}\text{O}^{12}\text{C}^{34}\text{S})/m_\text{u} = (8.7684)\times(R+R')^2 + (0.19366)\times(15.9949R^2 + 33.9679R'^2)$$

The spectral data provides the experimental values of the moments of inertia based on the relation $\tilde{\nu} = 2\tilde{B}(J+1)$ [12C.8a] with $\tilde{B} = \dfrac{\hbar}{4\pi cI}$ [12B.7]. These values are set equal to the above equations which are then solved for R and R'. The mean values of I obtained from the data are

$$I(^{16}\text{O}^{12}\text{C}^{32}\text{S}) = 1.37998\times10^{-45}\ \text{kg m}^2$$

$$I(^{16}\text{O}^{12}\text{C}^{34}\text{S}) = 1.41460\times10^{-45}\ \text{kg m}^2$$

Therefore, after conversion of the atomic mass units to kg, the equations we must solve are

$$1.37998\times10^{-45}\ \text{m}^2 = (1.4161\times10^{-26})\times(R+R')^2 + (5.3150\times10^{-27}R^2) \\ +(1.0624\times10^{-26}R'^2)$$

$$1.41460\times10^{-45}\ \text{m}^2 = (1.4560\times10^{-26})\times(R+R')^2 + (5.1437\times10^{-27}R^2) \\ +(1.0923\times10^{-26}R'^2)$$

These two equations may be solved for R and R'. They are tedious to solve by hand, but straightforward. Exercise 12B.4(b) illustrates the details of the solution. Readily available mathematical software can be used to quickly give the result. The outcome is $R = \boxed{116.28\ \text{pm}}$ and $R' = 155.97$ pm. These values may be checked by direct substitution into the equations.

Comment. The starting point of this problem is the actual experimental data on spectral line positions. Exercise 12B.4(b) is similar to this problem; its starting point is, however, given values of the rotational constants $\tilde{B}$, which were themselves obtained from the spectral line positions. So the results for R and R' are expected to be essentially identical and they are.

Question. What are the rotational constants calculated from the data on the positions of the absorption lines?

P12C.7 Plot frequency against J as in Fig. 12C.1.

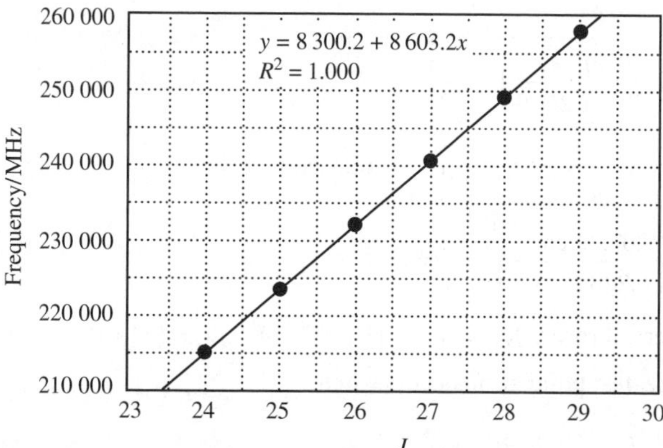

Figure 12C.1

The rotational constant is related to the wavenumbers of observed transitions by

$$\tilde{\nu} = 2\tilde{B}(J+1) = \frac{\nu}{c} \quad \text{so} \quad \nu = 2\tilde{B}c(J+1)$$

A plot of ν versus J, then, has a slope of $2\tilde{B}c$. From Fig. 12C.1, the slope is 8 603 MHz, so

$$\tilde{B} = \frac{8\,603 \times 10^6\ \mathrm{s^{-1}}}{2(2.988 \times 10^8\ \mathrm{m\ s^{-1}})} = \boxed{14.35\ \mathrm{m^{-1}}}$$

The most highly populated energy level is roughly

$$J_{\max} = \left(\frac{kT}{2hc\tilde{B}}\right)^{1/2} - \frac{1}{2}$$

$$\text{so } J_{\max} = \left(\frac{(1.381 \times 10^{-23}\ \mathrm{J\ K^{-1}}) \times (298\ \mathrm{K})}{(6.626 \times 10^{-34}\ \mathrm{J\ s}) \times (8\,603 \times 10^6\ \mathrm{s^{-1}})}\right)^{1/2} - \frac{1}{2} = \boxed{26} \text{ at 298 K}$$

$$\text{and } J_{\max} = \left(\frac{(1.381 \times 10^{-23}\ \mathrm{J\ K^{-1}}) \times (100\ \mathrm{K})}{(6.626 \times 10^{-34}\ \mathrm{J\ s}) \times (8\,603 \times 10^6\ \mathrm{s^{-1}})}\right)^{1/2} - \frac{1}{2} = \boxed{15} \text{ at 100 K}$$

P12C.9 $N \propto g\mathrm{e}^{-E/kT}$ [Boltzmann distribution, Chapter 15]

$N_J \propto g_J \mathrm{e}^{-E_J/kT} \propto (2J+1)\mathrm{e}^{-hc\tilde{B}J(J+1)/kT}$ [$g_J = 2J+1$ for a diatomic rotor]

The maximum population occurs when

$$\frac{\mathrm{d}}{\mathrm{d}J} N_J \propto \left\{2 - (2J+1)^2 \times \left(\frac{hc\tilde{B}}{kT}\right)\right\} \mathrm{e}^{-hc\tilde{B}J(J+1)/kT} = 0$$

and, since the exponential can never be zero at a finite temperature, when

$$(2J+1)^2 \times \left(\frac{hc\tilde{B}}{kT}\right) = 2$$

$$\text{or when } J_{\max} = \boxed{\left(\frac{kT}{2hc\tilde{B}}\right)^{1/2} - \frac{1}{2}}$$

For ICl, with $\dfrac{kT}{hc} = 207.22\ \mathrm{cm^{-1}}$ (inside front cover)

$$J_{\max} = \left(\frac{207.22\ \mathrm{cm^{-1}}}{0.2284\ \mathrm{cm^{-1}}}\right)^{1/2} - \frac{1}{2} = \boxed{30}$$

For a spherical rotor, $N_J \propto (2J+1)^2 \mathrm{e}^{-hc\tilde{B}J(J+1)/kT}$ [$g_J = (2J+1)^2$]

and the greatest population occurs when

$$\frac{\mathrm{d}N_J}{\mathrm{d}J} \propto \left(8J + 4 - \frac{hc\tilde{B}(2J+1)^3}{kT}\right)\mathrm{e}^{-hc\tilde{B}J(J+1)/kT} = 0$$

which occurs when

$$4(2J+1) = \frac{hc\tilde{B}(2J+1)^3}{kT}$$

$$\text{or at } J_{\max} = \boxed{\left(\frac{kT}{hc\tilde{B}}\right)^{1/2} - \frac{1}{2}}$$

$$\text{For CH}_4,\ J_{\max} = \left(\frac{207.22\ \mathrm{cm^{-1}}}{5.24\ \mathrm{cm^{-1}}}\right)^{1/2} - \frac{1}{2} = \boxed{6}$$

P12C.11 Temperature effects. At extremely low temperatures (10 K) only the lowest rotational states are populated. No emission spectrum is expected for the CO in the cloud and star light microwave absorptions by the CO in the cloud are from the lowest rotational states. At higher temperatures additional high-energy lines appear because higher energy rotational states are populated. Circumstellar clouds may exhibit infrared absorptions due to vibrational excitation as well as electronic transitions in the ultraviolet. Ultraviolet absorptions may indicate the photodissocation of carbon monoxide. High temperature clouds exhibit emissions.

Density effects. The density of an interstellar cloud may range from one particle to a billion particles per cm^{-3}. This is still very much a vacuum compared to the laboratory high vacuum of a trillion particles per cm^3. Under such extreme vacuum conditions the half-life of any quantum state is expected to be extremely long and absorption lines should be very narrow. At the higher densities the vast size of nebulae obscures distant stars. High densities and high temperatures may create conditions in which emissions stimulate emissions of the same wavelength by molecules. A cascade of stimulated emissions greatly amplifies normally weak lines—the maser phenomena of microwave amplification by stimulated emission of radiation.

Particle velocity effects. Particle velocity can cause Doppler broadening of spectral lines. The effect is extremely small for interstellar clouds at 10 K but is appreciable for clouds near high temperature stars. Outflows of gas from pulsing stars exhibit a red Doppler shift when moving away at high speed and a blue shift when moving toward us.

There will be many more transitions observable in circumstellar gas than in interstellar gas, because many more rotational states will be accessible at the higher temperatures. Higher velocity and density of particles in circumstellar material can be expected to broaden spectral lines compared to those of interstellar material by shortening collisional lifetimes. (Doppler broadening is not likely to be significantly different between circumstellar and interstellar material in the same astronomical neighbourhood. The relativistic speeds involved are due to large-scale motions of the expanding universe, compared to which local thermal variations are insignificant.) A temperature of 1 000 K is not high enough to significantly populate electronically excited states of CO; such states would have different bond lengths, thereby producing transitions with different rotational constants. Excited vibrational states would be accessible, though, and rovibrational transitions with P and R branches as detailed later in this chapter would be observable in circumstellar but not interstellar material. The rotational constant $\tilde{B}$ for $^{12}C^{16}O$ is 1.691 cm^{-1}. The first excited rotational energy level, $J=1$, with energy $J(J+1)hc\tilde{B}=2hc\tilde{B}$, is thermally accessible at about 6 K (based on the rough equation of the rotational energy to thermal energy kT). In interstellar space, only two or three rotational lines would be observable; in circumstellar space (at about 1 000 K) the number of transitions would be more like 20.

12D Vibrational spectroscopy of diatomic molecules

Answers to discussion questions

D12D.1 True harmonic oscillation implies the existence of a parabolic potential energy. For low vibrational energies, near the bottom of the potential well, the assumption of a parabolic potential energy is a very good approximation. However, molecular vibrations are always anharmonic to a greater or lesser extent. At high excitation energies, the parabolic approximation is poor, and it is totally wrong near the dissociation limit. An advantage of the parabolic potential energy is that it allows for a relatively straightforward solution of the Schrödinger equation for the vibrational motion. The Morse potential is a closer approximation to the true potential energy curve for molecular vibrations. It allows for the convergence of the energy levels at high values of the quantum numbers and for dissociation at large displacements. It fails at very short distances where it approaches a finite value. An advantage is that the Schrödinger equation can be solved exactly for the Morse potential.

D12D.3 Rotational constants $\tilde{B}_v$ differ from one vibrational level, v, to another. The method of combination differences as described in Section 12.4(b) can be used to determine two rotational constants individually. The procedure is illustrated in *Brief illustration* 12D.4 where the values of the rotational constants $\tilde{B}_0$ and $\tilde{B}_1$ are determined for $^1H^{35}Cl$ from the combination difference equations, eqns 12D.21a and b:

$$\Delta_0 = \tilde{\nu}_R(J-1) - \tilde{\nu}_P(J+1) = 4\tilde{B}_0(J+½), \text{ and}$$
$$\Delta_1 = \tilde{\nu}_R(J) - \tilde{\nu}_P(J) = 4\tilde{B}_1(J+½)$$

Substituting values for $\tilde{\nu}_R$ and $\tilde{\nu}_P$ from data that gives the wavenumbers of the transitions for $J=1,2,3,4$ etc. we calculate the values for Δ_0 and Δ_1. Then $\tilde{B}_0$ and $\tilde{B}_1$ are calculated from the simultaneous equations for Δ_0 and Δ_1.

Solutions to exercises

E12D.1(a) $\omega = 2\pi\nu = \left(\dfrac{k_f}{m}\right)^{1/2}$

$$k_f = 4\pi^2\nu^2 m = 4\pi^2 \times (2.0\,\mathrm{s^{-1}})^2 \times (1.0\,\mathrm{kg}) = 1.6\times10^2\,\mathrm{kg\,s^{-2}} = \boxed{1.6\times10^2\,\mathrm{N\,m^{-1}}}$$

E12D.2(a) $\omega = \left(\dfrac{k_f}{m_{\text{eff}}}\right)^{1/2}$ [12D.7]

The fractional difference is

$$\frac{\omega'-\omega}{\omega} = \frac{\left(\dfrac{k_f}{m'_{\text{eff}}}\right)^{1/2} - \left(\dfrac{k_f}{m_{\text{eff}}}\right)^{1/2}}{\left(\dfrac{k_f}{m_{\text{eff}}}\right)^{1/2}} = \frac{\left(\dfrac{1}{m'_{\text{eff}}}\right)^{1/2} - \left(\dfrac{1}{m_{\text{eff}}}\right)^{1/2}}{\left(\dfrac{1}{m_{\text{eff}}}\right)^{1/2}} = \left(\frac{m_{\text{eff}}}{m'_{\text{eff}}}\right)^{1/2} - 1$$

$$=\left(\frac{m(^{23}\text{Na})m(^{35}\text{Cl})\{m(^{23}\text{Na})+m(^{37}\text{Cl})\}}{\{m(^{23}\text{Na})+m(^{35}\text{Cl})\}m(^{23}\text{Na})m(^{37}\text{Cl})}\right)^{1/2}-1$$

$$=\left(\frac{m(^{35}\text{Cl})}{m(^{37}\text{Cl})}\times\frac{m(^{23}\text{Na})+m(^{37}\text{Cl})}{m(^{23}\text{Na})+m(^{35}\text{Cl})}\right)^{1/2}-1$$

$$=\left(\frac{34.9688}{36.9651}\times\frac{22.9898+36.9651}{22.9898+34.9688}\right)^{1/2}-1=-0.01077$$

Hence, the difference is $\boxed{1.077\text{ per cent}}$

E12D.3(a) $\omega=\left(\dfrac{k_\text{f}}{m_\text{eff}}\right)^{1/2}$ [12D.7]; $\omega=2\pi\nu=2\pi\left(\dfrac{c}{\lambda}\right)=2\pi c\tilde{\nu}$

Therefore, $k_\text{f}=m_\text{eff}\omega^2=4\pi^2 m_\text{eff}c^2\tilde{\nu}^2$, $m_\text{eff}=\tfrac{1}{2}m(^{35}\text{Cl})$

$$=(4\pi^2)\times\left(\frac{34.9688}{2}\right)\times(1.66054\times10^{-27}\text{ kg})\times[(2.997924\times10^{10}\text{ cm s}^{-1})\times(564.9\text{cm}^{-1})]^2$$

$$=\boxed{328.7\text{ N m}^{-1}}$$

E12D.4(a) $\omega=\left(\dfrac{k}{m_\text{eff}}\right)^{1/2}$ [12D.7], so $k=m_\text{eff}\omega^2=4\pi^2 m_\text{eff}c^2\tilde{\nu}^2$

$$m_\text{eff}=\frac{m_1m_2}{m_1+m_2}\ [43.3]$$

$$m_\text{eff}(\text{H}^{19}\text{F})=\frac{(1.0078)\times(18.9984)}{(1.0078)+(18.9984)}m_\text{u}=0.9570m_\text{u}$$

$$m_\text{eff}(\text{H}^{35}\text{Cl})=\frac{(1.0078)\times(34.9688)}{(1.0078)+(34.9688)}m_\text{u}=0.9796m_\text{u}$$

$$m_\text{eff}(\text{H}^{81}\text{Br})=\frac{(1.0078)\times(80.9163)}{(1.0078)+(80.9163)}m_\text{u}=0.9954m_\text{u}$$

$$m_\text{eff}(\text{H}^{127}\text{I})=\frac{(1.0078)\times(126.9045)}{(1.0078)+(126.9045)}m_\text{u}=0.9999m_\text{u}$$

We draw up the following table

	HF	HCl	HBr	HI
$\tilde{\nu}/\text{cm}^{-1}$	4141.3	2988.9	2649.7	2309.5
m_eff/m_u	0.9570	0.9697	0.9954	0.9999
$k/(\text{N m}^{-1})$	$\boxed{967.0}$	$\boxed{515.6}$	$\boxed{411.8}$	$\boxed{314.2}$

Note the order of stiffness HF > HCl > HBr > HI .

Question. Which ratio, $\dfrac{k}{B(\text{A–B})}$ or $\dfrac{\tilde{\nu}}{B(\text{A–B})}$, where B(A–B) are the bond energies, is the more nearly constant across the series of hydrogen halides? Why?

E12D.5(a) We write, with $N' = N$ (upper state) and $N = N$ (lower state)

$$\frac{N'}{N} = e^{-h\nu/kT} \text{ [from Boltzmann distribution]} = e^{-hc\tilde{\nu}/kT}$$

$$\frac{hc\tilde{\nu}}{k} = (1.4388\,\text{cm K}) \times (559.7\,\text{cm}^{-1}) \text{ [inside front cover]} = 805.3\,\text{K}$$

$$\frac{N(\text{upper})}{N(\text{lower})} = e^{-805.3\,\text{K}/T}$$

$$\textbf{(i)}\ \frac{N(\text{upper})}{N(\text{lower})} = e^{-805.3/298} = \boxed{0.067}\ (1{:}15) \quad \textbf{(ii)}\ \frac{N(\text{upper})}{N(\text{lower})} = e^{-805.3/500} = \boxed{0.20}\ (1{:}5)$$

E12D.6(a) Data on three transitions are provided. Only two are necessary to obtain the value of $\tilde{\nu}$ and x_e . The third datum can then be used to check the accuracy of the calculated values.

$$\Delta\tilde{G}(v = 1 \leftarrow 0) = \tilde{\nu} - 2\tilde{\nu}x_e = 1\,556.22\,\text{cm}^{-1}\ [12\text{D}.14]$$
$$\Delta\tilde{G}(v = 2 \leftarrow 0) = 2\tilde{\nu} - 6\tilde{\nu}x_e = 3\,088.28\,\text{cm}^{-1}\ [12\text{D}.15]$$

Multiply the first equation by 3, then subtract the second

$$\tilde{\nu} = (3)\times(1\,556.22\ \text{cm}^{-1}) - (3\,088.28\ \text{cm}^{-1}) = \boxed{1\,580.38\ \text{cm}^{-1}}$$

Then from the first equation

$$x_e = \frac{\tilde{\nu} - 1\,556.22\,\text{cm}^{-1}}{2\tilde{\nu}} = \frac{(1\,580.38 - 1\,556.22)\,\text{cm}^{-1}}{(2)\times(1\,580.38\,\text{cm}^{-1})} = \boxed{7.644\times10^{-3}}$$

x_e data are usually reported as $x_e\tilde{\nu}$ which is

$$x_e\tilde{\nu} = 12.08\ \text{cm}^{-1}$$

$$\Delta\tilde{G}(v = 3 \leftarrow 0) = 3\tilde{\nu} - 12\tilde{\nu}x_e$$
$$= (3) \times (1\,580.38\,\text{cm}^{-1}) - (12) \times (12.08\,\text{cm}^{-1}) = 4\,596.18\,\text{cm}^{-1}$$

which is very close to the experimental value.

E12D.7(a) $\Delta\tilde{G}_{v+1/2} = \tilde{\nu} - 2(v+1)x_e\tilde{\nu}$ [12D.14] where $\Delta\tilde{G}_{v+1/2} = \tilde{G}(v+1) - \tilde{G}(v)$

Therefore, since

$$\Delta\tilde{G}_{v+1/2} = (1 - 2x_e)\tilde{\nu} - 2vx_e\tilde{\nu}$$

a plot of $\Delta\tilde{G}_{v+1/2}$ against v should give a straight line which gives $(1-2x_e)\tilde{\nu}$ from the intercept at $v=0$ and $-2x_e\tilde{\nu}$ from the slope. We draw up the following table

v	0	1	2	3	4
$(\tilde{\nu})$ / cm^{-1}	1481.86	4367.50	7149.04	9826.48	12399.8
$\Delta\tilde{G}_{v+1/2}$ / cm^{-1}	2885.64	2781.54	2677.44	2573.3$\overline{4}$	

The points are plotted in Fig. 12D.1. The intercept lies at 2 885.6 and the slope is $\dfrac{-312.3}{3}=-104.1$; hence $x_e\tilde{\nu}=52.1\,\text{cm}^{-1}$.
Since $\tilde{\nu}-2x_e\tilde{\nu}=2\,885.6\,\text{cm}^{-1}$, it follows that $\tilde{\nu}=2\,989.8\,\text{cm}^{-1}$.

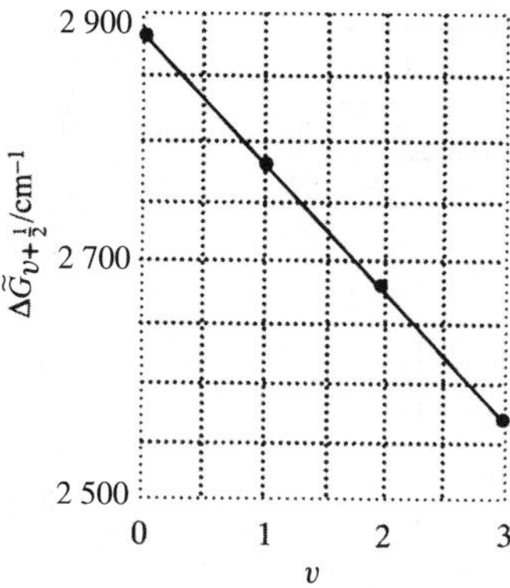

Figure 12D.1

The dissociation energy may be obtained by assuming that a Morse potential describes the molecule and that the constant $\tilde{D}_e$ in the expression for the potential is an adequate first approximation for it. Then

$$\tilde{D}_e=\frac{\tilde{\nu}}{4x_e}\,[12D.12]=\frac{\tilde{\nu}^2}{4x_e\tilde{\nu}}=\frac{(2\,989.8\,\text{cm}^{-1})^2}{(4)\times(52.1\,\text{cm}^{-1})}=42.9\times10^3\,\text{cm}^{-1}\ \text{or}\ 5.32\,\text{eV}$$

However, the depth of the potential well $\tilde{D}_e$ differs from $\tilde{D}_0$, the dissociation energy of the bond, by the zero-point energy; hence

$$\begin{aligned}\tilde{D}_0&=\tilde{D}_e-\tfrac{1}{2}\tilde{\nu}\ [\text{Fig. 12D.5 of text}]\\&=(42.9\times10^3\,\text{cm}^{-1})-(\tfrac{1}{2})\times(2\,989.8\,\text{cm}^{-1})\\&=41.4\times10^3\,\text{cm}^{-1}=\boxed{5.13\,\text{eV}}\end{aligned}$$

Solutions to problems

P12D.1 The Morse potential is $V(R)=hc\widetilde{D}_e\{1-e^{-a(R-R_e)}\}^2$ [12D.11]. Rewrite this expression as $V(R)=D\{1-e^{-ax}\}^2$ with $x=R-R_e$ and $D=hc\widetilde{D}_e$. The force constant is obtained from

$$k_f=\left(\frac{d^2V}{dx^2}\right)_{x=0} \text{ [12D.4]} =\{-2Da(ae^{-ax}-2ae^{-2ax})\}_{x=0} = \boxed{k_f=2Da^2.}$$

P12D.3 The wavenumbers of the transitions with $\Delta v=+1$ are

$$\Delta\widetilde{G}_{v+½}=\tilde{\nu}-2(v+1)x_e\tilde{\nu} \text{ [12D.14] and } \widetilde{D}_e=\frac{\tilde{\nu}^2}{4x_e\tilde{\nu}} \text{ [12D.12]}$$

A plot of $\Delta\widetilde{G}_{v+1/2}$ against $v+1$ should give a straight line with intercept $\tilde{\nu}$ at $v+1=0$ and slope $-2x_e\tilde{\nu}$.

Draw up the following table

$v+1$	1	2	3
$\Delta\widetilde{G}_{v+1/2}$ / cm^{-1}	284.50	283.00	281.502

The points are plotted in Fig. 12D.2.

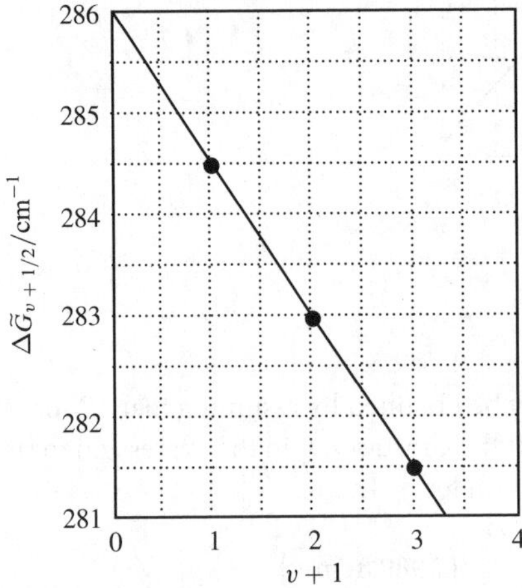

Figure 12D.2

The intercept is at 286.0, so $\tilde{\nu}=286\text{cm}^{-1}$. The slope is -1.50, so $x_e\tilde{\nu}=0.750\text{cm}^{-1}$. It follows that

$$\widetilde{D}_e=\frac{(286\text{cm}^{-1})^2}{(4)\times(0.750\text{cm}^{-1})}=27300\text{cm}^{-1}, \text{ or } 3.38 \text{ eV}$$

The zero-point level lies at $\boxed{142.81\,\text{cm}^{-1}}$ and so $D_0 = \boxed{3.36\text{ eV}}$. Since

$$m_{\text{eff}} = \frac{(22.99)\times(126.90)}{(22.99)+(126.90)}m_{\text{u}} = 19.46\bar{4}m_{\text{u}}$$

the force constant of the molecule is

$$\begin{aligned}k &= 4\pi^2 m_{\text{eff}}c^2\tilde{\nu}^2 \text{ [Exercise 12D.3(a)]}\\ &= (4\pi^2)\times(19.46\bar{4})\times(1.6605\times10^{-27}\text{ kg})\times[(2.998\times10^{10}\text{ cm s}^{-1})\times(286\text{ cm}^{-1})]^2\\ &= \boxed{93.8\text{ N m}^{-1}}\end{aligned}$$

P12D.5 $V(R) = hc\widetilde{D}_{\text{e}}\{1-\text{e}^{-a(R-R_{\text{e}})}\}^2$ [12D.11]

$$\tilde{\nu} = \frac{\omega}{2\pi c} = 936.8\text{ cm}^{-1} \quad x_{\text{e}}\tilde{\nu} = 14.15\text{ cm}^{-1}$$

$$a = \left(\frac{m_{\text{eff}}}{2hc\widetilde{D}_{\text{e}}}\right)^{1/2}\omega \quad x_{\text{e}} = \frac{\hbar a^2}{2m_{\text{eff}}\omega} \quad \widetilde{D}_{\text{e}} = \frac{\tilde{\nu}}{4x_{\text{e}}}$$

$$m_{\text{eff}}(\text{RbH}) \approx \frac{(1.008)\times(85.47)}{(1.008)+(85.47)}m_{\text{u}} = 1.654\times10^{-27}\text{ kg}$$

$$\widetilde{D}_{\text{e}} = \frac{\tilde{\nu}^2}{4x_{\text{e}}\tilde{\nu}} = \frac{(936.8\text{ cm}^{-1})^2}{(4)\times(14.15\text{ cm}^{-1})} = 15\,50\bar{5}\text{ cm}^{-1}\ (1.92\text{ eV})$$

$$\begin{aligned}a &= 2\pi\nu\left(\frac{m_{\text{eff}}}{2hc\widetilde{D}_{\text{e}}}\right)^{1/2}[12\text{D}.11] = 2\pi c\tilde{\nu}\left(\frac{m_{\text{eff}}}{2hc\widetilde{D}_{\text{e}}}\right)^{1/2}\\ &= (2\pi)\times(2.998\times10^{10}\text{ cm s}^{-1})\times(936.8\text{ cm}^{-1})\\ &\quad\times\left(\frac{1.654\times10^{-27}\text{ kg}}{(2)\times(15\,505\text{ cm}^{-1})\times(6.626\times10^{-34}\text{ J s})\times(2.998\times10^{10}\text{ cm s}^{-1})}\right)^{1/2}\\ &= 9.144\times10^{9}\text{ m}^{-1} = 9.14\text{ nm}^{-1} = \frac{1}{0.1094\text{ nm}}\end{aligned}$$

Therefore, $\dfrac{V(R)}{hc\widetilde{D}_{\text{e}}} = \{1-\text{e}^{-(R-R_{\text{e}})/(0.1094\text{ nm})}\}^2$

with $R_{\text{e}} = 236.7$ pm. We draw up the following table

R/pm	50	100	200	300	400	500	600	700	800
$V/(hc\widetilde{D}_{\text{e}})$	20.4	6.20	0.159	0.193	0.601	0.828	0.929	0.971	0.988

These points are plotted overleaf in Fig. 12D.3 as the line labelled $J = 0$.

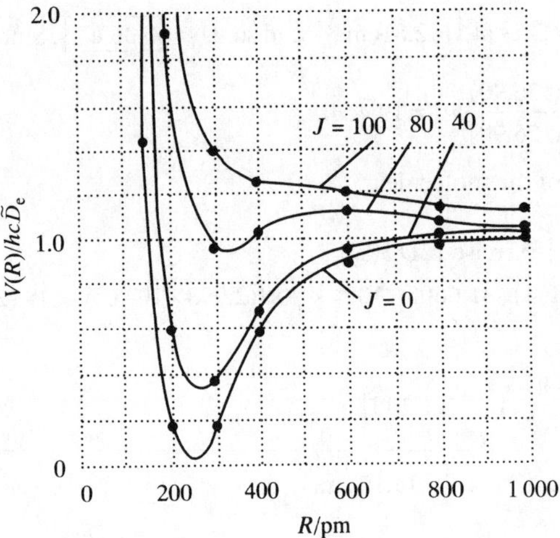

Figure 12D.3

For the second part, we note that $\tilde{B} \propto \dfrac{1}{R^2}$ and write

$$V_J^* = V + hc\tilde{B}_e J(J+1) \times \left(\frac{R_e^2}{R^2}\right)$$

with $\tilde{B}_e$ the equilibrium rotational constant, $\tilde{B}_e = 3.020\ \text{cm}^{-1}$.

We then draw up the following table using the values of V calculated above.

R/pm	50	100	200	300	400	600	800	1000
$\frac{R_e}{R}$	4.73	2.37	1.18	0.79	0.59	0.39	0.30	0.24
$\frac{V}{hc\tilde{D}_e}$	20.4	6.20	0.159	0.193	0.601	0.929	0.988	1.000
$\frac{V_{40}^*}{hc\tilde{D}_e}$	27.5	7.99	0.606	0.392	0.713	0.979	1.016	1.016
$\frac{V_{80}^*}{hc\tilde{D}_e}$	48.7	13.3	1.93	0.979	1.043	1.13	1.099	1.069
$\frac{V_{100}^*}{hc\tilde{D}_e}$	64.5	17.2	2.91	1.42	1.29	1.24	1.16	1.11

These points are also plotted in Fig. 12D.3.

P12D.7 The energy levels of a Morse oscillator, expressed as wavenumbers, are given by:

$$\tilde{G}(v) = (v+\tfrac{1}{2})\tilde{\nu} - (v+\tfrac{1}{2})^2 x_e\tilde{\nu} = (v+\tfrac{1}{2})\tilde{\nu} - (v+\tfrac{1}{2})^2\tilde{\nu}^2/4\tilde{D}_e.$$

States are bound only if the energy is less than the well depth, $\tilde{D}_e$, also expressed as a wavenumber:

$$\tilde{G}(v) < \tilde{D}_e \quad \text{or} \quad (v+\tfrac{1}{2})\tilde{\nu} - (v+\tfrac{1}{2})^2\tilde{\nu}^2/4\tilde{D}_e < \tilde{D}_e$$

Solve for the maximum value of v by making the inequality into an equality:

$$(v+\tfrac{1}{2})^2\tilde{\nu}^2/4\tilde{D}_e-(v+\tfrac{1}{2})\tilde{\nu}+\tilde{D}_e=0$$

Multiplying through by $4\tilde{D}_e$ results in an expression that can be factored by inspection into:

$$\left[(v+\tfrac{1}{2})\tilde{\nu}-2\tilde{D}_e\right]^2=0 \quad \text{so} \quad v+\tfrac{1}{2}=2\tilde{D}_e/\tilde{\nu} \quad \text{and} \quad v=\boxed{2\tilde{D}_e/\tilde{\nu}-\tfrac{1}{2}}.$$

Of course, v is an integer, so its maximum value is really the greatest integer less than this quantity.

P12D.9 $\tilde{B}=\dfrac{\hbar}{4\pi cI}$ [12B.7]; $I=m_{\text{eff}}R^2$; $R^2=\dfrac{\hbar}{4\pi cm_{\text{eff}}\tilde{B}}$

$$m_{\text{eff}}=\frac{m_Cm_O}{m_C+m_O}=\left(\frac{(12.0000m_u)\times(15.9949m_u)}{(12.0000m_u)+(15.9949m_u)}\right)\times(1.66054\times10^{-27}\ \text{kg}\,m_u^{-1})$$
$$=1.13852\times10^{-26}\ \text{kg}$$

$$\frac{\hbar}{4\pi c}=2.79932\times10^{-44}\ \text{kg m}$$

$$R_0^2=\frac{2.79932\times10^{-44}\ \text{kg m}}{(1.13852\times10^{-26}\ \text{kg})\times(1.9314\times10^2\ \text{m}^{-1})}=1.2730\bar{3}\times10^{-20}\ \text{m}^2$$

$$R_0=1.1283\times10^{-10}\ \text{m}=\boxed{112.83\ \text{pm}}$$

$$R_1^2=\frac{2.79932\times10^{-44}\ \text{kg m}}{(1.13852\times10^{-26}\ \text{kg})\times(1.6116\times10^2\ \text{m}^{-1})}=1.52565\times10^{-20}\ \text{m}^2$$

$$R_1=1.2352\times10^{-10}\ \text{m}=\boxed{123.52\ \text{pm}}$$

Comment. The change in internuclear distance is roughly 10 per cent, indicating that the rotations and vibrations of molecules are strongly coupled and that it is an oversimplification to consider them independently of each other.

P12D.11 Examination of the wavenumbers of the transitions allows the following identifications to be made as shown in the table below. The J value in the table is the J value of the rotational states in the $v=0$ vibrational level. The notation (R) and (P) after the J values refers to the R and P branches of the vibration–rotation spectrum.

$\tilde{\nu}$	2998.05	2981.05	2963.35	2944.99	2925.92	2906.25	2865.14	2843.63	2821.59	2799.00
J	5(R)	4(R)	3(R)	2(R)	1(R)	0(R)	1(P)	2(P)	3(P)	4(P)

The values of $\tilde{B}_0$ and $\tilde{B}_1$ are determined from the equations for the combination differences, eqns 12D.21a and b.

$$\Delta_0=\tilde{\nu}_R(J-1)-\tilde{\nu}_P(J+1)=4\tilde{B}_0(J+\tfrac{1}{2}),\ \text{and}$$
$$\Delta_1=\tilde{\nu}_R(J)-\tilde{\nu}_P(J)=4\tilde{B}_1(J+\tfrac{1}{2})$$

Substituting values for $\tilde{\nu}_R$ and $\tilde{\nu}_P$ from the above table for $J = 1, 2, 3$, and 4 we obtain the following values for Δ_0 and Δ_1.

J	1	2	3	4
$J+\frac{1}{2}$	3/2	5/2	7/2	9/2
Δ_0/cm^{-1}	62.62	104.33	145.99	
Δ_1/cm^{-1}	60.78	101.36	141.76	182.05
$\tilde{B}_0/\text{cm}^{-1}$	10.437	10.433	10.428	
$\tilde{B}_1/\text{cm}^{-1}$	10.130	10.136	10.126	10.113

An average of these values gives $\boxed{\tilde{B}_0 = 10.433\ \text{cm}^{-1}}$ and $\boxed{\tilde{B}_1 = 10.126\ \text{cm}^{-1}}$. Values for $\tilde{B}_0$ and $\tilde{B}_1$ may also be obtained by plotting the combination differences, Δ_0 and Δ_1, against $J+\frac{1}{2}$; the slopes give $4\tilde{B}_0$ and $4\tilde{B}_1$, respectively.

P12D.13 The virial theorem states that if the potential energy of a particle has the form $V = ax^b$ then its mean potential and kinetic energies are related by $2\langle E_k\rangle = b\langle V\rangle$ [8B.14]. For the harmonic oscillator potential energy $b=2$ and $\langle E_k\rangle = \frac{1}{2}E_v$ [8B.13c] $= \frac{1}{2}(v+\frac{1}{2})\hbar\omega$ [8B.4]. Hence, $\langle V\rangle = \langle E_K\rangle = \frac{1}{2}E_v = \frac{1}{2}(v+\frac{1}{2})\hbar\omega = \frac{1}{2}k_f\langle x^2\rangle$ and $\boxed{\langle x^2\rangle = \frac{1}{k_f}(v+\frac{1}{2})\hbar\omega.}$ We see that as v increases, $\langle x^2\rangle$ increases and $\langle R^2\rangle = R_e^2 + \langle x^2\rangle$ [Solution to Problem 2D.12] increases. As $\langle R^2\rangle$ increases, the moment of inertia I increases [Table 12.1]. As I increases the $\boxed{\text{rotational constant } B \text{ decreases}}$ [eqn 12B.7] as the oscillator is excited to higher quantum states. Anharmonicity results in greater average values of R^2 as v increases [Fig. 12.6]; hence, $\boxed{B \text{ decreases with increased anharmonicity}}$.

P12D.15 The set of peaks to the left of centre are the P branch, those to the right are the R branch. Within the rigid rotor approximation the two sets are separated by $4\tilde{B}$. The effects of the interactions between vibration and rotation and of centrifugal distortion are least important for transitions with small J values hence the separation between the peaks immediately to the left and right of centre will give good approximate values of $\tilde{B}$ and bond length.

(a) $\tilde{\nu}_Q(J) = \tilde{\nu}$ [12D.19b] $= \boxed{2143.26\ \text{cm}^{-1}}$

(b) The zero-point energy is $\frac{1}{2}\tilde{\nu} = 1071.63\text{cm}^{-1}$. The molar zero-point energy in J mol^{-1} is

$$N_A hc\times(1071.63\text{cm}^{-1}) = N_A hc\times(1.07163\times10^5\ \text{m}^{-1})$$
$$= 1.28195\times10^4\ \text{J mol}^{-1} = \boxed{12.8195\ \text{kJ mol}^{-1}}$$

(c) $k = 4\pi^2\mu c^2\tilde{\nu}^2$

$$\mu(^{12}\text{C}^{16}\text{O}) = \frac{m_C m_O}{m_C + m_O} = \left(\frac{(12.0000m_u)\times(15.9949m_u)}{(12.0000m_u)+(15.9949m_u)}\right)\times(1.66054\times10^{-27}\ \text{kg}\,m_u^{-1})$$
$$= 1.13852\times10^{-26}\ \text{kg}$$

$$k = 4\pi^2c^2\times(1.13852\times10^{-26}\ \text{kg})\times(2.14326\times10^5\ \text{m}^{-1})^2 = \boxed{1.85563\times10^3\ \text{N m}^{-1}}$$

(d) $4\tilde{B} \approx 7.655\,\text{cm}^{-1}$

$$\tilde{B} \approx \boxed{1.91\,\text{cm}^{-1}} \text{ [4 significant figures not justified]}$$

(e) $\tilde{B} = \dfrac{\hbar}{4\pi c I}$ [12B.7] $= \dfrac{\hbar}{4\pi c \mu R^2}$ [Table 12.1]

$$R^2 = \frac{\hbar}{4\pi c \mu \tilde{B}} = \frac{\hbar}{(4\pi c)\times(1.13852\times10^{-26}\,\text{kg})\times(191\,\text{m}^{-1})} = 1.28\overline{7}\times10^{-20}\,\text{m}^2$$

$$R = 1.13\times10^{-10}\,\text{m} = \boxed{113\,\text{pm}}$$

12E Vibrational spectroscopy of polyatomic molecules

Answers to discussion questions

D12E.1 The gross selection rules tell us which are the allowed spectroscopic transitions. For both microwave and infrared spectroscopy, the allowed transitions depend on the existence of an oscillating dipole moment which can stir the electromagnetic field into oscillation (and vice versa for absorption). For microwave rotational spectroscopy, this implies that the molecule must have a permanent dipole moment, which is equivalent to an oscillating dipole when the molecule is rotating. See Fig. 12C.1 of the text. In the case of infrared vibrational spectroscopy, the physical basis of the gross selection rule is that the molecule have a structure that allows for the existence of an oscillating dipole moment when the molecule vibrates. Polar molecules necessarily satisfy this requirement, but nonpolar molecules may also have a fluctuating dipole moment upon vibration. See Fig. 12D.3 of the text.

D12E.3 The exclusion rule applies to the benzene molecule because it has a centre of symmetry. Consequently, none of the normal modes of vibration of benzene can be both infrared and Raman active. If we wish to characterize all the normal modes we must obtain both kinds of spectra. See the solutions to Exercises 12E.7(a) and 12E.7(b) for specific illustrations of which modes are IR active and which are Raman active.

Solutions to exercises

E12E.1(a) See Section 12E.2. Select those molecules in which a vibration gives rise to a change in dipole moment. It is helpful to write down the structural formulas of the compounds. The infrared active compounds are

(b) HCl **(c)** CO_2 **(d)** H_2O

Comment. A more powerful method for determining infrared activity based on symmetry considerations is described in Section 12E.4(a). Also see Exercises 12E.5–12E.6.

E12E.2(a) The number of normal modes of vibration is given by (Section 12E.4(a))

$$N_{\text{vib}} = \begin{Bmatrix} 3N-5 \text{ for linear molecules} \\ 3N-6 \text{ for nonlinear molecules} \end{Bmatrix}$$

where N is the number of atoms in the molecule. Hence, since none of these molecules are linear,

(a) 3 **(b)** 6 **(c)** 12

Comment. Even for moderately sized molecules the number of normal modes of vibration is large and they are usually difficult to visualize.

E12E.3(a) This molecule is linear; hence the number of vibrational modes is $3N-5$. $N=44$ in this case; therefore, the number of vibrational modes is 127.

E12E.4(a) $\tilde{G}_q(v)=(v+\frac{1}{2})\tilde{\nu}_q \qquad \tilde{\nu}_q=\dfrac{1}{2\pi c}\left(\dfrac{k_q}{m_q}\right)^{1/2}$ [12E.1]

The lowest energy term is $\tilde{\nu}_2$ corresponding to the normal mode for bending. For this mode the oxygen atom may be considered to remain stationary and the effective mass is approximately $m_q=\dfrac{2m_{\text{H}}m_{\text{O}}}{2m_{\text{H}}+m_{\text{O}}}$. For the other modes the effective mass expression is more complicated and is beyond the scope of this text. However we know that in the ground vibrational state all normal modes have $v=0$. Thus, since H_2O has the three normal modes shown in text Fig. 12.3, the ground vibrational term is the sum of eqn 12.1 normal mode terms: $\tilde{G}_{\text{ground}}=\tilde{G}_1(0)+\tilde{G}_2(0)+\tilde{G}_3(0)=\boxed{\tfrac{1}{2}(\tilde{\nu}_1+\tilde{\nu}_2+\tilde{\nu}_3)}$

E12E.5(a) See Figs 12E.3 (H_2O, bent) and 12E.2 (CO_2, linear) of the text as well as the *Brief illustrations* in Section 12.4. Decide which modes correspond to (i) a changing electric dipole moment, (ii) a changing polarizability, and take note of the exclusion rule (Sections 12E.3 and 12E.4).

(a) Nonlinear: all modes both infrared and Raman active.

(b) Linear: the symmetric stretch is infrared inactive but Raman active.

The antisymmetric stretch is infrared active and (by the exclusion rule) Raman inactive. The two bending modes are infrared active and therefore Raman inactive.

E12E.6(a) The uniform expansion is depicted in Fig. 12E.1.

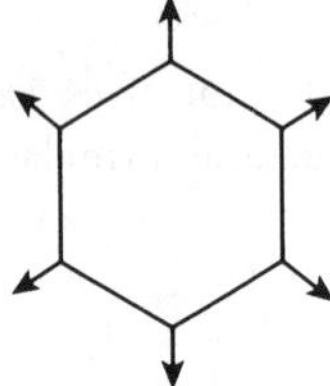

Figure 12E.1

Benzene is centrosymmetric, and so the exclusion rule applies (Section 12E.3). The mode is infrared inactive (symmetric breathing leaves the molelcular dipole moment unchanged at zero), and therefore the mode may be $\boxed{\text{Raman active}}$ (and is). In group theoretical terms, the breathing mode has symmetry A_{1g} in D_{6h}, which is the point group for benzene, and quadratic forms x^2+y^2 and z^2 have this symmetry (see the character table for C_{6h}, a subgroup of D_{6h}). Hence, the mode is Raman active.

E12E.7(a) Use the character table for the group C_{2v} (and see Example 12.3). The rotations span $A_2+B_1+B_2$. The translations span $A_1+B_1+B_2$. Hence the normal modes of vibration span the difference, $\boxed{4A_1+A_2+2B_1+2B_2}$

Comment. A_1, B_1 and B_2 are infrared active; all modes are Raman active.

E12E.8(a) See the comment in the solution to Exercise 12E.7(a).
$\boxed{A_1, B_1, \text{and } B_2 \text{ are infrared active}}$; $\boxed{\text{all modes are Raman active}}$.

Solutions to problems

P12E.1 A sketch of $V(h)=V_0\left(1-e^{-bh^4}\right)$ is presented in Fig. 12E.2.

(a) By analogy to the Morse potential and to the harmonic potential, we expect that $b^{1/4} \propto \nu \propto k^{1/2}$. Note that for small displacements from $h=0$ the potential energy function is flat (independent of h). Therefore the first and second derivatives of $V(h)$ near $h=0$ are zero and the force constant is zero. There is no restoring force for small displacements from the equilibrium position. The particle $\boxed{\text{cannot undergo simple harmonic motion}}$.

(b) Constructive interference of valence orbitals, shown in the adjacent sketch, should be the basic character of the ground-state wavefunction.

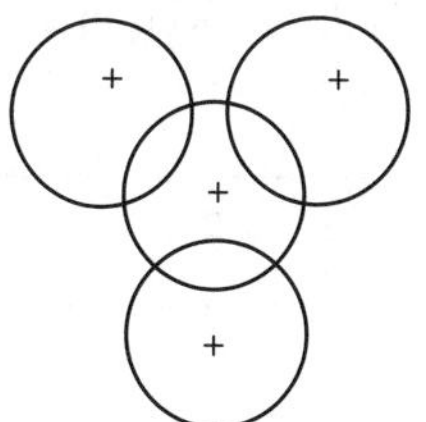

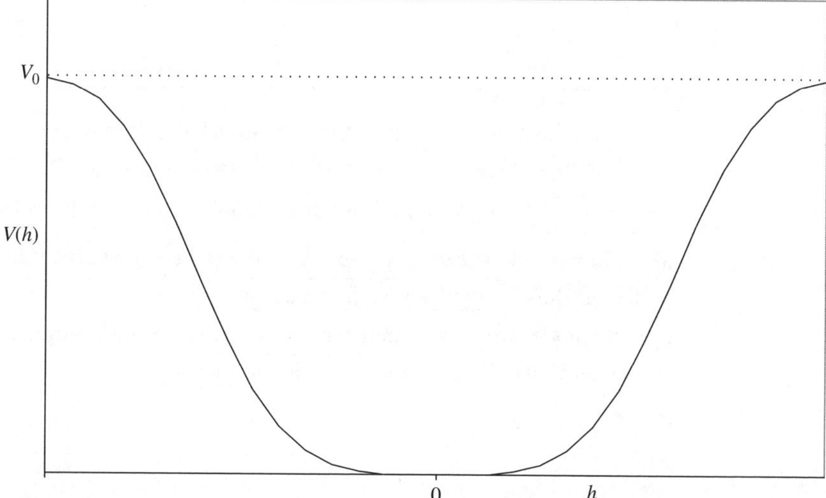

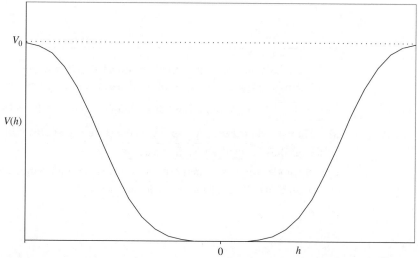

Figure 12E.2

P12E.3 **(a)** Follow the flow chart in Fig. 11E.7 of the text. CH_3Cl is not linear, it has a C_3 axis (only one), it does not have C_2 axes perpendicular to C_3, it has no σ_h, but does have 3 σ_v planes; so it belongs to $\boxed{C_{3v}}$.

(b) The number of normal modes of a nonlinear molecule is $3N-6$, where N is the number of atoms. So CH_3Cl has $\boxed{\text{nine}}$ normal modes

(c) To determine the symmetry of the normal modes, consider how the cartesian axes of each atom are transformed under the symmetry operations of the C_{3v} group; the 15 cartesian displacements constitute the basis here. All 15 cartesian axes are left unchanged under the identity, so the character of this operation is 15. Under a C_3 operation, the H atoms are taken into each other, so they do not contribute to the character of C_3. The z-axes of the C and Cl atoms, are unchanged, so they contribute 2 to the character of C_3; for these two atoms

$$x \to -\frac{x}{2} + \frac{3^{1/2}y}{2} \quad \text{and} \quad y \to -\frac{y}{2} + \frac{3^{1/2}x}{2},$$

so there is a contribution of $-1/2$ to the character from each of these coordinates in each of these atoms. In total, then $\chi=0$ for C_3. To find the character of σ_v, call one of the σ_v planes the yz plane; it contains C, Cl, and one H atom. The y and z coordinates of these three atoms are unchanged, but the x coordinates are taken into their negatives, contributing $6-3=3$ to the character for this operation; the other two atoms are interchanged, so contribute nothing to the character. To find the irreducible representations that this basis spans, we multiply its characters by the characters of the irreducible representations, also multiplying by the number of operations of each kind, sum those products, and divide the sum by the order h of the group (as in Section 11C). The table below illustrates the procedure.

	E	$2C_3$	$3\sigma_v$		E	$2C_3$	$3\sigma_v$	sum/h
tcbasis	15	0	3					
A_1	1	1	1	basis × A_1	15	0	3	4
A_2	1	1	−1	basis × A_2	15	0	−3	1
E	2	−1	0	basis × E	30	0	0	5

Of these 15 modes of motion, three are translations (an A_1 and an E) and three rotations (an A_2 and an E); we subtract these to leave the vibrations, which span

$\boxed{3A_1+3E}$ (three A_1 modes and three doubly-degenerate E modes).

(d) Any mode whose symmetry species is the same as that of x, y, or z is infrared active. Thus $\boxed{\text{all modes are infrared active}}$.

(e) Only modes whose symmetry species is the same as a quadratic form may be Raman active. Thus $\boxed{\text{all modes are Raman active}}$.

Integrated activities

I12.1 Refer to the flow chart in text Fig. 11A.7. Yes at the first question (linear?) leads to linear point groups and therefore linear rotors. If the molecule is not linear, then yes at the next question (two or more C_n with $n > 2$?) leads to cubic and icosahedral groups and therefore spherical rotors. If the molecule is not a spherical rotor, yes at the next question leads to symmetric rotors if the highest C_n has $n > 2$; if not, the molecule is an asymmetric rotor.

(a) CH_4: not linear, but more than two $C_n\,(n>2)$, so $\boxed{\text{spherical rotor}}$.

(b) CH_3CN: not linear, C_3 (only one of them), so $\boxed{\text{symmetric rotor}}$.

(c) CO_2: linear, so $\boxed{\text{linear rotor}}$.

(d) CH_3OH: not linear, no C_n, so $\boxed{\text{asymmetric rotor}}$.

(e) Benzene: not linear, C_6, but only one high-order axis, so $\boxed{\text{symmetric rotor}}$.

(f) Pyridine: not linear, C_2, is highest rotational axis, so $\boxed{\text{asymmetric rotor}}$.

I12.3 The H_3^+ ion has recently been found in the interstellar medium and in the atmospheres of Jupiter, Saturn, and Uranus. The rotational energy levels of H_3^+, an oblate symmetric rotor, are given by eqn 12B.13, with $\tilde{C}$ replacing $\tilde{A}$ when centrifugal distortion and other complications are ignored. Experimental values for vibrational–rotational constants are $\tilde{\nu}(\mathrm{E}')=2\,521.6\,\mathrm{cm^{-1}}$, $\tilde{B}=43.55\,\mathrm{cm^{-1}}$, and $\tilde{C}=20.71\,\mathrm{cm^{-1}}$. (a) Show that for a nonlinear planar molecule (such as H_3^+) that $I_C=2I_B$. The rather large discrepancy with the experimental values is due to factors ignored in eqn 12B.13. (b) Calculate an approximate value of the H–H bond length in H_3^+. (c) The value of R_e obtained from the best quantum mechanical calculations by J.B. Anderson (*J. Chem. Phys.* **96**, 3702 (1991)) is 87.32 pm Use this result to calculate the values of the rotational constants $\tilde{B}$ and $\tilde{C}$. (d) Assuming that the geometry and force constants are the same in D_3^+ and H_3^+, calculate the spectroscopic constants of D_3^+. The molecular ion D_3^+ was first produced by Shy et al. (*Phys. Rev. Lett.* **45**, 535 (1980)) who observed the band in the infrared.

(a) The problem states that the H_3^+ ion is an oblate symmetric rotor. Consequently, it must have the structure of an equilateral triangle with a C_3 principal axis. The rotational axis for $I_{\parallel}$, which is labelled I_C in this problem, is through the centre of the molecule, the location of the centre of mass, from which each atom is separated by the distance $R/\sqrt{3}$ where R is the equilibrium bond length. The rotational axis for $I_{\perp}$, which is labelled I_B in this problem, is on a line through one of the atoms and the centre of mass so the I_B axis bisects a H–H–H angle and two of the atoms are separated from the axis by the perpendicular distance $R/2$. The moment of inertia are given by

$$I_C=\sum_i m_H\times(\text{perpendicular distance from rotation axis})^2\quad[12B.1]$$

$$=3m_H\times(\text{perpendicular distance from rotation axis})^2=3m_H\times\left(R/\sqrt{3}\right)^2$$

$$=m_HR^2$$

$$I_B = \sum_i m_H \times (\text{perpendicular distance from rotation axis})^2 \quad [12B.1]$$
$$= 2m_H \times (\text{perpendicular distance from rotation axis})^2 = 2m_H \times (R/2)^2$$
$$= m_H R^2/2$$

Thus, $\boxed{I_C = 2I_B}$.

(b) $$\tilde{B} = \frac{\hbar}{4\pi c I_B} \; [12B.7] = \frac{2\hbar}{4\pi c m_H R^2} = \frac{\hbar}{2\pi c m_H R^2}$$

$$R = \left(\frac{\hbar}{2\pi c m_H \tilde{B}}\right)^{1/2} = \left(\frac{\hbar N_A}{2\pi c M_H \tilde{B}}\right)^{1/2}$$
$$= \left(\frac{(1.0546\times10^{-34}\ \text{J s})\times(6.0221\times10^{23}\ \text{mol}^{-1})}{2\pi\times(2.9979\times10^{10}\ \text{cm s}^{-1})\times(1.0079\times10^{-3}\ \text{kg mol}^{-1})\times(43.55\ \text{cm}^{-1})}\right)^{1/2}$$
$$= 87.64\ \text{pm}$$

Alternatively, the rotational constant $\tilde{C}$ can be used to calculate R.

$$\tilde{C} = \frac{\hbar}{4\pi c I_C} \; [12B.14] = \frac{\hbar}{4\pi c m_H R^2}$$

$$R = \left(\frac{\hbar}{4\pi c m_H \tilde{C}}\right)^{1/2} = \left(\frac{\hbar N_A}{4\pi c M_H \tilde{C}}\right)^{1/2}$$
$$= \left(\frac{(1.0546\times10^{-34}\ \text{J s})\times(6.0221\times10^{23}\ \text{mol}^{-1})}{4\pi\times(2.9979\times10^{10}\ \text{cm s}^{-1})\times(1.0079\times10^{-3}\ \text{kg mol}^{-1})\times(20.71\ \text{cm}^{-1})}\right)^{1/2}$$
$$= 89.87\ \text{pm}$$

The two calculations differ slightly so we approximate the bond length as the average:

$$R_{av} = \frac{(87.64+89.87)\ \text{pm}}{2} = \boxed{88.8\ \text{pm}}$$

(c) $$\tilde{B} = \frac{\hbar N_A}{2\pi c M_H R^2}$$
$$= \frac{(1.0546\times10^{-34}\ \text{J s})\times(6.0221\times10^{23}\ \text{mol}^{-1})}{2\pi\times(2.9979\times10^{10}\ \text{cm s}^{-1})\times(1.0079\times10^{-3}\ \text{kg mol}^{-1})\times(87.32\times10^{-12}\ \text{m})^2}$$
$$= \boxed{43.87\ \text{cm}^{-1}}$$

(d) $\dfrac{1}{m_{\text{eff},H_3^+}} = \sum_i \dfrac{1}{m_i} = \dfrac{3}{m_H}$ or $m_{\text{eff},H_3^+} = \dfrac{m_H}{3}$. Also, $m_D = 2\,m_H$ so $m_{\text{eff},D_3^+} = \dfrac{2m_H}{3}$.

Assuming that the bonding force constants are the same in D_3^+ and H_3^+, eqn 12D.8 relates the vibrational wavenumber to the effective mass.

$$\tilde{\nu}_{H_3^+} = \frac{1}{2\pi c}\left(\frac{k_f}{m_{\text{eff},H_3^+}}\right)^{1/2} \quad \text{and} \quad \tilde{\nu}_{D_3^+} = \frac{1}{2\pi c}\left(\frac{k_f}{m_{\text{eff},D_3^+}}\right)^{1/2}$$

Taking the ratio and solving for $\tilde{\nu}_{D_3^+}$ yields:

$$\tilde{\nu}_{D_3^+} = \left(\frac{m_{\text{eff},H_3^+}}{m_{\text{eff},D_3^+}}\right)^{1/2} \tilde{\nu}_{H_3^+} = \left(\frac{m_H/3}{2m_H/3}\right)^{1/2} \tilde{\nu}_{H_3^+} = \tilde{\nu}_{H_3^+}/\sqrt{2}$$
$$= \left(2\,521.6\ \text{cm}^{-1}\right)/\sqrt{2}$$
$$= \boxed{1783.0\ \text{cm}^{-1}}$$

Similarly,

$$\tilde{B}_{D_3^+} = \left(\frac{m_H}{m_D}\right)\tilde{B}_{H_3^+} = \tilde{B}_{H_3^+}/2$$
$$= \left(43.55\ \text{cm}^{-1}\right)/2$$
$$= \boxed{21.78\ \text{cm}^{-1}}$$

$$\tilde{C}_{D_3^+} = \left(\frac{m_H}{m_D}\right)\tilde{C}_{H_3^+} = \tilde{C}_{H_3^+}/2$$
$$= \left(20.71\ \text{cm}^{-1}\right)/2$$
$$= \boxed{10.36\ \text{cm}^{-1}}$$

I12.5 (a) SO_2 calculations are performed with Spartan '10 using the MP2 method with the 6-311++G** basis set. In a 6-311G basis set each atomic core basis function is expanded as a linear combination of six Gaussian functions. Valence orbitals are split into three basis set functions consisting of three, one, and one Gaussians. The 6-311++G basis set adds both an s-type and three p-type diffuse functions for each atom other than hydrogen and one s-type diffuse function for each hydrogen atom. The 6-311++G** basis set adds a set of five d-type polarization functions for each atom other than hydrogen and a set of three p-type polarization functions for each hydrogen atom.

The calculated equilibrium structure is shown in Fig. 12I.1 where we see that the two S–O bonds have equal length, thereby, justifying Lewis resonances structures; the Mulliken bond orders are found to be 1.47. The calculated bond length and angle agree with the experimental values of 143.21 pm and 119.54°. The calculated dipole is somewhat higher than the experimental value of 1.63 D. Greatly exaggerated stretches and bends of the three vibrational modes are also shown in Fig. I12.1.

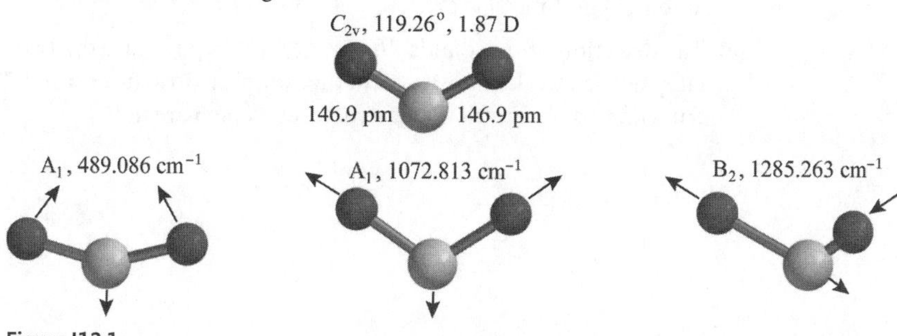

Figure I12.1

(b) The calculated values of the fundamental vibrational wavenumbers, $\boxed{489, 1073\text{, and } 1285\ \text{cm}^{-1}}$, are also reported in Fig. 12.2. They correlate well with experimental values but are about 35–80 cm^{-1} lower. SCF calculations often yield systematically lower or higher values than experiment while approximately paralleling the experimental to within an additive constant.

I12.7 (a) Resonance Raman spectroscopy is preferable to vibrational spectroscopy for studying the O–O stretching mode because such a mode would be $\boxed{\text{infrared inactive}}$, or at best only weakly active. (The mode is sure to be inactive in free O_2, because it would not change the molecule's dipole moment. In a complex in which O_2 is bound, the O–O stretch may change the dipole moment, but it is not certain to do so at all, let alone strongly enough to provide a good signal.)

(b) Assuming that the two isotopic oxygen molecules have identical O—O bond strengths when bound to haemerythrin, the vibrational wavenumber depends on the effective mass alone.

$$\tilde{\nu} \propto \left(\frac{1}{m_{\text{eff}}}\right)^{1/2} \quad [12\text{D}.8]$$

$$\frac{\tilde{\nu}(\text{bound }^{18}\text{O}_2)}{\tilde{\nu}(\text{bound }^{16}\text{O}_2)} = \left(\frac{m_{\text{eff}}(\text{bound }^{16}\text{O}_2)}{m_{\text{eff}}(\text{bound }^{18}\text{O}_2)}\right)^{1/2} = \left(\frac{16.0\,\text{u}}{18.0\,\text{u}}\right)^{1/2} = 0.943$$

and $\tilde{\nu}(\text{bound }^{18}\text{O}_2) = (0.943)\times(844\ \text{cm}^{-1}) = \boxed{796\ \text{cm}^{-1}}$

Note the assumption that the effective masses are proportional to the isotopic masses. This assumption is valid in the free molecule, where the effective mass of O_2 is equal to half the mass of the O atom; it is also valid if O_2 is strongly bound at one end, such that one atom is free and the other is essentially fixed to a very massive unit.

(c) The vibrational wavenumber is proportional to the square root of the force constant by eqn 12D.8. The force constant is itself a measure of the strength of the bond (technically of its stiffness, which correlates with strength), which in turn is characterized by bond order. Simple molecular orbital analysis of O_2, O_2–, and $O_2$2– results in bond orders of $\boxed{2, 1.5, \text{and } 1}$, respectively. Given decreasing bond order, one would expect decreasing vibrational wavenumbers (and vice versa).

(d) The wavenumber of the O—O stretch is very similar to that of the peroxide anion, suggesting $\boxed{Fe_2^{3+}O_2^{2-}}$.

(e) The detection of two bands due to $^{16}O^{18}O$ implies that the two O atoms occupy non-equivalent positions in the complex. Structures **6** and **7** are consistent with this observation, but structures **4** and **5** are not.

I12.9 (a) The molar absorption coefficient $\varepsilon(\tilde{\nu})$ is given by

$$\varepsilon(\tilde{\nu})=\frac{A(\tilde{\nu})}{L[CO_2]}=\frac{RTA(\tilde{\nu})}{Lx_{CO_2}p}$$ [Beer–Lambert law, perfect gas law, and Dalton's law]

where $T=298\,K$, $L=10\,cm$, $p=1\,bar$, and $x_{CO_2}=0.021$.

The absorption band originates with the $001 \leftarrow 000$ transition of the antisymmetric stretch vibrational mode at $2349\,cm^{-1}$ (text Fig. 12E.2). The band is very broad because of accompanying rotational transitions and lifetime broadening of each individual absorption (also called collisional broadening or pressure broadening). The spectra reveals that the Q branch is missing so we conclude that the transition $\boxed{\Delta J=0 \text{ is forbidden}}$ [12C.1] for the $D_{\infty h}$ point group of CO_2. The P-branch ($\Delta J=-1$) is evident at lower energies and the R-branch ($\Delta J=+1$) is evident at higher energies. See Figs I12.2(a) and (b).

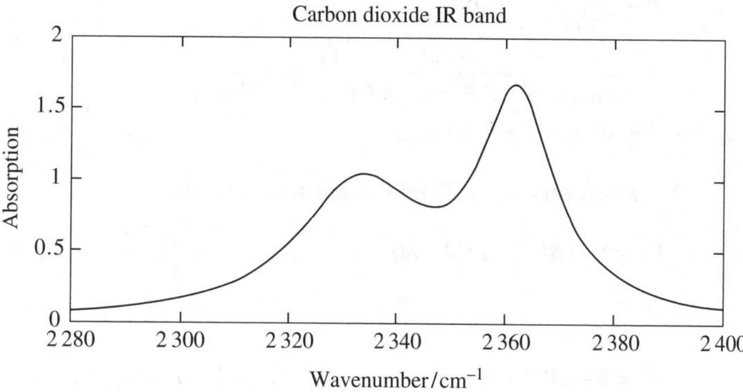

Figure I12.2(a)

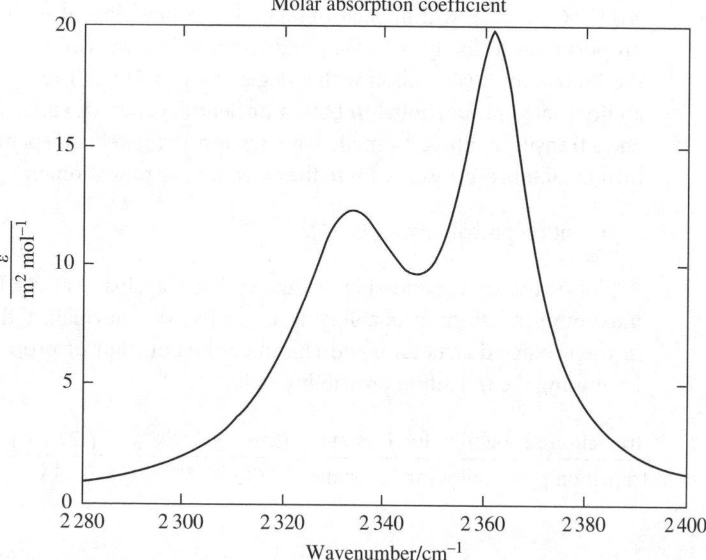

Figure I12.2(b)

(b) ^{16}O–^{12}C–^{16}O has two identical nuclei of zero spin so the CO_2 wavefunction must be symmetric with respect to nuclear interchange and it must obey Bose–Einstein nuclear statistics. Consequently, J takes on even values only for the $v=0$ vibrational state and odd values only for the $v=1$ state. The (v,J) states for this absorption band are $(1,J+1)\leftarrow(0,J)$ for $J=0,2,4,\ldots$. According to eqn 12D.18, the energy of the (0,J)) state is

$$\tilde{S}(0,J)=\tfrac{1}{2}\tilde{\nu}+\tilde{B}J(J+1) \text{ where } \tilde{\nu}=2349\ \text{cm}^{-1}$$

$$I=\frac{2M_{\text{O}}R^2}{N_{\text{A}}}$$
$$=\frac{2(0.01600\,\text{kg mol}^{-1})(116.2\times10^{-12}\,\text{m})^2}{6.022\times10^{23}\,\text{mol}^{-1}}$$
$$=7.175\times10^{-46}\,\text{kg m}^2 \quad \text{(Table 12B.1)}$$

$$\tilde{B}=\frac{h}{8\pi^2 cI} \quad [12\text{B}.7]$$
$$=\frac{6.626\times10^{-34}\,\text{J s}}{8\pi^2(2.998\times10^8\,\text{m s}^{-1})(7.175\times10^{-46}\,\text{kg m}^2)}$$
$$=39.02\,\text{m}^{-1}=0.3902\,\text{cm}^{-1}$$

The transitions of the P and R branches occur at

$$\tilde{\nu}_{\text{P}}=\tilde{\nu}-2\tilde{B}J \quad [12\text{D}.19\text{a}]$$

and

$$\tilde{\nu}_{\text{R}}=\tilde{\nu}+2\tilde{B}(J+1) \quad [12\text{D}.19\text{c}] \quad \text{where } J=0,2,4,6\ldots$$

The highest energy transition of the P branch is at $\tilde{\nu}-4\tilde{B}$; the lowest energy transition of the R branch is at $\tilde{\nu}+2\tilde{B}$. Transitions are separated by $4\tilde{B}$ ($1.5608\,\text{cm}^{-1}$) within each branch. The probability of each transition is proportional to the lower state population, which we assume to be given by the Boltzmann distribution with a degeneracy of $2J+1$. The transition probability is also proportional to both a nuclear degeneracy factor (eqn 12C.16) and a transition dipole moment, which is approximately independent of J. The former factors are absorbed into the constant of proportionality.

$$\text{transition probability} \propto (2J+1)\text{e}^{-S(0,J)hc/kT}$$

A plot of the right-hand-side of this equation against J at 298 K indicates a maximum transition probability at $J_{\text{max}}=16$. We 'normalize' the maximum in the predicted structure, and eliminate the constant of proportionality by examining the transition probability ratio:

$$\frac{\text{transition probability for } J^{\text{th}} \text{ state}}{\text{transition probability for } J_{\text{max}} \text{ state}}=\frac{(2J+1)\text{e}^{-\tilde{S}(0,J)hc/kT}}{33\text{e}^{-\tilde{S}(0,16)hc/RT}}=\left(\frac{2J+1}{33}\right)\text{e}^{-(J^2+J-272)\tilde{B}hc/kT}$$

A plot, Fig. I12.3, of the above ratio against predicted wavenumbers can be compared to the ratio $A(\tilde{\nu})/A_{max}$ where A_{max} is the observed spectrum maximum (1.677). It shows a fair degree of agreement between the experimental and simple theoretical band shapes.

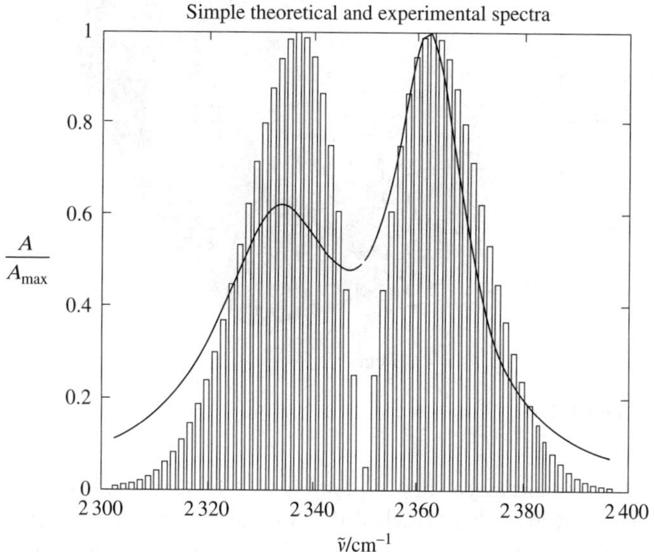

Figure I12.3

(c) Applying the Beer–Lambert law, we may write the relationship

$$A = \varepsilon(\tilde{\nu})\int_0^h [CO_2]dh$$

The strong absorption of the band suggests that h should not be a very great length and that $[CO_2]$ should be constant between the Earth's surface and h. Consequently, the integration gives

$$A = \varepsilon(\tilde{\nu})[CO_2]h$$
$$= \varepsilon(\tilde{\nu})h\left\{\frac{x_{CO_2}p}{RT}\right\} \quad \text{Dalton's law of partial pressures}$$

p and T are not expected to change much for modest values of h so we estimate that $p = 1$ bar and $T = 288$ K.

$$A = \varepsilon(\tilde{\nu})h\left\{\frac{(3.3\times10^{-4})(1\times10^5\ \text{Pa})}{(8.314\ \text{J K}^{-1}\ \text{mol}^{-1})(288\ \text{K})}\right\}$$
$$= (0.0138\ \text{m}^{-3}\ \text{mol})\times\varepsilon(\tilde{\nu})h$$

$$\text{Transmittance} = 10^{-A} = 10^{-(0.0138\,\text{m}^{-3}\,\text{mol})\times\varepsilon(\tilde{\nu})h}$$

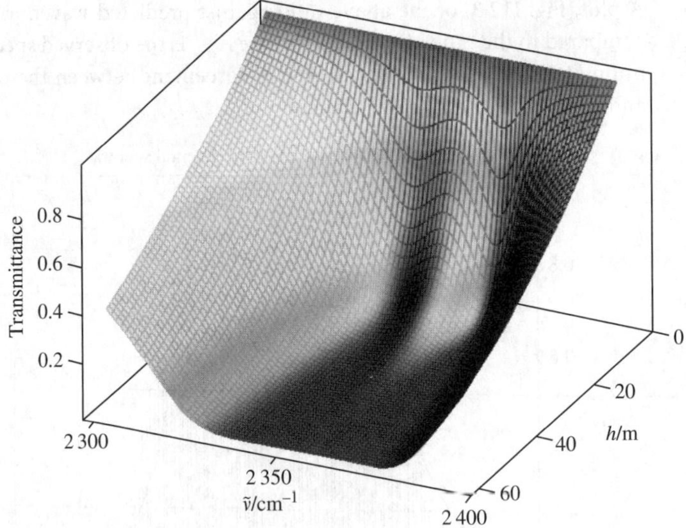

Figure I12.4

The transmittance surface plot, Fig. I12.4, clearly shows that before a height of about 30 m has been reached all of the Earth's IR radiation in the 2 320 – 2 380 cm^{-1} range has been absorbed by atmospheric carbon dioxide. A greyscale contour plot, shown in Fig. I12.5, is somewhat easier to read. We clearly see that in the wave-number range 2 325–2 375 cm^{-1} the transmittance has dropped below 0.1 at about 10 m. The transmittance extends to much greater heights on the band fringes. See C.A. Meserole, F.M. Mulcahy, J. Lutz, and H.A. Yousif, *J. Chem. Ed.*, *74*, 316 (1997).

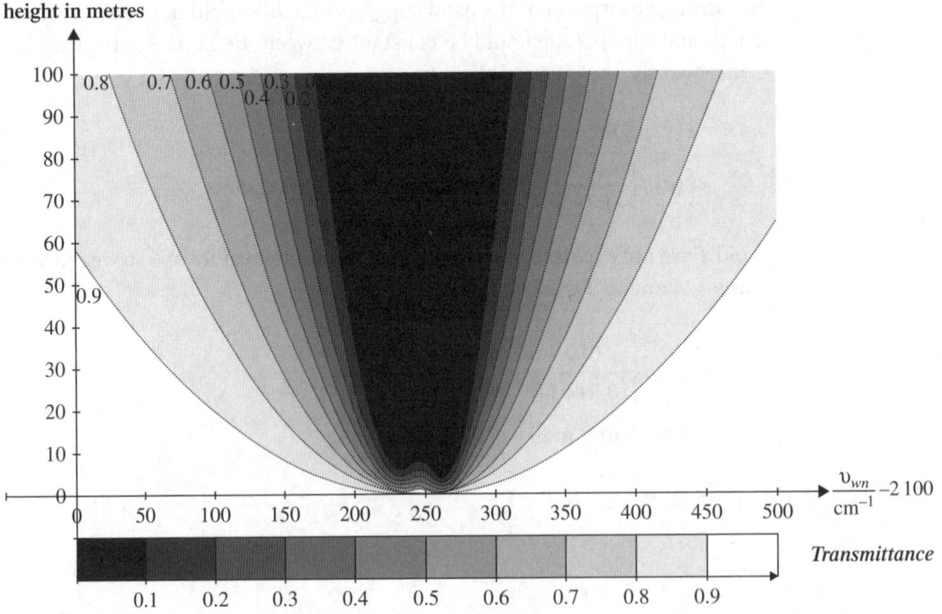

Figure I12.5

13 Electronic transitions

13A Electronic spectra

Answers to discussion questions

D13A.1 The ground electronic configuration of dioxygen, $1\sigma_g^2 1\sigma_u^2 2\sigma_g^2 1\pi_u^4 1\pi_g^2$, is discussed in Sections 10C.1(c) and 13A.1(a). The determination of the term symbol, ${}^3\Sigma_g^-$, is described in *Brief illustrations* 13A.1 and 13A.2. The term symbol is Σ to represent a total orbital angular momentum of zero about the internuclear axis. This happens because for every π orbital electron with $\lambda = +1$ there is a π orbital electron with $\lambda = -1$. For example, the two $1\pi_g$ electrons are, according to Hund's rules, in separate degenerate orbitals for which one orbital has $\lambda = +1$ and the other has $\lambda = -1$. Except for the two $1\pi_g$ electrons, electrons have paired α and β spins, which results in zero contribution to the total spin angular momentum. According to Hund's rules, the two $1\pi_g$ electrons (i.e. $1\pi_g^1 1\pi_g^1$) have parallel spins in the ground state. They provide a total spin angular momentum of $S = \frac{1}{2} + \frac{1}{2} = 1$ and a spin multiplicity of $2S + 1 = 3$, which appears as the left superscript 3. The term symbol indicates a gerade total symmetry because electrons are paired in the ungerade molecular orbitals and $u \times u = g$ and the resultant symmetry of electrons in different molecular orbitals must therefore be given by $g \times g = g$. The π orbitals change sign upon reflection in the plane that contains the internuclear axis. Consequently, the term symbol has the superscript—to indicate that the molecular wavefunction for O_2 changes sign upon reflection in the plane containing the nuclei.

D13A.3 A band head is the convergence of the frequencies of electronic transitions with increasing rotational quantum number, *J*. They result from the rotational structure superimposed on the vibrational structure of the electronic energy levels of the diatomic molecule. See Fig. 13A.10. To understand how a band head arises, one must examine the equations describing the transition energies (eqn 13A.8) where we see that convergence to an extreme in the energy absorption can only arise when terms in both $(B' - B)$ and $(B' + B)$ occur in the equation. The extreme occurs in the P branch when $B' > B$ but in is in the R branch when $B' < B$. Because only a term in $(B' - B)$ occurs for the Q branch, no band head can arise for that branch. The following Mathcad Prime 2 worksheet (in which v represents wavenumber) shows these branch heads by plotting $\Delta\tilde{\nu} \equiv \tilde{\nu}(J) - \tilde{\nu}$ [13.A.8a and c] against *J*.

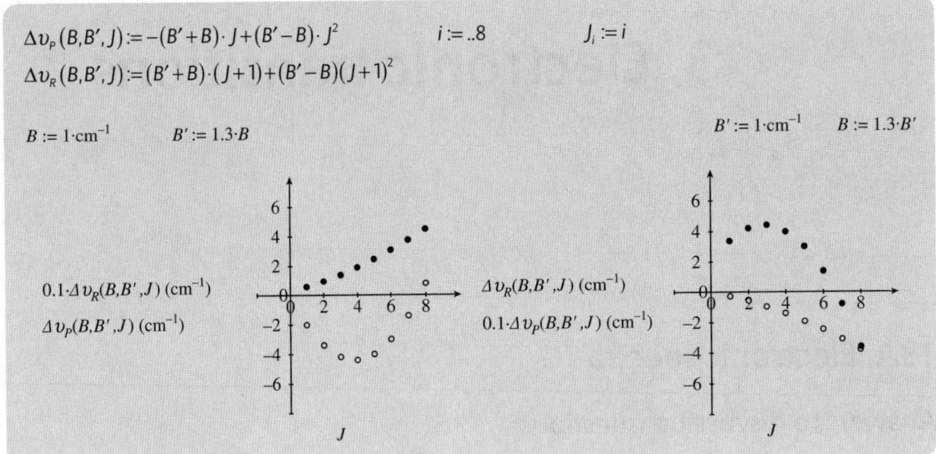

For the parameters chosen in the worksheet only the P branch shows an extreme Δv value at $J_{\text{extremum}} = 4$ when $B' > B$ and only the R branch shows a branch head at $J_{\text{extremum}} = 3$ when $B > B'$. We see that Δv values converge to the value at J_{extremum} from both low and high values of J as J approaches J_{extremum}. This is the branch head.

D13A.5 (a) The transition intensity is proportional to the square of the transition dipole moment. We initially suspect that the transition dipole moment should increase as the length L of the alternating carbon-to-carbon double/single/double/single···· bond sequence of the polyene is increased and, consequently, the transition intensity should also increase as L increases. To test this hypothesis, consider that the polyene has N pi electrons which fill the first $n = N/2$ quantum states of the particle in a one-dimensional box of length $L = Nd = 2nd$, where d is the average carbon-to-carbon bond length. (The length choice $L = Nd$ adds half a bond length at each end to the distance between the two end-carbon nuclei.) The transition dipole moment of this model is

$$\mu_x = \int_0^L \psi_{n_f} x \psi_{n_i} \, dx$$

We quickly find a selection rule for transitions with the substitution $x = f(x) + L/2$ where $f(x) = x - L/2$ is a function of ungerade symmetry w/r/t inversion through the centre of symmetry at $x = L/2$, and we note that the wavefunctions have alternating gerade and ungerade symmetry as n increases. Then,

$$\mu_x = \int_0^L \psi_{n_f} \{f(x) + L/2\} \psi_{n_i} \, dx = \int_0^L \psi_{n_f} f(x) \psi_{n_i} \, dx + (L/2) \int_0^L \psi_{n_f} \psi_{n_i} \, dx$$

The last integral vanishes because the wavefunctions of different energy levels are orthogonal which leaves

$$\mu_x = \int_0^L \psi_{n_f} f(x) \psi_{n_i} \, dx$$

Because the integrand factor $f(x)$ has ungerade symmetry, the product $\psi_{n_f}\psi_{n_i}$ of an allowed transition must also have ungerade symmetry so that the total symmetry of the integrand has u × u = g symmetry and the integral can be nonzero. Thus, the lowest energy transition is $n+1 \leftarrow n$ where $n = n_i$. It is now convenient to return to the original transition dipole moment integral and substitute the wavefunctions for the final and initial states:

$$\mu_x = \frac{2}{L}\int_0^L \sin\left(\frac{(n+1)\pi x}{L}\right) x \sin\left(\frac{n\pi x}{L}\right) \mathrm{d}x \quad [8\text{A}.7\text{a}]$$

We need not evaluate the integral exactly because the development of a method to improve the intensity of a dye requires only a knowledge of the approximate relationship between u_x and either n or L (proportional properties for a polyene) so we recognize that $n+1 \sim n$ for a large polyene and we make the estimate

$$\mu_x \propto \frac{1}{L}\int_0^L x \sin^2\left(\frac{n\pi x}{L}\right) \mathrm{d}x \quad \text{(The integral is found in standard mathematical tables.)}$$
$$\propto L$$

The model confirms the hypothesis that the transition dipole moment and, consequently, the transition intensity is increased by increasing the length of the polyene.

(b) Since $E_n = \dfrac{n^2h^2}{8m_eL^2}$ [8A.7b], $\Delta E = \dfrac{(2n+1)h^2}{8m_eL^2}[\Delta n = +1] \sim \dfrac{nh^2}{4m_eL^2}$ for large n.

But $L = 2nd$ is the length of the chain where d is the average carbon–carbon interatomic distance. Hence

$$\Delta E \propto \frac{1}{L}$$

Therefore, the transition moves toward the red as L is increased. When white light is used to illuminate the dyed object, the colour absorbed is the **complementary colour** to the reflected, observed colour. Newton's colour wheel, shown overleaf in Fig. 13A.1, usefully displays complementary and observed colours. For example, draw a line from complementary violet through the circle centre to find the observed colour of green-yellow. As the polyene length is increased, the complementary colour progresses from violet to indigo, to blue, to green, etc. while the observed colour progresses from green-yellow, to yellow-orange, to orange-red, to red-violet, to violet-indigo. We say that the apparent colour of the dye shifts towards blue.

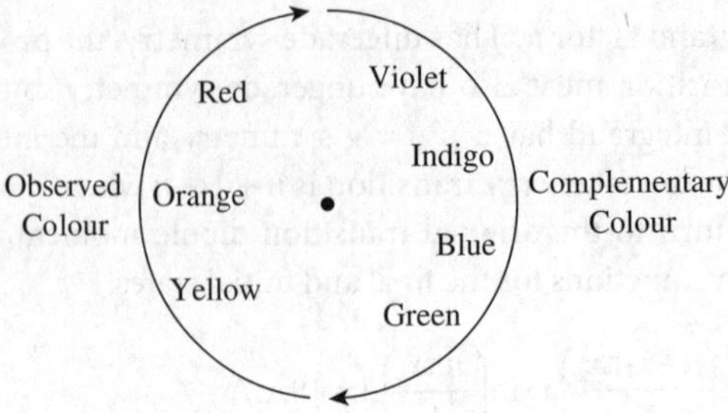

Figure 13A.1

Solutions to exercises

E13A.1(a) The $1\sigma_g^{\,2}1\sigma_u^{\,2}1\pi_u^{\,3}1\pi_g^{\,1}$ valence configuration has two unpaired electrons. Although Hund's rule does not apply to excited states, we examine the state of maximum spin multiplicity so $S = s_1 + s_2 = ½ + ½ = 1$ and the spin multiplicity is given by $2S+1=2(1)+1=\boxed{3}$. Because $\mathrm{u}\times\mathrm{u}=\mathrm{g}$ and $\mathrm{g}\times\mathrm{g}=\mathrm{g}$, the net parity of two electrons paired in an orbital is always gerade. Consequently, the overall parity is found by multiplying the parity of unpaired electrons. For this configuration, $\mathrm{u}\times\mathrm{g}=\boxed{\mathrm{u}}$.

E13A.2(a) The electronic spectrum selection rules concerned with changes in angular momentum are (eqn 13A.4): $\Delta\Lambda=0,\ \pm1\quad \Delta S=0\quad \Delta\Sigma=0\quad \Delta\Omega=0,\ \pm1$ where $\Omega=\Lambda+\Sigma$. Λ gives the total orbital angular momentum about the internuclear axis and Σ gives the total spin angular momentum about the internuclear axis. The ± superscript selection rule for reflection in the plane along the internuclear axis is $+\leftrightarrow+$ or $-\leftrightarrow-$ (i.e., $+\leftrightarrow-$ is forbidden). The **Laporte selection rule** states that for a centrosymmetric molecule (those with a centre of inversion) the only allowed transitions are transitions that are accompanied by a change of parity: $\mathrm{u}\leftrightarrow\mathrm{g}$.

(i) The changes in the transition $^2\Pi\leftrightarrow{}^2\Pi$ are $\Delta\Lambda=0$, $\Delta S=0$, $\Delta\Sigma=0$, and $\Delta\Omega=0$ so the transition is $\boxed{\text{allowed}}$.

(ii) The changes in the transition $^1\Sigma\leftrightarrow{}^1\Sigma$ are $\Delta\Lambda=0$, $\Delta S=0$, $\Delta\Sigma=0$, and $\Delta\Omega=0$ so the transition is $\boxed{\text{allowed}}$.

(iii) The changes in the transition $\Sigma\leftrightarrow\Delta$ are $\Delta\Lambda=2$ so the transition is $\boxed{\text{forbidden}}$.

(iv) The transition $\Sigma^+\leftrightarrow\Sigma^-$ is $\boxed{\text{forbidden}}$ because $+\leftrightarrow-$ is forbidden.

(v) The transition $\Sigma^+\leftrightarrow\Sigma^+$ is $\boxed{\text{allowed}}$ because $\Delta\Lambda=0$ and $+\leftrightarrow+$ is allowed.

E13A.3(a) We begin by evaluating the normalization constants N_0 and N_v.

$$N_0^2=\frac{1}{\int_{-\infty}^{\infty}\mathrm{e}^{-2ax^2}\,\mathrm{d}x}=\left(\frac{2a}{\pi}\right)^{1/2}\ \text{(standard integral)};\quad N_0=\left(\frac{2a}{\pi}\right)^{1/4}$$

Likewise, $$N_\Lambda^2=\frac{1}{\int_{-\infty}^{\infty}\mathrm{e}^{-2b(x-x_0)^2}\,\mathrm{d}x}=\left(\frac{2b}{\pi}\right)^{1/2};\quad N_\Lambda=\left(\frac{2b}{\pi}\right)^{1/4}$$

Furthermore, we can easily check that

$$ax^2+b(x-x_0)^2=z^2+\frac{ab}{a+b}x_0^2 \quad \text{where} \quad z=(a+b)^{1/2}x-\frac{b}{(a+b)^{1/2}}x_0 \quad \text{and} \quad \mathrm{d}x=\frac{1}{(a+b)^{1/2}}\mathrm{d}z$$

Then, the vibration overlap integral between the vibrational wavefunction in the upper and lower electronic states is:

$$S(v,0)=\langle v|0\rangle=N_0N_v\int_{-\infty}^{\infty}\mathrm{e}^{-ax^2}\mathrm{e}^{-b(x-x_0)^2}\mathrm{d}x=N_0N_v\int_{-\infty}^{\infty}\mathrm{e}^{-\{ax^2+b(x-x_0)^2\}}\mathrm{d}x$$

$$=\frac{N_0N_v}{(a+b)^{1/2}}\int_{-\infty}^{\infty}\mathrm{e}^{-\left\{z^2+\frac{ab}{a+b}x_0^2\right\}}\mathrm{d}z=\frac{N_0N_v}{(a+b)^{1/2}}\mathrm{e}^{-\frac{ab}{a+b}x_0^2}\int_{-\infty}^{\infty}\mathrm{e}^{-z^2}\mathrm{d}z=N_0N_v\left(\frac{\pi}{a+b}\right)^{1/2}\mathrm{e}^{-\frac{ab}{a+b}x_0^2}$$

$$=\left(\frac{2a}{\pi}\right)^{1/4}\left(\frac{2b}{\pi}\right)^{1/4}\left(\frac{\pi}{a+b}\right)^{1/2}\mathrm{e}^{-\frac{ab}{a+b}x_0^2}=(4ab)^{1/4}\left(\frac{1}{a+b}\right)^{1/2}\mathrm{e}^{-\frac{ab}{a+b}x_0^2}$$

For the case $b=a/2$, this simplifies to

$$S(v,0)=\frac{2}{\left(3\sqrt{2}\right)^{1/2}}\mathrm{e}^{-ax_0^2/3}$$

The Franck–Condon factor is

$$|S(v,0)|^2=\boxed{\frac{2\sqrt{2}}{3}\mathrm{e}^{-2ax_0^2/3}}$$

E13A.4(a) $\psi_0=\left(\frac{2}{L}\right)^{1/2}\sin\left(\frac{\pi x}{L}\right)$ for $0\le x\le L$ and 0 elsewhere.

$\psi_v=\left(\frac{2}{L}\right)^{1/2}\sin\left\{\frac{\pi}{L}\left(x-\frac{L}{4}\right)\right\}$ for $\frac{L}{4}\le x\le\frac{5L}{4}$ and 0 elsewhere.

$$S(v,0)=\langle v|0\rangle=\frac{2}{L}\int_{L/4}^{L}\sin\left(\frac{\pi x}{L}\right)\sin\left\{\frac{\pi}{L}\left(x-\frac{L}{4}\right)\right\}\mathrm{d}x$$

The above integral is recognized as the standard integral (see math handbook): $\int\sin(ax)\sin(ax+b)\mathrm{d}x=\frac{x}{2}\cos(b)-\frac{\sin(2ax+b)}{4a}$ with the transformations $a=\pi/L$ and $b=-\pi/4$. Thus,

$$S(v,0)=\frac{2}{L}\left[\frac{x}{2}\cos\left(-\frac{\pi}{4}\right)-\frac{\sin(2\pi x/L-\pi/4)}{4\pi/L}\right]_{x=L/4}^{x=L}=\frac{2}{L}\left[\frac{x}{2}\cos\left(\frac{\pi}{4}\right)-\frac{\sin(2\pi x/L-\pi/4)}{4\pi/L}\right]_{x=L/4}^{x=L}$$

$$=\left[x\cos\left(\frac{\pi}{4}\right)-\frac{\sin(2\pi x-\pi/4)}{2\pi}\right]_{x=1/4}^{x=1}$$

$$=\cos\left(\frac{\pi}{4}\right)-\frac{\sin(2\pi-\pi/4)}{2\pi}-\left[\frac{1}{4}\cos\left(\frac{\pi}{4}\right)-\frac{\sin(\pi/2-\pi/4)}{2\pi}\right]$$

$$=\frac{3}{4}\cos\left(\frac{\pi}{4}\right)-\frac{\sin(7\pi/4)}{2\pi}+\frac{\sin(\pi/4)}{2\pi}=\frac{3}{4}\cos\left(\frac{\pi}{4}\right)+\frac{\sin(\pi/4)}{2\pi}+\frac{\sin(\pi/4)}{2\pi}$$

$$=\frac{1}{4}\cos\left(\frac{\pi}{4}\right)\left\{3+\frac{4}{\pi}\right\}=\frac{\sqrt{2}}{8}\left\{3+\frac{4}{\pi}\right\}$$

The Franck–Condon factor is

$$|S(v,0)|^2 = \boxed{\frac{1}{32}\left(3+\frac{4}{\pi}\right)^2}$$

E13A.5(a) P branch ($\Delta J=-1$): $\tilde{\nu}_P(J)=\tilde{\nu}-(\tilde{B}'+\tilde{B})J+(\tilde{B}'-\tilde{B})J^2$ [13A.8a]

When the bond is shorter in the excited state than in the ground state, $\tilde{B}'>\tilde{B}$ and $\tilde{B}'-\tilde{B}$ is positive. In this case, the lines of the P branch appear at successively decreasing energies as J increases, begin to converge, go through a head at J_{head}, begin to increase with increasing J, and become greater than $\tilde{\nu}$ when $J>(\tilde{B}'+\tilde{B})/(B'-\tilde{B})$ (see Discussion question 13A.3); the quadratic shape of the $\tilde{\nu}_P$ against J curve is called the Fortrat parabola). This means that $\tilde{\nu}_P(J)$ is a minimum when $J=J_{\text{head}}$. It is reasonable to deduce that J_{head} is the closest integer to $\boxed{\tfrac{1}{2}(\tilde{B}'+\tilde{B})/(\tilde{B}'-\tilde{B})}$ because it takes twice as many J values to reach the minimum line of the P branch and to return to $\tilde{\nu}$. We can also find J_{head} by finding the minimum of the Fortrat parabola: $\mathrm{d}\tilde{\nu}_P/\mathrm{d}J=0$ when $J=J_{\text{head}}$.

$$\frac{\mathrm{d}\tilde{\nu}_P}{\mathrm{d}J}=\frac{\mathrm{d}}{\mathrm{d}J}\left\{\tilde{\nu}-(\tilde{B}'+\tilde{B})J+(\tilde{B}'-\tilde{B})J^2\right\}=-(\tilde{B}'+\tilde{B})+2(\tilde{B}'-\tilde{B})J$$

$$-(\tilde{B}'+\tilde{B})+2(\tilde{B}'-\tilde{B})J_{\text{head}}=0$$

$$J_{\text{head}}=\frac{(\tilde{B}'+\tilde{B})}{2(\tilde{B}'-\tilde{B})}$$

E13A.6(a) Since $\tilde{B}'<\tilde{B}$ and $\tilde{B}'-\tilde{B}$ is negative, the R branch shows a head at the closest integer to the value of $\boxed{\tfrac{1}{2}(\tilde{B}'+\tilde{B})/|(\tilde{B}'-\tilde{B})|-1}$ (see Exercise 13A.5(b)).

$$\frac{(\tilde{B}'+\tilde{B})}{2|(\tilde{B}'-\tilde{B})|}-1=\frac{(0.3101+0.3540)}{2|0.3101-0.3540|}-1=6.6$$

$$J_{\text{head}}=\boxed{7}$$

13A.7(a) When the R branch has a head, J_{head} is the closest integer to $\tfrac{1}{2}(\tilde{B}'+\tilde{B})/|(\tilde{B}'-\tilde{B})|-1$ (see Exercise 13A.5(b)). Thus, if we are only given that $J_{\text{head}}=1$ and $\tilde{B}=60.80\ \text{cm}^{-1}$, we know only that

$$0.5<\tfrac{1}{2}(\tilde{B}'+\tilde{B})/(\tilde{B}-\tilde{B}')-1<1.5$$

because the fractional value of a $\tfrac{1}{2}(\tilde{B}'+\tilde{B})/(\tilde{B}'-\tilde{B})-1$ calculation must be rounded-off to give the integer value J_{head}. Algebraic manipulation of the inequality yields

$$\frac{\{1+2(0.5)\}\tilde{B}}{\{3+2(0.5)\}}<\tilde{B}'<\frac{\{1+2(1.5)\}\tilde{B}}{\{3+2(1.5)\}}$$

$$\frac{\tilde{B}}{2}<\tilde{B}'<\frac{2\tilde{B}}{3}$$

$$\boxed{30.4\ \text{cm}^{-1}<\tilde{B}'<40.5\ \text{cm}^{-1}}$$

When $\tilde{B}' < \tilde{B}$, the bond length in the electronically excited state is $\boxed{\text{greater}}$ than that in the ground state.

Here's an alternative solution that gives the same answer with insight into the band head concept: At the head of an R band, $\tilde{\nu}_{J_{\text{head}}} > \tilde{\nu}_{J_{\text{head}}-1}$ where $\tilde{\nu}_{J_{\text{head}}-1}$ is the transition $J_{\text{head}} \leftarrow J = J_{\text{head}} - 1$. Substitution of eqn 13A.8(c) into this inequality yields the relation $\tilde{B}' > J_{\text{head}}\tilde{B}/(J_{\text{head}}+1)$. Similarly, $\tilde{\nu}_{J_{\text{head}}} > \tilde{\nu}_{J_{\text{head}}+1}$ where $\tilde{\nu}_{J_{\text{head}}+1}$ is the transition $J_{\text{head}}+2 \leftarrow J = J_{\text{head}}+1$. Substitution of eqn. 13A.8(c) into this inequality yields the relation $\tilde{B}' < (J_{\text{head}}+1)\tilde{B}/(J_{\text{head}}+2)$. Consequently, $J_{\text{head}}\tilde{B}/(J_{\text{head}}+1) < \tilde{B}' < (J_{\text{head}}+1)\tilde{B}/(J_{\text{head}}+2)$.

E13A.8(a) The transition wavenumber is $\tilde{\nu} = \dfrac{1}{\lambda} = \dfrac{1}{700\ \text{nm}} = 14\times10^3\ \text{cm}^{-1}$.

Water molecules are weak ligand field splitters, so we expect the d^5 electrons of Fe^{3+} to have the $t_{2g}^3e_g^2$ high spin ground state configuration in the octahedral $[Fe(OH_2)_6]^{3+}$ complex. The d-orbital electron spins are expected to be parallel with $S = 5/2$ and $2S + 1 = 6$ by Hund's maximum multiplicity rule. We also expect that $P > \Delta_O$ where P is the energy of repulsion for pairing two electrons in an orbital. A d–d transition to the $t_{2g}^4e_g^1$ octahedral excited state is expected to be both spin and parity forbidden and, therefore, have a very small molar absorption coefficient. This transition releases the energy Δ_O and requires the energy P needed to pair two electrons within a t_{2g} orbital. Thus, $\tilde{\nu} = P - \Delta_O$ and $\boxed{\Delta_O = P - \tilde{\nu}}$. Using the typical value $P \sim 28 \times 10^3\ \text{cm}^{-1}$ yields the estimate $\Delta_O \sim \boxed{14\times10^3\ \text{cm}^{-1}}$. See F.A. Cotton and G. Wilkinson, *Advanced Inorganic Chemistry*, 4th ed. (New York: Wiley-Interscience Publishers, 1980), 646, for electron-pairing energies.

E13A.9(a) The normalized wavefunctions are:

$$\psi_i = \left(\frac{1}{a}\right)^{1/2} \quad \text{for } 0 \le x \le a \text{ and 0 elsewhere.}$$

$$\psi_f = \left(\frac{1}{b-\tfrac{1}{2}a}\right)^{1/2} \quad \text{for } \tfrac{1}{2}a \le x \le b \text{ and 0 elsewhere.}$$

$$\int \psi_f x \psi_i\,dx = \left(\frac{1}{a}\right)^{1/2}\left(\frac{1}{b-\tfrac{1}{2}a}\right)^{1/2}\int_{\frac{1}{2}a}^{a} x\,dx = \left(\frac{1}{a}\right)^{1/2}\left(\frac{1}{b-\tfrac{1}{2}a}\right)^{1/2}\left.\frac{x^2}{2}\right|_{x=\frac{1}{2}a}^{x=a}$$

$$= \left(\frac{1}{a}\right)^{1/2}\left(\frac{1}{b-\tfrac{1}{2}a}\right)^{1/2}\left(\frac{3a^2}{8}\right) = \boxed{\frac{3}{8}\left(\frac{a^3}{b-\tfrac{1}{2}a}\right)^{1/2}}$$

E13A.10(a) The normalized wavefunctions are:

$$\psi_i = \left(\frac{1}{a\sqrt{\pi}}\right)^{1/2} e^{-x^2/2a^2} \quad \text{for } -\infty \le x \le \infty \text{ and width } a.$$

$$\psi_f = \left(\frac{1}{a\sqrt{\pi}}\right)^{1/2} e^{-(x-a/2)^2/2a^2} \quad \text{for } -\infty \le x \le \infty \text{ and width } a.$$

$$\int \psi_f x \psi_i\,dx = \left(\frac{1}{a\sqrt{\pi}}\right)\int_{-\infty}^{\infty} x\,e^{-x^2/2a^2}e^{-(x-a/2)^2/2a^2}\,dx = \left(\frac{1}{a\sqrt{\pi}}\right)\int_{-\infty}^{\infty} x\,e^{-\left\{x^2+(x-a/2)^2\right\}/2a^2}\,dx$$

Since $x^2+(x-a/2)^2=\left(2^{1/2}x-\frac{a}{2^{3/2}}\right)^2+\frac{a^2}{8}$, let $z=2^{1/2}x-\frac{a}{2^{3/2}}$ and $x=2^{-1/2}z+\frac{a}{4}$.

Then, $dx=2^{-1/2}\,dz$ and substitution gives:

$$\int\psi_f x\psi_i\,dx=\left(\frac{e^{-1/16}}{a\sqrt{2\pi}}\right)\int_{-\infty}^{\infty}\left(2^{-1/2}z+\frac{a}{4}\right)e^{-z^2/2a^2}\,dz$$

$$=\left(\frac{e^{-1/16}}{a\sqrt{2\pi}}\right)\left\{2^{-1/2}\int_{-\infty}^{\infty}ze^{-z^2/2a^2}\,dz+\frac{a}{4}\int_{-\infty}^{\infty}e^{-z^2/2a^2}\,dz\right\}$$

The factors within the first integral have ungerade and gerade symmetry. Because $u\times g=u$, the integrand has ungerade symmetry and the first integral is necessarily zero (the integral of an ungerade function over a symmetric interval equals zero).

$$\int\psi_f x\psi_i\,dx=\left(\frac{e^{-1/16}}{a\sqrt{2\pi}}\right)\left\{\frac{a}{4}\int_{-\infty}^{\infty}e^{-z^2/2a^2}\,dz\right\}=\left(\frac{e^{-1/16}}{4\sqrt{2\pi}}\right)(2a^2\pi)^{1/2}$$

$$=\boxed{\tfrac{1}{4}e^{-1/16}a}$$

E13A.11(a) π-electrons in polyenes may be considered as particles in a one-dimensional box. Applying the Pauli exclusion principle, the N conjugated electrons will fill the levels, two electrons at a time, up to the level $n=N/2$. Since N is also the number of alkene carbon atom, Nd is the length of the box, with d the carbon–carbon interatomic distance. Hence

$$E_n=\frac{n^2h^2}{8mN^2d^2}\quad[8A.7b]$$

For the lowest energy transition ($\Delta n=+1$)

$$\Delta E=h\nu=\frac{hc}{\lambda}=E_{N/2+1}-E_{N/2}=\frac{(N+1)h^2}{8mN^2d^2}$$

Therefore, the larger N, the larger λ. Hence the absorption at 243 nm is due to the diene and that at 192 nm to the butene.

Question. How accurate is the formula derived above in predicting the wavelengths of the absorption maxima in these two compounds?

An alternative analysis uses simple Hückel theory of π molecular orbitals. The substituted butene is modelled with the ethene π energies of text Fig. 110E.1 that shows a $\pi_{LUMO}\leftarrow\pi_{HOMO}$ transition energy of -2β (β is negative). The substituted 2,4-hexadiene is modelled with the butadiene π energies of text Fig. 10E.2 that shows a $\pi_{LUMO}\leftarrow\pi_{HOMO}$ transition energy of -1.24β. Thus, the transition energy of the conjugated diene is lower and has a longer wavelength. This is an example of the general principle that the difference between neighbouring energy levels becomes smaller as N becomes larger.

Solutions to problems

P13A.1 For a photon to induce a spectroscopic transition, the transition moment $\langle \mu \rangle$ must be nonzero, a requirement that leads to electronic spectrum selection rules concerned with changes in angular momentum. The rules for a homonuclear diatomic are (eqn 13A.4):

$$\Delta\Lambda = 0, \ \pm 1 \quad \Delta S = 0 \quad \Delta\Sigma = 0 \quad \Delta\Omega = 0, \ \pm 1 \qquad \text{where } \Omega = \Lambda + \Sigma$$

Λ gives the total orbital angular momentum about the internuclear axis and Σ gives the total spin angular momentum about the internuclear axis. The $\pm$ superscript selection rule for reflection in the plane along the internuclear axis is $+\leftrightarrow+$ or $-\leftrightarrow-$ (i.e. $+\leftrightarrow-$ is forbidden). The **Laporte selection rule** states that for a centrosymmetric molecule (those with a centre of inversion) the only allowed transitions are transitions that are accompanied by a change of parity: u$\leftrightarrow$g.

The electric-dipole transition $\boxed{{}^2\Sigma_g^+ \leftarrow {}^2\Sigma_u^+ \text{ is allowed}}$ because none of the above rules negate the possibility of this event. You may also wish to reach this conclusion by direct examination of the dipole transition moment integral, $\int \psi_f^* \mu \psi_i \mathrm{d}\tau$, where the dipole moment operator has components proportional to the Cartesian coordinates. The integral vanishes unless the integrand, or at least some part of it, belongs to the totally symmetric representation (A_{1g}; see text Sections 11C.1(a) and (c)). To find the symmetry species of the integrand, we multiply together the characters of its factors. Homonuclear diatomic molecules and ions belong to the $D_{\infty h}$ point group and the $D_{\infty h}$ character table tells us the symmetry species of each integrand factor.

$$\begin{aligned} &\psi_f && : A_{1g}\left(\Sigma_g^+\right) \\ &\mu_z && : A_{1u}\left(\Sigma_u^+\right) \\ &\psi_i && : A_{1u}\left(\Sigma_u^+\right) \\ &\text{Symmetry product}: A_{1g} \times A_{1u} \times A_{1u} = A_{1g} \times A_{1g} = A_{1g} \end{aligned}$$

Since the integrand spans A_{1g}, the transition ${}^2\Sigma_g^+ \leftarrow {}^2\Sigma_u^+$ is allowed.

An electric-dipole transition from a ${}^2\Sigma_u^+$ ground state to a Π_u excited state is forbidden by the Laporte selection rule.

Finally, we check the possibility of a transition from a ${}^2\Sigma_u^+$ ground state to a ${}^2\Pi_g$ excited state by finding whether or not the integrand of the transition integral spans the totally symmetric representation. The symmetry product for the μ_z component is: $E_{1g} \times A_{1u} \times A_{1u} = E_{1g} \times A_{1g} = E_{1g}$. Since the integrand does not span A_{1g}, the transition is forbidden for *z*-polarized light. The symmetry product for both the μ_x and μ_y components is: $E_{1g} \times E_{1u} \times A_{1u} = E_{1g} \times E_{1g}$. Since the species product $E_{1g} \times E_{1g}$ has an angular dependence and, therefore, does not contain the totally symmetric representation, the transition is forbidden for *x*- and *y*-polarized light. You may also wish to show this by application of the orthogonality theorem to find the coefficient of A_{1g} in the integrand.

$D_{\infty h}$	E	$\infty C_2'$	$2C_\phi$	i	$\infty\sigma_v$	$2S_\phi$
$\Sigma_u^+(A_{1u})$	1	−1	1	−1	1	−1
μ_x or $\mu_y(E_{1u})$	2	0	2 cos ϕ	−2	0	2 cos ϕ
$\Pi_g(E_{1g})$	2	0	2 cos ϕ	2	0	−2 cos ϕ
Integrand	4	0	4 cos^2 ϕ	4	0	4 cos^2 ϕ

The orthogonality theorem gives the coefficient of A_{1g} in the integrand as

$$c_{A_{1g}} = (1/h)\Sigma_C\, g(C)\chi(C) = [4+0+2(4\cos^2\phi)+4+0+2(4\cos^2\phi)]/h$$

Since the group order h, which equals infinity, does not cancel with a numerator factor, $c_{A_{1g}} = 0$.

P13A.3 Essential concepts associated with the vibration transition structure of an electronic absorption band are shown in text Fig. 13A.7 and Fig. 13A.2 below. The $X^3\Sigma_g^-$ ground state of O_2 dissociates into two $O(^3P)$ atoms while the $B^3\Sigma_u^-$ excited state dissociates into a $O(^3P)$ atom and a $O(^1D)$ atom. The electron absorption band for the $B^3\Sigma_u^- \leftarrow X^3\Sigma_g^-$ transition is characterized by the vibrational states (v' = 0, 1, 2,..., v_{max}) of the excited state because a great majority of the ground state molecules are in the lowest vibrational state. Thus, the vibrational transitions are 0 ← 0, 1 ← 0, 2 ← 0, ..., v' ← 0. It is important to recognize that the vibrational transitions near the dissociation limit are difficult to accurately identify among noise of comparable magnitude. To begin the data analysis, define $\Delta\tilde{\nu}_{v'}$ to be the difference between the wavenumber for the $v'+1 \leftarrow 0$ transition and the wavenumber for the $v \leftarrow 0$ transition:

$$\Delta\tilde{\nu}_{v'} = \tilde{\nu}_{v'+1} - \tilde{\nu}_{v'}$$

Examination of Fig. 13A.2 will confirm that $D_0\left(B^3\Sigma_u^-\right)$ equals the sum of all $\Delta\tilde{\nu}_v$, many of which must be determined with a Birge–Sponer extrapolation, plus a small energy δ needed to dissociate an excited molecule in the v'_{max} vibrational state. We take the latter energy to be given by the fractional quantum number obtained in the Birge–Sponer extrapolation.

$$D_0\left(B^3\Sigma_u^-\right) = \sum_{v=0} \Delta\tilde{\nu}_{v'} + \delta$$

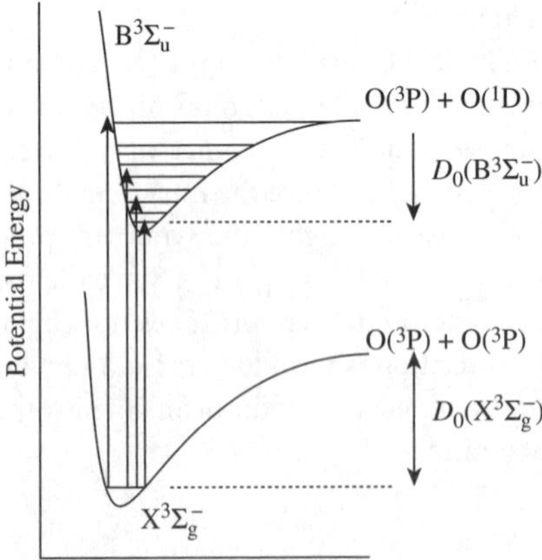

Figure 13A.2

We draw a table of $\Delta\tilde{\nu}_{\upsilon'}$ computations, prepare a plot of $\Delta\tilde{\nu}_{\upsilon'}$ against v', and calculate a quadratic regression fit to the curve, which provides the Birge–Sponer extrapolation to $\Delta\tilde{\nu}_{\upsilon'}=0$.

υ'	0	1	2	3	4	5	6	7
$\tilde{\nu}_{0\upsilon'}$ / cm⁻	50 062.6	50 725.4	51 369.0	51 988.6	52 579.0	53 143.4	53 679.6	54 177.0
$\Delta\tilde{\nu}_{\upsilon}$ / cm⁻	662.8	643.6	619.6	590.4	564.4	536.2	497.4	464.8

υ'	8	9	10	11	12	13	14
$\tilde{\nu}_{0\upsilon'}$ / cm⁻	54 641.8	55 078.2	55 460.0	55 803.1	56 107.3	56 360.3	56 570.6
$\Delta\tilde{\nu}_{\upsilon}$ / cm⁻	436.4	381.8	343.1	304.2	253.0	210.3	

The quadratic least squares fit to the $\Delta\tilde{\nu}_{\upsilon'}$ against v' plot is shown overleaf in Fig. 13A.3 provides a Birge–Sponer extrapolation which indicates that $v'_{\text{max}}=16$ and $v'_{\text{dissociation limit}}=16.8376$. Thus,

$$D_0\left(\mathrm{B}^3\Sigma_u^-\right)=\sum_{\upsilon'=0}^{15}\Delta\tilde{\nu}_{\upsilon'}+0.8376\times\Delta\tilde{\nu}_{16}$$

where the values of $\Delta\tilde{\nu}_{\upsilon'}$ are computed with the fit $y(v')$. Doing the sum, we find that

$$\begin{aligned}D_0\left(\mathrm{B}^3\Sigma_u^-\right)&=6\,767.908\ \mathrm{cm}^{-1}+0.8376\times\left(48.0992\ \mathrm{cm}^{-1}\right)\\&=\boxed{6\,808.2\ \mathrm{cm}^{-1}\ \text{or}\ 0.84411\ \mathrm{eV}}\end{aligned}$$

Examination of Fig. 13A.2 also indicates that

$$D_0\left(\mathrm{X}^3\Sigma_g^-\right)+E_{\mathrm{O(^1D)\leftarrow O(^3P)}}=\tilde{\nu}_{00}+D_0\left(\mathrm{B}^3\Sigma_u^-\right)$$

where $E_{\mathrm{O(^1D)\leftarrow O(^3P)}}$ is 190 kJ mol⁻¹ or 1.97 eV and $\tilde{\nu}_{00}$ equals 50 062.6 cm⁻¹ or 6.20697 eV.

$$\begin{aligned}D_0\left(\mathrm{X}^3\Sigma_g^-\right)&=\tilde{\nu}_{00}+D_0\left(\mathrm{B}^3\Sigma_u^-\right)-E_{\mathrm{O(^1D)\leftarrow O(^3P)}}\\&=(6.20697+0.84411-1.97)\ \mathrm{eV}\\&=\boxed{5.08\ \mathrm{eV}}\end{aligned}$$

This value of the ground state dissociation energy is in agreement with the accepted value.

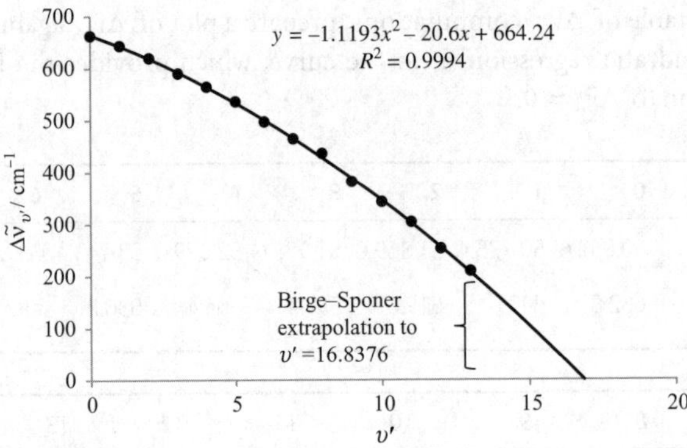

Figure 13A.3

P13A.5 (a) The H_2O^+ vibrational wavenumber (0.41 eV) of the photoelectron spectrum band at 12–13 eV corresponds to about 3 300 cm^{-1}, which is close to the 3 652 cm^{-1} symmetric stretching mode of the neutral ground state (see text Fig. 12E.3). This suggests loss of a non bonding electron. The absence of a long vibrational series in this band is compatible with an ionized equilibrium bond length that approximately equals that of the neutral ground state, thereby producing a large Franck–Condon factor for the adiabatic transition ($v' = 0 \leftarrow v = 0$) and smaller factors for other vibrational transitions ($v' = 1, 2, \ldots \leftarrow v = 0$). These observations are compatible with the loss of a non bonding electron because such a loss does not affect, or has little effect, on the bonding force constants, vibrational frequencies, or equilibrium bond lengths.

(b) The H_2O^+ vibrational wavenumber (0.125 eV) of the photoelectron spectrum band at 14–16 eV corresponds to about 1 000 cm^{-1}, which is very different than the 1 595 cm^{-1} bending mode of the neutral ground state (see text Fig. 12E.3). This suggests loss of a σ bonding electron that severely reduces bond order, a bond force constant, and a vibrational frequency. Consequently, we expect the bond length of the ionized state to be longer than that of the neutral ground state, an observation that is compatible with the long vibrational series of the band because this yields many vertical transitions with significant Franck-Condon factors.

P13A.7 We need to establish whether the transition dipole moments

$$\mu_{\rm fi} = \int \psi_{\rm f}^* \mu \psi_{\rm i} \, d\tau \quad [12\text{A}.10, \textit{Justification } 13\text{A}.2]$$

connecting the states $n = 1$ and 2 and the states $n = 1$ and 3 are zero or nonzero. The particle in a box wavefunctions are

$$\psi_n = (2/L)^{1/2} \sin(n\pi x/L) \quad [8\text{A}.7\text{a}]$$

Thus,

$$\mu_{2,1} \propto \int \sin\left(\frac{2\pi x}{L}\right) x \sin\left(\frac{\pi x}{L}\right) dx \propto \int x\left[\cos\left(\frac{\pi x}{L}\right) - \cos\left(\frac{3\pi x}{L}\right)\right] dx$$

and

$$\mu_{3,1} \propto \int \sin\left(\frac{3\pi x}{L}\right) x \sin\left(\frac{\pi x}{L}\right) dx \propto \int x\left[\cos\left(\frac{2\pi x}{L}\right) - \cos\left(\frac{4\pi x}{L}\right)\right] dx$$

having used $\sin\alpha\sin\beta = \frac{1}{2}\cos(\alpha-\beta) - \frac{1}{2}\cos(\alpha+\beta)$. Both of these integrals can be evaluated using the standard form

$$\int x(\cos ax)dx = \frac{1}{a^2}\cos ax + \frac{x}{a}\sin ax$$

$$\int_0^L x\cos\left(\frac{\pi x}{L}\right)dx = \frac{1}{(\pi/L)^2}\cos\left(\frac{\pi x}{L}\right)\Bigg|_0^L + \frac{x}{(\pi/L)}\sin\left(\frac{\pi x}{L}\right)\Bigg|_0^L = -2\left(\frac{L}{\pi}\right)^2 \neq 0$$

$$\int_0^L x\cos\left(\frac{3\pi x}{L}\right)dx = \frac{1}{(3\pi/L)^2}\cos\left(\frac{3\pi x}{L}\right)\Bigg|_0^L + \frac{x}{(3\pi/L)}\sin\left(\frac{3\pi x}{L}\right)\Bigg|_0^L = -2\left(\frac{L}{3\pi}\right)^2 \neq 0$$

Thus, $\mu_{2,1} \neq 0$. In a similar manner, $\mu_{3,1} = 0$.

Comment. A general formula for μ_{fi} applicable to all possible particle in a box transitions may be derived. The result is ($n = \mathrm{f}$, $m = i$)

$$\mu_{nm} = -\frac{eL}{\pi^2}\left[\frac{\cos(n-m)\pi - 1}{(n-m)^2} - \frac{\cos(n+m)\pi - 1}{(n+m)^2}\right]$$

For m and n both even or both odd numbers, $\mu_{\mathrm{nm}} = 0$; if one is even and the other odd, $\mu_{\mathrm{nm}} \neq 0$. Also, see Problem 13A.9.

Question. Can you establish the general relation for μ_{nm} above?

P13A.9 $S = \left[1 + \frac{R}{a_0} + \frac{1}{3}\left(\frac{R}{a_0}\right)^2\right]e^{-R/a_0}$ [10C.4]

$\mu = -eSR$ [given]

$$\mu/(-ea_0) = (R/a_0) \times S$$

$$= (R/a_0) \times \left[1 + \frac{R}{a_0} + \frac{1}{3}\left(\frac{R}{a_0}\right)^2\right]e^{-R/a_0}$$

We then draw up the following table:

R/a_0	0	1	2	3	4	5	6	7	8
$\mu/(-ea_0)$	0	0.858	1.173	1.046	0.757	0.483	0.283	0.155	0.081

These points are plotted overleaf in Fig. 13A.4. The dipole-transition moment rises from zero at $R = 0$ to a peak at $R \sim 2.1a_0$ after which it declines to zero at infinity. The transition moment is zero at $R = 0$, because the initial state ($1s_A$) and the final state ($1s_A$) are identical and, consequently, there is no dipole moment and no transition. As $R \to \infty$, the electron is confined to a single atom because its wavefunction does not extend to the other and, consequently, no transition is possible.

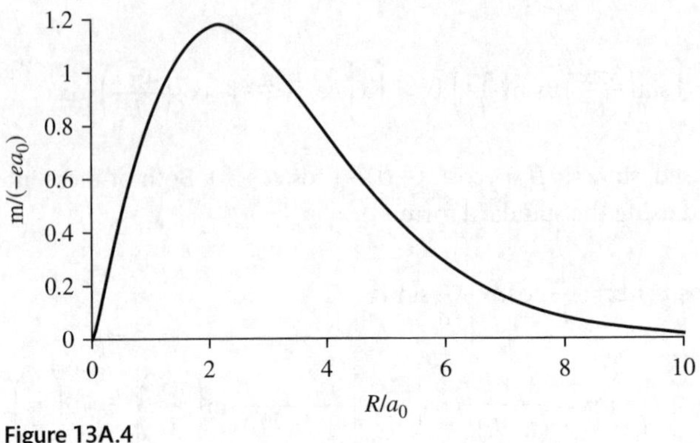

Figure 13A.4

13B Decay of excited states

Answer to discussion question

D13B.1 The overall process associated with fluorescence involves the following steps. The molecule is first promoted from the vibrational ground state of a lower electronic level to a higher vibrational–electronic energy level by absorption of energy from a radiation field. Because of the requirements of the Franck–Condon principle, the transition is to excited vibrational levels of the upper electronic state. See text Fig. 13B.2. Therefore, the absorption spectrum shows a vibrational structure characteristic of the upper state. The excited-state molecule can now lose energy to the surroundings through radiationless transitions and decay to the lowest vibrational level of the upper state. A spontaneous radiative transition now occurs to the lower electronic level and this fluorescence spectrum has a vibrational structure characteristic of the lower state. The fluorescence spectrum is not the mirror image of the absorption spectrum because the vibrational frequencies of the upper and lower states are different due to the difference in their potential energy curves.

Solutions to exercises

E13B.1(a) (a) Vibrational energy spacings of the $\boxed{\text{lower}}$ state are determined by the spacing of the peaks of fluorescence spectrum of benzophenone, which emission intensity A of text Fig. 13.3 shows to be $\boxed{\tilde{\nu} \approx 1\,800\ \text{cm}^{-1}}$.

(b) The fluorescence spectrum gives $\boxed{\text{no information}}$ about the spacing of the upper vibrational levels (without a detailed analysis of the intensities of the lines).

E13B.2(a) When the steeply repulsive section of the O_2 potential energy curve for the excited state lies slightly toward the short side of the equilibrium bond length of the ground state and the minimum of the excited state lies to the longer side (as shown in text Figs 13B.2 and 13B.7), a great many excited vibrational states overlap with the lowest energy vibration of the ground state, which makes the **Franck–Condon factor** appreciable for many vertical transitions. This, combined with continuous absorption above the dissociation limit, yields a relatively broad absorption band. Furthermore, predissociation to the unbound $^5\Pi_u$ state shortens the lifetime of excited vibrational states. This causes the high resolution lines of the corresponding vibrational-rotational transitions to be broad through the Heisenberg uncertainty principle $\Delta E \Delta t \geq \hbar / 2$.

Solution to problem

P13B.1 The anthracene vapour fluorescence spectrum gives the vibrational splitting of the lower state. The wavelengths stated correspond to the wavenumbers 22 730, 24 390, 25 640, 27 030 cm^{-1}, indicating spacings of 1 660, 1 250, and 1 390 cm^{-1}. The absorption spectrum spacing gives the separation of the vibrational levels of the upper state. The wavenumbers of the absorption peaks are 27 800, 29 000, 30 300, and 32 800 cm^{-1}. The vibrational spacings are therefore 1 200, 1 300, and 2 500 cm^{-1}. The data is compatible with the deactivation of the excited state vibrational modes before spontaneous emission returns the molecule to the ground electronic state. This produces a fluorescence band of lower energy than the absorption band. Furthermore, while the absorption band has a vibrational progression that depends upon vibrational modes of the excited state, the fluorescence band has a vibrational progression that depends upon vibrational modes of the ground state. The absorption and fluorescence spectra are not mirror images.

13C Lasers

Answer to discussion question

D13C.1 (a) Continuous-wave (CW) laser emission is possible when heat is easily dissipated and population inversion can be continuously maintained by pumping. The red laser pointer is an example of a CW laser. Typically, the light amplification by stimulated emission of radiation is continuous in the optical cavity and one of the two mirrors, the output coupler, at the ends of the cavity is partially transparent so that only a fraction of the cavity radiation can continuously escape. The gain medium is pumped to the excited state by electricity, a flash lamp, or another laser.

(b) The pulsed laser periodically emits a pulse of high peak power radiation, which is much higher than can be achieved with a CW laser because the average laser power is released in a pulse of short duration. Pulses may be achieved by Q-switching or mode locking. In Q-switching, the laser cavity resonance characteristics are modified to make the cavity conditions unfavourable for lasing, during which time a healthy population inversion is achieved; the cavity is then suddenly brought to resonance, releasing the radiation pulse. The electro-optical Pockels cell or a saturable absorber may be used as Q-switching devices that give pulses of about 5 ns duration. Picosecond pulses can be achieved by the technique of mode locking in which a range of resonant modes of different frequency are phase locked and superimposed. Interference of the modes gives rise to short, regular bursts of radiation. Mode locking is achieved by varying the Q-factor of the laser cavity periodically at the frequency $c/2L$. The modulation can be achieved by linking a prism in the cavity to a transducer driven by a radiofrequency source at a frequency $c/2L$. The transducer sets up standing-wave vibrations in the prism and modulates the loss it introduces into the cavity.

Solutions to exercises

E13C.1(a) Only an integral number of half-wavelengths fit into the cavity. These are the **resonant modes.**

$\lambda = 2L/n$ [13C.1] where n is an integer and L is the length of the cavity.

The resonant frequencies are given by $\nu = c/\lambda = nc/2L$. The lowest energy resonant modes ($n = 1$) in a 1.0 cm cavity are $\boxed{\lambda = 2.0\ \text{cm}\ (\nu = 15.0\ \text{GHz})}$.

E13C.2(a) Referring to Example 13C.1 of the text, we have

$$P_{\text{peak}} = E_{\text{pulse}}/t_{\text{pulse}} \quad \text{and} \quad P_{\text{average}} = E_{\text{total}}/t = E_{\text{pulse}} \times \nu_{\text{repetition}}$$

where $\nu_{\text{repetition}}$ is the pulse repetition rate.

$$t_{\text{pulse}} = E_{\text{pulse}}/P_{\text{peak}} = \frac{0.10\ \text{mJ}}{5.0\ \text{MW}} = \boxed{20\ \text{ps}}$$

$$\nu_{\text{repetition}} = P_{\text{average}}/E_{\text{pulse}} = \frac{7.0\ \text{kW}}{0.10\ \text{mJ}} = \boxed{70\ \text{MHz}}$$

Solution to problem

P13C.1 The laser is delivering photons of energy

$$E = h\nu = \frac{hc}{\lambda} = \frac{(6.626\times10^{-34}\ \text{J s})\times(2.998\times10^{8}\ \text{m s}^{-1})}{488\times10^{-9}\ \text{m}} = 4.07\times10^{-19}\ \text{J}$$

Since the laser is putting out 1.0 mJ of these photons every second, the rate of photon emission is:

$$r = \frac{1.0\times10^{-3}\ \text{J s}^{-1}}{4.07\times10^{-19}\ \text{J}} = 2.5\times10^{15}\ \text{s}^{-1}$$

The time it takes the laser to deliver 10^6 photons (and therefore the time the dye remains fluorescent) is

$$t=\frac{10^6}{2.5\times10^{15}\,\text{s}^{-1}}=\boxed{4\times10^{-10}\,\text{s}\quad\text{or}\quad 0.4\,\text{ns}}$$

Integrated activities

I13.1 The fluorescence spectrum of CN is sketched in Fig. I13.1.

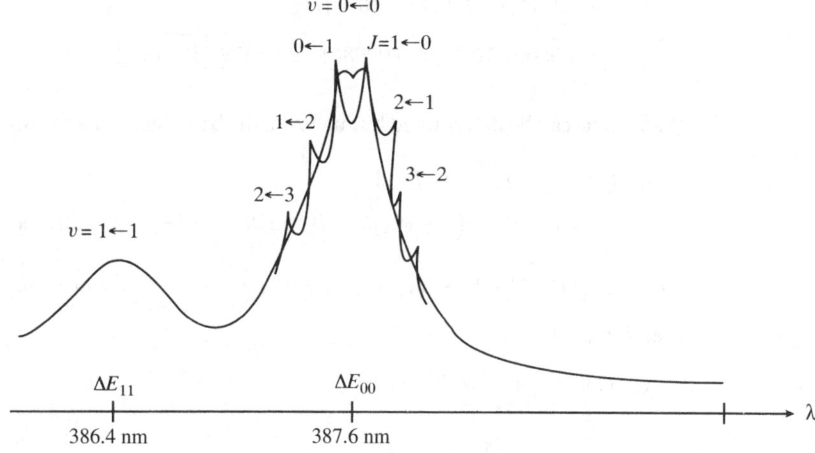

Figure I13.1

$$\Delta E_{11}=\frac{hc}{\lambda_{11}}=\frac{hc}{386.4\text{ nm}}=5.1409\times10^{-19}\text{ J}=3.2087\text{ eV}$$

and

$$\Delta E_{00}=\frac{hc}{\lambda_{00}}=\frac{hc}{387.6\text{nm}}=5.1250\times10^{-19}\text{ J}=3.1987\text{eV}$$

Energy of excited singlet, S_1: $E_1(\upsilon,J)=V_1+(\upsilon+\tfrac{1}{2})\tilde{\nu}_1hc+J(J+1)\tilde{B}_1hc$

Energy of ground singlet, S_0: $E_0(\upsilon,J)=V_0+(\upsilon+\tfrac{1}{2})\tilde{\nu}_0hc+J(J+1)\tilde{B}_0hc$

The midpoint of the 0–0 band corresponds to the forbidden Q branch ($\Delta J=0$) with $J=0$ and $\upsilon=0\leftarrow0$.

$$\Delta E_{00}=E_1(0,0)-E_0(0,0)=(V_1-V_0)+\tfrac{1}{2}(\tilde{\nu}_1-\tilde{\nu}_0)hc\quad(1)$$

The midpoint of the 1–1 band corresponds to the forbidden Q branch ($\Delta J=0$) with $J=0$ and $\upsilon=1\leftarrow1$.

$$\Delta E_{11}=E_1(1,0)-E_0(1,0)=(V_1-V_0)+\tfrac{3}{2}(\tilde{\nu}_1-\tilde{\nu}_0)hc\quad(1)$$

Multiplying eqn (1) by three and subtracting eqn (2) gives

$$3\Delta E_{00} - \Delta E_{11} = 2(V_1 - V_0)$$

$$\begin{aligned} V_1 - V_0 &= \tfrac{1}{2}(3\Delta E_{00} - \Delta E_{11}) \qquad (3) \\ &= \tfrac{1}{2}\{3(5.1250) - (5.1409)\} \times 10^{-19}\,\text{J} \\ &= 5.1171 \times 10^{-19}\,\text{J} = \boxed{3.1938\ \text{eV}} \end{aligned}$$

This is the potential energy difference between S_0 and S_1. Equations (1) and (3) may be solved for $\tilde{\nu}_1 - \tilde{\nu}_0$.

$$\begin{aligned} \tilde{\nu}_1 - \tilde{\nu}_0 &= 2\{\Delta E_{00} - (V_1 - V_0)\} \\ &= 2\{5.1250 - 5.1171\} \times 10^{-19}\,\text{J}/hc \\ &= 1.5800 \times 10^{-21}\ \text{J} = 0.0098615\ \text{eV} = \boxed{79.538\ \text{cm}^{-1}} \end{aligned}$$

The $\tilde{\nu}_1$ value can be determined by analyzing the band head data for which $J+1 \leftarrow J$.

$$\begin{aligned} \Delta E_{10}(J) &= E_1(0,J) - E_0(1,J+1) \\ &= V_1 - V_0 + \tfrac{1}{2}\left(\tilde{\nu}_1 - 3\tilde{\nu}_0\right)hc + J(J+1)\tilde{B}_1 hc - (J+1)\times(J+2)\tilde{B}_0 hc \end{aligned}$$

and $\Delta E_{00}(J) = V_1 - V_0 + \tfrac{1}{2}\left(\tilde{\nu}_1 - \tilde{\nu}_0\right)hc + J(J+1)\tilde{B}_1 hc - (J+1)\times(J+2)\tilde{B}_0 hc$

Therefore,

$$\Delta E_{00}(J) - \Delta E_{10}(J) = \tilde{\nu}_0 hc$$

$$\Delta E_{00}(J_{\text{head}}) = \frac{hc}{388.3\ \text{nm}} = 5.1158 \times 10^{-19}\,\text{J}$$

$$\Delta E_{10}(J_{\text{head}}) = \frac{hc}{421.6\ \text{nm}} = 4.7117 \times 10^{-19}\,\text{J}$$

$$\begin{aligned} \tilde{\nu}_0 &= \frac{\Delta E_{00}(J) - \Delta E_{10}(J)}{hc} \\ &= \frac{(5.1158 - 4.7117) \times 10^{-19}\ \text{J}}{hc} \\ &= \frac{4.0410 \times 10^{-20}\ \text{J}}{hc} = 0.25222\ \text{eV} = \boxed{20\,34.3\ \text{cm}^{-1}} \\ \tilde{\nu}_1 &= \tilde{\nu}_0 + 79.538\ \text{cm}^{-1} \\ &= (2\,034.3 + 79.538)\ \text{cm}^{-1} = \boxed{2\,113.8\ \text{cm}^{-1} = \frac{4.1990 \times 10^{-20}\ \text{J}}{hc}} \end{aligned}$$

$$\begin{aligned} \frac{I_{1\text{–}1}}{I_{0\text{–}0}} &\approx \frac{e^{-E_1(1,0)/kT_{\text{eff}}}}{e^{-E_1(0,0)/kT_{\text{eff}}}} = e^{-(E_1(1,0) - E_1(0,0))/kT_{\text{eff}}} \\ &\approx e^{-hc\tilde{\nu}_1/kT_{\text{eff}}} \end{aligned}$$

$$\ln\left(\frac{I_{1-1}}{I_{0-0}}\right) = -\frac{hc\tilde{\nu}_1}{kT_{\text{eff}}}$$

$$T_{\text{eff}} = \frac{hc\tilde{\nu}_1}{k\ln\left(\frac{I_{0-0}}{I_{1-1}}\right)} = \frac{4.1990\times10^{-20}\ \text{J}}{(1.38066\times10^{-23}\ \text{J K}^{-1})\ln(10)} = \boxed{1\,321\ \text{K}}$$

The relative population of the $\upsilon = 0$ and $\upsilon = 1$ vibrational states is the inverse of the relative intensities of the transitions from those states; hence $1/0.1 = \boxed{10}$.

It would seem that with such a high effective temperature more than eight of the rotational levels of the S_1 state should have a significant population. But the spectra of molecules in comets are never as clearly resolved as those obtained in the laboratory and that is most probably the reason why additional rotational structure does not appear in these spectra.

I13.3 (a) The calculations are perform with Spartan '10 using density functional theory (B3LYP) with the 6-31G* basis set. Equilibrium geometry characteristics for both the *trans* and *cis* form of (**4**) are summarized in the following table. The *trans* LUMO and HOMO are shown in Figs I13.2 and I13.3 while those of the *cis* form are shown in Figs I13.4 and I13.5.

Conformation	11-*trans* (4)	11-*cis* (4)
Total energy/eV	−23783.28	−23783.06
E_{LUMO}/eV	−5.78	−5.79
E_{HOMO}/eV	−7.91	−7.97
ΔE/eV	(a) 2.13	(b) 2.18
λ/nm	(a) 582	(b) 569
$C_5C_6C_7C_8$ torsion angle/°	170.97	170.26
$C_{11}C_{12}C_{13}C_{14}$ torsion angle/°	170.00	−179.73
C_1–C_2/pm	155.2	155.1
C_{11}–C_{12}/pm	139.0	139.6
C_{12}–C_{13}/pm	141.0	141.3
Dipole/D	13.58	12.27

The torsion angles are very close to the expected value of 180° with the exception of the torsion angle at C_{11}-*cis*, which is very close to the expected value of 0°. The aromatic character of the alternating π-bond system is evidenced by contrasting the computed bond lengths at a single bond away from the π-system (C_1–C_2), a double bond (C_{11}–C_{12}), and a single bond between doubles (C_{12}–C_{13}) within the Lewis structure. The former is a typical single C–C bond length. The latter two are approximately equal in both conformations and the value (~140 pm) is intermediate between a single and a double carbon bond, a characteristic of aromaticity. The dipole moment of the *cis* form is significantly smaller than that of the *trans* form and the difference may play an important role in the interaction of the chromophore with the surrounding protein matrix.

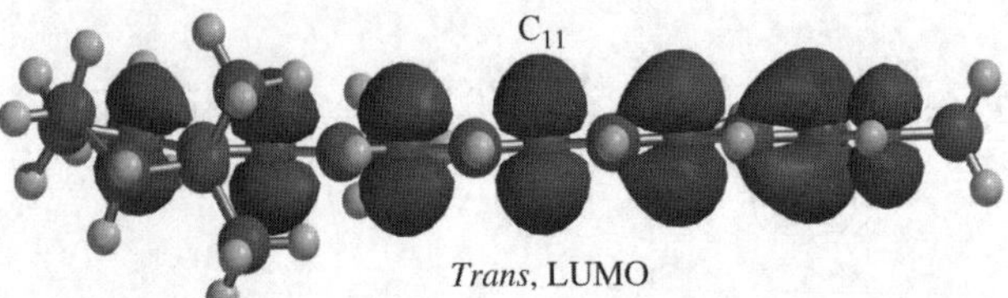

Figure I13.2

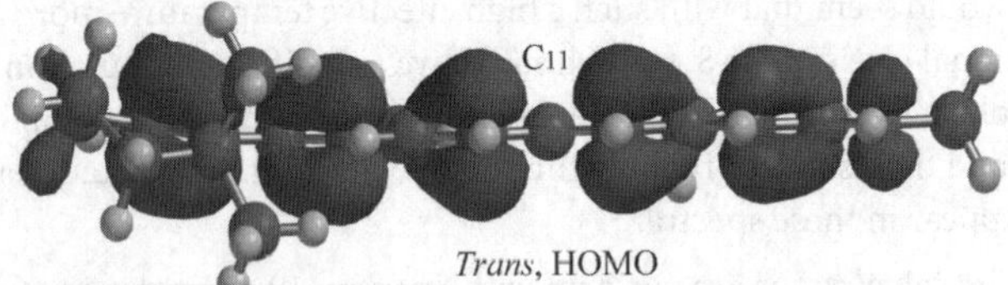

Figure I13.3

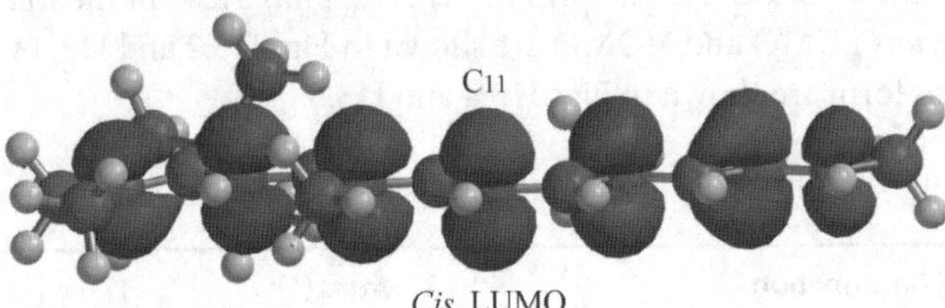

Figure I13.4

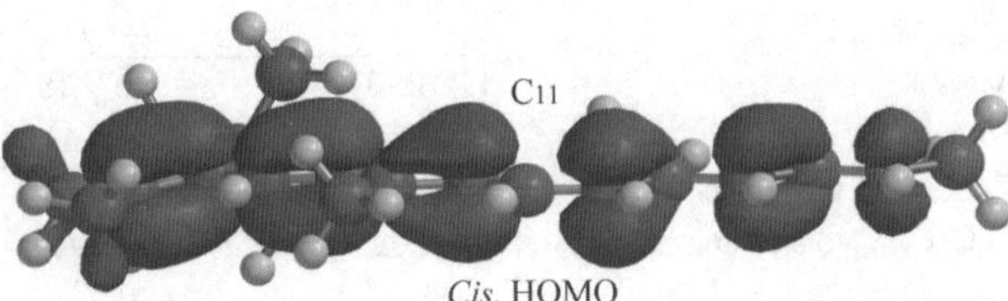

Figure I13.5

(c) The lowest energy $\pi^* \leftarrow \pi$ transition is from the HOMO to the LUMO occurs in the visible near yellow for both confirmations. The transition energy appears to be very slightly lower for the *trans* conformation and at a slightly longer wavelength. However, this difference is so small that it may be an artifact of the computation and not of physical significance.

Question. Figures I13.3 and I13.5 show that both the *trans* and *cis* conformations have HOMO nodes at C_{11}. Does this affect the transition between the two conformations?

I13.5 Use the Clebsch–Gordan series [9C.5 and 9C.6] to compound the two resultant angular momenta, and impose the conservation of angular momentum on the composite system.

(a) O_2 has $S=1$ (it is a spin triplet). The configuration of an O atom is [He]$2s^2\,2p^4$, which is equivalent to a Ne atom with two electron-like 'holes'. The atom may therefore exist as a spin singlet or as a spin triplet. Since $S_1 = 1$ and $S_2 = 0$, or $S_1 = 1$ and $S_2 = 1$ ($M_S = \pm 1, 0$) may each combine to give a resultant with $S = 1$, both may be the products of the reaction. Hence, atom multiplicities $\boxed{3+1}$ and $\boxed{3+3}$ are expected.

(b) N_2 has $S = 0$. The configuration of an N atom is [He]$2s^2 2p^3$. The atoms may have $S = {}^3/_2$ or ${}^1/_2$. Then we note that $S_1 = {}^3/_2$ and $S_2 = {}^3/_2$ ($M_S = \pm{}^3/_2, \pm{}^1/_2$) can combine to give $S=0$; $S_1 = {}^1/_2$ and $S_2 = {}^1/_2$ can also combine to give $S=0$ (but $S_1 = {}^3/_2$ and $S_2 = {}^1/_2$ cannot). Hence, atom multiplicities $\boxed{4+4}$ and $\boxed{2+2}$ may be expected.

14 Magnetic resonance

14A General principles

Answers to discussion questions

D14A.1 Refer to eqn 14A.6, which describes the condition for resonance. The resonance field and resonance frequency are directly proportional to each other. High magnetic fields imply high frequencies and vice versa. The intensity of an NMR transition depends on a number of factors, two of which are seen in eqn 14A.8a; one of these is the strength of the magnetic field itself and the other is the population difference between the energy levels of the α and β spins, which is also proportional to the magnetic field as shown by eqn 14A.8b. Hence the intensity of the signal is proportional to $\mathcal{B}_0^2$. Consequently an increase in external field results in an even larger increase in signal intensity. See Section 14A.1(b) and *Brief illustration* 14A.2 for a more detailed discussion. Chemical shifts as expressed in eqn 14B.1 are also seen to be proportional to $\mathcal{B}_0$. Finally, as discussed in Section 14B.3, the use of high magnetic fields simplifies the appearance of the complex spectra of macromolecules and allows them to be interpreted more readily. Although spin–spin splittings in NMR are not field dependent, strongly coupled (second order) spectra are simplified at high fields because the chemical shifts increase and individual groups of nuclei become identifiable again.

D14A.3 The Larmor frequency is the rate of procession of a magnetic moment (electron or nuclear) in a magnetic field. Resonance occurs when the frequency of the applied radiation matches the Larmor frequency.

Solutions to exercises

E14A.1(a) Work with eqn 14A.4c. $\gamma_N\hbar = g_I\mu_N$, $\hbar$ has unit J s, μ_N has units J T^{-1} therefore γ_N has units

$$\frac{\mathrm{J\,T^{-1}}}{\mathrm{J\,s}} = \boxed{\mathrm{s^{-1}\,T^{-1}}}$$

E14A.2(a) The magnitude of the angular momentum is given by $\{I(I+1)\}^{1/2}\hbar$. For a proton $I = 1/2$, hence magnitude $= \dfrac{\sqrt{3}}{2}\hbar = \boxed{9.133\times10^{-35}\ \mathrm{J\,s}}$. The components along the z–axis are $\pm\frac{1}{2}\hbar = \boxed{\pm 5.273\times10^{-35}\ \text{J s.}}$ The angles that the projections of the angular momentum make with the z-axis are

$$\theta = \pm\cos^{-1}\frac{\hbar/2}{\sqrt{3}\hbar/2} = \boxed{\pm 0.9553\ \text{rad} = \pm 54.74°}$$

E14A.3(a) The resonance frequency is equal to the Larmor frequency of the proton and is given by

$$\nu = \nu_{\text{L}} = \frac{\gamma_{\text{N}}\mathcal{B}_0}{2\pi}\,[14\text{A}.7]\ \text{with}\ \gamma_{\text{N}} = \frac{g_I\mu_{\text{N}}}{\hbar}\,[14\text{A}.4\text{c}]$$

$$\text{hence}\ \nu = \frac{g_I\mu_{\text{N}}\mathcal{B}_0}{h} = \frac{(5.5857)\times(5.0508\times10^{-27}\ \text{J T}^{-1})\times(13.5\ \text{T})}{6.626\times10^{-34}\ \text{J}\ s} = \boxed{574\ \text{MHz}}$$

E14A.4(a) $E_{m_I} = -\gamma_{\text{N}}\hbar\mathcal{B}_0 m_I\ [14\text{A}.4\text{c}] = -g_I\mu_{\text{N}}\mathcal{B}_0 m_I\ [14\text{A}.4\text{c}, \gamma_{\text{N}}\hbar = g_I\mu_{\text{N}}]$

$$m_I = \frac{3}{2}, \frac{1}{2}, -\frac{1}{2}, -\frac{3}{2}$$

$$E_{m_I} = (-0.4289)\times(5.051\times10^{-27}\ \text{J}\text{T}^{-1})\times(6.800\,\text{T}\times m_I) = \boxed{-1.473\times10^{-26}\ \text{J}\times m_I}$$

E14A.5(a) The energy level separation is

$$\Delta E = h\nu \quad \text{where}\ \nu = \frac{\gamma_{\text{N}}\mathcal{B}_0}{2\pi}\,[14\text{A}.6]$$

So

$$\nu = \frac{(6.73\times10^{7}\,\text{T}^{-1}\text{s}^{-1})\times15.4\,\text{T}}{2\pi} = 1.65\times10^{8}\ \text{s}^{-1}$$
$$= 1.65\times10^{8}\ \text{Hz} = \boxed{165\,\text{MHz}}$$

E14A.6(a) (i) By a calculation similar to that in Exercise 14A.3(a) we can show that a 600-MHz NMR spectrometer operates in a magnetic field of 14.1 T. Thus

$$\Delta E = \gamma_{\text{N}}\hbar\mathcal{B}_0 = h\nu_{\text{L}} = h\nu\,\text{at resonance}$$
$$= (6.626\times10^{-34}\ \text{J s})\times(6.00\times10^{8}\ \text{s}^{-1}) = \boxed{3.98\times10^{-25}\ \text{J}}$$

(ii) A 600-MHz NMR spectrometer means 600 MHz is the resonance frequency for protons for which the magnetic field is 14.1 T. In high-field NMRs, it is the field not the frequency which is fixed, so for the deuteron

$$\nu = \frac{g_I\mu_{\text{N}}\mathcal{B}_0}{h}\,[\text{Exercise 14A.3(a)}]$$
$$= \frac{(0.8575)\times(5.051\times10^{-27}\,\text{J}\text{T}^{-1})\times(14.1\text{T})}{6.626\times10^{-34}\,\text{J}\text{s}} = 9.21\bar{6}\times10^{7}\ \text{Hz} = 92.1\bar{6}\,\text{MHz}$$

$$\Delta E = h\nu = (6.626\times10^{-34}\ \text{J s})\times(9.21\bar{6}\times10^{7}\ \text{s}^{-1}) = \boxed{6.11\times10^{-26}\ \text{J}}$$

Thus the separation in energy is larger for the proton $\boxed{(a)}$.

E14A.7(a) The ground state has

$$m_I + \tfrac{1}{2} = \alpha\ \text{spin},\quad m_I = -\tfrac{1}{2} = \beta\ \text{spin}$$

Hence, with

$$\delta N = N_\alpha - N_\beta$$

$$\frac{\delta N}{N} = \frac{N_\alpha - N_\beta}{N_\alpha + N_\beta} = \frac{N_\alpha - N_\alpha e^{-\Delta E/kT}}{N_\alpha + N_\alpha e^{-\Delta E/kT}}\ [\textit{Justification}\ 14\text{A}.1]$$

$$= \frac{1-e^{-\Delta E/kT}}{1+e^{-\Delta E/kT}} \approx \frac{1-(1-\Delta E/kT)}{1+1} \approx \frac{\Delta E}{2kT} = \frac{g_I\mu_N\mathcal{B}_0}{2kT}\quad [\text{for } \Delta E \ll kT]$$

That is, $\dfrac{\delta N}{N} \approx \dfrac{g_I\mu_N\mathcal{B}_0}{2kT} = \dfrac{(5.5857)\times(5.0508\times10^{-27}\ \text{J T}^{-1})\times(\mathcal{B}_0)}{(2)\times(1.38066\times10^{-23}\ \text{J T}^{-1})\times(298\ \text{K})} \approx 3.43\times10^{-6}\ \mathcal{B}_0/\text{T}$

(i) $\mathcal{B}_0 = 0.3\ \text{T},\ \delta N/N = \boxed{1\times10^{-6}}$

(ii) $\mathcal{B}_0 = 1.5\ \text{T},\ \delta N/N = \boxed{5.1\times10^{-6}}$

(iii) $\mathcal{B}_0 = 10\ \text{T},\ \delta N/N = \boxed{3.4\times10^{-5}}$

E14A.8(a) $\delta N \approx \dfrac{Ng_I\mu_N\mathcal{B}_0}{2kT}[\text{Exercise 14A.7(a)}] = \dfrac{Nh\nu}{2kT}$

Thus, $\delta N \propto \nu$

$$\frac{\delta N(800\,\text{MHz})}{\delta N(60\,\text{MHz})} = \frac{800\,\text{MHz}}{60\,\text{MHz}} = \boxed{13}$$

This ratio is not dependent on the nuclide as long as the approximation $\Delta E \ll kT$ holds [Exercise 14A.7(a)].

E14A.9(a) (a) $\mathcal{B}_0 = \dfrac{h\nu}{g_I\mu_N} = \dfrac{(6.626\times10^{-34}\ \text{J Hz}^{-1})\times(9\times10^{9}\ \text{Hz})}{(5.5857)\times(5.051\times10^{-27}\ \text{J T}^{-1})} = \boxed{2\times10^{2}\ \text{T}}$

(b) $\mathcal{B}_0 = \dfrac{h\nu}{g_e\mu_B} = \dfrac{(6.626\times10^{-34}\ \text{J Hz}^{-1})\times(300\times10^{6}\ \text{Hz})}{(2.0023)\times(9.274\times10^{-24}\ \text{J T}^{-1})} = \boxed{10\,\text{mT}}$

Comment. Because of the sizes (very large and very small) of these magnetic fields neither experiment seems feasible.

Question. What frequencies are required to observe electron resonance in the magnetic field of a 300 MHz NMR magnet and nuclear resonance in the field of a 9 GHz ($g = 2.00$) ESR magnet? Are these experiments feasible?

Solutions to problems

P14A.1 $g_I = -3.8260$ (Table 14A.2)

$$\mathcal{B}_0 = \frac{h\nu}{g_I\mu_N} = \frac{(6.626\times10^{-34}\,\mathrm{J\,Hz^{-1}})\times\nu}{(-)(3.8260)\times(5.0508\times10^{-27}\,\mathrm{J\,T^{-1}})} = 3.429\times10^{-8}(\nu/\mathrm{Hz})\,\mathrm{T}$$

Therefore, with $\nu = 300$ MHz,

$$\mathcal{B}_0 = (3.429\times10^{-8})\times(300\times10^{6}\,\mathrm{T}) = \boxed{10.3\,\mathrm{T}}$$

$$\frac{\delta N}{N} \approx \frac{g_I\mu_N\mathcal{B}_0}{2kT} \quad \text{[Exercise 14A.7(a)]}$$

$$= \frac{(-3.8260)\times(5.0508\times10^{-27}\,\mathrm{J\,T^{-1}})\times(10.3\,\mathrm{T})}{(2)\times(1.381\times10^{-23}\,\mathrm{J\,K^{-1}})\times(298\,\mathrm{K})} = \boxed{2.42\times10^{-5}}$$

Since $g_I < 0$ (as for an electron, the magnetic moment is antiparallel to its spin), the $\boxed{\beta}$ state $\boxed{(m_I = -\tfrac{1}{2})}$ lies lower.

P14A.3 Assume that the radius of the disk is 1 unit. The volume of each slice is proportional to length of slice $\times\ \delta_z$ (see Fig. 14A.1).

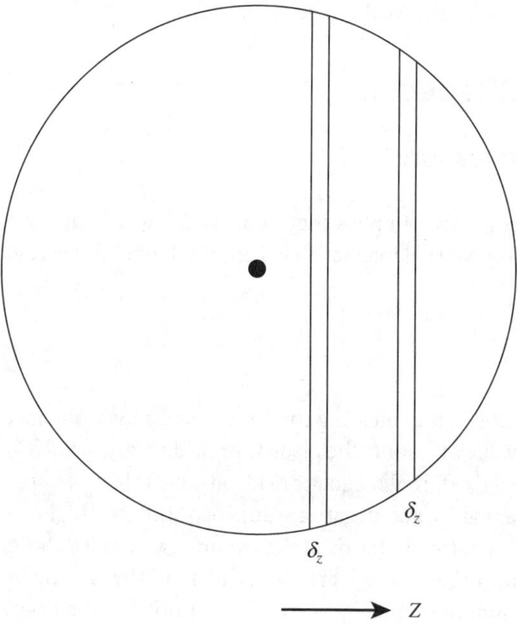

Figure 14A.1

length of slice at $z = 2\sin\theta$

$$z = \cos\theta$$

$$\theta = \arccos z$$

ranges from −1 to +1

length of slice at $z = 2\sin(\arccos z)$

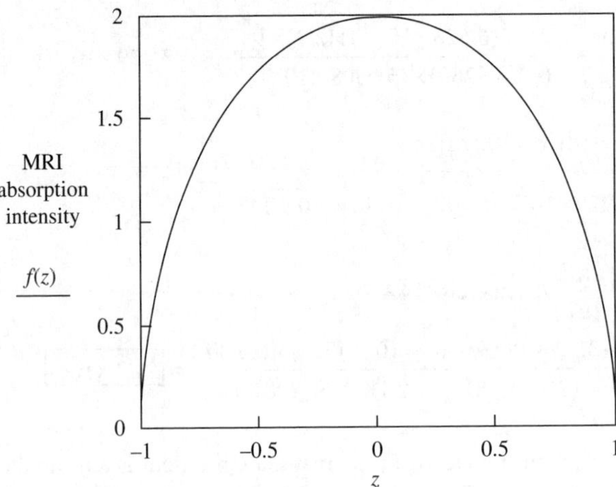

Figure 14A.2

Plot $f(z) = 2\sin(\arccos z)$ against z between the limits −1 and +1. The plot is shown above. The volume at each value of z is proportional to $f(z)$ and the intensity of the MRI signal is proportional to the volume, so the figure above represents the absorption intensity for the MRI image of the disk.

14B Features of NMR spectra

Answers to discussion questions

D14B.1 Examination of eqn 14B.3 implies that in an NMR spectrometer operating at a fixed frequency, the external magnetic field required to fulfill the resonance condition is given by

$$\mathcal{B}_0 = \frac{2\pi\nu_L}{\gamma(1-\sigma)}.$$

Thus, a positive value of the shielding constant, σ, shifts the resonance field to 'high field', a negative value of σ shifts the resonance field to 'low field'. Conversely, in a spectrometer at fixed external magnetic field, positive values of σ shift the resonant frequency to lower values and negative values of σ shift it to higher values. Chemical shifts are also reported in terms of the parameter, δ, called the 'chemical shift' and defined by eqn 14B.4; σ and δ are independent of the external magnetic field, but the chemical shift in frequency units, $n - n_0$, is not. See the *Brief illustration* in Section 14B.1.

D14B.3 See Section 14B.2(c), Figs 14B.18 and 14B.19, and *Justification* 14B.3 for a detailed explanation of this effect. As demonstrated in the *Justification*, spin–spin splittings between magnetically equivalent nuclei do not appear in the spectrum of groups of equivalent protons. For example, the protons of the methyl group in ethanol become magnetically equivalent because of rapid rotation about the C–C single bond. All three protons have the same time-averaged chemical environment and therefore the same resonance frequencies. If the rotation were slowed or stopped magnetic non-equivalence would re-emerge.

D14B.5 Spin–spin couplings in NMR are due to a polarization mechanism which is transmitted through bonds. The following description applies to the coupling between the protons in an H_X–C–H_Y group as is typically found in organic compounds. See Figs 14B.15–14B.17 of the text. On H_X, the Fermi contact interaction causes the spins of its proton and electron to be aligned antiparallel. The spin of the electron from C in the H_X–C bond is then aligned antiparallel to the electron from H_X due to the Pauli exclusion principle. The spin of the C electron in the bond with H_Y is then aligned parallel with the C electron from H_X because of Hund's rule. Finally the alignment is transmitted through the second bond in the same manner as the first. This progression of alignments (antiparallel × antiparallel × parallel × antiparallel × antiparallel) yields an overall energetically favourable parallel alignment of the two proton nuclear spins. Therefore, in this case the coupling constant, ${}^2J_{HH}$ is negative in sign.

Solutions to exercises

E14B.1(a) (i) $\delta = \dfrac{\nu - \nu^{\circ}}{\nu^{\circ}} \times 10^6$ [14B.4]

Since both ν and ν° depend upon the magnetic field in the same manner, namely

$$\nu = \frac{g_I \mu_N \mathcal{B}}{h} \quad \text{and} \quad \nu^{\circ} = \frac{g_I \mu_N \mathcal{B}_0}{h} \text{ [Exercise 14A.3(a)]}$$

δ is $\boxed{\text{independent}}$ of both $\mathcal{B}$ and ν.

(ii) Rearranging eqn 14B.4 we see $\nu - \nu^{\circ} = \nu^{\circ}\delta \times 10^{-6}$

and we see that the relative chemical shift is

$$\frac{\nu - \nu^{\circ}(800\text{ MHz})}{\nu - \nu^{\circ}(60\text{ MHz})} = \frac{(800\text{ MHz})}{(60\text{ MHz})} = \boxed{13}$$

Comment. This direct proportionality between $\nu - \nu^{\circ}$ and ν° is one of the major reasons for operating an NMR spectrometer at the highest frequencies possible.

E14B.2(a) $\mathcal{B}_{\text{loc}} = (1-\sigma)\mathcal{B}_0$ [14B.2]

$$|\Delta\mathcal{B}_{\text{loc}}| = |(\Delta\sigma)|\mathcal{B}_0 \approx |[\delta(\text{CH}_3) - \delta(\text{CHO})]| \times 10^{-6}\mathcal{B}_0 \quad \left[|\Delta\sigma| \approx \left|\frac{\nu - \nu^\circ}{\nu^\circ}\right|\right]$$

$$= |(2.20 - 9.80)| \times 10^{-6}\mathcal{B}_0 = 7.60 \times 10^{-6}\mathcal{B}_0$$

(i) $\mathcal{B}_0 = 1.5\text{ T}, \ |\Delta\mathcal{B}_{\text{loc}}| = 7.60 \times 10^{-6} \times 1.5\text{ T} = \boxed{11\ \mu\text{T}}$

(ii) $\mathcal{B}_0 = 15\text{T}, \ |\Delta\mathcal{B}_{\text{loc}}| = \boxed{110\ \mu\text{T}}$

E14B.3(a) $\nu - \nu^\circ = \nu^\circ\delta \times 10^{-6}$ [14B.4]

$$\begin{aligned} |\Delta\nu| &\equiv (\nu - \nu^\circ)(\text{CHO}) - (\nu - \nu^\circ)(\text{CH}_3) \\ &= \nu^\circ[\delta(\text{CHO}) - \delta(\text{CH}_3)] \times 10^{-6} \\ &= (9.80 - 2.20) \times 10^{-6}\nu^\circ = 7.60 \times 10^{-6}\nu^\circ \end{aligned}$$

(a) $\nu^\circ = 250\text{ MHz}, \ |\Delta\nu| = 7.60 \times 10^{-6} \times 250\text{ MHz} = 1.90\text{ kHz}$

The spectrum is shown in Fig. 14.B1 with the value of $|\Delta\nu|$ as calculated above.

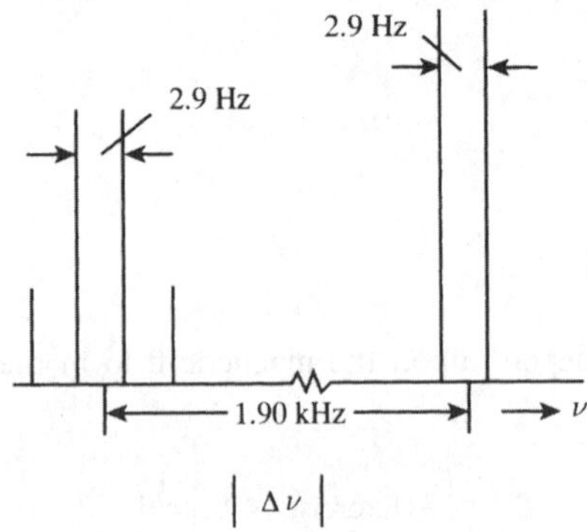

Figure 14B.1

(b) $\nu^\circ = 800\text{ MHz}, \ |\Delta\nu| = 6.08\text{ MHz}$

When the frequency is changed to 800 MHz, the $|\Delta\nu|$ changes to 6.08 kHz. The fine structure (the splitting within groups) remains the same as spin–spin splitting is unaffected by the strength of the applied field. However, the intensity of the lines increases by a factor of 3.2 because $\delta N/N \propto \nu$ (Exercise 14A.8(a)).

The observed splitting pattern is that of an AX_3 (or A_3X) species, the spectrum of which is described in Section 14B.3.

E14B.4(a) The four equivalent ^{19}F nuclei ($I = \frac{1}{2}$) give a single line. However, the ^{10}B nucleus ($I = 3$, 19.6 per cent abundant) splits this line into $2 \times 3 + 1 = 7$ lines and the ^{11}B nucleus ($I = \frac{3}{2}$, 80.4 per cent abundant) into $2 \times \frac{3}{2} + 1 = 4$ lines. The splitting arising from the ^{11}B nucleus will be larger than that arising from the ^{10}B nucleus (since its magnetic moment is larger, by a factor of 1.5, Table 14A.2). Moreover, the total intensity of the four lines due to the ^{11}B nuclei will be greater (by a factor of $80.4/19.6 \approx 4$) than the total intensity of the seven lines due to the ^{10}B nuclei. The individual line intensities will be in the ratio $\frac{7}{4} \times 4 = 7$ ($\frac{4}{7}$ the number of lines and about four times as abundant). The spectrum is sketched in Fig. 14.B2.

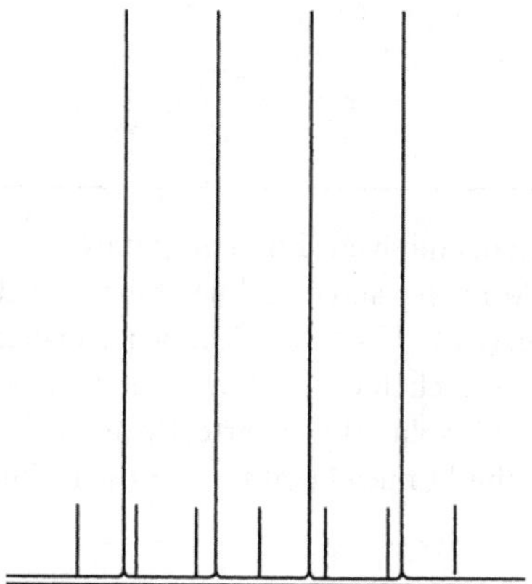

Figure 14B.2

E14B.5(a) $\nu = \dfrac{g_I \mu_N \mathcal{B}}{h}$ [Solution to Exercise 14A.3(a)]

Hence, $\dfrac{\nu(^{19}\text{F})}{\nu(^{1}\text{H})} = \dfrac{g(^{19}\text{F})}{g(^{1}\text{H})}$

or $\nu(^{19}\text{F}) = \dfrac{5.2567}{5.5857} \times 800\,\text{MHz} = \boxed{753\,\text{MHz}}$

The proton resonance consists of two lines $(2\times\frac{1}{2}+1)$ and the ^{19}F resonance of three lines $[2\times(2\times\frac{1}{2})+1]$. The intensities are in the ratio 1:2:1 (Pascal's triangle for two equivalent spin $\frac{1}{2}$ nuclei, Section 14B.3). The lines are spaced $\dfrac{5.5857}{5.2567} = 1.06$ times greater in the fluorine region than the proton region. The spectrum is sketched in Fig. 14.B3.

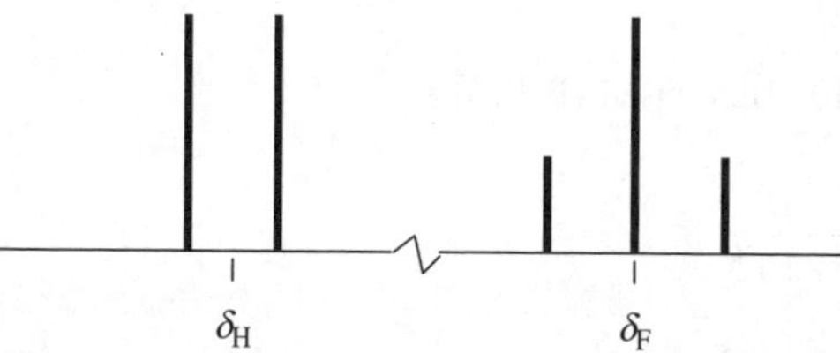

Figure 14B.3

E14B.6(a) See Section 14B.3(a), Example 14B.3, and Figs 14B.12 and 14B.13 for the approach to the solution to this exercise. Also see Example 14D.1 and Figs. 14D.4 and 14D.5. That latter example and those figures are applied specifically to EPR spectra, but the process of determining the intensity pattern in the fine structure of an NMR spectrum is the same. See the table below for the version of Pascal's triangle for up to 4 spin-3/2 nuclei. Each number in the table is the sum of the four ($I = 3/2$, $2I + 1 = 4$) numbers above it (2 to the right and 2 to the left).

1

1 1 1 1

1 2 3 4 3 2 1

1 3 6 10 12 12 10 6 3 1

1 4 10 20 31 40 44 40 31 20 10 4 1

E14B.7(a) The A, M, and X resonances lie in distinctively different groups. The A resonance is split into a 1:2:1 triplet by the M nuclei, and each line of that triplet is split into a 1:4:6:4:1 quintet by the X nuclei, (with $J_{AM} > J_{AX}$). The M resonance is split into a 1:3:3:1 quartet by the A nuclei and each line is split into a quintet by the X nuclei (with $J_{AM} > J_{MX}$). The X resonance is split into a quartet by the A nuclei and then each line is split into a triplet by the M nuclei (with $J_{AX} > J_{MX}$). The spectrum is sketched in Fig. 14.B4.

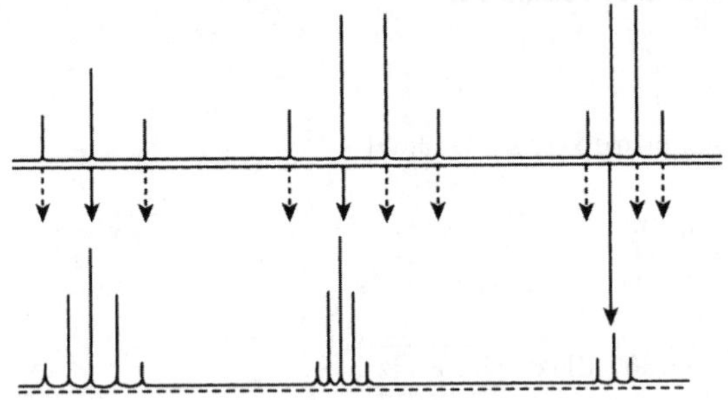

Figure 14B.4

E14B.8(a) (i) If there is rapid rotation about the axis, the H nuclei are both chemically and magnetically equivalent.

(ii) Since $J_{cis} \neq J_{trans}$, the H nuclei are chemically but not magnetically equivalent.

E14B.9(a) $\tau \approx \dfrac{\sqrt{2}}{\pi \Delta \nu}$ [14B.16, with $\delta\nu$ written as $\Delta\nu$]

$$\Delta\nu = \nu^{\circ}(\delta' - \delta) \times 10^{-6} \text{ [Exercise 14B.3(a)]}$$

Then $$\tau \approx \frac{\sqrt{2}}{\pi \nu_0 (\delta' - \delta) \times 10^{-6}}$$

$$\approx \frac{\sqrt{2}}{(\pi) \times (550 \times 10^{6}\,\text{Hz}) \times (4.8 - 2.7) \times 10^{-6}} \approx 3.9 \times 10^{-4}\,\text{s}$$

Therefore, the signals merge when the lifetime of each isomer is less than about $\boxed{0.39\,\text{ms}}$, corresponding to a conversion rate of about $\boxed{2.6 \times 10^{3}\,\text{s}^{-1}}$

Solutions to problems

P14B.1 We use $\nu = \dfrac{\gamma_N \mathcal{B}_{loc}}{2\pi} = \dfrac{\gamma_N}{2\pi}(1-\sigma)\mathcal{B}_0$ [14B.3]

where $\mathcal{B}_0$ is the applied field.

Because shielding constants are quite small (a few parts per million) compared to 1, we may write for the purposes of this calculation

$$\nu = \frac{\gamma_N \mathcal{B}_0}{2\pi}$$

$$\nu_L - \nu_R = 100\,\text{Hz} = \frac{\gamma_N}{2\pi}(\mathcal{B}_L - \mathcal{B}_R)$$

$$\mathcal{B}_L - \mathcal{B}_R = \frac{2\pi \times 100\ \text{s}^{-1}}{\gamma_N}$$

$$= \frac{2\pi \times 100\ \text{s}^{-1}}{26.752 \times 10^7\,\text{T}^{-1}\text{s}^{-1}} = 2.35 \times 10^{-6}\,\text{T}$$

$$= 2.35\ \mu\text{T}$$

The field gradient required is then

$$\frac{2.35\,\mu\text{T}}{0.08\,\text{m}} = \boxed{29\,\mu\text{T}\,\text{m}^{-1}}$$

Note that knowledge of the spectrometer frequency, applied field, and the numerical value of the chemical shift (because constant) is not required.

P14B.3 It seems reasonable to assume that only staggered conformations can occur. Therefore the equilibria are as shown in Fig. 14.B5.

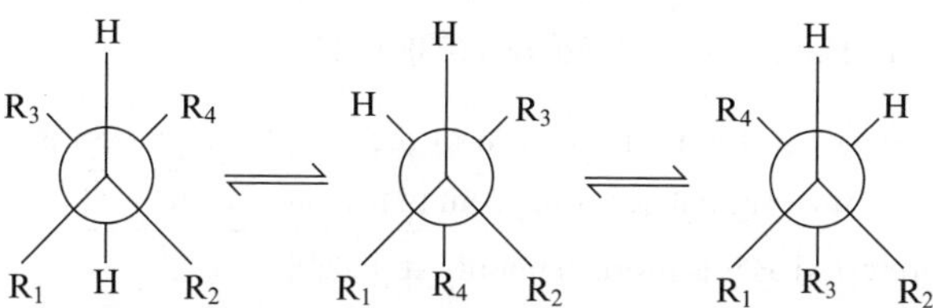

Figure 14B.5

When $R_3 = R_4 = H$, all three of the above conformations occur with equal probability:

$${}^3J_{HH}(\text{methyl}) = \tfrac{1}{3}\left({}^3J_t + 2\,{}^3J_g\right) \quad [\text{t} = \text{trans, g} = \text{gauche; CHR}_3\text{R}_4 = \text{methyl}]$$

The first conformation in the figure is trans, the second two are gauche.

Additional methyl groups will avoid being staggered between both R_1 and R_2. Therefore

$${}^3J_{HH}(\text{ethyl}) = \tfrac{1}{2}\left({}^3J_t + {}^3J_g\right) \quad [R_4 = H, R_3 = CH_3]$$

$${}^3J_{HH}(\text{isopropyl}) = {}^3J_t \qquad [R_3 = R_4 = CH_3]$$

We then have three simultaneous equations in two unknowns J_t and J_g.

$$\tfrac{1}{3}({}^3J_t + 2\,{}^3J_g) = 7.3\ \text{Hz} \quad (1)$$

$$\tfrac{1}{2}({}^3J_t + {}^3J_g) = 8.0\ \text{Hz} \quad (2)$$

$${}^3J_t = 11.2\ \text{Hz}$$

The two unknowns are overdetermined. The first two equations yield ${}^3J_t = 10.1$, ${}^3J_g = 5.9$. However, if we assume that ${}^3J_t = 11.2$ as measured directly in the isopropyl case then ${}^3J_g = 5.4$ (eqn 1) or 4.8 (eqn 2), with an average value of 5.1.

Using the original form of the Karplus equation

$${}^3J_t = A\cos^2(180°) + B = 11.2$$

$${}^3J_g = A\cos^2(60°) + B = 5.1$$

or

$$11.2 = A + B$$

$$5.1 = 0.25\,A + B$$

These simultaneous equations yield $A = 6.8$ Hz and $B = 4.8$ Hz. With these values of A and B, the original form of the Karplus equation fits the data exactly (at least to within the error in the values of ${}^3J_t$ and ${}^3J_g$ and in the measured values reported).

From the form of the Karplus equation in the text [14B.14] we see that those values of A, B and C cannot be determined from the data given, as there are three constants to be determined from only two values of J. However, if we use the values of A, B and C given in the text, then

$$J_t = 7\ \text{Hz} + 1\ \text{Hz}(\cos 180°) + 5\ \text{Hz}(\cos 360°) = 11\ \text{Hz}$$

$$J_g = 7\ \text{Hz} + 1\ \text{Hz}(\cos 60°) + 5\ \text{Hz}(\cos 120°) = 5\ \text{Hz}$$

The agreement with the modern form of the Karplus equation is excellent, but not better than the original version. Both fit the data equally well. But the modern version is preferred as it is more generally applicable.

P14B.5 $\quad {}^3J_{HH} = A + B\cos\phi + C\cos 2\phi \qquad [14B.14]$

$$\frac{d}{d\phi}({}^3J_{HH}) = -B\sin\phi - 2C\sin 2\phi = 0$$

This equation has a number of solutions:

$$\phi = 0,\ \phi = n\pi,\ \phi = \pi - \arccos\left(\frac{B}{4C}\right) = \arccos\left(\frac{B}{4C}\right)$$

The first two are trivial solutions.

If $\phi = \arccos\left(\dfrac{B}{4C}\right)$ then, $\sin\phi = \sqrt{1 - \dfrac{B}{16C^2}}$

$$\sin 2\phi = 2\sin\phi\cos\,\phi = 2\sqrt{1-\frac{B^2}{16C^2}}\left(\frac{B}{4C}\right)$$

$$B\sin\phi + 2C\sin 2\phi = B\sqrt{1-\frac{B^2}{16C^2}} + 4C\sqrt{1-\frac{B^2}{16C^2}}\left(\frac{B}{4C}\right) = 0$$

So $\dfrac{B}{4C} = \cos\phi$ clearly satisfies the condition for an extremum.

The second derivative is

$$\frac{d^2}{d\phi^2}({}^3J_{HH}) = -B\cos\phi - 4C\cos 2\phi = -B\cos\phi - 4C(2\cos^2\phi - 1)$$

$$= -B\left(\frac{B}{4C}\right) - 4C\left(2\frac{B^2}{16C^2} - 1\right) = -\frac{B^2}{4C} - \frac{2B^2}{4C} + 4C$$

This quantity is positive if $16C^2 > 3B^2$.

This is certainly true for typical values of B and C, namely $B = -1$ Hz and $C = 5$ Hz.

Therefore the condition for a minimum is as stated, namely, $\boxed{\cos\phi = B/4C}$.

14C Pulse techniques in NMR

Answers to discussion questions

D14C.1 Before the application of a pulse the magnetization vector, $\boldsymbol{M}$, points along the direction of the static external magnetic field $\mathcal{B}_0$. There are more α spins than β spins. When we apply a rotating magnetic field $\mathcal{B}_1$ at right angles to the static field, the magnetization vector as seen in the rotating frame begins to precess about the $\mathcal{B}_1$ field with angular frequency $\omega_1 = \gamma\mathcal{B}_1$. The angle through which M rotates is $\theta = \gamma\mathcal{B}_1 t$, where t is the time for which the $\mathcal{B}_1$ pulse is applied. When $t = \pi/2\gamma\mathcal{B}_1$, $\theta = \pi/2 = 90°$, and $\boldsymbol{M}$ has rotated into the xy plane. Now there are equal numbers of α and β spins. A 180° pulse applied for a time $\pi/\gamma\mathcal{B}_1$, rotates $\boldsymbol{M}$ antiparallel to the static field. Now there are more β spins than α spins. A population inversion has occurred.

D14C.3 For example, at room temperature, the tumbling rate of benzene, the small molecule, in a mobile solvent, may be close to the Larmor frequency, and hence its spin-lattice relaxation time will be short. As the temperature increases, the tumbling rate may increase well beyond the Larmor frequency, resulting in an increased spin–lattice relaxation time.

For the large molecule (like a polymer) at room temperature, the tumbling rate may be well below the Larmor frequency, but with increasing temperature it will approach the Larmor frequency due to the increased thermal motion of the molecule combined with the decreased viscosity of the solvent. Therefore, the spin-lattice relaxation time may decrease

D14C.5 The basic COSY experiment uses the simplest of all two-dimensional pulse sequences: a single 90° pulse to excite the spins at the end of the preparation period and a mixing period containing just a second 90° pulse.

The key to the COSY technique is the effect of the second 90° pulse, which can be illustrated by consideration of the four energy levels of an AX system (as shown in Figs 14C.15, 14C.16, and 14C.17 of the text). At thermal equilibrium, the population of the $\alpha_A\alpha_X$ level is the greatest, and that of $\beta_A\beta_X$ level is the smallest; the other two levels have the same energy and an intermediate population. After the first 90° pulse, the spins are no longer at thermal equilibrium. If a second 90° pulse is applied at a time t_1 that is short compared to the spin–lattice relaxation time T_1 the extra input of energy causes further changes in the populations of the four states. The changes in populations will depend on how far the individual magnetizations have precessed during the evolution period.

For simplicity, let us consider a COSY experiment in which the second 90° pulse is split into two selective pulses, one applied to X and one to A. Depending on the evolution time t_1, the 90° pulse that excites X may leave the population differences across each of the two X transitions unchanged, inverted, or somewhere in between. Consider the extreme case in which one population difference is inverted and the other unchanged. The 90° pulse that excites A will now generate an FID in which one of the two A transitions has increased in intensity, and the other has decreased. The overall effect is that precession of the X spins during the evolution period determines the amplitudes of the signals from the A spins obtained during the detection period. As the evolution time t_1 is increased, the intensities of the signals from A spins oscillate at rates determined by the frequencies of the two X transitions.

This transfer of information between spins is at the heart of two-dimensional NMR spectroscopy and leads to the correlation of different signals in a spectrum. In this case, information transfer tells us that there is a scalar coupling between A and X. If we conduct a series of experiments in which t_1 is incremented, Fourier transformation of the FIDs on t_2 yields a set of spectra $I(\nu_{1,}\ \nu_2)$ in which the A signal amplitudes oscillate as a function of t_1. A second Fourier transformation, this time on t_1, converts these oscillations into a two-dimensional spectrum $I(\nu_{1,}\ \nu_2)$. The signals are spread out in ν_1 according to their precession frequencies during the detection period. Thus, if we apply the COSY pulse sequence to our AX spin system, the result is a two-dimensional spectrum that contains four groups of signals centred on the two chemical shifts in ν_1 and ν_2. Each group will show fine structure, consisting of a block of four signals separated by J_{AX}. The diagonal peaks are signals centred on $(\delta_A\ \delta_A)$ and $(\delta_X\ \delta_X)$ and lie along the diagonal $\nu_1 = \nu_2$. They arise from signals that did not change chemical shift between t_1 and t_2. The cross peaks (or *off-diagonal peaks*) are signals centred on $(\delta_A\ \delta_X)$ and $(\delta_X\ \delta_A)$ and owe their existence to the coupling between A and X. Consequently, cross peaks in COSY spectra allow us to map the couplings between spins and to trace out the bonding network in complex molecules. Figure 14.1 of the text shows a simple example of a proton COSY spectrum of 1-nitropropane.

Solutions to exercises

E14C.1(a) Analogous to precession of the magnetization vector in the laboratory frame due to the presence of $\mathcal{B}_0$ that is

$$\nu_L = \frac{\gamma_N \mathcal{B}_0}{2\pi}\,[14A.7],$$

there is a precession in the rotating frame, due to the presence of $\mathcal{B}_1$, namely

$$\nu_L = \frac{\gamma_N \mathcal{B}_1}{2\pi} \quad \text{or} \quad \omega_1 = \gamma_N \mathcal{B}_1 \quad [\omega = 2\pi\nu]$$

Since ω is an angular frequency, the angle through which the magnetization vector rotates is

$$\theta = \gamma_N \mathcal{B}_1 t = \frac{g_I \mu_N}{\hbar} \mathcal{B}_1 t$$

So $\mathcal{B}_1 = \dfrac{\theta\hbar}{g_I \mu_N t} = \dfrac{(\pi)\times(1.0546\times10^{-34}\,\text{J s})}{(5.586)\times(5.0508\times10^{-27}\,\text{J T}^{-1})\times(12.5\times10^{-6}\,\text{s})} = \boxed{9.40\times10^{-4}\,\text{T}}$

a 90° pulse requires $\frac{1}{2}\times12.5\,\mu\text{s} = \boxed{6.25\,\mu\text{s}}$

E14C.2(a) The effective transverse relaxation time is given by

$$T_2^* = \frac{1}{\pi\Delta\nu_{1/2}}\,[14\text{C.7}] = \frac{1}{\pi\times1.5\,\text{s}^{-1}} = \boxed{0.21\,\text{s}}$$

E14C.3(a) The maximum enhancement is given by

$$\eta = \frac{\gamma_{^1\text{H}}}{2\times\gamma_{^{31}\text{P}}}[14\text{C.9}] = \frac{26.752\times10^7\,\text{T}^{-1}\text{s}^{-1}}{2\times10.840\times10^7\,\text{T}^{-1}\text{s}^{-1}}[\text{Table 14A.2}] = \boxed{1.234}$$

E14C.4(a) The proton COSY spectrum of 1-nitropropane shows that (a) the C_a–H resonance with $\delta = 4.3$ shares a cross-peak with the C_b–H resonance at $\delta = 2.1$ and (b) the C_b–H resonance with $\delta = 2.1$ shares a cross-peak with the C_c–H resonance at $\delta = 1.1$. Off diagonal peaks indicate coupling between H's on various carbons. Thus peaks at (4,2) and (2,4) indicate that the H's on the adjacent CH_2 units are coupled. The peaks at (1,2) and (2,1) indicate that the H's on CH_3 and central CH_2 units are coupled. See Fig. 14C.1.

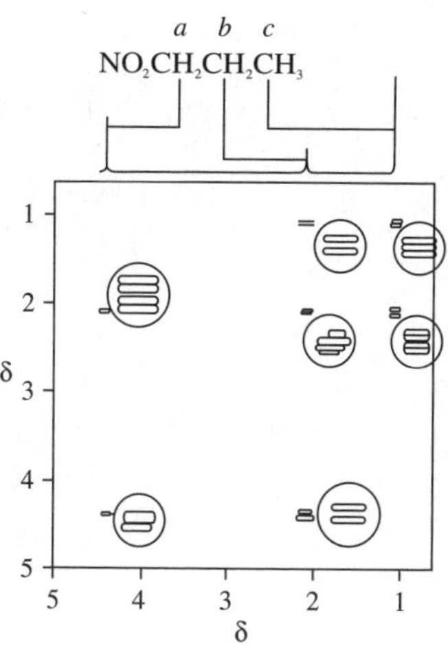

Figure 14C.1

Solutions to problems

P14C.1 The envelopes of maxima and minima of the curve are determined by T_2 through eqn 14C.1, but the time interval between the maxima of this decaying curve corresponds to the reciprocal of the frequency difference $\Delta\nu$ between the pulse frequency ν_0 and the Larmor frequency ν_L, that is $\Delta\nu = |\nu_0 - \nu_L|$

$$\Delta\nu = \frac{1}{0.12\,\text{s}} = 8.3\ \text{s}^{-1} = 8.3\,\text{Hz}$$

Therefore the Larmor frequency is $\boxed{400\times10^6\,\text{Hz}\pm8\,\text{Hz}}$

According to eqns 14C.1 and 14C.5, the intensity of the maxima in the FID curve decays exponentially as e^{-t/T_2}. Therefore T_2 corresponds to the time at which the intensity has been reduced to $1/e$ of the original value. In the text figure, this corresponds to a time slightly before the fourth maximum has occurred, or about $\boxed{0.29\,\text{s}}$

P14C.3 (a) The individual and total FIDs for the two nuclei are displayed in the following Mathcad worksheet. See Figs 14C.2(a), (b), (c), and (d).

Definitions:	$\mathbf{MHz} := 10^6\,\text{Hz}$	$\mathbf{\mu s} := 10^{-6}\,\text{s}$	$\mathbf{N} := 2^{12}$
	$\mathbf{m} := 0, 1..N-1$	$\mathbf{n} := 0, 1..\left(\frac{N}{2}-1\right)$	
Time domain:	$t_{max} := 1\,\mu\text{s}$	$t_m := \frac{m}{N} t_{max}$	
Frequency domain:	$\nu_n := \frac{n}{t_{max}}$		
FID of signal 1:	$S_1 := 1$	$\nu_1 = 50\,\text{MHz}$	$T_1 := 0.50\,\mu\text{s}$
	$F_{1_m} := S_1 \cos(2\pi\nu_1 t_m) e^{\frac{-t_m}{T_1}}$		
FID of signal 2:	$S_2 := 3$	$\nu_2 := 10\,\text{MHz}$	$T_2 := 1.0\,\mu\text{s}$
	$F_{2_m} := S_2 \cos(2\pi\nu_2 t_m) e^{\frac{-t_m}{T_2}}$		
Total FID signal:	$F_m := F_{1_m} + F_{2_m}$		

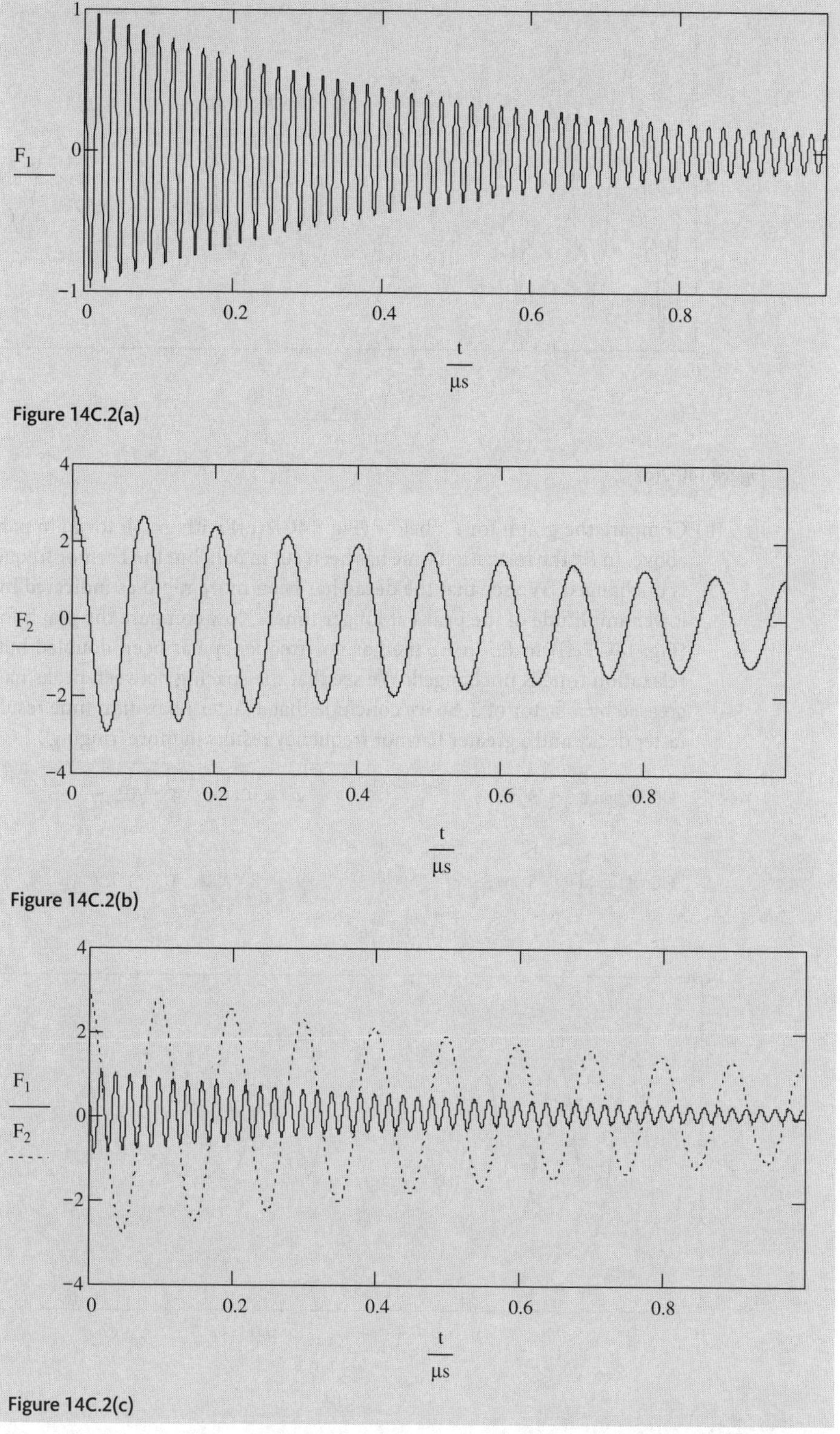

Figure 14C.2(a)

Figure 14C.2(b)

Figure 14C.2(c)

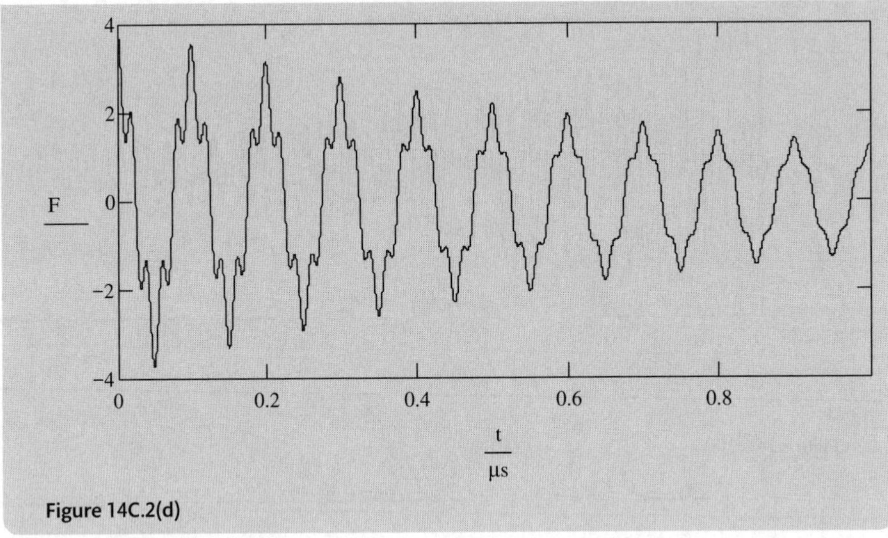

Figure 14C.2(d)

(b) Compare the graph for F_{22} below (Fig. 14C.2(e)) with graph for F_2 in part (a) above. In F_{22} the relaxation time has been cut in half, but the Larmor frequency is unchanged. We see that the decay has been more rapid as indicated by the lower amplitude of the peaks at longer times. Now compare the graph for F_{23} (Fig. 14C.2(f)) to F_2. In F_{23} the Larmor frequency has been doubled but the relaxation time is unchanged. We see that the spacing between peaks has decreased by a factor of 2. So we conclude that a faster relaxation time results in faster decay and a greater Larmor frequency results in more 'ringing'.

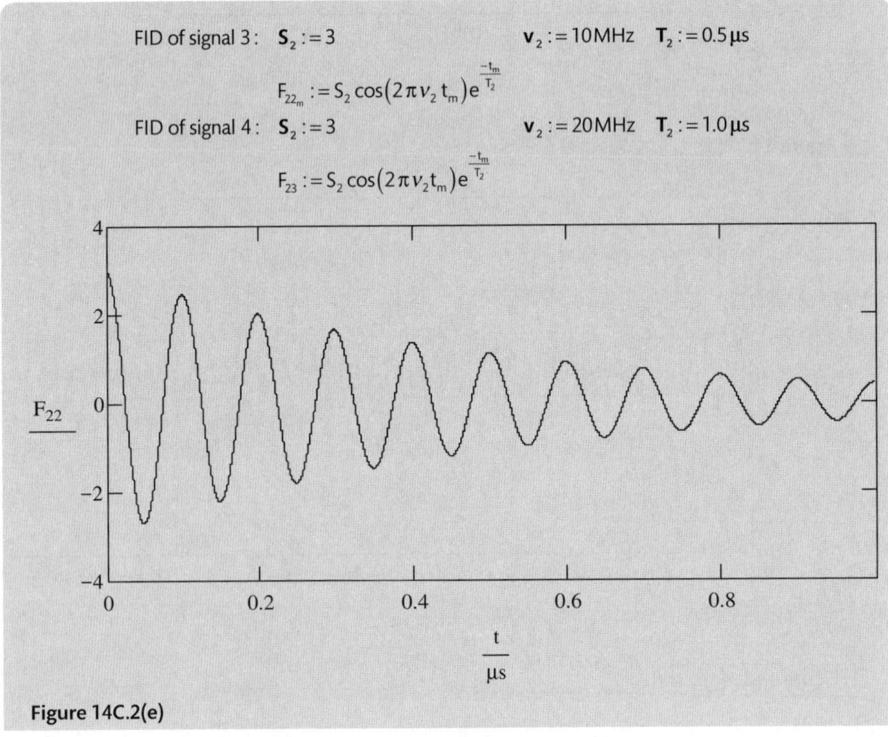

FID of signal 3: $S_2 := 3$
$\nu_2 := 10\,\text{MHz}$ $T_2 := 0.5\,\mu\text{s}$
$F_{22_m} := S_2 \cos(2\pi\nu_2 t_m)e^{\frac{-t_m}{T_2}}$
FID of signal 4: $S_2 := 3$
$\nu_2 := 20\,\text{MHz}$ $T_2 := 1.0\,\mu\text{s}$
$F_{23} := S_2 \cos(2\pi\nu_2 t_m)e^{\frac{-t_m}{T_2}}$
4
2
F_{22}
0
−2
−4
0
0.2
0.4
0.6
0.8
t
μs

Figure 14C.2(e)

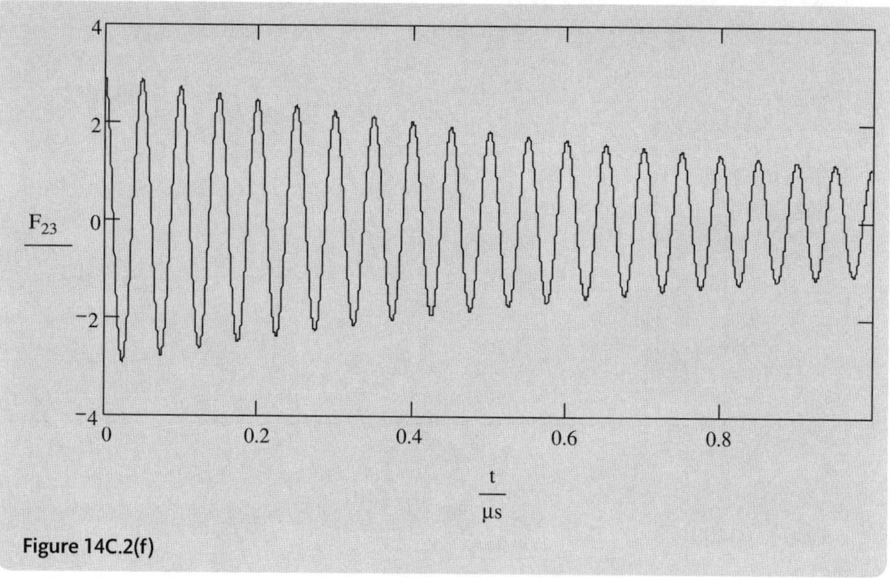

Figure 14C.2(f)

(c) The Fourier transforms of signals 1 and 2 are shown below in Figs 14C.2(g), (h), and (i).

Spectral intensities of separate and total FIDs:

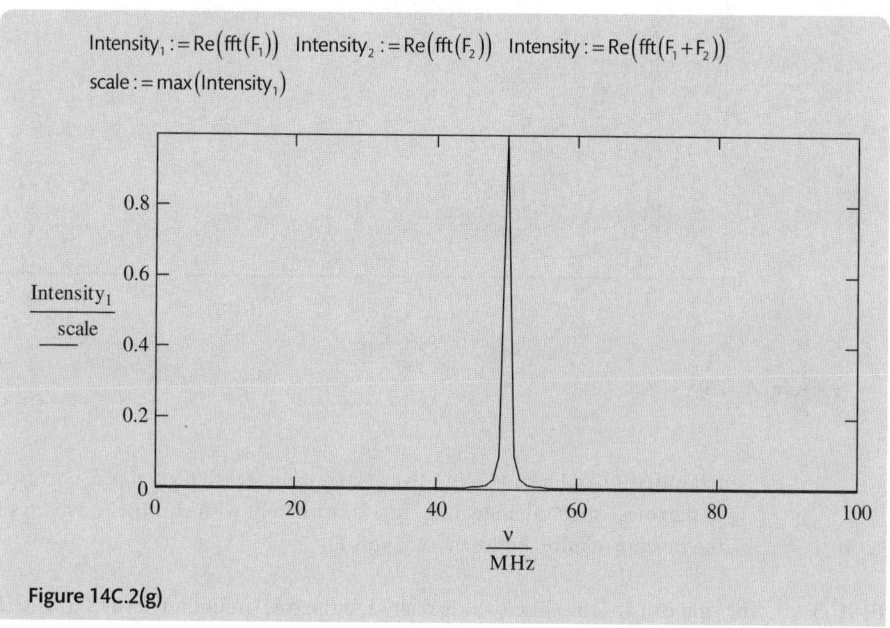

Figure 14C.2(g)

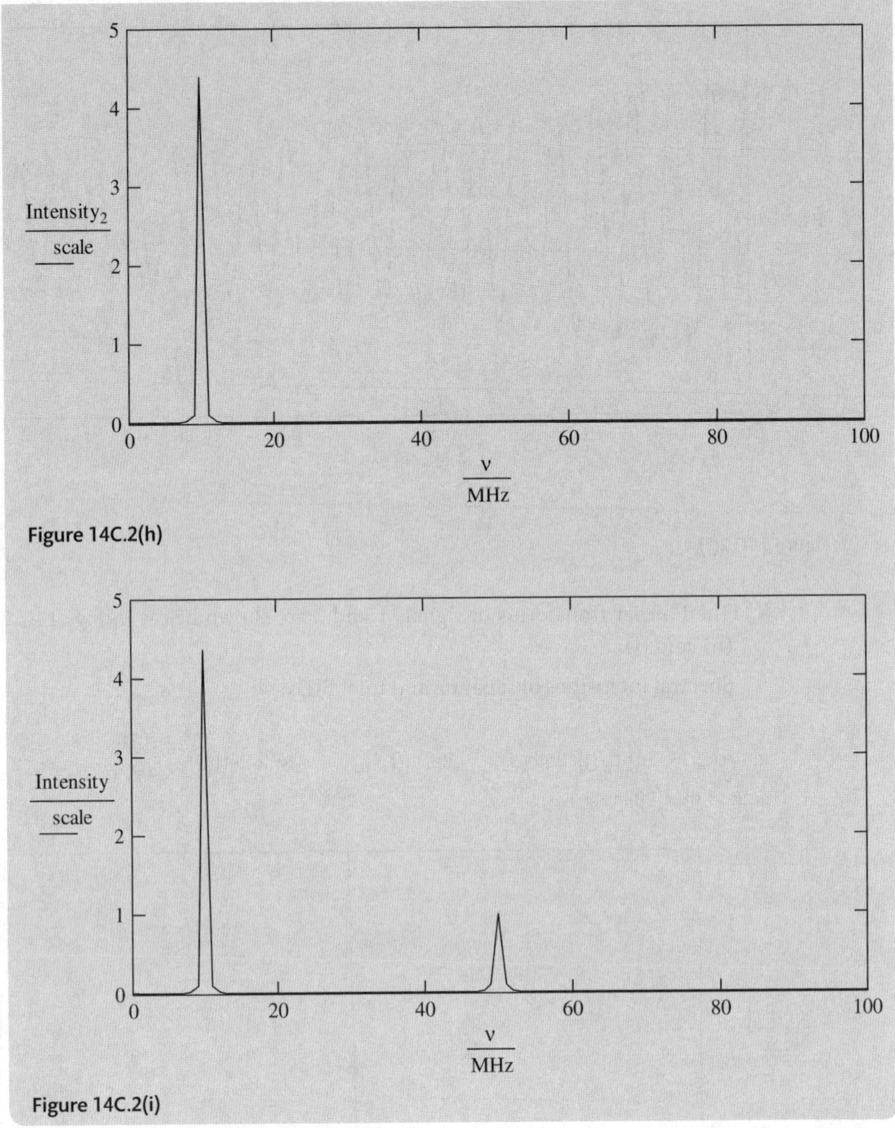

Figure 14C.2(h)

Figure 14C.2(i)

Examination of Figs 14C.2(g), (h), and (i) indicates that greater T_2 corresponds to narrower spectral linewidth. This is consistent with eqn 14C.7, which shows the inverse relation between $\Delta\nu_{1/2}$ and T_2.

P14C.5 The shape of spectral line $I(\omega)$ is related to the free induction decay signal $G(t)$ by

$$I(\omega) = a\,\mathrm{Re}\int_0^\infty G(t)\mathrm{e}^{i\omega t}\mathrm{d}t$$

where a is a constant and Re means take the real part of what follows. Calculate the lineshape corresponding to an oscillating, decaying function

$$G(t)=\cos\omega_0 t\,\mathrm{e}^{-t/\tau}$$

$$I(\omega)=a\,\mathrm{Re}\int_0^\infty G(t)\mathrm{e}^{\mathrm{i}\omega t}\,\mathrm{d}t$$

$$=a\,\mathrm{Re}\int_0^\infty \cos\omega_0 t e^{-t/\tau+\mathrm{i}\omega t}\,dt$$

$$=\frac{1}{2}a\,\mathrm{Re}\int_0^\infty (\mathrm{e}^{-\mathrm{i}\omega_0 t}+\mathrm{e}^{\mathrm{i}\omega_0 t})e^{-t/\tau+\mathrm{i}\omega t}\,\mathrm{d}t$$

$$=\frac{1}{2}a\,\mathrm{Re}\int_0^\infty \{\mathrm{e}^{\mathrm{i}(\omega_0+\omega+\mathrm{i}/\tau)t}+\mathrm{e}^{-\mathrm{i}(\omega_0-\omega-\mathrm{i}/\tau)t}\}dt$$

$$=-\frac{1}{2}a\,\mathrm{Re}\left[\frac{1}{\mathrm{i}(\omega_0+\omega+\mathrm{i}/\tau)}-\frac{1}{\mathrm{i}(\omega_0-\omega-\mathrm{i}/\tau)}\right]$$

when ω and ω_0 are similar to magnetic resonance frequencies (or higher), only the second term in brackets is significant (because $\frac{1}{(\omega_0+\omega)}\ll 1$ but $\frac{1}{(\omega_0-\omega)}$ may be large if $\omega\approx\omega_0$). Therefore

$$I(\omega)\approx\frac{1}{2}a\,\mathrm{Re}\frac{1}{\mathrm{i}(\omega_0-\omega)+1/\tau}$$

$$=\frac{1}{2}a\,\mathrm{Re}\frac{-\mathrm{i}(\omega_0-\omega)+1/\tau}{(\omega_0-\omega)^2+1/\tau^2}=\frac{1}{2}a\frac{1/\tau}{(\omega_0-\omega)^2+1/\tau^2}$$

$$=\boxed{\left(\frac{1}{2}\right)\frac{a\tau}{1+(\omega_0-\omega)^2\tau^2}}$$

which is a Lorentzian line centred on ω_0, of amplitude $\frac{1}{2}a\tau$ and width $\frac{2}{\tau}$ at half-height.

P14C.7 $$\mathcal{B}_{\text{nuc}}=-\frac{\gamma_{\mathrm{N}}\hbar\mu_0 m_{\mathrm{I}}}{4\pi R^3}(1-3\cos^2\theta)[\text{14C.10}]=\frac{g_{\mathrm{I}}\mu_{\mathrm{N}}\mu_0}{4\pi R^3}\quad\left[m_{\mathrm{I}}=+\frac{1}{2},\theta=0,\ \gamma_{\mathrm{N}}\hbar=g_{\mathrm{I}}\mu_{\mathrm{N}}\right]$$

which rearranges to

$$R=\left(\frac{g_I\mu_{\mathrm{N}}\mu_0}{4\pi\mathcal{B}_{\text{nuc}}}\right)^{1/3}=\left(\frac{(5.5857)\times(5.0508\times10^{-27}\,\mathrm{J\,T^{-1}})\times(4\pi\times10^{-7}\,\mathrm{T^2J^{-1}m^3})}{(4\pi)\times(0.715\times10^{-3}\,\mathrm{T})}\right)^{1/3}$$

$$=\left(3.946\times10^{-30}\ \mathrm{m^3}\right)^{1/3}=\boxed{158\ \mathrm{pm}}$$

P14C.9 $$\langle\mathcal{B}_{\text{nucl}}\rangle=\frac{-g_I\mu_{\mathrm{N}}\mu_0 m_I}{4\pi R^3}\times\frac{\int_0^{\theta_{\max}}(1-3\cos^2\theta)\sin\theta\,\mathrm{d}\theta}{\int_0^{\theta_{\max}}\sin\theta\,\mathrm{d}\theta}$$

The denominator is the normalization constant and ensures that the total probability of being between 0 and $\theta_{\max}$ is 1.

$$\langle\mathcal{B}_{\text{nucl}}\rangle=\frac{-g_I\mu_{\mathrm{N}}\mu_0 m_I}{4\pi R^3}\times\frac{\int_1^{x_{\max}}(1-3x^2)\mathrm{d}x}{\int_1^{x_{\max}}\mathrm{d}x}[x_{\max}=\cos\theta_{\max}]$$

$$\langle \mathcal{B}_{\text{nucl}} \rangle = \frac{-g_I \mu_N \mu_0 m_I}{4\pi R^3} \times \frac{x_{\max}(1-x_{\max}^2)}{x_{\max}-1} = \frac{+g_I \mu_N \mu_0 m_I}{4\pi R^3}(\cos^2\theta_{\max} + \cos\theta_{\max})$$

If $\theta_{\max} = 2\pi$ (complete rotation), $\cos\theta_{\max} = -1$ and $\langle \mathcal{B}_{\text{nucl}} \rangle = 0$. If $\theta_{\max} = 30°$,

$\cos^2\theta_{\max} + \cos\theta_{\max} = 1.616$ and

$$\langle \mathcal{B}_{\text{nucl}} \rangle = \frac{(5.5857)\times(5.0508\times10^{-27}\,\text{J T}^{-1})\times(4\pi\times10^{-7}\,\text{T}^2\,\text{J}^{-1}\text{m}^3)\times(1.616)}{(4\pi)\times(1.58\times10^{-10}\,\text{m})^3\times(2)}$$

$$= \boxed{0.58\,\text{mT}}$$

14D Electron paramagnetic resonance

Answer to discussion question

D14D.1 The hyperfine structure in the ESR spectrum of an atomic or molecular system is a result of two interactions: an anisotropic dipolar coupling between the net spin of the unpaired electrons and the nuclear spins and also an isotropic coupling due to the Fermi contact interaction. In solution, only the Fermi contact interaction contributes to the splitting as the dipolar contribution averages to zero in a rapidly tumbling system. In the case of π-electron radicals, such as $C_6H_6^-$, no hyperfine interaction between the unpaired electron and the ring protons might have been expected. The protons lie in the nodal plane of the molecular orbital occupied by the unpaired electron, so any hyperfine structure cannot be explained by a simple Fermi contact interaction which requires an unpaired electron density at the proton. However, an indirect spin polarization mechanism, similar to that used to explain spin–spin couplings in NMR, can account for the existence of proton hyperfine interactions in the ESR spectra of these systems. Refer to Fig. 14D.6 of the text. Because of Hund's rule, the unpaired electron and the first electron in the C–H bond (the one from the C atom), will tend to align parallel to each other. The second electron in the C–H bond (the one from H) will then align antiparallel to the first by the Pauli principle, and finally the Fermi contact interaction will align the proton and electron on H antiparallel. The net result (parallel × antiparallel × antiparallel) is that the spins of the unpaired electron and the proton are aligned parallel. The have effectively detected each other.

Solutions to exercises

E14D.1(a) $g = \dfrac{h\nu}{\mu_B \mathcal{B}_0}$ [14D.2]

We shall often need the value

$$\frac{h}{\mu_B} = \frac{6.62608\times10^{-34}\,\text{J Hz}^{-1}}{9.27401\times10^{-24}\,\text{J T}^{-1}} = 7.14478\times10^{-11}\,\text{T Hz}^{-1}$$

Then, in this case

$$g = \frac{(7.14478\times10^{-11}\,\mathrm{T\,Hz^{-1}})\times(9.2231\times10^{9}\,\mathrm{Hz})}{329.12\times10^{-3}\,\mathrm{T}} = \boxed{2.0022}$$

E14D.2(a) $a = \mathcal{B}(\text{line}\,3) - \mathcal{B}(\text{line}\,2) = \mathcal{B}(\text{line}\,2) - \mathcal{B}(\text{line}\,1)$

$$\left.\begin{aligned} \mathcal{B}_3 - \mathcal{B}_2 &= (334.8 - 332.5)\,\mathrm{mT} = 2.3\,\mathrm{mT} \\ \mathcal{B}_2 - \mathcal{B}_1 &= (332.5 - 330.2)\,\mathrm{mT} = 2.3\,\mathrm{mT} \end{aligned}\right\} a = \boxed{2.3\,\mathrm{mT}}$$

Use the centre line to calculate g

$$g = \frac{h\nu}{\mu_{\mathrm{B}}\mathcal{B}_0} = (7.14478\times10^{-11}\,\mathrm{T\,Hz^{-1}})\times\frac{9.319\times10^{9}\,\mathrm{Hz}}{332.5\times10^{-3}\,\mathrm{T}} = \boxed{2.002\overline{5}}$$

E14D.3(a) The centre of the spectrum will occur at $332.5\,\mathrm{mT}$. Proton 1 splits the line into two components with separation $2.0\,\mathrm{mT}$ and hence at $332.5\pm1.0\,\mathrm{mT}$. Proton 2 splits these two hyperfine lines into two, each with separation $2.6\,\mathrm{mT}$, and hence the lines occur at $332.5\pm1.0\pm1.3\,\mathrm{mT}$. The spectrum therefore consists of four lines of $\boxed{\text{equal intensity}}$ at the fields $\boxed{330.2\,\mathrm{mT}, 332.2\,\mathrm{mT}, 332.8\,\mathrm{mT}, 334.8\,\mathrm{mT}}$.

E14D.4(a) We construct Fig. 14D.1(a) for CH_3 and Fig. 14D.1(b) for CD_3. The predicted intensity distribution is determined by counting the number of overlapping lines of equal intensity from which the hyperfine line is constructed.

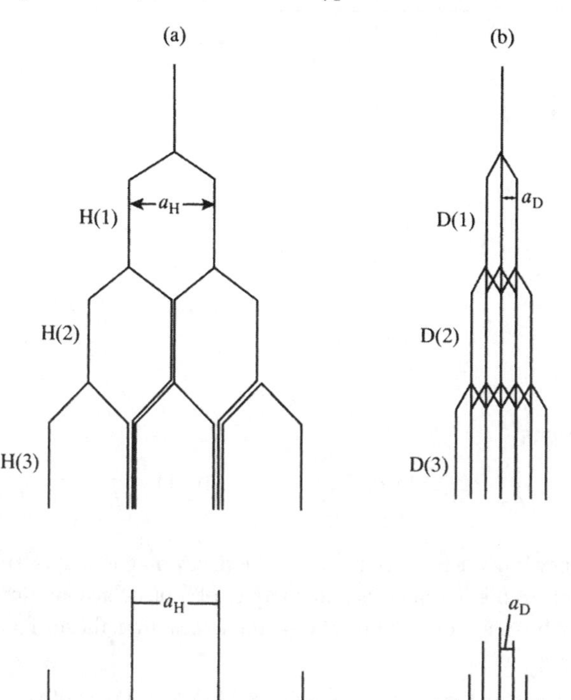

Figure 14D.1

E14D.5(a) $$\mathcal{B}_0 = \frac{h\nu}{g\mu_B} = \frac{7.14478\times10^{-11}}{2.0025}\,\mathrm{T\,Hz^{-1}}\times\nu\,[\text{Exercise 14D.1(a)}] = 35.68\,\mathrm{mT}\times(\nu/\mathrm{GHz})$$

(a) $\nu = 9.313\,\mathrm{GHz}$, $\mathcal{B}_0 = \boxed{332.3\,\mathrm{mT}}$

(b) $\nu = 33.80\,\mathrm{GHz}$, $\mathcal{B}_0 = 1206\,\mathrm{mT} = \boxed{1.206\,\mathrm{T}}$

E14D.6(a) Since the number of hyperfine lines arising from a nucleus of spin I is $2I+1$, we solve $2I+1=4$ and find that $\boxed{I=\frac{3}{2}}$.

Comment. Four lines of equal intensity could also arise from two inequivalent nuclei with $I=\frac{1}{2}$.

E14D.7(a) The X nucleus produces six lines of equal intensity. The pair of H nuclei in XH_2 split each of these lines into a 1:2:1 triplet (Fig. 14D.2(a)). The pair of D nuclei ($I=1$) in XD_2 split each line into a 1:2:3:2:1 quintet (Fig. 14D.2(b)). The total number of hyperfine lines observed is then $6\times3=18$ in XH_2 and $6\times5=30$ in XD_2.

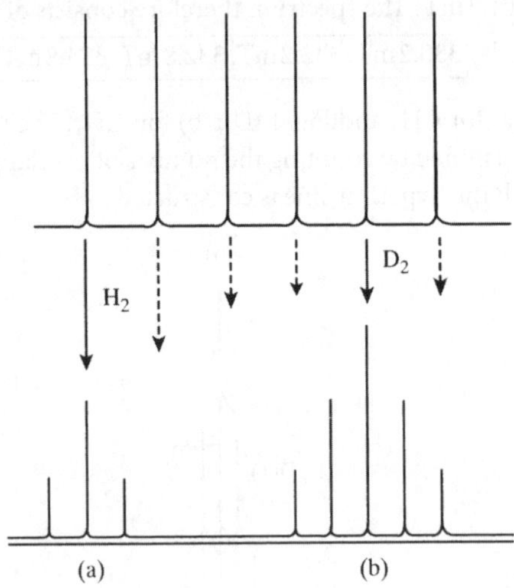

Figure 14D.2

Solutions to problems

P14D.1 $$\nu = \frac{g_e\mu_B\mathcal{B}}{h} = \frac{(2.00)\times(9.274\times10^{-24}\,\mathrm{J\,T^{-1}})\times(1.0\times10^{3}\,\mathrm{T})}{6.626\times10^{-34}\,\mathrm{J\,s}} = \boxed{2.8\times10^{13}\,\mathrm{Hz}}$$

This frequency is in the infrared region of the electromagnetic spectrum and hence is comparable to the frequencies and energies of molecular vibrations; it is much greater than those of molecular rotations, and far less than those of molecular electronic motion.

P14D.3 Refer to the figure in the solution to Exercise 14.D.4(a). The width of the CH_3 spectrum is $3a_H = \boxed{6.9\,\mathrm{mT}}$. The width of the CD_3 spectrum is $6a_D$. It seems reasonable to assume, since the hyperfine interaction is an interaction of the magnetic moments of the nuclei with the magnetic moment of the electron, that the strength of the interactions is proportional to the nuclear moments.

$$\mu = g_I\mu_N I \quad \text{or} \quad \mu_z = g_I\mu_N m_I [14A.4c]$$

and thus nuclear magnetic moments are proportional to the nuclear g-values; hence

$$a_D \approx \frac{0.85745}{5.5857} \times a_H = 0.1535 a_H = 0.35\,\text{mT}$$

Therefore, the overall width is $6a_D = \boxed{2.1\,\text{mT}}$

P14D.5 For $C_6H_6^-$, $a = Q\rho$ with $Q = 2.25\,\text{mT}$ [14D.5]. If we assume that the value of Q does not change from this value (a good assumption in view of the similarity of the anions), we may write

$$\rho = \frac{a}{Q} = \frac{a}{2.25\,\text{mT}}$$

Hence, we can construct the following maps

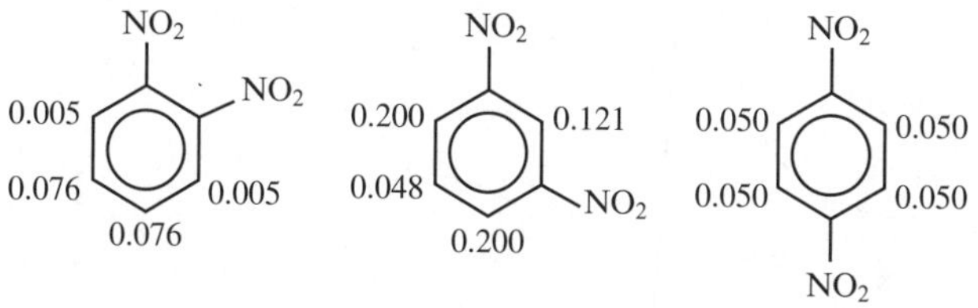

NO_2, NO_2; 0.005, 0.076, 0.076, 0.005

NO_2, NO_2; 0.200, 0.048, 0.200, 0.121

NO_2, NO_2; 0.050, 0.050, 0.050, 0.050

P14D.7 When spin label molecules approach to within 800 pm, orbital overlap of the unpaired electrons and dipolar interactions between magnetic moments cause an exchange coupling interaction between the spins. The electron exchange process occurs at a rate that increases as concentration increases. Thus the process has a lifetime that is too long at low concentrations to affect the 'pure' ESR signal. As the concentration increases, the linewidths increase until the triplet coalesces into a broad singlet. Further increase of the concentration decreases the exchange lifetime and therefore the linewidth of the singlet (see Fig. 14D.3 overleaf).

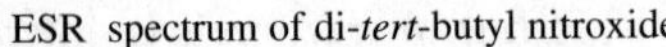

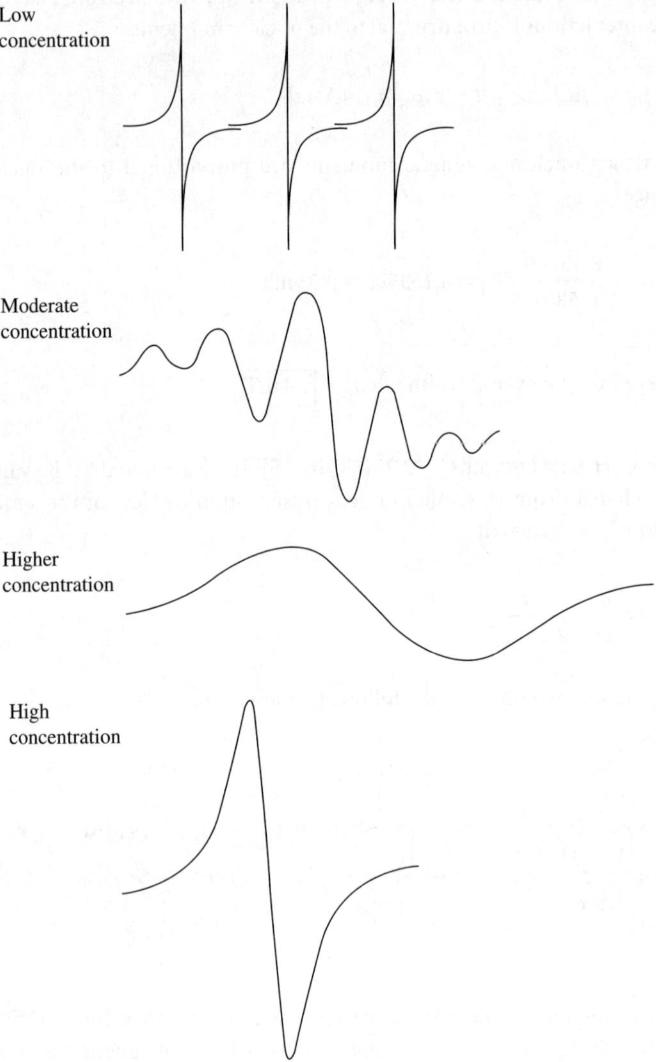

Figure 14D.3

When spin labels within biological membranes are highly mobile, they may approach closely and the exchange interaction may provide the ESR spectra with information that mimics the moderate and high concentration signals above.

Integrated activities

14.1 (a) The first table below displays experimental ^{13}C chemical shifts and computed* atomic charges on the carbon atom *para* to a number of substituents in substituted benzenes. Two sets of charges are shown, one derived by fitting the electrostatic potential and the other by Mulliken population analysis.

Substituent	CH_3	H	CF_3	CN	NO_2
δ	128.4	128.5	128.9	129.1	129.4
Electrostatic charge/e	−0.1273	−0.0757	−0.0227	−0.0152	−0.0541
Mulliken charge/e	−0.1089	−0.1021	−0.0665	−0.0805	−0.0392

*Semi-empirical, PM3 level, PC Spartan Pro™

In the table below we have recalculated net charges at a higher level than the semi-empirical PM3 level displayed above. This displays both the experimental and calculated* (HF-SCF/6-311G*) ^{13}C chemical shifts and computed* atomic charges on the carbon atom *para* to a number of substituents in substituted benzenes. Three sets of charges are shown, one derived by fitting the electrostatic potential, another by Mulliken population analysis, and the other by the method of 'natural' charges.

Substituent	CH_3	H	CF_3	CN	NO_2
δ_{exp}	128.4	128.5	128.9	129.1	129.4
δ_{calc}^{*}	134.4	133.5	132.1	138.8	141.8
Electrostatic charge*/e	−0.240	−0.135	−0.138	−0.102	−0.116
Mulliken charge*/e	−0.231	−0.217	−0.205	−0.199	−0.182
Natural charge*/e	−0.199	−0.183	−0.158	−0.148	−0.135

Spartan'06™; HF-SCF/6-311G

(b) Neither set of charges correlates well to the chemical shifts; however, some correlation is apparent, particularly for the Mulliken charges. See Fig. I14.1.

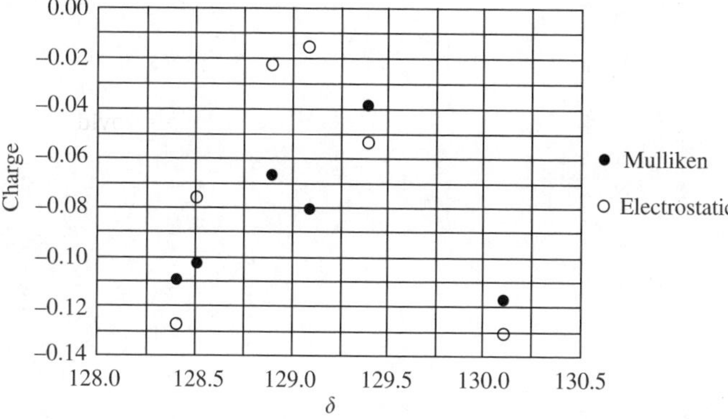

Figure I14.1

(c) & (d) The diamagnetic local contribution to shielding is roughly proportional to the electron density on the atom. The extent to which the *para*-carbon atom is affected by electron-donating or withdrawing groups on the other side of the benzene ring is reflected in the net charge on the atom. If the diamagnetic local contribution dominated, then the more positive the atom, the greater the deshielding and the greater the chemical shift δ would be. That no good correlation is observed leads to several possible hypotheses: for example, the diamagnetic local contribution is not the dominant contribution in these molecules (or not in all of these molecules), or the computation is not sufficiently accurate to provide meaningful atomic charges. See the solution to Problem F6.15 for additional discussion.

I14.3 When $\nu = 60$ MHz

$$\tau_J = \frac{\sqrt{2}}{\pi\delta\nu} \approx \frac{\sqrt{2}}{\pi\Delta\nu} = \frac{\sqrt{2}}{\pi \times \left((5.2-4.0)\times 10^{-6}\right) \times \left(60\times 10^{6}\,\text{Hz}\right)}$$

$\approx 6.25\,\text{ms}$, corresponding to a jump rate of $160\,\text{s}^{-1}$.

When $\nu = 300$ MHz

$$\tau_J = \frac{\sqrt{2}}{\pi\delta\nu} \approx \frac{\sqrt{2}}{\pi\Delta\nu} = \frac{\sqrt{2}}{\pi \times \left((5.2-4.0)\times 10^{-6}\right) \times \left(300\times 10^{6}\,\text{Hz}\right)}$$

$\approx 1.25\,\text{ms}$, corresponding to a jump rate of $8.0\times 10^{2}\,\text{s}^{-1}$.

Assume an Arrhenius-like jumping process (Chapter 20)

$$\text{rate} \propto e^{-E_a/RT}$$

Then, $\ln\left[\dfrac{\text{rate}(T')}{\text{rate}(T)}\right] = \dfrac{-E_a}{R}\left(\dfrac{1}{T'} - \dfrac{1}{T}\right)$

and therefore $E_a = \dfrac{R\ln(r'/r)}{\dfrac{1}{T} - \dfrac{1}{T'}} = \dfrac{8.314\,\text{J}\,\text{K}^{-1}\,\text{mol}^{-1} \times \ln\dfrac{800}{160}}{\dfrac{1}{280\,\text{K}} - \dfrac{1}{300\,\text{K}}} = \boxed{56\,\text{kJ}\,\text{mol}^{-1}}$

15 Statistical thermodynamics

15A The Boltzmann distribution

Answers to discussion questions

D15A.1 The **population** of a state is the number of molecules of a sample that are *in* that state, on average. The population of a state is the number of molecules in the sample times the probability of the state. The **configuration** of a system is a list of populations in order of the energy of the corresponding states. For example, $\{N-3, 2, 1, 0, \ldots\}$ is a possible configuration of a system of N molecules in which all but three molecules are in the ground state, two are in the next lowest state, one in the next state, etc. The **weight** of a configuration is the number of ways a given configuration can be comprised. In our example, the single molecule in the second excited state could be any of the system's N molecules, so there are N ways to arrange that state alone. The weight of the configuration $\{N_0, N_1, N_2, \ldots\}$ is

$$\mathcal{W} = \frac{N!}{N_0!N_1!N_2!\cdots} \quad [15\text{A}.1]$$

When N is large (as it is for any macroscopic sample), the most probable configuration is so much more probable than all other possible configurations that it is the system's dominant configuration. See Section 15A.1.

D15A.3 Because this chapter focuses on the application of statistics to the distribution of physical states in systems that contain a large number of atoms or molecules, we begin with a statistical answer: the thermodynamic temperature is the one quantity that determines the most probable populations of those states in systems at thermal equilibrium (Section 15A.1(b)). As a consequence, the temperature provides a necessary condition for thermal equilibrium; a system is at thermal equilibrium only if all of its sub-systems have the same temperature. Note that this is not a circular definition of temperature, for thermal equilibrium is not defined by uniformity of temperature: systems whose sub-systems can exchange energy tend toward thermal equilibrium. In this context, sub-systems can be different materials placed in contact (such as a block of copper in a beaker of water) or can be more abstract (such as rotational and vibrational modes of motion).

Finally, the equipartition theorem allows us to connect the temperature of statistical thermodynamics to the empirical concept of temperature developed long beforehand. Temperature is a measure of the intensity of thermal energy, directly proportional to the mean energy for each quadratic contribution to the energy (provided that the temperature is sufficiently high).

Solutions to exercises

E15A.1(a) The weight is given by

$$\mathcal{W}=\frac{N!}{N_0!N_1!N_2!\cdots}=\frac{16!}{0!1!2!3!8!0!0!0!0!2!}$$

This can be simplified by removing the common factor of 8! from the numerator and denominator and noting that 0! = 1! = 1:

$$\mathcal{W}=\frac{16\times15\times14\times13\times12\times11\times10\times9}{2\times(3\times2)\times2}=\boxed{21\,621\,600}$$

E15A.2(a) (i) $8!=8\times7\times6\times5\times4\times3\times2\times1=\boxed{40\,320}$ exactly.

(ii) According to Stirling's simple approximation [15A.2b],

$$\ln x!\approx x\ln x-x$$

$$\text{so}\ \ln 8!\approx 8\ln 8-8=8.636 \quad\text{and}\quad 8!\approx e^{8.636}=\boxed{5.63\times10^3}.$$

(iii) According to Stirling's better approximation [15A.2a],

$$x!\approx(2\pi)^{1/2}x^{x+1/2}e^{-x} \quad\text{so}\quad 8!\approx(2\pi)^{1/2}8^{8.5}e^{-8}=\boxed{3.99\times10^4}.$$

E15A.3(a) For two non-degenerate levels,

$$\frac{N_2}{N_1}=\frac{e^{-\beta\varepsilon_2}}{e^{-\beta\varepsilon_1}}=e^{-\beta(\varepsilon_2-\varepsilon_1)}=e^{-\beta\Delta\varepsilon}=e^{-\Delta\varepsilon/kT} \qquad \left[15A.7a \text{ with } \beta=\frac{1}{kT}\right]$$

Hence, as $T\to\infty$, $\dfrac{N_2}{N_1}=e^{-0}=\boxed{1}$. That is, the two levels would become equally populated.

E15A.4(a) For two non-degenerate levels,

$$\frac{N_2}{N_1}=\frac{e^{-\beta\varepsilon_2}}{e^{-\beta\varepsilon_1}}=e^{-\beta(\varepsilon_2-\varepsilon_1)}=e^{-\beta\Delta\varepsilon}=e^{-\Delta\varepsilon/kT} \qquad \left[15A.7a \text{ with } \beta=\frac{1}{kT}\right]$$

$$\text{so}\ \ln\frac{N_2}{N_1}=-\frac{\Delta\varepsilon}{kT} \quad\text{and}\quad T=-\frac{\Delta\varepsilon}{k\ln\dfrac{N_2}{N_1}}$$

$$\text{Thus}\ T=-\frac{6.626\times10^{-34}\ \text{J s}\times2.998\times10^{10}\ \text{cm s}^{-1}\times400\ \text{cm}^{-1}}{1.381\times10^{-23}\ \text{J K}^{-1}\times\ln(1/3)}=\boxed{524\ \text{K}}$$

E15A.5(a) See Example 15A.1. The ratio of populations of a particular *state* at the $J=5$ level to the population of the non-degenerate $J=0$ level is

$$\frac{N_2}{N_1}=\frac{e^{-\beta\varepsilon_2}}{e^{-\beta\varepsilon_1}}=e^{-\beta(\varepsilon_2-\varepsilon_1)}=e^{-\beta\Delta\varepsilon}=e^{-\Delta\varepsilon/kT} \qquad \left[15A.7\text{ a with } \beta=\frac{1}{kT}\right]$$

Because all of the states of a degenerate level are equally likely, the ratio of populations of a particular *level* is

$$\frac{N_5}{N_0}=\frac{g_5\mathrm{e}^{-\beta\varepsilon_5}}{g_0\mathrm{e}^{-\beta\varepsilon_0}}=\frac{g_5}{g_0}\mathrm{e}^{-(\varepsilon_5-\varepsilon_0)/kT}.$$

The degeneracy of linear rotor energy levels are

$$g_J=(2J+1)$$

and its energy levels are [12B.15]

$$\varepsilon_J=hc\tilde{B}J(J+1)$$

Thus, using $kT/hc=207.224\ \mathrm{cm^{-1}}$ at 298.15 K,

$$\frac{N_5}{N_0}=\frac{g_5}{g_0}\mathrm{e}^{-5(5+1)hc\tilde{B}/kT}=\frac{(2\times5+1)}{(2\times0+1)}\mathrm{e}^{-5(5+1)\times2.71\ \mathrm{cm^{-1}}/207.224\ \mathrm{cm^{-1}}}=\boxed{7.43}$$

E15A.6(a) For two non-degenerate levels,

$$\frac{N_2}{N_1}=\frac{\mathrm{e}^{-\beta\varepsilon_2}}{\mathrm{e}^{-\beta\varepsilon_1}}=\mathrm{e}^{-\beta(\varepsilon_2-\varepsilon_1)}=\mathrm{e}^{-\beta\Delta\varepsilon}=\mathrm{e}^{-\Delta\varepsilon/kT}\qquad\left[15\mathrm{A.7a\ with\ }\beta=\frac{1}{kT}\right]$$

so, assuming that other states (if any) are negligibly populated,

$$\ln\frac{N_2}{N_1}=-\frac{\Delta\varepsilon}{kT}\quad\text{and}\quad T=-\frac{\Delta\varepsilon}{k\ln\frac{N_2}{N_1}}$$

$$\text{Thus }T=-\frac{6.626\times10^{-34}\ \mathrm{J\,s}\times2.998\times10^{10}\ \mathrm{cm\,s^{-1}}\times540\ \mathrm{cm^{-1}}}{1.381\times10^{-23}\ \mathrm{J\,K^{-1}}\times\ln(10/90)}=\boxed{35\bar{4}\ \mathrm{K}}$$

Solutions to problems

P15A.1 (b) We draw up the following table

0	ε	2ε	3ε	4ε	5ε	$\mathcal{W}=\frac{N!}{n_1!n_2!\ldots}$ [15A.1]
4	0	0	0	0	1	5
3	1	0	0	1	0	20
3	0	1	1	0	0	20
2	2	0	1	0	0	30
2	1	2	0	0	0	30
1	3	1	0	0	0	20
0	5	0	0	0	0	1

The most probable configurations are $\boxed{\{2,2,0,1,0,0\}}$ and $\boxed{\{2,1,2,0,0,0\}}$.

(a) There is no configuration in which the molecules are distributed evenly over the available states. The distribution closest to uniform would have two states with two molecules each and one state with one molecule. Those are precisely the two configurations identified in part (b).

P15A.3 Listing all possible configurations for a 20-particle system would be very time-consuming and tedious indeed; however, listing representative configurations for a given total energy is manageable if done systematically. By 'representative', list only one configuration that has a given weight.

For example, consider systems that have a total energy of 10ε, where ε is the separation between adjacent energy levels. There are many distinct configurations that have 17 particles in the ground state and one particle in each of three different states, including configurations in which the singly-occupied levels are (ε, 2ε, 7ε), (2ε, 3ε, 5ε), and (ε, 4ε, 5ε). Any one of these configurations is 'representative' of all of them, however, because they all have the same weight, namely

$$\mathcal{W}=\frac{N!}{N_0!N_1!N_2!\cdots}[15A.1]=\frac{20!}{17!1!1!1!}=20\times19\times18=6\,840$$

So it is sufficient to enumerate just one of the configurations whose occupancy numbers are 17, 1, 1, and 1 (to make sure that one such configuration exists consistent with the desired total energy). A systematic way of keeping track of representative configurations is to lower maximum occupancy numbers. The next set of occupancy numbers to look at would be 16, 4; then 16, 3, 1; 16, 2, 2; 16, 2, 1, 1; and 16, 1, 1, 1, 1. We would eliminate 16, 4, because no such occupancy numbers yield a total energy of 10ε.

Set up a spreadsheet to generate one of each kind of representative configuration, while keeping the total number of particles constant at 20 and total energy constant (at 10ε for the moment). Again, the most systematic way to do this is to give the highest occupancy numbers to the lowest-energy available states. (This rule also generates the most 'exponential' configurations.) For example, out of several configurations corresponding to occupancy numbers of 16, 3, and 1, the one used is $N_0=16$, $N_1=3$, and $N_7=1$.

Next examine systems of total energy 10ε, 15ε, and 20ε. The most probable configurations and corresponding weights are:

$$10\varepsilon:\{N_0=12, N_1=6, N_2=2, N_3=0,\ldots\}\ \mathcal{W}=\frac{20!}{12!6!2!}=3\,527\,160$$

$$15\varepsilon:\{N_0=10, N_1=6, N_2=3, N_3=1, N_4=0,\ \ldots\}\ \mathcal{W}=\frac{20!}{10!6!3!1!}=155\,195\,040$$

$$20\varepsilon:\{N_0=10, N_1=4, N_2=3, N_3=2, N_4=1, N_5=0,\ \ldots\}$$

$$\mathcal{W}=\frac{20!}{10!4!3!2!1!}=2\,327\,925\,600$$

and $\{N_0=9, N_1=6, N_2=2, N_3=2, N_4=1, N_5=0,\ \ldots\}$

$$\mathcal{W}=\frac{20!}{9!6!2!2!1!}=2\,327\,925\,600$$

As the total energy increases, the most probable configuration has more occupied levels and occupancy of higher-energy levels.

The Boltzmann distribution would predict the following relative probabilities for equally spaced energy levels above the ground state:

$$\frac{p_j}{p_0} = e^{-\beta j\varepsilon} \text{ [15A.7a] so } \ln\frac{p_j}{p_0} = -\beta j\varepsilon$$

Thus, a plot of the natural log of relative probability vs. the ordinal number of the energy level (j) should be a straight line whose slope is $-\beta\varepsilon$. Using the nonzero occupancy numbers from the configuration, the plots are indeed roughly linear. Furthermore, the value of β decreases with increasing total energy, corresponding to an increase in temperature.

P15A.5 Look immediately after *Justification* 15A.1, to the expression for ln $\mathcal{W}$:

$$\ln\mathcal{W} = \ln N! - \sum_j \ln N_j!$$

Substitute the full version of Stirling's approximation:

$$\begin{aligned}\ln\mathcal{W} &= \ln\{(2\pi)^{1/2}N^{N+1/2}\mathrm{e}^{-N}\} - \sum_j \ln\{(2\pi)^{1/2}N_j^{N_j+1/2}\mathrm{e}^{-N_j}\}\\ &= \tfrac{1}{2}\ln(2\pi) + (N+\tfrac{1}{2})\ln N - N - \sum_j\left\{\tfrac{1}{2}\ln(2\pi) + (N_j+\tfrac{1}{2})\ln N_j - N_j\right\}\\ &= \tfrac{1}{2}\ln(2\pi) + (N+\tfrac{1}{2})\ln N - \sum_j\left\{\tfrac{1}{2}\ln(2\pi) + (N_j+\tfrac{1}{2})\ln N_j\right\}.\end{aligned}$$

With this expression in hand, we turn to Section 15A.2, to the derivation of the Boltzmann distribution. There the method of undetermined multipliers still yields

$$0 = \frac{\partial\ln\mathcal{W}}{\partial N_i} + \alpha - \beta\varepsilon_i \quad [15\text{A}.9]$$

Evaluating the derivative is similar with the more accurate expression for $\mathcal{W}$ as with the more approximate form. First of all, derivatives of the 2π terms vanish, leaving

$$\frac{\partial\ln\mathcal{W}}{\partial N_i} = \frac{\partial\{(N+\frac{1}{2})\ln N\}}{\partial N_i} - \sum_j \frac{\partial\{(N_j+\frac{1}{2})\ln N_j\}}{\partial N_i}$$

The first term is

$$\begin{aligned}\frac{\partial\{(N+\frac{1}{2})\ln N\}}{\partial N_i} &= \left(\frac{\partial(N+\frac{1}{2})}{\partial N_i}\right)\ln N + (N+\tfrac{1}{2})\left(\frac{\partial\ln N}{\partial N_i}\right)\\ &= \ln N + \left(\frac{N+\frac{1}{2}}{N}\right)\left(\frac{\partial N}{\partial N_i}\right) = \ln N + \left(\frac{N+\frac{1}{2}}{N}\right),\end{aligned}$$

which differs from the corresponding term in *Justification* 15A.2 only insomuch as $\frac{N+\frac{1}{2}}{N}$ differs from 1. In macroscopic samples, N is so large for this difference to be utterly negligible. Similarly

$$\sum_j \frac{\partial\{(N_j+\frac{1}{2})\ln N_j\}}{\partial N_i} = \ln N_i + \left(\frac{N_i+\frac{1}{2}}{N_i}\right)$$

Again, in macroscopic samples, the last term differs from unity only negligibly *for any states that have a reasonable probability of occupation.* For those samples, the remainder of the derivation is exactly as in Section 15A.2. Any deviations are limited to states of exceedingly low occupancy—the extreme tail of the distribution.

P15A.7 $$\frac{p(h)}{p(h_0)} = \frac{n(h)RT/V}{n(h_0)RT/V} = \mathrm{e}^{-\{(\varepsilon(h)-\varepsilon(h_0))/kT\}}[15\mathrm{A.7a}] = \mathrm{e}^{-mg(h-h_0)/kT}$$

Defining $p(h=0) \equiv p_0$, we obtain the desired barometric formula:

$$\frac{p(h)}{p_0} = \mathrm{e}^{-mgh/kT} = \boxed{\mathrm{e}^{-Mgh/RT}} = \frac{\mathcal{N}(h)}{\mathcal{N}(0)}$$

Note that the result depends on the temperature, and it assumes that the temperature does not vary with height. To proceed, we must pick a temperature, so we use the standard temperature of 298 K—which is reasonable for the surface of the earth—but not for 8.0 km altitude.

For oxygen at 8.0 km,

$$\frac{M(\mathrm{O_2})gh}{RT} = \frac{(0.0320\ \mathrm{kg\ mol^{-1}})\times(9.81\ \mathrm{m\ s^{-2}})\times(8.0\times10^3\ \mathrm{m})}{(8.3145\ \mathrm{J\ K^{-1}\ mol^{-1}})\times(298\ \mathrm{K})} = 1.01$$

$$\text{so}\ \frac{\mathcal{N}(8.0\ \mathrm{km})}{\mathcal{N}(0)} = \mathrm{e}^{-1.01} = \boxed{0.363}$$

For water,

$$\frac{M(\mathrm{H_2O})gh}{RT} = \frac{(0.0180\ \mathrm{kg\ mol^{-1}})\times(9.81\ \mathrm{m\ s^{-2}})\times(8.0\times10^3\ \mathrm{m})}{(8.3145\ \mathrm{J\ K^{-1}\ mol^{-1}})\times(298\ \mathrm{K})} = 0.57$$

$$\text{so}\ \frac{\mathcal{N}(8.0\,\mathrm{km})}{\mathcal{N}(0)} = \mathrm{e}^{-0.57} = \boxed{0.57}$$

15B Molecular partition functions

Answers to discussion questions

D15B.1 The molecular partition is roughly equal to the number of physically distinct states thermally accessible to a molecule at a given temperature. At low temperatures, very little energy is available, so only the lowest-energy states of a molecule are accessible; therefore, as the temperature approaches absolute zero, the partition function approaches the degeneracy of the molecule's ground state. The higher the temperature, the greater the Boltzmann weighting factor $e^{-\beta\varepsilon}$ and the more accessible a state of energy ε becomes. Thus the number of accessible states increases with temperature. (The partition function is only 'roughly equal' to the number of accessible states because at any nonzero temperature, each state is accessible with a finite probability proportional to $e^{-\beta\varepsilon}$. Thus, to state whether or not a given state is 'accessible' at a given temperature is somewhat arbitrary, so an exactly counted number of accessible states is also arbitrary. Because the partition function is the sum of all of these fractional probabilities, it is a good estimate of this number.) See Section 15B.1.

D15B.3 Distinct 'states' differ in one or more observable physical quantity such as energy or angular momentum. Thus, distinct states may have the same energy (may belong to the same 'energy level') if they differ in some other quantity; these are degenerate states. Each energy level has at least one state—more for degenerate levels. The partition function is a sort of count of **states**; it is a sum over all **states** of each state's Boltzmann factor. Because Boltzmann factors depend only on energy, the sum may practically be done as a sum over energy levels, as long as the degeneracy is included. That is, the sum over states within a given energy level is a sum of identical Boltzmann factors, equal to the degeneracy of the level times the Boltzmann factor.

Solutions to exercises

E15B.1(a) (i) The thermal wavelength is [15B.7b]

$$\Lambda=\frac{h}{(2\pi mkT)^{1/2}}$$

We need the molecular mass, not the molar mass:

$$m=\frac{150\times10^{-3}\ \text{kg mol}^{-1}}{6.022\times10^{23}\ \text{mol}^{-1}}=2.49\times10^{-25}\ \text{kg}$$

$$\text{So } \Lambda=\frac{6.626\times10^{-34}\ \text{J s}}{(2\pi\times2.49\times10^{-25}\ \text{kg}\times1.381\times10^{-23}\ \text{J K}^{-1}\times T)^{1/2}}=\frac{1.43\times10^{-10}\ \text{m}}{(T/\text{K})^{1/2}}$$

$$T=300\ \text{K:}\ \Lambda=\frac{1.43\times10^{-10}\ \text{m}}{(300)^{1/2}}=\boxed{8.23\times10^{-12}\ \text{m}}=\boxed{8.23\ \text{pm}}$$

$$T=3000\ \text{K:}\ \Lambda=\frac{1.43\times10^{-10}\ \text{m}}{(3000)^{1/2}}=\boxed{2.60\times10^{-12}\ \text{m}}=\boxed{2.60\ \text{pm}}$$

(ii) The translational partition function is

$$q^{\text{T}}=\frac{V}{\Lambda^3}\ [15\text{B}.10\text{b}]$$

$$T = 300\text{ K: } q^{\mathrm{T}} = \frac{1.00\text{ cm}^3}{(8.23\times10^{-12}\text{ m})^3} = \frac{(1.00\times10^{-2}\text{ m})^3}{(8.23\times10^{-12}\text{ m})^3} = \boxed{1.79\times10^{27}}$$

$$T = 3\,000\text{ K: } q^{\mathrm{T}} = \frac{(1.00\times10^{-2}\text{ m})^3}{(2.60\times10^{-12}\text{ m})^3} = \boxed{5.67\times10^{28}}$$

E15B.2(a) $q^{\mathrm{T}} = \dfrac{V}{\Lambda^3}$ [15B.10b], implying that $\dfrac{q}{q'} = \left(\dfrac{\Lambda'}{\Lambda}\right)^3$

However, $\Lambda = \dfrac{h}{(2\pi mkT)^{1/2}}$ [52.7b] $\propto \dfrac{1}{m^{1/2}}$ so $\dfrac{q}{q'} = \left(\dfrac{m}{m'}\right)^{3/2}$

Therefore, $\dfrac{q_{\mathrm{H_2}}}{q_{\mathrm{He}}} = \left(\dfrac{2\times1.008}{4.003}\right)^{3/2} = \boxed{0.3574}$

E15B.3(a) The high-temperature expression for the rotational partition function of a linear molecule is

$$q^{\mathrm{R}} = \frac{kT}{\sigma hc\tilde{B}}\text{ [15B.13b]},\quad \tilde{B} = \frac{\hbar}{4\pi cI}\text{ [12B.7]},\quad I = \mu R^2\text{ [Table 12B.1]}$$

Hence $q = \dfrac{8\pi^2 kTI}{\sigma h^2} = \dfrac{8\pi^2 kT\mu R^2}{\sigma h^2}$

For O_2, $\mu = \frac{1}{2}m(\mathrm{O}) = \frac{1}{2}\times16.00m_{\mathrm{u}} = 8.00m_{\mathrm{u}}$, and $\sigma = 2$; therefore

$$q = \frac{(8\pi^2)\times(1.381\times10^{-23}\text{ J K}^{-1})\times(300\text{ K})\times(8.00\times1.6605\times10^{-27}\text{ kg})\times(1.2075\times10^{-10}\text{ m})^2}{(2)\times(6.626\times10^{-34}\text{ J s})^2}$$

$$= \boxed{72.2}$$

E15B.4(a) The high-temperature expression for the rotational partition function of a non-linear molecule is [15B.14]

$$q^{\mathrm{R}} = \frac{1}{\sigma}\left(\frac{kT}{hc}\right)^{3/2}\left(\frac{\pi}{\tilde{A}\tilde{B}\tilde{C}}\right)^{1/2}$$

$$= \left(\frac{1.381\times10^{-23}\text{ J K}^{-1}\times T}{6.626\times10^{-34}\text{ J s}\times2.998\times10^{10}\text{ cm s}^{-1}}\right)^{3/2}\left(\frac{\pi}{3.1752\times0.3951\times0.3505\text{ cm}^{-3}}\right)^{1/2}$$

$$= 1.549\times(T/\text{K})^{3/2}$$

(i) At 25 °C, $q^{\mathrm{R}} = 1.549\times(298)^{3/2} = \boxed{7.97\times10^3}$

(ii) At 100 °C, $q^{\mathrm{R}} = 1.549\times(373)^{3/2} = \boxed{1.12\times10^4}$

E15B.5(a) The rotational partition function of a nonsymmetrical linear molecule is

$$q^{\mathrm{R}} = \sum_J (2J+1)\mathrm{e}^{-hc\tilde{B}J(J+1)/kT}\quad\left[15\text{B}.11\text{ with }\beta = \frac{1}{kT}\right]$$

Use $\dfrac{hc\tilde{B}}{k} = \dfrac{6.626\times10^{-34}\text{ J s}\times2.998\times10^{10}\text{ cm s}^{-1}\times1.931\text{ cm}^{-1}}{1.381\times10^{-23}\text{ J K}^{-1}} = 2.778\text{ K},$

so $q^{\mathrm{R}} = \sum_J (2J+1)\mathrm{e}^{-2.778\ \mathrm{K}\times J(J+1)/T}$

Use a spreadsheet or other mathematical software to evaluate the terms of the sum and to sum the terms until they converge. The high-temperature expression is

$$q^{\mathrm{R}} = \frac{kT}{hc\tilde{B}} = \frac{T}{2.778\ \mathrm{K}}$$

The explicit and high-temperature expressions are compared in Fig. 15B.1. The high-temperature expression reaches 95% of the explicit sum at $\boxed{18\mathrm{K}}$.

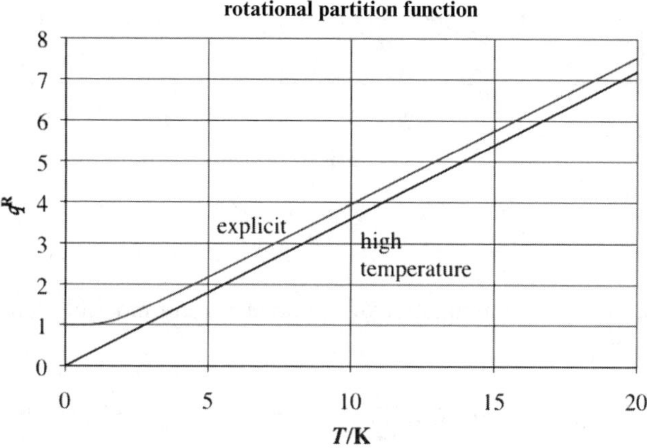

Figure 15B.1

E15B.6(a) The rotational partition function of a spherical rotor molecule, ignoring nuclear statistics, is

$$q^{\mathrm{R}} = \sum_J g_J \mathrm{e}^{-\varepsilon_J^{\mathrm{R}}/kT}\,[15\mathrm{B.1b}] = \sum_J (2J+1)^2\,\mathrm{e}^{-hc\tilde{B}J(J+1)/kT}\,[12\mathrm{B.8}]$$

Use $\dfrac{hc\tilde{B}}{k} = \dfrac{6.626\times10^{-34}\ \mathrm{J\,s}\times2.998\times10^{10}\ \mathrm{cm\,s^{-1}}\times5.241\ \mathrm{cm^{-1}}}{1.381\times10^{-23}\ \mathrm{J\,K^{-1}}} = 7.539\ \mathrm{K}$

so $q^{\mathrm{R}} = \sum_J (2J+1)^2\,\mathrm{e}^{-7.539\ \mathrm{K}\times J(J+1)/T}$

Use a spreadsheet or other mathematical software to evaluate the terms of the sum and to sum the terms until they converge. The high-temperature expression is eqn 15B.14, neglecting σ and with $\tilde{A} = \tilde{B} = \tilde{C}$:

$$q^{\mathrm{R}} = \pi^{1/2}\left(\frac{kT}{hc\tilde{B}}\right)^{3/2} = \pi^{1/2}\left(\frac{T}{7.539\ \mathrm{K}}\right)^{3/2}$$

The explicit and high-temperature expressions are compared in Fig. 15B.2. The high-temperature expression reaches 95% of the explicit sum at $\boxed{37\,\text{K}}$.

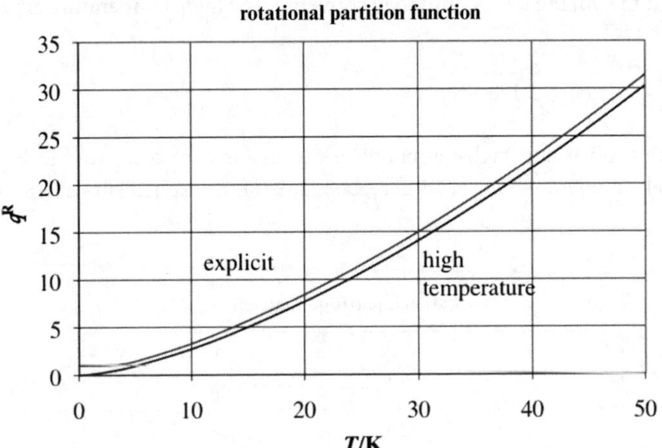

Figure 15B.2

E15B.7(a) The rotational partition function of a symmetric rotor molecule, ignoring nuclear statistics, is

$$q^R = \sum_{J,K} g_{J,K} e^{-\varepsilon^R_{J,K}/kT} [15B.1b] = \sum_{J=0} (2J+1)e^{-hc\tilde{B}J(J+1)/kT}\left(1+2\sum_{K=1}^{J} e^{-hc(\tilde{A}-\tilde{B})K^2/kT}\right) [12B.13]$$

Use $\dfrac{hc\tilde{B}}{k} = \dfrac{6.626\times10^{-34}\ \text{J s}\times 2.998\times10^{10}\ \text{cm s}^{-1}\times 0.443\ \text{cm}^{-1}}{1.381\times10^{-23}\ \text{J K}^{-1}} = 0.637\ \text{K}$, and

$$\frac{hc(\tilde{A}-\tilde{B})}{k} = \frac{6.626\times10^{-34}\ \text{J s}\times 2.998\times10^{10}\ \text{cm s}^{-1}\times(5.097-0.443)\ \text{cm}^{-1}}{1.381\times10^{-23}\ \text{J K}^{-1}} = 6.694\ \text{K}$$

so $q^R = \displaystyle\sum_{J=0} (2J+1)e^{-0.637\ \text{K}\times J(J+1)/T}\left(1+2\sum_{K=1}^{J} e^{-6.694\ \text{K}\times K^2/T}\right)$

Write a brief computer program or use other mathematical software to evaluate the terms of the sum and to sum the terms until they converge. Nested sums are straightforward to program in languages such as BASIC or FORTRAN, whereas spreadsheets are more unwieldy. Compare the results of the direct sum with the high-temperature expression, eqn 15B.14, with $\tilde{B}=\tilde{C}$:

$$q^R = \left(\frac{\pi}{\tilde{A}}\right)^{1/2}\left(\frac{kT}{hc}\right)^{3/2}\frac{1}{\tilde{B}}$$

The explicit and high-temperature expressions are compared in Fig. 15B.3. The high-temperature expression reaches 95% of the explicit sum at $\boxed{4.5\,\text{K}}$.

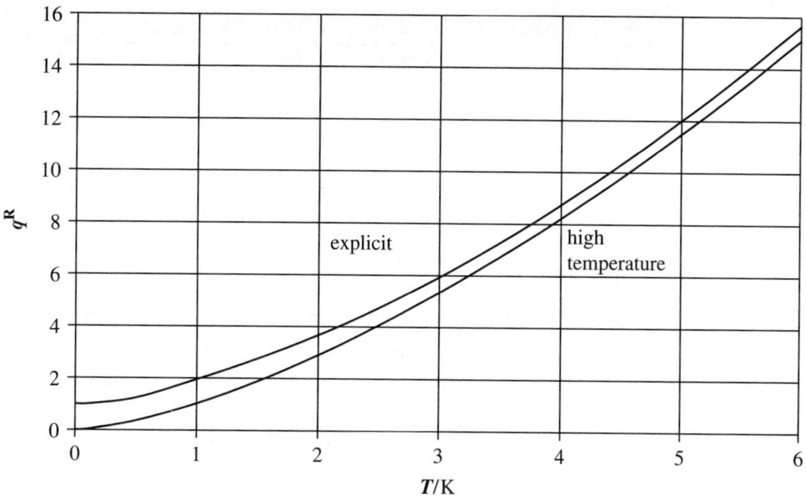

Figure 15B.3

E15B.8(a) The symmetry number is the order of the rotational subgroup of the group to which a molecule belongs (except for linear molecules, for which $\sigma = 2$ if the molecule has inversion symmetry and 1 otherwise). See Problem 15B.9.

(i) CO: Full group $C_{\infty v}$; subgroup C_1; hence $\sigma = \boxed{1}$

(ii) O_2: Full group $D_{\infty h}$; subgroup C_2; $\sigma = \boxed{2}$

(iii) H_2S: Full group C_{2v}; subgroup C_2; $\sigma = \boxed{2}$

(iv) SiH_4: Full group T_d; subgroup T; $\sigma = \boxed{12}$

(v) $CHCl_3$: Full group C_{3v}; subgroup C_3; $\sigma = \boxed{3}$

E15B.9(a) Ethene has four indistinguishable atoms that can be interchanged by rotations, so $\sigma = 4$. The high-temperature expression for the rotational partition function of a nonlinear molecule is [15B.14]

$$q^R = \frac{1}{\sigma}\left(\frac{kT}{hc}\right)^{3/2}\left(\frac{\pi}{\tilde{A}\tilde{B}\tilde{C}}\right)^{1/2}$$

$$= \frac{1}{4}\left(\frac{1.381\times10^{-23}\ \mathrm{J\,K^{-1}}\times 298.15\ \mathrm{K}}{6.626\times10^{-34}\ \mathrm{J\,s}\times 2.998\times10^{10}\ \mathrm{cm\,s^{-1}}}\right)^{3/2}\left(\frac{\pi}{4.828\times1.0012\times0.8282\ \mathrm{cm^{-3}}}\right)^{1/2}$$

$$= \boxed{660.6}$$

E15B.10(a) The partition function for a mode of molecular vibration is

$$q^V = \sum_v \mathrm{e}^{-vhc\tilde{\nu}/kT} = \frac{1}{1-\mathrm{e}^{-hc\tilde{\nu}/kT}} \quad [15B.15 \text{ with } \beta = 1/kT]$$

Use $\dfrac{hc\tilde{\nu}}{k} = \dfrac{6.626\times10^{-34}\,\mathrm{J\,s}\times 2.998\times10^{10}\,\mathrm{cm\,s^{-1}}\times 323.2\,\mathrm{cm^{-1}}}{1.381\times10^{-23}\,\mathrm{J\,K^{-1}}} = 464.9\,\mathrm{K}$,

so $q^{\mathrm{V}} = \sum_{v} \mathrm{e}^{-vhc\tilde{\nu}/kT} = \dfrac{1}{1-\mathrm{e}^{-464.9\,\mathrm{K}/T}}$

The high-temperature expression is

$$q^{\mathrm{V}} = \frac{kT}{hc\tilde{\nu}} = \frac{T}{464.9\,\mathrm{K}} \quad [15\mathrm{B}.16]$$

The explicit and high-temperature expressions are compared in Fig. 15B.4. The high-temperature expression reaches 95% of the explicit sum at $\boxed{4\,500\,\mathrm{K}}$.

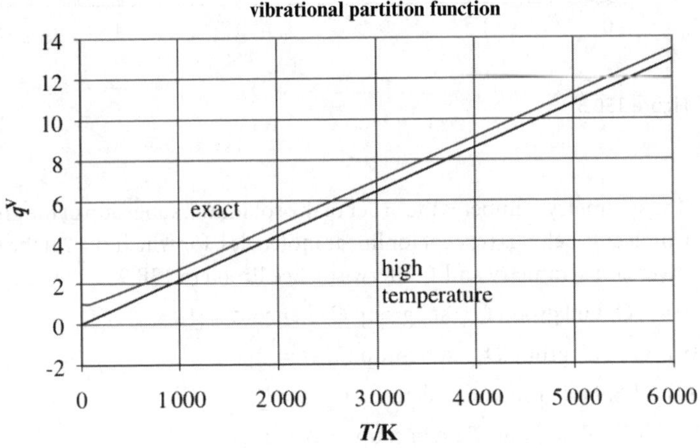

Figure 15B.4

E15B.11(a) The partition function for a mode of molecular vibration is

$$q^{\mathrm{V}} = \frac{1}{1-\mathrm{e}^{-hc\tilde{\nu}/kT}} \quad [15\mathrm{B}.15 \text{ with } \beta = 1/kT]$$

and the overall vibrational partition function is the product of the partition functions of the individual modes. (See Example 15B.2.) We draw up the following table:

Mode	1	2	3	4
$\tilde{\nu}/\mathrm{cm}^{-1}$	658	397	397	1535
$hc\tilde{\nu}/kT$	1.893	1.142	1.142	4.416
$q^{\mathrm{V}}_{\mathrm{mode}}$	1.177	1.469	1.469	1.012

The overall vibrational partition function is

$$q^{\mathrm{V}} = 1.177\times1.469\times1.469\times1.012 = \boxed{2.571}$$

E15B.12(a) The partition function for a mode of molecular vibration is

$$q^{V} = \frac{1}{1-e^{-hc\tilde{\nu}/kT}} \quad [15B.15 \text{ with } \beta = 1/kT]$$

and the overall vibrational partition function is the product of the partition functions of the individual modes. (See Example 15B.2.) We draw up the following table, including the degeneracy of each level:

Mode	1	2	3	4
$\tilde{\nu}$ / cm^{-1}	459	217	776	314
g_{mode}	1	2	3	3
$hc\tilde{\nu}/kT$	1.320	0.624	2.232	0.903
q^{V}_{mode}	1.364	2.15	1.120	1.681

The overall vibrational partition function is

$$q^{V} = 1.364\times 2.15^2 \times 1.120^3 \times 1.681^3 = \boxed{42.3}$$

E15B.13(a) $q = \sum_{\text{levels}} g_j e^{-\beta\varepsilon_j}[15\text{B.1b}] = \sum_{\text{levels}} g_j e^{-hc\tilde{\nu}_j/kT} = 4 + e^{-hc\tilde{\nu}_1/kT} + 2e^{-hc\tilde{\nu}_2/kT}$

where $\dfrac{hc\tilde{\nu}_j}{kT} = \dfrac{6.626\times 10^{-34}\,\text{J s}\times 2.998\times 10^{10}\,\text{cm s}^{-1}\times \tilde{\nu}_j}{1.381\times 10^{-23}\,\text{J K}^{-1}\times 1900\,\text{K}} = 7.571\times 10^{-4}\times(\tilde{\nu}_j/\text{cm}^{-1})$

Therefore

$$q = 4 + e^{-7.571\times 10^{-4}\times 2500} + 2e^{-7.571\times 10^{-4}\times 3500} = 4 + 0.151 + 2\times 0.0707 = \boxed{4.292}$$

The individual terms in the last expression are the relative populations of the levels, namely 2′0.0707 to 0.151 to 4 (second excited level to first to ground) or $\boxed{0.0353 \text{ to } 0.0377 \text{ to } 1}$

Solutions to problems

P15B.1 This problem can be carried out on a spreadsheet if care is taken with the layout. One may simply pick values of $\tilde{\nu}$; however, if one works in terms of the characteristic vibrational temperature, θ^{V}, one can employ more general dimensionless quantities as described below.

$$\theta^{V} = \frac{hc\tilde{\nu}}{k}$$

We note that the energy levels of the Morse oscillator can be written as

$$E_v = (v+\tfrac{1}{2})k\theta^{V} - (v+\tfrac{1}{2})^2 x_e k\theta^{V} = (v+\tfrac{1}{2})k\theta^{V}\{1-(v+\tfrac{1}{2})x_e\}$$

Thus, one can tabulate values of $E_v/k\theta^V$ without having to select a wavenumber. Similarly, one can employ the dimensionless temperature T/θ^V. As noted in the problem, Boltzmann factors require energies measured with respect to the ground state, so the energies in the exponents of the Boltzmann factors must be $E_v - E_0$. The energy expression for the Morse oscillator eventually reaches a maximum in v and then begins decreasing. Only the states up to and including the maximum energy are physically meaningful, so there are a finite number of Morse states. Thus, the partition function for a Morse oscillator is:

$$q = \sum_{v=0}^{v_{\max}} \exp\left(-\frac{E_v - E_0}{kT}\right) = \sum_{v=0}^{v_{\max}} \exp\left(-\frac{E_v - E_0}{k\theta^V} \times \frac{\theta^V}{T}\right)$$

To choose meaningful values of the anharmonicity, x_e, look up vibrational constants for some common diatomic molecules. x_e is about 0.03 for H_2 and about an order of magnitude smaller for I_2.

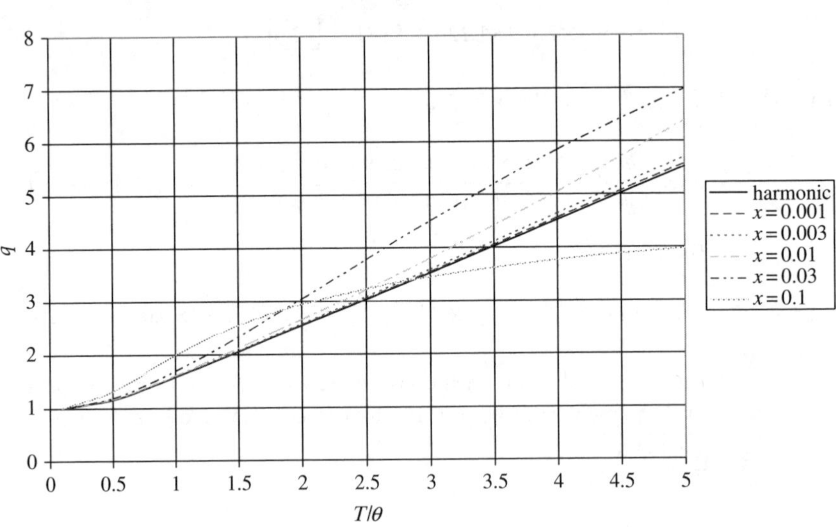

Figure 15B.5

A plot of partition functions with various anharmonicities is shown in Fig. 15B.5. For small values of x_e, the partition function closely resembles that of a harmonic oscillator [15B.15]. This provides a check on the calculation, for a Morse oscillator in the limit of small x_e **becomes** a harmonic oscillator. As x_e increases, the partition function gradually increases compared to the harmonic oscillator. This reflects the fact that the Morse oscillator energy levels become more closely spaced with increasing energy, so more levels are accessible at a given temperature. Eventually, however, the partition functions of highly anharmonic Morse oscillators fall below the harmonic curve at high temperatures. This reflects the fact that these Morse oscillators have a finite number of energy levels (indeed, a small number) so naturally a harmonic oscillator has more accessible levels at high temperature.

P15B.3 (a) $q=\sum_j g_j e^{-\beta\varepsilon_j}$[15B.1b]$=\sum_j g_j e^{-hc\beta\tilde{\nu}_j}$

We use

$$hc\beta=\frac{6.626\times10^{-34}\,\text{J s}\times2.998\times10^{10}\,\text{cm s}^{-1}}{1.381\times10^{-23}\,\text{J K}^{-1}\times298\,\text{K}}=\frac{1}{207\,\text{cm}^{-1}}\text{ at }298\,\text{K}$$

and $hc\beta=\dfrac{1}{3\,476\,\text{cm}^{-1}}$ at 5000 K. Therefore,

(i) $q=5+e^{-4707/207}+3e^{-4751/207}+5e^{-10559/207}$
$=(5)+(1.3\times10^{-10})+(3.2\times10^{-10})+(3.5\times10^{-22})=\boxed{5.00}$

(ii) $q=5+e^{-4707/3476}+3e^{-4751/3476}+5e^{-10559/3476}$
$=(5)+(0.26)+(0.76)+(0.24)=\boxed{6.26}$

(b) The proportion of atoms in energy level j is [15A.6, with degeneracy g_j included]

$$\frac{N_j}{N}=\frac{g_j e^{-\beta\varepsilon_j}}{q}=\frac{g_j e^{-hc\beta\tilde{\nu}_j}}{q}$$

Therefore, $\dfrac{N_0}{N}=\dfrac{5}{q}=\boxed{1.00}$ at 298 K and $\boxed{0.80}$ at 5000 K.

$$\frac{N_2}{N}=\frac{3e^{-4751/207}}{5.00}=\boxed{6.58\times10^{-11}}\text{ at }298\,\text{K}$$

$$\frac{N_2}{N}=\frac{3e^{-4751/3476}}{6.26}=\boxed{0.122}\text{ at }5000\,\text{K}$$

P15B.5 The partition function is

$$q=\sum_J g_J e^{-\beta\varepsilon_J}[15B.1b]=\sum_J g_J e^{-hc\beta\tilde{\nu}_J}=\sum_J g_J e^{-hc\tilde{\nu}_J/kT}=\sum_J(2J+1)e^{-hc\tilde{\nu}_J/kT}$$

where J is the level of the term (displayed as the subscript in the term symbol) and g_J is the degeneracy, given by

$g_J=2J+1.$

At 298 K, $\dfrac{hc\tilde{\nu}_j}{kT}=\dfrac{6.626\times10^{-34}\,\text{J s}\times2.998\times10^{10}\,\text{cm s}^{-1}}{1.381\times10^{-23}\,\text{J K}^{-1}\times298\,\text{K}}=4.83\times10^{-3}\times\tilde{\nu}_j/\text{cm}^{-1}$

so $q=1+3e^{-4.83\times10^{-3}\times557.1}+5e^{-4.83\times10^{-3}\times1410.0}+5e^{-4.83\times10^{-3}\times7125.3}+e^{-4.83\times10^{-3}\times16367.3}=\boxed{1.209}$.

At 1000 K, $\dfrac{hc\tilde{\nu}_j}{kT}=\dfrac{1.43877\,\text{cm K}\times\tilde{\nu}_j}{1000\,\text{K}}=1.439\times10^{-3}\times\tilde{\nu}_j/\text{cm}^{-1}$

so $q=1+3e^{-1.439\times10^{-3}\times557.1}+5e^{-1.439\times10^{-3}\times1410.0}+5e^{-1.439\times10^{-3}\times7125.3}+e^{-1.439\times10^{-3}\times16\,367.3}$
$=\boxed{3.004}$.

P15B.7 The partition function is

$$q=\sum_v e^{-\beta\varepsilon_v}[15B.1a]=\sum_v e^{-hc\tilde{\nu}_v/kT}$$

(a) At 100 K, $\dfrac{hc\tilde{\nu}_j}{kT}=\dfrac{6.626\times10^{-34}\ \text{J s}\times2.998\times10^{10}\ \text{cm s}^{-1}}{1.381\times10^{-23}\ \text{J K}^{-1}\times100\ \text{K}}=1.44\times10^{-2}\times\tilde{\nu}_j/\text{cm}^{-1}$,

so $q=1+e^{-0.0144\times213.30}+e^{-0.0144\times425.39}+e^{-0.0144\times636.27}+e^{-0.0144\times845.93}=\boxed{1.049}$

(b) At 298 K, $\dfrac{hc\tilde{\nu}_j}{kT}=\dfrac{6.626\times10^{-34}\ \text{J s}\times2.998\times10^{10}\ \text{cm s}^{-1}}{1.381\times10^{-23}\ \text{J K}^{-1}\times298\ \text{K}}=4.83\times10^{-3}\times\tilde{\nu}_j/\text{cm}^{-1}$

so $q=1+e^{-4.83\times10^{-3}\times213.30}+e^{-4.83\times10^{-3}\times425.39}+e^{-4.83\times10^{-3}\times636.27}+e^{-4.83\times10^{-3}\times845.93}=\boxed{1.548}$

The fractions of molecules at the various levels are

$$\frac{N_v}{N}=\frac{e^{-\beta\varepsilon_v}}{q}[15A.6]=\frac{e^{-hc\tilde{\nu}_v/kT}}{q}$$

So $\dfrac{N_0}{N}=\dfrac{1}{q}=$ (a) $\boxed{0.953}$, (b) $\boxed{0.645}$

$\dfrac{N_1}{N}=\dfrac{e^{-hc\tilde{\nu}_1/kT}}{q}=$ (a) $\boxed{0.044}$, (b) $\boxed{0.230}$

$\dfrac{N_2}{N}=\dfrac{e^{-hc\tilde{\nu}_2/kT}}{q}=$ (a) $\boxed{0.002}$, (b) $\boxed{0.083}$

Comment. Equation 15B.15 gives a closed-form expression for the vibrational partition function, based on an infinite ladder of harmonic oscillator states. The explicit sum will deviate slightly from this number because an actual molecule has a finite number of slightly anharmonic vibrational states below the bond dissociation energy.

P15B.9 (a) Ethene has four indistinguishable atoms that can be interchanged by rotations, so $\sigma=4$. The high-temperature expression for the rotational partition function of a nonlinear molecule is [15B.14]

$$q^{\text{R}}=\frac{1}{\sigma}\left(\frac{kT}{hc}\right)^{3/2}\left(\frac{\pi}{\tilde{A}\tilde{B}\tilde{C}}\right)^{1/2}$$
$$=\frac{1}{4}\left(\frac{1.381\times10^{-23}\ \text{J K}^{-1}\times298.15\ \text{K}}{6.626\times10^{-34}\ \text{J s}\times2.998\times10^{10}\ \text{cm s}^{-1}}\right)^{3/2}\left(\frac{\pi}{4.828\times1.0012\times0.8282\ \text{cm}^{-3}}\right)^{1/2}$$
$$=\boxed{660.6}$$

(b) Pyridine belongs to the C_{2v} group, the same as water, so $\sigma=2$.

$$q^{\text{R}}=\frac{1}{\sigma}\left(\frac{kT}{hc}\right)^{3/2}\left(\frac{\pi}{\tilde{A}\tilde{B}\tilde{C}}\right)^{1/2}$$
$$=\frac{1}{2}\left(\frac{1.381\times10^{-23}\ \text{J K}^{-1}\times298.15\ \text{K}}{6.626\times10^{-34}\ \text{J s}\times2.998\times10^{10}\ \text{cm s}^{-1}}\right)^{3/2}\left(\frac{\pi}{0.2014\times0.1936\times0.0987\ \text{cm}^{-3}}\right)^{1/2}$$
$$=\boxed{4.26\times10^4}$$

15C Molecular energies

Answer to discussion question

D15C.1 The equipartition theorem says that the average value of each quadratic contribution to the energy is the same and equal to $(1/2)kT = 1/(2\beta)$ This result agrees with the mean energy computed from a partition function, $\langle\varepsilon\rangle = -\frac{1}{q}\frac{\partial q}{\partial\beta}$ [15C.3], at sufficiently high temperatures. The partition function yields the general result for mean energy. Mathematically, evaluating $\partial q/\partial\beta$ involves summing over states. In mathematical terms, the equipartition theorem is valid when the temperature is high enough for the discrete sum to be well-approximated by a continuous integral. In physical terms, this corresponds to temperatures sufficiently high for the difference between energy levels to be small compared to kT.

Solutions to exercises

E15C.1(a) The mean energy is

$$\langle\varepsilon\rangle = \frac{1}{q}\sum_i \varepsilon_i \mathrm{e}^{-\beta\varepsilon_i}\,[15\text{C}.2] = \frac{\sum_i \varepsilon_i \mathrm{e}^{-\beta\varepsilon_i}}{\sum_i \mathrm{e}^{-\beta\varepsilon_i}} = \frac{\varepsilon \mathrm{e}^{-\beta\varepsilon}}{1+\mathrm{e}^{-\beta\varepsilon}} = \frac{\varepsilon}{1+\mathrm{e}^{\beta\varepsilon}},$$

where the last expression specializes to two non-degenerate levels. Substitute

$$\varepsilon = hc\tilde{\nu} = 6.626\times10^{-34}\,\text{J s}\times 2.998\times10^{10}\,\text{cm s}^{-1}\times 500\,\text{cm}^{-1} = 9.93\times10^{-21}\,\text{J},$$

and $\beta\varepsilon = \dfrac{\varepsilon}{kT} = \dfrac{9.93\times10^{-21}\,\text{J}}{1.381\times10^{-23}\,\text{J K}^{-1}\times 298\,\text{K}} = 2.41,$

so $\langle\varepsilon\rangle = \dfrac{9.93\times10^{-21}\,\text{J}}{1+\mathrm{e}^{2.41}} = \boxed{8.16\times10^{-22}\,\text{J}}$

E15C.2(a) The mean energy is

$$\langle\varepsilon\rangle = \frac{1}{q}\sum_{\text{states}} \varepsilon_i \mathrm{e}^{-\beta\varepsilon_i}\,[15\text{C}.2] = \frac{1}{q}\sum_{\text{levels}} g_i\varepsilon_i \mathrm{e}^{-\beta\varepsilon_i} = \frac{1}{q}\sum_J (2J+1)\varepsilon_J \mathrm{e}^{-\varepsilon_J/kT}$$

$$\varepsilon_J = hc\tilde{B}J(J+1) = 6.626\times10^{-34}\,\text{J s}\times 2.998\times10^{10}\,\text{cm s}^{-1}\times 1.931\,\text{cm}^{-1}\times J(J+1)$$
$$= J(J+1)\times 3.836\times10^{-23}\,\text{J}$$

and $\dfrac{\varepsilon_J}{k} = \dfrac{J(J+1)\times 3.836\times10^{-23}\,\text{J}}{1.381\times10^{-23}\,\text{JK}^{-1}} = J(J+1)\times 2.778\,\text{K}$

so $\langle\varepsilon\rangle = \dfrac{1}{q}\sum_J J(J+1)(2J+1)\times 3.836\times10^{-23}\,\text{J}\times \mathrm{e}^{-J(J+1)\times 2.778\,\text{K}/T}$

Use a spreadsheet or other mathematical software to evaluate the terms of the sum and to sum the terms until they converge. For the partition function, see Exercise 15B.5(a).

The equipartition value is simply kT (i.e. $kT/2$ for each rotational degree of freedom). The explicit and equipartition expressions are compared in Fig. 15C.1. The explicit sum reaches 95% of the equipartition value at about $\boxed{18.5\text{ K}}$.

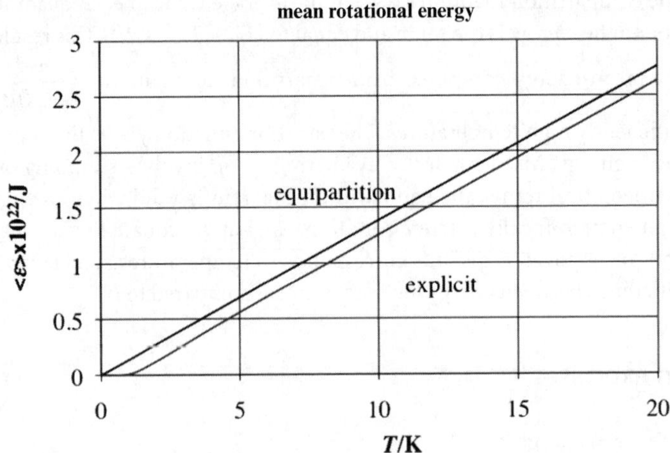

Figure 15C.1

E15C.3(a) The mean energy is

$$\langle\varepsilon\rangle=\frac{1}{q}\sum_{\text{states}}\varepsilon_i e^{-\beta\varepsilon_i}\,[15\text{C}.2]=\frac{1}{q}\sum_{\text{levels}}g_i\varepsilon_i e^{-\beta\varepsilon_i}=\frac{1}{\sigma q}\sum_J(2J+1)^2\varepsilon_J e^{-\varepsilon_J/kT}\quad[\text{Section 12B.2(b)}]$$

Note that the sum over levels is restricted by nuclear statistics; in order to avoid multiple counting, we sum over all J without restriction and divide the result by the symmetry number σ.

$$\varepsilon_J=hc\tilde{B}J(J+1)=6.626\times10^{-34}\text{ J s}\times2.998\times10^{10}\text{ cm s}^{-1}\times5.241\text{ cm}^{-1}\times J(J+1)$$
$$=J(J+1)\times1.041\times10^{-22}\text{ J}$$

and $\dfrac{\varepsilon_J}{k}=\dfrac{J(J+1)\times1.041\times10^{-22}\text{ J}}{1.381\times10^{-23}\text{ J K}^{-1}}=J(J+1)\times7.539\text{ K},$

so $\langle\varepsilon\rangle=\dfrac{1}{\sigma q}\sum_J J(J+1)(2J+1)^2\times1.041\times10^{-22}\text{ J}\times e^{-J(J+1)\times7.539\text{ K}/T}$

Use a spreadsheet or other mathematical software to evaluate the terms of the sum and to sum the terms until they converge. For σq, see Exercise 15B.6(a). The quantity evaluated explicitly in that exercise is σq, because we computed the partition function without taking the symmetry number into account; in effect, the sum evaluated here **and** the sum evaluated in the earlier exercise contain factors of σ, which cancel.

The equipartition value is simply $3kT/2$ (i.e., $kT/2$ for each rotational degree of freedom). The explicit and equipartition expressions are compared in Fig. 15C.2. The explicit sum reaches 95% of the equipartition value at about $\boxed{25\,\text{K}}$.

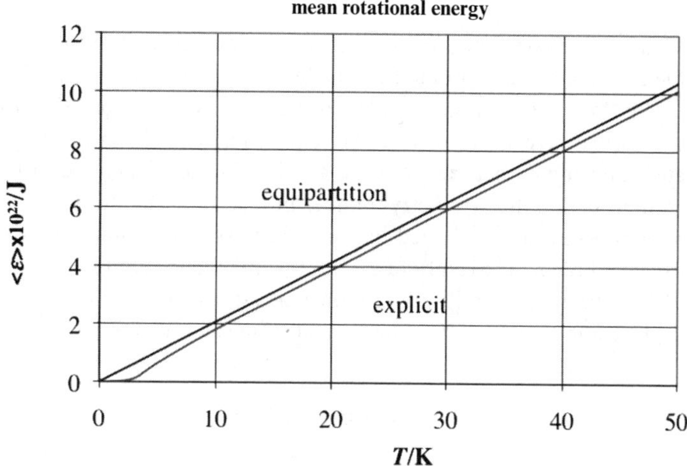

Figure 15C.2

E15C.4(a) The mean energy is

$$\langle\varepsilon\rangle = \frac{1}{q}\sum_{\text{states}} \varepsilon_i e^{-\beta\varepsilon_i} [15\text{C}.2] = \frac{1}{q}\sum_{\text{levels}} g_i\varepsilon_i e^{-\beta\varepsilon_i}$$

$$= \frac{1}{\sigma q}\sum_{J=0}(2J+1)e^{-hc\tilde{B}J(J+1)/kT}\left(\sum_{K=-J}^{J}\varepsilon_{J,K}e^{-hc(\tilde{A}-\tilde{B})K^2/kT}\right) [12\text{B}.13]$$

Note that the sum over levels is restricted by nuclear statistics; in order to avoid multiple counting, we sum over all J without restriction and divide the result by the symmetry number σ. (See Exercise 15C.3(a).)

$$\varepsilon_{J,K} = hc\left\{\tilde{B}J(J+1) + (\tilde{A}-\tilde{B})K^2\right\}$$

Use $hc\tilde{B} = 6.626\times10^{-34}\,\text{J\,s}\times2.998\times10^{10}\,\text{cm\,s}^{-1}\times0.443\,\text{cm}^{-1} = 8.80\times10^{-24}\,\text{J}$,

$$\frac{hc\tilde{B}}{k} = \frac{8.80\times10^{-24}\,\text{J}}{1.381\times10^{-23}\,\text{J\,K}^{-1}} = 0.637\,\text{K},$$

$$hc(\tilde{A}-\tilde{B}) = 6.626\times10^{-34}\,\text{J\,s}\times2.998\times10^{10}\,\text{cm\,s}^{-1}\times(5.097-0.443)\text{cm}^{-1}$$

$$= 9.245\times10^{-23}\,\text{J},$$

and $$\frac{hc(\tilde{A}-\tilde{B})}{k} = \frac{9.245\times10^{-23}\,\text{J}}{1.381\times10^{-23}\,\text{J\,K}^{-1}} = 6.694\,\text{K}$$

$$\text{so } \langle\varepsilon\rangle = \frac{1}{\sigma q}\sum_{J=0}(2J+1)e^{-0.637\ \mathrm{K}\times J(J+1)/T}$$
$$\times\sum_{K=-J}^{J}\{J(J+1)\times 8.80\times10^{-24}\ \mathrm{J} + K^2\times 9.245\times10^{-23}\ \mathrm{J}\}e^{-6.694\ \mathrm{K}\times K^2/T}$$

Write a brief computer program or use other mathematical software to evaluate the terms of the sum and to sum the terms until they converge. Nested sums are straightforward to program in languages such as BASIC or FORTRAN, whereas spreadsheets are more unwieldy. For σq, see Exercise 15B.7(a). The quantity evaluated explicitly in that exercise is σq, for there we computed the partition function without taking the symmetry number into account; in effect, the sum evaluated here **and** the sum evaluated in the earlier exercise contain factors of σ, which cancel. Compare the results of the direct sum with the equipartition value, namely $3kT/2$. The explicit and equipartition expressions are compared in Fig. 15C.3. The explicit sum reaches 95% of the equipartition value at about $\boxed{4.5\ \mathrm{K}}$.

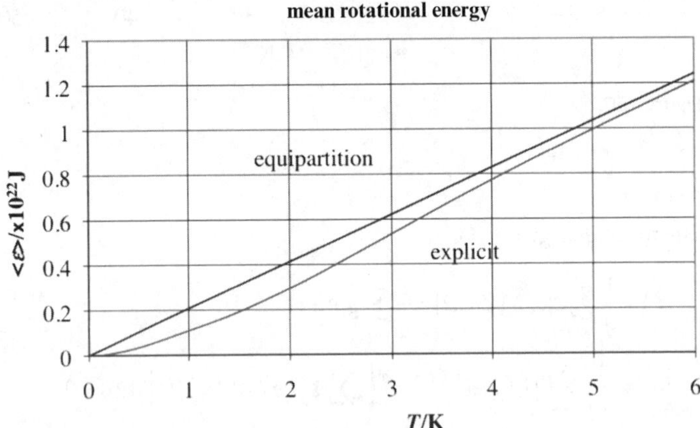

Figure 15C.3

E15C.5(a) The mean vibrational energy is

$$\langle\varepsilon^{\mathrm{V}}\rangle = \frac{hc\tilde{\nu}}{e^{hc\tilde{\nu}/kT}-1} \quad [15C.8 \text{ with } \beta = 1/kT].$$

Use $hc\tilde{\nu} = 6.626\times10^{-34}\ \mathrm{J\,s}\times 2.998\times10^{10}\ \mathrm{cm\,s^{-1}}\times 323.2\ \mathrm{cm^{-1}} = 6.420\times10^{-21}\ \mathrm{J}$

$$\text{and } \frac{hc\tilde{\nu}}{k} = \frac{6.420\times10^{-21}\ \mathrm{J}}{1.381\times10^{-23}\ \mathrm{J\,K^{-1}}} = 464.9\ \mathrm{K},$$

$$\text{so } \langle\varepsilon^{\mathrm{V}}\rangle = \frac{6.420\times10^{-21}\ \mathrm{J}}{e^{464.9\ \mathrm{K}/T}-1}$$

The equipartition value is simply kT for a single vibrational mode. The explicit and equipartition values are compared in Fig. 15C.4. The explicit expression reaches 95% of the equipartition value at $\boxed{4\,600\ \text{K}}$.

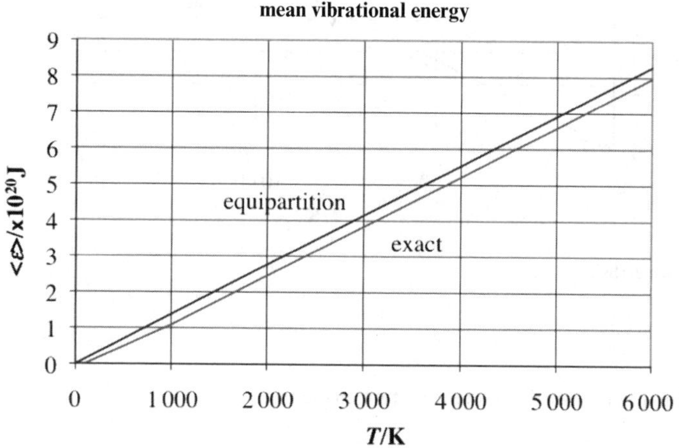

Figure 15C.4

E15C.6(a) The mean vibrational energy per mode is

$$\langle \varepsilon^{\mathrm{V}} \rangle = \frac{hc\tilde{\nu}}{e^{hc\tilde{\nu}/kT}-1} \quad [15\text{C}.8 \text{ with } \beta = 1/kT].$$

We draw up the following table

mode	1	2	3	4
$\tilde{\nu}$ / cm^{-1}	658	397	397	1535
$hc\tilde{\nu}$ / (10^{-21} J)	13.07	7.89	7.89	30.49
($hc\tilde{\nu}$ / k) / K	946	571	571	2208

$$\text{So } \langle \varepsilon^{\mathrm{V}} \rangle = \frac{1.307\times10^{-20}\ \text{J}}{e^{946\ \text{K}/T}-1} + 2\times\frac{7.89\times10^{-21}\ \text{J}}{e^{571\ \text{K}/T}-1} + \frac{3.049\times10^{-20}\ \text{J}}{e^{2208\ \text{K}/T}-1}$$

The equipartition value is simply $4kT$, that is, kT per vibrational mode. The explicit and equipartition values are compared in Fig. 15C.5. The explicit expression reaches 95% of the equipartition value at $\boxed{10\,500\ \text{K}}$.

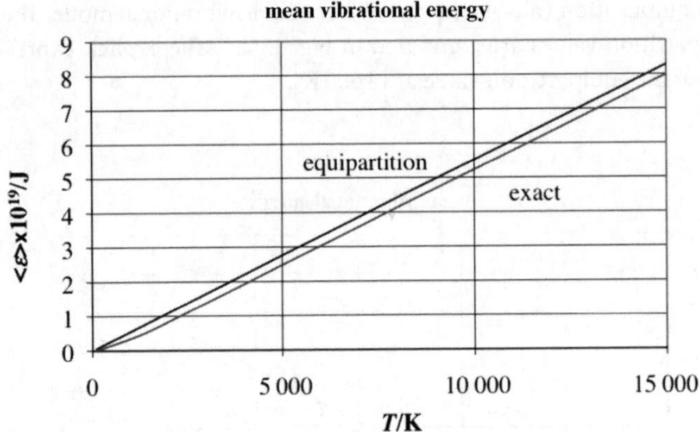

Figure 15C.5

E15C.7(a) The mean vibrational energy per mode is

$$\langle \varepsilon^{\mathrm{V}} \rangle = \frac{hc\tilde{\nu}}{\mathrm{e}^{hc\tilde{\nu}/kT}-1} \text{ [15C.8 with } \beta = 1/kT\text{]}.$$

We draw up the following table

mode	1	2	3	4
$\tilde{\nu}$ / cm^{-1}	459	217	776	314
degeneracy	1	2	3	3
$hc\tilde{\nu}$ / (10^{-21} J)	9.12	4.31	15.42	6.24
$(hc\tilde{\nu}/k)$ / K	660	312	1116	452

So $$\langle \varepsilon^{\mathrm{V}} \rangle = \frac{9.12\times10^{-21}\,\mathrm{J}}{\mathrm{e}^{660\,\mathrm{K}/T}-1} + 2\times\frac{4.31\times10^{-21}\,\mathrm{J}}{\mathrm{e}^{312\,\mathrm{K}/T}-1} + 3\times\frac{1.542\times10^{-20}\,\mathrm{J}}{\mathrm{e}^{1116\,\mathrm{K}/T}-1} + 3\times\frac{6.24\times10^{-21}\,\mathrm{J}}{\mathrm{e}^{452\,\mathrm{K}/T}-1}$$

The equipartition value is simply $9kT$, that is, kT per vibrational mode. The explicit and equipartition values are compared in Fig. 15C.6. The explicit expression reaches 95% of the equipartition value at $\boxed{6\,500\ \mathrm{K}}$.

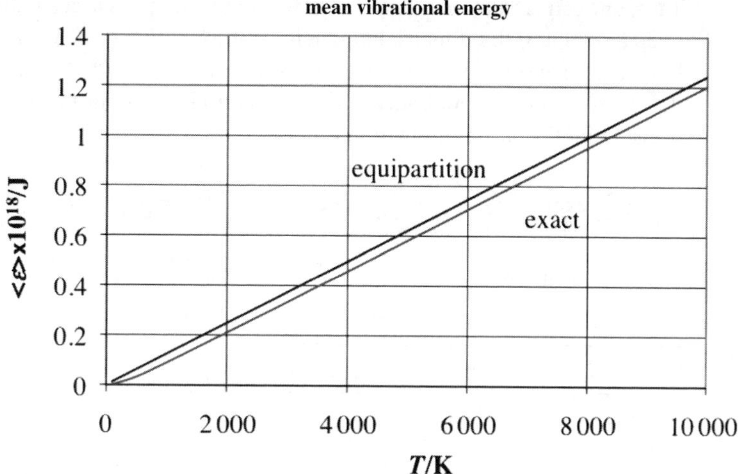

Figure 15C.6

E15C.8(a) $\langle\varepsilon\rangle=-\frac{1}{q}\frac{\partial q}{\partial\beta}[15C.3]=-\frac{1}{q}\frac{\partial}{\partial\beta}\left(4+e^{-\beta\varepsilon_1}+2e^{-\beta\varepsilon_2}\right)=-\frac{1}{q}\left(-\varepsilon_1 e^{-\beta\varepsilon_1}-2\varepsilon_2 e^{-\beta\varepsilon_2}\right)$

$$=\frac{hc}{q}\left(\tilde{\nu}_1 e^{-\beta hc\tilde{\nu}_1}+2\tilde{\nu}_2 e^{-\beta hc\tilde{\nu}_2}\right)=\frac{hc}{q}\left(\tilde{\nu}_1 e^{-hc\tilde{\nu}_1/kT}+2\tilde{\nu}_2 e^{-hc\tilde{\nu}_2/kT}\right)$$

Use $\frac{hc\tilde{\nu}_j}{kT}=7.571\times10^{-4}\times(\tilde{\nu}_j/\text{cm}^{-1})$ and $q=4.292$ [Exercise 15B.13a]

Thus $\langle\varepsilon\rangle=\frac{6.626\times10^{-34}\ \text{J s}\times2.998\times10^{10}\ \text{cm s}^{-1}}{4.292}$

$$\times\left(2\,500\ \text{cm}^{-1}\times e^{-7.571\times10^{-4}\times2\,500}+2\times3\,500\ \text{cm}^{-1}\times e^{-7.571\times10^{-4}\times3\,500}\right)$$

$$=\boxed{4.033\times10^{-21}\ \text{J}}$$

Solutions to problems

P15C.1 Follow the reasoning set out in Problem 15B.1 to employ more general dimensionless quantities. The energy divided by the Boltzmann constant has dimensions of temperature, so define the characteristic temperature for this system by

$$\theta\equiv\frac{\hbar^2}{2m_e R^2 k}$$

With this definition, note that the energy levels for this spherical well can be written as

$$E_{nl}=X_{nl}^2 k\theta$$

Thus, one can tabulate values of $E_{nl}/k\theta$ without having to select a radius. Likewise, one can employ the dimensionless temperature T/θ. As noted in the problem, Boltzmann factors require energies measured with respect to the ground state, so the energies **in the exponents of the Boltzmann factors** must be $E_{nl} - E_{10}$. The expression for the partition function is [15B.1]

$$q = \sum_{\text{states}} \exp\left(-\frac{E_{nl}-E_{10}}{kT}\right) = \sum_{\text{levels}} g_{nl}\exp\left(-\frac{E_{nl}-E_{10}}{kT}\right) = \sum_{\text{levels}} g_{nl}\exp\left(-\frac{E_{nl}-E_{10}}{k\theta}\times\frac{\theta}{T}\right)$$

Draw up the following table:

n	l	X_{nl}	$E_{nl}/k\theta$	$(E_{nl}-E_{10})/k\theta$	g_{nl}
1	0	3.142	9.872164	0	1
1	1	4.493	20.187049	10.314885	3
1	2	5.763	33.212169	23.340005	5
2	0	6.283	39.476089	29.603925	1
1	3	6.988	48.832144	38.95998	7
2	1	7.725	59.675625	49.803461	3

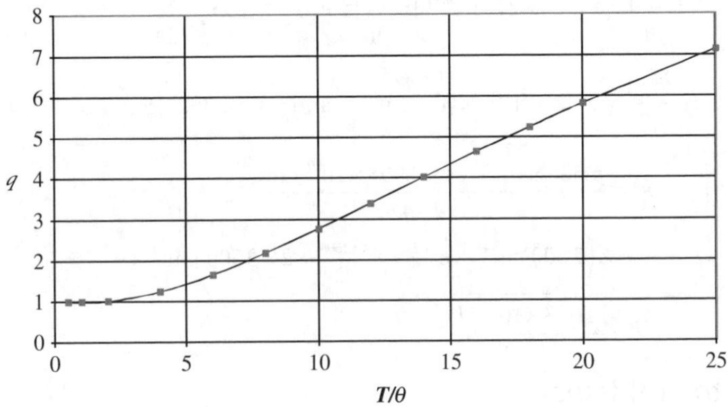

Figure 15C.7(a)

The partition function is plotted in Fig. 15C.7(a). Note that at $T/\theta = 25$, the highest state already contributes more than 5% of the six-level sum. Thus, by this temperature, truncating the sum after six levels is becoming inaccurate.

$$\frac{\langle\varepsilon\rangle}{k\theta} = \frac{1}{q}\sum_{\text{levels}} \frac{E_{nl}}{k\theta} g_{nl}\exp\left(-\frac{E_{nl}-E_{10}}{k\theta}\times\frac{\theta}{T}\right)$$

The average energy (average of $E_{nl}/k\theta$, not of $(E_{nl}-E_{10})/k\theta$) is plotted in Fig. 15C.7(b).

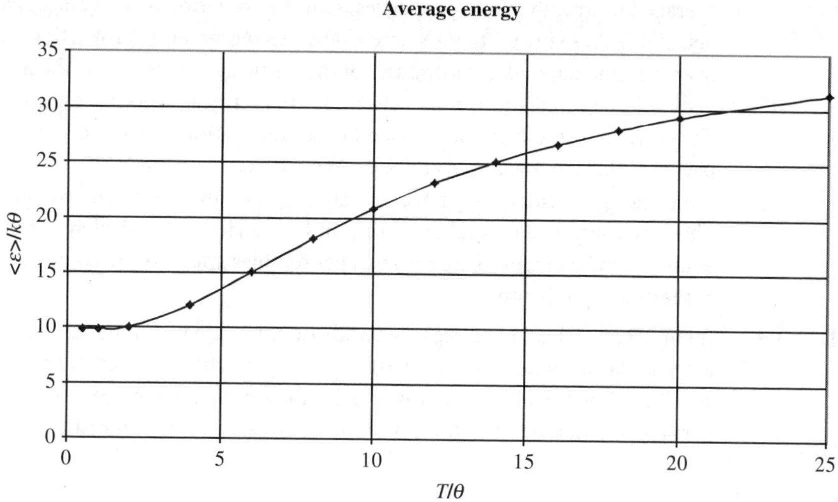

Figure 15C.7(b)

P15C.3 (a) $\langle\varepsilon\rangle = -\dfrac{1}{q}\dfrac{\partial q}{\partial\beta}$ [15C.3], with $q = \dfrac{1}{1-e^{-\beta\varepsilon}}$ [15B.15]

so $\langle\varepsilon\rangle = -\dfrac{1}{q}\dfrac{\partial q}{\partial\beta} = \dfrac{\varepsilon e^{-\beta\varepsilon}}{1-e^{-\beta\varepsilon}} = \dfrac{\varepsilon}{e^{\beta\varepsilon}-1} = a\varepsilon$

Hence, $e^{\beta\varepsilon} = \dfrac{1+a}{a}$, implying that $\beta = \dfrac{1}{\varepsilon}\ln\left(1+\dfrac{1}{a}\right)$

For a mean energy of ε, $a = 1$ and $\beta = \dfrac{\ln 2}{\varepsilon}$, implying that

$$T = \frac{\varepsilon}{k\ln 2} = (50\,\text{cm}^{-1})\times\left(\frac{hc}{k\ln 2}\right) = \boxed{104\,\text{K}}$$

(b) $q = \dfrac{1}{1-e^{-\beta\varepsilon}} = \dfrac{1}{1-\left(\dfrac{a}{1+a}\right)} = \boxed{1+a}$

15D The canonical ensemble

Answers to discussion questions

D15D.1 Ensembles are needed to treat systems of interacting particles (in contrast to the molecular partition functions that apply to independent particles). An ensemble is a set of a large number of imaginary replications of an actual physical (thermodynamic) system. Ensembles are useful in statistical thermodynamics because it is mathematically more tractable to perform an ensemble average to determine time averaged thermodynamic properties than it is to perform an average over time to determine these properties. (Macroscopic thermodynamic properties are averages over the time dependent properties of the particles that compose the macroscopic system.) The replications in an ensemble are identical in some respects but not in all respects. In the canonical ensemble, all replications have the same number of particles, the same volume, and the same temperature, but they need not have the same energy. Because they have the same temperature, they can exchange energy. Thus the canonical ensemble corresponds to a closed physical system at thermal equilibrium: averages using canonical ensembles apply to closed physical systems at thermal equilibrium

D15D.3 Identical particles can be regarded as distinguishable when they are localized as in a crystal lattice where we can assign a set of coordinates to each particle. Strictly speaking, it is the lattice site that carries the set of coordinates, but as long as the particle is tethered to the site, it too can be considered distinguishable.

Solution to exercise

E15D.1(a) Inclusion of a factor of $1/N!$ is necessary when considering indistinguishable particles. Because of their translational freedom, gases are collections of indistinguishable particles. Solids are collections of particles that are distinguishable by their positions. The factor must be included in calculations on (i) $\boxed{\text{He gas}}$, (ii) $\boxed{\text{CO gas}}$, and (iv) $\boxed{\text{H}_2\text{O vapour}}$, but not (iii) solid CO.

Solution to problem

P15D.1

$$p = kT\left(\frac{\partial \ln Q}{\partial V}\right)_{T,N} = kT\left(\frac{\partial \ln(q^N/N!)}{\partial V}\right)_{T,N}$$

$$= kT\left(\frac{\partial (N\ln q - \ln N!)}{\partial V}\right)_{T,N} = NkT\left(\frac{\partial \ln q}{\partial V}\right)_{T,N}$$

$$= NkT\left(\frac{\partial \ln(V/\Lambda^3)}{\partial V}\right)_{T,N}$$

$$= NkT\left(\frac{\partial (\ln V - \ln \Lambda^3)}{\partial V}\right)_{T,N} = NkT\left(\frac{\partial \ln V}{\partial V}\right)_{T,N}$$

$$= \frac{NkT}{V} = \frac{nRT}{V} \quad \text{or} \quad \boxed{pV = nRT}$$

15E The internal energy and the entropy

Answers to discussion questions

D15E.1 The temperature is always high enough for the mean translational energy to be $\frac{3}{2}kT$, the equipartition value (provided the gas is above its condensation temperature). Therefore, the molar constant-volume heat capacity for translation is $C_{V,\mathrm{m}}^{\mathrm{T}} = \frac{3}{2}R$.

Translation is the only mode of motion for a monatomic gas, so for such a gas $C_{V,\mathrm{m}} = \frac{3}{2}R = 12.47\ \mathrm{J\,K^{-1}\,mol^{-1}}$. This result is very reliable: helium, for example has this value over a range of 2 000 K.

When the temperature is high enough for the rotations of the molecules to be highly excited (when $T >> \theta_{\mathrm{R}}$) we can use the equipartition value kT for the mean rotational energy (for a linear rotor) to obtain $C_{\mathrm{V,m}} = R$. For nonlinear molecules, the mean rotational energy rises to $\frac{3}{2}kT$, so the molar rotational heat capacity rises to $\frac{3}{2}R$ when $T >> \theta_{\mathrm{R}}$. Only the lowest rotational state is occupied when the temperature is very low, and then rotation does not contribute to the heat capacity. We can calculate the rotational heat capacity at intermediate temperatures by differentiating the equation for the mean rotational energy (eqn 15C.26a for a linear molecule). Because the translational contribution is always present, we can expect the molar heat capacity of a gas of diatomic molecules ($C_{V,\mathrm{m}}^{\mathrm{T}} + C_{V,\mathrm{m}}^{\mathrm{R}}$) to change from $\frac{3}{2}R$ to $\frac{5}{2}R$ as the temperature is increased above θ_{R}.

Molecular vibrations contribute to the heat capacity, but only when the temperature is high enough for them to be significantly excited. For each vibrational mode, the equipartition mean energy is kT, so the maximum contribution to the molar heat capacity is R. However, it is very unusual for the vibrations to be so highly excited that equipartition is valid, and it is more appropriate to use the full expression for the vibrational heat capacity which is obtained by differentiating eqn 15.C8. The curve in Fig. 15E.1 of the text shows how the vibrational heat capacity depends on temperature. Note that even when the temperature is only slightly above the vibrational temperature, the heat capacity is close to its equipartition value.

The total heat capacity of a molecular substance is the sum of each contribution. When equipartition is valid (when the temperature is well above the characteristic temperature of the mode $T >> \theta^{\mathrm{M}}$) we can estimate the heat capacity by counting the numbers of modes that are active. In gases, all three translational modes are always active and contribute $\frac{3}{2}R$ to the molar heat capacity. If we denote the number of active rotational modes by $\nu^{\mathrm{R}*}$ (so for most molecules at normal temperatures $\nu^{\mathrm{R}*} = 2$ for linear molecules, and 3 for nonlinear molecules), then the rotational contribution is $\frac{1}{2}\nu^{\mathrm{R}*}R$. If the temperature is high enough for $\nu^{\mathrm{V}*}$ vibrational modes to be active the vibrational contribution to the molar heat capacity is $\nu^{\mathrm{V}*}R$. In most cases $\nu^{\mathrm{V}*} \approx 0$. It follows that the total molar heat capacity is [15E.6]

$$C_{V,\mathrm{m}} = \tfrac{1}{2}(3 + \nu^{\mathrm{R}*} + 2\nu^{\mathrm{V}*})R$$

D15E.3 Compare Section 15E.2, which introduces the statistical entropy, to Topic 3A, which introduced the thermodynamic entropy. The latter is defined by $dS = \frac{dq_{rev}}{T}$ [3A.1], that is, in terms of reversible heat transfer. The statistical entropy is defined by Boltzmann's formula, $S = k \ln \mathcal{W}$ [15E.7], in terms of the number of configurations or microstates consistent with a given total energy. Different as they appear on the face of them, the statistical and thermodynamic entropies are equivalent if the latter is taken to be zero at $T = 0$.

The concept of the number of microstates makes quantitative the qualitative concepts of 'disorder' and 'dispersal of matter and energy' that are often used to introduce the concept of entropy: a more 'disorderly' distribution of energy and matter corresponds to a greater number of microstates consistent with the same total energy. The more molecules that can participate in the distribution of energy, the more microstates there are for a given total energy and the greater the entropy than when energy is confined to a smaller number of molecules.

The molecular interpretation of entropy embodied in the Boltzmann formula also suggests the thermodynamic definition. At high temperatures where the molecules of a system can occupy a large number of available energy levels, a small additional transfer of energy as heat will cause only a small change in the number of accessible energy levels, whereas at low temperatures the transfer of the same quantity of heat will increase the number of accessible energy levels and microstates significantly. Hence the change in entropy upon heating will be greater when the energy is transferred to a cold body than when it is transferred to a hot body, as required by the thermodynamic definition.

D15E.5 The entropy of a monatomic perfect gas is given by the Sackur–Tetrode equation [15E.11a]:

$$S_m = R \ln\left(\frac{V_m e^{5/2}}{N_A \Lambda^3}\right) \quad \text{where} \quad \Lambda = \frac{h}{(2\pi mkT)^{1/2}} \quad [15B.7b]$$

Because the molar volume appears in the numerator, the molar entropy increases with the molar volume. In terms of the Boltzmann distribution, this is natural: large containers have more closely spaced energy levels than do small ones, so more states are thermally accessible. Temperature appears in the numerator of the expression (through the denominator of Λ), so the molar entropy increases with the temperature. Again, this is consistent with the Boltzmann distribution, because more states are accessible at higher temperatures than lower ones.

The fact that diatomic and polyatomic gases have rotational and vibrational modes of motion as well does not change the above arguments. The partition functions of those modes are independent of volume, so the volume dependence of the entropy is as described above. At most temperatures, rotational modes of motion are active and contribute to the partition function via the high-temperature expressions [15B.13 and 15B.14], which would only reinforce the temperature dependence described above. Finally, many vibrational modes contribute little if at all to the molecular partition function, and those that contribute significantly would also reinforce the temperature dependence described above.

Solutions to exercises

E15E.1(a) $C_{V,\mathrm{m}} = \frac{1}{2}(3+\nu^{\mathrm{R}*}+2\nu^{\mathrm{V}*})R$ [15E.6]

with a mode active if $T > \theta^{\mathrm{M}}$.

(i) $\mathrm{I_2}$: $\nu^{\mathrm{R}*}=2$, $\nu^{\mathrm{V}*}\approx 1$; hence $C_{V,\mathrm{m}} = \frac{1}{2}(3+2+2\times 1)R = \boxed{\frac{7}{2}R}$ [experimental = 3.4R]

Note that $\mathrm{I_2}$ has quite a low vibrational wavenumber, so

$$\theta^{\mathrm{V}} = \frac{hc\tilde{\nu}}{k} = \frac{(6.626\times10^{-34}\ \mathrm{J\,s})(2.998\times10^{10}\ \mathrm{cm\,s^{-1}})(214\ \mathrm{cm^{-1}})}{1.381\times10^{-23}\ \mathrm{J\,K^{-1}}} = 308\ \mathrm{K}$$

The temperature specified for this exercise is less than this, but only slightly. Looking at Fig. 15.E1 in the text suggests that the mode is closer to active than inactive when the temperature is even close to the vibrational temperature.

(ii) $\mathrm{CH_4}$: $\nu^{\mathrm{R}*}=3$, $\nu^{\mathrm{V}*}\approx 0$; hence $C_{V,\mathrm{m}} = \frac{1}{2}(3+3+0)R = \boxed{3R}$ [experimental = 3.2R]

(iii) $\mathrm{C_6H_6}$: $\nu^{\mathrm{R}*}=3$, $\nu^{\mathrm{V}*}\approx 4$; hence $C_{V,\mathrm{m}} = \frac{1}{2}(3+3+2\times 4)R = \boxed{7R}$ [experimental = 8.8R]

Note that data from the book by Herzberg (*Molecular Spectra and Molecular Structure II*) show low vibrational wavenumbers for four modes of benzene, and we have included these above. There are 26 more vibrational modes we have neglected, taking them to be inactive. Slight activity from these modes accounts for the difference of about 1.8R between the experimental value and our estimate.

E15E.2(a) Assuming that all rotational modes are active we can draw up the following table for $C_{V,\mathrm{m}}$, $C_{p,\mathrm{m}}$, and γ with and without active vibrational modes.

	$C_{V,\mathrm{m}}$	$C_{p,\mathrm{m}}$	γ	Expt	
$\mathrm{NH_3}$ ($\nu^{\mathrm{V}*}=0$)	$3R$	$4R$	1.33	1.31	$\boxed{\text{closer}}$
$\mathrm{NH_3}$ ($\nu^{\mathrm{V}*}=6$)	$9R$	$10R$	1.11		
$\mathrm{CH_4}$ ($\nu^{\mathrm{V}*}=0$)	$3R$	$4R$	1.33	1.31	$\boxed{\text{closer}}$
$\mathrm{CH_4}$ ($\nu^{\mathrm{V}*}=9$)	$12R$	$13R$	1.08		

The experimental values are obtained from Tables 2C.1 and 2C.2 assuming $C_{p,\mathrm{m}} = C_{V,\mathrm{m}} + R$. It is clear from the comparison in the above table that the vibrational modes are not active. This is confirmed by the experimental vibrational wavenumbers (see Herzberg, *Molecular Spectra and Molecular Spectra II*) all of which are much greater than kT at 298 K.

E15E.3(a) We assume based on the information given that Cl can effectively be treated as a two-level system. The partition function is then

$$q=\sum_j g_j e^{-\beta\varepsilon_j}$$

with degeneracies $g_j = 2J+1$ (where J is the subscript on the term symbol)

so $q=4+2e^{-\beta\varepsilon}\,[g(^2P_{3/2})=4, g(^2P_{1/2})=2]$

(i) $q=4+2e^{-\frac{hc\times 881\,\text{cm}^{-1}}{kT}}=4+2e^{-\frac{1.4388\,\text{cm K}\times 881\,\text{cm}^{-1}}{500\text{ K}}}=\boxed{4.158}$

(ii) $q=4+2e^{-\frac{hc\times 881\,\text{cm}^{-1}}{kT}}=4+2e^{-\frac{1.4388\,\text{cm K}\times 881\,\text{cm}^{-1}}{900\text{K}}}=\boxed{4.489}$

E15E.4(a) $C_{V,\text{m}}/R=f,\; f=\left(\frac{\theta^V}{T}\right)^2\times\left(\frac{e^{-\theta^V/2T}}{1-e^{-\theta^V/T}}\right)^2$ [15E.3]; $\theta^V=\frac{hc\tilde{\nu}}{k}$

We write $x=\frac{\theta^V}{T}=\frac{hc\tilde{\nu}}{kT}$; then $C_{V,\text{m}}/R=\frac{x^2e^{-x}}{(1-e^{-x})^2}$

This function is plotted in Fig. 15E.1. For the acetylene (ethyne) calculation, use the expression above for each mode. We draw up the following table using $kT/hc = 207\ \text{cm}^{-1}$ at 298 K and $348\ \text{cm}^{-1}$ at 500 K.

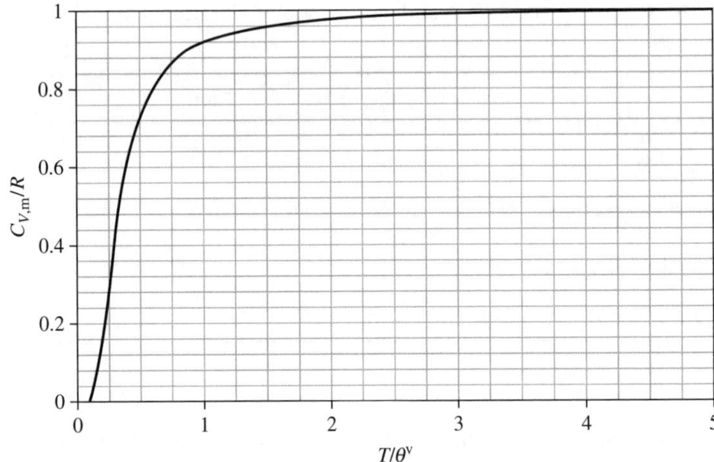

Figure 15E.1

$\tilde{\nu}/\text{cm}^{-1}$	x		$C_{V,m}/R$	
	298 K	500 K	298 K	500 K
612	2.96	1.76	0.505	0.777
612	2.96	1.76	0.505	0.777
729	3.52	2.09	0.389	0.704
729	3.52	2.09	0.389	0.704
1974	9.54	5.67	0.007	0.112
3287	15.88	9.45	3.2×10^{-5}	0.007
3374	16.30	9.70	2.2×10^{-5}	0.006

The heat capacity of the molecule is the sum of these contributions, namely

(i) $1.796R = \boxed{14.93\,\text{J K}^{-1}\,\text{mol}^{-1}}$ at 298 K and

(ii) $3.086R = \boxed{25.65\,\text{J K}^{-1}\,\text{mol}^{-1}}$ at 500 K.

E15E.5(a) $S_\text{m}^\ominus = R\ln\left(\dfrac{e^{5/2}kT}{p^\ominus \Lambda^3}\right)$ [15E.11b with $p = p^\ominus$]

(i) $$\Lambda = \frac{h}{(2\pi mkT)^{1/2}} = \frac{6.626\times10^{-34}\,\text{J s}}{\{(2\pi)(4.003)(1.6605\times10^{-27}\,\text{kg})(1.381\times10^{-23}\,\text{J K}^{-1})T\}^{1/2}}$$
$$= \frac{8.726\times10^{-10}\,\text{m}}{(T/\text{K})^{1/2}}$$

$$S_\text{m}^\ominus = R\ln\left(\frac{(e^{5/2})\times(1.381\times10^{-23}\,\text{J K}^{-1}T)}{(1.013\times10^5\,\text{Pa})\times(8.726\times10^{-10}\,\text{m})^3}\right)\times\left(\frac{T}{\text{K}}\right)^{3/2}$$
$$= R\ln\{2.499\times(T/\text{K})^{5/2}\} = (8.3145\,\text{J K}^{-1}\,\text{mol}^{-1})\ln\{2.499\times(298)^{5/2}\}$$
$$= \boxed{126\,\text{J K}^{-1}\,\text{mol}^{-1}}$$

(ii) $$\Lambda = \frac{h}{(2\pi mkT)^{1/2}} = \frac{6.626\times10^{-34}\,\text{J s}}{\{(2\pi)(131.29)(1.6605\times10^{-27}\,\text{kg})(1.381\times10^{-23}\,\text{J K}^{-1})T\}^{1/2}}$$
$$= \frac{1.524\times10^{-10}\,\text{m}}{(T/\text{K})^{1/2}}$$

$$S_\text{m}^\ominus = R\ln\left(\frac{(e^{5/2})\times(1.381\times10^{-23}\,\text{J K}^{-1}T)}{(1.013\times10^5\,\text{Pa})\times(1.524\times10^{-10}\,\text{m})^3}\right)\times\left(\frac{T}{\text{K}}\right)^{3/2}$$
$$= R\ln\{469.1\times(T/\text{K})^{5/2}\} = (8.3145\,\text{J K}^{-1}\,\text{mol}^{-1})\ln\{469.1\times(298)^{5/2}\}$$
$$= \boxed{169\,\text{J K}^{-1}\,\text{mol}^{-1}}$$

E15E.6(a) From the solution to Exercise 15E.5(a) we have

$$S_m^\ominus = R\ln\{2.499\times(T/\text{K})^{5/2}\} \text{ for helium}$$

and $S_m^\ominus = 169\ \text{J K}^{-1}\ \text{mol}^{-1}$ for xenon at 298 K

We solve for T:

$$T = \left\{\frac{1}{2.499}\times\exp\left(\frac{S_m^\ominus}{R}\right)\right\}^{2/5}\text{K} = \left\{\frac{1}{2.499}\times\exp\left(\frac{169\ \text{J K}^{-1}\ \text{mol}^{-1}}{8.3145\ \text{J K}^{-1}\ \text{mol}^{-1}}\right)\right\}^{2/5}\text{K}$$

$$= \boxed{2.35\times10^3\ \text{K}}$$

E15E.7(a) The high-temperature approximation to the rotational partition function of a non-linear molecule is (after substituting the numerical values of the constants in eqn 15B.14)

$$q^\text{R} = \frac{1.0270}{\sigma}\frac{(T/\text{K})^{3/2}}{(\tilde{A}\tilde{B}\tilde{C}/\text{cm}^{-3})^{1/2}} = \frac{1.0270\times298^{3/2}}{(2)\times(27.878\times14.509\times9.287)^{1/2}} = \boxed{43.1}$$

The high-temperature approximation is valid if $T > \theta^\text{R}$

and

$$\theta^\text{R} = \frac{hc(\tilde{A}\tilde{B}\tilde{C})^{1/3}}{k}$$

$$= \frac{(6.626\times10^{-34}\ \text{J s})(2.998\times10^{10}\ \text{cm s}^{-1})\{(27.878)(14.509)(9.287)\ \text{cm}^{-3}\}^{1/3}}{1.381\times10^{-23}\ \text{J K}^{-1}}$$

$$= \boxed{22.36\ \text{K}}$$

so it is valid in this case. All the rotational modes of water are fully active at 25°C; therefore

$$U_m^\text{R} - U_m^\text{R}(0) = E^\text{R} = \frac{3}{2}RT \text{, the equipartition value}$$

$$S_m^\text{R} = \frac{E^\text{R}}{T} + R\ln q^\text{R} = \frac{3}{2}R + R\ln 43.1 = \boxed{43.76\ \text{J K}^{-1}\ \text{mol}^{-1}}$$

Comment. Division of q^R by $N_\text{A}!$ is not required for the internal contributions; internal motions may be thought of as localized (distinguishable). It is the overall canonical partition function, which is a product of internal and external contributions, that is divided by $N_\text{A}!$.

E15E.8(a) We assume that the upper nine of the $(2\times\frac{9}{2}+1)=10$ spin–orbit states of the ion lie at an energy much greater than kT at 1 K; hence, since the spin degeneracy of Co^{2+} is 4 (the ion is a spin quartet), $q = 4$. The contribution to the entropy is

$$R\ln q = (8.3145\ \text{J K}^{-1}\ \text{mol}^{-1})\times(\ln 4) = \boxed{11.5\ \text{J K}^{-1}\ \text{mol}^{-1}}$$

E15E.9(a) The molar entropy of a collection of oscillators is given by

$$S_m = \frac{U_m - U_m(0)}{T} + k\ln Q \text{ [15E.8c]} = \frac{R(\theta^V/T)}{e^{\theta^V/T}-1} - R\ln(1-e^{-\theta^V/T}) \text{ [15E.14b]}$$

where $\theta^V = hc\tilde{\nu}/k$ is the vibrational temperature. The vibrational entropy of formic acid is the sum of contributions of this form from each of its nine normal modes. The table below shows results from a spreadsheet programmed to compute S_m/R at a given temperature for the normal-mode wavenumbers of formic acid.

		T = 298 K		T = 500 K	
$\tilde{\nu}$ / cm^{-1}	θ^V/K	T/θ^V	S_m/R	T/θ^V	S_m/R
638	918.0	0.324635	1.624247	0.544689	4.340205
1033	1486.3	0.200501	0.341767	0.33641	1.769583
625	899.3	0.331387	1.707377	0.556019	4.471129
1105	1589.9	0.187436	0.254961	0.31449	1.500608
1229	1768.3	0.168525	0.153044	0.28276	1.127318
1387	1995.6	0.149327	0.079118	0.250549	0.779719
1770	2546.7	0.117015	0.015427	0.196334	0.312684
2943	4234.4	0.070376	8.53E-05	0.118081	0.016529
3570	5136.5	0.058016	4.95E-06	0.097342	0.003239
			4.176032		14.32101

(i) At 298 K, $S_m = 4.176R = \boxed{34.72\text{ J K}^{-1}\text{ mol}^{-1}}$

(ii) At 500 K, $S_m = 14.32R = \boxed{119.06\text{ J K}^{-1}\text{ mol}^{-1}}$

Comment. These calculated values are the vibrational contributions to the standard molar entropy of formic acid. The total molar entropy would also include translational and rotational contributions, but without knowledge of the rotational constants the total molar entropy cannot be calculated.

Solutions to problems

P15E.1 $q^E = \sum_j g_j e^{-\beta\varepsilon_j} = 2 + 2e^{-\beta\varepsilon}, \quad \varepsilon = \Delta\varepsilon = 121.1\text{cm}^{-1}$

$$C_{V,m} = -k\beta^2\left(\frac{\partial U_m}{\partial \beta}\right)_V \text{ [15E.5]}$$

The molar internal energy is [15E.2a]

$$U_m - U_m(0) = -\frac{N_A}{q^E}\left(\frac{\partial q^E}{\partial \beta}\right)_V = \frac{N_A\varepsilon e^{-\beta\varepsilon}}{q^E}$$

Let $x = \beta\varepsilon$, then $\mathrm{d}\beta = \frac{1}{\varepsilon}\mathrm{d}x$

$$C_{V,\mathrm{m}} = -k\left(\frac{x}{\varepsilon}\right)^2 \varepsilon\frac{\partial}{\partial x}\left(\frac{N_A\varepsilon \mathrm{e}^{-x}}{1+\mathrm{e}^{-x}}\right) = -N_A k x^2 \times \frac{\partial}{\partial x}\left(\frac{\mathrm{e}^{-x}}{1+\mathrm{e}^{-x}}\right) = R\left(\frac{x^2\mathrm{e}^{-x}}{(1+\mathrm{e}^{-x})^2}\right)$$

We then draw up the following table

T/K	50	298	500
(kT/hc)/mol^{-1}	34.8	207	348
x	3.48	0.585	0.348
$C_{V,\mathrm{m}}/R$	0.351	0.079	0.029
$C_{V,\mathrm{m}}$/(J K^{-1} mol^{-1})	2.91	0.654	0.244

In order to plot the heat capacity against T we first evaluate $x = \beta\varepsilon = \frac{\varepsilon}{k}\left(\frac{1}{T}\right)$. After substituting values for ε and k we find $x = \frac{174.23}{T/\mathrm{K}}$. $C_{V,\mathrm{m}}/R \equiv c(T)$ is evaluated in the following Mathcad worksheet and plotted against T in Fig. 15E.2.

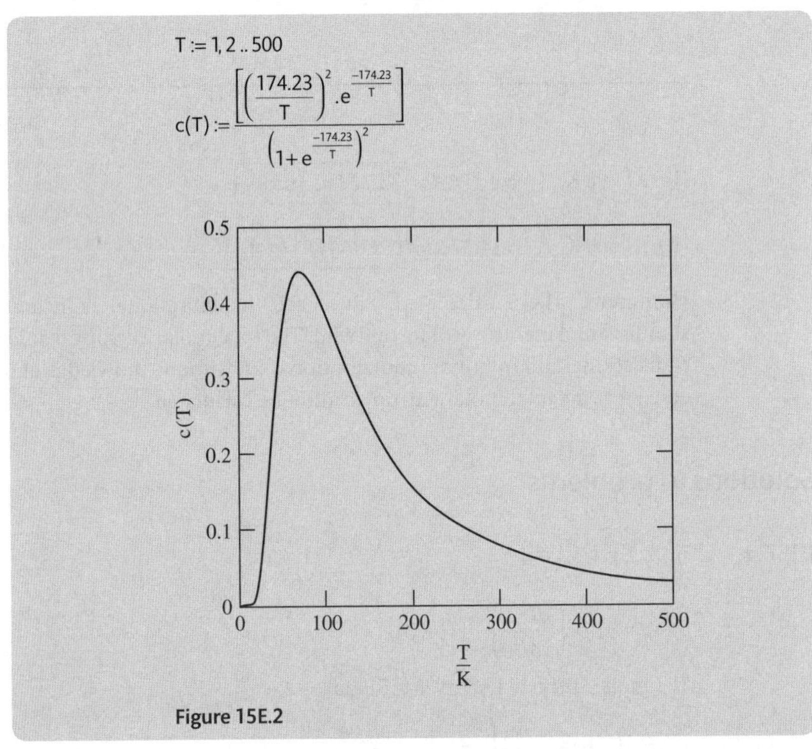

Figure 15E.2

Comment. Note that the double degeneracies do not affect the results because the two factors of 2 in q cancel when U is formed. In the range of temperature specified, the electronic contribution to the heat capacity decreases with increasing temperature.

P15E.3 The energy expression for a particle on a ring is

$$E_{m_l} = \frac{\hbar^2 m_l^2}{2I} \quad [8C.6b]$$

Therefore

$$q = \sum_{m=-\infty}^{\infty} e^{-m_l^2\hbar^2/2IkT} = \sum_{m=-\infty}^{\infty} e^{-\beta m_l^2\hbar^2/2I}$$

The summation may be approximated by an integration

$$q \approx \frac{1}{\sigma}\int_{-\infty}^{\infty} e^{-m_l^2\hbar^2/2IkT}\,dm_l = \frac{1}{\sigma}\left(\frac{2IkT}{\hbar^2}\right)^{1/2}\int_{-\infty}^{\infty} e^{-x^2}\,dx \approx \frac{1}{\sigma}\left(\frac{2\pi IkT}{\hbar^2}\right)^{1/2} = \frac{1}{\sigma}\left(\frac{2\pi I}{\hbar^2\beta}\right)^{1/2}$$

$$U - U(0) = -N\frac{\partial \ln q}{\partial \beta} = -N\frac{\partial}{\partial \beta}\ln\frac{1}{\sigma}\left(\frac{2\pi I}{\hbar^2\beta}\right)^{1/2} = \frac{N}{2\beta} = \frac{1}{2}NkT = \frac{1}{2}RT \quad (N = N_A)$$

$$C_{V,m} = \left(\frac{\partial U_m}{\partial T}\right)_V = \frac{1}{2}R = \boxed{4.2\ \mathrm{J\,K^{-1}\,mol^{-1}}}$$

$$\begin{aligned} S_m &= \frac{U_m - U_m(0)}{T} + R\ln q \\ &= \frac{1}{2}R + R\ln\frac{1}{\sigma}\left(\frac{2\pi IkT}{\hbar^2}\right)^{1/2} \\ &= \frac{1}{2}R + R\ln\frac{1}{3}\left(\frac{(2\pi)\times(5.341\times10^{-47}\ \mathrm{kg\,m^2})\times(1.381\times10^{-23}\ \mathrm{J\,K^{-1}})\times(298\ \mathrm{K})}{(1.055\times10^{-34}\ \mathrm{J\,s})^2}\right)^{1/2} \\ &= \frac{1}{2}R + 1.31R = 1.81R, \text{or } \boxed{15\ \mathrm{J\,K^{-1}\,mol^{-1}}} \end{aligned}$$

P15E.5 The vibrational temperature is defined by

$$k\theta^V = hc\tilde{\nu}$$

so a vibration with θ^V less than 1 000 K has a wavenumber less than

$$\tilde{\nu} = \frac{k\theta^V}{hc} = \frac{(1.381\times10^{-23}\ \mathrm{J\,K^{-1}})\times(1\,000\ \mathrm{K})}{(6.626\times10^{-34}\ \mathrm{J\,s})\times(2.998\times10^{10}\ \mathrm{cm\,s^{-1}})} = 695.2\ \mathrm{cm^{-1}}$$

There are seven such wavenumbers listed among those for C_{60}: two T_{1u}, a T_{2u}, a G_u, and three H_u. The number of *modes* involved, ν^{V*}, must take into account the degeneracy of these vibrational energies

$$\nu^{V*} = 2(3) + 1(3) + 1(4) + 3(5) = \boxed{28}$$

The molar heat capacity of a molecule is roughly

$$C_{V,\mathrm{m}} = \tfrac{1}{2}(3+\nu^{\mathrm{R}^*}+2\nu^{\mathrm{V}^*})R\ [15\mathrm{E}.6] = \tfrac{1}{2}(3+3+2\times 28)R = 31R$$
$$= 31(8.3145\ \mathrm{J\,mol^{-1}\,K^{-1}}) = \boxed{258\ \mathrm{J\,mol^{-1}\,K^{-1}}}$$

P15E.7 (a) $$U-U(0) = -\frac{N}{q}\frac{\partial q}{\partial \beta} = \frac{N}{q}\sum \varepsilon_j \mathrm{e}^{-\beta\varepsilon_j} = \frac{NkT}{q}\dot{q} = \boxed{nRT\left(\frac{\dot{q}}{q}\right)}$$

$$C_V = \left(\frac{\partial U}{\partial T}\right)_V = \frac{\partial \beta}{\partial T}\left(\frac{\partial U}{\partial \beta}\right)_V = -\frac{1}{kT^2}\frac{\partial}{\partial \beta}\left(\frac{N}{q}\sum_j \varepsilon_j \mathrm{e}^{-\beta\varepsilon_j}\right)$$
$$= \left(\frac{N}{kT^2}\right)\times\left[\frac{1}{q}\sum_j \varepsilon_j^2 \mathrm{e}^{-\beta\varepsilon_j} + \frac{1}{q^2}\left(\frac{\partial q}{\partial \beta}\right)\sum_j \varepsilon_j \mathrm{e}^{-\beta\varepsilon_j}\right]$$
$$= \left(\frac{N}{kT^2}\right)\times\left[\frac{1}{q}\sum_j \varepsilon_j^2 \mathrm{e}^{-\beta\varepsilon_j} - \frac{1}{q^2}\left(\sum_j \varepsilon_j \mathrm{e}^{-\beta\varepsilon_j}\right)^2\right]$$
$$= \left(\frac{N}{kT^2}\right)\times\left[\frac{k^2T^2\ddot{q}}{q} - \frac{k^2T^2}{q^2}\dot{q}^2\right]$$
$$= \boxed{nR\left\{\frac{\ddot{q}}{q} - \left(\frac{\dot{q}}{q}\right)^2\right\}}$$

$$S = \frac{U-U(0)}{T} + nR\ln\left(\frac{q}{N}+1\right) = \boxed{nR\left(\frac{\dot{q}}{q} + \ln\frac{\mathrm{e}q}{N}\right)}$$

(b) At 5 000 K, $\dfrac{kT}{hc} = 3\,475\ \mathrm{cm^{-1}}$. We form the sums

$$q = \sum_j g\mathrm{e}^{-\beta\varepsilon_j} = 1 + \mathrm{e}^{-21\,850/3\,475} + 3\mathrm{e}^{-21\,870/3\,475} + \cdots = 1.0167$$

$$\dot{q} = \sum_j \frac{\varepsilon_j}{kT}\mathrm{e}^{-\beta\varepsilon_j} = \frac{hc}{kT}\sum_j \tilde{\nu}_j \mathrm{e}^{-\beta\varepsilon_j}$$
$$= \left(\frac{1}{3\,475}\right)\times\{0 + 21\,850\mathrm{e}^{-21\,850/3\,475} + 3\times 21\,870\mathrm{e}^{-21\,870/3\,475} + \cdots\} = 0.1057$$

$$\ddot{q} = \sum_j \left(\frac{\varepsilon_j}{kT}\right)^2 \mathrm{e}^{-\beta\varepsilon_j} = \left(\frac{hc}{kT}\right)^2 \sum_j \tilde{\nu}_j^2 \mathrm{e}^{-\beta\varepsilon_j}$$
$$= \left(\frac{1}{3\,475}\right)^2 \times\{0 + 21\,850^2\,\mathrm{e}^{-21\,850/3\,475} + 3\times 21\,870^2\,\mathrm{e}^{-21\,870/3\,475} + \cdots\} = 0.6719$$

The electronic contribution to the molar constant-volume heat capacity is

$$C_{V,\mathrm{m}} = R\left\{\frac{\ddot{q}}{q} - \left(\frac{\dot{q}}{q}\right)^2\right\}$$
$$= 8.3145\ \mathrm{J\,K^{-1}\,mol^{-1}} \times \left\{\frac{0.6719}{1.0167} - \left(\frac{0.1057}{1.0167}\right)^2\right\} = \boxed{5.41\ \mathrm{J\,K^{-1}\,mol^{-1}}}$$

P15E.9 As the statement of this problem suggests, the derivation of

$$C_V = \frac{kN\beta^2}{2}\zeta(\beta)$$

given in Problem 15E.8 is completely general. That is, it makes no use of the fact that the energies and degeneracies in question were those of a linear rotor. The derivation and therefore the result can apply equally well to a nonlinear rotor, to electronic energy levels, or to the vibrational energy levels involved of this problem.

To evaluate contributions of individual excitations to the heat capacity, we re-write $\zeta(\beta)$ in notation associated with vibrational energy levels

$$\zeta(\beta)=\frac{1}{q^2}\sum_{v,v'}\{\varepsilon(v)-\varepsilon(v')\}^2 g(v)g(v')\mathrm{e}^{-\beta\{\varepsilon(v)+\varepsilon(v')\}}=\frac{1}{q^2}\sum_{v,v'}\{\varepsilon(v)-\varepsilon(v')\}^2\mathrm{e}^{-\beta\{\varepsilon(v)+\varepsilon(v')\}}$$

where the levels are non-degenerate, or at least are treated as such because vibrational modes are treated one by one. The energy levels are

$$\varepsilon(v)=hc\tilde{\nu}v=\theta^{\mathrm{V}}kv \quad \text{so} \quad \beta\varepsilon(v)=\theta^{\mathrm{V}}v/T.$$

The total heat capacity and the contributions of several transitions are plotted in Fig. 15E.3. For vibration, one can compute q and the total C_V/R analytically:

$$q^{\mathrm{V}}=\frac{1}{1-\mathrm{e}^{-\theta^{\mathrm{V}}/T}}\ [15B.15] \text{ and } \frac{C_{V,\mathrm{m}}}{R}=\left(\frac{\theta^{\mathrm{V}}}{T}\right)^2\frac{\mathrm{e}^{-\theta^{\mathrm{V}}/T}}{(1-\mathrm{e}^{-\theta^{\mathrm{V}}/T})^2}\ [15E.3]$$

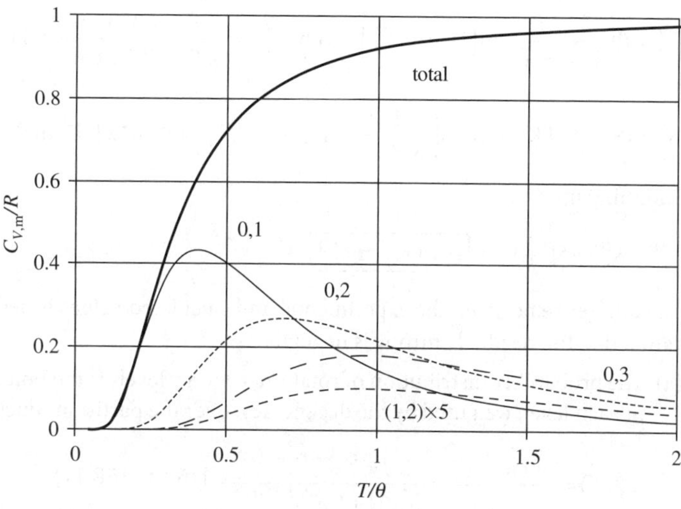

Figure 15E.3

P15E.11 The translational contribution to the entropy is given by the Sackur–Tetrode equation [15E.11a]:

$$S_m^{T\ominus} = R\ln\left(\frac{V_m e^{5/2}}{N_A \Lambda^3}\right) \quad \text{where} \quad \Lambda = \frac{h}{(2\pi mkT)^{1/2}} \quad [15B.7b]$$

After substituting values for the constants we obtain

$$\Lambda = \frac{h}{(2\pi mkT)^{1/2}} = \frac{6.626\times10^{-34}\ \text{J s}}{\{2\pi(28.02\times1.6605\times10^{-27}\ \text{kg})(1.381\times10^{-23}\ \text{JK}^{-1})(298\text{K})\}^{1/2}}$$
$$= 1.91\times10^{-11}\ \text{m}$$

and $S_m^{T\ominus} = (8.3145\ \text{J K}^{-1}\ \text{mol}^{-1})\ln\left(\frac{(2.479\times10^{-2}\ \text{m}^3)e^{5/2}}{(6.022\times10^{23}\ \text{mol}^{-1})(1.91\times10^{-11}\ \text{m})^3}\right)$

$$= 150\ \text{J K}^{-1}\ \text{mol}^{-1}$$

The rotational contribution is [15E.13a]

$$S_m^R = R\left(1 + \ln\frac{kT}{\sigma hc\tilde{B}}\right)$$

It will be useful to note that $\dfrac{kT}{hc} = \dfrac{(1.381\times10^{-23}\ \text{J K}^{-1})(298\ \text{K})}{(6.626\times10^{-34}\ \text{J s})(2.998\times10^{10}\ \text{cm s}^{-1})} = 207.2\ \text{cm}^{-1}$

Thus $S_m^R = (8.3145\ \text{J K}^{-1}\ \text{mol}^{-1})\left(1 + \ln\dfrac{207.2\ \text{cm}^{-1}}{(2)(1.9987\ \text{cm}^{-1})}\right) = 41.1\ \text{J K}^{-1}\ \text{mol}^{-1}$

The vibrational contribution is [15E.13a]

$$S_m^V = R\left(\frac{\theta^V/T}{e^{\theta^V/T}-1} - \ln\left(1-e^{-\theta^V/T}\right)\right) \quad \text{with} \quad \frac{\theta^V}{T} = \frac{hc\tilde{\nu}}{kT} = \frac{2\,358\ \text{cm}^{-1}}{207.2\ \text{cm}^{-1}} = 11.38$$

$$S_m^V = (8.3145\ \text{J K}^{-1}\ \text{mol}^{-1})\left(\frac{11.38}{e^{11.38}-1} - \ln\left(1-e^{-11.38}\right)\right) = 0.0012\ \text{J K}^{-1}\ \text{mol}^{-1}$$

Summing up

$$S_m^{\ominus} = S_m^{T\ominus} + S_m^R + S_m^V = \boxed{191\ \text{J K}^{-1}\ \text{mol}^{-1}}$$

The difference between the experimental and calculated values is negligible, indicating that the residual entropy is negligible.

P15E.13 (a) The probability distribution of rotational energy levels is the Boltzmann factor of each level, weighted by the degeneracy, over the partition function

$$p_J^R(T) = \frac{g(J)e^{-\varepsilon_J/kT}}{q^R} = \frac{(2J+1)e^{-hc\tilde{B}J(J+1)/kT}}{\sum_{J=0}(2J+1)e^{-hc\tilde{B}J(J+1)/kT}} \quad [15A.6, 15B.11]$$

It is conveniently plotted against J at several temperatures using mathematical software. This distribution at 100 K is shown in Fig. 15E.4(a) as both a bar plot and a line plot.

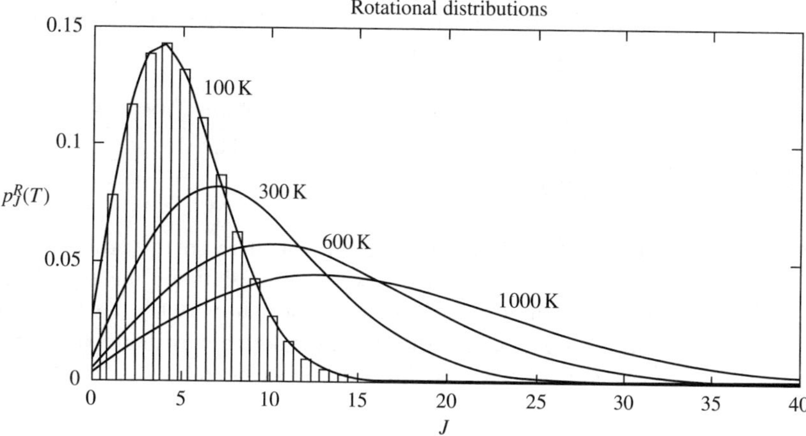

Figure 15E.4(a)

The plots show that higher rotational states become more heavily populated at higher temperature. Even at 100 K the most populated state has 4 quanta of rotational energy; it is elevated to 13 quanta at 1 000 K

Values of the vibrational state probability distribution,

$$p_v^{\mathrm{V}}(T)=\frac{\mathrm{e}^{-\varepsilon_j/kT}}{q^{\mathrm{V}}}=\mathrm{e}^{-vhc\tilde{\nu}/kT}(1-\mathrm{e}^{-hc\tilde{\nu}/kT})\ [15\mathrm{B}.15]$$

are conveniently tabulated against v at several temperatures. Computations may be discontinued when values drop below some small number like 10^{-7}.

	$p_v^{\mathrm{V}}(T)$			
v	**100 K**	**300 K**	**600 K**	**1 000 K**
0	1	1	0.095	0.956
1	2.77×10^{-14}	3.02×10^{-5}	5.47×10^{-3}	0.042
2		9.15×10^{-10}	3.01×10^{-5}	1.86×10^{-3}
3			1.65×10^{-7}	8.19×10^{-5}
4				3.61×10^{-6}
5				1.59×10^{-7}

Only the state $v=0$ is appreciably populated below 1 000 K and even at 1 000 K only 4% of the molecules have one quantum of vibrational energy.

(b) The classical (equipartition) rotational partition function is

$$q^{\mathrm{R}}_{\text{classical}}(T)=\frac{kT}{hc\tilde{B}}=\frac{T}{\theta^{\mathrm{R}}}\quad[15\mathrm{B}.13]$$

where θ^{R} is the rotational temperature. We would expect the partition function to be well approximated by this expression for temperatures much greater than the rotational temperature.

$$\theta^{\mathrm{R}}=\frac{hc\tilde{B}}{k}=\frac{(6.626\times10^{-34}\,\mathrm{J\,s})\times(2.998\times10^{10}\,\mathrm{cm\,s^{-1}})\times(1.931\,\mathrm{cm^{-1}})}{1.381\times10^{-23}\,\mathrm{J\,K^{-1}}}$$

$$=2.779\,\mathrm{K}$$

In fact $\theta^{\mathrm{R}} << T$ for all temperatures of interest in this problem (100 K or more). Agreement between the classical expression and the explicit sum is indeed good, as Fig. 15E.4(b) confirms. The figure displays the percentage deviation $(q^{\mathrm{R}}_{\text{classical}}-q^{\mathrm{R}})100/q^{\mathrm{R}}$. The maximum deviation is about –0.9% at 100 K and the magnitude decreases with increasing temperature.

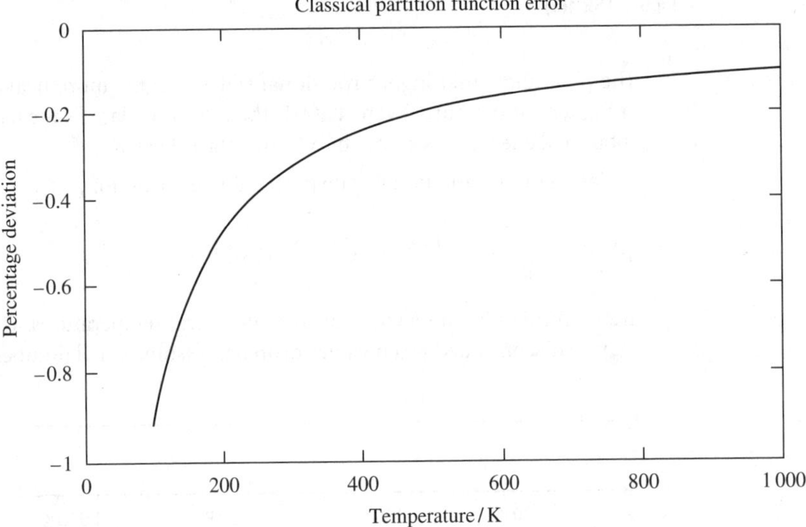

Figure 15E.4(b)

(c) The translational, rotational, and vibrational contributions to the total energy are specified by eqns 15C.5b, 15C.6b, and 15C.8, respectively, along with the realization that molar internal energy is $N_{\mathrm{A}}\langle\varepsilon\rangle$:

$$U^{\mathrm{T}}=\tfrac{3}{2}RT,\quad U^{\mathrm{R}}=RT,\quad U^{\mathrm{V}}=\frac{N_{\mathrm{A}}hc\tilde{\nu}}{e^{hc\tilde{\nu}/kT}-1}$$

The contributions to the difference in energy from its 100 K value are $\Delta U^{\mathrm{T}}(T)=U^{\mathrm{T}}(T)-U^{\mathrm{T}}(100\,\mathrm{K})$, etc. Figure 15E.4(c) shows the individual contributions to $\Delta U(T)$. Translational motion contributes 50% more than the rotational motion because it has three quadratic degrees of freedom compared to two quadratic degrees of freedom for rotation. Very little change occurs in the vibrational energy because very high temperatures are required to populate $v = 1, 2, \ldots.$ states (see part (a)).

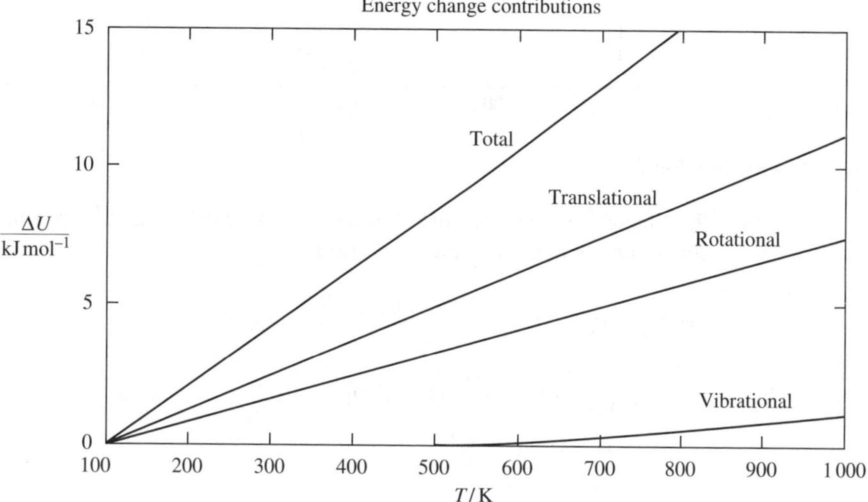

Figure 15E.4(c)

$$C_{V,\mathrm{m}}(T) = \left(\frac{\partial U(T)}{\partial T}\right)_V = \left(\frac{\partial}{\partial T}\right)_V (U^{\mathrm{T}} + U^{\mathrm{R}} + U^{\mathrm{V}})$$

$$= \frac{3}{2}R + R + \frac{\mathrm{d}U^{\mathrm{V}}}{\mathrm{d}T} = \frac{5}{2}R + \frac{\mathrm{d}U^{\mathrm{V}}}{\mathrm{d}T}$$

The derivative $\mathrm{d}U^{\mathrm{V}}/\mathrm{d}T$ may be evaluated numerically with numerical software (we advise exploration of the technique) or it may be evaluated analytically using eqn 15E.3:

$$C_{V,\mathrm{m}}^{\mathrm{V}} = \frac{\mathrm{d}U^{\mathrm{V}}}{\mathrm{d}T} = R\left\{\frac{\theta_{\mathrm{V}}}{T}\left(\frac{\mathrm{e}^{-\theta_{\mathrm{V}}/2T}}{1-\mathrm{e}^{-\theta_{\mathrm{V}}/T}}\right)\right\}^2$$

where $\theta^{\mathrm{V}} = hc\tilde{\nu}/k = 3122$ K. Figure 15E.4(d) shows the ratio of the vibrational contribution to the sum of translational and rotational contributions. Below 300 K, vibrational motion makes a small, perhaps negligible, contribution to the heat capacity. The contribution is about 10% at 700 K and grows with increasing temperature.

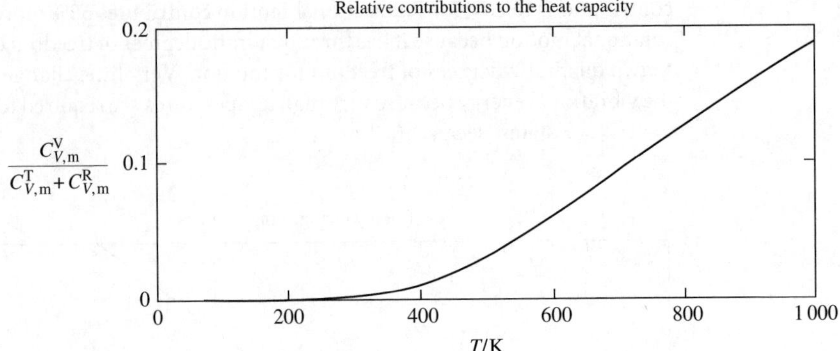

Figure 15E.4(d)

The change with temperature of molar entropy may be evaluated by numerical integration with mathematical software.

$$\Delta S(T)=S(T)-S(100\text{K})=\int_{100\text{K}}^{T}\frac{C_{p,\text{m}}(T)\text{d}T}{T}\quad[3\text{A}.19]$$

$$=\int_{100\text{K}}^{T}\frac{C_{V,\text{m}}(T)+R}{T}\text{d}T\quad[2\text{B}.9]$$

$$=\int_{100\text{K}}^{T}\frac{\frac{7}{2}R+C^{\text{V}}_{V,\text{m}}(T)}{T}\text{d}T$$

$$=\underbrace{\frac{7}{2}R\ln\left(\frac{T}{100\text{K}}\right)}_{\Delta S^{\text{T+R}}(T)}+\underbrace{\int_{100\text{K}}^{T}\frac{C^{\text{V}}_{V,\text{m}}(T)}{T}\text{d}T}_{\Delta S^{\text{V}}(T)}$$

Figure 15E.4(e) shows the ratio of the vibrational contribution to the sum of translational and rotational contributions. Even at the highest temperature the vibrational contribution to the entropy change is less than 2.5% of the contributions from translational and rotational motion. The vibrational contribution is negligible at low temperature.

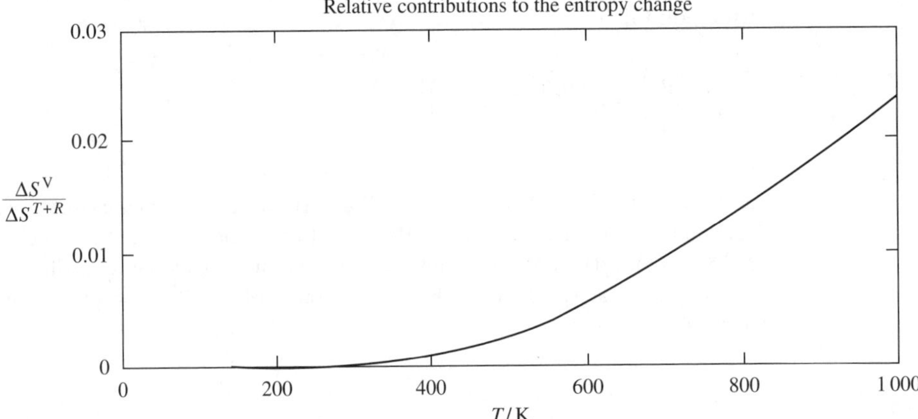

Figure 15E.4(e)

P15E.15 The partition function for a two-level system with energy separation ε is [*Brief illustration* 15C.1]

$$q=1+e^{-\varepsilon/kT}=1+e^{-\beta\varepsilon}$$

The entropy is calculated from

$$S=\frac{U-U(0)}{T}+Nk\ln q=k\beta\{U-U(0)\}+Nk\ln q \quad [15E.8a]$$

$$U-U(0)=-\frac{N}{q}\frac{dq}{d\beta}=\frac{N\varepsilon}{1+e^{\beta\varepsilon}} \quad [\textit{Brief illustration } 15C.1]$$

$$S=Nk\left[\frac{\beta\varepsilon}{1+e^{\beta\varepsilon}}+\ln(1+e^{-\beta\varepsilon})\right]$$

We now plot $\dfrac{S}{Nk}=\dfrac{S}{R}$ (for $N=N_A$) against $x\equiv\beta\varepsilon=\dfrac{\varepsilon}{kT}$ and against $y\equiv\dfrac{1}{x}=\dfrac{kT}{\varepsilon}$.

These plots (Figs 15E.5(a) and (b)) are shown in the following Mathcad worksheets.

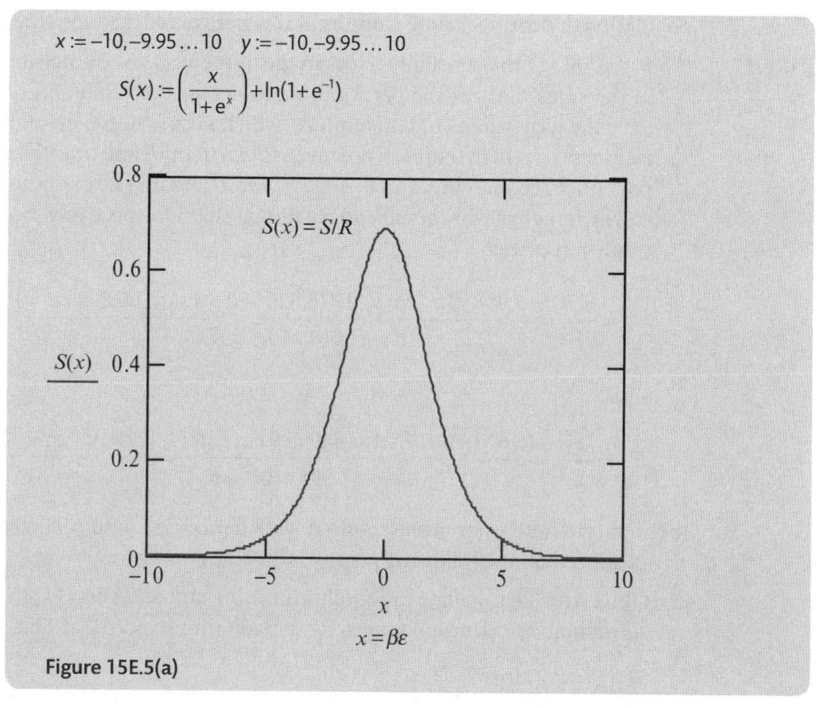

Figure 15E.5(a)

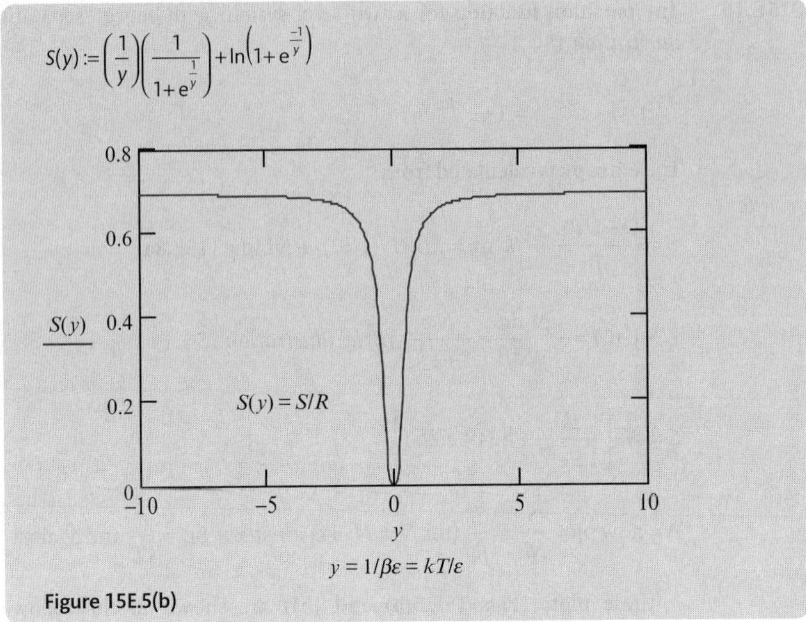

Figure 15E.5(b)

Physically, the increase in entropy for $T>0$ corresponds to the increasing accessibility of the upper state, and the decrease for $T<0$ corresponds to the shift towards population of the upper state alone as more energy is packed into the system.

P15E.17 (a) θ^{V} and θ^{R} are the constant factors in the numerators of the negative exponents in the sums that are the partition functions for vibration and rotation. They have the dimensions of temperature, which occurs in the denominator of the exponents. So high temperature means $T >> \theta^{\mathrm{V}}$ or θ^{R} and only then does the exponential become substantial. Thus θ^{V} and θ^{R} are measures of the temperature at which higher vibrational and rotational states, respectively, become significantly populated.

$$\theta^{\mathrm{R}} = \frac{hc\tilde{B}}{k} = \frac{(2.998\times10^{10}\,\mathrm{cm\,s^{-1}})\times(6.626\times10^{-34}\,\mathrm{J\,s})\times(60.864\,\mathrm{cm^{-1}})}{(1.381\times10^{-23}\,\mathrm{J\,K^{-1}})} = \boxed{87.55\,\mathrm{K}}$$

and

$$\theta^{\mathrm{V}} = \frac{hc\tilde{\nu}}{k} = \frac{(6.626\times10^{-34}\,\mathrm{J\,s})\times(4400.39\,\mathrm{cm^{-1}})\times(2.998\times10^{10}\,\mathrm{cm\,s^{-1}})}{(1.381\times10^{-23}\,\mathrm{J\,K^{-1}})} = \boxed{6\,330\,\mathrm{K}}$$

(b) and (c) These parts of the solution were performed with Mathcad and are reproduced on the following pages.

Objective: To calculate the equilibrium constant $K(T)$ and $C_p(T)$ for dihydrogen at high temperature for a system made with n mol H_2 at 1 bar.

$$H_2(g) \rightleftharpoons 2\,H(g)$$

At equilibrium the degree of dissociation, α, and the equilibrium amounts of H_2 and atomic hydrogen are related by the expressions

$$n_{H_2} = (1-\alpha)n \quad \text{and} \quad n_H = 2\alpha n$$

The equilibrium mole fractions are

$$x_{H_2} = (1-\alpha)n/\{(1-\alpha)n+2\alpha n\} = (1-\alpha)/(1+\alpha)$$

$$x_H = 2\alpha n/\{(1-\alpha)n+2\alpha n\} = 2\alpha/(1+\alpha)$$

The partial pressures are

$$p_{H_2} = (1-\alpha)p/(1+\alpha) \quad \text{and} \quad p_H = 2\alpha p/(1+\alpha)$$

The equilibrium constant is

$$K(T) = \frac{(p_H/p^\ominus)^2}{(p_{H_2}/p^\ominus)} = 4\alpha^2\frac{(p/p^\ominus)}{(1-\alpha^2)} = \frac{4\alpha^2}{(1-\alpha^2)} \quad \text{where } p = p^\ominus = 1\text{ bar}$$

The above equation is easily solved for α

$$\alpha = \left(\frac{K}{K+4}\right)^{1/2}$$

The heat capacity at constant volume for the equilibrium mixture is

$$C_V(\text{mixture}) = n_H C_{V,m}(H) + n_{H_2} C_{V,m}(H_2)$$

The heat capacity at constant volume per mole of dihydrogen used to prepare the equilibrium mixture is

$$C_V = C_V(\text{mixture})/n = \{n_H C_{V,m}(H) + n_{H_2} C_{V,m}(H_2)\}/n$$
$$= \boxed{2\alpha C_{V,m}(H) + (1-\alpha)C_{V,m}(H_2)}$$

The formula for the heat capacity at constant pressure per mole of dihydrogen used to prepare the equilibrium mixture (C_p) can be deduced from the molar relationship $C_{p,m} = C_{V,m} + R$ [2B.9]

$$C_p = \{n_H C_{p,m}(H) + n_{H_2} C_{p,m}(H_2)\}/n$$
$$= \frac{n_H}{n}\{C_{V,m}(H)+R\} + \frac{n_{H_2}}{n}\{C_{V,m}(H_2)+R\}$$
$$= \frac{n_H C_{V,m}(H) + n_{H_2} C_{V,m}(H_2)}{n} + R\left(\frac{n_H + n_{H_2}}{n}\right)$$
$$= C_V + R(1+\alpha)$$

Calculations

J = joule s = second kJ = 1 000 J

mol = mole g = gram bar = 1×10^5 Pa

$h = 6.62608\times10^{-34}$ J s $c = 2.9979\times10^8$ m s^{-1} $k = 1.38066\times10^{-23}$ J K^{-1}

$R = 8.31451$ J K^{-1} mol^{-1} $N_A = 6.02214\times10^{23}$ mol^{-1} $p^\ominus = 1$ bar

Molecular properties of

$\tilde{\nu} = 4400.39$ cm^{-1} $\tilde{B} = 60.864$ cm^{-1} $D = 432.1$ kJ mol^{-1}

$$m_H = \frac{1\,\text{g mol}^{-1}}{N_A} \qquad m_{H_2} = 2m_H$$

$$\theta^V = \frac{hc\tilde{\nu}}{k} \qquad \theta^R = \frac{hc\tilde{B}}{k}$$

Computation of $K(T)$ and $\alpha(T)$

$$N = 200 \quad i = 0,\ldots,N \quad T_i = 500\,\text{K} + \frac{i\times5\,500\,\text{K}}{N}$$

$$\Lambda_{Hi} = \frac{h}{(2\pi m_H kT_i)^{1/2}} \qquad \Lambda_{H_2i} = \frac{h}{(2\pi m_{H_2} kT_i)^{1/2}}$$

$$q_i^V = \frac{1}{1-e^{-(\theta^V/T_i)}} \qquad q_i^R = \frac{T_i}{2\theta^R}$$

$$\boxed{K_{eqi} = \frac{kT_i\left(\Lambda_{H_2i}\right)^3 e^{-(D/RT_i)}}{p^\ominus q_i^V q_i^R \left(\Lambda_{Hi}\right)^6}} \qquad \alpha_i = \left(\frac{K_{eqi}}{K_{eqi}+4}\right)^{1/2}$$

See Figs 15E.6(a) and (b).

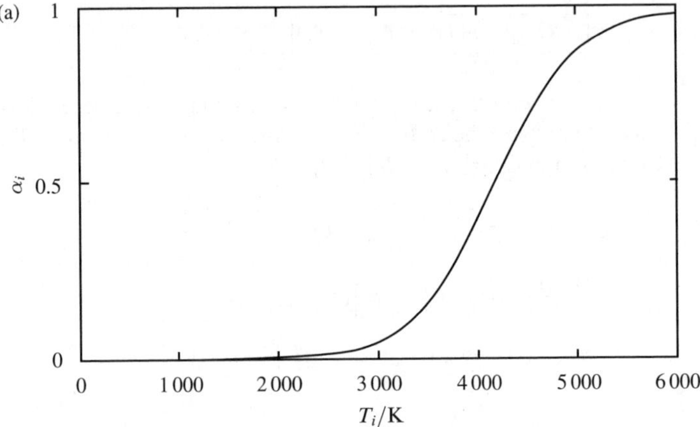

Figure 15E.6(a)

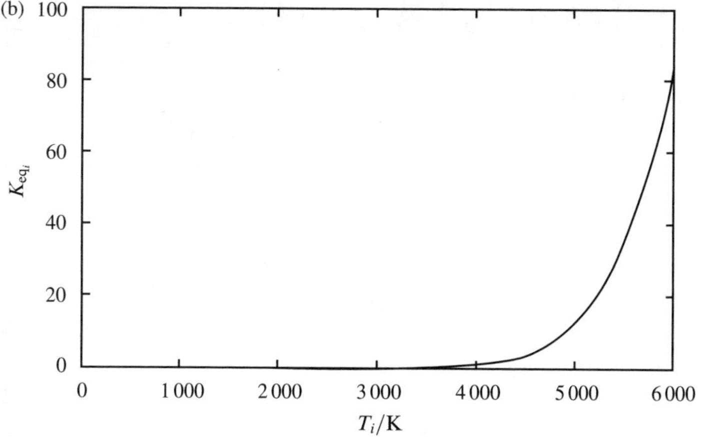

Figure 15E.6(b)

Heat capacity at constant volume per mole of dihydrogen used to prepare the equilibrium mixture is (see Fig. 15E.6(c))

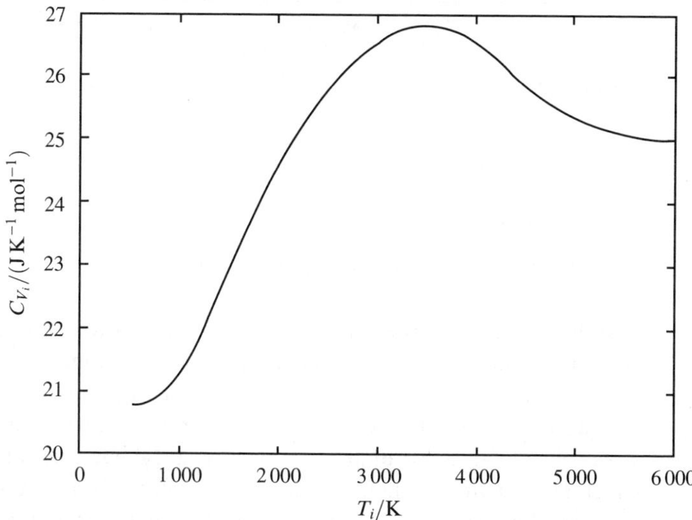

Figure 15E.6(c)

$C_{Vi}(\text{H}) = \boxed{1.5R}$

$$C_{Vi}(\text{H}_2) = \boxed{2.5R + \left[\frac{\theta^{\text{V}}}{T_i} \times \frac{\text{e}^{-(\theta^{\text{V}}/2T_i)}}{1-\text{e}^{\theta^{\text{V}}/T_i}}\right]^2 R} \quad C_{Vi} = 2\alpha_i C_{Vi}(\text{H}) + (1-\alpha_i) C_{Vi}(\text{H}_2)$$

The heat capacity at constant pressure per mole of dihydrogen used to prepare the equilibrium mixture is (see Fig. 15E.6(d))

$C_{pi} = C_{Vi} + R(1+\alpha_i)$

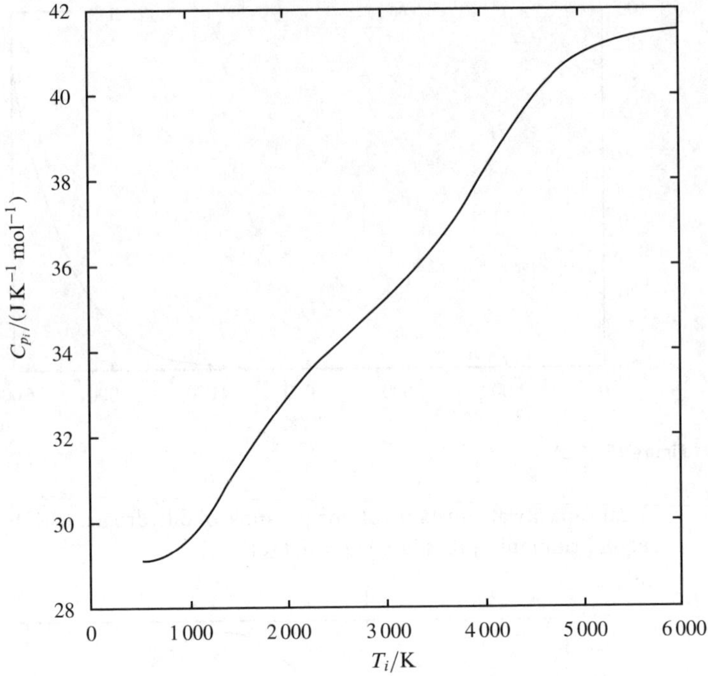

Figure 15E.6(d)

P15E.19 $S = k\ln \mathcal{W}$[15.27]

so $S = k\ln 4^N = Nk\ln 4$

$$= (5\times10^8)\times(1.38\times10^{-23}\,\mathrm{J\,K^{-1}})\times\ln 4 = \boxed{9.57\times10^{-15}\,\mathrm{J\,K^{-1}}}$$

Question. Is this a large residual entropy? The answer depends on what comparison is made. Multiply the answer by Avogadro's number to obtain the molar residual entropy, $5.76\times10^9\,\mathrm{J\,K^{-1}\,mol^{-1}}$, surely a large number—but then DNA is a macromolecule. The residual entropy per mole of base pairs may be a more reasonable quantity to compare to molar residual entropies of small molecules. To obtain that answer, divide the molecule's entropy by the number of base pairs before multiplying by N_A. The result is $11.5\,\mathrm{J\,K^{-1}\,mol^{-1}}$, a quantity more in line with examples discussed in Section 15E.2(e).

15F Derived functions

Answers to discussion questions

D15F.1 The relationship between pressure and the canonical partition function is [15F.3]

$$p = kT\left(\frac{\partial \ln Q}{\partial V}\right)_T$$

The derivative is a relationship between two non-negative quantities (Q and V). Furthermore, the derivative itself must be non-negative (since it is equal to the pressure times the non-negative quantity kT), so the derivative describes a monotonic relationship between Q and V: Q can only increase or remain constant with increasing V. The relationship might almost be interpreted anthropomorphically: to the extent that increasing the volume makes more states available (increasing Q), the system will push harder (higher p) to bring about that volume increase. Ultimately, it is a reflection of the Second Law of thermodynamics and the tendency toward increased disorder as quantified by the partition function.

D15F.3 The temperature dependence of the equilibrium constant in thermodynamic terms can be seen clearly in eqn 6A.18:

$$K = e^{\Delta_r S^\ominus/R} e^{-\Delta_r H^\ominus/RT}$$

In statistical thermodynamic terms, the equilibrium constant is [15F.10b]

$$K = \left\{\prod_J \left(\frac{q^\ominus_{J,m}}{N_A}\right)^{\nu_J}\right\} e^{-\Delta_r E_0/RT}$$

It is tempting to identify the entropic factor in the first expression with the partition function portion of the second and the enthalpic factor in the first with the energetic factor of the second, and such an identification would not be entirely misleading. A more comprehensive understanding, though, would recall that the standard reaction enthalpy at temperature T is not quite the same as the difference in ground state energies among the species involved. The enthalpies of the various species change with temperature by virtue of their heat capacities (Section 2B.2), which themselves can be related to partition functions (Section 15E.1(b)). So we can say that the temperature dependence of the equilibrium constant is largely but not entirely determined by the differences in ground-state energy levels of the species involved. Finally, recalling that raising the temperature of an equilibrium mixture tends to shift the reaction in the endothermic direction, we can recognize that it does so because raising the temperature has a greater effect on the number of accessible states of those species whose energy levels are higher; the species whose energy levels are lower already have significant population of several states at lower temperatures.

Solutions to exercises

E15F.1(a) In each case the contribution to G is given by

$G - G(0) = -nRT \ln q$ [15F.9 for non-translational modes]

Therefore, we first evaluate q^R and q^V.

$$q^R = \frac{kT}{\sigma hc\tilde{B}}[15B.13] = \frac{kT}{hc} \times \frac{1}{2(0.3902\ \mathrm{cm^{-1}})} = 265$$

The vibrational partition function is a product of three terms of the form of eqn 15B.15, one for each vibrational mode

$$q^{\mathrm{V}} = \left(\frac{1}{1-\mathrm{e}^{-a}}\right)\times\left(\frac{1}{1-\mathrm{e}^{-b}}\right)^2\times\left(\frac{1}{1-\mathrm{e}^{-c}}\right)$$

with

$$a = \frac{hc}{kT}\times 1388.2\ \mathrm{cm}^{-1} = 6.70\bar{2} \quad b = \frac{hc}{kT}\times 667.4\ \mathrm{cm}^{-1} = 3.22\bar{2}$$

and $c = \frac{hc}{kT}\times 2349.2\ \mathrm{cm}^{-1} = 11.3\bar{4}$

Hence

$$q^{\mathrm{V}} = \frac{1}{1-\mathrm{e}^{-6.702}}\times\left(\frac{1}{1-\mathrm{e}^{-3.222}}\right)^2\times\frac{1}{1-\mathrm{e}^{-11.34}} = 1.08\bar{6}$$

Therefore, the rotational contribution to the molar Gibbs energy is

$$-RT\ln q^{\mathrm{R}} = -8.314\ \mathrm{J\,K^{-1}\,mol^{-1}}\times 298\ \mathrm{K}\times\ln 265 = \boxed{-13.8\ \mathrm{kJ\,mol^{-1}}}$$

and the vibrational contribution is

$$-RT\ln q^{\mathrm{V}} = -8.314\ \mathrm{J\,K^{-1}\,mol^{-1}}\times 298\,\mathrm{K}\times\ln 1.08\bar{6} = \boxed{-0.20\ \mathrm{kJ\,mol^{-1}}}$$

E15F.2(a) See the solution to Exercise 15E.3(a).

(i) At 500 K

$$\beta\varepsilon = \frac{hc\times 881\ \mathrm{cm}^{-1}}{k\times 500\ \mathrm{K}} = 2.53\bar{5}$$

Therefore, the contribution to G_{m} is

$$G_{\mathrm{m}} - G_{\mathrm{m}}(0) = -RT\ln q \quad [15\mathrm{F}.9 \text{ for non-translational modes}]$$

$$G_{\mathrm{m}} - G_{\mathrm{m}}(0) = -(8.3145\,\mathrm{J K^{-1}\,mol^{-1}})\times(500\mathrm{K})\times\ln(4+2\times\mathrm{e}^{-2.535}) = \boxed{-6.42\,\mathrm{kJ mol^{-1}}}$$

(ii) At 900 K

$$\beta\varepsilon = \frac{hc\times 881\mathrm{cm}^{-1}}{k\times 500\mathrm{K}} = 1.40\bar{8}$$

Therefore, the contribution to G_{m} is

$$G_{\mathrm{m}} - G_{\mathrm{m}}(0) = -(8.3145\ \mathrm{J\,K^{-1}\,mol^{-1}})\times(900\ \mathrm{K})\times\ln(4+2\times\mathrm{e}^{-1.408}) = \boxed{-14.0\ \mathrm{kJ\,mol^{-1}}}$$

E15F.3(a) We use eqn 15F.11a with $\mathrm{X} = \mathrm{I}$, $\mathrm{X}_2 = \mathrm{I}_2$.

$$K = \left(\frac{(q^{\ominus}_{\mathrm{I,m}})^2}{q^{\ominus}_{\mathrm{I_2,m}}N_{\mathrm{A}}}\right)\mathrm{e}^{-\Delta_{\mathrm{r}}E_0/RT}$$

$$\Delta_{\mathrm{r}}E_0 = hc\tilde{D}_0 = hc\tilde{D}_{\mathrm{e}} - \frac{1}{2}hc\tilde{\nu}$$

$$= 1.5422\ \mathrm{eV}\times\frac{96\,485\ \mathrm{J\,mol^{-1}}}{1\ \mathrm{eV}} - \frac{hc}{2}\times 214.36\ \mathrm{cm}^{-1} = 1.475\times 10^5\ \mathrm{J\,mol^{-1}}$$

For I: $q^{\ominus}_{\mathrm{I,m}} = q^{\mathrm{T}\ominus}_{\mathrm{m}}(\mathrm{I})q^{\mathrm{E}}(\mathrm{I}), \quad q^{\mathrm{E}}(\mathrm{I}) = 4$

For I₂: $q^{\ominus}_{\mathrm{I_2,m}} = q^{\mathrm{T}\ominus}_{\mathrm{m}}(\mathrm{I_2})q^{\mathrm{R}}(\mathrm{I_2})q^{\mathrm{V}}(\mathrm{I_2})q^{\mathrm{E}}(\mathrm{I_2}), \quad q^{\mathrm{E}}(\mathrm{I_2}) = 1$

$$\frac{q^{\mathrm{T}\ominus}_{\mathrm{m}}(\mathrm{I_2})}{N_{\mathrm{A}}} = \frac{kT}{p^{\ominus}\Lambda^3} \qquad \Lambda = \frac{h}{(2\pi mkT)^{1/2}}$$

After substituting the values of the constants, we obtain

$$\frac{q^{\mathrm{T}\ominus}_{m}(\mathrm{I_2})}{N_{\mathrm{A}}} = 2.561\times10^{-2}(T/\mathrm{K})^{5/2}\times(M/\mathrm{g\,mol^{-1}})^{3/2}$$

$$= 2.561\times10^{-2}\times1\,000^{5/2}\times253.8^{3/2} = 3.27\times10^{9}$$

$$\frac{q^{\mathrm{T}\ominus}_{\mathrm{m}}(\mathrm{I})}{N_{\mathrm{A}}} = 2.561\times10^{-2}\times1\,000^{5/2}\times126.9^{3/2} = 1.16\times10^{9}$$

$$q^{\mathrm{R}} = \frac{kT}{\sigma hc\tilde{B}}[15\mathrm{B}.13] = \frac{k}{hc}\times\frac{1\,000\ \mathrm{K}}{2(0.0373\ \mathrm{cm^{-1}})} = 931\bar{6}$$

$$q^{\mathrm{V}} = \frac{1}{1-\mathrm{e}^{-\theta^{\mathrm{V}}/T}} \text{ where } \theta^{\mathrm{V}} = \frac{hc\tilde{\nu}}{k} = 1.4388\ \mathrm{K}\times(\tilde{\nu}/\mathrm{cm^{-1}})$$

$$q^{\mathrm{V}}(\mathrm{I_2}) = \frac{1}{1-\mathrm{e}^{-1.4388\times214.36/1000}} = 3.77$$

$$\text{so } K = \frac{(1.16\times10^{9}\times4)^2\mathrm{e}^{-17.741}}{(3.27\times10^{9})\times(9316)\times(3.77)} = \boxed{3.70\times10^{-3}}$$

Solutions to problems

P15F.1 The equilibrium constant is [15F.10, with $\Delta_{\mathrm{r}}E_0$ here defined as the molecular, not molar, energy difference; N_{A} factors cancel]

$$K = \frac{q^{\ominus}_{\mathrm{m}}(\mathrm{CHD_3})q^{\ominus}_{\mathrm{m}}(\mathrm{DCl})}{q^{\ominus}_{\mathrm{m}}(\mathrm{CD_4})q^{\ominus}_{\mathrm{m}}(\mathrm{HCl})}\mathrm{e}^{-\beta\Delta_{\mathrm{r}}E_0}$$

Use partition function expressions from Topic 15B.

The ratio of translational partition functions is

$$\frac{q^{\mathrm{T}}_{\mathrm{m}}(\mathrm{CHD_3})q^{\mathrm{T}}_{\mathrm{m}}(\mathrm{DCl})}{q^{\mathrm{T}}_{\mathrm{m}}(\mathrm{CD_4})q^{\mathrm{T}}_{\mathrm{m}}(\mathrm{HCl})} = \left(\frac{M(\mathrm{CHD_3})M(\mathrm{DCl})}{M(\mathrm{CD_4})M(\mathrm{HCl})}\right)^{3/2} = \left(\frac{19.06\times37.46}{20.07\times36.46}\right)^{3/2} = 0.964$$

The ratio of rotational partition functions is

$$\frac{q^{\mathrm{R}}(\mathrm{CHD_3})q^{\mathrm{R}}(\mathrm{DCl})}{q^{\mathrm{R}}(\mathrm{CD_4})q^{\mathrm{R}}(\mathrm{HCl})} = \frac{\sigma(\mathrm{CD_4})}{\sigma(\mathrm{CHD_3})}\frac{(\tilde{B}(\mathrm{CD_4}))^{3/2}\tilde{B}(\mathrm{HCl})}{(\tilde{A}(\mathrm{CHD_3})\tilde{B}(\mathrm{CHD_3})^2)^{1/2}\tilde{B}(\mathrm{DCl})}$$

$$= \frac{12}{3}\times\frac{2.63^{3/2}\times10.59}{(2.63\times3.28^2)^{1/2}\times5.445} = 6.24$$

The ratio of vibrational partition functions (call it f_V for convenience below) is

$$f_V = \frac{q^V(CHD_3)q^V(DCl)}{q^V(CD_4)q^V(HCl)} = \frac{q(2\,993)q(2\,142)q(1\,003)^3 q(1\,291)^2 q(1\,036)^2 q(2\,145)}{q(2\,109)q(1\,092)^2 q(2\,259)^3 q(996)^3 q(2\,991)}$$

where $q(x) = \dfrac{1}{1-e^{-hc\tilde{\nu}/kT}} = \dfrac{1}{1-e^{-1.4388x/(T/K)}}$

We also require $\Delta_r E_0$, which is equal to the difference in zero point energies

$$\begin{aligned}\frac{\Delta_r E_0}{hc} &= \frac{1}{2}\{(2\,993+2\,142+3\times1\,003+2\times1\,291+2\times1\,036+2\,145)\\ &\quad -(2\,109+2\times1\,092+3\times2\,259+3\times996+2\,991)\}\,\text{cm}^{-1}\\ &= -1\,053\ \text{cm}^{-1}\end{aligned}$$

So the exponent in the energy term is

$$-\beta\Delta_r E_0 = -\frac{\Delta_r E_0}{kT} = -\frac{hc}{k}\times\frac{\Delta_r E_0}{hc}\times\frac{1}{T} = -\frac{1.4388\times(-1\,053)}{T/K} = +\frac{1515}{T/K}$$

Hence,

$$K = 0.964\times6.24\times f_V e^{+1515/(T/K)} = 6.02 f_V e^{+1515/(T/K)}$$

We can now evaluate K and obtain the following values

T/K	300	400	500	600	700	800	900	1 000
K	945	273	132	83	61	49	42	37

The values of K are plotted in Fig. 15F.1.

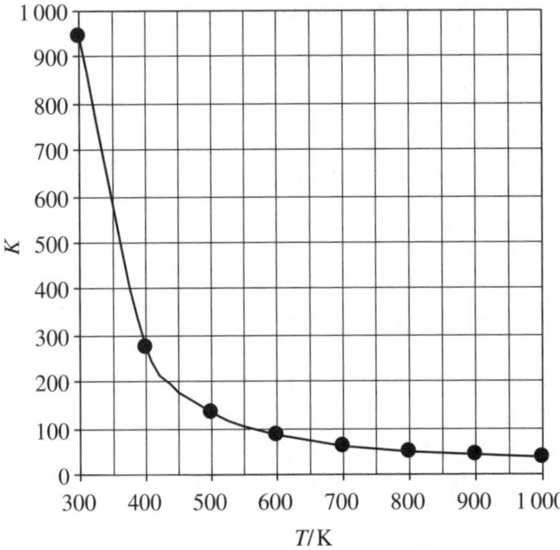

Figure 15F.1

P15F.3 All partition functions other than the electronic partition function of atomic I are unaffected by a magnetic field; hence the relative change in K is due to the relative change in q^{E}.

$$q^{\mathrm{E}}=\sum_{M_J}\mathrm{e}^{-g\mu_{\mathrm{B}}\beta\mathcal{B}M_J},\quad M_J=-\tfrac{3}{2},-\tfrac{1}{2},+\tfrac{1}{2},+\tfrac{3}{2};g=\tfrac{4}{3}$$

Since $g\mu_{\mathrm{B}}\beta\mathcal{B}<<1$ for normally attainable fields, we can expand the exponentials

$$q^{\mathrm{E}}=\sum_{M_J}\left\{1-g\mu_{\mathrm{B}}\beta\mathcal{B}M_J+\frac{1}{2}(g\mu_{\mathrm{B}}\beta\mathcal{B}M_J)^2+\dots\right\}$$

$$\approx 4+\frac{1}{2}(g\mu_{\mathrm{B}}\beta\mathcal{B})^2\sum_{M_J}M_J^2\left[\sum_{M_J}M_J=0\right]=4\left(1+\frac{10}{9}(\mu_{\mathrm{B}}\beta\mathcal{B})^2\right)\ \left[g=\frac{4}{3}\right]$$

This partition function appears squared in the numerator of the equilibrium constant expression. (See solution to Exercise 15F.3(a).) Therefore, if K is the equilibrium constant for an arbitrary field strength and K^0 is its value when $\mathcal{B}=0$, we write

$$\frac{K}{K^0}=\left(1+\frac{10}{9}(\mu_{\mathrm{B}}\beta\mathcal{B})^2\right)^2\approx 1+\frac{20}{9}\mu_{\mathrm{B}}^2\beta^2\mathcal{B}^2$$

For a shift of 1 per cent, we require

$$\frac{20}{9}\mu_{\mathrm{B}}^2\beta^2\mathcal{B}^2\approx 0.01,\quad\text{or}\quad \mu_{\mathrm{B}}\beta\mathcal{B}\approx 0.067$$

Hence

$$\mathcal{B}\approx\frac{0.067kT}{\mu_{\mathrm{B}}}=\frac{(0.067)\times(1.381\times10^{-23}\,\mathrm{J\,K^{-1}})\times(1\,000\,\mathrm{K})}{9.274\times10^{-24}\,\mathrm{J\,T^{-1}}}\approx\boxed{100\ \mathrm{T}}$$

P15F.5 The standard molar Gibbs energy is given by

$$G_{\mathrm{m}}^{\ominus}-G_{\mathrm{m}}^{\ominus}(0)=RT\ln\frac{q_{\mathrm{m}}^{\ominus}}{N_{\mathrm{A}}}\quad\text{where}\ \frac{q_{\mathrm{m}}^{\ominus}}{N_{\mathrm{A}}}=\frac{q_{\mathrm{m}}^{\mathrm{T}\ominus}}{N_{\mathrm{A}}}q^{\mathrm{R}}q^{\mathrm{V}}q^{\mathrm{E}}\ \ [15\mathrm{F}.9]$$

Translation:

$$\frac{q_{\mathrm{m}}^{\mathrm{T}\ominus}}{N_{\mathrm{A}}}=\frac{kT}{p^{\ominus}\Lambda^3}\quad \Lambda=\frac{h}{(2\pi mkT)^{1/2}}$$

After substituting the values of the constants, we obtain

$$\frac{q_{\mathrm{m}}^{\mathrm{T}\ominus}}{N_{\mathrm{A}}}=2.561\times10^{-2}(T/\mathrm{K})^{5/2}(M/\mathrm{g\,mol^{-1}})^{3/2}$$

$$=2.561\times10^{-2}\times(200.0)^{5/2}\times(102.9)^{3/2}=1.512\times10^{7}$$

Rotation of a nonlinear molecule:

$$q^{R}=\frac{1}{\sigma}\left(\frac{kT}{hc}\right)^{3/2}\left(\frac{\pi}{\tilde{A}\tilde{B}\tilde{C}}\right)^{1/2}=\frac{1.0270}{\sigma}\times\frac{(T/K)^{3/2}}{(\tilde{A}\tilde{B}\tilde{C}/\text{cm}^{-3})^{1/2}}$$

$$=\frac{1.0270}{2}\times\frac{[(200.0)\times(2.998\times10^{10}\,\text{cm}\,\text{s}^{-1})]^{3/2}}{[(13\,109.4)\times(2\,409.8)\times(2\,139.7)\times(10^{6}\,\text{s}^{-1})^{3}/\text{cm}^{-3}]^{1/2}}=2.900\times10^{4}$$

Vibration:

$$q^{V}=\frac{1}{1-\mathrm{e}^{-hc\tilde{\nu}/kT}}=\frac{1}{1-\exp\left(\frac{-1.4388(\tilde{\nu}/\text{cm}^{-1})}{T/\text{K}}\right)}$$

$$q_{1}^{V}=\frac{1}{1-\exp\left(\frac{-1.4388(\tilde{\nu}/\text{cm}^{-1})}{T/\text{K}}\right)}=\frac{1}{1-\exp\left(\frac{-1.4388(753)}{200.0}\right)}=1.004$$

$$q_{2}^{V}=\frac{1}{1-\exp\left(\frac{-1.4388(542)}{200.0}\right)}=1.021 \qquad q_{3}^{V}=\frac{1}{1-\exp\left(\frac{-1.4388(310)}{200.0}\right)}=1.120$$

$$q_{4}^{V}=\frac{1}{1-\exp\left(\frac{-1.4388(127)}{200.0}\right)}=1.670 \qquad q_{5}^{V}=\frac{1}{1-\exp\left(\frac{-1.4388(646)}{200.0}\right)}=1.010$$

$$q_{6}^{V}=\frac{1}{1-\exp\left(\frac{-1.4388(419)}{200.0}\right)}=1.052$$

$$q^{V}=\prod_{i=1}^{6}q_{i}^{V}=2.037$$

Putting it all together yields

$$G_{\text{m}}^{\ominus}-G_{\text{m}}^{\ominus}(0)=(8.3145\,\text{J}\,\text{mol}^{-1}\,\text{K}^{-1})\times(200.0\,\text{K})$$
$$\times\ln[(1.512\times10^{7})\times(2.900\times10^{4})\times(2.037)\times(1)]$$
$$=4.576\times10^{4}\,\text{J}\,\text{mol}^{-1}=\boxed{45.76\,\text{kJ}\,\text{mol}^{-1}}$$

Integrated activities

I15.1(a) The partition function is

$$q=\sum_{i}\mathrm{e}^{-\beta\varepsilon_{i}}=\sum_{i}\mathrm{e}^{-\varepsilon_{i}/kT}=1+\mathrm{e}^{-\varepsilon/kT}+\mathrm{e}^{-2\varepsilon/kt}$$

The partition function is plotted against kT/ε in Figs I15.1(a) and (b). As can be seen in both of these figures, the partition function rises from a value of 1 at low temperatures (where only the ground state is occupied) to approach a value of 3 at high temperatures (where all three states are nearly equally populated). When the horizontal axis is plotted on a logarithmic scale, the sigmoidal character of the function is more apparent: the slope of the plot approaches zero in the limits of both low and high temperature.

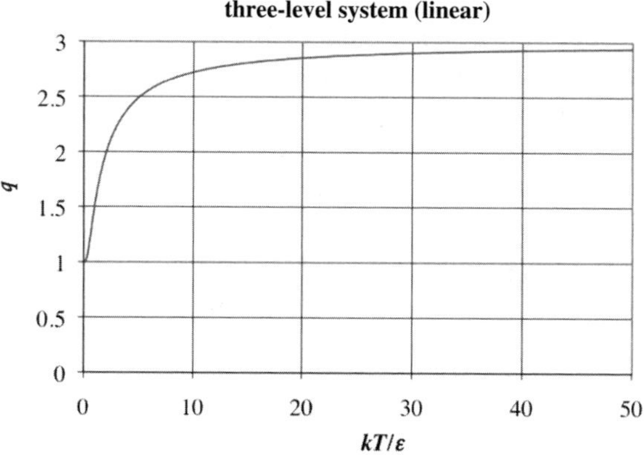

Figure I15.1(a)

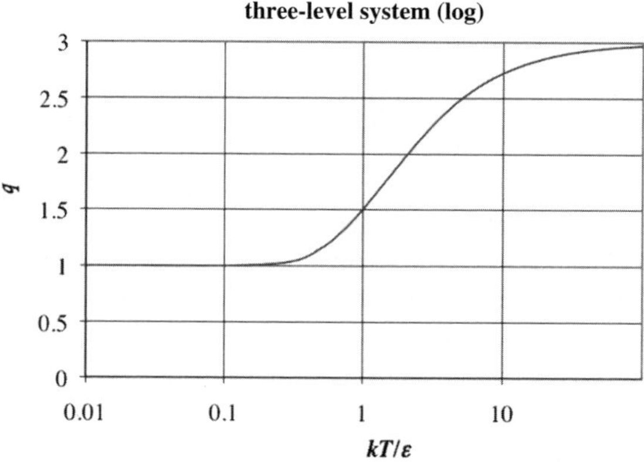

Figure I15.1(b)

I15.1(b) The statistical entropy is [15E.8a]

$$S=\frac{N\langle\varepsilon\rangle}{T}+Nk\ln q=Nk(\beta\langle\varepsilon\rangle+\ln q)$$

Substituting results from *Brief illustration* 15C.1 yields

$$S=Nk\left(\frac{\beta\varepsilon}{e^{\beta\varepsilon}+1}+\ln(1+e^{-\beta\varepsilon})\right)$$

So the temperature coefficient is

$$\frac{dS}{dT}=\frac{dS}{d\beta}\frac{d\beta}{dT}=Nk\left(\frac{\varepsilon}{e^{\beta\varepsilon}+1}-\frac{\beta\varepsilon^2e^{\beta\varepsilon}}{(e^{\beta\varepsilon}+1)^2}-\frac{\varepsilon e^{-\beta\varepsilon}}{1+e^{-\beta\varepsilon}}\right)\left(\frac{-1}{kT^2}\right)$$
$$=\frac{N}{T^2}\frac{\beta\varepsilon^2e^{\beta\varepsilon}}{(e^{\beta\varepsilon}+1)^2}=\left(\frac{\varepsilon}{kT}\right)^3\frac{k^2Ne^{\varepsilon/kT}}{\varepsilon(e^{\varepsilon/kT}+1)^2}$$

$\dfrac{\varepsilon}{k^2N}\dfrac{\mathrm{d}S}{\mathrm{d}T}$ is plotted (as 'dS/dT scaled') against $\dfrac{kT}{\varepsilon}$ on a logarithmic scale in Fig. I15.2. There is a maximum in the temperature coefficient of the entropy. The coefficient is non-negative everywhere, but it is quite close to zero at both very low and very high temperatures. This suggests that the entropy itself never decreases with increasing temperature, but that it increases only slowly at both very low and very high temperatures. In a two-level system, this is not surprising. At low temperatures, the lower state is nearly exclusively populated. The Boltzmann factor that represents upper-state population does not become nearly comparable with the lower-state population until the temperature is a significant fraction of ε/k. At the other extreme of temperature, the populations of the two states approach equality, this system's condition of maximum disorder. Near equality is reached when the temperature reaches a few times ε/k—well before it reaches $10\varepsilon/k$. Only in between those extremes is the growth of entropy with temperature rapid; only in between does dS/dT have a value near its maximum.

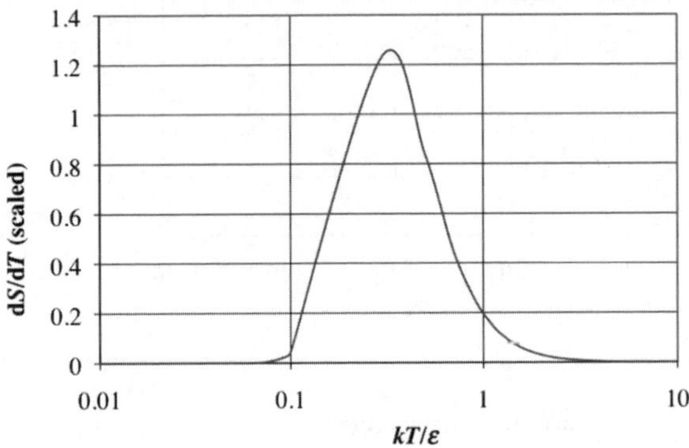

Figure I15.2

I15.1(c) One may simply pick values of the wavenumber $\tilde{\nu}$ and plot the partition function against temperature for the chosen wavenumbers. However, if one works in terms of the characteristic vibrational temperature

$$\theta^{\mathrm{V}} = \frac{hc\tilde{\nu}}{k}$$

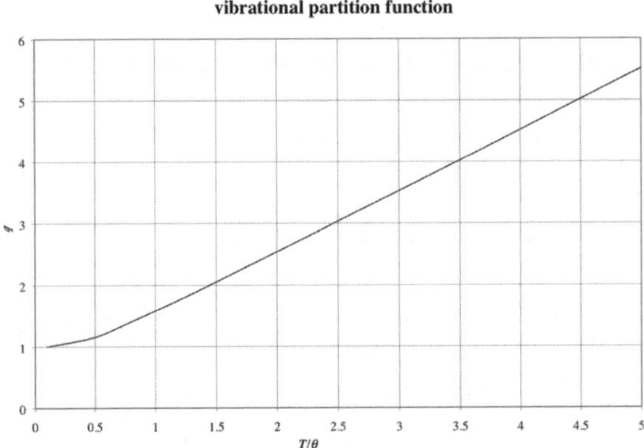

Figure I15.3

and a dimensionless temperature T/θ^V, one needs only a single plot, shown in Fig. I15.3.

The partition function is

$$q^V = \sum_v e^{-vhc\tilde{v}/kT} = \frac{1}{1-e^{-hc\tilde{v}/kT}} = \frac{1}{1-e^{-\theta^V/T}} \quad [15B.15]$$

As noted in Section 15B.2(c), at high temperatures, q approaches T/θ^V. In the figure, which extends only to moderately high dimensionless temperatures, it is apparent that the partition function varies very nearly linearly with temperature with a slope very slightly less than one. In fact, q appears to approach a limiting value of $T/\theta^V + 0.5$, and it approaches this value very quickly. Even at a dimensionless temperature of 1 (that is, at a temperature of $T = \theta^V$), $q = 1.582$, not very far from its limiting value of $T/\theta^V + 0.5 = 1.500$. As the temperature increases further, there are two distinct ways in which q approaches T/θ^V: one is that q continues slowly to get absolutely closer to $T/\theta^V + 0.5$, and the other is that 0.5 becomes *relatively* small compared to T/θ^V.

One might arbitrarily define the temperature at which the high-temperature limit is reached is the temperature at which q drops to within 10% of T/θ^V. By that criterion, the limit is reached at about $T = 5.2\theta^V$, because by that temperature, the 0.5 in the limiting value $T/\theta^V + 0.5$ becomes relatively small (by our definition) compared to T/θ^V. Perhaps a more telling, but still arbitrary criterion would define the high-temperature limit as the temperature at which q drops to $T/\theta^V + 0.55$ (that is, the temperature at which the difference between q and its limiting value T/θ^V + 0.5 drops to only 10% of the difference between its limiting value and T/θ^V). By that criterion, the limit is reached at about $\boxed{T = 1.7\theta^V}$. Indeed, by that temperature, the partition function curve is already very nearly a straight line.

16 Molecular interactions

16A Electric properties of molecules

Answers to discussion questions

D16A.1 A molecule with a permanent separation of electric charge has a **permanent dipole moment** and is said to be **polar molecule**. In molecules containing atoms of differing electronegativity, the bonding electrons may be displaced in such a way as to produce a net separation of charge. Separation of charge may also arise from a difference in atomic radii of the bonded atoms. The separation of charges in the bonds is usually, though not always, in the direction of the more electronegative atom but depends on the precise bonding and nuclear arrangement in the molecule. A heteronuclear diatomic molecule necessarily has a dipole moment if there is a difference in electronegativity between the atoms, but the situation in polyatomic molecules is more complex. A polyatomic molecule has a permanent dipole moment only if it belongs to the symmetry point group C_n, C_{nv}, or C_s as discussed in Topic 11A of the text.

An external electric field can distort the electron density in both polar and nonpolar molecules. This results in an **induced dipole moment** that is proportional to the field and the constant of proportionality is called the **polarizability**. Molecules with small HOMO–LUMO gaps typically have large polarizabilities.

D16A.3 Dipole moments are not measured directly, but are calculated from a measurement of the **relative permittivity**, $\varepsilon_r = \varepsilon/\varepsilon_0$ [16A.10, dielectric constant], of the medium as measured by comparing the capacitance of a capacitor with and without the sample present using $\varepsilon_r = C/C_0$. Equations 16A.11 (the **Debye equation**) and 16A.12,

$$\frac{\varepsilon_r - 1}{\varepsilon_r + 2} = \frac{\rho P_m}{M} \ [16A.11] \quad \text{where} \quad P_m = \frac{N_A}{3\varepsilon_0}\left(\alpha + \frac{\mu^2}{3kT}\right) \ [16A.12],$$

imply that the dipole moment can be determined from a measurement of ε_r as a function of temperature. This approach is illustrated in Example 16A.2. In another method, the relative permittivity of a solution of the polar molecule is measured as a function of concentration. The calculation is again based on the Debye equation, but in a modified form. The values obtained by this method are accurate only to about 10%. A third method is based on the relation between relative permittivity and refractive index, eqn 16A.14, and thus reduces to a measurement of the refractive index. Accurate values of the dipole moments of gaseous molecules can be obtained from the Stark effect in their microwave spectra.

Solutions to exercises

E16A.1(a) A molecule with a centre of symmetry may not be polar but molecules belonging to the groups C_n, C_{nv}, and C_s may be polar (Topic 11A). The C atom of CIF_3 is approximately sp^3 hybridized, which causes the molecule to belong to the C_{3v} point group. The highly electronegative F atoms cause the C–F bonds to be very polar and the average dipole of the three very polar C–F bonds is unbalanced by the less polar C–I bond. Therefore, $\boxed{CIF_3}$ is polar.

Ozone is a bent molecule that belongs to the C_{2v} point group. Lewis resonance structures for the molecule show a central atom with a double bond to an end oxygen atom and a single bond to oxygen at the other end. The central oxygen has a formal charge of +1 while the doubled bonded end oxygen has a formal charge of zero. The single bonded end oxygen has a −1 formal charge. The average position of the −1 formal charge of the two resonance structure predicts a small negative charge that lays half-way between the extremities and a fraction of a bond length away from the central oxygen, which is expected to have a small positive charge. Consequently, $\boxed{O_3}$ is polar.

Hydrogen peroxide belongs to the C_2 point group with each carbon atom being approximately sp^3 hybridized. The H-O-O bond angles are 96.87°. The H–O–O plane of one hydrogen atom is at a 93.85° angle with the O–O–H plane of the other hydrogen atom. Consequently, the dipole moments of the two polar O–H bonds do not cancel and the $\boxed{H_2O_2}$ molecule is polar.

E16A.2(a) $\mu_{res} = (\mu_1^2 + \mu_2^2 + 2\mu_1\mu_2\cos\theta)^{1/2}$ [16A.3a]

$$= [(1.5)^2 + (0.80)^2 + (2)\times(1.5)\times(0.80)\times(\cos 109.5°)]^{1/2}\,\text{D} = \boxed{1.4\ \text{D}}$$

E16A.3(a) The dipole moment is the vector sum shown in Fig. 16A.1.

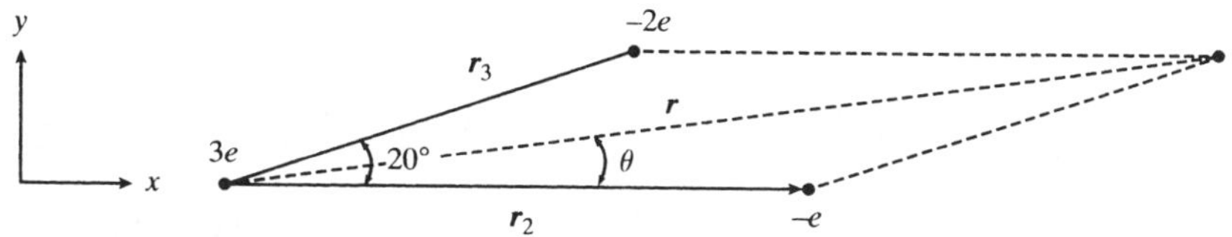

Figure16A.1

$$\mu = \sum_i Q_i r_i = 3e(0) - er_2 - 2er_3 \quad \text{where} \quad r_2 = \mathbf{i}x_2 \quad \text{and} \quad r_3 = \mathbf{i}x_3 + \mathbf{j}y_3$$

$$x_2 = +0.32\ \text{nm}$$

$$x_3 = |r_3|\cos 20° = (+0.23\ \text{nm})\times(0.94\bar{0}) = 0.21\bar{6}\ \text{nm}$$

$$y_3 = |r_3|\sin 20° = (+0.23\ \text{nm})\times(0.34\bar{2}) = 0.078\bar{7}\ \text{nm}$$

The components of the vector sum are the sums of the components.

$$\mu_x = -ex_2 - 2ex_3 = -e\times\{(0.32)+2\times(0.21\overline{6})\}\text{ nm} = -e\times(0.75\overline{2}\text{ nm})$$

$$\mu_y = -2ey_3 = -2e\times(0.078\overline{7}) = -e\times(0.15\overline{7}\text{ nm})$$

$$\mu = (\mu_x^2+\mu_y^2)^{1/2}\quad[16\text{A}.4\text{b}]$$

$$= e\times\{(0.75\overline{2}\text{ nm})^2+(0.15\overline{7}\text{ nm})^2\}^{1/2} = (1.602\times10^{-19}\text{ C})\times(0.76\overline{8}\times10^{-9}\text{ m})$$

$$= (1.2\overline{3}\times10^{-28}\text{ C m})\times\left(\frac{1\text{ D}}{3.33564\times10^{-30}\text{ C m}}\right) = \boxed{37\text{ D}}$$

The angle that μ makes with x-axis is given by

$$\cos\theta = \frac{|\mu_x|}{\mu} = \frac{0.75\overline{2}}{0.76\overline{8}}\quad\text{so}\quad \theta = \cos^{-1}\left(\frac{0.75\overline{2}}{0.76\overline{8}}\right) = \boxed{11.\overline{7}°}$$

E16A.4(a) Polarizability α, dipole moment μ, and molar polarization P_m are related by

$$P_\text{m} = \left(\frac{N_\text{A}}{3\varepsilon_0}\right)\times\left(\alpha+\frac{\mu^2}{3kT}\right)\quad[16\text{A}.12]$$

In order to solve for α, it is first necessary to obtain μ from the temperature variation of P_m.

$$\alpha+\frac{\mu^2}{3kT} = \frac{3\varepsilon_0 P_\text{m}}{N_\text{A}}$$

Therefore, $\left(\frac{\mu^2}{3k}\right)\times\left(\frac{1}{T}-\frac{1}{T'}\right) = \left(\frac{3\varepsilon_0}{N_\text{A}}\right)\times(P_\text{m}-P'_\text{m})$ [P_m at T, P'_m at T']

and hence

$$\mu^2 = \frac{9\varepsilon_0 k\times(P_\text{m}-P'_\text{m})}{N_\text{A}\times\left(\frac{1}{T}-\frac{1}{T'}\right)}$$

$$= \frac{9\times(8.854\times10^{-12}\text{ J}^{-1}\text{ C}^2\text{ m}^{-1})\times(1.381\times10^{-23}\text{ J K}^{-1})\times(70.62-62.47)\times10^{-6}\text{ m}^3\text{ mol}^{-1}}{(6.022\times10^{23}\text{ mol}^{-1})\times\left(\frac{1}{351.0\text{ K}}-\frac{1}{423.2\text{ K}}\right)}$$

$$= 3.06\overline{4}\times10^{-59}\text{ C}^2\text{ m}^2$$

$$\mu = 5.54\times10^{-30}\text{ C m}\times\left(\frac{1\text{ D}}{3.33564\times10^{-30}\text{ C m}}\right) = \boxed{1.66\text{ D}}$$

$$\alpha = \frac{3\varepsilon_0 P_\text{m}}{N_\text{A}} - \frac{\mu^2}{3kT}$$

$$= \frac{3\times(8.854\times10^{-12}\text{ J}^{-1}\text{ C}^2\text{ m}^{-1})\times(70.62\times10^{-6}\text{ m}^3\text{ mol}^{-1})}{6.022\times10^{23}\text{ mol}^{-1}} - \frac{3.06\overline{4}\times10^{-59}\text{ C}^2\text{ m}^2}{3\times(1.381\times10^{-23}\text{ J K}^{-1})\times(351.0\text{ K})}$$

$$= \boxed{1.01\times10^{-39}\text{ J}^{-1}\text{ C}^2\text{ m}^2}$$

Corresponding to $\alpha' = \dfrac{\alpha}{4\pi\varepsilon_0}$ [16A.6] $= \boxed{9.06\times10^{-30}\ \mathrm{m}^3}$

E16A.5(a) $M_{\mathrm{ClF_3}} = 92.45\ \mathrm{g\ mol^{-1}}$

$$\varepsilon_\mathrm{r} - 1 = \frac{\rho P_\mathrm{m}}{M}\times(\varepsilon_\mathrm{r}+2)\ \ [16\mathrm{A}.11]$$

$$\left(1-\frac{\rho P_\mathrm{m}}{M}\right)\varepsilon_\mathrm{r} = 1+\frac{2\rho P_\mathrm{m}}{M}$$

$$\begin{aligned}\varepsilon_\mathrm{r} &= \frac{M+2\rho P_\mathrm{m}}{M-\rho P_\mathrm{m}}\\ &= \frac{92.45\ \mathrm{g\ mol^{-1}} + 2\times(1.89\ \mathrm{g\ cm^{-3}})\times(27.18\ \mathrm{cm^3\ mol^{-1}})}{92.45\ \mathrm{g\ mol^{-1}} - (1.89\ \mathrm{g\ cm^{-3}})\times(27.18\ \mathrm{cm^3\ mol^{-1}})}\\ &= \boxed{4.75}\end{aligned}$$

E16A.6(a) $n_\mathrm{r} = (\varepsilon_\mathrm{r})^{1/2}$ [16A.14] and $\dfrac{\varepsilon_\mathrm{r}-1}{\varepsilon_\mathrm{r}+2} = \dfrac{\rho N_\mathrm{A}\alpha}{3M\varepsilon_0}$ [16A.13]

Therefore,

$$\begin{aligned}\alpha &= \frac{3M\varepsilon_0}{\rho N_\mathrm{A}}\left(\frac{n_\mathrm{r}^2-1}{n_\mathrm{r}^2+2}\right)\\ &= \frac{3\times(267.83\ \mathrm{g\ mol^{-1}})\times(8.854\times10^{-12}\ \mathrm{J^{-1}\ C^2\ m^{-1}})}{(3.32\times10^{6}\ \mathrm{g\ m^{-3}})\times(6.022\times10^{23}\ \mathrm{mol^{-1}})}\times\left(\frac{1.732^2-1}{1.732^2+2}\right)\\ &= \boxed{1.42\times10^{-39}\ \mathrm{J^{-1}\ C^2\ m^2}}\end{aligned}$$

E16A.7(a) $\alpha' = \dfrac{\alpha}{4\pi\varepsilon_0}$ [16A.6]

$$\alpha = 4\pi\varepsilon_0\alpha' = (1.11265\times10^{-10}\ \mathrm{J^{-1}\ C^2\ m^{-1}})\times(1.5\times10^{-30}\ \mathrm{m^3}) = 1.6\overline{7}\times10^{-40}\ \mathrm{J^{-1}\ C^2\ m^2}$$

Let

$$C = \frac{\rho N_\mathrm{A}\alpha}{3M\varepsilon_0} = \frac{(0.99707\times10^{6}\ \mathrm{g\ m^{-3}})\times(6.022\times10^{23}\ \mathrm{mol^{-1}})\times(1.6\overline{7}\times10^{-40}\ \mathrm{J^{-1}\ C^2\ m^2})}{3\times(18.02\ \mathrm{g\ mol^{-1}})\times(8.85419\times10^{-12}\ \mathrm{J^{-1}\ C^2\ m^{-1}})} = 0.20\overline{9}$$

and solve the Clausius–Mossotti equation [17A.13] for ε_r with which we calculate the refractive index.

$$\begin{aligned}\varepsilon_\mathrm{r} &= \frac{1+2C}{1-C}\\ &= \frac{1+2\times(0.20\overline{9})}{1-0.20\overline{9}}\\ &= 1.7\overline{9}\end{aligned}$$

$$n_\mathrm{r} = \varepsilon_\mathrm{r}^{1/2}\ [17.17] = (1.7\overline{9})^{1/2} = \boxed{1.34}$$

This result agrees with the experimental value of 1.33 because it lies within the uncertainty of the refractive index measurement.

E16A.8(a) $\mu = 1.57\ \text{D} = 5.23\bar{7}\times10^{-30}\ \text{C m}$ for chlorobenzene (112.55 g mol^{-1})

$$\alpha = 4\pi\varepsilon_0\alpha'\ [16\text{A}.6] = (1.11265\times10^{-10}\ \text{J}^{-1}\ \text{C}^2\ \text{m}^{-1})\times(1.23\times10^{-29}\ \text{m}^3)$$
$$= 1.36\bar{9}\times10^{-39}\ \text{J}^{-1}\ \text{C}^2\ \text{m}^2$$

$$P_\text{m} = \frac{N_\text{A}}{3\varepsilon_0}\left(\alpha + \frac{\mu^2}{3kT}\right)\ [16\text{A}.12]$$

$$= \frac{6.022\times10^{23}\ \text{mol}^{-1}}{3\times(8.85419\times10^{-12}\ \text{J}^{-1}\ \text{C}^2\ \text{m}^{-1})}\left(\begin{array}{l}(1.36\bar{9}\times10^{-39}\ \text{J}^{-1}\ \text{C}^2\ \text{m}^2)\\ +\dfrac{(5.23\bar{7}\times10^{-30}\ \text{C m})^2}{3\times(1.3807\times10^{-23}\ \text{J K}^{-1})\times(298.15\ \text{K})}\end{array}\right)$$

$$= 8.13\bar{8}\times10^{-5}\ \text{m}^3\ \text{mol}^{-1}$$

Let

$$C = \frac{\rho P_\text{m}}{M} = \frac{(1.173\times10^6\ \text{g m}^{-3})\times(8.13\bar{8}\times10^{-5}\ \text{m}^3\ \text{mol}^{-1})}{112.55\ \text{g mol}^{-1}} = 0.848\bar{1}$$

and solve the Debye equation [16A.11] for ε_r.

$$\frac{\varepsilon_\text{r}-1}{\varepsilon_\text{r}+2} = C\ [16\text{A}.11\text{, the Debye equation}]$$

$$\varepsilon_\text{r} = \frac{1+2C}{1-C}$$
$$= \frac{1+2\times(0.848\bar{1})}{1-0.848\bar{1}}$$
$$= \boxed{17.7}$$

Solutions to problems

P16A.1 Refer to Fig. 16A.1 of the text, and add moments vectorially using $\mu = 2\mu_1\cos\tfrac{1}{2}\theta$ [16A.3b].

(a) *p*-xylene: the resultant is zero, so $\mu = \boxed{0}$

(b) *o*-xylene: $\mu = (2)\times(0.4\ \text{D})\times\cos 30° = \boxed{0.7\ \text{D}}$

(c) *m*-xylene $\mu = (2)\times(0.4\ \text{D})\times\cos 60° = \boxed{0.4\ \text{D}}$

The *p*-xylene molecule to the group D_{2h}, and so it is necessarily nonpolar.

P16A.3 Individual acetic acid molecules have a nonzero dipole moment but the two dipoles of the dimer are exactly opposed and cancel. At low temperature a significant fraction of molecules are a part of a dimer, which means that their individual dipole moments will not be observed. However, as temperature is increased, hydrogen bonds of the dimer are broken, thereby, releasing individual molecules and the apparent dipole moment increases. This causes the molar polarization to increase by eqn 16A.12 and the relative permittivity to increase by eqn 16A.11.

P16A.5 The induced dipole moment of NH_3 is

$$\mu^*_{NH_3} = \alpha\mathcal{E}\ [16A.5a] = 4\pi\varepsilon_0\alpha'_{NH_3}\mathcal{E}\ [16A.6]$$

$$= 4\pi\times(8.854\times10^{-12}\ J^{-1}\ C^2\ m^{-1})\times(2.22\times10^{-30}\ m^3)\times(15.0\times10^3\ V\ m^{-1})$$

$$[1\ J = 1\ C\ V]$$

$$= (3.71\times10^{-36}\ C\ m)\times\left(\frac{1\ D}{3.33564\times10^{-30}\ C\ m}\right)$$

$$= \boxed{1.00\ \mu D}$$

Thus, we see that the induced dipole is much smaller than the permanent dipole moment of NH_3 (1.47 D).

P16A.7 The equations relating dipole moment and polarizability volume to the experimental quantities ε_r and ρ are

$$P_m = \left(\frac{M}{\rho}\right)\times\left(\frac{\varepsilon_r - 1}{\varepsilon_r + 2}\right)\ [16A.11]\ \text{and}\ P_m$$

$$= \frac{4\pi}{3}N_A\alpha' + \frac{N_A\mu^2}{9\varepsilon_0 kT}\quad [16A.12, \text{with } \alpha = 4\pi\varepsilon_0\alpha']$$

Therefore, we draw up the following table (with $M = 119.4\,g\,mol^{-1}$)

θ/°C	−80	−70	−60	−40	−20	0	20
T/K	193	203	213	233	253	273	293
$\frac{1000}{T/K}$	5.18	4.93	4.69	4.29	3.95	3.66	3.41
ε_r	3.1	3.1	7.0	6.5	6.0	5.5	5.0
$\frac{e_r - 1}{e_r + 2}$	0.41	0.41	0.67	0.65	0.63	0.60	0.57
ρ/g cm^{-3}	1.65	1.64	1.64	1.61	1.57	1.53	1.50
P_m/(cm^3 mol^{-1})	29.8	29.9	48.5	48.0	47.5	56.8	45.4

P_m is plotted against $\frac{1}{T}$ in Fig. 16A.2.

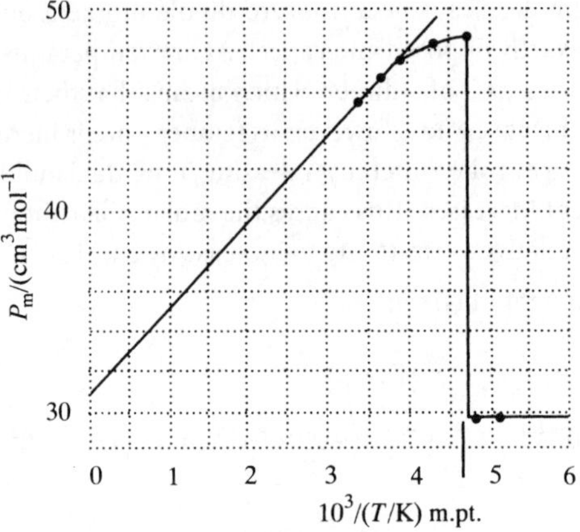

Figure 16A.2

The (dangerously unreliable) intercept is ≈ 30 $cm^3 mol^{-1}$ and the slope below the melting point is such that $\frac{dP_m}{d(1/T)} \approx 4.5 \times 10^3 cm^3 mol^{-1} K$. It follows that

$$\alpha' = \frac{3\times(30 \text{ cm}^3 \text{ mol}^{-1})}{4\pi\times(6.022\times10^{23} \text{ mol}^{-1})} = \boxed{1.2\times10^{-23} \text{ cm}^3}$$

$$\mu^2 = \frac{9\varepsilon_0 k}{N_A}\frac{dP_m}{d(1/T)}$$

$$= \left\{\frac{9\times(8.85419\times10^{-12} \text{ J}^{-1} \text{ C}^2 \text{ m}^{-1})\times(1.3807\times10^{-23} \text{ J K}^{-1})}{6.022\times10^{23} \text{ mol}^{-1}}\right\}\times(4.5\times10^{-3} \text{ m}^3 \text{ mol}^{-1} \text{ K})$$

$$= 8.2\times10^{-30} \text{ C}^2 \text{ m}^2$$

$$\mu = (8.2\times10^{-60} \text{ C}^2 \text{ m}^2)^{1/2}\times\left(\frac{1 \text{ D}}{3.33564\times10^{-30} \text{ C m}}\right) = \boxed{0.86 \text{ D}}$$

A sharp decrease in P_m occurs at the freezing point of chloroform (−64 °C), indicating that the dipole reorientation term no longer contributes. Note that P_m for the solid corresponds to the extrapolated, dipole-free, value of P_m, so the extrapolation is less hazardous than it looks.

P16A.9 $$P_m = \frac{4\pi}{3}N_A\alpha' + \frac{N_A\mu^2}{9\varepsilon_0 kT} \quad [16A.12, \text{with } \alpha = 4\pi\varepsilon_0\alpha']$$

Equation 16A.12 indicates that a plot of P_m against $1/T$ should be linear with a slope, $\frac{dP_m}{d(1/T)}$, equal to $\frac{N_A\mu^2}{9\varepsilon_0 k}$ and a $1/T = 0$ intercept that equals $\frac{4\pi}{3}N_A\alpha'$. Therefore, we draw up the following table and prepare a plot of P_m against $1/T$. If it is linear, we perform a linear least squares regression fit of the plot so as to acquire the slope and intercept from which we calculate α' and μ. A suitable plot is shown in Fig. 16A.3.

T/K	292.2	309.0	333.0	387.0	413.0	446.0
$\dfrac{1000}{T/\mathrm{K}}$	3.42	3.24	3.00	2.58	2.42	2.24
$P_m/(\mathrm{cm^3\ mol^{-1}})$	57.57	55.01	51.22	44.99	42.51	39.59

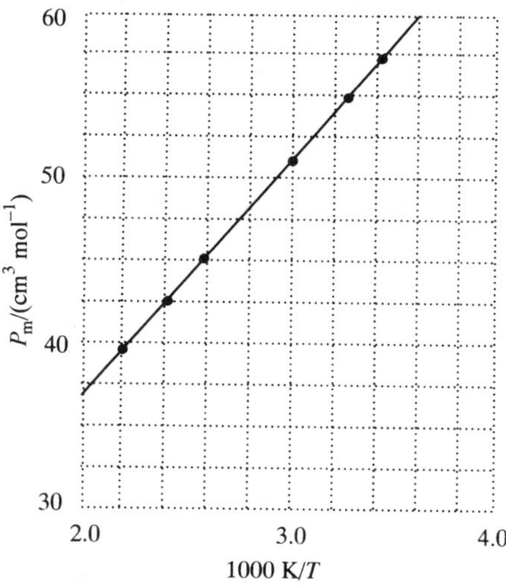

Figure 16A.3

The plot of P_m against $1/T$ is linear with a regression fit that gives an intercept of 5.65 cm^3 mol^{-1} (not shown in the figure), and the slope is such that $dP_m/d(1/T) = 1.52 \times 10^4$ cm^3 mol^{-1} K. It follows that

$$\alpha' = \frac{3P_m(\text{at intercept})}{4\pi N_A} = \frac{3\times(5.65\ \mathrm{cm^3\ mol^{-1}})}{4\pi\times(6.022\times10^{23}\ \mathrm{mol^{-1}})} = \boxed{2.24\times10^{-24}\ \mathrm{cm^3}}$$

$$\mu^2 = \frac{9\varepsilon_0 k}{N_A}\frac{dP_m}{d(1/T)}$$

$$= \left\{\frac{9\times(8.85419\times10^{-12}\ \mathrm{J^{-1}\ C^2\ m^{-1}})\times(1.3807\times10^{-23}\ \mathrm{J\,K^{-1}})}{6.022\times10^{23}\ \mathrm{mol^{-1}}}\right\}\times(1.52\times10^{-2}\ \mathrm{m^3\ mol^{-1}\ K})$$

$$= 2.78\times10^{-59}\ \mathrm{C^2\ m^2}$$

$$\mu = (2.78\times10^{-59}\ \mathrm{C^2\ m^2})^{1/2}\times\left(\frac{1\ \mathrm{D}}{3.33564\times10^{-30}\ \mathrm{C\,m}}\right) = \boxed{1.58\ \mathrm{D}}$$

The high-frequency contribution to the molar polarization, P'_m, at 273 K may be calculated from the refractive index

$$P'_m = \left(\frac{M}{\rho}\right)\times\left(\frac{\varepsilon_r - 1}{\varepsilon_r + 2}\right)[16A.11] = \left(\frac{M}{\rho}\right)\times\left(\frac{n_r^2 - 1}{n_r^2 + 2}\right)$$

Assuming that ammonia under these conditions (1.00 atm pressure assumed) can be considered a perfect gas, we have

$$\rho=\frac{pM}{RT} \text{ and } \frac{M}{\rho}=\frac{RT}{p}=\frac{(82.06\ \text{cm}^3\ \text{atm}\ \text{K}^{-1}\ \text{mol}^{-1})\times(273\ \text{K})}{1.00\ \text{atm}}=2.24\times10^4\ \text{cm}^3\ \text{mol}^{-1}$$

$$\text{Then } P'_\text{m}=2.24\times10^4\ \text{cm}^3\ \text{mol}^{-1}\times\left\{\frac{(1.000379)^2-1}{(1.000379)^2+2}\right\}=\boxed{5.66\ \text{cm}^3\ \text{mol}^{-1}}.$$

If we assume that the high-frequency contribution to P_m remains the same at 292.2 K then we have

$$\frac{N_\text{A}\mu^2}{9\varepsilon_0 kT}=P_\text{m}-P'_\text{m}=(57.57-5.66)\ \text{cm}^3\ \text{mol}^{-1}=51.91\ \text{cm}^3\ \text{mol}^{-1}=5.191\times10^{-5}\ \text{m}^3\ \text{mol}^{-1}$$

Solving for μ we have

$$\mu=\left(\frac{9\varepsilon_0 k}{N_\text{A}}\right)^{1/2}T^{1/2}(P_\text{m}-P'_\text{m})^{1/2}=4.275\times10^{-29}(\text{C}^2\ \text{mol}\ \text{K}^{-1}\ \text{m}^{-1})^{1/2}T^{1/2}(P_\text{m}-P'_\text{m})^{1/2}$$

$$=\left\{4.275\times10^{-29}\ (\text{C}^2\ \text{mol}\ \text{K}^{-1}\ \text{m}^{-1})^{1/2}\right\}\times(292.2\ \text{K})^{1/2}\times(5.191\times10^{-5}\ \text{m}^3\ \text{mol}^{-1})^{1/2}$$

$$=5.26\times10^{-30}\ \text{C m}=1.58\ \text{D}$$

The agreement between the two computational methods is exact.

P16A.11 $$P_\text{m}=\frac{M}{\rho}\times\left(\frac{\varepsilon_\text{r}-1}{\varepsilon_\text{r}+2}\right)\ [16\text{A}.11] \text{ and } P_\text{m}=\frac{N_\text{A}}{3\varepsilon_0}\left(\alpha+\frac{\mu^2}{3kT}\right)\ [16\text{A}.12]$$

The dipole moment μ contributes to the molar polarization of eqn 16A.12 in the low frequency limit of the applied electric field where molecular rotation can orientate quickly enough to follow the change in direction of the field. This corresponds to field oscillations of less than about 10^{12} Hz, a typical rotational frequency, which is the range of microwaves and radio waves. In the high frequency limit of 10^{14} Hz or higher (infrared, visible, and ultraviolet) molecules cannot orientate quickly enough to follow the change in direction of the field and the dipole moment does not contribute to eqn 16A.12. The limiting values of the molar polarization, calculated below, should be compared to the values reported and discussed in Problem 16A.8.

Low frequency limit of methanol:

$$P_\text{m}=\frac{N_\text{A}}{3\varepsilon_0}\left(\alpha+\frac{\mu^2}{3kT}\right)\ [16\text{A}.12]$$

$$=\frac{6.022\times10^{23}\ \text{mol}^{-1}}{3\times(8.85419\times10^{-12}\ \text{J}^{-1}\ \text{C}^2\ \text{m}^{-1})}\times\left(3.59\times10^{-40}\ \text{J}^{-1}\ \text{C}^2\ \text{m}^2+\frac{(5.70\times10^{-30}\ \text{C m})^2}{3\times(1.3807\times10^{-23}\ \text{J K}^{-1})\times(293.15\ \text{K})}\right)$$

$$=\boxed{68.8\ \text{cm}^3\ \text{mol}^{-1}}$$

Solving eqn 16A.11 for ε_r gives

$$\varepsilon_r = \frac{1+\dfrac{2\rho P_m}{M}}{1-\dfrac{\rho P_m}{M}} = \frac{M+2\rho P_m}{M-\rho P_m}$$

$$= \frac{32.0\text{ g mol}^{-1} + 2\times(0.7914\text{ g cm}^{-3})\times(68.8\text{ cm}^3\text{ mol}^{-1})}{32.0\text{ g mol}^{-1} - (0.7914\text{ g cm}^{-3})\times(68.8\text{ cm}^3\text{ mol}^{-1})}$$

$$= \boxed{4.40}$$

$$n_r = \varepsilon_r^{1/2}\ [17.17] = (4.40)^{1/2} = \boxed{2.10}$$

High frequency limit of methanol:

$$P_m = \frac{N_A\alpha}{3\varepsilon_0}$$

$$= \frac{(6.022\times10^{23}\text{ mol}^{-1})\times(3.59\times10^{-40}\text{ J}^{-1}\text{ C}^2\text{ m})}{3\times(8.85419\times10^{-12}\text{ J}^{-1}\text{ C}^2\text{ m}^{-1})}$$

$$= \boxed{8.14\text{ cm}^3\text{ mol}^{-1}}$$

Solving eqn 16A.11 for ε_r gives

$$\varepsilon_r = \frac{M+2\rho P_m}{M-\rho P_m}$$

$$= \frac{32.0\text{ g mol}^{-1} + 2\times(0.7914\text{ g cm}^{-3})\times(8.14\text{ cm}^3\text{ mol}^{-1})}{32.0\text{ g mol}^{-1} - (0.7914\text{ g cm}^{-3})\times(8.14\text{ cm}^3\text{ mol}^{-1})}$$

$$= \boxed{1.76}$$

$$n_r = \varepsilon_r^{1/2}\ [17.17] = (1.76)^{1/2} = \boxed{1.33}$$

P16A.13 Individual acetic acid molecules have a non-zero dipole moment but the two dipoles of the dimer are exactly opposed and cancel. At low temperature a significant fraction of molecules are a part of a dimer, which means that their individual dipole moments will not be observed. However, as temperature is increased, hydrogen bonds of the dimer are broken, thereby, releasing individual molecules and the apparent dipole moment increases. This causes the molar polarization to increase by eqn 16A.12 and the relative permittivity to increase by eqn 16A.11.

Isothermal dilution in benzene solution (a decrease in the preparation concentration c_{A_2} where A_2 is acetic acid dimer) has an effect that is analogous to heating. To see this consider equilibrium constant K for dimer dissociation and solve for the dissociation constant α. The equilibrium concentrations of A_2 and A are $(1-\alpha)c_{A_2}$ and $2\alpha c_{A_2}$, respectively. Thus, the equilibrium constant expression for

$$A_2(\text{benzene}) \rightarrow 2\text{ A(benzene)}$$

is

$$K=[\mathrm{A}]^2/[\mathrm{A}_2]=\left(2\alpha c_{\mathrm{A}_2}\right)^2/\left\{(1-\alpha)c_{\mathrm{A}_2}\right\}$$

$$4c_{\mathrm{A}_2}\alpha^2+K\alpha-K=0$$

$$\alpha=\tfrac{1}{8}\left\{-x+\sqrt{x^2+16x}\right\} \quad \text{with} \quad x=K/c_{\mathrm{A}_2} \quad \text{[quadratic equation solution]}$$

A plot of α against x shows that α increases as x increases. Thus, a decrease in c_{A_2} by isothermal dilution, which is equivalent to an increase in x, causes an increase in [A], an increase in the apparent dipole moment, an increase in the molar polarization, and an $\boxed{\text{increase in the relative permittivity}}$.

16B Interactions between molecules

Answers to discussion questions

D16B.1 (a) $V=-\dfrac{Q_2\mu_1}{4\pi\varepsilon_0 r^2}$ [16B.2]

V is the potential energy of interaction between a point dipole μ_1 and the point charge Q_2 at the separation r. The point charge lies on the axis of the dipole and the separation r is much larger than the separation of charge within the dipole so that the partial charges of the dipole seem to merge and cancel to create the so-called **point dipole**.

(b) $V=-\dfrac{Q_2\mu_1\cos\theta}{4\pi\varepsilon_0 r^2}$

V is the potential energy of interaction between a point dipole μ_1 and the point charge Q_2 at the separation r. The point charge lies at an angle θ to the axis of the dipole and the separation r is much larger than the separation of charge within the dipole so that the partial charges of the dipole seem to merge and cancel.

(c) $V=\dfrac{\mu_1\mu_2 f(\theta)}{4\pi\varepsilon_0 r^3}$ where $f(\theta)=1-3\cos^2\theta$ [16B.4]

V is the potential energy of interaction between the two point dipoles μ_1 and μ_2 at the separation r. The dipoles are parallel and the separation distance is at angle θ to the dipoles. The separation r is much larger than the separation of charge within the dipoles so that the partial charges of the dipoles seem to merge and cancel.

D16B.3 There are three van der Waals type interactions that depend upon distance as $1/r^6$; they are the Keesom interaction between rotating permanent dipoles, the permanent-dipole–induced-dipole interaction, and the induced-dipole–induced-dipole, or London dispersion, interaction. In each case, we can visualize the distance dependence of the potential energy as arising from the $1/r^3$ dependence of the field (and hence the magnitude of the induced dipole) and the $1/r^3$ dependence of the potential energy of interaction of the dipoles (either permanent or induced).

D16B.5 The increase in entropy of a solution when hydrophobic molecules or groups in molecules cluster together and reduce their structural demands on the solvent (water) is the origin of the hydrophobic interaction that tends to stabilize clustering of hydrophobic groups in solution. A manifestation of the hydrophobic interaction is the clustering together of hydrophobic groups in biological macromolecules. For example, the side chains of amino acids that are used to form the polypeptide chains of proteins are hydrophobic, and the hydrophobic interaction is a major contributor to the tertiary structure of polypeptides. At first thought, this clustering would seem to be a nonspontaneous process as the clustering of the solute results in a decrease in entropy of the solute. However, the clustering of the solute results in greater freedom of movement of the solvent molecules and an accompanying increase in disorder and entropy of the solvent. The total entropy of the system has increased and the process is spontaneous.

Solutions to exercises

E16B.1(a) The O–H bond length of a water molecule is 95.85 pm and the Li^+ cation is 100 pm from the dipole centre. Because these lengths are comparable, a calculation based on the assumption that the water dipole acts like a point dipole with a dipole length much shorter than the dipole–ion distance is unlikely to provide an accurate value of the dipole–ion interaction energy. However, such a calculation does provide an 'order-of-magnitude' estimate. The minimum value of the dipole–ion interaction occurs with the dipole pointing toward the cation.

$$V_{\text{min}} \sim -\frac{\mu_{H_2O}Q_{Li^+}}{4\pi\varepsilon_0 r^2}\ [16B.2] = -\frac{\mu_{H_2O}e}{4\pi\varepsilon_0 r^2}$$

$$\sim -\frac{(1.85\ \text{D})\times(3.336\times10^{-30}\ \text{C m D}^{-1})\times(1.602\times10^{-19}\ \text{C})}{(1.113\times10^{-10}\ \text{J}^{-1}\ \text{C}^2\ \text{m}^{-1})\times(100\times10^{-12}\ \text{m})^2}$$

$$\sim -8.88\times10^{-19}\ \text{J}$$

The interaction potential becomes a maximum upon flipping the dipole. This effectively changes the sign of the dipole in the previous calculation giving

$$V_{\text{max}} \sim 8.88\times10^{-19}\ \text{J}$$

The work w required to flip the dipole is the difference $V_{\text{max}}-V_{\text{min}}$.

$$w \sim V_{\text{max}} - V_{\text{min}} = 1.78\times10^{-18}\ \text{J}$$

$$w_{\text{m}} = wN_{\text{A}} \sim \boxed{1.07\times10^3\ \text{kJ mol}^{-1}}$$

E16B.2(a) The two linear quadrupoles are shown in Fig. 16B.1 with a collinear configuration.

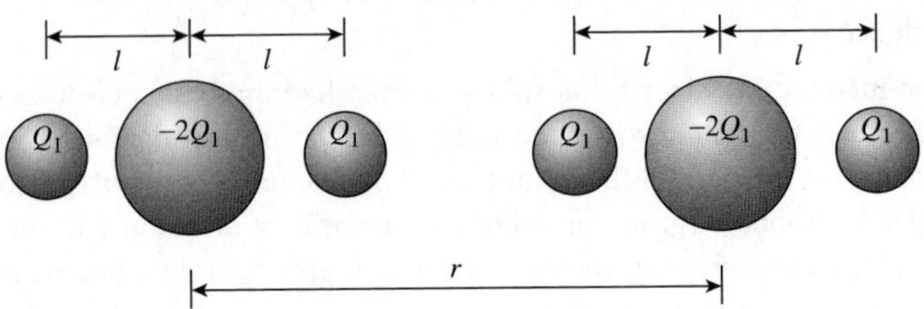

Figure 16B.1

The total potential energy of the interaction between the quadrupoles is:

$$4\pi\varepsilon_0 V = \frac{Q_1^2}{r} - \frac{2Q_1^2}{r+l} + \frac{Q_1^2}{r+2l} - \frac{2Q_1^2}{r-l} + \frac{4Q_1^2}{r} - \frac{2Q_1^2}{r+l} + \frac{Q_1^2}{r-2l} - \frac{2Q_1^2}{r-l} + \frac{Q_1^2}{r}$$

$$\frac{4\pi\varepsilon_0 rV}{Q_1^2} = 6 - \frac{4}{1+x} - \frac{4}{1-x} + \frac{1}{1+2x} + \frac{1}{1-2x} \quad \text{where} \quad x = \frac{l}{r}$$

With the point quadrupole condition that $x << 1$ the last four terms in the above expression can be expanded with the Taylor series:

$$(1+z)^{-1} = 1 - z + z^2 - z^3 + z^4 - \cdots \quad \text{and} \quad (1-z)^{-1} = 1 + z + z^2 + z^3 + z^4 + \cdots$$

where z is either $2x$ or x.

$$\begin{aligned}\frac{4\pi\varepsilon_0 rV}{Q_1^2} &= 6 - 4\{1 - x + x^2 - x^3 + x^4 - \cdots + 1 + x + x^2 + x^3 + x^4 + \cdots\} \\ &\quad + \{1 - (2x) + (2x)^2 - (2x)^3 + (2x)^4 - \cdots\} + \{1 + (2x) + (2x)^2 + (2x)^3 + (2x)^4 + \cdots\} \\ &= -8x^2 - 8x^4 + 8x^2 + 32x^4 + \cdots \\ &= 24x^4 + \text{higher order terms}\end{aligned}$$

In the limit of small x values the higher order terms are negligibly small, thereby, leaving

$$V = \frac{6x^4 Q_1^2}{\pi\varepsilon_0 r} = \boxed{\frac{6l^4 Q_1^2}{\pi\varepsilon_0 r^5}}$$

Thus, $V \propto \frac{1}{r^5}$ for the quadrupole–quadrupole interaction. See Discussion question 16B.2 and note that a quadrupole is n-pole array of charges with $n = 3$. So the above derivation demonstrates the general potential energy relation between an n-pole array and an m-pole array: $V \propto \frac{1}{r^{n+m-1}} = \frac{1}{r^{3+3-1}} = \frac{1}{r^5}$.

E16B.3(a) $$V_{\text{London}} = -\frac{3(\alpha'_{\text{He}} I_{\text{He}})^2}{2(I_{\text{He}}+I_{\text{He}})r^6}\ [16\text{B}.8] = -\frac{3(\alpha'_{\text{He}})^2 I_{\text{He}}}{4r^6}$$

$$= -\frac{3\times(0.20\times10^{-30}\ \text{m}^3)^2\times(2372.3\ \text{kJ mol}^{-1})}{4\times(1.0\times10^{-9}\ \text{m})^6}$$

$$= \boxed{0.071\ \text{J mol}^{-1}}$$

E16B.4(a) Using the partial charge presented in the table to the right, we estimate the partial charge on each hydrogen atom of a water molecule to be $Q_{\text{H}} = \delta e$ where $\delta = 0.42$. The electroneutrality of an H_2O molecule implies that the estimated partial charge on the oxygen atom is $Q_{\text{O}} = -2\delta e$. With a hydrogen bond length of 170 pm, the point charge model of the hydrogen bond in a vacuum estimates the potential of interaction to be

Partial charges in polypeptides (from *Quanta, Matter, and Change;* Atkins, de Paula, and Friedman, OUP, 1st ed. 2009)

Atom	Partial charge/*e*
C(=O)	+0.45
C(–CO)	+0.06
H(–C)	+0.02
H(–N)	+0.18
H(–O)	+0.42
N	–0.36
O	–0.38

$$V = \frac{Q_{\text{H}}Q_{\text{O}}}{4\pi\varepsilon_0 r} = -\frac{2(\delta e)^2}{4\pi\varepsilon_0 r}\ [16\text{B}.1]$$

$$= -\frac{2(0.42\times1.60\times10^{-19}\ \text{C})^2}{4\pi(8.85\times10^{-12}\ \text{J}^{-1}\ \text{C}^2\ \text{m}^{-1})\times(170\times10^{-12}\ \text{m})} = -4.8\times10^{-19}\ \text{J}$$

The molar energy required to break these bonds is

$$E_{\text{m}} = -N_{\text{A}}V = -(6.022\times10^{23}\ \text{mol}^{-1})\times(-4.8\times10^{-19}\ \text{J}) = \boxed{28\overline{9}\ \text{kJ mol}^{-1}}$$

Solutions to problems

P16B.1 The potential energy of the water dipole is $-\mu\mathcal{E}$ in the electric field $\mathcal{E}$ where $\mu = 1.85$ D $= 6.17\times10^{-30}$ C m. For the H_2O molecule to flip over requires a change in energy of $2\mu\mathcal{E}$.

$$V_{\text{flip}} = 2\mu\mathcal{E} = 2\times(6.17\times10^{-30}\ \text{C m})\times(1.00\times10^3\ \text{V m}^{-1}) = 1.23\times10^{-26}\ \text{J}$$

We want to find the distance r from the water dipole at which a polarizable argon atom ($a' = 1.66\times10^{-30}\ \text{m}^3$) gives an induced-dipole–dipole potential that has a magnitude equal to V_{flip}. Toward this end, we look ahead to eqn 16B.7, which describes the energy of interaction between a polar molecule and a polarizable atom.

$$2\mu\mathcal{E} = \frac{\mu^2\alpha'}{4\pi\varepsilon_0 r^6}\ [16\text{B}.7]$$

$$r^6 = \frac{\mu\alpha'}{8\pi\varepsilon_0\mathcal{E}} = \frac{(6.17\times10^{-30}\,\mathrm{C\,m})\times(1.66\times10^{-30}\,\mathrm{m^3})}{(8\pi)\times(8.854\times10^{-12}\,\mathrm{J^{-1}\,C^2\,m^{-1}})\times(1.0\times10^{3}\,\mathrm{V\,m^{-1}})} = 4.6\times10^{-53}\,\mathrm{m^6}$$

$$r = 1.9\times10^{-9}\,\mathrm{m} = \boxed{1.9\,\mathrm{nm}}$$

Comment. This distance is about 11 times the van der Waals radius of the Ar atom.

P16B.3 The interaction is a dipole–induced-dipole interaction. The energy is given by eqn 16B.7:

$$V = -\frac{\mu_1^2\alpha_2'}{4\pi\varepsilon_0 r^6} = -\frac{[(2.7\,\mathrm{D})(3.336\times10^{-30}\,\mathrm{C\,m\,D^{-1}})]^2(1.04\times10^{-29}\,\mathrm{m^3})}{4\pi(8.854\times10^{-12}\,\mathrm{J^{-1}\,C^2\,m^{-1}})(4.0\times10^{-9}\,\mathrm{m})^6}$$

$$V = \boxed{-1.8\times10^{-27}\,\mathrm{J} = -1.1\times10^{-3}\,\mathrm{J\,mol^{-1}}}.$$

Comment. This value seems exceedingly small. The distance suggested in the problem may be too large compared to typical values. A distance of 0.40 nm, yields $V = -1.1\ \mathrm{kJ\,mol^{-1}}$.

P16B.5 The potential of the London dispersion interaction is

$$V = -\frac{C}{r^6}\ [16\mathrm{B}.8] \quad \text{where} \quad C = \tfrac{3}{2}\alpha_1'\alpha_2'\frac{I_1 I_2}{I_1 + I_2}$$

$$F = -\frac{\mathrm{d}V}{\mathrm{d}r} = -\frac{\mathrm{d}}{\mathrm{d}r}\left(-\frac{C}{r^6}\right) = C\frac{\mathrm{d}}{\mathrm{d}r}\left(r^{-6}\right) = \boxed{\frac{-6C}{r^7}}$$

The negative sign indicates that the force is attractive.

P16B.7 (a) We want to construct the exponential-6 potential so that $V(r_0) = 0$ and the depth of the potential well is $-|\varepsilon|$. Consequently, we write

$$V(r) = A\varepsilon\left[Be^{-r/r_0} - \left(\frac{r_0}{r}\right)^6\right]$$

where the constants A and B are chosen to satisfy the conditions. For example, examine $V(r_0)$ to find an expression for B.

$$V(r_0) = A\varepsilon\left[Be^{-r_0/r_0} - \left(\frac{r_0}{r_0}\right)^6\right] = A\varepsilon\left[Be^{-1} - 1\right] = 0$$

So, $B = \mathrm{e}$ and the potential becomes $V(r) = A\varepsilon\left[e^{1-r/r_0} - \left(\frac{r_0}{r}\right)^6\right]$

The repulsive exponential term, the attractive term, and the potential (with $A\varepsilon = -1$) are sketched in Fig. 16B.2. The point at which the potential is a minimum is labelled as $x_e = r_e/r_0$.

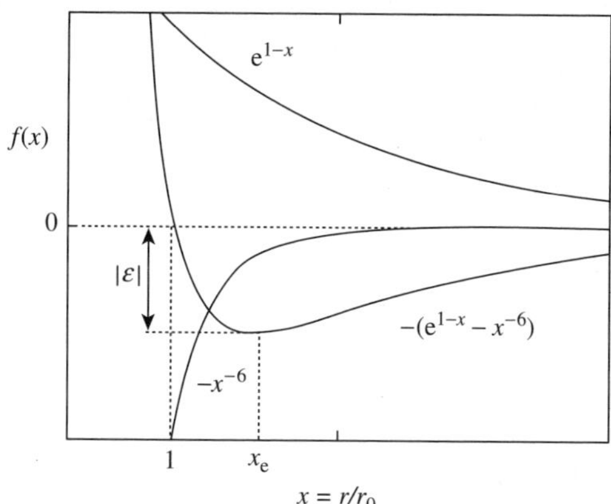

Figure 16B.2

(b) Let $x = r/r_0$, then $V(x) = A\varepsilon\left[e^{1-x} - \dfrac{1}{x^6}\right]$ and the potential minimum occurs when $\left.\dfrac{dV(x)}{dx}\right|_{x=x_e} = 0$

$$\frac{dV(x)}{dx} = A\varepsilon\left[-e^{1-x} + \frac{6}{x^7}\right]$$

$$\left.\frac{dV(x)}{dx}\right|_{x=x_e} = A\varepsilon\left[-e^{1-x_e} + \frac{6}{x_e^7}\right] = 0$$

Thus, the solution of the transcendental equation $e^{1-x_e} = 6x_e^{-7}$ gives the vallue x_e at which V is a minimum. x_e may be found as the intersection of the curves e^{1-x} and $6/x^7$ or it may be found using the numeric solver of a scientific calculator. Here's a short Mathcad Prime 2 worksheet solution for $x = x_e$.

Guess Values

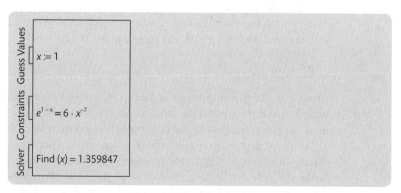

$x := 1$

Constraints

$e^{1-x} = 6 \cdot x^{-7}$

Solver

Find $(x) = 1.359847$

Thus, $\boxed{r_e = 1.3598 r_0}$.

Parameter A is determined with the criterion that $V(r_e) = \varepsilon$

$$V(x_e) = \varepsilon = A\varepsilon\left[e^{1-x_e} - \frac{1}{x_e^6}\right]$$

$$A = \left[e^{1-x_e} - \frac{1}{x_e^6}\right]^{-1} = \left[e^{1-1.3598} - \frac{1}{1.3598^6}\right]^{-1}$$

$$\boxed{A = 1.8531}$$

16C Liquids

Answer to discussion question

D16C.1 When the temperature of a thermodynamically stable, pure gaseous substance declines at constant pressure there comes a temperature at which the chemical potential of the liquid phase ($\mu_{\text{pure liquid}} = G_{\text{m,pure liquid}}$) is no longer greater than the chemical potential of the gas phase ($\mu_{\text{pure gas}} = G_{\text{m,pure gas}}$). That temperature, the boiling point, is part of the equilibrium vapour pressure curve for which $\mu_{\text{pure liquid}} = \mu_{\text{pure gas}}$ and the pressure is called the **vapour pressure of the liquid**. The relation between the vapour pressure p and the boiling point T, the liquid–vapour phase boundary, indicates that a plot of $\ln p$ against $1/T$ is linear with a slope equal to $-\Delta_{\text{vap}}H/R$:

$$p = p_{\text{ref}}e^{-\chi} \text{ where } \chi = \frac{\Delta_{\text{vap}}H}{R}\left(\frac{1}{T} - \frac{1}{T_{\text{ref}}}\right)$$

[4B.11, perfect gas, constant $\Delta_{\text{vap}}H$, experimental reference point $(p_{\text{ref}}, T_{\text{ref}})$]

Conditions below the boiling point are: $\mu_{\text{pure liquid}} < \mu_{\text{pure gas}}$, the liquid is the thermodynamically stable state, and condensation is spontaneous. Above the boiling point: $\mu_{\text{pure liquid}} > \mu_{\text{pure gas}}$, the gas is the thermodynamically stable state, and vaporization is spontaneous.

If an insoluble, inert gas is used to apply additional pressure ΔP on the liquid with the result that $P = p + \Delta P$, the vapour pressure becomes

$$p = p^* e^{V_m(l)\Delta P/RT} \quad [4B.3]$$

where p^* is the vapour pressure of the liquid in the absence of an additional pressure. The positive exponential factor indicates that the vapour pressure increases when the pressure acting on the condensed phase is increased.

The above discussion relates to bulk, macroscopic samples of matter in which surface effects are negligible. However, the condensation process can be explored further by considering the formation of small spherical droplets of liquid within the gas phase. The curvature of the droplet surface of radius r and surface tension γ plays a significant role because it gives rise to a positive pressure differential $\Delta P = 2\gamma/r$ between the droplet bulk and the gas phase. Equation 4B.3 becomes the **Kelvin equation**

$$p = p^* e^{2\gamma V_m(l)/rRT} \quad [16C.20]$$

The relation indicates that droplets with smaller values of r have greater vapour pressure. This presents a severe problem for the formation of liquid droplets from a pure gas state that has been cooled below its boiling point. We can imagine a swarm of gas phase molecules congregating into what would certainly be a very, very small droplet at best. The droplet, being exceedingly small, has an enhanced vapour pressure. Therefore, instead of growing it evaporates even though $\mu_{\text{pure liquid}} < \mu_{\text{pure gas}}$. This effect kinetically stabilizes the gas phase while liquid formation remains spontaneous. The rate of condensation is extremely slow and the direct mechanism of condensation is hindered. Thus, rapid formation of liquid–gas equilibrium depends on the presence of minute dust particles or other kinds of foreign matter to which the gas molecules can stick. These serve as **nucleation centres** for condensation by having larger radii with reasonably low vapour pressure.

Solutions to exercises

E16C.1(a) $V_m = \dfrac{M}{\rho} = \dfrac{18.02\ \text{g mol}^{-1}}{0.9982\ \text{g cm}^{-3}} = 18.05\ \text{cm}^3$

$$p = p^* e^{2\gamma V_m(\text{l})/rRT} \quad [\text{16C.20, the Kelvin equation}]$$

$$= (2.3\ \text{kPa}) \times \exp\left\{\frac{2\times(72.75\times10^{-3}\ \text{N m}^{-1})\times(18.05\times10^{-6}\ \text{m}^3\ \text{mol}^{-1})}{(10\times10^{-9}\ \text{m})\times(8.3145\ \text{J K}^{-1}\ \text{mol}^{-1})\times(293\ \text{K})}\right\}$$

$$= \boxed{2.6\ \text{kPa}}$$

E16C.2(a) $\gamma = \tfrac{1}{2}\rho g r h$ [16C.9]

$$= \tfrac{1}{2}\times(998.2\ \text{kg m}^{-3})\times(9.80665\ \text{m s}^{-2})\times(0.300\times10^{-3}\ \text{m})\times(4.96\times10^{-2}\ \text{m})$$

$$= 0.0728\bar{3}\ \text{kg s}^{-2} = \boxed{72.8\ \text{mN m}^{-1}}$$

E16C.3(a) $p_{\text{in}} - p_{\text{out}} = \dfrac{2\gamma}{r}$ [16C.7, the Laplace equation]

$$= \frac{2\times(72.75\times10^{-3}\ \text{N m}^{-1})}{200\times10^{-9}\ \text{m}} \quad [\text{Table 16C.1}]$$

$$= 7.28\times10^5\ \text{N m}^{-2} = \boxed{728\ \text{kPa}}$$

Pressure differentials for small droplets are quite large.

Solution to problem

P16C.1 A plot of surface tension γ against surfactant concentration [A] is shown in Fig. 16C.1. It is linear and the regression fit, summarized in the figure, gives a slope equal to -25.657×10^{-6} N m^2 mol^{-1}. Solving eqn 16C.19 for the surface excess $\Gamma_{[\text{A}]}$ indicates that the surface excess is proportional to the bulk concentration.

$$\Gamma_{[A]} = -\frac{[A]}{RT}\frac{d\gamma}{d[A]} \quad [16C.19]$$

$$= -\frac{(-25.657\times10^{-3}\ \text{N m}^{-1})}{(8.31447\ \text{J K}^{-1}\ \text{mol}^{-1})\times(293.15\ \text{K})}\times\left(\frac{[A]}{\text{mol dm}^{-3}}\right)$$

$$= (1.05\times10^{-5}\ \text{mol m}^{-2})\times\left(\frac{[A]}{\text{mol dm}^{-3}}\right)$$

This equation is used to calculate the surface excess for a given value of [A]. For example, when [A] = 0.10 mol dm^{-3}, $\Gamma_{[A]} = (1.05\times10^{-5}\ \text{mol m}^{-2})\times(0.10) = 1.05\times10^{-6}\ \text{mol m}^{-2}$.

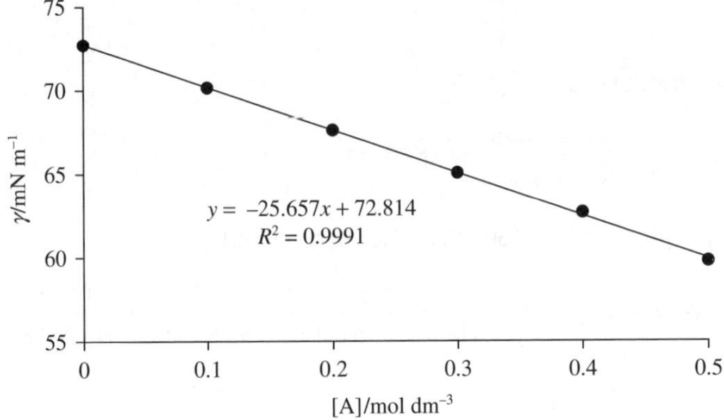

Figure 16C.1

Integrated activities

I16.1 Consider a single molecule surrounded by $N-1$ ($\approx N$) others in a container of volume V. The number of molecules in a spherical shell of thickness dR is $4\pi R^2\times\frac{N}{V}dR$. Molecules cannot approach more closely than the molecular diameter d so $R\geq d$ and, therefore, the pair interaction energy is

$$u = \int_d^\infty 4\pi R^2\times\left(\frac{N}{V}\right)\times\left(\frac{-C_6}{R^6}\right)dR = \frac{-4\pi NC_6}{V}\int_d^\infty\frac{dR}{R^4} = \left(\frac{4\pi NC_6}{3V}\right)\times\left(\frac{1}{\infty^3}-\frac{1}{d^3}\right)$$

$$= \frac{-4\pi NC_6}{3Vd^3}$$

The mutual pairwise interaction energy of all N molecules is $U=\frac{1}{2}Nu$ (the $\frac{1}{2}$ appears because each pair must be counted only once, i.e. A with B but not A with B and B with A). Therefore,

$$U = \boxed{\frac{-2\pi N^2C_6}{3Vd^3}}$$

For a van der Waals gas, $\dfrac{n^2a}{V^2}=\left(\dfrac{\partial U}{\partial V}\right)_T=\dfrac{2\pi N^2C_6}{3V^2d^3}$ and therefore $a=\boxed{\dfrac{2\pi N_A^2C_6}{3d^3}}$ $[N=nN_A]$

I16.3 (a) The left side molecule in Fig. I16.1 is methyladenine. Please note that we have taken the liberty of placing the methyl group in the position that would be occupied by a sugar in RNA and DNA. The wavefunction, structure, and atomic electrostatic charges (shown in the figure) calculation was performed with Spartan '10 using a Hartree–Fock procedure with a 6-31G* basis set. The atomic electrostatic charge (ESP) numerical method generates charges that reproduce the electrostatic field from the entire wavefunction. The right side molecule in Fig. I16.1 is methylthymine.

(b) The two molecules will hydrogen bond into a stable dimer in an orientation for which hydrogen bonding is linear, maximized, and steric hindrance is avoided. We expect hydrogen bonds of the type N–H·····O and N–H·····N with the N and O atoms having large negative electrostatic charges and the H atoms having large positive charges. These atoms are evident in the figure.

(c) Figure I16.1 shows one of three arrangements of hydrogen bonding between the two molecules. Another can be drawn by rotating methylthymine over the top of methyladenine and a third involves rotation to the bottom. The dashed lines show the alignments of two strong hydrogen bonds between the molecules, which the calculation show to have matching lengths of about 200 pm.

(d) The A-to-T base pairing shown in Fig. I16.1 has the largest charges in the most favourable positions for strong hydrogen bonding. Also the N-to-O distance of one hydrogen bond equals the N-to-N distance of the other, a favourable feature in RNA and DNA polymers where this pairing and alignment is observed naturally.

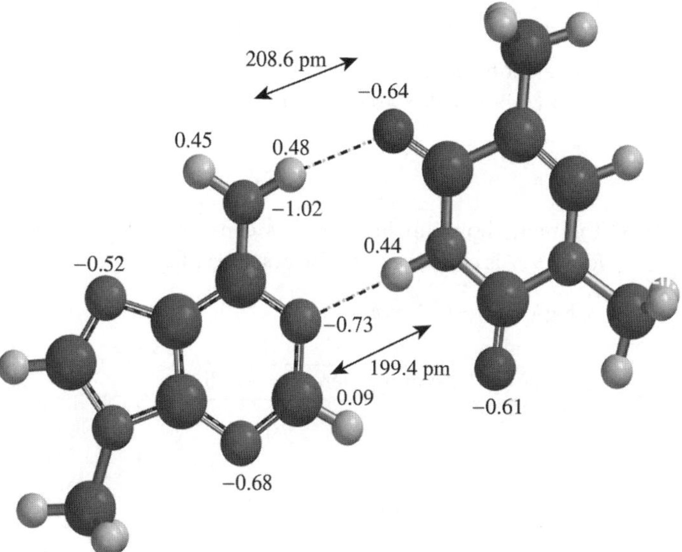

Figure I16.1

(e) The favourable orientation for hydrogen bonding between methylguanine (left side molecule) and methylcytosine (right side molecule) is shown in Fig. I16.2. Large counter charges align within distances that result in three strong hydrogen bonds with lengths that match those of Fig. I16.1, an essential feature of the DNA double helix.

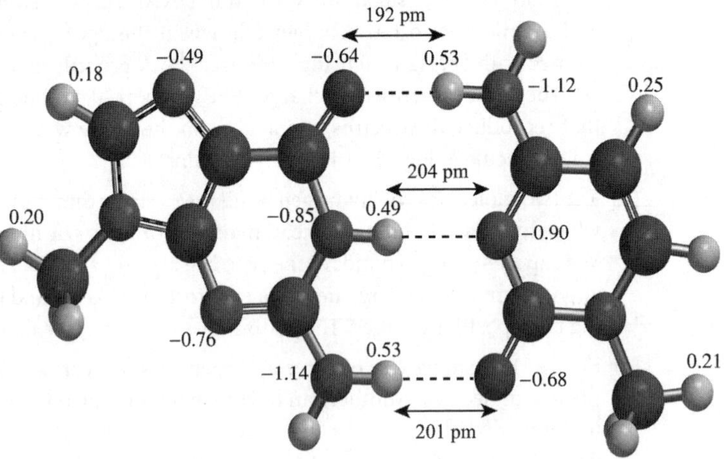

Figure I16.2

I16.5 (a) The hydrocarbons in question form a homologous series. They are straight-chain alkanes of the formula C_nH_{2n+2}, or R–H where $R = C_nH_{2n+1}$. Draw up the following table:

n	1	2	3	4	5
π	0.5	1.0	1.5	2.0	2.5

The relationship here is evident by inspection: $\pi = n/2$, so we predict for the seven-carbon hydrocarbon in question:

$$\pi = 7/2 = \boxed{3.5}$$

(b) The plot, shown in Fig. I16.3, is consistent with a linear relationship, for $R^2 = 0.997$ is close to unity. The best linear fit is:

$$\log K = -1.95 - 1.49\pi,$$

so $\boxed{\text{slope} = -1.49}$ and $\boxed{\text{intercept} = -1.95}$.

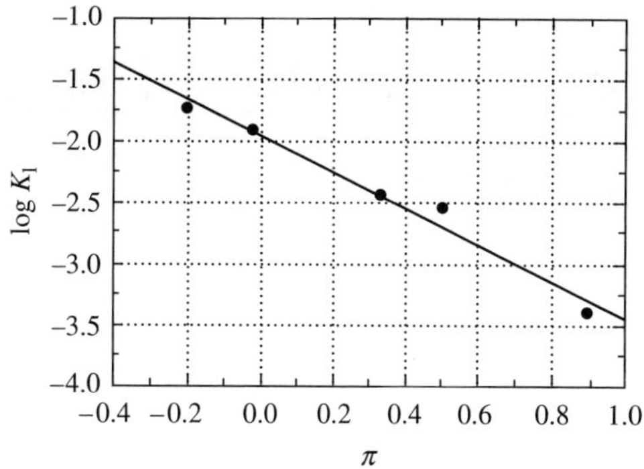

Figure I16.3

(c) If we know π for the substituent R = H, then we can use the linear SAR just derived. Our best estimate of π can be obtained by considering the zero-carbon 'alkane' H_2, whose radical H ought to have a hydrophobicity constant $\pi = 0/2 = 0$. This value yields

$$\log K = -1.95 - 1.49(0) = -1.95 \quad \text{so} \quad K = 10^{-1.95} = \boxed{1.12 \times 10^{-2}}$$

Note: the assumption that R = H is part of the homologous series of straight-chain alkanes is a reasonable but questionable one.

17 Macromolecules and self-assembly

17A The structures of macromolecules

Answers to discussion questions

D17A.1 The **primary structure** of a macromolecule is the sequence of small molecular residues making up the polymer whereas the **secondary structure** is the (often local) spatial arrangement of a chain. The **tertiary structure** is the overall three-dimensional structure of a single macromolecule. Two or more macromolecules may form an aggregation that yields an overall **quaternary structure**. Covalent bonding connects molecular residues of the primary structures. Higher order structures are created through hydrogen bonding, hydrophobic interactions, salt bridges, and disulfide links between residues that are widely separated along the polymer chain.

D17A.3 (a) **Contour length, $R_c = Nl$** [17A.3]: the length of the macromolecule measured along its backbone, the length of all its monomer units placed end to end. This is the stretched-out length of the macromolecule with bond angles maintained within the monomer units and 180° angles at unit links. It is proportional to the number of monomer units, N, and to the length of each unit.

(b) **Root mean square separation, $R_{rms} = N^{1/2}l$** [17A.4]: one measure of the average separation of the ends of a random coil. It is the square root of the mean value of R^2, where R is the separation of the two ends of the coil. R_{rms} is proportional to $N^{1/2}$ and the length of each unit.

(c) **Root mean square separation, $R_{rms} = (2N)^{1/2}l$** [17A.8]: a measure of the average separation of the ends of a tetrahedrally ($\theta = 109.5°$) constrained jointed chain. This model does not take into account the impossibility of two or more atoms occupying the same place.

(d) **Root mean square separation, $R_{rms} = N^{1/2}lF$** where $F = (1 - \cos\theta)^{1/2}/(1 + \cos\theta)^{1/2}$ [17A.7]: a measure of the average separation of the ends of a constrained jointed chain in which each successive individual bond is constrained to a single cone of angle θ relative to its neighbour.

(e) **Radius of gyration of a one-dimensional random coil, $R_g = N^{1/2}l$** [17A.5].

(f) **Radius of gyration of a three-dimensional random coil, $R_g = (N/6)^{1/2}l$** [17A.6]: radius of a hollow sphere that has the same moment of inertia (and therefore rotational characteristics) as the actual molecule of the same mass. $R_g^{\,2} = I/m_{tot}$. In general, it is not easy to visualize this distance geometrically. However, for the simple case of a molecule consisting of an unconstrained chain of identical atoms this quantity can be visualized as the root mean square distance of the atoms from the centre of mass. It also depends on $N^{1/2}$, but is smaller than the root mean square separation by a factor of $(1/6)^{1/2}$.

(g) **Radius of gyration of a tetrahedrally ($\theta = 109.5°$) constrained jointed chain, $R_g = (N/3)^{1/2}l$** [17A.8].

Solutions to problems

E17A.1(a) $R_{rms} = N^{1/2}l$ [17A.4] $= (700)^{1/2} \times (0.90\text{ nm}) = \boxed{27\text{ nm}}$

E17A.2(a) The repeating monomer unit of polyethylene is ($-CH_2-CH_2-$) which has a molar mass of 28 g mol^{-1}. The number of repeating units, N, is therefore

$$N = \frac{M_{\text{polymer}}}{M_{\text{monomer}}} = \frac{280\,000\text{ g mol}^{-1}}{28\text{ g mol}^{-1}} = 1.00 \times 10^4$$

$$l = 2R(\text{C–C}) \;[\text{Add half a bond length on either side of monomer.}]$$

$$R_c = Nl\;[17\text{A}.3] = 2 \times (1.00 \times 10^4) \times (154\text{ pm}) = 3.08 \times 10^6\text{ pm} = \boxed{3.08\text{ μm}}$$

$$R_{rms} = N^{1/2}l\;[17\text{A}.4] = 2 \times (1.00 \times 10^4)^{1/2} \times (154\text{ pm}) = 3.08 \times 10^4\text{ pm} = \boxed{30.8\text{ nm}}$$

E17A.3(a) For a one-dimensional random coil, the radius of gyration is

$$R_g = N^{1/2}l\;[17\text{A}.5] \text{ so the number of C–C links is } N = \left(\frac{R_g}{l}\right)^2 = \left(\frac{7.3\text{ nm}}{0.154\text{ nm}}\right)^2 = \boxed{2.4 \times 10^3}$$

E17A.4(a) The repeating monomer unit of polyethylene is ($-CH_2-CH_2-$) which has a molar mass of 28 g mol^{-1}. The number of repeating units, N, is therefore

$$N = \frac{M_{\text{polymer}}}{M_{\text{monomer}}} = \frac{65\,000\text{ g mol}^{-1}}{28\text{ g mol}^{-1}} = 2.3\bar{2} \times 10^3$$

$$l = 2R(\text{C–C}) = 2 \times 154\text{ pm} = 308\text{ pm}$$

$$[\text{Add half a bond length on either side of monomer.}]$$

In units of l the polymer ends are separated by the distance

$$n = \frac{10 \times 10^{-9}\text{ m}}{l} = \frac{10 \times 10^{-9}\text{ m}}{308 \times 10^{-12}\text{ m}} = 32.\bar{5}$$

Thus,

$$P = \left(\frac{2}{\pi N}\right)^{1/2} e^{-n^2/2N}\;[17\text{A}.1]$$

$$= \left(\frac{2}{\pi \times 2.3\bar{2} \times 10^3}\right)^{1/2} e^{-(32.\bar{5})^2/\{2 \times (2.3\bar{2} \times 10^3)\}} = \boxed{0.017}$$

E17A.5(a) We obtain (see Exercise 17A.4(a)) $N = 2.3\bar{2}\times10^3$ and $l = 308$ pm. Thus,

$$a = \left(\frac{3}{2Nl^2}\right)^{1/2} \quad [17A.2]$$

$$= \left(\frac{3}{2\times(2.3\bar{2}\times10^3)\times(308\times10^{-12}\text{ m})^2}\right)^{1/2} = 8.2\bar{6}\times10^7\text{ m}^{-1}$$

The 10.00 nm to 10.10 nm range of distances between the polymer ends is very small so we estimate that the distribution function $f(r)$ is the constant given by $f(r) = f(10.05\text{ nm})$. The probability that the polymer ends are in this range is

$$P = f(r)\Delta r = 4\pi\left(\frac{a}{\pi^{1/2}}\right)^3 r^2 e^{-a^2r^2}\Delta r \quad [17A.2]$$

$$= 4\pi\left(\frac{8.2\bar{6}\times10^7\text{ m}^{-1}}{\pi^{1/2}}\right)^3 \times(10.05\times10^{-9}\text{ m})^2 \times e^{-(8.2\bar{6}\times10^7\text{ m}^{-1})^2(10.05\times10^{-9}\text{ m})^2}$$

$$\times(10.1-10.00)\times10^{-9}\text{ m}$$

$$= \boxed{6.4\times10^{-3}}$$

E17A.6(a) The radius of gyration for a constrained chain that has successive individual bonds constrained to a single cone of angle $\theta = 109.5°$ (i.e. $\cos\theta = -⅓$) is

$$R_{\text{g,constrained coil}} = R_{\text{g,random coil}}F \quad \text{with} \quad F = \left(\frac{1-\cos\theta}{1+\cos\theta}\right)^{1/2} [17A.7] = \left(\frac{1+\frac{1}{3}}{1-\frac{1}{3}}\right)^{1/2} = \sqrt{2}$$

The percentage change in the radius of gyration upon application of the constraint is

$$\left(\frac{R_{\text{g,constrained coil}} - R_{\text{g,random coil}}}{R_{\text{g,random coil}}}\right)\times100\% = \left(\frac{R_{\text{g,constrained coil}}}{R_{\text{g,random coil}}} - 1\right)\times100\% = (F-1)\times100\%$$

$$= (\sqrt{2}-1)\times100\%$$

$$= \boxed{+41.42\%}$$

The percentage change in the volume upon application of the constraint is

$$\left(\frac{V_{\text{constrained coil}} - V_{\text{random coil}}}{V_{\text{random coil}}}\right)\times100\% = \left(\frac{R^3_{\text{g,constrained coil}} - R^3_{\text{g,random coil}}}{R^3_{\text{g,random coil}}}\right)\times100\%$$

$$= \left\{\left(\frac{R_{\text{g,constrained coil}}}{R_{\text{g,random coil}}}\right)^3 - 1\right\}\times100\% = (F^3-1)\times100\%$$

$$= (2^{3/2}-1)\times100\%$$

$$= \boxed{+182.8\%}$$

E17A.7(a) For a one-dimensional random coil: $R_g = R_{rms} = N^{1/2}l$ [17A.4 and 5]. Consequently, it seems reasonable to make the following analogy to eqn 17A.10.

$$R_g = R_{g,\text{random coil}}F = N^{1/2}lF \qquad \text{where} \qquad F = \left(\frac{2l_p}{l}-1\right)^{1/2}$$

Thus, the percentage increase in the radius of gyration when the polymer persistence length is changed from l to $0.050R_c$ with constant contour length is given by the expression

$$\left(\frac{R_g - R_{g,\text{random coil}}}{R_{g,\text{random coil}}}\right)\times 100\% = \left(\frac{R_g}{R_{g,\text{random coil}}}-1\right)\times 100\% = (F-1)\times 100\%$$

$$= \left\{\left(\frac{2l_p}{l}-1\right)^{1/2}-1\right\}\times 100\%$$

$$= \left\{\left(\frac{2\times(0.05R_c)}{l}-1\right)^{1/2}-1\right\}\times 100\%$$

$$= \left\{\left(\frac{2\times(0.05\times Nl)}{l}-1\right)^{1/2}-1\right\}\times 100\% \quad [17A.3]$$

$$= \left\{(0.10N-1)^{1/2}-1\right\}\times 100\%$$

$$= \boxed{+895\% \text{ when } N = 1\,000}$$

The percentage change in the volume as determined by the radius of gyration is

$$\left(\frac{V - V_{g,\text{random coil}}}{V_{g,\text{random coil}}}\right)\times 100\% = \left(\frac{R_g^3 - R_{g,\text{random coil}}^3}{R_{g,\text{random coil}}^3}\right)\times 100\% = (F^3-1)\times 100\%$$

$$= \left\{\left(\frac{2l_p}{l}-1\right)^{3/2}-1\right\}\times 100\%$$

$$= \left\{\left(\frac{2\times(0.05R_c)}{l}-1\right)^{3/2}-1\right\}\times 100\%$$

$$= \left\{\left(\frac{2\times(0.05\times Nl)}{l}-1\right)^{3/2}-1\right\}\times 100\% \quad [17A.3]$$

$$= \left\{(0.10N-1)^{3/2}-1\right\}\times 100\%$$

$$= \boxed{+9.84\times 10^4\% \text{ when } N = 1\,000}$$

E17A.8(a) By analogy to eqn 17A.10, we assume the following relation for a three-dimensional partially rigid polymer.

$$R_g = R_{g,\text{random coil}}F = \left(\frac{N}{6}\right)^{1/2} lF \ [17A.6] \quad \text{where} \quad F = \left(\frac{2l_p}{l} - 1\right)^{1/2} \ [17A.10]$$

Solving for l_p gives

$$l_p = \frac{l}{2}\left\{\frac{6R_g^2}{Nl^2} + 1\right\}$$

$$= \frac{150\ \text{pm}}{2}\left\{\frac{6\times(2\,100\ \text{pm})^2}{1\,000\times(150\ \text{pm})^2} + 1\right\} = \boxed{1.6\times10^2\ \text{pm}}$$

Solutions to problems

P17A.1 R_g is the radius of rotation of a point mass that has the same mass m and moment of inertia I as the object of interest. For an object that has a continuum of mass within its macroscopic boundaries:

$$I = mR_g^2 \quad \text{where} \quad I = \int \rho R^2\, d\tau \quad \text{and } R \text{ is the distance from the axis of rotation}$$

For a homogeneous object $\rho = m/V$ so

$$R_g^2 = V^{-1}\int R^2\, d\tau \quad \text{(The integrand is the square distance from the axis of rotation.)}$$

(a) For a solid sphere of radius a: $V = \tfrac{4}{3}\pi a^3$, $d\tau = r^2 \sin\theta\, dr\, d\theta\, d\phi$, and $R^2 = x^2 + y^2$.

$$R_g^2 V = \int_{\phi=0}^{2\pi}\int_{\theta=0}^{\pi}\int_{r=0}^{a} (x^2+y^2) r^2 \sin\theta\ dr\, d\theta\, d\phi$$

$$= \int_{\phi=0}^{2\pi}\int_{\theta=0}^{\pi}\int_{r=0}^{a} (r^2\sin^2\theta\cos^2\phi + r^2\sin^2\theta\sin^2\phi) r^2 \sin\theta\, dr\, d\theta\, d\phi$$

$$= \int_{\phi=0}^{2\pi}\int_{\theta=0}^{\pi}\int_{r=0}^{a} r^4 \sin^3\theta\ dr\, d\theta\, d\phi$$

$$= \left\{\int_0^{2\pi} d\phi\right\}\left\{\int_0^{\pi} \sin^3\theta\, d\theta\right\}\left\{\int_0^{a} r^4\, dr\right\}$$

$$= 2\pi\times\left\{-\cos\theta + \tfrac{1}{3}\cos^3\theta\right\}\Big|_{\theta=0}^{\theta=\pi} \times \left[\frac{r^5}{5}\right]_{r=0}^{r=a} = \frac{8\pi a^5}{15}$$

$$R_g^2 = \frac{8\pi a^5}{15V} = \frac{8\pi a^5}{15\times(\tfrac{4}{3}\pi a^3)} = \tfrac{2}{5}a^2$$

$$R_g = \boxed{\sqrt{\tfrac{2}{5}}\, a}$$

Comment. A common error involves using r^2 in place of the squared distance from the axis of rotation, which is actually equal to x^2+y^2 not to $x^2+y^2+z^2=r^2$. The common error gives the result $\sqrt{3/5}\,a$.

For a spherical macromolecule, the specific volume is:

$$v_s=\frac{V}{m}=\frac{4\pi a^3}{3}\times\frac{N_A}{M} \quad \text{so} \quad a=\left(\frac{3v_sM}{4\pi N_A}\right)^{1/3}$$

Therefore,

$$R_g=\left(\frac{2}{5}\right)^{1/2}\times\left(\frac{3v_sM}{4\pi N_A}\right)^{1/3}$$

$$=\left(\frac{2}{5}\right)^{1/2}\times\left(\frac{(3v_s/\text{cm}^3\,\text{g}^{-1})\times\text{cm}^3\,\text{g}^{-1}\times(M/\text{g}\,\text{mol}^{-1})\times\text{g}\,\text{mol}^{-1}}{4\pi\times(6.022\times10^{23}\,\text{mol}^{-1})}\right)^{1/3}$$

$$=(4.6460\times10^{-9}\,\text{cm})\times\{(v_s/\text{cm}^3\,\text{g}^{-1})\times(M/\text{g}\,\text{mol}^{-1})\}^{1/3}$$

$$R_g/\text{nm}=\boxed{0.046460\times\{(v_s/\text{cm}^3\,\text{g}^{-1})\times(M/\text{g}\,\text{mol}^{-1})\}^{1/3}}$$

Comment. The common error, discussed above, gives

$$0.056902\times\{(v_s/\text{cm}^3\,\text{g}^{-1})\times(M/\text{g}\,\text{mol}^{-1})\}^{1/3}.$$

When $M=100\,\text{kg}\,\text{mol}^{-1}$ and $v_s=0.750\,\text{cm}^3\,\text{g}^{-1}$,

$$R_g=0.046460\times\{(0.750)\times(100\times10^3)\}^{1/3}\,\text{nm}$$

$$=\boxed{1.96\,\text{nm}}$$

(b) For a solid rod of radius a and length l there are two moments of inertia. One about the axis of its length ($R_{g,\parallel}$, $R=r$) and another about the axis that is perpendicular to its length ($R_{g,\perp}$, $R=z$).

$$R_{g,\parallel}^2=(\pi a^2l)^{-1}\int r^2\,d\tau=(\pi a^2l)^{-1}\int_{r=0}^{a}\int_{z=0}^{l}\int_{\theta=0}^{2\pi}r^3\,d\theta\,dz\,dr \quad [\text{cylindrical coordinates}]$$

$$=(\pi a^2l)^{-1}\,2\pi l\int_0^a r^3\,dr$$

$$=2/a^2\left[\frac{r^4}{4}\right]_{r=0}^{r=a}$$

$$=\tfrac{1}{2}a^2$$

$$R_{g,\parallel}=\boxed{\sqrt{1/2}\,a}$$

$$R_{g,\perp}^2 = (\pi a^2 l)^{-1} \int z^2 \,\mathrm{d}\tau = (\pi a^2 l)^{-1} \int_{r=0}^{a} \int_{z=-l/2}^{l/2} \int_{\theta=0}^{2\pi} z^2 r \,\mathrm{d}\theta \,\mathrm{d}z \,\mathrm{d}r \quad \text{[cylindrical coordinates]}$$

$$= \tfrac{1}{12} l^2$$

$$R_{g,\perp} = \boxed{\sqrt{\tfrac{1}{12}}\, l}$$

For a rod-like macromolecule, the specific volume is:

$$\upsilon_s = \frac{V}{m} = \pi a^2 l \times \frac{N_A}{M} \quad \text{so} \quad l = \frac{\upsilon_s M}{\pi a^2 N_A}$$

Therefore,

$$R_{g,\parallel} = \sqrt{\tfrac{1}{2}} \times (0.50\ \text{nm}) = \boxed{0.35\ \text{nm}}$$

$$R_{g,\perp} = \sqrt{\tfrac{1}{12}} \left(\frac{\upsilon_s M}{\pi a^2 N_A} \right)$$

$$= \sqrt{\tfrac{1}{12}} \left\{ \frac{(0.750 \times 10^{-6}\ \text{m}^3\,\text{g}^{-1}) \times (100 \times 10^3\,\text{g}\,\text{mol}^{-1})}{\pi \times (0.50 \times 10^{-9}\ \text{m})^2 \times (6.022 \times 10^{23}\ \text{mol}^{-1})} \right\}$$

$$= \boxed{46\ \text{nm}}$$

P17A.3 The probability that the ends of a three-dimensional freely jointed chain lie in the range r to $r + \mathrm{d}r$ is $f(r)\mathrm{d}r$ where

$$f(r) = 4\pi \left(\frac{a}{\pi^{1/2}} \right)^3 r^2 \mathrm{e}^{-a^2 r^2} \qquad a = \left(\frac{3}{2Nl^2} \right)^{1/2} \quad [18.2]$$

The mean nth power of the end-to-end separation is

$$\langle r^n \rangle = \int_0^\infty r^n f(r) \mathrm{d}r$$

$$\langle r^2 \rangle = \int_0^\infty r^2 f(r) \mathrm{d}r$$

$$= 4\pi \left(\frac{a}{\pi^{1/2}} \right)^3 \int_0^\infty r^4 \mathrm{e}^{-a^2 r^2}\, \mathrm{d}r$$

$$= 4\pi \left(\frac{a}{\pi^{1/2}} \right)^3 \frac{\Gamma(\tfrac{5}{2})}{2a^5} \quad \text{where } \Gamma(\tfrac{5}{2}) = \tfrac{3}{4}\pi^{1/2} \quad \text{[standard integral]}$$

$$= \frac{3}{2a^2} = \frac{3}{2} \left(\frac{2Nl^2}{3} \right) = \boxed{Nl^2}$$

Note: The general form of the above standard integral is

$$\int_0^\infty r^m \mathrm{e}^{-a^2 r^2}\, \mathrm{d}r = \frac{\Gamma[(m+1)/2]}{2a^{m+1}} \quad \text{where } \Gamma \text{ is the gamma function found in maths handbooks}$$

P17A.5 R_g is the radius of rotation of a point mass that has the same mass m and moment of inertia I as the object of interest. For an object that has a continuum of mass within its macroscopic boundaries:

$$I = mR_g^2 \quad \text{where} \quad I = \int \rho R^2 \, d\tau \quad \text{and } R \text{ is the distance from the axis of rotation}$$

For a homogeneous object $\rho = m/V$ so

$$R_g^2 = V^{-1} \int R^2 \, d\tau \quad \text{(The integrand is the square distance from the axis of rotation.)}$$

(a) For a solid thin disk of radius a and thickness h there are two moments of inertia. One about the disk axis ($R_{g,\perp}$, $R = r$) and another about the axis along the disk diameter ($R_{g,\parallel}$, $R = y$).

$$R_{g,\perp}^2 = \left(\pi a^2 h\right)^{-1} \int r^2 \, d\tau = \left(\pi a^2 h\right)^{-1} \int_{r=0}^{a} \int_{z=0}^{h} \int_{\theta=0}^{2\pi} r^3 \, d\theta \, dz \, dr$$

[cylindrical coordinates]

$$= \left(\pi a^2 h\right)^{-1} \int_0^a r^3 \, dr \times \int_0^h dz \times \int_0^{2\pi} d\theta$$

$$= {}^{2}\!/_{a^2} \left[\frac{r^4}{4}\right]_{r=0}^{r=a}$$

$$= {}^{1}\!/_{2} a^2$$

$$R_{g,\perp} = \boxed{\sqrt{{}^{1}\!/_{2}}\, a}$$

$$R_{g,\parallel}^2 = \left(\pi a^2 h\right)^{-1} \int y^2 \, d\tau = \left(\pi a^2 h\right)^{-1} \int_{r=0}^{a} \int_{z=0}^{h} \int_{\theta=0}^{2\pi} (r \sin\theta)^2 \times r \, d\theta \, dz \, dr$$

[cylindrical coordinates]

$$= \left(\pi a^2 h\right)^{-1} \int_0^a r^3 \, dr \times \int_0^h dz \times \int_0^{2\pi} \sin^2\theta \, d\theta$$

$$= \left(\pi a^2 h\right)^{-1} \left\{ \left[\frac{r^4}{4}\right]_{r=0}^{r=a} \times h \times \left[\frac{\theta}{2} - \frac{\sin 2\theta}{4}\right]_{\theta=0}^{\theta=2\pi} \right\}$$

$$= {}^{1}\!/_{4} a^2$$

$$R_{g,\parallel} = \boxed{a/2}$$

(b) For a solid rod of radius a and length l there are two moments of inertia. One about the axis of its length ($R_{g,\parallel}$, $R = r$) and another about the axis that is perpendicular to its length ($R_{g,\perp}$, $R = z$).

$$R_{g,\parallel}^2 = (\pi a^2 l)^{-1}\int r^2\,d\tau = (\pi a^2 l)^{-1}\int_{r=0}^{a}\int_{z=0}^{l}\int_{\theta=0}^{2\pi} r^3\,d\theta\,dz\,dr \quad \text{[cylindrical coordinates]}$$

$$= (\pi a^2 l)^{-1}\,2\pi l\int_0^a r^3\,dr$$

$$= \tfrac{2}{a^2}\left[\frac{r^4}{4}\right]_{r=0}^{r=a}$$

$$= \tfrac{1}{2}a^2$$

$$R_{g,\parallel} = \boxed{\sqrt{\tfrac{1}{2}}a}$$

$$R_{g,\perp}^2 = (\pi a^2 l)^{-1}\int z^2\,d\tau = (\pi a^2 l)^{-1}\int_{r=0}^{a}\int_{z=-l/2}^{l/2}\int_{\theta=0}^{2\pi} z^2 r\,d\theta\,dz\,dr \quad \text{[cylindrical coordinates]}$$

$$= \tfrac{1}{12}l^2$$

$$R_{g,\perp} = \boxed{\sqrt{\tfrac{1}{12}}l}$$

(c) For a solid sphere of radius a: $V = \tfrac{4}{3}\pi a^3$, $d\tau = r^2\sin\theta\,dr\,d\theta\,d\phi$, and $R^2 = x^2 + y^2$.

$$R_g^2 V = \int_{\phi=0}^{2\pi}\int_{\theta=0}^{\pi}\int_{r=0}^{a}(x^2+y^2)r^2\sin\theta\;dr\,d\theta\,d\phi$$

$$= \int_{\phi=0}^{2\pi}\int_{\theta=0}^{\pi}\int_{r=0}^{a}(r^2\sin^2\theta\cos^2\phi + r^2\sin^2\theta\sin^2\phi)r^2\sin\theta\;dr\,d\theta\,d\phi$$

$$= \int_{\phi=0}^{2\pi}\int_{\theta=0}^{\pi}\int_{r=0}^{a} r^4\sin^3\theta\;dr\,d\theta\,d\phi$$

$$= \left\{\int_0^{2\pi}d\phi\right\}\left\{\int_0^{\pi}\sin^3\theta\,d\theta\right\}\left\{\int_0^a r^4\,dr\right\}$$

$$= 2\pi\times\left\{-\cos\theta + \tfrac{1}{3}\cos^3\theta\right\}\Big|_{\theta=0}^{\theta=\pi}\times\left[\frac{r^5}{5}\right]_{r=0}^{r=a} = \frac{8\pi a^5}{15}$$

$$R_g^2 = \frac{8\pi a^5}{15V} = \frac{8\pi a^5}{15\times(\tfrac{4}{3}\pi a^3)} = \tfrac{2}{5}a^2$$

$$R_g = \boxed{\sqrt{\tfrac{2}{5}}a}$$

Comment. A common error involves using r^2 in place of the squared distance from the axis of rotation, which is actually equal to x^2+y^2 not to $x^2+y^2+z^2=r^2$. The common error gives the result $\sqrt{\tfrac{3}{5}}\,a$.

P17A.7 The probability that the end separation equal nl is:

$$P = \frac{\text{number of polymers with } N_R \text{ bonds to the right}}{\text{total number of arrangements of bonds}}$$

$$= \frac{N!/N_R!(N-N_R)!}{2^N} = \frac{N!}{\{\tfrac{1}{2}(N+n)\}!\{\tfrac{1}{2}(N-n)\}!2^N}$$

The factorials are substituted with Stirling's formula:

$$\ln x! \approx \ln(2\pi)^{1/2} + \tfrac{1}{2}(2x+1)\ln x - x$$

$$\begin{aligned}
\ln P &= \ln N! - \ln\{\tfrac{1}{2}(N+n)\}! - \ln\{\tfrac{1}{2}(N-n)\}! - \ln 2^N \\
&= \ln(2\pi)^{1/2} + \tfrac{1}{2}(2N+1)\ln N - N - \ln 2^N \\
&\quad - \left\{\ln(2\pi)^{1/2} + \tfrac{1}{2}(N+n+1)\ln\{\tfrac{1}{2}(N+n)\} - \tfrac{1}{2}(N+n)\right\} \\
&\quad - \left\{\ln(2\pi)^{1/2} + \tfrac{1}{2}(N-n+1)\ln\{\tfrac{1}{2}(N-n)\} - \tfrac{1}{2}(N-n)\right\} \\
&= -\ln(2\pi)^{1/2} - \ln 2^N + \tfrac{1}{2}(2N+1)\ln N - N + \tfrac{1}{2}(N+n) + \tfrac{1}{2}(N-n) \\
&\quad - \tfrac{1}{2}(N+n+1)\ln\{\tfrac{1}{2}(N+n)\} - \tfrac{1}{2}(N-n+1)\ln\{\tfrac{1}{2}(N-n)\} \\
&= -\ln(2\pi)^{1/2} - \ln 2^N + \tfrac{1}{2}(2N+1)\ln N - \tfrac{1}{2}(N+n+1)\ln\{\tfrac{1}{2}\} - \tfrac{1}{2}(N-n+1)\ln\{\tfrac{1}{2}\} \\
&\quad - \tfrac{1}{2}(N+n+1)\ln\{N+n\} - \tfrac{1}{2}(N-n+1)\ln\{N-n\} \\
&= -\ln(2\pi)^{1/2} - \ln 2^N + \tfrac{1}{2}(2N+1)\ln N - (N+1)\ln\{\tfrac{1}{2}\} \\
&\quad - \tfrac{1}{2}(N+n+1)\ln\{N(1+\tfrac{n}{N})\} - \tfrac{1}{2}(N-n+1)\ln\{N(1-\tfrac{n}{N})\} \\
&= \ln(\tfrac{2}{\pi})^{1/2} + \tfrac{1}{2}(2N+1)\ln N - \tfrac{1}{2}(N+n+1)\ln N - \tfrac{1}{2}(N-n+1)\ln N \\
&\quad - \tfrac{1}{2}(N+n+1)\ln\{1+\tfrac{n}{N}\} - \tfrac{1}{2}(N-n+1)\ln\{1-\tfrac{n}{N}\} \\
&= \ln(\tfrac{2}{\pi})^{1/2} - \tfrac{1}{2}\ln N - \tfrac{1}{2}(N+n+1)\ln\{1+\tfrac{n}{N}\} - \tfrac{1}{2}(N-n+1)\ln\{1-\tfrac{n}{N}\} \\
&= \ln(\tfrac{2}{\pi N})^{1/2} - \tfrac{1}{2}(N+n+1)\ln\{1+\tfrac{n}{N}\} - \tfrac{1}{2}(N-n+1)\ln\{1-\tfrac{n}{N}\}
\end{aligned}$$

The final expression above for $\ln P$ is the desired relation. For a compact coil ($n/N << 1$) the expression simplifies using the relation $\ln\{1+\tfrac{n}{N}\} = \tfrac{n}{N} - \tfrac{1}{2}(\tfrac{n}{N})^2 \cdots$ where third and higher order terms are discarded as they are negligibly small.

$$\begin{aligned}
\ln P &= \ln(\tfrac{2}{\pi})^{1/2} - \tfrac{1}{2}\ln N - \tfrac{1}{2}(N+n+1)\{\tfrac{n}{N} - \tfrac{1}{2}(\tfrac{n}{N})^2\} - \tfrac{1}{2}(N-n+1)\{-\tfrac{n}{N} - \tfrac{1}{2}(\tfrac{n}{N})^2\} \\
&= \ln(\tfrac{2}{\pi})^{1/2} - \tfrac{1}{2}(N+n+1)\{\tfrac{n}{N}\} + \tfrac{1}{2}(N-n+1)\{\tfrac{n}{N}\} \\
&\quad + \tfrac{1}{2}(N+n+1)\{\tfrac{1}{2}(\tfrac{n}{N})^2\} + \tfrac{1}{2}(N-n+1)\{\tfrac{1}{2}(\tfrac{n}{N})^2\} \\
&= \ln(\tfrac{2}{\pi N})^{1/2} - \tfrac{n^2}{N} + \tfrac{1}{2}(N+1)(\tfrac{n}{N})^2 \\
&= \ln(\tfrac{2}{\pi N})^{1/2} - \tfrac{n^2}{2N} + \tfrac{1}{2}(\tfrac{n}{N})^2
\end{aligned}$$

Since $N >> 1$, the third term is negligibly small compared to the second term and the expression becomes

$$\ln P = \ln(\tfrac{2}{\pi N})^{1/2} - \tfrac{n^2}{2N} \qquad \text{or} \qquad P = (\tfrac{2}{\pi N})^{1/2}\, e^{-n^2/2N}$$

confirming eqn 17A.1.

P17A.9 As demonstrated in Problem 17A.1, the radius of gyration of solid sphere is given by:

(i) $R_g/\text{nm} = 0.046460 \times \left\{(v_s/\text{cm}^3\,\text{g}^{-1}) \times (M/\text{g}\,\text{mol}^{-1})\right\}^{1/3}$

but you may use

(ii) $R_g / \text{nm} = 0.056902 \times \{(v_s / \text{cm}^3\,\text{g}^{-1}) \times (M / \text{g}\,\text{mol}^{-1})\}^{1/3}$

because either of these very similar equations serves as a simple, rough model of a more, or less spherical macromolecule. These equations are not expected to do more than coarsely estimate the radius of gyration. If the calculated value approximates the observed value, that is evidence for a spherical-like macromolecule. We present calculated values of both eqns (i) and (ii) in the following table for your convenience only. Comparison of the calculated radius of gyration with the observed suggests that serum albumin and bushy stunt virus resemble solid globular spheres, but DNA does not.

	$M/(\text{g mol}^{-1})$	$\upsilon s/(\text{cm}^3\,\text{g}^{-1})$	$(Rg/\text{nm})_{\text{calc i}}$	$(Rg/\text{nm})_{\text{calc ii}}$	$(Rg/\text{nm})_{\text{expt}}$
Serum albumin	66×10^3	0.752	1.71	2.09	2.98
Bushy stunt virus	10.6×10^6	0.741	9.24	11.3	12.0
DNA	4×10^6	0.556	6.06	7.43	117.0

The large value of the DNA radius of gyration suggests that it may be rod-like. To explore this hypothesis we use the equation for $R_{g,\perp}$ of a solid rod (see Problem 17A.1(b)):

$$R_{g,\perp} = \sqrt{1/12}\left(\frac{v_s M}{\pi a^2 N_A}\right) \quad \text{[Problem 17A.1(b)]}$$

Using to DNA data to calculate a according to the rod-like model gives

$$a^2 = \sqrt{1/12}\left(\frac{v_s M}{\pi N_A R_{g,\perp}}\right)$$

$$= \sqrt{1/12}\left(\frac{(0.556\times10^{-6}\,\text{m}^3\,\text{g}^{-1})\times(4\times10^6\,\text{g}\,\text{mol}^{-1})}{\pi N_A(117.0\times10^{-9}\,\text{m})}\right) = 2.9\times10^{-18}\,\text{m}^2$$

$$a = 1.7\,\text{nm}$$

This is not too terribly different from the known radius of the DNA double-strand (~1 nm) so we cannot reject the rod-like hypothesis.

17B Properties of macromolecules

Answers to discussion questions

D17B.1 The **stress** on an object is the applied force divided by the area to which it is applied. The **strain** is the resulting distortion of the sample. The stress–strain curve in text Fig. 17B.2 shows how a material responds to stress.

D17B.3 Polymer melting occurs at a specific **melting temperature**, T_m, above which the crystallinity of polymers can be destroyed by thermal motion. Higher melting temperatures correspond to increased strength and number of intermolecular interactions in the material. Polymers undergo a transition from a state of high to low chain mobility at the **glass transition temperature**, T_g. There is sufficient energy available to an elastomer at normal temperatures for limited bond rotation to occur and the flexible chains writhe. At lower temperatures, the amplitudes of the writhing motion decrease until a specific temperature, T_g, is reached at which motion is frozen completely and the sample forms a glass.

Solutions to exercises

E17B.1(a) We obtain (see Exercise 17A.4(a)) $N = 2.3\bar{2}\times10^3$ and $l = 308$ pm. In units of l the polymer ends are moved apart by the distance

$$n = \frac{1.0\times10^{-9}\text{ m}}{l} = \frac{1.0\times10^{-9}\text{ m}}{308\times10^{-12}\text{ m}} = 3.2\bar{5}$$

Thus, the molar change in conformational entropy arising from the stretch is

$$\Delta S = -\tfrac{1}{2}RN\ln\{(1+\nu)^{1+\nu}(1-\nu)^{1-\nu}\}\ [18.7] \quad \text{with} \quad \nu = n/N = 3.2\bar{5}/2.3\bar{2}\times10^3 = 1.4\bar{0}\times10^{-3}$$

$$= -\tfrac{1}{2}\times\left(8.3145\text{ J mol}^{-1}\text{ K}^{-1}\right)\times\left(2.3\bar{2}\times10^3\right)\times\ln\left\{\left(1.0014\bar{0}\right)^{1.0014\bar{0}}\times(0.99860)^{0.99860}\right\}$$

$$= \boxed{-19\text{ mJ mol}^{-1}\text{ K}^{-1}}$$

E17B.2(a) Each polyethene chain bond has a length $l = 154$ pm. $= 0.154$ nm and the number N chain bonds is the polymer molar mass divided by the molar mass of the repeating CH_2 unit.

$$N = \frac{M_{\text{polymer}}}{M_{CH_2}} = \frac{65\,000\text{ g mol}^{-1}}{14\text{ g mol}^{-1}} = 4.6\bar{4}\times10^3$$

The restoring force of a freely jointed chain is

$$\mathcal{F} = \frac{kT}{2l}\ln\left(\frac{1+\nu}{1-\nu}\right) \qquad \nu = n/N\ [17\text{B.2a}]$$

n is the displacement from equilibrium in units of l so

$$\nu = \frac{n}{N} = \frac{(1.0\text{ nm})/(0.154\text{ nm})}{4.6\bar{4}\times10^3} = 1.4\times10^{-3}$$

Since $\nu << 1$, we use the simplified form of eqn 17B.2a.

$$\mathcal{F} = \frac{nkT}{Nl}\ [17\text{B.2b}]$$

$$= \frac{\left(1.4\times10^{-3}\right)\times\left(1.381\times10^{-23}\text{ J K}^{-1}\right)\times(293.15\text{ K})}{154\times10^{-12}\text{ m}} = \boxed{3.7\times10^{-14}\text{ N}}$$

Solutions to problems

P17B.1 The fundamental vibrational frequency ν is related to the harmonic oscillator force constant k_f by the expression

$$\nu=\frac{1}{2\pi}\left(\frac{k_f}{m}\right)^{1/2}\ [8B.4]=\frac{1}{2\pi}\left(\frac{\mathcal{F}}{mx}\right)^{1/2}\ [8B.1]$$

where $\mathcal{F}$ is the restoring force caused by displacement x. Substitution of $x = nl$ into the random coil model for the restoring force gives

$$\mathcal{F}=\frac{nkT}{Nl}\ [17B.2b, n/N \ll 1]=\frac{kTx}{Nl^2}$$

Thus, with the presumption that the effective mass of Hooke's law is $m = MN^{-1}N_A^{-1}$ we find that

$$\nu=\frac{1}{2\pi l}\left(\frac{kT}{Nm}\right)^{1/2}=\boxed{\frac{1}{2\pi l}\left(\frac{RT}{M}\right)^{1/2}}$$

For the case in which $l = 154$ pm., $M = 65\ \text{kg mol}^{-1}$, and $T = 293$ K the fundamental vibrational frequency is

$$\nu=\frac{1}{2\pi\times(154\times10^{-12}\ \text{m})}\left\{\frac{(8.3145\ \text{J K}^{-1}\ \text{mol}^{-1})\times(293\ \text{K})}{65\ \text{kg mol}^{-1}}\right\}^{1/2}=\boxed{6.3\ \text{GHz}}$$

The fundamental vibrational frequency has a $T^{1/2}$ dependence upon temperature which reflects the thermal energy needed to establish a displacement from equilibrium. The $M^{-1/2}$ dependence indicates that a larger chain mass requires more thermal energy to establish the same displacement.

P17B.3 The glass transition temperature T_g is the temperature at which internal bond rotations freeze. In effect, the easier such rotations are, the lower T_g. Internal rotations are more difficult for polymers that have bulky side chains than for polymers without such chains because the side chains of neighbouring molecules can impede each others' motion. Of the four polymers in this problem, polystyrene has the largest side chain (phenyl) and the largest T_g. The chlorine atoms in poly(vinyl chloride) interfere with each other's motion more than the smaller hydrogen atoms that hang from the carbon backbone of polyethylene. Poly(oxymethylene), like polyethylene, has only hydrogen atoms protruding from its backbone; however, poly(oxymethylene) has fewer hydrogen protrusions and a still lower T_g than polyethylene.

17C Self-assembly

Answers to discussion questions

D17C.1 We provide several IUPAC Gold Book definitions to supplement text discussion (see 2007 IUPAC, *Pure Appl. Chem.*, Vol. 79, No. 10, pp. 1801–1829, 2007).

Colloidal: State of subdivision, implying that the molecules or polymolecular particles dispersed in a medium have at least in one direction a dimension roughly between 1 nm and 1 μm, or that in a system discontinuities are found at distances of that order.

Colloidal sol: Fluid colloidal system of two or more components. Examples of colloidal sols are protein sols, gold sols, emulsions, and surfactant solutions above their critical micelle concentrations.

Aerosol: Sol in which the dispersed phase is a solid, a liquid, or a mixture of both and the continuous phase is a gas (usually air). Owing to their size, the particles of the dispersed phase have a comparatively small settling velocity and hence exhibit some degree of stability in the earth's gravitational field.

Particulate or polymeric sol: Sol in which the dispersed phase consists of solid particles or particles having a polymeric structure.

Emulsion: A fluid colloidal system in which liquid droplets and/or liquid crystals are dispersed in a liquid. The droplets often exceed the usual limits for colloids in size. An emulsion is denoted by the symbol O/W if the continuous phase is an aqueous solution and by W/O if the continuous phase is an organic liquid (an 'oil'). More complicated emulsions such as O/W/O (i.e. oil droplets contained within aqueous droplets dispersed in a continuous oil phase) are also possible.

Foam: A dispersion in which a large proportion of gas by volume in the form of gas bubbles, is dispersed in a liquid, solid or gel. The diameter of the bubbles is usually larger than 1μm, but the thickness of the lamellae between the bubbles is often in the usual colloidal size range. The term froth has been used interchangeably with foam. In particular cases froth may be distinguished from foam by the fact that the former is stabilized by solid particles and the latter by soluble substances.

D17C.3 Sterols, like cholesterol, prevent the hydrophobic chains of lipids bilayers from 'freezing' into a gel and, by disrupting the packing of the chains, spread the melting point of the membrane over a range of temperatures.

Solution to exercise

E17C.1(a) $\upsilon/(\mu\text{m s}^{-1}) = a + b(\text{pH}) + c(\text{pH})^2 + d(\text{pH})^3$ with $a = 0.50$, $b = -0.10$, $c = -3.0 \times 10^{-3}$, and $d = 5.0 \times 10^{-4}$

At the **isoelectric point** the pH is such that there is no net charge on the macromolecule, thereby, eliminating the normal protein velocity caused by an electric field. The following Mathcad Prime 2 worksheet sets the above speed expression equal to zero and numerically solves for the pH.

$a := 0.50 \quad b := -0.10 \quad c := -3.0 \cdot 10^{-3} \quad d := 5.0 \cdot 10^{-4}$

$$a + b \cdot pH + c \cdot pH^2 + c \cdot pH^3 \xrightarrow{solve, pH} \begin{bmatrix} 3.4328446224327753834 \\ -2.2164223112163876917 - 6.60591192008143i \\ -2.2164223112163876917 + 6.60591192008143 10554i \end{bmatrix}$$

The two imaginary solutions are non-physical so we conclude that the isoelectric point occurs at a pH of $\boxed{3.43}$.

Solution to problem

P17C.1 $N\,\text{M} \rightleftharpoons \text{M}_N$ where M is the monomer and M_N is the micelle that consists of N monomers.

$$K = \frac{[\text{M}_N]}{([\text{M}]_{\text{total}} - N[\text{M}_N])^N} \quad [17\text{C.5b}] \quad \text{or} \quad K([\text{M}]_{\text{total}} - N[\text{M}_N])^N = [\text{M}_N]$$

Figure 17C.6 of the text uses the definition: $f_N \equiv$ 'fraction present as micelles' $\equiv [\text{M}_N]/[\text{M}]_{\text{total}}$. The following Mathcad Prime 2 worksheet replicates Fig. 17C.6 using the symbolism M_N for $[\text{M}_N]$ and M_{total} for $[\text{M}]_{\text{total}}$. The worksheet establishes a solve block for numerically computing M_N as a function of N, M_{total}, and K. It plots the fraction present as micelles f_N against M_{total} for several N and K.

Guess Values: $M_N := -1$

Constraints: $K \cdot (M_{total} - N \cdot M_N)^N = M_N$

Solver: $M_N(N, M_{total}, K) := \text{Find}(M_N)$

$$f_N(N, M_{total}, K) := \frac{M_N(N, M_{total}, K)}{M_{total}}$$

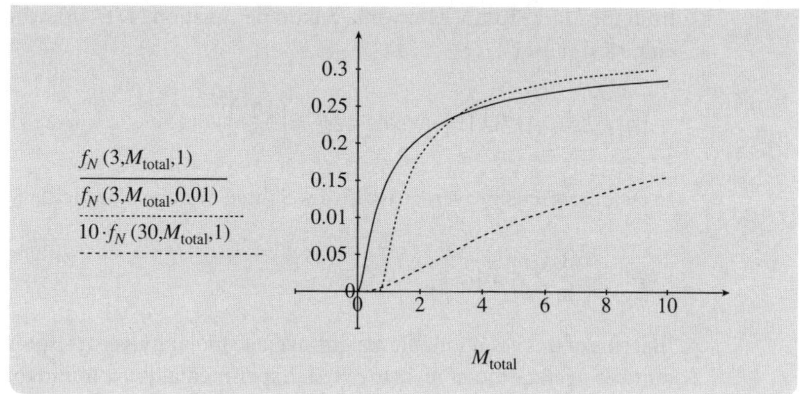

17D Determination of size and shape

Answer to discussion question

D17D.1 The **number-average molar mass** is the value obtained by weighting each molar mass by the number of molecules with that mass (eqn 17D.1).

$$\bar{M}_n = \frac{1}{N}\sum_i N_i M_i = \langle M \rangle \text{ [17D.1]}$$

In this expression, N_i is the number of molecules of molar mass M_i and N is the total number of molecules. Measurements of the osmotic pressures of macromolecular solutions yield the number-average molar mass.

The **weight-average molar mass** is the value obtained by weighting each molar mass by the mass of each one present (eqn 17D.2a).

$$\bar{M}_w = \frac{1}{m}\sum_i m_i M_i = \frac{\sum_i N_i M_i^2}{\sum_i N_i M_i} = \frac{\langle M^2 \rangle}{\langle M \rangle} \text{ [17D.2a and b]}$$

In this expression, m_i is the total mass of molecules with molar mass M_i and m is the total mass of the sample. Light scattering experiments give the weight-average molar mass.

The **Z-average molar mass** is defined through the formula (eqn 17D.2c).

$$\bar{M}_Z = \frac{\sum_i N_i M_i^3}{\sum_i N_i M_i^2} = \frac{\langle M^3 \rangle}{\langle M^2 \rangle} \text{ [17D.2c]}$$

The Z-average molar mass is obtained from sedimentation equilibria experiments.

The **viscosity-average molar mass** is the value obtained from measurements of the intrinsic viscosities $[\eta]$ of solutions of the macromolecule. It is obtained from the Mark–Kuhn–Houwink–Sakurada equation (17D.18), where K and a are empirical values:

$$[\eta]=K\bar{M}_{\mathrm{v}}^{a}\ [17\mathrm{D}.18] \qquad \text{so} \qquad \bar{M}_{\mathrm{v}}=\left(\frac{[\eta]}{K}\right)^{1/a}$$

In fact, the viscosity average is also a weighted average of individual molar masses:

$$\bar{M}_{\mathrm{v}}=\left(\frac{1}{m}\sum_i m_i M_i^{\,a}\right)^{1/a}$$

In terms of this equation, the weight-average molar mass corresponds to $a=1$, the number-average molar mass to $a=-1$. Experimentally, a is found to be in the range 0.5–1.0. Therefore, $\bar{M}_{\mathrm{v}}$ is closer to $\bar{M}_{\mathrm{w}}$ than to $\bar{M}_{\mathrm{n}}$.

Solutions to exercises

E17D.1(a) Equal amounts imply equal numbers of molecules. Hence the number average is (eqn 17D.1)

$$\bar{M}_{\mathrm{n}}=\frac{N_1M_1+N_2M_2}{N}=\frac{n_1M_1+n_2M_2}{n}=\tfrac{1}{2}(M_1+M_2)\ [n_1=n_2=\tfrac{1}{2}n]$$
$$=\tfrac{1}{2}(62+78)\ \mathrm{kg\,mol^{-1}}=\boxed{70\ \mathrm{kg\,mol^{-1}}}$$

and the weight average is (eqns 17D.2a and b)

$$\bar{M}_{\mathrm{w}}=\frac{m_1M_1+m_2M_2}{m}=\frac{n_1M_1^2+n_2M_2^2}{n_1M_1+n_2M_2}=\frac{M_1^2+M_2^2}{M_1+M_2}\ [n_1=n_2]$$
$$=\frac{62^2+78^2}{62+78}\ \mathrm{kg\,mol^{-1}}=\boxed{71\ \mathrm{kg\,mol^{-1}}}$$

E17D.2(a) (a) Osmometry gives the number-average molar mass, so

$$\bar{M}_{\mathrm{n}}=\frac{N_1M_1+N_2M_2}{N_1+N_2}\ [17\mathrm{D}.1]=\frac{n_1M_1+n_2M_2}{n_1+n_2}$$
$$=\frac{\left(\frac{m_1}{M_1}\right)M_1+\left(\frac{m_2}{M_2}\right)M_2}{\left(\frac{m_1}{M_1}\right)+\left(\frac{m_2}{M_2}\right)}=\frac{m_1+m_2}{\left(\frac{m_1}{M_1}\right)+\left(\frac{m_2}{M_2}\right)}$$
$$=\frac{100\ \mathrm{g}}{\left(\frac{30\ \mathrm{g}}{30\ \mathrm{kg\,mol^{-1}}}\right)+\left(\frac{70\ \mathrm{g}}{15\ \mathrm{kg\,mol^{-1}}}\right)}\ [\text{assume 100 g of solution}]=\boxed{18\ \mathrm{kg\,mol^{-1}}}$$

(b) Light scattering gives the mass-average molar mass, so

$$\bar{M}_{\rm w} = \frac{m_1 M_1 + m_2 M_2}{m_1 + m_2} [17D.2a] = \frac{(30)\times(30)+(70)\times(15)}{100}\ \mathrm{kg\,mol^{-1}} = \boxed{20\ \mathrm{kg\,mol^{-1}}}$$

E17D.3(a) Since the two spherical particles have different radii but identical densities, we immediately recognize that they have identical buoyancy and, therefore, their effective masses are proportional to the particle mass. The solute particle mass equals its volume multiplied by its density ρ_s. Therefore, the particle mass is proportional to the cube of the particle radius:

$$m_{\rm eff} \propto a^3 \rho_{\rm s}$$

According to eqn 17D.12, the Stokes frictional coefficient is proportional to a and eqn 17D.9 indicates that the sedimentation rate s is proportional to the effective mass and inversely proportional to the frictional coefficient. Thus,

$$\begin{aligned} s &\propto \frac{a^3 \rho_{\rm s}}{a} \\ &\propto a^2 \rho_{\rm s} \end{aligned}$$

The constant of proportionality cancels when taking ratios so the relative rates of sedimentation of the two different particles is

$$\frac{s_2}{s_1} = \frac{(10a_1)^2 \rho_2}{a_1^2 \rho_1} = 10^2 = \boxed{100} \text{ because } \rho_1 = \rho_2$$

E17D.4(a) $M = \bar{M}_{\rm n}$ and eqn 17D.12 gives

$$M = \frac{SfN_{\rm A}}{b} [17D.12] = \frac{SfN_{\rm A}}{1-\rho v_{\rm s}} [17D.8]$$

Assuming that the solution density equals that of water at 298 K (0.9969 g cm^{-3}) and substitution of the Stokes–Einstein relation, $f = kT/D$ [19B.19a], gives

$$\begin{aligned} M &= \frac{SRT}{(1-\rho v_{\rm s})D} \\ &= \frac{(4.48\ \mathrm{Sv})\times(10^{-13}\ \mathrm{s\,Sv^{-1}})(8.3145\ \mathrm{J\,mol^{-1}\,K^{-1}})\times(298.15\ \mathrm{K})}{\{1-(996.9\ \mathrm{kg\,m^{-3}})\times(0.749\times10^{-3}\ \mathrm{m^3\,kg^{-1}})\}\times(6.9\times10^{-11}\ \mathrm{m^2\,s^{-1}})} \\ &= \boxed{64\ \mathrm{kg\,mol^{-1}}} \end{aligned}$$

E17D.5(a) The net force acting upon the settling particle equals zero because of the balance between the gravitation pull, $m_{\text{eff}}g$, and the frictional force, fs where s is the drift speed and the frictional coefficient is given by Stokes' relation $f=6\pi a\eta$ [18.31] for a particle of radius a. Thus,

$$fs=m_{\text{eff}}g$$

$$s=\frac{m_{\text{eff}}g}{f}=\frac{bmg}{f}=\frac{(1-\rho/\rho_s)\times\{(\tfrac{4}{3}\pi a^3)\times\rho_s\}g}{6\pi a\eta}\quad[18.27\text{ and }18.31]$$

$$=\frac{2(\rho_s-\rho)a^2g}{9\eta}$$

$$=\frac{2\times(1750\,\text{kg m}^{-3}-1000\,\text{kg m}^{-3})\times(20\times10^{-6}\,\text{m})^2\times(9.8067\,\text{m s}^{-2})}{9\times(8.9\times10^{-4}\,\text{kg m}^{-1}\,\text{s}^{-1})}$$

$$=\boxed{0.73\,\text{mm s}^{-1}}$$

E17D.6(a) $\bar{M}_{\text{n}}=\dfrac{SfN_{\text{A}}}{b}\,[17\text{D}.11]=\dfrac{SfN_{\text{A}}}{1-\rho v_{\text{s}}}\,[17\text{D}.8]$

Substitution of the Stokes–Einstein relation, $f=kT/D$ [19B.19a], gives

$$\bar{M}_{\text{n}}=\frac{SRT}{(1-\rho v_{\text{s}})D}$$

$$=\frac{(3.2\,\text{Sv})\times(10^{-13}\,\text{s Sv}^{-1})(8.3145\,\text{J mol}^{-1}\,\text{K}^{-1})\times(293.15\,\text{K})}{\{1-(1.06\,\text{g cm}^{-3})\times(0.656\,\text{cm}^3\,\text{g}^{-1})\}\times(8.3\times10^{-11}\,\text{m}^2\,\text{s}^{-1})}$$

$$=\boxed{31\,\text{kg mol}^{-1}}$$

E17D.7(a) The number of solute molecules with energy E is proportional to $e^{-E/kT}$, hence

$$c\propto N\propto e^{-E/kT}\quad\text{where}\quad E=\tfrac{1}{2}m_{\text{eff}}r^2\omega^2$$

Therefore, $c\propto e^{Mb\omega^2r^2/2RT}$ $[m_{\text{eff}}=bm,\ M=mN_{\text{A}}]$ and

$$\ln c=\text{const.}+\frac{Mb\omega^2r^2}{2RT}\quad[b=1-\rho v_s]$$

This expression indicates that the slope of a plot of $\ln c$ against r^2 is equal to $Mb\omega^2/2RT$. Therefore

$$M=\frac{2RT\times\text{slope}}{b\omega^2}=\frac{2\times(8.3145\,\text{J K}^{-1}\,\text{mol}^{-1})\times(300\,\text{K})\times(729\times10^4\,\text{m}^{-2})}{(1-0.997\times0.61)\times\{(50\,000)\times(2\pi/60\,\text{s})\}^2}$$

$$=\boxed{3.4\times10^3\,\text{kg mol}^{-1}}$$

Solutions to problems

P17D.1 $dN = K\,e^{-(M-\bar{M})^2/2\gamma}\,dM$ where K is the constant of proportionality for the distribution. Evaluate K by requiring that $\int dN = N$.

Let $M-\bar{M} = (2\gamma)^{1/2}x$ so $dM = (2\gamma)^{1/2}\,dx$

and $N = \int_0^{\infty} K\,e^{-(M-\bar{M})^2/2\gamma}\,dM = K(2\gamma)^{1/2}\int_{-a}^{\infty} e^{-x^2}\,dx$ where $a = \dfrac{\bar{M}}{(2\gamma)^{1/2}}$

Note that the point $x = 0$ represents $M = \bar{M}$, and $x = -a$ represents $M = 0$. In a narrow distribution, the number of molecules with masses much different than the mean falls off rapidly as one moves away from the mean; therefore, $dN \approx 0$ at $M \le 0$ (that is, at $x \le -a$). Therefore

$$N \approx K(2\gamma)^{1/2}\int_{-\infty}^{\infty} e^{-x^2}\,dx = K(2\gamma)^{1/2}\pi^{1/2}$$

Hence, $K = \dfrac{N}{(2\pi\gamma)^{1/2}}$. It then follows from turning eqn 17D.1 into an integral that

$$\overline{M_n} = \frac{1}{N}\int M\,dN = \frac{1}{(2\pi\gamma)^{1/2}}\int_0^{\infty} M\,e^{-(M-\bar{M})^2/2\gamma}\,dM$$

$$= \frac{1}{(2\pi\gamma)^{1/2}}\int_0^{\infty}[(2\gamma)^{1/2}x+\bar{M}]e^{-x^2}(2\gamma)^{1/2}\,dx = \left(\frac{2\gamma}{\pi}\right)^{1/2}\int_{-a}^{\infty}\left(xe^{-x^2}+\frac{\bar{M}}{(2\gamma)^{1/2}}e^{-x^2}\right)dx$$

Once again extending the lower limit of integration to $-\infty$ adds negligibly to the integral, so

$$\overline{M_n} \approx \left(\frac{2\gamma}{\pi}\right)^{1/2}\times\left\{1+\left(\frac{\pi}{2\gamma}\right)^{1/2}\bar{M}\right\} = \boxed{\bar{M}+\left(\frac{2\gamma}{\pi}\right)^{1/2}}$$

P17D.3 (a) We seek an expression for a ratio of scattering intensities of a macromolecule in two different conformations, a rigid rod or a closed circle. The dependence on scattering angle θ is contained in the Rayleigh ratio $R(\theta)$. The definition of this quantity, in eqn 17D.4, gives an expression for the scattering intensity at scattering angle θ,

$$I(\theta,\phi,r) = \frac{R(\theta)I_0}{r^2},$$

I_0 is the intensity of incident light and r the distance between sample and detector. Thus, for any given scattering angle, the ratio of scattered intensity of two conformations is the same as the ratio of their Rayleigh ratios:

$$\frac{I_{\text{rod}}}{I_{\text{cc}}} = \frac{R_{\text{rod}}}{R_{\text{cc}}} = \frac{P_{\text{rod}}}{P_{\text{cc}}}.$$

The last equality stems from eqn 17D.5 with the inclusion of the structure factor $P(\theta)$ of eqn 17D.6, which relates the Rayleigh ratios to a number of angle-independent factors that would be the same for both conformations, and the structure factor that depends on both conformation and scattering angle. Finally, eqn 17D.6 gives an approximate value of the structure factor as a function of the macromolecule's radius of gyration R_g, the wavelength of light, and the scattering angle:

$$P(\theta) \approx 1 - \frac{16\pi^2 R_g^2 \sin^2(\frac{1}{2}\theta)}{3\lambda^2} = \frac{3\lambda^2 - 16\pi^2 R_g^2 \sin^2(\frac{1}{2}\theta)}{3\lambda^2}$$

The radius of gyration of a rod of length l is

$$R_{g,\perp} = \sqrt{1/12}\, l \text{ [See Problem 17A.1(b)]}$$

For a closed circle, the radius of gyration, which is the rms distance from the centre of mass [Problem 17A.8], is simply the radius of a circle whose circumference is l:

$$l = 2\pi R_{cc} \quad \text{so} \quad R_{cc} = \frac{l}{2\pi}.$$

The intensity ratio is:

$$\frac{I_{rod}}{I_{cc}} = \frac{3\lambda^2 - \frac{4}{3}\pi^2 l^2 \sin^2(\frac{1}{2}\theta)}{3\lambda^2 - 4l^2 \sin^2(\frac{1}{2}\theta)}.$$

Putting the numbers in yields:

$\theta/°$	20	45	90
I_{rod}/I_{cc}	0.976	0.876	0.514

(b) I would work at a detection angle at which the ratio is smallest, i.e. most different from unity, provided I had sufficient intensity to make accurate measurements. Of the angles considered in part (a), $\boxed{90°}$ is the best choice. With the help of a spreadsheet or symbolic mathematical program, the ratio can be computed for a large range of scattering angles and plotted (Fig. 17D.1).

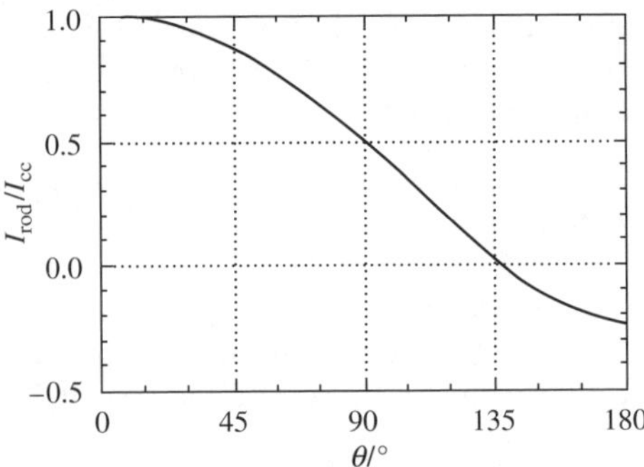

Figure 17D.1

A look at the results of such a calculation shows that both the intensity ratio and the intensities themselves decrease with increasing scattering angle from 0° through 180°, that of the closed circle conformation changing much more slowly than that of the rod.

Note: The approximation used above yields negative numbers for P_{rod} at large scattering angles; this is because the approximation, which depends on the molecule being much smaller than the wavelength, is shaky at best, particularly at large angles.

P17D.5 $$\bar{M}_w = \frac{2RT}{(r_2^2 - r_1^2)b\omega^2}\ln\frac{c_2}{c_1}\ [17\text{D}.14] = \frac{2RT}{(r_2^2 - r_1^2)b(2\pi\nu)^2}\ln\frac{c_2}{c_1}\ [\omega = 2\pi\nu]$$

Solving for the squared rotational frequency ν and substitution of $b = 1 - \rho v_s$ [17D.8] gives

$$\begin{aligned}\nu^2 &= \frac{RT\ln(c_2/c_1)}{2\pi^2\bar{M}_w \times (1-\rho v_s)\times(r_2^2 - r_1^2)} \\ &= \frac{(8.3145\ \text{J K}^{-1}\ \text{mol}^{-1})\times(298\ \text{K})\times\ln 5}{2\pi^2\times(1\times10^2\ \text{kg mol}^{-1})\times(1-0.75)\times(7.0^2 - 5.0^2)\times10^{-4}\ \text{m}^2} \\ &= 33\overline{67}\ \text{s}^{-2}\end{aligned}$$

$$\nu = (58\ \text{s}^{-1})\times(60\ \text{s min}^{-1}) = \boxed{3\,500\ \text{r.p.m.}}$$

P17D.7 $M = \bar{M}_n$ and eqn 17D.11 gives

$$M = \frac{SfN_A}{b}\ [17\text{D}.11] = \frac{SfN_A}{1-\rho v_s}\ [17\text{D}.8]$$

Substitution of the Stokes–Einstein relation, $f = kT/D$ [19B.19a], gives

$$\begin{aligned}M &= \frac{SRT}{(1-\rho v_s)D} \\ &= \frac{(4.5\ \text{Sv})\times(10^{-13}\ \text{s Sv}^{-1})(8.3145\ \text{J mol}^{-1}\ \text{K}^{-1})\times(293.15\ \text{K})}{\{1-(998\ \text{kg m}^{-3})\times(0.75\times10^{-3}\ \text{m}^3\ \text{kg}^{-1})\}\times(6.3\times10^{-11}\ \text{m}^2\ \text{s}^{-1})} \\ &= \boxed{69\ \text{kg mol}^{-1}}\end{aligned}$$

The Stokes' relation, $f = 6\pi a\eta$ [17D.12], and the Stokes–Einstein frictional coefficient, $f = RT/D$, are equated when estimating the effective radius a.

$$\begin{aligned}f_{\text{Stokes}} &= f_{\text{Einstein–Stokes}} \\ 6\pi a\eta &= kT/D\end{aligned}$$

$$a = \frac{kT}{6\pi D\eta}$$
$$= \frac{(1.381\times10^{-23}\ \mathrm{J K^{-1}})\times(293.15\ \mathrm{K})}{6\pi\times(6.3\times10^{-11}\ \mathrm{m^2\,s^{-1}})\times(1.00\times10^{-3}\ \mathrm{kg\,m^{-1}\,s^{-1}})}$$
$$= \boxed{3.4\ \mathrm{nm}}$$

P17D.9 $[\eta] = \lim_{c\to 0}\left(\frac{\eta/\eta_0 - 1}{c}\right)$ [17D.16]

We see that the y-intercept of a plot of the right-hand side against c, extrapolated to c = 0, gives $[\eta]$. We begin by constructing the following table using $\eta_0 = 0.985\ \mathrm{g\,m^{-1}\,s^{-1}}$.

$c/\mathrm{g\,dm^{-3}}$	1.32	2.89	5.73	9.17
$\left(\frac{\eta/\eta_0 - 1}{c}\right)\Big/\mathrm{dm^3\,g^{-1}}$	0.0731	0.0755	0.0771	0.0825

The points are plotted in Fig. 17D.2. The least-squares intercept is at 0.0716, so $[\eta] = \boxed{0.0716\ \mathrm{dm^3\,g^{-1}}}$

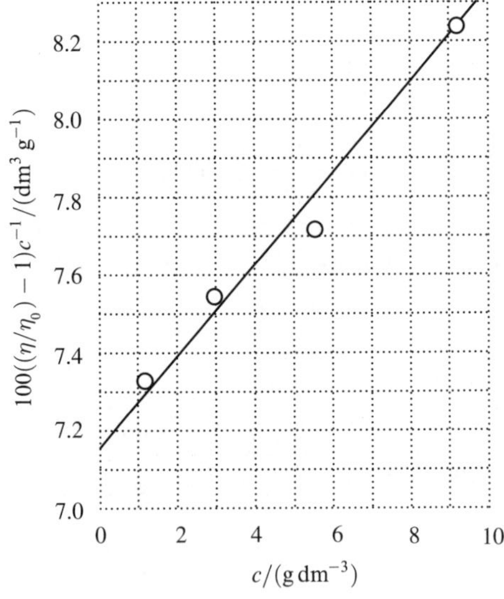

Figure 17D.2

P17D.11 We follow the procedure of Example 17D.5. Also compare to Problems 17D.9 and 17D.10.

$$[\eta] = \lim_{c\to 0}\left(\frac{\eta/\eta_0 - 1}{c}\right)\ [17D.16] \quad \text{and} \quad [\eta] = K\bar{M}_v^a\ [17D.18]$$

with K and a from Table 17D.3. We draw up the following table using $\eta_0 = 0.647 \times 10^{-3}\,\mathrm{kg\,m^{-1}\,s^{-1}}$, plot the points, and perform a linear regression fit as shown in Fig. 17D.3.

$100\,c/\mathrm{g\,cm^{-3}}$	0	0.2	0.4	0.6	0.8	1.0
$\eta/10^{-3}\,\mathrm{kg\,m^{-1}\,s^{-1}}$	0.647	0.690	0.733	0.777	0.821	0.865
$\left(\dfrac{\eta/\eta_0 - 1}{c}\right)\Big/100\,\mathrm{cm^3\,g^{-1}}$	–	$0.332\overline{3}$	$0.332\overline{3}$	$0.334\overline{9}$	$0.336\overline{2}$	$0.336\overline{9}$

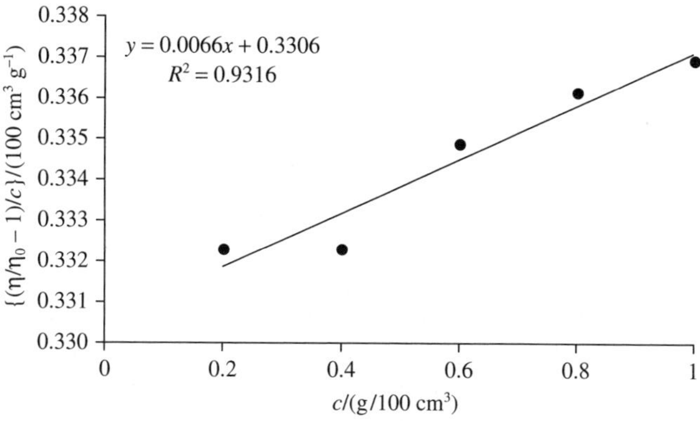

Figure 17D.3

The intercept of the linear regression fit gives $[\eta]$. Hence

$$[\eta] = (0.330\overline{6}) \times (100\,\mathrm{cm^3\,g^{-1}}) = 33.0\overline{6}\,\mathrm{cm^3\,g^{-1}}$$

and

$$\bar{M}_\mathrm{v} = \left(\frac{[\eta]}{K}\right)^{1/a} \mathrm{g\,mol^{-1}} = \left(\frac{33.0\overline{6}\,\mathrm{cm^3\,g^{-1}}}{8.3\times10^{-2}\,\mathrm{cm^3\,g^{-1}}}\right)^{1/0.50} \mathrm{g\,mol^{-1}} = \boxed{1.6\times10^5\,\mathrm{g\,mol^{-1}}}$$

Integrated activities

I17.1 See the discussion of radius of gyration in Section 17A.2(a). For a random coil $R_g \propto N^{1/2} \propto M^{1/2}$. For a rigid rod, the radius of gyration is proportional to the length of the rod, which is in turn proportional to the number of polymer units, N, and therefore also proportional to M. Therefore, poly(γ-benzyl-L-glutamate) in formamide is rod-like whereas polystyrene in butanone is a random coil.

I17.3 Computations are performed with Spartan 10 using the density functional B3LYP/6-31G* method. Starting and equilibrium dihedral angles along with differences in the equilibrium energies for the optimized geometries are reported in the following table. The reference energy for both the R = H and R = CH_3 case is that of the optimized geometry when starting with the torsion angles set as $\phi = 180°$ and $\psi = 180°$ (−456.536208 au and −495.852847 au, respectively). The ϕ(phi) dihedral angle involves the C(O)–N–C_α–C(O) bonds and the ψ(psi) dihedral angle involves the N–C_α–C(O)–N bonds.

	starting		optimized		
	ϕ/°	ψ/°	ϕ/°	ψ/°	$\Delta E/(10^{-3}$ au)
(a) R = H	+75	−65	+81.99	−69.22	−1.377
	+180	+180	+178.09	+178.80	0.000
	+65	+35	+120.56	−20.12	+2.849
(b) R = CH_3	+75	−65	+73.78	−56.79	+1.933
	+180	+180	−157.97	+163.43	0.000
	+65	+35	+66.91	+28.01	+9.892

None of the starting conformations relaxed to the same optimized conformation and the final conformations appear to represent local energy minima. It ought not to be surprising that there are several such minima in even a short peptide chain that contains several nearly free internal rotations. However, these results indicate that with so many local minima it is an intensive computational challenge to find the deepest energy wells of a Ramandran diagram. In a larger peptide it is certainly a daunting task to survey the energy contours that carry a polymer from a starting conformation to a global minimum energy conformation.

It is instructive to compare the all trans ($\phi = \psi = 180°$ shown in Fig. I17.1) initial conformation in the R = H and R = CH_3 cases. In the former, neither dihedral angle changes significantly, but the resulting structure is not the lowest-energy structure. In the latter, the methyl group appears to push the planes of the peptide link away from each other (albeit not far) due to steric effects; however, the resulting energy is lower than that of the other conformations examined.

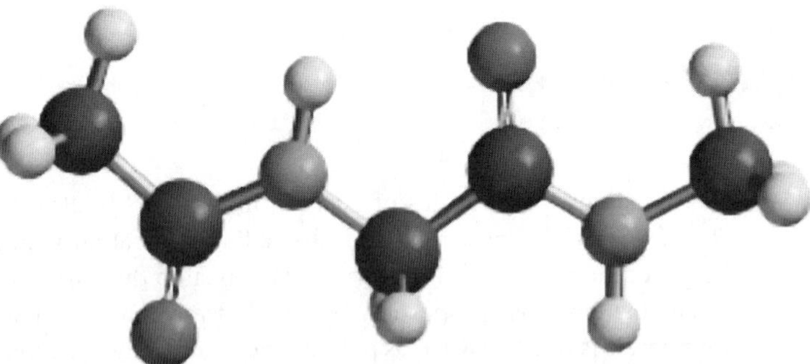

Figure I17.1

I17.5 $\eta=\eta_0(1+[\eta]c+k'[\eta]^2c^2)$

Assuming that $\eta \propto t_D$ [17D.17], we may write

$$t_D = t_{D,0}\left(1+[\eta]c+k'[\eta]^2 c^2\right)$$

(a) Perform a second-order polynomial, least squares regression fit of t_D against c. The regression coefficients a_i are related to solution properties: $a_0 = t_{D,0}$, $a_1 = t_{D,0}[\eta] = a_0\,[\eta]$, and $a_2 = t_{D,0}k'[\eta]^2 = a_1k'[\eta]$.

Regression fit in toluene: $t_D/\mathrm{s} = (8.371) + (0.7140) \times (c/\mathrm{g\,dm^{-3}}) + (0.02315) \times (c/\mathrm{g\,dm^{-3}})^2$ $R^2 = 0.999999$

$$[\eta]=\frac{a_1}{a_0}\,\mathrm{dm^3\,g^{-1}}=\frac{0.7140}{8.371}\,\mathrm{dm^3\,g^{-1}}$$
$$=\boxed{0.0853\,\mathrm{dm^3\,g^{-1}}}$$
$$k'=\frac{a_2\,\mathrm{dm^3\,g^{-1}}}{a_1[\eta]}=\frac{0.02315\,\mathrm{dm^3\,g^{-1}}}{0.7140\times\left(0.08529\,\mathrm{dm^3\,g^{-1}}\right)}$$
$$=\boxed{0.380}$$

Regression fit in cyclohexane: $t_D/\mathrm{s} = (8.320) + (0.3461)\times(c/\mathrm{g\,dm^{-3}}) + (0.004545) \times (c/\mathrm{g\,dm^{-3}})^2$ $R^2 = 0.999997$

$$[\eta]=\frac{a_1}{a_0}\,\mathrm{dm^3\,g^{-1}}=\frac{0.3461}{8.320}\,\mathrm{dm^3\,g^{-1}}$$
$$=\boxed{0.0416\,\mathrm{dm^3\,g^{-1}}}$$
$$k'=\frac{a_2\,\mathrm{dm^3\,g^{-1}}}{a_1[\eta]}=\frac{0.004545\,\mathrm{dm^3\,g^{-1}}}{0.3461\times\left(0.0416\,\mathrm{dm^3\,g^{-1}}\right)}$$
$$=\boxed{0.316}$$

(b) $[\eta]=K\bar{M}_v^{\,a}$ [17D.18] or $\overline{M_v}=\left(\frac{[\eta]}{K}\right)^{1/a}$

In toluene: $\bar{M}_v=\left(\frac{0.0853\,\mathrm{dm^2\,g^{-1}}}{1.15\times10^{-5}\,\mathrm{dm^3\,g^{-1}}}\right)^{(1/0.72)}\mathrm{g\,mol^{-1}}=\boxed{2.4\times10^2\ \mathrm{kg\,mol^{-1}}}$

In cyclohexane:

$$\bar{M}_v=\left(\frac{0.0416\,\mathrm{dm^3\,g^{-1}}}{8.2\times10^{-5}\ \mathrm{dm^3\,g^{-1}}}\right)^{(1/(1/2))}\mathrm{g\,mol^{-1}}=\boxed{2.6\times10^2\ \mathrm{kg\,mol^{-1}}}$$

(c) $[\eta]/(\mathrm{dm^3\,g^{-1}}) = \Phi\times(r_{\mathrm{rms}}/\mathrm{m})^3/(M\,\mathrm{mol\,g^{-1}})$ where $\Phi = 2.84\times10^{26}$

$$r_{\mathrm{rms}}=\left(\frac{[\eta]M\,\mathrm{mol}}{\Phi\,\mathrm{dm^3}}\right)^{1/3}\mathrm{m}$$

In toluene: $r_{\text{rms}} = \left(\dfrac{0.0853 \times 2.4 \times 10^5}{2.84 \times 10^{26}}\right)^{1/3} \text{m} = \boxed{42\ \text{nm}}$

In cyclohexane: $r_{\text{rms}} = \left(\dfrac{0.0416 \times 2.6 \times 10^5}{2.84 \times 10^{26}}\right)^{1/3} \text{m} = \boxed{34\ \text{nm}}$

(d) $M_{\text{styrene}} = 104\ \text{g mol}^{-1}$

The average number of monomeric units, $\langle n \rangle$ is $\langle n \rangle = \dfrac{\overline{M}_{\text{v}}}{M_{\text{styrene}}}$.

In toluene: $\langle n \rangle = \dfrac{2.4 \times 10^5\ \text{g mol}^{-1}}{104\ \text{g mol}^{-1}} = \boxed{2.3 \times 10^3}$

In cyclohexane: $\langle n \rangle = \dfrac{2.6 \times 10^5\ \text{g mol}^{-1}}{104\ \text{g mol}^{-1}} = \boxed{2.5 \times 10^3}$

(e) Consider the geometry in Fig. I17.2.

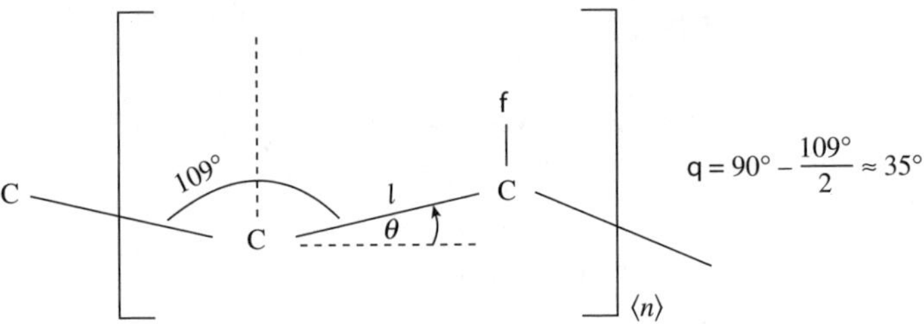

Figure I17.2

For a polymer molecule consisting of $\langle n \rangle$ monomers, the maximum molecular length, L_{max}, is

$$L_{\text{max}} = 2l\langle n \rangle \cos\theta = 2(0.154\ \text{nm})\langle n \rangle \cos 35° = (0.2507\ \text{nm})\langle n \rangle$$

In toluene: $L_{\text{max}} = (0.2507\ \text{nm}) \times (2.3 \times 10^3) = \boxed{5.8 \times 10^2\ \text{nm}}$

In cyclohexane: $L_{\text{max}} = (0.2507\ \text{nm}) \times (2.5 \times 10^3) = \boxed{6.3 \times 10^2\ \text{nm}}$

(f) $R_{\text{g}} = \left(\dfrac{\langle n \rangle}{3}\right)^{1/2} l\ [17\text{A}.6] = (0.0889\ \text{nm})\langle n \rangle^{1/2}$

Kirkwood–Riseman theory: $r_{\text{rms}}^{\text{KR}} = \left(\dfrac{[\eta] M\ \text{mol}}{\Phi\ \text{dm}^3}\right)^{1/3} \text{m}$ where $\Phi = 2.84 \times 10^{26}$

Constrained tetrahedral chain: $r_{\text{rms},109°} = (2\langle n \rangle)^{1/2} l\ [17\text{A}.8]$

Unconstrained chain: $r_{\text{rms,random}} = \langle n \rangle^{1/2} l\ [17\text{A}.4]$

Solvent	$\langle n \rangle$	R_g/nm	r_{rms}^{KR}/nm	$r_{rms,109°}$/nm	$r_{rms,random}$/nm
Toluene	2.3×10^3	4.3	42	10	7.4
Cyclohexane	2.5×10^3	4.4	33	11	7.7

R_g, $r_{rms,109°}$, and $r_{rms,random}$ are all much smaller than r_{rms}^{KR}. This is expected because they are based upon models that at best describe the lower limit of the polymer size.

(g) The manufacturer's claim that the polystyrene beads have an average molar mass of 250 kg mol^{-1} is consistent of the values of $\bar{M}_v$ found in part (b). Differences may be due to the use of a different solvent by the manufacturer or to the use of a non-viscosity technique for the molar mass measurement. The manufacturer's claim appears to be valid.

18 Solids

18A Crystal structure

Answers to discussion questions

D18A.1 A **spacc lattice** is the three-dimensional structural pattern formed by lattice points representing the locations of motifs which may be atoms, molecules, or groups of atoms, molecules, or ions within a crystal. All points of the space lattice have identical environments and they define the crystal structure. The **unit cell** is an imaginary parallelepiped from which the entire crystal structure can be generated, without leaving gaps, using translations of the unit cell alone. Each unit cell is defined in terms of lattice points and the unit cell is commonly formed by joining neighbouring lattice points by straight lines. The smallest possible unit cell is called the **primitive unit cell. Non-primitive unit cells** may exhibit lattice points within the cell, at the cell centre, on cell faces, or on cell edges.

D18A.3 We can use the Debye–Scherrer powder diffraction method with monochromic X-rays of wavelength λ and in particular look for systematic absences in the diffraction patterns. We can proceed through the following sequence:

1. Measure distances of the lines in the diffraction pattern from the centre.
2. From the known radius of the camera, convert the distances to incident angles θ for which the radiation exhibits constructive interference.
3. Calculate $\sin^2\theta$.
4. Based upon your experience, choose the cubic unit cell that you guess to be applicable and assign Miller indices (hkl) by $h^2 + k^2 + l^2$ values from lowest to highest to each incident angle. $h^2 + k^2 + l^2$ values equal to 7 and 15 are missing for a cubic P lattice (see *Brief illustration* 18A.4). When making the assignments, recognize that some indices will be absent depending upon the unit cell. For identical atoms in a cubic P (primitive cubic) arrangement, the hkl all-odd have zero intensity. For identical atoms in a cubic I (body-centred cubic) arrangement, the $h + k + l$ odd have zero intensity. For cubic F (face-centred cubic), only lines for which h, k, and l are either all even or all odd will be present, others will be absent. The Miller indices for observed spots are summarized in the following figure, which is a slight modification of text Fig. 18A.24.

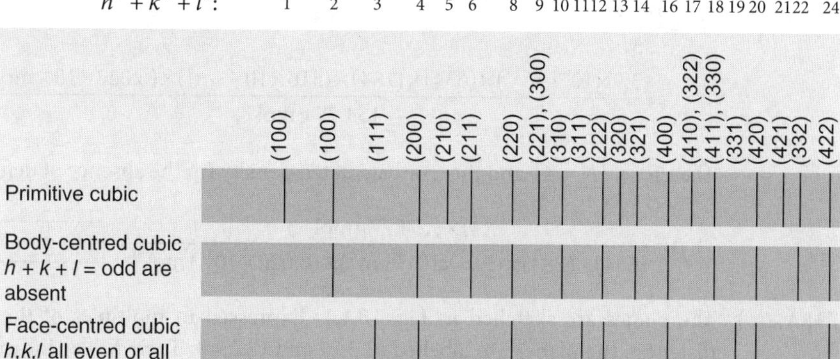

5. Find the factor $A = \lambda^2 / 4a^2 = \dfrac{\sin^2\theta}{h^2+k^2+l^2}$ for each spot [cubic unit cell, 18A.1b and 18A.2b].
6. If one or more A values disagree with the majority, select another cubic unit cell and repeat Steps 3 and 4.
7. If A values agree to within experimental uncertainty, solve $A = \lambda^2 / 4a^2$ for a.

D18A.5 A **systematic absence** is an accidently forbidden reflection in the powder diffraction pattern of a lattice. The absence is caused by destructive interference due to the positions of atoms or ions giving a structure factor, and an overall amplitude of a diffracted wave, equal to zero. This is discussed in text *Justification* 18A.3. The overall intensity of a diffracted wave from planes $\{hkl\}$ is determined from a calculation of the structure factor, F_{hkl}, which is a function of the positions (hence, of the Miller indices) and of the scattering factors of the atoms in the crystal (see eqns 18A.3 and 18A.4). If F_{hkl} is zero for $\{hkl\}$, that reflection is absent in the diffraction pattern. Recognition of systematic absences in a powder spectrum allows rapid identification a lattice structure. As summarized in Fig. 18A.24, a face-centred cubic lattice has a diffraction pattern for which h, k, and l are all even or all odd; other combinations are absent. A body-centred cubic lattice has systematic absences when the sum of h, k, and l is odd. Absences for which $(2a/\lambda)\times\sin\theta = 7^{1/2}$ or $15^{1/2}$ are characteristically absent in the pattern of a cubic P lattice (see *Brief illustration* 18A.4).

Solutions to exercises

E18A.1(a) $\rho = \dfrac{\text{mass of unit cell}}{\text{volume of unit cell}} = \dfrac{m}{V}$

$m = nM = \dfrac{N}{N_A}M$ [N is the number of formula units per unit cell]

Then, $\rho = \dfrac{NM}{VN_A}$

$$N=\frac{\rho VN_A}{M}$$
$$=\frac{(3.9\times10^6\ \text{g m}^{-3})\times(634)\times(784)\times(516\times10^{-36}\ \text{m}^3)\times(6.022\times10^{23}\ \text{mol}^{-1})}{154.77\ \text{g mol}^{-1}}=3.9$$

Therefore, $\boxed{N=4}$ and the true calculated density (in the absence of defects) is

$$\rho=\frac{(4)\times(154.77\ \text{g mol}^{-1})}{(634)\times(784)\times(516\times10^{-30}\ \text{cm}^3)\times(6.022\times10^{23})\ \text{mol}^{-1}}=\boxed{4.01\ \text{g cm}^{-3}}$$

E18A.2(a) The planes are sketched in Fig. 18A.1. Expressed in multiples of the unit cell distances the planes are labelled (2,3,2) and (2,2,∞). Their Miller indices are the reciprocals of these multiples with all fractions cleared, thus

$(2,3,2)\rightarrow\left(\frac{1}{2},\frac{1}{3},\frac{1}{2}\right)\rightarrow(3,2,3)$ [multiply by the lowest common denominator 6]

$(2,2,\infty)\rightarrow\left(\frac{1}{2},\frac{1}{2},0\right)\rightarrow(1,1,0)$ [multiply by the lowest common denominator 2]

Dropping the commas, the planes are written $\boxed{(323)\ \text{and}\ (110)}$.

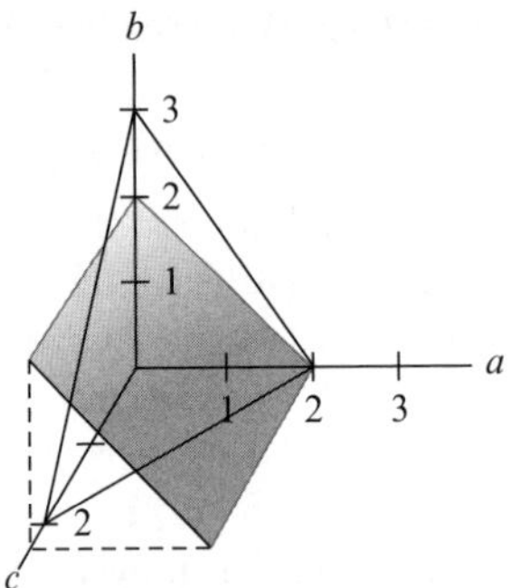

Figure 18A.1

E18A.3(a) For the cubic unit cell: $d_{hkl}=\dfrac{a}{\left(h^2+k^2+l^2\right)^{1/2}}$ [18A.1b]

$$d_{112}=\frac{a}{\left(1^2+1^2+2^2\right)^{1/2}}=\frac{562\ \text{pm}}{(6)^{1/2}}=\boxed{229\ \text{pm}}$$

$$d_{110}=\frac{a}{\left(1^2+1^2+0^2\right)^{1/2}}=\frac{562\ \text{pm}}{(2)^{1/2}}=\boxed{397\ \text{pm}}$$

$$d_{224}=\frac{a}{\left(2^2+2^2+4^2\right)^{1/2}}=\frac{562\ \text{pm}}{(24)^{1/2}}=\boxed{115\ \text{pm}}$$

E18A.4(a) $d_{hkl}=\left[\left(\dfrac{h}{a}\right)^2+\left(\dfrac{k}{b}\right)^2+\left(\dfrac{l}{c}\right)^2\right]^{-1/2}$ [18A.1c]

$$d_{321}=\left[\left(\frac{3}{812}\right)^2+\left(\frac{2}{947}\right)^2+\left(\frac{1}{637}\right)^2\right]^{-1/2}\ \text{pm}=\boxed{220\ \text{pm}}$$

E18A.5(a) $\lambda=2d\sin\theta$ [18A.2b] $=2\times(99.3\ \text{pm})\times\sin 20.85°=\boxed{70.7\ \text{pm}}$

E18A.6(a) Refer to Fig. 18A.24 of the text. Systematic absences correspond to $h+k+l=\text{odd}$. Hence the first three lines are from planes (110), (200), and (211).

$$\sin\theta_{hkl} = \frac{\lambda}{2d_{hkl}} \,[18\text{A.2b}] \quad d_{hkl} = \frac{a}{(h^2+k^2+l^2)^{1/2}} \,[18\text{A.1b}]$$

Then, $\sin\theta_{hkl} = (h^2+k^2+l^2)^{1/2} \times \left(\dfrac{\lambda}{2a}\right)$

In a bcc unit cell, the body diagonal of the cube is $4R$ where R is the atomic radius. The relationship of the side of the unit cell to R is therefore (using the Pythagorean theorem twice)

$$(4R)^2 = a^2 + 2a^2 = 3a^2 \quad \text{or} \quad a = \frac{4R}{3^{1/2}}$$

This can be seen from Fig. 18A.2.

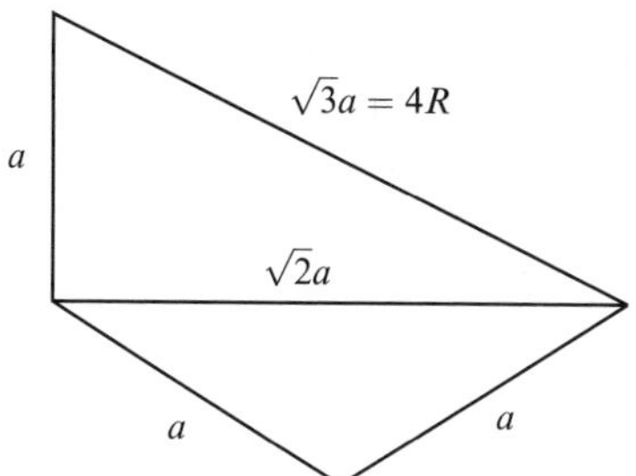

Figure 18A.2

$$a = \frac{4\times 126\text{ pm}}{3^{1/2}} = 291\text{ pm}$$

$$\frac{\lambda}{2a} = \frac{72\text{ pm}}{(2)\times(291\text{ pm})} = 0.12\bar{4}$$

$$\sin\theta_{110} = \sqrt{2}\times(0.12\bar{4}) = 0.17\bar{5} \qquad \theta_{110} = \boxed{10.1°}$$

$$\sin\theta_{200} = (2)\times(0.12\bar{4}) = 0.24\bar{8} \qquad \theta_{200} = \boxed{14.4°}$$

$$\sin\theta_{211} = \sqrt{6}\times(0.12\bar{4}) = 0.30\bar{4} \qquad \theta_{211} = \boxed{17.7°}$$

E18A.7(a) $\theta_{hkl} = \arcsin\dfrac{\lambda}{2d_{hkl}}$ [18A.2b]

$$= \arcsin\left\{\frac{\lambda}{2}\left[\left(\frac{h}{a}\right)^2 + \left(\frac{k}{b}\right)^2 + \left(\frac{l}{c}\right)^2\right]^{1/2}\right\} [18\text{A.1c}, \arcsin \equiv \sin^{-1}]$$

$$\theta_{100} = \arcsin\left\{\frac{154}{2}\times\left[\left(\frac{1}{542}\right)^2 + \left(\frac{0}{917}\right)^2 + \left(\frac{0}{645}\right)^2\right]^{1/2}\right\} = \boxed{8.16°}$$

$$\theta_{010} = \arcsin\left\{\frac{154}{2}\times\left[\left(\frac{0}{542}\right)^2 + \left(\frac{1}{917}\right)^2 + \left(\frac{0}{645}\right)^2\right]^{1/2}\right\} = \boxed{4.82°}$$

$$\theta_{111} = \arcsin\left\{\frac{154}{2}\times\left[\left(\frac{1}{542}\right)^2 + \left(\frac{1}{917}\right)^2 + \left(\frac{1}{645}\right)^2\right]^{1/2}\right\} = \boxed{11.7\bar{5}°}$$

E18A.8(a) $\theta = \arcsin\dfrac{\lambda}{2d}$ [18A.2b, $\arcsin \equiv \sin^{-1}$]

$$\Delta\theta = \arcsin\frac{\lambda_2}{2d} - \arcsin\frac{\lambda_1}{2d} = \arcsin\left(\frac{154.433\text{ pm}}{(2)\times(77.8\text{ pm})}\right) - \arcsin\left(\frac{154.051\text{ pm}}{(2)\times(77.8\text{ pm})}\right)$$

$$= 1.07° = 0.0187\text{ rad}$$

Consequently, the difference in the glancing angles (2θ) is $\boxed{2.14°}$.

E18A.9(a) *Justification* 18A.2 demonstrates that the scattering factor in the forward direction equals the number of electrons in the atom or simple ion. Consequently, $\boxed{f_{Br^-} = 36}$.

E18A.10(a) $F_{hkl} = \sum_j f_j e^{2\pi i(hx_j + ky_j + lz_j)}$ [18A.4] with $f_j = \frac{1}{8}f$ (each atom is shared by eight cells).

Therefore, $F_{hkl} = \frac{1}{8}f\{1 + e^{2\pi ih} + e^{2\pi ik} + e^{2\pi il} + e^{2\pi i(h+k)} + e^{2\pi i(h+l)} + e^{2\pi i(k+l)} + e^{2\pi i(h+k+l)}\}$

All the exponential terms have exponents that are a multiple of 2π because h, k, and l are all integers and a sum of any combination of these integers gives the integer n. Consequently, each exponential term has the form:

$$e^{in\times 2\pi} = \cos(n\times 2\pi) + i\sin(n\times 2\pi) = \cos(n\times 2\pi) = 1$$

Therefore, $F_{hkl} = \boxed{f}$

E18A.11(a) The structure factor is given by

$$F_{hkl} = \sum_i f_i e^{i\phi_i} \quad \text{where} \quad \phi_i = 2\pi(hx_i + ky_i + lz_i) \text{ [18A.4]}$$

All eight of the vertices of the cube are shared by eight cubes, so each vertex has a scattering factor of $f/8$ and their total contribution to the structure factor equals f (see Exercise 18A.10(a)). The two side-centred (C) ions of the unit cell are shared by two cells and in this problem each has a scattering factor of $2f$. Their coordinates are (½a, ½a, 0) and (½a, ½a, a). Thus, the contribution of the face-centred ions to the structure factor is

$$\sum_{i(\text{face-centred})} f_i e^{i\phi_i} = \tfrac{1}{2}(2f)e^{2\pi i(h/2+k/2)} + \tfrac{1}{2}(2f)e^{2\pi i(h/2+k/2+l)} = fe^{\pi i(h+k)}\left\{1 + e^{2\pi il}\right\}$$

But, according to Euler's identity ($e^{i\theta} = \cos\theta + i\sin\theta$),

$$e^{\pi i(h+k)} = \cos(\pi\{h+k\}) + i\sin(\pi\{h+k\}) = \cos(\pi\{h+k\}) = (-1)^{h+k}$$

and

$$e^{2\pi il} = \cos(2\pi l) + i\sin(2\pi l) = \cos(2\pi l) = 1$$

Thus, $\sum_{i(\text{face-centred})} f_i e^{i\phi_i} = f(-1)^{h+k}\{1+1\} = 2(-1)^{h+k}f$

and the structure factor becomes

$$F_{hkl} = \sum_{i(\text{vertices})} f_i e^{i\phi_i} + \sum_{i(\text{face-centred})} f_i e^{i\phi_i} = f + 2(-1)^{h+k} f = f\left\{1 + 2(-1)^{h+k}\right\}$$

$$F_{hkl} = \boxed{3f \text{ for } h+k \text{ even and } -f \text{ for } h+k \text{ odd}}$$

E18A.12(a) The electron density is given by

$$\rho(r) = \frac{1}{V}\sum_{hkl} F_{hkl} e^{-2\pi i(hx+ky+lz)} \quad [18\text{A}.5]$$

The component along the x direction is

$$\rho(x) = \frac{1}{V}\sum_{h} F_h e^{-2\pi i h x}$$

Using the data of this problem, we sum from $h = -9$ to $+9$ and use the relationship $F_{-h} = F_h$. The following Mathcad computation of $\rho(x)$ and plot shown in Fig. 18A.3 shows high electron density at the centre of a unit cell edge.

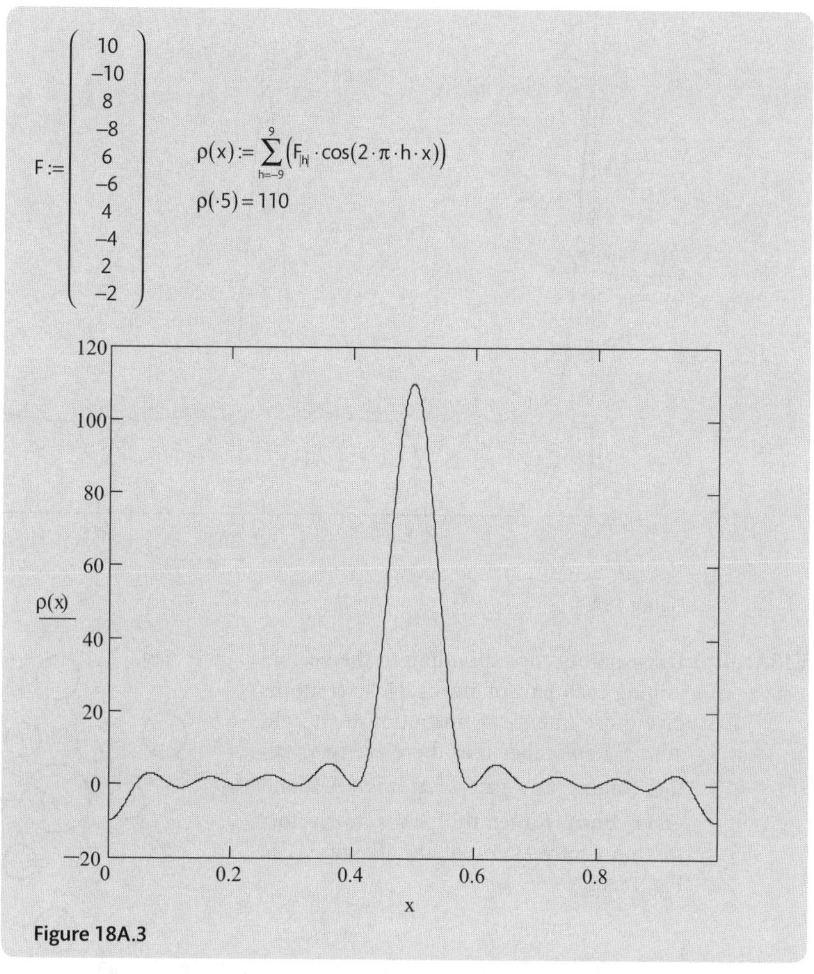

Figure 18A.3

E18A.13(a) Using the information of Exercise 18A.12(a), the Mathcad computation of $P(x)$ is performed with eqn 18A.6.

$$F := \begin{pmatrix} 10 \\ -10 \\ 8 \\ -8 \\ 6 \\ -6 \\ 4 \\ -4 \\ 2 \\ -2 \end{pmatrix} \qquad P(x) := \sum_{h=-9}^{9} \left[\left(F_{|h|}\right)^2 \cdot e^{-2\cdot\pi\cdot i\cdot h\cdot x} \right]$$

$$P(1.0) = 780$$

The Patterson synthesis $P(x)$ of Fig. 18A.4 shows that atoms represented by this data are separated by 1 a unit along the x axis.

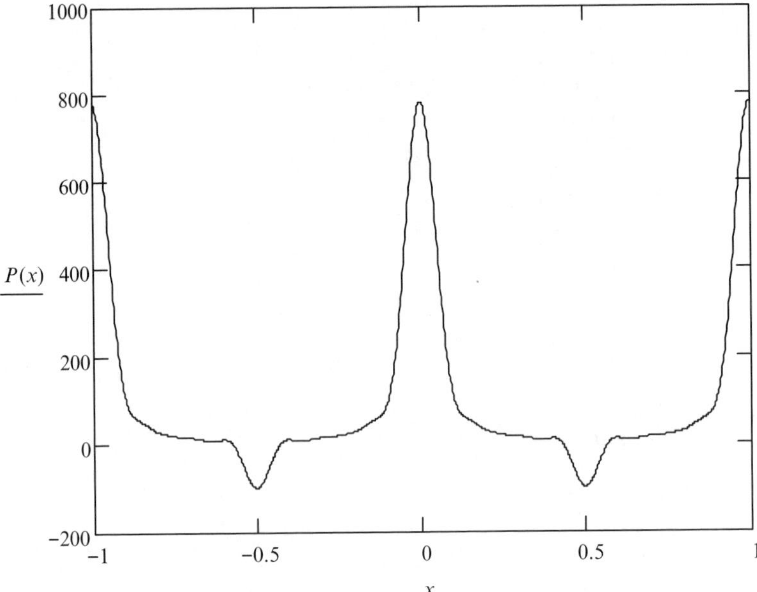

Figure 18A.4

E18A.14(a) Draw points corresponding to the vectors joining each pair of atoms. Heavier atoms give more intense contribution than light atoms. Remember that there are two vectors joining any pair of atoms ($\overrightarrow{AB}$ and $\overleftarrow{AB}$); don't forget the AA zero vectors for the centre point of the diagram. See Fig. 18A.5.

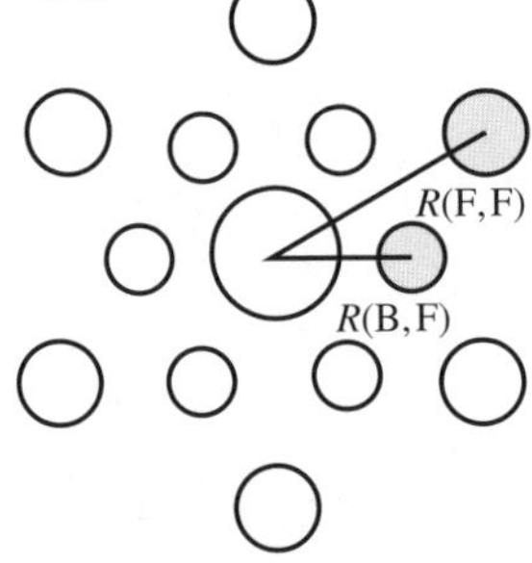

Figure 18A.5

E18A.15(a) $\lambda = \dfrac{h}{p} = \dfrac{h}{mv}$

Hence, $v = \dfrac{h}{m\lambda} = \dfrac{6.626\times10^{-34}\ \text{J s}}{(1.675\times10^{-27}\ \text{kg})\times(65\times10^{-12}\ \text{m})} = \boxed{6.1\ \text{km s}^{-1}}$

E18A.16(a) $\lambda = \dfrac{h}{p} = \dfrac{h}{mv}$ Hence, $v = \dfrac{h}{m\lambda}$.

Combine $E = \frac{1}{2}kT$ and $E = \frac{1}{2}mv^2 = \dfrac{h^2}{2m\lambda^2}$, to obtain

$$\lambda = \frac{h}{(mkT)^{1/2}} = \frac{6.626\times10^{-34}\ \text{J s}}{[(1.675\times10^{-27}\ \text{kg})\times(1.381\times10^{-23}\ \text{J K}^{-1})\times(350\text{K})]^{1/2}} = \boxed{233\ \text{pm}}$$

Solutions to problems

P18A.1 Although the crystallization of large biological molecules may not be as readily accomplished as that of small molecules, their crystal lattices are no different. Tobacco seed globulin forms face-centred cubic crystals with unit cell dimension of 12.3 nm and a mass density of 1.287 g cm^{-3}. Determine its molar mass.

The density of a face-centred cubic crystal is $4m/V$ where m is the mass of the unit hung on each lattice point and V is the volume of the unit cell. (The 4 comes from the fact that each of the cell's 8 vertices is shared by 8 cells, and each of the cell's 6 faces is shared by 2 cells.) So,

$$\rho = \frac{4m}{a^3} = \frac{4M}{N_A a^3} \text{ and } M = \tfrac{1}{4}\rho N_A a^3$$

$$M = \tfrac{1}{4}(1.287\ \text{g cm}^{-3})\times(6.022\times10^{23}\ \text{mol}^{-1})\times(12.3\times10^{-7}\ \text{cm})^3$$
$$= \boxed{3.61\times10^5\ \text{g mol}^{-1}}$$

P18A.3 The hexagonal unit cell is shown in Fig. 18A.6 from which we see that its volume V equals the value of c multiplied by the area of the hexagonal face. Since the area of the face equals 6 × (triangular area of Fig. 18A.6), we find that $\boxed{V = (3\sqrt{3}/2)a^2c}$.

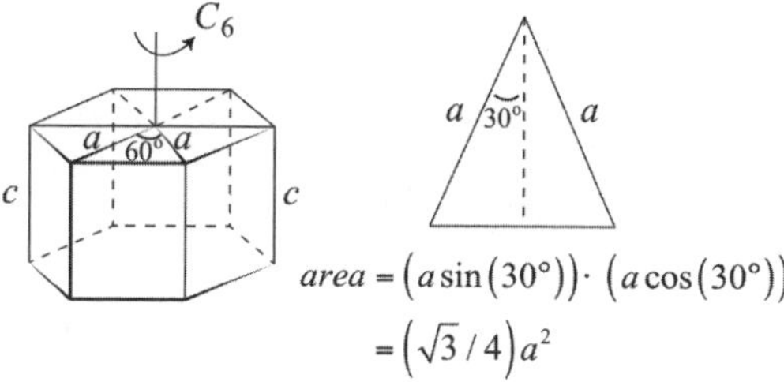

Figure 18A.6

P18A.5 $V = abc\sin\beta$ and the information given tells us that $a = 1.377b$, $c = 1.436b$, and $\beta = 122.82°$; hence

$$V = (1.377)\times(1.436\,b^3)\sin 122.82° = 1.662\,b^3$$

Since $\rho = \dfrac{NM}{VN_A} = \dfrac{2M}{1.662\,b^3N_A}$, we find that

$$b = \left(\frac{2M}{1.662\rho N_A}\right)^{1/3}$$

$$= \left(\frac{(2)\times(128.18\text{ g mol}^{-1})}{(1.662)\times(1.152\times10^6\text{ g m}^{-3})\times(6.022\times10^{23}\text{ mol}^{-1})}\right)^{1/3} = 605.8\text{ pm}$$

Therefore, $a = \boxed{834\text{ pm}}$, $b = \boxed{606\text{ pm}}$, and $c = \boxed{870\text{ pm}}$.

P18A.7 As shown in Problem 18A.2, the volume of a primitive monoclinic unit cell is

$$V = abc\,\sin\beta = (1.0427\text{ nm})\times(0.8876\text{ nm})\times(1.3777\text{ nm})\times\sin(93.254°)$$
$$= 1.2730\text{ nm}^3$$

The mass per unit cell is

$$m = \rho V = (2.024\text{ g cm}^{-3})\times(1.2730\text{ nm}^3)\times(10^{-7}\text{ cm nm}^{-1})^3 = 2.577\times10^{-21}\text{ g}$$

The repeating monomeric unit of the polymer is $[\{Cu(C_7H_9N_5O_2)(SO_4)\}{\cdot}2H_2O]_n$ (i.e. $CuC_7H_{13}N_5O_8S$), so its molar mass is

$$M = [63.546 + 7(12.011) + 13(1.008) + 5(14.007) + 8(15.999) + 32.066]\text{ g mol}^{-1}$$
$$= 390.82\text{ g mol}^{-1}$$

The number of monomer units, then, is the mass of the unit cell divided by the mass of the monomer

$$N = \frac{mN_A}{M} = \frac{(2.577\times10^{-21}\text{ g})\times(6.022\times10^{23}\text{ mol}^{-1})}{390.82\text{ g mol}^{-1}} = 3.97 \quad\text{or}\quad \boxed{4}$$

P18A.9 Consider for simplicity the two-dimensional lattice and planes shown in Fig. 18A.7.

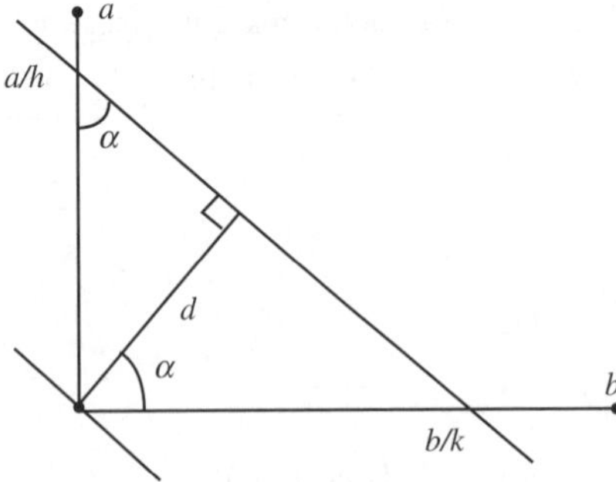

Figure 18A.7

The (hk) planes cut the a and b axes at a/h and b/k, and we have

$$\sin\alpha = \frac{d}{(a/h)} = \frac{hd}{a} \quad \text{and} \quad \cos\alpha = \frac{d}{(b/k)} = \frac{kd}{b}$$

(i.e., $\sin(90° - \alpha) = \sin 90° \cos\alpha - \cos 90° \sin\alpha = \cos\alpha$)

Then, since $\sin^2\alpha + \cos^2\alpha = 1$, we can write

$$\left(\frac{hd}{a}\right)^2 + \left(\frac{kd}{b}\right)^2 = 1$$

and therefore

$$\frac{1}{d^2} = \left(\frac{h}{a}\right)^2 + \left(\frac{k}{b}\right)^2$$

The same argument extends by analogy (or further trigonometry) to three dimensions, to give

$$\boxed{\frac{1}{d^2} = \left(\frac{h}{a}\right)^2 + \left(\frac{k}{b}\right)^2 + \left(\frac{l}{c}\right)^2} \quad [18A.1c]$$

P18A.11 Refer to *Self-test* 18A.6 of the text and Discussion questions 18A.3 and 18A.5 above. These references indicate systematic absences of diffraction lines from a cubic I lattice (body-centred cubic, bcc) corresponding to $h+k+l = \text{odd}$ thereby ruling out the presence of {100} planes in the diffraction pattern. These same planes are also ruled out of diffraction lines from a cubic F lattice (face-centred cubic, fcc) because the lines of this lattice have h, k, and l values which are either all even or all odd. These rules also indicate that {110} planes are not observed in the diffraction pattern cubic F lattices. Only the cubic P lattice exhibits diffraction lines from both {100} planes and {110} planes. The diffraction line from {111} planes of a cubic P lattice is expected to have low or zero intensity but, if observed, the separation between the sixth and seventh lines observed is larger than between the fifth and sixth lines because $\sin\theta_{hkl} \propto 1/d_{hkl} \propto (h^2+k^2+l^2)^{1/2}$, which corresponds to $5^{1/2}$, $6^{1/2}$, and $8^{1/2}$ for the 5th, 6th, and 7th observed lines. So we conclude that polonium has the $\boxed{\text{simple (primitive) cubic lattice}}$.

$$d_{hkl} = \frac{a}{(h^2+k^2+l^2)^{1/2}} [18\text{A}.1\text{b}] = \frac{\lambda}{2\sin\theta_{hkl}} [18\text{A}.2\text{b}] \quad \text{or} \quad a = \frac{\lambda}{2\sin\theta_{hkl}}(h^2+k^2+l^2)^{1/2}$$

Bragg reflection from {100}: $a = \dfrac{154\ \text{pm}}{(2)\times(0.225)} \times 1^{1/2} = 342\ \text{pm}$

from {110}: $a = \dfrac{154\ \text{pm}}{(2)\times(0.316)} \times 2^{1/2} = 345\ \text{pm}$

$$\text{from } \{111\}\text{: } a = \frac{154\,\text{pm}}{(2)\times(0.388)} \times 3^{1/2} = 344 \text{ pm}$$

Since these agree to within experimental uncertainty, we average to find that $\boxed{a = 344\,\text{pm}}$.

P18A.13 Therefore, $a = \dfrac{\lambda}{2\sin\theta_{100}}$ and

$$\frac{a(\text{KCl})}{a(\text{NaCl})} = \frac{\sin\theta_{100}(\text{NaCl})}{\sin\theta_{100}(\text{KCl})} = \frac{\sin 6.00°}{\sin 5.38°} = 1.115$$

Therefore, $a(\text{KCl}) = (1.115)\times(564\,\text{pm}) = \boxed{629\,\text{pm}}$.

The relative density ratio calculated from these unit cell dimensions is

$$\frac{\rho(\text{KCl})}{\rho(\text{NaCl})} = \left(\frac{M(\text{KCl})}{M(\text{NaCl})}\right)\times\left(\frac{a(\text{NaCl})}{a(\text{KCl})}\right)^3 = \left(\frac{74.55}{58.44}\right)\times\left(\frac{564\,\text{pm}}{629\,\text{pm}}\right)^3 = 0.920$$

whereas the ratio based upon macroscopic measurement of mass and volume is

$$\frac{\rho(\text{KCl})}{\rho(\text{NaCl})} = \frac{1.99\,\text{g}\,\text{cm}^{-3}}{2.17\,\text{g}\,\text{cm}^{-3}} = 0.917$$

The agreement of the two methods to within experiment uncertainties, $\boxed{\text{gave support}}$ to the X-ray method in the early period of its development.

P18A.15 The four values of $hx + ky + lz$ that occur in the exponential functions of F have the values 0, 5/2, 3, and 7/2, and so

$$F_{114} = f_{\text{I}}\sum_j e^{i\phi_{114}(\text{I}_j)}\ [18\text{A}.4] = f_{\text{I}}\sum_j e^{2i\pi(x_j + y_j + 4z_j)}$$
$$= f_{\text{I}}\{1 + e^{5i\pi} + e^{6i\pi} + e^{7i\pi}\} = f_{\text{I}}\{1 - 1 + 1 - 1\} = \boxed{0}$$

P18A.17 (a) When there is only one pair of identical atoms, the Wierl equation reduces to

$$I(\theta) = f^2\,\frac{\sin sR}{sR} \quad \text{where} \quad s = \frac{4\pi}{\lambda}\sin\tfrac{1}{2}\theta$$

Using θ_{ex} and sR_{ex}, meaning $(sR)_{\text{ex}}$, to represent θ and sR values when $I(sR)$ is an extremum, we find an analytic formula for the extrema by taking the derivative of $I(sR)$ w/r/t sR, set it equal to zero, and solve for sR_{ex}. We find

$$sR_{\text{ex}} = \frac{\sin sR_{\text{ex}}}{\cos sR_{\text{ex}}} = \tan sR_{\text{ex}}$$

This equation may be solved numerically, as shown in the following Mathcad Prime 2 worksheet, to give the extrema. Alternatively, we can simply plot

$I_{ff} \equiv I(sR)/f^2$ against sR to find sR values at which I_{ff} is a maximum or minimum. The graphical method is shown in Fig. 18A.8.

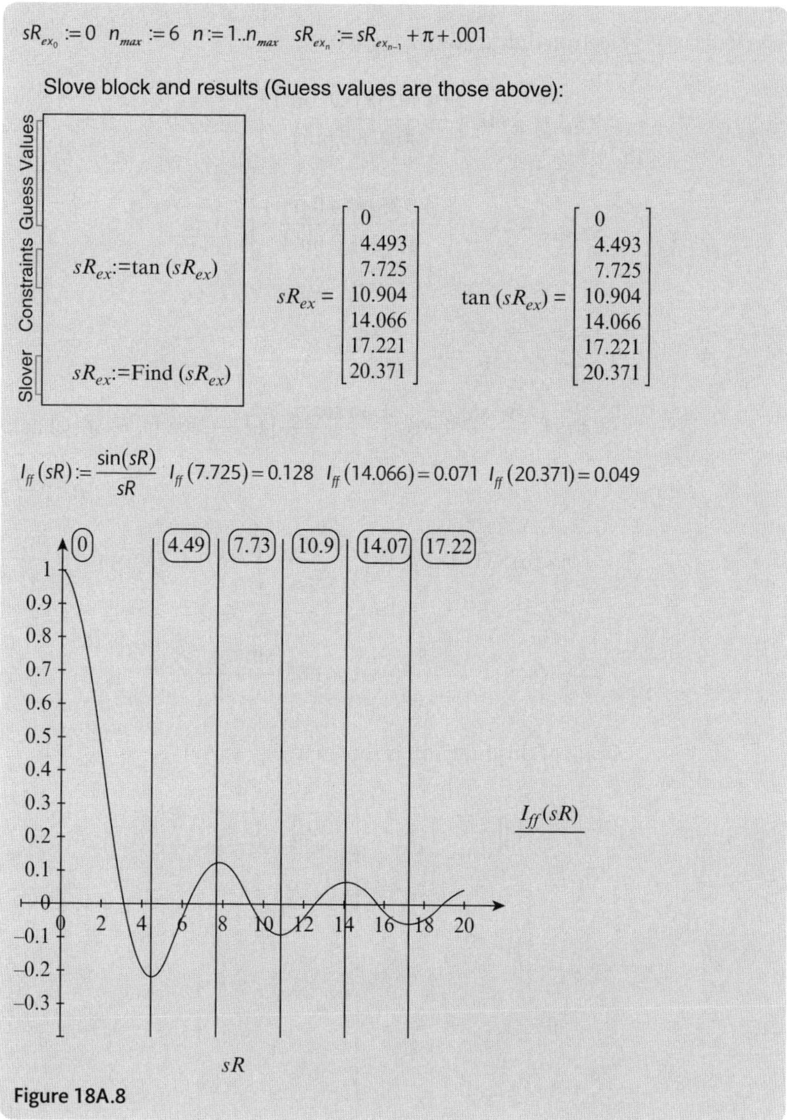

$sR_{ex_0} := 0 \quad n_{max} := 6 \quad n := 1..n_{max} \quad sR_{ex_n} := sR_{ex_{n-1}} + \pi + .001$
Slove block and results (Guess values are those above):
Slover Constraints Guess Values
$sR_{ex} := \tan(sR_{ex})$
$sR_{ex} := \text{Find}(sR_{ex})$
$sR_{ex} = [0, 4.493, 7.725, 10.904, 14.066, 17.221, 20.371]$
$\tan(sR_{ex}) = [0, 4.493, 7.725, 10.904, 14.066, 17.221, 20.371]$
$I_{ff}(sR) := \frac{\sin(sR)}{sR} \quad I_{ff}(7.725) = 0.128 \quad I_{ff}(14.066) = 0.071 \quad I_{ff}(20.371) = 0.049$
0
4.49
7.73
10.9
14.07
17.22
$I_{ff}(sR)$
sR

Figure 18A.8

The angles of extrema are calculated using the Br_2 bond length of 228.3 pm, the equation $\theta_{ex} = 2\sin^{-1}\left(\frac{(sR)_{ex}\lambda}{4\pi R}\right)$ and sR extrema.

Neutron diffraction: $\theta_{1st\ max} = 0$

$$\theta_{1st\ min} = 2\sin^{-1}\left(\frac{(4.493)\times(78\ \text{pm})}{4\pi(228.3\ \text{pm})}\right) = \boxed{14.0°}$$

$$\theta_{\text{2nd max}} = 2\sin^{-1}\left(\frac{(7.725)\times(78\text{ pm})}{4\pi(228.3\text{ pm})}\right) = \boxed{24.2°}$$

Electron diffraction: $\theta_{\text{1st max}} = 0$

$$\theta_{\text{1st min}} = 2\sin^{-1}\left(\frac{(4.493)\times(4.0\text{ pm})}{4\pi(228.3\text{ pm})}\right) = \boxed{0.72°}$$

$$\theta_{\text{2nd max}} = 2\sin^{-1}\left(\frac{(7.725)\times(4.0\text{ pm})}{4\pi(228.3\text{ pm})}\right) = \boxed{1.23°}$$

(b) $I = \sum_{i,j} f_i f_j \frac{\sin sR_{i,j}}{sR_{i,j}}, \quad s = \frac{4\pi}{\lambda}\sin\frac{1}{2}\theta$

$$= 4 f_{\text{C}} f_{\text{Cl}} \frac{\sin sR_{\text{CCl}}}{sR_{\text{CCl}}} + 6 f_{\text{Cl}}^2 \frac{\sin sR_{\text{ClCl}}}{sR_{\text{ClCl}}} \quad [4\text{ C–Cl pairs, } 6\text{ Cl–Cl pairs, } R_{\text{ClCl}} = (\tfrac{8}{3})^{1/2} R_{\text{CCl}}]$$

$$= (4)\times(6)\times(17)\times(f^2)\times\left(\frac{\sin sR_{\text{CCl}}}{sR_{\text{CCl}}}\right) + (6)\times(17)^2\times(f^2)\times\frac{\sin(\tfrac{8}{3})^{1/2} sR_{\text{CCl}}}{(\tfrac{8}{3})^{1/2} sR_{\text{CCl}}}$$

$$\frac{I}{f^2} = (408)\times\frac{\sin sR_{\text{CCl}}}{sR_{\text{CCl}}} + (1061.85)\frac{\sin(\tfrac{8}{3})^{1/2} sR_{\text{CCl}}}{sR_{\text{CCl}}}$$

A plot of this function is shown in Fig. 18A.9.

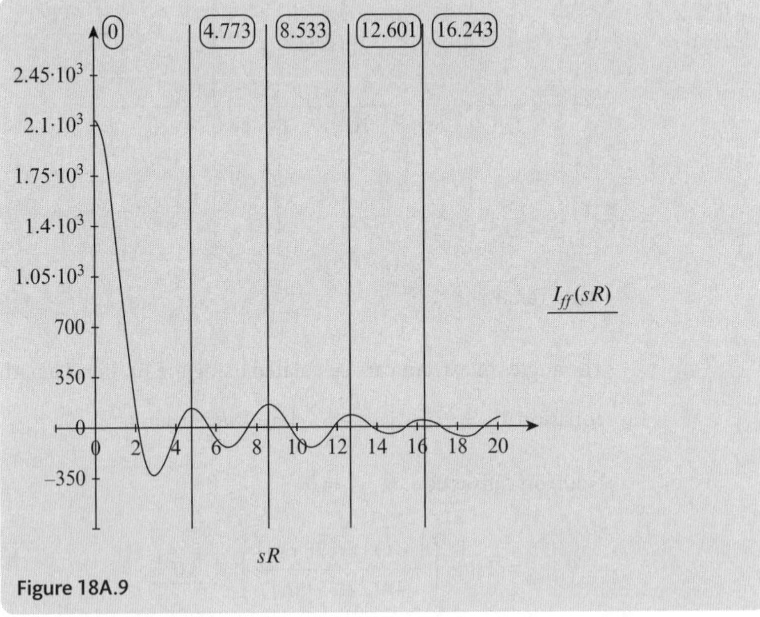

Figure 18A.9

We find sR_{ex} from the graph, and s_{ex} from the data. Then, we take the ratio $(sR/s)_{ex}$ to find the bond length R_{CCl}. The calculation of s requires the wavelength of the 10.0 keV electron beam.

$$\frac{p^2}{2m_e}=eV \quad \text{or} \quad p=(2m_eeV)^{1/2}$$

From the de Broglie relation,

$$\lambda=\frac{h}{p}=\frac{h}{(2m_eeV)^{1/2}}$$

$$=\frac{6.626\times10^{-34}\ \text{J s}}{2\times(9.109\times10^{-31}\ \text{kg})\times(1.609\times10^{-19}\ \text{C})\times(1.00\times10^{4}\ \text{V})}$$

$$=12.2\ \text{pm}$$

We draw up the following table.

	Maxima			Minima			
θ_{ex}(exp)	3.17°	5.37°	7.90°	1.77°	4.10°	6.67°	9.17°
s_{ex}/pm^{-1}	0.0285	0.0483	0.0710	0.0156	0.0368	0.0599	0.0823
$(sR)_{ex}$	4.773	8.533	12.60	2.931	6.549	10.62	14.45
$(sR/s)_{ex}$/pm	167	177	177	173	178	177	176

We reject the first value of 167 pm and average the others to find R_{CCl} = 176 pm and R_{ClCl} = 289 pm. The agreement of the $(sR/s)_{ex}$ values confirms the tetrahedral geometry.

18B Bonding in solids

Answer to discussion question

D18B.1 The majority of metals crystallize in structures which can be interpreted as the closest packing arrangements of hard spheres. These are the cubic close-packed (ccp) and hexagonal close-packed (hcp) structures. In these models, 74% of the volume of the unit cell is occupied by the atoms (packing fraction = 0.74). Most of the remaining metallic elements crystallize in the body-centred cubic (bcc) arrangement which is not too much different from the close-packed structures in terms of the efficiency of the use of space (packing fraction 0.68 in the hard sphere model). Polonium is an exception; it crystallizes in the simple cubic structure which has a packing fraction of 0.52. (See the solution to Exercise 38.2(a) for a derivation of all the packing fractions in cubic systems.) If atoms were truly hard spheres, we would expect that all metals would crystallize in either the ccp or hcp close-packed structures. The fact that a significant number crystallize in other structures is proof that a simple hard sphere model is an inaccurate representation of the interactions between the atoms. Covalent bonding between the atoms may influence the structure.

Solutions to exercises

E18B.1(a) The hatched area in Fig. 18B.1 is $h\times 2R = 3^{1/2}R\times 2R = 2\sqrt{3}R^2$ where $h = 2R\cos 30°$. The net number of cylinders in a hatched area is 1, and the area of the cylinder's base is πR^2. The volume of the prism (of which the hatched area is the base) is $2\sqrt{3}R^2L$, and the volume occupied by the cylinders is $\pi R^2 L$. Hence, the packing fraction is

$$f = \frac{\pi R^2 L}{2\sqrt{3}R^2 L} = \frac{\pi}{2\sqrt{3}} = \boxed{0.9069}$$

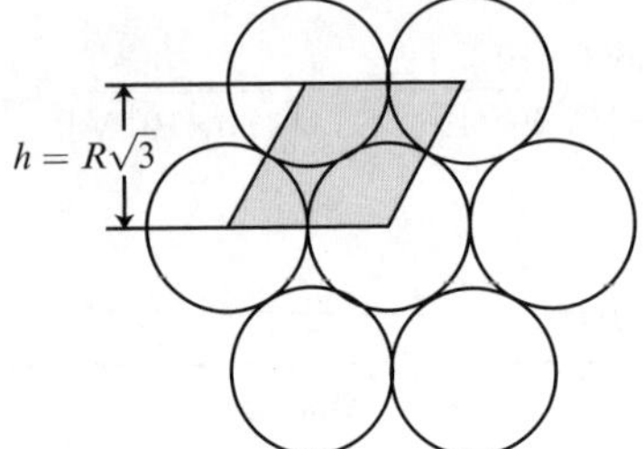

Figure 18B.1

E18B.2(a) $f = \dfrac{NV_a}{V_c}$ where N is the number of atoms in each unit cell, V_a their individual volumes, and V_c the volume of the unit cell itself. Refer to Fig. 18B.2 for a view of the primitive unit cell, the bcc unit cell, and the fcc unit cell.

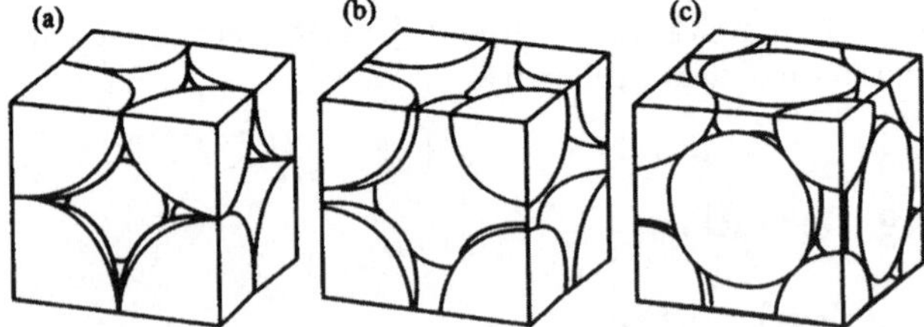

Figure 18B.2

(i) For a primitive unit cell: $N = 1$, $V_a = \frac{4}{3}\pi R^3$, and $V_c = (2R)^3$

$$f = \frac{\left(\frac{4}{3}\pi R^3\right)}{(2R)^3} = \frac{\pi}{6} = \boxed{0.5236}$$

(ii) For the bcc unit cell:

$$N = 2,\quad V_a = \frac{4}{3}\pi R^3,\quad \text{and}\quad V_c = \left(\frac{4R}{\sqrt{3}}\right)^3 \quad [\text{body diagonal of a unit cube is } 4R]$$

$$f = \frac{2\times\frac{4}{3}\pi R^3}{\left(\frac{4R}{\sqrt{3}}\right)^3} = \frac{\pi\sqrt{3}}{8} = \boxed{0.6802}$$

(iii) For the fcc unit cell: $N=4, \quad V_a=\frac{4}{3}\pi R^3, \quad \text{and} \quad V_c=(2\sqrt{2}R)^3$

$$f=\frac{4\times\frac{4}{3}\pi R^3}{(2\sqrt{2}R)^3}=\frac{\pi}{3\sqrt{2}}=\boxed{0.7405}$$

E18B.3(a) (i) For sixfold coordination see Fig. 18B.3.

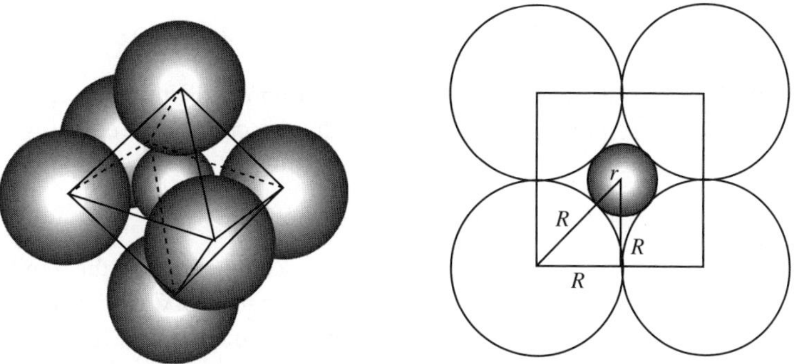

Figure 18B.3

We assume that the larger spheres of radius R touch each other and that they also touch the smaller interior sphere. Hence, by the Pythagorean theorem

$$(R+r)^2=2(R)^2 \quad \text{or} \quad \left(1+\frac{r}{R}\right)^2=2$$

Thus, $\frac{r}{R}=0.414$ and $r=0.414R=(0.414)\times(181\text{ pm})=\boxed{74.9\text{ pm}}$

(ii) For eightfold coordination see Fig. 18B.4.

The body diagonal of a cube is $a\sqrt{3}$. Hence

$$a\sqrt{3}=2R+2r \quad \text{or} \quad \sqrt{3}R=R+r \quad [a=2R]$$

$$\frac{r}{R}=0.732 \quad \text{and} \quad r=0.732R=(0.732)\times(181\text{ pm})=\boxed{132\text{ pm}}$$

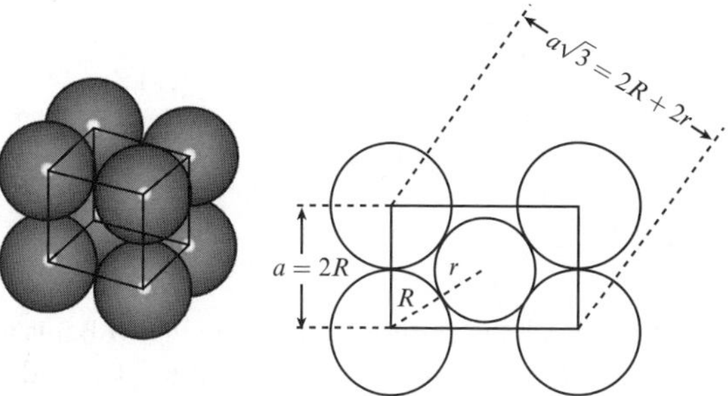

Figure 18B.4

E18B.4(a) The volume change is a result of two partially counteracting factors: (1) different packing fraction (f), and (2) different radii.

$$\frac{V(\text{bcc})}{V(\text{hcp})}=\frac{f(\text{hcp})}{f(\text{bcc})}\times\frac{\nu(\text{bcc})}{\nu(\text{hcp})} \quad \text{where } \nu \text{ is atomic volume}$$

$$f(\text{hcp})=0.7405,\quad f(\text{bcc})=0.6802 \quad [\text{Exercise 18B.2(a)}]$$

$$\frac{V(\text{bcc})}{V(\text{hcp})}=\frac{0.7405}{0.6802}\times\frac{(142.5)^3}{(145.8)^3}=1.016$$

Hence there is an $\boxed{\text{expansion}}$ of 1.6% in a transformation from hcp to bcc.

E18B.5(a) The lattice enthalpy is the difference in enthalpy between an ionic solid and the corresponding isolated ions. In this exercise, it is the enthalpy corresponding to the process

$$\text{CaO(s)}\rightarrow \text{Ca}^{2+}\text{(g)}+\text{O}^{2-}\text{(g)}$$

The standard lattice enthalpy can be computed from the standard enthalpies given in the exercise by considering the formation of CaO(s) from its elements as occurring through the following steps: sublimation of Ca(s), removing two electrons from Ca(g), atomization of O_2(g), two-electron attachment to O(g), and formation of CaO(s) lattice from gaseous ions. The formation reaction of CaO(s) is

$$\text{Ca(s)}+\tfrac{1}{2}\text{O}_2\text{(g)}\rightarrow\text{CaO(s)}$$

$$\begin{aligned}\Delta_f H^{\ominus}(\text{CaO,s})=&\ \Delta_{\text{sub}}H^{\ominus}(\text{Ca,s})+\Delta_{\text{ion}}H^{\ominus}(\text{Ca,g})\\ &+\frac{1}{2}\Delta_{\text{bond diss}}H^{\ominus}(\text{O}_2,\text{g})+\Delta_{\text{eg}}H^{\ominus}(\text{O,g})\\ &+\Delta_{\text{eg}}H^{\ominus}(\text{O}^-,\text{g})-\Delta_{\text{L}}H^{\ominus}(\text{CaO,s})\end{aligned}$$

So the lattice enthalpy is

$$\begin{aligned}\Delta_{\text{L}} H^{\ominus}(\text{CaO,s})=&\ \Delta_{\text{sub}}H^{\ominus}(\text{Ca,s})+\Delta_{\text{ion}}H^{\ominus}(\text{Ca,g})\\ =&\ \frac{1}{2}\Delta_{\text{bond diss}}H^{\ominus}(\text{O}_2,\text{g})+\Delta_{\text{eg}}H^{\ominus}(\text{O,g})\\ &+\Delta_{\text{eg}}H^{\ominus}(\text{O}^-,\text{g})-\Delta_{\text{f}}H^{\ominus}(\text{CaO,s})\end{aligned}$$

$$\begin{aligned}\Delta_L H^{\ominus}(\text{CaO,s})&=[178+1735+\frac{1}{2}(497)-141+844+635]\ \text{kJ mol}^{-1}\\ &=\boxed{3500.\ \text{kJ mol}^{-1}}\end{aligned}$$

Solutions to problems

P18B.1 Text Fig. 18B.15 shows the diamond structure. Fig. 18B.5 is an easier to visualize form of the structure which shows the unit cell of diamond. The number of carbon atoms in the unit cell is $(8\times\frac{1}{8})+(6\times\frac{1}{2})+(4\times1)=8$ ($\frac{1}{8}$ for a corner atom, $\frac{1}{2}$ for a face-centred atom, and 1 for an atom entirely in the cell). The positions of the atoms are $(0,0,0)$, $(\frac{1}{2},\frac{1}{2},0)$, $(\frac{1}{2},0,\frac{1}{2})$, $(0,\frac{1}{2},\frac{1}{2})$, $(\frac{1}{4},\frac{1}{4},\frac{1}{4})$, $(\frac{1}{4},\frac{3}{4},\frac{3}{4})$, $(\frac{3}{4},\frac{1}{4},\frac{3}{4})$, $(\frac{3}{4},\frac{3}{4},\frac{1}{4})$, as indicated in Fig. 18B.6.

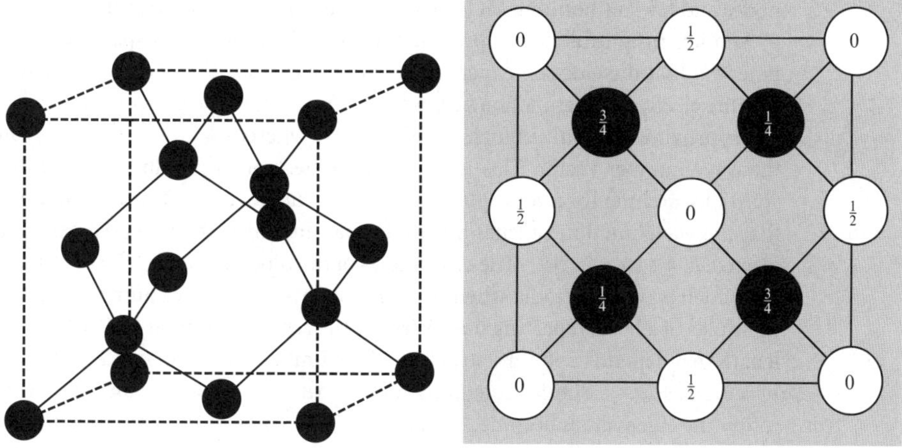

Figure 18B.5

Figure 18B.6

The fractions in Fig. 18B.6 denote height above the base in units of the cube edge, a. Two atoms that touch lie along the body diagonal at (0,0,0) and $(\frac{1}{4},\frac{1}{4},\frac{1}{4})$. Hence the distance $2r$ is one-fourth of the body diagonal which is $\sqrt{3}a$ in a cube. That is $2r = \sqrt{3}a/4$

The packing fraction is $\dfrac{\text{volume of atoms}}{\text{volume of unit cell}} = \dfrac{8V_a}{a^3} = \dfrac{(8)\times\frac{4}{3}\pi r^3}{\left(8r/\sqrt{3}\right)^3} = \boxed{0.340}$

P18B.3 As demonstrated in Example 18B.1 of the text and Exercise 18B.2(a), close-packed spheres (cubic F or face-centred cubic) fill 0.7404 of the total volume of the crystal. Therefore 1 cm^3 of close-packed carbon atoms would contain

$$\frac{0.74040\ \text{cm}^3}{\left(\frac{4}{3}\pi r^3\right)} = 3.838\times10^{23}\ \text{atoms}$$

$$\left(r = \left(\frac{154.45}{2}\right)\text{pm} = 77.225\ \text{pm} = 77.225\times10^{-10}\ \text{cm}\right)$$

Hence the close-packed density would be

$$\rho = \frac{\text{mass in 1 cm}^3}{1\ \text{cm}^3} = \frac{(3.838\times10^{23}\ \text{atom})\times(12.01\ \text{u/atom})\times(1.6605\times10^{-24}\ \text{g u}^{-1})}{1\ \text{cm}^3}$$

$$= \boxed{7.654\ \text{g cm}^{-3}}$$

The diamond structure is a very open structure which is dictated by the tetrahedral bonding of the carbon atoms. As a result many atoms that would be touching each other in a normal fcc structure do not in diamond; for example, the C atom in the centre of a face does not touch the C atoms at the corners of the face. This reduces the density to the much lower 3.516 g cm^{-1}.

P18B.5 Permitted states at the low energy edge of the band must have a relatively long characteristic wavelength while the permitted states at the high energy edge of the band must have a relatively short characteristic wavelength. There are few wavefunctions that have these characteristics so the density of states is lowest at the edges. This is analogous to the MO picture that shows a few bonding MOs that lack nodes and few antibonding MOs that have the maximum number of nodes.

Another insightful view is provided by consideration of the spatially periodic potential that the electron experiences within a crystal. The periodicity demands that the electron wavefunction be a periodic function of the position vector $\vec{r}$. We can approximate it with a Bloch wave: $\psi \propto e^{i\vec{k}\cdot\vec{r}}$ where $\vec{k} = k_x\hat{i} + k_y\hat{j} + k_z\hat{k}$ is called the wave number vector. This is a bold, 'free' electron approximation and in the spirit of searching for a conceptual explanation, not an accurate solution, suppose that the wavefunction satisfies a Hamiltonian in which the potential can be neglected: $\hat{H} = -(\hbar^2/2m)\nabla^2$. The eigenvalues of the Bloch wave are: $E = \hbar^2|\vec{k}|^2/2m$. The Bloch wave is periodic when the components of the wave number vector are multiples of a basic repeating unit. Writing the repeating unit as $2\pi/L$ where L is a length that depends upon the structure of the unit cell, we find: $k_x = 2n_x\pi/L$ where $n_x = 0, \pm1, \pm2, \ldots$. Similar equations can be written for k_y and k_z and with substitution the eigenvalues become: $E = (1/2m)(2\pi\hbar/L)^2(n_x^2 + n_y^2 + n_z^2)$. This equation suggests that the density of states for energy level E can be visually evaluated by looking at a plot of permitted n_x, n_y, n_z values as shown in the following graph of Fig. 18B.7. The number of n_x, n_y, n_z values within a thin, spherical shell around the origin equals the density of states which have energy E. Three shells, labelled 1, 2, and 3, are shown in the graph. All have the same width but their energies increase with their distance from the origin. It is obvious that the low energy shell 1 has a much lower density of states than the intermediate energy shell 2. The sphere of shell 3 has been cut into the shape determined by the periodic potential pattern of the crystal and, because of this phenomena, it also has a lower density of states than the intermediate energy shell 2. The general concept is that the low energy and high energy edges of a band have lower density of states than that of the band centre.

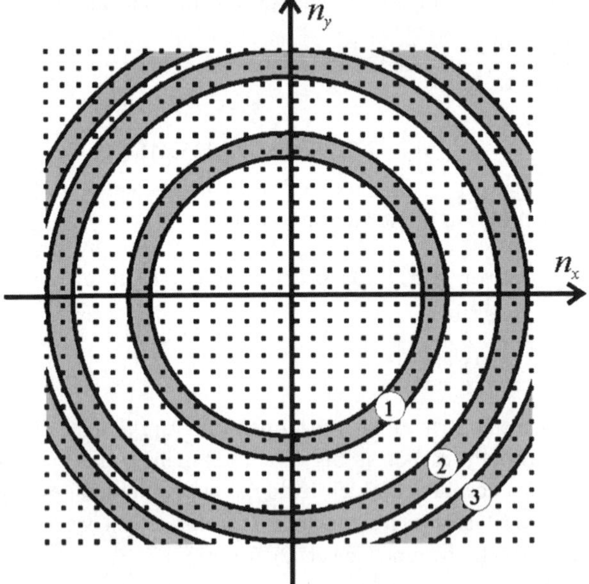

Figure 18B.7

P18B.7 (a) Consider the sixfold coordination shown in Fig. 18B.8 where we take the smallest distance between the larger hard-spheres to be $2R$ and the closest approach of a large and small hard-sphere to be $R + r$. This lattice gives the minimum value of the **radius ratio**, $\gamma = r/R$ [18B.3]. By the Pythagorean theorem

$$(R+r)^2 = 2R^2 \quad \text{or} \quad \left(1+\frac{r}{R}\right)^2 = 2$$

$$\gamma = \sqrt{2}-1 = \boxed{0.41421}$$

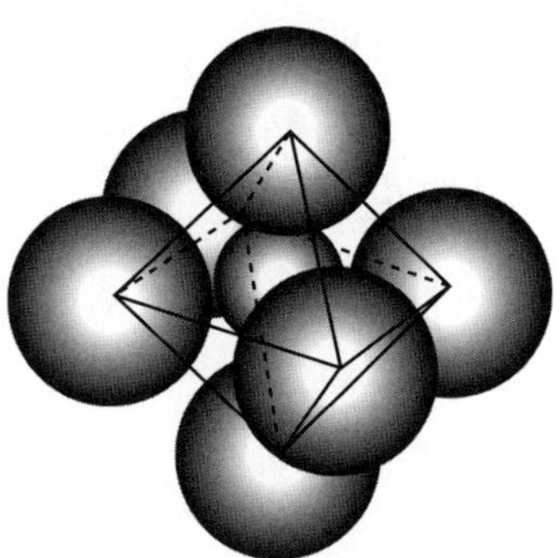

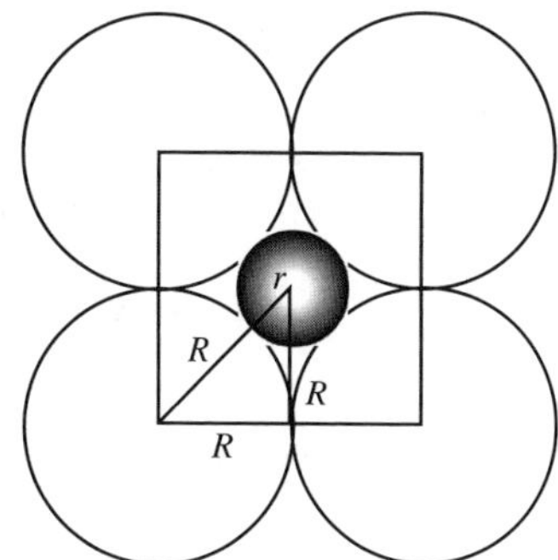

Figure 18B.8

(b) Consider the eightfold coordination shown in Fig. 18B.9 where we take the smallest distance between the larger hard-spheres at corners to be $2R$ and the closest approach of a large and small hard-sphere to be $R + r$. This lattice gives the minimum value of the **radius ratio**, $\gamma = r/R$ [18B.3]. The body diagonal of the cube is $a\sqrt{3}$. Hence

$$a\sqrt{3} = 2R+2r \quad \text{or} \quad \sqrt{3}\,R = R+r \;\; [a=2R] \quad \text{or} \quad \sqrt{3} = 1+\gamma$$

$$\gamma = \sqrt{3}-1 = \boxed{0.73205}$$

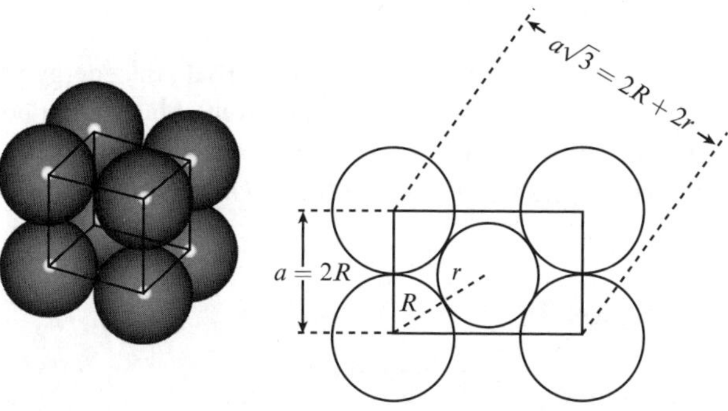

Figure 18B.9

P18B.9 $E_p = -A \times \dfrac{|z_1 z_2| N_A e^2}{4\pi\varepsilon_0 d}$ [18B.4] and $E_p^* = N_A C' e^{-d/d^*}$ [18B.5]

The sum of these two expressions gives the total potential energy and the minimum in this quantity is obtained by differentiation with respect to d.

$$\frac{d(E_P+E_P^*)}{dd}=0=\frac{A|z_1z_2|N_Ae^2}{4\pi\varepsilon_0 d^2}-N_AC'\left(\frac{1}{d^*}\right)e^{-d/d^*}$$

N_A can be eliminated and we can write

$$C'e^{-d/d^*}=\frac{A|z_1z_2|e^2}{4\pi\varepsilon_0 d}\left(\frac{d^*}{d}\right)$$

Then after substitution into eqn 18B.5 we obtain

$$\begin{aligned}E_{p,\min}&=-A\frac{|z_1z_2|N_Ae^2}{4\pi\varepsilon_0 d}+AN_A\frac{|z_1z_2|e^2}{4\pi\varepsilon_0 d}\left(\frac{d^*}{d}\right)\\&=-AN_A\left(\frac{|z_1z_2|e^2}{4\pi\varepsilon_0 d}\right)\left(1-\frac{d^*}{d}\right)\end{aligned}$$

which is eqn 18B.6.

18C Mechanical, electrical, and magnetic properties of solids

Answer to discussion question

D18C.1 The **Fermi–Dirac distribution** is a version of the Boltzmann distribution that takes into account the effect of the Pauli exclusion principle. It can therefore be used to calculate the probability of occupancy of a state of energy E, $f(E)$, in a many-electron system at a temperature T:

$$f(E)=\frac{1}{e^{(E-\mu)/kT}+1}\quad[18C.6a]$$

In this expression, μ is the '**chemical potential**', the energy of the level for which $f=½$ provided that $T>0$. For energies well above μ, the population resembles a Boltzmann distribution, decaying exponentially with increasing energy, i.e. the higher the temperature, the longer the exponential tail.

Solutions to exercises

E18C.1(a) Poisson's ratio: $\nu_P=\dfrac{\text{transverse strain}}{\text{normal strain}}$ [18C.2]

$$=0.45$$

We note that the transverse strain is usually a contraction and that it is usually evenly distributed in both transverse directions. That is, if $(\Delta L/L)_z$ is the normal strain, then the transverse strains, $(\Delta L/L)_x$ and $(\Delta L/L)_y$, are equal. In this case of a 1.0% uniaxial stress:

$$\left(\frac{\Delta L}{L}\right)_z = +0.010, \quad \left(\frac{\Delta L}{L}\right)_x = \left(\frac{\Delta L}{L}\right)_y = -0.010 \times 0.45$$
$$= -0.0045 \text{ [a contraction of widths]}$$

Application of the stress to 1 cm^3 cube of polyethene results in a volume equal to

$$(1-0.0045)^2 \times (1+0.010) \times 1\,\text{cm}^3 = 1.00093\,\text{cm}^3$$

The change in volume is $\boxed{9.3\times10^{-4}\,\text{cm}^3}$.

E18C.2(a) n-type; the dopant, arsenic, belongs to Group 15 whereas germanium belongs to Group 14.

E18C.3(a) $m = g_e\{S(S+1)\}^{1/2}\mu_B$ [18C.9, with S in place of s]

Therefore, because $m = 3.81\,\mu_B$ and $g_e \approx 2$,

$$S(S+1) = (\tfrac{1}{4}) \times (3.81)^2 = 3.63$$

implying that $S = 1.47$. Because $S \approx \tfrac{3}{2}$, there must be $\boxed{\text{three unpaired spins}}$.

E18C.4(a) $\chi_m = \chi V_m$[18C.8]$= \chi M/\rho = (-7.2\times10^{-7})\times(78.11\,\text{g mol}^{-1})/(0.879\,\text{g cm}^{-3})$

$$= \boxed{-6.4\times10^{-5}\,\text{cm}^3\text{mol}^{-1}} = \boxed{-6.4\times10^{-11}\,\text{m}^3\,\text{mol}^{-1}}$$

E18C.5(a) The molar susceptibility is given by

$$\chi_m = \frac{N_A g_e^2 \mu_0 \mu_B^2 S(S+1)}{3kT} \text{ [18C.10a]} \quad \text{so} \quad S(S+1) = \frac{3kT\chi_m}{N_A g_e^2 \mu_0 \mu_B^2}$$

$$S(S+1) = \frac{3(1.381\times10^{-23}\,\text{J K}^{-1})\times(294.53\,\text{K})\times(0.1463\times10^{-6}\,\text{m}^3\,\text{mol}^{-1})}{(6.022\times10^{23}\,\text{mol}^{-1})\times(2.0023)^2\times(4\pi\times10^{-7}\,\text{T}^2\,\text{J}^{-1}\,\text{m}^3)\times(9.27\times10^{-24}\,\text{J T}^{-1})^2}$$
$$= 6.841$$

$$S^2 + S - 6.841 = 0 \quad \text{and} \quad S = \frac{-1+\sqrt{1+4(6.841)}}{2} = 2.163$$

corresponding to $\boxed{4.326}$ effective unpaired spins. The theoretical number is $\boxed{5}$ corresponding to the $3d^5$ electronic configuration of Mn^{2+}.

Comment. The discrepancy between the two values is accounted for by an antiferromagnetic interaction between the spins which alters χ_m from the form of eqn 18C.10a.

E18C.6(a) $\chi_m = (6.3001\times10^{-6})\times\left(\dfrac{S(S+1)}{T/\text{K}}\,\text{m}^3\text{mol}^{-1}\right)$ [See text Example 18C.1]

Since Cu(II) is a d^9 species, it has one unpaired spin, and so $S = s = \tfrac{1}{2}$. Therefore,

$$\chi_m = \frac{(6.3001\times10^{-6})\times(\tfrac{1}{2})\times(\tfrac{3}{2})}{298}\,\text{m}^3\text{mol}^{-1} = \boxed{+1.6\times10^{-8}\,\text{m}^3\,\text{mol}^{-1}}$$

Solutions to problems

P18C.1 According to eqn 18C.3:

$$G = \frac{E}{2(1+\nu_P)} \text{ and } K = \frac{E}{3(1-2\nu_P)}$$

Substituting the Lamé-constant expressions for E and ν_P into the right-hand side of these relationships yields:

$$G = \frac{\dfrac{\mu(3\lambda+2\mu)}{\lambda+\mu}}{2\left(1+\dfrac{\lambda}{2(\lambda+\mu)}\right)} \text{ and } K = \frac{\dfrac{\mu(3\lambda+2\mu)}{\lambda+\mu}}{3\left(1-\dfrac{\lambda}{\lambda+\mu}\right)}.$$

Expanding leads to:

$$G = \frac{\dfrac{\mu(3\lambda+2\mu)}{\lambda+\mu}}{2\left(\dfrac{2\lambda+2\mu+\lambda}{2(\lambda+\mu)}\right)} = \frac{\mu(3\lambda+2\mu)}{3\lambda+2\mu} = \boxed{\mu}$$

$$\text{and,} \quad K = \frac{\dfrac{\mu(3\lambda+2\mu)}{\lambda+\mu}}{3\left(\dfrac{\lambda+\mu+\lambda}{(\lambda+\mu)}\right)} = \frac{\mu(3\lambda+2\mu)}{3\mu} = \boxed{\frac{3\lambda+2\mu}{3}},$$

as the problem asks us to prove.

P18C.3 $f(E) = \dfrac{1}{e^{(E-\mu)/kT}+1}$ [18C.6a] where E is measured from in minimum value of $E_{min} = 0$

In the case for which $E < \mu$, $\lim_{T\to 0}(E-\mu)/kT = \lim_{T\to 0}(-|E-\mu|/kT) = -\infty$.
Thus,

$$\lim_{T\to 0} f(E) = \frac{1}{e^{-\infty}+1} = \frac{1}{0+1} = 1$$

Conclusion: $\boxed{\lim_{T\to 0} P(E) = 1 \text{ when } E < \mu}$

In the case for which $E > \mu$, $\lim_{T\to 0}(E-\mu)/kT = \lim_{T\to 0}(|E-\mu|/kT) = +\infty$
Thus,

$$\lim_{T\to 0} f(E) = \frac{1}{e^{+\infty}+1} = \frac{1}{\infty+1} = 0$$

Conclusion: $\boxed{\lim_{T\to 0} f(E) = 0 \text{ when } E > \mu}$

Since $N_e = \int_0^\infty \rho(E)f(E)\mathrm{d}E$ [18C.5] $= \int_0^\mu \rho(E)f(E)\mathrm{d}E + \int_\mu^\infty \rho(E)f(E)\mathrm{d}E$, we find μ at $T=0$ by substitution of $f=1$ in the first integral and $f=0$ in the second integral.

$$N_e = \int_0^{\mu(0)} \rho(E)\times 1\,\mathrm{d}E + \int_{\mu(0)}^{\infty} \rho(E)\times 0\,\mathrm{d}E = \int_0^{\mu(0)} \rho(E)\,\mathrm{d}E$$

$$= \int_0^{\mu(0)} \rho(E)\,\mathrm{d}E = C\int_0^{\mu(0)} E^{1/2}\mathrm{d}E \quad \text{where} \quad C = 4\pi V\left(2m_e/h^2\right)^{3/2}$$

$$= \left.\frac{2CE^{3/2}}{3}\right|_{E=0}^{E=\mu(0)} = \frac{2C\mu(0)^{3/2}}{3}$$

Solving for $\mu(0)$ yields

$$\mu(0) = \left(\frac{3N_e}{2C}\right)^{2/3} = \boxed{(3\mathcal{N}/8\pi)^{2/3}\left(h^2/2m_e\right)} \quad \text{where} \quad \mathcal{N} = N_e/Vw$$

Because a sodium atom contributes a single valence electron and sodium has a density of about 0.97 g cm^{-3},

$$\mathcal{N} = N_e/V = N_A/V_m = N_A \times \text{mass density / molar mass}$$
$$= \left(6.022\times10^{23}\ \text{mol}^{-1}\right)\times\left(0.97\ \text{g cm}^{-3}\right)/\left(23\ \text{g mol}^{-1}\right) = 2.5\times10^{28}\ \text{m}^{-3}$$

and

$$\mu(0) = \left(\frac{3\left(2.5\times10^{28}\ \text{m}^{-3}\right)}{8\pi}\right)^{2/3}\frac{\left(6.63\times10^{-34}\ \text{J s}\right)^2}{2\left(9.11\times10^{-31}\ \text{kg}\right)} = 5.0\times10^{-19}\ \text{J} = \boxed{3.1\ \text{eV}}$$

P18C.5 $G = G_0 e^{-E_g/2kT}$

So $\ln(G/\text{S}) = \ln(G_0/\text{S}) - \left(\frac{E_g}{2k}\right)\times\frac{1}{T}$ with conductance G in the siemens unit (1 S = 1 Ω^{-1}). Thus, the slope of a $\ln(G)$ against $1/T$ plot equals $-E_g/2k$. The data has minimal uncertainty so the slope can be calculated by the two-point difference method. Alternatively, a linear regression fit of $(1/T, \ln(G/\text{S}))$ data points gives the slope.

$$\text{slope} = \frac{\Delta\ln(G/\text{S})}{\Delta\left(\frac{1}{T}\right)} = \frac{\ln(0.0847) - \ln(2.86)}{\frac{1}{312\ \text{K}} - \frac{1}{420\ \text{K}}} = -4\,270\ \text{K}$$

$$E_g = -2k\times(\text{slope}) = -2\times\left(1.381\times10^{-23}\ \text{J K}\right)\times(-4\,270)$$
$$= 1.18\times10^{-19}\ \text{J}$$

This is equivalent to 71.0 kJ mol^{-1} or $\boxed{0.736\ \text{eV}}$

P18C.7 The molar magnetic susceptibility is given by [see text Example 18C.1]:

$$\chi_m = \frac{N_A g_e^2 \mu_0 \mu_B^2 S(S+1)}{3kT} = (6.3001\times10^{-6})\times\frac{S(S+1)}{T/\text{K}}\ \text{m}^3\ \text{mol}^{-1}$$

$$\text{For } S=2,\ \chi_m = \frac{(6.3001\times10^{-6})\times(2)\times(2+1)}{298}\ \text{m}^3\ \text{mol}^{-1} = \boxed{0.127\times10^{-6}\ \text{m}^3\ \text{mol}^{-1}}$$

$$\text{For } S=3,\ \chi_m = \frac{(6.3001\times10^{-6})\times(3)\times(3+1)}{298}\ \text{m}^3\text{mol}^{-1} = \boxed{0.254\times10^{-6}\ \text{m}^3\ \text{mol}^{-1}}$$

$$\text{For } S=4,\ \chi_m = \frac{(6.3001\times10^{-6})\times(4)\times(4+1)}{298}\ \text{m}^3\ \text{mol}^{-1} = \boxed{0.423\times10^{-6}\ \text{m}^3\ \text{mol}^{-1}}$$

Instead of a single value of S we use an average weighted by the Boltzmann factor

$$\exp\left(\frac{-50\times10^3\ \text{J mol}^{-1}}{(8.3145\ \text{J mol}^{-1}\ \text{K}^{-1})\times(298\ \text{K})}\right) = 1.7\times10^{-9}$$

Thus the $S=2$ and $S=4$ forms are present in negligible quantities compared to the $S=3$ form. The compound's susceptibility, then, is that of the $S=3$ form, namely $\boxed{0.254\ \text{cm}^3\ \text{mol}^{-1}}$.

P18C.9 If the unit cell volume does not change upon substitution of Ca for Y, then, the density of the superconductor and that of the Y-only compound will be proportional to their molar masses.

$$M_{super} = [2(200.59)+2(137.327)+(1-x)\times(88.906)+x(40.078)$$
$$+2(63.546)+7.55(15.999)]\ \text{g mol}^{-1}$$

$$M_{super}/(\text{g mol}^{-1}) = 1012.6 - 48.828x$$

The molar mass of the Y-only compound is $1012.6\ \text{g mol}^{-1}$, and the ratio of their densities is

$$\frac{\rho_{\text{super}}}{\rho_{\text{Y-only}}} = \frac{1012.6-48.828x}{1012.6} = 1-0.04822x \text{ so } x = \frac{1}{0.04822}\left(1-\frac{\rho_{\text{super}}}{\rho_{\text{Y-only}}}\right)$$

The density of the Y-only compound is its mass over its volume. The volume is

$$V_{\text{Y-only}} = a^2c = (0.38606\ \text{nm})^2\times(2.8915\ \text{nm}) = 0.43096\ \text{nm}^3 = 0.43096\times10^{-21}\ \text{cm}^3$$

so the density is

$$\rho_{\text{Y-only}} = \frac{2M}{N_A V} = \frac{2(1012.6\ \text{g mol}^{-1})}{(6.022\times10^{23}\ \text{mol}^{-1})\times(0.43096\times10^{-21}\ \text{cm}^3)} = 7.804\ \text{g cm}^{-3}$$

The extent of Ca substitution is

$$x = \frac{1}{0.04822}\left(1-\frac{7.651}{7.804}\right) = \boxed{0.41}$$

Comment. The precision of this method depends strongly on just how constant the lattice volume really is.

18D The optical properties of solids

Answer to discussion question

D18D.1 The Davydov splitting in the exciton bands of a crystal can be understood by considering the allowed transitions of interacting dipoles of neighbouring molecules. As shown in text Fig. 18D.2, the parallel alignment of transition dipoles is energetically unfavourable, and the exciton absorption is shifted to higher energy (blue shifted) than in the isolated molecules. However, a head-to-tail alignment is energetically favourable, and the transition occurs at a lower frequency (red shifted) than in the isolated molecules. Transition dipoles are rarely in a single parallel or head-to-tail alignment. As shown in text Fig. 18D.3, transition moments may often be either energetically favoured or energetically unfavoured. These crystals exhibit both the blue shifted and the red shifted band and the separation of the bands is the **Davydov** or **exciton splitting**.

Solution to exercise

E18D.1(a) $E_g = h\nu_{min} = \dfrac{hc}{\lambda_{max}} = \dfrac{(6.626\times10^{-34}\,\text{J s})(2.998\times10^{8}\,\text{m s}^{-1})}{350\times10^{-9}\,\text{m}}\left(\dfrac{1\,\text{eV}}{1.602\times10^{-19}\,\text{J}}\right) = \boxed{3.54\,\text{eV}}$

Solutions to problems

P18D.1 Figure 18D.1 depicts the single transition of the monomer with the transition dipole moment μ_{mon}. It also defines the excited states and transitions of the dimer.

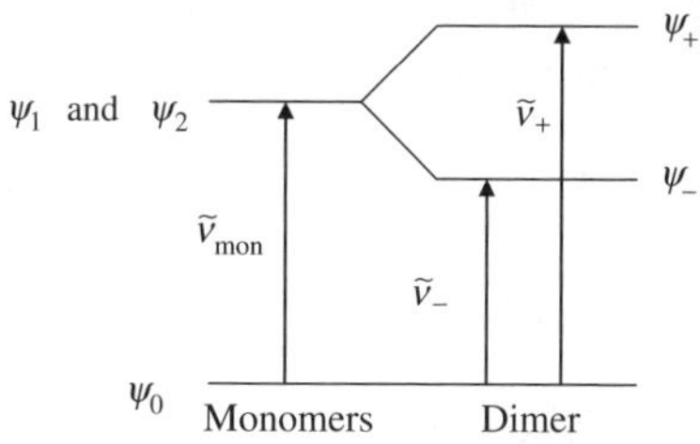

Figure 18D.1

$$\mu_{+or-} = \langle\psi_{+or-}|\mu|\psi_0\rangle = \langle c_{+or-,1}\psi_1 + c_{+or-,2}\psi_2|\mu|\psi_0\rangle$$
$$= c_{+or-,1}\langle\psi_1|\mu|\psi_0\rangle + c_{+or-,2}\langle\psi_2|\mu|\psi_0\rangle$$

But $\langle\psi_1|\mu|\psi_0\rangle = \langle\psi_2|\mu|\psi_0\rangle = \mu_{mon}$ so $\mu_{+or-} = (c_{+or-,1} + c_{+or-,2})\mu_{mon}$

P18D.3 If a substance responds nonlinearly to an electric field E, then it induces a dipole moment:

$$\mu = \alpha E + \beta E^2.$$

If the electric field is oscillating at two frequencies, we can write the electric field as

$$E = E_1\cos\omega_1 t + E_2\cos\omega_2 t,$$

and the nonlinear response as

$$\beta E^2 = \beta(E_1 \cos\omega_1 t + E_2 \cos\omega_2 t)^2,$$

$$\beta E^2 = \beta(E_1^2 \cos^2 \omega_1 t + E_2^2 \cos^2 \omega_2 t + 2E_1 E_2 \cos\omega_1 t \cos\omega_2 t).$$

Application of trigonometric identities allows a product of cosines to be rewritten as a sum:

$$\cos A \cos B = \tfrac{1}{2}\cos(A-B) + \tfrac{1}{2}\cos(A+B).$$

Using this result (a special case of which applies to the $\cos^2$ terms), yields:

$$\beta E^2 = \frac{1}{2}\beta[E_1^2(1+\cos 2\omega_1 t) + E^2(1+\cos 2\omega_2 t) + 2E_1 E_2(\cos(\omega_1+\omega_2)t + \cos(\omega_1-\omega_2)t].$$

This expression includes responses at twice the original frequencies as well as at the sum and difference frequencies.

Integrated activities

I18.1 $\theta_{111}(100\text{ K}) = 22.0403°,\ \ \theta_{111}(300\text{ K}) = 21.9664°$

$$\sin\theta_{111}(100\text{ K}) = 0.37526,\ \ \sin\theta_{111}(300\text{ K}) = 0.37406$$

$$\frac{\sin\theta(300\text{ K})}{\sin\theta(100\text{ K})} = 0.99681 = \frac{a(100\text{ K})}{a(300\text{ K})}$$ [18A.2b, Bragg's law; see Problem 18A.13]

$$a(300\text{ K}) = \frac{\lambda\sqrt{3}}{2\sin\theta_{111}} = \frac{(154.0562\text{ pm})\times\sqrt{3}}{(2)\times(0.37406)} = 356.67\text{ pm}$$

$$a(100\text{ K}) = (0.99681)\times(356.67\text{ pm}) = 355.53\text{ pm}$$

$$\frac{\delta a}{a} = \frac{356.67-355.53}{355.53} = 3.206\times10^{-3}$$

$$\frac{\delta V}{V} = \frac{356.67^3-355.53^3}{355.53^3} = 9.650\times10^{-3}$$

$$\alpha_{\text{volume}} = \frac{1}{V}\frac{\delta V}{\delta T} = \frac{9.560\times10^{-3}}{200\text{ K}} = \boxed{4.8\times10^{-5}\text{ K}^{-1}}$$

$$\alpha_{\text{linear}} = \frac{1}{a}\frac{\delta a}{\delta T} = \frac{3.206\times10^{-3}}{200\text{ K}} = \boxed{1.6\times10^{-5}\text{ K}^{-1}}$$

I18.3 1s hydrogenic radial wavefunction: $R_{1s}(\rho) = 2(Z/a_o)e^{-\rho/2}$ where $\rho = (2Z/a_o)r$

We start by constructing a Gaussian function, $G_{1s}(\rho) = Ne^{-b\rho^2}$, that 'fits' or 'matches' $R_{1s}(\rho)$. Finding the normalization constant N requires evaluation of the normalization integral. This begins with the transformation $r \to \rho$.

$$\int_0^\infty G_{1s}(\rho)^2 d\tau = N^2\int_0^\infty e^{-2b\rho^2} r^2 dr\ [18.5b] = N^2(a_o/2Z)^3\int_0^\infty e^{-2b\rho^2}\rho^2 d\rho = 1$$

$$N = \left(\frac{(2Z/a_o)^3}{\int_0^\infty e^{-2b\rho^2}\rho^2 d\rho}\right)^{1/2}$$

The integral in the denominator of the above expression is found in any math handbook. Here is the integration in the Mathcad Prime 2 worksheet used throughout this problem.

$$N(Z,b,a_0):=\left(\frac{2\cdot Z}{a_0}\right)^{\frac{3}{2}}\cdot\left(\frac{1}{\int_0^{\infty}\left(e^{-b\cdot\rho^2}\right)^2\cdot\rho^2\,d\rho}\right)^{\frac{1}{2}}\rightarrow\left(\frac{2\cdot Z}{a_0}\right)^{\frac{3}{2}}\cdot\sqrt{\frac{8\cdot\sqrt{2}\cdot b^{\frac{3}{2}}}{\sqrt{\pi}}}$$

To find the constant b, we calculate R_{1s} at 200 points in the range $0 \le r \le 3a_0$, setup the worksheet to calculate G_{1s} at each of these points, subtract the two, square the difference, and sum the result over all points. This gives the sum of squares (SS) for which we wish to vary b so as to minimize the sum, the total difference error (E). We call the sum *SSE* in the worksheet and Mathcad performs the minimization process with the 'minerr()' function within a solve block.

$$\text{pm}:=10^{-12}\cdot\text{m}\quad a_0:=52.9177\cdot\text{pm}\quad \rho(r,Z):=\frac{2\cdot Z\cdot r}{a_0}$$

$$R_{1s}(r,Z):=2\cdot\left(\frac{Z}{a_0}\right)^{\frac{3}{2}}\cdot e^{-2^{-1}\cdot\rho(r,Z)}\qquad G_{1s}(r,Z,b):=N(Z,b,a_0)\cdot e^{-b\cdot\rho(r,Z)^2}$$

$$M:=200\quad i:=0..M\quad r_{max}:=3\cdot a_0\quad r_i:=\frac{i}{M}\cdot r_{max}$$

$$SSE(Z,b):=\sum_i\left(R_{1s}(r_i,Z)-G_{1s}(r_i,Z,b)\right)^2$$

$Z:=1$

Solver — Guess Values: $b:=0.1$

Constraints: $SSE(Z,b)=0$

Solver: $b:=\text{minerr}(b)$

Computed b values:
$Z=1,\ b=0.127952$
$Z=2,\ b=0.128470$
$Z=3,\ b=0.129258$

The following worksheet plot compares G_{1s} with R_{1s} over the range of r for $Z = 1$. The Gaussian wavefunction reduces the electron density at the nucleus and removes the sharp peak. The Gaussian function effectively replaces the cusp of the exponential decay near $r = 0$ with a broader, flatter function while simultaneously decreasing the long tail of the exponential. Physically, this is analogous to contraction of the orbital by increasing the nuclear charge Z. At intermediate r the Gaussian increases electron density but at large r the Gaussian approaches zero more rapidly. This comparison also applies for larger values of Z.

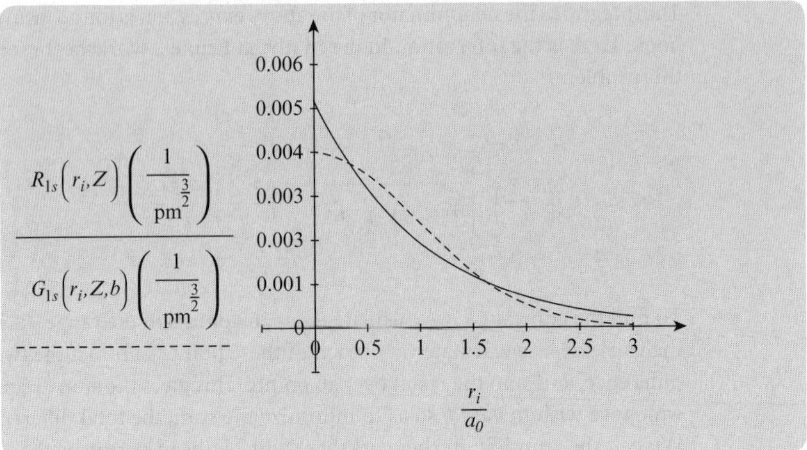

We can now proceed to a comparison of scattering factors for the two wavefunction over a range of $\xi \equiv \sin(\theta)/\lambda$ values.

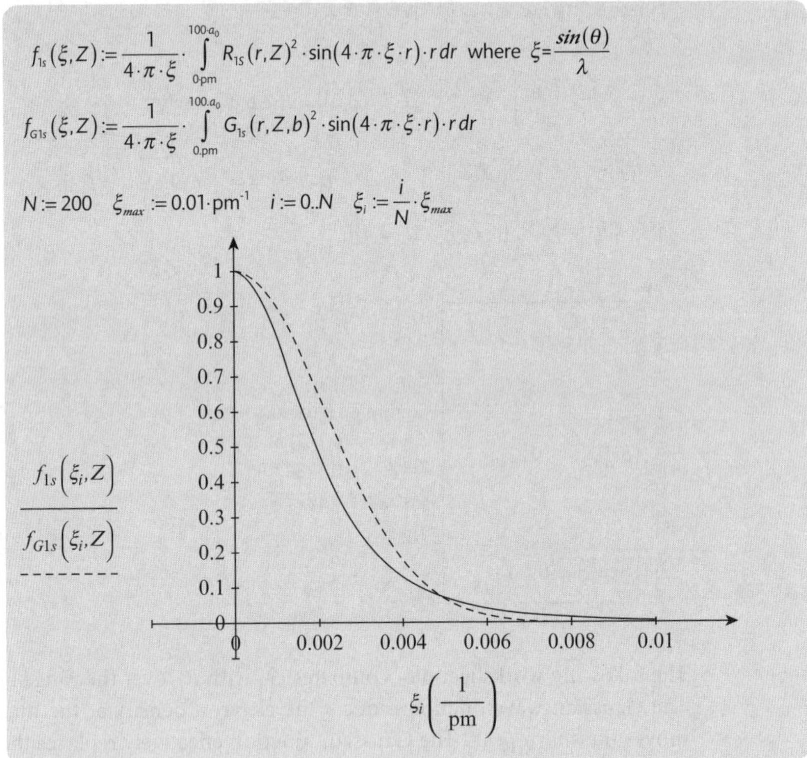

The above plot clearly shows that the replacement of the hydrogenic 1s wavefunction with a Gaussian wavefunction shifts the reflection angle away from the forward direction ($\theta = 0$). In the context of the discussion of Integrated activity 18.2 this is equivalent to the effect of increasing the atomic number, Z.

19 Molecules in motion

19A Transport properties of a perfect gas

Answers to discussion questions

D19A.1 Fick's first law, $J_z = -D\dfrac{d\mathcal{N}}{dz}$ [19A.3], arises in a non-equilibrium gas in which the number density $\mathcal{N}$ is a smooth, continuous function of the spatial coordinates. As discussed in Section 19A.2(a), the net flux of matter from one infinitesimally small volume of space to an adjacent volume is the difference the matter flux from density to low density and the matter flux from the low density to the high density. This difference is non zero and proportional to the density (concentration) gradient in the absence of equilibrium. The constant of proportionality is the diffusion constant D and the net flux is always positive down the concentration gradient. That is, there is a net flow of mass from high to low concentration.

D19A.3 As a hypothesis, we speculate that gaseous molecular attraction effectively reduces the magnitude of the collision cross-section σ. Since the transport properties of diffusion, thermal energy transport, and viscosity in the approximate regime $\sigma << \lambda <<$ *container size* are all inversely proportional to σ, the hypothesis suggests an increase in these transport properties when molecular attraction becomes relevant. Stronger molecular repulsions effectively increase the molecular size parameter σ, thereby, decreasing the magnitude of the transport coefficients, which decreases the gaseous transport properties.

Solutions to exercises

E19A.1(a) For a perfect argon gas: $\lambda = \dfrac{kT}{\sigma p}$ [1B.13] $= \dfrac{RT}{\sigma N_A p} = \dfrac{1}{\sigma N_A[\text{Ar}]}$. Thus,

$$\kappa = \tfrac{1}{3}\lambda \upsilon_{\text{mean}} C_{V,\text{m}}[\text{Ar}] \ [19\text{A}.13\text{b}] = \tfrac{1}{3}\left(\frac{1}{\sigma N_A[\text{Ar}]}\right)\upsilon_{\text{mean}} C_{V,\text{m}}[\text{Ar}] = \tfrac{1}{3}\left(\frac{C_{V,\text{m}}}{\sigma N_A}\right)\upsilon_{\text{mean}}$$

$$= \tfrac{1}{3}\left(\frac{C_{V,\text{m}}}{\sigma N_A}\right)\times\left(\frac{8RT}{\pi M}\right)^{1/2} \ [1\text{B}.8]$$

$$= \tfrac{1}{3}\left(\frac{12.5\ \text{J K}^{-1}\ \text{mol}^{-1}}{(0.36\times10^{-18}\ \text{m}^2)\times(6.022\times10^{23}\ \text{mol}^{-1})}\right)\times\left(\frac{8\times(8.3145\ \text{J K}^{-1}\ \text{mol}^{-1})\times(298\ \text{K})}{\pi\times(39.95\times10^{-3}\ \text{kg mol}^{-1})}\right)^{1/2}$$

$$= \boxed{7.6\times10^{-3}\ \text{J K}^{-1}\ \text{m}^{-1}\ \text{s}^{-1}}$$

Comment. This calculated value does not agree well with the value of κ listed in Table 19A.1.

Question. Can the differences between the calculated and experimental values of κ be accounted for by the difference in temperature (298 K here, 273 K in Table 79.1)? If not, what might be responsible for the difference?

E19A.2(a) $D=\tfrac{1}{3}\lambda v_{\text{mean}}$ [19A.10]

$$=\tfrac{1}{3}\left(\frac{v_{\text{rel}}}{z}\right)v_{\text{mean}}\ [1\text{B}.12]=\tfrac{1}{3}\left(\frac{v_{\text{rel}}}{\sigma v_{\text{rel}}p/kT}\right)v_{\text{mean}}\ [1\text{B}.11\text{b}]$$

$$=\tfrac{1}{3}\left(\frac{kT}{\sigma p}\right)\times\left(\frac{8RT}{\pi M}\right)^{1/2}\ [1\text{B}.8]$$

$$=\tfrac{1}{3}\left(\frac{(1.381\times10^{-23}\ \text{J K}^{-1})\times(293.15\ \text{K})}{(0.36\times10^{-18}\ \text{m}^2)}\right)\times\left(\frac{8\times(8.3145\ \text{J K}^{-1}\ \text{mol}^{-1})\times(293.15\ \text{K})}{\pi\times(39.95\times10^{-3}\ \text{kg mol}^{-1})}\right)^{1/2}$$

$$\times\frac{1}{p/\text{Pa}}$$

$$=(1.4\bar{8}\ \text{m}^2\ \text{s}^{-1})\times\frac{1}{p/\text{Pa}}$$

$$\frac{J_z}{N_\text{A}}=-\frac{D}{N_\text{A}}\frac{\text{d}\mathcal{N}}{\text{d}z}[19\text{A}.3]=-\frac{D}{N_\text{A}}\frac{\text{d}}{\text{d}z}\left(\frac{N_\text{A}p}{RT}\right)$$

$$=-\left(\frac{D}{RT}\right)\frac{\text{d}p}{\text{d}z}$$

$$=-\left(\frac{1}{(8.3145\ \text{J K}^{-1}\ \text{mol}^{-1})\times(293.15\ \text{K})}\right)\times\left(\frac{1.4\bar{8}\ \text{m}^2\ \text{s}^{-1}}{p/\text{Pa}}\right)\times(1.0\times10^5\ \text{Pa m}^{-1})$$

$$=-\frac{60.\bar{7}\ \text{mol m}^{-2}\ \text{s}^{-1}}{p/\text{Pa}}$$

(a) $p=1.00$ Pa, $\boxed{D=1.5\ \text{m}^2\ \text{s}^{-1}, J_z/N_A=-61\ \text{mol m}^{-2}\ \text{s}^{-1}}$

(b) $p=100$ kPa, $\boxed{D=1.5\times10^{-5}\ \text{m}^2\ \text{s}^{-1}, J_z/N_A=-6.1\times10^{-4}\ \text{mol m}^{-2}\ \text{s}^{-1}}$

(c) $p=10.0$ MPa, $\boxed{D=1.5\times10^{-7}\ \text{m}^2\ \text{s}^{-1}, J_z/N_\text{A}=-6.1\times10^{-6}\ \text{mol m}^{-2}\ \text{s}^{-1}}$

E19A.3(a) For a perfect argon gas: $\lambda=\dfrac{kT}{\sigma p}\ [1\text{B}.13]=\dfrac{RT}{\sigma N_\text{A}p}=\dfrac{1}{\sigma N_\text{A}[\text{Ar}]}$. Thus,

$$\kappa=\tfrac{1}{3}\lambda v_{\text{mean}}C_{V,\text{m}}[\text{Ar}]\ [19\text{A}.13\text{b}]=\tfrac{1}{3}\left(\frac{1}{\sigma N_\text{A}[\text{Ar}]}\right)v_{\text{mean}}C_{V,\text{m}}[\text{Ar}]=\tfrac{1}{3}\left(\frac{C_{V,\text{m}}}{\sigma N_\text{A}}\right)v_{\text{mean}}$$

$$=\tfrac{1}{3}\left(\frac{C_{V,\text{m}}}{\sigma N_\text{A}}\right)\times\left(\frac{8RT}{\pi M}\right)^{1/2}\ [1\text{B}.8]$$

$$=\tfrac{1}{3}\left(\frac{12.5\ \text{J K}^{-1}\ \text{mol}^{-1}}{(0.36\times10^{-18}\ \text{m}^2)\times(6.022\times10^{23}\ \text{mol}^{-1})}\right)\times\left(\frac{8\times(8.3145\ \text{J K}^{-1}\ \text{mol}^{-1})\times(280\ \text{K})}{\pi\times(39.95\times10^{-3}\ \text{kg mol}^{-1})}\right)^{1/2}$$

$$=7.4\times10^{-3}\ \text{J K}^{-1}\ \text{m}^{-1}\ \text{s}^{-1}$$

$$J_z = -\kappa\frac{\mathrm{d}T}{\mathrm{d}z}[19\text{A}.4]$$
$$= -(7.4\times10^{-3}\,\text{J K}^{-1}\,\text{m}^{-1}\,\text{s}^{-1})\times(10.5\,\text{K m}^{-1})$$
$$= \boxed{-0.078\,\text{J m}^{-2}\,\text{s}^{-1}}$$

E19A.4(a) For a perfect neon gas: $C_{V,\text{m}} = C_{p,\text{m}} - R = 20.786\,\text{J K}^{-1}\,\text{mol}^{-1} - R = 12.472\,\text{J K}^{-1}\,\text{mol}^{-1}$

$$\kappa = \tfrac{1}{3}\lambda v_{\text{mean}} C_{V,\text{m}}[\text{Ne}][19\text{A}.13\text{b}] = \tfrac{1}{3}\left(\frac{1}{\sigma N_\text{A}[\text{Ne}]}\right)v_{\text{mean}} C_{V,\text{m}}[\text{Ne}] = \tfrac{1}{3}\left(\frac{C_{V,\text{m}}}{\sigma N_\text{A}}\right)v_{\text{mean}}$$
$$= \tfrac{1}{3}\left(\frac{C_{V,\text{m}}}{\sigma N_\text{A}}\right)\times\left(\frac{8RT}{\pi M}\right)^{1/2}[1\text{B}.8]$$

Solve for σ.

$$\sigma = \tfrac{1}{3}\left(\frac{C_{V,\text{m}}}{\kappa N_\text{A}}\right)\times\left(\frac{8RT}{\pi M}\right)^{1/2}$$
$$= \tfrac{1}{3}\left(\frac{12.472\,\text{J K}^{-1}\,\text{mol}^{-1}}{(46.5\times10^{-3}\,\text{J K}^{-1}\,\text{m}^{-1}\,\text{s}^{-1})\times(6.022\times10^{23}\,\text{mol}^{-1})}\right)$$
$$\times\left(\frac{8\times(8.3145\,\text{J K}^{-1}\,\text{mol}^{-1})\times(273\text{K})}{\pi\times(20.18\times10^{-3}\,\text{kg mol}^{-1})}\right)^{1/2}$$
$$= 7.95\times10^{-20}\,\text{m}^2 = \boxed{0.0795\,\text{nm}^2}$$

The value reported in Table 1B.1 is $0.24\,\text{nm}^2$.
Question. What approximations inherent in the equation used in the solution to this exercise are likely to cause a factor of 3 difference?

E19A.5(a) The thermal energy flux ('heat' flux) is described by: $J_z(\text{energy}) = -\kappa\frac{\mathrm{d}T}{\mathrm{d}z}[19\text{A}.4]$ where the negative sign indicates flow toward lower temperature. This is the rate of energy transfer per unit area. The total rate of energy transfer across area A is

$$\frac{\mathrm{d}E}{\mathrm{d}t} = AJ_z(\text{energy}) = -\kappa A\frac{\mathrm{d}T}{\mathrm{d}z}$$

To calculate the temperature gradient with the given data, we assume that the gradient is in a steady-state. Then, recognizing that temperature differences have identical magnitude in Celsius or Kelvin units,

$$\frac{\mathrm{d}T}{\mathrm{d}z} = \frac{\Delta T}{\Delta z} = \frac{\{(-15)-(28)\}\text{K}}{1.0\times10^{-2}\,\text{m}} = -4.3\times10^{3}\,\text{K m}^{-1}.$$

We now assume that the coefficient of thermal conductivity of the gas between the window panes is comparable to that of nitrogen given in Table 19A.1: $\kappa \approx 0.0240\ \mathrm{J\,K^{-1}\,m^{-1}\,s^{-1}}$. Then, the rate of outward energy transfer is

$$\frac{\mathrm{d}E}{\mathrm{d}t} \approx -\left(0.0240\,\mathrm{J K^{-1} m^{-1} s^{-1}}\right)\times\left(1.0\,\mathrm{m^2}\right)\times\left(-4.3\times10^{3}\,\mathrm{K m^{-1}}\right)$$

$$\approx 10\bar{3}\,\mathrm{J s^{-1}} \quad \text{or} \quad \boxed{10\bar{3}\,\mathrm{W}}$$

A $10\bar{3}$ W heater is needed to balance this rate of heat loss.

E19A.6(a) $\eta = \dfrac{pMD}{RT}\ [19\text{A}.16\text{c}] = \left(\dfrac{pM}{RT}\right)\times\left(\dfrac{kT}{3\sigma p}\right)\times\left(\dfrac{8RT}{\pi M}\right)^{1/2}$

$$= \left(\frac{k}{3\sigma}\right)\times\left(\frac{8MT}{\pi R}\right)^{1/2}$$

Solve for σ.

$$\sigma = \left(\frac{k}{3\eta}\right)\times\left(\frac{8MT}{\pi R}\right)^{1/2}$$

$$= \tfrac{1}{3}\left(\frac{1.381\times10^{-23}\ \mathrm{J\,K^{-1}}}{298\times10^{-6}\times10^{-1}\ \mathrm{kg\,m^{-1}\,s^{-1}}}\right)\times\left(\frac{8\times\left(20.18\times10^{-3}\,\mathrm{kg\,mol^{-1}}\right)\times\left(273\,\mathrm{K}\right)}{\pi\times\left(8.3145\,\mathrm{J\,K^{-1}\,mol^{-1}}\right)}\right)^{1/2}$$

$$= 2.01\times10^{-19}\ \mathrm{m^2} = \boxed{0.201\,\mathrm{nm^2}}$$

E19A.7(a) $\eta = \dfrac{pMD}{RT}[19\text{A}.16\text{c}] = \left(\dfrac{pM}{RT}\right)\times\left(\dfrac{kT}{3\sigma p}\right)\times\left(\dfrac{8RT}{\pi M}\right)^{1/2}[19\text{A}.10] = \left(\dfrac{k}{3\sigma}\right)\times\left(\dfrac{8MT}{\pi R}\right)^{1/2}$

$$= \tfrac{1}{3}\left(\frac{1.381\times10^{-23}\ \mathrm{J\,K^{-1}}}{0.40\times10^{-18}\,\mathrm{m^2}}\right)\times\left(\frac{8\times\left(29.0\times10^{-3}\,\mathrm{kg\,mol^{-1}}\right)\times T}{\pi\times\left(8.3145\ \mathrm{J\,K^{-1}\ mol^{-1}}\right)}\right)^{1/2}$$

$$= (1.08\times10^{-6}\,\mathrm{kg\ m^{-1}\ s^{-1}})\times(T/\mathrm{K})^{1/2} = (1.08\times10^{-5}\,\mathrm{P})\times(T/\mathrm{K})^{1/2}$$

(a) At 273 K, $\boxed{\eta = 178\,\mu\mathrm{P}}$

(b) At 298 K, $\boxed{\eta = 186\,\mu\mathrm{P}}$

(c) At 1 000 K, $\boxed{\eta = 342\,\mu\mathrm{P}}$

E19A.8(a) $A = \left(2.5\,\mathrm{mm}\right)\times\left(3.0\,\mathrm{mm}\right) = 7.5\times10^{-6}\ \mathrm{m^2}$

The collision frequency of the Ar gas molecules with surface area A equals $Z_\mathrm{W}A$.

$$Z_\mathrm{W}A = \frac{pAN_\mathrm{A}}{(2\pi MRT)^{1/2}}[19\text{A}.17]$$

$$= \frac{\left(90\ \mathrm{Pa}\right)\times\left(6.022\times10^{23}\ \mathrm{mol^{-1}}\right)\times\left(7.5\times10^{-6}\ \mathrm{m^2}\right)}{\left\{2\pi\left(39.95\times10^{-3}\ \mathrm{kg\,mol^{-1}}\right)\times\left(8.3145\ \mathrm{J\ K^{-1}\ mol^{-1}}\right)\times\left(500\ \mathrm{K}\right)\right\}^{1/2}}$$

$$= 1.2\bar{6}\times10^{19}\ \mathrm{s^{-1}}$$

The number of argon molecule collisions within A in time interval t equals $Z_W At$ if p does not change significantly during the period t.

$$Z_W At = (1.2\bar{6}\times10^{19}\,\text{s}^{-1})\times(15\,\text{s}) = \boxed{1.9\times10^{20}}$$

E19A.9(a) The mass loss equals the effusion mass loss multiplied by the time period t: m_{loss} = (rate of effusion) × t × m = (rate of effusion) × t × M/N_A.

$$\begin{aligned} m_{\text{loss}} &= \left(\frac{pA_0N_A}{(2\pi MRT)^{1/2}}\right)\times\left(\frac{Mt}{N_A}\right)[19\text{A}.17] \\ &= pA_0t\times\left(\frac{M}{2\pi RT}\right)^{1/2} \\ &= (0.835\,\text{Pa})\times\{\pi(1.25\times10^{-3}\,\text{m})^2\}\times(2\times3600\,\text{s}) \\ &\quad\times\left(\frac{0.260\,\text{kg mol}^{-1}}{2\pi(8.3145\,\text{J mol}^{-1}\text{K}^{-1})\times(400\,\text{K})}\right)^{1/2} \\ &= 1.04\times10^{-4}\,\text{kg} = \boxed{104\,\text{mg}} \end{aligned}$$

Question. For the same solid shaped in the form of a sphere of radius 0.050m and suspended in a vacuum, what will be the mass loss in 2.00 h? *Hint*. Make any reasonable approximations.

E19A.10(a) The mass loss equals the effusion mass loss multiplied by the time period t: m_{loss} = (rate of effusion) × t × m = (rate of effusion) × t × M/N_A.

$$m_{\text{loss}} = \left(\frac{pA_0N_A}{(2\pi MRT)^{1/2}}\right)\times\left(\frac{Mt}{N_A}\right)[19\text{A}.17] = pA_0t\times\left(\frac{M}{2\pi RT}\right)^{1/2}$$

Solving for p gives

$$\begin{aligned} p &= \frac{m_{\text{loss}}}{A_0t}\left(\frac{2\pi RT}{M}\right)^{1/2} \\ &= \frac{285\times10^{-6}\,\text{kg}}{\{\pi(0.25\times10^{-3}\,\text{m})^2\}\times(400\,\text{s})}\left(\frac{2\pi(8.3145\,\text{J mol}^{-1}\text{K}^{-1})\times(673\,\text{K})}{0.100\,\text{kg mol}^{-1}}\right)^{1/2} \\ &= \boxed{2.15\times10^3\,\text{Pa}} \end{aligned}$$

E19A.11(a) The pressure of this exercise changes significantly during time period t so it is useful to spend a moment finding an expression for $p(t)$. Mathematically, the rate of effusion is the derivative $-\mathrm{d}N/\mathrm{d}t$. Substitution of the perfect gas law for N, $N = pVN_A/RT$ where V and T are constants, reveals that the rate of effusion can be written as $(-N_AV/RT)\mathrm{d}p/\mathrm{d}t$. This formulation of the rate of effusion, along with eqn 19A.17, is used to find $p(t)$.

$$-\left(\frac{N_A V}{RT}\right)\frac{dp}{dt}=\frac{pA_0N_A}{(2\pi MRT)^{1/2}}\ [19A.17]$$

$$\frac{dp}{dt}=-\frac{pA_0}{V}\left(\frac{RT}{2\pi M}\right)^{1/2}$$

$$\frac{dp}{p}=-\frac{dt}{\tau}\quad \text{where}\quad \tau=\frac{V}{A_0}\left(\frac{2\pi M}{RT}\right)^{1/2}$$

$$\int_{p_0}^{p}\frac{dp}{p}=-\frac{1}{\tau}\int_0^t dt\quad \text{where } p_0 \text{ is the initial pressure}$$

$$\ln\frac{p}{p_0}=-\frac{t}{\tau}\quad \text{or}\quad p(t)=p_0\mathrm{e}^{-t/\tau}$$

The nitrogen gas and carbon dioxide gas data can be used to determine the relaxation time, τ, for each.

$$\tau_{N_2}=\frac{t_{N_2}}{\ln(p_0/p)_{N_2}}=\frac{42\ \text{s}}{\ln(75/50)}=104\ \text{s}$$

$$\tau_{CO_2}=\frac{t_{CO_2}}{\ln(p_0/p)_{CO_2}}=\frac{52\ \text{s}}{\ln(75/50)}=128\ \text{s}$$

The above definition of τ shows that it is proportional to $M^{1/2}$. Since the ratio of the relaxation times cancels the constant of proportionality,

$$\left(\frac{M_{CO_2}}{M_{N_2}}\right)^{1/2}=\frac{\tau_{CO_2}}{\tau_{N_2}}$$

$$M_{CO_2}=\left(\frac{\tau_{CO_2}}{\tau_{N_2}}\right)^2 M_{N_2}$$

$$=\left(\frac{128}{104}\right)^2\times(28.02\ \text{g mol}^{-1})=\boxed{42.4\ \text{g mol}^{-1}}$$

E19A.12(a) In Exercise 19A.11(a) it is shown that

$$\ln\frac{p}{p_0}=-\frac{t}{\tau}\quad \text{or}\quad p(t)=p_0\mathrm{e}^{-t/\tau}\quad \text{where}\quad \tau=\frac{V}{A_0}\left(\frac{2\pi M}{RT}\right)^{1/2}$$

The relaxation time, τ, of oxygen is calculated with the data.

$$\tau=\left(\frac{3.0\ \text{m}^3}{\pi(0.10\times10^{-3}\ \text{m})^2}\right)\times\left\{\frac{2\pi(32.00\times10^{-3}\ \text{kg mol}^{-1})}{(8.3145\ \text{J mol}^{-1}\,\text{K}^{-1})\times(298\ \text{K})}\right\}^{1/2}$$

$$=8.6\times10^5\ \text{s}=10.\bar{0}\ \text{days}$$

The time required for the specified pressure decline is calculated with the above eqn.

$$t = \tau \ln(p_0/p) = (10.\bar{0}\ \text{days}) \times \ln(80/70) = \boxed{1.3\ \text{days}}$$

Solutions to problems

P19A.1 Viscosity is independent of pressure in simple molecular kinetic theory.

$$\eta = \frac{pMD}{RT}[\text{19A.16c}] = \left(\frac{pM}{RT}\right) \times \left(\frac{kT}{3\sigma p}\right) \times \left(\frac{8RT}{\pi M}\right)^{1/2} \quad [\text{19A.10}]$$
$$= \left(\frac{k}{3\sigma}\right) \times \left(\frac{8MT}{\pi R}\right)^{1/2}$$

Solve for σ.

$$\sigma = \left(\frac{k}{3\eta}\right) \times \left(\frac{8MT}{\pi R}\right)^{1/2}$$
$$= \tfrac{1}{3}\left(\frac{1.381\times10^{-23}\ \text{J K}^{-1}}{(10^{-7}\ \text{kg m}^{-1}\ \text{s}^{-1})\times(\eta/\mu\text{P})}\right) \times \left(\frac{8\times(17.03\times10^{-3}\ \text{kg mol}^{-1})\times\text{K}}{\pi\times(8.3145\ \text{J K}^{-1}\ \text{mol}^{-1})}\right)^{1/2} \times (T/\text{K})^{1/2}$$
$$= (3.325\times10^{-18}\ \text{m}^2)\times\frac{(T/\text{K})^{1/2}}{\eta/\mu\text{P}} = (3.325\ \text{nm}^2)\times\frac{(T/\text{K})^{1/2}}{\eta/\mu\text{P}}$$

(a) At 270 K, $\eta = 90.8\ \mu\text{P}$, $\boxed{\sigma = 0.602\ \text{nm}^2}$, $\boxed{d = (\sigma/\pi)^{1/2} = 438\ \text{pm}}$

(b) At 490 K, $\eta = 174.9\ \mu\text{P}$, $\boxed{\sigma = 0.421\ \text{nm}^2}$, $\boxed{d = (\sigma/\pi)^{1/2} = 366\ \text{pm}}$

The smaller molecular diameter at higher temperature is consistent with the idea that, at higher temperatures, more forceful collisions contract a molecule's perimeter.

P19A.3 $\lambda = kT/(\sigma p)$ [1B.13] or, since $T/p = 1/k\mathcal{N}$ where $\mathcal{N}$ is the number density, $\lambda = 1/\sigma\mathcal{N}$

$$D = \tfrac{1}{3}\lambda v_{\text{mean}}\ [\text{19A.10}]$$
$$= \tfrac{1}{3}\left(\frac{1}{\sigma\mathcal{N}}\right)v_{\text{mean}}[\text{1B.13}] = \tfrac{1}{3}\left(\frac{1}{\pi(2a_0)^2\mathcal{N}}\right)\times\left(\frac{8RT}{\pi M}\right)^{1/2} \quad [\text{1B.8}]$$
$$= \tfrac{1}{3}\left(\frac{1}{\pi\times(2\times5.29\times10^{-11}\ \text{m})^2\times(1.0\times10^{6}\ \text{m}^{-3})}\right)$$
$$\times\left(\frac{8\times(8.3145\ \text{J K}^{-1}\ \text{mol}^{-1})\times(1.00\times10^4\ \text{K})}{\pi\times(1.0079\times10^{-3}\ \text{kg mol}^{-1})}\right)^{1/2}$$
$$= \boxed{2.37\times10^{17}\ \text{m}^2\ \text{s}^{-1}}$$

$$\kappa = \tfrac{1}{3}\lambda v_{\text{mean}} C_{V,\text{m}}[\text{H}]\ [19\text{A}.13\text{b}] = \tfrac{1}{3}\left(\frac{1}{\sigma N_\text{A}[\text{H}]}\right) v_{\text{mean}} C_{V,\text{m}}[\text{H}] = \tfrac{1}{3}\left(\frac{C_{V,\text{m}}}{\sigma N_\text{A}}\right) v_{\text{mean}}$$

$$= \tfrac{1}{3}\left(\frac{\tfrac{3}{2}R}{\sigma N_\text{A}}\right) \times \left(\frac{8RT}{\pi M}\right)^{1/2} [\text{1B.8 and equipartition theorem}]$$

$$= \tfrac{1}{2}\left(\frac{k}{\pi(2a_0)^2}\right) \times \left(\frac{8RT}{\pi M}\right)^{1/2}$$

$$= \tfrac{1}{2}\left(\frac{1.381\times10^{-23}\,\text{J K}^{-1}}{\pi\times(2\times5.29\times10^{-11}\,\text{m})^2}\right) \times \left(\frac{8\times(8.3145\,\text{J K}^{-1}\,\text{mol}^{-1})\times(1.00\times10^{4}\,\text{K})}{\pi\times(1.0079\times10^{-3}\,\text{kg mol}^{-1})}\right)^{1/2}$$

$$= \boxed{2.85\,\text{J K}^{-1}\,\text{m}^{-1}\,\text{s}^{-1}}$$

The validity of these calculations is in doubt because the kinetic theory of gases assumes the Maxwell–Boltzmann distribution, essentially an equilibrium distribution. In such a dilute medium, the timescales on which particles exchange energy by collision make an assumption of equilibrium unwarranted. It is especially dubious considering that atoms are more likely to interact with photons from stellar radiation than with other atoms.

P19A.5 The atomic current is the number of atoms emerging from the slit per second, which is $Z_\text{W}A$ with $A = 1.0\times10^{-7}\ \text{m}^2$. We use

$$Z_\text{W} = \frac{p}{(2\pi mkT)^{1/2}}\ [19\text{A}.6]$$

$$= \frac{p/\text{Pa}}{\left[(2\pi)\times(M/\text{g mol}^{-1})\times(1.6605\times10^{-27}\,\text{kg})\times(1.381\times10^{-23}\,\text{J K}^{-1})\times(380\,\text{K})\right]^{1/2}}$$

$$= (1.35\times10^{23}\,\text{m}^{-2}\,\text{s}^{-1}) \times \left(\frac{p/\text{Pa}}{(M/\text{g mol}^{-1})^{1/2}}\right)$$

(a) Cadmium:

$$Z_\text{W}A = (1.35\times10^{23}\,\text{m}^{-2}\,\text{s}^{-1})\times(1.0\times10^{-7}\,\text{m}^2)\times\left(\frac{0.13}{(112.4)^{1/2}}\right) = \boxed{1.7\times10^{14}\,\text{s}^{-1}}$$

(b) Mercury:

$$Z_\text{W}A = (1.35\times10^{23}\,\text{m}^{-2}\,\text{s}^{-1})\times(1.0\times10^{-7}\,\text{m}^2)\times\left(\frac{12}{(200.6)^{1/2}}\right) = \boxed{1.1\times10^{16}\,\text{s}^{-1}}$$

19B Motion in liquids

Answers to discussion questions

D19B.1 The **ionic radius**, as assigned according to the distances between ions in a crystal, is a measure of ion size. The **hydrodynamic radius** (or **Stokes radius**) of an ion is its effective radius in solution taking into account all the water molecules it carries in its **hydration shell**. A hydrodynamic radius of a small ion is typically much larger than the ionic radius. This happens because small ions give rise to stronger electric fields than large ones so the small ions are more extensively solvated than big ones. Thus, an ion of small ionic radius may have a large hydrodynamic radius because it drags many solvent molecules through the solution as it migrates.

D19B.3 Equation 19B.12 indicates that, for the passage of a solvated ion sphere through a continuous media, ion mobility is inversely proportional to viscosity. To test the thought that this transport mechanism explains observed differences between proton mobility in water and in liquid ammonia, we look up the respective viscosities and find that the viscosity of ammonia is 0.276×10^{-3} kg m^{-1} s^{-1} at $-40\,°$C and the viscosity of water is 0.891×10^{-3} kg m^{-1} s^{-1} at $25\,°$C. This mechanism predicts a larger proton mobility in ammonia, the reverse of fact. This, and the fact that proton mobility in water is extraordinarily large, forces us to look for another mechanism of proton transport. Thus, we turn to text Fig. 19B.2.

Text Fig. 19B.2 is a schematic illustration of fast proton transfer along a chain of hydrogen bonded water molecules. Each step along the chain of events has the basic form illustrated in Fig. 19B.1.

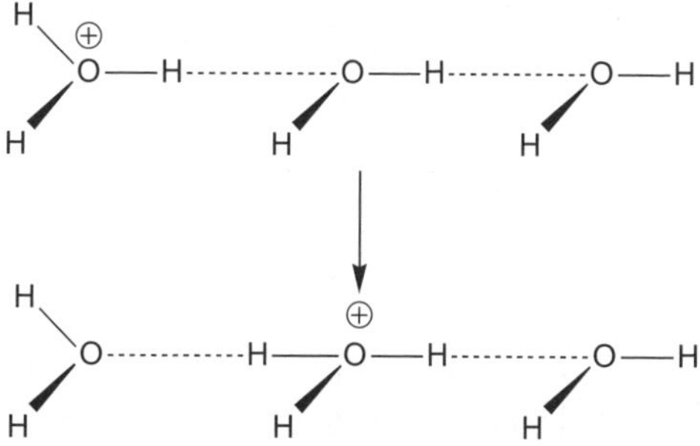

Figure 19B.1

There is a strong hydrogen bond in liquid water and the speed of proton transfer suggests a low activation energy for the proton jump. This activation energy is readily available via molecule thermal motions. Proton mobility in liquid ammonia may occur by a similar mechanism but, whereas the protonated species in water is viewed as the simplified H_3O^+ species, the protonated species in ammonia is NH_4^+. The hydrogen bond between ammonia molecules is weaker than the hydrogen bond in water but the lower charge mobility suggests a larger ratio of activation energy to available thermal energy for the proton jump. The collision frequency is comparable between water and ammonia; however, fewer of these thermal collisions have the requisite energy to promote the proton jump in liquid ammonia.

Solutions to exercises

E19B.1(a) We take the natural logarithm of eqn 19B.2 and solve for the activation energy, E_a.

$$\eta = \eta_0 e^{E_a/RT} \quad [19B.2]$$
$$\ln\eta = \ln\eta_0 + E_a/RT$$
$$\ln\eta_{T_1} - \ln\eta_{T_2} = \frac{E_a}{R}\times\left(\frac{1}{T_1}-\frac{1}{T_2}\right)$$

Therefore,

$$E_a = \frac{R\ln(\eta_{T_1}/\eta_{T_2})}{\left(\frac{1}{T_1}-\frac{1}{T_2}\right)}$$
$$= \frac{(8.3145\,\mathrm{J\,K^{-1}\,mol^{-1}})\times\ln(1.002/0.7975)}{\left(\frac{1}{293\,\mathrm{K}}-\frac{1}{303\,\mathrm{K}}\right)} = \boxed{16.8\,\mathrm{J\,mol^{-1}}}$$

E19B.2(a) Molar ionic conductivity is related to mobility by

$$\lambda = zuF \quad [19B.13]$$
$$= 1\times(7.91\times10^{-8}\,\mathrm{m^2\,s^{-1}\,V^{-1}})\times(96\,485\,\mathrm{C\,mol^{-1}})$$
$$= \boxed{7.63\times10^{-3}\,\mathrm{S\,m^2\,mol^{-1}}}$$

E19B.3(a) $s = u\mathcal{E}$ [19B.11] and $\mathcal{E} = \frac{\Delta\phi}{l}$ [19B.7]

Therefore,

$$s = u\left(\frac{\Delta\phi}{l}\right)$$
$$= (7.92\times10^{-8}\,\mathrm{m^2\,s^{-1}\,V^{-1}})\times\left(\frac{25.0\,\mathrm{V}}{7.00\times10^{-3}\,\mathrm{m}}\right)$$
$$= 2.83\times10^{-4}\,\mathrm{m\,s^{-1}} \quad \text{or} \quad \boxed{283\,\mathrm{\mu m\,s^{-1}}}$$

E19B.4(a) The basis for the solution is the law of independent migration of ions (eqn 19B.6). Switching counterions does not affect the mobility of the remaining other ion at infinite dilution.

$$\Lambda_m^{\circ} = \nu_+\lambda_+ + \nu_-\lambda_- \quad [19B.6]$$

$$\Lambda_m^{\circ}(\mathrm{NaI}) = \lambda(\mathrm{Na^+}) + \lambda(\mathrm{I^-}) = 12.69\,\mathrm{mS\,m^2\,mol^{-1}}$$

$$\Lambda_m^\circ(\text{NaNO}_3)=\lambda(\text{Na}^+)+\lambda(\text{NO}_3^-)=12.16\text{ mS m}^2\text{ mol}^{-1}$$

$$\Lambda_m^\circ(\text{AgNO}_3)=\lambda(\text{Ag}^+)+\lambda(\text{NO}_3^-)=13.34\text{ mS m}^2\text{ mol}^{-1}$$

Hence,

$$\Lambda_m^\circ(\text{AgI})=\Lambda_m^\circ(\text{AgNO}_3)+\Lambda_m^\circ(\text{NaI})-\Lambda_m^\circ(\text{NaNO}_3)$$
$$=(13.34+12.69-12.16)\text{mS m}^2\text{ mol}^{-1}=\boxed{13.87\text{ mS m}^2\text{ mol}^{-1}}$$

Question. How well does this result agree with the value calculated directly from the data of Table 19B.2?

E19B.5(a) $u=\dfrac{\lambda}{zF}$ [19B.13]; $z=1$; $1\text{S}=1\Omega^{-1}=1\text{C V}^{-1}\text{ s}^{-1}$

$$u(\text{Li}^+)=\frac{3.87\text{ mS m}^2\text{ mol}^{-1}}{9.6485\times10^4\text{ C mol}^{-1}}=4.01\times10^{-5}\text{ mS C}^{-1}\text{ m}^2=\boxed{4.01\times10^{-8}\text{ m}^2\text{ V}^{-1}\text{ s}^{-1}}$$

$$u(\text{Na}^+)=\frac{5.01\text{ mS m}^2\text{ mol}^{-1}}{9.6485\times10^4\text{ C mol}^{-1}}=\boxed{5.19\times10^{-8}\text{ m}^2\text{ V}^{-1}\text{ s}^{-1}}$$

$$u(\text{K}^+)=\frac{7.35\text{ mS m}^2\text{ mol}^{-1}}{9.6485\times10^4\text{ C mol}^{-1}}=\boxed{7.62\times10^{-8}\text{ m}^2\text{ V}^{-1}\text{ s}^{-1}}$$

E19B.6(a) $a=\dfrac{kT}{6\pi\eta D}$ [19B.19b]; $1\text{ P}=10^{-1}\text{ kg m}^{-1}\text{ s}^{-1}$

$$a=\frac{(1.381\times10^{-23}\text{ J K}^{-1})\times(298\text{ K})}{6\pi\times(1.00\times10^{-3}\text{ kg m}^{-1}\text{ s}^{-1})\times(5.2\times10^{-10}\text{ m}^2\text{ s}^{-1})}=4.2\times10^{-10}\text{ m}\quad\text{or}\quad\boxed{420\text{ pm}}$$

E19B.7(a) $D=\dfrac{uRT}{zF}$ [19B.16]; $z=1$; $1\text{ C V}=1\text{ J}$

$$D=\frac{(7.40\times10^{-8}\text{ m}^2\text{ V}^{-1}\text{ s}^{-1})\times(8.3145\text{ J K}^{-1}\text{ mol}^{-1})\times(298\text{ K})}{1\times9.6485\times10^4\text{ C mol}^{-1}}=\boxed{1.90\times10^{-9}\text{ m}^2\text{ s}^{-1}}$$

Solutions to problems

P19B.1 We take the natural logarithm of eqn 19B.2

$$\eta\propto e^{E_a/RT}\ [19B.2]$$
$$\ln\eta=\text{constant}+E_a/RT$$

and recognize that a plot of ln η against $1/T$ has a slope equal to E_a/R. Thus, a linear regression fit of ln η against $1/T$, shown in Fig. 19B.2, yields the slope from which we calculate E_a with the expression $E_a=\text{slope}\times R$.

$$E_a=(1\,222\text{ K})\times(8.3145\text{ J K}^{-1}\text{ mol}^{-1})=\boxed{10.2\text{ kJ mol}^{-1}}$$

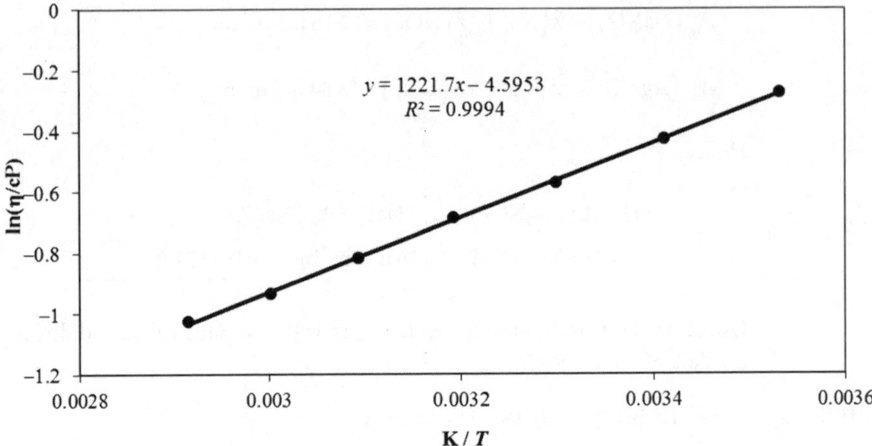

Figure 19B.2

P19B.3 The molar conductivity, Λ_m, is related to the conductivity, κ, by $\Lambda_m = \kappa/c$ [19B.4] and the **Kohlrausch law** [19B.5] indicates that molar conductivity is linear in $c^{1/2}$.

$$\Lambda_m = \Lambda_m^\circ - \mathcal{K}c^{1/2} \text{ [19B.5]}$$

We draw a data table and calculate values for a plot of Λ_m against $c^{1/2}$:

$c/(\text{mol dm}^{-3})$	1.334	1.432	1.529	1.672	1.725
$\chi/(\text{mS cm}^{-1})$	131	139	147	156	164
$c^{1/2}/(\text{mol dm}^{-3})^{1/2}$	1.155	1.197	1.237	1.293	1.313
$\Lambda_m/(\text{mS m}^2\text{ mol}^{-1})$	9.82	9.71	9.61	9.33	9.51

The plot, shown in Fig. 19B.3, is linear and the linear regression fit yields the Kohlrausch parameters:

$$\Lambda_m^\circ = \boxed{12.78 \text{ mS m}^2 \text{ mol}^{-1}}$$

$$\mathcal{K} = \boxed{2.57 \text{ mS m}^2 (\text{mol dm}^{-1})^{-3/2}}$$

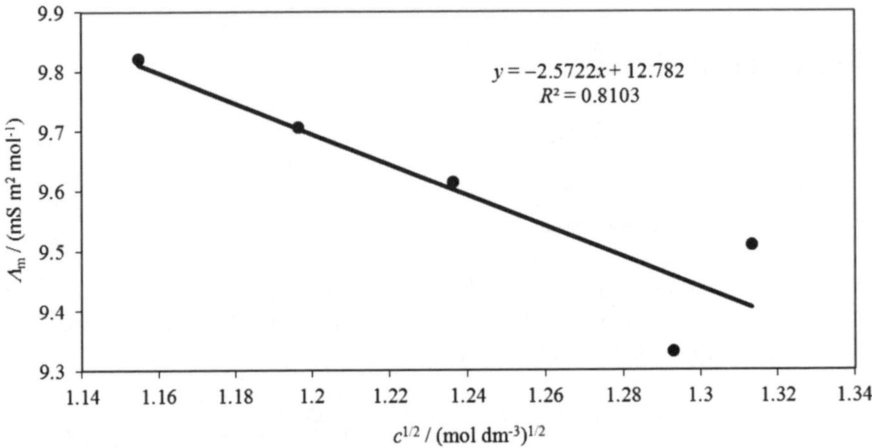

Figure 19B.3

P19B.5 The molar conductivity, Λ_m, is related to the conductivity, κ, by $\Lambda_m = \kappa/c$ [19B.4] = C/Rc where the cell constant is $C = 0.2063\ \text{cm}^{-1}$. The Kohlrausch law [19B.5] indicates that molar conductivity is linear in $c^{1/2}$.

$$\Lambda_m = \Lambda_m{}^\circ - \mathcal{K}c^{1/2}\ [19B.5]$$

We draw a data table and calculate values for a plot of Λ_m against $c^{1/2}$:

$c/(\text{mol dm}^{-3})$	0.00050	0.0010	0.0050	0.010	0.020	0.050
R/Ω	3314	1669	342.1	174.1	89.08	37.14
$c^{1/2}/(\text{mol dm}^{-3})^{1/2}$	0.0224	0.0316	0.0707	0.100	0.141	0.224
$\Lambda_m/(\text{mS m}^2\ \text{mol}^{-1})$	12.45	12.36	12.06	11.85	11.58	11.11

The plot, shown in Fig. 19B.4, is linear and the linear regression fit yields the intercept and slope. The intercept is the limiting molar conductivity and the slope is the negative of the Kohlrausch parameter K:

$$\Lambda_m^\circ = \boxed{12.6\ \text{mS m}^2\ \text{mol}^{-1}}$$

$$\mathcal{K} = \boxed{6.66\ \text{mS m}^2\ (\text{mol dm}^{-1})^{-3/2}}$$

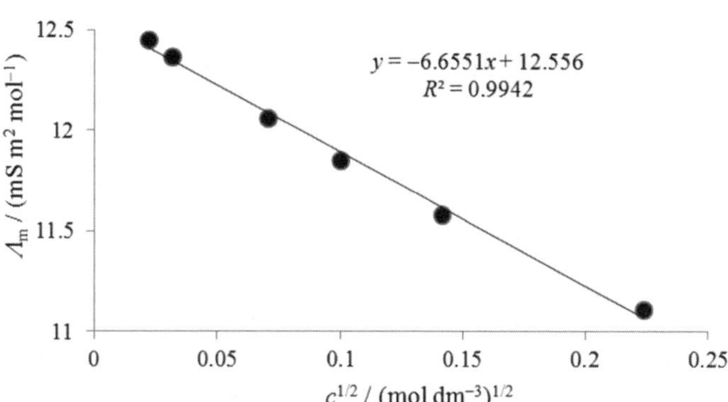

Figure 19B.4

(a) $\Lambda_m = \left((5.01 + 7.68) - 6.66 \times (0.010)^{1/2}\right) \text{mS m}^2\ \text{mol}^{-1}$
$= \boxed{12.02\ \text{mS m}^2\ \text{mol}^{-1}}$

(b) $\kappa = c\Lambda_m = (10\ \text{mol m}^{-3}) \times (12.02\ \text{mS m}^2\ \text{mol}^{-1}) = 120\ \text{mS m}^2\ \text{m}^{-3} = \boxed{120\ \text{mS m}^{-1}}$

(c) $R = \dfrac{C}{\kappa} = \dfrac{20.63\text{m}^{-1}}{120\ \text{mS m}^{-1}} = \boxed{172\Omega}$

P19B.7 The diffusion constant of a C_{60}^- ion in solution is related to the mobility of the ion and to its hydrodynamic radius a in eqn 19B.12, the Stokes formula:

$$a=\frac{ze}{6\pi\eta u}\,[19B.12]$$
$$=\frac{(1)\times(1.602\times10^{-19}\,\text{C})}{6\pi(0.93\times10^{-3}\,\text{kg m}^{-1}\,\text{s}^{-1})\times(1.1\times10^{-8}\,\text{m}^2\,\text{V}^{-1}\,\text{s}^{-1})}=8.3\times10^{-10}\,\text{m}=\boxed{0.83\,\text{nm}}$$

This is substantially larger than the 0.5 nm van der Waals radius of a Buckminsterfullerene (C_{60}) molecule because the anion attracts a considerable hydration shell through the London dispersion attraction to the nonpolar solvent molecules and through the ion–induced dipole interaction. The Stokes radius reflects the larger effective radius of the combined anion and its solvation shell.

P19B.9 Since $D\propto\eta^{-1}$ [19B.16 and 19B.12] and $\eta\propto e^{E_a/RT}$ [19B.2], we expect that $D\propto e^{-E_a/RT}$

Therefore, we take the ratio of D at two different temperatures so as to eliminate the constant of proportionality, and solve for E_a to find:

$$E_a=-\frac{R\ln(D_{T_1}/D_{T_2})}{\left(\frac{1}{T_1}-\frac{1}{T_2}\right)}=-\frac{(8.3145\,\text{J K}^{-1}\text{mol}^{-1})\times\ln(2.89/2.05)}{\frac{1}{298\,\text{K}}-\frac{1}{273\,\text{K}}}=9.3\,\text{kJ mol}^{-1}$$

That is, the activation energy for diffusion is $\boxed{9.3\,\text{kJ mol}^{-1}}$

19C Diffusion

Answer to discussion question

D19C.1 The **thermodynamic force** F is defined in eqn 19C.1.

$$\mathcal{F}=-\left(\frac{\partial\mu}{\partial x}\right)_{p,T}\quad[19C.1]$$

This expression is a summary of the second thermodynamic law that molecules move in the direction that minimizes the chemical potential of the molecules when p and T are local constants. F is not one of the 'real' forces such as gravity or electromagnetism; it is the negative gradient of the chemical potential, which has a balance of terms involving enthalpy and entropy:

$$\mu_J=\left(\frac{\partial G}{\partial n_J}\right)_{p,T,n'}[5A.4]=\left(\frac{\partial}{\partial n_J}\right)_{p,T,n'}(H-TS)$$

Thus, the thermodynamic force moves molecules so as to minimize the enthalpy, to which molecular interactions provide a great contribution, while simultaneously attempting to maximize entropy. Often times one, or the other, of these tendencies predominate. For an ideal solution the gradient of the molar enthalpy is zero so the force represents the spontaneous tendency for molecules to disperse so that entropy is maximized.

Solutions to exercises

E19C.1(a) Equation 19C.14, $\langle x^2\rangle = 2Dt$, gives the mean square distance travelled in any one dimension in time t. We need the distance travelled from a point in any direction. The distinction here is the distinction between the one-dimensional and three-dimensional diffusion. The mean square three dimensional distance can be obtained from the one dimensional mean square distance since motions in the three directions are independent.

$$r^2 = x^2 + y^2 + z^2 \text{[Pythagorean theorem]},$$

$$\begin{aligned}\langle r^2\rangle &= \langle x^2\rangle + \langle y^2\rangle + \langle z^2\rangle = 3\langle x^2\rangle \text{[independent motion]}\\ &= 3\times 2Dt \text{ [19C.14 for } \langle x^2\rangle]\\ &= 6Dt\end{aligned}$$

Therefore, $t = \dfrac{\langle r^2\rangle}{6D} = \dfrac{(5.0\times10^{-3}\,\text{m})^2}{(6)\times(6.73\times10^{-10}\,\text{m}^2\text{s}^{-1})} = \boxed{6.2\times10^3\,\text{s}}$

E19C.2(a) The diffusion equation solution for these boundary conditions is provided in eqn 19C.11 with

$$n_0 = (20.0\text{ g})\times\left(\frac{1\text{ mol sucrose}}{342.30\text{g}}\right) = 5.84\times10^{-2}\text{ mol sucrose},$$

$A = 5.0\text{ cm}^2$, $D = 5.216\times10^{-9}\text{ m}^2\text{ s}^{-1}$, and $x = 10$ cm.

$$c(x,t) = \frac{n_0}{A(\pi Dt)^{1/2}}\text{e}^{-x^2/4Dt} \quad [19\text{C}.11]$$

$$\begin{aligned}c(10\text{ cm},t) &= \frac{5.84\times10^{-2}\text{ mol}}{(5.0\times10^{-4}\text{ m}^2)\times\{\pi(5.216\times10^{-9}\text{ m}^2\text{ s}^{-1})\}^{1/2}t^{1/2}}\text{e}^{-(0.10\text{ m})^2/4\times(5.216\times10^{-9}\text{ m}^2\text{ s}^{-1})\times t}\\ &= (9.12\times10^5\text{ mol m}^{-3})\times(t/\text{s})^{-1/2}\text{e}^{-(4.79\times10^5)/(t/\text{s})}\end{aligned}$$

(a) $t = 10$ s:

$$c(10\text{ cm},10\text{ s}) = (9.12\times10^5\text{ mol m}^{-3})\times(10)^{-1/2}\text{e}^{-(4.79\times10^5)/(10)} = \boxed{0.00\text{ mol dm}^{-3}}$$

(b) $t = 24\times3\,600\text{ s} = 8.64\times10^4$ s:

$$\begin{aligned}c(10\text{ cm}, 8.64\times10^4\text{ s}) &= (9.12\times10^5\text{ mol m}^{-3})\times(8.64\times10^4)^{-1/2}\text{e}^{-(4.79\times10^5)/(8.64\times10^4)}\\ &= 12.1\text{ mol m}^{-3} = \boxed{0.0121\text{ mol dm}^{-3}}\end{aligned}$$

E19C.3(a) The linear decay in concentration c has the form $c(x)=c_0-Ax$ where c_0 is the concentration at $x=0$ and the constant A is determined by the condition that $c(10\text{ cm})=\tfrac{1}{2}c_0$.

$$A=\frac{c_0-c(x)}{x}=\frac{c_0-\tfrac{1}{2}c_0}{10\text{ cm}}=(0.050\text{ cm}^{-1})\times c_0=(5.0\text{ m}^{-1})\times c_0$$

The thermodynamic force is determined with eqn 19C.2b.

$$\mathcal{F}(x)=-\frac{RT}{c}\frac{\text{d}c}{\text{d}x}\ [19\text{C}.2\text{b}]=-\frac{RT}{c}\frac{\text{d}}{\text{d}x}(c_0-Ax)=\frac{ART}{c}$$
$$=\frac{(5.0\text{m}^{-1})\times(8.3145\text{ J K}^{-1}\text{ mol}^{-1})\times(298\text{K})}{1-(0.050\text{ cm}^{-1})\times x}$$
$$=\frac{12.\bar{4}\text{ kN mol}^{-1}}{1-(0.050\text{ cm}^{-1})\times x}$$

$$\mathcal{F}(10\text{ cm})=\frac{12.\bar{4}\text{ kN mol}^{-1}}{1-(0.050\text{ cm}^{-1})\times(10\text{ cm})}=\boxed{25\text{ kN mol}^{-1}}$$

$\mathcal{F}(20\text{ cm})=\boxed{\infty}$ because $c=0$ when $x\geq 20$ cm.

E19C.4(a) The Gaussian decay in concentration c has the form $c(x)=c_0\text{e}^{-Ax^2}$ where c_0 is the concentration at $x=0$ and the constant A is determined by the condition that $c(5\text{ cm})=\tfrac{1}{2}c_0$.

$$A=\frac{\ln(c_0/c)}{x^2}=\frac{\ln 2}{(5\text{ cm})^2}=(0.040\text{ cm}^{-2})\times\ln 2=(4.00\text{ cm}^{-1}\text{ m}^{-1})\times\ln 2$$

The thermodynamic force is determined with eqn 19C.2b.

$$\mathcal{F}(x)=-\frac{RT}{c}\frac{\text{d}c}{\text{d}x}[19\text{C}.2\text{b}]=-\frac{RT}{c}\frac{\text{d}}{\text{d}x}c_0\text{e}^{-Ax^2}=2ARTx$$
$$=2\times(4.00\text{m}^{-1})\times\ln 2\times(8.3145\text{JK}^{-1}\text{ mol}^{-1})\times(293\text{K})\times(x/\text{cm})$$
$$=(13.5\text{ kN mol}^{-1})\times(x/\text{cm})$$

$$\mathcal{F}(5\text{ cm})=(13.5\text{ kN mol}^{-1})\times(5)=\boxed{67.5\text{ kN mol}^{-1}}$$

E19C.5(a) Equation 19C.14, $\langle x^2\rangle=2Dt$, gives the mean square distance travelled in any one dimension in time t. We need the distance travelled from a point in any direction. The distinction here is the distinction between the one-dimensional and three dimensional diffusion. The mean square three dimensional distance can be obtained from the one dimensional mean square distance since motions in the three directions are independent. Since $r^2=x^2+y^2+z^2$ [Pythagorean theorem],

$$\langle r^2\rangle = \langle x^2\rangle + \langle y^2\rangle + \langle z^2\rangle = 3\langle x^2\rangle \text{[independent motion]}$$

$$= 3\times 2Dt\,[19\text{C}.14] = 6Dt.$$

Therefore, $t = \dfrac{\langle r^2\rangle}{6D} = \dfrac{(5.0\times 10^{-3}\,\text{m})^2}{6\times(3.17\times 10^{-9}\,\text{m}^2\,\text{s}^{-1})} = \boxed{1.3\times 10^3\,\text{s}}$

E19C.6(a) $a = \dfrac{kT}{6\pi\eta D}$[19B.19b]; $1\,\text{P} = 10^{-1}\,\text{kg}\,\text{m}^{-1}\,\text{s}^{-1}$

$$a = \frac{(1.381\times 10^{-23}\,\text{J}\,\text{K}^{-1})\times(298\,\text{K})}{6\pi\times(1.00\times 10^{-3}\,\text{kg}\,\text{m}^{-1}\,\text{s}^{-1})\times(5.2\times 10^{-10}\,\text{m}^2\,\text{s}^{-1})} = 4.2\times 10^{-10}\,\text{m} = \boxed{0.42\,\text{nm}}$$

E19C.7(a) The Einstein–Smoluchowski equation [19C.16] relates the diffusion constant to the jump distance λ and time τ required for a jump.

$$D = \frac{\lambda^2}{2\tau}\,[19\text{C}.16] \quad \text{so} \quad \tau = \frac{\lambda^2}{2D}$$

If the jump distance is about one molecular diameter, or two effective molecular radii, then the jump distance can be obtained by use of the Stokes–Einstein equation[19B.19b].

$$\lambda = 2a = 2\left(\frac{kT}{6\pi\eta D}\right)[19\text{B}.19\text{b}] = \frac{kT}{3\pi\eta D}$$

$$\tau = \frac{1}{2D}\times\left(\frac{kT}{3\pi\eta D}\right)^2$$

$$= \frac{1}{2\times(2.13\times 10^{-9}\,\text{m}^2\,\text{s}^{-1})}\times\left\{\frac{(1.381\times 10^{-23}\,\text{J}\text{K}^{-1})\times(298\text{K})}{3\pi\times(0.601\times 10^{-3}\,\text{kg}\text{m}^{-1}\,\text{s}^{-1})\times(2.13\times 10^{-9}\,\text{m}^2\,\text{s}^{-1})}\right\}^2 \quad \text{[Table19B.1]}$$

$$= 2.73\times 10^{-11}\,\text{s} = \boxed{27.3\,\text{ps}}$$

Comment. In the strictest sense we are dealing with three dimensional diffusion. However, since we are assuming that only one jump occurs, it is probably an adequate approximation to use an equation derived for one dimensional diffusion. For three dimensional diffusion the equation analogous to eqn 19C.16 is $\tau = \lambda^2/6D$.

Question. Can you derive the equation? Use an analysis similar to that described in the solution to Exercise 19C.5(a).

E19C.8(a) For three-dimensional diffusion we use an equation analogous to eqn 19C.16 derived in Exercise 19C.5(a).

$$\langle r^2\rangle = 6Dt$$

For iodine in benzene [data from Table 19B.3]

$$\langle r^2\rangle^{1/2} = [6\times(2.13\times10^{-9}\ \mathrm{m^2\,s^{-1}})\times(1.0\ \mathrm{s})]^{1/2} = \boxed{113\ \mu\mathrm{m}}$$

For sucrose in water

$$\langle r^2\rangle^{1/2} = [6\times(0.522\times10^{-9}\ \mathrm{m^2\,s^{-1}})\times(1.0\ \mathrm{s})]^{1/2} = \boxed{56\ \mu\mathrm{m}}$$

Solutions to problems

P19C.1 $\mathcal{F} = -\dfrac{RT}{c}\times\dfrac{\mathrm{d}c}{\mathrm{d}x}$ [19C.2b]

$$\frac{\mathrm{d}c}{\mathrm{d}x} = \frac{(0.05-0.10)\ \mathrm{mol\,dm^{-3}}}{0.10\,\mathrm{m}} = -0.50\ \mathrm{mol\,dm^{-3}\,m^{-1}}\ \text{[linear gradation]}$$

$$RT = 2.48\times10^{3}\ \mathrm{J\,mol^{-1}} = 2.48\times10^{3}\ \mathrm{N\,m\,mol^{-1}}$$

(a) $$\mathcal{F} = -\left(\frac{2.48\times10^{3}\ \mathrm{N\,m\,mol^{-1}}}{0.10\ \mathrm{mol\,dm^{-3}}}\right)\times(-0.50\ \mathrm{mol\,dm^{-3}\,m^{-1}})$$

$$= \boxed{12\ \mathrm{kN\,mol^{-1}}}\ \text{or}\ \boxed{2.0\times10^{-20}\ \mathrm{N\,molecule^{-1}}}$$

(b) $$\mathcal{F} = -\left(\frac{2.48\times10^{3}\ \mathrm{N\,m\,mol^{-1}}}{0.075\ \mathrm{mol\,dm^{-3}}}\right)\times(-0.50\ \mathrm{mol\,dm^{-3}\,m^{-1}})$$

$$= \boxed{16.\bar{5}\ \mathrm{kN\,mol^{-1}}}\ \text{or}\ \boxed{2.7\times10^{-20}\ \mathrm{N\,molecule^{-1}}}$$

(c) $$\mathcal{F} = -\left(\frac{2.48\times10^{3}\ \mathrm{N\,m\,mol^{-1}}}{0.050\ \mathrm{mol\,dm^{-3}}}\right)\times(-0.50\ \mathrm{mol\,dm^{-3}\,m^{-1}})$$

$$= \boxed{24.\bar{8}\ \mathrm{kN\,mol^{-1}}}\ \text{or}\ \boxed{4.1\times10^{-20}\ \mathrm{N\,molecule^{-1}}}$$

P19C.3 $\mathcal{F} = -\dfrac{RT}{c}\times\dfrac{\mathrm{d}c}{\mathrm{d}x}$ [19 C.2b] with the axis origin at the centre of the tube

$$RT = 2.48\times10^{3}\ \mathrm{J\,mol^{-1}} = 2.48\times10^{3}\ \mathrm{N\,m\,mol^{-1}}$$

$$c = c(x) = c_0\left(1-\mathrm{e}^{-ax^2}\right)\ \text{where}\ c_0 = 0.100\ \mathrm{mol\,dm^{-3}}\ \text{and}\ a = 0.10\ \mathrm{cm^{-2}}$$

$$\frac{\mathrm{d}c}{\mathrm{d}x} = 2xac_0\mathrm{e}^{-ax^2}$$

$$\mathcal{F} = 2aRTx\mathrm{e}^{-ax^2}\left(1-\mathrm{e}^{-ax^2}\right)^{-1} = 2aRTx\left(1-\mathrm{e}^{-ax^2}\right)^{-1} = \left(50\ \mathrm{kN\,cm^{-1}\,mol^{-1}}\right)x\left(1-\mathrm{e}^{ax^2}\right)^{-1}$$

A plot of the force per mole against x is shown in Fig. 19C.1. It demonstrates that mass is pushed by the thermodynamic force toward the centre of the tube where the concentration is lowest; a negative force pushes toward the left, positive force pushes toward the right. The force per molecule is calculated with the equation:

$$\mathrm{F} = 2aRTx\left(1-\mathrm{e}^{ax^2}\right)^{-1}/N_\mathrm{A} = \left(8.2\times10^{-23}\ \mathrm{kN\,cm^{-1}\ molecule^{-1}}\right)x\left(1-\mathrm{e}^{ax^2}\right)^{-1}$$

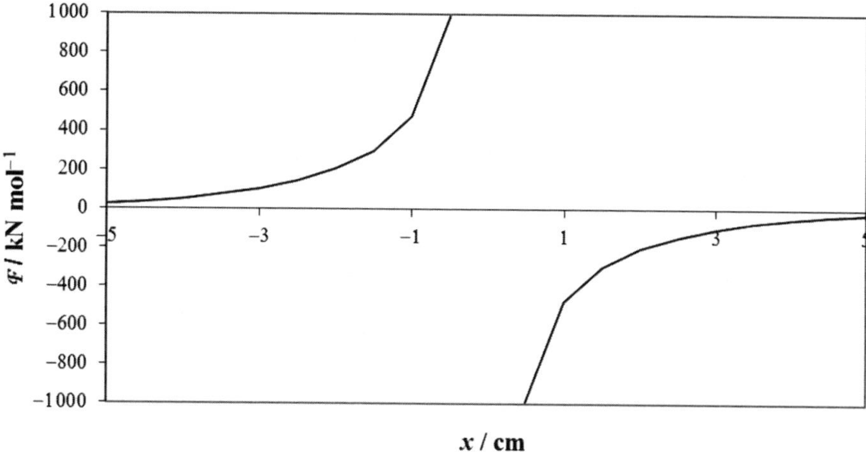

Figure 19C.1

P19C.5 The diffusion equation is: $\dfrac{\partial c}{\partial t} = D\dfrac{\partial^2 c}{\partial x^2}$ [19C.7].

We confirm that $c(x,t) = \dfrac{a}{t^{1/2}}\mathrm{e}^{-bx^2/t}$ [19C.11], where $a = \dfrac{n_0}{A(\pi D)^{1/2}}$ and $b = \frac{1}{4D}$, is a solution to the diffusion equation by taking both the partial derivative w/r/t time and the second partial w/r/t position to find whether they are proportional.

$$\frac{\partial c}{\partial t} = -\left(\frac{1}{2}\right)\times\left(\frac{a}{t^{3/2}}\right)\mathrm{e}^{-bx^2/t} + \left(\frac{a}{t^{1/2}}\right)\times\left(\frac{bx^2}{t^2}\right)\mathrm{e}^{-bx^2/t} = -\frac{c}{2t} + \frac{bx^2}{t^2}c$$

$$\frac{\partial c}{\partial x} = \left(\frac{a}{t^{1/2}}\right)\times\left(\frac{-2bx}{t}\right)\mathrm{e}^{-bx^2/t}$$

$$\frac{\partial^2 c}{\partial x^2} = -\left(\frac{2b}{t}\right)\times\left(\frac{a}{t^{1/2}}\right)\mathrm{e}^{-bx^2/t} + \left(\frac{a}{t^{1/2}}\right)\times\left(\frac{2bx}{t}\right)^2\mathrm{e}^{-bx^2/t} = -\left(\frac{2b}{t}\right)c + \left(\frac{2bx}{t}\right)^2 c$$

$$= -\left(\frac{1}{2Dt}\right)c + \left(\frac{bx^2}{Dt^2}\right)c$$

$$= \frac{1}{D}\frac{\partial c}{\partial t}\ \text{as required.}$$

Initially the material is concentrated at $x=0$. Note that $c=0$ for $x>0$ when $t=0$ on account of the very strong exponential factor $\left(\mathrm{e}^{-bx^2/t} \to 0 \text{ more strongly than } \frac{1}{t^{1/2}} \to \infty\right)$. When $x=0$, $\mathrm{e}^{-x^2/4Dt}=1$. We confirm the correct behaviour by noting that $\langle x\rangle = 0$ and $\langle x^2\rangle = 0$ at $t=0$ [19C.13 and 19C.14], and so all the material must be at $x=0$ at $t=0$.

P19C.7 As discussed in Example 19C.1 of the text, the probability density that a diffusing molecule has travelled the distance x from the origin in time t is given by the expression

$$P(x)=\frac{1}{(\pi Dt)^{1/2}}\mathrm{e}^{-x^2/4Dt}$$

Thus, to find the mean value of any function of x, we needed only weigh the function of x with $P(x)$ and integrate over all possible x.

$$\begin{aligned}\langle x^2\rangle &=\frac{1}{(\pi Dt)^{1/2}}\int_0^\infty x^2\mathrm{e}^{-x^2/4Dt}\,\mathrm{d}x\\ &=\frac{1}{(\pi Dt)^{1/2}}\times\{2\sqrt{\pi}(Dt)^{3/2}\}\ \text{[standard integral]}\\ &=2Dt\end{aligned}$$

$$\begin{aligned}\langle x^4\rangle &=\frac{1}{(\pi Dt)^{1/2}}\int_0^\infty x^4\mathrm{e}^{-x^2/4Dt}\,\mathrm{d}x\\ &=\frac{1}{(\pi Dt)^{1/2}}\times\{12\sqrt{\pi}(Dt)^{5/2}\}\ \text{[standard integral]}\\ &=12(Dt)^2\end{aligned}$$

Thus, $\langle x^4\rangle/\langle x^2\rangle^2=3$ or $\boxed{\langle x^4\rangle^{1/4}/\langle x^2\rangle^{1/2}=3^{1/4}}$.

P19C.9 $P(n\lambda)=N!/(N-N_R)!N_R!2^N$

Since $N=N_R+N_L$ and $n=N_R-N_L$, we make the substitution $N_R=(N+n)/2$ to arrive at the working equation:

$$P(n)=\frac{N!}{\left(\frac{N-n}{2}\right)!\left(\frac{N+n}{2}\right)!\,2^N}$$

(a) The pair $N=4$ and $n=6$ is physically impossible so $P(6)=\boxed{0}$.

(b) $N=6$ and $n=6$: $P(6)=\dfrac{6!}{\left(\frac{6-6}{2}\right)!\left(\frac{6+6}{2}\right)!2^6}=\boxed{0.0156}$

(c) $N=12$ and $n=6$: $P(6)=\dfrac{12!}{\left(\frac{12-6}{2}\right)!\left(\frac{12+6}{2}\right)!2^{12}}=\boxed{0.0537}$

P19C.11 Using the definitions $N=N_R+N_L$ and $n=N_R-N_L$, solve for N_R and N_L to find

$$N_R=½(N+n)\quad\text{and}\quad N_L=½(N-n)$$

Following the discussion of *Justification* 19C.2, we then have

$$\begin{aligned}P(n\lambda)&=\frac{\text{number of paths with } N_R \text{ steps to the right}}{\text{total number of paths}}=\frac{N!}{(N_L)!N_R!2^N}\\ &=\frac{N!}{\{½(N+n)\}!\{½(N-n)\}!2^N}\end{aligned}$$

The intermediate mathematical manipulations of *Justification* 19C.2 begin with the above expression. Simplification of the expression proceeds by taking the natural logarithm of the expression, applying Stirling's approximation to each term that has the ln(x!) form, checking for term cancellations, and simplification using basic logarithm properties.

Stirling's approximation: $\ln x! = \ln(2\pi)^{1/2} + (x+\tfrac{1}{2})\ln x - x$

Basic logarithm properties: $\ln(x \times y) = \ln x + \ln y$

$\ln(x/y) = \ln x - \ln y$

$\ln(x^y) = y\ln x$

Taking the natural logarithm and applying Stirling's formula gives

$$\begin{aligned}\ln P &= \ln\left\{\frac{N!}{\{\tfrac{1}{2}(N+n)\}!\{\tfrac{1}{2}(N-n)\}!2^N}\right\}\\ &= \ln N! - \ln(\{\tfrac{1}{2}(N+n)\}!) - \ln(\{\tfrac{1}{2}(N-n)\}!) - \ln 2^N\\ &= \cancel{\ln(2\pi)^{1/2}} + (N+\tfrac{1}{2})\ln N - \cancel{N}\\ &\quad -\left[\cancel{\ln(2\pi)^{1/2}} + \{\tfrac{1}{2}(N+n)+\tfrac{1}{2}\}\ln\{\tfrac{1}{2}(N+n)\} - \tfrac{1}{2}(\cancel{N}+\cancel{n})\right]\\ &\quad -\left[\ln(2\pi)^{1/2} + \{\tfrac{1}{2}(N-n)+\tfrac{1}{2}\}\ln\{\tfrac{1}{2}(N-n)\} - \tfrac{1}{2}(\cancel{N}-\cancel{n})\right] - \ln 2^N\end{aligned}$$

$$\begin{aligned}\ln P &= (N+\tfrac{1}{2})\ln N - \ln(2\pi)^{1/2} - \ln 2^N\\ &\quad -\{\tfrac{1}{2}(N+n)+\tfrac{1}{2}\}\ln\{\tfrac{1}{2}(N+n)\} - \{\tfrac{1}{2}(N-n)+\tfrac{1}{2}\}\ln\{\tfrac{1}{2}(N-n)\}\\ &= \ln\left\{\frac{(N/2)^{N+\frac{1}{2}}}{\pi^{1/2}}\right\} - \tfrac{1}{2}\{N+n+1\}\ln\left\{\frac{N}{2}\left(1+\frac{n}{N}\right)\right\} - \tfrac{1}{2}\{N-n+1\}\ln\left\{\frac{N}{2}\left(1-\frac{n}{N}\right)\right\}\\ &= \ln\left\{\frac{(N/2)^{N+\frac{1}{2}}}{\pi^{1/2}}\right\} - \tfrac{1}{2}\{N+n+1\}\left\{\ln\left(\frac{N}{2}\right)+\ln\left(1+\frac{n}{N}\right)\right\} - \tfrac{1}{2}\{N-n+1\}\left\{\ln\left(\frac{N}{2}\right)+\ln\left(1-\frac{n}{N}\right)\right\}\\ &= \ln\left\{\frac{(N/2)^{N+\frac{1}{2}}}{\pi^{1/2}}\right\} - \tfrac{1}{2}\{N+\cancel{n}+1\}\ln\left(\frac{N}{2}\right) - \tfrac{1}{2}\{N+n+1\}\ln\left(1+\frac{n}{N}\right)\\ &\quad -\tfrac{1}{2}\{N-\cancel{n}+1\}\ln\left(\frac{N}{2}\right) - \tfrac{1}{2}\{N-n+1\}\ln\left(1-\frac{n}{N}\right)\\ &= \ln\left\{\frac{(N/2)^{N+\frac{1}{2}}}{\pi^{1/2}}\right\} - \{N+1\}\ln\left(\frac{N}{2}\right) - \tfrac{1}{2}\{N+n+1\}\ln\left(1+\frac{n}{N}\right) - \tfrac{1}{2}\{N-n+1\}\ln\left(1-\frac{n}{N}\right)\\ &= \ln\left\{\frac{(N/2)^{N+\frac{1}{2}}}{(N/2)^{N+1}\pi^{1/2}}\right\} - \tfrac{1}{2}\{N+n+1\}\ln\left(1+\frac{n}{N}\right) - \tfrac{1}{2}\{N-n+1\}\ln\left(1-\frac{n}{N}\right)\\ \ln P &= \ln\left(\frac{2}{\pi N}\right)^{\frac{1}{2}} - \tfrac{1}{2}\{N+n+1\}\ln\left(1+\frac{n}{N}\right) - \tfrac{1}{2}\{N-n+1\}\ln\left(1-\frac{n}{N}\right)\end{aligned}$$

We expect that for a reasonably large value of N the number of right steps should nearly equal the number of left steps so the value of n/N should be much less than 1. Thus, it is appropriate to expand $\ln\left(1\pm\frac{n}{N}\right)$ in a Taylor series around $n/N = 0$ and discard the third and higher order terms.

$$\ln\left(1\pm\frac{n}{N}\right)=\pm\frac{n}{N}-½\left(\frac{n}{N}\right)^2 \quad \text{for} \quad \frac{n}{N}\ll 1$$

$$\ln P=\ln\left(\frac{2}{\pi N}\right)^{½}-½\{N+n+1\}\left(\frac{n}{N}-½\left(\frac{n}{N}\right)^2\right)-½\{N-n+1\}\left(-\frac{n}{N}-½\left(\frac{n}{N}\right)^2\right)$$

$$=\ln\left(\frac{2}{\pi N}\right)^{½}-½\left(\frac{n}{N}\right)^2(N-1)$$

$$=\ln\left(\frac{2}{\pi N}\right)^{½}-½\left(\frac{n^2}{N}\right) \quad \text{for } N\gg 1$$

Or, $\boxed{P=\left(\frac{2}{\pi N}\right)^{½}\mathrm{e}^{-n^2/2N}}$ where $x=n\lambda$ because $n\equiv N_\mathrm{R}-N_\mathrm{L}$

P19C.13 $c(x,t)=c_0+(c_\mathrm{s}-c_0)\{1-\mathrm{erf}(\xi)\}$ where $\xi(x,t)=x/(4Dt)^{1/2}$ and $\mathrm{erf}(\xi)=1-\frac{2}{\pi^{1/2}}\int_{\xi}^{\infty}e^{y^2}\,\mathrm{d}y$

In order for $c(x,t)$ to be the correct solution of this diffusion problem it must satisfy the boundary condition, the initial condition, and the diffusion equation (eqn 19C.7). At the boundary $x=0$, $\xi=0$, and $\mathrm{erf}(0)=1-\frac{2}{\pi^{1/2}}\int_0^{\infty}e^{-y^2}\,\mathrm{d}y=1-\left(\frac{2}{\pi^{1/2}}\right)\times\left(\frac{\pi^{1/2}}{2}\right)=0$

Thus, $c(0,t)=c_0+(c_\mathrm{s}-c_0)\{1-0\}=c_\mathrm{s}$. The boundary condition is satisfied. At the initial time $(t=0)$, $\xi(x,0)=\infty$ and $\mathrm{erf}(\infty)=1$. Thus, $c(x,0)=c_0+(c_\mathrm{s}-c_0)\{1-1\}=c_0$. The initial condition is satisfied. We must find the analytical forms for $\partial c/\partial t$ and $\partial^2 c/\partial x^2$. If they are proportional with a constant of proportionality equal to D, then $c(x,t)$ satisfies the diffusion equation.

$$\frac{\partial c(x,t)}{\partial x}=D\left[\frac{1}{2}\frac{(c_\mathrm{s}-c_0)x}{\sqrt{\pi}(Dt)^{3/2}}\mathrm{e}^{-x^2/4Dt}\right]$$

$$\frac{\partial^2 c(x,t)}{\partial x^2}=\left[\frac{1}{2}\frac{(c_\mathrm{s}-c_0)x}{\sqrt{\pi}(Dt)^{3/2}}\mathrm{e}^{-x^2/4Dt}\right]$$

The constant of proportionality between the partials equals D and we conclude that the suggested solution satisfies the diffusion equation. Diffusion through alveoli sites (about 1 cell thick) of oxygen and carbon dioxide between lungs and blood capillaries (also about 1 cell thick) occurs through about 0.075 mm (the diameter of a red blood cell). So we will examine diffusion profiles for $0\le x\le 0.1$ mm. The largest distance suggests that the longest time that must be examined is estimated with eqn 19C.13.

$$t_\mathrm{max}\simeq\frac{\pi x_\mathrm{max}^2}{4D}=\frac{\pi(1\times10^{-4}\,\mathrm{m})^2}{4(2.10\times10^{-9}\,\mathrm{m^2\,s^{-1}})}=3.74\,\mathrm{s}$$

Figure 19C.2 shows oxygen concentration distributions for times between 0.01 s and 4.0 s. We set c_0 equal to zero and calculate c_s with Henry's law [5A.23].

$$b_{\mathrm{O_2}}=\frac{p_{\mathrm{O_2}}}{K_{\mathrm{O_2}}}[5\mathrm{A}.23]=\frac{21\,\mathrm{kPa}}{7.9\times10^4\,\mathrm{kPa\,kg\,mol^{-1}}}[\text{Table 5A.1}]=2.9\times10^{-4}\,\mathrm{mol\,kg^{-1}}$$

So, $c_\mathrm{s}=2.9\times10^{-4}$ mol dm^{-3}

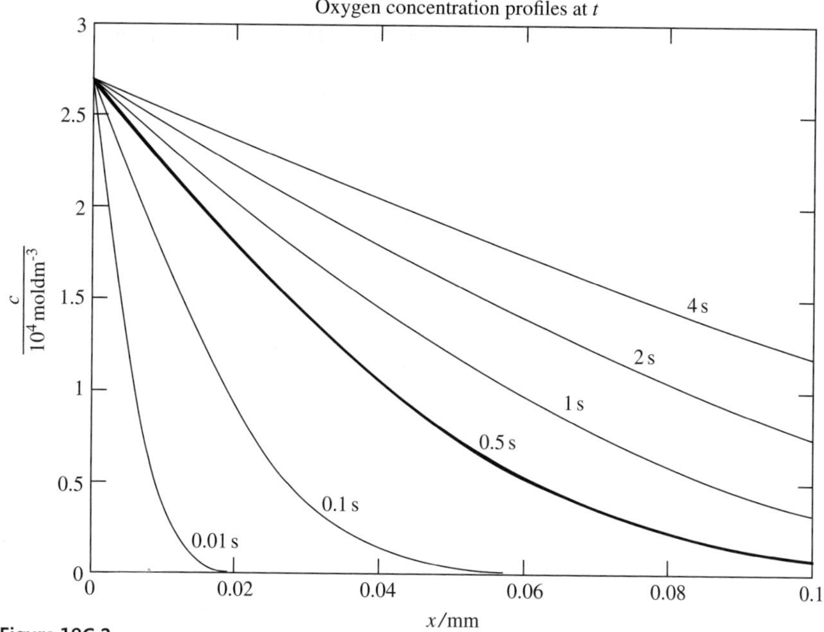

Figure 19C.2

Integrated activities

I19.1 Here's a Mathcad Prime 2 worksheet for generating the curves.

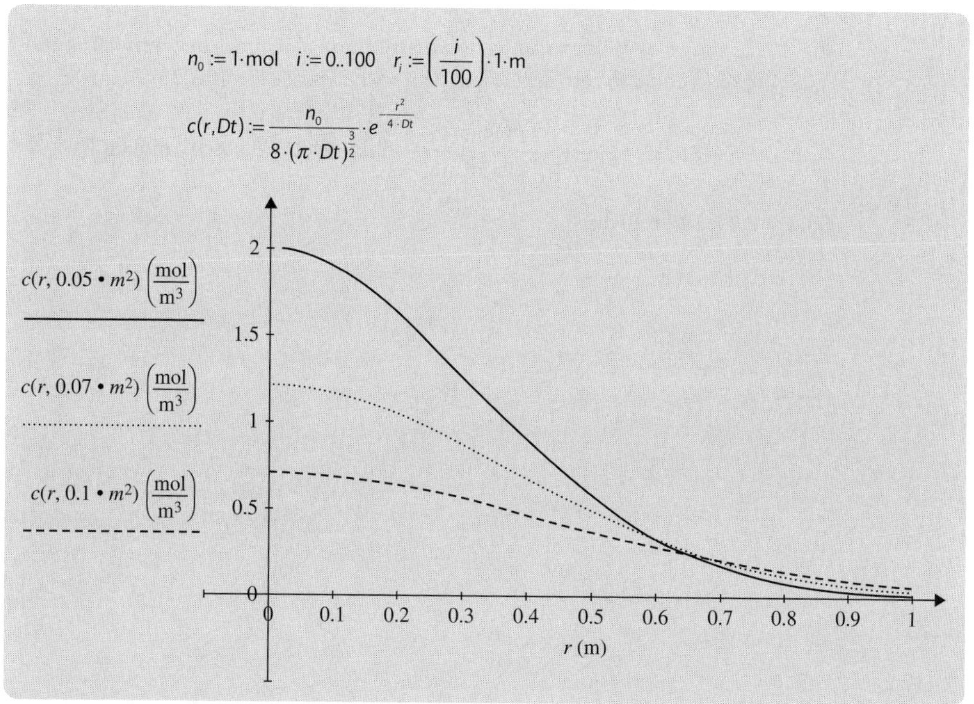

I19.3 $c(x,t)=c_0+(c_s-c_0)\{1-\text{erf}(\xi)\}$ where $\xi(x,t)=x/(4Dt)^{1/2}$ and $\text{erf}(\xi)=1-\dfrac{2}{\pi^{1/2}}\int_\xi^\infty e^{y^2}\,dy$

In order for $c(x,t)$ to be the correct solution of this diffusion problem it must satisfy the boundary condition, the initial condition, and the diffusion equation (eqn 19C.7). At the boundary $x=0$, $\xi=0$, and $\text{erf}(0)=1-\dfrac{2}{\pi^{1/2}}\int_0^\infty e^{-y^2}\,dy=1-\left(\dfrac{2}{\pi^{1/2}}\right)\times\left(\dfrac{\pi^{1/2}}{2}\right)=0$ Thus, $c(0,t)=c_0+(c_s-c_0)\{1-0\}=c_s$. The boundary condition is satisfied. At the initial time $(t=0)$, $\xi(x,0)=\infty$ and $\text{erf}(\infty)=1$. Thus, $c(x,0)=c_0+(c_s-c_0)\{1-1\}=c_0$. The initial condition is satisfied. We must find the analytical forms for $\partial c/\partial t$ and $\partial^2 c/\partial x^2$. If they are proportional with a constant of proportionality equal to D, then $c(x,t)$ satisfies the diffusion equation.

$$\frac{\partial c(x,t)}{\partial x}=D\left[\frac{1}{2}\frac{(c_s-c_0)x}{\sqrt{\pi}(Dt)^{3/2}}e^{-x^2/4Dt}\right]$$

$$\frac{\partial^2 c(x,t)}{\partial x^2}=\left[\frac{1}{2}\frac{(c_s-c_0)x}{\sqrt{\pi}(Dt)^{3/2}}e^{-x^2/4Dt}\right]$$

The constant of proportionality between the partials equals D and we conclude that the suggested solution satisfies the diffusion equation. Diffusion through alveoli sites (about 1 cell thick) of oxygen and carbon dioxide between lungs and blood capillaries (also about 1 cell thick) occurs through about 0.075 mm (the diameter of a red blood cell). So we will examine diffusion profiles for $0\le x\le 0.1$ mm. The largest distance suggests that the longest time that must be examined is estimated with eqn 19C.13.

$$t_{\text{max}}\simeq\frac{\pi x_{\text{max}}^2}{4D}=\frac{\pi(1\times10^{-4}\text{ m})^2}{4(2.10\times10^{-9}\text{ m}^2\text{ s}^{-1})}=3.74\text{ s}$$

Fig. 19C.2 shows oxygen concentration distributions for times between 0.01 s and 4.0 s. We set c_0 equal to zero and calculate c_s with Henry's law [5A.23].

$$b_{O_2}=\frac{p_{O_2}}{K_{O_2}}[5A.23]=\frac{21\text{ kPa}}{7.9\times10^4\text{ kPa kg mol}^{-1}}[\text{Table 5A.1}]=2.9\times10^{-4}\text{ mol kg}^{-1}$$

So, $c_s=2.9\times10^{-4}$ mol dm^{-3}

20 Chemical kinetics

20A The rates of chemical reactions

Answer to discussion questions

D20A.1 As far as this topic is concerned, the order of a reaction with respect to a species (a product or a reactant) is the power to which the concentration of that species is raised in a power-law type of rate law (eqn 20A.7). The **overall order** of a reaction is the sum of the orders with respect to all of the species that appear in the rate law. Reaction orders are empirical quantities, and they need not be integers or half-integers. In Topic 20E, we will see that integer reaction orders are common, though, and we will see how they come about.

The relevance of a pseudofirst-order reaction or of any pseudo reaction order for that matter, is not so much what the order appears to be but rather the strategy involved that allows for the empirical determination of the rate law. In the technique called the **isolation method**, reaction conditions can often be manipulated such that the concentrations of all reactants but one are in such excess that their concentrations can be considered constant over the course of a kinetics experiment. In that case, a rate law of the form

$$v = k_r[\mathrm{A}]^a[\mathrm{B}]^b\ldots$$

would appear to simplify to

$$v = k_r'[\mathrm{A}]^a$$

where the apparent rate constant, k_r' includes the actual rate constant and nearly constant concentration terms:

$$k_r' = k_r[\mathrm{B}]_0^b\ldots$$

A pseudofirst-order reaction, then, is a reaction that is first-order with respect to one reactant run under conditions such that the overall order of the reaction appears to be first order. (That is, in the example above, $a = 1$.)

Topic 20B examines the mathematical properties of various rate laws in some detail. We will consider only reactions whose rates depend on the concentration of a single reactant (so that the overall order is the order with respect to that one reactant).

In a first-order reaction, the rate of reaction is directly proportional to the concentration of the reactant:

$$v = k_r[A] \quad \text{and} \quad \ln[A] = \ln[A]_0 - k_r t [20B.1]$$

A plot of the logarithm of reactant concentration against time is a straight line.

In a zero-order reaction, the rate of reaction is constant, independent of the reactant concentration:

$$v = k_r \quad \text{and} \quad [A] = [A]_0 - k_r t$$

A plot of the reactant concentration itself against time is a straight line.

In a second-order reaction, the rate of reaction is proportional to the square of the reactant concentration:

$$v = k_r[A]^2 \quad \text{and} \quad \frac{1}{[A]} = \frac{1}{[A]_0} - k_r t [20B.4]$$

A plot of the reciprocal of reactant concentration against time is a straight line.

D20A.3 Because rate laws are differential equations, we must integrate them if we want to find the concentrations as a function of time. Even the most complex rate laws may be integrated numerically. However, in a number of simple cases analytical solutions are easily obtained and prove to be very useful. In order to determine the rate law, one plots the left hand side of the integrated rate laws shown in eqns 20B.1 and 20B.4, for example, against t in order to see which of them results in a straight line through the origin. The one that does is the correct rate law.

Solutions to exercises

E20A.1(a) Let the initial amount of ICl be n_{ICl} and the initial amount of H_2 be n_H; the initial amounts of I_2 and HCl are assumed to be zero. Thus, the initial total quantity of gas is $n_{ICl} + n_H$. Let the amount of I_2 formed at any given time be n. In that case, the amount of HCl is $2n$, that of H_2 is $n_H - n$, and the amount of ICl is $n_{ICl} - 2n$. At any given time, then, the total quantity of gas is

$$n_{total} = n_{ICl} - 2n + n_H - n + n + 2n = n_{ICl} + n_H = n_{initial}$$

Thus there is no change in the amount of gas during the course of the reaction. Since there is no change in volume or temperature either, there is $\boxed{\text{no change in pressure}}$

Comment. Measuring the pressure would **not** be a practical way of monitoring the progress of this reaction.

E20A.2(a) $$v = \frac{1}{\nu_J}\frac{d[J]}{dt} \ [20A.3b] \quad \text{so} \quad \frac{d[J]}{dt} = \nu_J v$$

Rate of formation of $C = 3v = \boxed{8.1 \text{ mol dm}^{-3}\,\text{s}^{-1}}$

Rate of formation of $D = v = \boxed{2.7 \text{ mol dm}^{-3}\,\text{s}^{-1}}$

Rate of consumption of A $= v = \boxed{2.7\ \text{mol dm}^{-3}\,\text{s}^{-1}}$

Rate of consumption of B $= 2v = \boxed{5.4\ \text{mol dm}^{-3}\,\text{s}^{-1}}$

E20A.3(a) $v = \dfrac{1}{\nu_\text{J}}\dfrac{\text{d[J]}}{\text{d}t}\ [\text{20A.3b}] = \dfrac{1}{2}\dfrac{\text{d[C]}}{\text{d}t} = \tfrac{1}{2}\times(2.7\,\text{mol dm}^{-3}\,\text{s}^{-1}) = \boxed{1.3\bar{5}\,\text{mol dm}^{-3}\,\text{s}^{-1}}$

Rate of formation of D $= 3v = \boxed{4.0\bar{5}\ \text{mol dm}^{-3}\,\text{s}^{-1}}$

Rate of consumption of A $= 2v = \boxed{2.7\ \text{mol dm}^{-3}\,\text{s}^{-1}}$

Rate of consumption of B $= v = \boxed{1.3\bar{5}\ \text{mol dm}^{-3}\,\text{s}^{-1}}$

E20A.4(a) The rate is expressed in mol dm^{-3} s^{-1}; therefore

$$\text{mol dm}^{-3}\,\text{s}^{-1} = [k_\text{r}]\times(\text{mol dm}^{-3})\times(\text{mol dm}^{-3}),$$

where $[k_\text{r}]$ denotes units of k_r, requires the units to be $\boxed{\text{dm}^3\,\text{mol}^{-1}\,\text{s}^{-1}}$

(i) Rate of consumption of A $= v = \boxed{k_\text{r}[\text{A}][\text{B}]}$

(ii) Rate of formation of C $= 3v = \boxed{3k_\text{r}[\text{A}][\text{B}]}$

E20A.5(a) Given $\dfrac{\text{d[C]}}{\text{d}t} = k_\text{r}[\text{A}][\text{B}][\text{C}]$,

the rate of reaction is [20A.3b]

$$v = \frac{1}{\nu_\text{J}}\frac{\text{d[J]}}{\text{d}t} = \frac{1}{2}\frac{\text{d[C]}}{\text{d}t} = \boxed{\frac{1}{2}k_\text{r}[\text{A}][\text{B}][\text{C}]}$$

The units of k_r, $[k_\text{r}]$, must satisfy

mol dm^{-3} s^{-1} $= [k_\text{r}] \times$ (mol dm^{-3}) $\times$ (mol dm^{-3}) $\times$ (mol dm^{-3})

Therefore, $[k_\text{r}] = \boxed{\text{dm}^6\,\text{mol}^{-2}\,\text{s}^{-1}}$

E20A.6(a) (i) For a second-order reaction, denoting the units of k_r by $[k_\text{r}]$

mol dm^{-3} s^{-1} $= [k_\text{r}] \times$ (mol dm^{-3})^2; therefore $\boxed{[k_\text{r}] = \text{dm}^3\,\text{mol}^{-1}\,\text{s}^{-1}}$

For a third-order reaction

mol dm^{-3} s^{-1} $= [k_\text{r}] \times$ (mol dm^{-3})^3; therefore $\boxed{[k_\text{r}] = \text{dm}^6\,\text{mol}^{-2}\,\text{s}^{-1}}$

(ii) For a second-order reaction

kPa s^{-1} $= [k_\text{r}] \times$ kPa^2; therefore $\boxed{[k_\text{r}] = \text{kPa}^{-1}\,\text{s}^{-1}}$

For a third-order reaction

kPa s^{-1} $= [k_\text{r}] \times$ kPa^3; therefore $\boxed{[k_\text{r}] = \text{kPa}^{-2}\,\text{s}^{-1}}$

Solutions to problems

P20A.1 The rate law is

$$v = k_\text{r}[\text{A}]^a \propto p_\text{A}^a = \{p_{\text{A},0}(1-f)\}^a$$

where f is the fraction reacted. Thus

$$\frac{v_1}{v_2}=\frac{p_{A,1}^a}{p_{A,2}^a}=\left(\frac{1-f_1}{1-f_2}\right)^a$$

Taking logarithms

$$\ln\left(\frac{v_1}{v_2}\right)=a\ln\left(\frac{1-f_1}{1-f_2}\right)$$

$$\text{so}\quad a=\frac{\ln\left(\frac{v_1}{v_2}\right)}{\ln\left(\frac{1-f_1}{1-f_2}\right)}=\frac{\ln\left(\frac{9.71}{7.67}\right)}{\ln\left(\frac{0.90}{0.80}\right)}=2.0$$

The reaction is $\boxed{\text{second order}}$.

Comment. Knowledge of the initial pressure is not required for the solution to this exercise. The ratio of pressures was computed using fractions of the initial pressure.

P20A.3 Begin with the supposition that the rate law has the form $v_0=k_r[\text{complex}]_0^a[\text{Y}]_0^b$ or, taking logarithms,

$\log(v_0)=\log(k_r)+a\log([\text{complex}]_0)+b\log([\text{Y}]_0)=\log(k_{r,\text{eff}})+a\log([\text{complex}]_0)$

where $\log(k_{r,\text{eff}})=\log(k_r)+b\log([\text{Y}]_0)$.

Thus, if the supposition is correct, a plot of $\log(v_0)$ against $\log([\text{complex}]_0)$ at fixed $[\text{Y}]_0$ will be linear with a slope equal to a and an intercept equal to $\log(k_{r,\text{eff}})$. A subsequent plot of $\log(k_{r,\text{eff}})$ against $\log([\text{Y}]_0$ will have a slope equal to b and an intercept equal to $\log(k_r)$. Plots of $\log(v_0)$ against $\log([\text{complex}]_0)$ for $[\text{Y}]_0$ = 2.7 mmol dm^{-3} and 6.1 mmol dm^{-3} are shown in Fig. 20A.1

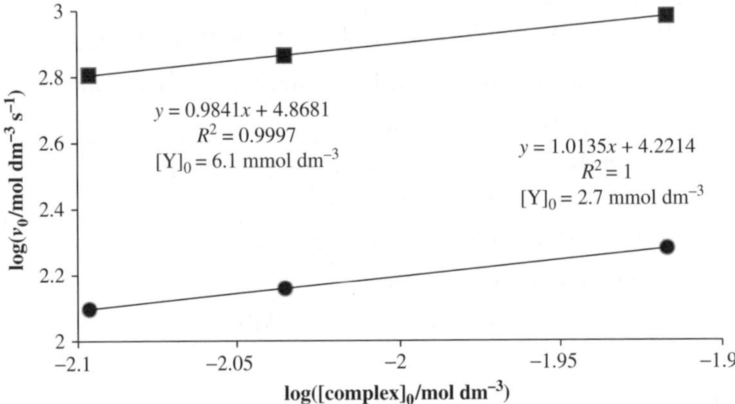

Figure 20A.1

(a) The plots of Fig. 20A.1, being linear, confirm the postulated form of the rate law and the inserts display the linear least squares regression fits of the data. Since both plots have slopes very close to 1 with small deviations due to experimental error, we conclude that the reaction order w/r/t the complex is $\boxed{1}$. Rather than preparing a two-point plot of $\log(k_{r,\text{eff}})$ against $\log([Y]_0)$ for the determination of b, we will simply use the mathematic definition of a slope and the intercepts of Fig. 20A.1 in a computation.

$$b = slope = \frac{d(\log k_{r,\text{eff}})}{d(\log[Y]_0)} = \frac{\Delta(\log k_{r,\text{eff}})}{\Delta(\log[Y]_0)} = \frac{\log k_{r,\text{eff}\,1} - \log k_{r,\text{eff}\,2}}{\log[Y]_{0,1} - \log[Y]_{0,2}} = \frac{\log k_{r,\text{eff}\,1} - \log k_{r,\text{eff}\,2}}{\log([Y]_{0,1}/[Y]_{0,2})}$$

$$= \frac{4.2214 - 4.8681}{\log(2.7/6.1)}$$

$$= 1.83$$

The computed value of b is close to 2 with deviation due to experimental error so we conclude that the reaction order w/r/t Y is $\boxed{2}$ and the overall reaction order is $\boxed{3}$.

(b) We have not prepared a two-point plot of $\log(k_{r,\text{eff}})$ against $\log([Y]_0$ so we calculate k_r using the definition that $k_r = k_{r,\text{eff}}[Y]_0^{-b}$ where $b = 2$.

(i) For the $[Y]_0 = 2.7\ \text{mmol dm}^{-3}$ data,

$$k_r = (10^{4.2214}\ \text{s}^{-1}) \times (2.7 \times 10^{-3}\ \text{mol dm}^{-3})^{-2} = 2.3 \times 10^9\ \text{mol}^{-2}\,\text{dm}^6\,\text{s}^{-1}.$$

(ii) For the $[Y]_0 = 6.1\ \text{mmol dm}^{-3}$ data,

$$k_r = (10^{4.8681}\ \text{s}^{-1}) \times (6.1 \times 10^{-3}\ \text{mol dm}^{-3})^{-2} = 2.0 \times 10^9\ \text{mol}^{-2}\,\text{dm}^6\,\text{s}^{-1}.$$

Since these values reasonably agree, we conclude that on the average $k_r = \boxed{2.2 \times 10^9\ \text{mol}^{-2}\,\text{dm}^6\,\text{s}^{-1}}$.

20B Integrated rate laws

Answer to discussion question

D20B.1 The determination of a rate law is simplified by the isolation method in which the concentrations of all the reactants except one are in large excess. If B is in large excess, for example, then to a good approximation its concentration is constant throughout the reaction. Although the true rate law might be $v = k_r[A][B]$, we can approximate [B] by $[B]_0$ and write

$$v = k_r'[A], \text{ where } k_r' = k_r[B]_0$$

which has the form of a first-order rate law. Because the true rate law has been forced into first-order form by assuming that the concentration of B is constant, it is called a pseudo first-order rate law. The dependence of the rate on the concentration of each of the reactants may be found by isolating them in turn (by having all the other substances present in large excess), and so constructing the overall rate law.

In the method of initial rates, which is often used in conjunction with the isolation method, the rate is measured at the beginning of the reaction for several different initial concentrations of reactants. We shall suppose that the rate law for a reaction with A isolated is $v = k_r[A]^a$; then its initial rate, v_0, is given by the initial values of the concentration of A, and we write $v_0 = k_r[A]_0^a$. Taking logarithms gives

$$\log v_0 = \log k_r + a\log[A]_0$$

For a series of initial concentrations, a plot of the logarithms of the initial rates against the logarithms of the initial concentrations of A should be a straight lime with slope *a*.

The method of initial rates might not reveal the full rate law, for the products may participate in the reaction and affect the rate. For example, products participate in the synthesis of HBr, where the full rate law depends on the concentration of HBr. To avoid this difficulty, the rate law should be fitted to the data throughout the reaction. The fitting may be done, in simple cases at least, by using a proposed rate law to predict the concentration of any component at any time, and comparing it with the data.

Because rate laws are differential equations, we must integrate them if we want to find the concentrations as a function of time. Even the most complex rate laws may be integrated numerically. However, in a number of simple cases analytical solutions are easily obtained and prove to be very useful. These are summarized in Table 20B.3. In order to determine the rate law, one plots the right hand side of the integrated rate laws shown in the table against *t* in order to see which of them results in a straight line through the origin. The one that does is the correct rate law.

Solutions to exercises

E20B.1(a) Table 20B.3 gives a general expression for the half-life of a reaction of the type $A \rightarrow P$ for orders other than 1:

$$t_{1/2} = \frac{2^{n-1}-1}{(n-1)k_r[A]_0^{n-1}} \propto [A]_0^{1-n} \propto p_0^{1-n}$$

where the proportionality constants may be functions of the reaction order, the rate constant, or even the temperature, but not of the concentration. Form a ratio of the half-lives at different initial pressures:

$$\frac{t_{1/2}(p_{0,1})}{t_{1/2}(p_{0,2})} = \left(\frac{p_{0,1}}{p_{0,2}}\right)^{1-n} = \left(\frac{p_{0,2}}{p_{0,1}}\right)^{n-1}$$

Hence $\ln\left(\dfrac{t_{1/2}(p_{0,1})}{t_{1/2}(p_{0,2})}\right)=(n-1)\ln\left(\dfrac{p_{0,2}}{p_{0,1}}\right)$

or $(n-1)=\dfrac{\ln\left(\dfrac{410\ \text{s}}{880\ \text{s}}\right)}{\ln\left(\dfrac{169\ \text{Torr}}{363\ \text{Torr}}\right)}=0.999\approx 1$

Therefore, $\boxed{n=2}$.

E20B.2(a) $2\,N_2O_5 \rightarrow 4\,NO_2+O_2 \qquad v=k_r[N_2O_5]$

Therefore, rate of consumption of $N_2O_5=2v=2k_r[N_2O_5]$

$$\frac{d[N_2O_5]}{dt}=-2k_r[N_2O_5] \quad \text{so} \quad [N_2O_5]=[N_2O_5]_0\,e^{-2k_rt}$$

Solve this for t:

$$t=\frac{1}{2k_r}\ln\frac{[N_2O_5]_0}{[N_2O_5]}$$

Therefore, the half-life is:

$$t_{1/2}=\frac{1}{2k_r}\ln 2=\frac{\ln 2}{(2)\times(3.38\times10^{-5}\ \text{s}^{-1})}=\boxed{1.03\times10^4\ \text{s}}$$

Since the partial pressure of N_2O_5 is proportional to its concentration

$$p(N_2O_5)=p_0(N_2O_5)e^{-2k_rt}$$

(i) $p(N_2O_5)=(500\,\text{Torr})\times\left(e^{-(2\times3.38\times10^{-5}/\text{s})\times(50\ \text{s})}\right)=\boxed{498\,\text{Torr}}$

(ii) $p(N_2O_5)=(500\,\text{Torr})\times\left(e^{-(2\times3.38\times10^{-5}/\text{s})\times(20\times60\ \text{s})}\right)=\boxed{461\,\text{Torr}}$

Comment. The half-life formula in Table 20B.3 is based on a rate constant for the rate of change of the reactant, that is, based on the assumption that

$$-\frac{d[A]}{dt}=k_r[A]$$

Our expression for the rate of consumption has $2k_r$ instead of k_r, and our expression for $t_{1/2}$ does likewise.

E20B.3(a) The integrated second-order rate law for a reaction of the type A + B → products is

$$k_rt=\frac{1}{[B]_0-[A]_0}\ln\left(\frac{[B]/[B]_0}{[A]/[A]_0}\right) \quad [20B.8]$$

Introducing $[B]=[B]_0-x$ and $[A]=[A]_0-x$ and rearranging we obtain

$$k_r t = \left(\frac{1}{[B]_0 - [A]_0}\right) \ln\left(\frac{[A]_0([B]_0 - x)}{([A]_0 - x)[B]_0}\right)$$

Solving for x yields, after some rearranging,

$$x = \frac{[A]_0[B]_0\left(e^{k_r([B]_0-[A]_0)t} - 1\right)}{[B]_0 e^{([B]_0-[A]_0)k_r t} - [A]_0} = \frac{(0.060)\times\left(0.110\,\text{mol dm}^{-3}\right)\times\left(e^{(0.110-0.060)\times 0.11\times t/\text{s}} - 1\right)}{(0.110)\times e^{(0.110-0.060)\times 0.11\times t/\text{s}} - 0.060}$$

$$= \frac{\left(0.060\,\text{mol dm}^{-3}\right)\times\left(e^{0.0055\,t/\text{s}} - 1\right)}{e^{0.0055\,t/\text{s}} - 0.55}$$

(i) After 20 s

$$x = \frac{(0.060\,\text{mol dm}^{-3})\times(e^{0.0055\times 20} - 1)}{e^{0.0055\times 20} - 0.55} = 0.0122\,\text{mol dm}^{-3}$$

which implies that

$$[CH_3COOC_2H_5] = (0.110 - 0.0122)\,\text{mol dm}^{-3} = \boxed{0.098\,\text{mol dm}^{-3}}$$

(ii) After 15 min = 900 s,

$$x = \frac{(0.060\,\text{mol dm}^{-3})\times(e^{0.0055\times 900} - 1)}{e^{0.0055\times 900} - 0.55} = 0.060\,\text{mol dm}^{-3}$$

so $[CH_3COOC_2H_5] = (0.110 - 0.060)\,\text{mol dm}^{-3} = \boxed{0.050\,\text{mol dm}^{-3}}$

E20B.4(a) The rate of consumption of A is

$$-\frac{d[A]}{dt} = 2v = 2k_r[A]^2 \quad [\nu_A = -2]$$

which integrates to $\frac{1}{[A]} - \frac{1}{[A]_0} = 2k_r t$ [20B.4 with k_r replaced by $2k_r$]

Therefore, $t = \frac{1}{2k_r}\left(\frac{1}{[A]} - \frac{1}{[A]_0}\right)$

$$t = \left(\frac{1}{2\times 4.30\times 10^{-4}\,\text{dm}^3\,\text{mol}^{-1}\,\text{s}^{-1}}\right)\times\left(\frac{1}{0.010\,\text{mol dm}^{-3}} - \frac{1}{0.210\,\text{mol dm}^{-3}}\right)$$

$$= \boxed{1.11\times 10^5\,\text{s}} = \boxed{1.28\text{ days}}$$

Solutions to problems

P20B.1 The concentration of A varies with time as

$$[A] = [A]_0 e^{-k_r t} \quad [20B.1]$$

Dimensionless concentrations $[A]/[A]_0$ and $[B]/[A]_0$ are plotted against the dimensionless time $k_r t$ in Fig. 20B.1(a). The same variables are plotted against a logarithmic horizontal axis in Fig. 20B.1(b). The latter graph shows that most of the change happens in the interval $0.3 < k_r t < 3$.

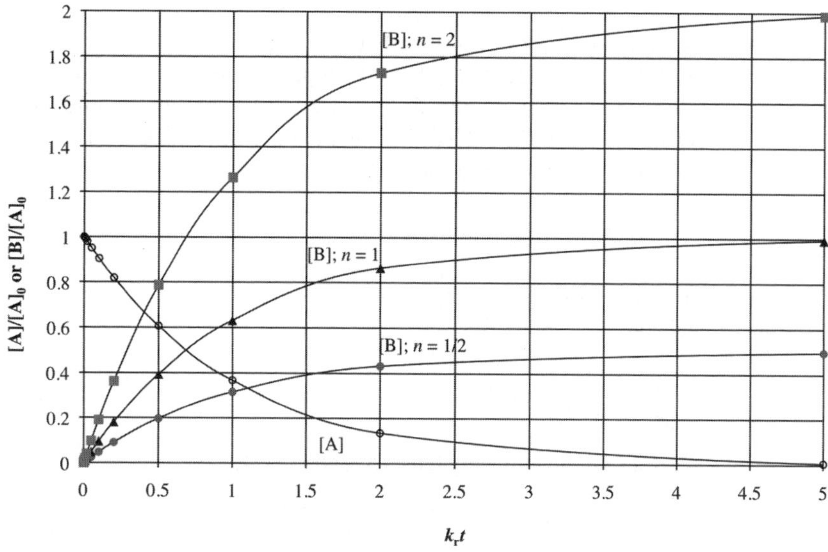

Figure 20B.1(a)

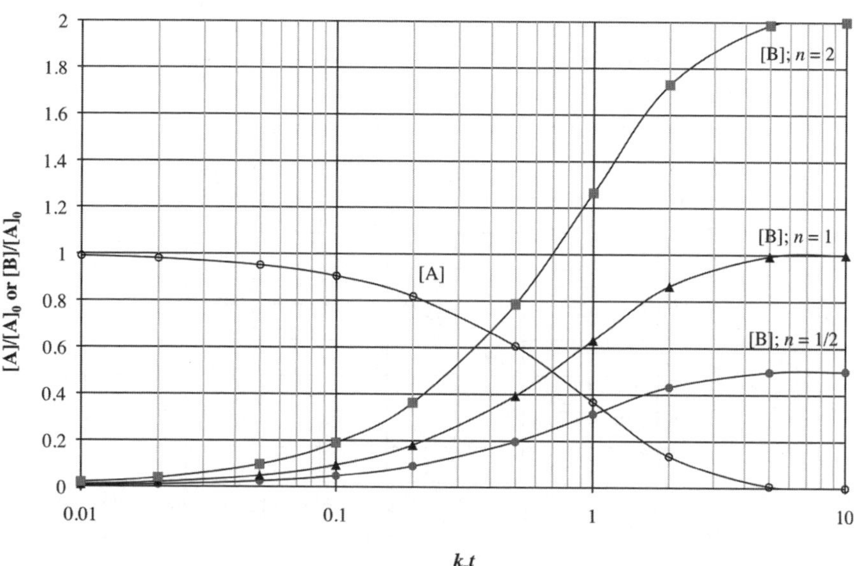

Figure 20B.1(b)

P20B.3 A simple but practical approach is to make an initial guess at the order by observing whether the half-life of the reaction appears to depend on concentration. If it does not, the reaction is first-order; if it does, refer to Table 20B.3 for an expression for the half-life of a reaction of the type $A \rightarrow P$ for orders other than 1:

$$t_{1/2} = \frac{2^{n-1}-1}{(n-1)k_r[A]_0^{n-1}} \propto [A]_0^{1-n}$$

Examination of the data shows that the first half-life is roughly 45 minutes and the second is about double the first. (Compare the 0–50 minute data to the 50–150 minute data.) That is, the half-life starting from **half** of the initial concentration is about **twice** the initial half-life, suggesting that the half-life is inversely proportional to initial concentration:

$$t_{1/2} \propto [A]_0^{-1} = [A]_0^{1-n} \text{ with } n = 2$$

Confirm this suggestion by plotting 1/[A] against time. A second-order reaction will obey

$$\frac{1}{[A]} = k_r t + \frac{1}{[A]_0} \quad [20B.4]$$

We draw up the following table ($A = NH_4CNO$)

t/min	0	20.0	50.0	65.0	150
m(urea)/g	0	7.0	12.1	13.8	17.7
m(A)/g	22.9	15.9	10.8	9.1	5.2
[A]/(mol dm^{-3})	0.381	0.265	0.180	0.152	0.0866
[A]$^{-1}$/(dm^3 mol^{-1})	2.62	3.78	5.56	6.60	11.5

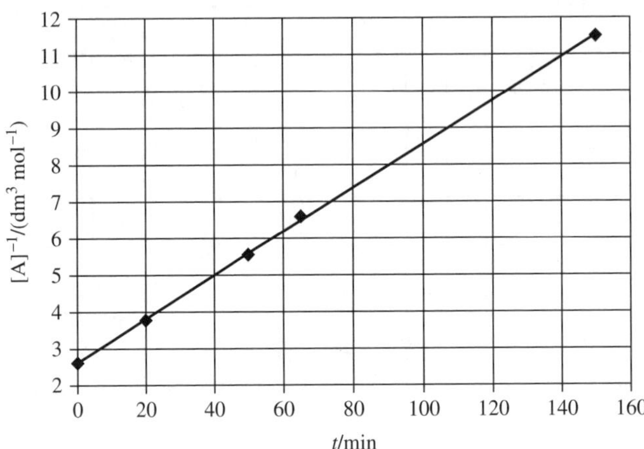

Figure 20B.2

The data are plotted in Fig. 20B.2 and fit closely to a straight line. Hence, the reaction is indeed $\boxed{\text{second-order}}$. The rate constant is the slope: $\boxed{k_r = 0.059\overline{4}\,\text{dm}^3\,\text{mol}^{-1}\,\text{min}^{-1}}$. To find [A] at 300 min, use eqn 20B.4:

$$[\mathrm{A}]=\frac{[\mathrm{A}]_0}{1+k_{\mathrm{r}}t[\mathrm{A}]_0}=\frac{0.381\,\mathrm{mol\,dm^{-3}}}{1+(0.059\bar{4})\times(300)\times(0.381)}=0.048\bar{9}\,\mathrm{mol\,dm^{-3}}$$

The mass of NH_4CNO left after 300 minutes is

$$m=(0.048\bar{9}\,\mathrm{mol\,dm^{-3}})\times(1.00\,\mathrm{dm^3})\times(60.06\,\mathrm{g\,mol^{-1}})=\boxed{2.94\ \mathrm{g}}$$

P20B.5 Use the procedure adopted in the solutions to Problems 20B.3 and 20B.4: is the half-life (or any other similarly defined 'fractional life') constant, or does it vary over the course of the reaction? The data are not quite so clear-cut. The half-life appears to be approximately constant at about 10 minutes: in the interval 0–10 the initial concentration drops by just over half, while in the interval 2–12 the concentration drops by slightly less than half. Another measure would compare the fractional consumption in two equal time intervals. The fractional consumption in the 0–2 interval is about 1/6, while that in the 10–12 interval is about 1/10—suggesting that the fractional consumption is **not** constant over the time the reaction was monitored. We draw up the following table (A = nitrile) in order to examine both first-order and second-order plots. The former (see Fig. 20B.3(a)) is a plot of $\ln\left(\frac{[\mathrm{A}]}{[\mathrm{A}]_0}\right)$ against time (eqn 20B.1); the latter (see Fig. 20B.3(b)) is a plot of $1/[\mathrm{A}]$ against time.

$t/(10^3\ \mathrm{s})$	0	2.00	4.00	6.00	8.00	10.00	12.00
$[\mathrm{A}]/(\mathrm{mol\,dm^{-3}})$	1.50	1.26	1.07	0.92	0.81	0.72	0.65
$\frac{[\mathrm{A}]}{[\mathrm{A}]_0}$	1.00	0.840	0.713	0.613	0.540	0.480	0.433
$\ln\left(\frac{[\mathrm{A}]}{[\mathrm{A}]_0}\right)$	0	−0.174	−0.338	0.489	−0.616	−0.734	−0.836
$1/[\mathrm{A}]$	0.667	0.794	0.935	1.09	1.23	1.39	1.54

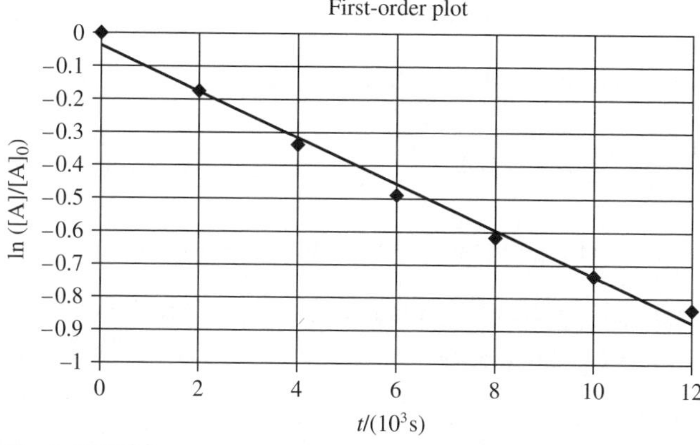

Figure 20B.3(a)

The first-order plot is not bad: the correlation coefficient is 0.991. The corresponding first-order rate constant is $k_{\mathrm{r}}=-\mathrm{slope}=\boxed{7.0\times10^{-5}\ \mathrm{s^{-1}}}$

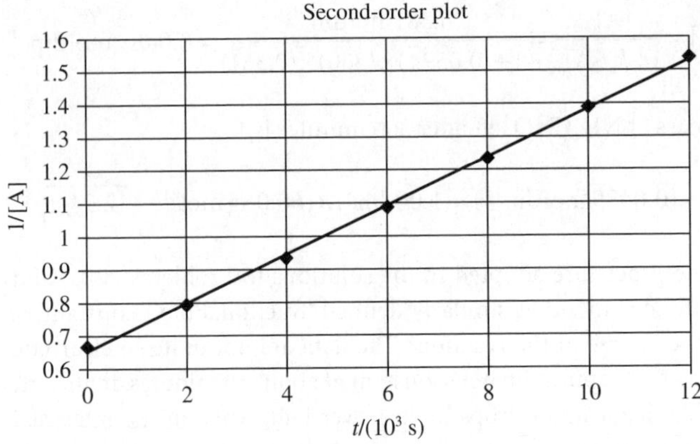

Figure 20B.3(b)

The second-order plot looks even better: the correlation coefficient is 0.999. The corresponding second-order rate constant is k_r = slope = $\boxed{7.3\times10^{-5}\ \text{dm}^3\,\text{mol}^{-1}\,\text{s}^{-1}}$

Comment. Based on the given data, the reaction appears to be closer to second-order than to first-order. (The reaction order need not be an integer.) This conclusion is a tenuous one, though, based on the assumption that the data contain little experimental error. The best course of action for an investigator seeking to establish the reaction order would be further experimentation, following the reaction over a wider range of concentrations.

P20B.7 The initial rate is

$$\begin{aligned} v_0 &= (3.6\times10^{6}\ \text{dm}^9\,\text{mol}^{-3}\,\text{s}^{-1})\times(5\times10^{-5}\,\text{mol}\,\text{dm}^{-3})^2\times(10^{-5.6}\ \text{mol}\,\text{dm}^{-3})^2 \\ &= \boxed{6\times10^{-14}\,\text{mol}\,\text{dm}^{-3}\,\text{s}^{-1}} \end{aligned}$$

The half-life for a second-order reaction is

$$t_{1/2} = \frac{1}{k_{\text{eff}}[\text{HSO}_3^-]_0} \quad [20\text{B}.5]$$

where k_{eff} is the rate constant in the expression

$$-\frac{\text{d}[\text{HSO}_3^-]}{\text{d}t} = 2v = k_{\text{eff}}[\text{HSO}_3^-]^2$$

Comparison to the given rate law and rate constant shows

$$\begin{aligned} k_{\text{eff}} &= 2k_r[\text{H}^+]^2 = 2(3.6\times10^{6}\ \text{dm}^9\,\text{mol}^{-3}\,\text{s}^{-1})\times(10^{-5.6}\ \text{mol}\,\text{dm}^{-3})^2 \\ &= 4.5\times10^{-5}\,\text{dm}^3\,\text{mol}^{-1}\,\text{s}^{-1} \end{aligned}$$

and $$t_{1/2} = \frac{1}{(4.5\times10^{-5}\ \text{dm}^3\,\text{mol}^{-1}\,\text{s}^{-1})\times(5\times10^{-5}\,\text{mol}\,\text{dm}^{-3})} = \boxed{4.\bar{4}\times10^{8}\ \text{s} = 14\ \text{yr}}$$

P20B.9 Examination of the data shows that the half-life remains constant at about 2 minutes. Therefore, the reaction is $\boxed{\text{first-order}}$. This can be confirmed by fitting any two pairs of data to the integrated first-order rate law, solving for the rate constant from each pair, and checking to see that they are the same to within experimental error.

$$\ln\left(\frac{[A]}{[A]_0}\right) = -k_r't \ \left[20B.1, A = N_2O_5\right]$$

Note: k_r' is the rate constant in the differential equation

$$-\frac{d[A]}{dt} = k_r'[A]$$

Because of the stoichiometry of the reaction, the rate of the reaction is half the rate of consumption of N_2O_5:

$$v = -\frac{1}{2}\frac{d[A]}{dt} = k_r[A] = \frac{k_r'}{2}[A]$$

Solving for k_r',

$$k_r' = \frac{\ln\left(\frac{[A]_0}{[A]}\right)}{t}$$

At $t = 1.00$ min, $[A] = 0.705$ mol dm^{-3} and

$$k_r' = \frac{\ln\left(\frac{1.000}{0.705}\right)}{1.00\,\text{min}} = 0.350\,\text{min}^{-1} = 5.83\times10^{-3}\,\text{s}^{-1}$$

wAt $t = 3.00$ min, $[A] = 0.349$ mol dm^{-3} and

$$k_r' = \frac{\ln\left(\frac{1.000}{0.349}\right)}{3.00\,\text{min}} = 0.351\,\text{min}^{-1} = 5.85\times10^{-3}\,\text{s}^{-1}$$

Values of k_r' may be determined in a similar manner at all other times. The average value of k_r' obtained is $\boxed{5.84\times10^{-3}\,\text{s}^{-1}}$ (which makes $\boxed{k_r = 2.92\times10^{-3}\,\text{s}^{-1}}$). The constancy of k_r', which varies only between 5.83 and 5.85 × 10^{-3} s^{-1} confirms that the reaction is $\boxed{\text{first-order}}$. A linear regression of $\ln[A]$ against t yields the same result. The half-life is (eqn 20B.2)

$$t_{1/2} = \frac{\ln 2}{k_r'} = \frac{0.693}{5.84\times10^{-3}\,\text{s}^{-1}} = 118.\bar{7}\,\text{s} = \boxed{1.98\ \text{min}}.$$

P20B.11 The data for this experiment do not extend much beyond one half-life. Therefore the half-life method of predicting the order of the reaction as described in the solutions to Problems 20B.3 and 20B.4 cannot be used here. However, a similar method based on 'three-quarters lives' will work. For a first-order reaction, we may write (analogous to the derivation of eqn 20B.2)

$$k_r t_{3/4} = -\ln\frac{\frac{3}{4}[A]_0}{[A]_0} = -\ln\frac{3}{4} = \ln\frac{4}{3} = 0.288 \quad \text{or} \quad t_{3/4} = \frac{0.288}{k_r}$$

Thus the three-quarters life (or any given fractional life) is also independent of concentration for a first-order reaction. Examination of the data shows that the first three-quarters life (time to $[A] = 0.237\,\text{mol dm}^{-3}$) is about 80 min and by interpolation the second (time to $[A] = 0.178\,\text{mol dm}^{-3}$) is also about 80 min. Therefore the reaction is first order and the rate constant is approximately

$$k_r = \frac{0.288}{t_{3/4}} \approx \frac{0.288}{80\,\text{min}} = 3.6\times10^{-3}\,\text{min}^{-1}$$

A least-squares fit of the data to the first-order integrated rate law [20B.1] gives the slightly more accurate result, $k_r = \boxed{3.65\times10^{-3}\ \text{min}^{-1}}$. The half-life is

$$t_{1/2} = \frac{\ln 2}{k_r} = \frac{\ln 2}{3.65\times10^{-3}\,\text{min}^{-1}} = \boxed{190\ \text{min}}$$

The average lifetime is calculated from

$$\frac{[A]}{[A]_0} = e^{-k_r t} \quad [20B.1]$$

which has the form of a distribution function in the sense that the ratio $\frac{[A]}{[A]_0}$ is the fraction of sucrose molecules which have lived to time t. The average lifetime is then

$$\langle t\rangle = \frac{\int_0^\infty t e^{-k_r t}\,dt}{\int_0^\infty e^{-k_r t}\,dt} = \frac{1}{k_r} = \boxed{274\ \text{min}}$$

The denominator ensures normalization of the distribution function.

Comment. The average lifetime is also called the relaxation time. Note that the average lifetime is not the half-life. The latter is 190 minutes. Also note that $2\times t_{3/4} \neq t_{1/2}$.

P20B.13 The data do not extend much beyond one half-life; therefore, we cannot see whether the half-life is constant over the course of the reaction as a preliminary step in guessing a reaction order. In a first-order reaction, however, not only the half-life but any other similarly defined fractional lifetime remains constant. (See Problem 20B.11.) In this problem, we can see that the ⅔-life is **not** constant. (It takes less than 1.6 ms for [ClO] to drop from the first recorded value ($8.49\,\mu\text{mol dm}^{-3}$) by more than ⅓ of that value (to $5.79\,\mu\text{mol dm}^{-3}$); it takes more than 4.0 more ms for the concentration to drop by not even ⅓ of **that** value (to $3.95\,\mu\text{mol dm}^{-3}$). So our working assumption is that the reaction is not first-order but second-order. Draw up the following table:

t/ms	[ClO]/(μmol dm^{-3})	(1/[ClO])/(dm^3 μmol^{-1})
0.12	8.49	0.118
0.62	8.09	0.124
0.96	7.10	0.141
1.60	5.79	0.173
3.20	5.20	0.192
4.00	4.77	0.210
5.75	3.95	0.253

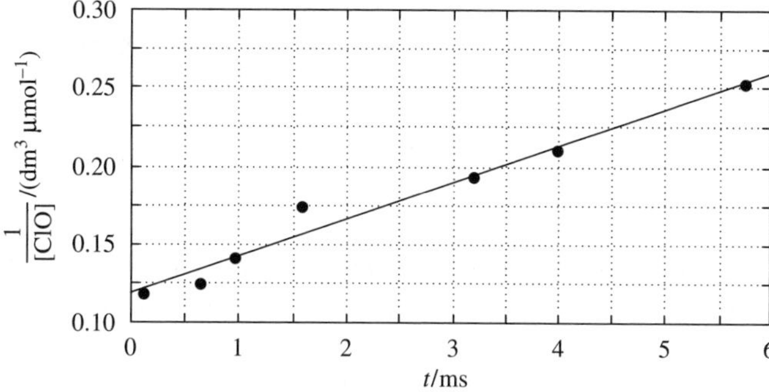

Figure 20B.4

The plot of 1/[ClO] vs. t (Fig. 20B.4) yields a reasonable straight line; the linear least-squares fit is:

$$(1/[\text{ClO}])/(\text{dm}^3\,\mu\text{mol}^{-1}) = 0.118 + 0.0237(t/\text{ms}) \quad R^2 = 0.974$$

The rate constant is equal to the slope

$$k_r' = 0.0237\,\text{dm}^3\,\mu\text{mol}^{-1}\,\text{ms}^{-1} = \boxed{2.37\times10^7\ \text{dm}^3\,\text{mol}^{-1}\,\text{s}^{-1}}$$

Note: k_r' is the rate constant in the differential equation

$$-\frac{\text{d[ClO]}}{\text{d}t} = k_r'[\text{ClO}]$$

Because of the stoichiometry of the reaction, the rate of the reaction is half the rate of consumption of ClO:

$$v = -\frac{1}{2}\frac{\text{d[ClO]}}{\text{d}t} = k_r[\text{ClO}] = \frac{k_r'}{2}[\text{ClO}]$$

so $$\boxed{k_r = 1.18\times10^7\ \text{dm}^3\,\text{mol}^{-1}\,\text{s}^{-1}}$$

The half-life depends on the initial concentration (eqn 20B.5):

$$t_{1/2}=\frac{1}{k_r[\text{ClO}]_0}=\frac{1}{(2.37\times10^7\ \text{dm}^3\ \text{mol}^{-1}\ \text{s}^{-1})(8.47\times10^{-6}\ \text{mol dm}^{-3})}=\boxed{4.98\times10^{-3}\ \text{s}}$$

P20B.15 $\text{A}+\text{B}\rightarrow\text{P},\quad \frac{\text{d[P]}}{\text{d}t}=k_r[\text{A}]^m[\text{B}]^n$

and for a short interval δt,

$$\delta[\text{P}]\approx k_r[\text{A}]^m[\text{B}]^n\,\delta t$$

Therefore, since $\delta[\text{P}]=[\text{P}]_t-[\text{P}]_0=[\text{P}]_t$

$$\frac{[\text{P}]}{[\text{A}]}=k_r[\text{A}]^{m-1}[\text{B}]^n\,\delta t$$

$\frac{[\text{Chloropropane}]}{[\text{Propene}]}$ is independent of [Propene], implying that $m=1$.

$$\frac{[\text{Chloropropane}]}{[\text{HCl}]}=\begin{cases} p(\text{HCl}) & 10 \quad 7.5 \quad 5.0 \\ & 0.05 \quad 0.03 \quad 0.01 \end{cases}$$

These results suggest that the ratio is roughly proportional to $p(\text{HCl})^2$, and therefore that $m=3$ when A is identified with HCl. The rate law is therefore

$$\frac{\text{d[Chloropropane]}}{\text{d}t}=k_r[\text{Propene}][\text{HCl}]^3$$

and the reaction is $\boxed{\text{first-order}}$ in propene and $\boxed{\text{third-order}}$ in HCl.

P20B.17 $v=\frac{d[\text{P}]}{dt}=k_r[\text{A}][\text{B}]$

Let the initial concentrations be $[\text{A}]_0=A_0$, $[\text{B}]_0=B_0$, and $[\text{P}]_0=0$. Then, when P is formed in concentration x, the concentration of A changes to A_0-2x and that of B changes to B_0-3x. Therefore

$$\frac{\text{d[P]}}{\text{d}t}=\frac{\text{d}x}{\text{d}t}=k_r(A_0-2x)(B_0-3x)\quad\text{with } x=0 \text{ at } t=0$$

$$\int_0^t k_r\text{d}t=\int_0^x\frac{\text{d}x}{(A_0-2x)\times(B_0-3x)}$$

Apply partial fractions decomposition to the integrand on the right.

$$\int_0^t k_r\text{d}t=\int_0^x\left(\frac{6}{2B_0-3A_0}\right)\times\left(\frac{1}{3(A_0-2x)}-\frac{1}{2(B_0-3x)}\right)\text{d}x$$

$$=\left(\frac{-1}{(2B_0-3A_0)}\right)\times\left(\int_0^x\frac{\text{d}x}{x-(1/2)A_0}-\int_0^x\frac{\text{d}x}{x-(1/3)B_0}\right)$$

$$k_r t = \left(\frac{-1}{(2B_0-3A_0)}\right)\times\left[\ln\left(\frac{x-\frac{1}{2}A_0}{-\frac{1}{2}A_0}\right)-\ln\left(\frac{x-\frac{1}{3}B_0}{-\frac{1}{3}B_0}\right)\right]$$

$$=\left(\frac{-1}{2B_0-3A_0}\right)\ln\left(\frac{(2x-A_0)B_0}{A_0(3x-B_0)}\right)$$

$$=\boxed{\left(\frac{1}{3A_0-2B_0}\right)\ln\left(\frac{(2x-A_0)B_0}{A_0(3x-B_0)}\right)}$$

P20B.19 The rate law $\frac{d[A]}{dt}=-k_r[A]^n$ for $n\neq 1$ integrates to

$$k_r t = \left(\frac{1}{n-1}\right)\times\left(\frac{1}{[A]^{n-1}}-\frac{1}{[A]_0^{n-1}}\right) \text{[Table 20B.3]}$$

At $t=t_{1/2}$, $$kt_{1/2}=\left(\frac{1}{n-1}\right)\left[\left(\frac{2}{[A]_0}\right)^{n-1}-\left(\frac{1}{[A]_0}\right)^{n-1}\right] \text{[Table 20B.3]}$$

At $t=t_{3/4}$, $$kt_{3/4}=\left(\frac{1}{n-1}\right)\left[\left(\frac{4}{3[A]_0}\right)^{n-1}-\left(\frac{1}{[A]_0}\right)^{n-1}\right]$$

Hence, $$\frac{t_{1/2}}{t_{3/4}}=\boxed{\frac{2^{n-1}-1}{\left(\frac{4}{3}\right)^{n-1}-1}}$$

20C Reactions approaching equilibrium

Answer to discussion question

D20C.1 A temperature-jump experiment makes use of the fact that large quantities of energy can be delivered to a system in a very short time by electrical discharge or laser, with the result that a system's temperature can be changed in a very short time. Changing the temperature of an equilibrium mixture suddenly disturbs the equilibrium, since equilibrium constants and therefore equilibrium compositions depend on temperature. By changing the temperature suddenly, a system that was at equilibrium at its initial temperature is no longer at equilibrium. Monitoring the concentration of one or more components of the equilibrium as the concentrations adjust to establish equilibrium at the new temperature can yield information on the rate of reaction at the new temperature. 'Relaxation' (adjustment toward equilibrium) occurs with a first-order rate constant equal to the sum of the forward and reverse rate constants of the equilibrium reaction. See Section 20C.2 and Example 20C.1.

Solution to exercise

E20C.1(a) The reactions whose rate constants are sought are the forward and reverse reactions in the following equilibrium.

$$NH_3(aq)+H_2O(l) \underset{k_r'}{\overset{k_r}{\rightleftharpoons}} NH_4^+(aq)+OH^-(aq)$$

The rate constants are related by

$$K_b = \frac{k_r}{k_r'} = \frac{[NH_4^+][OH^-]}{[NH_3]} = 1.78\times10^{-5}\ \text{mol dm}^{-3}$$

where the concentrations are equilibrium concentrations. (We assign units to K_b, which technically is a pure number, to help us keep track of units in the rate constants. Keeping track of the units makes us realize that k_r is a pseudo first-order protonation of NH_3 in excess water, for water does not appear in the above expression.) We need one more relationship between the constants, which we can obtain by proceeding as in Example 20C.1:

$$\frac{1}{\tau} = k_r + k_r'([NH_4^+]+[OH^-])$$

Substitute into this expression

$$k_r = K_b k_r' \quad \text{and} \quad [NH_4^+]=[OH^-]=(K_b[NH_3])^{1/2}$$

hence $\frac{1}{\tau} = K_b k_r' + 2k_r'(K_b[NH_3])^{1/2} = k_r'\{K_b + 2(K_b[NH_3])^{1/2}\}$

So the reverse rate constant is

$$k_r' = \frac{1}{\tau\{K_b + 2(K_b[NH_3])^{1/2}\}}$$

$$= \frac{1}{7.61\times10^{-9}\ \text{s}\{1.78\times10^{-5}\ \text{mol dm}^{-3} + 2(1.78\times10^{-5}\times0.15)^{1/2}\ \text{mol dm}^{-3}\}}$$

$$= \boxed{4.0\times10^{10}\ \text{dm}^{-3}\ \text{mol s}^{-1}}$$

and the forward constant is

$$k_r = K_b k_r' = 1.78\times10^{-5}\ \text{mol dm}^{-3}\times4.0\times10^{10}\ \text{dm}^3\ \text{mol}^{-1}\ \text{s}^{-1} = \boxed{7.1\times10^5\ \text{s}^{-1}}$$

Recall that k_r is the rate constant in the pseudo first-order rate law

$$-\frac{d[NH_3]}{dt} = k_r[NH_3]$$

Let us call k_2 the rate constant in the overall second-order rate law

$$-\frac{d[NH_3]}{dt}=k_2[NH_3][H_2O]$$

Setting these two expressions equal to each other yields

$$k_2=\frac{k_r}{[H_2O]}=\frac{7.1\times10^5\ s^{-1}}{(1\,000\ g\,dm^{-3})/(18.02\ g\,mol^{-1})}=\boxed{1.28\times10^4\ dm^3\ mol^{-1}\ s^{-1}}$$

Solutions to problems

P20C.1 Differentiate eqn 20C.4:

$$\frac{d[A]}{dt}=\frac{d}{dt}\left(\frac{k_r'+k_r e^{-(k_r+k_r')t}}{k_r'+k_r}\right)[A]_0$$

The only time dependence in the expression is in the exponential, so

$$\frac{d[A]}{dt}=\frac{[A]_0 k_r}{k_r'+k_r}\frac{d}{dt}e^{-(k_r+k_r')t}=-[A]_0 k_r e^{-(k_r+k_r')t}$$

According to eqn 20C.3, this expression should be equal to $(k_r+k_r'[A]+k_r'[A]_0$, so we expand the latter by substituting eqn 20C.4 for A:

$$-(k_r+k_r')[A]+k_r'[A]_0=-(k_r+k_r')\left(\frac{k_r'+k_r e^{-(k_r+k_r')t}}{k_r'+k_r}\right)[A]_0+k_r'[A]_0$$

$$=-(k_r'+k_r e^{-(k_r+k_r')t})[A]_0+k_r'[A]_0=-k_r e^{-(k_r+k_r')t}[A]_0$$

Indeed, this is equal to the result of the differentiation above.

P20C.3 $A \rightleftharpoons B$

$$\frac{d[A]}{dt}=-k_r[A]+k_r'[B] \quad \text{and} \quad \frac{d[B]}{dt}=-k_r'[B]+k_r[A]$$

At all times, $[A]+[B]=[A]_0+[B]_0$

Therefore, $[B]=[A]_0+[B]_0-[A]$

$$\frac{d[A]}{dt}=-k_r[A]+k_r'\{[A]_0+[B]_0-[A]\}=-(k_r+k_r')[A]+k_r'([A]_0+[B]_0)$$

To solve, one must integrate

$$\int\frac{d[A]}{(k_r+k_r')[A]-k_r'([A]_0+[B]_0)}=-\int dt$$

The solution is $[\mathrm{A}]=\boxed{\dfrac{k_\mathrm{r}'([\mathrm{A}]_0+[\mathrm{B}]_0)+(k_\mathrm{r}[\mathrm{A}]_0-k_\mathrm{r}'[\mathrm{B}]_0)\mathrm{e}^{-(k_\mathrm{r}+k_\mathrm{r}')t}}{k_\mathrm{r}+k_\mathrm{r}'}}$

The final composition is found by setting $t=\infty$

$$[\mathrm{A}]_\infty=\boxed{\left(\frac{k_\mathrm{r}'}{k_\mathrm{r}+k_\mathrm{r}'}\right)\times([\mathrm{A}]_0+[\mathrm{B}]_0)}$$

and $[\mathrm{B}]_\infty=[\mathrm{A}]_0+[\mathrm{B}]_0-[\mathrm{A}]_\infty=\boxed{\left(\dfrac{k_\mathrm{r}}{k_\mathrm{r}+k_\mathrm{r}'}\right)\times([\mathrm{A}]_0+[\mathrm{B}]_0)}$

Note that $\boxed{\dfrac{[\mathrm{B}]_\infty}{[\mathrm{A}]_\infty}=\dfrac{k_\mathrm{r}}{k_\mathrm{r}'}}$

P20C.5 (a) First, find an expression for the relaxation time, using Example 20C.1 as a model:

$$\frac{\mathrm{d}[\mathrm{A}]}{\mathrm{d}t}=-2k_\mathrm{a}[\mathrm{A}]^2+2k_\mathrm{a}'[\mathrm{A}_2]$$

Rewrite the expression in terms of a difference from equilibrium values, $[\mathrm{A}]=[\mathrm{A}]_\mathrm{eq}+x$

$$\frac{\mathrm{d}[\mathrm{A}]}{\mathrm{d}t}=\frac{\mathrm{d}([\mathrm{A}]_\mathrm{eq}+x)}{\mathrm{d}t}=\frac{\mathrm{d}x}{\mathrm{d}t}=-2k_\mathrm{a}([\mathrm{A}]_\mathrm{eq}+x)^2+2k_\mathrm{a}'([\mathrm{A}_2]_\mathrm{eq}-\tfrac{1}{2}x)$$

$$\frac{\mathrm{d}x}{\mathrm{d}t}=-2k_\mathrm{a}[\mathrm{A}]_\mathrm{eq}^2-4k_\mathrm{a}[\mathrm{A}]_\mathrm{eq}x-2k_\mathrm{a}x^2+2k_\mathrm{a}'[\mathrm{A}_2]_\mathrm{eq}-k_\mathrm{a}'x$$

Neglect powers of x greater than x^1, and use the fact that at equilibrium the forward and reverse rates are equal,

$$k_\mathrm{a}[\mathrm{A}]_\mathrm{eq}^2=k_\mathrm{a}'[\mathrm{A}_2]_\mathrm{eq}$$

to obtain

$$\frac{\mathrm{d}x}{\mathrm{d}t}\approx-(4k_\mathrm{a}[\mathrm{A}]_\mathrm{eq}+k_\mathrm{a}')x \quad \text{so} \quad \frac{1}{\tau}\approx4k_\mathrm{a}[\mathrm{A}]_\mathrm{eq}+k_\mathrm{a}'$$

To get the desired expression, square the reciprocal relaxation time,

(i) $\dfrac{1}{\tau^2}\approx16k_\mathrm{a}^2[\mathrm{A}]_\mathrm{eq}^2+8k_\mathrm{a}k_\mathrm{a}'[\mathrm{A}]_\mathrm{eq}+(k_\mathrm{a}')^2$

introduce $[\mathrm{A}]_\mathrm{tot}=\mathrm{A}]_\mathrm{eq}+2[\mathrm{A}_2]_\mathrm{eq}$ into the middle term,

$$\frac{1}{\tau^2}\approx16k_\mathrm{a}^2[\mathrm{A}]_\mathrm{eq}^2+8k_\mathrm{a}k_\mathrm{a}'([\mathrm{A}]_\mathrm{tot}-2[\mathrm{A}_2]_\mathrm{eq})+(k_\mathrm{a}')^2$$

$$\approx16k_\mathrm{a}^2[\mathrm{A}]_\mathrm{eq}^2+8k_\mathrm{a}k_\mathrm{a}'[\mathrm{A}]_\mathrm{tot}-16k_\mathrm{a}k_\mathrm{a}'[\mathrm{A}_2]_\mathrm{eq}+(k_\mathrm{a}')^2=\boxed{8k_\mathrm{a}k_\mathrm{a}'[\mathrm{A}]_\mathrm{tot}+(k_\mathrm{a}')^2}$$

and use the equilibrium condition again to see that the remaining equilibrium concentrations cancel each other.

Comment. Introducing $[A]_{tot}$ into just one term of eqn (i) above is a permissible step but not a very systematic one. It is worth trying because of the resemblance between eqn (i) and the desired expression: We would be finished if we could get $[A]_{tot}$ into the middle term and somehow get the first term to disappear! A more systematic but messier approach would be to express $[A]_{eq}$ in terms of the desired $[A]_{tot}$ by using the equilibrium condition and $[A]_{tot} = [A]_{eq} + 2[A_2]_{eq}$: Solve both of those equations for $[A_2]_{eq}$, set the two resulting expressions equal to each other, solve for $[A]_{eq}$ in terms of the desired $[A]_{tot}$, and substitute **that** expression for $[A]_{eq}$ everywhere in eqn (i).

(b) Plot $\dfrac{1}{\tau^2}$ vs. $[A]_{tot}$

The resulting curve should be a straight line whose y-intercept is $(k_a')^2$ and whose slope is $8k_a k_a'$.

(c) Draw up the following table:

$[A]_{tot}/(\text{mol dm}^{-3})$	0.500	0.352	0.251	0.151	0.101
τ / ns	2.3	2.7	3.3	4.0	5.3
$1/(\tau/\text{ns})^2$	0.189	0.137	0.092	0.062	0.036

The plot is shown in Fig. 20C.1

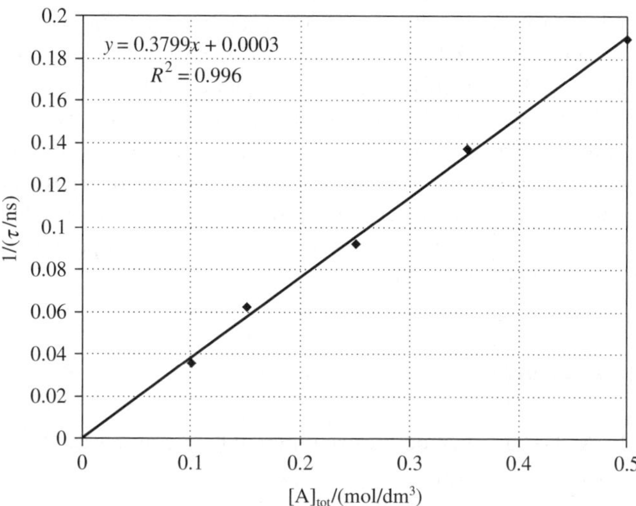

Figure 20C.1

The y-intercept is 0.0003 ns^{-2} and the slope is 0.38 $\text{ns}^{-2}\,\text{dm}^3\,\text{mol}^{-1}$, so

$$k_a' = \{3\times10^{-4}\times(10^{-9}\text{ s})^{-2}\}^{1/2} = (3\times10^{14}\text{ s}^{-2})^{1/2} = \boxed{1.\overline{7}\times10^{7}\text{ s}^{-1}}$$

$$k_a = \frac{0.38\times(10^{-9}\text{ s})^{-2}\text{ dm}^3\text{ mol}^{-1}}{8\times(1.\overline{7}\times10^{7}\text{ s}^{-1})} = \boxed{2.\overline{7}\times10^{9}\text{ dm}^3\text{ mol}^{-1}\text{ s}^{-1}}$$

and $$K = \frac{k_a / \text{dm}^3\,\text{mol}^{-1}\,\text{s}^{-1}}{k'_a / \text{s}^{-1}} = \frac{2.\bar{7}\times10^9}{1.\bar{7}\times10^7} = \boxed{1.\bar{6}\times10^2}$$

Comment. The data define a good straight line, as the correlation coefficient $R^2 =$ 0.996 shows. That straight line appears to go through the origin, but the best-fit equation gives a small nonzero y-intercept. Inspection of the plot shows that several of the data points lie about as far from the fit line as the y-intercept does from zero. This suggests that y-intercept has a fairly high relative uncertainty and so do the rate constants.

20D The Arrhenius equation

Answer to discussion question

D20D.1 In eqn. 20D.1, the Arrhenius equation, k_r is the **rate constant,** A the **pre-exponential factor** (a name that makes more sense in light of the form of the Arrhenius equation given in eqn 20D.4), and E_a the **activation energy**. The pre-exponential factor is sometimes called the frequency factor or even the Arrhenius A-factor. The Arrhenius equation is both an empirical expression that often provides a good summary of the temperature dependence of reaction rates and an expression whose parameters have sound physical significance. As an empirical expression, the equation is good, but neither unique nor universal. Over relatively small temperature ranges, other expressions can fit rate data; for example, plots of $\ln k_r$ vs. $\ln T$ or vs. T are often linear. Over large temperature ranges, plots of $\ln k_r$ vs. $1/T$ sometimes exhibit curvature, and kineticists may then turn to three- or four-parameter fits for greater accuracy. Equation 20D.4 bears some resemblance to the Boltzmann distribution, and the activation energy is the minimum energy required for an encounter between reactant molecules to lead to reaction. The pre-exponential factor can be interpreted in terms of the frequency of collisions between reactant molecules that have the right orientation for reaction to take place. More detailed interpretation of the Arrhenius parameters will be seen in Chapter 21, Reaction dynamics.

Solutions to exercises

E20D.1(a) The Arrhenius equation for two different temperatures can be rearranged to yield the activation energy:

$$E_a = \frac{R\ln\dfrac{k_{r,2}}{k_{r,1}}}{\left(\dfrac{1}{T_1}-\dfrac{1}{T_2}\right)} = \frac{(8.3145\ \text{J K}^{-1}\,\text{mol}^{-1})\ln\dfrac{2.67\times10^{-2}}{3.80\times10^{-3}}}{\left(\dfrac{1}{(273+35)\ \text{K}}-\dfrac{1}{(273+50)\ \text{K}}\right)}$$

$$= \boxed{1.08\times10^5\ \text{J mol}^{-1}} = \boxed{108\ \text{kJ mol}^{-1}}$$

With the activation energy in hand, the pre-exponential factor can be found from either rate constant by rearranging eqn 20D.4.

$$A = k_r e^{E_a/RT} = (3.80\times10^{-3}\ \text{dm}^3\ \text{mol}^{-1}\ \text{s}^{-1})e^{1.08\times10^5\ \text{J mol}^{-1}/(8.3145\ \text{J K}^{-1}\ \text{mol}^{-1})(273+35)\ \text{K}}$$

$$A = \boxed{6.50\times10^{15}\ \text{dm}^3\ \text{mol}^{-1}\ \text{s}^{-1}}.$$

Computing A from both provides a useful check on the calculation.

$$A = k_r e^{E_a/RT} = (2.67\times10^{-2}\ \text{dm}^3\ \text{mol}^{-1}\ \text{s}^{-1})e^{1.08\times10^5\ \text{J mol}^{-1}/(8.3145\ \text{J K}^{-1}\ \text{mol}^{-1})(273+50)\ \text{K}}$$

$$A = \boxed{6.50\times10^{15}\ \text{dm}^3\ \text{mol}^{-1}\ \text{s}^{-1}}.$$

E20D.2(a) Proceed as in Exercise 20D.1(a):

$$E_a = \frac{R\ln\left(\dfrac{k_r(T_2)}{k_r(T_1)}\right)}{\left(\dfrac{1}{T_1}-\dfrac{1}{T_2}\right)} = \frac{(8.3145\ \text{J K}^{-1}\ \text{mol}^{-1})\times\ln 3}{\dfrac{1}{297\ \text{K}}-\dfrac{1}{322\ \text{K}}} = \boxed{35\ \text{kJ mol}^{-1}}$$

Solutions to problems

P20D.1 Equation 20D.3 defines the activation energy whether or not that quantity is a constant with temperature:

$$E_a = RT^2\left(\frac{d\ln k_r}{dT}\right)$$

Multiply both sides by dT and divide by RT^2:

$$\frac{E_a dT}{RT^2} = d\ln k_r$$

Now integrate both sides, taking E_a to be constant:

$$\int\frac{E_a dT}{RT^2} = \frac{E_a}{R}\int\frac{dT}{T^2} = -\frac{E_a}{RT}+c = \int d\ln k_r = \ln k_r$$

This is eqn 20D.1 once we identify the constant of integration c as $\ln A$.

P20D.3 Subtract eqn 20D.1 at T_1 from the same equation at T_2:

$$\ln k_2 - \ln k_1 = \frac{E_a}{RT_1}-\frac{E_a}{RT_2} \quad \text{so} \quad E_a = R\ln\left(\frac{k_1}{k_2}\right)\left(\frac{1}{T_2}-\frac{1}{T_1}\right)^{-1}$$

T_1/K	300.3	300.3	341.2
T_2/K	341.2	392.2	392.2
$10^{-7}\ k_1/(\mathrm{dm^3\ mol^{-1}\ s^{-1}})$	1.44	1.44	3.03
$10^{-7}\ k_2/(\mathrm{dm^3\ mol^{-1}\ s^{-1}})$	3.03	6.9	6.9
$E_a/(\mathrm{kJ\ mol^{-1}})$	15.5	16.7	18.0

The mean is $\boxed{16.7\ \mathrm{kJ\,mol^{-1}}}$. Compute A from each rate constant, using the mean E_a and $A = k_r e^{E_a/RT}$

T/K	300.3	341.2	392.2
$10^{-7}\ k_r/(\mathrm{dm^3\ mol^{-1}\ s^{-1}})$	1.44	3.03	6.9
E_a/RT	6.69	5.89	5.12
$10^{-10}\ A/(\mathrm{dm^3\ mol^{-1}\ s^{-1}})$	1.16	1.10	1.16

The mean is $\boxed{1.14\times10^{10}\ \mathrm{dm^3\ mol^{-1}\ s^{-1}}}$

P20D.5 (a) The rate of reaction is

$$v = k_r[\mathrm{CH_4}][\mathrm{OH}]$$
$$= (1.13\times10^9\ \mathrm{dm^3\ mol^{-1}\ s^{-1}})\times\exp\left(\frac{-14.1\times10^3\ \mathrm{J\,mol^{-1}}}{(8.3145\ \mathrm{J\,K^{-1}\ mol^{-1}})\times(263\ \mathrm{K})}\right)$$
$$\times(4.0\times10^{-8}\,\mathrm{mol\,dm^{-3}})\times(3.0\times10^{-15}\,\mathrm{mol\,dm^{-3}}) = \boxed{2.1\times10^{-16}\,\mathrm{mol\,dm^{-3}\ s^{-1}}}.$$

(b) The mass is the amount consumed (in moles) times the molar mass; the amount consumed is the rate of consumption times the volume of the 'reaction vessel' times the time.

$$m = MvVt = (0.01604\ \mathrm{kg\,mol^{-1}})\times(2.1\times10^{-16}\ \mathrm{mol\,dm^{-3}\ s^{-1}})$$
$$\times(4\times10^{21}\ \mathrm{dm^3})\times(365\times24\times3\,600\ \mathrm{s})$$
$$= \boxed{4.3\times10^{11}\ \mathrm{kg\ or\ 430\ Tg}}.$$

20E Reaction mechanisms

Answers to discussion questions

D20E.1 The overall reaction order is the sum of the powers of the concentrations of all of the substances appearing in the *experimental* rate law for the reaction (eqn 20A.7); hence, it is the sum of the individual orders (exponents) associated with a given reactant (or, occasionally, product). Reaction order is an experimentally determined, *not theoretical*, quantity, although theory may attempt to predict or explain it. Molecularity is the number of reactant molecules participating in an *elementary* reaction. Molecularity has meaning only for an elementary reaction, but reaction order applies to any reaction. In general, reaction order bears no necessary relation to the stoichiometry of the reaction, with the exception of elementary reactions, where the order of the reaction corresponds to the number of molecules participating in the reaction; that is, to its molecularity. Thus for an elementary reaction (but *only* for an elementary reaction), overall order and molecularity are the same and are determined by the stoichiometry.

D20E.3 The steady-state approximation is the assumption that the rate of change of the concentrations of intermediates in consecutive chemical reactions is negligibly small. It is a good approximation when at least one of the reaction steps involving the intermediate is very fast, that is, has a large rate constant relative to other steps. A pre-equilibrium approximation is similar in that it is a good approximation when the rate of formation of the intermediate from the reactants and the rate of its reversible decay back to the reactions are both very fast in comparison to the rate of formation of the product from the intermediate. This results in the intermediate being in approximate equilibrium with the reactants over relatively long time periods (though short compared to the overall time scale of the reaction). Thus, the concentration of the intermediate remains approximately constant over the time period that the equilibrium can be considered to be maintained. This allows one to relate the rate constants and concentrations to each other through a constant (the pre-equilibrium constant). The approximations differ in ***how*** the intermediate concentration is kept nearly constant. In the steady-state case, it is because formation of the equilibrium is slow, limiting the reactions that come afterward. In the pre-equilibrium case, it is because formation of the intermediate and dissociation back to reactants are both fast compared to the reactions that follow.

D20E.5

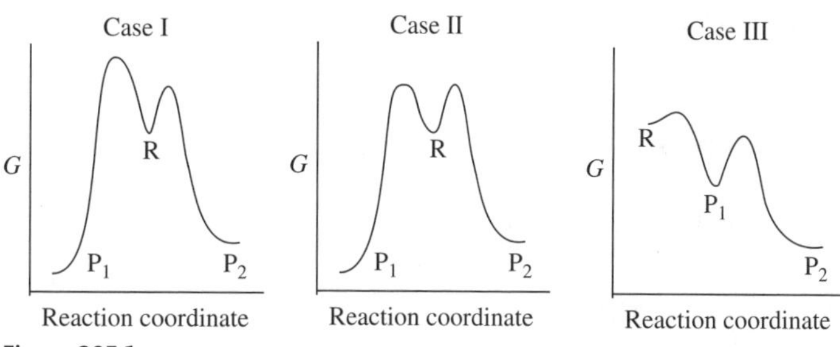

Figure 20E.1

Simple diagrams of Gibbs energy against reaction coordinate are useful for distinguishing between kinetic and thermodynamic control of a reaction. For the simple parallel reactions R → P_1 and R → P_2, shown in Fig. 20E.1 as Cases I and II, the product P_1 is thermodynamically favoured because the Gibbs energy decreases to a greater extent for its formation. However, the rate at which each product appears does not depend upon thermodynamic favourability. Rate constants depend upon activation energy. In Case I the activation energy for the formation of P_1 is much larger than that for formation of P_2. At low or moderate temperatures the large activation energy may not be readily available and P_1 either cannot form or forms at a slow rate. The much smaller activation energy for P_2 formation is available and, consequently, P_2 is produced even though it is not the thermodynamically favoured product. This is kinetic control. In this case, $[P_2]/[P_1] = k_2/k_1 > 1$ [20E.14].

The activation energies for the parallel reactions are equal in Case II and, consequently, the two products appear at identical rates. If the reactions are irreversible, $[P_2]/[P_1] = k_2/k_1 = 1$ at all times. The results are very different for reversible reactions. The activation energy for $P_1 \rightarrow R$ is much larger than that for $P_2 \rightarrow R$ and P_1 accumulates as the more rapid $P_2 \rightarrow R \rightarrow P_1$ occurs. Eventually the ratio $[P_2]/[P_1]$ approaches the equilibrium value

$$\left(\frac{[P_2]}{[P_1]}\right)_{eq} = e^{-(\Delta G_2 - \Delta G_1)/RT} < 1$$

This is thermodynamic control.

Case III above represents an interesting consecutive reaction series $R \rightarrow P_1 \rightarrow P_2$. The first step has relatively low activation energy and P_1 rapidly appears. However, the relatively large activation energy for the second step is not available at low and moderate temperatures. By using low or moderate temperatures and short reaction times it is possible to produce more of the thermodynamically less favourable P_1. This is kinetic control. High temperatures and long reaction times will yield the thermodynamically favoured P_2.

The ratio of reaction products is determined by relative reaction rates in kinetic controlled reactions. Favourable conditions include short reaction times, lower temperatures, and irreversible reactions. Thermodynamic control is favoured by long reaction times, higher temperatures, and reversible reactions. The ratio of products depends on the relative stability of products for thermodynamically controlled reactions.

Solutions to exercises

E20E.1(a) The rate of the overall reaction is

$$v = \frac{d[P]}{dt} = k_2[A][B]$$

however, we cannot have the concentration of an intermediate in the overall rate law.

(i) Assume a pre-equilibrium with

$$K = \frac{[A]^2}{[A_2]}, \quad \text{implying that} \quad [A] = K^{1/2}[A_2]^{1/2}$$

and $$v = \boxed{k_2K^{1/2}[A_2]^{1/2}[B]} = k_{eff}[A_2]^{1/2}[B]$$

where $k_{eff} = k_2K^{1/2}$

(ii) Apply the steady-state approximation:

$$\frac{d[A]}{dt} \approx 0 = 2k_1[A_2] - 2k_1'[A]^2 - k_2[A][B]$$

This is a quadratic equation in [A]

$$[A] = \frac{-b \pm \sqrt{b^2 - 4ac}}{2a} = \frac{k_2[B] \pm \sqrt{k_2^2[B]^2 + 16k_1'k_1[A_2]}}{-4k_1'}$$

$$= \frac{k_2[B]}{4k_1'}\left(\sqrt{1 + \frac{16k_1'k_1[A_2]}{k_2^2[B]^2}} - 1\right)$$

(In the last step, choose the sign that gives a positive quantity for [A].) Thus the rate law is

$$v = k_2[A][B] = \boxed{\frac{k_2^2[B]^2}{4k_1'}\left(\sqrt{1 + \frac{16k_1'k_1[A_2]}{k_2^2[B]^2}} - 1\right)}$$

This is a perfectly good rate law, albeit a complicated one. It is not in typical power-law form, but it is a function of reactant concentrations only, with no intermediates. This law simplifies under certain conditions. If $16k_1k_1'[A_2] >> k_2^2[B]^2$, then

$$\sqrt{1 + \frac{16k_1'k_1[A_2]}{k_2^2[B]^2}} - 1 \approx \frac{4\sqrt{k_1'k_1[A_2]}}{k_2[B]}$$

and $v \approx \frac{k_2^2[B]^2}{4k_1'} \times \frac{4\sqrt{k_1'k_1[A_2]}}{k_2[B]} = k_2[B]\sqrt{\frac{k_1[A_2]}{k_1'}} = \boxed{k_2K^{1/2}[A_2]^{1/2}[B]}$ recovering the pre-equilibrium rate law. If, on the other hand, $16k_1k_1'[A_2] << k_2^2[B]^2$, we expand the square root

$$\sqrt{1 + \frac{16k_1'k_1[A_2]}{k_2^2[B]^2}} \approx 1 + \frac{8k_1'k_1[A_2]}{k_2^2[B]^2}$$

and $v \approx \frac{k_2^2[B]^2}{4k_1'}\left(1 + \frac{8k_1'k_1[A_2]}{k_2^2[B]^2} - 1\right) = \boxed{2k_1[A_2]}$

Comment. If the equilibrium is 'fast', the latter condition will not be fulfilled. In fact, this special case amounts to having the first step rate-limiting. Note that the full (messy) steady-state approximation is less severe than either the pre-equilibrium or the rate-limiting step approximations, for it includes both as special cases.

E20E.2(a) Let the steps be

$$A + B \rightleftharpoons I \quad (\text{fast: } k_a, k_a')$$

and $I \rightarrow P \quad (k_b)$

Then the rate of reaction is

$$v = \frac{d[P]}{dt} = k_b[I]$$

Applying the pre-equilibrium approximation yields

$$\frac{[I]}{[A][B]} = K = \frac{k_a}{k'_a} \quad \text{so} \quad [I] = \frac{k_a[A][B]}{k'_a}$$

and $$v = \frac{k_a k_b [A][B]}{k'_a} = k_r[A][B] \quad \text{with} \quad k_r = \frac{k_a k_b}{k'_a}$$

Thus $E_a = E_{a,a} + E_{a,b} - E'_{a,a}[20E.13] = (25+10-38)\ \text{kJ mol}^{-1} = \boxed{-3\ \text{kJ mol}^{-1}}$

Comment. Activation energies are rarely negative; however, some composite reactions are known to have small, negative activation energies.

Solutions to problems

P20E.1 Using spreadsheet software to evaluate the equation after eqn 20E.4a., one can draw up a plot like that in Fig. 20E.2. The curves in this plot represent the concentration of the intermediate [I] as a function of time. They are labelled with the ratio k_a/k_b, where $k_b = 1\ \text{s}^{-1}$ for all curves and k_a varies. The thickest curve, labelled 10, corresponds to $k_a = 10\ \text{s}^{-1}$, as specified in part (a) of the problem. As the ratio k_a/k_b gets smaller (or, as the problem puts it, the ratio k_b/k_a gets larger), the concentration profile for I becomes lower, broader, and flatter; that is, [I] becomes more nearly constant over a longer period of time. This is the nature of the $\boxed{\text{steady-state approximation}}$, which becomes more and more valid as consumption of the intermediate becomes fast compared with its formation.

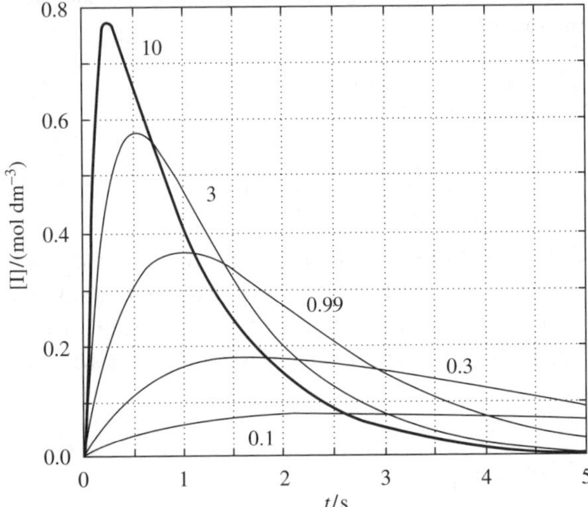

Figure 20E.2

P20E.3 The rate equations are

$$\frac{d[A]}{dt} = -k_a[A] + k_a'[B]$$

$$\frac{d[B]}{dt} = k_a[A] - k_a'[B] - k_b[B] + k_b'[C]$$

$$\frac{d[C]}{dt} = k_b[B] - k_b'[C]$$

These equations are a set of coupled differential equations. Although it is not immediately apparent, they have a closed-form general solution; however, we are looking for the particular circumstances under which the mechanism reduces to the second form given. Since the reaction involves an intermediate, let us explore the result of applying the steady-state approximation to it. Then

$$\frac{d[B]}{dt} = k_a[A] - k_a'[B] - k_b[B] + k_b'[C] \approx 0$$

and $[B] \approx \dfrac{k_a[A] + k_b'[C]}{k_a' + k_b}$

Therefore, $\dfrac{d[A]}{dt} = -\dfrac{k_a k_b}{k_a' + k_b}[A] + \dfrac{k_a' k_b'}{k_a' + k_b}[C]$

This rate expression may be compared to that given in the text [Topic 20C] for a first-order reaction approaching equilibrium

$$A \underset{k_r'}{\overset{k_r}{\rightleftarrows}} C$$

Here $k_r = \dfrac{k_a k_b}{k_a' + k_b}$ and $k_r' = \dfrac{k_a' k_b'}{k_a' + k_b}$

The solutions are $[A] = \left(\dfrac{k_r' + k_r e^{-(k_r'+k_r)t}}{k_r' + k_r}\right) \times [A]_0$ [20C.4]

and $[C] = [A]_0 - [A]$

Thus, the conditions under which the first mechanism given reduces to the second are the conditions under which the steady-state approximation holds, namely, when B can be treated as a $\boxed{\text{steady-state intermediate}}$.

P20E.5 $2\,HCl \rightleftharpoons (HCl)_2 \quad K_1 \qquad [(HCl)_2] = K_1[HCl]^2$

$HCl + CH_3CH{=}CH_2 \rightleftharpoons \text{complex} \quad K_2[\text{complex}] = K_2[HCl][CH_3CH{=}CH_2]$

$(HCl)_2 + \text{complex} \rightarrow CH_3CHClCH_3 + 2\,HCl \quad k_r$

$$v = \frac{d[CH_3CHClCH_3]}{dt} = k_r[(HCl)_2][\text{complex}]$$

Both $(HCl)_2$ and the complex are intermediates, so substitute for them using equilibrium expressions:

$$v = k_r[(HCl)_2][\text{complex}] = k_r(K_1[HCl]^2)(K_2[HCl][CH_3CH{=}CH_2])$$
$$= \boxed{k_r K_1 K_2 [HCl]^3[CH_3CH{=}CH_2]},$$

which is third-order in HCl and first-order in propene. One approach to experimental verification is to look for evidence of proposed intermediates, using infrared spectroscopy to search for $(HCl)_2$, for example.

20F Examples of reaction mechanisms

Answer to discussion question

D20F.1 The expression $k = k_a k_b[A]/(k_b + k_a'[A])$ for the effective rate constant of a unimolecular reaction $A \rightarrow P$ is based on the validity of the assumption of the existence of the pre-equilibrium $A + A \underset{k_a'}{\overset{k_a}{\rightleftharpoons}} A^* + A$. This can be a good assumption if both k_a and k_a' are much larger than k_b. The expression for the effective rate-constant, k, can be rearranged to

$$\frac{1}{k} = \frac{k_a'}{k_a k_b} + \frac{1}{k_a[A]}$$

A test of the theory is to plot $1/k$ against $1/[A]$, and to expect a straight line. Another test is based on the prediction from the Lindemann–Hinshelwood mechanism that as the concentration (and therefore the partial pressure) of A is reduced, the reaction should switch to overall second-order kinetics. Whereas the mechanism agrees in general with the switch in order of unimolecular reactions, it does not agree in detail. A typical graph of $1/k$ against $1/[A]$ has a pronounced curvature, corresponding to a larger value of k (a smaller value of $1/k$) at high pressures (low $1/[A]$) than would be expected by extrapolation of the reasonably linear low pressure (high $1/[A]$) data.

Solutions to exercises

E20F.1(a) $\dfrac{1}{k_r} = \dfrac{k_a'}{k_a k_b} + \dfrac{1}{k_a p_A}$ [analogous to 20F.8]

Therefore, for two different pressures we have

$$\frac{1}{k_r(p_1)} - \frac{1}{k_r(p_2)} = \frac{1}{k_a}\left(\frac{1}{p_1} - \frac{1}{p_2}\right)$$

$$\text{so}\quad k_a = \frac{\dfrac{1}{p_1} - \dfrac{1}{p_2}}{\dfrac{1}{k_r(p_1)} - \dfrac{1}{k_r(p_2)}} = \frac{\dfrac{1}{12\ \text{Pa}} - \dfrac{1}{1.30\times10^3\ \text{Pa}}}{\dfrac{1}{2.10\times10^{-5}\ \text{s}^{-1}} - \dfrac{1}{2.50\times10^{-4}\ \text{s}^{-1}}} = \boxed{1.9\times10^{-6}\ \text{Pa}^{-1}\ \text{s}^{-1}}$$

or $\boxed{1.9\ \text{MPa}^{-1}\ \text{s}^{-1}}$

E20F.2(a) The degree of polymerization is [20F.12]

$$\begin{aligned}\langle N\rangle &= 1+k_r t[\mathrm{A}]_0 \\ &= 1+(1.39\ \mathrm{dm^3\,mol^{-1}\,s^{-1}})\times 5.00\ \mathrm{h}\times 3\,600\ \mathrm{s\,h^{-1}}\times 1.00\times 10^{-2}\ \mathrm{mol\,dm^{-3}} \\ &= \boxed{251}\end{aligned}$$

The fraction condensed is related to the degree of polymerization by

$$\langle N\rangle = \frac{1}{1-p} \quad \text{so} \quad p=\frac{\langle N\rangle -1}{\langle N\rangle}=\frac{251-1}{251}=\boxed{0.996}$$

E20F.3(a) The kinetic chain length varies with concentration as

$$\nu = k_r[\mathrm{M}][\mathrm{In}]^{-1/2}\ [20\mathrm{F}.13]$$

so the ratio of kinetic chain lengths under different concentrations is

$$\frac{\nu_2}{\nu_1}=\frac{[\mathrm{M}]_2}{[\mathrm{M}]_1}\times\left(\frac{[\mathrm{In}]_1}{[\mathrm{In}]_2}\right)^{1/2}=\frac{1}{4.2}\times\left(\frac{1}{3.6}\right)^{1/2}=\boxed{0.125}$$

Solutions to problems

P20F.1 $\dfrac{1}{k_r}=\dfrac{k_a'}{k_a k_b}+\dfrac{1}{k_a p}$ [analogous to 20F.8]

We expect a straight line when $\dfrac{1}{k_r}$ is plotted against $\dfrac{1}{p}$. We draw up the following table:

p/Torr	84.1	11.0	2.89	0.569	0.120	0.067
$1/(p/\mathrm{Torr})$	0.012	0.091	0.346	1.76	8.33	14.9
$10^4\ k_r/\mathrm{s^{-1}}$	2.98	2.23	1.54	0.857	0.392	0.303
$10^{-4}/(k_r/\mathrm{s^{-1}})$	0.336	0.448	0.629	1.17	2.55	3.30

These points are plotted in Fig. 20F.1. There are marked deviations at low pressures, indicating that the Lindemann theory is deficient in that region.

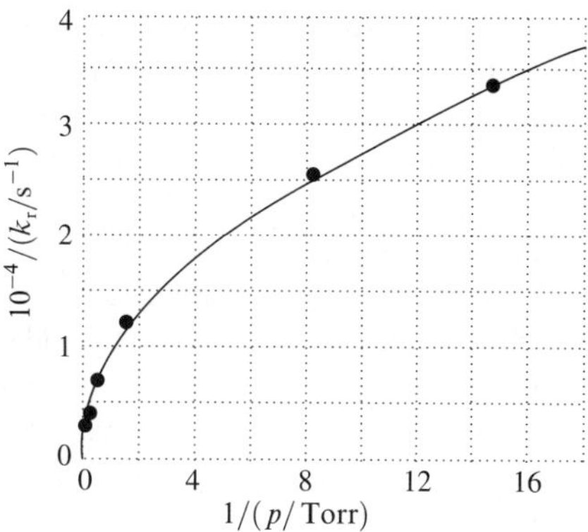

Figure 20F.1

P20F.3 $$\frac{d[A]}{dt} = -k_r[A]^2[OH] = -k_r[A]^3 \text{ because } [A] = [OH]$$

$$\frac{d[A]}{[A]^3} = -k_r dt \text{ and } \int_{[A]_0}^{[A]} \frac{d[A]}{[A]^3} = -k_r \int_0^t dt = -k_r t$$

Since $\int \frac{dx}{x^3} = \frac{-1}{2x^2}$, the equation becomes

$$\frac{1}{[A]^2} - \frac{1}{[A]_0^2} = 2k_r t \quad \text{or} \quad [A] = [A]_0(1 + 2k_r t[A]_0^2)^{-1/2}$$

By eqn 20F.12 the degree of polymerization, $\langle N \rangle$, is given by

$$\langle N \rangle = \frac{[A]_0}{[A]} = \boxed{\left(1 + 2k_r t[A]_0^2\right)^{1/2}}$$

20G Photochemistry

Answer to discussion question

D20G.1 The time scales of atomic processes are rapid indeed: according to the following table, a nanosecond is an eternity. Note that the times given here are in some way typical values for times that may vary over two or three orders of magnitude. For example, vibrational wavenumbers can range from about 4400 cm^{-1} (for H_2) to 100 cm^{-1} (for I_2) and even lower, with a corresponding range of associated times. Radiative decay rates of electronic states can vary even more widely: Times associated with phosphorescence can be in the millisecond and even second range. A large number of time scales for physical, chemical, and biological processes on the atomic and molecular scale are reported in Figure 2 of A.H. Zewail, 'Femtochemistry: Atomic-Scale Dynamics of the Chemical Bond,' *Journal of Physical Chemistry A* **104**, 5660 (2000).

Radiative decay of excited electronic states can range from about 10^{-9} s to 10^{-4} s—even longer for phosphorescence involving 'forbidden' decay paths. Molecular rotational motion takes place on a scale of 10^{-12} s to 10^{-9} s. Molecular vibrations are faster still, about 10^{-14} s to 10^{-12} s. Proton transfer reactions occur on a timescale of about 10^{-10} s to 10^{-9} s, although protons can hop from molecule to molecule in water even more rapidly (1.5×10^{-12} s, Topic 19B). Harvesting of light during plant photosynthesis involves very fast time scales of several energy-transfer and electron-transfer steps in photosynthesis. Initial energy transfer (to a nearby pigment) has a time scale of around 10^{-13} s to 5×10^{-12} s, with longer-range transfer (to the reaction centre) taking about 10^{-10} s. Immediate electron transfer is also very fast (about 3 ps), with ultimate transfer (leading to oxidation of water and reduction of plastoquinone) taking from 10^{-10} s to 10^{-3} s. The mean time between collisions in liquids is similar to vibrational periods, around 10^{-13} s. One can estimate collision times in liquids very roughly by applying the expression for collisions in gases (Topic 19A) to liquid conditions.

process	t/ns	reference
radiative decay of electronic excited state	1×10^{1}	Topic 13B
molecular rotational motion	3×10^{-2}	$B\approx 1\ \text{cm}^{-1}$
molecular vibrational motion	3×10^{-5}	$\tilde{\nu}\approx 1\,000\ \text{cm}^{-1}$
proton transfer	0.3	Zewail 2000
initial chemical reaction of vision*	2×10^{-4}	
energy transfer in photosynthesis†	1×10^{-3}	
electron transfer in photosynthesis	3×10^{-3}	
collision frequency in liquids	4×10^{-4}	Topic 19B‡

*photoisomerisation of retinal from 11-cis to all-trans
†time from absorption until electron transfer to adjacent pigment
‡Use formula for gas collision frequency at 300 K, parameters for benzene from data section, and density of liquid benzene

Solutions to exercises

E20G.1(a) Number of photons absorbed = ϕ^{-1} × number of molecules that react [20G.1a]. Therefore,

$$\text{Number absorbed} = \frac{(2.28\times10^{-3}\ \text{mol}/2)\times(6.022\times10^{23}\ \text{einstein}^{-1})}{2.1\times10^{2}\ \text{mol einstein}^{-1}} = \boxed{3.3\times10^{18}}$$

E20G.2(a) The Stern–Volmer equation (eqn 20G.8) relates the ratio of fluorescence quantum yields in the absence and presence of quenching

$$\frac{\phi_{\text{F},0}}{\phi_{\text{F}}} = 1+\tau_0 k_{\text{Q}}[\text{Q}] = \frac{I_{\text{F},0}}{I_{\text{F}}}$$

The last equality reflects the fact that fluorescence intensities are proportional to quantum yields. Solve this equation for [Q]:

$$[\text{Q}] = \frac{(I_{\text{F},0}/I_{\text{F}})-1}{\tau_0 k_{\text{Q}}} = \frac{2-1}{(6.0\times10^{-9}\ \text{s})\times(3.0\times10^{8}\ \text{dm}^3\ \text{mol}^{-1}\ \text{s}^{-1})} = \boxed{0.56\ \text{mol dm}^{-3}}$$

Solutions to problems

P20G.1 The quantum yield is defined as the amount of reacting molecules n_{A} divided by the amount of photons absorbed n_{abs}. The fraction of photons absorbed f_{abs} is one minus the fraction transmitted f_{trans}; and the amount of photons emitted n_{photon} can be inferred from the energy of the light source (power P times time t) and the energy of per photon (hc/λ).

$$\phi = \frac{n_{\text{A}}}{n_{\text{abs}}} = \frac{n_{\text{A}}hcN_{\text{A}}}{(1-f_{\text{trans}})\lambda Pt}$$

$$= \frac{(0.324\ \text{mol})\times(6.626\times10^{-34}\ \text{J s})\times(2.998\times10^{8}\ \text{m s}^{-1})\times(6.022\times10^{23}\ \text{mol}^{-1})}{(1-0.257)\times(320\times10^{-9}\ \text{m})\times(87.5\ \text{W})\times(28.0\ \text{min})\times(60\ \text{s min}^{-1})}$$

$$= \boxed{1.11}.$$

P20G.3 (a) The fluorescence intensity is proportional to the concentration of fluorescing species, so

$$\frac{I_F}{I_0}=\frac{[S]}{[S]_0}=e^{-t/\tau_0}\ [20G.4] \quad \text{so} \quad \ln\left(\frac{I_F}{I_0}\right)=-\frac{t}{\tau_0}$$

A plot of $\ln(I_F/I_0)$ against t should be linear with a slope equal to $-1/\tau_0$ (i.e. $\tau_0 = -1/\text{slope}$) and an intercept equal to zero. See Fig. 20G.1. The plot is linear, with slope $-0.150\ \text{ns}^{-1}$, so

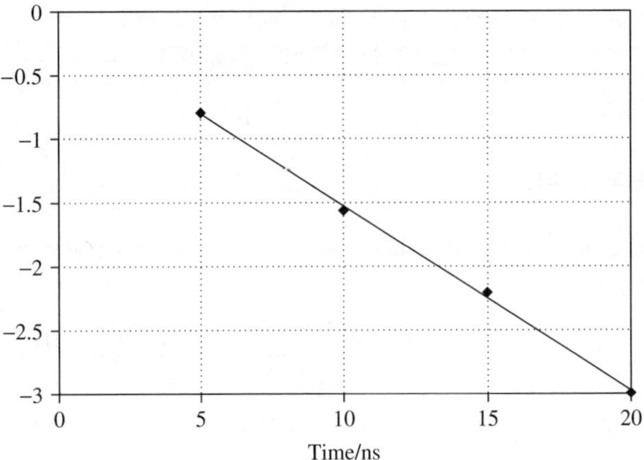

Figure 20G.1

$$\tau_0=-1/(-0.15\overline{0}\ \text{ns}^{-1})=\boxed{6.7\ \text{ns}}$$

Alternatively, average the experimental values of $\frac{1}{t}\ln\left(\frac{I_F}{I_0}\right)$ and check that the standard deviation is a small fraction of the average (it is). The average equals $-1/\tau_0$ (i.e. $\tau_0=-1/\text{average}$).

(b) The quantum yield for fluorescence is related to the rate constants for the various decay mechanisms of the excited state by

$$\phi_F=\frac{k_F}{k_F+k_{ISC}+k_{IC}}\ [20G.6]=k_F\tau_0\ [20G.7]$$

$$\text{so}\quad k_F=\phi_F/\tau_0=0.70/(6.7\ \text{ns})=\boxed{0.10\overline{5}\ \text{ns}^{-1}}$$

P20G.5 Proceed as in Problem 20G.3. In the absence of a quencher, a plot of $\ln I_f/I_0$ against t should be linear with a slope equal to $-1/\tau_0$. The plot is in fact linear with a best-fit slope of $-0.100\overline{4}\ \mu\text{s}^{-1}$. (See Fig. 20G.2)

$$\tau_0=\frac{-1}{-0.100\overline{4}\ \mu\text{s}^{-1}}=9.96\ \mu\text{s}$$

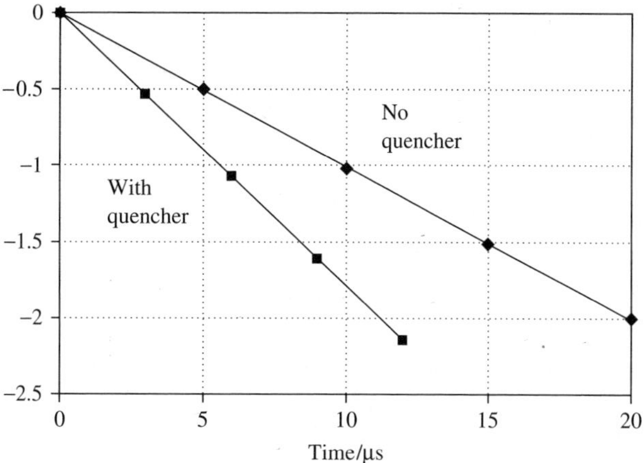

Figure 20G.2

In the presence of a quencher, a graph of $\ln I_F/I_0$ against t is still linear but with a slope equal to $-1/\tau$. This plot is found to be linear with a regression slope equal to $-1.78\bar{8}\ \mu s^{-1}$

$$\tau = \frac{-1}{-0.178\bar{8}\ \mu s^{-1}} = 5.59 \mu s$$

The rate constant for quenching (i.e. for energy transfer to the quencher) can be obtained from

$$\frac{1}{\tau} = \frac{1}{\tau_0} + k_Q[Q] \qquad \text{[Example 20G.2]}$$

Thus $$k_Q = \frac{\tau^{-1} - \tau_0^{-1}}{[N_2]} = \frac{RT(\tau^{-1} - \tau_0^{-1})}{p_{N_2}}$$

$$= \frac{(0.08206\ \text{dm}^3\ \text{atm}\ \text{K}^{-1}\ \text{mol}^{-1})(300\ \text{K})(0.1788 - 0.1004) \times (10^{-6}\ \text{s})^{-1}}{9.74 \times 10^{-4}\ \text{atm}}$$

$$= \boxed{1.98 \times 10^9\ \text{dm}^3\ \text{mol}^{-1}\ \text{s}^{-1}}$$

P20G.7 $$\eta_T = \frac{R_0^6}{R_0^6 + R^6} \quad \text{or} \quad \frac{1}{\eta_T} = 1 + \left(\frac{R}{R_0}\right)^6 \quad \text{[20G.10]}$$

A plot of η_T^{-1} vs. R^6 ought to be linear with a slope of $(R_0/\text{nm})^{-6}$.

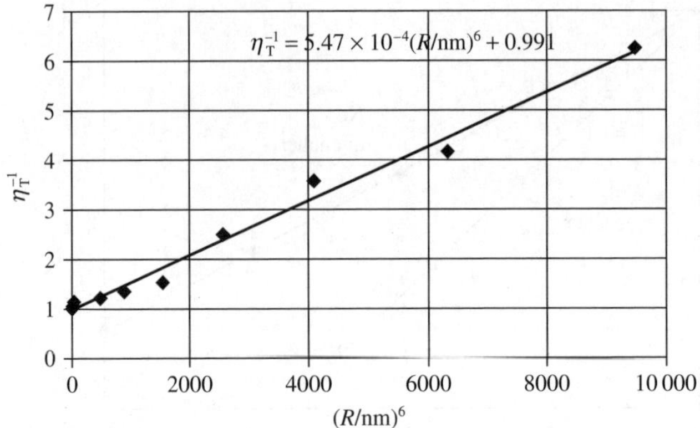

Figure 20G.3

The plot (Fig. 20G.3) appears to be linear with an intercept equal to 1, so we conclude that eqn 20G.10 adequately describes the data. The slope of the best-fit line is $(R_0/\text{nm})^{-6}$, so $R_0 = (5.47\times10^{-4})^{-1/6}\ \text{nm} = \boxed{3.5\ \text{nm}}$

20H Enzymes

Answers to discussion questions

D20H.1 The **Michaelis–Menten mechanism** of enzyme activity models the enzyme with one active site, that weakly and reversibly, binds a substrate in homogeneous solution. It is a three-step mechanism. The first and second steps are the reversible formation of the enzyme–substrate complex (ES). The third step is the decay of the complex into the product. The steady-state approximation is applied to the concentration of the intermediate (ES) and its use simplifies the derivation of the final rate expression. However, the justification for the use of the approximation with this mechanism is suspect, in that both rate constants for the reversible step may not be as large, in comparison to the rate constant for the decay to products, as they need to be for the approximation to be valid. The mechanism clearly indicates that the simplest form of the rate law, $v = v_{\max} = k_b[\text{E}]_0$, occurs when $[\text{S}]_0 \gg K_M$ and the general form of the rate law does seem to match the principal experimental features of enzyme catalysed reactions. It provides a mechanistic understanding of both the turnover number and catalytic efficiency. The model may be expanded to include multisubstrate reactions and inhibition.

D20H.3 Text Fig. 20H.6 summarizes the important characteristics of the three major modes of enzyme inhibition: competitive inhibition, uncompetitive inhibition, and non-competitive inhibition. Mathematical models for inhibition, which are the analogues of the Michaelis–Menten and Lineweaver–Burk equations (20H.3a and 20H.3b), are presented in eqns 20H.7 and 20H.8.

$$\frac{1}{v}=\frac{\alpha'}{v_{max}}+\left(\frac{\alpha K_M}{v_{max}}\right)\frac{1}{[S]_0} \quad [20H.8] \text{ where } \alpha=1+[I]/K_I \text{ and } \alpha'=1+[I]/K_I'$$

$$K_I=[E][I]/[EI],\ K_I'=[ES][I]/[ESI]$$

In **competitive inhibition** the inhibitor binds only to the active site of the enzyme and thereby inhibits the attachment of the substrate. This condition corresponds to $\alpha>1$ and $\alpha'=1$ (because ESI does not form). The slope of the Lineweaver–Burk plot increases by a factor of α relative to the slope for data on the uninhibited enzyme ($\alpha=\alpha'=1$). The y-intercept does not change as a result of competitive inhibition.

In **uncompetitive inhibition** the inhibitor binds to a site of the enzyme that is removed from the active site, but only if the substrate is already present. The inhibition occurs because ESI reduces the concentration of ES, the active type of the complex. In this case $\alpha=1$ (because EI does not form) and $\alpha'>1$. The y- intercept of the Lineweaver–Burk plot increases by a factor of α' relative to the y-intercept for data on the uninhibited enzyme, but the slope does not change.

In **non-competitive inhibition** (also called **mixed inhibition**) the inhibitor binds to a site other than the active site, and its presence reduces the ability of the substrate to bind to the active site. Inhibition occurs at both the E and ES sites. This condition corresponds to $\alpha>1$ and $\alpha'>1$. Both the slope and y-intercept of the Lineweaver–Burk plot increase upon addition of the inhibitor. Figure 20H.6c shows the special case of $K_I=K_I'$ and $\alpha=\alpha'$, which results in intersection of the lines at the x-axis.

In all cases, the efficiency of the inhibitor may be obtained by determining K_M and v_{max} from a control experiment with uninhibited enzyme and then repeating the experiment with a known concentration of inhibitor. From the slope and y-intercept of the Lineweaver–Burk plot for the inhibited enzyme (eqn 20H.8), the mode of inhibition, the values of α or α', and the values of K_I, or K_I' may be obtained.

Solutions to exercises

E20H.1(a) The fast, reversible step suggests the pre-equilibrium approximation:

$$K=\frac{[BH^+][A^-]}{[AH][B]} \quad \text{and} \quad [A^-]=\frac{K[AH][B]}{[BH^+]}.$$

Thus, the rate of product formation is

$$\frac{d[P]}{dt}=k_b[AH][A^-]=\boxed{\frac{k_bK[AH]^2[B]}{[BH^+]}}.$$

The application of the steady-state approximation to $[A^-]$ gives a similar, but significantly different, result:

$$\frac{d[A^-]}{dt}=k_a[AH][B]-k_a'[A^-][BH^+]-k_b[A^-][AH]=0.$$

Therefore, $[A^-]=\dfrac{k_a[AH][B]}{k_a'[BH^+]+k_b[AH]}$

and the rate of formation of product is

$$\frac{d[P]}{dt}=k_b[AH][A^-]=\frac{k_ak_b[AH]^2[B]}{k_a'[BH^+]+k_b[AH]}.$$

E20H.2(a) Since $v=\dfrac{v_{max}}{1+K_M/[S]_0}$ [20H.3a],

$$\begin{aligned}v_{max}&=\left(1+K_M/[S]_0\right)v\\&=(1+0.046/0.105)\times\left(1.04\ \text{mmol}\,\text{dm}^{-3}\,\text{s}^{-1}\right)\\&=\boxed{1.50\ \text{mmol}\,\text{dm}^{-3}\,\text{s}^{-1}}\end{aligned}$$

E20H.3(a) Equation 20H.7 describes competitive inhibition as the case for which $\alpha=1+[I]/K_I$ and $\alpha'=1$. Thus,

$$v=\frac{v_{max}}{1+\alpha K_M/[S]_0}$$

By setting the ratio v ([I] = 0)/v ([I]) equal to 2 and solving for α, we can subsequently solve for the inhibitor concentration that reduces the catalytic rate by 50%.

$$\frac{v([I]=0)}{v([I])}=\frac{1+\alpha K_M/[S]_0}{1+K_M/[S]_0}=2$$

$$\begin{aligned}\alpha&=\frac{2\left(1+K_M/[S]_0\right)-1}{K_M/[S]_0}\\&=\frac{2(1+3.0/0.10)-1}{3.0/0.10}=2.0\overline{3}\\ [I]&=(\alpha-1)K_I\\&=1.0\overline{3}\times\left(2.0\times10^{-5}\ \text{mol}\,\text{dm}^{-3}\right)=\boxed{2.0\times10^{-5}\ \text{mol}\,\text{dm}^{-3}}\end{aligned}$$

Solutions to problems

P20H.1 Michaelis–Menten mechanism: $E+S \rightleftarrows ES \quad k_a, k_a'$

$ES\rightarrow P+E \quad k_b$

Assuming a rapid pre-equilibrium of E, S, and ES for the reversible step implies that

$$K=\frac{k_a}{k_a'}=\frac{[ES]}{[E][S]}\quad\text{and}\quad [ES]=K[E][S].$$

But the law of mass balance demands that $[E]=[E]_0-[ES]$ so $[ES]=K([E]_0-[ES])[S]$ and, solving for [ES], we find that

$$[\mathrm{ES}]=\frac{[\mathrm{E}]_0}{1+\dfrac{1}{K[\mathrm{S}]_0}}$$

where the free substrate concentration has been replaced by $[\mathrm{S}]_0$ because the substrate is typically in large excess relative to the enzyme. Now substitute the latter expression into the Michaelis–Menten expression for the rate of product P formation.

$$v=k_\mathrm{b}[\mathrm{ES}]=\frac{k_\mathrm{b}[\mathrm{E}]_0}{1+\dfrac{1}{K[\mathrm{S}]_0}}\quad \text{where } v_\mathrm{max}=k_\mathrm{b}[\mathrm{E}]_0$$

$$\boxed{v=\frac{v_\mathrm{max}}{1+\dfrac{1}{K[\mathrm{S}]_0}}\qquad \text{Rate law based on rapid pre-equilibrium approximation}}.$$

With $K_\mathrm{M}=(k_\mathrm{a}'+k_\mathrm{b})/k_\mathrm{a}$ the steady-state approximation (*Justification* 20H.1) gives:

$$v=\frac{v_\mathrm{max}}{1+\dfrac{K_\mathrm{M}}{[\mathrm{S}]_0}}\qquad \text{Rate law based on steady-state approximation.}$$

Inspection reveals that the two approximations are identical when $K_\mathrm{M}=1/K$, which implies that

$$(k_\mathrm{a}'+k_\mathrm{b})/k_\mathrm{a}=k_\mathrm{a}'/k_\mathrm{a} \text{ or } \boxed{k_\mathrm{a}'\gg k_\mathrm{b}}.$$

P20H.3 Assume that the steady-state approximation is appropriate for both intermediates ([ES] and [ES′]).

For [ES]:

$$\frac{\mathrm{d}[\mathrm{ES}]}{\mathrm{d}t}=k_\mathrm{a}[\mathrm{E}][\mathrm{S}]-k_\mathrm{a}'[\mathrm{ES}]-k_\mathrm{b}[\mathrm{ES}]=0 \quad\text{and}\quad [\mathrm{ES}]=\left(\frac{k_\mathrm{a}}{k_\mathrm{a}'+k_\mathrm{b}}\right)[\mathrm{E}][\mathrm{S}].$$

For [ES′]:

$$\frac{\mathrm{d}[\mathrm{ES}']}{\mathrm{d}t}=k_\mathrm{b}[\mathrm{ES}]-k_\mathrm{c}[\mathrm{ES}']=0 \quad\text{and}\quad [\mathrm{ES}']=\left(\frac{k_\mathrm{b}}{k_\mathrm{c}}\right)[\mathrm{ES}].$$

We now have two equations in the three unknowns [E], [ES], and [ES′]. A third is provided by the mass balance expression $[\mathrm{E}]_0=[\mathrm{E}+[\mathrm{ES}]+[\mathrm{ES}']$. These three equations may be solved to give expressions for each of the three unknowns in terms of the rate constants, $[\mathrm{E}]_0$, and [S]. (For practical purposes the free substrate concentration is replaced by $[\mathrm{S}]_0$ because the substrate is typically in large excess relative to the enzyme.) The expression found for [ES′] is

$$[\mathrm{ES'}]=\frac{v_{\max}/k_c}{1+K_M/[\mathrm{S}]_0}\quad\text{where}\quad v_{\max}=\left(\frac{k_bk_c}{k_b+k_c}\right)[\mathrm{E}]_0\quad\text{and}\quad K_M=\frac{k_c(k_a'+k_b)}{k_a(k_b+k_c)}.$$

Substitution into the rate expression for product formation yields the desired equation.

$$v=k_c[\mathrm{ES'}]=\frac{v_{\max}}{1+K_M/[\mathrm{S}]_0}.$$

P20H.5 We draw up the table below, which includes data rows required for a Lineweaver-Burk plot ($1/v$ against $1/[\mathrm{S}]_0$). The linear regression fit is summarized in the Fig. 20H.1 plot.

[ATP]/(μmol dm^{-3})	**0.60**	**0.80**	**1.4**	**2.0**	**3.0**
v/(μmol dm^{-3} s^{-1})	0.81	0.97	1.30	1.47	1.69
1/{[ATP]/(μmol dm^{-3})}	1.67	1.25	0.714	0.500	0.333
1/{v/(μmol dm^{-3} s^{-1})}	1.23	1.03	0.769	0.680	0.592

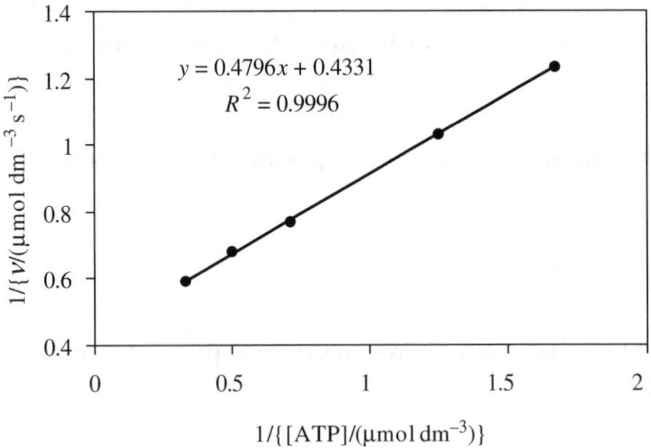

Figure 20H.1

$1/v_{\max}=intercept$ [20H.3b]

$$v_{\max}=1/intercept=1/(0.433\ \mu\mathrm{mol\,dm^{-3}\,s^{-1}})=\boxed{2.31\ \mu\mathrm{mol\,dm^{-3}\,s^{-1}}}$$

$$k_b=v_{\max}/[E]_0\,[20\mathrm{H.4}]=(2.31\ \mu\mathrm{mol\,dm^{-3}\,s^{-1}})/(0.020\ \mu\mathrm{mol\,dm^{-3}})=\boxed{115\ \mathrm{s^{-1}}}$$

$$k_{\mathrm{cat}}=k_b\,[20\mathrm{H.4}]=\boxed{115\ \mathrm{s^{-1}}}$$

$$K_M=v_{\max}\times slope[20\mathrm{H.3b}]=(2.31\ \mu\mathrm{mol\,dm^{-3}\,s^{-1}})\times(0.480\ \mathrm{s})=\boxed{1.11\ \mu\mathrm{mol\,dm^{-3}}}$$

$$\eta=k_{\mathrm{cat}}/K_M\,[20\mathrm{H.5}]=(115\ \mathrm{s^{-1}})/(1.11\ \mu\mathrm{mol\,dm^{-3}})=\boxed{104\ \mathrm{dm^3\,\mu mol^{-1}\,s^{-1}}}$$

Integrated activities

I20.1 (a) $A + P \rightarrow P + P$ autocatalytic step, $v = k_r[A][P]$

Let $[A] = [A]_0 - x$ and $[P] = [P]_0 + x$.

We substitute these definitions into the rate expression, simplify, and integrate.

$$v = -\frac{d[A]}{dt} = k_r[A][P]$$

$$-\frac{d([A]_0 - x)}{dt} = k_r([A]_0 - x)([P]_0 + x)$$

$$\frac{dx}{([A]_0 - x)([P]_0 + x)} = k_r\, dt$$

$$\frac{1}{[A]_0 + [P]_0}\left(\frac{1}{[A]_0 - x} + \frac{1}{[P]_0 + x}\right)dx = k_r\, dt$$

$$\frac{1}{[A]_0 + [P]_0}\int_0^x \left(\frac{1}{[A]_0 - x} + \frac{1}{[P]_0 + x}\right)dx = k_r\int_0^t dt$$

$$\frac{1}{[A]_0 + [P]_0}\left\{\ln\left(\frac{[A]_0}{[A]_0 - x}\right) + \ln\left(\frac{[P]_0 + x}{[P]_0}\right)\right\} = k_r t$$

$$\ln\left\{\left(\frac{[A]_0}{[P]_0}\right)\left(\frac{[P]_0 + x}{[A]_0 - x}\right)\right\} = k_r([A]_0 + [P]_0)t$$

$$\ln\left\{\left(\frac{[A]_0}{[P]_0}\right)\left(\frac{[P]}{[A]_0 + [P]_0 - [P]}\right)\right\} = k_r([A]_0 + [P]_0)t$$

$$\ln\left\{\left(\frac{1}{b}\right)\left(\frac{[P]}{[A]_0 + [P]_0 - [P]}\right)\right\} = at \quad \text{where} \quad a = k_r([A]_0 + [P]_0) \quad \text{and} \quad b = \frac{[P]_0}{[A]_0}$$

$$\frac{[P]}{[A]_0 + [P]_0 - [P]} = be^{at}$$

$$[P] = ([A]_0 + [P]_0)be^{at} - be^{at}[P]$$

$$(1 + be^{at})[P] = [P]_0\left(1 + \frac{[A]_0}{[P]_0}\right)be^{at} = [P]_0\left(1 + \frac{1}{b}\right)be^{at} = [P]_0(b+1)e^{at}$$

$$\boxed{\frac{[P]}{[P]_0} = (b+1)\frac{e^{at}}{1 + be^{at}}}$$

(b) See Fig. I20.1(a).

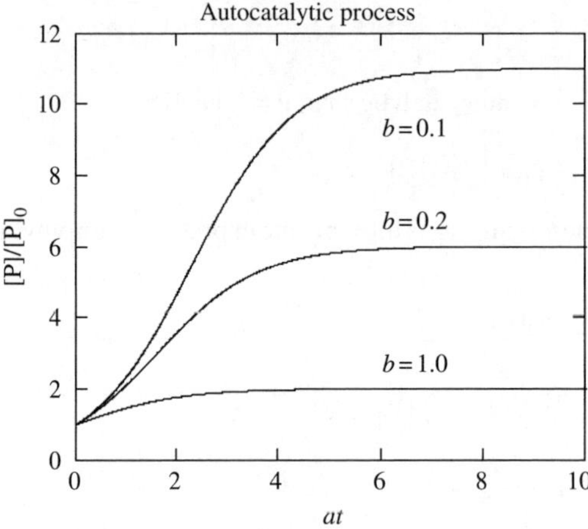

Figure I20.1(a)

The growth to [P] reaches a maximum at very long times. As $t\rightarrow\infty$, the exponential term in the denominator of $[\mathrm{P}]/[\mathrm{P}]_0 = (b+1)\mathrm{e}^{at}/\left(1+b\mathrm{e}^{at}\right)$ becomes so large that the denominator becomes $b\mathrm{e}^{at}$. Thus, $\left([\mathrm{P}]/[\mathrm{P}]_0\right)_{\max} = (b+1)\mathrm{e}^{at}/\left(b\mathrm{e}^{at}\right) = (b+1)/b$ where $b = [\mathrm{P}]_0/[\mathrm{A}]_0$ and this maximum occurs as $t\rightarrow\infty$.

The autocatalytic curve $[\mathrm{P}]/[\mathrm{A}]_0 = (b+1)e^{at}/(1+be^{at})$ has a shape that is very similar to that of the first-order process $[\mathrm{P}]/[\mathrm{A}]_0 = 1-\mathrm{e}^{-k_r t}$. However, $[\mathrm{P}]_{\max} = [\mathrm{A}]_0$ at $t\rightarrow\infty$ for the first-order process whereas $[\mathrm{P}]_{\max} = (1 + 1/b)[\mathrm{P}]_0$ for the autocatalytic mechanism. In a series of experiments at fixed $[\mathrm{A}]_0$ and assorted $[\mathrm{P}]_0$, only the autocatalytic mechanism will show variation in $[\mathrm{P}]_{\max}$. Another difference is that the autocatalytic curve is initially concave up, which gives an overall sigmoidal curve, whereas the first-order curve is concave down. See Fig. I20.1(b).

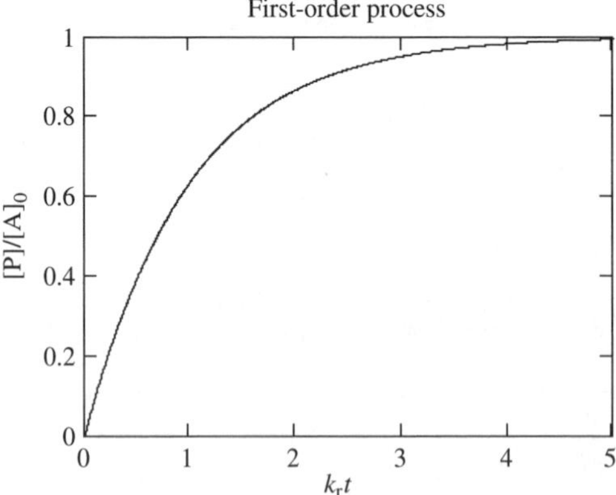

Figure I20.1(b)

(c) Let $[\mathrm{P}]_{v_{\max}}$ be the concentration of P at which the reaction rate is a maximum and let $t_{\max}$ be the corresponding time.

$$v = k_r[\mathrm{A}][\mathrm{P}] = k_r([\mathrm{A}]_0 - x)([\mathrm{P}]_0 + x)$$
$$= k_r\{[\mathrm{A}]_0[\mathrm{P}]_0 + ([\mathrm{A}]_0 - [\mathrm{P}]_0)x - x^2\}$$
$$\frac{\mathrm{d}v}{\mathrm{d}t} = k_r([\mathrm{A}]_0 - [\mathrm{P}]_0 - 2x)$$

The reaction rate is a maximum when $\mathrm{d}v/\mathrm{d}t = 0$. This occurs when

$$x = [\mathrm{P}]_{v_{\max}} - [\mathrm{P}]_0 = \frac{[\mathrm{A}]_0 - [\mathrm{P}]_0}{2} \quad \text{or} \quad \frac{[\mathrm{P}]_{v_{\max}}}{[\mathrm{P}]_0} = \frac{b+1}{2b}$$

Substitution into the final equation of part (a) gives:

$$\frac{[\mathrm{P}]_{v_{\max}}}{[\mathrm{P}]_0} = \frac{b+1}{2b} = (b+1)\frac{\mathrm{e}^{at_{\max}}}{1+b\mathrm{e}^{at_{\max}}}$$

Solving for $t_{\max}$:

$$1 + b\mathrm{e}^{at_{\max}} = 2b\mathrm{e}^{at_{\max}}$$
$$\mathrm{e}^{at_{\max}} = b^{-1}$$
$$at_{\max} = \ln(b^{-1}) = -\ln(b)$$
$$t_{\max} = \boxed{-\frac{1}{a}\ln(b)}$$

(d) $\dfrac{\mathrm{d}[\mathrm{P}]}{\mathrm{d}t} = k_r[\mathrm{A}]^2[\mathrm{P}]$

$$[\mathrm{A}] = A_0 - x, \quad [\mathrm{P}] = P_0 + x, \quad \frac{\mathrm{d}[\mathrm{P}]}{\mathrm{d}t} = \frac{\mathrm{d}x}{\mathrm{d}t} = k_r(A_0 - x)^2(P_0 + x)$$

$$\int_0^x \frac{\mathrm{d}x}{(A_0 - x)^2(P_0 + x)} = k_r t$$

Solve the integral by partial fractions.

$$\frac{1}{(A_0 - x)^2(P_0 + x)} = \frac{\alpha}{(A_0 - x)^2} + \frac{\beta}{A_0 - x} + \frac{\gamma}{P_0 + x}$$
$$= \frac{\alpha(P_0 + x) + \beta(A_0 - x)(P_0 + x) + \gamma(A_0 - x)^2}{(A_0 - x)^2(P_0 + x)}$$

$$\left.\begin{aligned} P_0\alpha + A_0P_0\beta + A_0^2\gamma &= 1 \\ \alpha + (A_0 - P_0)\beta - 2A_0\gamma &= 0 \\ -\beta + \gamma &= 0 \end{aligned}\right\}$$

This set of simultaneous equations solves to

$$\alpha = \frac{1}{A_0 + P_0}, \quad \beta = \gamma = \frac{\alpha}{A_0 + P_0}$$

Therefore,

$$k_r t = \left(\frac{1}{A_0 + P_0}\right)\int_0^x \left[\left(\frac{1}{A_0 - x}\right)^2 + \left(\frac{1}{A_0 + P_0}\right)\left(\frac{1}{A_0 - x} + \frac{1}{P_0 - x}\right)\right] dx$$

$$= \left(\frac{1}{A_0 + P_0}\right)\left\{\left(\frac{1}{A_0 - x}\right) - \left(\frac{1}{A_0}\right) + \left(\frac{1}{A_0 + P_0}\right)\left[\ln\left(\frac{A_0}{A_0 - x}\right) + \ln\left(\frac{P_0 + x}{P_0}\right)\right]\right\}$$

$$= \left(\frac{1}{A_0 + P_0}\right)\left[\left(\frac{x}{A_0(A_0 - x)}\right) + \left(\frac{1}{A_0 + P_0}\right)\ln\left(\frac{A_0(P_0 + x)}{(A_0 - x)P_0}\right)\right].$$

Therefore with $y = \frac{x}{A_0}$ and $p = \frac{P_0}{A_0}$,

$$A_0(A_0 + P_0)kt = \boxed{\left(\frac{y}{1-y}\right) + \left(\frac{1}{1-p}\right)\ln\left(\frac{p+y}{p(1-y)}\right)}$$

The maximum rate occurs at

$$\frac{dv_P}{dt} = 0, \quad v_P = k_r[A]^2[P]$$

and hence at the solution of

$$2k_r\left(\frac{d[A]}{dt}\right)[A][P] + k_r[A]^2\frac{d[P]}{dt} = 0$$

$$-2k_r[A][P]v_P + k_r[A]^2 v_P = 0 \qquad [\text{as } v_A = -v_P]$$

$$k_r[A]([A] - 2[p])v_P = 0$$

This is, the rate is a maximum when $[A] = 2[P]$, which occurs at

$$A_0 - x = 2P_0 + 2x, \text{ or } x = \tfrac{1}{3}(A_0 - 2P_0); \quad y = \tfrac{1}{3}(1 - 2p)$$

Substituting this condition into the integrated rate law gives

$$A_0(A_0 + P_0)k_r t_{max} = \left(\frac{1}{1+p}\right)\left(\frac{1}{2}(1-2p) + \ln\frac{1}{2p}\right)$$

or $\boxed{(A_0 + P_0)^2 kt_{max} = \tfrac{1}{2} - p - \ln 2p}$.

(e) $$\frac{d[P]}{dt} = k_r[A][P]^2$$

$$\frac{dx}{dt} = k_r(A_0 - x)(P_0 + x)^2 \quad [x = P - P_0]$$

$$kt = \int_0^x \frac{dx}{(A_0 - x)(P_0 + x)^2}$$

Integrate by partial fractions (as in part (d)).

$$k_r t = \left(\frac{1}{A_0 + P_0}\right) \int_0^x \left\{ \left(\frac{1}{P_0 + x}\right)^2 + \left(\frac{1}{A_0 + P_0}\right) \left[\frac{1}{P_0 + x} + \frac{1}{A_0 - x}\right] \right\} dx$$

$$= \left(\frac{1}{A_0 + P_0}\right) \left\{ \left(\frac{1}{P_0} - \frac{1}{P_0 + x}\right) + \left(\frac{1}{A_0 + P_0}\right) \left[\ln\left(\frac{P_0 + x}{P_0}\right) + \ln\left(\frac{A_0}{A_0 - x}\right) \right] \right\}$$

$$= \left(\frac{1}{A_0 + P_0}\right) \left[\left(\frac{x}{P_0(P_0 + x)}\right) + \left(\frac{1}{A_0 + P_0}\right) \ln\left(\frac{(P_0 + x)A_0}{P_0(A_0 - x)}\right) \right]$$

Therefore, with $y = \frac{x}{[\mathrm{A}]_0}$ and $p = \frac{P_0}{A_0}$,

$$A_0(A_0 + P_0)k_r t = \boxed{\left(\frac{y}{p(p+y)}\right) + \left(\frac{1}{1+p}\right) \ln\left(\frac{p+y}{p(1-y)}\right)}$$

The rate is maximum when

$$\frac{dv_\mathrm{P}}{dt} = 2k_r[\mathrm{A}][\mathrm{P}]\left(\frac{d[\mathrm{P}]}{dt}\right) + k_r\left(\frac{d[\mathrm{A}]}{dt}\right)[\mathrm{P}]^2$$

$$= 2k_r[\mathrm{A}][\mathrm{P}]v_\mathrm{P} - k_r[\mathrm{P}]^2 v_\mathrm{P} = k_r[\mathrm{P}](2[\mathrm{A}] - [\mathrm{P}])v_\mathrm{P} = 0.$$

That is, at $[\mathrm{A}] = \frac{1}{2}[\mathrm{P}]$.

On substitution of this condition into the integrated rate law, we find

$$A_0(A_0 + P_0)k_r t_{\max} = \left(\frac{2-p}{2p(1+p)}\right) + \left(\frac{1}{1+p}\right) \ln\frac{2}{p}$$

or $(A_0 + P_0)^2 k_r t_{\max} = \boxed{\frac{2-p}{2p} + \ln\frac{2}{p}}$

I20.3 (a) The rates of change are

(1) $\frac{d[\mathrm{F_2O}]}{dt} = -k_a[\mathrm{F_2O}]^2 - k_b[\mathrm{F}][\mathrm{F_2O}]$

(2) $\frac{d[\mathrm{F}]}{dt} = k_a[\mathrm{F_2O}]^2 - k_b[\mathrm{F}][\mathrm{F_2O}] + 2k_c[\mathrm{OF}]^2 - 2k_d[\mathrm{F}]^2[\mathrm{F_2O}] \approx 0$ [steady-state]

(3) $\frac{d[\mathrm{OF}]}{dt} = k_a[\mathrm{F_2O}]^2 + k_b[\mathrm{F}][\mathrm{F_2O}] - 2k_c[\mathrm{OF}]^2 \approx 0$ [steady-state]

Applying the steady-state approximation to both [F] and [OF] and adding eqns (2) and (3) gives

$$
\begin{array}{ll}
k_a[F_2O]^2 - k_b[F][F_2O] + 2k_c[OF]^2 - 2k_d[F]^2[F_2O] = 0 \\
k_a[F_2O]^2 + k_b[F][F_2O] - 2k_c[OF]^2 \qquad\qquad = 0 \\
\hline
2k_a[F_2O]^2 \qquad\qquad - 2k_d[F]^2[F_2O] = 0
\end{array}
$$

Solving for [F] gives

$$[F] = \left(\frac{k_a}{k_d}[F_2O]\right)^{1/2}$$

Substituting this result into (1) gives

$$\frac{d[F_2O]}{dt} = -k_a[F_2O]^2 - k_b\left(\frac{k_a}{k_d}\right)^{1/2}[F_2O]^{3/2}$$

$$\text{or}\quad -\frac{d[F_2O]}{dt} = \boxed{k_a[F_2O]^2 + k_b\left(\frac{k_a}{k_d}\right)^{1/2}[F_2O]^{3/2}}$$

Comparison with the experimental rate law reveals that they are consistent when we make the following identifications.

$$k_r = k_a = 7.8\times10^{13}\,e^{-E_1/RT}\,dm^3\,mol^{-1}\,s^{-1}$$

$$E_a = (19\,350\text{ K})R = 160.9\text{ kJ mol}^{-1} = \text{activation energy for step (1),}$$

$$k_r' = k_b\left(\frac{k_a}{k_d}\right)^{1/2} = 2.3\times10^{10}\,e^{-E_a'/RT}\,dm^3\,mol^{-1}\,s^{-1}$$

$$E_a' = (16\,910\text{ K})R = 140.6\text{ kJ mol}^{-1}$$

(b) $\tfrac{1}{2}O_2 + F_2 \rightarrow F_2O \qquad \Delta_f H(F_2O) = 24.41\text{ kJ mol}^{-1}$

$2F \rightarrow F_2 \qquad \Delta H = -D(F\text{–}F) = -160.6\text{ kJ mol}^{-1}$

$O \rightarrow \tfrac{1}{2}O_2 \qquad \Delta H = -\tfrac{1}{2}D(O\text{–}O) = -249.1\text{ kJ mol}^{-1}$

$2F + O \rightarrow F_2O$

$$
\begin{aligned}
\Delta H(\text{FO–F}) + \Delta H(\text{O–F}) &= -\left[\Delta_f H(F_2O) - D(\text{F–F}) - \tfrac{1}{2}D(\text{O–O})\right] \\
&= -(24.41 - 160.6 - 249.1)\text{ kJ mol}^{-1} \\
&= 385.3\text{ kJ mol}^{-1}
\end{aligned}
$$

We estimate that $\Delta H(\text{FO–F}) \approx E_1 = \boxed{160.9\text{ kJ mol}^{-1}}$

Then $\Delta H(\text{O–F}) \approx (385.3 - 160.9)\text{ kJ mol}^{-1} \approx \boxed{224.4\text{ kJ mol}^{-1}}$

In order to determine the activation energy of reaction (2) we assume that each rate be expressed in Arrhenius form, then

$$\ln k_r' = \ln k_b + \tfrac{1}{2}\ln k_a - \tfrac{1}{2}\ln k_b$$

and $E_a' = E_2 + \tfrac{1}{2}E_1 - \tfrac{1}{2}E_4 = 140.6\ \text{kJ mol}^{-1}$

Hence $E_2 - \tfrac{1}{2}E_4 = E_a' - \tfrac{1}{2}E_1 = (140.6 - 80.4)\ \text{kJ mol}^{-1} = 60.2\ \text{kJ mol}^{-1}$

E_4 is expected to be small since reaction (4) is a radical–radical combination, so we set $E_4 \approx 0$; then $E_2 \approx \boxed{60\ \text{kJ mol}^{-1}}$

I20.5 (a) To find mean cube and mean square molar masses, we need an expression for the probability P_N that a polymer consists of N monomers. In stepwise polymerization, P_N is the probability that the polymer has $N-1$ reacted end groups and one unreacted end group. In terms of the fraction p of total end groups in the mixture that have reacted, the former probability is p^{N-1}; the latter $1-p$. Therefore, the total probability of finding an N-mer is

$$P_N = p^{N-1}(1-p)$$

Since the molar mass of an N-mer is N times the monomer mass, M_1, the mean square molar mass is

$$\langle M^2\rangle = \sum_N (M_1N)^2 P_N = M_1^2\sum_N N^2P_N = M_1^2\langle N^2\rangle = M_1^2(1-p)\sum_N N^2p^{N-1}$$

$$= M_1^2(1-p)\frac{\mathrm{d}}{\mathrm{d}p}p\frac{\mathrm{d}}{\mathrm{d}p}\sum_N p^N = M_1^2(1-p)\frac{\mathrm{d}}{\mathrm{d}p}p\frac{\mathrm{d}}{\mathrm{d}p}(1-p)^{-1} = \frac{M_1^2(1+p)}{(1-p)^2}.$$

Similarly the mean cube molar mass is

$$\langle M^3\rangle = \sum_N (M_1N)^3 P_N = M_1^3(1-p)\sum_N N^3p^{N-1}$$

$$= M_1^3(1-p)\frac{\mathrm{d}}{\mathrm{d}p}\sum_N N^2p^N = M_1^3(1-p)\frac{\mathrm{d}}{\mathrm{d}p}p\frac{\mathrm{d}}{\mathrm{d}p}p\frac{\mathrm{d}}{\mathrm{d}p}\sum_N p^N$$

$$= M_1^3(1-p)\frac{\mathrm{d}}{\mathrm{d}p}p\frac{\mathrm{d}}{\mathrm{d}p}p\frac{\mathrm{d}}{\mathrm{d}p}(1-p)^{-1} = \frac{M_1^3(1+4p+p^2)}{(1-p)^3}$$

Therefore, $\dfrac{\langle M^3\rangle}{\langle M^2\rangle} = \boxed{\dfrac{M_1(1+4p+p^2)}{1-p^2}}$

(b) $\langle N\rangle = \dfrac{1}{1-p}$ [20F.12a], so $p = 1 - \dfrac{1}{\langle N\rangle}$

I20.7 Consider the following mechanism:

$Cl_2 + h\nu \rightarrow 2\,Cl$	I_a
$Cl + CHCl_3 \rightarrow CCl_3 + HCl$	k_2
$CCl_3 + Cl_2 \rightarrow CCl_4 + Cl$	k_3
$2\,CCl_3 + Cl_2 \rightarrow 2\,CCl_4$	k_4

(i) $\dfrac{\mathrm{d}[\mathrm{CCl}_4]}{\mathrm{d}t} = 2k_4[\mathrm{CCl}_3]^2[\mathrm{Cl}_2] + k_3[\mathrm{CCl}_3][\mathrm{Cl}_2]$

(ii) $\dfrac{\mathrm{d}[\mathrm{CCl}_3]}{\mathrm{d}t} = k_2[\mathrm{Cl}][\mathrm{CHCl}_3] - k_3[\mathrm{CCl}_3][\mathrm{Cl}_2] - 2k_4[\mathrm{CCl}_3]^2[\mathrm{Cl}_2] = 0$

(iii) $\dfrac{\mathrm{d}[\mathrm{Cl}]}{\mathrm{d}t} = 2I_\mathrm{a} - k_2[\mathrm{Cl}][\mathrm{CHCl}_3] + k_3[\mathrm{CCl}_3][\mathrm{Cl}_2] = 0$

(iv) $\dfrac{\mathrm{d}[\mathrm{Cl}_2]}{\mathrm{d}t} = -I_\mathrm{a} - k_3[\mathrm{CCl}_3][\mathrm{Cl}_2] - k_4[\mathrm{CCl}_3]^2[\mathrm{Cl}_2]$

Therefore, $I_\mathrm{a} = k_4[\mathrm{CCl}_3]^2[\mathrm{Cl}_2]$ [Add eqns (ii) and (iii).]

which implies that

$$[\mathrm{CCl}_3] = \left(\frac{I_\mathrm{a}}{k_4[\mathrm{Cl}_2]}\right)^{1/2}$$

Then, with (i),

$$\frac{\mathrm{d}[\mathrm{CCl}_4]}{\mathrm{d}t} = 2I_\mathrm{a} + \frac{k_3 I_\mathrm{a}^{1/2}[\mathrm{Cl}_2]^{1/2}}{k_4^{1/2}}$$

When the pressure of chlorine is high, and the initiation rate is slow (in the sense that the lowest powers of I_a dominate), the second term dominates the first, giving

$$\frac{\mathrm{d}[\mathrm{CCl}_4]}{\mathrm{d}t} = \frac{k_3 I_\mathrm{a}^{1/2}}{k_4^{1/2}}[\mathrm{Cl}_2]^{1/2} = \boxed{k_\mathrm{r} I_\mathrm{a}^{1/2}[\mathrm{Cl}_2]^{1/2}}$$

with $k_\mathrm{r} = \dfrac{k_3}{k_4^{1/2}}$. It seems necessary to suppose that Cl + Cl recombination (which needs a third body) is unimportant.

21 Reaction dynamics

21A Collision theory

Answers to discussion questions

D21A.1 Collision theory expresses a rate of reaction as a fraction of the rate of collision, on the assumption that reaction happens only between colliding molecules, and then only if the collision has enough energy and the proper orientation. So the rate of reaction is directly proportional to the rate of collision, and the expression of this rate comes directly from kinetic-molecular theory. The fraction of collisions energetic enough for reaction also comes from kinetic-molecular theory via the Boltzmann distribution of energy.

D21A.3 The harpoon mechanism accounts for the large steric factor of reactions of the kind $K+Br_2 \rightarrow KBr+Br$ in beams. It is supposed that an electron hops across from K to Br_2 when they are within a certain distance, and then the two resulting ions are drawn together by their mutual Coulombic attraction.

Solutions to exercises

E21A.1(a) The collision frequency is [1B.11a]

$$z = \sigma v_{\text{rel}} \mathcal{N}$$

where $v_{\text{rel}} = \left(\frac{16kT}{\pi m}\right)^{1/2}$ [1B.10a & 1B.9], $\sigma = \pi d^2 = 4\pi R^2$, and $\mathcal{N} = \frac{p}{kT}$

$$\text{Therefore, } z = \sigma\mathcal{N}\left(\frac{16kT}{\pi m}\right)^{1/2} = 16pR^2\left(\frac{\pi}{mkT}\right)^{1/2}$$

$$= 16\times(120\times10^3\ \text{Pa})\times(190\times10^{-12}\ \text{m})^2$$

$$\times\left(\frac{\pi}{17.03\ m_{\text{u}}\times1.661\times10^{-27}\ \text{kg}\ m_{\text{u}}^{-1}\times1.381\times10^{-23}\ \text{J K}^{-1}\times303\ \text{K}}\right)^{1/2}$$

$$= \boxed{1.13\times10^{10}\ \text{s}^{-1}}$$

The collision density is [*Justification* 21A.1]

$$Z=\frac{z\mathcal{N}_A}{2}=\frac{z}{2}\left(\frac{p}{kT}\right)=\frac{1.13\times10^{10}\ \mathrm{s^{-1}}}{2}\left(\frac{120\times10^{3}\ \mathrm{Pa}}{1.381\times10^{-23}\ \mathrm{J\,K^{-1}}\times303\ \mathrm{K}}\right)$$

$$=\boxed{1.62\times10^{35}\ \mathrm{s^{-1}\ m^{-3}}}$$

For the percentage increase at constant volume, note that $\mathcal{N}$ is constant at constant volume, so the only constant-volume temperature dependence on z (and on Z) is in the speed factor.

$$z\propto T^{1/2}\quad\text{so}\quad \frac{1}{z}\left(\frac{\partial z}{\partial T}\right)_V=\frac{1}{2T}\quad\text{and}\quad \frac{1}{Z}\left(\frac{\partial Z}{\partial T}\right)_V=\frac{1}{2T}$$

Therefore $\dfrac{\delta z}{z}=\dfrac{\delta T}{Z}\approx\dfrac{\delta T}{2T}=\dfrac{1}{2}\left(\dfrac{10\ \mathrm{K}}{303\ \mathrm{K}}\right)=0.017$

so both z and Z increase by about $\boxed{1.7\text{ per cent}}$.

E21A.2(a) The fraction of collisions having at least E_a along the line of flight may be inferred by dividing out of the collision-theory rate constant (eqn 21A.9) those factors that can be identified as belonging to the steric factor or collision rate: $f=e^{-E_a/RT}$

(i) (1) $\dfrac{E_a}{RT}=\dfrac{20\times10^{3}\ \mathrm{J\,mol^{-1}}}{(8.3145\ \mathrm{J\,K^{-1}\ mol^{-1}})\times(350\ \mathrm{K})}=6.9$ so $f=e^{-6.9}=\boxed{1.04\times10^{-3}}$

(2) $\dfrac{E_a}{RT}=\dfrac{20\times10^{3}\ \mathrm{J\,mol^{-1}}}{(8.3145\ \mathrm{J\,K^{-1}\ mol^{-1}})\times(900\ \mathrm{K})}=2.67$ so $f=e^{-2.67}=\boxed{0.069}$

(ii) (1) $\dfrac{E_a}{RT}=\dfrac{100\times10^{3}\ \mathrm{J\,mol^{-1}}}{(8.3145\ \mathrm{J\,K^{-1}\ mol^{-1}})\times(350\ \mathrm{K})}=34.4$ so $f=e^{-34.4}=\boxed{1.19\times10^{-15}}$

(2) $\dfrac{E_a}{RT}=\dfrac{100\times10^{3}\ \mathrm{J\,mol^{-1}}}{(8.3145\ \mathrm{J\,K^{-1}\ mol^{-1}})\times(900\ \mathrm{K})}=13.4$ so $f=e^{-13.4}=\boxed{1.57\times10^{-6}}$

E21A.3(a) A straightforward approach would be to compute $f=e^{-E_a/RT}$ at the new temperature and compare it to that at the old temperature. An approximate approach would be to note that f changes from $f_0=e^{-E_a/RT}$ to $\exp\left(\dfrac{-E_a}{RT(1+x)}\right)$, where x is the fractional increase in the temperature. If x is small, the exponent changes from $-E_a/RT$ to approximately $-E_a(1-x)/RT$ and f changes from f_0 to

$$f\approx e^{-E_a(1-x)/RT}=e^{-E_a/RT}\left(e^{-E_a/RT}\right)^{-x}=f_0f_0^{-x}$$

Thus the new fraction is the old one times a factor of f_0^{-x}. The increase in f expressed as a percentage is

$$\frac{f-f_0}{f_0}\times100\%=\frac{f_0f_0^{-x}-f_0}{f_0}\times100\%=(f_0^{-x}-1)\times100\%$$

(i) (1) $f_0^{-x} = (1.04\times10^{-3})^{-10/350} = 1.22$ and the percentage change is $\boxed{22\%}$

(2) $f_0^{-x} = (0.069)^{-10/900} = 1.03$ and the percentage change is $\boxed{3\%}$

(ii) (1) $f_0^{-x} = (1.19\times10^{-15})^{-10/350} = 2.7$ and the percentage change is $\boxed{170\%}$

(2) $f_0^{-x} = (1.57\times10^{-6})^{-10/900} = 1.16$ and the percentage change is $\boxed{16\%}$

E21A.4(a) $k_r = P\sigma\left(\frac{8kT}{\pi\mu}\right)^{1/2} N_A e^{-E_a/RT}$ [21A.9]

$$k_r = 0.36\times(10^{-9}\,\text{m})^2\times\left(\frac{8\times(1.381\times10^{-23}\,\text{J K}^{-1})\times(650\,\text{K})}{\pi\times(3.32\times10^{-27}\,\text{kg})}\right)^{1/2}\times(6.022\times10^{23}\,\text{mol}^{-1})$$

$$\times\exp\left(\frac{-171\times10^3\,\text{J}}{(8.3145\,\text{J K}^{-1}\,\text{mol}^{-1})\times(650\,\text{K})}\right)$$

$$= \boxed{1.03\times10^{-5}\,\text{m}^3\,\text{mol}^{-1}\,\text{s}^{-1}} = \boxed{1.03\times10^{-2}\,\text{dm}^3\,\text{mol}^{-1}\,\text{s}^{-1}}$$

E21A.5(a) The steric factor, P, is

$$P = \frac{\sigma^*}{\sigma}\ \text{[Section 21A.1(c)]}$$

The mean collision cross-section is $\sigma = \pi d^2$ with $d = (d_A + d_B)/2$

Get the diameters from the collision cross-sections:

$$d_A = (\sigma_A/\pi)^{1/2} \quad \text{and} \quad d_B = (\sigma_B/\pi)^{1/2},$$

$$\text{so}\quad \sigma = \frac{\pi}{4}\left\{\left(\frac{\sigma_A}{\pi}\right)^{1/2} + \left(\frac{\sigma_B}{\pi}\right)^{1/2}\right\}^2 = \frac{(\sigma_A^{1/2}+\sigma_B^{1/2})^2}{4} = \frac{\{(0.95\,\text{nm}^2)^{1/2}+(0.65\,\text{nm}^2)^{1/2}\}^2}{4}$$

$$= \boxed{0.79\,\text{nm}^2}$$

Therefore, $P = \frac{9.2\times10^{-22}\,\text{m}^2}{0.79\times(10^{-9}\,\text{m})^2} = \boxed{1.16\times10^{-3}}$

E21A.6(a) According to RRK theory, the steric P-factor is given by eqn 21A.10a

$$P = \left(1-\frac{E^*}{E}\right)^{s-1}$$

where s is the number of vibrational modes in the reacting molecule. For a non-linear molecule composed of N atoms, the number of modes is [Section 12E.1]

$$s = 3N - 6 = 3\times5 - 6 = 9.$$

Rearranging eqn 21A.10a yields

$$\frac{E^*}{E} = 1 - P^{\frac{1}{s-1}} = 1-(3.0\times10^{-5})^{\frac{1}{8}} = \boxed{0.73}$$

E21A.7(a) According to RRK theory, the steric P-factor is given by eqn 21A.10a

$$P=\left(1-\frac{E^*}{E}\right)^{s-1}=\left(1-\frac{200\ \text{kJ mol}^{-1}}{250\ \text{kJ mol}^{-1}}\right)^{10-1}=\boxed{5.1\times10^{-7}}$$

Solutions to problems

P21A.1 Comparing the Arrhenius equation [20D.1] to eqn 21A.9, with $\sigma^*=P\sigma$ yields

$$A=N_A\sigma^*\left(\frac{8kT}{\pi\mu}\right)^{1/2}$$

where $\mu=m(CH_3)/2=(15.03\ m_u/2)\times1.6605\times10^{-27}\ \text{kg}\ m_u^{-1}=1.248\times10^{-26}\ \text{kg}$

$$\text{So}\quad A=\sigma^*\times(6.022\times10^{23}\ \text{mol}^{-1})\times\left(\frac{(8)\times(1.381\times10^{-23}\ \text{J K}^{-1})\times(298\ \text{K})}{(\pi)\times(1.248\times10^{-26}\ \text{kg})}\right)^{1/2}$$

$$=(5.52\times10^{26})\times(\sigma^*)\ \text{mol}^{-1}\ \text{m s}^{-1}$$

$$\text{(a)}\quad \sigma^*=\frac{2.4\times10^{10}\ \text{mol}^{-1}\ \text{dm}^3\ \text{s}^{-1}}{5.52\times10^{26}\ \text{mol}^{-1}\ \text{m s}^{-1}}=\frac{2.4\times10^{7}\ \text{mol}^{-1}\ \text{m}^3\ \text{s}^{-1}}{5.52\times10^{26}\ \text{mol}^{-1}\ \text{m s}^{-1}}=\boxed{4.3\bar{5}\times10^{-20}\ \text{m}^2}$$

(b) Take $\sigma\approx\pi d^2$ and estimate d as twice the bond length; therefore

$$\sigma=(\pi)\times(154\times2\times10^{-12}\ \text{m})^2=3.0\times10^{-19}\ \text{m}^2$$

$$\text{Hence}\quad P=\frac{\sigma^*}{\sigma}=\frac{4.3\bar{5}\times10^{-20}}{3.0\times10^{-19}}=\boxed{0.15}$$

P21A.3 For radical recombination it has been found experimentally that $E_a\approx0$. The maximum rate of recombination is obtained when $P=1$ (or more), and then (see Problem 21A.1)

$$k_r=A=\sigma^*N_A\left(\frac{8kT}{\pi\mu}\right)^{1/2}=4\sigma^*N_A\left(\frac{kT}{\pi m}\right)^{1/2}\quad[\mu=\tfrac{1}{2}m]$$

$$\sigma^*\approx\pi d^2=\pi\times(308\times10^{-12}\ \text{m})^2=3.0\times10^{-19}\ \text{m}^2$$

$$k_r=(4)\times(3.0\times10^{-19}\ \text{m}^2)\times(6.022\times10^{23}\ \text{mol}^{-1})$$

$$\text{Hence}\quad\times\left(\frac{(1.381\times10^{-23}\ \text{J K}^{-1})\times(298\ \text{K})}{(\pi)\times(15.03\ m_u)\times(1.6605\times10^{-27}\ \text{kg}/m_u)}\right)^{1/2}$$

$$=1.7\times10^{8}\ \text{m}^3\ \text{mol}^{-1}\ \text{s}^{-1}=\boxed{1.7\times10^{11}\ \text{mol}^{-1}\ \text{dm}^3\ \text{s}^{-1}}$$

The rate constant is for the rate law

$$v=k_r[CH_3]^2$$

$$\text{Therefore}\quad -\frac{d[CH_3]}{dt}=2k_r[CH_3]^2$$

and its solution is $\frac{1}{[CH_3]}-\frac{1}{[CH_3]_0}=2k_r t$

For 90% recombination, $[CH_3]=0.10\times[CH_3]_0$, which occurs when

$$2k_r t=\frac{9}{[CH_3]_0} \quad \text{or} \quad t=\frac{9}{2k_r[CH_3]_0}$$

The mole fraction of CH_3 radicals when 10 mol % of ethane is dissociated is

$$\frac{(2)\times(0.10)}{1+0.10}=0.18$$

The initial partial pressure of CH_3 radicals is thus

$$p_0=0.18p=1.8\times10^4\ \text{Pa}$$

and $[CH_3]_0=\frac{1.8\times10^4\ \text{Pa}}{RT}$

Therefore $t=\frac{9RT}{(2k_r)\times(1.8\times10^4\ \text{Pa})}=\frac{(9)\times(8.3145\ \text{J K}^{-1}\ \text{mol}^{-1})\times(298\text{K})}{(1.7\times10^8\ \text{m}^3\ \text{mol}^{-1}\ \text{s}^{-1})\times(3.6\times10^4\ \text{Pa})}$.

$$=\boxed{3.6\,\text{ns}}$$

P21A.5 Linear regression analysis of ln(rate constant) against $1/T$ yields the following results:

$$\ln(k_r/22.4\ \text{dm}^3\ \text{mol}^{-1}\ \text{min}^{-1})=C+B/T$$

where $C=34.36$ (standard deviation 0.36),

$B=-23\,227$ K (standard deviation 252 K),

and $R=0.99976$ (indicating a good fit).

$$\ln(k_r'/22.4\text{dm}^3\ \text{mol}^{-1}\ \text{min}^{-1})=C'+B'/T$$

where $C'=28.30$ (standard deviation = 0.84),

$B'=-21\,065$K (standard deviation = 582K),

and $R=0.99848$ (indicating a good fit).

The regression parameters can be used in the calculation of the pre-exponential factor (A) and the activation energy (E_a) using $\ln k_r=\ln A-E_a/RT$.

$$\ln A=C+\ln(22.4)=37.47$$

so $A=1.87\times10^{16}\text{dm}^3\ \text{mol}^{-1}\ \text{min}^{-1}=\boxed{3.12\times10^{14}\text{dm}^3\ \text{mol}^{-1}\ \text{s}^{-1}}$

$$E_a=-RB=-(8.3145\ \text{J K}^{-1}\text{mol}^{-1})\times(-23227\ \text{K})\times\left(\frac{1\ \text{kJ}}{10^3\ \text{J}}\right)=\boxed{193\ \text{kJ mol}^{-1}}$$

$$\ln A' = C' + \ln(22.4) = 31.41$$

so $A' = 4.37\times10^{13}\ \text{dm}^3\,\text{mol}^{-1}\,\text{min}^{-1} = \boxed{7.29\times10^{11}\ \text{dm}^3\,\text{mol}^{-1}\,\text{s}^{-1}}$

$$E_a' = -RB' = -(8.3145\ \text{J K}^{-1}\text{mol}^{-1})\times(-21\,065\ \text{K})\times\left(\frac{1\ \text{kJ}}{10^3\ \text{J}}\right) = \boxed{175\ \text{kJ mol}^{-1}}$$

To summarize:

	$A/(\text{dm}^3\ \text{mol}^{-1}\text{s}^{-1})$	$E_a/(\text{kJ mol}^{-1})$
forward reaction	3.12×10^{14}	193
reverse reaction	7.29×10^{11}	175

Both sets of data, k and k', fit the Arrhenius equation very well. To the extent that the Arrhenius equation [20D.1] is consistent with collision theory (whose rate constant is given by eqn. 21A.9), these data are also consistent with collision theory. The numerical values for A' and A may be compared to the results of *Brief illustration* 21A.3. The pre-exponential factor for the reverse reaction is comparable to the one estimated in that problem based on collision density; however, the pre-factor for the forward reaction appears to be much larger than collision density. Because they imply an excessively high collision density, these data are not really compatible with collision theory.

P21A.7 $$P = \frac{n!(n-n^*+s-1)!}{(n-n^*)!(n+s-1)!}$$

Take the natural logarithm of both sides, turning the log of a product (quotient) into a sum (difference) of logs:

$$\ln P = \ln n! + \ln(n-n^*+s-1)! - \ln(n-n^*)! - \ln(n+s-1)!$$

Now apply Stirling's approximation to the logs of factorials:

$$\ln P \approx n(\ln n - 1) + (n-n^*+s-1)\{\ln(n-n^*+s-1)-1\} - (n-n^*)\{\ln(n-n^*)-1\} - (n+s-1)\{\ln(n+s-1)-1\}$$

Replace $n-n^*+s-1$ by $n-n^*$ and $n+s-1$ by n inside log terms:

$$\ln P \approx n(\ln n - 1) + (n-n^*+s-1)\{\ln(n-n^*)-1\} - (n-n^*)\{\ln(n-n^*)-1\} - (n+s-1)(\ln n - 1)$$

Collect terms and simplify:

$$\ln P \approx (n-n-s+1)(\ln n - 1) + (n-n^*+s-1-n+n^*)\{\ln(n-n^*)-1\}$$

$$\ln P \approx (s-1)\{\ln(n-n^*)-1-\ln n+1\} = (s-1)\ln\left(\frac{n-n^*}{n}\right) = \ln\left(\frac{n-n^*}{n}\right)^{s-1}$$

Take the antilog of both sides to get the desired result:

$$P \approx \left(\frac{n-n^*}{n}\right)^{s-1}$$

21B Diffusion-controlled reactions

Answer to discussion question

D21B.1 A reaction in solution can be regarded as the outcome of two stages: one is the encounter of two reactant species; this is followed by their reaction in the second stage, if they acquire their activation energy. If the rate-determining step is the former, then the reaction is said to be diffusion controlled. If the rate-determining step is the latter, then the reaction is activation controlled. For a reaction of the form $A+B \rightarrow P$ that obeys the second-order rate law $v = k_r[A][B]$, in the diffusion-controlled regime,

$$k_r = k_d = 4\pi R^* D N_A$$

where D is the sum of the diffusion coefficients of the two reactant species and R^* is the distance at which reaction occurs. A further approximation is that each molecule obeys the Stokes–Einstein relation and Stokes' law, and then

$$k_d \approx \frac{8RT}{3\eta} \quad [21B.4]$$

where η is the viscosity of the medium. The result suggests that k_d is independent of the radii of the reactants. It also suggests that the rate constant depends only weakly on temperature, so the activation energy (eqn 20D.3) is small.

Solutions to exercises

E21B.1(a) The rate constant for a diffusion-controlled bimolecular reaction is

$$k_d = 4\pi R^* D N_A \quad [21B.3]$$

where $D = D_A + D_B = 2\times(6\times10^{-9}\,\text{m}^2\,\text{s}^{-1}) = 1.2\times10^{-8}\,\text{m}^2\,\text{s}^{-1}$

$$k_d = 4\pi\times(0.5\times10^{-9}\,\text{m})\times(1.2\times10^{-8}\,\text{m}^2\,\text{s}^{-1})\times(6.022\times10^{23}\,\text{mol}^{-1})$$

$$k_d = \boxed{4.\overline{5}\times10^{7}\,\text{m}^3\,\text{mol}^{-1}\,\text{s}^{-1}} = \boxed{4.\overline{5}\times10^{10}\,\text{dm}^3\,\text{mol}^{-1}\,\text{s}^{-1}}$$

E21B.2(a) The rate constant for a diffusion-controlled bimolecular reaction is

$$k_d = \frac{8RT}{3\eta}\ [21A.4] = \frac{8\times(8.3145\,\text{J K}^{-1}\,\text{mol}^{-1})\times(298\,\text{K})}{3\eta} = \frac{6.61\times10^{3}\,\text{J mol}^{-1}}{\eta}$$

(i) For water, $\eta = 1.00\times10^{-3}\,\text{kg m}^{-1}\,\text{s}^{-1}$

$$k_d = \frac{6.61\times10^3\ \text{J mol}^{-1}}{1.00\times10^{-3}\ \text{kg m}^{-1}\ \text{s}^{-1}} = \boxed{6.61\times10^6\ \text{m}^3\ \text{mol}^{-1}\ \text{s}^{-1}} = \boxed{6.61\times10^9\ \text{dm}^3\ \text{mol}^{-1}\ \text{s}^{-1}}$$

(ii) For pentane, $\eta = 2.2\times10^{-4}\ \text{kg m}^{-1}\ \text{s}^{-1}$

$$k_d = \frac{6.61\times10^3\ \text{J mol}^{-1}}{2.2\times10^{-4}\ \text{kg m}^{-1}\ \text{s}^{-1}} = \boxed{3.0\times10^7\ \text{m}^3\ \text{mol}^{-1}\ \text{s}^{-1}} = \boxed{3.0\times10^{10}\ \text{dm}^3\ \text{mol}^{-1}\ \text{s}^{-1}}$$

E21B.3(a) The rate constant for a diffusion-controlled bimolecular reaction is [21B.4]

$$k_d = \frac{8RT}{3\eta} = \frac{8\times(8.3145\ \text{J K}^{-1}\ \text{mol}^{-1})\times(320\ \text{K})}{3\times(0.89\times10^{-3}\ \text{kg m}^{-1}\ \text{s}^{-1})}$$

$$= \boxed{8.0\times10^6\ \text{m}^3\ \text{mol}^{-1}\ \text{s}^{-1}} = \boxed{8.0\times10^9\ \text{dm}^3\ \text{mol}^{-1}\ \text{s}^{-1}}$$

Since this reaction is elementary bimolecular it is second-order; hence

$$t_{1/2} = \frac{1}{2k_d[\text{A}]_0}\ \left[\text{Table 20B.3, with } k_r = 2k_d \text{ because 2 atoms are consumed}\right]$$

$$\text{so}\ \ t_{1/2} = \frac{1}{2\times(8.0\times10^9\ \text{dm}^3\ \text{mol}^{-1}\ \text{s}^{-1})\times(1.5\times10^{-3}\ \text{mol dm}^{-3})} = \boxed{4.2\times10^{-8}\ \text{s}}$$

E21B.4(a) Since the reaction is diffusion controlled, the rate-limiting step is bimolecular and therefore second-order; hence

$$\frac{d[\text{P}]}{dt} = k_d[\text{A}][\text{B}]$$

where

$$k_d = 4\pi R^* D N_A [21B.3] = 4\pi N_A R^*(D_A + D_B)$$

$$= 4\pi N_A \times (R_A + R_B)\times \frac{kT}{6\pi\eta}\left(\frac{1}{R_A}+\frac{1}{R_B}\right)[19B.19b] = \frac{2RT}{3\eta}(R_A+R_B)\times\left(\frac{1}{R_A}+\frac{1}{R_B}\right)$$

$$k_d = \frac{2\times(8.3145\ \text{J K}^{-1}\ \text{mol}^{-1})\times(313\ \text{K})}{3\times(2.93\times10^{-3}\ \text{kg m}^{-1}\text{s}^{-1})}\times(655+1820)\times\left(\frac{1}{655}+\frac{1}{1820}\right)$$

$$= 3.04\times10^6\ \text{m}^3\,\text{mol}^{-1}\,\text{s}^{-1} = 3.04\times10^9\ \text{dm}^3\,\text{mol}^{-1}\,\text{s}^{-1}$$

Therefore, the initial rate is

$$\frac{d[\text{P}]}{dt} = (3.04\times10^9\ \text{dm}^3\ \text{mol}^{-1}\ \text{s}^{-1})\times(0.170\,\text{mol dm}^{-3})\times(0.350\,\text{mol dm}^{-3})$$

$$= \boxed{1.81\times10^8\ \text{mol dm}^{-3}\ \text{s}^{-1}}$$

Comment. If the approximation of eqn 21B.4 is used, $k_d = 2.37\times10^9\ \text{dm}^3\ \text{mol}^{-1}\ \text{s}^{-1}$. In this case the approximation results in a difference of about 25% compared to the expression used above.

Solutions to problems

P21B.1 We are to show that eqn 21B.8

$$[\mathrm{J}]^* = [\mathrm{J}]\mathrm{e}^{-k_r t}$$

is a solution of eqn 21B.7,

$$\frac{\partial[\mathrm{J}]^*}{\partial t} = D\frac{\partial^2[\mathrm{J}]^*}{\partial x^2} - k_r[\mathrm{J}]^*$$

provided that [J] is a solution of

$$\frac{\partial[\mathrm{J}]}{\partial t} = D\frac{\partial^2[\mathrm{J}]}{\partial x^2}$$

Evaluate the derivatives of [J]*:

$$\frac{\partial[\mathrm{J}]^*}{\partial t} = \frac{\partial[\mathrm{J}]}{\partial t}\mathrm{e}^{-k_r t} - k_r[\mathrm{J}]\mathrm{e}^{-k_r t} \quad \text{and} \quad \frac{\partial^2[\mathrm{J}]^*}{\partial x^2} = \frac{\partial^2[\mathrm{J}]}{\partial x^2}\mathrm{e}^{-k_r t}$$

Use the fact that [J] is a solution in the absence of reaction

$$D\frac{\partial^2[\mathrm{J}]^*}{\partial x^2} = D\frac{\partial^2[\mathrm{J}]}{\partial x^2}\mathrm{e}^{-k_r t} = \frac{\partial[\mathrm{J}]}{\partial t}\mathrm{e}^{-k_r t} = \frac{\partial[\mathrm{J}]^*}{\partial t} + k_r[\mathrm{J}]\mathrm{e}^{-k_r t}$$

which gives us back eqn 21B.7, as required.

P21B.3 We are to show that

$$[\mathrm{J}]^* = k_r\int_0^t [\mathrm{J}]\mathrm{e}^{-k_r t}\,\mathrm{d}t + [\mathrm{J}]\mathrm{e}^{-k_r t}$$

is a solution of eqn 21B.7,

$$\frac{\partial[\mathrm{J}]^*}{\partial t} = D\frac{\partial^2[\mathrm{J}]^*}{\partial x^2} - k_r[\mathrm{J}]^*,$$

(subject to the stated initial condition and boundary condition) provided that [J] is a solution of

$$\frac{\partial[\mathrm{J}]}{\partial t} = D\frac{\partial^2[\mathrm{J}]}{\partial x^2}.$$

First evaluate the derivatives of [J]*:

$$\frac{\partial[\mathrm{J}]^*}{\partial t} = k_r[\mathrm{J}]\mathrm{e}^{-k_r t} + \left(\frac{\partial[\mathrm{J}]}{\partial t}\right)\mathrm{e}^{-k_r t} - k_r[\mathrm{J}]\mathrm{e}^{-k_r t} = \left(\frac{\partial[\mathrm{J}]}{\partial t}\right)\mathrm{e}^{-k_r t}$$

and $$\frac{\partial^2[\mathrm{J}]^*}{\partial x^2} = k_r\int_0^t\left(\frac{\partial^2[\mathrm{J}]}{\partial x^2}\right)\mathrm{e}^{-k_r t}\,\mathrm{d}t + \left(\frac{\partial^2[\mathrm{J}]}{\partial x^2}\right)\mathrm{e}^{-k_r t}.$$

Thus, $$D\frac{\partial^2[\mathrm{J}]^*}{\partial x^2} = k_r\int_0^t\left(\frac{\partial[\mathrm{J}]}{\partial t}\right)\mathrm{e}^{-k_r t}\,\mathrm{d}t + \left(\frac{\partial[\mathrm{J}]}{\partial t}\right)\mathrm{e}^{-k_r t} = k_r\int_0^t\left(\frac{\partial[\mathrm{J}]^*}{\partial t}\right)\mathrm{d}t + \frac{\partial[\mathrm{J}]^*}{\partial t}.$$

The integral on the right is

$$\int_0^t \left(\frac{\partial[\mathrm{J}]^*}{\partial t}\right)\mathrm{d}t = [\mathrm{J}]^* - [\mathrm{J}]_0^* = [\mathrm{J}]^*,$$

where we have used the condition that requires [J] and therefore also [J]* to be zero initially.

$$\text{Thus,}\quad D\frac{\partial^2[\mathrm{J}]^*}{\partial x^2} = k_\mathrm{r}[\mathrm{J}]^* + \frac{\partial[\mathrm{J}]^*}{\partial t},$$

which is equivalent to eqn 21B.7.

21C Transition-state theory

Answers to discussion questions

D21C.1 The Eyring equation (eqn 21C.10) results from activated complex theory which is an attempt to account for the rate constants of bimolecular reactions of the form $\mathrm{A+B \rightleftharpoons C^{\ddagger} \rightarrow P}$ in terms of the formation of an activated complex. In the formulation of the theory, it is assumed that the activated complex and the reactants are in equilibrium, and the concentration of activated complex is calculated in terms of an equilibrium constant, which in turn is calculated from the partition functions of the reactants and a postulated form of the activated complex. It is further supposed that one normal mode of the activated complex, the one corresponding to displacement along the reaction coordinate, has a very low force constant. Displacement along this mode leads to products, provided that the complex enters a certain configuration of its atoms, known as the transition state. The derivation of the equilibrium constant from the partition functions leads to eqn 21C.9 and in turn to eqn 21C.10, the Eyring equation.

D21C.3 In the kinetic salt effect, the rate of a reaction in solution is changed by modification of the ionic strength of the medium. If the reactant ions have the same charge sign (as in cation/cation or anion/anion reactions), then an increase in ionic strength increases the rate constant. If the reactant ions have opposite signs (as in cation/anion reactions), then an increase in ionic strength decreases the rate constant. In the former case, the effect can be traced to the denser ionic atmosphere that forms round the newly formed and highly charged ion that constitutes the activated complex and the stronger interaction of that ion with the atmosphere. In the latter case, the ion corresponding to the activated complex has a lower charge than the reactants and hence it has a more diffuse ionic atmosphere and interacts with it more weakly. In the limit of low ionic strength the rate constant can be expected to follow the relation (eqn 21C.18)

$$\log k_\mathrm{r} = \log k_\mathrm{r}^\circ + 2Az_\mathrm{A}z_\mathrm{B}I^{1/2}$$

Solutions to exercises

E21C.1(a) The entropy of activation for a bimolecular solution reaction is [Section 21C.2(a) footnote]

$$\Delta^{\ddagger}H = E_a - RT = (8681\,\text{K} - 303\,\text{K}) \times 8.3145\,\text{J mol}^{-1}\,\text{K}^{-1} = \boxed{+69.7\,\text{kJ mol}^{-1}}$$

$$k_r = Be^{\Delta^{\ddagger}S/R}e^{-\Delta^{\ddagger}H/RT}, \quad B = \left(\frac{kT}{h}\right)\times\left(\frac{RT}{p^{\ominus}}\right) = \frac{kRT^2}{hp^{\ominus}}$$

$$= Be^{\Delta^{\ddagger}S/R}e^{-E_a/RT}e = Ae^{-E_a/RT}$$

Therefore, $A = eBe^{\Delta^{\ddagger}S/R}$, implying that $\Delta^{\ddagger}S = R\left(\ln\frac{A}{B} - 1\right)$

$$B = \frac{(1.381\times10^{-23}\,\text{J K}^{-1})\times(8.3145\,\text{J K}^{-1}\,\text{mol}^{-1})\times(303\,\text{K})^2}{6.626\times10^{-34}\,\text{J s}\times10^5\,\text{Pa}}$$

$$= 1.59\times10^{11}\,\text{m}^3\,\text{mol}^{-1}\,\text{s}^{-1} = 1.59\times10^{14}\,\text{dm}^3\,\text{mol}^{-1}\,\text{s}^{-1}$$

and hence $\Delta^{\ddagger}S = R\left[\ln\left(\frac{2.05\times10^{13}\,\text{dm}^3\,\text{mol}^{-1}\,\text{s}^{-1}}{1.59\times10^{14}\,\text{dm}^3\,\text{mol}^{-1}\,\text{s}^{-1}}\right) - 1\right]$

$$= 8.3145\,\text{J K}^{-1}\,\text{mol}^{-1}\times(-3.05) = \boxed{-25\,\text{J K}^{-1}\,\text{mol}^{-1}}$$

E21C.2(a) The enthalpy of activation for a bimolecular solution reaction is [Section 21C.2(a) footnote]

$$\Delta^{\ddagger}H = E_a - RT = 8.3145\,\text{J K}^{-1}\text{mol}^{-1}\times(9134\,\text{K} - 303\,\text{K}) = \boxed{+73.4\,\text{kJ mol}^{-1}}$$

The entropy of activation is [Exercise 21C.1(a)]

$$\Delta^{\ddagger}S = R\left(\ln\frac{A}{B} - 1\right)$$

with $B = \dfrac{kRT^2}{hp^{\ominus}} = 1.59\times10^{14}\,\text{dm}^3\,\text{mol}^{-1}\,\text{s}^{-1}$

Therefore, $\Delta^{\ddagger}S = 8.3145\,\text{J K}^{-1}\,\text{mol}^{-1}\times\left[\ln\left(\frac{7.78\times10^{14}}{1.59\times10^{14}}\right) - 1\right] = +4.9\,\text{J K}^{-1}\,\text{mol}^{-1}$

Hence, $\Delta^{\ddagger}G = \Delta^{\ddagger}H - T\Delta^{\ddagger}S = \{73.4 - (303)\times(4.9\times10^{-3})\}\,\text{kJ mol}^{-1} = \boxed{+71.9\,\text{kJ mol}^{-1}}$

E21C.3(a) Use eqn 21C.15(a) to relate a bimolecular gas-phase rate constant to activation energy and entropy:

$$k_r = e^2Be^{\Delta^{\ddagger}S/R}e^{-E_a/RT}$$

where $B = \left(\frac{kT}{h}\right)\times\left(\frac{RT}{p^{\ominus}}\right)$ [21C.14]

$$= \frac{(1.381\times10^{-23}\,\text{J K}^{-1})\times(338\,\text{K})^2\times(8.3145\,\text{J mol}^{-1}\,\text{K}^{-1})}{(6.626\times10^{-34}\,\text{J s})\times(10^5\,\text{Pa})}$$

$$= 1.98\times10^{11}\,\text{m}^3\,\text{mol}^{-1}\,\text{s}^{-1}$$

Solve for the entropy of activation:

$$\Delta^{\ddagger}S = R\left(\ln\frac{k_r}{B} - 2\right) + \frac{E_a}{T}$$

The derivations in Section 21C.2(a) are based on a k_r that contains concentration units, whereas the rate constant given here has pressure units. So k_r in the above expressions is not the rate constant given in the exercise, but

$$k_r = 7.84\times10^{-3}\,\text{kPa}^{-1}\text{s}^{-1}\times RT$$

$$k_r = 7.84\times10^{-3}\times\left(10^3\,\text{Pa}\right)^{-1}\text{s}^{-1}\times 8.3145\,\text{J K}^{-1}\text{mol}^{-1}\times 338\,\text{K} = 0.0220\,\text{m}^3\text{mol}^{-1}\text{s}^{-1}$$

Hence

$$\Delta^{\ddagger}S = 8.3145\,\text{J K}^{-1}\,\text{mol}^{-1}\times\left(\ln\frac{0.0220\,\text{m}^3\,\text{mol}^{-1}\,\text{s}^{-1}}{1.98\times10^{11}\,\text{m}^3\,\text{mol}^{-1}\,\text{s}^{-1}} - 2\right) + \frac{58.6\times10^3\,\text{J mol}^{-1}}{338\,\text{K}}$$

$$= \boxed{-91\,\text{J K}^{-1}\,\text{mol}^{-1}}$$

E21C.4(a) For a bimolecular gas-phase reaction [Exercise 21C.3(a)],

$$\Delta^{\ddagger}S = R\left(\ln\frac{k_r}{B} - 2\right) + \frac{E_a}{T} = R\left(\ln\frac{A}{B} - \frac{E_a}{RT} - 2\right) + \frac{E_a}{T} = R\left(\ln\frac{A}{B} - 2\right)$$

where $B = \dfrac{kRT^2}{hp^{\ominus}}$

For two structureless particles, the rate constant is the same as that of collision theory

$$k_r = N_A\sigma^*\left(\frac{8kT}{\pi\mu}\right)^{1/2} e^{-\Delta E_0/RT} \quad \text{[Example 21C.1]}$$

The activation energy is [20D.3]

$$E_a = RT^2\frac{d\ln k_r}{dT} = RT^2\frac{d}{dT}\left(\ln N_A\sigma^* + \frac{1}{2}\ln\frac{8k}{\pi\mu} + \frac{1}{2}\ln T - \frac{\Delta E_0}{RT}\right)$$

$$= RT^2\left(\frac{1}{2T} + \frac{\Delta E_0}{RT^2}\right) = \Delta E_0 + \frac{RT}{2},$$

so the prefactor is

$$A = k_r e^{E_a/RT} = N_A\sigma^*\left(\frac{8kT}{\pi\mu}\right)^{1/2} e^{-\Delta E_0/RT}\left(e^{\Delta E_0/RT}e^{1/2}\right) = N_A\sigma^*\left(\frac{8kT}{\pi\mu}\right)^{1/2} e^{1/2}$$

Hence $\Delta^{\ddagger}S = R\left\{\ln N_A\sigma^*\left(\dfrac{8kT}{\pi\mu}\right)^{1/2} + \dfrac{1}{2} - \ln\dfrac{kRT^2}{p^{\ominus}h} - 2\right\} = R\left\{\ln\dfrac{\sigma^* p^{\ominus}h}{(kT)^{3/2}}\left(\dfrac{8}{\pi\mu}\right)^{1/2} - \dfrac{3}{2}\right\}$

For identical particles,

$$\mu = m/2 = (65m_u)(1.661\times10^{-27}\,\text{kg}\,m_u^{-1})/2 = 5.4\times10^{-26}\,\text{kg},$$

and hence

$$\Delta^{\ddagger}S = 8.3145\ \mathrm{J\,K^{-1}\,mol^{-1}}$$
$$\times\left\{\ln\frac{0.35\times(10^{-9}\ \mathrm{m})^2\times10^5\ \mathrm{Pa}\times6.626\times10^{-34}\ \mathrm{J\,s}}{(1.381\times10^{-23}\ \mathrm{J\,K^{-1}}\times300\ \mathrm{K})^{3/2}}\left(\frac{8}{\pi\times5.4\times10^{-26}\ \mathrm{kg}}\right)^{1/2}-\frac{3}{2}\right\}$$
$$= \boxed{-74\ \mathrm{J\,K^{-1}\,mol^{-1}}}$$

E21C.5(a) At low pressure, the reaction can be assumed to be bimolecular. (The rate constant is certainly second-order.)

(a) $\Delta^{\ddagger}S = R\left(\ln\frac{A}{B}-2\right)$ [Exercise 21C.4(a)]

where

$$B=\frac{kRT^2}{hp^{\ominus}}\ \text{[Exercise 21C.4(a)]}=\frac{1.381\times10^{-23}\ \mathrm{J\,K^{-1}}\times8.3145\ \mathrm{J\,K^{-1}\,mol^{-1}}\times(298\ \mathrm{K})^2}{6.626\times10^{-34}\ \mathrm{J\,s}\times10^5\ \mathrm{Pa}}$$
$$=1.54\times10^{11}\ \mathrm{m^3\,mol^{-1}\,s^{-1}}=1.54\times10^{14}\ \mathrm{dm^3\,mol^{-1}\,s^{-1}}$$

Hence $\Delta^{\ddagger}S = 8.3145\ \mathrm{J\,K^{-1}\,mol^{-1}}\times\left(\ln\frac{4.6\times10^{12}\ \mathrm{dm^3\,mol^{-1}\,s^{-1}}}{1.54\times10^{14}\ \mathrm{dm^3\,mol^{-1}\,s^{-1}}}-2\right)$

$$=\boxed{-46\ \mathrm{J\,K^{-1}\,mol^{-1}}}$$

(b) The enthalpy of activation for a bimolecular gas-phase reaction is [Section 21C.2(a) footnote]

$$\Delta^{\ddagger}H = E_{\mathrm{a}}-2RT = 10.0\,\mathrm{kJ\,mol^{-1}}-2\times8.3145\,\mathrm{J\,mol^{-1}\,K^{-1}}\times298\ \mathrm{K}=\boxed{+5.0\,\mathrm{kJ\,mol^{-1}}}$$

(c) The Gibbs energy of activation at 298 K is

$$\Delta^{\ddagger}G=\Delta^{\ddagger}H-T\Delta^{\ddagger}S = 5.0\,\mathrm{kJ\,mol^{-1}}-(298\ \mathrm{K})\times(-46\times10^{-3}\,\mathrm{kJ\,K^{-1}\,mol^{-1}})$$
$$\Delta^{\ddagger}G=\boxed{+18.7\,\mathrm{kJ\,mol^{-1}}}$$

E21C.6(a) Use eqn 21C.18 to examine the effect of ionic strength on a rate constant:

$$\log k_{\mathrm{r}} = \log k_{\mathrm{r}}^{\circ} + 2A|z_{\mathrm{A}}z_{\mathrm{B}}|I^{1/2}$$

Hence

$$\log k_{\mathrm{r}}^{\circ} = \log k_{\mathrm{r}} - 2A|z_{\mathrm{A}}z_{\mathrm{B}}|I^{1/2} = \log 12.2-2\times0.509\times|1\times(-1)|\times(0.0525)^{1/2}=0.85$$

and $k_{\mathrm{r}}^{\circ} = \boxed{7.1\ \mathrm{dm^6\,mol^{-2}\,min^{-1}}}$

Solutions to problems

P21C.1 Draw up the following table for an Arrhenius plot:

θ/°C	−24.82	−20.73	−17.02	−13.00	−8.95
T/K	248.33	252.42	256.13	260.15	264.20
10^3 K/T	4.027	3.962	3.904	3.844	3.785
$10^4\,k_r$/s^{-1}	1.22	2.31	4.39	8.50	14.3
ln (k_r/s^{-1})	−9.01	−8.37	−7.73	−7.07	−6.55

The points are plotted in Fig. 21C.1.

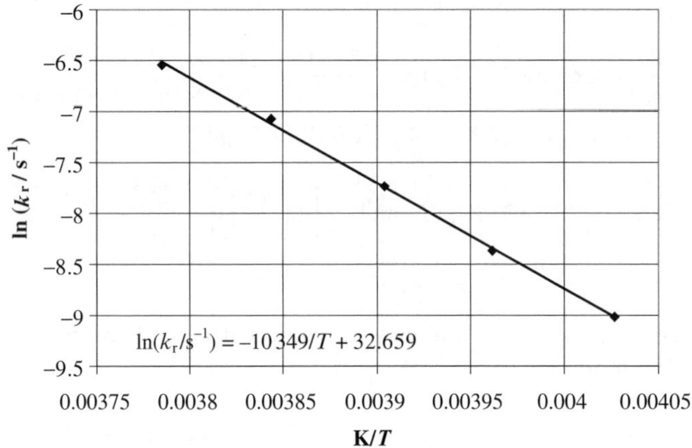

Figure 21C.1

A least squares fit of the data yields the intercept + 32.7 at $1/T = 0$, which implies that $\ln\left(\dfrac{A}{\text{s}^{-1}}\right) = 32.7$, and hence that $A = 1.53\times10^{14}\ \text{s}^{-1}$. The slope is

$$-1.035\times10^4\ \text{K} = -E_a/R,\ \text{and hence}\ \boxed{E_a = 86.0\ \text{kJ mol}^{-1}}$$

In solution $\Delta^{\ddagger}H = E_a - RT$ [Section 21C.2(a)], so at −20 °C

$$\Delta^{\ddagger}H = 86.0\ \text{kJ mol}^{-1} - (8.3145\ \text{J mol}^{-1}\ \text{K}^{-1})\times(253\ \text{K}) = \boxed{+83.9\ \text{kJ mol}^{-1}}$$

We assume that the reaction is first-order for which, by analogy to Section 21C.1(d)

$$K^{\ddagger} = \frac{kT}{h\nu}\bar{K}^{\ddagger} \quad \text{and} \quad k_r = k^{\ddagger}K^{\ddagger} = \nu\times\frac{kT}{h\nu}\times\bar{K}^{\ddagger}$$

with $\Delta^{\ddagger}G = -RT\ln\bar{K}^{\ddagger}$

Therefore, $k_r = A\,\text{e}^{-E_a/RT} = \dfrac{kT}{h}\text{e}^{-\Delta^{\ddagger}G/RT} = \dfrac{kT}{h}\text{e}^{\Delta^{\ddagger}S/R}\,\text{e}^{-\Delta^{\ddagger}H/RT}$

We can identify $\Delta^{\ddagger}S$ by writing

$$k_r = \frac{kT}{h}\text{e}^{\Delta^{\ddagger}S/R}\,\text{e}^{-E_a/RT}\,\text{e} = A\,\text{e}^{-E_a/RT}$$

and hence obtain

$$\Delta^{\ddagger}S = R\left[\ln\left(\frac{hA}{kT}\right)-1\right]$$

$$= 8.3145\ \mathrm{J\,K^{-1}\,mol^{-1}}\times\left[\ln\left(\frac{(6.626\times10^{-34}\ \mathrm{J\,s})\times(1.53\times10^{14}\ \mathrm{s^{-1}})}{(1.381\times10^{-23}\ \mathrm{J\,K^{-1}})\times(253\ \mathrm{K})}\right)-1\right]$$

$$= \boxed{+19.6\ \mathrm{J\,K^{-1}\,mol^{-1}}}.$$

Therefore,

$$\Delta^{\ddagger}G = \Delta^{\ddagger}H - T\Delta^{\ddagger}S = 83.9\ \mathrm{kJ\,mol^{-1}} - 253\ \mathrm{K}\times 19.6\ \mathrm{J\,K^{-1}\,mol^{-1}} = \boxed{+79.0\ \mathrm{kJ\,mol^{-1}}}$$

P21C.3 In effect, we are asked to simplify the expression in the middle of Example 21C.1:

$$k_r = \kappa\frac{kT}{h}\frac{RT}{p^{\ominus}}\left(\frac{N_A\Lambda_A^3\Lambda_B^3}{\Lambda_{C^\ddagger}^3V_m^{\ominus}}\right)\frac{2IkT}{\hbar^2}e^{-\Delta E_0/RT}$$

Before we substitute for the thermal wavelengths, we notice that the expression contains several related constants, such as k and $R = N_Ak$ as well as h and $\hbar = h/2\pi$. Let us replace R (except in the exponent) and $\hbar$, as well as noting

$$p^{\ominus}V_m^{\ominus} = RT = N_AkT$$

so

$$k_r = \kappa\frac{kT}{h}\frac{N_AkT}{N_AkT}\left(\frac{N_A\Lambda_A^3\Lambda_B^3}{\Lambda_{C^\ddagger}^3}\right)\frac{2IkT(2\pi)^2}{h^2}e^{-\Delta E_0/RT} = \kappa\frac{2^3\pi^2N_Ak^2T^2}{h^3}\left(\frac{\Lambda_A^3\Lambda_B^3}{\Lambda_{C^\ddagger}^3}\right)Ie^{-\Delta E_0/RT}$$

The thermal wavelengths are

$$\Lambda_J = \frac{h}{(2\pi m_JkT)^{1/2}} = \frac{h}{(2\pi kT)^{1/2}m_J^{1/2}}$$

where we separate the one piece of the expression (the mass) that differs for different species. Thus the factor that involves the thermal wavelengths can be written as

$$\frac{\Lambda_A^3\Lambda_B^3}{\Lambda_{C^\ddagger}^3} = \frac{h^3(m_A+m_B)^{3/2}}{(2\pi kT)^{3/2}m_A^{3/2}m_B^{3/2}} = \frac{h^3}{(2\pi kT)^{3/2}}\left(\frac{m_A+m_B}{m_Am_B}\right)^{3/2} = \frac{h^3}{(2\pi kT)^{3/2}\mu^{3/2}}$$

Seeing the reduced mass μ reminds us to substitute for the moment of inertia $I = \mu r^2$. So

$$k_r = \kappa\frac{2^3\pi^2N_Ak^2T^2}{h^3}\left(\frac{h^3}{(2\pi kT)^{3/2}\mu^{3/2}}\right)\mu r^2e^{-\Delta E_0/RT} = \kappa r^2\frac{2^{3/2}\pi^{1/2}N_Ak^{1/2}T^{1/2}}{\mu^{1/2}}e^{-\Delta E_0/RT}$$

Identifying the reactive cross-section $\sigma^* = \kappa\pi r^2$ and collecting the remaining terms gives the expression at the end of the Example:

$$k_r = \sigma^*N_A\left(\frac{8kT}{\pi\mu}\right)^{1/2}e^{-\Delta E_0/RT}$$

P21C.5 For a bimolecular gas-phase reaction

$$\Delta^{\ddagger}H = E_a - 2RT \text{ [Section 21C.2(a) footnote]}$$

so $\Delta^{\ddagger}H = 65.43\,\text{kJ mol}^{-1} - 2\times 8.3145\,\text{J mol}^{-1}\,\text{K}^{-1}\times 300\,\text{K} = \boxed{+60.44\,\text{kJ mol}^{-1}}$

$$\Delta^{\ddagger}H = \Delta^{\ddagger}U + \Delta^{\ddagger}(pV)$$

so $\Delta^{\ddagger}U = \Delta^{\ddagger}H - \Delta^{\ddagger}(pV) = \Delta^{\ddagger}H - RT\Delta\nu_{\text{gas}}$

$$\Delta^{\ddagger}U = 60.44\text{ kJ mol}^{-1} - (8.3145\text{ J K}^{-1}\text{mol}^{-1})\times(300\text{ K})\times(-1) = \boxed{+62.9\text{ kJ mol}^{-1}}$$

$$\Delta^{\ddagger}S = R\left(\ln\frac{A}{B} - 2\right) \text{ [Exercise 21C.4(b)]}$$

where $B = \dfrac{kRT^2}{hp^{\ominus}} = \dfrac{(1.381\times10^{-23}\,\text{J K}^{-1})\times(300\,\text{K})^2\times(8.3145\text{ J K}^{-1}\text{ mol}^{-1})}{(6.626\times10^{-34}\,\text{J s})\times(10^5\,\text{Pa})}$

$$= 1.56\times10^{11}\text{ m}^3\text{ mol}^{-1}\text{ s}^{-1} = 1.56\times10^{14}\text{ dm}^3\text{ mol}^{-1}\text{ s}^{-1}$$

so $\Delta^{\ddagger}S = (8.3145\text{ J K}^{-1}\text{ mol}^{-1})\times\left(\ln\dfrac{4.07\times10^5\text{ dm}^3\text{ mol}^{-1}\text{ s}^{-1}}{1.56\times10^{14}\text{ dm}^3\text{ mol}^{-1}\text{ s}^{-1}} - 2\right)$

$$= \boxed{-181\text{ J K}^{-1}\text{ mol}^{-1}}$$

$$\Delta^{\ddagger}G = \Delta^{\ddagger}H - T\Delta^{\ddagger}S = 60.44\,\text{kJ mol}^{-1} - (300\text{ K})\times(-181\,\text{J K}^{-1}\text{ mol}^{-1})$$

$$\Delta^{\ddagger}G = \boxed{+114.7\,\text{kJ mol}^{-1}}$$

P21C.7 $\text{A} + \text{B} \rightarrow \text{C}^{\ddagger} \rightarrow \text{P}$

$$k_r = \left(\kappa\frac{kT}{h}\right)\times\bar{K}_c^{\ddagger}\text{[21C.10]} = \left(\kappa\frac{kT}{h}\right)\times\left(\frac{N_A RT}{p^{\ominus}}\right)\frac{q_{C^{\ddagger}}^{\ominus}}{q_A^{\ominus}q_B^{\ominus}}e^{-\Delta E_0/RT}\text{ [21C.9]}$$

We assume that the only factor that changes between the atomic and molecular case is the ratio of the partition functions.

For collisions between atoms

$$q_A^{\ominus} = q_A^{T} \approx 10^{26} \approx q_B^{\ominus} = q_B^{T}$$

$$q_C^{\ominus} = (q_C^{R})^2 q_C^{V} q_C^{T} \approx (10^{1.5})^2\times(1)\times(10^{26}) = 10^{29}$$

$$k_r(\text{atoms}) \propto \frac{10^{29}}{10^{26}\times10^{26}} = 10^{-23}$$

For collisions between nonlinear molecules

$$q_A^{\ominus} = (q_A^{R})^3(q_A^{V})^{3N-6}(q_A^{T}) \approx (10^{1.5})^3\times(1)\times(10^{26}) = 3\times10^{30} \approx q_B^{\ominus}$$

$$q_C^{\ominus} = (q_C^{R})^3(q_C^{V})^{3(N+N')-6}(q_C^{T}) \approx (10^{1.5})^3\times(1)\times(10^{26}) = 3\times10^{30}$$

$$k_r(\text{molecules}) \propto \frac{3\times10^{30}}{1\times10^{61}} = 3\times10^{-31}$$

$$\text{Therefore}\quad k_r(\text{atoms})/k_r(\text{molecules}) \approx \frac{10^{-23}}{3\times10^{-31}} = \boxed{3\times10^{7}}$$

P21C.9 According to the Debye–Hückel limiting law, the logarithms of ionic activity coefficients, γ, are proportional to $I^{1/2}$ (where I is ionic strength). The ionic strength in a 2:1 electrolyte solution is three times the molal concentration. If the limiting law holds, then a plot of $\log k_r$ versus $I^{1/2}$ should give a straight line whose y-intercept is $\log k_r^\circ$ and whose slope is $2Az_Az_B$, where z_A and z_B are charge numbers of the component ions of the activated complex (eqn 21C.18).

The line based on the limiting law appears curved. The zero-ionic-strength rate constant based on it is

$$k_r^\circ = 10^{-0.690}\,\text{dm}^{3/2}\,\text{mol}^{-1/2}\,\text{s}^{-1} = 0.204\,\text{dm}^{3/2}\,\text{mol}^{-1/2}\,\text{s}^{-1}$$

The slope is positive, so the complex must overcome repulsive interactions. The product of charges, however, works out to be 0.5, not easily interpretable in terms of charge numbers.

The extended Debye–Hückel law (eqn 5F.11a) has $\log\gamma$ proportional to $\left(\dfrac{I^{1/2}}{1+BI^{1/2}}\right)$, so it requires plotting $\log k_r$ versus $\left(\dfrac{I^{1/2}}{1+BI^{1/2}}\right)$, and it also has a slope of $2Az_Az_B$ and a y-intercept of $\log k_r^\circ$. B is an empirical parameter, though, and we don't know its value. (In fact, the extended law reduces to the limiting law with $B=0$.) But it is easy to try a few values, and the resulting curves are plotted in Fig. 21C.2.

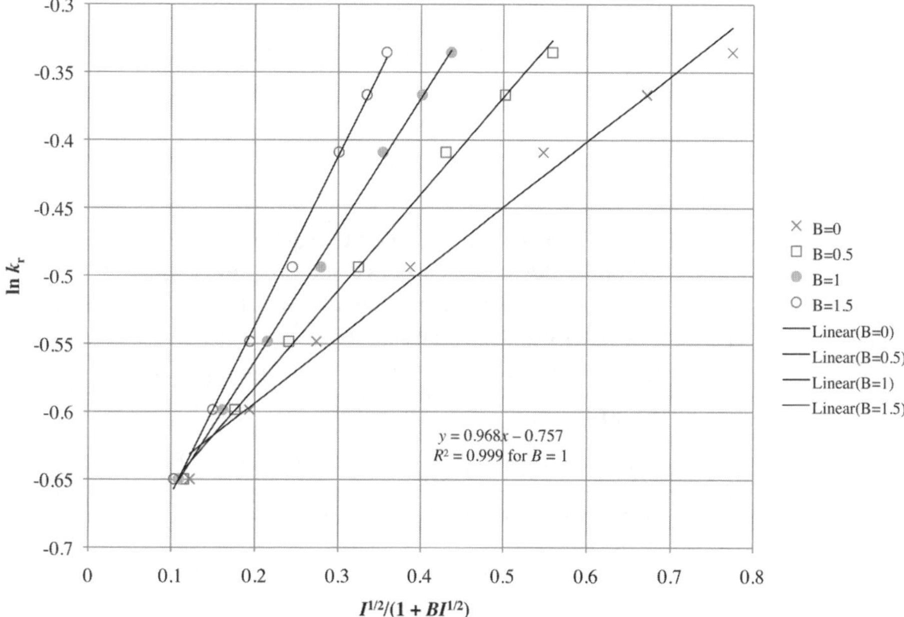

Figure 21C.2

The line based on the extended law with $B = 1$ appears straighter and has a better correlation coefficient than the limiting law; it also has a better correlation coefficient than $B = 0.5$ or $B = 1.5$ curves. The zero-ionic-strength rate constant based on $B = 1$ is

$$k_r{}^\circ = 10^{-0.757}\ \text{dm}^{3/2}\ \text{mol}^{-1/2}\ \text{s}^{-1} = 0.175\,\text{dm}^{3/2}\ \text{mol}^{-1/2}\ \text{s}^{-1}$$

The product of charges works out to be 0.9, nearly 1, interpretable in terms of a complex of [two univalentions of the same sign].

The transformed data are in the following table.

$[Na_2SO_4]/(\text{mol kg}^{-1})$	0.2	0.15	0.1	0.05	0.025	0.0125	0.005
$k_r\,/\,(\text{dm}^{3/2}\,\text{mol}^{-1/2}\,\text{s}^{-1})$	0.462	0.430	0.390	0.321	0.283	0.252	0.224
$I^{1/2}$	0.775	0.671	0.548	0.387	0.274	0.194	0.122
$I^{1/2}/(1 + I^{1/2})$	0.436	0.401	0.354	0.279	0.215	0.162	0.109
$\log k_r$	−0.335	−0.367	−0.409	−0.493	−0.548	−0.599	−0.650

P21C.11 (a) Assuming that the rate-determining step is the scission of a C–H bond, the ratio of rate constants for the tritiated versus protonated reactant should be

$$\frac{k_r(\text{C–T})}{k_r(\text{C–H})} = e^{-\zeta},\ \text{where}\ \zeta = \left(\frac{\hbar k_f^{1/2}}{2kT}\right)\times\left(\frac{1}{\mu_{\text{CH}}^{1/2}} - \frac{1}{\mu_{\text{CT}}^{1/2}}\right)\ [\text{21C.20 with}\ \hbar\omega = \hbar(k_f/\mu)^{1/2}]$$

The reduced masses are roughly m_u and $3m_u$, respectively, for the protons and ^{3}H nuclei are far lighter than the rest of the molecule to which they are attached. So

$$\zeta = \left(\frac{(1.0546\times10^{-34}\ \text{J s})(450\ \text{N m}^{-1})^{1/2}}{2(1.381\times10^{-23}\ \text{J K}^{-1})(298\ \text{K})}\right)$$
$$\times\left(\frac{1}{m_u^{1/2}} - \frac{1}{(3m_u)^{1/2}}\right)(1.6605\times10^{-27}\ \text{kg}\ m_u^{-1})^{-1/2}$$
$$\approx 2.8$$

so $$\frac{k_r(\text{C–T})}{k_r(\text{C–H})} \approx e^{-2.8} = \boxed{0.06}$$

(b) The analogous expression for ^{16}O and ^{18}O requires reduced masses for C–^{16}O and C–^{18}O bonds. The reduced masses could vary rather widely depending on the size of the whole molecule, but in no case will they be terribly different for the two isotopes. Take ^{12}CO, for example:

$$\mu_{16} = \frac{(16.0m_u)\times(12.0m_u)}{(16.0+12.0)m_u} = 6.86m_u \quad\text{and}\quad \mu_{18} = \frac{(18.0m_u)\times(12.0m_u)}{(18.0+12.0)m_u} = 7.20m_u$$

$$\zeta = \left(\frac{(1.0546\times10^{-34}\ \text{J s})(1750\ \text{N m}^{-1})^{1/2}}{2(1.381\times10^{-23}\ \text{J K}^{-1})(298\ \text{K})}\right)$$
$$\times\left(\frac{1}{(6.86m_u)^{1/2}} - \frac{1}{(7.20m_u)^{1/2}}\right)(1.6605\times10^{-27}\ \text{kg}\ m_u^{-1})^{-1/2}$$
$$\approx 0.12$$

$$\text{so} \quad \frac{k_r(\mathrm{C^{18}O})}{k_r(\mathrm{C^{16}O})} \approx \mathrm{e}^{-0.12} = \boxed{0.89}$$

At the other extreme, the O atoms could be attached to heavy fragments such that the reduced mass of the relevant vibration approximates the mass of the oxygen isotope. That is, $\mu_{16} \approx 16m_u$ and $\mu_{18} \approx 18m_u$, so $\zeta \approx 0.19$ and

$$\frac{k_r(\mathrm{C^{18}O})}{k_r(\mathrm{C^{16}O})} \approx \mathrm{e}^{-0.19} = \boxed{0.83}$$

Note that ζ is temperature dependent, so the rate-constant ratio would decrease (that is, get further from unity) with decreasing temperature. That is, the reactions become more isotopically selective the lower the temperature, because at high temperatures there is enough energy to react at an appreciable rate even with the less favourable heavier isotopes.

21D The dynamics of molecular collisions

Answers to discussion questions

D21D.1 *Infrared chemiluminescence.* Chemical reactions may yield products in excited states. The emission of radiation as the molecules decay to lower energy states is called chemiluminescence. If the emission is from vibrationally excited states, then it is infrared chemiluminescence. The vibrationally excited product molecule in the example of Fig. 21D.11 in the text is CO. By studying the intensities of the infrared emission spectrum, the populations of the vibrational states in the product CO may be determined and this information allows us to determine the relative rates of formation of CO in these excited states.

Laser-induced fluorescence (LIF) uses a laser to excite product molecules from a particular rotational-vibrational state and then 'counts' those molecules by detecting fluorescence from the excited molecules. By tuning the exciting laser to different rotational-vibrational states, one can gather information about the distribution of product molecules among different states.

Multi-photon ionization (MPI). Multi-photon absorption is the absorption of two or more photons by the molecule in its transition to a higher electronic state. The frequencies of the photons satisfy the condition

$$\Delta E = h\nu_1 + h\nu_2 + \cdots$$

which is similar to the frequency condition for one-photon absorption. However, multi-photon selection rules are different from one-photon selection rules. Therefore, multi-photon processes allow examination of energy states that otherwise could not be reached. In multi-photon ionization, the second or third photon takes the molecule into the energy continuum above its highest lying energy state. This technique is especially useful for the study of weakly fluorescing molecules.

Resonant multi-photon ionization (REMPI). This is a variant of MPI described above, in which one or more photons promote a molecule to an electronically excited state and then additional photons generate ions from the excited state. The power of this method in the study of chemical reactions is its selectivity. In a chemically reacting system, individual reactants and products can be chosen by tuning the frequency of the laser generating the radiation to the electronic absorption band of specific molecules.

Reaction product imaging. In this technique, product ions are accelerated by an electric field toward a phosphorescent screen and the light emitted from the screen is imaged by a charge-coupled device. The significance of this experiment to the study of chemical reactions is that it allows for a detailed analysis of the angular distribution of products.

D21D.3 The Rb atom must hit the I side of CH_3I in order to produce RbI + CH_3. The orientation of CH_3I can be controlled by exciting rotations about the CI axis with linearly polarized light; the optimal orientation aims the I side of CH_3I at the direction of approach of the beam of Rb atoms. Two possible alignments of the reactant beams are shown in Fig. 21D.1. In the top depiction, the beams are antiparallel, thereby maximizing the likelihood of collision and the volume within which collision can occur (but also putting each beam source in the path of the other beam). In the lower depiction, the beam paths are at right angles, thereby minimizing the region in which the beams collide, but facilitating the study of that well-defined collision volume by a 'probe' laser at right angles to both beams.

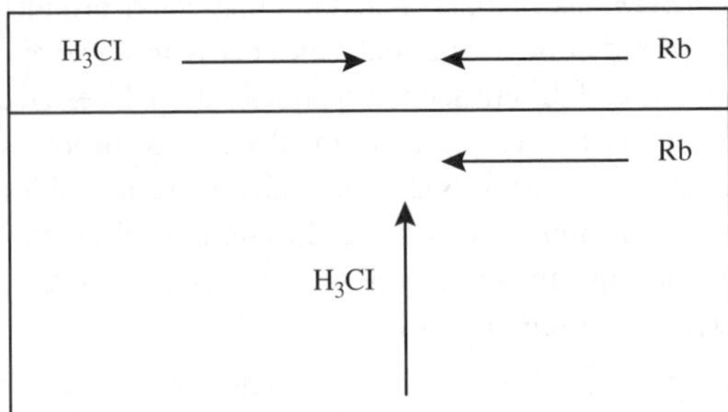

Figure 21D.1

D21D.5 Molecular beams may be used to prepare molecules in specific rotational and vibrational states and then to examine the results of collisions between such precisely prepared species. Section 21D.1(a) describes how molecular beams are prepared such that the molecules in them have a very narrow range of velocities and therefore relatively few collisions to redistribute their energies. Molecules in such beams can be prepared in specific vibrational states, for example, by having lasers excite vibrations. Crossing two molecular beams allows collisions to be staged between two sets of precisely characterized molecules. Detectors can then be used to study the results of those collisions, recording what molecules in what state are scattered where as a result.

Solutions to exercises

E21D.1(a) Refer to Fig. 21D.19 of the main text, which shows an attractive potential energy surface as well as trajectories of both a successful reaction and an unsuccessful one. The trajectories begin in the lower right, representing reactants. The successful trajectory passes through the transition state (marked by a circle with the symbol ‡ near it). That trajectory is fairly straight from the lower right through the transition state, indicating little or no vibrational excitation in the reactant. Therefore most of its energy is in translation. Since it has enough total energy to reach the transition state, we can say that the $\boxed{\text{reactant is high in translational energy and low in vibrational energy}}$. That successful trajectory moves from side to side along the valley representing products, so the $\boxed{\text{product is high in vibrational energy and relatively lower in translational energy}}$. The unsuccessful trajectory, by contrast, has a reactant high in vibrational energy; it moves from side to side in the reactant valley without reaching the transition state.

E21D.2(a) The numerator of eqn 21D.6 is

$$\int_0^\infty \bar{P}(E)\mathrm{e}^{-E/RT}\,\mathrm{d}E = \bar{P}\int_0^\infty \mathrm{e}^{-E/RT}\,\mathrm{d}E = -\frac{\bar{P}}{RT}\left(\mathrm{e}^{-E/RT}\right)\Big|_{E=0}^{E=-\infty} = \frac{\bar{P}}{RT}$$

Thus, if the cumulative reaction probability were independent of energy (could be taken outside the integral), then the numerator would decrease with increasing temperature and so, then, would the rate constant. (In fact, the rate constant would decrease with increasing temperature even faster, because the denominator of eqn. 21D.6 would increase with increasing temperature.)

Solutions to problems

P21D.1 The change in intensity of the beam, dI, is proportional to the number of scatterers per unit volume, $\mathcal{N}$, the intensity of the beam, I, and the path length dL. The constant of proportionality must be the collision cross-section σ, the 'target area' of each scatterer. Thus, σdL is the volume of scatterers to be encountered within the beam, and

$$\mathrm{d}I = -\sigma\mathcal{N}I\mathrm{d}L \quad \text{or} \quad \mathrm{d}\ln I = -\sigma\mathcal{N}I\mathrm{d}L$$

If the incident intensity (at $L = 0$) is I_0 and the emergent intensity is I, we can write

$$\ln\frac{I}{I_0} = -\sigma\mathcal{N}L \quad \text{or} \quad \boxed{I = I_0\mathrm{e}^{-\sigma\mathcal{N}L}}$$

P21D.3 Refer to Fig. 21D.2.

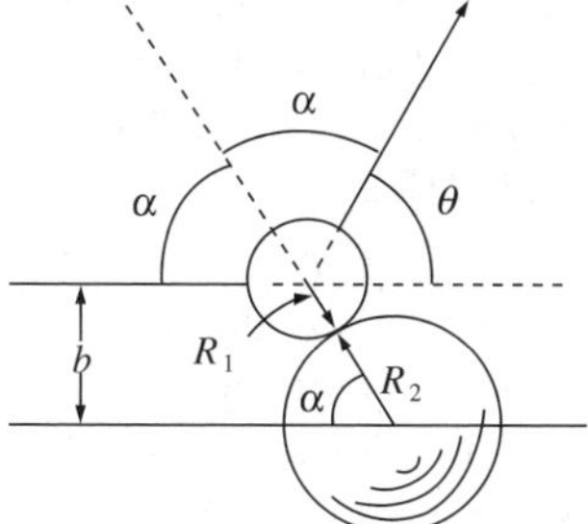

Figure 21D.2

The scattering angle is $\theta = \pi - 2\alpha$ if specular reflection occurs in the collision (angle of impact equal to angle of departure from the surface). For $b \le R_1 + R_2$, $\sin\alpha = \dfrac{b}{R_1 + R_2}$

$$\theta = \begin{cases} \pi - 2\arcsin\left(\dfrac{b}{R_1+R_2}\right) & b \le R_1 + R_2 \\ 0 & b > R_1 + R_2 \end{cases}$$

The function is plotted in Fig. 21D.3.

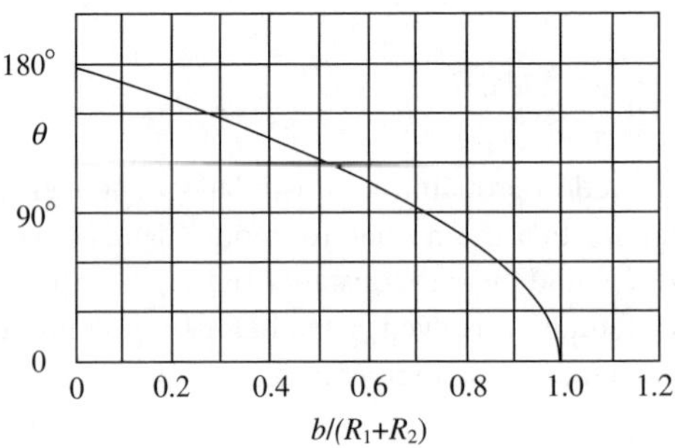

Figure 21D.3

21E Electron transfer in homogeneous systems

Answers to discussion questions

D21E.1 An expression of the rate constant for electron transfer is given by eqn. 21E.5:

$$k_{\text{et}} = CH_{\text{et}}(d)^2 \, \text{e}^{-\Delta^{\ddagger}G/RT}$$

This rate constant depends on the distance between donor and acceptor (d) through the function $H_{\text{et}}(d)^2$, which is given by eqn 21E.4; k_{et} decays exponentially with increasing d. The parameter β in eqns 21E.4 and 21E.8 determines how strongly k_{et} depends on distance. This parameter changes with the transfer medium. The standard Gibbs energy of the electron-transfer process ($\Delta_r G^{\ominus}$) affects the rate through the activation Gibbs energy and a quantity called the reorganization energy (ΔE_{R}), as shown in eqn 21E.6. In systems where the reorganization energy is constant, the dependence of $\ln k_{\text{et}}$ on $\Delta_r G^{\ominus}$ is given by eqn 21E.9; it is an inverted parabola in which the maximum rate occurs when $\Delta_r G^{\ominus} = \Delta E_{\text{R}}$. A more thorough discussion can be found in Section 21E.2.

D21E.3 The inverted region is discussed just before *Brief illustration* 21E.2 in the main text. The phenomenon in question refers to the fact that the rate constant for the reaction can decrease even as the reaction becomes more thermodynamically spontaneous (that is, as it becomes more exergonic). Equation 21E.6 explains this phenomenon by relating the activation Gibbs energy ($\Delta^{\ddagger}G$), which affects reaction rate, and the standard reaction Gibbs energy ($\Delta_r G^{\ominus}$), which affects thermodynamic stability. Perhaps the best way to understand this relationship is with reference to Fig. 21E.3 of the main text. The activation energy is determined by the crossing point between Gibbs energy curves for the product P and reactant R. One can see from the figure that as the parabola representing P is lowered, the energy of the crossing point decreases until the minimum in curve R is reached, and then the energy of the crossing point increases as the crossing point travels up the other side of curve R.

Solutions to exercises

E21E.1(a) The rate constant for electron transfer is

$$k_{\text{et}} = C\{H_{\text{et}}(d)\}^2 \mathrm{e}^{-\Delta^{\ddagger}G/RT} \quad [21\text{E}.5]$$

The reorganization energy, ΔE_{R}, appears in two of these factors:

$$\Delta^{\ddagger}G = \frac{(\Delta_r G^{\ominus} + \Delta E_{\text{R}})^2}{4\Delta E_{\text{R}}} \quad [21\text{E}.6] \quad \text{and} \quad C = \frac{1}{h}\left(\frac{\pi^3}{RT\Delta E_{\text{R}}}\right)^{1/2} \quad [21\text{E}.7],$$

$$\text{so} \quad k_{\text{et}} = \frac{\{H_{\text{et}}(d)\}^2}{h}\left(\frac{\pi^3}{RT\Delta E_{\text{R}}}\right)^{1/2} \exp\left(\frac{-(\Delta_r G^{\ominus} + \Delta E_{\text{R}})^2}{4RT\Delta E_{\text{R}}}\right)$$

$$= \frac{\{H_{\text{et}}(d)\}^2}{h}\left(\frac{\pi^3}{kT\Delta E_{\text{R}}}\right)^{1/2} \exp\left(\frac{-(\Delta_r G^{\ominus} + \Delta E_{\text{R}})^2}{4kT\Delta E_{\text{R}}}\right)$$

depending on whether the energies are expressed in molar units or molecular units. The only unknown in this equation is ΔE_{R}. Isolating ΔE_{R} analytically is not possible; however, one can solve for it numerically using the root-finding command of a symbolic mathematics package, or graphically by plotting the right-hand side vs. the (constant) left-hand side and finding the value of ΔE_{R} at which the two lines cross. Before we put in numbers, we must make sure to use compatible units. We recognize that $H_{\text{et}}(d)$ and $\Delta_r G^{\ominus}$ are both given in molecular units, but that the former is really a wavenumber rather than an energy. So we choose to express all energies per molecule in joules:

$$\frac{H_{\text{et}}^2}{h}\left(\frac{\pi^3}{kT}\right)^{1/2} = \frac{(8\times10^{-25}\ \text{J})^2}{6.626\times10^{-34}\ \text{J s}}\left(\frac{\pi^3}{1.381\times10^{-23}\ \text{J K}^{-1}\times298\ \text{K}}\right)^{1/2} = 8\times10^{-5}\ \text{J}^{0.5}\ \text{s}^{-1}$$

$$\Delta_r G^{\ominus} = -0.185\ \text{eV}\times1.602\times10^{-19}\ \text{J eV}^{-1} = -2.96\times10^{-20}\ \text{J}$$

$$\text{and} \quad 4kT = 4\times(1.381\times10^{-23}\ \text{J K}^{-1})\times(298\text{K}) = 1.65\times10^{-20}\ \text{J}$$

$$\text{Thus} \quad 37.5 = 8\times10^{-5}\left(\frac{\text{J}}{\Delta E_{\text{R}}}\right)^{1/2} \exp\left(\frac{-(-2.96\times10^{-20}\ \text{J} + \Delta E_{\text{R}})^2}{\Delta E_{\text{R}}\times1.65\times10^{-20}\ \text{J}}\right)$$

where $\Delta E_{\text{R}} = \boxed{4\times10^{-21}\ \text{J}}$ or $\boxed{2\ \text{kJ mol}^{-1}}$.

E21E.2(a) For the same donor and acceptor at different distances, eqn 21E.8 applies:

$$\ln k_{\text{et}} = -\beta d + \text{constant}$$

The slope of a plot of k_{et} versus d is $-\beta$. The slope of a line defined by two points is:

$$\text{slope} = \frac{\Delta y}{\Delta x} = \frac{\ln k_{\text{et},2} - \ln k_{\text{et},1}}{d_2 - d_1} = -\beta = \frac{\ln 4.51\times10^4 - \ln 2.02\times10^5}{(1.23-1.11)\ \text{nm}}$$

so $\beta = \boxed{12.\bar{5}\ \text{nm}^{-1}}$

Solutions to problems

P21E.1 First, assume that electron transfer is rate limiting, and not diffusion. In that case, the rate constant is (Section 21E.1)

$$k_{\text{r}} = Kk_{\text{et}} = K\kappa\nu^{\ddagger}\text{e}^{-\Delta^{\ddagger}G/RT}$$

(a) $$\Delta^{\ddagger}G = \frac{(\Delta_{\text{r}}G^{\ominus} + \Delta E_{\text{R}})^2}{4\Delta E_{\text{R}}}\ [21\text{E}.6]$$

so $$\Delta^{\ddagger}G_{\text{DD}} = \frac{(0+\Delta E_{\text{R,DD}})^2}{4\Delta E_{\text{R,DD}}} = \frac{\Delta E_{\text{R,DD}}}{4} \quad \text{and} \quad \Delta^{\ddagger}G_{\text{AA}} = \frac{\Delta E_{\text{R,AA}}}{4}$$

and $$\Delta^{\ddagger}G_{\text{DA}} = \frac{(\Delta_{\text{r}}G^{\ominus} + \Delta E_{\text{R,DA}})^2}{4\Delta E_{\text{R,DA}}} = \frac{(\Delta_{\text{r}}G^{\ominus})^2 + 2\Delta_{\text{r}}G^{\ominus}\Delta E_{\text{R,DA}} + (\Delta E_{\text{R,DA}})^2}{4\Delta E_{\text{R,DA}}}$$

(b) If $|\Delta_{\text{r}}G^{\ominus}| \ll \Delta E_{\text{R,DA}}$, then $\Delta^{\ddagger}G_{\text{DA}} \approx \dfrac{\Delta_{\text{r}}G^{\ominus}}{2} + \dfrac{\Delta E_{\text{R,DA}}}{4}$

Assume $$\Delta E_{\text{R,DA}} = \frac{\Delta E_{\text{R,AA}} + \Delta E_{\text{R,DD}}}{2} = 2(\Delta^{\ddagger}G_{\text{AA}} + \Delta^{\ddagger}G_{\text{DD}})$$

Hence $$\Delta^{\ddagger}G_{\text{DA}} \approx \frac{\Delta_{\text{r}}G^{\ominus} + \Delta^{\ddagger}G_{\text{AA}} + \Delta^{\ddagger}G_{\text{DD}}}{2}$$

(c) From the above expression, the rate constants for the self-exchange reactions are

$$k_{\text{AA}} = K_{\text{AA}}\kappa\nu^{\ddagger}\text{e}^{-\Delta^{\ddagger}G_{\text{AA}}/RT} \quad \text{and} \quad k_{\text{DD}} = K_{\text{AA}}\kappa\nu^{\ddagger}\text{e}^{-\Delta^{\ddagger}G_{\text{DD}}/RT}$$

(d) Compare these results to the rate constant for the reaction of interest:

$$k_{\text{r}} = K_{\text{DA}}\kappa\nu^{\ddagger}\text{e}^{-\Delta^{\ddagger}G_{\text{DA}}/RT} \approx K_{\text{DA}}\kappa\nu^{\ddagger}\text{e}^{-\Delta_{\text{r}}G/2RT}\text{e}^{-\Delta^{\ddagger}G_{\text{AA}}/2RT}\text{e}^{-\Delta^{\ddagger}G_{\text{DD}}/2RT}$$

(e) Thus $$k_{\text{r}} = (k_{\text{AA}}k_{\text{DD}})^{1/2}\frac{K_{\text{DA}}}{(K_{\text{AA}}K_{\text{DD}})^{1/2}}\text{e}^{-\Delta_{\text{r}}G/2RT}$$

The constants K_{DA} etc. are equilibrium constants for diffusive pairing, i.e. for steps like eqn 21E.2a. It is reasonable to expect that $K_{\text{DA}} \approx (K_{\text{AA}}K_{\text{DD}})^{1/2}$, eliminating them from the expression. Finally, the equilibrium constant for the overall reaction is

$$e^{-\Delta_r G/RT} = K$$

so the exponential term in our expression is its square root. Therefore

$$\boxed{k_r \approx (k_{AA}k_{DD}K)^{1/2}}$$

P21E.3 For a series of reactions with a fixed edge-to-edge distance and reorganization energy, the logarithm of the rate constant depends quadratically on the reaction free-energy:

$$\ln k_{et} = -\frac{(\Delta_r G^\ominus)^2}{4kT\Delta E_R} - \frac{\Delta_r G^\ominus}{2kT} + \text{constant [21E.9 in molecular units]}$$

Draw up the following table:

$\Delta_r G^\ominus$ / eV	$k_{et}/(10^6\ s^{-1})$	$\ln(k_{et}/s^{-1})$
−0.665	0.657	13.4
−0.705	1.52	14.2
−0.745	1.12	13.9
−0.975	8.99	16.0
−1.015	5.76	15.6
−1.055	10.1	16.1

and plot $\ln k_{et}$ vs. $\Delta_r G^\ominus$ (Fig. 21E.1).

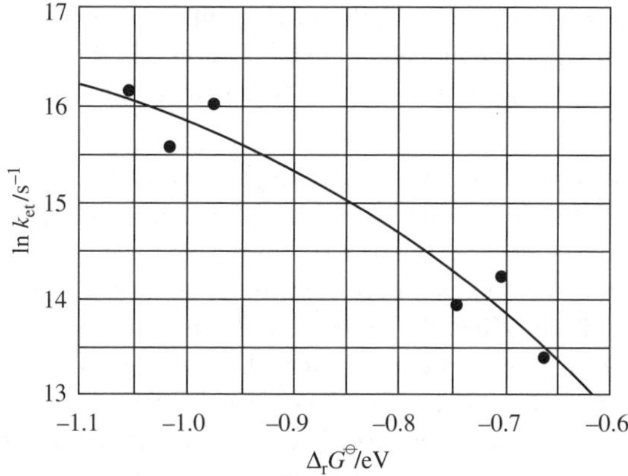

Figure 21E.1

The least squares quadratic fit equation is

$$\ln k_{et}/s^{-1} = 3.23 - 21.1(\Delta_r G^\ominus/eV) - 8.48(\Delta_r G^\ominus/eV)^2, \quad r^2 = 0.938$$

The coefficient of the quadratic term is

$$-\frac{1}{4\lambda kT} = -\frac{8.48}{eV^2}$$

$$\text{so}\quad \Delta E_\text{R} = \frac{(eV)^2}{4(8.48)kT} = \frac{(1.602\times10^{-19}\ \text{J eV}^{-1})(eV)^2}{2(8.48)(1.381\times10^{-23}\ \text{J K}^{-1})(298\ \text{K})} = \boxed{1.15\ \text{eV}}$$

P21E.5 The theoretical treatment of Topic 21E applies only at relatively high temperatures. At temperatures above 130 K, the reaction in question is observed to follow a temperature dependence consistent with eqn 21E.5, namely increasing rate with increasing temperature. Below 130 K, the temperature dependent terms in eqn 21E.5 are replaced by Franck–Condon factors (Topic 13A); that is, temperature-dependent terms are replaced by temperature-independent wavefunction overlap integrals.

21F Processes at electrodes

Answers to discussion questions

D21F.1 Models of the interface between an electrode and an electrolyte solution are described in Section 21F.1. The most primitive is the **electrical double layer** model, in which the electrode and electrolyte are represented by sheets of opposite electrical charge (positive in the electrode and negative in the solution or vice versa as appropriate). In reality, however, the transition between the charged surface of an electrode and the electrically neutral bulk solution is more gradual, and other models attempt to describe this more gradual transition. The **Helmholtz layer** model represents the solution by solvated ions attracted to the electrode but held at some distance from it by their hydration spheres. The **Gouy–Chapman** model adds thermal disorder to the Helmholtz model. Still more sophisticated models like the **Stern** and **Grahame** models combine features of a rigid Helmholtz layer close to the electrode and a more disordered Gouy–Chapman-like description further from it.

D21F.3 In cyclic voltammetry, the current at a working electrode is monitored as the applied potential difference is changed back and forth at a constant rate between pre-set limits (Fig. 21F.12 of the main text). As the potential difference approaches $E^\ominus$ for a solution that contains the reduced component (Red), current begins to flow as Red is oxidized. When the potential difference is swept beyond $E^\ominus$, the current passes through a maximum and then falls as all the Red near the electrode $E^\ominus$ is consumed and converted to Ox, the oxidized form. When the direction of the sweep is reversed and the potential difference passes through $E^\ominus$, current flows in the reverse direction. This current is caused by the reduction of the Ox formed near the electrode on the forward sweep. It passes through the maximum as Ox near the electrode is consumed. The forward and reverse current maxima bracket $E^\ominus$, so the species present can be identified. Furthermore, the forward and reverse peak currents are proportional to the concentration of the couple in the solution, and vary with the sweep rate. If the electron transfer at the electrode is rapid, so that the ratio of the concentrations of Ox and Red at the electrode surface have their equilibrium values for the applied potential (that is, their relative concentrations are given by the Nernst equation), the voltammetry is said to be *reversible*.

If homogeneous chemical reactions accompany the oxidation or reduction of the couple at the electrode, the shape of the voltammogram changes, and the observed changes give valuable information about the kinetics of the reactions as well as the identities of the species present.

Solutions to exercises

E21F.1(a) The conditions are in the limit of large, positive overpotentials, so eqn 21F.5b applies:

$$\ln j = \ln j_0 + (1-\alpha)f\eta$$

where $f = \dfrac{F}{RT} = \dfrac{96\,845\ \mathrm{C\,mol^{-1}}}{(8.3145\ \mathrm{J\,K^{-1}\,mol^{-1}})\times(298\ \mathrm{K})} = 38.9\ \mathrm{V^{-1}}$

Subtracting this equation from the same relationship between another set of currents and overpotentials, we have

$$\ln\frac{j'}{j} = (1-\alpha)f(\eta'-\eta)$$

which rearranges to

$$\eta' = \eta + \frac{\ln(j'/j)}{(1-\alpha)f} = (125\times10^{-3}\ \mathrm{V}) + \frac{\ln(75/55.0)}{(1-0.39)\times(38.9\ \mathrm{V})^{-1}} = \boxed{0.138\ \mathrm{V}}$$

E21F.2(a) Use eqn 21F.5a; then

$$j_0 = j\,\mathrm{e}^{-(1-\alpha)\eta f} = (55.0\,\mathrm{mA\,cm^{-2}})\times \mathrm{e}^{-(1-0.39)\times0.125\ \mathrm{V}\times38.9/\mathrm{V}} = \boxed{2.82\,\mathrm{mA\,cm^{-2}}}.$$

E21F.3(a) For a large positive overpotential we use eqn 21F.5b

$$\ln j = \ln j_0 + (1-\alpha)f\eta,$$

which implies $\ln\dfrac{j'}{j} = (1-\alpha)f(\eta'-\eta) = (1-0.5)\times(0.60-0.40)\ \mathrm{V}\times 38.9\ \mathrm{V^{-1}} = 3.\bar{9}$

and $j' = j\mathrm{e}^{3.\bar{9}} = (1.0\ \mathrm{mA\ cm^{-2}})\times(4\bar{9}) = 4\bar{9}\ \mathrm{mA\ cm^{-2}}$

Hence, the anodic current density $\boxed{\text{increases}}$ by roughly a $\boxed{\text{factor of 50}}$ with a corresponding increase in gas evolution.

E21F.4(a) From Table 21F.1, $j_0 = 6.3\times10^{-6}\ \mathrm{A\ cm^{-2}}$ and $\alpha = 0.58$

(i) According to the Butler–Volmer equation [21F.1]

$$j = j_0(\mathrm{e}^{(1-\alpha)f\eta} - \mathrm{e}^{-\alpha f\eta})$$

so $\dfrac{j}{j_0} = \mathrm{e}^{(1-0.58)\times(38.9/\mathrm{V})\times0.20\ \mathrm{V}} - \mathrm{e}^{-0.58\times(38.9/\mathrm{V})\times0.20\ \mathrm{V}} = (26.\bar{3}-0.011)\approx 26$

$$j = (26)\times(6.3\times10^{-6}\,\mathrm{A\,cm^{-2}}) = \boxed{1.7\times10^{-4}\,\mathrm{A\,cm^{-2}}}.$$

(ii) Equation 21F.5a (also known as the Tafel equation) corresponds to the neglect of the second exponential above, which is very small for an overpotential of 0.2 V. (Even when it was kept, in part (a), it was negligible.) Hence

$$j = \boxed{1.7\times10^{-4}\ \mathrm{A\,cm^{-2}}}.$$

The validity of the Tafel equation increases with higher overpotentials, but decreases at lower overpotentials. A plot of j against η becomes linear (non-exponential) as $\eta \to 0$.

E21F.5(a) $\dfrac{j}{j_0} = \mathrm{e}^{(1-\alpha)f\eta} - \mathrm{e}^{-\alpha f\eta}$ [21F.1] $= \mathrm{e}^{(1/2)f\eta} - \mathrm{e}^{-(1/2)f\eta}$ $[\alpha = 0.5]$

$$= 2\sinh\left(\frac{f\eta}{2}\right) \quad \left[\sinh x = \frac{\mathrm{e}^{x} - \mathrm{e}^{-x}}{2}\right]$$

Use $\dfrac{f\eta}{2} = \dfrac{(38.9\ \mathrm{V^{-1}})\eta}{2} = 19.5(\eta/\mathrm{V}) = 0.0195(\eta/\mathrm{mV})$

Thus $j = 2j_0 \sinh\left(\tfrac{1}{2}f\eta\right) = (1.58\ \mathrm{mA\ cm^{-2}})\times\sinh\left(\dfrac{0.0195\,\eta}{\mathrm{mV}}\right)$

(a) If $\eta = 10\,\mathrm{mV}$, then $j = (1.58\,\mathrm{mA\,cm^{-2}}) \times \sinh(0.195) = \boxed{0.31\,\mathrm{mA\,cm^{-2}}}$.

(b) If $\eta = 100\,\mathrm{mV}$, then $j = (1.58\,\mathrm{mA\,cm^{-2}}) \times \sinh(1.95) = \boxed{5.44\,\mathrm{mA\,cm^{-2}}}$.

(c) If $\eta = -50\,\mathrm{V}$, then $j = (1.58\,\mathrm{mA\,cm^{-2}}) \times \sinh(-97.5) = \boxed{-2\times10^{42}\ \mathrm{mA\,cm^{-2}}}$.

Comment. The extraordinarily large magnitude of the answer to (c) strongly suggests that the Butler–Volmer equation is not applicable at such large negative overpotentials!

E21F.6(a) The current density of electrons is j_0/e because each one carries a charge of magnitude e. Look up j_0 values in Table 21F.1, and recall that $1\ \mathrm{A} = 1\ \mathrm{C\,s^{-1}}$.

(a) $\mathrm{Pt}\,|\,\mathrm{H_2}\,|\,\mathrm{H^+}$ $\quad j_0 = 7.9\times10^{-3}\ \mathrm{A\ cm^{-2}}$

$$\frac{j_0}{e} = \frac{0.79\times10^{-3}\ \mathrm{A\ cm^{-2}}}{1.602\times10^{-19}\ \mathrm{C}} = \boxed{4.9\times10^{15}\ \mathrm{cm^{-2}\ s^{-1}}}$$

(b) $\mathrm{Pt}\,|\,\mathrm{Fe^{3+}, Fe^{2+}}$ $\quad j_0 = 2.5\times10^{-3}\ \mathrm{mA\ cm^{-2}}$

$$\frac{j_0}{e} = \frac{2.5\times10^{-3}\ \mathrm{A\ cm^{-2}}}{1.602\times10^{-19}\ \mathrm{C}} = \boxed{1.6\times10^{16}\ \mathrm{cm^{-2}\ s^{-1}}}$$

(c) $\mathrm{Pb}\,|\,\mathrm{H_2}\,|\,\mathrm{H^+}$ $\quad j_0 = 5.0\times10^{-12}\ \mathrm{A\ cm^{-2}}$

$$\frac{j_0}{e} = \frac{2.5\times10^{-3}\ \mathrm{A\ cm^{-2}}}{1.602\times10^{-19}\ \mathrm{C}} = \boxed{3.1\times10^{7}\ \mathrm{cm^{-2}\ s^{-1}}}$$

There are approximately $\dfrac{(1.0\times10^{-2}\ \mathrm{m})^2}{(280\times10^{-12}\ \mathrm{m})^2} = 1.3\times10^{15}$ atoms in each square centimeter of surface. The numbers of electrons per atom are therefore $\boxed{3.9\ \mathrm{s^{-1}}}$, $\boxed{12\ \mathrm{s^{-1}}}$, and $\boxed{24\times10^{-8}\ \mathrm{s^{-1}}}$, respectively. The last corresponds to less than one event per year.

E21F.7(a) When the overpotential is small, its relation to the current density is [21F.4]

$$\eta = \frac{RTj}{Fj_0} = \frac{j}{fj_0}$$

which implies that the current through surface area S is

$$I = Sj = Sj_0 f\eta.$$

An ohmic resistance r obeys $\eta = Ir$, and so we can identify the resistance as

$$r = \frac{\eta}{I} = \frac{1}{Sj_0 f} = \frac{1}{1.0\ \text{cm}^2 \times 38.9\ \text{V}^{-1} \times j_0} = \frac{2.57\times10^{-2}\ \Omega}{(j_0/\text{A cm}^{-2})}\ [1\text{V} = 1\text{A}\Omega]$$

(a) $\text{Pt}\,|\,\text{H}_2\,|\,\text{H}^+$ $\quad j_0 = 7.9\times10^{-4}\ \text{A cm}^{-2}$

$$r = \frac{2.57\times10^{-2}\ \Omega}{7.9\times10^{-4}} = \boxed{33\ \Omega}$$

(b) $\text{Hg}\,|\,\text{H}_2\,|\text{H}^+$ $\quad j_0 = 7.9\times10^{-13}\ \text{A cm}^{-2}$

$$r = \frac{2.57\times10^{-2}\ \Omega}{7.9\times10^{-13}} = \boxed{3.3\times10^{10}\ \Omega} = 33\ \text{G}\Omega$$

E21F.8(a) We assume $\alpha \approx 0.5$. We look up the standard potential (zero-current, unit activity) of the Zn^{2+}, Zn couple; it is –0.76 V. Zinc will deposit from a solution of unit activity when the potential is below –0.76 V. The hydrogen ion current toward the zinc electrode is then

$$j(\text{H}^+) = j_0 e^{-\alpha f\eta}\ [21\text{F.6a}] = (50\times10^{-12}\ \text{A cm}^{-2}) \times e^{-0.5\times(38.9/\text{V})\times(-0.76\ \text{V})} = 1.3\times10^{-4}\ \text{A cm}^{-2}.$$

This is a borderline negligible current. If we consider a current of 1 mA cm^{-2} or greater to be significant (i.e. to produce a significant rate of evolution of H_2), then $\boxed{\text{one can (barely) deposit zinc}}$ under these conditions without a lot of hydrogen evolution.

Solutions to problems

P21F.1 The high-overpotential version of the Butler–Volmer equation predicts a linear dependence of j on η:

$$\ln j = \ln j_0 + (1-\alpha)f\eta\ [21\text{F.5b}]$$

Draw up the following table for regression analysis:

η/mV	50	100	150	200	250
ln(j/mA cm^{-2})	0.98	2.19	3.40	4.61	5.81

The points are plotted overleaf in Fig. 21F.1.

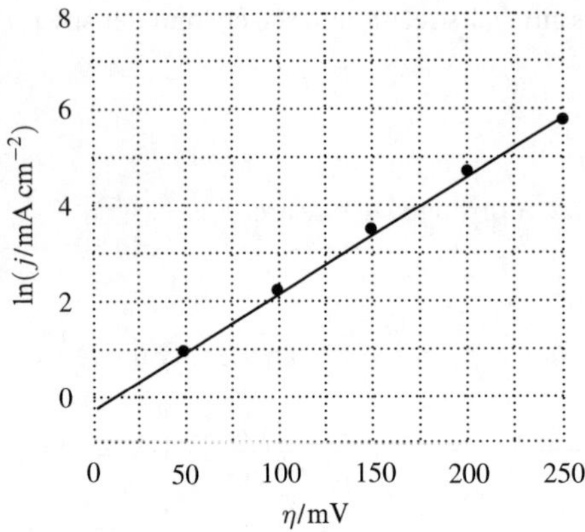

Figure 21F.1

The plot is linear over the entire range of data. The y-intercept is at –0.25, and so

$$j_0\,/\,(\text{mA cm}^{-2}) = \text{e}^{-0.25} = \boxed{0.78}.$$

The slope is 0.0243, and so $(1-\alpha)f = 0.0243\,\text{mV}^{-1} = 24.3\,\text{V}^{-1}$. Since $f = 38.9\ \text{V}^{-1}$ at 25 °C [Exercise 21F.1(a)], it follows that $1-\alpha = 0.62$, and $\alpha = \boxed{0.38}$. If η were large but negative,

$$|j| \approx j_0\text{e}^{-\alpha f\eta}\,[21\text{F.6a}] = (0.78\,\text{mA cm}^{-2})\,\text{e}^{-0.38\eta\times(0.0389/\text{mV})} = (0.78\,\text{mA cm}^{-2})\text{e}^{-0.015(\eta/\text{mV})}$$

and we draw up the following table

η/mV	–50	–100	–150	–200	–250
$\lvert j\rvert$/mA cm^{-2}	1.65	3.50	7.40	15.7	33.2

P21F.3 $\text{Fe}^{2+} + 2\,\text{e}^- \rightarrow \text{Fe}$ $\nu = 2;\ E^\ominus = -0.447\ \text{V}$ [*CRC Handbook*]

(a) The zero-current potential is given by the Nernst equation [6C.4]:

$$E_0 = E^\ominus - \frac{RT}{\nu F}\ln Q$$

$$= E^\ominus - \frac{RT}{\nu F}\ln\frac{1}{[\text{Fe}^{2+}]} \quad \text{assuming } \gamma_{\text{Fe}^{2+}} = 1$$

$$= -0.447\ \text{V} - \frac{1}{2\times(38.9\ \text{V}^{-1})}\ln\left(\frac{1}{1.70\times10^{-6}}\right) = \boxed{-0.618\ \text{V}} \quad \left[\frac{F}{RT} = 38.9\ \text{V}^{-1}\right]$$

$$\eta = E' - E_0\ [21\text{F.2}]$$

η values are reported in the table below.

(b) The current density is related to the reaction rate

$$j = \frac{\nu F}{A}\frac{\text{d}n_{\text{Fe}}}{\text{d}t} = \frac{2\times(96\,845\ \text{C mol}^{-1})}{9.1\ \text{cm}^2}\frac{\text{d}n_{\text{Fe}}}{\text{d}t}$$

The same quantity is given by the Butler-Volmer equation

$$j = j_0(\text{e}^{(1-\alpha)f\eta} - \text{e}^{-\alpha f\eta})[21\text{F.1}] = j_0\text{e}^{-\alpha f\eta}(\text{e}^{f\eta}-1) = j_\text{c}(\text{e}^{f\eta}-1)[21\text{F.6a}]$$

so $\quad j_c = \dfrac{j}{e^{f\eta}-1} = \dfrac{j}{e^{(38.9/V)\eta}-1}$

j_c values are reported in the following table

$\dfrac{dn_{Fe}}{dt}/(10^{-12}\ \text{mol s}^{-1})$	$-E'/\text{mV}$	$-\eta/\text{mV}$	$j/\mu\text{A cm}^{-2}$	$j_c/\mu\text{A cm}^{-2}$
1.47	702	84	0.0313	0.0325
2.18	727	109	0.0464	0.0471
3.11	752	134	0.0662	0.0666
7.26	812	194	0.155	0.155

(c) $j_c = j_0 e^{-\alpha f\eta}$ [21F.6a]

so $\ln j_c = \ln j_0 - \alpha f\eta$

The data are plotted in Fig. 21F.2. Performing a linear regression analysis of the ln j_c versus η data, we find

$$\ln j_0/\mu\text{A cm}^{-2} = -4.612 \quad \text{so} \quad j_0 = 9.94\times10^{-3}\ \mu\text{A cm}^{-2}$$

and $\alpha f = 0.0141\ \text{mV}^{-1}$ so $\alpha = 0.363$

$R^2 = 0.9999$, indicating that the plot provides an excellent description of the data

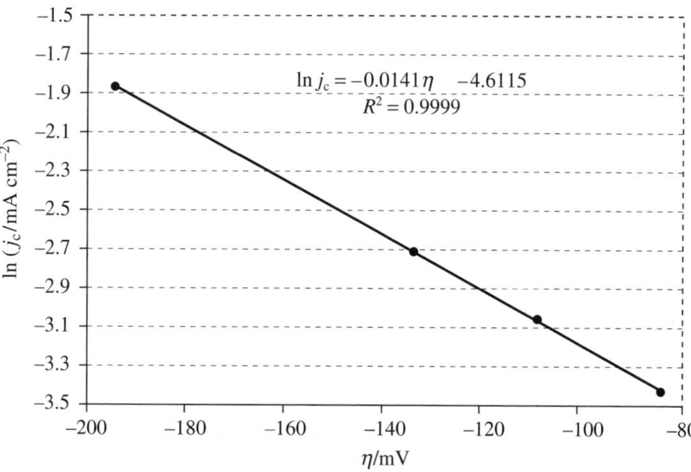

Figure 21F.2

P21F.5 At large positive values of the overpotential the current density is anodic.

$$j = j_0(e^{(1-\alpha)f\eta} - e^{-\alpha f\eta})[21F.1] \approx j_0 e^{(1-\alpha)f\eta}\ [21F.5a]$$

$$\ln j = \ln j_0 + (1-\alpha)f\eta\ [\text{anode}]$$

Performing a linear regression analysis of $\ln j$ against η, we find

$$\text{slope} = (1-\alpha)f = 19.550\ \text{V}^{-1},\ \text{standard deviation} = 0.355$$

$$y\text{-intercept} = \ln\{j_0/(\text{mA m}^{-2})\} = -10.826,\quad \text{standard deviation} = 0.287$$

$$R = 0.99901$$

$$j_0 = \text{e}^{-10.826}\ \text{mA m}^{-2} = \boxed{2.00\times10^{-5}\ \text{mA m}^{-2}}.$$

$$\alpha = 1-\frac{19.550\ \text{V}^{-1}}{f} = 1-\frac{19.550\ \text{V}^{-1}}{38.9\ \text{V}^{-1}} = \boxed{0.498}$$

The linear regression explains 99.90 per cent of the variation in a $\ln j$ against η plot and standard deviations are low. There are essentially $\boxed{\text{no}}$ deviations from the Tafel equation/plot.

P21F.7 Let η oscillate between η_+ and η_- around a mean value η_0. Then η_- is large and positive and η_+ still larger

$$j = j_0\text{e}^{(1-\alpha)\eta f}\ [21\text{F.5a}] = j_0\text{e}^{\eta f/2}\ [\alpha = 1/2]$$

where $f = \dfrac{F}{RT}$ and η varies as depicted in Fig. 21F.3(a)

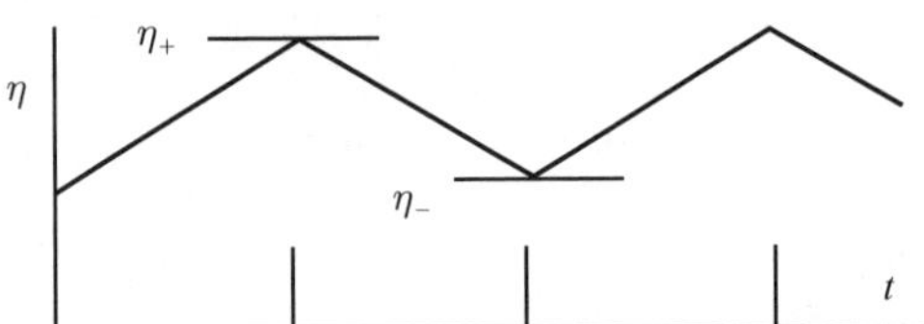

Figure 21F.3(a)

Therefore, j is a chain of increasing and decreasing exponential functions. During the increasing phase,

$$j = j_0\text{e}^{(\eta_-+\gamma t)f/2} \propto \text{e}^{t/\tau}$$

where $\tau = \dfrac{2RT}{\gamma F}$ and γ is a constant that characterizes the rate at which η rises. During the decreasing phase,

$$j = j_0\text{e}^{(\eta_+-\gamma t)f/2} \propto \text{e}^{-t/\tau}.$$

The variation of j is depicted in Fig. 21F.3(b).

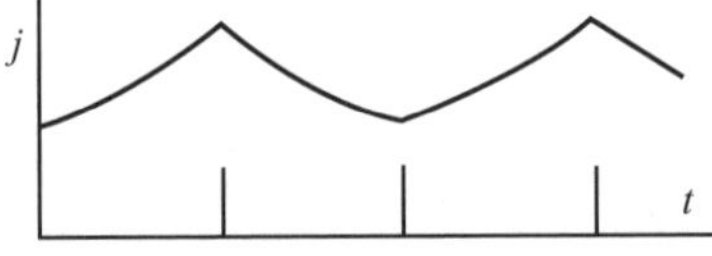

Figure 21F.3(b)

Integrated activities

I21.1 The standard molar partition function is the molecular partition function at standard pressure. See the Checklist of equations from Topic 15B for the partition function factors.

$$\frac{q_{\rm m}^{\ominus\rm T}}{N_{\rm A}}=\frac{V_{\rm m}^{\ominus}}{N_{\rm A}\Lambda^3}=\frac{RT}{N_{\rm A}p^{\ominus}\Lambda^3}=\frac{kT}{p^{\ominus}}\frac{(2\pi mkT)^{3/2}}{h^3}=\frac{(kT)^{5/2}}{p^{\ominus}h^3}\left(\frac{2\pi M}{N_{\rm A}}\right)^{3/2}$$

For $T\approx 300\,\rm K$, $M\approx 50\,\rm g\,mol^{-1}$,

$$\frac{q_{\rm m}^{\ominus\rm T}}{N_{\rm A}}=\frac{\{(1.381\times10^{-23}\ \rm J\,K^{-1})(300\ K)\}^{5/2}}{(10^5\ \rm Pa)(6.626\times10^{-34}\ J\,s)^3}\left(\frac{2\pi(50\times10^{-3}\ \rm kg\,mol^{-1})}{6.022\times10^{23}\ \rm mol^{-1}}\right)^{3/2}\approx\boxed{1.4\times10^7}$$

$$q^{\rm R}(\text{nonlinear})=\frac{1}{\sigma}\left(\frac{kT}{hc}\right)^{3/2}\left(\frac{\pi}{\tilde{A}\tilde{B}\tilde{C}}\right)^{1/2}$$

For a non linear molecule, $T\approx 300\,\rm K, \tilde{A}\approx\tilde{B}\approx\tilde{C}\approx 2\,cm^{-1}, \sigma\approx 2$

$$q^{\rm R}=\frac{1}{2}\left(\frac{(1.381\times10^{-23}\ \rm J\,K^{-1})(300\ K)}{(6.626\times10^{-34}\ \rm J\,s)(2.998\times10^{10}\ cm\,s^{-1})}\right)^{3/2}\left(\frac{\pi}{(2\ \rm cm^{-1})^3}\right)^{1/2}=\boxed{900}$$

$$q^{\rm R}(\text{linear})=\frac{kT}{\sigma hc\tilde{B}}$$

For a linear molecule, $T\approx 300\,\rm K$, $\tilde{B}\approx 1\,\rm cm^{-1}$, $\sigma\approx 1$

$$q^{\rm R}(\text{linear})=\frac{(1.381\times10^{-23}\ \rm J\,K^{-1})(300\ K)}{(1)(6.626\times10^{-34}\ \rm J\,s)(2.998\times10^{10}\ cm\,s^{-1})(1\ cm^{-1})}\approx\boxed{200}$$

Energies of most excited vibrational states in small molecules and of most excited electronic states are high enough to make $q^{\rm V}\approx q^{\rm E}\approx\boxed{1}$

$$k_{\rm r}=\frac{\kappa kT}{h}\bar{K}_{\rm c}^{\ddagger}\,[21\text{C}.10]=\left(\frac{\kappa kT}{h}\right)\times\left(\frac{RT}{p^{\ominus}}\right)\times\left(\frac{N_{\rm A}\bar{q}_{\rm C^{\ddagger}}^{\ominus}}{q_{\rm A}^{\ominus}q_{\rm B}^{\ominus}}\right){\rm e}^{-\Delta E_0/RT}\,[21\text{C}.9]=A{\rm e}^{-E_{\rm a}/RT}$$

If A and B are structureless molecules, then we use estimates from above to evaluate

$$\frac{q_{\rm A}^{\ominus}}{N_{\rm A}}=\frac{q_{\rm A}^{\ominus\rm T}}{N_{\rm A}}\approx1.4\times10^7\approx\frac{q_{\rm B}^{\ominus}}{N_{\rm A}}=\frac{q_{\rm B}^{\ominus\rm T}}{N_{\rm A}}$$

$$\frac{\bar{q}_{\rm C^{\ddagger}}^{\ominus}}{N_{\rm A}}=\frac{q_{\rm C^{\ddagger}}^{\ominus\rm T}q^{\rm R}(\text{linear})}{N_{\rm A}}\approx(2^{3/2})\times(1.4\times10^7)\times(200)=8\times10^9$$

(The factor of $2^{3/2}$ comes from $m_{\rm C}=m_{\rm A}+m_{\rm B}\approx 2m_{\rm A}$ and $q^{\rm T}\propto m^{3/2}$.)

$$\frac{RT}{p^{\ominus}}=\frac{(8.3145\ \rm J\,K^{-1}\ mol^{-1})\times(300\ K)}{10^5\ \rm Pa}=2.5\times10^{-2}\ \rm m^3\ mol^{-1}$$

and $\dfrac{\kappa kT}{h} \approx \dfrac{kT}{h} = \dfrac{(1.381\times10^{-23}\ \mathrm{J\,K^{-1}})\times(300\ \mathrm{K})}{6.626\times10^{-34}\ \mathrm{J\,s}} = 6.25\times10^{12}\ \mathrm{s^{-1}}$

Strictly speaking, E_a is not exactly the same as ΔE_0, but they are approximately equal—close enough for the purposes of estimating the order of magnitude of the pre-exponential factor. So once we identify $E_a \approx \Delta E_0$, we identify the pre-exponential factor with everything *other than* the exponential term. Therefore, the pre-exponential factor

$$A \approx \frac{(6.25\times10^{12}\ \mathrm{s^{-1}})\times(2.5\times10^{-2}\ \mathrm{m^3\,mol^{-1}})\times(8\times10^{9})}{(1.4\times10^{7})^2}$$

$$\approx 6.3\times10^{6}\ \mathrm{m^3\,mol^{-1}\,s^{-1}} = \boxed{6.3\times10^{9}\ \mathrm{dm^3\,mol^{-1}\,s^{-1}}}.$$

According to collision theory [21A.9]

$$A = P\sigma\left(\frac{8kT}{\pi\mu}\right)^{1/2} N_A$$

Take $\sigma \approx 0.5\ \mathrm{nm^2} = 5\times10^{-19}\ \mathrm{m^2}$ as a typical value for small molecules:

$$A = P\times5\times10^{-19}\ \mathrm{m^2}\times\left(\frac{8\times(1.381\times10^{-23}\ \mathrm{J\,K^{-1}})\times(300\ \mathrm{K})}{\pi\times(25\,m_u)\times(1.661\times10^{-27}\ \mathrm{kg}\,m_u^{-1})}\right)^{1/2}\times6.022\times10^{23}\ \mathrm{mol^{-1}}$$

$$= 1.5\times10^{8}\ \mathrm{m^3\,mol^{-1}\,s^{-1}}\times P.$$

The values are quite consistent, for they imply $P \approx 0.04$, which is certainly a plausible value.

If A and B are non linear triatomics, then

$$\frac{q_A^{\ominus}}{N_A} \approx (1.4\times10^{7})\times(900) = 1.3\times10^{10} \approx \frac{q_B^{\ominus}}{N_A}$$

$$\frac{q_{C^\ddagger}^{\ominus}}{N_A} \approx (2^{3/2})\times(1.4\times10^{7})\times(900) = 3.6\times10^{10}$$

and $A \approx \dfrac{(6.25\times10^{12}\,\mathrm{s^{-1}})\times(2.5\times10^{-2}\ \mathrm{m^3\,mol^{-1}})\times(3.6\times10^{10})}{(1.3\times10^{10})^2}$

$$\approx 33\ \mathrm{m^3\,mol^{-1}\,s^{-1}} = \boxed{3.3\times10^{4}\ \mathrm{dm^3\,mol^{-1}\,s^{-1}}}.$$

Comparison to the expression from collision theory implies $\boxed{P = 2\times10^{-7}}$

I21.3 $E^{\ominus} = \dfrac{-\Delta_r G^{\ominus}}{\nu F}$

(a) $H_2 + \frac{1}{2}O_2 \rightarrow H_2O$; $\Delta_r G^{\ominus} = -237\ \mathrm{kJ\,mol^{-1}}$

Since $\nu = 2$,

$$E^{\ominus} = \frac{-(-237\ \mathrm{kJ\,mol^{-1}})}{2\times(96.48\ \mathrm{kC\,mol^{-1}})} = \boxed{+1.23\ \mathrm{V}}$$

(b) $CH_4 + 2\,O_2 \rightarrow CO_2 + 2\,H_2O$

$$\Delta_r G^\ominus = 2\Delta_f G^\ominus(H_2O) + \Delta_f G^\ominus(CO_2) - \Delta_f G^\ominus(CH_4)$$
$$= \{(2)\times(-237.1) + (-394.4) - (-50.7)\}\ \text{kJ mol}^{-1} = -817.9\ \text{kJ mol}^{-1}$$

The reaction corresponds to the transfer of eight electrons. (The carbon goes from – 4 to + 4 while each of the four oxygen atoms goes from 0 to –2.) It follows that, for the species in their standard states,

$$E^\ominus = \frac{-(-817.9\,\text{kJ mol}^{-1})}{8\times(96.48\ \text{kC mol}^{-1})} = \boxed{+1.06\ \text{V}}$$

(c) $C_3H_8 + 5\,O_2 \rightarrow 3\,CO_2 + 4\,H_2O$

$$\Delta_r G^\ominus = 4\Delta_f G^\ominus(H_2O) + 3\Delta_f G^\ominus(CO_2) - \Delta_f G^\ominus(C_3H_8)$$
$$= \{4\times(-237.1) + 3\times(-394.4) - (-23.5)\}\ \text{kJ mol}^{-1} = -2\,108.1\ \text{kJ mol}^{-1}$$

The reaction corresponds to the transfer of 20 electrons. (Each of the 10 oxygen atoms goes from 0 to –2 while the carbon atoms go from –8 collectively (2 at –3 and one at –2) to + 12 collectively (+ 4 each).) It follows that, for the species in their standard states,

$$E^\ominus = \frac{-(-2108.1\ \text{kJ mol}^{-1})}{20\times(96.48\ \text{kC mol}^{-1})} = \boxed{+1.093\ \text{V}}$$

22 Processes on solid surfaces

22A An introduction to solid surfaces

Answer to discussion question

D22A.1 (a) Topographical features of a clean surface include terraces, steps, and dislocations such as edge and screw dislocations. A **terrace** is a flat layer of atoms on a surface. There can be more than one terrace on a surface, each at a different height. Steps are the joints between the terraces; the height of the step can be constant or variable.

(b) **Dislocations**, or discontinuities in the regularity of a crystal lattice, result in surface steps and terraces. The edge dislocation can be envisioned by imagining small clumps of crystalline matter sticking together from either a melt or a solution. The lowest energy pattern sticks them together with valence requirements satisfied or atoms in a close-packed arrangement. This process is expected to form surface terraces because terraces yield the maximum possible number of nearest neighbours at a surface and the lowest possible surface energy. However, the very process of small clumps of matter rapidly sticking is very unlikely to always produce a perfect space-filled, crystalline structure. Crystal defects such as the half-plane of atoms shown in Fig. 22A.1 may form near the surface of the growing crystal. This **edge dislocation** distorts adjacent planes into a high energy configuration that is inherently unstable but thermal agitations of the growth process cause the dislocation to propagate to the surface, thereby, forming a step (see Fig. 22A.2).

A **screw dislocation** is shown in Fig. 22A.3. Imagine a cut in the crystal, with the atoms to the left of the cut pushed up through a distance of one unit cell. The unit cells now form a continuous spiral around the end of the cut, which is called the **screw axis**. A path encircling the screw axis spirals up to the top of the crystal, and where the dislocation breaks through to the surface it takes the form of a spiral ramp.

The surface defect formed by a screw dislocation is a step, possibly with kinks, where growth can occur. The incoming particles lie in ranks on the ramp, and successive ranks reform the step at an angle to its initial position. As deposition continues the step rotates around the screw axis, and is not eliminated. Growth may therefore continue indefinitely. Several layers of deposition may occur, and the edges of the spirals might be cliffs several atoms high. Propagating spiral edges can also give rise to flat terraces. Terraces are formed if growth occurs simultaneously at neighbouring left- and right-handed screw dislocations. Successive tables of atoms may form as counter-rotating defects collide on successive circuits, and the terraces formed may then fill up by further deposition at their edges to give flat crystal planes.

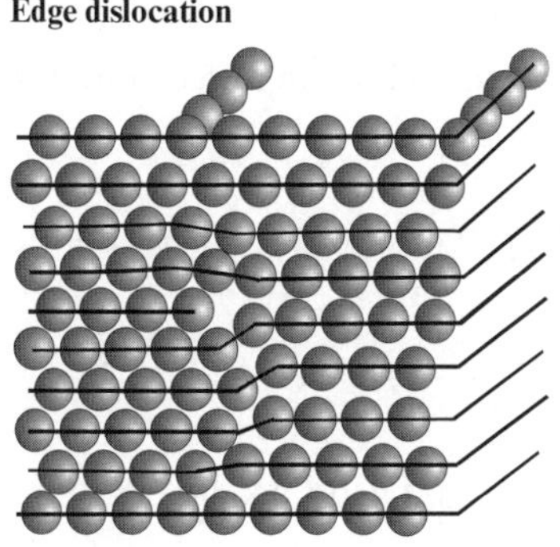

Figure 22A.1

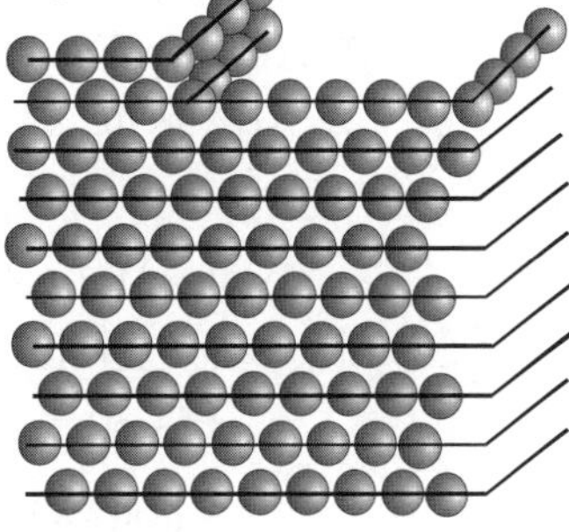

Figure 22A.2

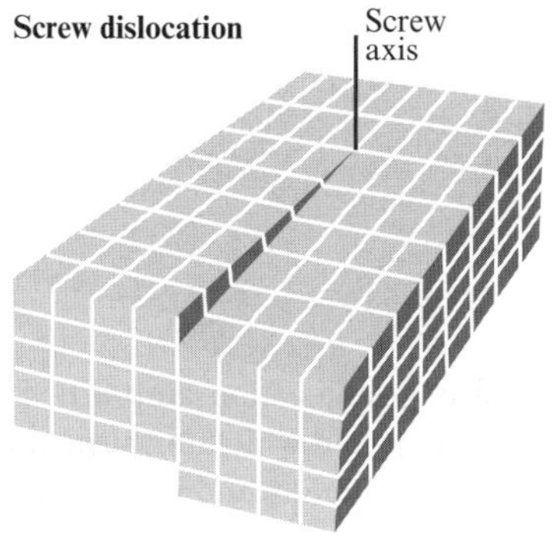

Figure 22A.3

Solutions to exercises

E22A.1(a) The collision frequency, Z_W, of gas molecules with an ideally smooth surface area is given by eqn 22A.1.

$$p = 0.10\,\mu\text{Torr} = 1.33\times10^{-5}\ \text{Pa}$$

$$Z_W = \frac{p}{(2\pi MkT/N_A)^{1/2}}\ [22\text{A}.1;\ m = M/N_A]$$

$$= \frac{(1.33\times10^{-5}\ \text{Pa})\times(\text{m}/10^2\ \text{cm})^2}{\{2\pi\times(1.381\times10^{-23}\ \text{J K}^{-1})\times(298.15\ \text{K})\times(\text{kg mol}^{-1})/(6.022\times10^{23}\ \text{mol}^{-1})\}^{1/2}\{M/(\text{kg mol}^{-1})\}^{1/2}}$$

$$= \frac{6.42\times10^{12}}{\{M/(\text{kg mol}^{-1})\}^{1/2}}\ \text{cm}^{-2}\ \text{s}^{-1} \quad \text{at } 25\ ^\circ\text{C and } p = 0.10\,\mu\text{Torr}$$

(i) Hydrogen ($M = 0.002016$ kg mol^{-1}), $Z_W = \boxed{1.4\times10^{14}\ \text{cm}^{-2}\ \text{s}^{-1}}$

(ii) Propane ($M = 0.04410$ kg mol^{-1}), $Z_W = \boxed{3.1\times10^{13}\ \text{cm}^{-2}\ \text{s}^{-1}}$

E22A.2(a) $A = \pi d^2/4 = \pi(1.5\,\text{mm})^2/4 = 1.77\times10^{-6}\ \text{m}^2$

The collision frequency of the Ar gas molecules with surface area A is $Z_W A = 4.5\times10^{20}\ \text{s}^{-1}$.

$$Z_W A = \frac{p}{(2\pi MkT/N_A)^{1/2}} A \quad [22\text{A}.1;\ m = M/N_A]$$

$$\begin{aligned} p &= (Z_W A)\times(2\pi MkT/N_A)^{1/2}/A \\ &= (4.5\times10^{20}\ \text{s}^{-1}) \\ &\quad \times\left\{\begin{matrix} 2\pi(39.95\times10^{-3}\ \text{kg mol}^{-1})\times(1.381\times10^{-23}\ \text{J K}^{-1}) \\ \times(425\ \text{K})/(6.022\times10^{23}\ \text{mol}^{-1}) \end{matrix}\right\}^{1/2} \\ &\quad /(1.77\times10^{-6}\ \text{m}^2) \\ &= 1.3\times10^4\ \text{Pa} = \boxed{0.13\ \text{bar}} \end{aligned}$$

Solutions to problems

P22A.1 Figure 22A.4(a) shows a dark univalent probe cation atop a two-dimensional square ionic lattice of grey univalent cations and white univalent anions. Let $d_0 = 200$ pm be the distance between nearest neighbours and let V_0 be the Coulombic interaction between nearest neighbours.

$$V_0 = -\frac{e^2}{4\pi\varepsilon_0 d_0} = -\frac{(1.602\times10^{-19}\ \text{C})^2}{(1.113\times10^{-10}\ \text{J}^{-1}\ \text{C}^2\ \text{m}^{-1})\times(200\times10^{-12}\ \text{m})} = -1.153\times10^{-18}\ \text{J}$$

The symmetry of the lattice with respect to the probe cation consists of one region like that of Fig. 22A.4(b) and two regions like that of Fig. 22A.4(c) so we calculate

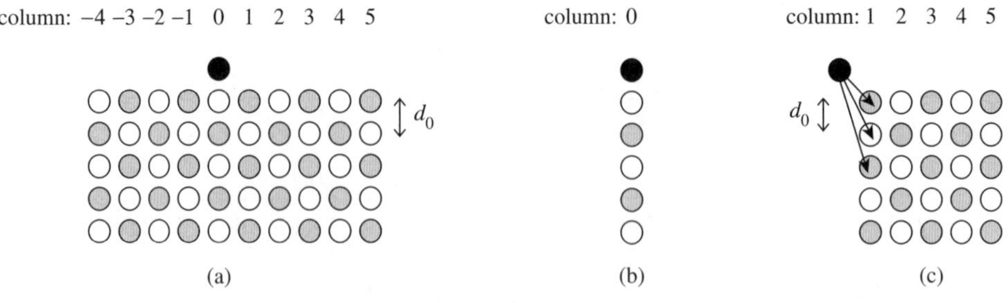

Figure 22A.4

the total Coulombic interaction of the probe with the lattice, $V_{\text{total Fig 95.2(a)}}$, by adding the interaction within Fig. 22A.4(b) to twice the interaction within Fig. 22A.4(c).

$$V_{\text{total Fig 95.2(a)}} = V_{\text{Fig 95.2(b)}} + 2V_{\text{Fig 95.2(c)}}$$

The calculation is pursued one column at a time and the column interactions with the probe cation are summed.

Probe-to-column 0 interaction:

$$\begin{aligned} V_{\text{Fig 22A.4(b)}} &= V_0 \times \left(1 - \frac{1}{2} + \frac{1}{3} - \frac{1}{4} + \frac{1}{5} \cdots\right) = V_0 \ln 2 \\ &= 0.6931\, V_0 \end{aligned}$$

Probe-to-column 1 interaction using the Pythagorean theorem for the probe-ion distance:

$$V_{\text{column 1}} = -V_0 \times \left(\frac{1}{2^{1/2}} - \frac{1}{5^{1/2}} + \frac{1}{10^{1/2}} - \frac{1}{17^{1/2}} \cdots\right) = -V_0 \sum_{n=1}^{\infty} \frac{(-1)^{n+1}}{\left(n^2 + 1^2\right)^{1/2}}$$

Similarly, the probe-to-column m interaction, using the Pythagorean theorem for the probe-ion distance, is

$$V_{\text{column } m} = -V_0 \sum_{n=1}^{\infty} \frac{(-1)^{n+m}}{\left(n^2 + m^2\right)^{1/2}} \quad \text{for } 1 \le m < \infty$$

The total interaction for the region shown in Fig. 22A.4(c) is the sum of the above expression over all m columns

$$\begin{aligned} V_{\text{Fig 22A.4(c)}} &= -V_0 \sum_{m=1}^{\infty} \sum_{n=1}^{\infty} \frac{(-1)^{n+m}}{\left(n^2 + m^2\right)^{1/2}} \quad \text{(the sum is performed with a calculator or software)} \\ &= -0.2893\, V_0 \end{aligned}$$

$$\begin{aligned} \text{Thus, } V_{\text{total Fig 22A.4(a)}} &= (0.6931 - 2 \times 0.2893) V_0 \\ &= 0.1145\, V_0 = (0.1145) \times \left(-1.1153 \times 10^{-18}\, \text{J}\right) \\ &= -1.2770 \times 10^{-19}\, \text{J or } \boxed{-76.9\, \text{kJ mol}^{-1}} \end{aligned}$$

Now consider the probe at the corner formed by a step and the terrace shown overleaf in Fig. 22A.5. The symmetry of the lattice with respect to the probe cation consists of two regions like that of Fig. 22A.4(b) and three regions like that of Fig. 22A.4(c) so we calculate the total Coulombic interaction of the probe with the lattice, $V_{\text{total Fig 22A.5}}$, by adding twice the interaction within Fig. 22A.4(b) to thrice the interaction within Fig. 22A.4(c).

$$\begin{aligned}V_{\text{total Fig 22A.5}} &= 2V_{\text{Fig 22A.4(b)}} + 3V_{\text{Fig 22A.4(c)}}\\ &= \big(2\times(0.6931) - 3\times(0.2893)\big)V_0\\ &= 0.5183\,V_0 = 0.5183\times\big(-1.1153\times10^{-18}\ \text{J}\big)\\ &= -5.7806\times10^{-19}\ \text{J or } \boxed{-348.1\ \text{kJ mol}^{-1}}\end{aligned}$$

Figure 22A.5

The potential energy of the probe cation is much lower at the corner formed by a step and a terrace than it is upon a simple terrace so the $\boxed{\text{corner is the likely settling point}}$.

P22A.3 Refer to Fig. 22A.6.

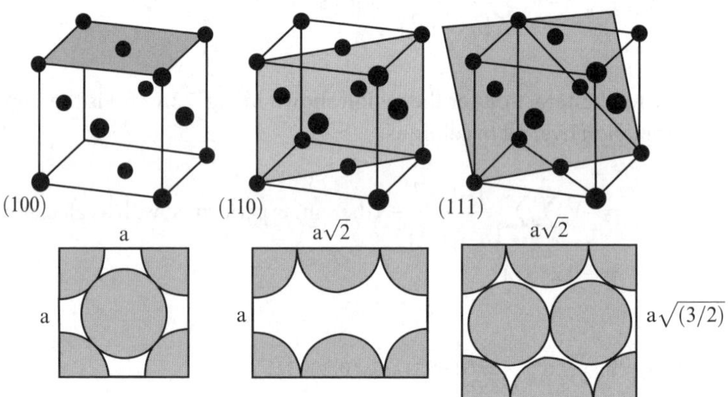

Figure 22A.6

The (100) and (110) faces each expose two atoms, and the (111) face exposes four. The areas of the faces of each cell are (a) $(352\ \text{pm})^2 = 1.24\times10^{-15}\ \text{cm}^2$, (b) $\sqrt{2}\times(352\ \text{pm})^2 = 1.75\times10^{-15}\ \text{cm}^2$, and (c) $\sqrt{3}\times(352\ \text{pm})^2 = 2.15\times10^{-15}\ \text{cm}^2$. The numbers of atoms exposed per square centimetre, the surface number density, are therefore

(a) $$\frac{2}{1.24\times10^{-15}\ \text{cm}^2} = \boxed{1.61\times10^{15}\ \text{cm}^{-2}}$$

(b) $\dfrac{2}{1.75\times10^{-15}\ \text{cm}^2} = \boxed{1.14\times10^{15}\ \text{cm}^{-2}}$

(c) $\dfrac{4}{2.15\times10^{-15}\ \text{cm}^2} = \boxed{1.86\times10^{15}\ \text{cm}^{-2}}$

The collision frequency, Z_W, of gas molecules with a surface is given by eqn 22A.1.

$$Z_W = \frac{p}{(2\pi MkT/N_A)^{1/2}} \quad [22\text{A}.1;\ m = M/N_A]$$

$$= \frac{p\times\{(\text{kg m}^{-1}\ \text{s}^{-2})/\text{Pa}\}\times(10^{-4}\ \text{m}^2/\text{cm}^2)}{\{2\pi\times(1.381\times10^{-23}\ \text{J K}^{-1})\times(298.15\ \text{K})\times(\text{kg mol}^{-1})/(6.022\times10^{23}\ \text{mol}^{-1})\}^{1/2}\{M/(\text{kg mol}^{-1})\}^{1/2}}$$

$$= 4.825\times10^{17}\left(\frac{p/\text{Pa}}{\{M/(\text{kg mol}^{-1})\}^{1/2}}\right)\text{cm}^{-2}\ \text{s}^{-1} \quad \text{at } 25\ ^\circ\text{C}$$

(a) Hydrogen ($M = 0.002016\ \text{kg mol}^{-1}$)
 (i) $p = 100\ \text{Pa}$, $Z_W = 1.07\times10^{21}\ \text{cm}^{-2}\ \text{s}^{-1}$
 (ii) $p = 0.10\ \mu\text{Torr} = 1.33\times10^{-5}\ \text{Pa}$, $Z_W = 1.4\times10^{14}\ \text{cm}^{-2}\ \text{s}^{-1}$

(b) Propane ($M = 0.04410\ \text{kg mol}^{-1}$)
 (i) $p = 100\ \text{Pa}$, $Z_W = 2.30\times10^{20}\ \text{cm}^{-2}\ \text{s}^{-1}$
 (ii) $p = 0.10\ \mu\text{Torr} = 1.33\times10^{-5}\ \text{Pa}$, $Z_W = 3.1\times10^{13}\ \text{cm}^{-2}\ \text{s}^{-1}$

The frequency of collision per surface atom, Z, is calculated by dividing Z_W by the surface number densities for the different planes. We can therefore draw up the following table:

	Hydrogen		Propane	
$Z/(\text{atom}^{-1}\text{s}^{-1})$	100 Pa	10^{-7} Torr	100 Pa	10^{-7} Torr
(100)	6.6×10^5	8.7×10^{-2}	1.4×10^5	1.9×10^{-2}
(110)	9.4×10^5	1.2×10^{-1}	2.0×10^5	2.7×10^{-2}
(111)	5.8×10^5	7.5×10^{-2}	1.2×10^5	1.7×10^{-2}

22B Adsorption and desorption

Answer to discussion question

D22B.1 The characteristic conditions of the **Langmuir isotherm** are:

1. Adsorption cannot proceed beyond monolayer coverage.
2. All sites are equivalent and the surface is uniform.
3. The ability of a molecule to adsorb at a given site is independent of the occupation of neighbouring sites.

For the **BET isotherm** condition number 1 above is removed and the isotherm applies to multi-layer coverage.

For the **Temkin isotherm** condition number 2 is removed and it is assumed that the energetically most favourable sites are occupied first. The Temkin isotherm corresponds to supposing that the adsorption enthalpy changes linearly with pressure.

The **Freundlich isotherm** removes condition 2 but this isotherm corresponds to a logarithmic change in the adsorption enthalpy with pressure.

Solutions to exercises

E22B.1(a) $\theta = \frac{V}{V_\infty} = \frac{V}{V_{\text{mon}}} = \frac{\alpha p}{1+\alpha p}$ [22B.2]

This rearranges to [Example 22B.1]

$$\frac{p}{V} = \frac{p}{V_{\text{mon}}} + \frac{1}{\alpha V_{\text{mon}}}$$

Hence, $\frac{p_2}{V_2} - \frac{p_1}{V_1} = \frac{p_2}{V_{\text{mon}}} - \frac{p_1}{V_{\text{mon}}}$

Solving for V_{mon}

$$V_{\text{mon}} = \frac{p_2 - p_1}{(p_2/V_2 - p_1/V_1)} = \frac{(760-145.4)\ \text{Torr}}{(760/1.443 - 145.4/0.286)\ \text{Torr cm}^{-3}} = \boxed{33.6\ \text{cm}^3}$$

E22B.2(a) The enthalpy of adsorption is typical of $\boxed{\text{chemisorption}}$ (Table 22A.2) for which $\tau_0 \approx 10^{-14}$ s because the adsorbate-substrate bond is stiff (see Brief illustration 22B.2). The half-life for remaining on the surface is

$$t_{1/2} = \tau_0 e^{E_{\text{a,des}}/RT}\ [22B.13] \approx (10^{-14}\ \text{s}) \times (e^{120\times10^3/(8.3145\times400)})\ [E_{\text{d}} \approx -\Delta_{\text{ad}}H] \approx \boxed{50\ \text{s}}$$

E22B.3(a) $\frac{m_1}{m_2} = \frac{\theta_1}{\theta_2} = \frac{p_1}{p_2} \times \frac{1+\alpha p_2}{1+\alpha p_1}$ [22B.2 and $V \propto m/p$]

which solves to

$$\alpha = \frac{(m_1 p_2 / m_2 p_1) - 1}{p_2 - (m_1 p_2 / m_2)} = \frac{(m_1/m_2)\times(p_2/p_1) - 1}{1-(m_1/m_2)} \times \frac{1}{p_2}$$
$$= \frac{(0.44/0.19)\times(3.0/26.0) - 1}{1-(0.44/0.19)} \times \frac{1}{3.0\ \text{kPa}} = 0.19\ \text{kPa}^{-1}$$

Therefore,

$$\theta_1 = \frac{(0.19\,\text{kPa}^{-1})\times(26.0\,\text{kPa})}{(1)+(0.19\,\text{kPa}^{-1})\times(26.0\,\text{kPa})} = \boxed{0.83}\quad [22B.2]\quad \text{and}$$

$$\theta_2 = \frac{(0.19)\times(3.0)}{(1)+(0.19)\times(3.0)} = \boxed{0.36}$$

E22B.4(a) $\theta=\dfrac{Kp}{1+Kp}$ [22B.2, $\alpha=K=k_a/k_d$], which implies that $p=\left(\dfrac{\theta}{1-\theta}\right)\dfrac{1}{K}$.

(a) $p=(0.15/0.85)/0.75\,\text{kPa}^{-1}=\boxed{0.24\,\text{kPa}}$

(b) $p=(0.95/0.05)/0.75\,\text{kPa}^{-1}=\boxed{25\,\text{kPa}}$

E22B.5(a)

$$\left(\frac{\partial\ln(p/p^{\ominus})}{\partial(1/T)}\right)_{\theta}=\frac{\Delta_{ad}H^{\ominus}}{R}$$

$$=-\frac{\Delta_{des}H^{\ominus}}{R}\quad[\text{Example 22B.2}, \Delta_{ad}H=-\Delta_{des}H=-10.2\,\text{J}/1.00\,\text{mmol}]$$

Assuming that $\Delta_{des}H$ is independent of temperature, integration and evaluation gives

$$\ln\frac{p_2}{p_1}=-\frac{\Delta_{des}H}{R}\left(\frac{1}{T_2}-\frac{1}{T_1}\right)=-\left(\frac{10.2\text{ kJ mol}^{-1}}{8.3145\text{ J K}^{-1}\text{ mol}^{-1}}\right)\times\left(\frac{1}{313\text{ K}}-\frac{1}{298\text{ K}}\right)=0.197$$

which implies that $p_2=(12\text{ kPa})\times(e^{0.197})=\boxed{15\,\text{kPa}}$.

E22B.6(a)

$$\left(\frac{\partial\ln(p/p^{\ominus})}{\partial(1/T)}\right)_{\theta}=\frac{\Delta_{ad}H^{\ominus}}{R}\quad[\text{Example 22B.2}]$$

Assuming that $\Delta_{ad}H$ is independent of temperature, integration and evaluation gives

$$\ln\frac{p_2}{p_1}=\frac{\Delta_{ad}H}{R}\left(\frac{1}{T_2}-\frac{1}{T_1}\right)$$

$$\Delta_{ad}H=R\left(\frac{1}{T_2}-\frac{1}{T_1}\right)^{-1}\ln\frac{p_2}{p_1}$$

$$=(8.3145\text{ J K}^{-1}\text{ mol}^{-1})\times\ln\left(\frac{3.2\times10^3\text{ kPa}}{490\text{ kPa}}\right)\times\left(\frac{1}{250\text{ K}}-\frac{1}{190\text{ K}}\right)^{-1}$$

$$=\boxed{-12.\bar{4}\text{ kJ mol}^{-1}}$$

E22B.7(a) The desorption time for a given volume is proportional to the half-life of the absorbed species and, consequently, the ratio of desorption times at two different temperatures is given by:

$$t(2)/t(1)=t_{1/2}(2)/t_{1/2}(1)=e^{E_{a,des}/RT_2}/e^{E_{a,des}/RT_1}\ [22B.13]=e^{E_{a,des}(1/T_2-1/T_1)/R}$$

Solving for the activation energy for desorption, $E_{a,des}$, gives:

$$E_{a,des}=R\ln\{t(2)/t(1)\}(1/T_2-1/T_1)^{-1}$$

$$=(8.3145\text{ J K}^{-1}\text{ mol}^{-1})\times\ln\left(\frac{2.0\text{ min}}{27\text{ min}}\right)\times\left(\frac{1}{1978\text{ K}}-\frac{1}{1856\text{ K}}\right)^{-1}$$

$$=\boxed{65\bar{1}\text{ kJ mol}^{-1}}$$

The desorption time, t, for the same volume at temperature t is given by:

$$t = t(1)\mathrm{e}^{E_{\mathrm{a,des}}(1/T - 1/T_1)/R}$$

$$= (27\ \mathrm{min})\exp\left\{(65\bar{1}\times10^3\ \mathrm{J\,mol^{-1}})\times\left(\frac{1}{T} - \frac{1}{1856\ \mathrm{K}}\right)/(8.3145\ \mathrm{J\,K^{-1}\,mol^{-1}})\right\}$$

$$= (27\ \mathrm{min})\exp\left\{(78.3)\times\left(\frac{1}{T/1\,000\ \mathrm{K}} - \frac{1}{1.856}\right)\right\}.$$

(i) At 298 K, $t = \boxed{1.6\times10^{97}\ \mathrm{min}}$, which is about forever.

(ii) At 3 000 K, $t = \boxed{2.8\times10^{-6}\ \mathrm{min}}$

E22B.8(a) The average time of molecular residence is proportional to the half-life of the absorbed species and, consequently, the ratio of average residence times at two different temperatures is given by:

$$t(2)/t(1) = t_{1/2}(2)/t_{1/2}(1) = \mathrm{e}^{E_{\mathrm{a,des}}/RT_2}/\mathrm{e}^{E_{\mathrm{a,des}}/RT_1}\ [22B.13] = \mathrm{e}^{E_{\mathrm{a,des}}(1/T_2 - 1/T_1)/R}$$

Solving for the activation energy for desorption, $E_{\mathrm{a,des}}$, gives:

$$E_{\mathrm{a,des}} = R\ln\{t(2)/t(1)\}(1/T_2 - 1/T_1)^{-1}$$

$$= (8.3145\ \mathrm{J\,K^{-1}\,mol^{-1}})\times\ln\left(\frac{3.49\ \mathrm{s}}{0.36\ \mathrm{s}}\right)\times\left(\frac{1}{2\,362\ \mathrm{K}} - \frac{1}{2\,548\ \mathrm{K}}\right)^{-1}$$

$$= \boxed{61\bar{1}\ \mathrm{kJ\,mol^{-1}}}$$

E22B.9(a) At 400 K: $t_{1/2} = \tau_0 \mathrm{e}^{E_{\mathrm{a,des}}/RT}\ [22B.13] = (0.10\,\mathrm{ps})\times\mathrm{e}^{0.301E_{\mathrm{a,des}}/\mathrm{kJ\,mol^{-1}}}$

At 1 000 K: $t_{1/2} = \tau_0 \mathrm{e}^{E_{\mathrm{a,des}}/RT}\ [22B.13] = (0.10\,\mathrm{ps})\times\mathrm{e}^{0.120E_{\mathrm{a,des}}/\mathrm{kJ\,mol^{-1}}}$

(i) $E_{\mathrm{a,des}} = 15\,\mathrm{kJ\,mol^{-1}}$

$$t_{1/2}(400\,\mathrm{K}) = (0.10\,\mathrm{ps})\times\mathrm{e}^{0.301\times15} = \boxed{9.1\,\mathrm{ps}},\ t_{1/2}(1\,000\,\mathrm{K}) = (0.10\,\mathrm{ps})\times\mathrm{e}^{0.120\times15}$$

$$= \boxed{0.60\,\mathrm{ps}}$$

(ii) $E_{\mathrm{a,des}} = 150\,\mathrm{kJ\,mol^{-1}}$

$$t_{1/2}(400\,\mathrm{K}) = (0.10\,\mathrm{ps})\times\mathrm{e}^{0.301\times150} = \boxed{4.1\times10^6\ \mathrm{s}},\ t_{1/2}(1\,000\,\mathrm{K}) = (0.10\,\mathrm{ps})\times\mathrm{e}^{0.120\times150}$$

$$= \boxed{6.6\,\mu\mathrm{s}}$$

E22B.10(a) Rate of desorption $= k_\mathrm{r}\theta = \dfrac{k_\mathrm{r}\alpha p}{1+\alpha p}$ [22B.1b, 22B.2]

(a) On gold, $\theta \approx 1$, and $k_\mathrm{r}\theta \approx$ constant, a $\boxed{\text{zeroth-order}}$ reaction.

(b) On platinum, $\theta \approx \alpha p$ (as $\alpha p << 1$), so the rate of desorption $= k_\mathrm{r}Kp$ and the reaction is $\boxed{\text{first-order}}$.

Solutions to problems

P22B.1 We use Mathcad Prime 2 to study the functional dependences of this problem. Parameters within the worksheets are easily changed to view their effect.

(a) Inversion of eqn 22B.2, $\dfrac{1}{\theta}=1+\dfrac{1}{\alpha p}$, shows that for Langmuir adsorption isotherms without dissociation $1/\theta$ is linear in $1/p$ with an intercept of 1 and slope of $1/\alpha$.

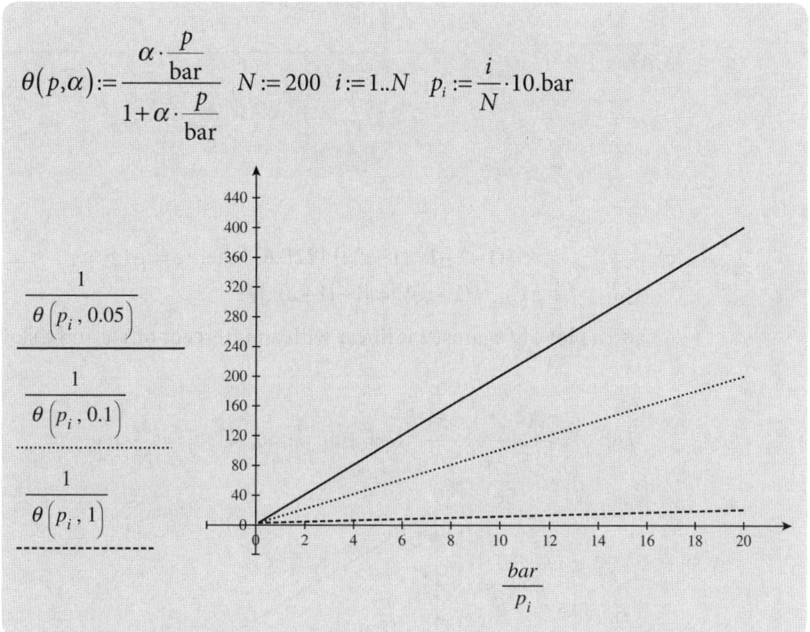

(b) Inversion of eqn 22B.4, $\dfrac{1}{\theta}=1+\dfrac{1}{(\alpha p)^{1/2}}$, shows that for Langmuir adsorption isotherms with dissociation $1/\theta$ is linear in $1/p^{1/2}$ with an intercept of 1 and slope of $1/\alpha^{1/2}$. Change the parameter 'power' in the following worksheet so that it equals one, you'll see that the curves become non linear. Thus, should an experimental data plot of $1/\theta_{exp}$ against $1/p_{exp}$ be non linear while a data plot of $1/\theta_{exp}$ against $(1/p_{exp})^{1/2}$ is linear, we suspect dissociation.

$$\theta(p,\alpha):=\frac{\left(\alpha\cdot\dfrac{p}{\text{bar}}\right)^{\frac{1}{2}}}{1+\left(\alpha\cdot\dfrac{p}{\text{bar}}\right)^{\frac{1}{2}}}\quad N:=200\quad i:=1..N\quad p_i:=\frac{i}{N}\cdot 10.\text{bar}$$

$$power:=\frac{1}{2}$$

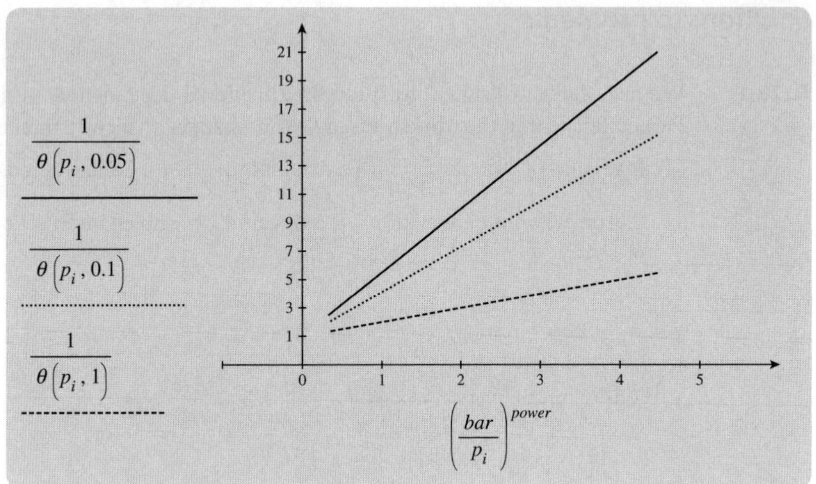

(c) $V/V_{\text{mon}} = cz/(1-z)\{1-(1-c)z\}$ [22B.6] where $z = p/p^*$

Thus, $f \equiv zV_{\text{mon}}/(1-z)V = \{1-(1-c)z\}/c$

and a plot of f against z is linear with an intercept of $1/c$ and a slope of $1 - 1/c$.

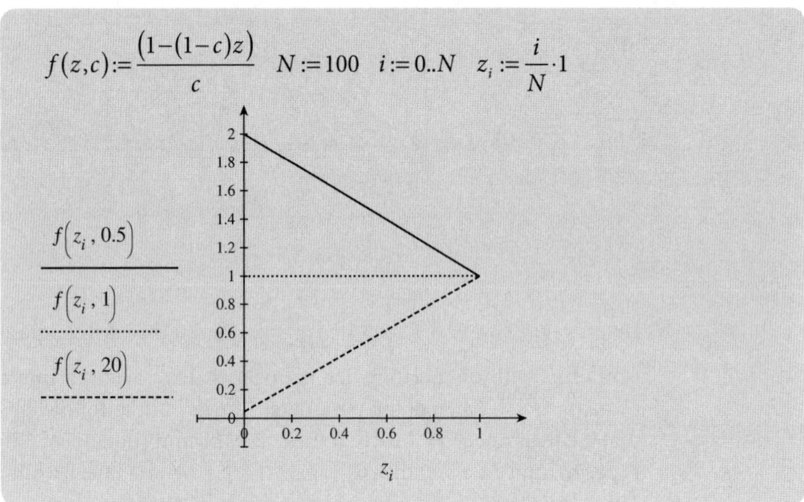

P22B.3 $$\frac{V}{V_{\text{mon}}} = \frac{cz}{(1-z)\{1-(1-c)z\}} \quad \text{with} \quad z = \frac{p}{p^*} \quad [22B.6]$$

This rearranges to

$$\frac{z}{(1-z)V} = \frac{1}{cV_{\text{mon}}} + \frac{(c-1)z}{cV_{\text{mon}}}$$

A plot of the left-hand side, $z/(1-z)V$, against z should result in a straight line if the data obeys the BET isotherm. Should it be linear, a linear regression fit of the plot yields values for the intercept and slope, which are related to c and V_{mon} by

$$1/cV_{\text{mon}} = \text{intercept} \quad \text{and} \quad (c-1)/cV_{\text{mon}} = \text{slope}.$$

Solving for c and V_{mon} yields

$$c = 1 + \text{slope}/\text{intercept} \quad \text{and} \quad V_{\text{mon}} = 1/(c \times \text{intercept}).$$

We draw up the following tables:

(a) $0\,°\text{C}, p^* = 429.6\,\text{kPa}$

p/kPa	**14.0**	**37.6**	**65.6**	**79.2**	**82.7**	**100.7**	**106.4**
$10^3 z$	32.6	87.5	152.7	184.4	192.4	234.3	247.7
$\dfrac{10^3 z}{(1-z)(V/\text{cm}^3)}$	3.03	7.11	12.1	14.1	15.4	17.7	20.0

(b) $18\,°\text{C}, p^* = 819.7\,\text{kPa}$

p/kPa	**5.3**	**8.4**	**14.4**	**29.2**	**62.1**	**74.0**	**80.1**	**102.0**
$10^3 z$	6.5	10.2	17.6	35.6	75.8	90.3	97.8	124.4
$\dfrac{10^3 z}{(1-z)(V/\text{cm}^3)}$	0.70	1.06	1.74	3.27	6.35	7.58	8.08	10.1

The $z/(1-z)V$ against z points are plotted in Fig. 22B.1. It is apparent that the plots are linear so we conclude that the data fits the BET isotherm. The linear regression fits are summarized in the figure.

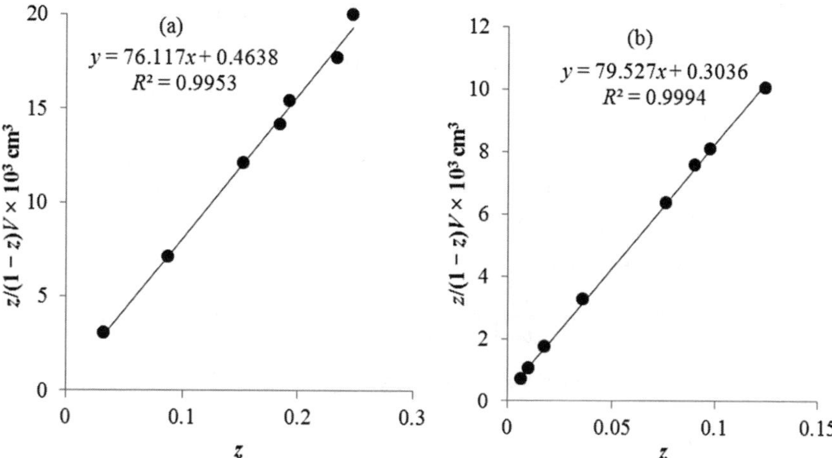

Figure 22B.1

(a) $\text{intercept} - 0.4638\times10^{-3}\ \text{cm}^{-3}$ and $\text{slope} = 76.12\times10^{-3}\ \text{cm}^{-3}$

$c = 1 + \text{slope}/\text{intercept} = 1 + 76.12/0.4638 = \boxed{165}$

$V_{\text{mon}} = 1/(c\times\text{intercept}) = 1/(165\times0.4638\times10^{-3}\ \text{cm}^{-3}) = \boxed{13.1\,\text{cm}^3}$

(b) $\text{intercept} = 0.3036\times10^{-3}\ \text{cm}^{-3}$ and $\text{slope} = 79.53\times10^{-3}\ \text{cm}^{-3}$

$c = 1 + \text{slope}/\text{intercept} = 1 + 79.53/0.3036 = \boxed{263}$

$V_{\text{mon}} = 1/(c\times\text{intercept}) = 1/(263\times0.3036\times10^{-3}\ \text{cm}^{-3}) = \boxed{12.5\,\text{cm}^3}$

P22B.5 The Langmuir isotherm (eqn 22B.2 with $\alpha = K$) is

$$\theta = \frac{Kp}{1+Kp} = \frac{n}{n_\infty} \text{ so } n(1+Kp) = n_\infty Kp \text{ and } \frac{p}{n} = \frac{p}{n_\infty} + \frac{1}{Kn_\infty}$$

So a plot of p/n against p should be a straight line with slope $1/n_\infty$ and y-intercept $1/Kn_\infty$. The transformed data and plot (Fig. 22B.2) follow.

p/kPa	31.00	38.22	53.03	76.38	101.97	130.47	165.06	182.41	205.75	219.91
$n/(\text{mol kg}^{-1})$	1.00	1.17	1.54	2.04	2.49	2.90	3.22	3.30	3.35	3.36
$\frac{p/n}{\text{kPa mol}^{-1}\,\text{kg}}$	31.00	32.67	34.44	37.44	40.95	44.99	51.26	55.28	61.42	65.45

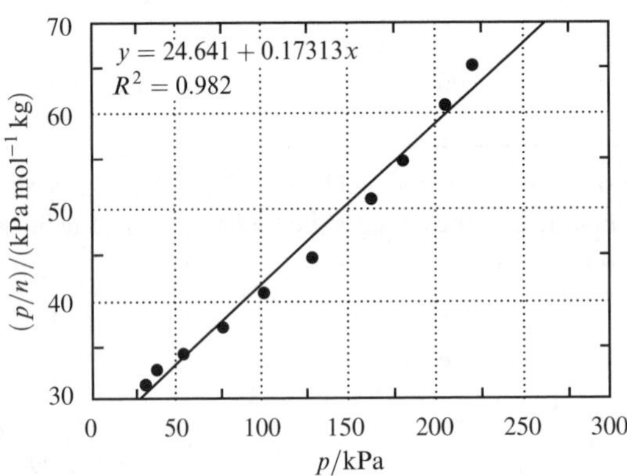

Figure 22B.2

$$n_\infty = \frac{1}{0.17313\ \text{mol}^{-1}\ \text{kg}} = \boxed{5.78\ \text{mol kg}^{-1}}$$

The y-intercept is

$$b = \frac{1}{Kn_\infty} \text{ so } K = \frac{1}{bn_\infty} = \frac{1}{(24.641\ \text{kPa mol}^{-1}\ \text{kg})\times(5.78\ \text{mol kg}^{-1})}$$

$$K = 7.02\times10^{-3}\ \text{kPa}^{-1} = \boxed{7.02\ \text{Pa}^{-1}}$$

P22B.7 Application of the van 't Hoff equation (eqn 22B.5, $K = \alpha p^{\ominus}$) to adsorption equilibria yields

$$\left(\frac{\partial \ln K}{\partial T}\right)_{\theta} = \frac{\Delta_{ad}H^{\ominus}}{RT^2} \text{ or } \left(\frac{\partial \ln K}{\partial(1/T)}\right)_{\theta} = \frac{-\Delta_{ad}H^{\ominus}}{R}$$

A plot (Fig. 22B.3) of $\ln K$ against $1/T$ should be a straight line with slope $-\Delta_{ad}H^{\ominus}/R$. The transformed data and plot follow:

T/K	28.3	298	308	318
10^{-11} K	2.642	2.078	1.286	1.085
1 000 K/T	3.53	3.36	3.25	3.14
ln K	26.30	26.06	25.58	25.41

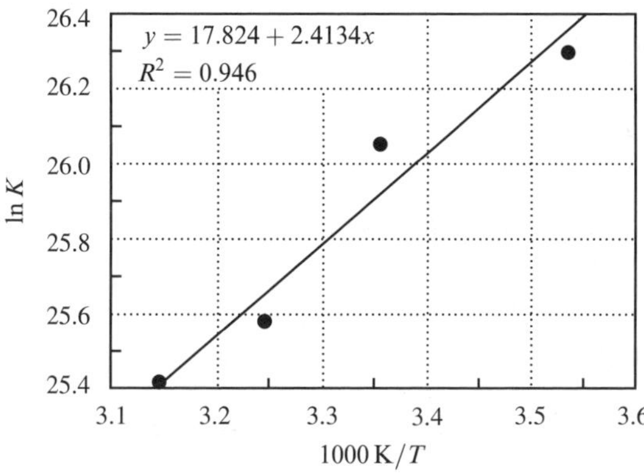

Figure 22B.3

$$\Delta_{ad}H^{\ominus} = -R\times\text{slope} = -(8.3145\text{ J mol}^{-1}\text{K}^{-1})\times(2.41\times10^3\text{ K})$$
$$= -20.0\times10^3\text{ J mol}^{-1} = \boxed{-20.0\text{ kJ mol}^{-1}}$$

The Gibbs energy for absorption is

$$\Delta_{ad}G^{\ominus} = \Delta_{ad}H^{\ominus} - T\Delta_{ad}S^{\ominus} = -20.2\text{ kJ mol}^{-1} - (298\text{ K})\times(0.146\text{ kJ mol}^{-1}\text{ K}^{-1})$$
$$= \boxed{-63.5\text{ kJ mol}^{-1}}$$

P22B.9 $$\frac{1}{q_{\text{VOC,RH=0}}} = \frac{1+bc_{\text{VOC}}}{abc_{\text{VOC}}} = \frac{1}{abc_{\text{VOC}}} + \frac{1}{a}$$

Parameters of regression fit:

θ/°C	1/a	1/ab	R	a	b/ppm^{-1}
33.6	9.07	709.8	0.9836	0.110	0.0128
41.5	10.14	890.4	0.9746	0.0986	0.0114
57.4	11.14	1 599	0.9943	0.0898	0.00697
76.4	13.58	2 063	0.9981	0.0736	0.00658
99	16.82	4 012	0.9916	0.0595	0.00419

The linear regression fit is generally good at all temperatures with

$\boxed{R \text{ values in the range } 0.975 \text{ to } 0.991}$

(b) $\ln a = \ln k_{\rm a} - \dfrac{\Delta_{\rm ad}H}{R}\dfrac{1}{T}$ and $\ln b = \ln k_{\rm b} - \dfrac{\Delta_{\rm b}H}{R}\dfrac{1}{T}$

Linear regression analysis of $\ln a$ versus $1/T$ gives the intercept $\ln k_{\rm a}$ and slope $-\Delta_{\rm ad}H/R$ while a similar statement can be made for a $\ln b$ versus $1/T$ plot. The temperature must be in Kelvin.

For $\ln a$ versus $1/T$:

$$\ln k_{\rm a} = -5.605, \quad \text{standard deviation} = 0.197$$

$$-\Delta_{\rm ad}H/R = 1\,043.2\ \text{K}, \quad \text{standard deviation} = 65.4\ \text{K}$$

$$R = 0.9942 \quad [\text{good fit}]$$

$$k_{\rm a} = \mathrm{e}^{-5.605} = \boxed{3.68\times10^{-3}}$$

$$\Delta_{\rm ad}H = -(8.31451\ \text{J}\,\text{K}^{-1}\text{mol}^{-1})\times(1\,043.2\,\text{K})$$

$$= \boxed{-8.67\,\text{kJ}\,\text{mol}^{-1}}$$

For $\ln b$ versus $1/T$:

$$\ln\left(k_b/(\text{ppm}^{-1})\right) = -10.550, \quad \text{standard deviation} = 0.713$$

$$-\Delta_{\rm b}H/R = 1\,895.4\,\text{K}, \quad \text{standard deviation} = 236.8$$

$$R = 0.9774 \quad [\text{good fit}]$$

$$k_{\rm b} = \mathrm{e}^{-10.550}\ \text{ppm}^{-1} = \boxed{2.62\times10^{-5}\ \text{ppm}^{-1}}$$

$$\Delta_{\rm b}H = -(8.31451\,\text{J}\,\text{K}^{-1}\text{mol}^{-1})\times(1\,895.4\ \text{K})$$

$$\boxed{\Delta_{\rm b}H = -15.7\,\text{kJ}\,\text{mol}^{-1}}$$

(c) $k_{\rm a}$ may be interpreted to be the maximum adsorption capacity at an adsorption enthalpy of zero, while $k_{\rm b}$ is the maximum affinity in the case for which the adsorbant–surface bonding enthalpy is zero.

P22B.11 Taking the natural logarithm of the isotherm $c_{\rm ads} = Kc_{\rm sol}^{1/n}$ gives

$$\ln c_{\rm ads} = \ln K + (\ln c_{\rm sol})/n$$

so a plot of $\ln c_{\rm ads}$ versus $\ln c_{\rm sol}$ would have a slope of $1/n$ and a y-intercept of $\ln K$. The transformed data and plot are shown in Fig. 22B.4.

$c_{sol}/(\text{mg g}^{-1})$	8.26	15.65	25.43	31.74	40.00
$c_{ads}/(\text{mg g}^{-1})$	4.4	19.2	35.2	52.0	67.2
$\ln c_{sol}$	2.11	2.75	3.24	3.46	3.69
$\ln c_{ads}$	1.48	2.95	3.56	3.95	4.21

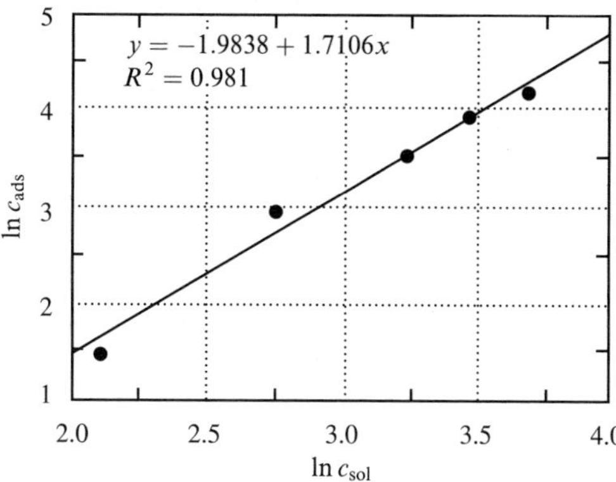

Figure 22B.4

$$K = e^{-1.9838}\ \text{mg g}^{-1} = \boxed{0.138\ \text{mg g}^{-1}} \quad \text{and} \quad n = 1/1.71 = \boxed{0.58}$$

In order to express this information in terms of fractional coverage, the amount of adsorbate corresponding to monolayer coverage must be known. This saturation point, however, has no special significance in the Freundlich isotherm (i.e. it does not correspond to any limiting case).

P22B.13 (a) $q_{\text{water}} = k(\text{RH})^{1/n}$

With a power law regression analysis we find

$\boxed{k = 0.2289}$, standard deviation = 0.0068

$1/n = 1.6182$, standard deviation = 0.0093; $\boxed{n = 0.6180}$

$R = 0.999508$

A linear regression analysis may be performed by transforming the equation to the following form by taking the logarithm of the Freundlich type equation.

$$\ln q_{\text{water}} = \ln k + \frac{1}{n}\ln(\text{RH})$$

$\ln k = -1.4746$, standard deviation = 0.0068; $\boxed{k = 0.2289}$

$\frac{1}{n} = 1.6183$, standard deviation = 0.0093; $\boxed{n = 0.6180}$

$R = 0.999508$

The two methods give exactly the same result because the software package for performing the power law regression performs the transformation to linear form for you. Both methods are actually performing a linear regression. The correlation coefficient indicates that 99.95 per cent of the data variation is explained with the Freundlich type isotherm. The Freundlich fit hypothesis looks very good.

(b) The Langmuir isotherm model describes adsorption sites that are independent and equivalent. This assumption seems to be valid for the VOC case in which molecules interact very weakly. However, water molecules interact much more strongly through forces such as hydrogen bonding and multilayers may readily form at the lower temperatures. The intermolecular forces of water apparently cause adsorption sites to become non-equivalent and dependent. In this particular case the Freundlich type isotherm becomes the better description.

(c) $r_{\text{VOC}} = 1 - q_{\text{water}}$ where $r_{\text{VOC}} \equiv q_{\text{VOC}} / q_{\text{VOC,RH=0}}$

$$r_{\text{VOC}} = 1 - k(\text{RH})^{1/n}$$

$$1 - r_{\text{VOC}} = k(\text{RH})^{1/n}$$

To determine the goodness-of-fit values for k and n, we perform a power law regression fit of $1 - r_{\text{VOC}}$ against RH. Results are:

$\boxed{k = 0.5227}$, standard deviation $= 0.0719$

$\frac{1}{n} = 1.3749$, standard deviation $= 0.0601$; $\boxed{n = 0.7273}$

$R = 0.99620$

Since 99.62 per cent of the variation is explained by the regression, we conclude that the hypothesis that $r_{\text{VOC}} = 1 - q_{\text{water}}$ may be very useful. The values of R and n differ significantly from those of part (a). It may be that water is adsorbing to some portions of the surface and VOC to others.

22C Heterogeneous catalysis

Answer to discussion question

D22C.1 In the Langmuir–Hinshelwood mechanism of surface-catalysed reactions, the reaction takes place by encounters between molecular fragments and atoms already adsorbed on the surface. We therefore expect the rate law to be second-order in the extent of surface coverage:

$$\text{A} + \text{B} \rightarrow \text{P} \qquad v = k_{\text{r}}\theta_{\text{A}}\theta_{\text{B}} \quad [22\text{C}.2]$$

Insertion of the appropriate isotherms for A and B then gives the reaction rate in terms of the partial pressures of the reactants. For example, if A and B follow Langmuir isotherms (Example 22B.1), and adsorb without dissociation, then it follows that the rate law is

$$v = \frac{k_r \alpha_A \alpha_B p_A p_B}{(1+\alpha_A p_A + \alpha_B p_B)^2} \quad \text{[Example 22C.1]}$$

The parameters α in the isotherms and the rate constant k_r are all temperature dependent, so the overall temperature dependence of the rate may be strongly non-Arrhenius (in the sense that the reaction rate is unlikely to be proportional to $\exp(-E_a/RT)$.

In the Eley–Rideal mechanism (ER mechanism) of a surface-catalysed reaction, a gas phase molecule collides with another molecule already adsorbed on the surface. The rate of formation of product is expected to be proportional to the partial pressure, p_B of the non-adsorbed gas B and the extent of surface coverage, θ_A, of the adsorbed gas A. It follows that the rate law should be

$$A + B \rightarrow P \qquad v = k_r p_B \theta_A \quad [22C.3]$$

The rate constant, k, might be much larger than for the uncatalysed gas-phase reaction because the reaction on the surface has a low activation energy and the adsorption itself is often not activated.

If we know the adsorption isotherm for A, we can express the rate law in terms of its partial pressure, p_A. For example, if the adsorption of A follows a Langmuir isotherm in the pressure range of interest, then the rate law would be

$$v = \frac{k_r \alpha p_A p_B}{1+\alpha p_A} \quad [22C.4]$$

If A were a diatomic molecule that adsorbed as atoms, we would substitute the isotherm given in eqn 22B.4 instead.

According to eqn 22B.6, when the partial pressure of A is high (in the sense $\alpha p_A \gg 1$), there is almost complete surface coverage, and the rate is equal to $k_r p_B$. Now the rate-determining step is the collision of B with the adsorbed fragments. When the pressure of A is low $(\beta p_A \ll 1)$, perhaps because of its reaction, the rate is equal to $k_r \alpha p_A p_B$. Now the extent of surface coverage is important in the determination of the rate.

Solutions to exercises

E22C.1(a) Let us assume that the nitrogen molecules are close-packed, as shown in Fig. 22C.1 as spheres, in the monolayer. Then, one molecule occupies the parallelogram area of $2\sqrt{3}\,r^2$ where r is the radius of the adsorbed molecule. Furthermore, let us assume that the collision cross-section of Table 1B.1 ($\sigma = 0.43\ \text{nm}^2 = 4\pi r^2$) gives a reasonable estimate of r: $r = (\sigma/4\pi)^{1/2}$. With these assumptions the surface area occupied by one molecule is:

$$\begin{aligned} A_{\text{molecule}} &= 2\sqrt{3}(\sigma/4\pi) = \sqrt{3}\,\sigma/2\pi \\ &= \sqrt{3}\left(0.43\ \text{nm}^2\right)/2\pi = 0.12\ \text{nm}^2 \end{aligned}$$

In this model the surface area per gram of the catalyst equals $A_{\text{molecule}}N$ where N is the number of adsorbed molecules. N can be calculated with the 0 °C data, a temperature that is so high compared to the boiling point of nitrogen that all molecules are likely to be desorbed from the surface as perfect gas.

$$N = \frac{pV}{kT} = \frac{(760\ \text{Torr})\times(133.3\ \text{Pa/Torr})\times(3.86\times10^{-6}\ \text{m}^3)}{(1.381\times10^{-23}\ \text{J K}^{-1})\times(273.15\ \text{K})} = 1.04\times10^{20}$$

$$A_{\text{molecule}}N = (0.12\times10^{-18}\ \text{m}^2)\times(1.04\times10^{20}) = \boxed{12\ \text{m}^2}$$

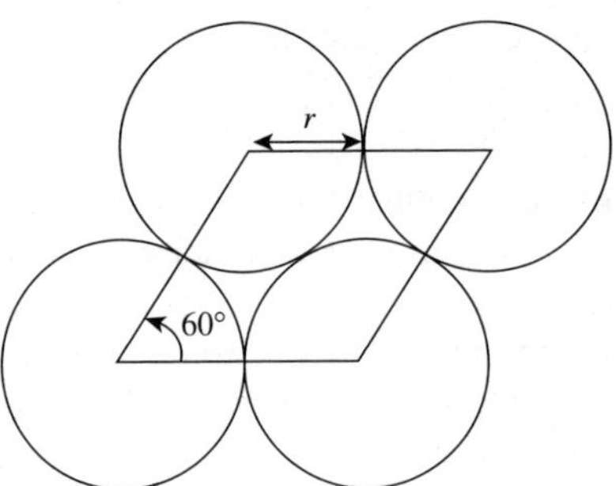

Figure 22C.1

Solution to problem

P22C.1 Proposed mechanism:

$$\text{NH}_3(\text{g}) + \text{M}_{\text{surface}} \xrightarrow{k_a} \text{NH}_3\cdot\text{M}_{\text{surface}} \rightarrow \xrightarrow{\text{rapid}} \rightarrow \tfrac{1}{2}\text{N}_2(\text{g})$$
$$+\tfrac{3}{2}\text{H}_2\cdot\text{M}_{\text{surface}} \xrightleftharpoons{K=k_{\text{a,H}_2}/k_{\text{d,H}_2}} \tfrac{3}{2}\text{H}_2(\text{g}) + \text{M}_{\text{surface}}$$

This mechanism postulates that the decomposition rate is proportional both to p_{NH_3} and to the fraction of adsorption sites that remain unoccupied by the strongly adsorbed H_2, $1-\theta$. We assume that gaseous hydrogen and adsorbed hydrogen are in equilibrium at all $t \geq 30$ s. Then,

$$\theta = \frac{Kp_{\text{H}_2}}{1+Kp_{\text{H}_2}}\ [22\text{B}.2,\ \alpha = K = k_{\text{a,H}_2}/k_{\text{d,H}_2}]\ \text{ and }\ 1-\theta = \frac{1}{1+Kp_{\text{H}_2}}$$

For a strongly adsorbed hydrogen species, $Kp_{\text{H}_2} \gg 1$ and $1-\theta = 1/Kp_{\text{H}_2}$. Since the reaction rate is proportional to the pressure of ammonia and the fraction of sites left uncovered by the strongly adsorbed hydrogen product, we can write

$$\frac{\text{d}p_{\text{NH}_3}}{\text{d}t} = -k_r p_{\text{NH}_3}(1-\theta) \approx \boxed{-\frac{k_r}{K}\frac{p_{\text{NH}_3}}{p_{\text{H}_2}}}$$

According to the reaction stoichiometry

$$p_{H_2} = \tfrac{3}{2}\{p_{0,NH_3} - p_{NH_3}\}\ [NH_3 \rightarrow \tfrac{1}{2}N_2 + \tfrac{3}{2}H_2]$$

from which it follows that, with $p = p_{NH_3}$

$$\frac{-dp}{dt} = \frac{k_c p}{p_0 - p} \quad \text{where } k_c = \frac{2k_r}{3K}$$

This equation integrates as follows.

$$\int_{p_0}^{p}\left(1 - \frac{p_0}{p}\right)dp = k_c\int_0^t dt \quad \text{so} \quad \boxed{k_c = \frac{p - p_0}{t} - \frac{p_0}{t}\ln\frac{p}{p_0}}$$

Thus, we prepare a table in which $\frac{p - p_0}{t} - \frac{p_0}{t}\ln\frac{p}{p_0}$ is calculated at each data pair and examine these calculated k_c values. The mean and standard deviation are also reported in the table.

t/s	0	30	60	100	160	200	250	mean	std. dev.
p_{NH_3} / kPa	13.3	11.7	11.2	10.7	10.3	9.9	9.6		
$k_c/10^{-3}$ kPa s^{-1}		3.49	3.09	2.93	2.50	2.63	2.54	3.02	0.67

The standard deviation is seen to be rather large and it appears that calculated k_c values are high at early times, decline, and reach a constant value at times larger than about 160 s. The Fig. 22C.2 plot of k_c against t appears to provide confirm of this trend so we accept the average of large time values as the best approximation of k_c:

$$\boxed{k_c = 2.5 \times 10^{-3}\ \text{kPa s}^{-1}}$$

The variation of k_c values may indicate that equilibrium between gaseous and adsorbed hydrogen is not achieved until the latter times.

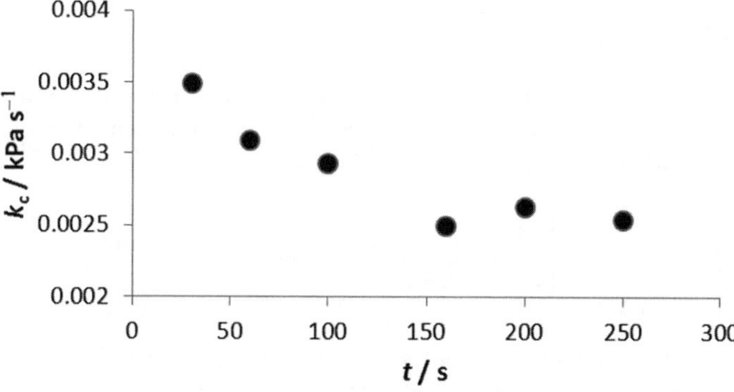

Figure 22C.2

Integrated activities

I22.1 Our model will treat the solid argon (1.784 g mol^{-1}, face-centred ccp, 525.6 pm sides unit cell) as a continuum of matter that has the number density:

$$\mathcal{N} = \rho N_A / M = 2.689\times10^{28}\ \mathrm{m}^{-3}$$

Referring to Fig. I22.1 for the definitions of R, z, and r, the Lennard-Jones (6,12)-potential [16B.14] for the interaction between an Ar atom and a point within the solid at the distance ξ from the atom is

$$V(R,z,r) = 4\varepsilon\left\{\left(\frac{r_0}{\xi}\right)^{12} - \left(\frac{r_0}{\xi}\right)^{6}\right\} \quad \text{where} \quad \xi = \left[(R+z)^2 + r^2\right]^{1/2}$$

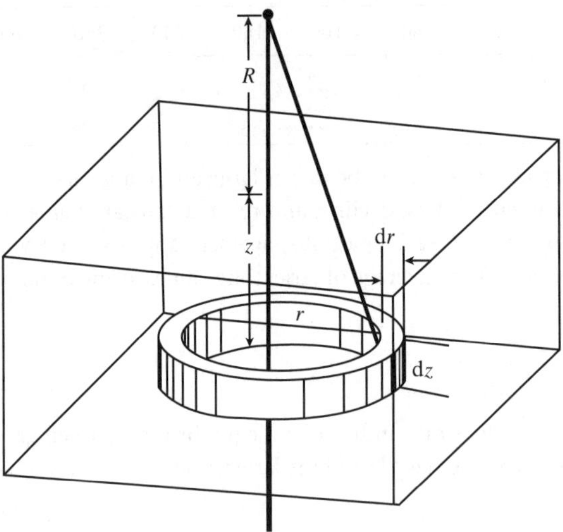

Figure I22.1

The differential number of atoms in the annulus between r and $r + \mathrm{d}r$ and thickness $\mathrm{d}z$ at a depth z below the surface is $2\pi\mathcal{N}r\,\mathrm{d}r\,\mathrm{d}z$ so the interaction energy between the adsorbate atom and the atoms of the annulus is

$$\mathrm{d}U = 8\pi\varepsilon\mathcal{N}\times\left\{\left(\frac{r_0}{\xi}\right)^{12} - \left(\frac{r_0}{\xi}\right)^{6}\right\}\times r\ \mathrm{d}r\ \mathrm{d}z$$

To get the total interaction energy when the adsorbate is at the distance R from the solid surface we need to integrate $\mathrm{d}U$ over the ranges $0 \le z \le \infty$ and $0 \le r \le \infty$.

$$U = 8\pi\varepsilon\mathcal{N}\int_0^\infty\left(\int_0^\infty\left\{\left(\frac{r_0}{\left[(R+z)^2+r^2\right]^{1/2}}\right)^{12} - \left(\frac{r_0}{\left[(R+z)^2+r^2\right]^{1/2}}\right)^{6}\right\}r\ \mathrm{d}r\right)\mathrm{d}z$$

Using the standard integral $\int_0^\infty \frac{r\,\mathrm{d}r}{(a^2+r^2)^{n/2}} = \frac{1}{(n-2)a^{n-2}}$ for $n=12$ or 6, we perform the inner integral.

$$U = 8\pi\varepsilon\mathcal{N}\int_0^\infty \left(\frac{r_0^{12}}{10\times(R+z)^{10}} - \frac{r_0^6}{4\times(R+z)^4}\right)\mathrm{d}z$$

For the last integral use the standard integral $\int_0^\infty \frac{1}{(R+z)^{n-2}}\,\mathrm{d}z = \frac{1}{(n-3)R^{n-3}}$.

$$\boxed{U = \tfrac{4}{3}\pi\varepsilon r_0^3\mathcal{N}\left\{\frac{1}{15}\left(\frac{r_0}{R}\right)^9 - \frac{1}{2}\left(\frac{r_0}{R}\right)^3\right\}}$$

(a) The negative term of the above expression is the attractive interaction and we see that $\boxed{U_{\text{attraction}} \sim R^{-3}}$.

(b) For the position of equilibrium, we look for the value of R for which $\mathrm{d}U/\mathrm{d}R = 0$.

$$\left.\frac{\mathrm{d}U}{\mathrm{d}R}\right|_{R=R_{\text{eq}}} = \tfrac{4}{3}\pi\varepsilon r_0^3\mathcal{N}\left\{-\frac{9r_0^9}{15R_{\text{eq}}^{10}} + \frac{3r_0^3}{2R_{\text{eq}}^4}\right\} = 0$$

Solving for R_{eq}, we find $\boxed{R_{\text{eq}} = (2/5)^{1/6}\,r_0}$.

From Table 16B.2 we have $\varepsilon = 128$ kJ mol^{-1} and $r_0 = 342$ pm Consequently, this model predicts that the adsorbate equilibrium interaction with the solid has the properties

$$R_{\text{eq}} = (2/5)^{1/6}\times(342\,\text{pm}) = \boxed{294\,\text{pm}}$$

and

$$U_{\min} = \tfrac{4}{3}\pi\times(128\text{ kJ mol}^{-1})\times(342\times10^{-12}\text{ m})^3\times\left(\frac{2.689\times10^{28}\text{ m}^{-3}}{N_\text{A}}\right)$$
$$\times\left\{\frac{1}{15}\left(\frac{1}{(2/5)^{1/6}}\right)^9 - \frac{1}{2}\left(\frac{1}{(2/5)^{1/6}}\right)^3\right\}$$
$$= \boxed{-5.05\times10^{-22}\text{ kJ}}\text{ or } -304\text{ kJ per mol adsorbate atoms}$$

I22.3 The Coulombic force is

$$F = -\frac{\mathrm{d}V}{\mathrm{d}r} = -\frac{\mathrm{d}}{\mathrm{d}r}\left(\frac{Q_1Q_2}{4\pi\varepsilon_0 r}\right) = \frac{Q_1Q_2}{4\pi\varepsilon_0 r^2}$$
$$= \frac{(1.602\times10^{-19}\text{ C})^2}{(1.113\times10^{-10}\text{ J}^{-1}\text{ C}^2\text{ m}^{-1})\times(2.00\times10^{-9}\text{ m})^2}$$
$$= \boxed{5.77\times10^{-11}\text{ N}}$$